AF606158

Ullmann's Encyclopedia of Industrial Chemistry

Volume B 2

Unit Operations I

Ullmann's Encyclopedia of Industrial Chemistry

Volumes A 1 – A 28: alphabetically arranged articles
Volumes B 1 – B 8: basic knowledge

Ullmann's Encyclopedia of Industrial Chemistry

Fifth, Completely Revised Edition

Volume B2:
Unit Operations I

Executive Editor: Wolfgang Gerhartz

Editors: Barbara Elvers, Michael Ravenscroft, James F. Rounsaville, Gail Schulz

Numerical data, descriptions of methods or equipment, and other information presented in this book have been carefully checked for accuracy. Nevertheless, authors and publishers do not assume any liability for misprints, faulty statements, or other kinds of errors. Persons intending to handle chemicals or to work according to information derived from this book are advised to consult the original sources as well as relevant regulations in order to avoid possible hazards.

Production Director: Maximilian Montkowski
Production Manager: Myriam Nothacker

Library of Congress Card No. 84-25-829

Deutsche Bibliothek, Cataloguing-in-Publication Data:

Ullmann's encyclopedia of industrial chemistry / executive ed.: Wolfgang Gerhartz. Ed.: Barbara Elvers ... [Ed. advisory board Hans-Jürgen Arpe ...]. — Weinheim ; Basel (Switzerland) ; Cambridge ; New York, NY : VCH
Teilw. mit d. Erscheinungsorten Weinheim, Deerfield Beach, Fl. —
Teilw. mit d. Erscheinungsorten Weinheim, New York, NY
Bis 4. Aufl. u. d. T.: Ullmanns Encyklopädie der technischen Chemie
NE: Gerhartz, Wolfgang [Hrsg.]; Encyclopedia of industrial chemistry

Vol. B. Basic knowledge.
2. Unit operations. — 1. — 5., completely rev. ed. — 1988
ISBN 3-527-20132-7 (Weinheim ...) Pp.
ISBN 0-89573-537-7 (Cambridge ...) Pp.

British Library Cataloguing in Publication Data

Ullmann's encyclopedia of industrial chemistry.
— 5th completely rev. ed.
Vol. B 2: Unit operations, I
I. Ullmann, Fritz, **1875–1939** II. Gerhartz, Wolfgang III. [Ullmanns Enzyklopädie der technischen Chemie]. **English**
661'.003'21

ISBN 3-527-20132-7

Distribution
VCH Verlagsgesellschaft, P.O. Box 12 60/12 80. D-6940 Weinheim (Federal Republic of Germany)
Switzerland: VCH Verlags-AG, P.O. Box, CH-4020 Basel (Switzerland)
Great Britain and Ireland: VCH Publishers (UK) Ltd., 8 Wellington Court, Wellington Street, Cambridge CB1 1HW (Great Britain)
USA and Canada: VCH Publishers, Suite 909, 220 East 23rd Street, New York NY 10010-4606 (USA)

Cover design: Wolfgang Schmidt
Composition, printing, and bookbinding: Graphischer Betrieb Konrad Triltsch, D-8700 Würzburg
Printed in the Federal Republic of Germany

Foreword

The concept of unit operations simplifies the understanding of industrial chemical processes. By learning the fundamentals of the pertinent unit operations, the chemist or chemical engineer need not study each chemical process as a separate entity. Instead, he or she need only study the individual building blocks, i.e., the unit operations. These deal with the physical changes that occur in a process but usually not with the specific chemical reactions.

For example, in the gasification of coal, which also involves combustion reactions, the important unit operations are the *solids handling* of the coal, including *mixing* and *sizing*, plus *heat transfer* and treatment of the product gases by *adsorption, absorption,* and *distillation.*

The various unit operations may overlap with each other. For example, heat and mass transfer are involved in *evaporation, distillation, absorption, combustion, drying,* and *crystallization.* This allows the different unit operations to be grouped under major headings, thus simplifying the presentation and discussion of unit operations in this encyclopedia. The major headings used are

Volume B2	Particle Technology Separation and Classification Mixing
Volume B3	Diffusional Separation Processes Separations in Biotechnology Energy Management and Heat Generation Low-Temperature Technology Vacuum Technology

Particle Technology. Solids handling and treatment are the very core of process systems that have streams of solids as major components. Particle technology is broken down in this encyclopedia into the unit operations of *particle-size analysis, crystallization and precipitation, drying of solid materials, size reduction, spraying and atomizing, size enlargement,* and *solids handling.*

Separation and Classification. Separations are grouped by the pair of phases involved:

Solid–Liquid Separation
Filtration
Centrifugation and Hydrocyclone Separation
Sedimentation

Solid–Gas Separation
Dust Separation

Solid–Solid Separation
Screening
Elutriation
Air Classifying
Mineral Sorting
Magnetic Separation
Electrostatic Separation
Gravity Concentration
Dense-Medium Separation
Flotation

Mixing. The well-developed principles of *stirring* and stirring equipment are clearly explained, followed by a discussion of the special problems involved in the *mixing of highly viscous media.* The mixing discussion is completed by the unit operation of *mixing of solids.*

Diffusional Separation. The design and operation of *heat exchangers* is simply the effective application of the principles of heat transfer. Because of the importance of heat exchange, this topic occupies a major portion of this section. Directly related to heat exchange is *evaporation,* an evaporator merely being a special type of heat exchanger used to accomplish mass transfer by vaporization of a liquid.

The unit operation *distillation and rectification* is of paramount importance in many industrial processes and as a result distillation and rectification are given the same emphasis as heat exchange. Its treatment as a unit operation requires the use of basic principles of energy and material balances. Inclusion of equilibrium relationships allows separation values to be established, the equipment to be designed, and operation to be optimized, with the resultant improvement in product quality.

Mass transfer, heat transfer, and equipment design are also important factors to be considered in *sublimation, liquid–liquid extraction, liquid–solid extraction, absorption,* and *adsorption.* Each of these operations is discussed in the encyclopedia with emphasis on the basic principles and practices.

Separations in Biotechnology. *Chromatography* is an important tool for separating components in a mixture and is widely used for chemical analysis as well as for industrial-scale separation. Separation of components of a mixture can also be accomplished by means of membranes, see *Membranes and Membrane Separation* in the A Series. These two separation methods have become increasingly important in recent years, and a strong background of basic operating principles has been developed along with a considerable amount of information on the equipment. The new *separation methods in biochemistry,* which are currently being developed the world over, are also discussed.

Energy Management and Heat Generation. The general technology of energy use, including both heating and cooling, is a central aspect of industrial chemistry. The concepts developed under the heading of diffusional separation processes form the basis of efficient energy use. This includes the survey topic *energy management,* which is of special importance in the chemical industry. A new method, *pinch technology,* has shown promising results in reducing the overall energy requirement of chemical processes.

The technology of *combustion,* which must, of course, include redox reactions, is also a basic unit operation important in many industrial processes. The *electrical generation of heat, radiation heating, evaporation cooling,* and *heating with circulating heat carriers* are all applications of basic unit operations and, as such, are treated in the encyclopedia as applications of fundamental principles of heat and mass transfer.

Low-Temperature Technology. Low-temperature systems and processes are a special application of heat and mass transfer and receive detailed treatment as *refrigeration technology* and *cryogenic technology.* These two topics have a long history of development, and much essential information on principles and practice is presented.

Vacuum Technology. The last section of unit operations deals with *vacuum technology.* Operations under vacuum are widely used in industrial processes. Consideration of this technology as a unit operation permits presentation of basic principles and equipment operation.

Taken together, the information provided by these two volumes of the encyclopedia on unit operations should give the reader the necessary background to understand and apply the principles to all types of industrial chemical processes. The information on the individual unit operations provides the necessary detail for that particular operation. A fully operational process is achieved when the unit operations are integrated into one smoothly functioning unit, which then produces the desired product under optimum conditions.

Max S. Peters
University
of Colorado
Boulder, Colorado
United States

Wolfgang Gerhartz
VCH Verlagsgesellschaft
Weinheim/Bergstraße
Federal Republic
of Germany

Contents

Cross References

* A related topic, Continuous Mixing of Fluids, is treated in Volume B4.

Symbols and Units

Symbols and units agree with SI standards (for conversion factors see pp. X–XI). The following list gives the most important symbols used in the encyclopedia. Articles with many specific units and symbols have a similar list as front matter.

Symbol	Unit	Physical Quantity
a_B		activity of substance B
A_r		relative atomic mass (atomic weight)
A	m^2	area
c_B	mol/m^3, mol/L (M)	concentration of substance B
C	C/V	electric capacity
c_p, c_v	$J\,kg^{-1}K^{-1}$	specific heat capacity
d	cm, m	diameter
d		relative density (ϱ/ϱ_{water})
D	m^2/s	diffusion coefficient
D	Gy (= J/kg)	absorbed dose
e		elementary charge
E	J	energy
E	V/m	electric field strength
E	V	electromotive force
E_A	J	activation energy
f		activity coefficient
F	C/mol	Faraday constant
F	N	force
g	m/s^2	acceleration due to gravity
G	J	Gibbs free energy
h	m	height
h	$W \cdot s^2$	Planck constant
H	J	enthalpy
I	A	electric current
I	cd	luminous intensity
k	(variable)	rate constant of a chemical reaction
k	J/K	Boltzmann constant
K	(variable)	equilibrium constant
l	m	length
m	g, kg, t	mass
M_r		relative molecular mass (molecular weight)
n_D^{20}		refractive index (sodium D-light, 20 °C)
n	mol	amount of substance
N_A	mol^{-1}	Avogadro constant (6.023×10^{23} mol^{-1})
p	Pa; bar *	pressure
Q	J	quantity of heat
r	m	radius
R	$J\,K^{-1}mol^{-1}$	gas constant
R	Ω	electric resistance
S	J/K	entropy
t	s, min, h, d, month, a	time
t	°C	temperature
T	K	absolute temperature
u	m/s	velocity

* The official unit of pressure is the Pascal (Pa).

Symbols and units (continued from p. IX)

Symbol	Unit	Physical Quantity
U	V	electric potential
U	J	internal energy
V	m^3, L, mL	volume
w		mass fraction
W	J	work
x_B		mole fraction of substance B
α		cubic expansion coefficient
α	$W m^{-2} K^{-1}$	heat-transfer coefficient (heat-transfer number)
α		degree of dissociation of electrolyte
$[\alpha]$	$10^{-2} deg\ cm^2 g^{-1}$	specific rotation
η	Pa · s	dynamic viscosity
θ	°C	temperature
$\varkappa$		c_p/c_v
λ	$W m^{-1} K^{-1}$	thermal conductivity
λ	nm, m	wavelength
μ		chemical potential
ν	Hz; s^{-1}	frequency
ν	m^2/s	kinematic viscosity (η/ϱ)
π	Pa	osmotic pressure
ϱ	g/cm^3	density
σ	N/m	surface tension
τ	Pa (N/m^2)	shear stress
φ		volume fraction
χ	Pa^{-1} (m^2/N)	compressibility

Conversion Factors

SI unit	Non-SI unit	From SI to non-SI multiply by
Mass		
kg	pound (avoirdupois)	2.205
kg	ton (long)	9.842×10^{-4}
kg	ton (short)	1.102×10^{-3}
Volume		
m^3	cubic inch	6.102×10^4
m^3	cubic foot	35.315
m^3	gallon (U.S., liquid)	2.642×10^2
m^3	gallon (Imperial)	2.200×10^2
Temperature		
°C	°F	°C × 1.8 + 32
Force		
N	dyne	1.0×10^5

Conversion factors (continued from p. IX)

SI unit	Non-SI unit	From SI to non-SI multiply by
Energy, Work		
J	Btu (int.)	9.480×10^{-4}
J	cal (int.)	2.389×10^{-1}
J	eV	6.242×10^{18}
J	erg	1.0×10^{7}
J	kW · h	2.778×10^{-7}
J	kp · m	1.020×10^{-1}
Pressure		
MPa	at	10.20
MPa	atm	9.869
MPa	bar	10
kPa	mbar	10
kPa	mm Hg	7.502
kPa	psi	0.145
kPa	torr	7.502

Prefixes for Powers of Ten

T (tera) 10^{12}	k (kilo) 10^{3}	d (deci) 10^{-1}	μ (micro) 10^{-6}
G (giga) 10^{9}	h (hecto) 10^{2}	c (centi) 10^{-2}	n (nano) 10^{-9}
M (mega) 10^{6}		m (milli) 10^{-3}	p (pico) 10^{-12}

Abbreviations

The following is a list of the abbreviations used in the text. Common terms, the names of publications and institutions, and legal agreements are included along with their full identities. Other abbreviations will be defined wherever they first occur in an article. For further abbreviations, see p. IX (Symbols and Units), p. XVI (Companies and Country Codes in Patent References). The names of periodical publications are abbreviated exactly as done by Chemical Abstracts Service.

abs. absolute
a.c. alternating current
ACGIH American Conference of Governmental Industrial Hygienists
ACS American Chemical Society
ADI acceptable daily intake
ADN accord européen relatif au transport international des marchandises dangereuses par voie de navigation interieure (European agreement concerning the international transportation of dangerous goods by inland waterways)
ADNR ADN par le Rhin (regulation concerning the transportation of dangerous goods on the Rhine and all national waterways of the countries concerned)
ADR accord européen relatif au transport international des marchandises dangereuses par route (European agreement concerning the international transportation of dangerous goods by road)
AEC Atomic Energy Commission (United States)
AIChE American Institute of Chemical Engineers
AIME American Institute of Mining, Metallurgical, and Petroleum Engineers

APhA	American Pharmaceutical Association
ASTM	American Society for Testing and Materials
BAM	Bundesanstalt für Materialprüfung (Federal Republic of Germany)
BAT	Biologischer Arbeitsstoff-Toleranzwert (biological tolerance value for a working material, established by MAK commission, see MAK)
Beilstein	Beilstein's Handbook of Organic Chemistry, Springer, Berlin-Heidelberg-New York
BET	Brunauer-Emmett-Teller
BGBl.	Bundesgesetzblatt (Federal Republic of Germany)
BIOS	British Intelligence Objectives Subcommitee Report (see also FIAT)
BOD	biological oxygen demand
bp	boiling point
B.P.	British Pharmacopeia
BS	British Standard
ca.	circa
calcd.	calculated
CAS	Chemical Abstracts Service
cat.	catalyst; catalyzed
cf.	compare
CFR	Code of Federal Regulations (United States)
Chap.	chapter
ChemG	Chemikaliengesetz (Federal Republic of Germany)
C.I.	Colour Index
CIOS	Combined Intelligence Objectives Subcommitee Report (see also FIAT)
CNS	Central Nervous System
Co.	Company
COD	chemical oxygen demand
conc.	concentrated
const.	constant
Corp.	Corporation
crit.	critical
CTFA	The Cosmetic, Toiletry and Fragrance Association (United States)
DAB 9	Deutsches Arzneibuch, 9th ed., Deutscher Apotheker-Verlag, Stuttgart 1986
d.c.	direct current
decomp.	decompose, decomposition
DFG	Deutsche Forschungsgemeinschaft (German Science Foundation)
dil.	dilute, diluted
DIN	Deutsche Industrie Norm (Federal Republic of Germany)
DOE	Department of Energy (United States)
DOT	Department of Transportation – Materials Transportation Bureau (United States)
DTA	differential thermal analysis
ed.	editor, editors, edition, edited
EEC	European Economic Community
e.g.	for example
emf	electromotive force
EPA	Environmental Protection Agency (United States)
EPR	electron paramagnetic resonance
Eq.	equation
ESCA	electron spectroscopy for chemical analysis
esp.	especially
ESR	electron spin resonance
Et	ethyl substituent ($-C_2H_5$)
et al.	and others
etc.	et cetera
EVO	Eisenbahnverkehrsordnung (Federal Republic of Germany)
exp (...)	$e^{(...)}$, mathematical exponent
FAO	Food and Agriculture Organization (United Nations)
FDA	Food and Drug Administration (United States)
FD & C	Food, Drug and Cosmetic Act (United States)
FHSA	Federal Hazardous Substances Act (United States)
FIAT	Field Information Agency, Technical (United States reports on the chemical industry in Germany, 1945)
Fig.	figure
fp	freezing point
Friedländer	P. Friedländer, Fortschritte der Teerfarbenfabrikation und verwandter Industriezweige, Vol. 1–25, Springer, Berlin 1888–1942
FT	Fourier transform
(g)	gas, gaseous
GC	gas chromatography
GGVE	Verordnung in der Bundesrepublik Deutschland über die Beförderung gefährlicher Güter mit der Eisenbahn (regulation in the Federal Republic of Germany concerning the transportation of dangerous goods by rail)

GGVS Verordnung in der Bundesrepublik Deutschland über die Beförderung gefährlicher Güter auf der Straße (regulation in the Federal Republic of Germany concerning the transportation of dangerous goods by road)
GGVSee Verordnung in der Bundesrepublik Deutschland über die Beförderung gefährlicher Güter mit Seeschiffen (regulation in the Federal Republic of Germany concerning the transportation of dangerous goods by sea-going vessels)
GLC gas-liquid chromatography
Gmelin Gmelin's Handbook of Inorganic Chemistry, 8th ed., Springer, Berlin-Heidelberg-New York
GRAS generally recognized as safe
Hal halogen substituent (−F, −Cl, −Br, −I)
Houben-Weyl Methoden der organischen Chemie, 4th ed., Georg Thieme Verlag, Stuttgart
HPLC high performance liquid chromatography
IARC International Agency for Research on Cancer, Lyon, France
IAEA International Atomic Energy Agency
IATA-DGR International Air Transport Association, Dangerous Goods Regulations
i.e. that is
i.m. intramuscular
IMDG International Maritime Dangerous Goods Code
IMO Inter-Governmental Maritime Consultive Organization (in the past: IMCO)
Inst. Institute
i.p. intraperitoneal
IR infrared
ISO International Organization for Standardization
IUPAC International Union of Pure and Applied Chemistry
i.v. intravenous
Kirk-Othmer Encyclopedia of Chemical Technology, 3rd ed., J. Wiley & Sons, New York-Chichester-Brisbane-Toronto 1978–1984
(l) liquid
Landolt-Börnstein Zahlenwerte u. Funktionen aus Physik, Chemie, Astronomie, Geophysik u. Technik, Springer, Heidelberg 1950–1980; Zahlenwerte und Funktionen aus Naturwissenschaften und Technik, Neue Serie, Springer, Heidelberg, since 1961
LC_{50} lethal concentration
LCLo lowest published lethal concentration
LD_{50} lethal dose
LDLo lowest published lethal dose
ln logarithm (base e)
LNG liquefied natural gas
log logarithm (base 10)
LPG liquefied petroleum gas
M mol/L
M metal (in chemical formulas)
MAK Maximale Arbeitsplatz-Konzentration (maximum concentration at the workplace in the Federal Republic of Germany); cf. Deutsche Forschungsgemeinschaft (ed.): Maximale Arbeitsplatzkonzentrationen (MAK) und Biologische Arbeitsstoff-Toleranzwerte (BAT), VCH Verlagsgesellschaft, Weinheim (published annually)
max. maximum
MCA Manufacturing Chemists Association (United States)
Me methyl substituent ($-CH_3$)
Methodicum Chimicum Methodicum Chimicum, Georg Thieme Verlag, Stuttgart
MIK maximale Immissionskonzentration (maximum immission concentration)
min. minimum
mp melting point
MS mass spectrum, mass spectrometry
NAS National Academy of Sciences (United States)
NASA National Aeronautics and Space Administration (United States)
NBS National Bureau of Standards (United States)
NCTC National Collection of Type Cultures (United States)
NIH National Institutes of Health (United States)

NIOSH	National Institute for Occupational Safety and Health (United States)
NMR	nuclear magnetic resonance
no.	number
NRC	Nuclear Regulatory Commission (United States)
NRDC	National Research Development Corporation (United States)
NSC	National Service Center (United States)
NSF	National Science Foundation (United States)
NTSB	National Transportation Safety Board (United States)
OECD	Organization for Economic Cooperation and Development
OSHA	Occupational Safety and Health Administration (United States)
p., pp.	page, pages
Patty	G. D. Clayton, F. E. Clayton (ed.): Patty's Industrial Hygiene and Toxicology, 3rd ed., Wiley-Interscience, New York
PB report	Publication Board Report (U.S. Department of Commerce, Scientific and Industrial Reports)
PEL	permitted exposure limit
Ph	phenyl substituent ($-C_6H_5$)
Ph. Eur.	European Pharmacopoeia, 2nd. ed., Council of Europe, Strasbourg 1981 –
phr	part per hundred rubber (resin)
q. v.	which see (quod vide)
ref.	refer, reference
resp.	respectively
R_f	retention factor (TLC)
R. H.	relative humidity
RID	règlement international concernant le transport des marchandises dangereuses par chemin de fer (international convention concerning the transportation of dangerous goods by rail)
rpm	revolutions per minute
RTECS	Registry of Toxic Effects of Chemical Substances, edited by the National Institute of Occupational Safety and Health (United States)
(s)	solid
SAE	Society of Automotive Engineers (United States)
s.c.	subcutaneous
SI	International System of Units
SIMS	secondary ion mass spectrometry
STEL	Short Term Exposure Limit (see TLV)
STP	standard temperature and pressure (0° C, 101.325 kPa)
T_g	glass transition temperature
TA Luft	Technische Anleitung zur Reinhaltung der Luft (clean air regulation in Federal Republic of Germany)
TA Lärm	Technische Anleitung zum Schutz gegen Lärm (low noise regulation in Federal Republic of Germany)
tan	tangent
TDLo	lowest published toxic dose
TLC	thin layer chromatography
TLV	Threshold Limit Value (TWA and STEL); published annually by the American Conference of Governmental Industrial Hygienists (ACGIH), Cincinnati, Ohio
TOD	total oxygen demand
TRK	Technische Richtkonzentration (lowest technically feasible level)
TSCA	Toxic Substances Control Act (United States)
TWA	Time Weighted Average
Ullmann	Ullmanns Encyklopädie der Technischen Chemie, 4th ed., Verlag Chemie, Weinheim 1972–1984; 3rd ed., Urban und Schwarzenberg, München 1951–1970
USAEC	United States Atomic Energy Commission
USAN	United States Adopted Names
USD	United States Dispensatory
USDA	United States Department of Agriculture
U.S.P.	United States Pharmacopeia
UV	ultraviolet
UVV	Unfallverhütungsvorschriften der Berufsgenossenschaft (workplace safety regulations in the Federal Republic of Germany)
VbF	Verordnung in der Bundesrepublik Deutschland über die Errichtung und den Betrieb von Anlagen zur Lagerung, Abfüllung und Beförderung brennbarer Flüssigkeiten (regulation in the Federal Republic of Germany concerning the con-

	struction and operation of plants for storage, filling, and transportation of flammable liquids; classification according to the flash point of liquids, recently in accordance with the classification in the United States)
VDE	Verband Deutscher Elektroingenieure (Federal Republic of Germany)
VDI	Verein Deutscher Ingenieure (Federal Republic of Germany)
VO	Verordnung
vol	volume
vol.	volume (of a series of books)
vs.	versus
WHO	World Health Organization (United Nations)
Winnacker-Küchler	Chemische Technologie, Carl Hanser Verlag, München
wt	weight
$	U.S. dollar, unless otherwise stated

Abbreviations for the Names of Frequently Cited Companies

Air Products	Air Products and Chemicals
Akzo	Algemene Koninklijke Zout Organon
Alcoa	Aluminum Company of America
Allied	Allied Corporation
Amer. Cyanamid	American Cyanamid Company
BASF	BASF Aktiengesellschaft
Bayer	Bayer AG
BP	British Petroleum Company
Celanese	Celanese Corporation
Daicel	Daicel Chemical Industries
Dainippon	Dainippon Ink and Chemicals Inc.
Dow Chemical	The Dow Chemical Company
DSM	Dutch Staats Mijnen
Du Pont	E.I. du Pont de Nemours & Company
Exxon	Exxon Corporation
FMC	Food Machinery & Chemical Corporation
GAF	General Aniline & Film Corporation
W.R. Grace	W.R. Grace & Company
Hoechst	Hoechst Aktiengesellschaft
IBM	International Business Machines Corporation
ICI	Imperial Chemical Industries
INCO	International Nickel Company
3M	Minnesota Mining and Manufacturing Company
Mitsubishi Chemical	Mitsubishi Chemical Industries
Monsanto	Monsanto Company
Nippon Shokubai	Nippon Shokubai Kagaku Kogyo
PCUK	Pechiney Ugine Kuhlmann
PPG	Pittsburg Plate Glass Industries
Searle	G.D. Searle & Company
SKF	Smith Kline & French Laboratories
SNAM	Societá Nazionale Metandotti
Sohio	Standard Oil of Ohio
Stauffer	Stauffer Chemical Company
Sumitomo	Sumitomo Chemical Company
Toray	Toray Industries Inc.
UCB	Union Chimique Belge
Union Carbide	Union Carbide Corporation
UOP	Universal Oil Products Company
VEBA	Vereinigte Elektrizitäts- und Bergwerks-AG
Wacker	Wacker Chemie GmbH

Country Codes

The following list contains a selection of standard country codes used in the patent references.

AT	Austria
AU	Australia
BE	Belgium
BG	Bulgaria
BR	Brazil
CA	Canada
CH	Switzerland
CS	Czechoslovakia
DD	German Democratic Republic
DE	Federal Republic of Germany (and Germany before 1949)*
DK	Denmark
ES	Spain
FI	Finland
FR	France
GB	United Kingdom
GR	Greece
HU	Hungary
ID	Indonesia
IL	Israel
IT	Italy
JP	Japan*
LU	Luxembourg
MA	Morocco
NL	Netherlands*
NO	Norway
NZ	New Zealand
PL	Poland
PT	Portugal
SE	Sweden
SU	Soviet Union
US	United States of America
YU	Yugoslavia
ZA	South Africa
EP	European Patent Office*
WO	World Intellectual Property Organization

* For Europe, Federal Republic of Germany, Japan, and the Netherlands, the type of patent is specified: EP (patent), EP-A (application), DE (patent), DE-OS (Offenlegungsschrift), DE-AS (Auslegeschrift), JP (patent), JP-Kokai (Kokai tokkyo koho), NL (patent), and NL-A (application).

Periodic Table of Elements

1A ("European" group designation according to old IUPAC recommendation)
1 (group designation according to 1985 IUPAC proposal)
IA ("American" group designation, also used by the Chemical Abstracts Service until the end of 1986)

1A 1 IA	2A 2 IIA	3A 3 IIIB	4A 4 IVB	5A 5 VB	6A 6 VIB	7A 7 VIIB	8 8 VIII	8 9 VIII	8 10 VIII	1B 11 IB	2B 12 IIB	3B 13 IIIA	4B 14 IVA	5B 15 VA	6B 16 VIA	7B 17 VIIA	0 18 VIIIA
1.0079 $_{1}$**H**																	4.0026 $_{2}$**He**
6.941 $_{3}$**Li**	9.0122 $_{4}$**Be**											10.811 $_{5}$**B**	12.011 $_{6}$**C**	14.007 $_{7}$**N**	15.9994 $_{8}$**O**	18.998 $_{9}$**F**	20.180 $_{10}$**Ne**
22.990 $_{11}$**Na**	24.305 $_{12}$**Mg**											26.982 $_{13}$**Al**	28.086 $_{14}$**Si**	30.974 $_{15}$**P**	32.066 $_{16}$**S**	35.453 $_{17}$**Cl**	39.948 $_{18}$**Ar**
39.098 $_{19}$**K**	40.078 $_{20}$**Ca**	44.956 $_{21}$**Sc**	47.88 $_{22}$**Ti**	50.942 $_{23}$**V**	51.996 $_{24}$**Cr**	54.938 $_{25}$**Mn**	55.847 $_{26}$**Fe**	58.933 $_{27}$**Co**	58.69 $_{28}$**Ni**	63.546 $_{29}$**Cu**	65.39 $_{30}$**Zn**	69.723 $_{31}$**Ga**	72.61 $_{32}$**Ge**	74.922 $_{33}$**As**	78.96 $_{34}$**Se**	79.904 $_{35}$**Br**	83.80 $_{36}$**Kr**
85.468 $_{37}$**Rb**	87.62 $_{38}$**Sr**	88.906 $_{39}$**Y**	91.224 $_{40}$**Zr**	92.906 $_{41}$**Nb**	95.94 $_{42}$**Mo**	98.906 $_{43}$**Tc** *	101.07 $_{44}$**Ru**	102.91 $_{45}$**Rh**	106.42 $_{46}$**Pd**	107.87 $_{47}$**Ag**	112.41 $_{48}$**Cd**	114.82 $_{49}$**In**	118.71 $_{50}$**Sn**	121.75 $_{51}$**Sb**	127.60 $_{52}$**Te**	126.90 $_{53}$**I**	131.29 $_{54}$**Xe**
132.91 $_{55}$**Cs**	137.33 $_{56}$**Ba**		178.49 $_{72}$**Hf**	180.95 $_{73}$**Ta**	183.85 $_{74}$**W**	186.21 $_{75}$**Re**	190.2 $_{76}$**Os**	192.22 $_{77}$**Ir**	195.08 $_{78}$**Pt**	196.97 $_{79}$**Au**	200.59 $_{80}$**Hg**	204.38 $_{81}$**Tl**	207.2 $_{82}$**Pb**	208.98 $_{83}$**Bi**	208.98 $_{84}$**Po** *	209.99 $_{85}$**At** *	222.02 $_{86}$**Rn** *
223.02 $_{87}$**Fr** *	226.03 $_{88}$**Ra** *																

138.91 $_{57}$**La**	140.12 $_{58}$**Ce**	140.91 $_{59}$**Pr**	144.24 $_{60}$**Nd**	146.92 $_{61}$**Pm** *	150.36 $_{62}$**Sm**	151.97 $_{63}$**Eu**	157.25 $_{64}$**Gd**	158.93 $_{65}$**Tb**	162.50 $_{66}$**Dy**	164.93 $_{67}$**Ho**	167.26 $_{68}$**Er**	168.93 $_{69}$**Tm**	173.04 $_{70}$**Yb**	174.97 $_{71}$**Lu**
227.03 $_{89}$**Ac** *	232.04 $_{90}$**Th** *	231.04 $_{91}$**Pa** *	238.03 $_{92}$**U** *	237.05 $_{93}$**Np** *	244.06 $_{94}$**Pu** *	243.06 $_{95}$**Am** *	247.07 $_{96}$**Cm** *	247.07 $_{97}$**Bk** *	251.08 $_{98}$**Cf** *	252.08 $_{99}$**Es** *	257.10 $_{100}$**Fm** *	258.10 $_{101}$**Md** *	259.10 $_{102}$**No** *	260.11 $_{103}$**Lr** *

* Elements with unstable isotopes; mass of most important isotope given

1. Solids Technology, Introduction

KLAUS BORHO, FRIEDRICH RICHARD FAULHABER, REINHARD POLKE, PETER THOMA,
BASF Aktiengesellschaft, Ludwigshafen, Federal Republic of Germany

1. Objectives of Solids Technology

Solids technology is concerned with the production of disperse solid products. Characteristically, the properties of solid products depend not only on chemical composition but also on the state of dispersion. The production processes are composed of individual unit operations which include all necessary activities from preparation of the solid to filling of the product for sale. The articles on unit operations in volumes B 2 and B 3 of this encyclopedia deal with scientific principles, the types of apparatus used, and the scale-up methods of various operations.

This introduction is intended to show, on the basis of principles underlying unit operations, how to design a solids process. This involves optimizing the interaction of various unit operations to fulfill the objective of solids technology, namely the economical preparation of products having specific characteristics with regards to production and application.

The principles of solids technology were only developed in this century. According to these principles, the state of dispersion of a product is related to its properties. RUMPF in particular realized that a product's properties are closely related to its state of dispersion. This resulted in the concept of the "property function" in the scientific discussion of solids technology.

The production of disperse solid products is characterized by the generation and repeated modification of disperse states. A typical example of a solids processing chain includes the unit operations of (1) crystallization, (2) filtration, (3) drying, and (4) size conversion (Fig. 1 A) Every unit operation in the processing chain modifies the physical state of the product. The state changes from solution to suspension in crystallization (→ 3. Crystallization and Precipitation), from suspension to filter cake during filtration (→ 10. Filtration), from filter cake to powder or lumpy product on drying (→ 4. Drying of Solid Materials), and from powder to granulate or from lumpy product to powder by size conversion (→ 5. Size Reduction, → 7. Size Enlargement) (Fig. 1 B).

The state of dispersion of the solid changes during the process. A suspension contains individual particles or loose flakes which are considerably compacted in the course of filtration to form moist, loose agglomerates. Drying results in considerably stronger agglomerates, which can be formed into larger secondary agglomerates or pulverized at the size conversion stage (Fig. 1 B).

Production Properties. The state of dispersion affects production properties (Fig. 1 C). Thus, only specific processing operations may be appropriate, and the cost of their implementation depends on the required state of dispersion of the product. Typical production properties are:

crystallizability
filterability
washability

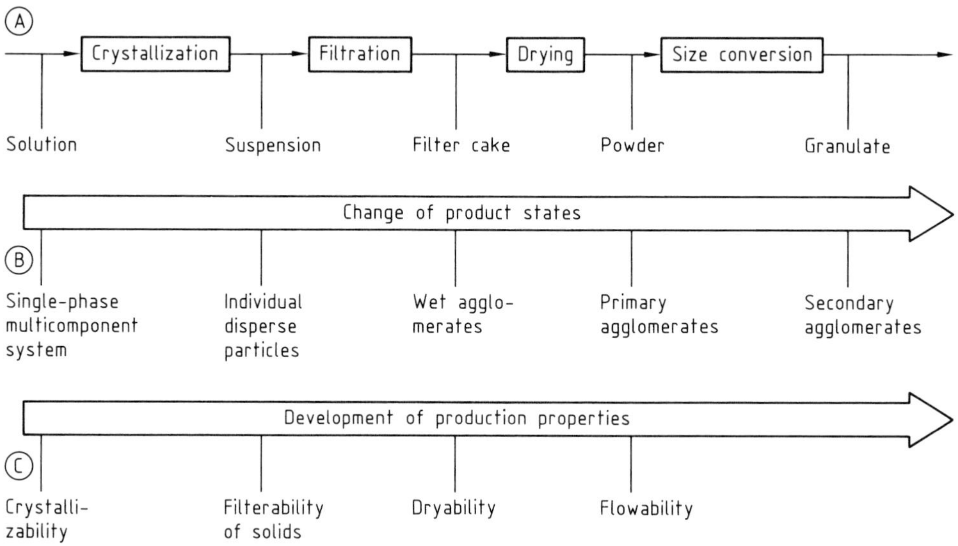

Figure 1. A) A typical solids technology processing chain; B) Change of product states during the processing chain; C) Development of production properties during the processing chain

dryability
suitability for pumping
bin storage capacity

Application Properties. Users of solid chemicals demand that the end products have clearly defined properties, called application properties, which depend on the state of dispersion of the material. Application properties, on which the performance of a dye, medicine, or fertilizer may depend, include:

dispersibility
dissolution rate
floating capacity
abrasion resistance
freedom from dust

Modifying the state of dispersion throughout the solids processing chain leads to a product with the desired application properties. However, the processing chain affects the application properties not only in predetermined ways but also in many cases, accidentally. Application properties become apparent only in the end product, although they originate at various stages throughout the production process.

The concept of production and application properties may seem somewhat abstract: it can be clarified by considering the following examples:

a tablet which does not disintegrate and, as a result, fails to release its active agent
a paint whose pigment adheres to the bottom of the can to form a solid layer
powdered, soluble coffee extract which forms lumps on the spoon
crystals that cannot be filtered
a caking powder that sticks to the inside walls of bins

Products with these properties clearly do not satisfy the requirements of either the process engineer or the end user.

2. Stages of Implementation

Figure 2 shows the various stages in the development of a process. In the *first stage*, the property profile is established, a concept of the relationship between product properties and dispersion values is formulated, and working hypotheses are deduced from the concept. In the *second stage* the working hypotheses are tested experimentally in the laboratory. In the *third stage*, unit operations are selected and designed to achieve an optimal processing chain. In the *fourth stage*, pilot-scale operations are carried out, if necessary, to ensure that the process has been designed correctly.

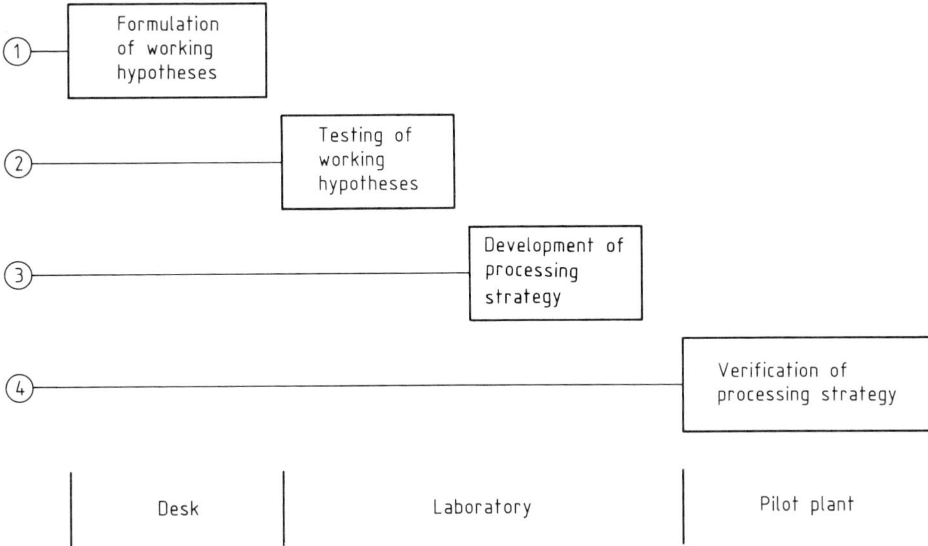

Figure 2. Four stages in process development and scale of experimental work

2.1. Formulation of Working Hypotheses

The formulation of working hypotheses involves several steps (Fig. 3). In developing a process, the first is to specify the objective, that is, the *profile of properties* of the end product. At first sight this may seem simple; however, in practice, the profile of properties is often difficult to finalize at such an early stage of process development because the required product properties represent—in almost all cases—a compromise between what is possible in terms of process technology and what is acceptable in terms of manufacturing costs. For this reason, the final profile of properties can only be established at a later stage.

If the profile of properties is available, step 2 involves formulating a *model* to indicate what the state of the end product must be if it is to display the required properties. This state, as described by physical properties and particularly by dispersion data, must be achieved by the processing operations. Often, several models are

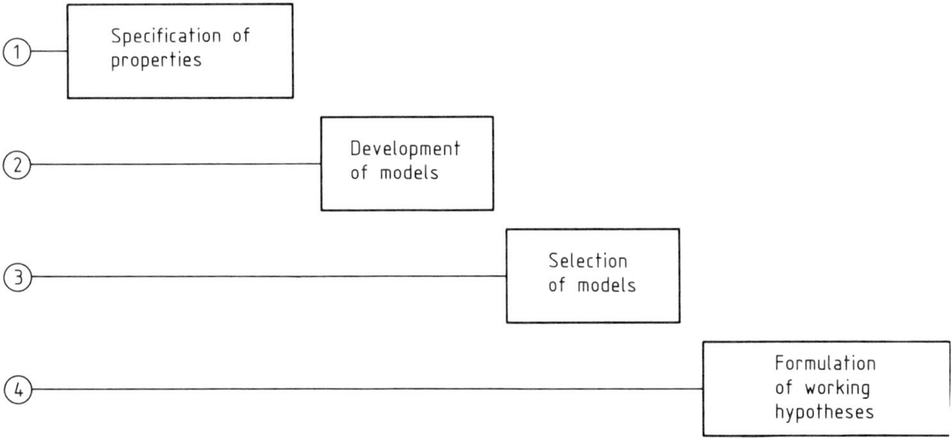

Figure 3. Four steps in the formulation of working hypotheses

worked out, along with a number of physically justified parameters that describe them.

Testing models by experimental measurement of all the parameters involved is far too expensive. Hence, those physical properties must be selected which are likey to exert most influence on the required product properties. In this way, various working hypotheses are formulated and then tested experimentally. The cycle (i.e., models, working hypothesis, test) may have to be repeated several times before a working hypothesis can be accepted.

The accepted hypothesis provides only a postulated connection between product properties and physical variables; it does not confirm the correctness of the underlying model. Indeed, the sequence of steps described is not intended to confirm models but to quickly establish parameters that are physically justified and provide a solution to the specific problem. The choice of a correct model is, therefore, of decisive importance. The process engineer must be trained in working out models and must be fully conversant with the models available. In practice, the sequence of steps can be shortened. In the most favorable case, the relationship between product and physical properties is known, or existing conceptual models can be used.

In the following example, a single, fixed product property is involved: initial laboratory testing of a new product shows that it flows very poorly. The flow properties must be improved, and the first action is to specify a profile of properties (step 1, Fig. 3).

Flow properties depend on the structure of the aggregate and the adhesion between particles. Therefore, a conceptual model for the first processing stage already exists (step 2, Fig. 3). Only the adhesion mechanisms can be affected by the process to a significant extent, and they may be classified in various groups:

1) Adhesion mechanisms without material bridges
   a) Van der Waals forces
   b) Electrostatic forces
   c) Mechanical interlocking

2) Adhesion mechanisms with solid bridges
   a) Crystallization of substances in solution
   b) Recrystallization of substances in solution
   d) Sintering
   d) Chemical reactions

3) Adhesion mechanisms with liquid bridges

In the following discussion, two specific adhesion mechanisms have been selected (step 3, Fig. 3): van der Waals forces (1 a) and crystallization of substances in solution (2 a). Physically relevant variables for these two mechanisms are listed below.

1 a) *Van der Waals forces*
   - Particle-size distribution
   - Particle form
   - Distance between particles
   - Interaction constant
   - Hardness

2 a) *Crystallization of substances in solution*
   - Particle-size distribution
   - Solubility of product and byproducts
   - Moisture transport

In step 4 (Fig. 3), the variables must be determined which can be influenced by the process and which control the required properties. Each of these variables can, if the assumption is correct, improve flow properties. This is a working hypothesis in accordance with step 4 of Figure 3. In this example, two working hypotheses emerge:

1) Poor flow properties are caused by van der Waals forces. By producing relatively large particles, the flow properties can be improved.

2) Poor flow properties are caused by interactions at the particle level resulting from the crystallization of dissolved impurities. Flow properties can be improved by reducing the concentration of byproducts.

The next stage must determine which hypotheses should be retained for further process development.

2.2. Testing of Working Hypotheses

In this stage, samples must be produced for experimental work, whose major parameters vary over a wide range. The required product properties should then be represented as a function of these physical variables. This presupposes, on the one hand, a test method that permits the production of such samples at a realistic cost and, on the other hand, efficient measurements capable of relating product properties to physical quantities. This stage can only be carried out in the laboratory because of the large number of samples required. Samples should not weigh more than 1 kg. The laboratory apparatus must set or modify physically relevant product variables. In this context, the laboratory apparatus need not be a scaled-down version of production equipment.

In the example, the first working hypothesis (van der Waals forces, particle size) can be tested by production of samples with different particle sizes. This can be achieved by screening or by controlling particle size during crystallization. In this way, a link between flow properties and particle size is established. The second working hypothesis (solid-substance bridges, byproducts) can be tested by washing the filter cake more thoroughly or by lowering the level of byproducts in the mother liquor.

As a result of these experiments, hypothesis 1 may be accepted and hypothesis 2 discarded. The studies carried out to test hypothesis 1 also yield information on the minimum particle size necessary to achieve the required flow properties.

2.3. Development of Processing Strategy

The unit operations which lead to the required particle-size distribution must now be determined. Some data are already available from experimental testing of the working hypotheses. The preparation of samples in the laboratory provides some information on production properties. The range of feasible technologies is theoretically very large. It can be decreased considerably by further measuring the values characteristic of specific substances and bearing in mind the boundary conditions. Enough data must be produced to ensure that the required product properties can be attained economically. The overall optimum is not necessarily the sum of individually optimized unit operations but the entire process chain must be optimized. This may require parallel study of several possible process chains or repeated implementation of the same process chain. In extreme cases new working hypotheses must be introduced or the objectives modified at this stage; i.e., the development stages already described must be repeated.

In addition to scaleup values, the data obtained must show how sensitive the process is when various unit operations are changed. A provisional process flow chart is eventually prepared.

Quick and economical testing methods are required. They must be sufficiently sophisticated to ensure that test products have the same properties as those produced industrially and that the values they yield permit accurate scaleup. This cannot be achieved on a pilot scale plant because of the time and cost involved. Therefore, these tasks must also be performed in the laboratory. To continue the example of improving poor flow properties: crystallization, filtration, and drying may constitute the most promising process chain. Better flow properties should be attained once the minimum particle size, previously determined, has been achieved during crystallization and then maintained during further process stages. The crystallization operation may involve intermittent or continuous cooling and vacuum or evaporation crystallization; these matters must be decided along with the components and size of the apparatus, and the way the system should be operated. This work yields data on the required particle-size distribution and permissible slurry concentration.

The remaining unit operations are developed along similar lines. In this example, product handling in and between process stages must be observed, because abrasion or size reduction may hinder the realization of the process objective.

Optimization involves the selection of the technology and process conditions for every unit operation so as to achieve the required flow characteristics reliably and economically. Sensitivity analysis is used to determine tolerance limits of the required operation states.

2.4. Verification of Processing Strategy

Not all process operations can be simulated reliably in laboratory experiments. Such operations can be examined on a pilot scale to ensure that the process has been designed correctly. This completes the actual process development.

3. Measurement Technology

Any development in solids technology requires effective measuring techniques. The variables to be measured can be divided into product properties and process parameters.

A discussion of *product properties* must include the *dispersion data* that describe these properties. Essential *process parameters* must be established if the plant is to operate in a repeatable and reliable manner; set values and tolerance limits must be determined for these parameters. In addition, *product-specific data* which characterize the product behavior within each unit are required for the selection and design of equipment.

When new processes are developed or when product properties deviate from set values, the demands on measurement technology are likely to be greater than those on production control, where measurement of auxiliary variables or production properties is often sufficient.

At present, in solids technology the armoury of sensors available for measuring both the product properties and the process parameters is rather limited. Some of these gaps can be closed by methods based on mathematical models. With these models, variables that are not amenable to direct measurement can be determined indirectly from more easily measured parameters. As computers become less expensive, these methods will increase in importance.

3.1. Product Properties

Properties of products from either unit operations or the entire process must meet predetermined requirements. Production and application properties summarize the effects of several variables. For example, the Jenike shear test assays flow capacity, and the measurement of activity determines the effectiveness of a catalyst.

Standardized test methods are used to determine certain product properties such as the strength of concrete, the color intensity of pigments, or the brightness of delustering agents. Other properties such as odor or settling behavior cannot be indicated by a metric scale. Frequently, application-specific measuring devices must be used to determine the same property for different products. For example, the dispersibility of tablets is assessed differently from the dispersibility of crop-protection agents or pigments.

3.2. Dispersion Data

Describing the disperse state, the most important variables are particle size and particle-size distribution (→ 2. Particle Size Analysis and Characterization of a Classification Process, p. **2**-4). They must be assessed as they occur in the process, i.e., particle-size distribution must not be changed by sampling, preparation, or measurement. For instance, in the segregation of particles from gases, measurements must be carried out with the particles suspended in the gas; if deposits were formed, they might agglomerate, whereas dispersion would destroy agglomerates.

3.3. Process Parameters

To design and verify a processing strategy, process parameters must be measured and maintained. Even simple measurements such as pressure or temperature can be difficult, for example, because of incrustation on the probes. For measuring more complicated parameters of systems charged with solids (e.g., viscosity, throughput, or concentration), only a few suitable sensors exist. For the reliable transfer of information from one stage to another, measurements carried out during the development phase should make clear which data are, or should be, available for quality assurance and with what accuracy.

3.4. Specific Data

The most important substance- or process-specific variables must be characterized for every unit to enable the selection of appropriate apparatus and operating parameters. In crystallization, for example, solubility and metastable zone width are the measured variables; in drying, sorption isotherms as well as softening and melting points must be determined. Safety-related and toxicological data are also required.

4. Outlook

Solids technology is a relatively new industrial science. It is based on a large number of fundamental and application-related studies. In addition, the research and development work is characterized by intensive interaction between research organizations in universities and in industry. University research is devoted primarily to determining fundamental relationships, whereas industrial studies are concerned mainly with the application of models evolved at the university. Constructive collaboration between academia and industry is the only possible way of obtaining further advances in solids technology. Many examples of such advances are available but are not discussed here for two reasons. Firstly, in view of the many research workers engaged in this field, selection of representative examples is extremely difficult. Secondly, detailed descriptions of the relevant operations are given in the following articles.

2. Particle Size Analysis and Characterization of a Classification Process

KURT LESCHONSKI, Technische Universität Clausthal, Clausthal, Federal Republic of Germany

In addition to the symbols defined in the front matter of this volume, the following symbols are defined:

a	acceleration
A	Cunningham factor
A	area
A_1	fine particles in the feed material (Eq. 49)
A_2	coarse particles in the feed material (Eq. 50)
A_3	fine particles in the coarse fraction (Eq. 51)
A_4	coarse particles in the coarse fraction (Eq. 52)
A_5	fine particles in the fine fraction (Eq. 53)
A_6	coarse particles in the fine fraction (Eq. 54)
C_V	volume concentration
C_N	number concentration
D	characteristic fractal dimension
E	expected value
E_M	Ecart of Mayer [22], [23]
E_T	Ecart Terra
f	focal length
g	gravity constant
H	height
I	imperfection

I intensity of light
I_0 intensity of incident light
k constant
k_S surface area shape factor
k_V volume shape factor
K_0 corrected cumulative distribution of the feed material
K_1 corrected cumulative distribution of the fine fraction
K_2 corrected cumulative distribution of the coarse fraction
L length
m mass
m_∞ mass of particles at time $t = \infty$
m spread parameter of the power function
m_0 mass of the feed material
m_1 mass of the fine fraction
m_2 mass of the coarse fraction
m_s mass of solids
m_t mass of particles at time t
$\dot{m}$ mass flow rate
$\dot{m}_0$ mass flow rate of the feed material
$\dot{m}_1$ mass flow rate of the fine fraction
$\dot{m}_2$ mass flow rate of the coarse fraction
n spread parameter of the RRSB function
n number of intervals
q density distribution
q_0 number density distribution
q_0 density distribution of the feed material
q_1 density distribution of the fine fraction
q_2 density distribution of the coarse fraction
Q cumulative distribution
Q_0 cumulative distribution of the feed material
Q_1 cumulative distribution of the fine fraction
Q_2 cumulative distribution of the coarse fraction
Q_3 mass or volume cumulative distribution
r radius
r_i radius of the liquid surface
R residue
Re Reynolds number
s standard deviation
s_g logarithmic standard deviation

S surface area
S_m mass-related surface area
S_V volume-related surface area
t first derivation of T
t student factor
t time
T selectivity curve
T_{tot} overall classification efficiency
u_p peripheral velocity
U perimeter
v_a velocity of air
v_r radial velocity
v_1 relative amount of the fine fraction, fine yield
v_2 relative amount of the coarse fraction, coarse yield
V volume
w settling rate
w_{at} centrifugal settling rate
w_g stationary settling rate
x' position parameter of the RRSB function
x_a analytical cut size
x_{min} minimum diameter
x_{min2} minimum diameter of the coarse fraction
x_{max} maximum diameter
x_{max1} maximum diameter of the fine fraction
x_O cut size with $1-Q_1(x_O) = Q_2(x_O)$
x_A equivalent projected area diameter
x_S equivalent surface area diameter
x_t cut size
x_V equivalent volume diameter
x_w equivalent settling rate diameter
x_{Am} equivalent medial position projected area diameter
x_{As} equivalent stable position projected area diameter
x_{50} median diameter
$x_{50,r}$ median of the $Q_r(x)$ distribution
z dimensionless variable of the error function

α angle
α dimensionless particle size $= \pi x/\lambda$
α weighted sum of variances
Δx interval of diameter
ε size-dependent error
η sharpness of cut parameter
η dynamic viscosity
η_i ordinate for evaluation of $v_1 = \eta_i/\xi_i$
Θ scattering angle
κ sharpness of cut
λ stride length
λ wavelength of light
λ mean free path
ξ integration variable for size
ξ_i abscissa for evaluation of $v_1 = Q_{1i} - Q_{2i}$
ϱ_d suspension density
ϱ_{fl} fluids density
ϱ_s solids density
σ^2 variance
τ dead flux
φ ratio of weighted density distributions
ψ sphericity
ω angular velocity

1. Introduction

Particle size analysis or particle characterization comprises the measurement and the quantitative description of physical properties of particulate matter. Therefore, it characterizes on the one hand the properties of single stationary or moving particles and on the other hand those of a bulk material.

This article describes the main areas of particle characterization, the representation of particle size analysis data, the characterization of a classification or separation process, and the methods of particle size analysis.

Currently, only a limited number of different measuring techniques dominate instrumentation; therefore, this article does not describe all possible techniques, but concentrates on the most important ones.

Any unit operation of particle technology changes certain physical properties of solid or liquid particles coarser than ca. 10^{-7} m. In heat and mass transfer operations, the state of the process is described adequately by average values of the molecular state. In all particle technology processes, however, the distribution of certain characteristic particle properties must be analyzed. The main task of particle size analysis is to measure these distributions.

Disperse or particulate systems consist of a finely divided particulate phase in a uniform carrier or dispersion medium [1]. The particulate phase and the carrier medium may be solid, liquid, or gaseous. Therefore, particle size analysis predominantly investigates such systems as suspensions, emulsions, aerosols, or aerodispersions. However, the particulate phase may also consist of bubbles or pores.

The range of particle sizes to be measured covers more than five to six orders of ten, with smallest particle sizes 10^{-9} to 10^{-7} m. Many physical properties of particulate matter, e.g., tendency to agglomerate, settling characteristics, solubility, and permeability, depend strongly on particle size.

2. Graphic Representation [2]

The individual elements of particulate matter always form a distribution, the elements of which can be classified according to their size. The distribution is, therefore, characterized by a quantity, e.g., the number or the mass of particles present in individual size classes. The elements can be solid particles, droplets, bubbles, or pores.

Many methods can be used to analyze particulate matter. Each group of measurement techniques based on the same physical principle provides, in general, different distribution curves. The reason for this must partly be attributed to the fact that in particle size analysis every possible physical principle has been used to determine these distributions. For several reasons, the results are generally not the same [3]. One of the main problems still encountered with most methods in use is their unknown absolute accuracy. However, apart from this the distributions differ because (1) different physical properties are used to characterize the "size" of individual particles, and (2) different types and different measures of quantity are chosen.

2.1. Particle Characteristics and Types and Measures of Quantity

In a graphic representation of particle size analysis data, the independent variable plotted on the abscissa describes the physical property chosen to characterize the size of the particles (Table 1). The dependent variable characterizes the type and measure of quantity and is plotted on the ordinate.

Different types of quantity are

- $r = 0$: number
- $r = 1$: length
- $r = 2$: area, surface area
- $r = 3$: volume, mass

Measures of quantity are

- Q_r: cumulative distribution
- q_r: density distribution

The density distribution is obtained from the cumulative distribution by differentiation:

$$q_r(x) = \frac{dQ_r(x)}{dx}$$

If particles having the same physical property are counted, the *type of quantity* used is a number; if the particles are weighed, the type of

Table 1. Types and measures of quantity

Type of quantity		Measure of quantity	
		density distribution	cumulative distribution
Number:	$r = 0$	$q_0(x)$	$Q_0(x)$
Length:	$r = 1$	$q_1(x)$	$Q_1(x)$
Area:	$r = 2$	$q_2(x)$	$Q_2(x)$
Volume:	$r = 3$	$q_3(x)$	$Q_3(x)$
General description:	$r = r$	$q_r(x)$	$Q_r(x)$

quantity is a mass. Particles laid end to end form a line of a certain length, but they also represent a certain projected area. Thus, length and area are other ways of quantifying particle size.

Therefore, the following types of quantity are distinguished: number, length, area, volume, or—at constant particle density—mass. This is the decisive difference from ordinary statistics, where the type of quantity that is used is always a number. Since number and volume distributions of the same powder may differ considerably, the type of quantity measured or chosen when representing the data must be taken into account.

However, a distinction must also be made between two *measures of quantity*. This is necessary because in a counting method either all particles smaller than a certain size may be counted, or only those found in certain size classes. The different relative amounts of particles measured in certain size intervals form the so-called *density distribution*, $q_r(x)$, which represents the first derivative of the *cumulative distribution*, $Q_r(x)$. In this notation x characterizes a physical property uniquely related to particle size. The subscript, r, indicates the type of quantity chosen.

Table 1 shows the types of distribution that can be distinguished:

The physical properties that characterize particles are as follows:

1) linear dimension
2) surface area, and projected area
3) volume
4) mass
5) settling rate
6) response of electrical, optical. or acoustical field

These are the one-, two-, or three-dimensional particle properties, such as linear dimensions, chords of different definition, the projected area of particles, their surface area, or their volume. Other characteristics are the mass of particles, their settling rate or terminal velocity in a viscous fluid, or the response of an electrical or an optical field to the presence of particles.

Each of these properties can be used to characterize the size of a particle. In principle, however, the properties can be applied usefully to the characterization of particles only if a unique relationship exists between the physical property used and a one-dimensional property unequivocally defining "size." With irregular particles, this can be achieved only approximately.

So-called *equivalent diameters* are defined, which are diameters of spheres that yield the same value of a certain physical property when analyzed under the same conditions as the irregularly-shaped particle. The relationship between the physical property chosen and the equivalent particle diameter must be either evaluated theoretically or measured experimentally.

Therefore, an equivalent diameter represents the diameter of a sphere that has the same physical property as the irregularly-shaped particle. One distinguishes, for instance, between

x_A: equivalent projected area diameter
x_S: equivalent surface area diameter
x_{Am}: equivalent projected area diameter for mean particle position
x_{As}: equivalent projected area diameter for stable particle position
x_V: equivalent volume diameter
x_w: equivalent settling rate diameter

For a given sphere, all equivalent diameters are the same; they differ from each other if the particles are irregularly-shaped. Under certain assumptions and after a shape factor has been defined that describes the deviation in shape from a sphere, the equivalent diameters given above can be intercorrelated using the shape factor *sphericity*, as defined by WADELL [4]:

$$\Psi = x_V^2 / x_S^2 \tag{1}$$

If Cauchy's theorem [5] is used and a force balance is applied to a particle settling in a viscous fluid under laminar conditions, the following is obtained:

$$\begin{aligned}
x_S &= x_{Am} &&= x_w \psi^{-3/4} = x_V \psi^{-1/2} \\
x_{Am} &= x_S &&= x_w \psi^{-3/4} = x_V \psi^{-1/2} \\
x_w &= x_S \psi^{3/4} = x_{Am} \psi^{3/4} &&= x_V \psi^{1/4} \\
x_V &= x_S \psi^{1/2} = x_{Am} \psi^{1/2} &&= x_w \psi^{-1/4}
\end{aligned}$$

In principle, the following inequality holds:

$$x_{As} > x_{Am} > x_S > x_V > x_w$$

2.2. Cumulative and Density Distributions

The graphic representation of particle size analysis data is shown in Figures 1–3. The term

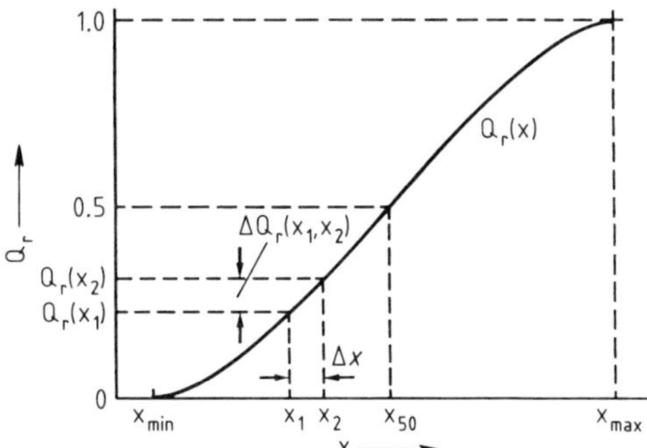

Figure 1. Cumulative distribution $Q_r(x)$

x represents an arbitrary equivalent diameter, the dimension of which is a length such as µm.

Figure 1 shows the normalized cumulative distribution $Q_r(x)$. Each individual point of $Q_r(x)$ specifies the relative amount of particles smaller than or equal to a certain equivalent diameter. The kind of quantity chosen is indicated by the subscript r. The normalized curve $Q_r(x)$ has values between 0 and 1:

$$Q_r(x = x_{\min}) = 0$$
$$Q_r(x = x_{\max}) = 1$$
(condition of normalization)

Figure 2 shows the normalized *histogram* of a density distribution, $\bar{q}_r(x_1, x_2)$. The difference ΔQ_r, of two values of the cumulative distribution, $Q_r(x)$, between two equivalent diameters x_1 and $x_2 = x_1 + \Delta x$ is represented by

$$\Delta Q_r(x_1, x_2) = Q_r(x_2) - Q_r(x_1) \tag{2}$$

The histogram can then be obtained from

$$\bar{q}_r(x_1, x_2) = \frac{\Delta Q_r(x_1, x_2)}{\Delta x} = \frac{Q_r(x_2) - Q_r(x_1)}{x_2 - x_1} \tag{3}$$

The value $\bar{q}_r(x_1, x_2)$ then represents the constant height of the rectangular shaded area of Figures 2 and 3 of width Δx. Therefore, the shaded area is given by

$$\bar{q}_r(x_1, x_2) \cdot \Delta x = \Delta Q_r(x_1, x_2) \tag{4}$$

and represents the relative amount of particles within the size interval, Δx.

If the curve $Q_r(x)$, as plotted in Figure 1, can be differentiated, the density distribution, $q_r(x)$, is obtained from

$$q_r(x) = \frac{\mathrm{d}Q_r(x)}{\mathrm{d}x} \tag{5}$$

This function, $q_r(x)$, is plotted in Figure 3. It represents the first derivative or the slope of $Q_r(x)$. It should be noted that $q_r(x)$ has the dimension of an inverse length. Irrespective of the type of quantity plotted, the overall area underneath $q_r(x)$ equals unity or 100% (condition of normalization).

$$\int_{x_{\min}}^{x_{\max}} q_r(x)\,\mathrm{d}x = Q_r(x_{\max}) - Q_r(x_{\min}) = 1 \tag{6}$$

Each of the three curves shown in Figures 1–3 represents the information needed for further evaluation of the data. If the cumulative distribution, $Q_r(x)$, has been obtained by multiple analyses, statistical methods allow the most probable distribution curve to be found. If certain assumptions with regard to an analytical function that fits $Q_r(x)$ are possible, regression analysis using least squares methods yield the most probable parameters of the analytical function. The main problems in evaluating more data

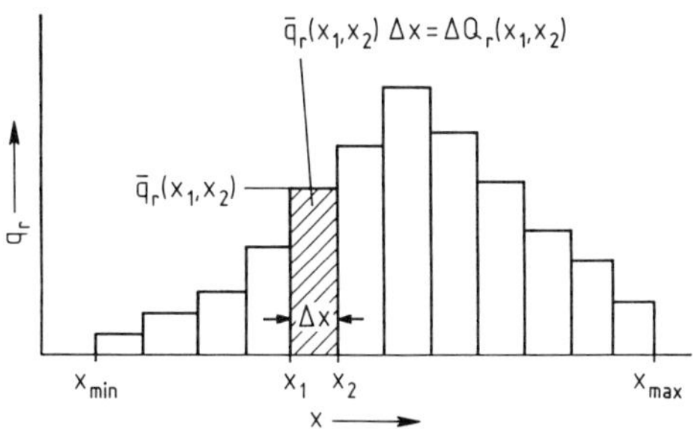

Figure 2. Histogram of density distribution $q_r(x)$

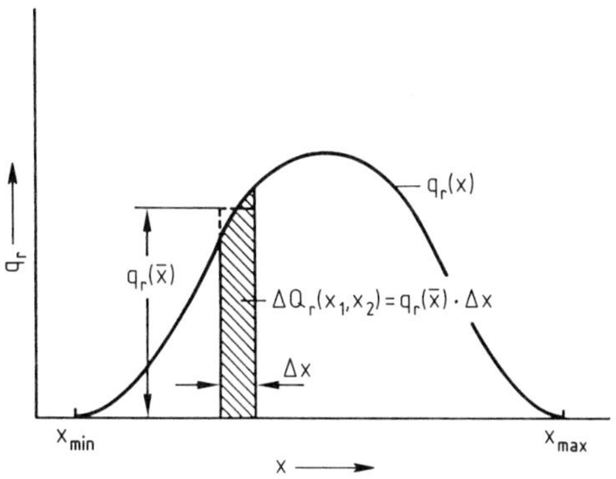

Figure 3. Density distribution $q_r(x)$

from these curves originate mainly from two sources:

1) The distribution curve, as plotted, has been analyzed with several methods of size analysis using different types of quantity and different physical properties defining "size." Therefore, the values plotted on both the ordinate and the abscissa must be recalculated and reduced to the same form. It may be necessary, for instance, to transform a number distribution of volumes into a volume distribution of equivalent sizes.

2) The distribution curves obtained must be correlated to other material properties, such as strength, solubility, or pore volume. Usually one figure, or at the most two figures, must then represent the size distribution adequately. The problem of defining an average particle size (a position parameter) and another parameter describing the spread of the distribution (a spread parameter) then arises. Some of the existing possibilities are discussed in [2].

2.3. Distribution Functions

Several analytical particle size distributions have been found to approximately describe empirically-determined particle size distributions. Special graph papers have been developed in which cumulative distributions following these functions form a straight line. Three of the best known analytical distribution functions are the Gaudin–Schumann distribution, the RRSB function, and the Gaussian distribution.

1) The power function

$$Q_3(x) = (x/x_{\max})^m \tag{7}$$

is also known as the *Gaudin–Schumann distribution* [6], [7]. It contains a position parameter, $x_{\max}$, and a spread parameter, m. The latter parameter indicates the slope of the straight line, which is obtained when the above function is plotted on a log Q_3–log x paper.

2) The *Rosin, Rammler, Sperling, Bennett (RRSB) function* has the following equation [8]–[12]:

$$1 - Q_3(x) = \exp(-(x/x')^n) \tag{8}$$

x' represents the position parameter where the function $1 - Q_3(x') = e^{-1} = 0.368$. The spread parameter, n, represents the slope of the straight line obtained in a special graph paper when $\log\log(1/R)$ is plotted on the ordinate and $\log x$ is plotted on the abscissa. The straight line is represented by

$$\log\log(1/R) = n\log x - n\log x' + \log\log e \tag{9}$$

3) The third analytical function is the *normal distribution* or *Gaussian distribution*, given here in its general form [11]:

$$q_r^*(z) = \frac{1}{\sqrt{2\pi}} \exp(-z^2/2) \tag{10}$$

with z being a dimensionless variable describing particle size. The cumulative distribution is obtained from

$$Q_r^*(z) = \int_{-\infty}^{z} q_r^*(\xi)\,\mathrm{d}\xi \tag{11}$$

Normal distributions differ by the definition of z and the distinction between (1) the Gaussian distribution with linear abszissa and (2) the logarithmic normal distribution.

In the *Gaussian distribution with linear abscissa*, z is defined as

$$z = (x - x_{50,r})/s \tag{12}$$

where $x_{50,r}$ is the median of the $Q_r(x)$ distribution and s represents the standard deviation.

The quantity $Q_r^*(z)$ is tabulated for certain z values; therefore, the standard deviation, s, can be calculated from

$$z = 1{:} \quad Q_r^*(1) = 0.841; \quad s = x_{84,r} - x_{50,r} \tag{13}$$

$$z = -1{:} \quad Q_r^*(-1) = 0.159; \quad s = -(x_{16,r} - x_{50,r}) \tag{14}$$

In the logarithmic normal distribution, the following substitution is made:

$$z = \ln(x/x_{50,r})/s \tag{15}$$

or according to Herdan [11]:

$$z = \log(x/x_{50,r})/\log s_g = \ln(x/x_{50,r})/\ln s_g \tag{16}$$

By comparison of Equations (15) and (16), Equation (17) is obtained:

$$s = \ln s_g \quad \text{or} \quad s_g = e^s \tag{17}$$

The standard deviation can be calculated from

$$s = \ln(x_{84,r}/x_{50,r}) = \ln(x_{50,r}/x_{16,r}) = \ln s_g \tag{18}$$

2.4. Calculation of Specific Surface Area

Specific surface area may be defined in two ways:

1) The ratio of surface area, S, to volume, V, is defined as the volume-related surface area, S_V:

$$S_V = S/V \tag{19}$$

2) The ratio of surface area, S, to mass, m, is defined as the mass-related surface area, S_m:

$$S_m = S/m \tag{20}$$

S_m and S_V differ by the solids density, ϱ_s:

$$S_V = S_m \varrho_s \tag{21}$$

The volume-related surface area S_V of particles of a size interval $\Delta x = x_2 - x_1$ can be calculated from a number distribution $q_0(x)$:

$$S_V(x_1, x_2) = \int_{x_1}^{x_2} k_S(x)\, x^2 q_0(x)\,\mathrm{d}x \Big/ \int_{x_1}^{x_2} k_V(x)\, x^3 q_0(x)\,\mathrm{d}x \tag{22}$$

where $k_S(x)$ = surface area shape factor and $k_V(x)$ = volume shape factor

In general, k_S and k_V are assumed to be independent of size, and a new shape factor, f, is introduced according to HEYWOOD which takes into account deviations in shape from spherical particles [13]. If Equation (23) is introduced

$$6f = k_S(x)/k_V(x), \tag{23}$$

Equation (24) is obtained:

$$S_V(x_1, x_2) = 6f \frac{\int_{x_1}^{x_2} x^2 q_0(x)\,\mathrm{d}x}{\int_{x_1}^{x_2} x^3 q_0(x)\,\mathrm{d}x} \tag{24}$$

3. Characterization of a Classification or Separation Process [14], [15]

Classification is a basic unit operation of particle technology. As shown in Figure 4, the feed material consists of particular matter of size, settling rate, density, or shape distribution $q_0(\xi)$; this is separated in a classification process into a fine or light $q_1(\xi)$ and coarse or heavy fraction $q_2(\xi)$.

The subscripts 0, 1, and 2 should not be mistaken for the subcripts used in Chapter 2. Actually $q_{r,0}(\xi)$ or $q_{r,1}(\xi)$ should have been employed; however, because the quantity used in the calculation of the mass balances of Section 3.1.2 is always assumed to be the same, the first subscript, r, has been omitted.

The parameter ξ represents any kind of physical property that is uniquely correlated to the "size" of a particle. The "size" of a particle is usually represented by the diameter of a sphere equivalent in physical behavior to the irregularly-shaped particle (equivalent diameter; see Section 2.2). The equivalent spherical diameter is called x in this and the following chapters. The parameter ξ may, however, also represent a settling rate in a gas or a liquid, a density, or even a shape-related piece of information.

According to Figure 4, the feed material is separated into at least two parts, with one part mainly containing particles below a specific value, and the other part containing particles above that value. Irrespective of the classification methods used, the classification itself can be described completely by using the density and/or the cumulative distribution curves of the feed material and the product fractions [15]. Therefore, it is important to determine these curves as accurately as possible. In Section 3.1 the characterization of a classification is described under the assumption that the distribution curves de-

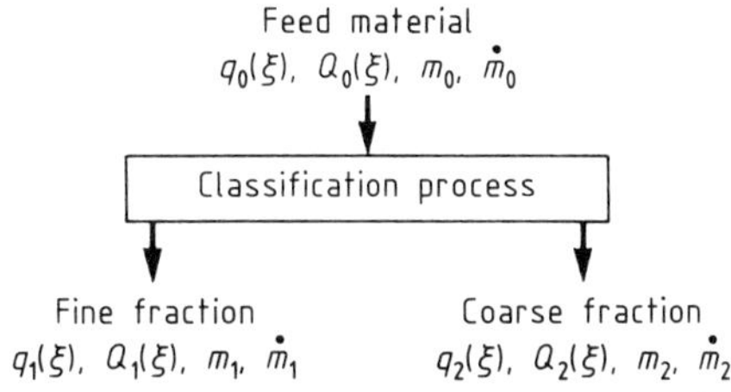

Figure 4. Fraction and distributions produced in a one-step classification process

scribing the feed material and the products, as well as the overall mass balance, are free of errors. This is the traditional approach; it has been applied in the way described throughout the world for many years. Only recently, the errors involved in particle size analysis have been included in the characterization of a classification process by LESCHONSKI and HERMANN [16]; Section 3.2 describes this method and its improvements.

Section 3.3 deals with systematic changes of the feed material in the classifier, for instance, due to comminution of the feed material in the classification zone. Systematic changes lead to abnormal grade efficiency or classifier selectivity curves, $T(x)$, which no longer extend between 0 and 1.

The general description of a classification process, based on error-free distribution curves and an error-free overall mass balance, was first attempted in the processing of minerals or coal; it has been the subject of many papers [17]. In the following chapters, the description of a separation process is explained by using classification as an example. However, the same curves and equations are valid for separations based on density, shape, or any other, e.g., size-dependent physical property.

In classification processes, the characteristic physical property of a single particle is described by its equivalent settling rate diameter, x_w, plotted on the abscissa. It has the dimension of a length and is usually given in 10^{-6} m.

3.1. Classification Based on Error-Free Distribution Curves and Mass Balances

3.1.1. Density and Cumulative Distribution Curves

In all classification processes a given feed material (subscript 0) is classified into at least two parts, which are called the fine (subscript 1) and the coarse (subscript 2) fractions. If an ideal classification were possible, the fine fraction would contain particles below or equal to a certain size, x_t, the so-called *cut size*, and the coarse fraction would contain all particles above that size. Such a classification process is shown in Figure 5.

The shaded areas beneath the weighted density distributions of the fine and the coarse product represent the overall relative amount of the

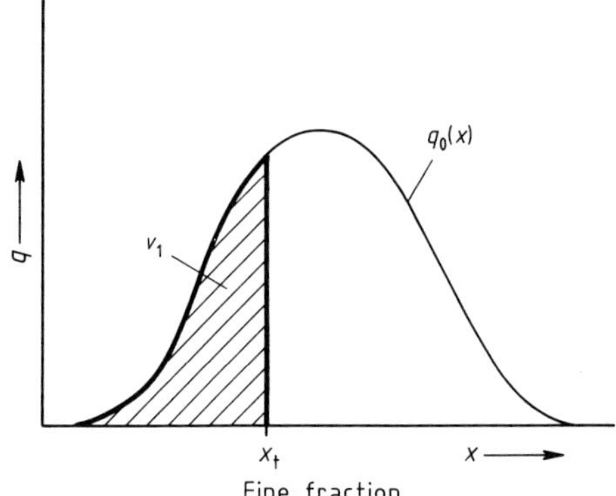

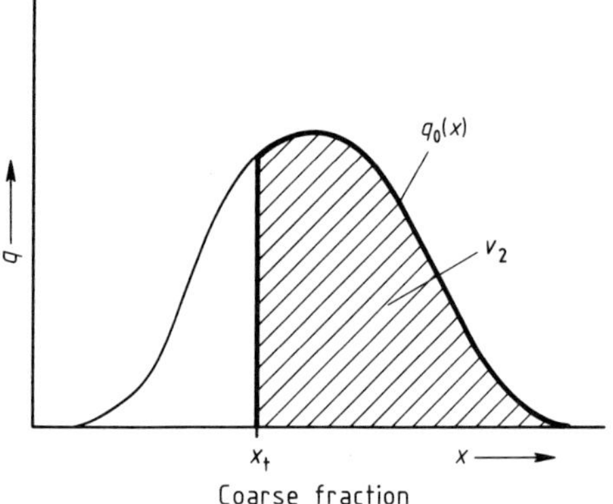

Figure 5. Weighted density distributions of feed, fine, and coarse fractions of an ideal classification

fine fraction, v_1, also known as *fine yield*, and of the coarse fraction, v_2, also known as *coarse yield*, the sum of which equals 100% or 1.

In reality, however, in a certain range of sizes ($x_{\min 2} \leq x \leq x_{\max 1}$), particles of the same size, x, are present in both the fine and the coarse fraction, as shown in Figure 6.

If the density distribution curves of the fine, $q_1(x)$, and the coarse fraction, $q_2(x)$, are multiplied by their relative amounts, v_1 and v_2 (see Eqs. 27 and 28), the weighted density distribution curves $v_1 q_1(x)$ and $v_2 q_2(x)$ are obtained. These curves overlap and intersect each other in the size range

$$x_{\min 2} \leq x \leq x_{\max 1}$$

The point of intersection is attributed to the so-called cut size, x_t. The particles below and above x_t in the coarse or fine fraction, respectively, have been incorrectly classified. These parti-

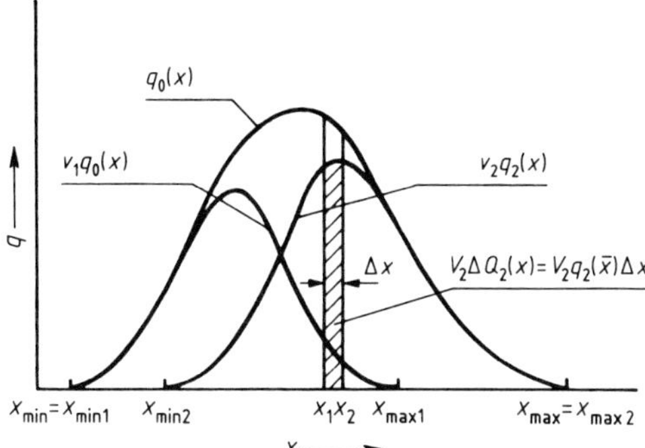

Figure 6. Weighted density distributions of the feed material $q_0(x)$, the fine fraction $[v_1\, q_1(x)]$, and the coarse fraction $(v_2 q_2(x))$

cles are commonly called *misplaced particles* or *misplaced material*. The less precise a classification process has been, the greater the areas representing the misplaced materials are. The amount of misplaced material in the coarse fraction is represented by the integral:

$$v_2 \int_{x_{min2}}^{x_t} q_2(x)\,dx = v_2\,Q_2(x_t) \qquad (25)$$

The amount of misplaced material in the fine fraction can be calculated from Equation (26):

$$v_1 \int_{x_t}^{x_{max1}} q_1(x)\,dx = v_1\,Q(x_{max1}) - v_1 Q_1(x_t) = v_1[1 - Q_1(x_t)] \qquad (26)$$

The classification, as described, can also be represented by using the cumulative distribution curves $Q_0(x)$, $Q_1(x)$, and $Q_2(x)$, as shown in Figure 7.

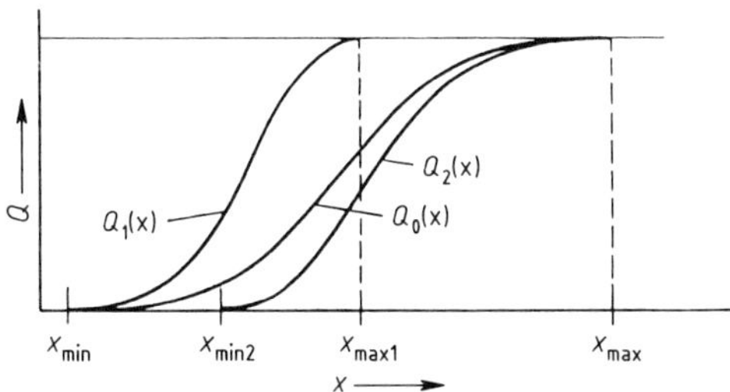

Figure 7. Cumulative distribution curves of the feed material $Q_0(x)$ and the fine fraction $Q_1(x)$ and coarse fraction $Q_2(x)$

3.1.2. Mass Balances

In the representation of density distribution curves, the relative mass is often used as the quantity plotted on the ordinate. In principle, other quantities such as number, length, area, or volume may be used, without change in the grade efficiency or classifier selectivity curves. Therefore, the distribution curves and all other quantities are based on particle mass. It should be repeated that in the equations to follow the density and cumulative distribution curves are assumed to be free of errors.

Mass Balance in the Size Range from x_{min} to x_{max}. In the classification process, the mass, m_0, or the mass flow rate, $\dot{m}_0$, of the feed material is split into the mass, m_1, or the mass flow rate, $\dot{m}_1$, of the fine material and the mass, m_2, or the mass flow rate, $\dot{m}_2$, of the coarse material.

Therefore, the relative amount of fine or coarse material, v_1 and v_2, is obtained from

$$v_1 = m_1/m_0 = \dot{m}_1/\dot{m}_0 \qquad (27)$$
$$v_2 = m_2/m_0 = \dot{m}_2/\dot{m}_0 \qquad (28)$$

Because of

$$m_0 = m_1 + m_2 \quad \text{or} \quad \dot{m}_0 = \dot{m}_1 + \dot{m}_2 \qquad (29)$$

the following is obtained:

$$v_1 + v_2 = 1 \qquad (30)$$

In Figures 5 and 6, v_1 and v_2 are represented by the areas beneath the weighted density distribution curves of the fine $[v_1\, q_1(x)]$ and the coarse $[v_2\, q_2(x)]$ material. The area beneath the density distribution curve of the feed material $q_0(x)$ equals unity. For example:

$$\int_{x_{min}}^{x_{max1}} v_1\, q_1(x)\,dx = v_1[Q_1(x_{max1}) - Q_1(x_{min})] = v_1 \qquad (31)$$

Mass Balance in a Small Size Range. In the size range $x_{min\,2} \le x \le x_{max1}$ the quantity

$$dQ_0(x) = q_0(x)\,dx \qquad (32)$$

is split into two fractions:

$$dQ_0(x) = v_1\,dQ_1(x) + v_2\,dQ_2(x) \qquad (33)$$

Replacing $dQ(x)$ by $q(x)\,dx$ yields

$$q_0(x) = v_1\,q_1(x) + v_2\,q_2(x) \qquad (34)$$

If only two density distributions and the relative amount of the fine or the coarse material are known, Equation (34) may be used to determine the missing third density distribution or to check and control values of this density distribution, if also measured. Equation (34) is also used to construct the set of density distribution curves in Figure 6. For instance, if the density distributions of the feed and the coarse material have been chosen or measured, the density distribution of the fine material can be calculated or constructed from

$$v_1 q_1(x) = q_0(x) - v_2 q_2(x)$$

This must be taken into account when diagrams are plotted, as shown in Figure 6.

Mass Balance below a Certain Size. Integrating Equation (34) between x_{min} and x yields

$$Q_0(x) = v_1 Q_1(x) + v_2 Q_2(x) \tag{35}$$

3.1.3. Indirect Evaluation of v_1 and v_2

In many cases of practical application, v_1 and v_2 cannot be calculated from the relevant masses or mass flow rates because these values are not available, are difficult to measure, etc. However, if representative samples of the three product streams have been measured, Equation (30), (34), and (35) may be used to calculate v_1 or v_2. Introducing, for instance, Equation (30) into Equation (35) and solving with respect to v_1 yields

$$\begin{aligned} v_1 &= 1 - v_2 \\ &= [Q_0(x) - Q_2(x)]/[Q_1(x) - Q_2(x)] \end{aligned} \tag{36}$$

If the cumulative distributions $Q_0(x)$, $Q_1(x)$, and $Q_2(x)$ are free of errors (i.e., the mass balance according to Equation (36) leaves no remainder), then v_1 or v_2 will be constant and independent of size x. In reality, however, the calculated values of v_1 will vary due to errors of size analysis and sampling when measuring $Q_0(x)$, $Q_1(x)$, and $Q_2(x)$. In this case, average values of v_1 and v_2 must be calculated by using the procedures suggested in Section 3.2.1.

3.1.4. Cut Size

In principle, any value of x between x_{min1} and x_{max2} (Fig. 6) can be used as cut size, x_t. However, its definition should be correlated as closely with the field of application of the product as possible.

Reference Cut Size. In many cases of practical application, the grade efficiency curve (see Section 3.1.5) is not determined and the fine product is available for analysis only. Then a so-called reference cut size is defined which represents a particle size, x_y, at y% weight undersize [$Q_3(x_y) = y$%]. The value y is, for instance, taken to be 97% or even 99%. It should be noted that x_y does not in the true sense represent a cut size of a classifier because the other products and the relative amounts of the fine or the coarse material have not been taken into account.

Three definitions of cut size are commonly used: (1) the median of the grade efficiency curve x_{50}, (2) the analytical cut size x_a, and (3) the overlap cut size x_0.

Median of the Grade Efficiency Curve. In Figure 6 the weighted density distribution curves intersect at a certain size x_t. This is the cut size.

$$x_t = x_{50} \tag{37}$$

where

$$v_1 q_1(x_{50}) = v_2 q_2(x_{50}) \tag{38}$$

Particles of size x_{50} are equally present in the coarse and the fine fractions.

The grade efficiency curve, or classifier selectivity curve, $T(x)$, as defined in Section 3.1.5, has a value of 0.5 or 50% at this particular size. Therefore, the cut size, x_{50}, also represents the median value of the grade efficiency curve.

Analytical Cut Size. The analytical cut size x_a represents the value of x at which a vertical line

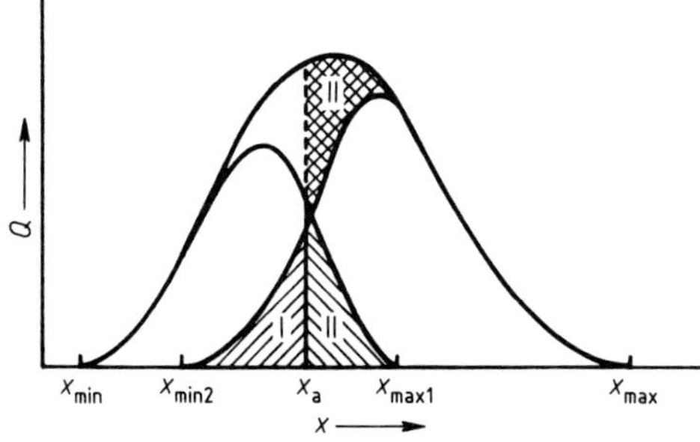

Figure 8. Definition and representation of the analytical cut size x_a

divides the shaded areas I and II in Figure 8 beneath the $v_1 q_1(x)$ and $v_2 q_2(x)$ curves into two equal parts. At this cut size, the coarse and fine material contains equal quantities of misplaced material.

The integration yields

$$v_2 \int_{x_{min2}}^{x_a} q_2(x)\,dx = v_1 \int_{x_a}^{x_{max1}} q_1(x)\,dx = \int_{x_a}^{x_{max1}} (q_0(x) - v_2 q_2(x))\,dx \quad (39)$$

or

$$v_1 = 1 - v_2 = Q_0(x_a) \quad (40)$$

An analytical air classifier can be seen as a black box, similar to the situation shown in Figure 4, operated under defined conditions. A certain mass, m_0, of the feed material is fed to the classifier, and the mass, m_2, of the coarse product only is quantitatively obtained. Therefore, it is possible to calculate $m_2/m_0 = v_2$.

The relative amount, v_2, of the coarse material as determined by the experiment is usually understood to equal the so-called residue R, that is, the amount of particles coarser than x_t. Equation (40) shows that the cut size x_t must equal the analytical cut size, x_a, if $R = v_2$. Then v_2 represents one point of the size distribution of the feed material to be measured.

Overlap Cut Size. The value x_0 is taken as the cut size of technical air classifiers, where v_1 or v_2 are unknown and where $Q_1(x)$ and $Q_2(x)$ can be obtained relatively easily. The value x_0 represents a cut size, x_t, where $Q_2(x_0)$ equals the oversize fraction of the fine product: $1 - Q_1(x_0)$

$$1 - Q_1(x_0) = Q_2(x_0) \quad (41)$$

The determination of x_0 is shown in Figure 9.

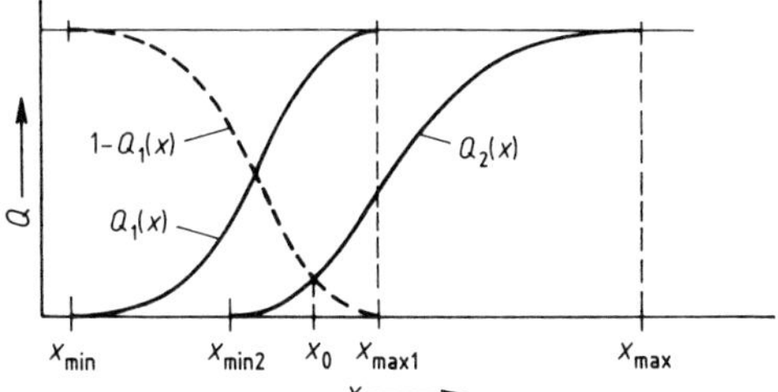

Figure 9. Determination of x_0

3.1.5. Grade Efficiency Curve (Tromp's Curve)

To describe the efficiency of a classification process, the so-called grade efficiency curve, $T(x)$, is normally derived from the density distribution curves of Figure 6 [17].

The *grade efficiency* or *classifier selectivity*, T, represents the ratio of the relative amount of material of a certain size present in the coarse material ($v_2 q_2(x)\,dx$), to the relative amount of the same size initially present in the feed material: $q_0(x)\,dx$. Therefore, the grade efficiency is equal to

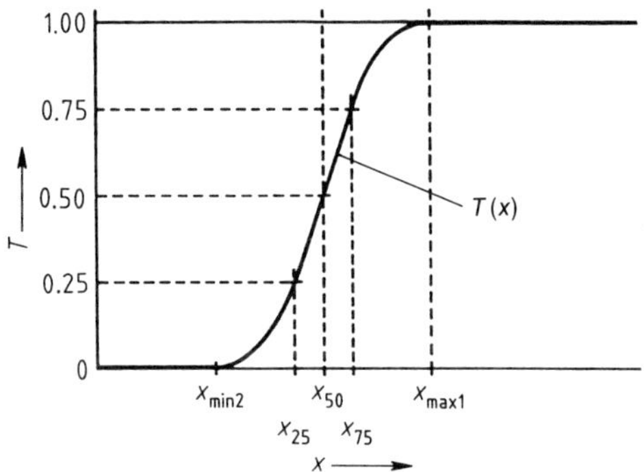

Figure 10. Grade efficiency or classsifier selectivity curve, $T(x)$

$$T(x) = v_2 q_2(x)/q_0(x) \quad (42)$$

The grade efficiency $T(x)$ can be obtained from the ratio of the ordinates of the coarse and the feed materials, as shown in Figure 6. It is equal to 0 at particle sizes below x_{min2}; it is equal to 1 at particle sizes above x_{max1}, and is 0.5 at $x = x_{50}$.

If T is plotted against particle size, x, with x being the equivalent settling rate diameter, the resulting curve is the *grade efficiency curve*, or *classifier selectivity curve* (Fig. 10). It should start at 0 and remain there between x_{min1} and x_{min2}, and it should reach the value of 1 at particle sizes equal to or coarser than x_{max2}.

Attempts have often been made to fit analytical functions to grade efficiency curves. In most cases these attempts have not been very successful. A universally applicable function does not seem to be available as yet. TROMP [17], MAYER [18] and EDER [19] succeeded in special cases; TRAWINSKI [20] also published an analysis of useful functions to fit grade efficiency curves.

3.1.6. Sharpness of Cut and Other Parameters

In principle, the sharpness or the quality of a classification process is better when the overlapping size range $x_{min2} \leq x \leq x_{max1}$ is smaller, or when the amounts of misplaced material are smaller. A considerable number of parameters have been defined to indicate quantitatively the sharpness or the lack of sharpness of a cut in a classification process.

These parameters can be used meaningfully only when they are selected with regard to the technical application for which the classification process has been used. Therefore, one parameter alone is in many cases not as adequate for a complete description of the classification process as a series or even a combination of different parameters. Most parameters that are given quantify only a part or parts of the information obtainable from the grade efficiency curve. In 1972, LESCHONSKI formed three groups of parameters which suffice to include the existing parameters: (1) parameters formed with characteristic particle sizes, (2) parameters derived from cumulative distribution curves, and (3) parameters derived from the overall course of the grade efficiency curve.

Parameters from Characteristic Particle Sizes. Parameters of this type indicate a difference or a ratio of characteristic particle sizes taken from the grade efficiency curve. Some examples follow:

1) Ecart Terra [21]: $E_T = (x_{75} - x_{25})/2$ (43)
2) F. W. MAYER [22], [23]: $E_M = x_{90} - x_{10}$ (44)
3) Imperfection [24]: $I = (x_{75} - x_{25})/2x_{50}$ (45)
4) K. GRUMBRECHT [25]: $\kappa_{75/25} = x_{75}/x_{25}$ (46)
5) TH. EDER [24]: $\kappa_{25/75} = x_{25}/x_{75}$ (47)

$$\kappa_{35/65} = x_{35}/x_{65} \quad (48)$$

where x subscript y indicates the value of size x where the grade efficiency curve has a value of $T = y\%$.

Parameters from Cumulative Distribution Curves [7]. These parameters can be determined from the cumulative distribution curves $Q_0(x)$, $Q_1(x)$, and $Q_2(x)$ and the relative amounts of the fine material, v_1 or v_2. The parameters have the advantage that the grade efficiency curve is not required for their determination. In principle, a distinction is made between six different areas beneath the three density distribution curves, as shown in Figure 11.

From these areas a number of characteristic parameters can be derived:

1) Parameters composed of the areas $A_1 - A_6$, as shown in Figure 11. These parameters indicate either the amount of fine or coarse particles below or above a certain size in the feed,

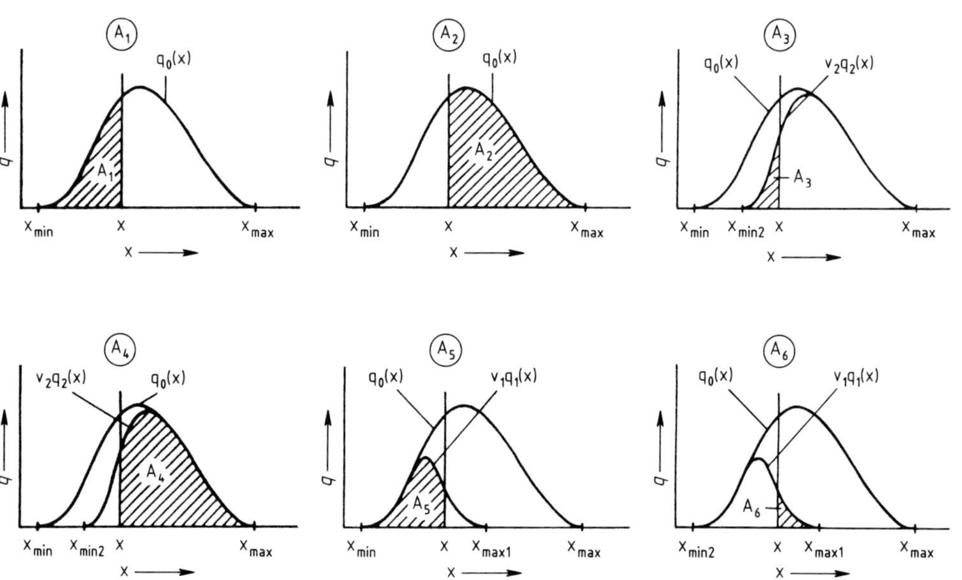

Figure 11. Definition and representation of six areas beneath the three density distribution curves

the fine or the coarse fractions. The following distinctions are made:

– Fine particles in the feed material:

$$A_1 = Q_0(x) = \int_{x_{min}}^{x} q_0(\xi)\,d\xi \tag{49}$$

– Coarse particles in the feed material:

$$A_2 = 1 - Q_0(x) = \int_{x}^{x_{max}} q_0(\xi)\,d\xi \tag{50}$$

– Fine particles in the coarse fraction:

$$A_3 = v_2 Q_2(x) = \int_{x_{min}}^{x} v_2 q_2(\xi)\,d\xi \tag{51}$$

– Coarse particles in the coarse fraction:

$$A_4 = v_2[1 - Q_2(x)] = \int_{x}^{x_{max}} v_2 q_2(\xi)\,d\xi \tag{52}$$

– Fine particles in the fine fraction:

$$A_5 = v_1 Q_1(x) = \int_{x_{min}}^{x} v_1 q_1(\xi)\,d\xi \tag{53}$$

– Coarse particles in the fine fraction:

$$A_6 = v_1[1 - Q_1(x)] = \int_{x}^{x_{max}} v_1 q_1(\xi)\,d\xi \tag{54}$$

The parameters A_1–A_6 vary with size; A_3 and A_6 are plotted vs. x in Figure 12. The misplaced coarse material in the fine fraction, A_6 drops from v_1 at small sizes to 0 at x_{max1}; the misplaced fine material in the coarse fraction, A_3, on the other hand, increases from 0 at x_{min2} to v_2. The sum of A_3 and A_6 represents the total misplaced material; it forms a minimum at $x = x_{50}$ [25].

2) Relative parameters may also be formed from ratios of the areas A_1 to A_6, as for instance:

— the *retrieval* or *recovery* of fine particles related to those initially present in the feed material (or related to the feed undersize)

$$A_5/A_1 = v_1 Q_1(x)/Q_0(x) \tag{55}$$

— the *retrieval* or *recovery* of coarse particles related to those initially present in the feed material (or related to the feed oversize)

$$A_4/A_2 = v_2(1 - Q_2(x))/(1 - Q_0(x)) \tag{56}$$

3) More complicated relative parameters have been suggested by MADEL [26], ROSIN and RAMMLER [27], and WHITE [28]; they are derived from differences of relative parameters according to paragraph 2:

$$\eta = \frac{A_4}{A_2} - \frac{A_3}{A_1} = \frac{A_5}{A_1} - \frac{A_6}{A_2} = \frac{v_2[Q_0(x) - Q_2(x)]}{Q_0(x)(1 - Q_0(x))} = \frac{v_1[Q_1(x) - Q_0(x)]}{Q_0[(x)(1 - Q_0(x))]} \tag{57}$$

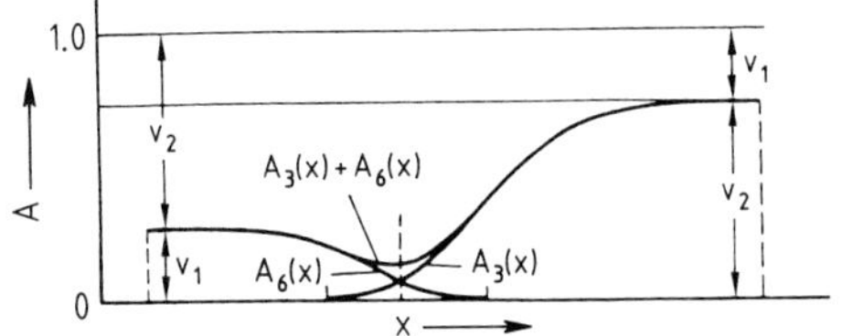

Figure 12. Course of $A_6(x)$, $A_3(x)$, and $(A_3(x) + A_6(x))$

One or more of the given parameters or other parameters must be used to quantify the quality of the process, depending on the separation or classification process to be characterized. The parameters should be chosen from those that best characterize the aim of the separation or classification process.

Parameters from the Overall Course of the Grade Efficiency Curve. The most important parameter of this group is the so-called *overall classification or separation (collection) efficiency*, T_{tot}. This parameter is generally used to describe the quality of dust removal systems (overall collection efficiency, → 13. Dust Separation, p. 13-4). However, it can also be used for classification processes. It corresponds to the already defined relative amount of coarse material, v_2, and may be calculated from the grade efficiency curve $T(x)$ and the density distribution curve of the feed material, $q_0(x)$, as follows:

$$T_{tot} = v_2 = \int_{x_{min}}^{x_{max}} T(x)\,q_0(x)\,dx = \int_0^1 T(x)\,dQ_0(x) \tag{58}$$

If the density distribution $q_0(x)$ is available as a histogram, the integration may be replaced by a summation:

$$T_{tot} = v_2 = \sum_{i=1}^{n} T_i(\bar{x}_i)\,\Delta Q_{0i} \tag{59}$$

The use of variance and the expected value of the grade efficiency curves for its characterization have been suggested [29], [30]:

If $t(x)$ is the first derivative of $T(x)$ with regard to x

$$t(x) = dT(x)/dx \tag{60}$$

then the expected value $E(x) = x$ and the variance, s^2, of the grade efficiency curve may be determined in the usual way as follows:

$$E(x) = \bar{x} = \int_{x_{min}}^{x_{max}} x\,t(x)\,dx = \int_0^1 x\,dT(x) \tag{61}$$

$$s^2 = \int_{x_{min}}^{x_{max}} (x - \bar{x})^2\,t(x)\,dx$$

$$= \int_0^1 (x - \bar{x})^2\,dT(x) \tag{62}$$

Further parameters have been suggested by MAYER [31]–[33].

3.2. Influence of Stochastic Errors

The mass balances given in Section 3.1.2 and the evaluation of the grade efficiency curve as described in Section 3.1.5 are valid only under the assumption that the relevant Equations (30) and (33)–(35) leave no remainder. However, in reality Equation (34) does not hold with, e.g., the density distributions as obtained by conventional particle size analysis for the feed, fine, and coarse material. A size-dependent error $E(x)$ is observed, and Equation (34) must be rewritten as follows:

$$q_0(x) - v_1 q_1(x) - v_2 q_2(x) = E(x) \tag{63}$$

The same applies to Equation (35), which yields

$$Q_0(x) - v_1 Q_1(x) - v_2 Q_2(x) = \varepsilon(x) \tag{64}$$

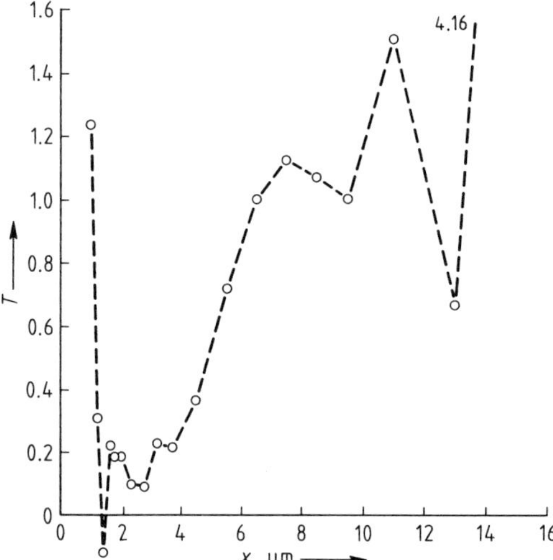

Figure 13. Uncorrected grade efficiency curves, $T(x_i)$

The size-dependent errors $E(x)$ and $\varepsilon(x)$ are due to errors of particle size analysis when either $q_0(x)$, $q_1(x)$, and $q_2(x)$ or $Q_0(x)$, $Q_1(x)$, and $Q_2(x)$ are measured. Further errors may arise when v_1 or v_2 are measured directly.

Errors of particle size analysis may be caused not only by erroneous sampling and sample splitting, but also (and sometimes mainly) by errors produced by handling the instruments of particle size analysis.

If the average grade efficiency, T, of a small size class, Δx_i, is calculated, as usual, with $x_{i+1} = x_i + \Delta x_i$ and the average size $\bar{x}_i$:

$$\bar{x}_i = (x_i + x_{i+1})/2 \tag{65}$$

from

$$T(x_i) = \frac{v_2 [Q_2(x_{i+1}) - Q_2(x_i)]}{Q_0(x_{i+1}) - Q_0(x_i)} \tag{66}$$

results are obtained as shown in Figure 13 [34], which reveals a remarkable difference to the grade efficiency curve shown in Figure 10. The results shown in Figure 13 have been obtained with the Acucut B18-classifier from Donaldson Co. Inc., Minneapolis, Minnesota, USA, in a modified form suggested and investigated by LESCHONSKI and BOECK [34]. A fine limestone has been used as feed material. The cumulative size distributions of the feed, fine and coarse material have been obtained with the X-ray sedimentometer, Sedigraph 5000, of Micromeritics Instr. Corp., Norcross, Georgia, USA. The points calculated for different average sizes, $\bar{x}_i$, scatter considerably, and a reliable average grade efficiency curve cannot be obtained.

Grade efficiency curves, as shown in Figure 13, are not unusual. Published grade efficiency curves generally follow a rather smooth course because either a graphic or numerical regression technique has been used. HERRMANN and LESCHONSKI [16] have suggested a statistically sound correcting method that yields the best possible answer, i.e., the best estimate of the grade efficiency curve from the experimental evidence as obtained by size analysis.

3.2.1. Indirect Evaluation of v_1 and v_2

In many practical cases, the relative amounts of the fine and the coarse materials, v_1 and v_2, cannot be obtained from the solids mass flow rates directly, as suggested in Equation (9) and (10). However, if sampling is possible from the three streams of the feed, the fine, and the coarse material, the cumulative size distributions [$Q_0(x)$, $Q_1(x)$, and $Q_2(x)$] of the feed material and the two products may be obtained. As shown in Section 3.1.3, v_1 or v_2 may be calculated from Equation (36). According to this equation, v_1 and/or v_2 must be constant and independent of x if the cumulative distributions $Q_0(x)$, $Q_1(x)$, and $Q_2(x)$ are free of errors. In reality this is not true, which consequently leads to values of v_1 and v_2 that vary with size x (Figure 14). Therefore, suitable equations to calculate an average value v_1 from $v_1(x)$ have been suggested by a number of authors [25], [35]–[38].

Equation (36) may be interpreted as a linear equation, if it is rewritten as follows:

$$Q_0(x) - Q_2(x) = v_1[Q_1(x) - Q_2(x)] \qquad (67)$$

If $\eta_i = Q_{0i} - Q_{2i}$ is plotted on the ordinate against $\xi_i = Q_{1i} - Q_{2i}$ on the abscissa, a series of points is obtained that approximates a straight line through the origin (Fig. 15).

$$\eta_i = v_1 \cdot \xi_i \qquad (68)$$

Regression analysis yields the average slope, for instance: $\bar{v}_1$, for n values of ξ_i and η_i, from

$$\bar{v}_1 = \sum_{i=1}^{n} \xi_i \eta_i \Big/ \sum_{i=1}^{n} \xi_i^2 \qquad (69)$$

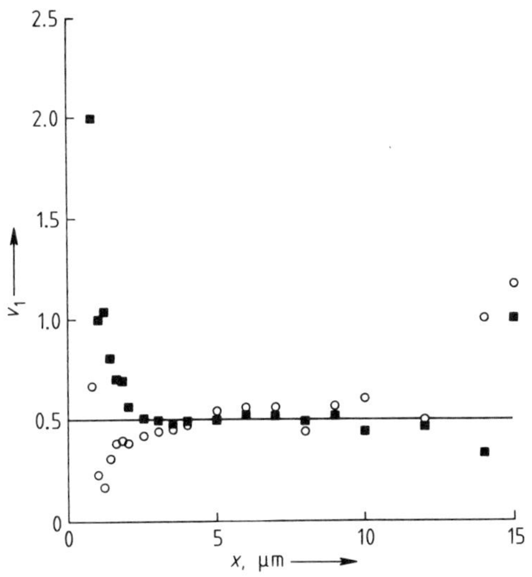

Figure 14. $v_1(x)$ as calculated from erroneous cumulative size distributions $Q_0(x)$, $Q_1(x)$, $Q_2(x)$
■ $\bar{v}_1 = 0.502$
○ $\bar{v}_1 = 0.506$

Equation (69) coincides with the solution suggested by GRUMBRECHT [35], obtained when $\bar{v}_1$ is calculated under the assumption that

$$\sum_{i=1}^{n} \varepsilon_i^2 \overset{!}{=} \min \qquad (70)$$

and the constraint that $v_1 + v_2 = 1$ holds. Equation (68) is found when Equation (64) is substituted into Equation (69) and this equation is solved by using Lagrange's method.

However, the method has the general disadvantage that Equation (68) is valid only under the assumption that the errors of particle size analysis are attributed only to the η_i values. The ξ_i values are taken to be free of errors. However, strictly speaking, neither η_i nor ξ_i are free of errors, and the minimum of the squared errors should be calculated perpendicular to the straight line drawn through the measured points under the angle α (Fig. 16).

The most probable $\bar{\alpha}$, is obtained with $\tan\bar{\alpha} = \bar{v}_1$, from

$$\sum z_i^2 = \sum(\xi_i \sin\alpha - \eta_i \cos\alpha)^2 \overset{!}{=} \min \qquad (70\,a)$$

The solution yields

$$\tan 2\bar{\alpha} = \frac{2\sum_{i=1}^{n} \xi_i \eta_i}{\sum_{i=1}^{n} \xi_i^2 - \sum_{i=1}^{n} \eta_i^2} = a \qquad (71)$$

with

$$\bar{\alpha} = \frac{\text{arc tan } a}{2} \qquad (72)$$

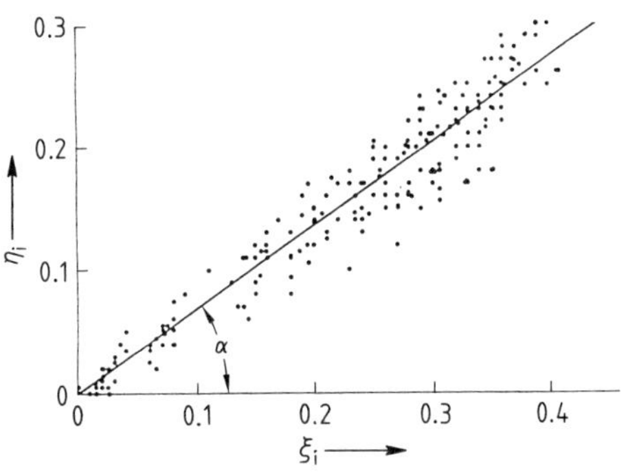

Figure 15. $\eta_i = v_1 \cdot \xi_i$ as obtained from multiple analysis of a hydrocyclone cut [38]
$\tan\alpha = 0.67991$

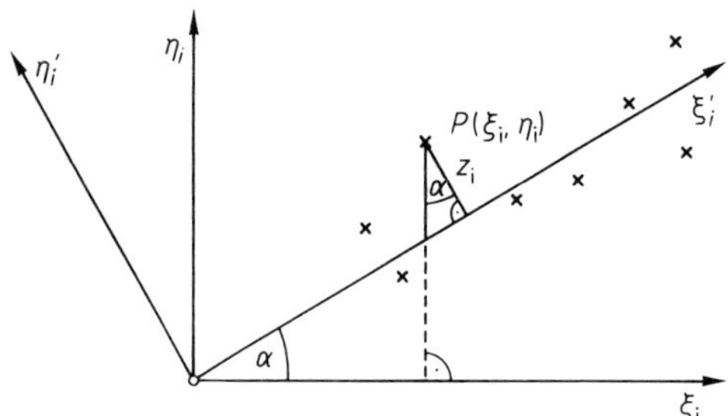

Figure 16. Definition of the deviation z_i

Numerical values of $\bar{v}_1$ calculated by using Equation (68) or (70) for the same set of results differ very little. Therefore, in most cases of application it is advisable and legitimate to use the much simpler Equation (68).

Regression analysis not only yields average values of $\bar{v}_1$, but also its standard deviation or variance, as well as its confidence interval. In the case of Equation (68), the variance, s^2, may be calculated from Equation (73) [39]:

$$s^2 = \frac{1}{(n-1)\sum \xi_i^2}\left(\sum \eta_i^2 - \frac{(\sum \xi_i \eta_i)^2}{\sum \xi_i^2}\right) \tag{73}$$

with the confidence interval being defined in the usual way by

$$\pm t \cdot s/\sqrt{n} \tag{74}$$

where t equals the student factor, i.e., approximately 1.96 with a probability of 95% and $n > 25$. Equation (73) is, in fact, a well-known equation in regression analysis, but unfortunately has not been used in the characterization of a classification process. It should be remembered that

$$1/v_1 = \dot{m}_0/\dot{m}_1 \tag{75}$$

with $1/v_1$ being the so-called recirculating load. Therefore, in a continuous classification process, $1/v_1$ represents the average ratio of the mass throughput of the feed, $\dot{m}_0$, to the mass throughput of the fine material, $\dot{m}_1$.

3.2.2. Evaluation of the Grade Efficiency Curve, $T(x)$

As shown in Figure 13, erroneous cumulative size distributions $Q_0(x)$, $Q_1(x)$, and $Q_2(x)$ lead to nonideal grade efficiency curves. Therefore, HERRMANN and LESCHONSKI suggested a statistical correction of the measured values that not only reduces the error as obtained, but also allows the calculation of the best statistical estimate of the grade efficiency curve, $T(x)$, based on the given data of particle size analysis [16]. According to this, corrected cumulative distribution values $K_0(x)$, $K_1(x)$, and $K_2(x)$ are derived from the measured cumulative distribution values $Q_0(x)$, $Q_1(x)$, and $Q_2(x)$, for which the mass balance holds:

$$K_0(x) - v_1 K_1(x) - v_2 K_2(x) = 0 \tag{76}$$

A least squares method was applied with $1/\sigma^2$ as weight for the square of the deviation $(Q - K)^2$ to derive Equations (77)–(80) for the correction, where σ^2 represents the variance or its estimated value s^2, as calculated from q_1 values, obtained by multiple analysis of the same product. The statistical correction yields the following equations:

$$K_0(x) = Q_0(x) - \frac{\sigma_0^2(x)}{\alpha(x)}\varepsilon(x) \tag{77}$$

$$K_1(x) = Q_1(x) - \frac{v_1 \sigma_1^2(x)}{\alpha(x)}\varepsilon(x) \tag{78}$$

$$K_2(x) = Q_2(x) - \frac{v_2 \sigma_2^2(x)}{\alpha(x)}\varepsilon(x) \tag{79}$$

with

$$\alpha(x) = \sigma_0^2(x) + v_1^2 \sigma_1^2(x) + v_2^2 \sigma_2^2(x) \tag{80}$$

The correction terms on the right-hand side of Equations (77)–(79) depend not only on the error variance, σ^2, of each individual measured value relative to the weighted sum of the variances, $\alpha(x)$ (Eq. 80), but also on the error $\varepsilon(x)$ (Eq. 64). When the suggested correction method is applied, the error distributions σ^2 or s^2 must be known for the particular case of application and the particle size analysis instruments used for the determination of $Q_0(x)$, $Q_1(x)$, and $Q_2(x)$. These curves can be obtained by repeated analyses of the feed and the fine and coarse products. As shown in Figure 17, the error distributions, $s(x)$, may differ for the three streams.

In Figure 17 the average error, $\bar{s}$, as obtained from 21 measurements of $Q_0(x)$, $Q_1(x)$, and $Q_2(x)$, is plotted against the equivalent settling rate diameter, $x = x_w$. On a preliminary basis,

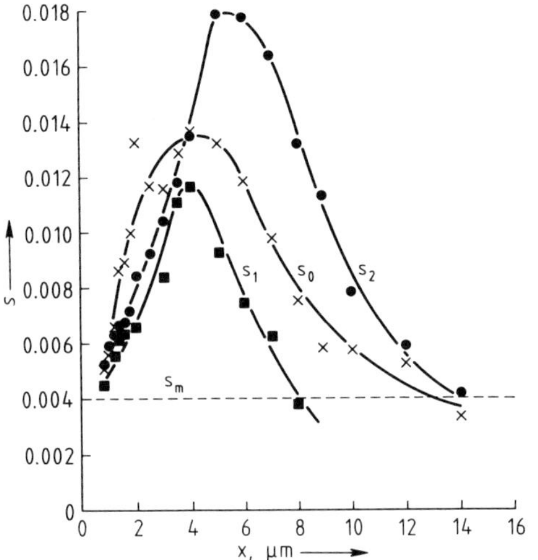

Figure 17. Error distributions $s_0(x)$, $s_1(x)$, and $s_2(x)$

the three differing curves may be described by Equation (81):

$$\sigma^2(x) = s_m^2 + C(x_{50})Q(x)[1 - Q(x)] \qquad (81)$$

If the suggested correction is applied to grade efficiency curves, as shown in Figure 13, the solid lines of Figure 18 are obtained.

Apart from the error distributions $s^2(x)$, only one set of results of particle size analysis has been used in this presentation. The suggested correction smooths the grade efficiency curves especially well in the coarse size range.

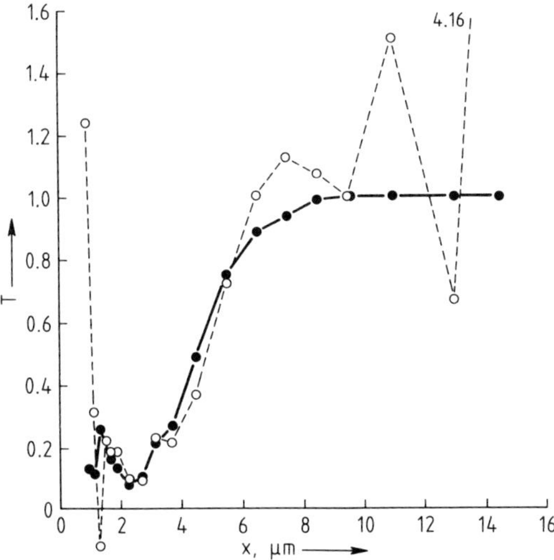

Figure 18. Corrected grade efficiency curves using $\sigma^2(x)$ but only one set of results

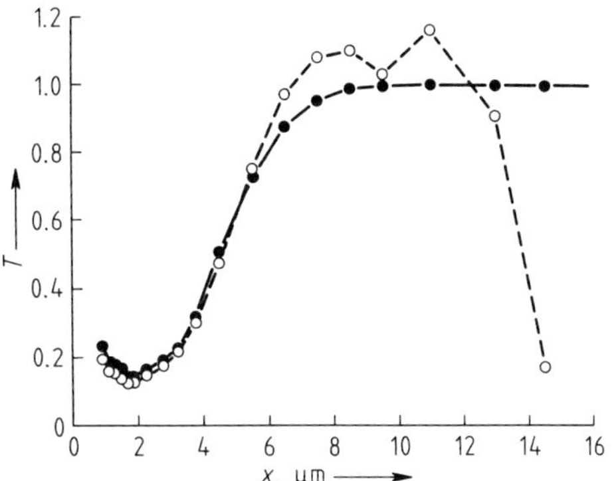

Figure 19. Average and average corrected grade efficiency curves

From the results obtained, the obvious question can be answered, whether the use of Equations (77)–(80) can be dropped altogether, if an average grade efficiency curve is calculated from average cumulative size distributions of the products. The answer is no, even though each product had been analyzed 21 times in this case.

The dotted line in Figure 19 shows the average grade efficiency curve. While the fine end seems to be right, the coarse end neither reaches unity nor does it seem to follow the right course. The best answer is obtained with multiple analyses and an application of the suggested correction. The full line in Figure 19 indicates the difference.

This example clearly shows the necessity for multiple analyses. These can be performed on an economical basis only if particle size distributions are analyzed quickly and, if possible, online. Therefore, the development and use of rapid-response methods or on-line methods of particle size analysis are more important than generally recognized and acknowledged. A complete theory of the correction method and further statistical measures have been published [40], [41].

3.3. Influence of Systematic Errors

Various effects can cause systematic deviations from the ideal course of a grade efficiency curve, as shown in Figures 20–22. Such effects are

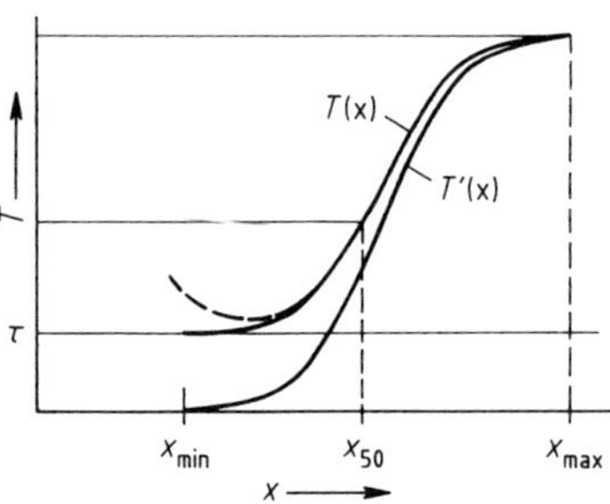

Figure 20. Influence of a splitting process in the classifier on $T(x)$

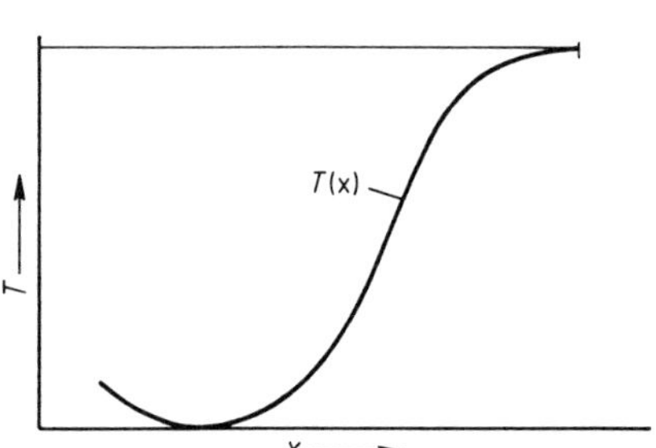

Figure 21. Deviation from the ideal grade efficiency curve caused by incomplete dispersion of the feed material in the classifier

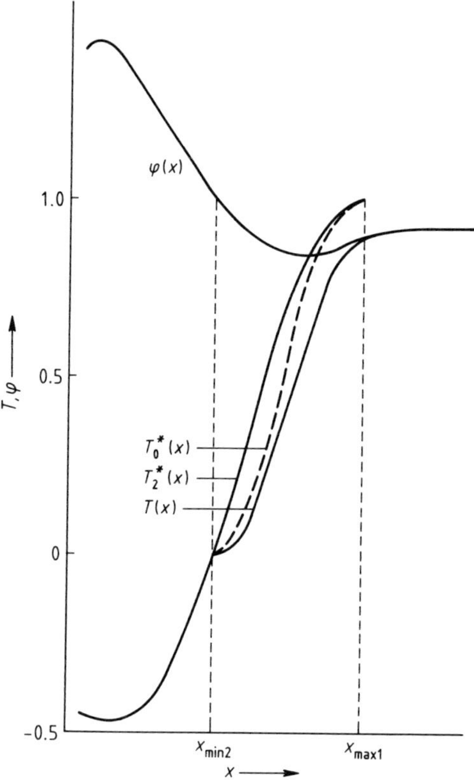

Figure 22. $\varphi(x)$ and $T(x)$ if comminution occurs in a classification process

1) systematic analytical errors of sampling and sample splitting
2) superposition of classification and splitting within the classifier
3) undispersed agglomerated fine particles which are transferred to the coarse product
4) comminution of the feed material in the classifier.

Under the assumption that sampling from the product streams and the splitting of the samples of the three products have been performed with greatest care, the error listed under 1) may be disregarded. The others, however, must be taken into account.

3.3.1. Splitting in the Classifier

If the grade efficiency curve, as shown in Figure 20, does not drop to 0 at small particle sizes, but ends parallel to the abscissa at a certain grade efficiency $T = \tau$, a splitting process is superimposed on the classification. A modified grade efficiency curve $T'(x)$ may then be calculated [18]:

$$T'(x) = (T(x) - \tau)/(1 - \tau) \tag{82}$$

In a hydrocyclone classification, the value of τ describes the so-called *dead flux*. $T'(x)$ resembles the ideal grade efficiency curve of Figure 10; it extends between 0 and 1. When this technique is used, it should be remembered, the effective grade efficiency curve is represented by $T(x)$ only; however, the use of $T'(x)$ and τ may simplify the description of the grade efficiency curve and may give additional information. In many technical applications, it is usually advisable to minimize τ.

3.3.2. Incomplete Dispersion of the Feed Material

If the feed material is not properly dispersed before it enters the classification zone, or even within the classification zone, coarse agglomerates of fine particles may be taken by the classifier as coarse particles and may consequently appear in the coarse product. On the other hand if the samples are well dispersed, e.g., in a sedimentation liquid, during the analysis of the three products, the agglomerates disappear. The analysis then detects more fine particles than the classifier did, and the calculated grade efficiency curve deviates from the ideal one, as shown in Figure 21.

The classification process itself is not to blame for this result. Better dispersion of the feed material, if possible before entering the classification zone, reduces this effect. Therefore, the feed material should pass a separate dispersion zone before it enters the classifier.

3.3.3. Comminution in the Classifier

If the feed material is comminuted in the classification zone, i.e., if the classifier partly acts as a grinding machine, the fine feed material differs from the one obtained from sampling the original feed. New fine material is produced and an equivalent amount of coarse particles disappears.

This error has already been described and is defined in Equation (63):

$$q_0(x) - v_1 q_1(x) - v_2 q_2(x) = E(x) \tag{63}$$

It can be used to formulate a new variable, $\varphi(x)$, which indicates whether the deviations observed are either stochastic or systematic and which course they follow [42]:

$$\varphi(x) = \frac{v_1 q_1(x) + v_2 q_2(x)}{q_0(x)} = 1 - \frac{E(x)}{q_0(x)} \tag{83}$$

If the sum of the weighted density distributions of the fine and the coarse material equals the density distribution of the feed, then $E(x) = 0$ and $\varphi(x) = 1$. Stochastic deviations of $\varphi(x)$ from a value of 1 indicate stochastic errors, $E(x)$, in the mass balance of Equation (63). The suggested error correction and multiple analysis should then be used in the evaluation of the grade efficiency curve, as described in Section 3.2. However, if $\varphi(x)$ deviates systematically from 1, then comminution of the feed material in the classifier might be responsible. A typical example is given in Figure 22.

Table 2 shows the relevant equations and an explanation of the subscripts chosen. If $q_2^*(x)$ and $q_0{}^*(x)$ are introduced into the equation for $T_2^*(x)$ and $T_0^*(x)$ (Table 2), the following is obtained:

$$\begin{aligned} T_2^*(x) &= T(x) + [1 - \varphi(x)] \\ &= 1 - v_1 q_1(x)/q_0(x) \end{aligned} \tag{90}$$

$$T_0^*(x) = T(x)/\varphi(x) = 1\bigg/\left(\frac{v_1 q_1(x)}{v_2 q_2(x)} + 1\right) \tag{91}$$

While $T_0^*(x)$ can be calculated from the fine and the coarse materials, $T_2^*(x)$ depends on the fine and the feed materials. The function $T_0^*(x)$ covers the size range between $x_{\min 2}$ and $x_{\max 1}$ and is not influenced by the course of $\varphi(x)$ as soon as $T(x) = 0$. Therefore, neither $T(x)$ nor $T_0^*(x)$ gives evidence of the increased amount of fine material at small particle sizes due to comminution. This information can be obtained only from $T_2^*(x)$. Therefore, $T_2^*(x)$ should be calculated as soon as $T(x)$ remains smaller than 1 at coarse particle sizes.

Table 2. Definition of $\varphi(x)$ and various grade efficiencies

Measured values		Calculated values	Equation
Feed material:	$q_0(x)$	$q_0^*(x) = v_1 q_1(x) + v_2 q_2(x) = q_0(x) - E(x)$	(84)
Fine material:	$q_1(x)$, v_1		
Coarse material:	$q_2(x)$, v_2	$q_2^*(x) = \frac{q_0(x) - v_1 q_1(x)}{v_2} = q_2(x) - \frac{E(x)}{v_2}$	(85)
		$\varphi(x) = \frac{q_0^*(x)}{q_0(x)} = 1 - \frac{E(x)}{q_0(x)}$	(86)
Grade efficiencies:		$T(x) = \frac{v_2 q_2(x)}{q_0(x)}$	(87)
		$T_2^*(x) = \frac{v_2 q_2^*(x)}{q_0(x)} = T(x) + (1 - \varphi(x))$	(88)
		$T_0^*(x) = \frac{v_2 q_2(x)}{q_0^*(x)} = \frac{T(x)}{\varphi(x)}$	(89)

4. Sample Splitting

Particle size analysis may be performed only if a suitable sample of particulate matter to be analyzed has been taken from the bulk material in the factory or delivered to the laboratory. The sample mass may vary from less than 1 mg to more than several kilograms depending on the method of analysis or the particular instrument chosen, and the average size of the particles and their spread. The analysis sample must be taken from the parent population of particulate matter. Its composition should equal the composition of the parent population in all aspects; it should be a so-called *representative sample*. Since the overall mass of the parent population may be of arbitrary size, the mass of the analysis sample must be obtained by repetitive splitting of the original sample. The answer obtained after the analysis sample has been analyzed may represent the particle size distribution, the surface area, etc., of the parent population more or less accurately, depending on the quality of the splitting process.

Differences may occur due both to statistical variations of the composition of the analysis samples and to systematic deviations in the composition caused by the splitting process.

Statistical variations of the samples can generally be decreased by using an appropriate amount of particulate material for the analysis, if the instrument that is applied permits this. Systematic and statistical errors can also be reduced if the analysis sample consists of a large number of much smaller samples.

The methods applied in sample splitting may be subdivided into two groups, those for powders and those for suspensions and emulsions. Sample splitting of powder is achieved by *coning* and *quartering*, or by repeatedly splitting the laboratory sample (10–100 kg) in stationary or rotating rifflers. Sample splitting from suspensions or emulsions can be achieved by withdrawing an appropriate small amount of suspension or emulsion from the bulk material after the laboratory sample has been mixed thoroughly.

5. Sedimentation Analysis

Sedimentation methods use the effect that single particles settle, for instance, under the influence of gravity in a static liquid medium with constant velocity after a short and negligible period of acceleration. The settling rates that are measured may be used to calculate the size of a sphere, e.g., the equivalent settling rate diameter x_w, travelling with the same speed in the same environment.

In most cases the analysis starts with a *homogeneous suspension* in which the particles are evenly, or better, randomly distributed throughout the liquid volume. The individual solid particles then start to settle, with settling rates depending on the particles' size, shape, and density. The suspension density, which was constant at the beginning of the experiment, starts to change, forming an image of the cumulative distribution of settling rates with respect to the height of the sedimentation column, which can be used to analyze the sample. The physical methods used to measure the time- and height-dependent change of suspension densities define and distinguish the different instruments of sedimentation analysis.

5.1. Calculation of the Equivalent Settling Rate Diameter, x_w

The liquid used in a sedimentation analysis should neither physically nor chemically affect the particles to be analyzed. Furthermore, it should not cause reagglomeration of the solid particles, which in most cases is guaranteed by adding small amounts of suitable surfactants. The liquid density must be smaller than the solids density, and its viscosity should match the size distribution under investigation, avoiding, for instance, extremely long measuring times for the smallest particles [44].

The stationary motion of a single spherical particle in a static, infinite liquid under the influence of gravity can be calculated from the force balance: weight minus buoyancy force equals the drag force. With spherical particles and laminar flow conditions, i.e., under the validity of Stokes law, the drag coefficient c_d equals

$$c_d = 24/Re \qquad (92)$$

The dependency between equivalent settling rate diameter, x_w, and stationary settling rate, w_g, in the gravity field can then be calculated from

$$x_w^2 = \frac{18\eta\, w_g}{(\varrho_s - \varrho_{fl})\, g} \qquad (93)$$

or

$$v_g = \frac{(\varrho_s - \varrho_{fl})\, g\, x_w^2}{18\eta} \qquad (94)$$

The Reynolds number, *Re*, as used in Equation (92), equals

$$Re = w_g \, x_w \, \varrho_{\mathrm{fl}}/\eta \tag{95}$$

The equivalent settling rate diameter, x_w, of particles settling in a centrifugal field with the acceleration

$$a = r\,\omega^2 \tag{96}$$

where the centrifuge rotates with constant angular velocity, ω, can be obtained from

$$x_w^2 = \frac{18\,\eta}{(\varrho_s - \varrho_{\mathrm{fl}})} \cdot \frac{\ln r/r_i}{\omega^2\, t} \tag{97}$$

where r_i represents the radius of the liquid surface. The equivalent settling rate diameter, x_w, as calculated using Equations (93) and (97) describes the settling behavior of a sphere in an infinite static liquid. In reality, however, settling takes place in the neighborhood of 10^6 other particles. Furthermore, the particles are in most cases not spherical and the vessels used have limited dimensions. Walls and the bottom of the vessel decrease settling rate, but for practical applications the influence of both may be neglected.

The settling rate increases under certain conditions with increasing solids volume concentrations C_V; it reaches a maximum at approximately $C_V = 1\,\%$ and then decreases to 0 [45]–[47]. The settling rate maximum disappears if fine particles form the suspension and if the distribution of settling rates is widely spread. The settling rate is decreased from the stationary settling rate, w_g, at increasing solids concentration by the increased liquid flow that is caused by the settling of the particles into a confined liquid volume. The settling rate only increases, if hydrodynamically-linked particles form clusters that settle faster than neighboring single particles [45]. Therefore, there is an upper limit to solids concentration in sedimentation analysis.

The solids volume concentration should be kept smaller than $C_V = 2 \times 10^{-3} = 0.2\,\%$. An upper limit, where deviations in settling rate from w_g may still be small, is $C_V = 1\,\%$.

With particles smaller than 10^{-6} m and settling in liquids, the Brownian motion is no longer negligible. This diffusion process produces wider size distributions than are present in reality. It can be avoided only if the analysis is repeated in a centrifuge.

Settling of small particles in air is influenced by the *slip velocity*. If the mean free path, $\bar{\lambda}$, of the air molecules is of the order of the average size of the particles, the settling rate is increased and the *Cunningham correction* must be taken into account [48]:

$$w/w_g = 1 + A\,\bar{\lambda}/x \tag{98}$$

$A = 1.73$ for oil droplets in air,
$A = 0.7$ for rough sphere in air

In addition, the settling rate may be strongly influenced by *density convection currents*. They may be caused either by temperature gradients in the liquid or density differences in the suspension. Therefore, convection currents can be avoided only if the density of the suspension increases toward the bottom of the vessel. Cooling or evaporation of liquid at the top of the liquid column should be avoided, as well as temperature differences greater than 0.01 °C/min caused by the temperature control of thermostats. Differences in suspension density are caused mainly by the inclined walls of the sedimentation vessel and solid obstacles, e.g., a weighing pan or a with-drawal tube immersed in the suspension.

5.2. Methods of Sedimentation Analysis

The methods and instrumentation used in sedimentation analysis can be subdivided into two principal groups:

1) *Suspension techniques*: at the beginning of the analysis the suspension is evenly distributed throughout the liquid volume
2) *Superimposed layer techniques*: at the beginning of the analysis the suspension is superimposed in a thin layer on top of a solids-free liquid column.

Both methods may be applied in a gravity or in a centrifugal field. Furthermore, the measurement of the amount of particles present in a certain size class, or smaller than that certain size, is either performed in a measuring volume, which is small in comparison to the overall settling height, or includes all particles above or below the plane of measurement. The first group comprises the *incremental methods*, the second one the *cumulative methods*. Therefore, eight principal possibilities are distinguished:

Suspension	Gravity field	Incremental Cumulative
	Centrifugal field	Incremental Cumulative
Superimposed layer	Gravity field	Incremental Cumulative
	Centrifugal field	Incremental Cumulative

The measurements may be performed at a fixed height with respect to time, or after a fixed time with respect to settling height; in this way, the number of possible methods may be at least doubled. The number of instruments is much larger than the number of principal methods because different physical methods are used to measure the concentration, mass, or volume of particles in a thin suspension layer or below or above the height of measurement. The measurements are performed gravimetrically; they use the scattering and absorption of electromagnetic radiation, as well as buoyancy, hydrodynamic pressure, and other physical variables that depend on solids concentration, surface, or number.

5.2.1. Suspension Techniques

Pipette Method. One of the best known incremental gravitational suspension methods is the pipette method [49]–[52]. With this method, the cumulative weight or mass undersize, Q_3, is determined from the ratio of two volume concentrations measured at a fixed height at the beginning of the analysis and after varying times t, (Eq. 99).

$$Q_{3\,t} = C_{V\,t}/C_{V\,0} \tag{99}$$

Fixed suspension volumes are withdrawn, the solids contents of which are determined, e.g., by weighing after the liquid is evaporated.

Diver Method. BERG developed the diver method [53]. A small spherical diver, the density of which lies between the density of the solids-free liquid and the density of the homogeneous suspension, is submersed in a suspension that has already formed its settling-rate-dependent concentration gradient. The diver then settles to the suspension layer of the same density and its position is measured. The suspension density and Q_3 are correlated according to Equation (100):

$$Q_3 = (\varrho_d - \varrho_{fl})/[(\varrho_s - \varrho_{fl})\, C_{V\,0}] \tag{100}$$

Therefore, the missing settling rate is determined from the diver position and the length of time the experiment has taken.

Optical Measurement. Other incremental techniques use the attenuation of a beam either of white light or of γ-rays by the particles. Lambert–Beer's law of attenuation yields the theoretical background for photosedimentometers [54].

Sedimentation Balance. The best known example for cumulative suspension methods is the sedimentation balance first suggested by ODÉN in 1916 [52], [55]. This method determines $Q_3(w_g)$ from the mass of particles, m_t, collected on a submerged balance pan at the lower end of a suspension column. The quantity $Q_3(w_g)$ can be calculated from the curve m_t vs. t by using Equation (101):

$$m_\infty Q_{3\,t} = m_t - t\, \mathrm{d}m_t/\mathrm{d}t \tag{101}$$

The methods mentioned have been applied in both gravitational and centrifugal applications. Centrifugal methods, which are generally used for particles smaller than ca. 5×10^{-6} m, apply the same techniques as gravitational ones; however, the mathematical evaluation of the data is more sophisticated.

5.2.2. Superimposed Layer Techniques

The superimposed layer technique was first applied in 1930 by MARSHALL [56]. The superposition of a thin layer of suspension on top of a solids-free liquid free from disturbances is the basic problem of the superimposed or stratified layer technique.

In most cases the analyst experiences instabilities at the interface between the suspension and the solids-free liquid during or after the superposition. These instabilities depend on a number of variables [57]. To achieve a stable superposition free of instabilities, the *density gradient technique* must be applied [58].

Incremental concentration measurements in gravitational or centrifugal instruments and the direct measurement of a density distribution can be performed best with conventional photometers. Examples for cumulative gravity methods have been the *Micromerograph* [59], where the superposition is performed in air, the *Woods–Hole–Rapid–Sediment Analyzer* [60], and other methods [61], [62].

The methods are applicable mainly to size distributions with particles coarser than ca. 60×10^{-6} m.

Modern centrifuges no longer use beakers, but disk-shaped rotors. A typical example is the *Joyce–Loebl centrifuge*. A perspex rotor is filled with solids-free liquid at the beginning of the experiment, and the suspension sample to be analyzed is introduced near the axis of rotation with a hypodermic syringe. While adhering to the rotating wall, the suspension is radially accelerated and finally forms a thin suspension ring on top of the solids-free liquid, from which the particles start to settle. Instabilities can be avoided only if the density of the solids-free liquid surpasses at any radial position the suspension density formed by the particles and the surrounding liquid at slightly smaller radii.

6. Classification Methods

With all classification methods of particle size analysis, a given sample of particulate matter is separated into at least two size or settling rate classes, one comprising particles smaller and the other coarser than a certain size. The characterization and a general description have already been given in Chapter 3. Sieve analysis and air classification are the most important classification techniques used in particle characterization.

6.1. Sieve Analysis

The principle of sieve analysis is generally well known. Several sieves, one on top of each other, form a set, with the size of the sieve openings decreasing from top to bottom. A representative sample of the material to be analyzed is fed to the coarsest sieve, and the sieve column is vibrated either mechanically or electro-mechanically. The particles pass through the sieve openings, and the coarser ones are retained on the sieves. As soon as the particles cease to pass through the openings, the vibration is stopped. The amount of particles on each individual sieve is weighed, and the histogram of the density distribution is obtained by weight.

The mass of particles retained between two sieves can be measured relatively simply and with little error. The problem of sieve analysis lies in the determination of the cut sizes of the individual sieves. In most cases of practical application, the nominal width of a sieve opening is taken to represent its cut size. However, since the size of the openings of a sieve follows a distribution, the cut size does not necessarily coincide with the nominal width of the sieve openings, and a calibration procedure becomes necessary for high-quality sieve analyses.

Sieve cloths are standardized in most countries and internationally:

Federal Republic of Germany:

DIN 4188	woven wire test sieves
DIN 4187	perforated plate test sieves

United Kingdom:

BS 410	woven wire test sieves

United States:

ASTME 11	woven wire test sieves
ASTME 161	micromesh sieves (electroformed sieves)

The range of the widths of sieve openings varies from 5.5 μm (ASTM 161) or 25 μm (DIN 4188) to 125 mm, thus covering more than four powers of ten.

The particles can pass through the openings of a sieve cloth only if the sieve cloth and the particles move relative to each other. The relative motion may be produced either by a motion of the sieve cloth itself or when the particles are transported to the sieve opening with an air or liquid flow. Therefore, the quality of sieving depends on how often a particle is presented to a free sieve opening within a given period of time.

The sieving process also depends on further variables, which are correlated with the material to be sieved and the sieving principle. (→ 15. Screening) The latter includes intensity and frequency of relative motion, sieving aids, and the aperture width distributions at the sieves. Variables that depend on the material to be sieved are the solids loading on each individual sieve, the size distribution that is mainly the amount of particles, the size of which is similar to the width of the sieve openings, and particle shape and their tendency to clog the openings of the sieve. Furthermore, sieving depends on the tendency of fine particles to agglomerate, i.e., on the absolute size and the moisture content of the material to be sieved.

Many authors have investigated theoretically and experimentally how these variables influence sieving results [63]–[68]. However, until now a universal formula has not been found, describing the course of sieving with time, which takes into account the above variables.

Cut size, sharpness of cut, and the amount of material that has passed through a sieve depend on and increase with sieving time. As soon as a reasonable sharpness of cut has been obtained, cut size and undersize weight should be determined independently. A simple and reasonably accurate method to measure an equivalent to a cut size has been suggested [70], [71]. A counting and weighing technique is applied to the fraction of particles passing through the sieve just after stopping the shaking or sieving action. The average volume diameter, x_V, of this fraction is taken to be the cut size of the sieve [71]. It can be calculated from the mass, m_s, of n particles collected:

$$x_V = \sqrt[3]{\frac{6\,m_s}{\pi\,\varrho_s\,n}}$$

A subcommittee of the Bureau Commutative de Référence (BCR) has prepared a number of quartz samples for calibration purposes, using the above technique [71 a].

Results obtained for a coarse quartz are given in Figure 23. The cumulative distribution by mass shows an extremely low scatter. When this sample is used as calibration material, it is fed to a stack of sieves and sieving is performed under the same conditions as those chosen for the calibration. The over- or undersize weight values of each size are determined and the analytical cut sizes, x_a, can easily be obtained from the cumulative distribution curve of the calibration material (see Figure 23).

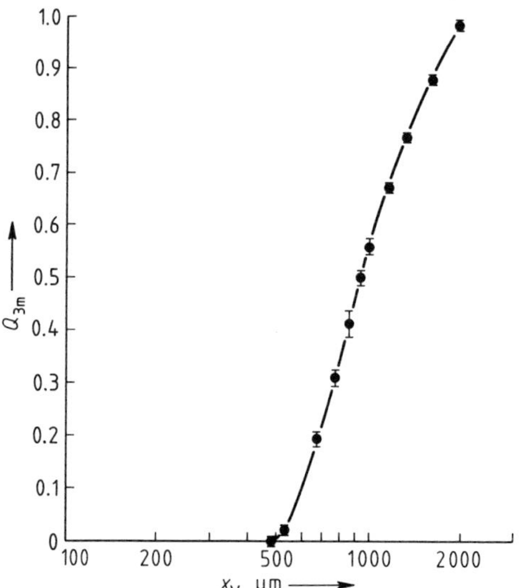

Figure 23. BCR certificate with extremely low scatter. BCR sample 131 was used and regression analysis performed.

6.2. Air Classification

Particles introduced into the flow field of the classification zone of an air classifier travel on trajectories to different points or areas of the classification zone, and are separated into at least two fractions, a fine and a coarse one. These trajectories depend on size, or more accurately, on settling rate. A distinction is made between the *counterflow* and the *cross flow principle*, which may be applied independently or in combination. The terms counterflow and cross flow have been used to describe the movement of the coarse particles either against or partly perpendicular to the main air flow (→ 17. Air Classifying).

Air classification as applied in particle size analysis should at least fulfill the following conditions:

1) It should be possible, either to precalculate the cut size from the main variables or to use a calibration curve for its determination.
2) Sharpness of cut should be as high as possible, and/or the amount of misplaced material should be the same in the fine and in the coarse fraction.
3) The coarse fraction should be recovered without any losses.

6.2.1. Cross-flow Classifiers

Cross-flow classifiers have been used mainly for the sampling of fine particles from aerosols and their size-dependent deposition on a carrier medium suitable for further size analysis. The principle is explained in Figure 24.

A cross-flow classifier basically consists of a horizontal tube of rectangular cross section. Solids-free air enters it from one side, and the particles to be classified are fed into the airstream through a small slit in the top wall. If plug flow and a negligible particle entry velocity, v_0, are assumed, the particle trajectories are straight lines, the slope of which is equal to the ratio of

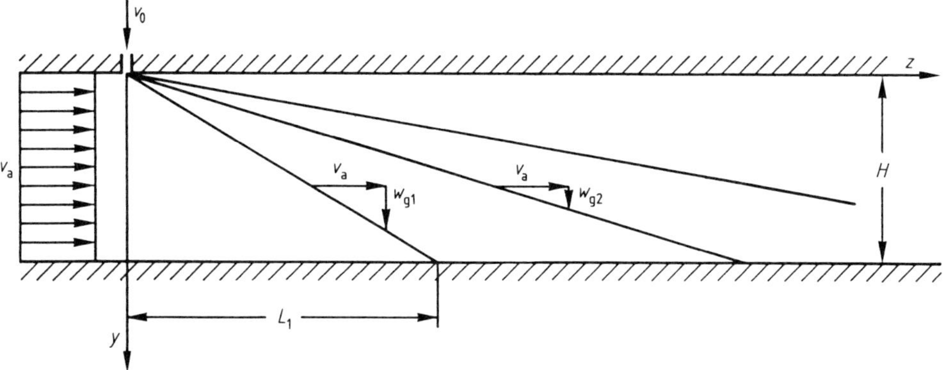

Figure 24. Principle of a cross flow classifier

settling rate, w_g, to the average air velocity, v_a, if the particles settle perpendicular to the flow under the influence of gravity. If the assumed plug flow is replaced with a laminar flow profile of the same average velocity, the particle trajectory changes, but the position L_1 at which a particle of settling rate $w_{g\,1}$ meets the bottom remains the same. The straight lines, i.e., the particle trajectories in Figure 24, are represented by Equation (103):

$$y = z \cdot w_g / v_a \tag{103}$$

and the point of deposition is equal to:

$$L = v_a H / w_g \tag{104}$$

Thus, the particles are spread at the bottom in a pattern that depends on settling rate. The particles may be removed, for instance, by adherence to a transparent, sticky tape and analyzed microscopically with respect to their geometrical size distribution. Typical examples for gravity applications have been suggested [72]–[74].

The above cross-flow principle may also be used with forces other than gravity that cause the particles to move perpendicular to the flow. These are centrifugal forces, electrical forces as in electrostatic precipitators, and thermal forces as used in thermal precipitators.

Aerosol Centrifuge. The first aerosol centrifuge using the cross flow principle was the *Conifuge* [75]–[78]. Its design has been changed to a spiral channel in a plane plate [79], which has later been improved [80]. A comprehensive description of the development of aerosol centrifuges is given in [81].

Impactor. For several years the use of impactors has become increasingly important. These instruments can be used for the analysis of solid particles or droplets. Its principle is shown in Figure 25. One stage of the impactor consists of a jet of diameter d arranged at a distance s perpendicular to a plate of diameter D. The air flow passing the jet impinges on the plate and bypasses it, forming a so-called stagnation point flow. Some of the coarse particles entrained in the air flow cannot follow the flow lines because of their inertia; they collide with the plate surface and are retained there. The impact depends not only on the velocity of the air and the particle when leaving the jet, but also on the particle's

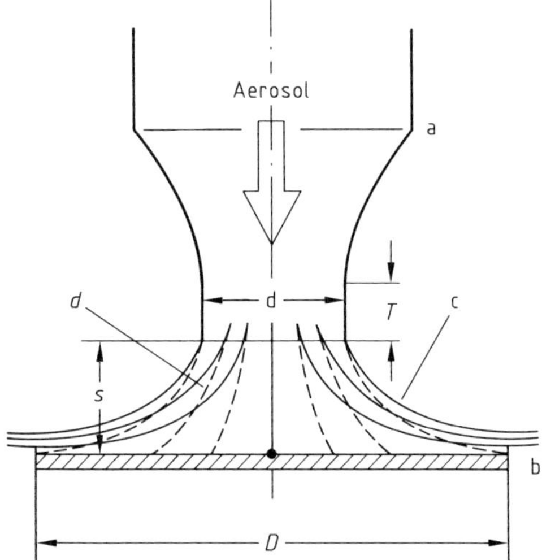

Figure 25. Principle of an impactor
a) Nozzle; b) Plate; c) Stream line; d) Particle trajectory

starting position in the jet orifice and the other variables already mentioned when the principle was described. A size-dependent grade efficiency curve is obtained, a great number of which have been calculated [82], [83]. The first impactor was described in 1945 [84]. If several impactor stages are connected in series with the cut size of each stage continuously decreasing, a *cascade impactor* is obtained.

6.2.2. Counterflow Classifiers

A distinction is made between counterflow classifiers in the gravity or centrifugal field.

Conventional *gravity counterflow classifiers* consist of a long, cylindrical tube of circular or rectangular cross section arranged vertically. The tube is tapered at its bottom, leading into a receptacle for the coarse material. Air is introduced at the bottom end, dispersing the particles which are then introduced into the rising flow. Fine particles follow the flow to the top end, while the coarse ones settle [85]–[88].

Centrifugal counterflow classifiers are commonly called *spiral classifiers*. The classification takes place in a flat, rotationally symmetrical classification zone. Air is introduced tangentially at its periphery, passing through it on spiral flow lines. Fine particles leave the classification zone with the air through a central orifice, while the coarse particles circle at the outer periphery and are removed from there continuously. The cut size of the classification process can be calculated from a radial force balance, which yields the following:

$$v_r = w_{at} \qquad (105)$$

The radial velocity component of the air, v_r, equals the centrifugal settling rate, w_{at}, of the particle, obtained with the acceleration a:

$$a = u_p^2/r \qquad (106)$$

Spiral classifiers were first investigated by RUMPF [89]; a laboratory spiral classifier has also been developed [90].

7. Counting Methods

In all counting methods, the measure of quantity is a number. To characterize the "size" of an irregularly shaped particle, many different physical particle properties are used. Number distributions determined by counting methods can be obtained only if the size of each particle is measured from its image. The physical properties used to measure size are as follows:

1) characteristic linear dimensions, chord lengths, the circumference, or the area of the projected areas of the particles
2) particle volume
3) the disturbance of an electromagnetic field caused by the particle passing through it; the most important techniques use the disturbance of an electrical field (Coulter principle) or of a light beam (optical methods).

The measurement may be performed with the particle ifself or its image; hence, the distinction is made between *direct* and *indirect counting methods*, respectively. Proper dispersion of a representative sample to be analyzed is one of the basic assumptions of all counting methods. Each single particle must be presented to the measuring zone of the instrument used. Since this can only be achieved at a negligible solids concentration, an error caused by the simultaneous presence of two or more particles in the measuring zone must be accepted. If a Poisson process is assumed to govern the particle approach to the measuring zone, and the probability for two particles being present in the zone is 5 %, the number concentration, C_N, per volume, V, of fluid may be calculated from Equation (107):

$$C_N = 0.05/V \qquad (107)$$

In most cases the sample size is fairly small and statistical errors may arise because only a limited number of particles are measured. In addition, all counting methods have a lower limit of recognition, because, e.g., optical microscopes cannot produce an image if the particle size drops below the wavelength of light. In direct counting methods, among other reasons, a lower size limit exists when the signal created by the particle is of the same order as the electronic noise of the amplifier. The total number of all particles passing through the measuring zone is, therefore, unknown in most cases, and the resulting number distributions are shifted toward coarser sizes.

7.1. Image Analysis

Image analysis is an indirect counting method. The scale may vary from 1:1000 and more (image reduction) to 10^5:1 (image magnifica-

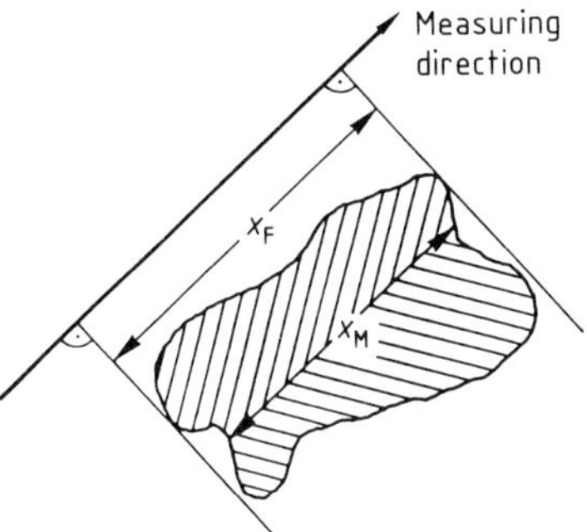

Figure 26. Feret and Martin diameter of a particle projection

Figure 27. Demonstration of fractal construction

tion), depending on the optical method used, (camera or optical or electron microscope). The problems and methods used in sample preparation are described in [91].

Of all particle size analysis methods, image analysis presumably yields the most definite size and shape description. However, if high accuracy is required, it is one of the most time-consuming and labor-intensive methods in use. Particle size is determined from the projected, hence two-dimensional, image of the particle. Figure 26 shows the sharp outline of the projected area of a particle. The so-called *Feret diameter*, x_F, is represented by the distance of two tangents to the particle contour perpendicular to the direction of measurement. The so-called *Martin diameter*, x_M, drawn parallel to the direction of measurement bisects the projected area. With an irregularly-shaped particle outline, each of these definitions yields different particle sizes if the direction of measurement changes. A distribution of *Feret* diameters is obtained, and an average value, $\bar{x}_F$, if the direction of measurement is rotated over 360°. The average *Feret* diameter, $\bar{x}_F$, is equal to the equivalent circumferential diameter, x_U, that is, the diameter of a circle of the same perimeter, U [92]:

$$\bar{x}_F = U/\pi = x_U \qquad (108)$$

Fractals. MANDELBROT introduced the use of fractals into the description of particle images [93]. With a pair of compasses set at a certain stride length, λ, estimation of the perimeter of the profile is made by walking the compasses around it to construct a polygon, the perimeter of which becomes the estimate of the profile at resolution, λ [94]. If this procedure is repeated at different stride lengths, λ, Figure 27 is obtained.

Equation (109) describes the relationship between perimeter, U, and stride length, λ:

$$U = k\lambda^{1-D} \qquad (109)$$

where D is the characteristic fractal dimension.

Fourier Analysis. The particle outline may also be described with Fourier analysis [95].

The deviation of irregularly-shaped particles from spheres can quantitatively be described by shape factors, ψ, which can be defined as the ratio of two different equivalent spherical diameters, x_α and x_β, with α and β representing two different physical properties measured on the same particle or the same fraction of particles.

$$\psi_{\alpha,\beta} = x_\alpha/x_\beta \qquad (110)$$

One of the best known definitions of a shape factor is the *sphericity*, ψ [4]. The sphericity ψ represents the ratio of the surface of the sphere of the same volume to the surface of the particle.

$$\psi = \pi x_V^2/\pi x_S^2 = (x_V/x_S)^2 = \psi_{V,S}^2 \qquad (111)$$

Shape factors may be used for substituting the abscissa of a size distribution, or more generally, for replacing one equivalent diameter by another.

Automatic Image Analysis. Image analysis can also be performed automatically [96]. A video camera is then used to evaluate the particle image. The cathode ray of the camera scans the image and produces brightness-dependent voltage information. The electrical analog or digitized signal is then, for instance, transformed into a better image, rich in contrast. Further mathematical algorithms yield other image modifications, such as shading, erosion, or dilatation. The

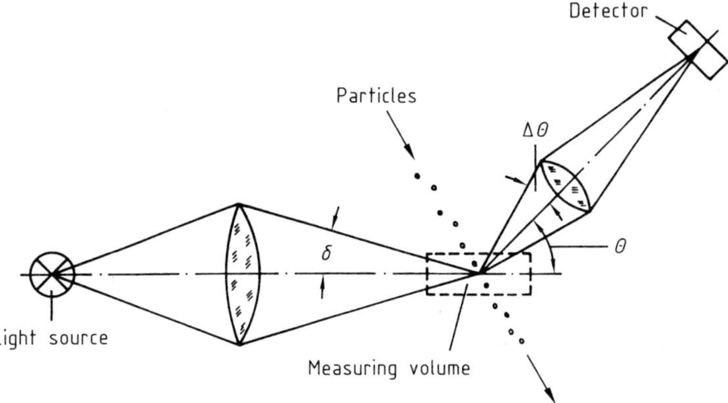

Figure 28. Optical setup for measuring scattered light

corrected data are evaluated with respect to different chord lengths, equivalent diameters, or projected area.

7.2. Electrical Sensing Zone Method

In electrical sensing, a homogeneous electrical field is set up in a cylindrical orifice. A particle passing through the orifice changes its electrical resistance and a voltage impulse is created, the height of which is directly proportional to the volume of the particle. With one particle after the other passing through the orifice, the number density distribution of particle volumes can be measured [97].

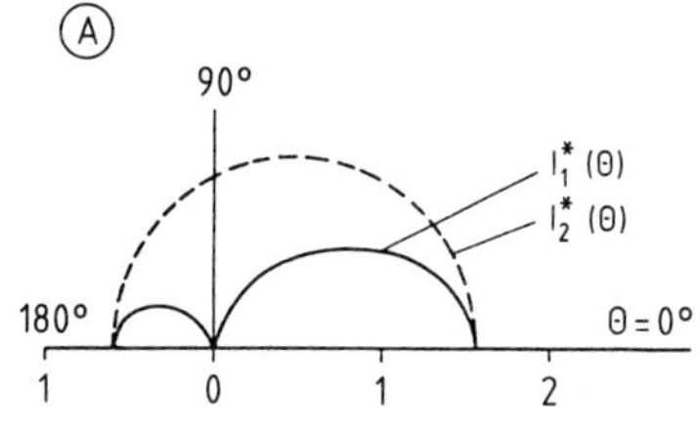

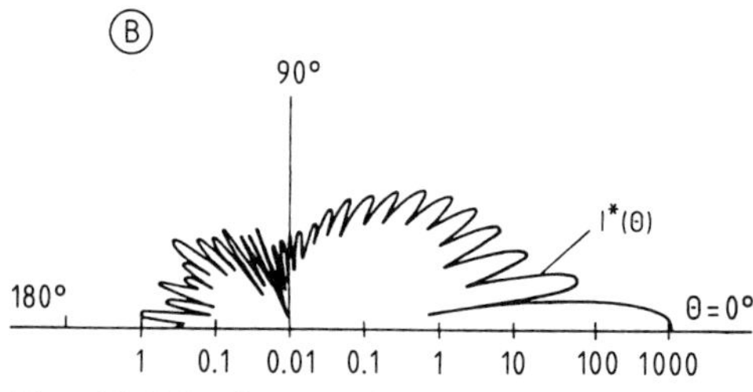

Figure 29. Polar diagrams of scattered light for two values of α

A) $\alpha = 1$, $n = 1.46$

B) $\alpha = 30$, $n = 1.33$

7.3. Measurement by Light Scattering or Absorption

Small particles scatter and absorb electromagnetic radiation. If a planar electromagnetic wave interferes with a homogeneous, isotropic sphere, it is scattered three-dimensionally. Scattering represents the deflection of a light beam by refraction, reflection, and diffraction. Absorption attenuates the light by converting it into different kinds of energy.

The schematic setup used for scattered light methods is shown in Figure 28. A small measur-

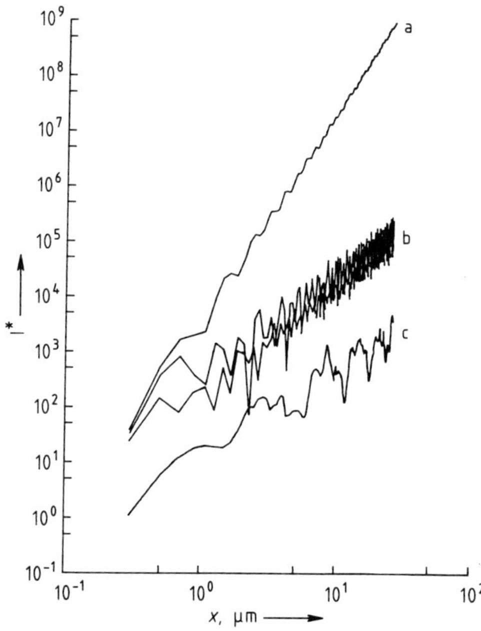

Figure 30. Scattered intensities of monochromatic light at different angles ($\lambda = 0.4$ μm, $n = 1.46$)

a) $\theta = 0°$; b) $\theta = 15°$; c) $\theta = 30°$; d) $\theta = 90°$

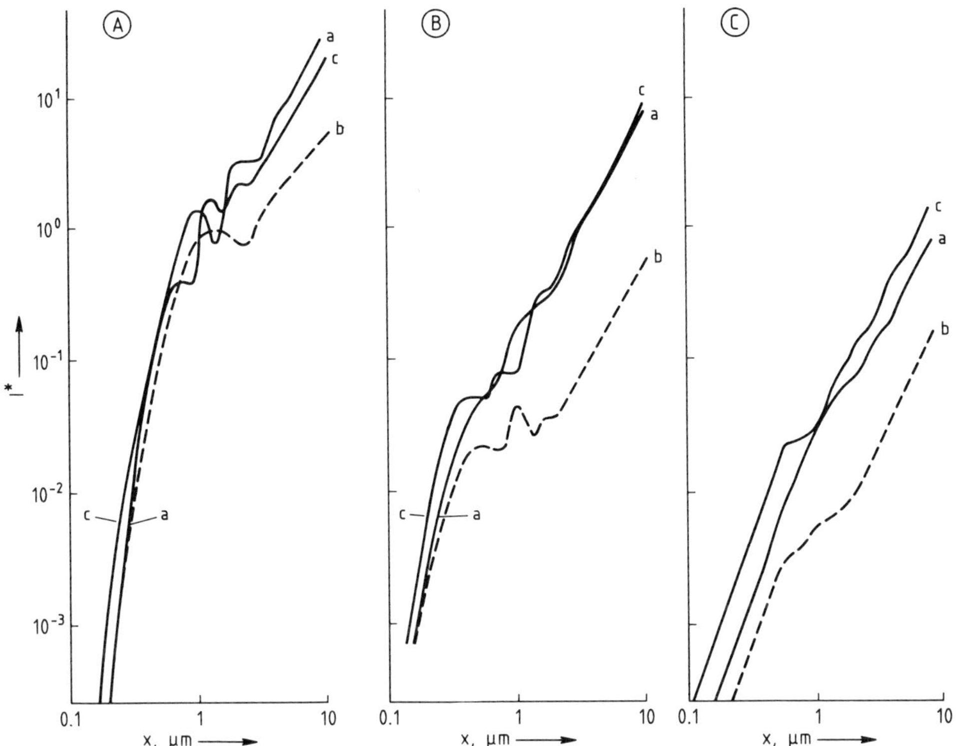

Figure 31. Scattered intensities of white light at $\theta = 17°$ (A), $\theta = 47°$ (B), and $\theta = 90°$ (C) and different refractive indices of spheres
a) $n = 1.46$; b) $n = 1.46 - 0.15\,i$; c) $n = 1.8$

ing volume receives the incident light beam from a light source and a suitable optical arrangement. An individual particle, passed through the measuring volume by a gas or a liquid stream, creates scattered light which is transmitted by a second optical system to a photoelectrical detector. The detector signal depends on the aperture, $\Delta\Theta$, of the second optical system and its position (angle Θ) with respect to the forward direction of light, where $\Theta = 0$.

The theory of light scattering was first described in 1908 [98]. This theory yields the distribution of scattered light in space for a sphere from the Maxwell field equations. Results are given in Figure 29. This figure shows the polar diagram of the scattered light. The two graphs indicate the differences obtained in the scattered light, with respect to the dimensionless particle size, $\alpha = \pi x/\lambda$. The value x represents the diameter of the sphere, and λ the wavelength of light. Figure 29 A shows dipole scattering in the so-called Rayleigh range. At higher α, the number of the maxima and minima of the polar diagram increases, as well as the intensity of light in the forward direction.

To use light scattering for the characterization of particles, a monotonic dependency is needed between the intensity of the scattered light, I^*, and particle size, x. Figure 30 [99] shows the theoretical dependency for monochromatic light with $\lambda = 0.4$ μm for different values of Θ. It is realized that an approximate monotonic dependency is obtained only at $\Theta = 0°$. At $x < 0.7$ μm; i.e., in the Rayleigh range, the scattered light drops proportionally to x^6.

Figure 31 shows similar curves for white light [100].

Dependencies close to monotonic ones are obtained only for $\Theta = 90°$ (Figure 31 C). Contrary to monochromatic light, however, the curves strongly depend on the refractive index, i.e., the material to be measured.

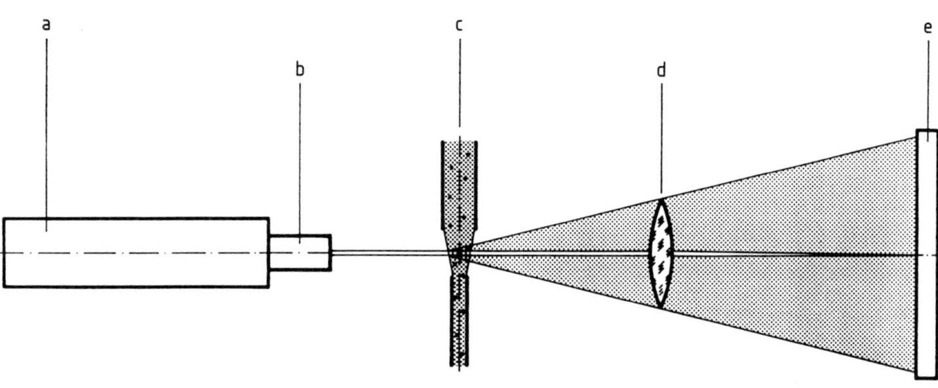

Figure 32. Optical setup for Fraunhofer diffraction pattern analysis
a) Laser; b) Beam expander; c) Aerosol beam; d) Lens; e) Detector

8. Fraunhofer Diffraction

In 1817 FRAUNHOFER proposed an optical setup for the generation of diffraction patterns, which was similar to that shown in Figure 32 [101]. It consists of a monochromatic light source, optical means to expand the diameter of the beam, a measuring zone, a lens, and a suitable detector. Spherical monosized particles passing through the measuring zone, dispersed in a gas or a liquid, create a radially symmetrical diffraction pattern, which consists of an extremely bright central spot, followed by a succession of dark and bright rings. This intensity distribution was first described in 1835 with an equation similar to Equation (111) which is generally called the Fraunhofer approximation [102]. With $\Theta < 8°$, $\tan \Theta = r/f$ and $k = \pi/\lambda f$, the following is obtained:

$$I(r, x)/I_0 = (k x^2/2)^2 (J_1(k r x)/k r x)^2 \quad (112)$$

where f represents the focal length, and $J_1(k r x)$ is the first kind, first-order Bessel function.

If the particles passing the measuring zone form a number density distribution, $q_0(x)$, the diffraction patterns are superimposed and a radially symmetrical, diffuse diffraction pattern with the following intensity distribution is obtained:

$$I(r) = \int_{x_{\min}}^{x_{\max}} N_0 q_0(x) I(r, x) \, dx \quad (113)$$

Equation (112) represents a first kind, first-order *Fredholm integral equation.* Its solution with respect to $q_0(x)$ was first given in 1955 [103], and then in 1956 [104], using the Mellin transformation [105]. The above solution has been used to develop a new diffraction pattern analyser (HELOS, Sympatec, Federal Republic of Germany). At present, three other instruments are available on the world market, manufactured by CILAS, France; Leeds and Northrup, United States; and Malvern, Great Britain.

9. References

[1] W. H. Westphal: *Physikalisches Wörterbuch*, Springer-Verlag, Berlin-Göttingen-Heidelberg 1952, pp. 238–239.

[2] K. Leschonski: "Representation and Evaluation of Particle Size Analysis Data," *Part. Charact.* **1** (1984) 89–95.

[3] K. Leschonski: "Particle Characterization, Present State and Possible Future Trends," *Prepr. 1. World Congress Particle Technology*, Part I, Particle Characterization, Nürnberg 1986, pp. 1–16.

[4] H. Wadell: "Volume, Shape and Roundness of Rock Particles," *J. Geol.* **40** (1932) 443–451.

[5] A. Cauchy: "Nôte sur divers theorèmes relatifs à la rectification des courbes," *C. R. Hebd. Seances Acad. Sci.* **13** (1841) 1060–1065.

[6] A. M. Gaudin: "An Investigation of Crushing Phenomena," *Trans. Am. Inst. Min. Metall. Pet. Eng.* **73** (1926) 253–316.

[7] R. Schumann: "Principles of Communition I–Size Distribution and Surface Calculations," Amer. Inst. of Mining and Metallurgical Engineers, Techn. Publ. No. 1189 (1940), pp. 1–11.

[8] P. Rosin, E. Rammler, U. Sperling: "Korngrößenprobleme des Kohlenstaubs," *Bericht C 52 des Reichskohlenrates*, Berlin 1933, pp. 1–33.

[9] E. Rammler: "Gesetzmäßigkeiten in der Kornverteilung zerkleinerter Stoffe," *Z. Ver. Dtsch. Ing., Beih. Verfahrenstech.* **5** (1937) 101–168.

[10] I. G. Bennett: "Broken Coal," *J. Inst. Fuel.* **10** (1936) 23–39.

[11] G. Herdan: *Small Particle Statistics*, Butterworths, London 1960.
[12] M. Langemann: "Zur Berechnung der spezifischen Oberflächen von Dispersionen," *Chem. Ing. Tech.* **27** (1955) 27–32.
[13] H. Heywood: "Numerical Definitions of Particle Size and Shape," *Chem. Ind.* **15** (1937) 149–154.
[14] K. Leschonski: "Classification of Particles in Gases," *IFPRI-Report*, 1981 K. Leschonski, Clausthal.
[15] *Ullmann*, 4th ed., **2**, 35–42.
[16] K. Leschonski, H. Herrmann: "Einfluß und Berücksichtigung von Fehlern der Partikelgrößenanalyse bei der Ermittlung von Trennkurven," *Proc. 2. Europ. Symp. "Partikelmeßtechnik,"* Nürnberg 1979, pp. 41–58.
[17] F. K. Tromp: "Neue Wege in der Beurteilung von Steinkohlen," *Glückauf* **73** (1937) 125–131; 151–156.
[18] F. W. Mayer: "Die Teilungskurve und ihre wichtigsten Anomalien," *Bergakademie* **9** (1957) 1–13.
[19] Th. Eder: "Probleme der Trennschärfe," *Aufbereit. Techn.* **2** (1961) 104–109; 136–148; 443–446; 484–495.
[20] H. Trawinski: "Die mathematische Formulierung der Tromp-Kurve," *Aufbereit. Tech.* **17** (1976) 248–254; 449–459.
[21] H. Heidenreich: *Die Erfolgsrechnung im Aufbereitungsbetrieb*, Verlag Glückauf, Essen 1954.
[22] F. W. Mayer: "Die Trennschärfe von Sichtern," *Zem. Kalk-Gips* **19** (1966) 259–268.
[23] F. W. Mayer: "Die Trennschärfe von im Mahlkreislauf arbeitenden Sichtern anhand von Tromp-Kurven," *Tech. Mitt.* **59** (1966) 246–249.
[24] Th. Eder: "Zur einheitlichen Kennzeichnung der Trennschärfe," *Montanzeitung, Zeitschrift für Bergbau und Hüttenwesen* **67** (1951) 163–165.
[25] K. Grumbrecht, F. W. Mayer: "Die Lage des Trennschnittes bei Trennvorgängen in der Aufbereitung," *Glückauf* **96** (1960) 186–188.
[26] H. Madel: "Über den Trennungsgrad von Windsichtern," *Zement* **19** (1930) 958–961.
[27] P. Rosin, E. Rammler: "Windsichter und ihre Untersuchung," *Glückauf* **68** (1932) 529–537.
[28] J. White: *J. Chem. Metall. Min. Soc. S. Africa* (1925).
[29] H. Rumpf: "Über die Feinheitsbestimmung von technischen Stäuben," *Staub* **25** (1965) 15–22.
[30] S. Pethö, E. Tompos: "Über die statistische Auswertung der Tromp-Kurven," *Bergakademie* **21** (1969) 430–433.
[31] F. W. Mayer: "Die Grundlagen der Trennschärfenermittlung von Aufbereitungsvorgängen mit Hilfe der Tromp'schen Verteilungszahlen," *Glückauf* **88** (1952) 4–13.
[32] F. W. Mayer: "Allgemeine Grundlagen der T-Kurven," *Aufbereit. Tech.* **8** (1967) 429–440; 673–678; **9** (1968) 14–23.
[33] F. W. Mayer: "Berechnung des neuen Trennschärfe-Kennwertes," *Aufbereit. Tech.* **12** (1971) 82–90; 203–211.
[34] K. Leschonski, Th. Boeck: "Die Bedeutung des Dispergierens und Dosierens bei Feinsttrennung in Windsichtern," *Prepr. Europ. Symp. Particle Technology 1980*, vol. B, Amsterdam 1980, pp. 746–761.
[35] K. Grumbrecht: "Über die Genauigkeit der Ermittlung von Mengenausbringen nach graphischen Methoden in der Steinkohleaufbereitung," *Glückauf* **89** (1953) 175–178.
[36] P. Moiset: "Über die Bestimmung der funktionellen Eigenschaften einer Klassiermaschine," *Aufbereit. Tech.* **2** (1961) 395–403.
[37] K. Koulen, H. Schneider: "Zur Berechnung des Gewichtsausbringes bei der Sichtung," *Aufbereit. Tech.* **6** (1965) 586–589.
[38] K. Leschonski: "Bewertung des Trenneffektes bei der Trennung von Suspensionen," *Proc. 17. Diskussionstagung Mechanische Flüssigkeitsabtrennung*, 21.10.80, TH Dresden, DDR.
[39] H. Heidenreich: "Zur Berechnung des Gewichtsausbringens in der Kohlenaufbereitung," Bergbau-Arch. **24** (1963) 27–34.
[40] H. Herrmann: "Ein allgemeines Verfahren zur Verbesserung des Meßfehlers in der Partikelgrößenanalyse," *Chem. Ing. Tech.* **51** (1979) 1140–1141.
[41] H. Herrmann: "Zur Ermittlung der Trennkurve—Fehleranalyse und Fehlerverbesserung," *Chem. Ing. Tech.* **52** (1980) 466–467.
[42] H. J. Smigerski: "Die Bilanzierung von Haufwerkstrennungen mit Hilfe mikroskopischer Korngrößenmessungen," *VI Verfahrenstech.* **11** (1977) 546–548.
[43] *Ullmann*, 4th ed., **5**, 725–753.
[44] DIN 66111, "Grundlagen der Sedimentationsanalyse".
[45] B. H. Kaye, R. P. Boardman: "Cluster Formation in Dilute Suspensions," *Proc. 3rd. Congr. Europ. Tech. Chem. Eng. London* 1962, pp. A 17–21.
[46] R. Johne: "Einfluß der Konzentration einer monodispersen Suspension auf die Sinkgeschwindigkeit ihrer Teilchen," *Fortschr. Ber. VDI Z., Reihe 9*, **11** (1966).
[47] B. Koglin: "Untersuchung zur Sedimentationsgeschwindigkeit von Einzelteilchen," *Freiberg. Forschungsh. A*, **A 484** (1971) 35–44.
[48] E. Cunningham: "On the Velocity of Steady Fall of Spherical Particles through Fluid Medium," *Proc. R. Soc. London Ser. A*, **83** (1909/10) 357–365.
[49] D. S. Jennings, M. D. Thomas, W. Gardner: "A New Method of Mechanical Analysis of Soils," *Soil Sci.* **14** (1922) 485–499.
[50] G. W. Robinsion: "A New Method for the Mechanical Analysis of Soils and other Dispersion," *J. Agric. Sci.* **12** (1922) 306–321.
[51] D. H. M. Andreasen: "Zur Kenntnis des Mahlgutes. Kap. D: Untersuchung über zulässige Meßunsicherheit nebst Berichten über die rechnerische Behandlung der Meßresultate," *Kolloidchem. Beih.* **27** (1928) 384–409.
[52] K. Leschonski: "Vergleichende Untersuchungen der Sedimentationsanalyse," *Staub* **22** (1962) 475–486.
[53] S. Berg: "Studies on Particle-Size Distribution," *Ingenioervidensk. Skr.* **2** (1940) 5–243. Akad. Tekn. Videnskaber u. Dansk Ing. Sorening, Kopenhagen.
[54] H. E. Rose, H. B. Lloyd: "On the Measurement of the Size Characteristics of Powders Photoextinction Methods, Part 1," *J. Soc. Chem. Ind. London* **65** (1946) 52–58.
[55] S. Odin: "Eine neue Methode zur Bestimmung der Körnerverteilung in Suspensionen," *Kolloid Z.* **28** (1916) 33–47.
[56] C. E. Marshall: "A New Method for Determining the Distribution Curve of Polydisperse Colloidal Systems," *Proc. Roy. Soc.* (*London*) *Ser* A **126** (1930) 427–439.

[57] H. Rumpf, W. Alex, R. Johne, K. Leschonski: "Korngrößenanalyse feiner Teilchen, eine kritische Betrachtung der Methoden," Ber. Bunsenges. Phys. Chemie **71** (1967) 253–270.
[58] *An Introduction to Density Gradient Centrifugation,* Beckmann Instruments Inc., Spinco Div., Palo Alto, Cal. USA 1960.
[59] F. S. Eadin, R. E. Payne: "Particle Size Distribution Analysed Quickly, Accurately," *Iron Age* **174** (1954) 99–102.
[60] J. M. Zeigler, G. G. I. R. Whitney, C. R. Hayes: "Woods Hole Rapid Sediment Analyser," *J. Sediment. Petrol.* **30** (1960) 490–495.
[61] B. Bienek, H. Huffmann, H. Meder: "Korngrößenanalyse mit Hilfe von Sedimentationswaagen," *Erdöl Kohle Erdgas Petrochem.* **18** (1965) 509–513.
[62] J. Brezina: "Sedimentation of Stratified Suspensions above the Stokes Range and its use for Grain Size Analysis," *2nd Int. Symp. Part. Size Anal.,* Bradford 9./11.09.1970.
[63] K. T. Whitby: *The mechanics of fine sieving,* Ph.D. Thesis. Univ. of Minnesota 1945.
[64] F. A. Shergold: "Effect of Sieve Loading on the Results of Sieve Analysis of Natural Sands," *Trans. Soc. Chem. Ind.* **65** (1946) 245–249.
[65] F. G. Carpenter, V. R. Deitz: "Methods of Sieve Analysis with Particular Reference to Bone Char," *J. Res. Natl. Bur. Stand. U.S.* **45** (1950) 328–346.
[66] W. Batel: "Kritische Betrachtungen zur Teilchengrößenbestimmung durch Siebanalyse, Windsichten, Sedimentieren und den Blaine-Test," Chem. Ing. Tech. **29** (1957) 581–589.
[67] B. H. Kaye: *Physical Problems of Particle Size Analysis,* Ph.D. Thesis, London Univ. 1962.
[68] H. E. Rose: "The Derivation of an Equation for the Performance of a Screen, " *DECHEMA Monogr.* **79** (1976) 351–369.
[69] A. H. M. Andreasen: "Eine selbstregistrierende Sedimentationswaage, die die Sedimentationskurve direkt aufzeichnet," *Sprechsaal* **60** (1927) 869–871.
[70] J. Andersen: "Über Maschenweiten und Korngrößen," *Zement* **20** (1931) 224–226, 242–245.
[71] K. Leschonski: "Sieve Analysis, the Cinderella of Particle Size Analysis Methods," *Powder Technol.* **24** (1979) 115–124.
[71a] Subcommittee 4.13.1 of Bureau Commutative de Référence, Rue de la Loi 200, B-1049 Bruxelles, Belgium.
[72] H. Polley, K. D. Friedberg: " Zur gravimetrischen Konzentrationsbestimmung von Fraktionen des Schwebestaubes mit einem Spaltkanalgerät," *Silikosebericht NRW, Arbeitsgemeinschaft Staub- und Silikosebekämpfung,* vol. 7, Verlag H. Bösmann GmbH, Detmold 1969, pp. 65–71.
[73] H. Boose: "Gerät zur fraktionierten Abscheidung von Staub für Korngrößenanalysen," *Staub Reinhalt. Luft* **22** (1962) 109–112.
[74] V. Timbrell: "The Terminal Velocity and Size of Airborne Dust Particles," *Br. J. Appl. Phys.* **5** (1954) Suppl. **3,** 86–90.
[75] K. F. Sawyer, W. H. Walton: "The Conifuge – A Size Separating Sampling Device for Airborne Particles," *J. Sci. Instrum.* **27** (1950) 272–276.
[76] C. H. Keith, J. C. Derrick: "Measurement of the Particle Size Distribution and Concentration of Cigarette Smoke by the "Conifuge" ", *J. Colloid Sci.* **15** (1960) 340–356.
[77] A. Goetz, J. R. Stevenson, O. Preining: "The Design and Performance of the Aerosol Spectrometer," *J. Air Pollut. Control. Assoc.* **10** (1960) 378–383, 414, 416.
[78] W. Stöber, U. Zessack: "Zur Theorie einer konischen Aerosolzentrifuge," *Staub Reinhalt. Luft* **24** (1964) 295–305.
[79] W. Kast: "Staubmeßgerät zur Schnellbestimmung der Staubkonzentration," *Staub Reinhalt. Luft* **21** (1961) 215–223.
[80] W. Stöber, H. Flachsbart: "High Resolution Aerodynamic Size Spectrometry of Quasi-Monodisperse Latex Spheres with a Spiral Centrifuge,"*J. Aerosol Sci.* **2** (1971) 103–116.
[81] W. Stöber: "Spiralenzentrifugen als Aerosolmeßgeräte," *Proc. 2. Europ. Symp. "Partikelmeßtechnik,"* Nürnberg, 24./26.09.1979, 91–148. Engl. Fassung: *Proc. Symp. "Fine Particles,"* US Env. Prot. Agency, Minneapolis, 28./30. May, 1975, Academic Press, New York 1976, pp. 351–397.
[82] V. A. Marple: *A Fundamental Study of Inertial Impactors,* Ph. D. Thesis, Univ. of Minnesota, Particle Technology Laboratory, Publ. No. 144 (1970).
[83] V. A. Marple, B. Y. H. Liu: "Characteristics of Laminar Jet Impactors," *Environ. Sci. Techn.* **8** (1974) 648–654.
[84] K. R. May: "The Cascade Impactor; An Instrument for Sampling Coarse Aerosols," *J. Sci. Instrum.* **22** (1945) 187–195.
[85] H. W. Gonell: "Ein Windsichtverfahren zur Bestimmung der Kornzusammensetzung staubförmiger Stoffe," *VDI Z.* **72** (1928) 945–950.
[86] P. S. Roller: "Measurement of Particle Size with an Accurate Air Analyzer. The Fineness and Particle Size Distribution of Portland Zement," *Proc. Am. Soc. Test. Mater.* **1932,** no. 59, 607–628.
[87] M. Weilbacher: *Untersuchungen zur Schwerkraft- und Fliehkraftwindsichtung für Teilchengrößenanalysen,* Diss. Universität Karlsruhe 1968.
[88] K. Leschonski, H. Rumpf: "Principle and Construction of Two New Air Classifiers for Size Analysis," *Powder Technol.* **2** (1968/69) 175–185.
[89] H. Rumpf: *Über die bei der Bewegung von Pulvern in spiraligen Luftströmungen auftretende Sichtwirkung,* Diss. TH Karlsruhe 1939.
[90] K. A. Gustavson: "Stoftcentrifug för Kornstorleksanalyse," *Tek. Tidskr.* **78** (1948) 667–670.
[91] W. Alex: "Prinzipien and Systematik der Zählverfahren in der Teilchengrößenanalyse," *Aufbereit. Tech.* **13** (1972) 105–111, 168–182, 639–652, 723–732.
[92] C. N. Davies: "Measurement of Particles," *Nature (London)* **195** (1962) 768–770.
[93] B. B. Mandelbrot: *"Fractals, Form, Chance, and Dimension,"* W. H. Freeman and Comp., San Francisco 1977.
[94] B. H. Kaye: "Multifractal Description of a Rugged Fineparticle Profile," *Part. Charact.* **1** (1984) 14–21.
[95] R. Weichert, D. Huller: "Volumenbestimmung und Formerkennung unregelmäßig geformter Partikeln mittels dreidimensionaler Bildanalyse," *Proc.2. Europ. Symp. "Partikelmeßtechnik,"* NMA, Nürnberg 24./26.09.79, pp. 600–615.

[96] K. Hermes, U. Kesten: "Die Bildanalyse als modernes Verfahren der Partikelmeßtechnik," *Chem. Ing. Tech.* **53** (1981) 780–786.

[97] Coulter-Counter: Equipment Description.

[98] G. Mie: "Beiträge zur Optik trüber Medien, speziell kolloidaler Metallösungen," *Ann. Phys.* 4. Folge 25 (1908) 377–445.

[99] Th. Boeck: *Entwicklung eines photometrischen online Oberflächenmeßverfahrens für trockene, disperse Feststoffe,* Diss. TU Clausthal 1983.

[100] R. Broßmann: *Die Lichtstreuung an kleinen Teilchen als Grundlage einer Teilchengrößenbestimmung,* Diss. TH Karlsruhe 1966.

[101] J. Fraunhofer: "Bestimmung des Brechungs- und Farbzerstreuungsvermögens verschiedener Glasarten," *Gilberts Ann. Phys.* **56** (1817) 193–226.

[102] G. B. Airy: "On the Diffraction of an Object Glass with Circular Aperture," *Trans. Cambridge Philos. Soc.* **5** (1835) 283–290.

[103] J. H. Chin, C. M. Sliepcevich, M. Tribus: "Determination of Particle Size Distributions in Polydispersed Systems," *J. Phys. Chem.* **59** (1955) 845–848.

[104] K. S. Shifrin: *Tr. VZLTI* (*Leningrad*) **2** (1956) 153–162.

[105] E. C. Titchmarsh: *Introduction to the Theory of Fourier Integrals,* Claredon Press. Oxford 1924.

[106] M. Heuer, K. Leschonski: "Erfahrungen mit einem neuen Gerät zur Messung von Partikelgrößenverteilungen aus Beugungsspektren," *3. Europ. Symp. "Partikelmeßtechnik,"* NMA Nürnberg, 9./11.05.84, pp. 515–537.

3. Crystallization and Precipitation

John W. Mullin, University College London, Torrington Place, London WC1E 7JE, United Kingdom

In addition to the standard symbols defined in the front matter of this volume, the following symbols are used:

a	activity
a^*	activity of a saturated solution
$a_{\pm}$	mean ionic activity
A	particle surface area, preexponential factor, m^2
b	kinetic order of secondary nucleation
B	secondary nucleation rate, $m^{-3}\,s^{-1}$
c	solution concentration
c^*	equilibrium saturation concentration
C	specific heat capacity
D	diffusivity, m^2/s
D	homogeneous distribution coefficient
g	kinetic order of growth
G	crystal growth rate, m/s
i	relative kinetic order ($= b/g$)
J	nucleation rate, $m^{-3}\,s^{-1}$
k	Boltzmann constant
K	rate constant, solubility product
L	crystal size, m
M	relative molecular mass
M_T	total magma density, kg/m^3
n	kinetic order of primary nucleation
n°	population density of nuclei ($= B/G$), m^{-4}
N	impeller rotational speed, Hz
q	heat of crystallization, J/kg
r_c	size of critical nucleus, m
R	ratio of molecular masses, mass deposition rate
S	supersaturation ratio ($= c/c^*$)
S'	activity supersaturation
v	molar volume, m^3/mol
$\bar{v}$	mean linear growth velocity, m/s
V	water lost by evaporation, kg/kg
W	initial mass of water, kg
x	mole fraction
Y	crystal yield, kg
z	valence
α	surface roughness factor, volume shape factor
β	surface shape factor
γ	interfacial tension, J/m^2
$\gamma_{\pm}$	mean ionic activity coefficient
λ	latent heat, heterogeneous distribution coefficient
ν	number of moles of ions
σ	relative supersaturation
τ	induction period, residence time, batch time, s

1. Introduction

Few branches of the chemical and process industries do not, at some stage, employ crystallization or precipitation for production or separation purposes. Vast quantities of crystalline substances are manufactured commercially: sodium chloride and sucrose, for example, have worldwide production rates exceeding 10^8 t/a, while annual rates for fertilizer chemicals such as ammonium nitrate, potassium chloride, ammonium phosphates, and urea exceed 10^6 t. Although crystalline products of the pharmaceutical, organic fine chemical, and dye industries are produced in relatively low tonnages, they still represent a valuable and important industrial sector. Crystallization is also a key operation in, for example, the desalination of seawater, the concentration of fruit juices, and the removal of unwanted materials or recovery of valuable constituents in many industrial processes (e.g., sodium sulfate removal from viscose spin-bath liquors, recovery of metal salts from electroplating baths). Crystallization is increasingly employed in the production of materials for the electronics industry; applications of precipitation have been extended to biotechnology, e.g., for processing proteins.

The unit operation of crystallization is governed by very complex, interacting variables. It is a simultaneous heat- and mass-transfer process with a strong dependence on fluid and particle mechanics. Crystallization occurs in a multiphase, multicomponent system and involves particulate solids whose size and size distribution are incapable of definition and vary with time. The solids are suspended in a solution which can fluctuate between a so-called metastable equilibrium and a labile state that is unstable and prone to change; the solution composition can also vary with time. Nucleation and growth kinetics, the key processes in this operation, are often influenced profoundly by traces of impurities: a few parts per million may alter the product beyond recognition.

Five main types of information are generally required to design a crystallization process:

1) Solubility and phase relationships
2) Metastability limits
3) Nucleation characteristics
4) Crystal growth characteristics
5) Hydrodynamics of crystal suspensions

Solubility and phase relationships influence the choice of crystallizer and method of operation. These data must be obtained by using the materials to be encountered in the plant because traces of impurity often have a considerable effect on phase relationships. Metastable limits define acceptable operating conditions for the minimization of uncontrolled nucleation and encrustation of heat-exchange surfaces. The processes of nucleation and growth are both exceedingly complex; they are influenced greatly by temperature, supersaturation, and impurities. A knowledge of these system characteristics is essential in design. Crystal suspension velocities must also be known so that liquor circulation rates in fluidized-bed crystallizers and agitation rates in stirred vessels can be specified. Because the crystals are present in large quantities, settling is hindered; further complications can arise if their shapes are irregular.

Many different methods are available for crystallization. Crystals can be grown from the liquid (solution or melt) or vapor phase (desublimation, see → 5. Sublimation, **B3**), but in all cases the state of supersaturation must first be achieved. The method used to obtain supersaturation depends on the characteristics of the crystallizing system: some solutes are readily deposited from solution when cooled, whereas others may crystallize only after some solvent has been removed. The addition of another substance to the system to alter the equilibrium conditions is used frequently in precipitation processes. Supersaturation is sometimes achieved as a result of a chemical reaction between two or more substances and one of the reaction products is precipitated.

Comprehensive accounts of industrial crystallization have been presented in several books [1]–[4], review articles [5]–[8] and symposium proceedings [9]–[13].

2. System Properties

2.1. Saturation and Supersaturation

A saturated solution is a solution that is in thermodynamic equilibrium with the solid phase of its solute at a specified temperature. However, solutions frequently contain more dissolved solute than that given by the equilibrium saturation value and are then said to be supersaturated. The degree of supersaturation can be expressed by the concentration difference Δc:

$$\Delta c = c - c^* \quad (1)$$

where c is the actual solution concentration and c^* the equilibrium saturation value. Other common expressions are the supersaturation ratio S and the relative supersaturation σ which are both dimensionless:

$$S = c/c^* \quad (2)$$

$$\sigma = \Delta c/c^* = S - 1 \quad (3)$$

Solution concentration may be expressed in a variety of units, but for general mass balance calculations, units such as kilograms of anhydrate per kilogram of solvent or kilograms of hydrate per kilogram of free solvent are most convenient. The former avoids complications if different phases (e.g., anhydrates and hydrates) can crystallize over the temperature range considered. The latter simplifies yield calculations when a single hydrate phase crystallizes.

Although the terms S and σ are dimensionless, their magnitudes depend on the units used to express solution concentration. For example, a supersaturated solution of sucrose at 20 °C contains 2.45 kg of sucrose per kilogram of water, and the corresponding value for c^* is 2.04; therefore, the value of S (= c/c^*) is 1.20. However, if the composition is expressed as kilograms of sucrose per kilogram of solution ($c = 0.710$, $c^* = 0.671$), the value of S becomes 1.06 [3].

None of the above expressions represents exactly the true thermodynamic supersaturation. The fundamental driving force for crystallization is the difference between the chemical potential of a given substance in the transferring and the transferred state, i.e., in solution (state 1) and in the crystal (state 2). For an unsolvated solute crystallizing from a binary solution, this may be written as

$$\Delta\mu = \mu_1 - \mu_2 \quad (4)$$

The chemical potential μ is defined in terms of the standard potential μ_0 and the activity a by

$$\mu = \mu_0 + RT \ln a \quad (5)$$

The fundamental dimensionless driving force for crystallization may, therefore, be expressed as

$$\Delta\mu/RT = \ln (a/a^*) = \ln S' \quad (6)$$

where a^* is the activity of a saturated solution, S' the activity supersaturation, R the gas constant, and T the absolute temperature; i.e.,

$$S' = \exp(\Delta\mu/RT) \quad (7)$$

For electrolyte solutions, use of the mean ionic activity $a_\pm$ is more appropriate; this is defined by

$$a = a_\pm^\nu \quad (8)$$

where ν is the number of moles of positive and negative ions in one mole of solute ($\nu = \nu_+ + \nu_-$). Therefore

$$\Delta\mu/RT = \nu \ln S_a \quad (9)$$

where

$$S_a = a_\pm/a_\pm^* \quad (10)$$

Alternatively, supersaturation can be expressed as

$$\sigma_a = S_a - 1 \quad (11)$$

and Equation (9) becomes

$$\Delta\mu/RT = \nu \ln(1 + \sigma_a) \quad (12)$$

For low supersaturations ($\sigma_a < 0.1$), the following approximation is valid:

$$\Delta\mu/RT \cong \nu\sigma_a \quad (13)$$

In practice, however, supersaturation is generally expressed directly in terms of solution concentration (Eqs. 1–3). The relationship between concentration-based supersaturation and fundamental (activity-based) supersaturation can be expressed by means of the relevant concentration-dependent activity coefficient ratios; for details see [14].

Metastability. The state of supersaturation is an essential feature of all crystallization operations. OSTWALD [15] introduced the terms *labile* and *metastable* supersaturation to describe supersaturated solutions in which spontaneous (primary) nucleation (see Section 4.1) would or would not occur, respectively. MIERS and ISAAC [16] proposed a representation of the metastable zone by means of a solubility–supersolubility diagram (Fig. 1). The lower, continuous, equilibrium solubility curve can be determined accurately, but the position of the upper, broken, supersolubility curve is less certain because it is influenced considerably by factors such as the rate at which supersaturation is generated, the intensity of agitation, and the presence of crystals or impurities. The width of the metastable zone is usually expressed as a temperature difference $\Delta\theta$, which is related to the corresponding concentration difference Δc by the local slope of the solubility curve, $dc^*/d\theta$:

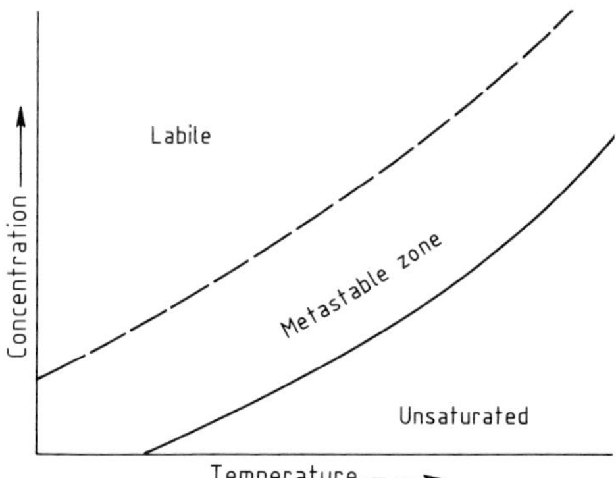

Figure 1. Solubility–supersolubility diagram

$$\Delta c = (dc^*/d\theta)\Delta\theta \quad (14)$$

The measurement of metastable zone width is discussed in Section 4.1.3.

Supersaturation Measurement. The supersaturation of a solution can be calculated simply from Equations (1)–(3) if the actual solution concentration and the corresponding equilibrium saturation concentration at a given temperature are known. Collections of solubility data for two- and three-component systems have been compiled [17]–[19]. A relationship of the form

$$\log x = A + (B/T) + C \log T \quad (15)$$

where x is the solute concentration expressed as a mole fraction, T is the absolute temperature, and A, B, and C are empirical constants, is generally suitable for correlating solubility data.

Many ways exist for measuring supersaturation (i.e., concentration), but all of these are not readily applicable to industrial crystallization. If chemical analysis is difficult, measurement of a concentration-dependent property of the system (e.g., density or refractive index) may be possible. Both of these properties can usually be measured with high precision on a sample transferred under controlled laboratory conditions.

On the other hand, an in situ, preferably continuous method of concentration measurement is usually required for a crystallizer operating under laboratory or pilot-plant conditions. Although the above properties are temperature-dependent, they can often be measured with sufficient accuracy for supersaturation determination. In industrial crystallization, temperature and feedstock concentration can fluctuate and thus make the assessment of supersaturation difficult. Under these conditions, a crude method based on a mass balance coupled with feedstock and exit-liquor concentrations and crystal production rates averaged over several hours may be adequate.

2.2. Crystal Size and Solubility

If the solute particles dispersed in a solution are small enough, the solute concentration may greatly exceed the normal equilibrium saturation value. The relationship between particle size and solubility, first derived for liquid–vapor systems by THOMSON (1870), utilized by GIBBS (1890), and applied to solid–liquid systems by OSTWALD (1900) and FREUNDLICH (1909), may be expressed in the form

$$\ln\left[\frac{c(r)}{c^*}\right] = \frac{2\,M\gamma}{\nu RT\varrho r} \qquad (16)$$

where $c(r)$ is the solubility of particles with radius r, c^* the normal equilibrium solubility of the substance, R the gas constant, T the absolute temperature, ϱ the density of the solid, M the relative molecular mass of the solute in solution, and γ the interfacial tension of the crystallization surface in contact with its solution. The quantity ν represents the number of moles of ions formed from one mole of electrolyte. For a nonelectrolyte, $\nu = 1$.

For most inorganic salts in water, the solubility increase becomes significant only for particle sizes smaller than ca. 1 μm. For example, for barium sulfate at 25 °C: $T = 298$ K, $M = 0.233$ kg/mol, $\nu = 2$, $\varrho = 4500$ kg/m^3, $\gamma = 0.13$ J/m^2, $R = 8.3$ J mol^{-1} K^{-1}. Thus for a 1-μm crystal ($r = 5 \times 10^{-7}$ m), $c/c^* = 1.005$ (i.e., 0.5% increase); for 0.1 μm, $c/c^* = 1.06$ (i.e., 6% increase); and for 0.01 μm, $c/c^* = 1.72$ (i.e., 72% increase). For a soluble organic compound such as sucrose ($M = 0.342$ kg/mol, $\nu = 1$, $\varrho = 1590$ kg/m^3, $\gamma = 0.01$ J/m^2), the effect is much more marked: 1 μm (4% increase); 0.01 μm (3000%). These numbers are, however, merely approximations since, for illustrative purposes, the values of interfacial tension γ have only been estimated.

2.3. Effect of Impurities

Industrial solutions invariably contain dissolved impurities which can increase or decrease the solubility of the prime solute considerably. Therefore, the solubility data used to design crystallization processes must relate to the actual system used. Impurities can also have profound effects on other crystallization characteristics such as nucleation and growth (cf. Chap. 4).

3. Phase Equilibria

3.1. One-Component Systems

Temperature and pressure are the two variables that can affect the phase equilibria in a one-component system. The phase diagram in Figure 2 depicts equilibria between the solid, liquid, and vapor states of water: ice, water, and steam, respectively. At the *triple point* B (0.6 kPa, 0.01 °C), all three phases are in equilibrium.

The *sublimation curve* AB records the vapor pressure of ice, the *vaporization curve* BC records the vapor pressure of liquid water, and the *fusion curve* BD records the effect of pressure on the

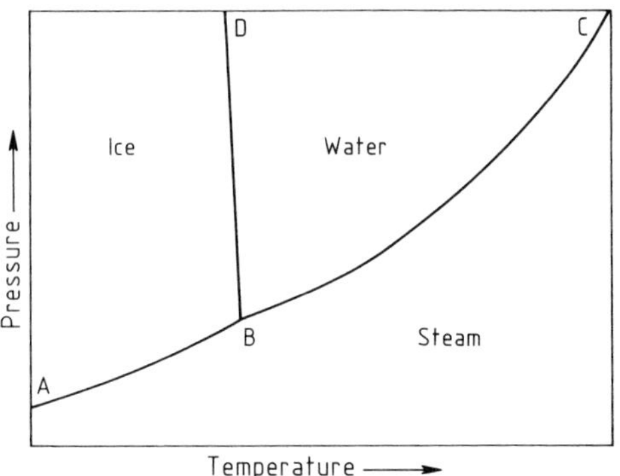

Figure 2. Phase diagram for water

melting point of ice. The fusion curve for ice is very unusual because an increase of pressure increases the melting point of most other one-component systems.

Polymorphs. A single substance can crystallize in more than one of seven crystal systems, 32 classes, or 230 point groups. Because these crystalline forms differ in their lattice arrangement, they can exhibit not only different basic shapes but also different physical properties. A substance capable of crystallizing in more than one different crystalline form is said to exhibit *polymorphism,* and the different forms are called *polymorphs.* Carbon, for example, has two polymorphs: graphite (hexagonal) and diamond (regular). Calcium carbonate has three: calcite (hexagonal), aragonite (tetragonal), and vaterite (trigonal).

Each polymorph is composed of a common component but constitutes a separate phase. Since only one polymorph is thermodynamically stable at a specified temperature and pressure, all the other polymorphs are potentially capable of being transformed into the stable polymorph. Some polymorphic transformations are rapid and reversible, others are not.

Polymorphs may be *enantiotropic* (interconvertible) or *monotropic* (incapable of transformation). Graphite and carbon, for example, are monotropic at ambient temperature and pressure, whereas ammonium nitrate has five enantiotropic polymorphs over the temperature range − 18 to 125 °C:

$$\text{Regular} \underset{125\,°C}{\rightleftharpoons} \text{Trigonal} \underset{84\,°C}{\rightleftharpoons} \text{Orthorhombic(I)}$$

$$\underset{32\,°C}{\rightleftharpoons} \text{Orthorhombic(II)} \underset{-18\,°C}{\rightleftharpoons} \text{Tetragonal}$$

Figure 3 A shows the phase reactions exhibited by two enantiotropic forms, α and β, of the same substance. AB is the vapor pressure curve for the α form; BC, for the β form; and CD, for the liquid. Point B, where the vapor pressure curves of the two solids intersect, is the transition point at which the two forms can coexist in equilibrium at the specified temperature and pressure. Point C is a triple point at which vapor, liquid, and β solid can coexist; it can be considered the melting point of the β form. If the α solid is heated slowly, it changes into the β solid and finally melts: vapor pressure curve ABC is followed. Conversely, if the liquid is cooled slowly, the β form crystallizes first and then changes to the α form. Rapid heating or cooling, however,

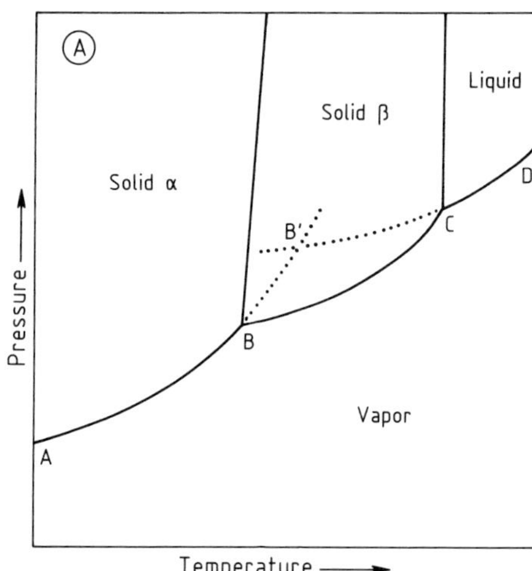

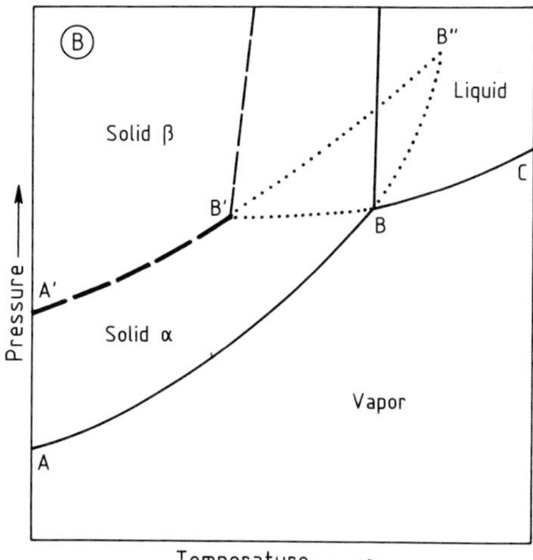

Figure 3. Phase diagrams for polymorphic substances exhibiting enantiotropy (A) and monotropy (B)

can result in different behavior. The vapor pressure of the α form changes along curve BB′, a continuation of AB, the α form being metastable in section BB′. Similarly, the liquid vapor pressure changes along curve CB′, a continuation of DC, the liquid being metastable in sector B′C. Point B′, therefore, is a metastable triple point at which the liquid, vapor, and α solid can coexist in metastable equilibrium.

Figure 3 B shows the pressure–temperature curves for a monotropic substance: AB and BC are the vapor pressure curves for the α solid and liquid, respectively, and A′B′ is that for the β solid. The vapor pressure curves of the α and β

forms do not intersect, so no transition point occurs. The solid form with the higher vapor pressure at any given temperature (β in this case) is the metastable form. Curves BB′ and BB″ are vapor pressure curves for the liquid and metastable α solid, respectively; thus B′ is a metastable triple point. If this system were to exhibit a true transition point, it would lie at point B″, but because this represents a temperature higher than the melting point of the solid, it cannot exist.

3.2. Two-Component Systems

Three variables (temperature, pressure, and concentration) can affect the phase equilibria in a two-component (binary) system. The effect of pressure, which is usually negligible, is generally ignored so that the relevant data can be shown on a two-dimensional temperature–concentration plot.

Three important basic types of binary system — eutectics, solid solutions, and systems with compound formation — are discussed. They are also referred to in the discussion of melt crystallization (Chap. 6). Although the terminology used is specific to melt systems, the types of behavior described may also be exhibited by aqueous solutions of salts, for example. In fact, no fundamental difference exists between a melt and a solution [3].

3.2.1. Eutectics

An example of a binary eutectic system AB is shown schematically in Figure 4. A eutectic (Greek: *eu tektos* = easily melted) is the mixture of components that has the lowest crystallization temperature in the system. The concentration of component B in system AB is plotted as the abscissa and temperature as the ordinate. Point A is the crystallization temperature (freezing point) of pure component A, and point B that of component B. Curves AE and EB represent the crystallization temperatures of all mixtures of the two components A and B. Above the two curves, all mixtures of A and B exist only in the liquid state, i.e., as a melt. If a melt represented by point X is cooled along the vertical line XZ, crystals will start to be deposited at point Y (with proper initiation); theoretically, these crystals should be pure component B. On further cooling, more

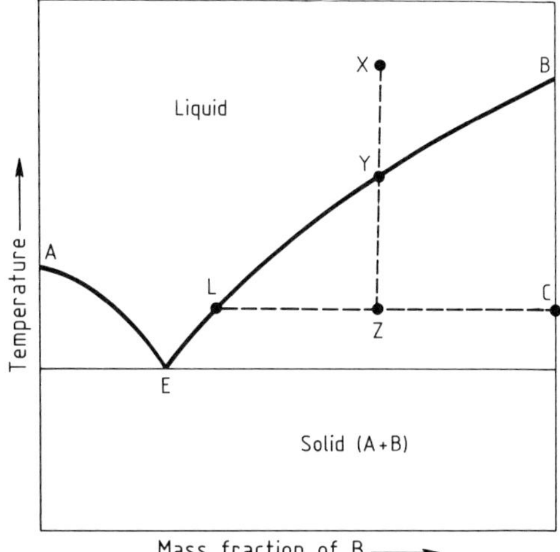

Figure 4. Phase diagram for a binary eutectic system AB For explanation of symbols, see text.

crystals of pure component B will be deposited until, at the eutectic point E, the system solidifies completely. The horizontal line passing through point E is called the *solidus*. The liquid-phase composition during cooling changes continuously along curve BE, which is called the *liquidus*. For example, if mixture X is cooled to point Z, the crystals C will be pure component B and the liquid L will be a mixture of A and B. The mass proportion of solid phase (crystal) to liquid phase (residual melt) at temperature Z is quantified by the ratio of the lengths LZ to CZ. This relationship is known by various names such as the *mixture rule* or the *lever rule*. Similar reasoning can be applied to mixtures located in the region of curve AE, in which case the crystals would be pure component A.

The eutectic point E is common to both curves. A liquid of this composition cooled to the eutectic temperature crystallizes with unchanged composition, and continues to deposit such crystals until the whole system solidifies. Although a eutectic in a given system has a fixed composition, it is not a chemical compound but simply a physical mixture of the individual components. The component crystals are often clearly visible under a low-power microscope.

3.2.2. Solid Solutions

The second common type of binary system is the continuous series of solid solutions. The term *solid solution* or *mixed crystal* refers to an intimate mixture, on the molecular scale, of two or more components. The components of a solid-solution system cannot be separated as easily as those of a eutectic system, as shown in Figure 5. Points A and B represent the crystallization temperatures (freezing points) of pure components A and B, respectively. The upper curve (the liquidus) represents the temperature at which mixtures of A and B begin to crystallize on cooling. The lower curve (the solidus) represents temperatures at which mixtures begin to melt on heating. A melt of composition X begins to crystallize on reaching temperature Y. At temperature Z, the system consists of a mixture of crystals with composition C and a liquid with composition L. The ratio of crystals to liquid is again given by the mixture rule (see Section 3.2.1). However, the crystals do not consist of a single pure component (as in a simple eutectic system) but are an intimate mixture (a solid solution) of components A and B. To purify the crystals further, they must first be melted; the resulting melt must then be cooled and recrystallized. This sequence may have to be repeated many times to achieve the desired purity.

In brief, a simple eutectic system may be purified in a single-stage crystallization operation, whereas a solid-solution system always needs a multistage operation.

3.2.3. Compound Formation

The solute and solvent of a binary system can combine to form one or more different compounds, for example, hydrates in aqueous solutions. If the compound can coexist in stable equilibrium with a liquid phase of the same composition, it is said to have a *congruent* melting point; i.e., melting occurs without change in composition (Fig. 6). If it cannot, the melting point is said to be *incongruent* (Fig. 7).

In Figure 6, the heating–cooling cycle follows the vertical line through point D; i.e., melting and crystallization occur without any change of composition. In Figure 7, however, compound D decomposes at a temperature T_1 below its theoretical melting point T_2. Thus, if compound D is heated, melting begins at temperature T_1, but it is not complete. At temperature T_1, the system of overall composition D contains crystals of pure component B in a melt of composition C. If this mixture is then cooled, a solid mixture of B and C is obtained. Subsequent heating and cooling cycles result in further decomposition of compound D.

A considerable amount of interest currently exists in the use of inorganic salt hydrates as heat-storage materials, particularly for storage of solar heat in domestic and industrial space heating. Ideally, the hydrate should have a congruent melting point so that sequences of crystallization–

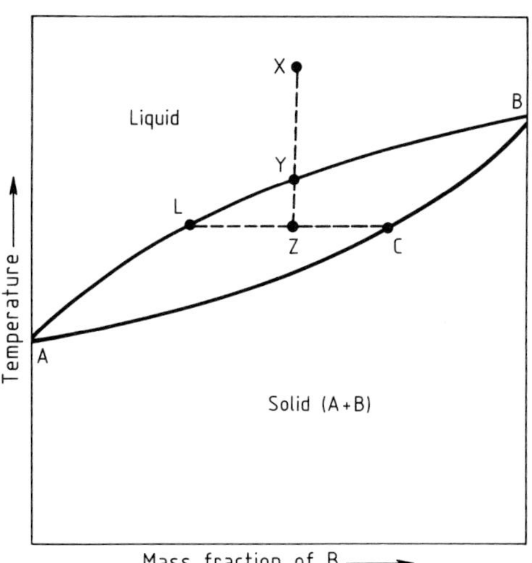

Figure 5. Phase diagram for a binary system AB composed of a continuous series of solid solutions
For explanation of symbols, see text.

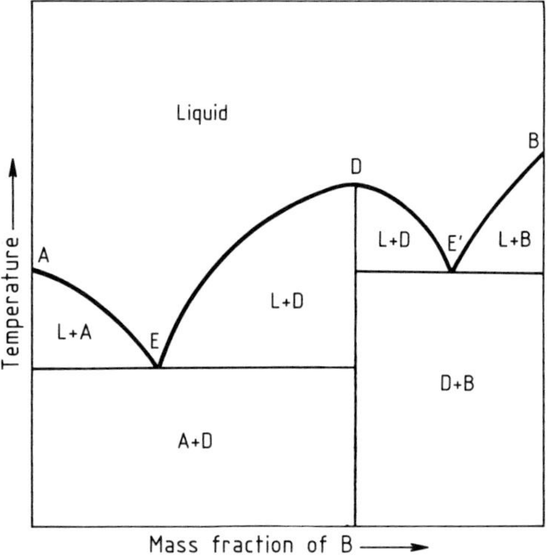

Figure 6. Phase diagram of a binary system AB that forms a compound D with a congruent melting point
Symbols indicate the phases coexisting in equilibrium: L = liquid; E, E′ = eutectics.

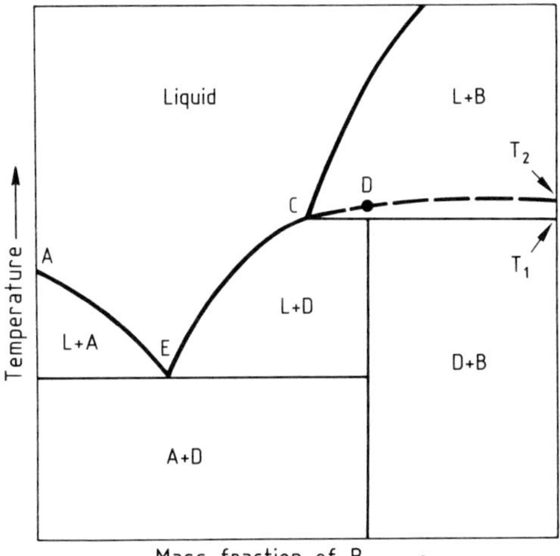

Figure 7. Phase diagram of a binary system AB that forms a compound D with an incongruent melting point
Symbols indicate the phases coexisting in equilibrium: L = liquid; E = eutectic; T_1 = decomposition temperature of D; T_2 = theoretical melting point of D.

melting–crystallization can be repeated indefinitely. Incongruently melting hydrate systems tend to stratify on repeated temperature cycling, with a consequent loss of efficiency. This occurs because melting yields a liquid phase that contains crystals of a lower hydrate or the anhydrous salt, which settle to the bottom of the container and fail to redissolve on subsequent heating.

Calcium chloride hexahydrate, although not a true congruently melting hydrate, appears to be one of the most promising materials [20], [21]. Sodium sulfate decahydrate, sodium acetate trihydrate, and sodium thiosulfate pentahydrate, which have incongruent melting points, are also worthy of mention [22], [23].

3.3. Three-Component Systems

The phase equilibria in three-component (ternary) systems can be affected by four variables: temperature, pressure, and the compositions of any two of the three components. The effect of pressure (usually negligible in normal ranges) is generally ignored, and the phase equilibria are plotted on an isothermal triangular diagram.

Equilibrium relationships in three-component systems can be represented on a temperature–concentration space model as shown in Figure 8 A. The ternary system 2-, 3-, and 4-nitrophenol (*o*-, *m*-, and *p*-nitrophenol), in which compound formation does not occur, has been chosen for illustrative purposes. The three components are referred to as O, M, and P, respectively. Points O′, M′, and P′ on the vertical edges of the model represent the melting points of the pure components *o*- (45 °C), *m*- (97 °C) and *p*-nitrophenol (114 °C). The vertical faces of the prism represent temperature–concentration diagrams for the three binary eutectic systems O–M, O–P, and M–P, which are all similar to that shown in Figure 4.

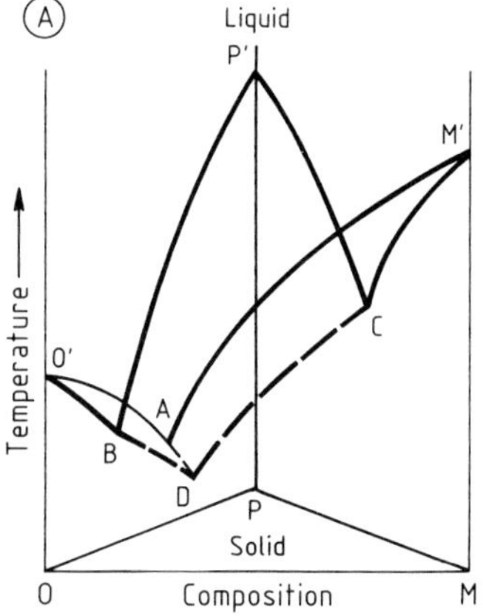

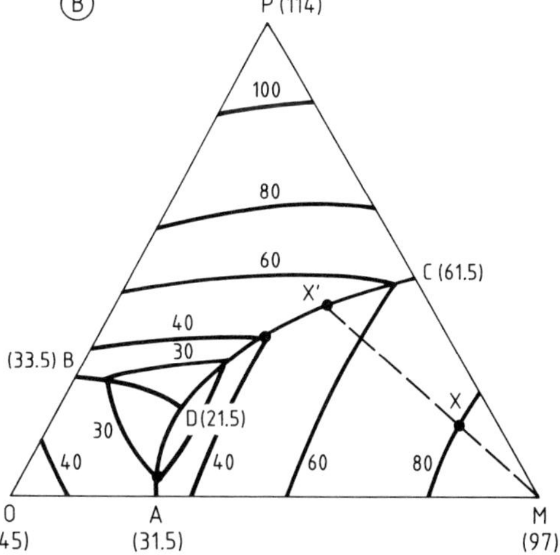

Figure 8. Eutectic formation in the ternary system *o*-, *m*- and *p*-nitrophenol
A) Temperature–concentration space model; B) Projection on a triangular diagram
Numerical values represent temperatures in °C.

The binary eutectics are represented by points A (31.5 °C; 72.5 % O, 27.5 % M), B (33.5 °C; 75.5 % O, 24.5 % P), and C (61.5 °C; 54.8 % M, 45.2 % P). Curve AD within the prism represents the effect of addition of component P to the O–M binary eutectic A. Similarly, curves BD and CD denote the lowering of freezing points of the binary eutectics B and C, respectively, upon addition of the third component. Point D, indicating the lowest temperature at which solid and liquid phases can coexist in equilibrium in this system, is a ternary eutectic point (21.5 °C; 57.7% O, 23.2% M, 19.1% P). At this temperature and concentration, the liquid freezes to form a solid mixture of the three components. The section of the space model above the freezing point surfaces formed by the liquidus curves represents the homogeneous liquid phase. The section below these surfaces down to a temperature represented by point D denotes solid and liquid phases in equilibrium. The section of the model below this temperature represents a completely solidified system.

Figure 8 B is the projection of the curves AD, BD, and CD in Figure 8 A onto the triangular base of the prism. The apexes of the triangle represent pure components O, M, and P, and their melting points are indicated in parentheses. Points A, B, and C on the sides of the triangle indicate the three binary eutectic points; point D is the ternary eutectic point. The projection diagram is divided by curves AD, BD, and CD into three regions which denote the three liquidus surfaces in the space model. The temperature falls from the apexes and sides of the triangle toward the eutectic point D; several isotherms showing points on the liquidus surfaces are also drawn. The phase reactions occurring when a given ternary mixture is cooled can be traced. A molten mixture with a composition X starts to solidify when the temperature is reduced to 80 °C. Point X lies in the region ADCM, so pure *m*-nitrophenol is deposited when the temperature is decreased. The composition of the remaining melt changes along line MX′ in the direction away from point M which represents the deposited solid phase (the mixture rule). At X′, where line MX′ intersects curve CD, the temperature is about 50 °C and *p*-nitrophenol also starts to crystallize. On further cooling, both *m*- and *p*-nitrophenol are deposited, and the composition of the liquid phase changes in the direction X′D. When the melt composition and temperature reach point D, the third component (*o*-nitrophenol) also crystallizes, and the system solidifies without further change in composition. Similar reasoning can be applied to the cooling or melting of systems represented by points in other regions of the diagram.

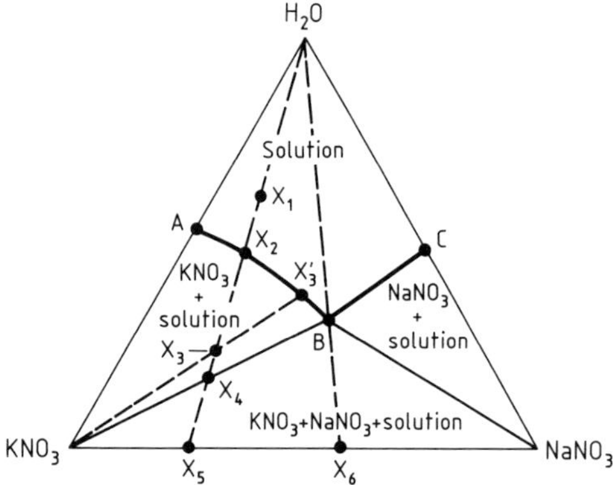

Figure 9. Phase diagram for the ternary system KNO_3–$NaNO_3$–H_2O at 50 °C

Many different types of phase behavior are encountered in ternary systems that consist of water and two solid solutes. One simple case is considered here: the system KNO_3–$NaNO_3$–H_2O at 50 °C (Fig. 9). The salts do not form hydrates nor do they combine chemically. Point A represents the solubility of KNO_3 in water at 50 °C (46.2 g per 100 g of solution) and point C the solubility of $NaNO_3$ (53.2 g per 100 g of solution). Curve AB indicates the composition of saturated ternary solutions in equilibrium with solid KNO_3 and curve BC those in equilibrium with solid $NaNO_3$. The upper area enclosed by ABC represents the region of unsaturated homogeneous solutions. Three other triangular areas are constructed by drawing straight lines from point B to the two remaining apexes of the triangle; the compositions of the phases within these regions are marked on the diagram. At point B, the solution is saturated with both KNO_3 and $NaNO_3$.

If water is evaporated isothermally from an unsaturated solution represented by point X_1, the solution concentration increases along the line X_1X_2. Pure KNO_3 is deposited when the concentration reaches point X_2. If more water is evaporated to give a system of composition X_3, the solution composition is represented by point X'_3 on the saturation curve AB, and by point B when composition X_4 is reached; further removal of water causes deposition of $NaNO_3$. All solutions in contact with solid will thereafter have a constant composition B, which is referred to as the *eutonic point or drying-up point* of the system. After complete evaporation of water, the composition of the solid residue is indicated by point X_5 on the base line. Similarly, if an unsaturated solution, represented by a point located to the right of B in the diagram, is evaporated isothermally, only $NaNO_3$ is deposited until the solution composition reaches point B. The salt KNO_3 is then also deposited, and the solution composition remains constant until evaporation is complete. If water is removed isothermally from a solution of composition B, the composition of deposited solid is given by point X_6 on the base line, and it remains unchanged throughout the evaporation process.

3.4. Multicomponent Systems

The more components a system contains, the more complex are the phase equilibria and the more difficult it is to represent phase reactions graphically. Specific descriptions of multicomponent solid–liquid diagrams and their uses are dealt with comprehensively in several books and monographs [3], [24]–[27]. Techniques for predicting multicomponent solid–liquid phase equilibria have also been described [26]–[30].

3.5. Phase Transformations

Metastable crystalline phases frequently crystallize prior to an expected, more stable phase, in accordance with Ostwald's rule of stages (cf. Section 8.2). The more common types of phase transformation that occur in crystallizing and precipitating systems include those between polymorphs (Section 3.1) and solvates. Transformations can occur in the solid state, particularly at temperatures near the melting point of the crystalline solid, or due to the intervention of a solvent.

A stable phase has a lower solubility than a metastable phase, as indicated by the solubility curves in Figures 10 A and 10 B for enantiotropic and monotropic systems, respectively. Transformation cannot occur between the metastable (I) and stable (II) phases in the monotropic system in the temperature range shown, but it is possible above the transition temperature in an enantiotropic system.

The occurrence of polymorphic transformation adds complexity to a phase diagram. For example, Figures 11 A and 11 B both show simple cases in which solids A and B form a eutectic E, but solid B has two enantiomorphs (α and β).

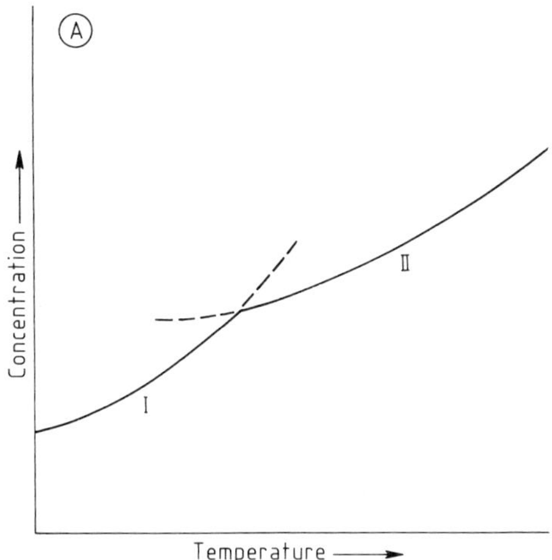

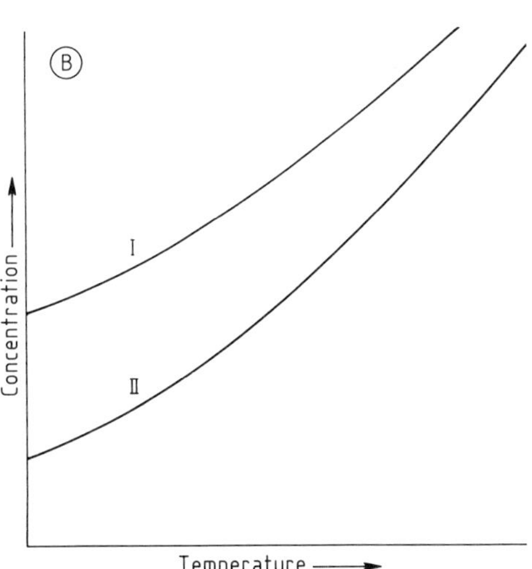

Figure 10. Solubility curves for substances with two polymorphs I and II
A) Enantiotropic polymorphs; B) Monotropic polymorphs
Broken lines in A indicate the metastable solubility curves of the two enantiomorphs.

In Figure 11 A, the polymorphic transformation temperature is higher than the eutectic melting point, as would be the case for a liquid-phase-mediated transformation. In Figure 11 B the transformation occurs at a temperature below the eutectic point, i.e., a case of solid-state transformation.

NANCOLLAS and coworkers [31], [32] have studied dissolution–recrystallization transformations in hydrate systems. CARDEW et al. [33], [34] presented theoretical analyses of both solid-state and solvent-mediated transformations in an attempt to predict their kinetics.

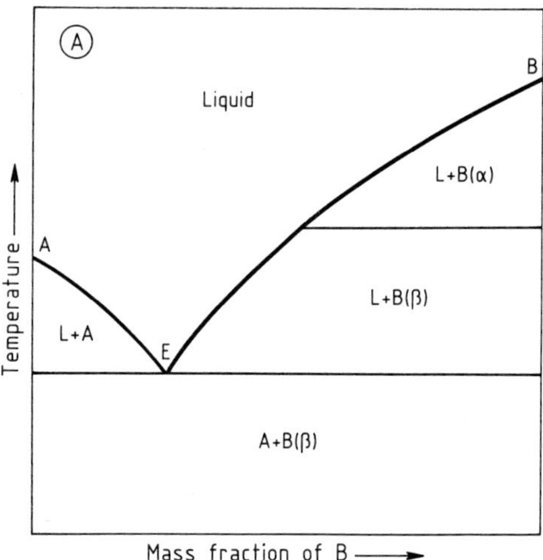

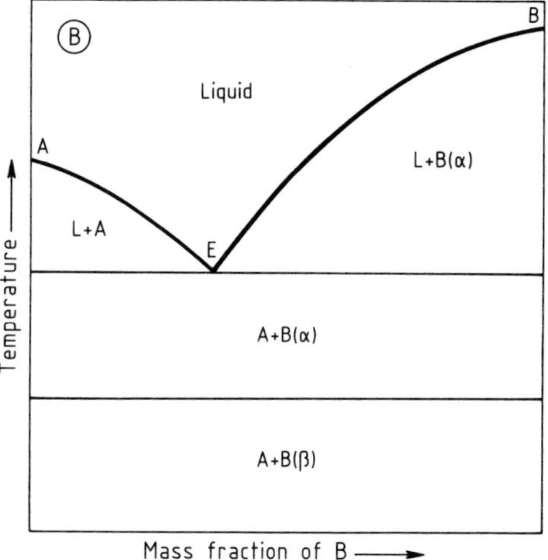

Figure 11. Phase diagrams for a binary system AB showing polymorphic transformations that occur at higher (Fig. A) or lower (Fig. B) temperature than the eutectic melting point
Symbols indicate the phases coexisting in equilibrium: L = liquid; E = eutectic; α, β = enantiomorphs of component B.

4. Kinetics and Mechanisms of Crystallization

4.1. Crystal Nucleation

Nucleation, i.e., the creation of crystalline bodies within a supersaturated fluid, is a complex, often ill-defined event, and nuclei may be generated by many different mechanisms. Numerous nucleation classification schemes have been proposed; most distinguish between two basic modes:

1) *Primary nucleation* — in the absence of crystals
2) *Secondary nucleation* — in the presence of crystals

Several comprehensive reviews of general nucleation phenomena [40], [41] and secondary nucleation [42] are available.

4.1.1. Primary Nucleation

Classical theories of primary nucleation are based on sequences of bimolecular collisions and interactions in a supersaturated fluid, which result in the buildup of lattice-structured bodies that may or may not achieve thermodynamic stability. This type of primary nucleation is referred to as *homogeneous,* although the terms *spontaneous* and *classical* have also been used.

Ample experimental evidence demonstrates that ordered solute clustering can occur in supersaturated solutions prior to the onset of homogeneous nucleation [35]–[39]; the presence of quasi-solid-phase species has even been detected in some unsaturated solutions [37]. Concentration gradients develop readily in supersaturated solutions of citric acid [38] under the influence of gravity; recent theoretical analysis of this phenomenon estimates the size of the clusters at 4–10 nm [39].

Primary nucleation may also be initiated by suspended particles of foreign substances (e.g., dust or other debris). This mechanism is generally referred to as *heterogeneous* nucleation.

Most primary nucleation in industrial crystallization is almost certainly heterogeneous rather than homogeneous; i.e., it is induced by the foreign solid particles that are invariably present in working solutions. The mechanism of heterogeneous nucleation is not well understood, but it probably begins with adsorption of the crystallizing species on the surface of solid particles, thus creating apparently crystalline bodies larger than the critical nucleus size. These stable particles then grow into macrocrystals.

Homogeneous Nucleation. Consideration of the energy involved in solid-phase formation and in creation of the surface of an arbitrary spherical crystal of radius r in a supersaturated fluid leads to the relationship

$$\Delta G = 4\,\pi r^2 \gamma + (4\,\pi/3)\, r^3 \Delta G_v \qquad (17)$$

where ΔG is the overall excess free energy associated with the formation of the crystalline body, γ is the interfacial tension between the crystal and its surrounding supersaturated fluid, and ΔG_v is the free energy change associated with the phase change. The first term on the right-hand side of Equation (17), representing the surface contribution, is positive and proportional to r^2. The second term, representing the volume contribution, is negative and proportional to r^3. The overall dependence of ΔG on r is shown in Figure 12. Any crystal smaller than the critical nucleus size r_c is unstable and will tend to dissolve. Any crystal larger than r_c is stable and will tend to grow.

Combination of Equations (16) and (17) and expression of the rate of nucleation J in the form of an Arrhenius reaction rate equation give

$$J = A \exp\left[\frac{-16\,\pi \gamma^3 v^2}{3 k^3 T^3 (\ln S)^2}\right] \qquad (18)$$

where A is a preexponential factor, γ is interfacial tension, v is molar volume, k is the Boltzmann constant, T is absolute temperature, and S is the supersaturation ratio. Equation (18) not only demonstrates the powerful effect of supersaturation on homogeneous nucleation, predicting an explosive increase in the nucleation rate beyond some so-called critical value of S, but also indicates the possibility of nucleation at any level of supersaturation.

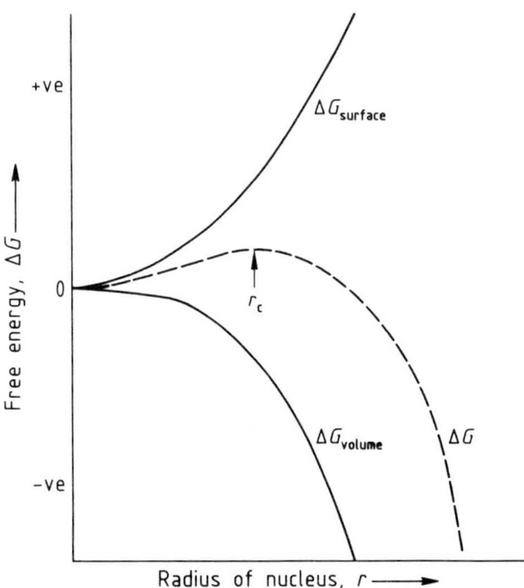

Figure 12. Free energy diagram for homogeneous nucleation demonstrating the critical nucleus size

Heterogeneous Nucleation. The presence of foreign particles (heteronuclei) enhances the nucleation rate of a given solution. Equations similar to that for homogeneous nucleation (Eq. 18) have been proposed to express this enhancement. However, the result is simply a displacement of the nucleation rate vs. supersaturation curve as shown in Figure 13, indicating that nucleation occurs more readily, i.e., at lower supersaturation.

For primary nucleation in industrial crystallization, classical relationships like those based on Equation (18) have little practical use. All that can be justified is a simple empirical relationship such as

$$J = K_n \Delta c^n \qquad (19)$$

which relates the primary nucleation rate J to the supersaturation Δc (Eq. 1). The primary nucleation rate constant K_n and the order of the nucleation process n depend on the physical properties and hydrodynamics of the system. Values of n are usually greater than 2.

4.1.2. Secondary Nucleation

Secondary nucleation, by definition, can take place only if crystals of the species under consideration are already present. Since this is mainly the case in working crystallizers, secondary nu-

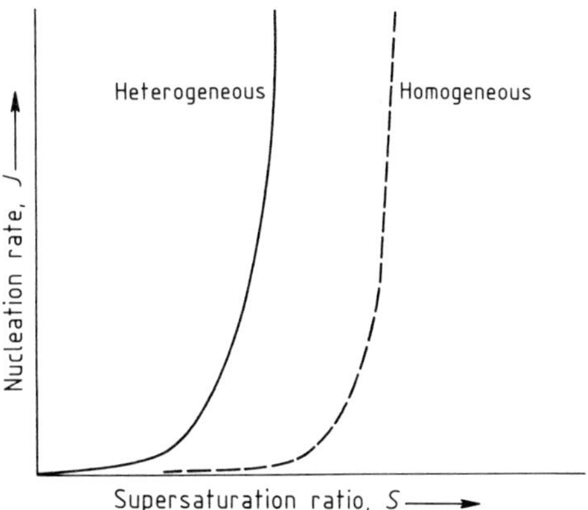

Figure 13. Effect of supersaturation on the rates of homogeneous and heterogeneous nucleation

cleation has a profound influence on virtually all industrial crystallization processes.

Apart from deliberate or accidental introduction of tiny seed crystals to the system, and productive interactions between existing crystals and quasicrystalline embryos or clusters in solution, the most influential mode of new crystal generation in an industrial crystallizer is *contact* secondary nucleation. The contact in this case can be between the existing crystals themselves, between crystals and the walls or other internal parts of the crystallizer, or between crystals and the mechanical agitator.

Secondary nucleation rates in industrial crystallizers are most commonly correlated by empirical relationships such as

$$B = K_b M_T^j N^l \Delta c^b \tag{20}$$

where B is the rate of secondary nucleation (birthrate), K_b is the birthrate constant, M_T is the slurry concentration (magma density), N is a term that gives some measure of the intensity of agitation in the system (e.g., the rotational speed of an impeller), and Δc is the supersaturation.

The exponents j, l, and b vary according to the operating conditions.

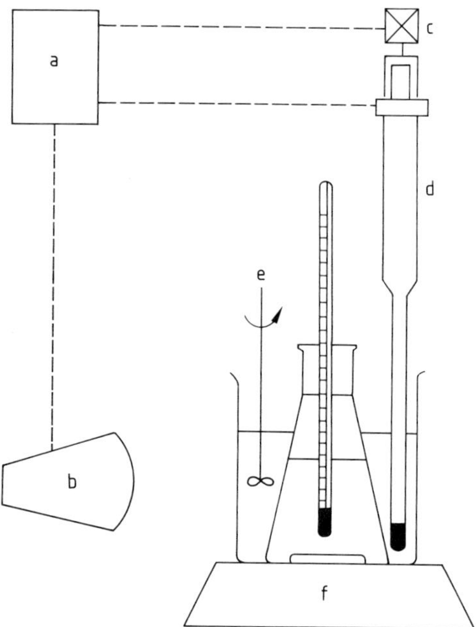

Figure 14. Simple apparatus for measuring metastable zone widths [43]
a) Control unit; b) Hot–cold air blower; c) Rotating magnet; d) Contact thermometer; e) Water bath stirrer; f) Magnetic stirrer

4.1.3. Nucleation Measurements

One of the earliest attempts to derive meaningful nucleation kinetics for solution crystallization was proposed by NÝVLT [2], [43]. The method is based on the measurement of metastable zone widths (Section 2.2) by using a simple apparatus (Fig. 14) that consists of a 50-mL flask fitted with a thermometer and a magnetic stirrer, located in an external cooling bath. Nucleation is detected visually. Both primary and secondary nucleation can be studied this way. The typical results [3] shown in Figure 15 demonstrate that seeding has a considerable influence on the nucleation process. The difference between the slopes of the two lines indicates that primary and secondary nucleation occur by different mechanisms.

Solution turbulence also affects nucleation. In general, agitation reduces the metastable zone width. For example, the metastable zone width for gently agitated potassium sulfate solutions is about 12 °C; vigorous agitation reduces this to ca. 8 °C. The presence of crystals also induces secondary nucleation at supercooling around 4 °C. The relation between supercooling $\Delta\theta$ and supersaturation Δc is given by Equation (14).

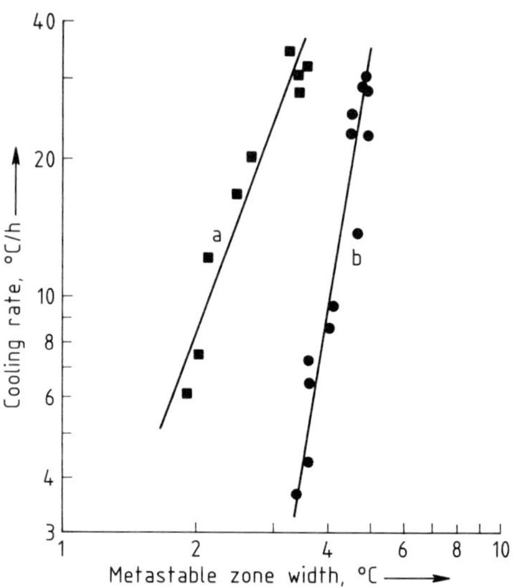

Figure 15. Metastable zone width of aqueous ammonium sulfate solutions as a function of cooling rate [3]
a) Seeded solution; b) Unseeded solution

Useful information on secondary nucleation kinetics for crystallizer operation and design can be determined only from model experiments that employ techniques such as those developed for MSMPR (mixed-suspension mixed-product removal) crystallizers (see Section 5.8). In a real crystallizer, however, both nucleation and growth proceed together and interact with other system parameters in a complex manner [2]–[4], [43], [44].

4.1.4. Induction Periods

A delay occurs between achievement of supersaturation and detection of the first newly created crystals in a solution. This so-called induction period τ is a complex quantity that involves both nucleation and growth components. However, if the assumption is made that τ is essentially concerned with nucleation (i.e., τ is inversely proportional to J), then from the classical nucleation equation (Eq. 6) it follows that [3]

$$\log(\tau^{-1}) \sim \gamma^3/T^3 (\log S)^2 \qquad (21)$$

Thus, for a given temperature T, a plot of log τ vs. $(\log S)^{-2}$ should yield a straight line which, if the data truly represent homogeneous nucleation, will allow calculation of the interfacial tension γ. The effect of temperature on γ can also be evaluated. An attempt has been made to derive a general correlation between interfacial tension and the solubility of inorganic salts (Fig. 16) [45]. However, the success of this method for evaluating interfacial tension depends on precise measurement of the induction period τ, which presents problems if τ is less than a few seconds.

Short induction periods can be determined by a stopped-flow technique that detects rapid changes in the conductivity of a supersaturated solution [46] (e.g., during a precipitation reaction). Typical results for $CaCO_3$, $SrCO_3$, and $BaCO_3$, produced by mixing an aqueous solution of Na_2CO_3 with a solution of the appropriate chloride, are shown in Figure 17. The slopes of the linear (high-supersaturation) regions are used to calculate the interfacial tensions (80–120 mJ/m²), which compare reasonably well with the values predicted from the interfacial tension–solubility relationship in Figure 15.

Interfacial tensions evaluated from experimentally measured induction periods can be jus-

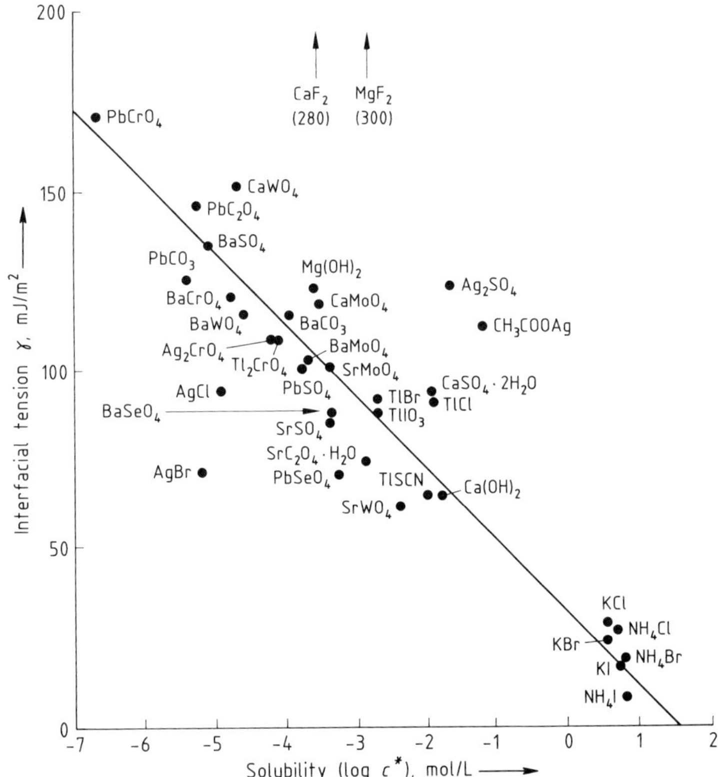

Figure 16. Interfacial tension as a function of solubility [45]

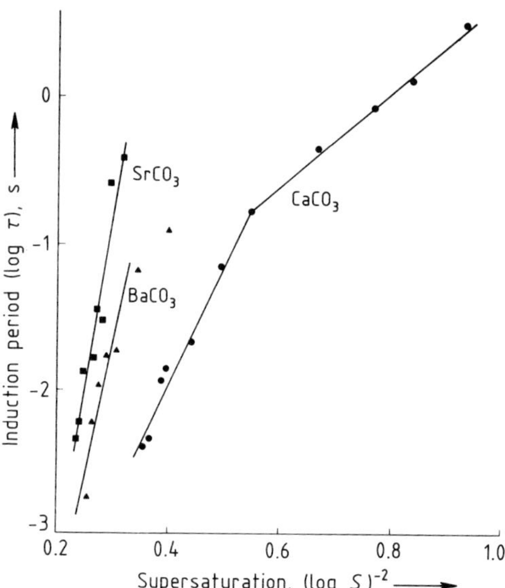

Figure 17. Induction period as a function of initial supersaturation [46]

tifiably argued to be unreliable. Nevertheless, measurements of the induction period can provide useful information on other crystallization phenomena, particularly the effect of impurities.

4.2. Crystal Growth

(For a detailed discussion of crystal growth, see → Crystal Growth, treated in the A series.)

Once a stable nucleus has been created in a solution, it is capable of growing into a crystal. In its simplest form, crystal growth may be considered a two-step process involving (1) mass transport, either by diffusion or convection from the bulk solution to the crystal face, followed by (2) a surface reaction in which the growth units are integrated into the crystal lattice. Either step may control the overall growth process, although convective mass transport is unimportant for crystals smaller than ca. 10 μm because these crystals are scarcely affected by turbulent eddies and diffusional mass transport predominates.

For diffusion-controlled growth, the theoretical linear crystal growth rate G for a nonionic species may be calculated from

$$G = Dv\Delta c / L \qquad (22)$$

where D is the diffusion coefficient, v the molar volume, Δc the supersaturation, and L the crystal size; i.e., the diffusion-controlled growth rate is directly proportional to the degree of supersaturation [3].

However, most crystal growth processes are not simply diffusion-controlled because the surface reaction (integration) step generally plays a contributing role. The two most common surface processes are referred to as spiral and polynuclear growth. *Spiral growth* (also known as BCF growth because it was first postulated by Burton, Cabrera, and Frank [47]) stems from the emergence of screw dislocations on a crystal face, as illustrated in Figure 18. Growth proceeds at relatively low supersaturation at a rate proportional to the square of the supersaturation.

Polynuclear growth develops from surface (monolayer) nucleation on the edges, corners, and faces of a crystal. These surface nuclei spread across the crystal face, and further nuclei develop on them. This mechanism, illustrated in Figure 19, is sometimes referred to as the B + S (birth and spread) model. Polynuclear growth is related to supersaturation by a complex exponential relationship [48].

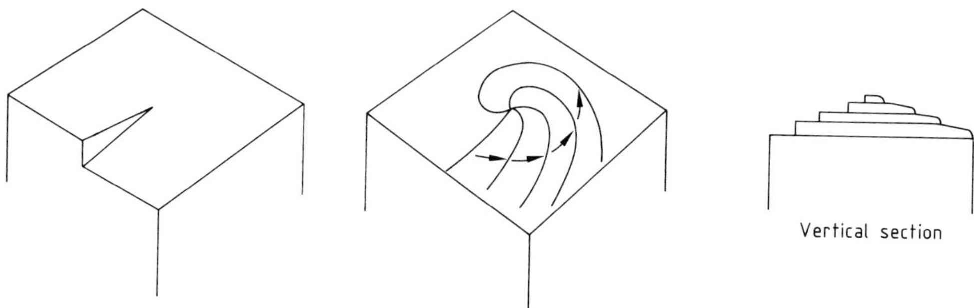

Figure 18. Spiral (BCF) crystal growth beginning at a screw dislocation

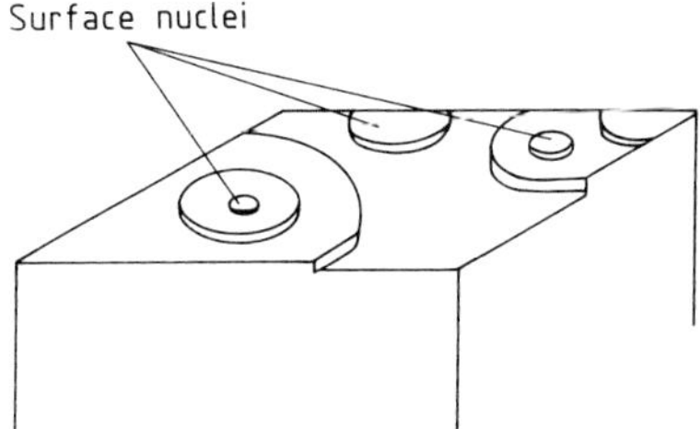

Figure 19. Polynuclear (B + S) crystal growth initiated by surface nucleation

At first sight, therefore, identifying the mechanism of growth by examining the relationship between the experimentally determined growth rate and supersaturation might appear to be possible. In practice, however, this is not easy because of the many errors inherent in growth rate measurement and, consequently, the relatively poor reproducibility of the data.

Several comprehensive reviews of modern theories of crystal growth are available [7], [48]–[51] (see also → Crystal Growth, treated in the A series).

4.2.1. Measurement of Growth Rate

A variety of methods have been used to measure crystal growth rate; they are divided into two main categories: (1) direct measurement of the linear growth rate of a chosen crystal face and (2) indirect estimation of an overall linear growth rate from mass deposition rates measured either on individual crystals or on groups of freely suspended crystals [2]–[4], [52], [53].

Face Growth Rates. Different crystal faces grow at different rates. In general, high-index faces grow faster than low-index faces. Changes in growth environment (temperature, supersaturation, pH, impurities, etc.) can also have a profound effect on growth. Differences in individual face growth rates give rise to habit (shape) changes in crystals (see Section 4.4).

Equipment for precise measurement of individual crystal face growth rates is depicted in Figure 20. A fixed crystal in a glass cell is observed with a traveling microscope. Solution temperature, supersaturation, and velocity are precisely controlled [3].

The solution velocity past the fixed crystal is frequently an important growth-determining parameter. It is sometimes responsible for the so-

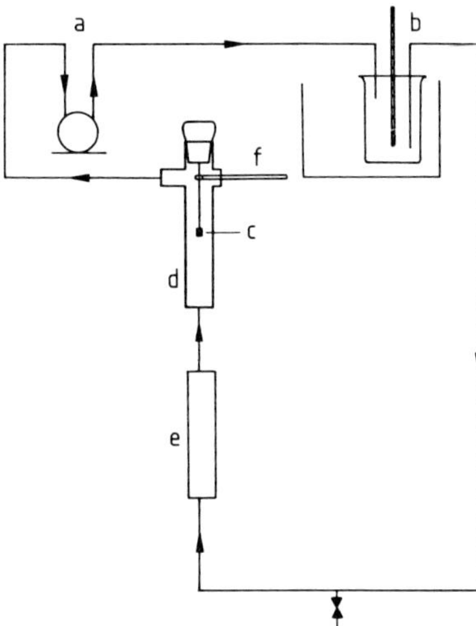

Figure 20. Equipment for measuring the growth of a single crystal
a) Pump; b) Saturator; c) Single crystal fixed on a wire; d) Glass cell; e) Rotameter; f) Thermometer

called size-dependent growth effect often observed in agitated vessel and fluidized-bed crystallizers. Large crystals have higher settling velocities than small crystals and, if their growth is diffusion-controlled, they tend to grow faster. Other reasons for size-dependent growth rates are discussed in Section 4.2.3. Examples of salts that exhibit solution-velocity-dependent growth rates include the alums, nickel ammonium sulfate, and potassium sulfate. However, salts such as ammonium sulfate and ammonium or potassium dihydrogen phosphate are not affected by solution velocity.

Overall Growth Rates. In the laboratory, growth rate data for crystallizer design can be measured in fluidized beds or agitated vessels. A typical laboratory fluidized-bed crystallizer is shown in Figure 21. Crystal growth rates are measured by growing large numbers of carefully sized seeds in fluidized suspension under strictly controlled conditions. A warm undersaturated solution of known concentration is circulated in the crystallizer and supersaturated by cooling to the working temperature. About 5 g of closely sized seed crystals with a narrow size distribution and a mean size of ca. 500 μm is introduced into the crystallizer, and the upward solution velocity is adjusted so that the crystals are maintained in a reasonably uniform fluidized state in the growth zone. The crystals are allowed to grow at

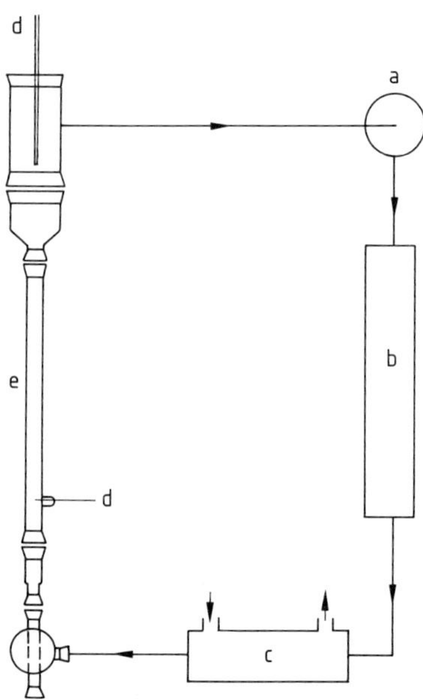

Figure 21. Laboratory-scale fluidized-bed crystallizer
a) Pump; b) Heater; c) Cooler; d) Thermometer; e) Crystal bed

a constant temperature until their total mass is ca. 10 g. They are then removed from the crystallizer, washed, dried, and weighed. The final solution concentration is measured, and the mean of the initial and final supersaturations is taken as the average for the run. This assumption does not involve any significant error because the solution concentration is usually not allowed to change by more than about 1 % during a run. The overall crystal growth rate is then calculated in terms of mass deposited per unit area per unit time at a specified supersaturation.

4.2.2. Expression of Growth Rate

Crystal growth rates are commonly expressed in three different ways:

1) As mass deposition rate R, e.g., $\mathrm{kg\,m^{-2}\,s^{-1}}$
2) As mean linear growth velocity $\bar{v}$ ($= \mathrm{d}r/\mathrm{d}t$), m/s
3) As overall linear growth rate G ($= \mathrm{d}L/\mathrm{d}t$), m/s

The relationships between these quantities are

$$R = K_g \Delta c^g = \frac{1}{A} \cdot \frac{\mathrm{d}m}{\mathrm{d}t} = \frac{3\alpha\varrho G}{\beta} = \frac{6\,\alpha\varrho}{\beta} \cdot \frac{\mathrm{d}r}{\mathrm{d}t} = \frac{6\alpha\varrho\bar{v}}{\beta} \tag{23}$$

where L is some characteristic dimension of the crystal (e.g., the equivalent sieve aperture size), r is the radius of the equivalent sphere, and ϱ is the density of the crystal. The order of the growth process, g, is generally between 1 and 2. The volume and surface shape factors (α and β, respectively) are related to particle mass m and surface area A, respectively, by

$$m = \alpha\varrho L^3 \tag{24}$$

and

$$A = \beta L^2 \tag{25}$$

For spheres and cubes, $6\alpha/\beta = 1$.

4.2.3. Dependence of Growth Rate on Crystal Size

A considerable amount of experimental evidence indicates that crystal growth kinetics often depend on crystal size. Apart from cases of solution velocity dependence (Section 4.2.1), this condition may result from the crystal size dependence of the surface integration kinetics. Different crystals of the same size can also have different growth rates, for example, because of differences in surface structure or perfection. Furthermore, small crystals ($< 50\ \mu\mathrm{m}$) of many substances grow much more slowly than larger crystals, and some do not grow at all [54]–[57].

The behavior of very small crystals has considerable influence on the performance of continuously operated industrial crystallizers because new crystals with a size of 1–10 µm are constantly generated by secondary nucleation. These subsequently grow to populate the full crystal size distribution. Therefore, the ability to predict the growth rates of small crystals is useful in assessing the performance of crystallizers.

4.3. Growth–Nucleation Interactions

Crystal nucleation and growth in an industrial crystallizer cannot be considered in isolation because they interact with one another and with other system parameters in a complex manner. These factors are summarized in Figure 22.

For a complete description of the crystal size distribution of the product in a continuously operated crystallizer, both the nucleation and the growth processes must be quantified, and the

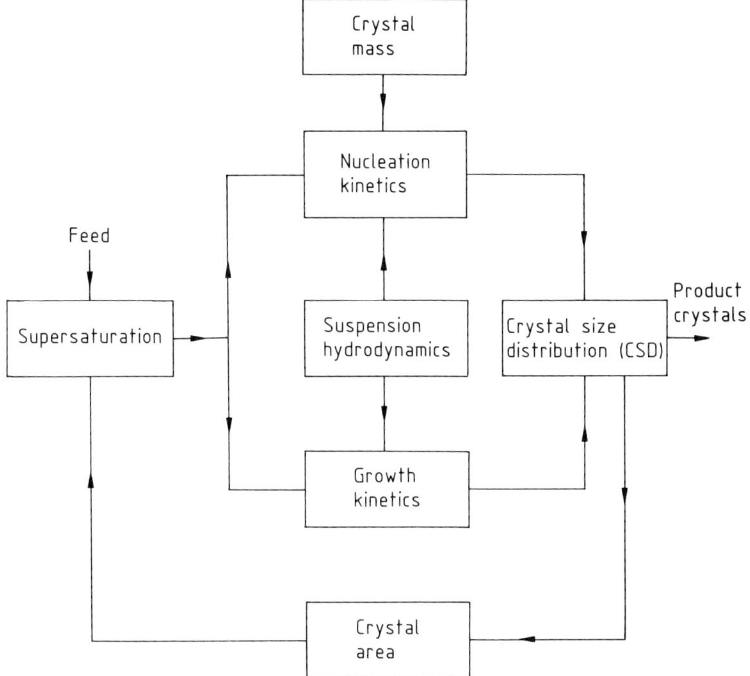

Figure 22. Complex interactions in a continuous crystallization process

laws of conservation of mass, energy, and crystal population must be applied. The importance of population balance, in which all particles are accounted for, was first stressed in the pioneering work of RANDOLPH and LARSON [44] (cf. Section 5.8.1).

4.4. Crystal Habit Modification

Changes in the face growth rates of crystals give rise to changes in their habit (shape). The growth kinetics of individual crystal faces usually depend to various extents on supersaturation, so that crystal habit can sometimes be controlled by adjusting operating conditions. The most common cause of habit modification, however, is the presence of impurities. The crystallizing solution may already contain impurities (e.g., raffinose which induces crystallization of characteristic flat sucrose crystals from beet sugar feedstocks), or they can be added deliberately.

In some cases, mere traces (ca. 1 mg/kg) of an impurity can cause significant changes in the crystal habit, as is the case for $Fe(CN)_6^{4-}$ which causes sodium chloride to crystallize in dendritic form. On the other hand, habit modification sometimes occurs only in the presence of large quantities of impurity, e.g., > 5 wt% of biuret is needed to change the shape of urea crystals from needles to bricks. Transition-metal ion species, such as complexes of Cr(III) and Fe(III), are particularly active impurities at < 10 mg/kg for a variety of inorganic salts in aqueous solution. Surface-active agents can also change crystal habit: anionic surfactants (e.g., alkyl sulfates; alkyl and arylalkyl sulfonates) and cationic surfactants (e.g., quaternary ammonium salts) are commonly used. Low molecular mass organic phosphonates and high molecular mass polyelectrolytes, polyacrylamides, poly(vinyl alcohols), etc., also find specific application. A change of solvent frequently results in a change of crystal habit because the solvent represents a massive impurity in a crystallizing system.

In the simple Kossel model of crystal growth, three basic sites may be considered for the adsorption of an impurity species: at a kink (Fig. 23 A), at a step (layer front, Fig. 23 B), or on a ledge (face) between steps (Fig. 23 C). Adsorption at kinks, the key sites at which growth units are incorporated into the lattice, effectively

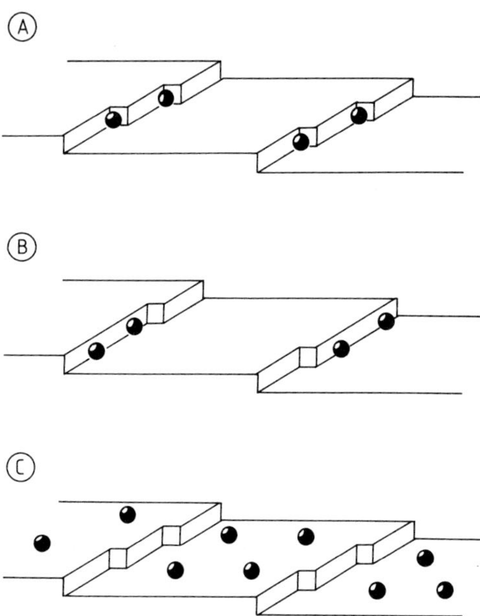

Figure 23. Kossel model of crystal growth showing sites for impurity adsorption on a crystal surface
A) At a kink; B) At a step; C) On a ledge (face)

reduces both their number and the velocity at which the step advances. Consequent increases in interkink distance diminish the importance of surface diffusion in the growth process and cause the step velocity to depend on kink density; this results in polygonization of the growth layers [58]. Adsorption at a step also reduces both the number of sites available for growth and the step velocity. However, unless the adsorbed units are closer together than the diameter of a critical two-dimensional nucleus, the advancing step can still squeeze between impurity species and progress relatively unimpeded. Step adsorption, therefore, implies a critical concentration of dissolved impurity, below which the step velocity is unaffected and above which it decreases rapidly. Adsorption of an impurity on a face is likely to be effective in cases where surface processes, particularly diffusion, play an important role in crystal growth. The result is a decrease in the surface flux of growth units to the step and, thus, in the step velocity.

In addition to physical obstruction, the effects of impurity adsorption and change of solvent on crystal growth may be explained in terms of changes in crystal surface properties. These changes can be quantified by the α factor, also called the surface entropy factor or surface roughness factor, which is related to the interfacial tension γ between the growing crystal and its solution [59].

In brief, at low values of α, i.e., at low interfacial tension γ, the crystal surface is inherently rough, on the molecular scale; under these conditions, growth is controlled only by the rate of diffusion of growth units to the crystal face. At high values of α and γ, the surface is inherently smooth and growth cannot proceed at low supersaturation unless growth centers (e.g., emergent screw dislocations) exist on the face. Growth under these conditions may then proceed by the spiral (BCF) mechanism. At intermediate values of α and γ, growth may proceed by the polynuclear (B + S) mechanism (see Section 4.2).

The interfacial tension γ can be changed by additives. Adsorption on a crystal face, for example, reduces γ and could, therefore, change the growth mechanism. The dependence of growth rate on supersaturation thus changes, and if this occurs differently on different faces of the crystal, the habit can consequently change. Similar reasoning can be applied to the removal of a particular impurity from the crystallizing solution. Furthermore, since γ is a solvent-dependent property (it increases as solubility decreases), a change of solvent can have the same effect on crystal habit as an additive.

Some form of habit modification is employed in a large proportion of all industrial crystallization and precipitation operations to control the type of crystal produced and to improve the rheological properties of the slurry, downstream processes such as filtration or washing and the handling properties of the dried product and its stability on storage. This may be done by controlling the rate of crystallization (e.g., by adjusting the rate of cooling or evaporation, the degree of supersaturation, or the temperature at which crystallization occurs). Alternative methods involve choosing an appropriate solvent, adjusting the pH of the solution, or deliberately adding (or perhaps removing) some habit-modifying impurity to (or from) the system. A combination of several of these methods may have to be used.

The search for a habit modifier is complex, and the results of small-scale laboratory investigations may not always be useful for large-scale industrial application; in some cases, they can even be misleading. Realistic pilot-plant trials on batches greater than about 100 L, however, generally yield useful information. Several comprehensive reviews [3], [60], [61] report industrial applications of habit modification and discuss factors that must be considered in selecting a suitable habit modifier.

4.5. Inclusions in Crystals

Inclusions are small pockets of solid, liquid, or gaseous impurities entrapped in crystals. They usually occur randomly throughout the crystal, but sometimes a regular pattern may be observed.

A simple technique for observing inclusions is to immerse the crystal in an inert liquid of similar refractive index or, alternatively, in its own saturated solution. In the latter case, the inclusion can be identified under the microscope by raising the temperature slightly to dissolve the crystal. If the inclusion is a liquid, concentration streamlines will be seen as the two fluids meet; if it is a vapor, a bubble will be released [3].

Crystals produced industrially may contain significant amounts of included mother liquor, in extreme cases up to 1 wt%, which can significantly affect product purity. Stored crystals may cake because of liquid seepage from inclusions in broken crystals, which leads to subsequent recrystallization (see Section 4.6). To minimize inclusions, the crystallizing system should be free of dirt and other solid debris to prevent its incorporation into the crystals. Vigorous agitation or boiling should be avoided because it can lead to the formation of air or vapor inclusions. Ultrasonic irradiation may suppress adherence of bubbles or particles to a growing crystal face and, hence, reduce inclusion formation. Fast crystal growth is probably the most common cause of inclusion formation; this means that, in general, high supersaturation should be avoided.

Several comprehensive reviews and accounts of crystal inclusions are available [62]–[65].

4.6. Caking of Crystals

Crystalline materials frequently cake (i.e., cement together) on storage. The size of the crystals, as well as their shape, moisture content, and storage conditions (time, temperature and humidity fluctuations, pressure, etc.), can all contribute to the caking tendency.

In general, caking is caused initially by dampening of the crystal surfaces in a storage container, e.g., because of inefficient drying or an increase in atmospheric humidity above some critical value that depends on both substance and temperature. For example, at 15 °C, crystals of $Na_2SO_4 \cdot 10\,H_2O$ become damp when atmospheric humidity exceeds ca. 93 % saturation; the same applies to NaCl at 78 % saturation and $CaCl_2 \cdot 6\,H_2O$ at 32% saturation. The presence of a hygroscopic trace impurity in the crystals, therefore, can greatly influence their tendency to absorb atmospheric moisture. Moisture may also be released from inclusions if crystals fracture under storage conditions (Section 4.5). If crystal surface moisture later evaporates, e.g., because of atmospheric temperature or humidity changes, adjacent crystals become firmly joined together with a cement of recrystallized solute.

Precautions to reduce caking include efficient drying, packaging in airtight containers, and avoiding compaction on storage. In addition, crystals may be coated with an inert dust that acts as a moisture barrier; e.g., crystals of table salt are sometimes coated with finely powdered magnesium carbonate. However, the crystals themselves play a dominant role in caking. Small crystals are more prone to cake than large crystals because of the greater number of contact points per unit mass, but actual size is less important than size distribution and shape. The narrower the size distribution and the more granular the shape, the lower is the tendency of crystals to cake. Crystal size distribution can be controlled by adjusting operating conditions of the crystallizer (Section 5.4), and crystal shape may be influenced by the use of habit modifiers (Section 4.4). A comprehensive account of the inhibition of caking by trace additives is given in [66].

5. Crystallization from Solutions

Solution crystallizers are generally classified according to the method by which supersaturation is achieved, e.g., cooling, evaporation, vacuum (adiabatic cooling), reaction, salting out. The designation *controlled* denotes supersaturation control; *classifying* refers to classification of product size. The term mixed-suspension mixed-product removal is abbreviated as MSMPR (see Section 5.8).

5.1. Cooling Crystallizers

5.1.1. Nonagitated Vessels

The simplest type of cooling crystallizer is the unstirred tank: a hot feedstock solution is charged to the open vessel where it is allowed to cool, often for several days, by natural convection. Metallic rods may be suspended in the solution so that large crystals can grow on them and reduce the amount of product that sinks to the bottom of the crystallizer. The product is removed by hand.

Because cooling is slow, large interlocked crystals are usually obtained and retention of mother liquor is unavoidable. As a result, the dried crystals are generally impure. Because of the uncontrolled nature of the process, product crystals range from a fine dust to large agglomerates.

Labor costs are generally high, but the method is economical for small batches because capital, operating, and maintenance costs are low. However, the productivity of this type of equipment is low and space requirements are high.

5.1.2. Agitated Vessels

Installation of an agitator in an open-tank crystallizer generally results in smaller, more uniform crystals and reduced batch time. The final product tends to have a higher purity because less mother liquor is retained by the crystals after filtration and more efficient washing is possible. Water jackets are usually preferred to coils for cooling because the latter often become encrusted with crystals and cease to operate efficiently. Where possible, the internal surfaces of the crystallizer should be smooth and flat to suppress encrustation.

An agitated cooler is more expensive to operate than a simple tank crystallizer, but it has a much higher productivity. Labor costs for product handling may still be rather high. The design of tank crystallizers varies from shallow pans to large cylindrical tanks.

The large agitated cooling crystallizer shown in Figure 24 A has an upper conical section which slows down the upward velocity of liquor and prevents the crystalline product from being swept out with the spent liquor. An agitator located in the lower region of a draft tube circulates the crystal slurry (magma) through the growth zone of the crystallizer. If required, cooling surfaces may be provided inside the crystallizer.

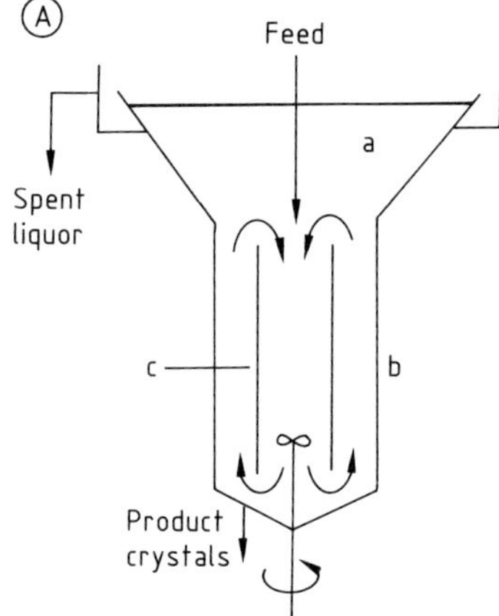

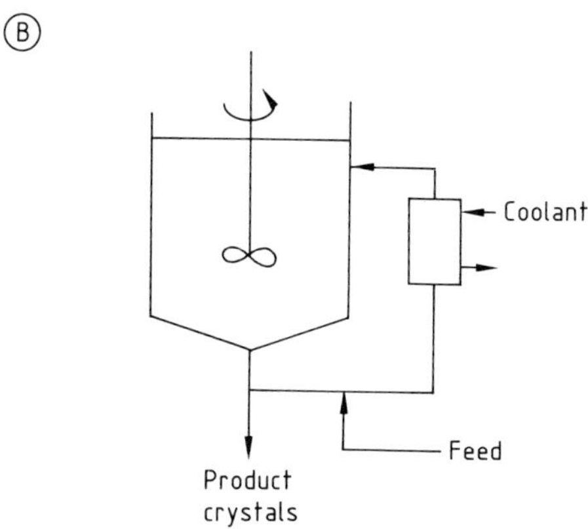

Figure 24. Cooling crystallizers
A) Internal circulation through a draft tube; B) External circulation through a heat exchanger
a) Calming section; b) Growth zone; c) Draft tube

Use of external circulation allows good mixing inside the crystallizer and high rates of heat transfer between the liquor and coolant (Fig. 24 B). An internal agitator may be installed in the crystallization tank if needed. The liquor velocity in the tubes is high; therefore, small temperature differences are usually adequate for cooling purposes and encrustation on heat-transfer surfaces can be reduced considerably. The unit shown may be used for batch or continuous operation.

5.1.3. Direct-Contact Cooling

The use of a conventional heat exchanger and the problems caused by crystal encrustation can be avoided by employing direct-contact cooling (DCC) in which supersaturation is achieved by allowing the process liquor to come into contact with a cold heat-transfer medium. Other potential advantages of DCC over conventional indirect-contact methods include better heat transfer and smaller cooling load. However, problems are also associated with DCC crystallization; these include product contamination from the coolant and the cost of extra processing required to recover the coolant for further use.

A solid, liquid, or gaseous coolant can be used; heat exchange may occur via the transfer of sensible or latent heat. The coolant may or may not boil during the operation, and it can be miscible or immiscible with the process liquor. Thus, several types of DCC crystallization are possible:

1) immiscible, boiling, solid or liquid coolant: heat is removed mainly by transfer of latent heat of sublimation or vaporization;
2) immiscible, nonboiling, solid, liquid, or gaseous coolant: mainly sensible heat transfer;
3) miscible, boiling, liquid coolant: mainly latent heat transfer; and
4) miscible, nonboiling, liquid coolant: mainly sensible heat transfer.

Crystallization processes employing DCC have been used successfully in recent years in the dewaxing of lubricating oils [67], the desalination of water [68], and the production of inorganic salts from aqueous solution [69].

5.2. Evaporating Crystallizers

If the solubility of a solute in a solvent is not appreciably decreased by lowering the temperature, the appropriate degree of solution supersaturation can be achieved by evaporating some of the solvent. Evaporation techniques have been used for centuries to crystallize salts; the simplest method — utilization of solar energy — is still employed commercially throughout the world [70]. Common salt is produced widely from brine in steam-heated evaporators, and similar evaporating crystallizers, often in multiple-effect series, are used in sugar refining. Many types of forced-circulation evaporating crystallizers are now in large-scale use [1], [3], [71].

Evaporating crystallizers are normally operated under reduced pressure to aid in solvent removal, minimize heat consumption, or decrease the operating temperature of the solution; they are best described as "reduced-pressure evaporating crystallizers."

5.3. Vacuum (Adiabatic Cooling) Crystallizers

A vacuum crystallizer operates on a slightly different principle from the reduced-pressure evaporating crystallizer described in the previous section.

Supersaturation is achieved in a vacuum crystallizer by simultaneous evaporation and adiabatic cooling of the feedstock. A hot, saturated solution is fed into an insulated vessel maintained under reduced pressure. If the feed liquor temperature is higher than the boiling point of the solution under the low pressure existing in the vessel, the liquor cools adiabatically to this temperature. The sensible heat and any heat of crystallization liberated by the solution evaporate some of the solvent and concentrate the solution.

5.4. Continuous Crystallizers

Many different continuously operated crystallizers are available, but the majority can be divided into three basic types: forced-circulation, fluidized-bed (Oslo), and draft-tube agitated units. A small selection of the large number of commercial types available is described.

Forced-Circulation Crystallizers. A *Swenson forced-circulation crystallizer* that operates under reduced pressure is shown in Figure 25. A high recirculation rate through the external heat exchanger is used to provide good heat transfer and minimize encrustation. The crystal magma is circulated from the lower conical section of the evaporator body through the vertical tubular heat exchanger and reintroduced tangentially into the evaporator below the liquor level to create a swirling action and prevent flashing (sudden evaporation). Feedstock enters on the pump inlet side of the circulation system. Product crystal magma is removed below the conical section.

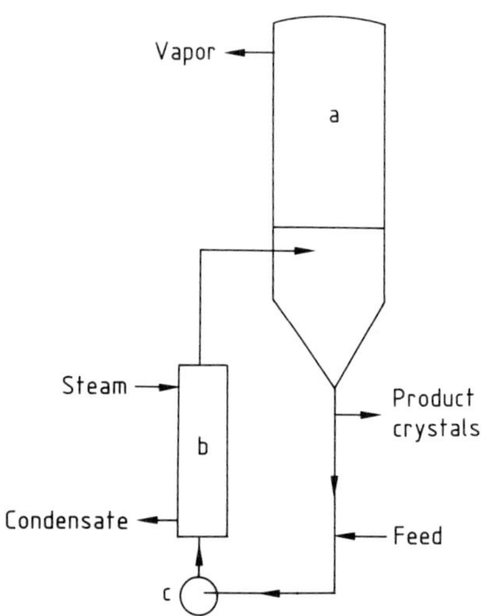

Figure 25. Forced-circulation Swenson crystallizer
a) Evaporator; b) Heat exchanger; c) Pump

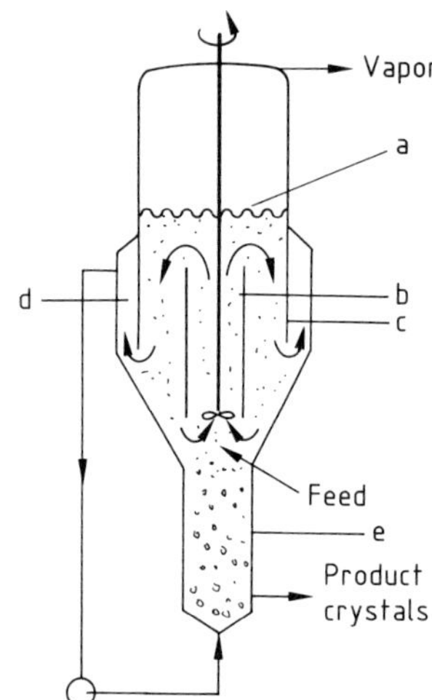

Figure 27. Swenson draft-tube-baffled (DTB) crystallizer
a) Boiling surface; b) Draft tube; c) Baffle; d) Settling zone; e) Elutriating leg

Fluidized-Bed Crystallizers. In an *Oslo fluidized-bed crystallizer,* a bed of crystals is suspended in the vessel by the upward flow of supersaturated liquor in the annular region surrounding a central downcomer (Fig. 26). Although originally designed as classifying crystallizers, fluidized-bed Oslo units are now frequently operated in a mixed-suspension mode to improve productivity, although this reduces product crystal size. A cooling-type Oslo crystallizer operates in the classifying mode as follows. The hot, concentrated feed solution is fed into the vessel at a point directly above the inlet to the circulation pipe. Saturated solution from the upper regions of the crystallizer, together with the small amount of feedstock, is circulated through the tubes of the heat exchanger, which is cooled by forced circulation of water or brine. On cooling, the solution becomes supersaturated, but not enough for spontaneous nucleation to occur; great care, in fact, is taken to prevent this. Product crystal magma is removed from the lower regions of the vessel.

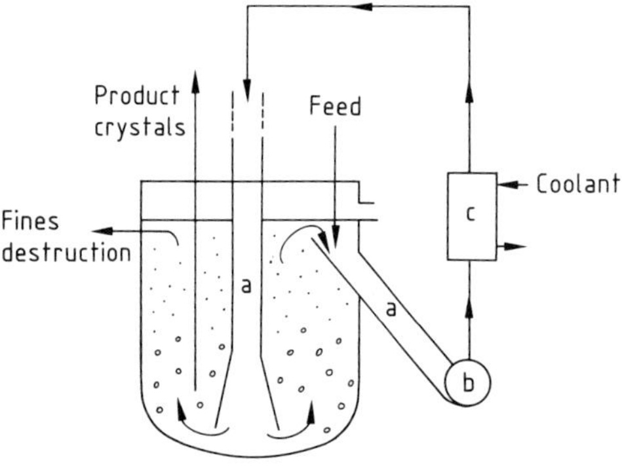

Figure 26. Oslo cooling crystallizer
a) Downcomer; b) Pump; c) Heat exchanger

Draft-Tube Agitated Vacuum Crystallizers. A *Swenson draft-tube-baffled* (*DTB*) *vacuum unit* is shown in Figure 27. A relatively slow-speed propeller agitator is located in a draft tube which extends to a few inches below the liquor level in the crystallizer. Hot, concentrated feedstock enters at the base of the draft tube. The steady movement of magma and feedstock up to the surface of the liquor produces a gentle, uniform boiling action over the whole cross-sectional area of the crystallizer. The degree of supercooling thus produced is very low ($< 1\,°C$), and in the absence of violent vapor flashing, both excessive nucleation and salt buildup on the inner walls are minimized. The internal baffle in the crystallizer forms an annular space in which agitation effects are absent. This provides a settling zone that permits regulation of the magma density and control of the removal of excess nuclei. An integral elutriating leg may be installed underneath the crystallization zone (as depicted in Fig. 27) to effect some degree of product classification.

The *Standard–Messo turbulence crystallizer* (Fig. 28) is another successful draft-tube vacuum

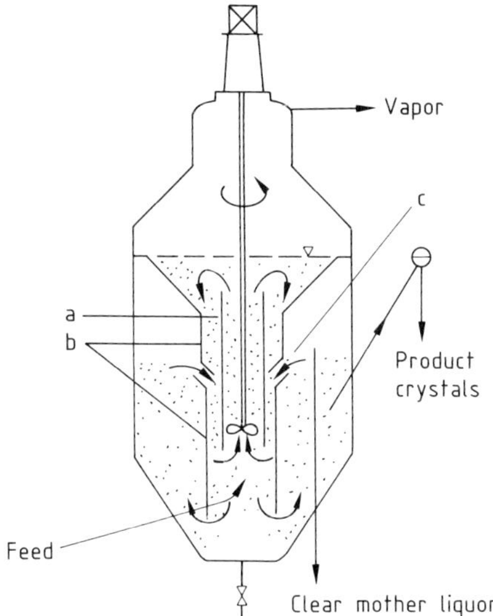

Figure 28. Standard–Messo turbulence crystallizer
a) Draft tube; b) Downcomer; c) Circumferential slot

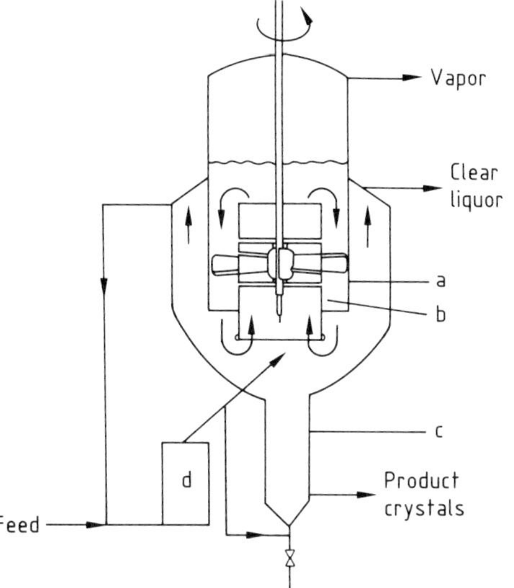

Figure 29. Escher–Wyss Tsukishima double-propeller (DP) crystallizer
a) Baffle; b) Draft tube; c) Elutriating leg; d) Heat exchanger

unit. Two liquor flow circuits are created by concentric pipes: an outer ejector tube with a circumferential slot and an inner guide tube. Circulation is effected by a variable-speed agitator in the guide tube. The principle of the Oslo crystallizer is utilized in the growth zone; partial classification occurs in the lower regions, and fine crystals segregate in the upper regions. The primary circuit is created by a fast upward flow of liquor in the guide tube and a downward flow in the annulus; liquor is thus drawn through the slot between the ejector tube and the baffle, and a secondary flow circuit is formed in the lower region of the vessel. Feedstock is introduced into the guide tube and passes into the vaporizer section where flash evaporation takes place. Nucleation, therefore, occurs in this region, and the nuclei are swept into the primary circuit. Mother liquor can be drawn off via a control valve, thus providing a means of controlling crystal slurry density.

The *Escher–Wyss Tsukishima* double-propeller (*DP*) *crystallizer* (Fig. 29) is essentially a draft-tube agitated crystallizer with some novel features. The DP unit contains an annular baffled zone and a double-propeller agitator which maintains a steady upward flow inside the draft tube and a downward flow in the annular region. Very stable suspension characteristics are claimed.

5.5. Crystal Yield

The crystal yield for simple cooling or evaporating crystallization can be estimated from the solubility characteristics of the solution. For aqueous solutions, the following general equation applies:

$$Y = \frac{WR\,[c_1 - c_2(1 - V)]}{1 - c_2(R - 1)} \tag{26}$$

where c_1 is the initial solution concentration, kg anhydrous salt per kg water; c_2 is the final solution concentration, kg anhydrous salt per kg water; W is the initial mass of water, kg; V is the water lost by evaporation, kg per kg of original water present; R is the ratio of molecular masses of hydrated to anhydrous salts; and Y is the crystal yield, kg.

The actual yield may differ slightly from that calculated by Equation (26). For example, if the crystals are washed with fresh solvent on the filter, losses may occur through dissolution. On the other hand, if mother liquor is retained by the crystals, an extra quantity of crystalline material will be deposited on drying. Furthermore, published solubility data usually refer to pure solvents and solutes. Because pure systems are rarely encountered industrially, solubilities should always be checked on the actual working liquors.

Before Equation (26) can be applied to vacuum (adiabatic cooling) crystallization, the quantity V must be estimated:

$$V = \frac{q R (c_1 - c_2) + C (t_1 - t_2)(1 + c_1)[1 - c_2 (R - 1)]}{\lambda [1 - c_2 (R - 1)] - q R c_2} \tag{27}$$

where λ is the latent heat of evaporation of the solvent, J/kg; q is the heat of crystallization of the product, J/kg; t_1 is the initial temperature of the solution, °C; t_2 is the final temperature of the solution, °C; C is the specific heat capacity of the solution, J kg^{-1} K^{-1}; and c_1 and c_2 have the same meaning as in Equation (26).

5.6. Controlled Crystallization

Carefully selected seed crystals are sometimes added to a crystallizer to control the final product crystal size. The effect of rapid cooling on an unseeded solution is shown in Figure 30 A. The solution cools at constant concentration until the limit of the metastable zone is reached, where nucleation occurs. The temperature increases slightly due to the release of latent heat of crystallization, but cooling reduces it and more nucleation occurs. The temperature and concentration subsequently fall as indicated. In such a process, nucleation and growth cannot be controlled.

Figure 30 B demonstrates the slow cooling of a seeded solution in which temperature and solution composition are controlled within the metastable zone throughout the cooling cycle. Crystal growth occurs at a controlled rate only on the added seeds; spontaneous nucleation is avoided because the system is never allowed to

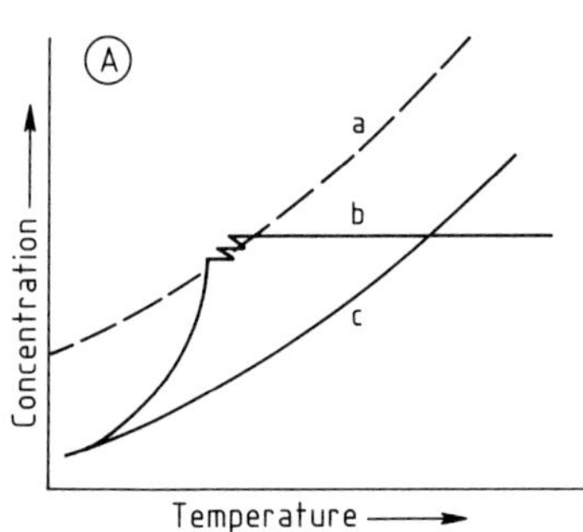

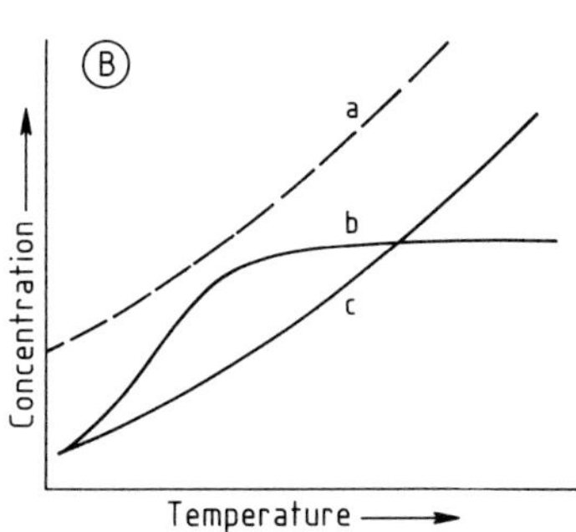

Figure 30. Effect of seeding on cooling crystallization
A) Rapid cooling of an unseeded solution; B) Slow cooling of a seeded solution
a) Supersolubility curve of the solute; b) Cooling curve of the solution; c) Solubility curve of the solute

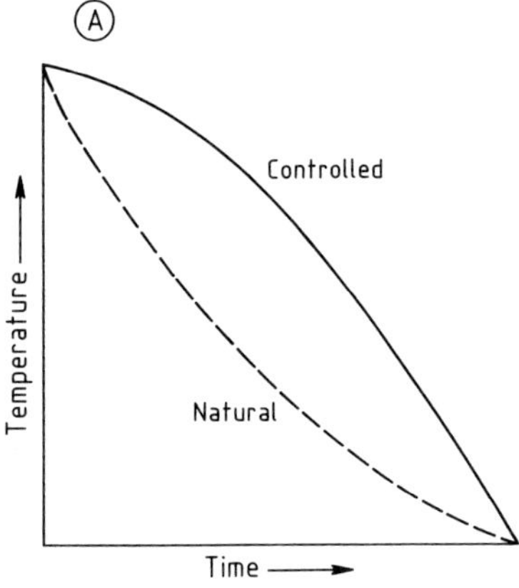

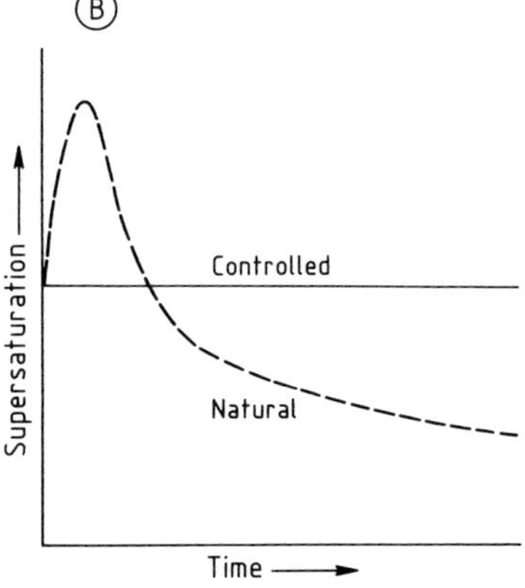

Figure 31. Natural and controlled cooling in batch crystallization
A) Temperature profile;
B) Supersaturation profile

become labile. This batch operating method is known as *controlled crystallization*; many modern large-scale crystallizers operate on this principle.

If crystallization occurs only on the added seeds, the mass M_s of seeds of size L_s that can be added to a crystallizer depends on the required crystal yield Y (Eq. 26) and the product crystal size L_p:

$$M_s = YL_s^3/(L_p^3 - L_s^3) \qquad (28)$$

The product crystal size from a batch crystallizer can also be controlled by adjusting cooling or evaporation rates. Natural cooling (Fig. 31 A), for example, produces a supersaturation peak in the early stages of the process when rapid, uncontrolled heavy nucleation inevitably occurs. However, nucleation can be controlled within acceptable limits by following a cooling path that maintains a constant low level of supersaturation (Fig. 31 B).

The calculation of optimum cooling curves for different operating conditions is complex [72], but the following simplified relationship is usually adequate for general application:

$$\theta_t = \theta_0 - (\theta_0 - \theta_f)(t/\tau)^3 \qquad (29)$$

where θ_0, θ_f, and θ_t are the temperatures at the beginning, end, and any time t during the process, respectively, and τ is the overall batch time.

5.7. Comparison of Batch and Continuous Crystallization

Although continuous, steady-state operation is often regarded as the ideal procedure for much processing plant equipment, this is not always true for crystallization processes. Batch operation often offers considerable advantages, such as simplicity of equipment and minimization of encrustation on heat-exchanger surfaces. In many cases, only a batch crystallizer can produce the required crystal form, size distribution, or purity. On the other hand, the operating costs of a batch system can be significantly higher than those of a comparable continuous unit, and problems of product variation from batch to batch may be encountered.

The particular attraction of a continuous crystallizer is its built-in flexibility for control of temperature, supersaturation, nucleation, crystal growth, and all the other parameters that influence crystal size distribution. A continuous crystallizer, however, does not discharge its product under equilibrium conditions (a batch unit can, if the batch time is adjusted appropriately) so the product slurry may have to be passed to a holdup tank to allow equilibrium to be reached. Omission of this step may cause problems due to further crystallization occurring in other parts of the plant (e.g., unwanted deposition in pipelines and effluent tanks). A holdup (aging) tank may also be necessary in a continuous system if the product exists in a phase (polymorph, hydrate, etc.) that differs from the one which appears initially.

A distinct advantage of batch crystallization, widely acknowledged in the pharmaceutical industry, is that the crystallizer can be cleaned thoroughly at the end of each batch to prevent contamination (seeding) of the next charge with any undesirable phase that might have arisen from transformation, rehydration, dehydration, air oxidation, etc., during the batch cycle. Continuous crystallization systems often self-seed undesirably after a certain operating time, which necessitates frequent shutdown and washout.

Semicontinuous crystallization processes often combine the best features of both batch and continuous operation. For example, a rapid mixer with a short residence time (possibly an on-line device) can discharge its product slurry into an agitated residence tank. In many cases, a series of tanks may have to be installed, which can then be operated as individual units or arranged in cascade [2], [44], [73], [74].

5.8. Crystallizer Modeling and Design

5.8.1. Population Balance

As described in Section 4.3, the processes of growth and nucleation interact in a crystallizer, and both contribute to the final crystal size distribution (CSD) of the product. In such assessments, the utility of the population balance [44] is widely acknowledged.

Application of the population balance is described most easily with reference to the simple, idealized case of an MSMPR (mixed-suspension mixed-product removal) crystallizer operated continuously in the steady state. The assumptions are made that no crystals are present in the feed stream, that all crystals are of the same shape, that crystals do not break down by attrition, and that crystal growth rate is independent of crystal size.

The relationship between crystal size L and population density n (number of crystals per unit size per unit volume of the system), derived directly from the population balance over the system [44], is

$$n = n^0 \exp(-L/G\tau) \quad (30)$$

where n^0 is the population density of nuclei (zero-sized crystals) and τ is the residence time. Equation (30) describes the crystal size distribution for steady-state operation. Rates of nucleation B and growth G ($= \mathrm{d}L/\mathrm{d}t$) are conventionally written in terms of supersaturation as

$$B = k_1 \Delta c^b \quad (31)$$

and

$$G = k_2 \Delta c^g \quad (32)$$

These empirical expressions can be combined to give

$$B = k_3 G^i \quad (33)$$

where

$$i = b/g \quad (34)$$

in which b and g are the kinetic orders of nucleation and growth, respectively, and i is the relative kinetic order. The relationship between nucleation and growth may be expressed as

$$B = n^0 G \quad (35)$$

or

$$n^0 = k_4 G^{i-1} \quad (36)$$

Experimental measurement of crystal size distribution (recorded on a number basis) in a steady-state MSMPR crystallizer can thus be used to quantify nucleation and growth rates. A plot of log n vs. L should give a straight line of slope $-(G\tau)^{-1}$ with an intercept at $L = 0$ equal to n^0 (Eq. 30 and Fig. 32 A); if the residence time τ is known, the crystal growth rate G can be calculated. Similarly, a plot of log n^0 vs. log G should give a straight line of slope $i - 1$ (Eq. 36 and Fig. 32 B); if the order g of the growth process is known (Eq. 34), the order of nucleation b can be calculated.

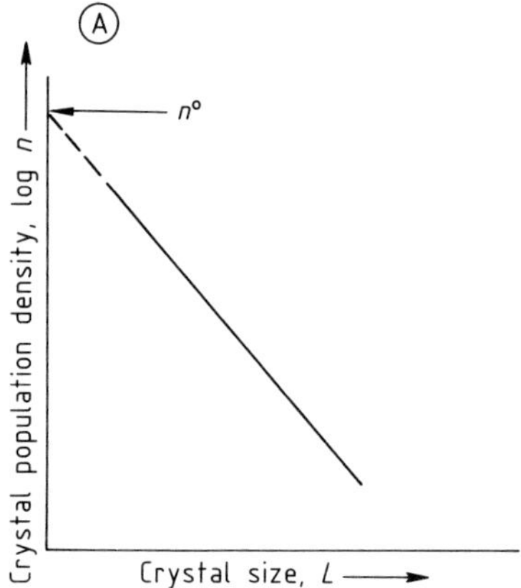

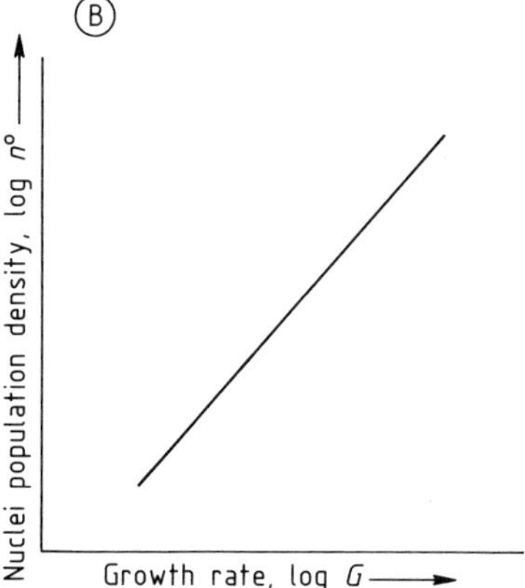

Figure 32. Population plots for a continuous mixed-suspension mixed-product removal (MSMPR) crystallizer
A) Crystal size distribution, slope $= -(G\tau)^{-1}$; B) Nucleation and growth kinetics, slope $= i - 1$
For explanation of symbols, see text.

The total mass M_T of crystals in the system, the so-called magma density (mass of crystals per unit volume of the system), is given by

$$M_T = 6\,\alpha\,\varrho\,n^0 (G\tau)^4 \quad (37)$$

where α is the volume shape factor (Eq. 24) and ϱ is the crystal density.

The peak of the mass distribution, the dominant size L_D of the CSD, is given by

$$L_D = 3\,G\tau \quad (38)$$

and can be related to the crystallization kinetics by

$$L_D \sim \tau^{(i-1)/(i-3)} \quad (39)$$

This interesting relationship [44] enables the effect of changes in residence time to be evaluated. For example, if $i = 2$ (a typical value for many inorganic salt systems), a doubling of the residence time would increase the dominant product crystal size by only ca. 15%. To double the residence time, however, either the crystallizer volume would have to be doubled or the volumetric feed rate, and hence the production rate, would have to be halved. Therefore, residence time adjustment is usually not a very effective means of controlling product crystal size.

Population-balance-based CSD modeling can be applied to crystallizer configurations other than MSMPR [44] and has become a distinct, self-contained branch of reaction engineering. The ability to simulate process configurations, however, presently exceeds the capabilities for measuring CSD on-line, predicting growth and nucleation rates, maintaining crystal magmas in a well-mixed state, and measuring supersaturation [75].

5.8.2. Scaleup Problems

Industrial crystallizers are commonly designed by using data measured on laboratory-scale (1 – 10 L) or pilot-scale (50 – 250 L) units. In difficult cases, information may have to be obtained for both scales of operation. One of the main problems in crystallizer scaleup is characterization of particle-fluid hydrodynamics and assessment of their effects on the kinetics of nucleation and crystal growth.

In fluidized-bed crystallizers, for example, evaluation of the crystal suspension velocity is necessary. This parameter is related to crystal size, size distribution, and shape, as well as bed voidage and other system properties such as density differences and viscosity. Possible ways of estimating crystal suspension velocity are discussed in [3] – [7]. In practice, however, determination of suspension velocities on actual crystal samples by simple experimental techniques is often advisable [3].

In agitated vessels, the "just-suspended" agitator speed N_{JS} must be established, i.e., the minimum rotational speed necessary to keep all crystals in suspension. Not only do the crystals have to be kept in suspension, but the development of "dead spaces" in the vessel must also be avoided because they are unproductive zones and regions of high supersaturation in which vessel surfaces can become encrusted. Fluid and crystal properties, together with vessel and agitator geometries, are important in establishing N_{JS} values [3] – [7].

Agitated vessel crystallizers are often scaled up successfully on the crude basis of either constant power input per unit volume or constant agitator tip speed. A refinement of the latter criterion, for draft-tube agitated vessels, is to maintain the quantity $(TS)^2/(TO)$ constant, where TS is the agitator tip speed and TO is the turnover time, i.e., TO is the vessel volume divided by the volumetric circulation rate [76].

6. Crystallization from Melts

Melt is the common name given to a liquid or a liquid mixture at a temperature near its freezing point. Melt crystallization is the process of separating the components of a liquid mixture by cooling until a quantity of crystallized solid is deposited from the liquid phase.

The basic requirement of melt crystallization is that the composition of the crystallized solid differ from that of the liquid mixture from which it is deposited. The ease or difficulty of separating one component from a multicomponent mixture by crystallization is best represented by a phase diagram as in Figures 4 and 5, both of which depict binary systems: the former shows a eutectic, and the latter shows a continuous series of solid solutions. These two systems behave quite differently on freezing; as described previously, a eutectic system can deposit a pure component (Section 3.2.1), whereas a solid solution can only deposit a mixture of components (Section 3.2.2).

6.1. Basic Techniques

Two basic techniques of melt crystallization are

1) gradual deposition of a crystalline layer on a chilled surface in a static or laminar flowing melt and
2) fast crystallization of discrete crystals in the body of an agitated vessel.

An example of category 1 is found in the Proabd refiner [77] which is essentially a batch cooling process. A static liquid feedstock is progressively crystallized onto extensive cooling surfaces (e.g., fin-tube heat exchangers supplied with a cold heat-transfer fluid) located inside a crystallization tank. As crystallization proceeds, the remaining liquid becomes increasingly impure and, in some cases, crystallization may be continued until virtually the entire charge has solidified. The crystallized mass is then slowly melted by circulating a hot fluid through the heat exchanger. The impure fraction melts first and drains out of the tank. As melting proceeds, the melt runoff becomes progressively richer in the desired component, and fractions may be taken off during the melting stage if required. A typical flow diagram, based on a scheme for the purification of naphthalene, is shown in Figure 33. In this case, the circulating fluid is usually cold water which is heated during the melting stage by steam injection.

Another example in the first category is the rotary drum crystallizer which usually consists of a horizontally mounted cylinder that is partially immersed in the melt or supplied with feedstock in some other way. The coolant enters and leaves the inside of the hollow drum through trunnions. As the drum rotates, a crystalline layer forms on the cold surface and is subsequently removed with a scraper knife. Two feed and discharge arrangements are shown in Figure 34. Rotary drum behavior and design have been discussed by GELPERIN [78] and PONOMARENKO [79].

Analysis of general layer freezing processes shows that the structure and impurity levels of growing crystal layers are determined primarily by mass-transfer effects at the layer front [80]. Effective distribution coefficients described by a single parameter that combines growth velocity, mass-transfer coefficient, and concentration of mother liquor can be determined by simple laboratory tests and employed to facilitate process scaleup.

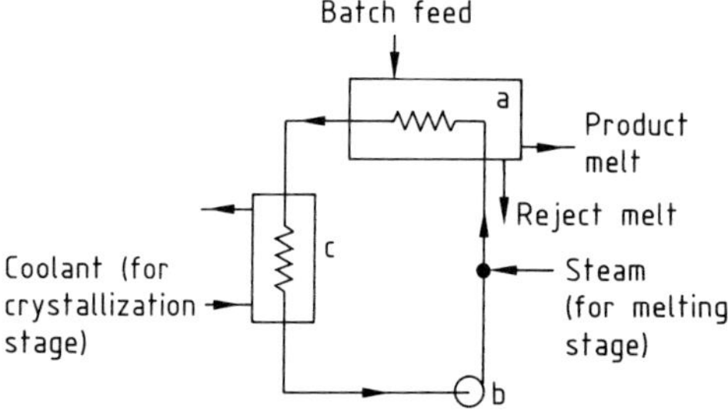

Figure 33. Batch cooling crystallization of melts: flow diagram for the Proabd refiner
a) Crystallizer; b) Pump; c) Heat exchanger

An example of melt crystallization in category 2 is the scraped-surface heat exchanger, which is basically a cylindrical tube surrounded by a cylindrical heat-exchange jacket. The tube is fitted with close-clearance scraper blades and rotates at relatively low speed. Two basic types are available: the large (> 200 mm in diameter, > 3 m long) slow-speed (< 10 rpm) unit and the small (< 150 mm in diameter, < 1.5 m long) high-speed (> 500 rpm) machine. Both types, but especially the latter, can handle viscous magmas, and operate at temperatures as low as − 80 °C. They are widely used, for example, in the manufacture of margarine (crystallization of triglycerides), dewaxing of lubricating oils (crystallization of higher *n*-alkanes), and large-scale processing of many organic substances (naphthalene, *p*-xylene, chlorobenzenes, etc.). The magma emerging from a scraped-surface crystallizer generally contains very small crystals (often < 10 μm) which can cause separation and subsequent reprocessing problems unless the crystals are first grown to a larger size, e.g., in a separate holdup tank.

6.2. Multistage Processes

The single-stage crystallization described in Section 6.1 might not always be sufficient to achieve the required purity of the final product, in which case further separation, melting, washing, or refining may be required. Two basic models can be considered:

1) a repeating sequence of crystallization, melting, and recrystallization steps (cf. Section 9.2);
2) a single crystallization step followed by countercurrent contacting of the crystals with a relatively pure liquid stream.

Scheme 1 is preferred if the concentration of impurities in the feedstock is high; it is essential if the system forms a continuous series of solid solutions. Scheme 2 is applicable if the concentration of impurities is low. Some industrial operations, however, require a combination of both systems. For a comparative analysis of different types of multistage crystallization schemes, see [81].

A number of industrial melt crystallization processes have been developed [3], [76], [82], but interest in this technique is currently being stimulated by the energy-saving potential in large-

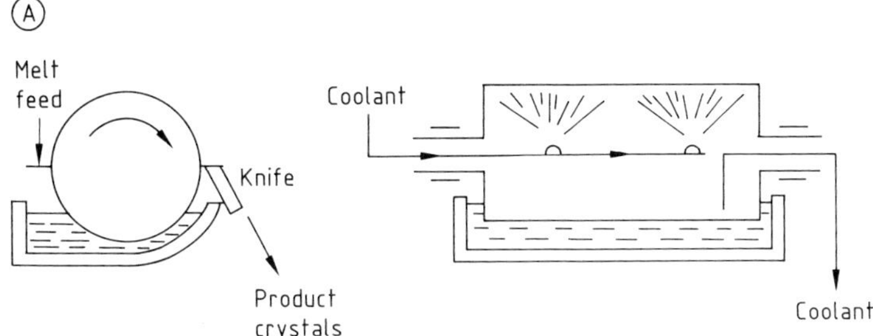

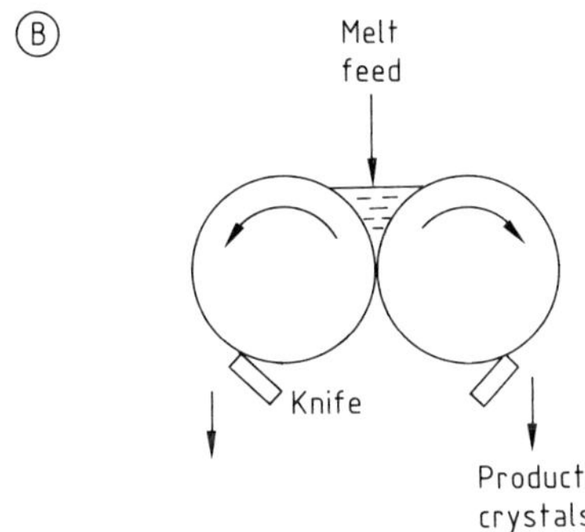

Figure 34. Two feeding modes (A) and (B) for drum crystallizers

scale processing, compared with distillation, e.g., for the separation of close-boiling organic substances.

For example, in the *Newton–Chambers process* [77], benzene is produced from a coal-tar benzole fraction by contacting the impure feedstock with an immiscible refrigerant (brine). The slurry is centrifuged, yielding benzene crystals (*fp* 5.4 °C) and a mixture of brine and mother liquor which is then allowed to settle; the brine is returned for refrigeration and the mother liquor is reprocessed for motor fuel. Process efficiency depends to a large extent on the efficient removal of impure mother liquor that adheres to the benzene crystals. Several modes of operation are possible. In the thaw–melt method (Fig. 35), benzene crystals are washed in the centrifuge with brine at a temperature > 6 °C. Some of the benzene crystals partially melt (thaw), which helps wash away the adhering mother liquor. The thaw liquor can be recycled. Multistage operation can be employed: the first crop of crystals is removed as product and the second, from the liquor, is melted for recycle. Crystals from the second stage should not be less pure than the original feedstock.

The *Sulzer MWB process* [83] is a recent development that operates by crystallization on a

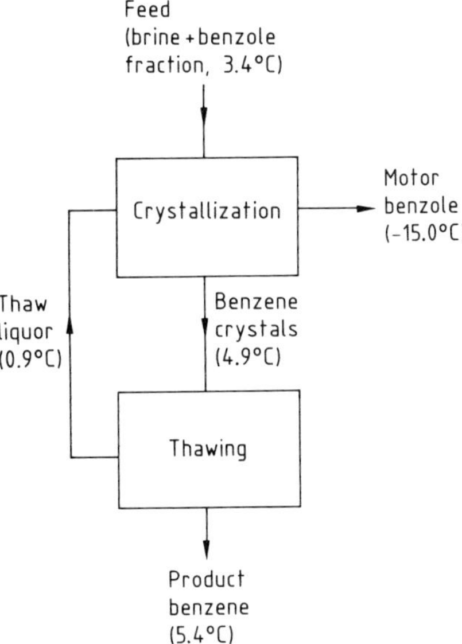

Figure 35. Newton–Chambers process for purification of benzene

cold surface; it also has features that permit it to operate effectively as a multistage separation device. Consequently, it can be used to purify solid solutions. An effective multistage countercurrent scheme is illustrated for four-stage oper-

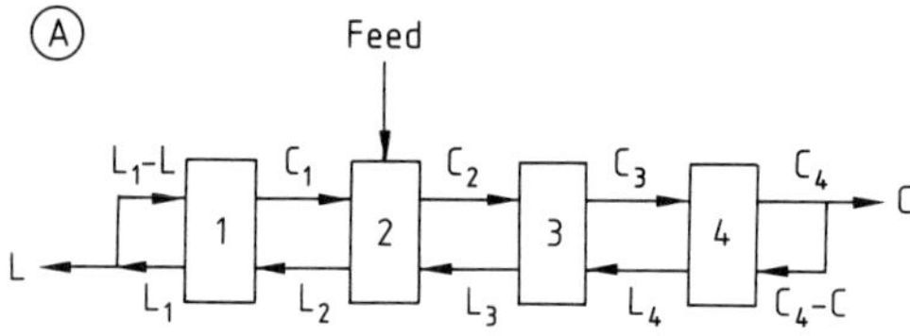

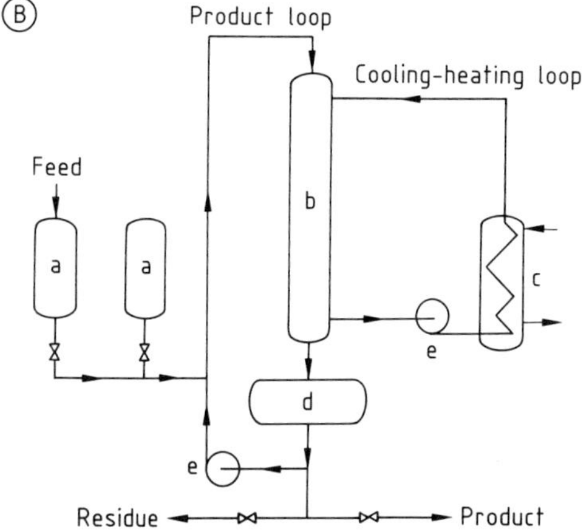

Figure 36. Sulzer MWB process
A) Multistage flow diagram
Symbols: C = crystals, L = liquor
B) Schematic of practical layout
a) Residue melt storage tanks; b) Crystallizer; c) Heat exchanger; d) Collecting tank; e) Pump

ation in Figure 36 A. Stage 1 is fed with melt L_2 and recycle liquor $L_1 - L$ (L denotes the impure reject liquor stream). Crystals C_1 are deposited in stage 1, and after melting, they are mixed with melt L_3 and fresh feedstock. Crystallization of this mixture yields crystals C_2 and melt L_2 in stage 2. Similar patterns are followed in stages 3 and 4, and the final high-purity stream C_4 is remelted and split into product C and recycle melt $C_4 - C$. Only one crystallizer, a vertical multitube heat exchanger, is required in this scheme. The crystals do not have to be transported; they remain deposited on the internal heat-exchange surfaces in the vessel, until they are melted for further processing. The intermediate storage tanks and crystallizer are linked by a control system consisting of a program timer, actuating valves, pumps, and the cooling loop (Fig. 36 B). The process has been used on a large scale in the purification of a range of organic substances (e.g., chloro- and chloronitrobenzenes, nitrotoluenes, cresols, and xylenols) and in the separation of fatty acids.

6.3. Column Crystallizers

Because melt feedstock components can form both eutectic and solid-solution systems with one another, sequences of washing, partial or complete melting, and recrystallization are often necessary to produce one of the components in near-pure form. However, the operation of a sequence of melt crystallization steps can be time-consuming and costly, especially if the liquid feedstock has to be cooled until it crystallizes and the crystals have to be separated from the residual melt, washed, and then remelted before the cycle can be repeated. Many attempts have been made to effect some of these events in a single unit, such as the column crystallizer developed by SCHILDKNECHT in the late 1950s.

The basic features of a *Schildknecht column* are shown in Figure 37. Liquid feedstock enters the column continuously at an arbitrary midpoint. Freezing at the bottom of the column is accomplished with a suitable refrigerant fluid, melting at the top, with a hot fluid or an electrical heating element. The crystals and liquid pass through the column countercurrently, the solid phase being transported downward by a helical conveyor fixed on a central shaft. The purification zone is usually operated at a virtually constant temperature between those of the freezing and melting sections. Crystals are formed mainly in the freezing section, but they can also deposit on the inner surface of the column from which

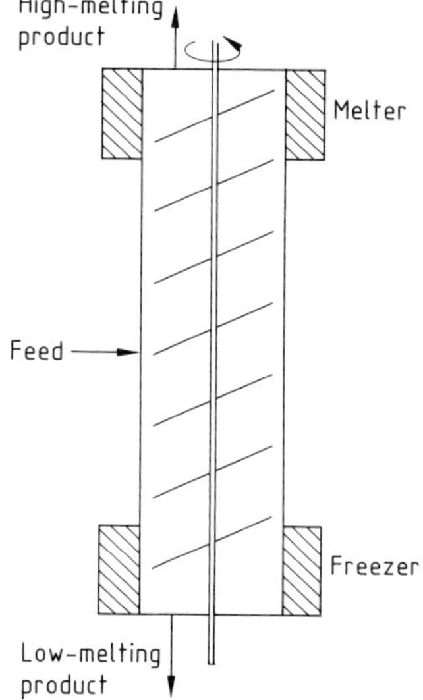

Figure 37. Schildknecht column

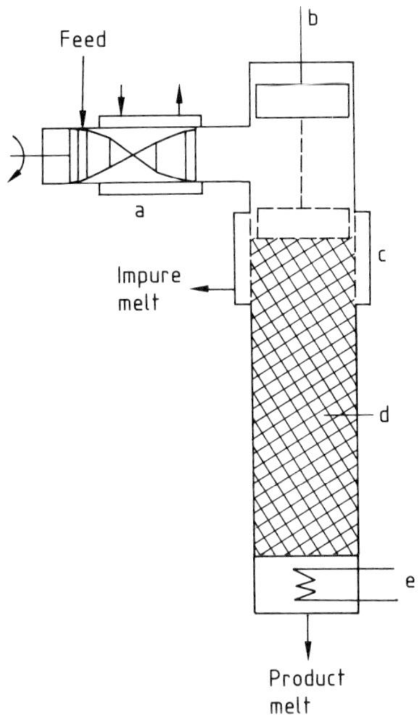

Figure 38. Phillips pulsed-column crystallizer
a) Scraped-surface chiller; b) Piston; c) Wall filter; d) Crystal bed; e) Heater

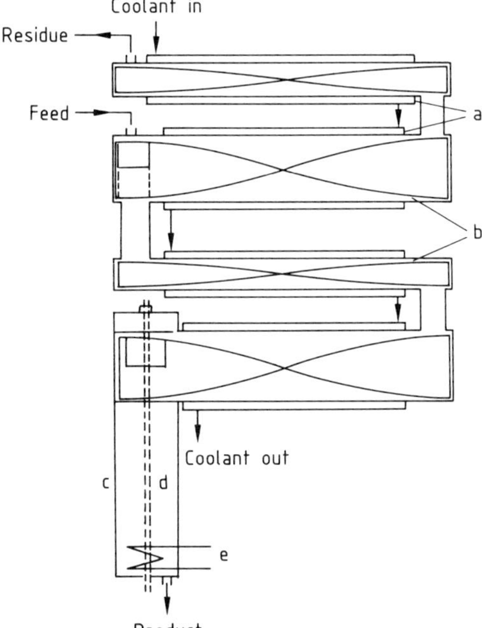

Figure 39. Brodie purifier
a) Cooling jackets; b) Scraper conveyors; c) Purifying column; d) Slow agitator; e) Heater

the helical conveyor removes them by scraping. During conveyance, crystals come into contact with the counterflowing liquid melt and are thus subjected to surface washing. The reverse mode of operation has also been used, i.e., upwardly flowing liquid in contact with crystals being conveyed downward. In this case, the locations of the freezer and melter are the reverse of those depicted in Figure 37. Comprehensive studies of the modeling of column crystallizers appear in [84], [85].

Although successful, the Schildknecht column is basically a laboratory apparatus; no large-scale industrial applications have been reported. A melt crystallizer of the wash column type, however, was developed by Phillips Petroleum Company in the 1960s for large-scale production of *p*-xylene [86]. The key features of the *Phillips pulsed-column crystallizer* are shown in Figure 38. A cold slurry feed, produced in a scraped-surface chiller, enters at the top of the column. Crystals in the vertical bed are pulsed downward by a piston, and impure mother liquor leaves through a wall filter. The upwardly flowing wash liquor is generated by the bottom heater which melts pure crystals before they are removed from the column.

The *Brodie purifier* [87] developed in the late 1960s has several features of the column crystallizers described above, but it also has the potential to deal effectively with solid-solution systems. It is essentially a center-fed column which can convey crystals from one end to the other (Fig. 39). As the crystals are conveyed through the unit, their temperature is gradually increased by a deliberately imposed temperature gradient along the flow path; they are thus subjected to partial melting which encourages the release of low-melting impurities. The interconnected scraped-surface heat exchangers are of progressively smaller diameter to maintain reasonably constant axial flow velocities and prevent backmixing. The vertical purifying column acts as a countercurrent washer in which falling, near-pure crystals meet an upflow of pure melt. The Brodie purifier has been used in the large-scale production of high-purity 1,4-dichlorobenzene and naphthalene.

The *Tsukishima Kikai* (*TSK*) *countercurrent cooling crystallization process* [88] is, in effect, a development of Brodie technology. The flow sheet in Figure 40 shows three conventional cooling crystallizers connected in series. Feed enters the first-stage vessel and partially crystallizes.

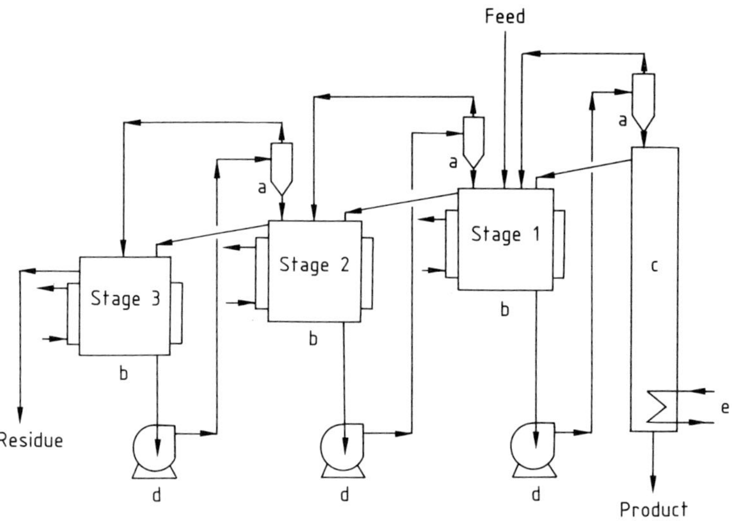

Figure 40. Tsukishima Kikai (TSK) countercurrent cooling crystallization process
a) Hydrocyclone; b) Conventional crystallizer; c) Brodie purifying column; d) Pump; e) Heater

The slurry is then concentrated in a hydrocyclone before passing into a Brodie purifying column. After passage through a settling zone in the crystallizer, clear liquid overflows to the next stage. Slurry pumping and overflow of clear liquid in each stage result in a countercurrent flow of liquid and solid. The process has been applied in the large-scale production of *p*-xylene.

Finally, brief reference may be made to *high-pressure crystallization* [89] in which an impure liquid feedstock is subjected to pressures up to 300 MPa in a relatively small (1.5-L) chamber under adiabatic conditions. As the pressure and temperature of the charge increase, fractional crystallization ensues and the impurities are concentrated in the liquid phase, which is then discharged from the pressure chamber. At the end of the cycle, further purification is possible because residual impurities in the compressed crystalline plug may be "sweated out" when the pressure is released. A single-cycle operation lasting less than 5 min is claimed to be sufficient to effect substantial purification in a wide range of organic binary melt systems [89]. The technique shows potential interest for specialized commercial exploitation.

6.4. Prilling and Granulation

Prilling is a melt spray crystallization process that results in the formation of solid spherical granules. It is employed widely in the manufacture of fertilizer chemicals such as ammonium nitrate and urea. In the ammonium nitrate prilling process [90], a very concentrated solution, containing ca. 5% water, is sprayed at 140 °C into the top of a 30-m-high, 6-m-diameter tower in which the droplets fall countercurrently to an upwardly flowing air stream that enters the base of the tower at 20 °C. The solidified droplets (prills), which leave the tower at 80 °C, contain ca. 4% water and must be dried to an acceptable moisture content at < 80 °C to prevent the occurrence of polymorphic transitions (see Section 3.1).

In a recent development of the melt granulation technique for urea [91], molten urea is sprayed at 148 °C onto cascading granules in a rotary drum and seed granules (< 0.5 mm) are thereby built up to product size (2–3 mm). Heat released by the solidifying melt is removed primarily by evaporation of a fine mist of water sprayed into air that is passed through the granulation drum.

7. Crystallization from Vapors

In the rapidly expanding field of single-crystal growth, a tendency exists nowadays to classify the various processes of growth from the vapor according to the manner in which vapor is generated [92], e.g., sublimation, chemical vapor transport (CVT), chemical vapor deposition (CVD) (→ Crystal Growth, **A8**, pp. 133–134).

Sublimation processes are characterized by vapor production resulting from heating the solid phase and subsequent crystallization of the sublimate by condensation under conditions below the triple point. Substances such as zinc selenide and gallium arsenide are grown by this technique. Sublimation–desublimation processes are also used on a large scale by the chemical industry to produce a wide range of organic and inorganic substances (→ 5. Sublimation, **B3**).

8. Precipitation

Precipitation is widely used in the laboratory for chemical analysis and in industry for the manufacture of paints, pigments, pharmaceutical and photographic chemicals, etc. In the production of ultrafine crystalline powders, precipitation is often considered an attractive alternative to comminution, particularly for heat-labile substances (i.e., substances that are unstable when heated).

No generally accepted, unambiguous definition of the term precipitation exists; it may refer simply to very fast crystallization, although precipitation is often an irreversible process, i.e., many precipitates are virtually insoluble substances produced by a chemical reaction. The products of conventional crystallization, on the other hand, can often be redissolved when the original conditions of temperature and concentration are restored. Nevertheless, precipitation and crystallization have much in common and are governed by the same laws.

8.1. Solubility Products

The solubility of a sparingly soluble electrolyte in water may be expressed in terms of the concentration solubility product K_c. For example, if such an electrolyte dissociates in solution into x cations and y anions according to

$$M_xA_y \rightleftharpoons x\,M^{z+} + y\,A^{z-} \qquad (40)$$

where z^+ and z^- are the valences of the metal cation M and the anion A, respectively, then for a saturated solution

$$(c_+)^x\,(c_-)^y = \text{constant} = K_c \qquad (41)$$

where c_+ and c_- are the ionic concentrations.

For 1–1, 2–2, etc., electrolytes (i.e., $x = y = 1$, $c_+ = c_- = c^*$, the equilibrium solubility), Equation (41) becomes

$$c^* = (K_c)^{1/2} \qquad (42)$$

In general,

$$c^* = (K_c/x^x y^y)^{1/(x+y)} \qquad (43)$$

For a 2–1 electrolyte, therefore,

$$c^* = (K_c/4)^{1/3} \qquad (44)$$

A more fundamental approach to the solubility product involves the use of activities rather than concentrations; the activity solubility product K_a is then defined by

$$(a_+)^x\,(a_-)^y = \text{constant} = K_a \qquad (45)$$

where a_+ and a_- are the ionic activities, or by

$$(c_+\gamma_+)^x\,(c_-\gamma_-)^y = K_a \qquad (46)$$

where γ is the ionic activity coefficient. Therefore,

$$K_a = K_c\,(\gamma_\pm)^\nu \qquad (47)$$

where $\gamma_\pm$ is the mean ionic activity coefficient and ν $(= x + y)$ is the number of moles of ions produced by one mole of electrolyte.

In practice, K_a and K_c may be assumed equal for concentrations up to about 10^{-3} mol/L, but above this, significant differences can occur. The activity of an ion depends on the concentration of all the other ions in solution, so the presence of a foreign electrolyte can greatly influence the value for $\gamma_\pm$ of a sparingly soluble salt.

A number of cases that appear anomalous when the simple solubility product is used can be explained when activity coefficients are considered [93]. For instance, the addition of a common ion generally decreases the solubility of a given salt, but cases are known in which addition of a common ion increases the solubility of the salt. This is because a large increase in ionic concentration can bring about a reduction in the activity coefficients. Thus, from Equation (50), an increase in c_- will result in a decrease in c_+ (i.e., precipitation of the sparingly soluble salt) if γ_+ and γ_- remain fairly constant; however, an increase in c_- to a value that reduces both γ_+ and γ_- must result in an increase in c_+ if K_a is to remain constant. The addition of a salt without a common ion often increases the solubility; this again occurs because the increased ionic concentration reduces the activity coefficients.

8.2. Ostwald's Rule of Stages

Ostwald's rule of stages, or rule of successive transformations, was originally stated as follows [94]: "An unstable chemical system does not spontaneously transform directly into that state which, under the given conditions, is the most stable of all the possible states, but into that which most closely resembles its own, i.e., into the state whose formation from the original is accompanied by the smallest loss in free energy."

Although many exceptions to Ostwald's rule have been recorded and little theoretical support has been found, it provides a useful guide to the possible behavior of precipitating systems. The rule appears, in fact, to be a manifestation of the frequently noted behavior of chemical systems that, if more than one reaction is thermodynamically possible, the resulting reaction is not the one which is thermodynamically most likely, but the one with the fastest rate. In other words, kinetics are often more important than thermodynamics.

Therefore, a metastable solid phase may be precipitated first and then, at some later stage, transformed into a more stable phase. Transformations such as those of one polymorph to another, of one hydrate to another hydrate or to an anhydrous form, and of an amorphous precipitate to a crystalline phase are quite common (Section 3.5).

8.3. Development of Precipitates

Precipitation, like all crystallization processes, consists of three basic steps: (1) the creation of supersaturation, followed by (2) the generation of nuclei and (3) the subsequent growth of these nuclei to visible size. The kinetics of nucleation and growth are described in Sections 4.1 and 4.2, respectively. The agglomeration of small crystals into clusters and changes in size distribution caused by ripening can also play important roles in the development of a precipitate.

8.3.1. Ripening

When solid particles are dispersed in their own saturated solution, the smaller particles tend to dissolve and the resulting solute is then deposited on the larger particles. Thus, the small particles disappear, the large ones grow larger, and theoretically, the particle-size distribution should ultimately become monodisperse. The reason for this behavior is that the solid phase in the system adjusts itself so as to achieve a minimum total surface free energy. This process of particle coarsening is called ripening.

The driving force for ripening is the difference in solubility between small and large particles, as given by the size–solubility (Gibbs–Thomson–Ostwald–Freundlich) relationship (Eq. 16), although the effect only becomes significant for particle sizes $< 1\ \mu m$ (see Section 2.2). If mass transport occurs between the particles in a polydisperse precipitate and if the growth kinetics are diffusion controlled, all particles of size

$$r' = 2\, v\gamma c^* / \nu RT(c - c^*) \quad (48)$$

are in equilibrium with the bulk solution ($dr/dt = 0$) [3]; v is molar volume, γ interfacial tension and ν the number of ions. All particles smaller than r' will dissolve ($dr/dt < 0$) and all particles larger than r' will grow ($dr/dt > 0$).

The speed at which ripening occurs depends to a large extent on particle size r and solubility. For diffusion-controlled growth kinetics, the linear growth velocity may be expressed approximately as

$$dr/dt \cong \gamma v^2 D c^* / 3\, \nu RT r^2 \quad (49)$$

where D is the diffusion coefficient. However, because ripening generally occurs at very low supersaturation, it is more likely to be controlled by surface reaction (p. 17) than by diffusion; under these circumstances, ripening could be considerably slower than indicated by Equation (49).

Analyses of ripening mechanisms have been made [95]; experimental studies have also been reported [96], [97].

Ripening changes the particle-size distribution of a precipitate over a period of time, even in an isothermal system, but the change can be accelerated by controlled temperature fluctuation. This process, known as temperature cycling, has been utilized to alter the physical characteristics of precipitates [98], [99].

8.3.2. Agglomeration

Small particles in liquid suspension tend to agglomerate into clusters. A theoretical basis for the rate of agglomeration of colloidal particles in suspension was first proposed by SMOLUCHOWSKI [100]. Two types of behavior may be distinguished: *perikinetic* (static fluid with particles in Brownian motion) and *orthokinetic* (agitated dispersions), which can both be important in precipitation processes. In agitated precipitators, orthokinetic agglomeration becomes more important as the particle size and the shear rate increases.

Double-layer repulsion forces and van der Waals attraction forces operate independently in disperse systems. Repulsion forces decrease exponentially over a distance corresponding to the thickness of the ionic double layer, whereas attraction forces decrease, over a larger distance from the particle surface, as an inverse power of the distance. Consequently, attraction normally predominates at very small and very large distances, and repulsion over intermediate distances [101].

When assessing the effects of agitation on particle agglomeration, the breakup of agglomerates under laminar and turbulent shear must also be considered [102]–[104].

8.3.3. Precipitate Morphology

The morphological development of a precipitate is a complex combination of a variety of processes including nucleation, habit modification, phase transformation, ripening, and agglomeration. The most influential system parameters are supersaturation and the concentration of active impurities, although pH can also exert a profound effect in some aqueous systems.

The dominant influence of supersaturation on the particle-size characteristics of a precipitate has been summed up in the so-called Weimarn laws of precipitation [105] which, while open to theoretical criticism, provide very useful guidelines for batch precipitation behavior. They are illustrated in Figure 41:

1) As the concentration of reactants increases, the median particle size of the precipitate (determined at a given time after mixing the reactants) increases to a maximum and then decreases. As the time interval increases, the maximum is displaced to lower initial supersaturation and higher median particle size.

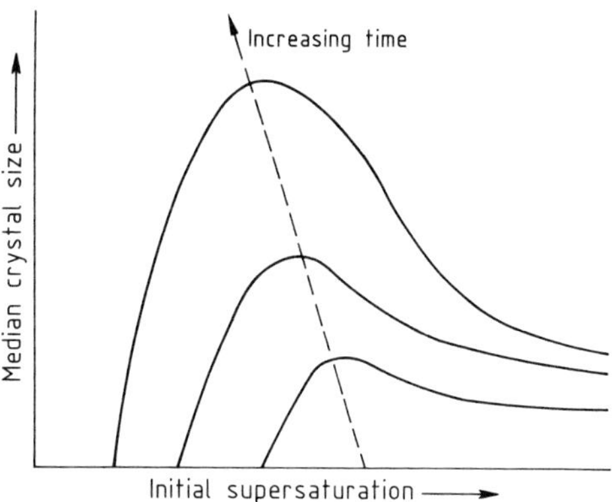

Figure 41. Illustration of Weimarn's laws of precipitation

2) For a completed precipitation, the median size of the precipitate crystals decreases as the initially created supersaturation S is increased.

For a given system, maximum agglomeration often appears to occur at a certain level of supersaturation; this behavior has been linked to the character of the adsorption layer surrounding a growing crystal which consists of loosely bonded, partially integrated groups of the crystallizing species [3].

An oversimplified but graphic example of the interactive effects of supersaturation, nucleation, and growth in the development of precipitated particles is given by WALTON [106], who considered the homogeneous nucleation of three different systems at an arbitrary value of supersaturation $S = 100$ (Eq. 2). If the number of particles nucleated is $10^6/cm^3$, the maximum precipitated particle sizes expected for three different solubilities c^* of 10^{-7}, 10^{-4}, and 10^{-1} mol/L, were concluded to be 1, 10, and 100 μm, respectively.

Alternatively, the approximate particle size can be estimated if the above three solutions are all assumed to have the same concentration, e.g., 1 mol/L. The first ($S = 10^7$) would nucleate homogeneously, forming approximately 1-nm particles and producing a colloidal system (gel) which could remain stable for long periods before the primary particles agglomerated. The second ($S = 10^4$) would also nucleate homogeneously, forming primary particles around 0.1 μm which would agglomerate easily and develop rapidly into a conventional precipitate. The third solution ($S = 10$) would probably nucleate heterogeneously, yielding approximately 10-μm crystals which could, under favorable conditions, remain discretely dispersed.

8.3.4. Coprecipitation

All precipitates are contaminated to some extent with materials originally present in the mother liquor. The general term coprecipitation may be used to describe the many different types of impurity incorporation that can occur, including surface adsorption and lattice entrapment of foreign ions and solvent molecules, as well as physical inclusion of pockets of mother liquor.

The adsorption of salts having an ion in common with the precipitate roughly obeys the Paneth–Fajans–Hahn adsorption rule which postulates that the less soluble the salt, the more easily is it incorporated into a precipitate. For example, barium chloride is more readily adsorbed by barium sulfate than is barium iodide, which is more soluble than the chloride. The dissociability of the adsorbed salt is also important; adsorption decreases as the degree of dissociation of the adsorbed salt increases.

The distribution of an impurity between solid (i.e., solid solution) and liquid phases may be represented by the Chlopin [107] equation:

$$\frac{x}{y} = D\left(\frac{a-x}{b-y}\right) \tag{50}$$

where a and b are the amounts of components A and B in the original solid, x and y are the amounts of A and B in the crystallized solid, and $a-x$ and $b-y$ are the amounts of A and B retained in the solution. D is a distribution coefficient. Alternatively, the logarithmic Doerner–Hoskins [108] equation may be used:

$$\ln(a/x) = \lambda \ln(b/y) \tag{51}$$

The constant λ has been called a heterogeneous distribution coefficient to distinguish it from the homogeneous distribution coefficient D in Equation (50). Under ideal conditions $D = \lambda$.

If component A is the impurity, $\lambda > 1$ indicates that the impurity will be enriched in the precipitate; conversely, if $\lambda < 1$ it will be depleted. The effect of precipitation rate on λ is shown in Figure 42 [106]. In enrichment systems, $\lambda \to \lambda_e = D_e$ as the precipitation rate tends to zero. For fast rates of precipitation $\lambda \to 1$. In depletion systems, an analogous situation exists, with $\lambda \to \lambda_d = D_d$ for very low precipitation rates and $\lambda \to 1$ for rapid precipitation.

Both the Chlopin and the Doerner–Hoskins relationships have been widely used to correlate the results of fractional precipitation schemes (see Chap. 9) although neither is entirely satisfactory from a theoretical point of view.

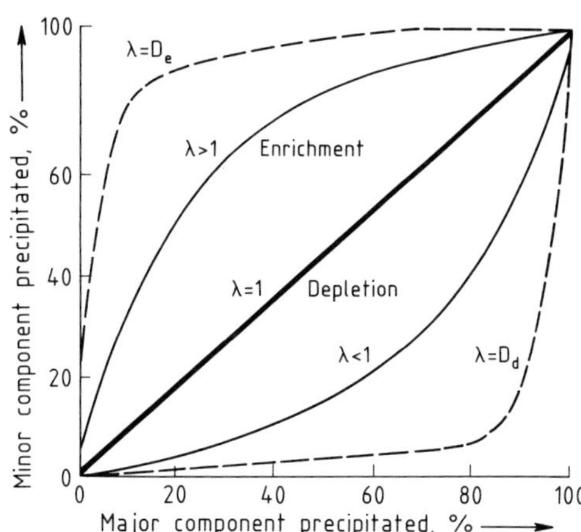

Figure 42. Effect of precipitation rate on the heterogeneous distribution coefficient λ [106]
Broken lines denote very slow precipitation; thickened line, very rapid precipitation.

8.4. Precipitation Techniques

8.4.1. Reaction Precipitation

A common method for producing a precipitate is to mix two reacting solutions together quickly to create a highly supersaturated system. One practical difficulty, however, is to maintain reasonably uniform conditions throughout the reaction vessel. The choice of method used to mix the reactants is, therefore, very important because zones of excessive supersaturation should not be allowed to develop. The sequence of reactant mixing can also be of critical importance: addition of A to B to produce a precipitate C often yields a very different product from the addition of B to A. Factors such as the development of local pockets of reactants in nonstoichiometric ratios and undesirable pH levels can have highly detrimental effects on precipitation.

Primary nucleation does not necessarily commence as soon as the reactants are mixed, even when the level of supersaturation is very high. The mixing stage is followed by a time lag (induction period) before nuclei appear, which depends on the temperature, supersaturation, efficiency of mixing, state of agitation, and presence of im-

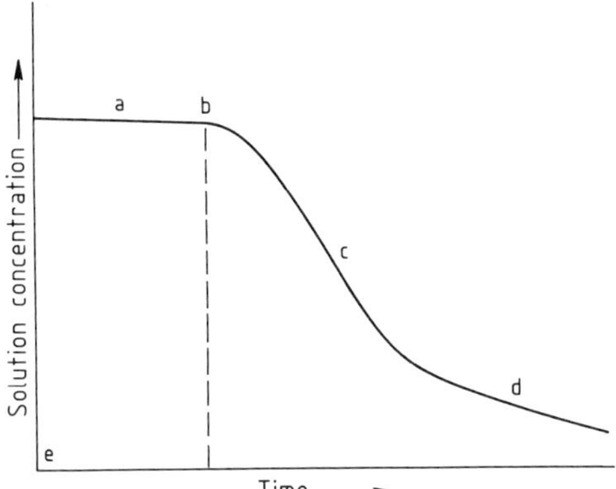

Figure 43. A typical desupersaturation curve
a) Development of nuclei; b) Induction period; c) Growth of crystals; d) Ripening of precipitate; e) Equilibrium saturation concentration

purities (see Section 4.1.4). After the induction period, rapid desupersaturation ensues (Fig. 43), during which primary (homogeneous and heterogeneous) and secondary nucleation may occur together. The predominant process at this stage, however, is growth of the nuclei. Later, if sufficient time is allowed to elapse, particle coarsening may occur as a result of ripening or agglomeration.

Precipitation of solid particles resulting from a chemical reaction between gases or liquids is a standard method for the preparation of many industrial chemicals. Precipitation occurs because the gaseous or liquid phase becomes highly supersaturated with respect to the solid phase. A crude precipitation operation, therefore, can be transformed into a crystallization process by careful control of the degree of supersaturation.

Reaction precipitation–crystallization is used widely in industries where waste gases containing valuable or noxious substances are produced. For instance, ammonia can be recovered from coke-oven gases by converting it into ammonium sulfate via reaction with sulfuric acid. In this case, agitation within the vessel is effected by a combination of the vigorous nature of the exothermic reaction and air sparging. The heat of reaction is removed by the evaporation of water added to the reaction zone [71].

Few publications address the problems of precipitation plant design. The design and operation of a continuous pilot plant for precipitation of ammonium diuranate from uranyl nitrate solutions by addition of ammonia have been described [109]. A practical approach to design theories for sparingly soluble substances is given in [110].

Precipitation from Homogeneous Solution. For the purpose of gravimetric analysis, when a solid must be efficiently separated from a liquid, precipitation is generally carried out slowly from dilute solution. However, substances such as the hydroxides and basic salts of aluminum, iron, and tin demand extremely high dilution and excessively long time for coarse filterable particles to be produced. Precipitation from homogeneous solution (PFHS) offers a useful way of overcoming these difficulties [111].

Briefly, the technique consists of slowly generating the precipitating agent homogeneously within a well-mixed solution by means of a chemical reaction. Undesirable concentration effects are thus eliminated, a dense granular precipitate is formed, and coprecipitation is minimized. For example, silver chloride crystals can be produced from an aqueous solution of silver nitrate by reaction with allyl chloride (3-chloropropene):

$$CH_2{=}CHCH_2Cl + H_2O \longrightarrow Cl^- + CH_2{=}CHCH_2OH + H^+$$

$$Cl^- + Ag^+ \longrightarrow AgCl$$

An example of a PFHS reaction in nonaqueous solution is the precipitation of silver iodide in ethanol:

$$2\,C_2H_5I + 2\,AgNO_3 + C_2H_5OH \longrightarrow 2\,AgI + (C_2H_5)_2O + HNO_3 + C_2H_5NO_3$$

Other PFHS methods used to produce crystalline precipitates by controlled generation of the required anions in an appropriate aqueous solution include the hydrolysis of dimethyl oxalate ($C_2O_4^{2-}$), triethyl phosphate (PO_4^{3-}), dimethyl sulfate (SO_4^{2-}), and thioacetamide (S^{2-}).

Precipitation from homogeneous solution plays an important role in modern analytical chemistry. It is also used to investigate coprecipitation and nucleation because the slow, controlled precipitation allows a close approach to equilibrium between the solid and the solution. Industrial applications of PFHS have so far been limited, but it appears to be a promising technique. It generally improves fractional precipitation methods and has been applied to the difficult separation of radium and barium used in the production of carriers for radioactive materials.

A further use is in the preparation of monodisperse suspensions of pigments and polishing agents [111], [112].

8.4.2. Salting Out

A solution can be made supersaturated with respect to a given solute by addition of a substance generally referred to as the precipitant, which reduces the solubility of the solute in the solvent. The precipitant may be a liquid, solid, or gas. This operation is known by a variety of terms, salting out being the most common. The term watering out is used in the pharmaceutical industry to describe the precipitation–crystallization of organic substances from water-miscible organic solvents by the controlled addition of water. The term solventing out has been applied to the use of water-miscible organic solvents to precipitate electrolytes from aqueous solution [113].

The properties required of a precipitant are that it be miscible with the solvent of the original solution, that the solute be relatively insoluble in it, and that the final solvent–precipitant mixture be easily separable if it contains valuable components.

Salting out has many advantages. For example, highly concentrated initial solutions can be made by dissolving an impure crystalline material in a suitable solvent. If the solute is very soluble in the chosen solvent, dissolution may be effected at low temperature, which is advantageous when heat-labile substances are processed. Solute recovery is usually high. Purification is often better than in straightforward crystallization because the mixed mother liquor often retains more undesirable impurities than the original solvent does. On the other hand, salting out has the disadvantage that a recovery unit may be needed to handle fairly large quantities of mother liquor in order to separate valuable solvents and precipitants.

Several potential large-scale applications of salting out have been reported [3]. Examples are the production of pure inorganic salts with liquid organic precipitants (particularly anhydrous salts from aqueous solution at ambient temperature when a hydrated species is the thermodynamically stable phase [114]), the treatment of seawater with alcohols to recover fertilizer-grade double salts (e.g., hydrated potassium magnesium sulfate) [115], and the separation of inorganic salt mixtures [113], [116].

Gases or solids may be used as precipitants provided they are soluble in the original solvent and do not react with the solute to be precipitated. Ammonia can assist in the production of potassium sulfate by the reaction of calcium sulfate and potassium chloride: in a pure aqueous medium the yield is low, but in aqueous ammonia the yield is greatly improved. Hydrazine acts in a similar manner [117].

An example of the use of a solid precipitant is the addition of sodium chloride crystals to salt out organic dyes from aqueous solution. The sodium chloride acts in the solution phase, i.e., it must dissolve in the water present before it can act as a precipitant; its precise mode of action is probably quite complex.

Crystalline salts can be added to solutions to precipitate other salts, for example, as a result of the formation of a stable salt pair. This behavior is encountered when two solutes AX and BY, usually without a common ion, react in solution and undergo double decomposition (metathesis).

$$AX + BY \rightleftharpoons AY + BX$$

The four salts AX, BY, AY, and BX constitute a reciprocal salt pair. One of these pairs, AX – BY or AY – BX, is stable and is composed of compatible salts which can coexist in solution; the other salt pair is unstable and composed of incompatible salts which cannot coexist. This principle is used in the large-scale production of potassium sulfate by addition of solid glaserite ($3\,K_2SO_4 \cdot Na_2SO_4$) to a concentrated aqueous solution of potassium chloride [118]. Conversion to a stable salt pair occurs; the Na_2SO_4 and KCl remain in solution and K_2SO_4 precipitates.

8.5. Precipitation Methods and Equipment

Although continuous, steady-state operation is often regarded as ideal for process plant equipment, this is not always true for crystallization [3] and perhaps even less so for precipitation. The advantages and disadvantages discussed in Section 5.7 also apply to precipitation.

Most industrial precipitation units are constructed simply. The main aim is usually to mix reacting fluids rapidly and allow the development of a precipitate with certain desirable physical and chemical characteristics. An efficient mixing device, therefore, is of prime importance and may be a conventional agitated vessel or a small in-line mixer.

9. Fractional Crystallization

A single crystallization operation performed on a solution or melt can fail to produce a pure crystalline product for a variety of reasons. For example, the impurity may have solubility characteristics similar to those of the desired pure component, and both substances consequently cocrystallize. Alternatively, the impurity may be present in such large amounts that the crystals inevitably become contaminated. Furthermore, a pure substance cannot be produced in a single crystallization stage if the impurity and the required substance form a solid solution (see Section 3.2.2). Recrystallization (i.e., repeated crystallization steps) from a solution or melt is, therefore, widely employed to increase crystal purity.

9.1. Recrystallization from Solutions

Most of the impurities from a crystalline mass can often be removed by dissolving the crystals in a small amount of fresh hot solvent and cooling the solution to produce a fresh crop of purer crystals. However, the impurities must then be more soluble in the solvent than the main product is. Recrystallization may have to be repeated many times before crystals of the desired purity are obtained. A simple recrystallization scheme is

$$\begin{array}{ccccccc} S & & S & & S & & \\ \downarrow & & \downarrow & & \downarrow & & \\ AB & \longrightarrow & X_1 & \longrightarrow & X_2 & \longrightarrow & X_3 \\ \downarrow & & \downarrow & & \downarrow & & \\ L_1 & & L_2 & & L_3 & & \end{array}$$

An impure crystalline mass AB (A is the less soluble, desired component) is dissolved in the minimum amount of hot solvent S and then cooled. The first crop of crystals X_1 will contain less impurity B than the original mixture, and B is concentrated in the liquor L_1. To achieve a higher degree of crystal purity, the procedure can be repeated.

In such a sequence, losses of the desired component A can be considerable, and the final amount of "pure" crystals may easily be a small fraction of the starting mixture AB. Many schemes have been designed to increase both the yield and the separation efficiency of fractional recrystallization. The choice of solvent depends on the characteristics of the required substance A and the impurity B. Ideally, B should be very soluble in the solvent at the lowest temperature employed and A should have a high temperature coefficient of solubility, so that high yields of A can be obtained from operation within a small temperature range.

9.2. Recrystallization from Melts

Melts can be fractionally recrystallized by schemes similar to those described in Section 9.1 for solutions, although a solvent is not normally added. Usually, simple sequences of heating (melting) and cooling (partial crystallization) are followed by separation of the purified crystals from the residual melt. Selected melt fractions may be mixed at intervals according to the type of scheme employed, and fresh feedstock may be added at different stages if necessary. Several such schemes have been proposed for purification of fats and waxes [119].

As described in Section 3.2, eutectic systems can theoretically be purified by single-stage crystallization, whereas solid solutions always require multistage operations. Countercurrent fractional crystallization processes in column crystallizers are described in Section 6.3.

9.3. Recrystallization Schemes

A number of fractional crystallization schemes have been devised [3], [111], the triangular scheme shown in Figure 44 [120] demonstrates their essential features. Individual fractions, represented by the circles, are designated by a horizontal row number and a diagonal column letter. Consider the separation of two components A and B: if a constant fraction of A is precipitated at each operation and fractions are combined as in the triangular scheme, then the fraction x of the original A which appears at any given point in the triangular scheme is given by the binominal expansion

$$[x + (1 - x)]^n = 1 \qquad (52)$$

Similarly, for component B, where a constant fraction y is precipitated at each step, the distri-

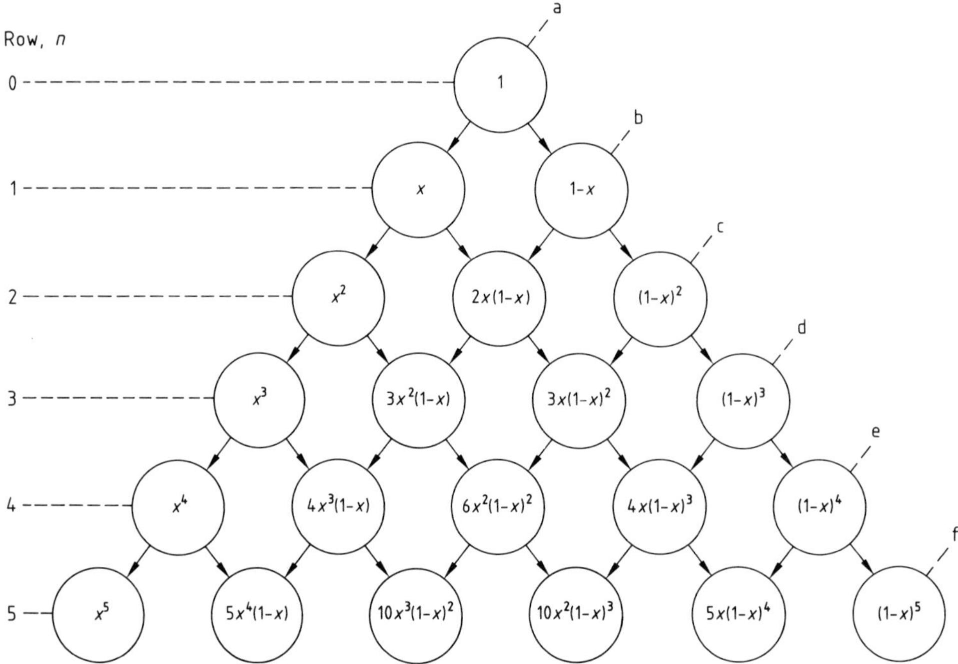

Figure 44. Triangular fractional crystallization scheme [120]

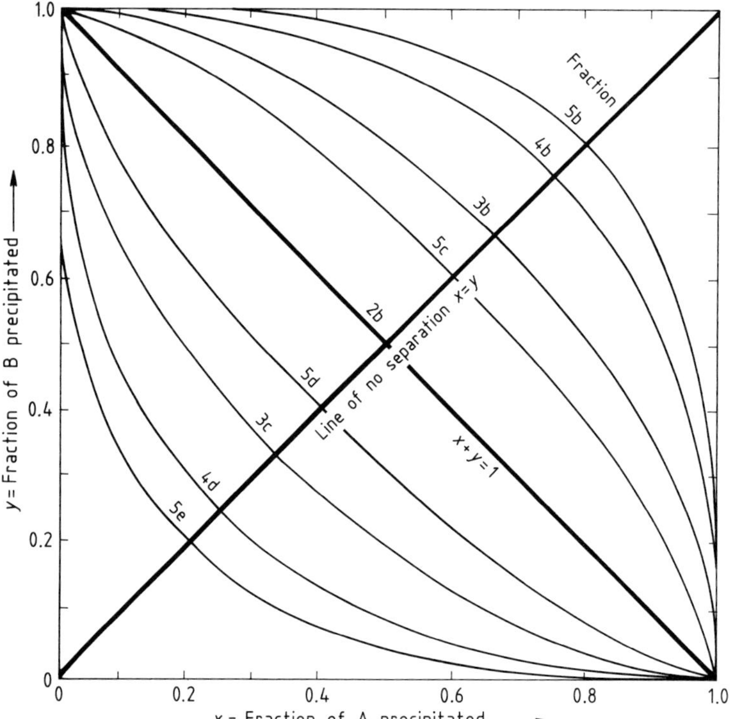

Figure 45. Operating curves for systems with repeating composition [120]
Curves were calculated by assuming that $\lambda = 7.21$; λ is the heterogeneous distribution, coefficient defined in Equation (51), p. **3**-38.

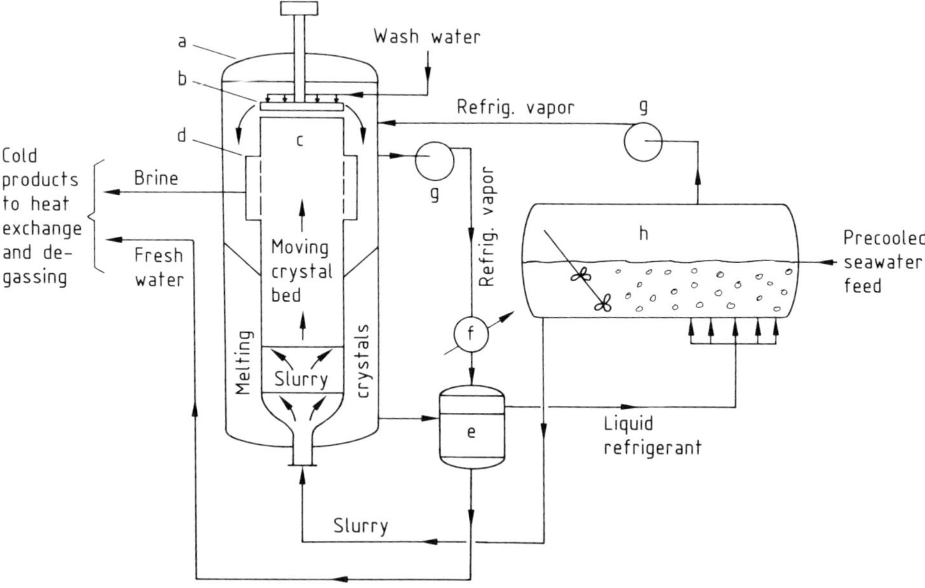

Figure 46. Desalination of seawater by freezing [67]
a) Washer–melter; b) Scraper; c) Wash column; d) Screens; e) Decanter; f) Heat exchanger; g) Compressor; h) Crystallizer

bution of B in various fractions is given by the expansion

$$[y + (1 - y)]^n = 1 \quad (53)$$

Fractionation schemes are simplified when fractions of repeating composition recur at regular intervals. The first fraction in the above scheme that can have the same composition (i.e., A:B ratio) as the original mixture is 2b. In this case, every fraction has the same composition as the fraction vertically above it in the scheme. Fraction 2b will also be identical in composition to the original if the fraction of A which reaches that point is the same as the fraction of B reaching that point, i.e.,

$$2x(1 - x) = 2y(1 - y) \quad (54)$$

This equation has two solutions ($x = y$ and $x + y = 1$). The first represents a line of no separation which, when drawn as in Figure 45, divides the diagram into two zones: enrichment of component B in the precipitate occurs in the upper, left-hand zone (enrichment being expressed by the ratio y/x) and enrichment of A in the lower, right-hand zone. The second solution to the equation ($x + y = 1$), represented by the other diagonal, gives repeating compositions for fractions in the same vertical columns and also gives enrichment of $(y/x)^n$ in the end fractions (column a in Fig. 44). This second diagonal may be called the operating line for a repeating composition in fraction 2b. Operating lines can be calculated in a similar manner for repeating compositions in the other fractions, as depicted in Figure 45 [120]. Examples of the use of such a diagram are described in [120], [121].

10. Freeze Crystallization

Crystallization by freezing, more commonly known as freeze crystallization, is a process in which heat is removed from a solution to form crystals of the solvent rather than the solute. Subsequent steps include separation of crystals from the concentrated solution, washing the crystals with near-pure solvent, and finally melting them to generate visually pure solvent. The product of freeze crystallization can be either the melted crystals (near-pure solvent), as in water desalination, or the concentrated solution, as in the concentration of fruit juice or coffee extract. Freeze crystallization is applicable in principle to a variety of solvents and solutions; however, because it is most commonly applied to aqueous systems, the following account refers exclusively to the freezing of water.

One of the more obvious advantages of freezing over evaporation for removal of water from solutions is the potential for saving heat energy: the enthalpy of crystallization of ice (334 kJ/kg) is only one-seventh of the enthalpy of vaporization of water (2260 kJ/kg). Process energy consumption, however, may be reduced below that predicted by the phase-change enthalpy by utilizing energy recycle methods, such as multiple-effect or vapor compression, commonly employed in evaporation. In freeze crystallization plants operating by direct heat exchange, vapor compression has been employed to recover refrigeration energy by using the crystals to condense the refrigerant evaporated in the crystallizer.

Another advantage of freeze crystallization, important in many food applications, is that the volatile flavor components normally lost during conventional evaporation can be retained in the freeze-concentrated product. In fact, freeze crystallization is used mainly in the food industry. Despite earlier enthusiasm, large-scale applications in desalination, effluent treatment, dilute liquor concentration, solvent recovery, etc., have not yet emerged.

All freeze separation processes depend on the formation of pure solvent crystals from solution, as described for eutectic systems in Section 3.2.1; this allows single-stage operation. Solid-solution systems, which require multistage operation, are not usually handled. Several types of freeze crystallization can be designated according to the kind of refrigeration system used.

In *indirect-contact freezing,* for instance, the liquid feedstock is crystallized in a scraped-surface heat exchanger (Section 6.2, Brodie purifier) fitted with internal rotating scraper blades and an external heat-transfer jacket through which a liquid refrigerant flows. The resulting ice–brine slurry passes to a wash column where the ice crystals are separated and washed before melting. One of a number of commercial systems based on this type of freezing process is described in [122].

Direct-contact freezing processes utilizing inert, immiscible refrigerants have been investigated widely for desalination purposes. A typical scheme is shown in Figure 46 [68]. Seawater, precooled close to its freezing point, is fed continuously into the crystallization vessel where it comes in direct contact with a liquid refrigerant (e.g., *n*-butane) which vaporizes and causes ice crystals to form due to the exchange of latent heat. The ice–brine slurry is fed to a wash column where it is washed countercurrently with fresh water. The emerging brine-free ice is melted by the enthalpy of vapor condensation released from the compressed refrigerant. Energy input is required for the compressors.

Vacuum freezing processes do not require a conventional heat exchanger, and the problems of scale formation on heat-transfer surfaces are avoided. Cooling is effected by flash-evaporating some of the solvent when the liquid feedstock enters a crystallization vessel maintained at reduced pressure. Vacuum freezing is potentially attractive for aqueous systems (7 kg of ice can be produced for every kilogram of water evaporated) but has not yet achieved commercial success.

A general review of freeze concentration as an industrial separation process is given in [123]. The status of freeze desalination [67] and the potential of freeze crystallization in the recycling and reuse of wastewater [124] have also been reviewed. The kinetics of ice crystallization in aqueous sugar solutions and fruit juice are described in [125].

11. References

[1] G. Matz: *Kristallisation,* 2nd ed., Springer Verlag, Berlin 1969.
[2] J. Nývlt: *Industrial Crystallization from Solutions.* Butterworths, London 1971.
[3] J. W. Mullin: *Crystallization,* 2nd ed., Butterworths, London 1972.
[4] S. J. Jančić, P. A. M. Grootscholten: *Industrial Crystallization,* Reidel, Dordecht 1984.
[5] A. Mersmann, *Int. Chem. Eng.* **24** (1984) 401–418.
[6] E. J. de Jong, *Int. Chem. Eng.* **24** (1984) 419–431.
[7] J. Garside, *Chem. Eng. Sci.* **40** (1984) 3–26.
[8] J. S. Wey, *Chem. Eng. Commun.* **35** (1985) 231–252.
[9] J. W. Mullin (ed.): *Industrial Crystallization 75,* Plenum Publishing, London 1976.
[10] E. J. de Jong, S. J. Jančić (eds.): *Industrial Crystallization 78,* North-Holland, Amsterdam 1979.
[11] S. J. Jančić, E. J. de Jong (eds.): *Industrial Crystallization 81,* North-Holland, Amsterdam 1982.
[12] S. J. Jančić, E. J. de Jong (eds.): *Industrial Crystallization 84,* Elsevier, Amsterdam 1984.
[13] *AIChE Symp. Ser.* (a) **65** (1969) no. 95; (b) **67** (1971) no. 110; (c) **68** (1972) no. 121; (d) **72** (1976) no. 153; (e) **76** (1980) no. 193; (f) **78** (1982) no. 215; (g) **80** (1984) no. 240.
[14] J. W. Mullin, O. Söhnel, *Chem. Eng. Sci.* **32** (1977) 683–686.
[15] Wilhelm Ostwald, *Z. Phys. Chem.* (*Leipzig*) **34** (1900) 493–503.
[16] H. A. Miers, F. Isaac, *J. Chem. Soc.* **89** (1906) 413–454.

[17] A. Seidell in W. F. Linke (ed.): *Solubilities of Inorganic and Metal Organic Compounds,* 4th ed., vol. 1, Van Nostrand, New York 1958, vol. 2, Am. Chem. Soc., Washington, D.C., 1965.
[18] H. Stephen, T. Stephen: *Solubilities of Inorganic and Organic Compounds,* Pergamon, London 1963.
[19] M. Broul, J. Nývlt, O. Söhnel: *Solubilities in Binary Aqueous Solutions,* Academia, Prague 1981.
[20] H. Feilchenfeld, S. Sarig, *Ind. Eng. Chem. Process Des. Dev.* **24** (1985) 130–133.
[21] H. Kimura, *J. Cryst. Growth* **73** (1985) 53–62.
[22] F. Grønvold, K. K. Meisingset, *J. Chem. Thermodyn.* **14** (1982) 1083–1098.
[23] H. Kimura, K. Junjiro, *Sol. Energy* **35** (1985) 527–534.
[24] A. Findlay, A. N. Campbell: *The Phase Rule and its Applications,* 9th ed., Longmans, London 1951.
[25] J. E. Ricci: *The Phase Rule and Heterogeneous Equilibrium,* Van Nostrand, New York 1951.
[26] H. R. Null: *Phase Equilibrium in Process Design,* Wiley-Interscience, New York 1970.
[27] J. Nývlt: *Solid-Liquid Phase Equilibrium,* Academia, Prague, Elsevier, Amsterdam 1977.
[28] H. Hörmeyer, *Ger. Chem. Eng. (Engl. Transl.)* **6** (1983) 277–281.
[29] C. L. Kusik, H. P. Meissner, E. L. Field, *AIChE J.* **25** (1979) 759–762.
[30] B. Sander, P. Rasmussen, A. Fredenslund, *Chem. Eng. Sci.* **41** (1986) 1197–1202.
[31] G. H. Nancollas, M. M. Reddy, F. Tsai, *J. Cryst. Growth* **20** (1973) 125–134.
[32] G. H. Nancollas, M. M. Reddy, *Soc. Pet. Eng. J.* April 1974, 117–123.
[33] P. T. Cardew, R. J. Davey, A. J. Ruddik, *J. Chem. Soc. Faraday Trans. 2,* **80** (1984) 659–668.
[34] P. T. Cardew, R. J. Davey, *Proc. R. Soc. London Ser. A* **398** (1985) 415–428.
[35] A. R. Ubbelohde: *Melting and Crystal Structure,* OUP, Oxford 1965.
[36] V. A. Garten, R. B. Head, *Philos. Mag.* **8** (1963) 1793–1803 and **14** (1966) 1243–1253.
[37] K. A. Berglund et al., in Reference 12, pp. 21–26, 229–236.
[38] J. W. Mullin, C. L. Leci, *Philos. Mag.* **19** (1969) 1075–1077.
[39] M. A. Larson, J. Garside, *Chem. Eng. Sci.* **41** (1986) 1285–1289.
[40] R. F. Strickland-Constable: *Kinetics and Mechanism of Crystallization,* Academic Press, London 1968.
[41] A. C. Zettlemoyer (ed.): *Nucleation,* Marcel Dekker, New York 1969.
[42] J. Garside, R. J. Davey, *Chem. Eng. Commun.* **4** (1980) 393–424.
[43] J. Nývlt, O. Söhnel, M. Matuchová, M. Broul: *The Kinetics of Industrial Crystallization,* Academia, Prague 1985.
[44] A. D. Randolph, M. A. Larson: *Theory of Particulate Processes,* Academic Press, New York 1971.
[45] A. E. Nielsen, O. Söhnel, *J. Cryst. Growth* **11** (1971) 233–242.
[46] O. Söhnel, J. W. Mullin, *J. Cryst. Growth* **44** (1978) 377–382.
[47] W. K. Burton, N. Cabrera, F. C. Frank, *Philos. Trans. R. Soc. London Ser. A.* **243** (1951) 299–358.
[48] A. E. Nielsen, *J. Cryst. Growth* **67** (1984) 289–310.
[49] B. R. Pamplin (ed.): *Crystal Growth,* 2nd ed., Pergamon Press, Oxford 1980.
[50] E. Kaldis, H. J. Scheel: *Crystal Growth and Materials,* North-Holland, Amsterdam 1977.
[51] P. Bennema, in Reference 9, pp. 91–112.
[52] T. Tengler, A. Mersmann, *Ger. Chem. Eng. (Engl. Transl.)* **7** (1984) 248–259.
[53] W. Wöhlk: "Meßanordnungen zur Bestimmung von Kristallwachstumsgeschwindigkeiten," *Fortschr. Ber. VDI Z., Reihe 3,* (1982) no. 71.
[54] E. T. White, L. L. Bendig, M. A. Larson, in Reference 13d, pp. 41–47.
[55] A. G. Jones, J. W. Mullin, *Chem. Eng. Sci.* **29** (1974) 105–118.
[56] A. H. Janse, E. J. de Jong, in Reference 9, pp. 145–154.
[57] J. Garside, S. J. Jančić, *AIChE J.* **22** (1976) 887–894.
[58] G. H. Gilmer, P. Bennema, *J. Cryst. Growth* **13/14** (1972) 142–153.
[59] J. R. Bourne, R. J. Davey, *J. Cryst. Growth* **36** (1976) 278–286.
[60] G. D. Botsaris, in Reference 11, pp. 109–116.
[61] R. J. Davey, in Reference 10, pp. 169–184.
[62] G. Deicha: *Lacunes des Cristeaux et leurs Inclusions Fluides,* Masson, Paris 1955.
[63] H. E. C. Powers, *Sugar Technol. Rev.* **1** (1970) 85–190.
[64] W. R. Wilcox, V. H. S. Kuo, *J. Cryst. Growth* **19** (1973) 221–229.
[65] D. Senol, A. S. Myerson, in Reference 13f, pp. 37–41.
[66] L. Phoenix, *Brit. Chem. Eng.* **11** (1966) 33–38.
[67] J. D. Bushnell, J. F. Eagen, *Oil Gas J.* **73** (1975) no. 42, 80–84.
[68] A. J. Barduhn, *Chem. Eng. Prog.* **71** (1975) no. 11, 80–87.
[69] J. W. Mullin, J. R. Williams, *Chem. Eng. Res. Des.* **62** (1984) 296–302.
[70] E. Finkelstein, *J. Heat Recovery Syst.* **3** (1983) 431–437.
[71] A. W. Bamforth: *Industrial Crystallization,* Leonard Hill, London 1965.
[72] J. W. Mullin, J. Nývlt, *Chem. Eng. Sci.* **26** (1971) 369–377.
A. G. Jones, J. W. Mullin, *Chem. Eng. Sci.* **29** (1974) 105–118.
[73] J. Robinson, J. E. Roberts, *Can. J. Chem. Eng.* **35** (1957) 105–112.
[74] C. F. Abbeg, N. S. Balakrishnam, in Reference 136, pp. 88–94.
[75] A. D. Randolph, in Reference 13g, pp. 14–22.
[76] R. C. Bennett, H. Fiedelman, A. D. Randolph, *Chem. Eng. Prog.* **69** (1973) no. 7, 86–93.
[77] J. G. D. Molinari in M. Zeif, W. R. Wilcox (eds.): *Fractional Solidification,* Dekker, New York 1967, pp. 393–400.
[78] N. I. Gelperin, G. A. Nosov, *Sov. Chem. Ind. (Engl. Transl.)* **9** (1977) 713–717.
[79] V. G. Ponomorenko, G. F. Potebnya, V. I. Bei, *Theor. Found. Chem. Eng. (Engl. Transl.)* **13** (1980) 724–729.
[80] K. Wintermantel, *Chem. Ing. Tech.* **58** (1986) no. 6, 498–499.
[81] G. R. Atwood, in Reference 13a, pp. 112–121.
[82] S. Rittner, R. Steiner, *Chem. Ing. Tech.* **57** (1985) no. 2, 91–102.
[83] O. Fischer, S. J. Jančić, K. Saxer, in Reference 12, pp. 153–157.
[84] W. C. Gates, J. E. Powers, *AIChE J.* **16** (1970) 648–655.

[85] J. D. Henry, J. E. Powers, *AIChE J.* **16** (1970) 1055–1063.
[86] D. L. McKay, in Reference 77, pp. 427–439.
[87] J. A. Brodie, *Mech. Chem. Trans. Inst. Engrs. Australia* **7** (1971) no. 1, 37–44.
[88] K. Takegami, N. Nakamaru, M. Morita, in Reference 12, pp. 143–146.
[89] M. Moritoki, *Int. Chem. Eng.* **20** (1980) 394–401.
[90] W. H. Shearon, W. B. Dunwoody, *Ind. Eng. Chem.* **45** (1953) 496–504.
[91] L. M. Nunelly, F. T. Cartney, *Ind. Eng. Chem. Prod. Res. Dev.* **21** (1982) 617–620.
[92] M. M. Faktor, I. Garrett: *Growth of Crystals from the Vapor,* Chapman and Hall, London 1974.
[93] S. Lewin: *The Solubility Product Principle,* Pitman, London 1960.
[94] Wilhelm Ostwald: *Lehrbuch Allgemeine Chemie,* vol. 2, Engleman, Leipzig 1896, p. 401.
[95] E. Hanitzch, M. Kahlweit in *Symposium on Industrial Crystallization,* Inst. Chem. Engrs., London 1969, pp. 130–141.
G. Matz, *Ger. Chem. Eng.* (*Engl. Transl.*) **8** (1985) 255–265.
[96] A. Winzer, H.-H. Emons, in Reference 9, pp. 163–171.
[97] T. Nakai, H. Nakamura, in Reference 10, pp. 75–79.
[98] J. Nývlt, in A. L. Smith (ed.): *Particle Growth in Suspensions,* Academic Press, London 1973, pp. 131–159.
[99] J. Skřivánek, S. Žáček, J. Hostomský, V. Vacek, in Reference 9, pp. 173–185.
[100] M. von Smoluchowski, *Z. Phys. Chem.* (*Leipzig*) **92** (1918) 129–168.
[101] T. F. Tadros, *Adv. Colloid Interface. Sci.* **12** (1980) 141–150.
[102] A. J. Karabelas, *AIChE J.* **22** (1976) 765–769.
[103] I. Patterson, M. R. Kamal, *Can. J. Chem. Eng.* **54** (1976) 623–625.
[104] C. G. J. Baker, M. A. Bergougnou, *Can. J. Chem. Eng.* **52** (1974) 246–250.
[105] P. P. von Weimarn, *Chem. Rev.* **2** (1926) 217–235.
[106] A. G. Walton: *The Formation and Properties of Precipitates,* Interscience, New York 1967.
[107] V. Chlopin, *Zeit. Anorg. Allg. Chem.* **143** (1925) 97–117.
[108] H. A. Doerner, W. M. Hoskins, *J. Am. Chem. Soc.* **47** (1925) 662–675.
[109] D. G. Stevenson, *Trans. Inst. Chem. Eng.* **42** (1964) 316–322.
[110] K. Toyokura, K. Ono, J. Ueno, in Reference 10, pp. 357–365.
[111] L. Gordon, M. L. Salutsky, H. H. Willard: *Precipitation from Homogeneous Solution,* Wiley-Interscience, New York 1959.
[112] P. F. S. Cartwright, E. J. Newman, D. D. Wilson, *Analyst* (*London*) **92** (1967) 663–679.
[113] Z. B. Alfassi, *Sep. Sci. Technol.* **14** (1979) 155–161.
[114] H. Hoppe, *Chem. Process Eng.* (*London*) **49** (1968) no. 12, 61–63.
[115] J. R. Fernandez-Lozano, *Ind. Eng. Chem. Process Des. Dev.* **15** (1976) 445–449.
[116] Z. B. Alfassi, S. Mosseri, *AIChE J.* **30** (1984) 874–876.
[117] J. R. Fernandez-Lozano, A. Wint, *Chem. Eng. London* **349** (1979) 688–690.
[118] D. E. Garrett, *Chem. Eng. Prog.* **54** (1958) no. 12, 65–68.
[119] A. E. Bailey: *Solidification of Fats and Waxes,* Interscience, New York 1950.
[120] E. F. Joy, J. H. Payne, *Ind. Eng. Chem.* **47** (1955) 2157–2161.
[121] M. L. Salutsky, J. G. Sites, *Ind. Eng. Chem.* **47** (1955) 2162–2166.
[122] W. H. van Pelt, *Food Eng.* (1975) Nov., 77–79.
[123] H. A. C. Thijssen, in A. Spicer (ed.): *Advances in Preconcentration and Dehydration of Foods,* Wiley-Interscience, New York 1974.
[124] J. A. Heist, *AIChE Symp. Ser.* **77** (1984) no. 201, 259–272.
[125] A. M. Omran, C. J. King, *AIChE J.* **20** (1974) 799–801.

4. Drying of Solid Materials

Evangelos Tsotsas, Volker Gnielinski, Ernst-Ulrich Schlünder, Institut für Thermische Verfahrenstechnik, University of Karlsruhe, Karlsruhe, Federal Republic of Germany

In addition to the standard symbols defined in the front matter of this volume, the following symbols are used.

A surface area of the drying solid, m^2
C excess air factor in Eq. (38)
C radiation coefficient, $W m^{-2} K^{-4}$
c specific heat, $J kg^{-1} K^{-1}$
d particle diameter, m
h enthalpy, J/kg
L dryer length, m
l modified mean free path, m
M mass, kg
$\dot{M}$ mass flow rate, kg/s
$\tilde{M}$ molecular mass, kg/kmol
$\dot{m}$ drying rate, $kg m^{-2} s^{-1}$
N number of moles, kmol
P pressure, Pa
$\dot{Q}$ heat transfer rate, W
$\dot{q}$ heat flux, W/m^2
R gas constant, $J kg^{-1} K^{-1}$
s width of sample, m
T temperature, K
t time, s
u velocity of drying agent, m/s
v velocity of liquid in capillaries, m/s
$\dot{W}$ power, W
X moisture content in solid, kg moisture/kg dry solid
$\tilde{x}$ mole fraction in the liquid phase
Y humidity of the drying agent, kg moisture/kg drying agent

$\tilde{y}$ mole fraction in the gas phase
z axial coordinate in a dryer, m
z distance from the free surface of sample, m
α heat-transfer coefficient, $Wm^{-2}K^{-1}$
β mass-transfer coefficient, m/s
γ accommodation coefficient
δ diffusion coefficient, m^2/s
δ surface roughness, m
ζ dimensionless axial coordinate in a dryer
Λ mean free path, m
λ thermal conductivity, $Wm^{-1}K^{-1}$
μ ratio of molecular masses
$\dot{v}$ normalized drying rate, Eq. (23)
ξ normalized moisture content, Eq. (24)
ϱ density, kg/m^3
τ dimensionless time
Φ dimensionless number, Eq. (29)
φ relative humidity
φ surface coverage
ψ porosity

Subscripts

a ambient
bed bed
crit at the critical point
eq in equilibrium
f final, at the outlet
g gas (drying agent)
h maximum hygroscopic moisture content
heat from the heater
i initial, at the inlet
l liquid (moisture)
lost heat losses to the surroundings
max maximum
p at constant pressure
r radiation
s dry solid
v vapor
vent ventilator
W wall
w wetting
1 component index (isopropyl alcohol)
I in the first period of drying

Superscripts

0 effective coefficient, Eqs. (15), (17)
* saturated
~ molar quantities

1. Theoretical Fundamentals of the Drying Process

1.1. Concepts, Definitions

Drying denotes the separation of volatile liquids from solid materials by vaporizing the liquid and removing the vapor. The liquid that is to be removed is usually water, but it could also be a solvent such as alcohol or acetone, or a mixture of such solvents. The solid material that is to be dried can be a natural product such as wood, a semifinished or a finished good (such as paper).

The removal of water from other fluids such as refrigerants or from gases, such as natural gas, is also considered a drying process, but this topic is not treated in this article.

The vaporization of liquids requires the supply of heat. Accordingly, drying can be considered a *thermal separation process*. The removal of liquids from solids without the application of heat, which is the case in a centrifuge, does not come under the strict definition of drying.

The product that is to be dried is denoted as the moist solid, or simply as the solid. The substance that carries the necessary heat is called the drying agent. This substance could be air, an inert gas, or superheated steam. Heat could also be supplied by radiation, by hot surfaces, or by microwaves. The moisture content of the solid is denoted by X and measured in kg liquid per kg of dry solid. For the humidity of a gaseous drying agent the symbol Y is used (in kg vapor per kg of dry gas). The saturation humidity is denoted by Y^*. The mass flux of the vapor leaving the surface of the solid per unit time is called the drying rate, and is denoted by the symbol $\dot{m}$ (in $kgm^{-2}s^{-1}$). The drying rate is usually determined by measuring the change of moisture content with time dX/dt. It follows that

$$\dot{m} = -\frac{M_s}{A}\frac{dX}{dt}$$

M_s is the mass of the dry solid, and A is the portion of its surface area that is in contact with the drying agent. The drying rate $\dot{m}$ depends on the conditions of drying and on the moisture content X. The drying conditions are specified by factors such as the air pressure, temperature, and humidity, the radiator temperature, the temperature of a heating surface, or the strength of the microwaves. The relationship of the drying rate

$\dot{m}$ and the moisture content X under constant drying conditions—this is $\dot{m}(X)$—is called the *drying rate curve*.

1.2. Characteristics of Moist Solids

Almost all industrial products have to be dried one or more times in the course of their manufacture. Consequently, the variety and characteristics of moist solids are manifold. However, only two of these characteristics have a direct influence on the drying rate curve $\dot{m}(X)$. They relate to the questions:

1) Is the liquid in the solid freely mobile or is it bonded to the solid by sorption?
2) Does vaporization take place at the surface or in the interior of the solid?

The first characteristic is described by the thermodynamic equilibrium (sorption isotherms), the second by the kinetics of liquid migration in the interior of the solid (capillarity, diffusion). Both topics will be treated thoroughly subsequently. In any case it should be borne in mind that drying is not only a thermal separation process; it is also a means to manufacture specified products and to influence their quality. Paper is an example of such a specified product whose quality is controlled by the choice of drying conditions. Further, the risk of possible product damage during drying must be reduced. This is particularly important for sensitive products such as foodstuffs. In this context, a large number of product characteristics differing from material to material should be accounted for. Some remarks on the handling of temperature-sensitive materials are given in Section 3.1.

Free and Bonded Moisture in Solids. If free (unbound) liquid is in contact with its own vapor, then the vapor pressure is equal to the saturation pressure for the respective temperature $P_v^*(T)$, e.g., $P_v^* = 0.1$ MPa for water at 100 °C. In the presence of an inert gas (e.g., air), the total pressure at the surface of the liquid P is equal to the sum of the saturation pressure P_v^* and of the partial pressure of the inert gas P_g:

$$P = P_v^*(T) + P_g \tag{1}$$

In most practical cases the total pressure P is prescribed. The partial pressure of the inert gas can be obtained from Eq (1):

$$P_g = P - P_v^*(T) \tag{2}$$

The ratio of the partial pressure to the total pressure is equal to the mole fraction of the respective component. It follows:

$$\tilde{y}_v^*(T) = \frac{N_v}{N_v + N_g} = \frac{P_v^*}{P} \tag{3}$$

and correspondingly

$$\tilde{y}_g = \frac{N_g}{N_v + N_g} = \frac{P_g}{P} \tag{4}$$

The number of moles N (expressed in kmol), is related to the mass of the respective component M (expressed in kg) by the molecular mass $\tilde{M}$ (which is given in kg/kmol) by

$$M_v = N_v \tilde{M}_v \quad M_g = N_g \tilde{M}_g \tag{5a,b}$$

($\tilde{M}_v = 18.01$ kg/kmol for water and $\tilde{M}_g =$ 28.96 kg/kmol for air)
From Eqs. (1)–(5) the saturation humidity of the inert gas [$Y^*(T)$ in kg vapor per kg inert gas] can be obtained:

$$Y^*(T) = \frac{\tilde{M}_v}{\tilde{M}_g} \frac{P_v^*(T)}{P - P_v^*(T)} \tag{6}$$

The humidity of the unsaturated inert gas is given by the relationship

$$Y = \frac{\tilde{M}_v}{\tilde{M}_g} \frac{\varphi P_v^*(T)}{P - \varphi P_v^*(T)} \tag{7}$$

where φ is the relative humidity, defined as

$$\varphi = P_v / P_v^*(T) \tag{8}$$

The limits of φ are 0 and 1, which correspond to completely dry air and vapor-saturated air, respectively.

In summary, the thermodynamic equilibrium between the unbound moisture within a solid that is in contact with a gas–vapor atmosphere is characterized by $\varphi = 1$ and $Y = Y^*(T)$. This is the case regardless of the magnitude of the moisture content X.

If the moisture is bonded (bound) to the solid material, the air humidity under conditions of thermodynamic equilibrium (denoted by Y_{eq}) depends not only on the temperature T, but also on the moisture content X_{eq}, i.e., $Y_{eq} = Y_{eq}(T, X_{eq})$. The parameter Y_{eq} is always smaller than Y^* and consequently, the partial vapor pressure in the gas–vapor mixture $P_{v,eq}$ is always smaller than

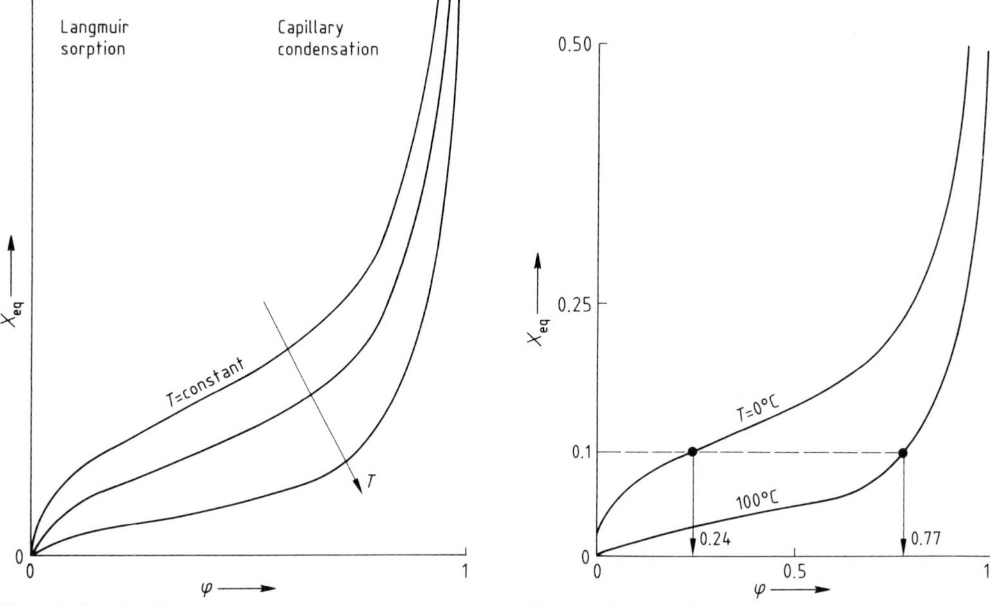

Figure 1. Sorption isotherms $X_{eq}(\varphi, T)$ (qualitative graph)

Figure 2. Sorption isotherms for potatoes [1]

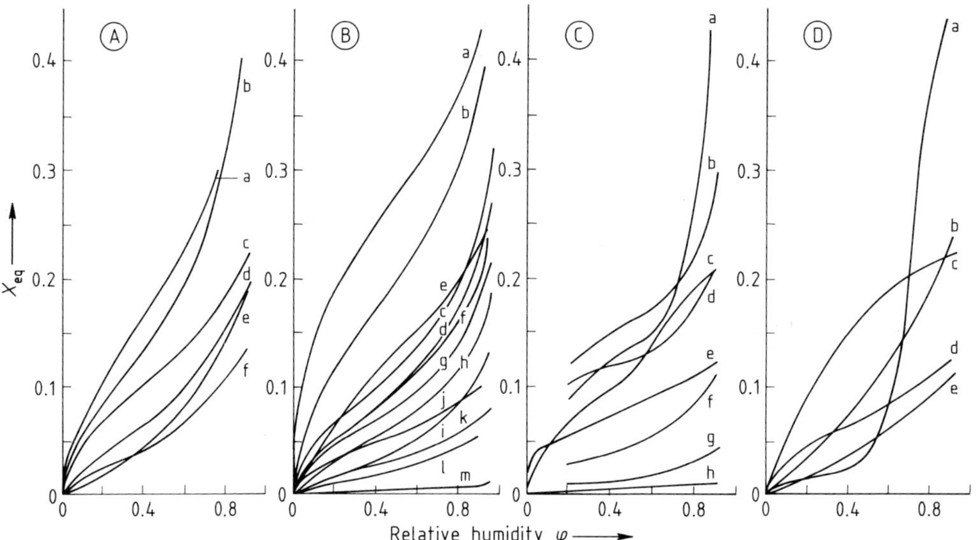

Figure 3. Sorption isotherms for various materials [1]
A) Cereal foods and tobacco
a) Tobacco leaves; b) Tobacco; c) Macaroni; d) Flour; e) Bread; f) Crackers
B) Textile fibers
a) Beryllium alginate fiber (25 °C); b) Calcium alginate fiber (25 °C); c) Nitrate fiber (25 °C), copper fiber (25 °C), viscose fiber (25 °C), wool (worsted, 25 °C); d) Casein fiber (20 °C), wool (35.6 °C); e) Jute; f) Cotton (mercerized, 20 °C), silk; g) Flax (30 °C), hemp; h) Cotton, fluffed (20 °C); i) Acetate fiber (25 °C); j) Linen; k) Perlon fiber (25 °C), nylon fiber (25 °C); l) Cellulose acetate fiber; m) Pe-Ce fiber (20 °C)
C) Leather, rubber, catgut, feathers
a) Sheepskin; b) Leather (sole, oak tanned); c) Catgut; d) Gold beater skin; e) Feathers; f) Latex dipped cord; g) Reclaimed rubber; h) Rubber
D) Binding agents, adsorption agents, soaps
a) Activated charcoal (Novitkol, 400 kg/m^3); b) Soap; c) Silicagel; d) Glue, starch; e) Gelatin

the saturation vapor pressure $P_v^*(T)$. The diminishing of the vapor pressure at the surface of materials with a high moisture content is associated with the phenomenon called *capillary condensation*. With a low moisture content (down to the limit $X \to 0$), the intermolecular attraction between the fluid and the solid plays a dominant role. One speaks of *Langmuir sorption*. For practical purposes the equilibrium moisture content of a solid X_{eq} is depicted as a function of the relative humidity of the air φ at various temperatures T, i.e., $X_{eq} = X_{eq}(\varphi, T)$. Such curves are called *sorption isotherms* (Fig. 1).

Figure 2 shows two sorption isotherms for potatoes. To dry potatoes to a residual moisture content of $X_{eq} = 0.10$, for example, cold air with a temperature of $T = 0$ °C and a low relative humidity of $\varphi = 0.24$, or hot air with a temperature of 100 °C and a high relative humidity of $\varphi = 0.77$ could be used.

Sorption isotherms also aid in choosing the drying conditions when using steam as the drying agent. Taking the potatoes as an example, if the steam temperature is 100 °C, then its pressure should not exceed 77 kPa in order to attain a residual moisture of $X_{eq} = 0.10$.

Solids that contain a considerable amount of bonded moisture at normal conditions are usually called hygroscopic. Sorption isotherms for many such solids are to be found in [1]; some are presented in Figure 3.

Liquids that are bound to solids by sorption are in a state of lower free energy than those that are unbound. Hence, in the drying of hygroscopic substances, the enthalpy of wetting Δh_w must be included in addition to the vaporization enthalpy Δh_v. The so-called differential heat of wetting can be calculated from the relationship

$$\Delta h_w = -R_v \left[\frac{d \ln \varphi}{d(1/T)} \right]_{X_{eq} = \text{constant}} \tag{9}$$

where R_v is the gas constant of the vapor. When the logarithm of the relative humidity is plotted

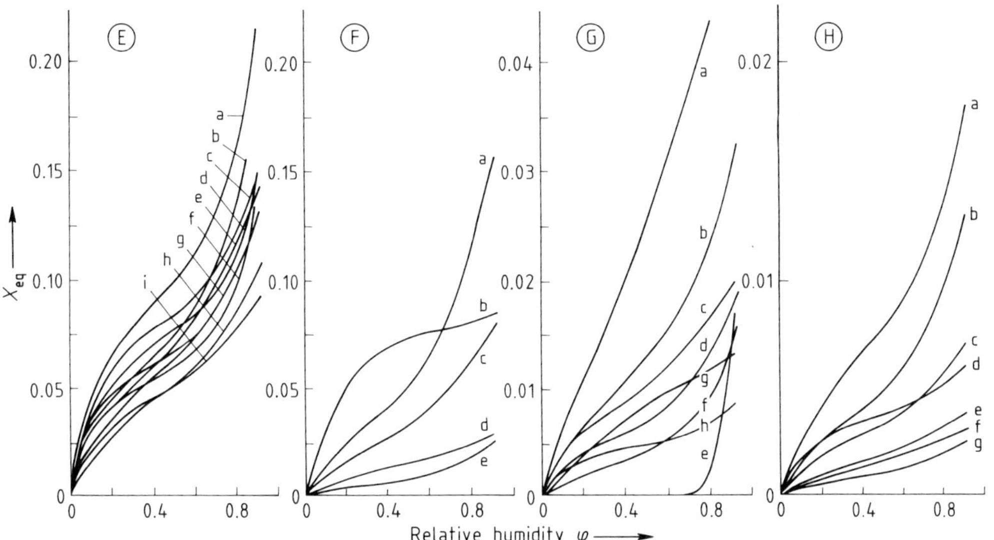

E) Paper and wood shavings
a) Wood shavings; b) Manila paper; c) Kraft paper; d) Cellulose paper and printing paper from sulfite pulp; e) Writing paper; f) Book paper; g) Writing paper (high quality, white); h) Newsprint, filter paper; i) Brazil wood paper, offset press paper
F) Plastics and carbon black
a) Poly(vinyl alcohol) powder (25 °C); b) Carbon black (25 °C); c) 6-Polyamide (20 °C); d) Polyacrylonitrile powder (50 °C); e) Mixed polymer powder (85% poly(vinyl chloride), 14% poly(vinyl acetate), 1% maleic acid (50 °C)
G) Building materials, soils, asbestos
a) Cement mortar 2040 kg/m³; b) Diatomaceous earth; c) Concrete 2300 kg/m³; d) Lime mortar 1800 kg/m³; e) Gypsum 1340 kg/m³; f) Lime stucco 1600 kg/m³; g) Kaolin h) Asbestos
H) Plastics
a) Polystyrene, polymerized in blocks (40 °C); b) Poly(vinyl chloride) powder (50 °C); c) Mixed polymer powders (85% poly(vinyl chloride), 15% poly(vinyl acetate) (50°C); d) Polyethylene powder (25 °C); e) Polytrifluoroethylene powder (50 °C); f) Polyethylene granules with carbon black (25 °C); g) Polyethylene granules (25 °C)

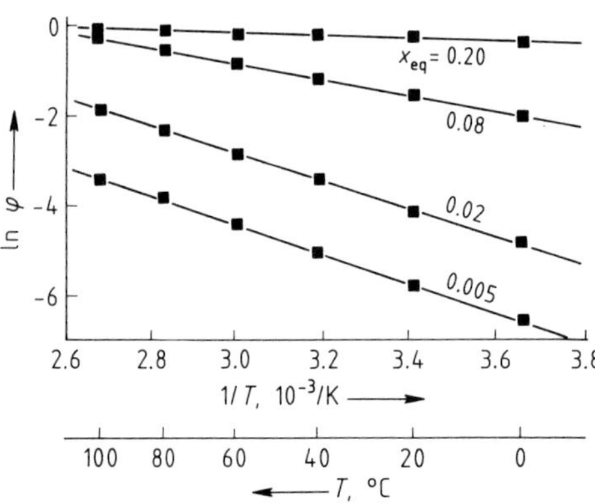

Figure 4. Sorption isosteres for potatoes
The graph can be used for the determination of the differential heat of wetting.

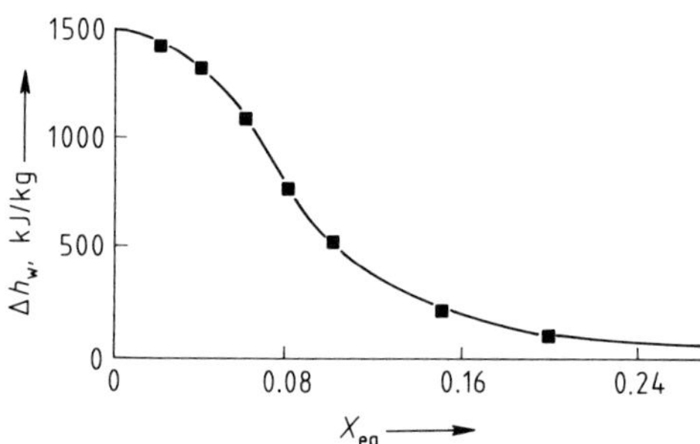

Figure 5. Differential heat of wetting Δh_w for potatoes as a function of their equilibrium moisture content X_{eq} [1] The points for $X_{eq} = 0.20$, 0.08, and 0.02 can be calculated from the slope of the corresponding straight lines in Figure 4, according to Eq. (9).

versus the reciprocal value of the absolute temperature at constant equilibrium moisture content, straight lines are obtained (*sorption isosteres*). The sorption isosteres and the differential heat of wetting for potatoes are plotted in Figures 4 and 5, respectively. To calculate the amount of heat needed to accomplish drying to a final moisture content $X_{eq,f}$ from an initial moisture content $X_{eq,i}$, the integral heat of wetting is needed.

$$\overline{\Delta h_w} = \frac{1}{X_{eq,i} - X_{eq,f}} \int_{X_{eq,f}}^{X_{eq,i}} \Delta h_w \, dX_{eq} \tag{10}$$

The total enthalpy change Δh_{total} is the sum of the vaporization enthalpy and the integral enthalpy of wetting:

$$\Delta h_{total} = \Delta h_v + \overline{\Delta h_w} \tag{11}$$

1.3. Drying Rate Curves for Convection Drying

In most cases, drying rate curves are experimentally determined using air as the drying agent. During each experiment of this kind the air conditions, that is the total pressure P, the temperature T, the relative humidity φ, and the air velocity u are kept constant (Fig. 6). The sample is weighed at specified intervals of time Δt, the reduction of mass ΔM is calculated ($\Delta M = M_s \Delta X$), and the drying rate is obtained:

$$\dot{m} = -\frac{1}{A}\frac{\Delta M}{\Delta t} = -\frac{M_s}{A}\frac{\Delta X}{\Delta t} \tag{12}$$

For most moist solids, especially those having capillary porosity, the drying rate $\dot{m}$ depends upon the moisture content X in a manner similar to that shown in Figure 7 [1], [2].

In the first drying period, the drying rate remains practically constant. During this period, unbound liquid is vaporized from the surface of the solid and carried away by the drying agent. When the moisture content is reduced below a critical value X_{crit}, the surface of the solid dries out, and further evaporation takes place in the

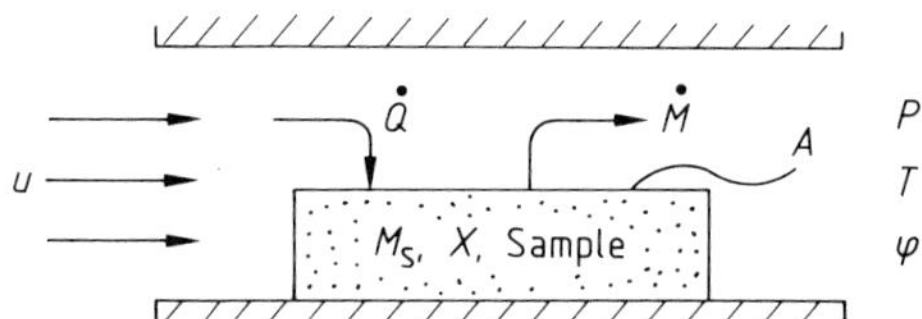

Figure 6. Schematic of a drying tunnel to measure the drying rate curve

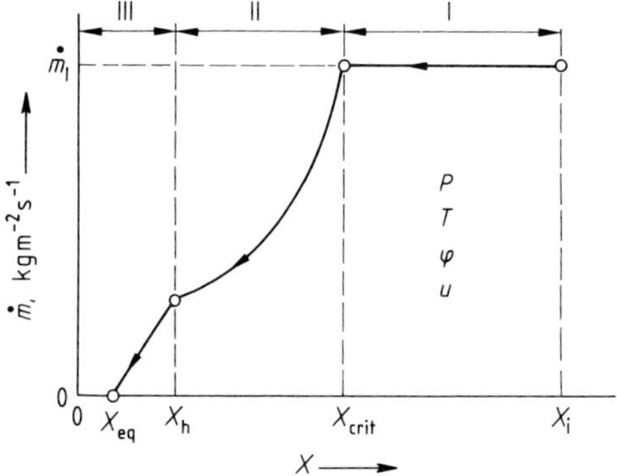

Figure 7. Typical drying rate curve $\dot{m}(X)$, showing the first, second, and third period of drying

interior of the porous solid. The drying rate decreases with decreasing moisture content. This is called the second drying period. The residual moisture in the solid is bound to it by sorption. The drying rate decreases rapidly with decreasing moisture content and tends to zero as the hygroscopic equilibrium moisture content $X_{eq}(T, \varphi)$ is approached. The regime between the maximum hygroscopic moisture content X_h and the equilibrium value of X_{eq} is designated as the third drying period.

First Drying Period. The drying rate during the first period (denoted by $\dot{m}_I$) is dependent only upon the conditions of drying and not upon the characteristics of the solid: there is only evaporation from the surface of the solid. The temperature of the solid, denoted by T_I, is established so that the heat flux from the drying agent to the solid $\dot{q}$ is equal to the product of the drying rate and the vaporization enthalpy,

$$\dot{q} = \dot{m}_I \, \Delta h_v \tag{13}$$

The drying agent, usually air, is completely saturated at the surface of the solid,

$$Y \text{ (at the surface of the solid)} = Y^*(T_I) \tag{14}$$

Assuming that the humidity in the bulk of the drying agent is equal to Y, where $Y < Y^*(T_I)$, the drying rate $\dot{m}_I$ is proportional to the humidity difference $Y^*(T_I) - Y$. The proportionality constant is the effective mass-transfer coefficient β_g^0. Hence,

$$\dot{m}_I = \varrho_g \beta_g^0 [Y^*(T_I) - Y] \tag{15}$$

In this expression ϱ_g is the density of air (or other drying agent). The coefficient β_g^0 is related to the real mass-transfer coefficient β_g by

$$\beta_g^0 = \beta_g \frac{\ln \dfrac{1 + \mu Y^*(T_I)}{1 + \mu Y}}{\mu [Y^*(T_I) - Y]} \tag{16}$$

in which

$$\mu = \tilde{M}_g / \tilde{M}_v \tag{16a}$$

is the ratio of the molecular masses of the inert gas and the vapor ($\mu = 28.96/18.01 = 1.606$ for air and water vapor).

For low levels of humidity ($Y < Y^*(T_I) < 0.05$), Equation 16 reduces to

$$\beta_g^0 = \beta_g \tag{16b}$$

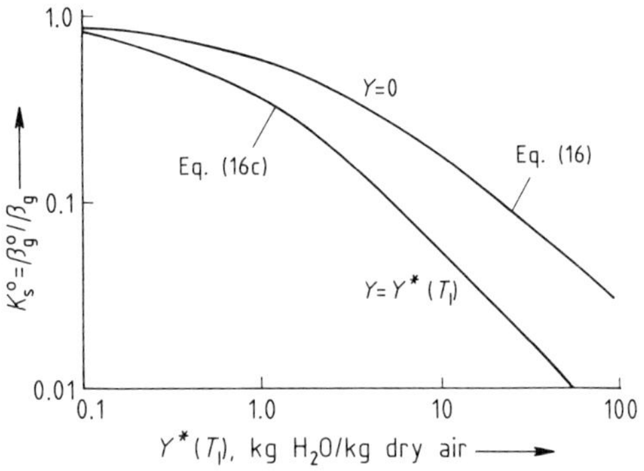

Figure 8. Stefan correction K_S^0 as a function of the saturation humidity $Y^*(T_I)$ for the limiting cases $Y = 0$ and $Y = Y^*(T_I)$
Curves are calculated for water vapor and air, $\mu = 1.608$.

For the case $Y \to Y^*(T_I)$,

$$\beta_g^0 = \beta_g \frac{1}{1 + \mu Y^*(T_I)} \tag{16c}$$

is obtained.

The effective mass-transfer coefficient β_g^0 is always smaller than the real one β_g. Only at sufficiently low humidities Y are the two coefficients equal. This fact is illustrated in Figure 8, where the dependence of the so-called *Stefan correction* $K_s^0 = \beta_g^0/\beta_g$ on the saturation humidity is depicted.

The saturation humidity $Y^*(T_I)$ can be found from the energy balance given by Equation 13, where the heat flux $\dot{q}$ is given by

$$\dot{q} = \alpha^0 (T - T_I) \tag{17}$$

In this equation T_I is the solid temperature during the first drying period, T is the temperature in the bulk of the drying agent, and α^0 is the effective heat-transfer coefficient. The latter is related to the real heat-transfer coefficient α by the following relationship:

$$\alpha^0 = \alpha \cdot \frac{\Delta h_v}{c_{pv}(T - T_I)} \ln \left[1 + \frac{c_{pv}(T - T_I)}{\Delta h_v} \right] \tag{18}$$

In this equation Δh_v is the vaporization enthalpy of the liquid and c_{pv} is the heat capacity of the vapor. The value of α^0 is smaller than the value of α. If the temperature difference $(T - T_I)$ is sufficiently small, these two heat-transfer coefficients are practically the same. Figure 9 shows the relationship between $\alpha^0/\alpha = K_A$ and $c_{pv}(T - T_I)/\Delta h_v$. The former is sometimes called

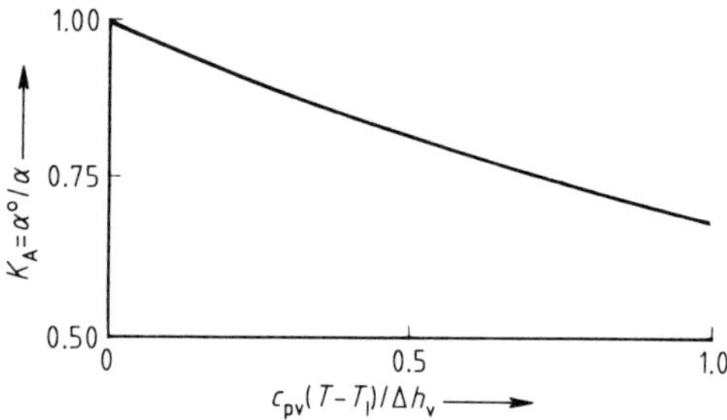

Figure 9. Ackermann correction K_A as a function of the phase change number $c_{pv}(T-T_I)/\Delta h_v$

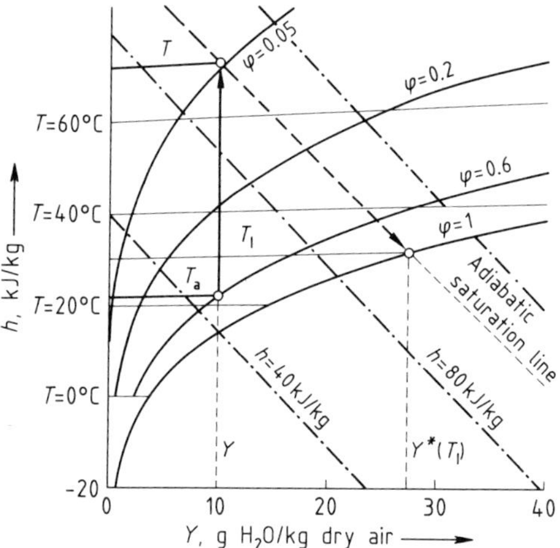

Figure 10. Mollier diagram for moist air
Evaluation of the solid's temperature T_I in the first period of drying

the *Ackermann correction*, while the latter is termed the *phase change number*.

The temperature that the solid reaches in the first drying period T_I can be found by inserting the equation for the mass transfer (Eq. 15) and the equation for the heat transfer (Eq. 17) into the energy balance (Eq. 13). The result is

$$\alpha^0 (T - T_I) = \varrho_g \beta_g^0 [Y^*(T_I) - Y] \Delta h_v \qquad (19)$$

For solids wetted by water that are to be dried by air, the approximation

$$\frac{\alpha^0}{\beta_g^0} \frac{1}{\varrho_g} \approx c_{pg} + Y c_{pv} \qquad (20)$$

is useful provided that the saturation humidity $Y^*(T_I)$ is not too high. The quantities c_{pg} and c_{pv} are the heat capacities of air and water vapor, respectively. Whenever Equation (20) is valid, Equation (19) can be transformed into one that is written in the coordinates of the Mollier diagram:

$$\frac{h^*(T_I) - h}{Y^*(T_I) - Y} = c_{pl}\, T_I \qquad (21)$$

In this relationship c_{pl} is the heat capacity of the liquid and h is the enthalpy of the moist air

$$h = c_{pg}\, T + Y(\Delta h_v^0 + c_{pv}\, T) \qquad (22)$$

where Δh_v^0 is the vaporization enthalpy at 0 °C. Equation (21) is the locus of points of the *adiabatic saturation line*, crossing the curve with $\varphi = 1$ at the point (h^*, Y^*) (Fig. 10).

If the temperature of the solid T_I and either the heat-transfer coefficient α^0 or the mass-transfer coefficient β_g^0 are known, then the drying rate can be determined numerically either with Equations (17) and (13) or with Equation (15).

Numerical Example. Consider the drying agent to be ambient air ($T_a = 22$ °C, $\varphi = 0.60$) that has been heated to $T = 72$ °C. The heat-transfer coefficient at the air–solid interface is 20 $W m^{-2} K^{-1}$ ($= \alpha^0$), the moisture is water. What is the drying rate during the first drying period?

The temperature of the solid during the first drying period T_I can be evaluated with the help of the Mollier diagram (Fig. 10). Condition of ambient air: $T_a = 22$ °C, $\varphi = 0.60 \Rightarrow Y =$ 0.010 kg H_2O/kg dry air. Condition of preheated air: $T = 72$ °C, $Y = 0.010$ kg H_2O/kg dry air $\Rightarrow \varphi = 0.05$. Following the adiabatic saturation line, a value of $T_I = 30$ °C is obtained. The corresponding vaporization enthalpy of water is $\Delta h_v = 2\,430\,300$ J/kg (tabulated value). Using Equations (13) and (17), the drying rate during the first drying period $\dot{m}_I = 1.24$ $kg m^{-2} h^{-1}$ is obtained.

Second and Third Drying Periods. The first drying period is completed when the solid's moisture content X has reached the critical value X_{crit} (cf. Fig. 7). After this point the capillary forces are no longer sufficient to transport the liquid to the surface of the solid. The liquid–vapor interface (drying front) moves inside the solid. The dried portion of the solid near the surface thermally insulates the moist inner portions of the solid. At the same time these dried regions impede the transport of the vapor to the bulk of the drying agent. These are the reasons

Table 1. Critical moisture content of solids when dried in air [3]

Solid	Layer thickness, cm	Critical moisture content X_{crit}
Sulfite pulp	0.6–1.9	0.6–0.8
Paper, white eggshell	0.02	0.41
Paper, fine book	0.0125	0.33
Paper, coated	0.01	0.34
Paper, newsprint		0.6–0.7
Beaverboard	0.43	>1.2
Poplar wood	0.42	1.2
Wool fabric, worsted		0.31
Wool, undyed serge		0.08
Sole leather	0.63	> 0.9
Chrome leather	0.1	1.26
Sand (50–150 mesh)	5	0.05
Sand (200–325 mesh)	5	0.1
Sand (through 325 mesh)	5	0.21
Sea sand (on trays)	0.63	0.03
	1.27	0.047
	2.5	0.059
	5.0	0.06
Brick clay	1.6	0.14
Kaolin		0.14
Barium nitrate (crystals)	2.5	0.07
Carbon pigment	1	0.4
Copper carbonate	2.5–3.8	0.6
Iron blue pigment	0.63–1.9	1.1
Lithopone press cake	0.63	0.064
	1.27	0.08
	1.9	0.12
	2.5	0.16
Prussian blue pigment		0.4
Gelatin (X_i = 4.0)	0.25–0.5 (moist)	3.0
White lead		0.11
Rock salt	2.5	0.07

why the drying rate decreases while the solid's temperature rises after the completion of the first drying period. The critical moisture content X_{crit} must usually be determined experimentally; Table 1 gives representative values for various materials.

In Figure 11, measurements of the moisture content within a paper sample during drying are depicted. The width of the sample was $s = 30$ mm, its depth—measured from the surface which is in contact with the drying agent—is denoted by z. After 7 or 8 h, drying is still taking place in the first drying period. The surface of the sample is moist, the moisture content is almost evenly distributed over the entire depth of the sample. About 9 h after the beginning of the experiment, the critical point is reached. The surface of the sample dries out and the second drying period begins. After 65 or 95 h, drying takes place in the third drying period. All moisture is bonded, the drying rate is very small. The drying rate curve, associated with the experiment in Figure 11 is depicted in Figure 12 (Curve d). Additionally, three other curves for thinner samples are shown. The thinner samples dry more rapidly in the second and third drying periods than the thicker ones.

The history of the temperature at various positions within a layer of powder is presented in Figure 13. In the first drying period the entire sample takes on the temperature T_I that can be

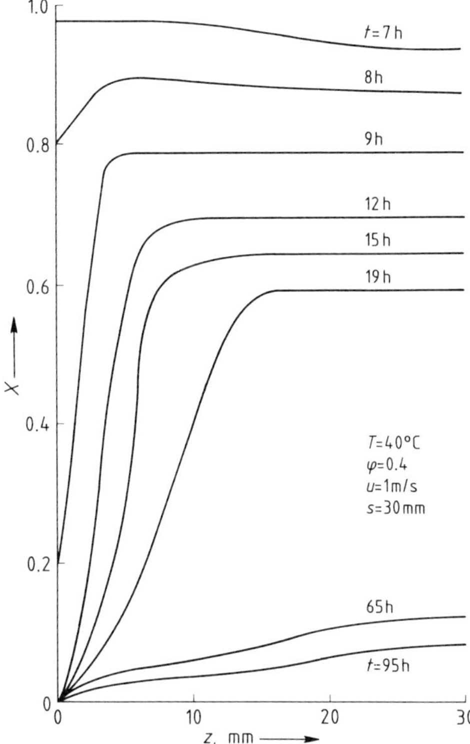

Figure 11. Moisture profile in paper stock as a function of the duration of drying [1]

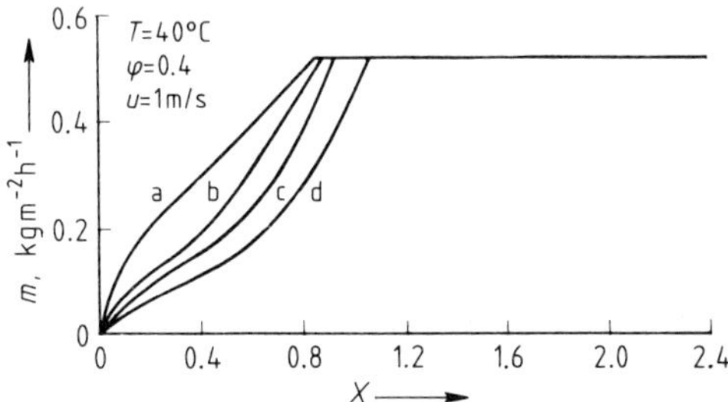

Figure 12. Drying rate curves for paper stock of various thicknesses [1]
Sample thickness s in mm a) 10; b) 15; c) 20; d)30

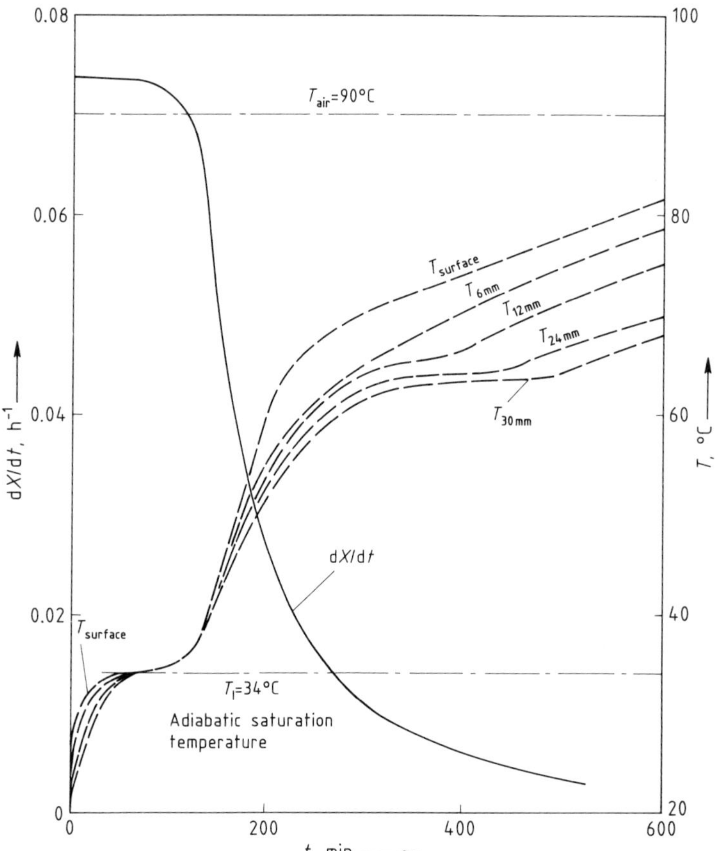

Figure 13. Drying rate curve and temperatures at various depths of the sample during the drying of powdered $CaCO_3$ [1]

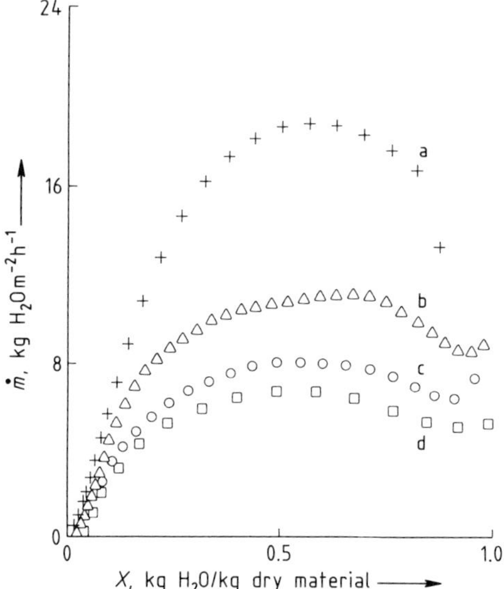

Figure 14. Drying rate curves for a textile fabric dried by passing air through it [4]
a) $T = 71.7\,°C$; b) $T = 39.4\,°C$; c) $T = 25.3\,°C$; d) $T = 22.8\,°C$

read from a Mollier diagram. This temperature is usually referred to as the adiabatic saturation temperature. In the second and third drying periods, the temperature rises continuously in the various internal layers of the solid with the result that they approach the temperature of the drying agent. Large differences in moisture content and temperature can occur in the solid after the first drying period has been completed. Such differences can produce fissures in the solid, as well as scaling and discoloration.

Drying rate curves $\dot{m}(X)$ are usually determined experimentally. If the process is sufficiently slow, the sample can be periodically weighed. In rapid processes, such as the drying of paper by a transverse flow of air, which is completed in a matter of a few seconds, the humidity of the exhaust air can be determined continuously with an infrared spectrometer. Figure 14 shows drying rate curves that were obtained in this way for a textile fabric through which air was blown [4].

As already discussed, drying rate curves are obtained from laboratory tests that are conduct-

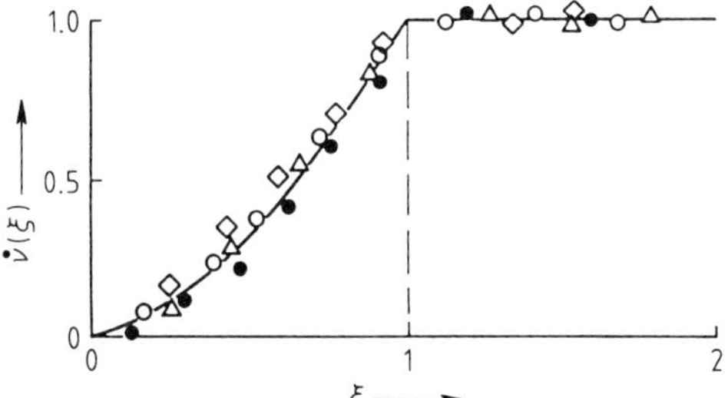

Figure 15. Normalized drying rate curve for pottery clay The air temperature T and the relative humidity φ were varied [1]:
○ $T = 45\,°C$, $\varphi = 0.537$; Δ$T = 15\,°C$, $\varphi = 0.537$
● $T = 25\,°C$, $\varphi = 0.187$; ◇$T = 25\,°C$, $\varphi = 0.758$

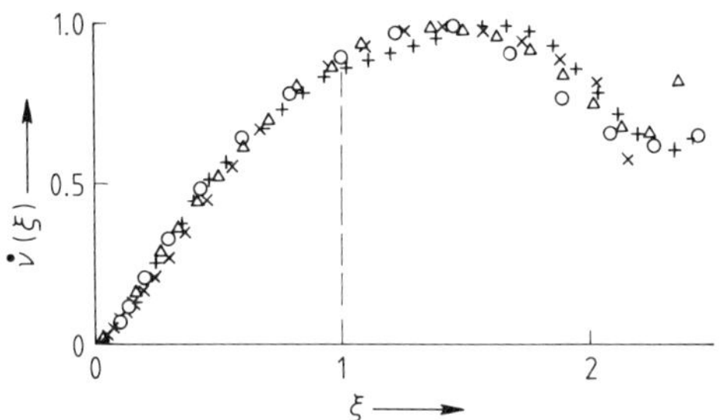

Figure 16. Normalized drying rate curve calculated from the data of Figure 14
× $T = 71.7\,°C$; + $T = 39.4\,°C$; Δ $T = 25.3\,°C$; ○ $T = 22.8\,°C$

ed under constant drying agent conditions. In reality, the condition of the drying agent changes with time and location in any commercial drying apparatus. Because it is seldom possible to simulate all of the conditions that can occur in a production dryer, methods of interpolating the laboratory data are needed. In this context, the use of a drying rate curve is very helpful. The normalized drying rate is defined by

$$\dot{\nu} = \dot{m}/\dot{m}_{\mathrm{I}} \tag{23}$$

and the normalized moisture content by

$$\xi = \frac{X - X_{\mathrm{eq}}}{X_{crit} - X_{\mathrm{eq}}} \tag{24}$$

When $\dot{\nu}$ is plotted versus ξ, most drying rate curves measured in air under different conditions coincide (Figs. 15 and 16). In such cases, only one drying rate curve has to be determined by experiment.

1.4. Drying Rate Curves for Contact Drying

In contact drying the heat necessary to vaporize the moisture in the solid is transferred by direct contact with a heated surface. The process of drying can take place in an atmosphere containing only the liquid's vapor, or air can also be present. Vacuum dryers are an example of the former, tube dryers for peat an example of the latter. Figure 17 shows a laboratory tray dryer equipped with an agitator. The entire experimental setup is placed on a balance and can be used in order to measure drying rate curves of porous, granular materials in vacuum. During each experiment the temperature of the heating surface T_{w} and the chamber pressure P are held constant. The same apparatus can be used in order to measure drying rate curves in the presence of inert gas.

The drying rate curves shown in Figure 18 were obtained with the apparatus of Figure 17 during vacuum contact drying of moist granular aluminum silicate; the particle diameter d was varied. Figure 19 gives similar results for peat in the presence of air (at normal pressure). In both cases, the drying rate depends upon the efficiency of the heat transfer between the heating surface and the granular bed, as well as upon the intensity of mechanical mixing. The reduction of the drying rate with decreasing moisture content is a consequence of the increased blocking of the heating surface by the particles that have already been dried. Consequently, the drying rate curve appears to be a property of the agitated bed as a whole, and not of the individual porous particles.

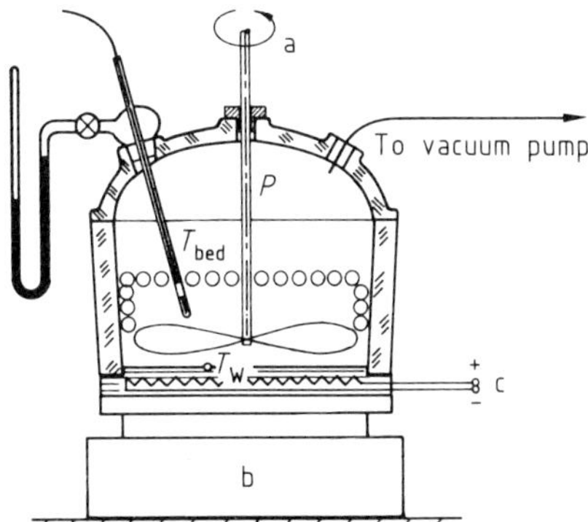

Figure 17. Laboratory tray dryer to determine the drying rate during contact drying
a) Mixer; b) Balance; c) Heater

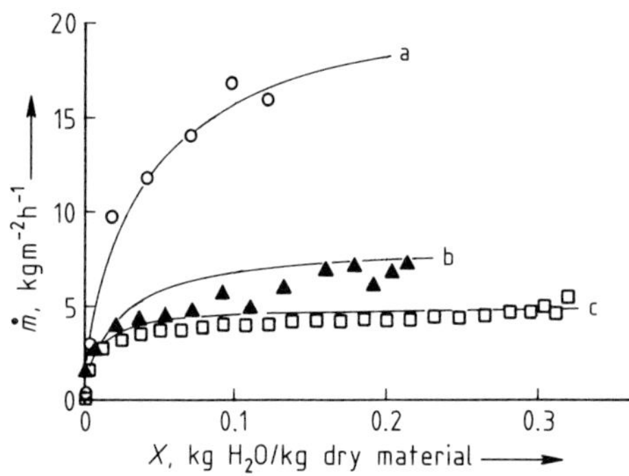

Figure 18. Drying rate curves for nonhygroscopic, vacuum-dried aluminum silicate particles
Conditions: $P = 2.63$ kPa, $T_w = 80\,°C$, and mixer speed = 45 rpm
a) $d = 0.83$ mm; b) $d = 3.25$ mm; c) $d = 6.60$ mm

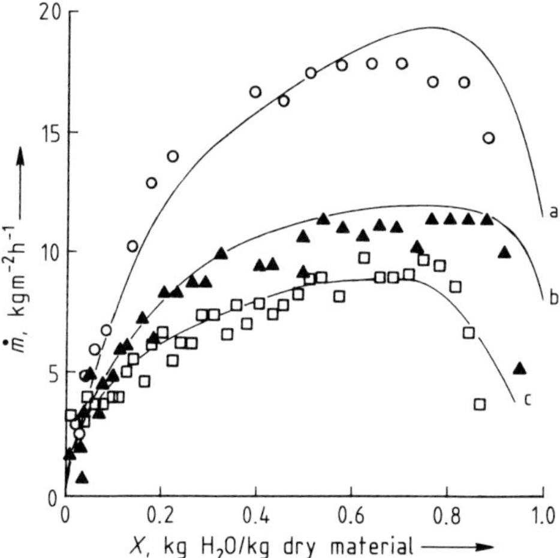

Figure 19. Drying rate curves for hygroscopic peat in the tray dryer of Figure 17
Conditions: $P = 0.1$ MPa (air), $T_w = 130\,°C$, and mixer speed = 40 rpm
a) $d = 0.75$ mm; b) $d = 3.0$ mm; c) $d = 6.0$ mm

Contact drying becomes faster as the temperature difference between the heated wall and the bed ($T_w - T_{bed}$) increases or the mixing speed is raised. For coarse-grained materials ($d \approx 10$ mm), the drying rate is directly proportional to the temperature difference, for fine materials ($d \approx 100$ µm) it depends approximately upon the square root of this difference $(T_w - T_{bed})^{0.5}$. In contrast, the mixing intensity has a stronger influence on the drying rate of fine-grained materials than on the drying rate of coarse-grained ones. The physical explanation for this behavior can be found in [5]. The maximum drying rate that can be expected is given by

$$\dot{m}_{max} = \alpha_{max}(T_w - T_{bed})/\Delta h_v \tag{25}$$

The maximum possible heat-transfer coefficient, α_{max}, can be found from

$$\alpha_{max} = \alpha_r + \varphi_w \frac{4\lambda}{d}\left[\left(1 + \frac{2(l+\delta)}{d}\right)\ln\left(1 + \frac{d}{2(l+\delta)}\right) - 1\right] \tag{26}$$

with

$$l = 2\Lambda\left(\frac{2}{\gamma} - 1\right) \tag{27}$$

and

$$\alpha_r = 4C_{w,\,bed}\left(\frac{T_w + T_{bed}}{2}\right)^3 \tag{28}$$

λ	molecular thermal conductivity of vapor or inert gas
d	mean particle diameter
δ	surface roughness of the particles (in most cases $0 < \delta < 10$ µm [5], [6])
φ_w	fraction of the heated surface that is covered ($\varphi_w \approx 0.8$ for random beds of spherical particles)
Λ	mean free path of the vapor or inert gas molecules
γ	accommodation coefficient ($\gamma \approx 0.9$ for air or water vapor at moderate temperatures)
$C_{w,\,bed}$	coefficient for thermal radiation between the heated surface and the bed

The radiant heat-transfer coefficient α_r has a value of approximately 5 $Wm^{-2}\,K^{-1}$ at room temperature. For high vacuum, the molecular conduction, the second term in Equation (26), is insignificant so that $\alpha_{max} = \alpha_r$ is obtained. If this limit is reached, the drying rate can only be increased if the temperature difference ($T_w - T_{bed}$) is made larger; in this instance, the degree of mixing has no effect on the drying process.

1.5. Drying Rate and Moisture-Composition Curves for Solids Wetted by Liquid Mixtures

A large variety of solids are moistened with liquid mixtures containing water and other substances such as alcohol, acids, or esters. In the course of drying such solids, the composition of the moisture generally changes because different liquids evaporate at different rates. This is termed *selective drying*, to denote the preferential removal of individual components from the mixture. However, under certain circumstances the composition of the moisture remains constant during drying (*unselective drying*). The properties of the dried item are often very strongly dependant upon the composition of the residual moisture. Moreover, there are accepted industry and health standards that the final moisture content must meet. Pharmaceutical items must not contain any toxic residues; foodstuffs should be dried to remove their water but not their (highly volatile) aromas. Although lacquers should dry with a constant composition, no fixed rules can be given because the requirements in individual cases differ.

The drying rate as well as the selectivity of the drying process are controlled by five interrelated physical mechanisms:

Phase equilibrium between the gaseous, liquid, and solid phases
Diffusion in the gaseous phase
Diffusion in the liquid phase
Capillary transport of moisture
Heat transfer

The combined influence of these five mechanisms on selectivity of drying has been theoretically and experimentally examined using simple liquid mixtures [7], [8]. According to the results of such investigations, the influence of vapor–liquid equilibria and gas-phase diffusion on the one hand, and the influence of capillary transport and liquid-phase diffusion on the other hand act together.

Furthermore, it is necessary to differentiate between liquid mixtures containing only volatile components and those containing at least one nonvolatile component. A mixture of alcohol, water, and glycerol is an example of the second kind. In all drying processes with a sufficiently high drying rate and a large thickness of the solid (large samples, coarse-grained products), the selectivity is determined by the capillary motion and the diffusion of the liquid constituents. This is the case when the dimensionless number

$$\Phi_l = \frac{\delta_l}{v_l (d/2)} \quad (29)$$

is considerably greater than unity. This parameter is equal to the ratio of two lengths: the penetration depth of the concentration profile, $\varepsilon = \delta_l / v_l$, and the depth of the sample, $d/2$. The velocity of the capillary flow v_l is directly proportional to the drying rate $\dot{m}$, i.e., $v_l = \dot{m}/(\varrho_l \cdot \psi)$. (Note: ϱ_l is the density of the liquid phase, δ_l is the diffusion coefficient in this phase, and ψ is the porosity of the sample.)

If the solution within the pores of the solid contains only volatile components, such as water, alcohol, or acetone, then for $\Phi_l \ll 1$ "sharp" drying proceeds unselectively (in sharp drying the drying agent has a high temperature, and the drying rate is high). But should the liquid in the pores contain one or more nonvolatile components, then the volatile component that diffuses most quickly through the nonvolatile constituent of the mixture disappears preferentially. An example is the water in a solution with glycerol and alcohol. The relative volatility of the individual components, as well as the speed of diffusion into the gaseous drying agent, is inconsequential in this case. This situation is important because of the possibility of supressing the loss of aroma during drying of foodstuffs.

If the dimensionless number Φ_l is considerably larger than unity (low temperature of drying agent, low drying rate), then the selectivity is solely determined by the relative volatility and by the velocity of diffusion into the gaseous phase. The more volatile components disappear first, as long as "dynamic azeotropy" does not occur. The appearence of such azeotropic points is due to the interaction of relative volatility and gaseous diffusion. For the solution of isopropyl alcohol and water, the (static) azeotropic mole fraction of isopropyl alcohol is about 0.65. However, using dry air as drying agent a dynamic azeotropic mole fraction of 0.41 can be attained; the reason for this behavior is that water diffuses into the air much faster than alcohol. At this composition the drying process would be unselective for all values of the parameter Φ_l.

Figures 20 and 21 clarify the drying behavior of solids that are wetted by solutions with only volatile components. The experimental drying rate and composition curves for a clay cylinder are shown in Figure 20. The clay has been soaked in a solution of isopropyl alcohol and water. The drying agent is dry air ($T = 60\,°C$, $u = 0.2$ m/s). Because of the relatively large diameter of the sample, the parameter Φ_l remains much lower than unity during the first drying period. Consequently, the process is controlled by the diffusion in the liquid phase. In the absence of an inert liquid component the drying process occurs unselectively, i.e., the mole fraction of the alcohol, $\tilde{x}_1$, remains constant, as illustrated in Figure 20 B. This situation changes in the course of the second drying period: with decreasing moisture content, the liquid–vapor interface retreats inside the solid, so that the drying rate and the width of the moist region diminish. The result is that Φ_l increases. Towards the end of the drying process a pronounced change in the composition of the remaining moisture is recorded. The extent and trend of this selectivity are dictated by the diffusion occuring in the gaseous phase, and by the thermodynamic equilibrium. For an initial mole fraction of alcohol $\tilde{x}_{1,i} > 0.41$, the criterion for dynamic azeotropy, the solution becomes depleted of water, and for $\tilde{x}_{1,i} < 0.41$, alcohol is lost.

The region of selective drying can be considerably extended if the thickness of the sample is decreased. In this

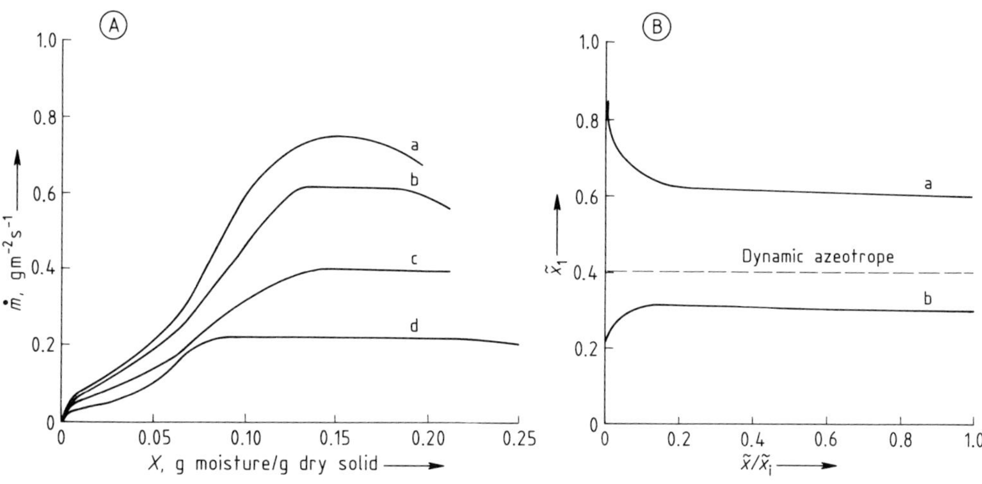

Figure 20. Drying rate (A) and composition curves (B) for a clay cylinder wetted by isopropyl alcohol ('1') and water
Conditions: diameter of sample = 39.2 mm, length = 95.7 mm; drying agent: dry air, $T = 60\,°C$, $u = 0.2$ m/s [7]
A) a) $\tilde{x}_{1,i} = 1.0$; b) $\tilde{x}_{1,i} = 0.6$; c) $\tilde{x}_{1,i} = 0.3$; d) $\tilde{x}_{1,i} = 0.0$
B) a) $\tilde{x}_{1,i} = 0.6$; b) $\tilde{x}_{1,i} = 0.3$

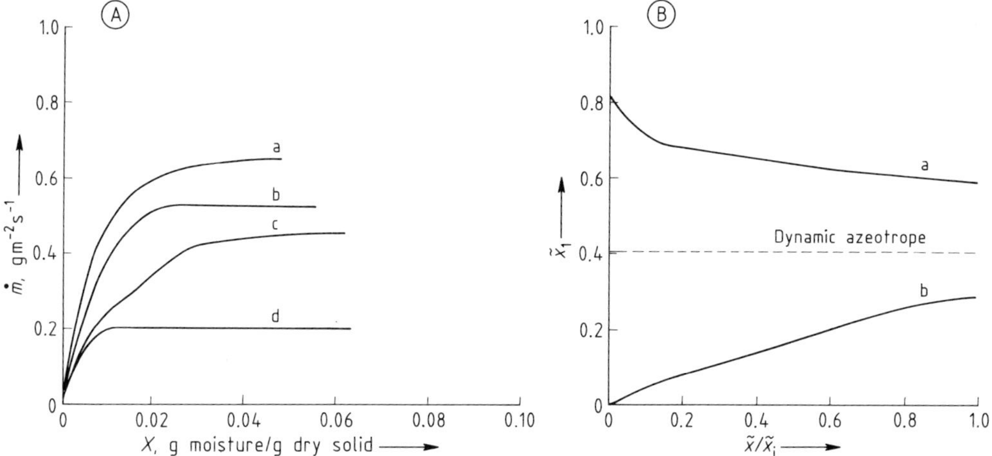

Figure 21. Drying rate (A) and composition curves (B) for a hollow cylinder of sintered bronze wetted by isopropyl alcohol ('1') and water
Conditions: outer diameter of sample = 31.0 mm; inner diameter of sample = 30.0 mm; length = 95.7 mm; drying agent: dry air, $T = 60\,°C$ and $u = 0.2$ m/s [7]
A) *a)* $\tilde{x}_{1,i} = 1.0$; b) $\tilde{x}_{1,i} = 0.6$; c) $\tilde{x}_{1,i} = 0.3$; d) $\tilde{x}_{1,i} = 0.0$
B) a) $\tilde{x}_{1,i} = 0.6$; b) $\tilde{x}_{1,i} = 0.3$

manner, the drying of a thin-walled hollow cylinder made of sintered bronze with dry air ($T = 60\,°C$ and $v = 0.2$ m/s), occurs selectively from the beginning (see Fig. 21).

2. Drying Methods and Dryer Types

In this chapter the most important methods of drying are arranged according to the way the heat is transferred. In a convection dryer the liquid is vaporized by the heat that is transferred from the drying agent. In the case of a contact dryer the heat is conducted from a heated surface to the solid. The solid may be transported over the heat-transfer surface, or it may rest upon it. Radiant heating, in which the heat is supplied from a radiation source that is remote from the surface of the solid but with an unobstructed view of it, is also used. Special methods of drying include dielectric drying and freeze drying. The myriad of dryer types is a consequence of the different behavior that the solid exhibits during drying, the particular product needs, and many economic considerations.

2.1. Convection Drying

The methods of convection drying differ from one another in the manner by which the moist solid contacts the drying agent, which is usually hot air.

2.1.1 Flowing Gas

Drying with a flowing gas is particularly suitable for materials which should not be mechanically stressed during drying.

Drying Oven (Kiln). With small quantities of moist solids, the simplest, cheapest dryer is a drying oven similar to that shown in Figure 22. The solid is placed upon racks or in trays that are mounted on a cart. A fan circulates warm air through a heater and then through the drying racks or trays. Flow dividers ensure that the air is uniformly distributed. Drying proceeds uniformly and without overdrying portions of the material. The drying process for forced convection dryers is controlled by the amount of exhaust and intake air, as well as its velocity and temperature. Such dryers are employed for processing sensitive materials that require long drying times, i.e., gentle drying. The kilns for some grades of lumber can be 200 m^3 or larger. The atmosphere in such chambers can best be regulated by a feedback system with suitable sensors.

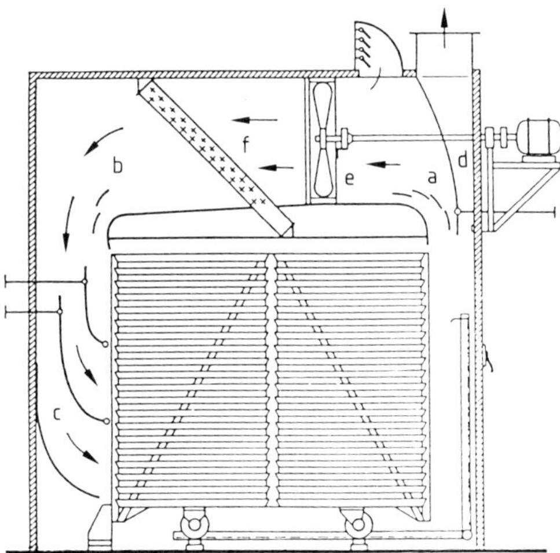

Figure 22. Compartment dryer
(Courtesy of Babcock BSH)
a)–c) Turning vanes; d) Air-exhaust duct with damper; e) Fan; f) Heaters

Tunnel Dryers. Large quantities of materials are dried in a tunnel in a continuous process. The cart is placed within the tunnel at its entrance and conveyed, usually continuously, but sometimes step by step, along the tunnel to its exit. The moving air can be cocurrent or countercurrent with the direction of the solids.

Sometimes the flow of the drying agent is changed, and it is blown perpendicular to the solid's direction of motion. This can lessen the excessive drying of edges for materials such as plasterboard, and permits different drying conditions in different portions of the tunnel (Fig. 23). Applications of this drying method are freshly lacquered chassis parts and glass plates that have a layer of leather glued to them; they are carried on moving hooks through the drying tunnels. Goods that are in the form of large sheets, such as plasterboard or wood particle board, are moved by a conveyor system consisting of many parallel rollers that are driven by an interconnecting chain. Many layers of such a roller system can form decks to use the tunnel's space effectively.

Spiral Belt Dryer. A dryer of this type (Fig. 24) is often ideal for materials that require a long, undisturbed drying time. The moist solid is placed upon a circulating belt at a position outside the dryer. The solid remains undisturbed while it dries. The belt enters at the top of the dryer and moves in a spiral fashion toward the bottom. Several blowers rotate about the vertical axis of the dryer and service a particular elevation of it. The air is blown past the solid, onto the heating pipes, and then sucked back across the solid. It is possible to regulate both the temperature and air speed of each vertical drying zone separately.

Segmented Rotating Tray (Wiped-Tray) Dryer. The continuous drying of large quantities of crystalline, granular, or pasty materials that must be dried gently can be effectively done with a segmented rotating tray dryer of the type shown in Figure 25. The solid can have a broad grain-size spectrum. A lazy Susan that slowly rotates around the central axis of the dryer is its

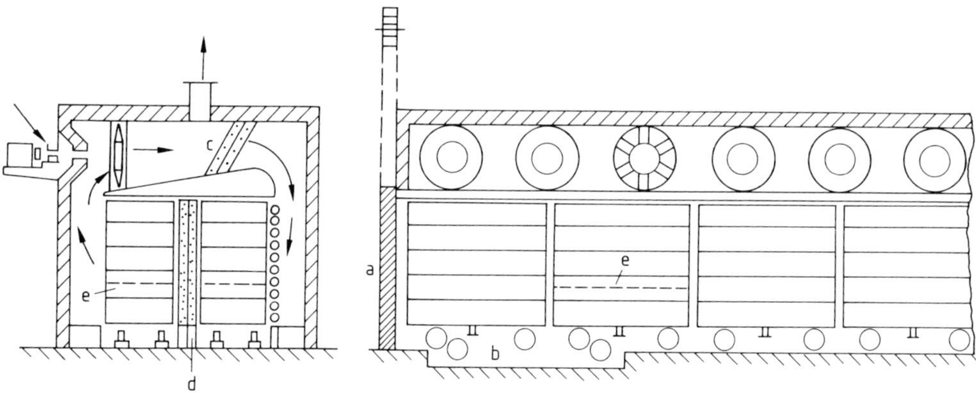

Figure 23. Double-truck dryer for two rows of carts
a) Loading door; b) Cart pool; c) Heater; d) Intermediate heater; e) Racks for loading

main component. Each tray is subdivided into pie-shaped sections by radial slots—60–120 mm wide—through which the material can fall. The moist solid is placed upon the topmost tray in an even layer. A stationary wiping arm comes into contact with the dried surface material and pushes it circumferentially along the disk toward the radial slots. The material that falls through a slot lands upon the disk below, where it initially forms a heap or ball. An arm is used to spread the material into a uniform layer, and the procedure continues until the dried material reaches the bottom of the dryer, where it is removed. Pasty and sticky materials can form clumps in the early stages of drying. In order to achieve uniform drying, such clumps are subdivided between each disk by an appropriate device.

In the dryer shown in Figure 25, the blowers are centrally located and force the air over the material on one disk and back again over another disk after the air passes through a heating unit. The dryer can have several zones for the circulating air, each having a different temperature. The dimensions for these types of dryers range from 1.2 m in diameter with 8 m^2 of useful drying surface to 10 m in diameter with 1500 m^2. The rotational speed of the dryer is specified by the drying time and the number of vertical stages.

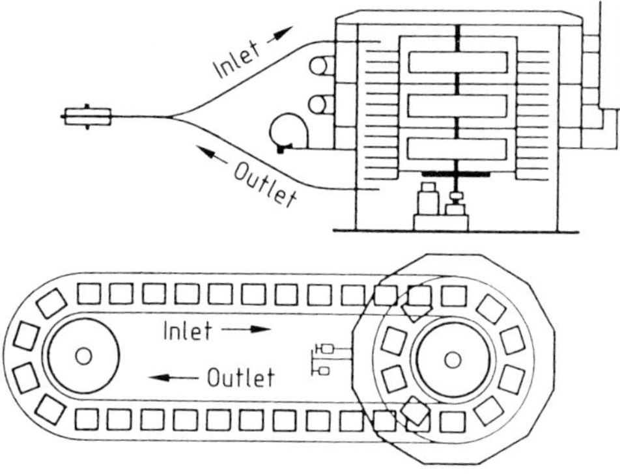

Figure 24. Spiral belt dryer (Courtesy of Babcock BSH)

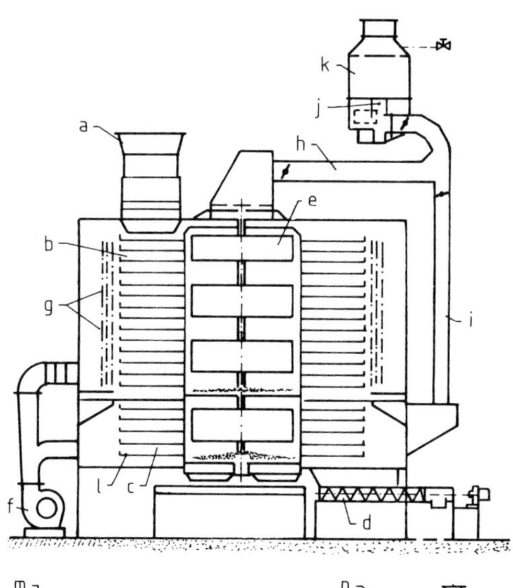

Figure 25. Segmented rotating tray (wiped-tray) dryer
Top: vertical, axial section
Bottom: Detail of the wiping and distributing method
a) Loading device; b) Drying zone; c) Cooling zone; d) Unloading conveyor; e) Turbo fans; f) Fresh air fan; g) Heaters; h) Exhaust duct; i) Cooling air duct; j) Wet washer; k) Demister; l) Segmented tray; m) Segment; n) Wiper; o) Distributor

Disking Dryers. A circulating disking dryer is advantageous for materials that must be constantly turned over. Like the wiped-tray dryer, it contains a set of vertically stacked, circular trays. In addition, fixed disks of different diameters, in effect, disk-harrow the material on the trays. In this way, the material that is to be dried is both agitated and transported. Attached to the drying chamber is a separate chamber for the fans and heaters to supply and heat the air. An important application of this kind of dryer is for materials moistened by volatile solvents.

Jet Dryers. Dryers of the type shown in Figure 26 blow hot air out of slots or circular holes at a high speed onto the surface of the moist solid. As a consequence of the high air speeds resulting from large rates of circulated air, very large mass-transport coefficients and high drying rates of the liquid near the surface, i.e., the first stage of drying, are achieved. This minimizes the redistribution of the moisture within the solid because of the increased exchange of mass. Jet dryers are mostly used to dry flat materials such as wood veneer, cardboard, foils, textiles, or photographic papers.

Continuous sheets of materials, such as coated papers, foils after printing, or photographic film, are transported through the dryer on a cushion of air that can be over 100 m long. Pieces of veneer are placed onto a conveyor belt; extruded pellets, for example, either dry or wet feed, are transported on a continuous stainless steel belt. The speed of the drying air must be adjusted so that the pieces are not blown away. A tentering frame dryer has hooks or clamps for holding the cloth that are attached to a chain that moves through the dryer. The material is laterally constrained in this way, with the result that the fabric undergoes a favorable tensile treatment while drying.

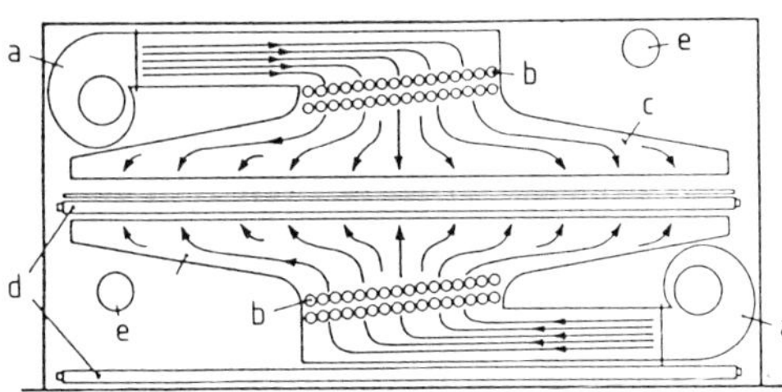

Figure 26. Impinging jet dryer
(Courtesy of Deutsche Babcock Anlage)
a) Fresh air fan; b) Heaters; c) Nozzle boxes; d) Roller system loaded with strip material; e) Exhaust

2.1.2. The Solid is Aerated

If air can be blown through a layer of solids, or if the solid can be processed to make this possible, aeration is the most efficient form of convection drying. Because nearly all of the surface of the moist particles is exposed to the drying agent, maximum heat and mass transfer are achieved for the available conditions. High drying rates are possible even under relatively mild thermal and mechanical conditions.

Through-Circulation Batch Dryer. Agricultural products are often dried by this device. Grain or hay are placed into a container that has a perforated bottom, or onto a rack, and warm air is blown through it. Uniformity of the drying within a pile of cereal can be facilitated with stirring forks that simultaneously transport the material.

More than 100 t/h of grain can be continuously dried in inclined tubular dryers that also aerate the material. There are dryers that resemble silos, and these are filled from the top. The grain slides past roof-like inserts (Fig. 27), as it moves toward the bottom where it is removed from a funnel-shaped collector.

The gas that is used as the drying agent flows through a duct and into the layer of grain via the roof-shaped forms. The gas is subsequently sucked back through similar vee-shaped openings that are placed somewhat higher. The uppermost portion of the dryer is not ventilated and is called the sweating zone. Here the material slides past heaters that are usually filled with hot water. The sweating of the grain considerably reduces the necessary drying time. In the lower portion of the duct, a cooling zone is often provided. Fodder is dried with air diluted with (warm) exhaust fumes, but legumes and grains used for food or producing oil are dried with air that is indirectly heated to prevent contamination.

Aerated, Stationary Rack Kiln. Small quantities of fragile materials can be placed on racks that can be put into a kiln. Air is blown through the material. Such ovens are loaded and unloaded by hand.

Aerated, Moving Rack Kiln (Simplizior Dryer). The Simplizior dryer (Fig. 28) is a semicontinuously operating rack dryer that is mainly used by the food industry to treat vegetable products such as parsley, carrots, spinach, mush-

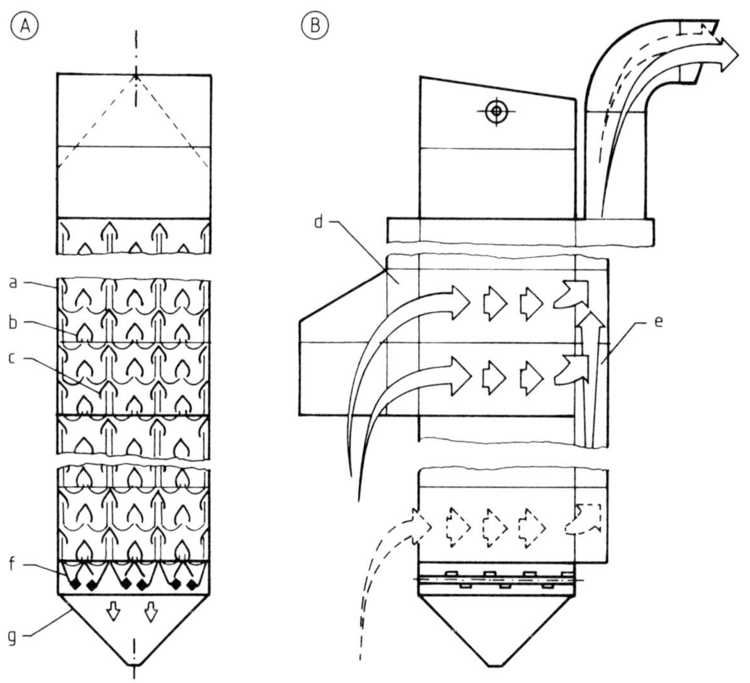

Figure 27. Vertical gravity bed continuous flow dryer
A) Construction of the dryer; B) Air flow (view perpendicular to that in A)
a) Chute; b) Fresh air; c) Exhaust; d) Inlet air duct; e) Exhaust duct; f) Unloading device; g) Funnel

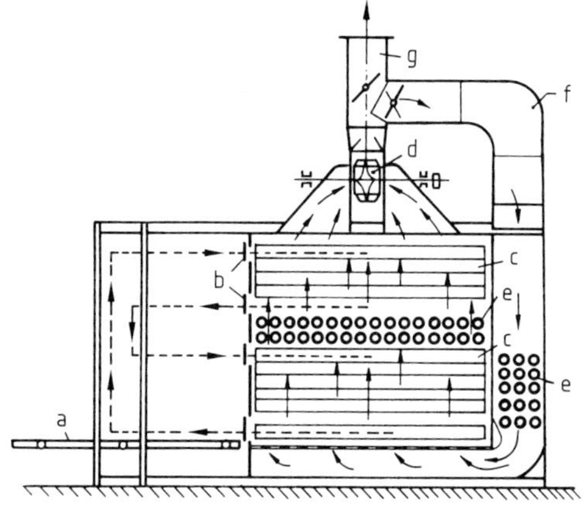

Figure 28. Through-circulation floating rack compartment dryer
Simplizior (Courtesy of Deutsche Babcock Anlage)
a) Elevator; b) Flaps; c) Rack piles; d) Fan; e) Heaters; f) Recirculation duct; g) Exhaust duct
---→ Path of solid; ——→ Path of air

rooms, as well as spices. Because the moisture content of the drying air increases on its way through the dryer, the racks are moved through the dryer countercurrently to the air. They are loaded with the moist goods and installed in the upper part of the dryer. Then they move toward the bottom, where they are unloaded. The racks are finally loaded with new moist goods, and sent back to the top of the dryer to begin another drying round. The motive power can be supplied by a hand crank in small units, while large installations have power systems that are automatically controlled. In this way the material in all of the racks reaches the same final condition, and the process approaches the efficiency of a fully continuous process.

The dryer that is illustrated in Figure 28 requires removal of the racks above the intermediate heating unit and their reinsertion below it. Intermediate heating of the air is necessary because the air cools after passing through the lower racks, and because very high drying temperatures cannot be used due to the sensitivity of the goods. (The Simplizior dryer is also produced without an intermediate heater.) Useable surface areas up to 60 m^2 are common.

Belt Dryers. In belt dryers a loading device that is especially designed for the product is used to place the moist solid on the surface of a belt which passes through a drying chamber that resembles a tunnel. At the end of this chamber the material falls from the belt into a chute for further processing. In some installations the material falls onto another belt that moves in the opposite direction to the first one. Depending upon the characteristics of the material to be dried, multiple passages through the dryer are possible. In this way the material is mixed while drying, and new surfaces are exposed to the drying agent. A shorter, more gentle drying process is the result.

Centrifugal or axial flow blowers are used to aerate the moist materials. The air stream can enter the solid from below or above. The drying agent in belt dryers is commonly supplied laterally so that they are operated as convection dryers with partly recirculating air. The drying agent is heated indirectly with steam or hot water. Fuel oil or natural gas can also be used to indirectly or directly heat the drying agent.

Stepwise heating of the solid is possible with a unidirectional belt dryer, and this allows further control of the drying process. Belt dryers are ideal for friable, molded, granular, or crystalline products. They are utilized in all branches of industry.

Aerated Perforated Drum Dryers. This type of machine is employed to dry solids that can form porous layers on curved surfaces. Examples of such materials are cellulose fibers, wool, and cotton. The dryer consists of a series of closely spaced drums with diameters ranging up to 2 m and lengths of 6 m, which turn about a horizontal axis (Fig. 29). The cylindrical surfaces of these drums are made of perforated sheet metal. The material to be dried is transported to the top of one drum and is transferred to the bottom of the next; it moves to the top of the following drum, etc. It is held against the drum by a flow of high velocity air that is sucked into the drums through guides and hemicylindrical baffles at their periphery.

Air-permeable paper is produced on rotating cylinder dryers that consist of only one drum with a surface made of metal screening. The efficacy of these dryers is dependent upon the porosity of the layer of moist solid, the maximum permissible air temperature, and the vacuum that can be achieved within the drum which, in turn, determines the quantity of air that passes through the paper. It takes only seconds to dry the paper in this way.

Rotating Drum Dryers. The drum of a rotating drum dryer (Fig. 30) is longer in relation to its diameter than that of the perforated drum dryers and rotates about a horizontal axis that is slightly inclined. Consequently, the material that is loaded at the upper end slowly wanders toward the lower end, where it usually falls into a loading hopper. Longitudinal protrusions (flights) inside the kiln distribute the material more or less uniformly over the peripheral surface for drying and promote transport in some instances. These protrusions are often shaped like crosses or quadrants and reduce the drying time in two ways: (1) they constantly stir the solid and (2) intermittently cause it to fall away from the wall as a shower of fine particles through the drying agent, which streams axially along the kiln. Some of these lifting flights do not perform well when the solid is quite dry because they create too much dust, but are very effective for pasty or

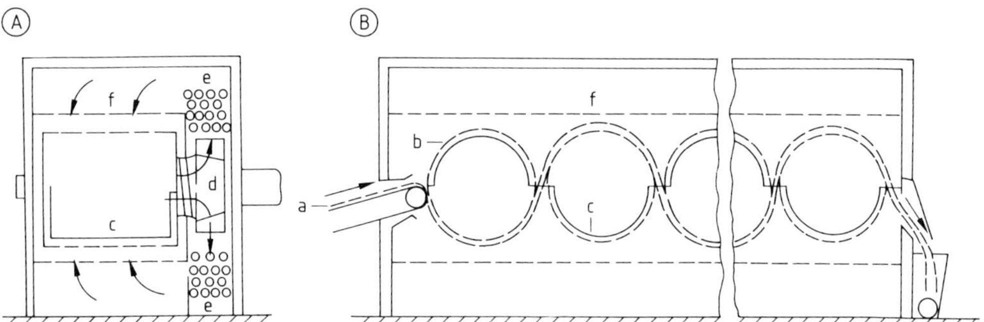

Figure 29. Through-circulation perforated drum dryer for loose, fibrous materials (Courtesy of Fleißner Egelsbach)
A) End view; B) Side view
a) Loading conveyor; b) Perforated drum; c) Cover plates; d) Fan; e) Heaters; f) Air distributor

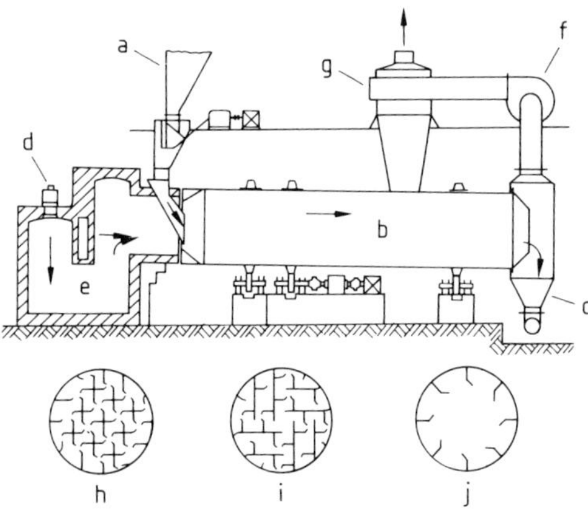

Figure 30. Cocurrent direct-heat rotary dryer and various flight arrangements
a) Feed chute; b) Drum; c) Discharge; d) Burner; e) Combustion chamber; f) Exhaust fan; g) Dust separator; h) Cross-shaped flights; i) Quadrant flights; j) 45° lip flights

sludge-like materials. If the drum must be cleaned often, or blockage of the drum can occur, flights that have a relatively simple shape are preferable.

The moist material and the drying agent flow either cocurrently or countercurrently. This type of kiln is often direct-fired. They have diameters between 0.3 and 6 m and are used principally for granular and crumbly materials. Pasty substances and slurries are often transformed into crumbs in a short time after they come in contact with the hot drying agent. Some liquid or semiliquid materials form small clumps when mixed with an already dried portion of the same materials. Then they can be dried with this kind of dryer. To increase the residence time in rotary drum dryers, circumferential baffles are used.

Roto-Louvre Dryer. This particular type of drum dryer has internal guide vanes (Fig. 31) that convey the drying medium in crosscurrent to the moist solid in very close proximity. The solid is simultaneously well stirred. The channels for the drying gas are tapered toward the rear of the dryer to achieve a good flow distribution. This results in a conical dryer, which does not need to be inclined.

Aerated, Double-Screw Dryer. The moisture in clumps of materials and coarsely fibered or flaky granules can be extracted in an aerated double-screw dryer (Fig. 32). Two parallel screws that rotate closely together convey the

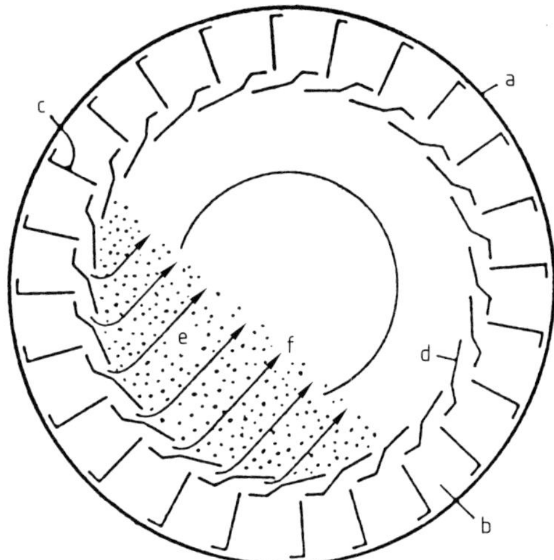

Figure 31. Cross section through a Roto-Louvre dryer (Courtesy of Dunford & Ellioth, London)
a) Rotating drum; b) Inlet ducts for hot air; c) Radial guide vanes; d) Tangential guide vanes; e) Moist solid; f) Air flow through the solid

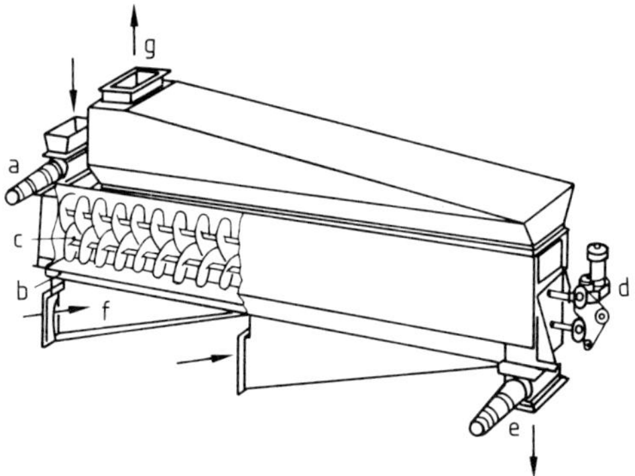

Figure 32. Twin-screw dryer
(Courtesy of Werner Pfleiderer Maschinenfabrik, Stuttgart)
a) Product feed valve; b) Fine screen plate; c) Conveyor screws; d) Screw drive; e) Product discharge valve; f) Fresh air distributor duct; g) Exhaust air connector

solid over a perforated floor through which the drying agent flows upward. The dryer can be completely sealed so that the drying agent flows in a closed circuit to facilitate the recovery of volatile solvents.

2.1.3. Large-Scale Agitation of the Solid

Spouting Bed Dryers. These units are noted for their high drying efficiency. One model is

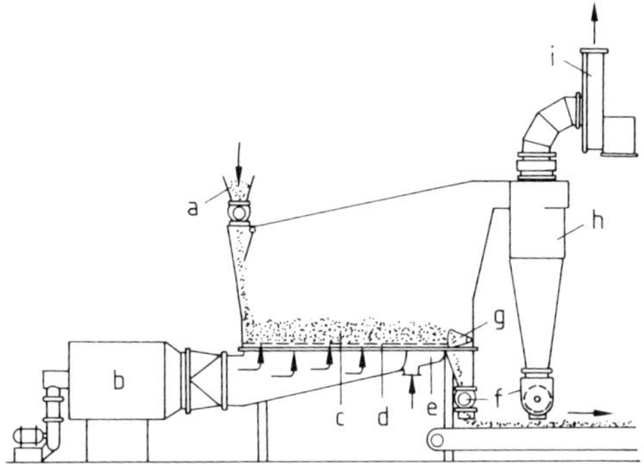

Figure 33. Fluidized bed dryer
(Courtesy of Büttner-Schilde-Haas)
a) Fluidized bed; b) Heater; c) Solids feeder; d) Gas distributor; e) Cooling zone; f) Discharge; g) Control baffle; h) Dust separator; i) Fan

illustrated in Figure 33. The solid moves horizontally in a chute and the drying agent flows vertically through a perforated floor to fluidize the solid. These machines can operate continuously, because the solid that enters the dryer via an adjustable opening is transported through it while suspended in the drying agent. The throughput can be increased by belts fitted with scoops to collect and move the material.

To achieve fluidization of the solid with minimum air velocity, vibrating fluidized beds are used. The dryer is mounted on a chute that vibrates under the action of a shaking device. The lengthwise oscillations of the chute serve to transport the material. The residence time of the solid within the dryer can be modified by adjusting the amount of vibration. If the airflow is divided into two sequential streams, the first can be used to dry the solid, while the second can cool it for further processing or packaging. Versions of these kinds of dryers that have a closed circuit for the drying agent can be used to recover solvents.

Materials that can be suspended in the drying agent are usually powders, crystals, and granular or short-fibered products that remain finely divided or have only a slight tendency to stick or cake. Pastes and slurries are mixed with previously dried material and are then easily fluidized. Solutions and suspensions can be dried by spraying them into the swirling layer with nozzles. The minute solid particles that are present are then covered by a thin film of liquid, which drys quickly, primarily due to the surface evaporation that characterizes the first period of drying. The small particles grow in size by conglomerating with others. Particles are added to the dryer when it is operating continuously to act as seeds. Small dried particles are recirculated; large particles are pulverized, classified, and recirculated. Particles between 0.5 and 5 mm in diameter can be produced from various solutions in this way.

Spin-Flash Dryer. Pastes and high viscosity liquids are often dried best in the type of dryer that is pictured in Figure 34. The dryer is loaded by a screw conveyor or a pump. The material to be dried is first stirred and broken up by a multi-armed paddle that turns at 50–500 rpm. Air is fed in tangentially at the bottom. A kind of fluidized bed is produced in the region of the stirrer. The air flow entrains the smaller, drier particles, and the larger, wetter ones fall back into the stirred region, where they are reduced in size.

Centrifugal Dryer. Figure 35 shows a centrifugal dryer whose motion serves to subdivide the moist solid and bring it into effective contact with the drying agent. The dryer consists of a stationary closed vessel that has at its base a centrifuge, or possibly two if the unit is large. They whirl the moist solid around in the dryer where the drying agent, usually air mixed with the combustion products of coal, oil, or natural gas, can perform its drying function. The fumes are diluted with enough fresh air to attain the needed inlet temperature and to increase the gas volume sufficiently so that the humidity does not become too high.

There are also dryers that combine the mechanical and aerodynamical methods to entrain

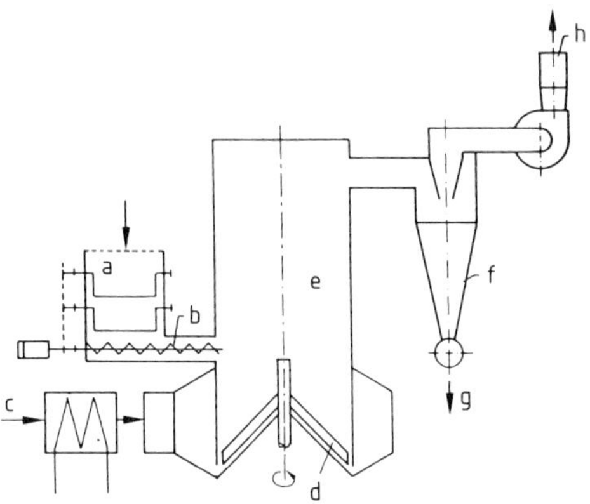

Figure 34. Spin-flash dryer
(Courtesy of Anhydro AS, Soborg-Kopenhagen)
a) Moist solid bin; b) Screw feeder; c) Air inlet; d) Stirrer; e) Main drying chamber; f) Material separator; g) Product discharge; h) Exhaust exit

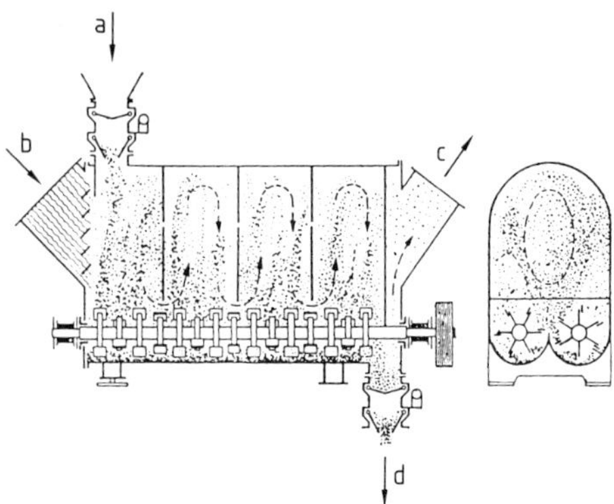

Figure 35. Centrifugal dryer
(Courtesy of Hazemag, Münster/Westfalen)
a) Loading zone; b) Hot gas inlet; c) Exhaust outlet; d) Material discharge

the solid in the vortical motion of the drying agent. Machines that swirl the solid are used to dry granules of plastics, salts, coal, and other chemicals.

Centrifugal-Impact Dryer. The unit pictured in Figure 36 functions as a centrifugal dryer and an impact pulverizer. Materials that have an edge length of less than 500 mm and a moisture content of less than 30% are reduced to a grain size between 0–10 mm and dried in gases that reach 900 °C. Within the dryer a rotating wheel with impact bars catapults the particles onto hardened plates, which shatter the clumps of moist solid. The pulverized pieces of the moist solid are dried in a stream of hot gases. The smallest particles are entrained in the stream and pass out of the dryer with the drying agent and must be recovered. The coarser particles fall to the bottom of the dryer, where they are removed. Dryers of this type are often used in the sand and gravel industry.

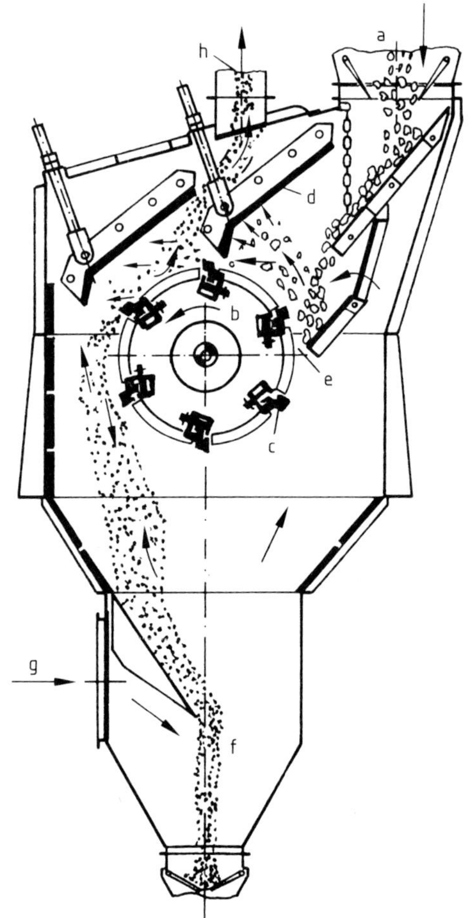

Figure 36. Centrifugal-impact dryer
(Courtesy of Hazemag mbH, Münster/Westfalen)
a) Moist solid loading; b) Impact rotor; c) Impact bar; d) Striker plate; e) Adjustable slit; f) Discharge collector; g) Hot-air inlet; h) Exhaust gas exit

2.1.4. The Solid Moves in the Drying Agent

Pneumatic Conveyor Dryers. Materials that can be pneumatically transported can be dried simultaneously. The simplest form for a dryer encompassing this dual role for the drying agent consists of a vertical tube in which granular or pulverized materials are dried while suspended in a gas or air stream (Fig. 37). The available drying time is only a few seconds: only fine materials, with their high rates of heat and mass transfer, or coarse products, with only surface moisture to be removed, are used in such dryers. Solids that contain internal moisture can only be dried to a limited extent by this method. Sometimes such materials can be dried in a multistaged gas-lift dryer.

The drying agent and the solid move cocurrently in air-lift dryers. The moist solid comes into contact with the hottest drying gases when it first enters the dryer. Because the solid is completely submerged in the hot drying gas, the conditions for high heat- and mass-transfer rates are good. The solid has the wet bulb temperature as long as surface evaporation is dominant. Only when this phase of drying is completed does the temperature of the solid begin to rise. This

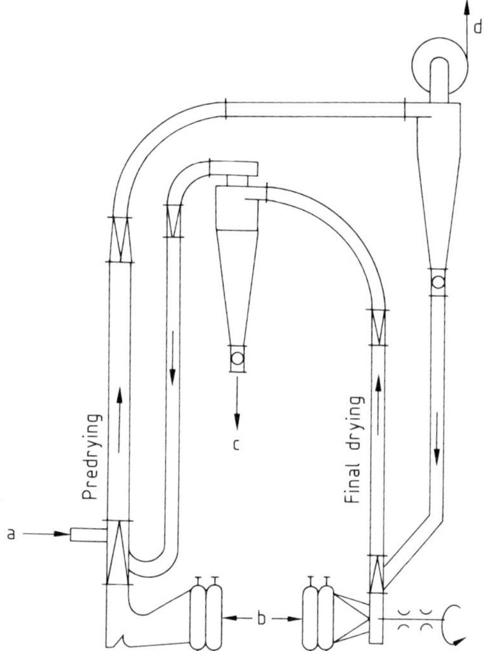

Figure 37. Two-stage pneumatic conveyor dryer
(Courtesy of H. Orth, Böhl/Pfalz)
a) Wet feed; b) Fresh air; c) Material discharge; d) Exhaust

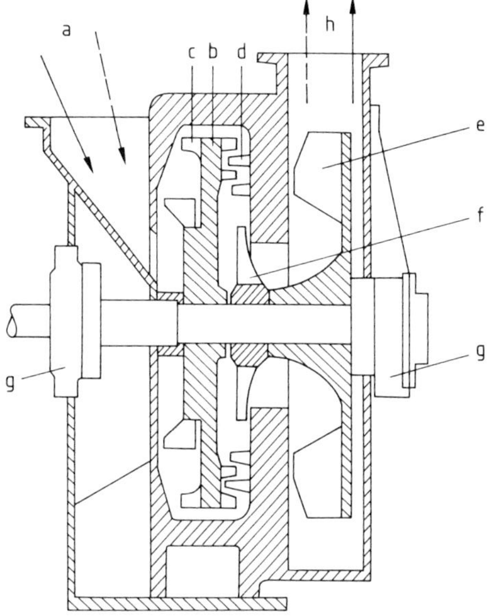

Figure 38. Atritor mill dryer
(Courtesy of Alfred Herbertg, Coventry, UK)
a) Moist solid and heating gas; b) Rotor; c) Hammer segments; d) Stator; e) Fan; f) Finger separator; g) Bearing; h) To the separator
---→ Heating gas; ⟶ Moist solid

warming process is limited, however, because the drying agent has been cooled by evaporating the moisture. Therefore, sensitive materials can be dried without damage in such dryers, even though the gas inlet temperatures are relatively high. The limited residence time is also a factor. Both organic and inorganic salts that have some moisture after having been centrifuged or filtered can be dried effectively with pneumatic conveyor dryers. Plastic powders, granules, foodstuffs, fodder, wood chips, sand, and quartz can also be dried in this way.

Mill Dryers. As the name of the dryer implies, in mill dryers (Fig. 38) the moist solid is simultaneously ground and dried. Although the energy consumption of these devices is relatively large, even that used for grinding the solid ultimately appears in the form of the heat that is necessary for drying. Hot gases also flow through the dryer to achieve the desired drying times. The mill shown in Figure 38 has a loading chute in which the moist material slides to the front surface of a rotating grinding wheel. The particles that this disk produces are carried away by the hot gases into a second grinding chamber of the dryer. Here the solid and entraining gas pass through a series of stationary and rotating pegs that further reduce the particle size. The gases and solids are ultimately forced out of the dryer by the blower.

This kind of dryer can produce powder from granular, caked, or partially liquid materials that can have a moisture content up to 80% and a representative size of 50 mm. Mill dryers are used to dry and comminute peat clumps at power plants.

2.1.5. The Material is Sprayed

Spray drying is used for the drying of pastes, suspensions, or solutions. The moist material is sprayed into the drying agent, and is converted into a powder that is entrained by the gas stream. The volatile liquids vaporize quickly. The gas and the dry powder can be separated at the exit of the dryer.

In this kind of drying process, a uniform fog of moist material should be produced within the dryer. Each type of solid affects the design used to produce a uniform product. The most important types are the disk atomizer (Fig. 39 A) and

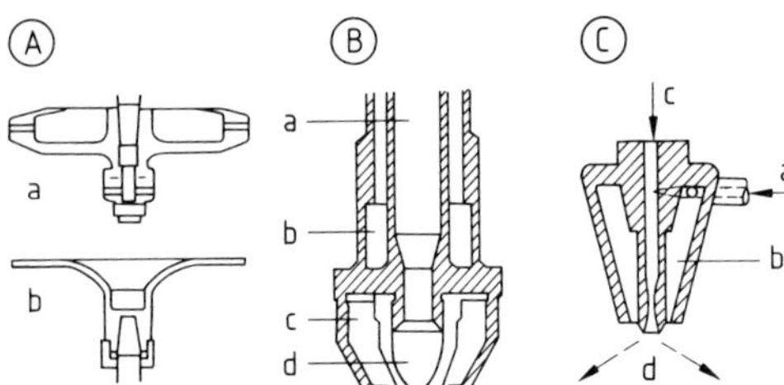

Figure 39. Devices for spraying
A) Disk atomizer
a) Suspended installation; b) Standing installation
B) Pressure nozzle
a) Fluid delivery; b) Heating or cooling jacket; c) Nozzle support; d) Nozzle orifice
C) Two-fluid nozzle
a) Tangential high-pressure air supply; b) Conical expansion and swirl chamber; c) Liquid supply line; d) Spray zone

pressure nozzle (Fig. 39 B, C). (→ 6. Spraying and Atomizing).

Centrifugal disks atomize liquids by extending them into thin sheets, which are discharged at high speeds from the periphery of the rapidly rotating (4000–15 000 rpm), specially designed disks (diameter 50–350 mm). The speed depends upon the desired size of the particles. This type of atomizer is used for pastes and suspensions because these materials would damage nozzles by abrasion, or even clog them. Very thick pastes can be treated in disk atomizers but they require a high-pressure pumping system to introduce them into the dryer.

Powders are produced from a spray nozzle, either a simple nozzle operating under sufficient liquid pressure or a two-fluid nozzle that concurrently sprays streams of liquid and gas—usually air. The resulting mist contains individual particles with diameters between 20 and 300 μm.

Figure 40 shows a two-fluid nozzle that exploits the fact that the speed of sound (critical speed) in a gas–liquid mixture is much smaller than in the gas or liquid phase. The two fluids leave the mixing chamber, which is at a low pressure, and enter an atmosphere where the mixture suddenly expands. The shock wave that occurs at the end of the mixing chamber causes the liquid—with suspended solids in some cases—to be distributed into a fine mist with a narrow size spectrum.

Because the liquid is transformed into minute drops, a very large surface area is placed in contact with the drying agent. Very short drying times—from fractions of a second to a maximum of a few seconds—are the result.

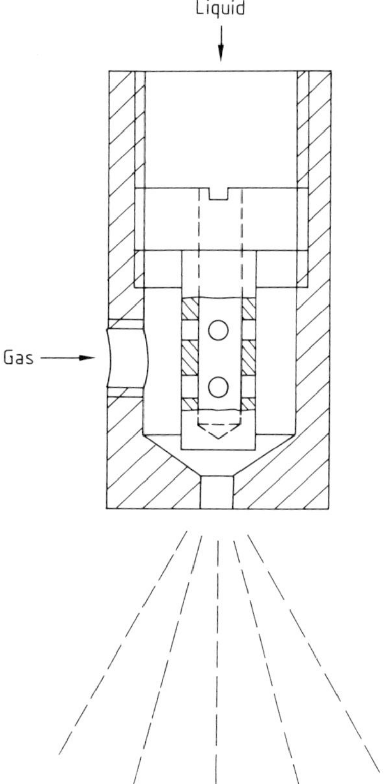

Figure 40. Two-fluid sonic nozzle
(Courtesy of Caldyn, Ettingen, FRG)

One or more powderizing systems can be installed at the top of a cylindrical structure, which is often as high as 20 m and several meters in diameter. Hot gases are introduced into the drying tower at the bottom or the top. The final product falls into a conical hopper at the bottom of the dryer, where it is unloaded.

Spray dryers are particularly suitable for drying solids that are temperature sensitive. Examples are milk products, baby foods, eggs, blood and blood plasma, pharmaceutical products, chemicals, dyes, plastics, glues, tannin, and soaps. An advantage of these dryers is the high solubility of the powders that are produced.

2.2. Contact Drying

In contact drying, the heat is directly transferred to the solid from a heated surface upon which it rests. The solid may be either stationary or be continually transferred from one hot surface to another.

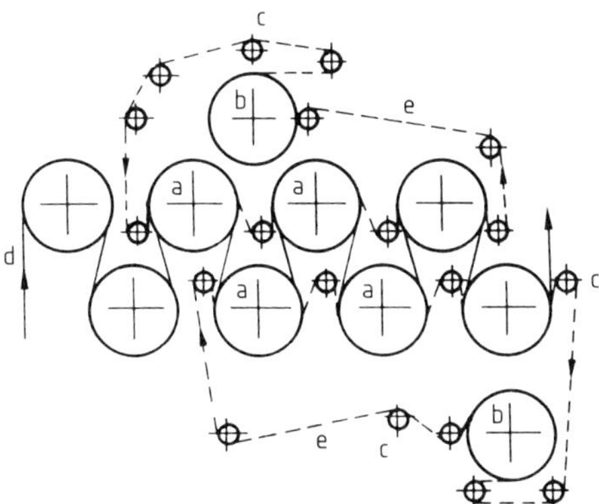

Figure 41. Schematic of a multicylinder dryer with guiding belt
a) Drying cylinders for the moist solid; b) Drying cylinders for the guiding belt; c) Guide rollers; d) Moist material; e) Guide belt

2.2.1. Flat and Strip Materials

Materials that are flat or in strips, such as textiles, paper, or cardboard, are dried with a drum dryer. The moist material is wrapped around a series of rotating, horizontal cylinders that are usually heated internally with steam. The steam and its condensate enter and leave the cylinder through hollow-journals, often the same journal. Materials that have a low strength when moist, e.g., felt and cotton, move between the drums on endless belts. The belts press the solid against the drying drums and simultaneously absorb some of the moisture. This moisture must be subsequently removed from the belts (Fig. 41). The possibility of introducing air between the drums and the moist material in rapidly rotating drum dryers is reduced by the pressure that the belts exert against the material. Consequently, the contact resistance to heating is reduced, and the heat transfer between the drum and product is improved.

2.2.2. Low-Viscosity Materials

Roller Dryers. In order to contact dry solutions of organic or inorganic material, a continuously operating roller dryer is almost always used (Fig. 42). They are usually selected because of their effective utilization of heat.

The material flows onto the drying roll as a thin layer. The liquid is vaporized by the heat coming from the heated rollers. Shortly before the dried solid reaches the end of the dryer, it is peeled away from the hot surface by a scraper to produce a film, coarse or fine flakes, or a powder. Some of the methods of applying the moist material to the rollers are shown in Figure 43. Each one has its particular uses, depending upon the adhesion and consistency of the moist solid.

The rollers are usually heated internally with steam, although hot water or heat carrier oils are sometimes used. Induction-heated drying cylinders are also available; they can produce controllable surface temperatures between 40 and 400 °C.

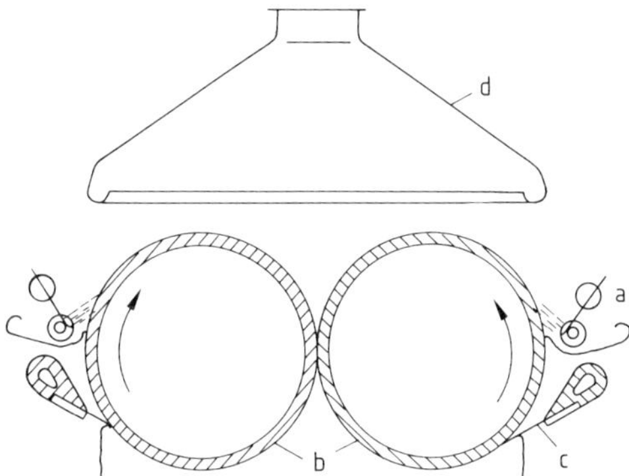

Figure 42. Double-roller spray dryer
(Courtesy of Escher Wyss)
a) Spraying apparatus; b) Drying rollers; c) Scraping knife; d) Moisture exhaust

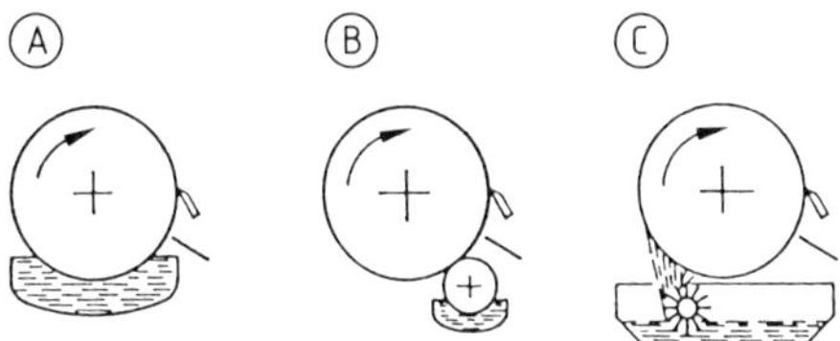

Figure 43. Methods for feeding the moist solid for a roller dryer
A) Submerged roller; B) Distributing roller; C) Barbed roller

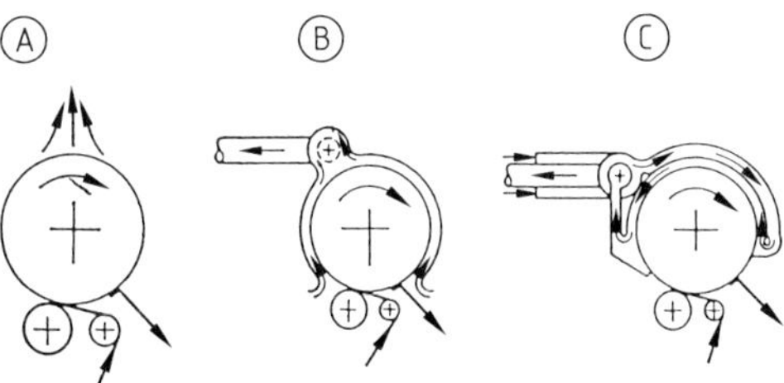

Figure 44. Aerating a single-roller dryer
A) Dryer without a hood; B) Dryer with hood and exhaust; C) Dryer with hood, warm air supply, and exhaust

Hoods can be placed over the dryer to reduce the ambient pressure and facilitate drying. The dryer can also be completely encased to make it airtight and dustproof. In some applications where increased drying efficiency is needed, the roller is additionally heated with a stream of hot air or exposed to a radiant heater (Fig. 44).

2.2.3. Pasty Materials

Roller dryers are also effective for drying pasty or creamy materials if they can be satisfactorily applied to the rollers. A proven single-roller device is shown in Figure 45.

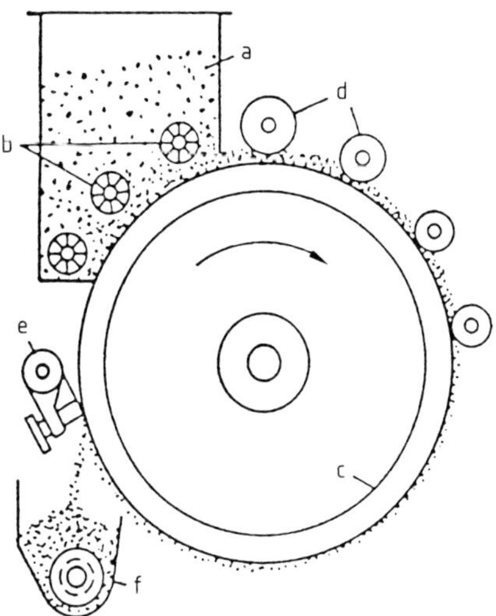

Figure 45. Single-roller dryer for mealy materials
a) Feed container; b) Stirrers; c) Drying roller; d) Pressurizing rollers; e) Scraping knife; f) Discharge

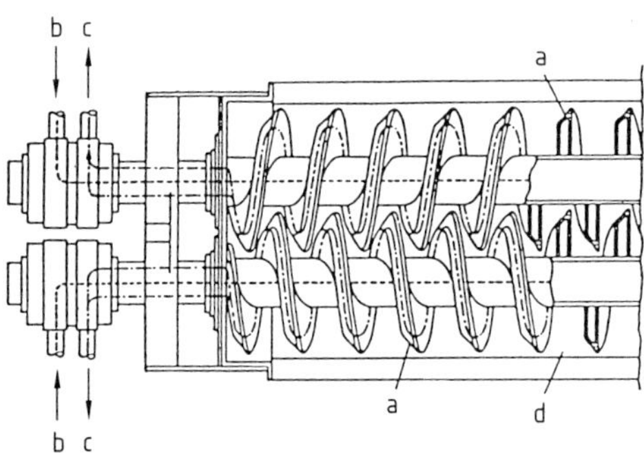

Figure 46. Hollow-screw heat exchanger
(Courtesy of Lurgi)
a) Hollow-screw conveyors; b) Inlets for the heating agent; c) Outlets for the heating agent; d) Trough

Grooved Roller Dryer. A special type of roller dryer reduces the moisture of pastes by 8–10 % and produces short segments for further drying steps. The paste is not rolled out into a thin layer, instead it is forced by a special roller into the grooves of the drying roller. As the roller turns, the material is pressed and thickened. Finally, a serrated scraper removes the dried material. To dry the solid completely a belt dryer is added. The drying of materials such as stearates, carbonates, clay, kaolin, white lead, or titanium dioxide are some special applications of this kind of dryer.

Hollow-Screw Heat Exchanger. Hollow-screw heat exchangers (Fig. 46) have proved useful for the continuous drying of paste. This dryer consists of a trough that contains one or two hollow screws. The screws turn in the same or the opposite directions, and partially overlap. Screws turning in the same direction are preferred for pastes because of the greater overlap and the self-cleaning action. Hot water, saturated steam, or hot organic liquids serve to provide the heat.

List AP Dryer (All Phase Dryer). The dryer shown in Figure 47 can be used for materials that can range from fluid or pasty to putty-like or

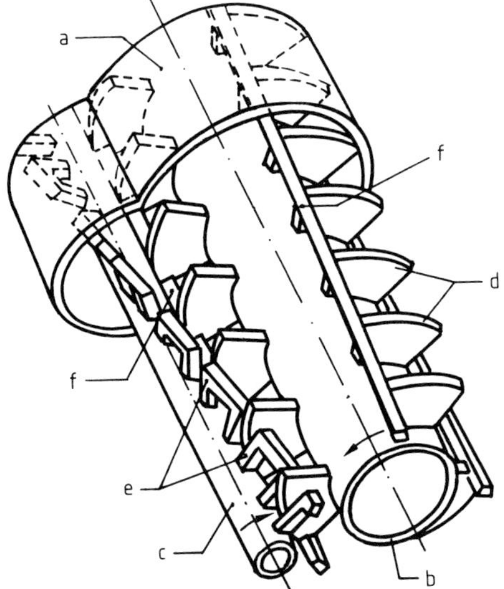

Figure 47. Contact dryer with stirrer and kneader
(Courtesy of List, Prattein)
a) Double-walled container; b) Main shaft; c) Cleaning shaft; d) Cleaning disk; e) Square-hoop stirrer; f) Kneading bars

granular during the drying process. It is also suitable for substances that are sticky or form crusts. A suitably designed horizontal housing surrounds the main roller, which has disks attached to it. This roller rotates in the opposite direction to a cleaner roller that has rectangular hoops that stir the material. The cleaner roller's speed is four times that of the main roller, whose surface is cleaned by it. Kneadable solids adhere to the disk protrusions on the main roller and scrape against the housing to scour and clean it. The throughput of the device can be controlled by the orientation of the rollers. The housing, its lid, the hollow main roller, and the disk elements are heated with steam, hot water, or hot oil.

2.2.4. Granular Materials

Rotary Dryer. The heat necessary for drying is transferred through the peripheral walls of a rotary dryer in contact drying. Only a small amount of air is necessary to carry off the moisture that is taken from the solid. Accordingly, the air velocity in these units is quite low. This is advantageous when drying materials that dust easily or form dust during drying. If necessary, ducting can be installed to introduce gases into the drum where the solid is drying.

Rotary dryers are preferred for drying washed ores, coal, and cement. Such materials are constantly turned over and mixed by lifting flights attached inside the drum.

Rotary contact dryers are produced in two main types. The first is constructed with integral heating pipes within the drum. The solid falls in a shower past these hot pipes once during each revolution of the drum. In the second type, the moist solid is in tubes that are attached to the walls of a drum into which steam is fed. The steam condenses on the surface of the tubes, heating the moist solid within. The drum is inclined to the horizontal so that gravity transports the material down the tubes, but some assistance is provided by the flow of vapor that ensues in the tubes.

Screw Conveyor Dryers. Inside the drum, screw conveyor dryers (Fig. 48) have a rotor that is fitted with many small paddles that constantly force the moist solid into contact with the hot wall, as well as move it along the axis. The residence time of the solid in the dryer is regulated by changing the angle of the paddles.

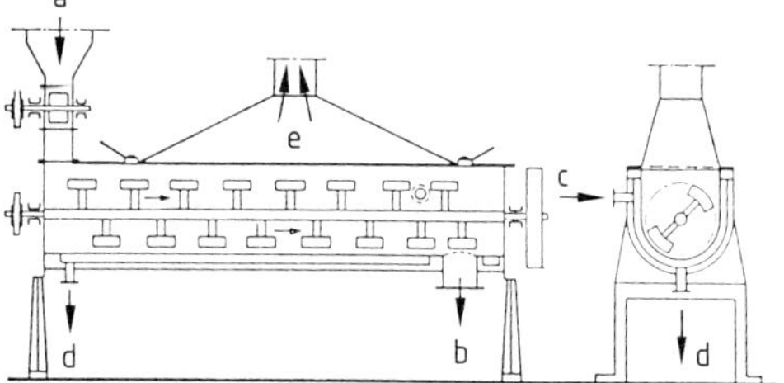

Figure 48. Screw conveyor dryer
(Courtesy of Büttner-Schilde-Haas)
a) Moist solid loading; b) Dried solid discharge; c) Inlet for heating agent; d) Outlet for heating agent; e) Solvent vapors

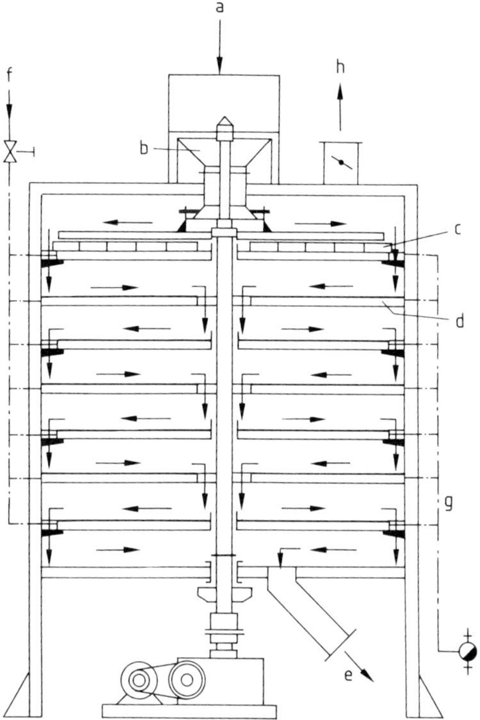

Figure 49. Tray dryer
(Courtesy of Büttner-Schilde-Haas)
a) Moist solid loading; b) Rotating tray feeder; c) Rakes (only drawn for topmost level); d) Heated tray; e) Product discharge; f) Steam; g) Condensate; h) Exhaust

Cone-Worm Dryer. A cone-worm dryer operates along the inclined wall of a vertical funnel-shaped container. The chamber is sealed and equipped with a heating jacket. The screw serves to transport the material to the top and to produce a gentle stirring of the moist solid.

Tray Dryers. When contact drying is employed, the trays (Fig. 49) are heated. Different

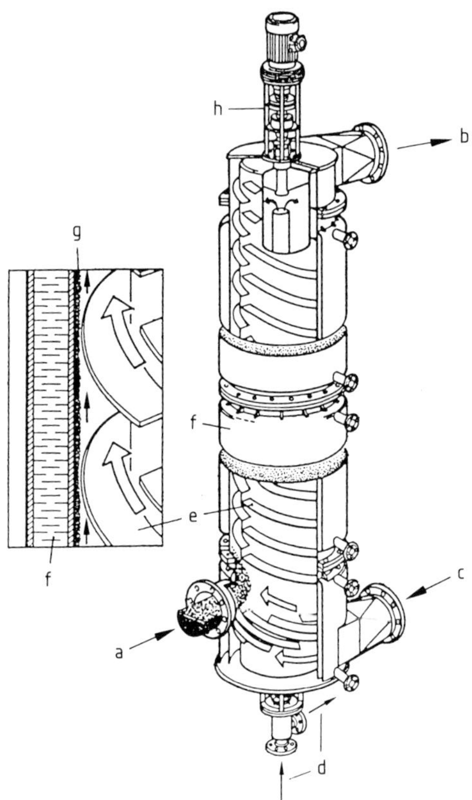

Figure 50. Spiral tube dryer (Ruhrchemie system, greatly shortened)
(Courtesy of Werner & Pfleiderer)
a) Moist solid; b) Dry solid; c) Transporting air; d) Heating liquid; e) Air baffles; f) Heating jacket; g) Product path; h) Compactor drive

heating intensities on the different trays provide the optimum conditions for drying. In this way, for example, the material can be cooled in the lowest levels of such dryers. The vapor is removed by warm air flowing countercurrently.

Spiral Tube Dryers. The drying of solids that range from granular to powdery is facilitated by spiral tube dryers (Fig. 50). The material is only momentarily in contact with the heated surface of the tube and is transported pneumatically. Residence time in the dryer is but a few seconds. These dryers require only a small amount of gas. Therefore, the drying of solids wetted by solvents can be done very economically with inert gases flowing in a complete circuit that includes a solvent recovery system. Spiral tube dryers find application in processing plastic granules, starches, bread crumbs, and copper powder.

2.3. Radiant Heat or Infrared Drying

Very short drying times are possible with infrared radiation sources that transmit large amounts of energy per unit of surface area to the solid in the form of heat. This method of drying is costly because of the electrical energy. It is only economical for drying long runs of similar materials that have a sufficiently thin surface film.

The intensity and emisson spectrum of the radiation must be optimized with respect to the moist solid to keep the energy costs low and to prevent excessive heating of the solid. Ovens with conveyor systems are often divided into zones to achieve this optimization; the individual radiation sources can be adjusted to match the conditions of the solid (→ 16. Radiation Heating, **B 3**).

The main use of this method is the drying of painted surfaces because warming occurs rapidly, forcing the solvents within the paint to move to the surface, which inhibits the formation of an undesirable surface skin. This method of drying is also important for drying thin-walled ceramic products, which can be dried quickly but without cracks by this method.

A combination of radiant and contact drying is effective if the drying time or the size of the dryer must be reduced. An example of such a combination dryer is shown in Figure 51. The dryer is provided with a means for jet drying a woven belt and a radiant heater to predry the material. The rapid drying of the fabric surface hinders the undesirable migration of dyes and other chemicals to the surface.

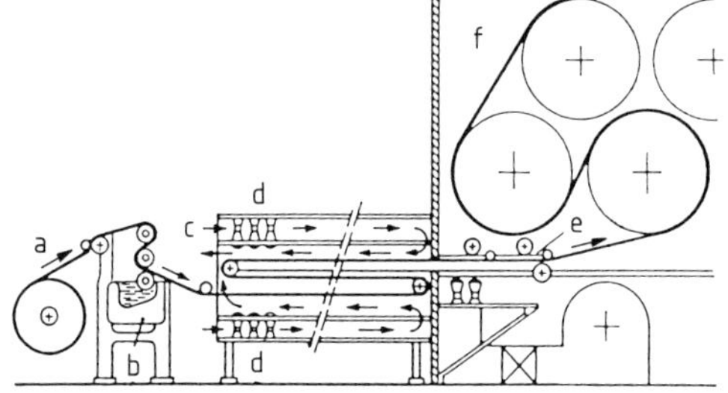

Figure 51. Predryer for textiles
(Courtesy of Philips)
a) Fabric; b) Impregnating trough; c) Air inlet; d) Infrared radiators; e) Sheet-metal reflector; f) Main dryer

2.4. Dielectric Heating

Drying with high-frequency heating has an advantage over other methods of drying: the temperature within the moist solid rapidly rises. The internal temperature can be maintained at a prescribed value regardless of the surface temperature. When wood is dried by this means, the moisture distribution can be adjusted so that the surface is always more moist than the interior. As a result, compressive stresses are induced in the surface fibers, strengthening the product, whereas the other means of drying tend to weaken the material and produce cracks. Only a few hours are needed to dry wood and ceramic goods with high frequency (→ 15. Electrically Generated Heat, **B3**).

Relatively high energy and installation costs are associated with this method of drying so that it is only cost effective to dry particularly valuable materials such as special hardwoods, large ceramic pieces, or temperature-sensitive foodstuffs and gourmet items.

2.5. Vacuum and Freeze Drying

Almost all dryers that operate at atmospheric pressure and supply the necessary heat by conduction or radiation can be converted to a vacuum dryer. The principal difference in the vacuum dryers are their seals and the means to produce the vacuum. Continuously operating vacuum dryers require, in addition, special devices for loading and unloading. Drying under reduced pressure is advantageous for materials that are temperature sensitive or easily decomposed because the vaporization temperature is reduced. Also, drying times in vacuum dryers at the maximum temperature are shorter. When drying materials containing organic solvents in vacuum dryers, the solvents are recovered more economically than with convection dryers because the humid drying agent need only be cooled to the condensation temperature of the solvent at the reduced pressure. Vacuum dryers are most often used to dry pharmaceutical products and foodstuffs. If the substance to be dried is poisonous or potentially explosive, vacuum drying lessens the emission of poisonous gas or vapor into the environment.

The simplest form of a vacuum dryer for batch drying is the vacuum shelf dryer. The moist solid lies on a heated plate. Improved heat transfer with higher efficiency is obtained in the vacuum tumble dryer (Fig. 52), in which the moist solid is constantly agitated and mixed. An example of a continuously operating vacuum dryer is a roller dryer with one or two rollers. Air-lift dryers that operate at reduced pressures are effective when large amounts of materials must be dried at low temperature and only a relatively small amount of moisture must be removed.

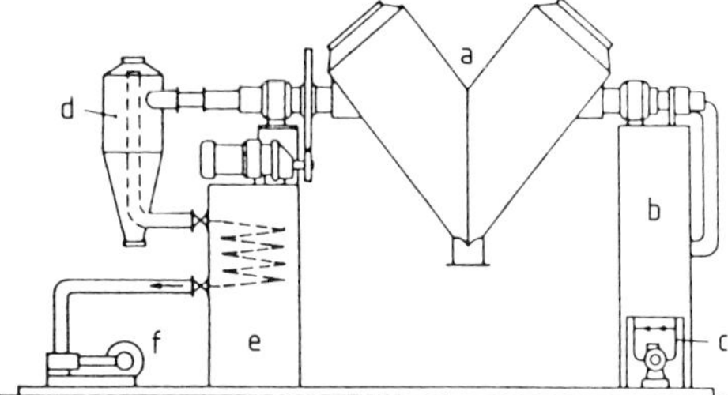

Figure 52. Vacuum tumble dryer
(Courtesy of Patterson-Kelley)
a) Rotating container; b) Heating boiler; c) Pump; d) Dust separator; e) Condensor; f) Vacuum pump

Freeze Drying. Freeze drying is characterized by the removal of water (sublimation of ice) from the solid at temperatures below 0 °C and reduced pressure. Freezing halts nearly all the chemical and biological processes in the material so that it is biochemical, physiological, and therapeutic characteristics remain essentially unchanged. Accordingly, this method of drying is used to process high-value foods, fruit juices, spices, tea, coffee, pharmaceuticals, virus and bacteria cultures, vaccines, and preparations containing protein.

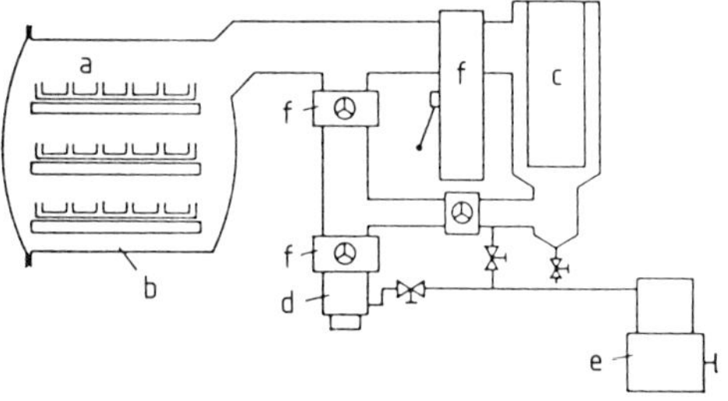

Figure 53. Oven vacuum drying
(Courtesy of E. Leybolds)
a) Heated trays with moist solid; b) Drying oven; c) Condenser; d) Diffusion pump; e) Booster pump; f) Shutoff valves

A schematic diagram of a freeze dryer is shown in Figure 53. Usually the freezer is a separate first stage and then the solid is vacuum-dried, but in some systems one apparatus performs both functions. Shallow pans that contain the solid are heated by conduction in the dryer. The vapors condense as ice upon reaching the cooling surfaces. These surfaces must be thawed at the end of the process when the condenser is isolated from the drying chamber by closing a valve. The dryer is then evacuated with the aid of a diffusion pump.

The continuous operation of a freeze dryer that processes foodstuffs is accomplished with a vacuum disk dryer. Such a unit is characterized by considerably shorter drying times than a vacuum chamber dryer (→ 19. Refrigeration Technology, **B 3**, p. **19**-33).

3. Selecting, Sizing, and Energy Requirements of Dryers

3.1. Choosing the Type of Dryer

The choice of a particular dryer type is mainly determined by the characteristics and production rate of the material that requires drying. Often several types would be suitable so that operating and economic aspects become the decisive criteria.

3.1.1. The Role of the Material Properties of the Solid

The mechanical properties of the moist solid when it arrives at the dryer determine the ways the material can be loaded into and transported through the dryer. Lumpy, granular, and crumbly materials can remain stationary (as in simple drying chambers, Simplizior, tunnel, or screw dryers), or they may be moving in respect to their supporting elements (as in fully-automated rack dryers, belt dryers, air-lift dryers, drum dryers, cyclones, and spiral tube dryers). Liquefied and pasty materials can be converted into fine drops (spray dryer), or thin layers (roller dryer), or they can be transformed into a crumbly material by partial drying (grooved roller dryer).

The thermal sensitivity of the material determines the temperature at which the heat can be transferred safely and the residence time in the dryer. If low temperatures are necessary, moisture removal can be achieved with a vacuum dryer. Small residence times are possible in forced convection dryers, spiral tube dryers, and spray dryers.

Materials that are chemically insensitive can be dried in direct contact with the hot drying agent, either air or combustion products. If the moist material is chemically sensitive, an inert gas must be used as the drying agent or some other heating method must be applied.

In the event that an organic solvent is present, it may be economical to recover it. In some cases the solid or the moistening liquid may be flammable or toxic. In such cases, completely enclosed dryers, such as vacuum dryers or special contact dryers, are to be recommended.

3.1.2. Production Rate

The throughput of the solid determines whether the dryer is to be operated in a continuous or batch mode. Batch operation requires a high energy consumption and worker effort per unit mass of dry product, but installation costs are relatively low. Nevertheless, this method of drying (a simple drying oven or chamber) is often economically viable for products with a small throughput or often changing properties.

When it is necessary to dry a well-specified product at a high production rate, a continuous dryer is advantageous. Even though the initial costs of such installations are high, the standardized physical and geometric characteristics of the product permit optimization of the energy requirements for this mode of operation. Moreover, the manpower necessary to operate the system is quite small. Most of the common dryer types are able to operate in the continuous mode. The efficiency of a dryer is largely determined by the amount of moisture that can be removed per unit area or volume of the dryer, and the amount of heat that is necessary to vaporize the liquid on a unit mass basis. Table 2 gives information about these parameters.

3.1.3. Dryer Ventilation

The ventilation of a dryer is important, and the various possibilities are given according to a scheme by Görling [12] in Figure 54.

Table 2. Vaporizing effectiveness and heat consumption of dryers

Dryer	Vaporizing effectiveness, kg H_2O m^{-2} h^{-1}*	Heat consumption, kJ/m^3h vaporized H_2O
Air-lift dryer	$d = 0.5$ mm: 100 kg m^3 h^{-1}	
(d = particle diameter)	$d = 1$ mm: 20 kg m^3 h^{-1}	3700–8800
	$d = 5$ mm: 4 kg m^3 h^{-1}	
Belt dryer		4000–5000
Centrifugal dryer	90 kg m^3 h^{-1}	3300–3500
Compartment dryer	0.1–12 (ventilated)	
	0.1–15 (flown over)	5600–13 000
Cylinder dryer	7–25	2900–5700
Drum dryer	25–50 kg m^3 h^{-1}	4000–8000
Kiln	20	5000–5800
Roller dryer	5–60	2900–5700
Rotating steam tube contact dryer		3800–6300
Roto-Louvre dryer	35–75	3500–10 000
Screw conveyor dryer	5–15 (high humidity)	
	0.5–2.5 (low humidity)	3400–5600
Spiral tube dryer	500 kg H_2O m^3 h^{-1}	3400
Spray dryer	1.5–48 kg m^3 h^{-1}	4600–11 000
Tray dryer	4–8	
Vacuum chamber dryer	0.15–1	2900–4600
Vacuum paddle dryer	10-15	2900–4200
	0.4–0.6**	10 000
Wiped tray dryer	3.5–4	3500–9200

* If not noted otherwise. ** With intensive terminal drying.

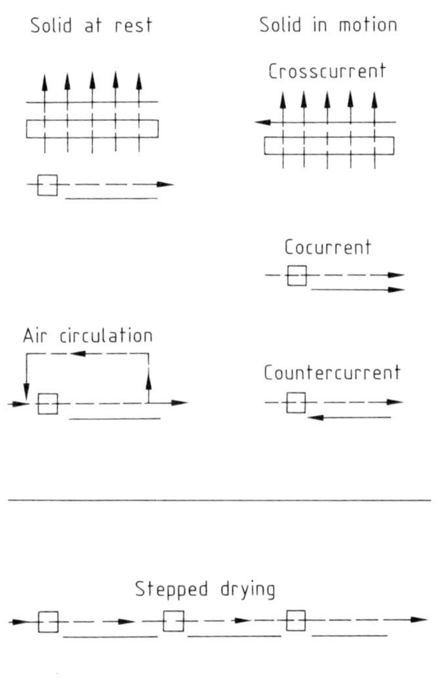

Figure 54. Various arrangements between solid, heaters, and drying agent
—, ⟶ Solid; ---→ Drying agent; □ Heaters

Stationary Solids. Solids that remain stationary in a convection dryer usually encounter a flow that is parallel to the surface on which they rest. However, it is possible to blow the air through a layer of granular material. This requires more energy to realize a given amount of air flow, but shorter drying times are possible so that the total energy consumption can remain satisfactory.

The usual manner of using air as a drying agent is to constantly circulate it past the moist solid while adding only a fraction of fresh air to the system. This is economical and minimizes the possibility of uneven drying or overheating the solid, which otherwise usually occurs near the air inlet duct. Through such a circulating system, the drying process can be satisfactorily matched to the characteristics of the solid by adjusting the temperature and speed of the air. Moreover, the influence of the daily atmospheric conditions are easy to compensate with this system.

Moving Solids. If the solid is moving in the dryer, the stream of air can be flowing in the same, opposite, or perpendicular direction to produce cocurrent, countercurrent, or crosscurrent drying, respectively.

In *cocurrent drying* the hot, dry air encounters the solid in its most moist condition at the dryer's entrance. As a consequence of the large temperature and humidity differences between the solid and the air at the beginning of the drying process, the drying rate is high. However, as a result of increasing humidity and decreasing temperature of the air along its path, the drying rate decreases continuously and becomes quite small near the end of the dryer. Thus with cocurrent drying, a small residual moisture content in the solid is not realizable with economical amounts of drying agent. Cocurrent drying is economically attractive if the dry solid is sensitive to high temperatures, the moisture removal does not have to include an extensive part of the hyroscopic moisture, and the initial high temperature in the dryer is not deleterious to the properties of the solid.

In countercurrent dryers the hot, fresh air comes in contact with the solid when it is at the end of its drying process; the cooled, moist air flows past the fresh, moist solid at the beginning. At first the solid is dried slowly, which is beneficial for products such as clay, but at the end of the drying process the solid is exposed to very high temperatures. Countercurrent dryers are not suitable for temperature-sensitive products.

A matching of the drying air's temperature with the requirements of the material to be dried is not possible in one-stage *crosscurrent dryers*. For this reason the solid must not be sensitive to high temperatures. The advantage of the method is that short drying times can be achieved. Using crosscurrent dryers which are subdivided in zones with different air temperatures, the conditions in cocurrent as well as in countercurrent dryers can be approximated.

As with stationary solids, moving solids can be dried more economically and uniformly when a circulating system for the drying agent is used. A further refinement consists of heating the air at several places along its path through the dryer. In this manner, the dryer is subdivided into several zones, and the temperature of the solid in each zone can be easily regulated. However, the change of humidity and the velocity of the drying agent are prescribed. In order to freely adjust the air velocity as well, the zone construction can be combined with the circulation of the drying agent. In each zone of this type of dryer, the air is circulated in large amounts and can be kept at the desired temperature by individual heaters. The humidity is adjusted by adding small amounts of air from the next zone, which is then mixed with the primary air. This method can be used for stationary or moving solid.

3.2. Sizing the Dryer

After the type of dryer that is needed has been determined, then its size (major dimensions) must be specified, which cannot always be done theoretically. The engineer often needs to turn to values obtained from existing units or experimental tests.

3.2.1. Batch Dryers

Drying with Constant Air Properties. The conditions of the air are practically constant during the entire drying process when large amounts of air are used or a large surplus of fresh air is circulated. If the normalized drying rate curve $\dot{v}(\xi)$ is known, then the drying time t can be calculated:

$$\frac{A\,\dot{m}_{\mathrm{I}}}{M_{\mathrm{s}}(X_{\mathrm{crit}}-X_{\mathrm{eq}})}\,t=\int_0^{\xi_i}\frac{\mathrm{d}\xi}{\dot{v}(\xi)} \tag{30}$$

A is the surface area and M_s is the mass of the dry solid. The drying rate in the first period of drying is $\dot{m}_{\mathrm{I}}$ and is determined from Equation (15):

$$\dot{m}_{\mathrm{I}}=\varrho_{\mathrm{g}}\beta_{\mathrm{g}}^{0}\,[Y^{*}(T_{\mathrm{I}})-Y]$$

Drying with Variable Air Properties. If the fresh air surplus is small or the amount of circulating air is limited, the conditions of the drying agent change considerably as it flows in the dryer. The normalized moisture content of the solid ξ is not only a function of time but also depends upon the path length of the drying agent z. In the analysis of the drying process, it is useful to introduce the dimensionless drying time τ

$$\tau=\frac{\varrho_{\mathrm{g}}\beta_{\mathrm{g}}^{0}\,A\,[Y^{*}(T_{\mathrm{I}})-Y(z=0)]}{M_{\mathrm{s}}(X_{\mathrm{crit}}-X_{\mathrm{eq}})} \tag{31 a}$$

and the dimensionless path length ζ

$$\zeta=\frac{\varrho_{\mathrm{g}}\beta_{\mathrm{g}}^{0}\,A}{\dot{M}_{\mathrm{g}}}\cdot\frac{z}{L} \tag{31 b}$$

in which $\dot{M}_g$ is the mass flow rate of the drying agent and L is the length of the dryer. The fundamental equation for the temporal and spatial distribution of the solid's moisture $\xi(\zeta,\tau)$ is

$$\frac{\partial^{2}\xi}{\partial\tau\,\partial\zeta}+\dot{v}(\xi)\frac{\partial\xi}{\partial\tau}-\frac{1}{\dot{v}(\xi)}\cdot\frac{\mathrm{d}\dot{v}}{\mathrm{d}\xi}\cdot\frac{\partial\xi}{\partial\zeta}\cdot\frac{\partial\xi}{\partial\tau}=0 \tag{32}$$

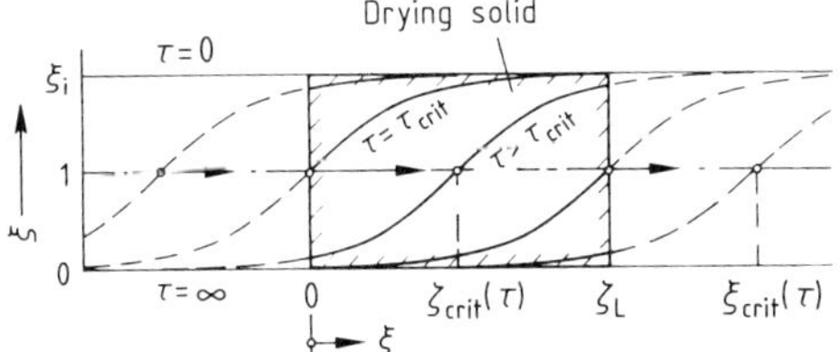

Figure 55. Moisture distribution in a solid as a function of dimensionless time τ

This equation assumes that the partial pressure of the vapor is small with respect to the total pressure, and that the normalized drying rate curve $\dot{v}(\xi)$ is independent of the condition of the drying agent. In addition, the critical moisture content X_{crit} must be independent of the drying rate. Solutions to this equation are given in the literature [9]. A qualitative picture of the temporal and spatial moisture distribution in the solid is given in Figure 55. The initial moisture content is ξ_i, and $\xi = 1$ corresponds to the critical moisture content. At $\tau = \tau_{crit}$, the solid's critical moisture content is achieved at the entrance of the dryer, $\zeta = 0$. Subsequently, the location of the critical moisture content of the solid $\zeta_{crit}(\tau)$ is shifted through the dryer. Whenever $\zeta_{crit} > \zeta_L$, the entire solid would be in the second or possibly third stage of drying.

In order to estimate the extent of drying, it is often sufficient to approximate the normalized drying rate curve $\dot{v}(\xi)$ with the linear approximation

$$\dot{v} = \xi \quad \text{for} \quad 0 \leq \xi \leq 1$$

$$\dot{v} = 1 \quad \text{for} \quad 1 \leq \xi \leq \xi_i$$

In this case Equation 32 has an analytical solution. For

$$\tau_{crit} = \xi_i - 1 \tag{33}$$

in the region $\zeta \geq \zeta_{crit}$ and $0 \leq \tau/\tau_{crit} < \infty$

$$\xi(\zeta, \tau) = \xi_i - (\xi_i - 1) \exp[-(\zeta - \zeta_{crit})] \tag{34}$$

is obtained. In the region

$\zeta \leq \zeta_{crit}$ and $0 \leq \tau/\tau_{crit} < \infty$

the result is

$$\xi(\zeta, \tau) = \frac{\xi_i}{1 + (\xi_i - 1) \exp[-\xi_i(\zeta - \zeta_{crit})]} \tag{35}$$

The dimensionless location of the critical moisture content ζ_{crit} can be calculated by

$$\zeta_{crit} = \ln(\tau/\tau_{crit}) \tag{36}$$

for $\tau/\tau_{crit} < 1$, and by

$$\zeta_{crit} = \frac{1}{\xi_i} \ln \frac{\xi_i \exp[(\xi_i - 1)(\tau/\tau_{crit} - 1)] - 1}{\xi_i - 1} \tag{37}$$

for $\tau/\tau_{crit} \geq 1$.

Even though Equations (36) and (37) encompass all values of ζ_{crit} between $\pm\infty$, only the values between 0 and ζ_L have a physical meaning.

3.2.2. Continuous Dryers

Useful suggestions and short-cut methods for the sizing of drum, convection, and spray dryers can be found in [10]. The following theoretical analysis utilizes the same simplifying assumptions that were used in the analysis of batch drying (see Section 3.2.1).

Drying with Constant Air Properties. The drying time t is determined by Equation (30). This corresponds to the residence time of the solid in the dryer. If u is the speed of the conveyor, then the necessary dryer length is $ut = L$.

Drying with Variable Air Properties. In contrast to batch drying, the location of the solid's critical moisture content within the dryer is independent of time. The basic equation for the spatial variation of the moisture in a cocurrent dryer is

$$\frac{d\xi}{d\zeta} + (\xi + C)\,\dot{v}(\xi) = 0 \tag{38}$$

In Equation (38) C is the so-called *excess air factor*, which is defined as

$$C = \frac{[Y^*(T_1) - Y(\zeta = 0)]\,\dot{M}_g - (X_i - X_{eq})\,\dot{M}_s}{(X_{crit} - X_{eq})\,\dot{M}_s} \tag{39}$$

If $C = 0$, the air exactly reaches saturation when the solid has reached its equilibrium moisture content. In practice, $C > 0$ is used.

To obtain an order of magnitude, Equation (38) can be integrated with a simple function for the normalized drying rate curve. For this purpose the linear approximation $\dot{v}(\xi) = 1$ for $\xi \geq 1$ and $\dot{v}(\xi) = \xi$ for $\xi \leq 1$ is used. For the region $\xi \geq 1, \zeta \leq \zeta_{\text{crit}}$ the result is

$$\xi = (1 + C) \exp[-(\zeta - \zeta_{\text{crit}})] - C \tag{40}$$

For the region $\xi \leq 1, \zeta \geq \zeta_{\text{crit}}$

$$\xi = \frac{C \exp[-(\zeta - \zeta_{\text{crit}}) C]}{1 + C - \exp[-(\zeta - \zeta_{\text{crit}}) C]} \tag{41}$$

is obtained. In both cases

$$\zeta_{\text{crit}} = \ln \frac{\xi_i + C}{1 + C} \tag{42}$$

holds.

Countercurrent drying can be analyzed with the following basic equation

$$\frac{d\xi}{d\zeta} + (C^* - \xi) \dot{v}(\xi) = 0 \tag{43}$$

in which $C^* = C + \xi_i + \xi_f$; the subscripts denote the initial and final moisture content of the solid. Integration of Equation (43) for the region $\xi \geq 1, \zeta \leq \zeta_{\text{crit}}$ yields

$$\xi = (1 - C^*) \exp(\zeta - \zeta_{\text{crit}}) + C^* \tag{44}$$

and for the region $\xi \leq 1, \zeta \geq \zeta_{\text{crit}}$

$$\xi = \frac{C^* \exp[-(\zeta - \zeta_{\text{crit}}) C^*]}{C^* - 1 + \exp[-(\zeta - \zeta_{\text{crit}}) C^*]} \tag{45}$$

The dimensionless location of the critical moisture content is given by

$$\zeta_{\text{crit}} = \ln \frac{C^* - 1}{C^* - \xi_i} \tag{46}$$

Figure 56 is an example of the variation of the moisture content ξ along the length of a dryer for cocurrent and countercurrent operation. For the calculation the values $\xi_i = 2, \xi_f = 0.2$, and $C = 0$ have been used. The cocurrent dryer must be 1.7 times longer than the countercurrent unit.

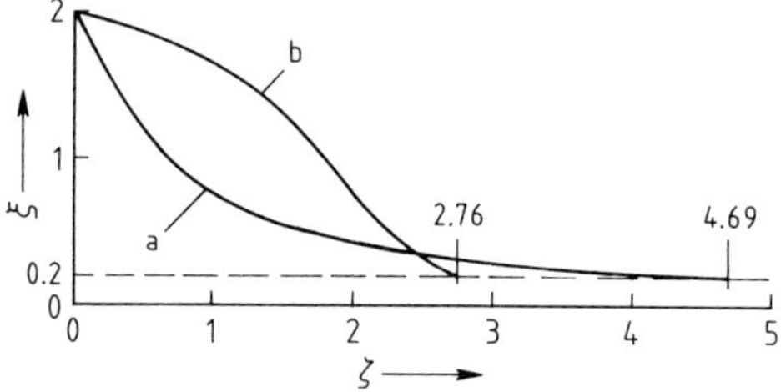

Figure 56. Dimensionless moisture content versus dimensionless location ζ in a dryer with
a) Cocurrent; b) Countercurrent flow ($\xi_i = 2, \xi_f = 0.2$, and $C = 0$)

3.3. Heat and Driving Power Requirements

The heat consumption of dryers is determined by

1) the heat needed to vaporize or desorb the moisture from the solid
2) the heat needed to warm the solid up to the drying temperature and to heat the vapor up to its exit temperature
3) the heat lost to the surroundings

The heat consumption is determined from an energy and mass balance of the system. Figure 57 shows the situation for a batch dryer. Mass balance is

$$\dot{M}_g (Y_i - Y_f) = M_s \frac{dX}{dt} \tag{47}$$

Energy balance is

$$\dot{Q}_{\text{Heat}} + \dot{W}_{\text{Vent}} = \dot{M}_g (h_f - h_i) + \dot{M}_s \frac{d(h_s + X h_l)}{dt} + \dot{Q}_{\text{lost}} \tag{48}$$

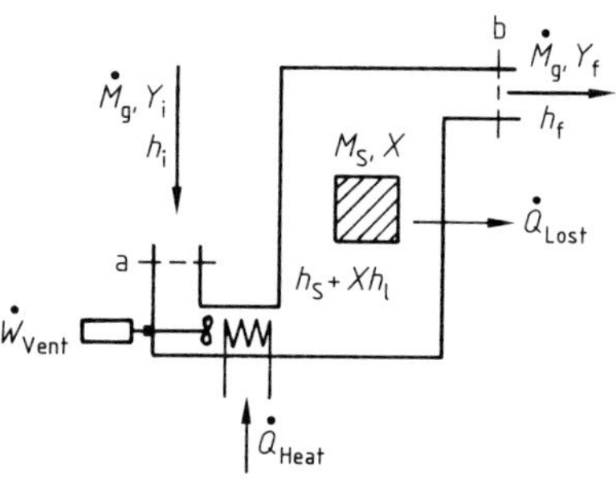

Figure 57. Schematic drawing of a batch dryer for the derivation of overall mass and energy balances
a) Inlet; b) Outlet

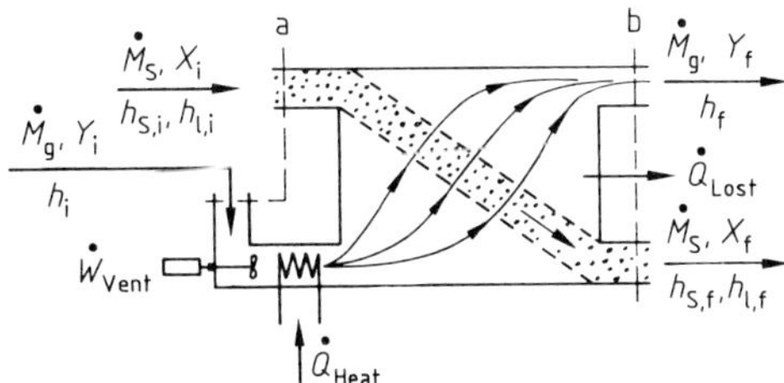

Figure 58. Schematic drawing of a continuous dryer for the derivation of overall mass and energy balances
a) Inlet; b) Outlet

In these equations, h_i and h_f are the entering (initial) and exiting (final) enthalpies of the moist drying agent, respectively; h_s is the enthalpy of the dry solid and h_l the enthalpy of the liquid. $\dot{Q}_{heat}$ is the heat requirement from the heaters; $\dot{W}_{vent}$ is the power required by the ventilation unit; and $\dot{Q}_{lost}$ is the heat lost to the surroundings.

The corresponding equations for a continuous dryer, shown schematically in Figure 58, are

$$\dot{M}_g(Y_f - Y_i) = \dot{M}_s(X_i - X_f) \qquad (49)$$

for the mass balance and

$$\dot{Q}_{Heat} + \dot{W}_{Vent} = \dot{M}_g(h_f - h_i) + \dot{M}_s[(h_{sf} - h_{si}) + (X_f h_{lf} - X_i h_{li})] + \dot{Q}_{lost} \qquad (50)$$

for the energy balance.

Approximate values for the heat consumption of the most important types of dryers are given in Table 2.

Driving Power Consumption. Not only is energy needed as heat to the dryer, also both the solid and the drying agent must be transported. Most of the power needed to move these materials in a convection dryer is consumed by the fans or blowers. This power has already been accounted for in the overall balances of Equation (50). The vacuum pump absorbs most of the power in a vacuum dryer. A guide to the driving power requirements of the various kinds of dryers is given in [10].

4. References

[1] O. Krischer, W. Kast: *Die wissenschaftlichen Grundlagen der Trocknungstechnik*, 3rd ed., Springer, Berlin-Heidelberg-New York 1978.

[2] R. B. Keey: *Drying: Principles and Practice*, Pergamon Press, Oxford 1972.

[3] R. H. Perry, D. Green (eds.): *Perry's Chemical Engineer's Handbook*, 6th ed. McGraw-Hill, New York-Toronto-London, 1984, pp. 20-1–20-58.

[4] P. Gummel: *Durchströmungstrocknung: Experimentelle Bestimmung und Analyse der Trocknungsgeschwindigkeit und des Druckverlustes luftdurchströmter Textilien und Papiere*, Dissertation, Karlsruhe 1977.

[5] E.-U. Schlünder, N. Mollekopf: "Vacuum contact drying of free flowing mechanically agitated particulate material," *Chem. Eng. Process* **18** (1984) 93–111.

[6] E.-U. Schlünder, "Heat transfer to packed and stirred beds from the surface of immersed bodies," *Chem. Eng. Process.* **18** (1984) 31–53.

[7] F. Thurner, E.-U. Schlünder: "Progress towards understanding the drying of porous materials wetted with binary mixtures," *Chem. Eng. Process.* **20** (1986) 9–25.

[8] J. Schwarzbach, E.-U. Schlünder: "Microconvection due to Marangoni-type instabilities in porous media – Effect on pervaporation and selective drying," *Proc. of 3rd World Congress of Chemical Engineering*, Tokyo 1986, 560–563.

[9] D. A. van Meel: "Adiabatic convection batch drying with recirculation of air," *Chem. Eng. Sci.* **9** (1958) 36–44.

[10] K. Kröll: *Trockner und Trocknungsverfahren*, 2nd ed., Springer, Berlin-Heidelberg-New York 1978.

5. Size Reduction

SIEGFRIED BERNOTAT, Institut für Mechanische Verfahrenstechnik der Technischen Universität Braunschweig, Federal Republic of Germany (Chap. 2)

KLAUS SCHÖNERT, Institut für Aufbereitung und Veredelung der Technischen Universität Clausthal, Federal Republic of Germany (Chap. 1)

In addition to the standard symbols defined in the front matter of this volume, the following symbols are used:

a	half the crack length, half the diameter of the contact surface
A	cross sectional area of specimen
A_B	fracture surface area
b_{ij}	breakage distribution coefficient of fragments
B	breakage function, sample thickness
B_f	breakage fraction
c	specific heat
c_v	volume concentration of solid
d	particle size, diameter of ball or disk
d_L	diameter of perforations
D	mill diameter, roller diameter
E	energy, Young's modulus
E_B	energy of fracturing
E_M	energy applied relative to mass,
E_u	energy utilization
F	force
F_{sp}	specific contact pressure
g	acceleration due to gravity
G	energy release rate
h	stroke
H	height of the crusher chamber
k_{opt}	optimal recirculation factor
l	crack length
L	length of cylindrical specimen, mill length, roller length
$\dot{m}$	mass throughput
m_i	mass fraction of interval i
M	mass
M_0	total particle mass
M_p	mass of fragments below lower limit of size range
n_c	critical rotational speed
p_m	maximum compressive stress
P	power
q_A	relative area of the ascending zone
q_L	relative area of the perforations
Q	heat produced during fracturing, particle-size distribution
r	radius of curvature
r_s	amplitude of oscillation
R	crack resistance
s	gap width
S	newly created surface area
S_M	specific surface area

t	time
u	peripheral speed of roller
u_p	velocity of particle in gap
U_{el}	elastic energy of stress field
T	driving torque
v	velocity
V	volume
$\dot{V}$	volumetric flow rate
w_i	specific rate of breakage of fraction i
W	work
W_{ad}	specific work (or energy) required in additional process steps
W_B	energy of fracturing
W_i	Bond index
W_M	specific work (or energy) of comminution
W_t	total work
x	particle size
x_F	particle size of feed
x_P	particle size of product
x_V	equivalent diameter of a sphere with same volume as particle
z	displacement
α	crushing angle
β	angle of force action
δ	length of fracture zone
ν	Poisson number
ΔS	newly created surface area
ΔT_m	maximum temperature increase
Δx	particle-size range
ϱ	density
ϱ_s	density of solid
σ	tensile stress in unperturbed state
σ_B	breaking stress
σ_m	maximum tensile stress
σ_y	yield stress
τ_m	maximum shear stress
ω	angular velocity

Size reduction, also known as comminution, is defined as the mechanical breakdown of solids into smaller particles without changing their state of aggregation. Processes such as the atomization of liquids or the separation of gas into bubbles are not considered in this article.

Products that have undergone size reduction surround us daily. Some examples are foods such as flour, sugar, and spices; construction materials such as cement, lime, plaster and sand; and pigment particles in lacquer and paint. Throughputs in industrial size reduction range from a few kilograms to hundreds of tons per hour. Particles ranging in size from blocks with edge lengths of 1–2 m down to very fine particles with a diameter of 1 μm are subjected to size reduction.

The two main objectives of size reduction are:

1) *Production of a specified particle-size distribution or a specific surface area.* The maximum permitted amount of residual coarse material at the upper particle-size distribution limit or the maximum permitted amount of fines at the lower limit is also stipulated. These requirements are always applied if the dispersity properties of the material have a pronounced effect on product quality (e.g., pigment color intensity, taste of chocolate, cement strength) and properties (e.g., agglomeration, miscibility, and solubility [1], [2]).
2) *Treatment of multicomponent materials.* In the case of inhomogeneous materials such as ores, the content of a valuable substance in a particle frequently depends on the particle size. As a result, the size must be reduced sufficiently so that the components are obtained as particles of maximum purity which can then be separated.

Size-reduction devices have been in use ever since people began to eat cereal crops. The exploitation of water, wind, and steam power led to the development of size-reduction equipment. However, a thorough treatment of size reduction from a scientific and engineering viewpoint began only in the late 1950s. The work of Rumpf and coworkers in the Federal Republic of Germany deserves special mention [3]. Additional research groups now exist in the German Democratic Republic, the United States, and Japan [4]. The European Symposia on Size Reduction provide an international insight into this field [4]–[6]. They cover research and progress in the basic understanding and mathematical description of size reduction as well as developments in machinery, processes, and protection against wear.

A comprehensive survey of size-reduction equipment is given in [7]. Reviews of size reduction are given in [8]–[18].

1. Fundamentals

Areas covered in this chapter include the basic mechanism of particle breakdown, the comminution results obtained on a laboratory scale using particles under defined loading conditions, and a description of size-reduction processes.

1.1. Particle Breakdown

Particle breakdown describes the sequence of events that leads to particle fracture:

1) Loading of the particle
2) Particle deformation
3) Buildup of a nonuniform stress field
4) Occurrence or nonoccurrence of fracturing

Division of the process into the above steps facilitates understanding of the point at which the many variable parameters exert their effects. Examples of such parameters are particle shape, particle size, material properties (e.g., elasticity), temperature, and type of stress. Stress may be applied between two surfaces (*compression*) or at a single surface (*impacting*). Figure 1 summarizes the various parameters and their effects, which are discussed below.

The *state of stress* plays a key role in particle breakdown and determines whether inelastic deformation or fracturing occurs. The latter two phenomena are termed the particle reaction. The particle reaction, in turn, affects the state of stress; this is especially important in inelastic deformation. Furthermore, the state of stress depends on the loading force, type of comminution tool used (flat plates or cutters), particle shape, and mechanical properties of the material.

In size-reduction equipment, the particles are stressed by *contact forces*. Generally, the cause of the stress (i.e., contact between neighboring particles or between particles and a moving part of the machine) is unimportant. The conditions in various types of equipment (e.g., crushers, ball mills, or impact mills) differ with respect to the number and direction of contact forces and the loading velocity.

Deformation and stress are related by the *mechanical properties of the material*, which are described by three limiting cases: elastic, plastic, and viscous behavior. Plastic and viscous behavior are both inelastic. In practice, however, the behavior of materials subjected to loads sufficient to cause breakdown seldom corresponds to only one of these cases.

Elastic behavior (described in the simplest case by Hook's law) is characterized by reversibility and time independence. Because of the latter, the stress in the contact region depends solely on the contact force and not on the loading velocity. Therefore, impacting and slower com-

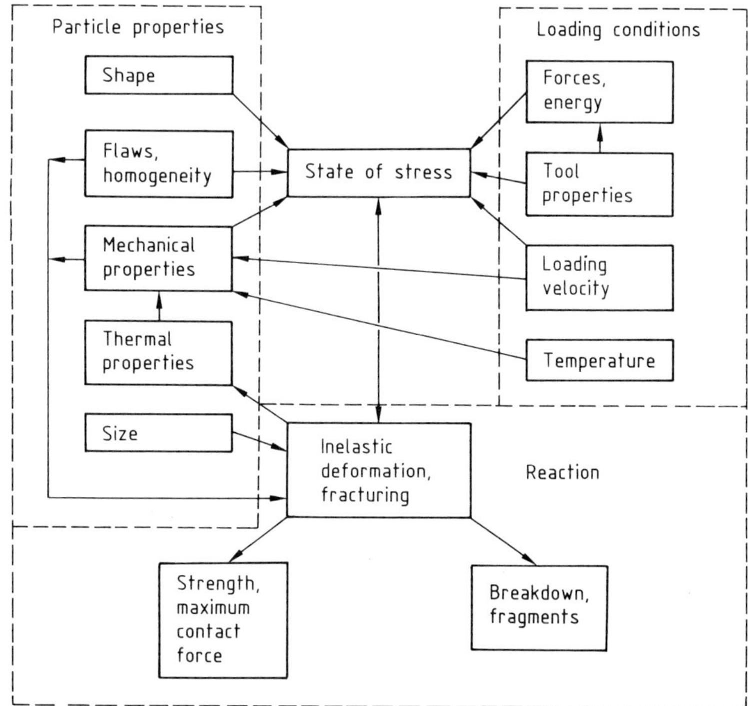

Figure 1. Factors affecting particle breakdown in size reduction

pression have the same effects. Elastic waves of significant intensity occur only at impact velocities above 300 m/s. Such values are seldom achieved in mills.

The limiting case of *plastic behavior* is characterized by irreversibility, as well as time independence. However, for plastic materials encountered in practice, a slight time dependence exists but can be neglected. Plastic deformation produces internal stresses and embrittlement. As a result, particles that undergo this type of deformation (e.g., metal powders) must usually be stressed repeatedly to break them. Because the loading velocity has little or no effect on the stress, only a slight advantage is obtained by using impact instead of compression.

Viscous behavior is distinguished from the two other limiting cases by its strong dependence on strain velocity and the related temperature. Plastics such as polyethylene and polystyrene are typical examples of this class of materials. The instantaneous stress is a function of the previous course of deformation. Therefore, under certain conditions, the maximum stress in a sphere can occur during removal of the load. The loading velocity plays an important role in materials exhibiting viscous behavior. In high-pressure polyethylene, a very rapid deformation lasting a few microseconds produces a four- to fivefold higher stress than a slow deformation lasting a few seconds. Consequently, in this case, impacting is superior to other methods of stress application. Cooling reduces the viscous component of deformation. Regions undergoing plastic deformation experience an increase in temperature, which in turn increases viscous behavior. Because the particles are generally stressed many times before breaking, intermittent cooling is advantageous. In cases of highly viscous behavior, application of stress with a cutting blade is more successful than impact stress. However, particle fineness is limited by equipment design.

The deformation of *irregularly shaped particles* and the resultant stresses cannot be calculated. However, to a close approximation, the contact region of the particle can be considered spherical. Solutions obtained by using the Hertz–Huber theory are valid only for linear elastic behavior and small deformations. Nevertheless, important information results from them and is discussed in the following.

Beneath the contact surface, a region exists in which only compressive and shear forces are found. The maximum compressive stress p_m occurs at the center of the contact surface, whereas the maximum tensile stress σ_m occurs at its perimeter, and the maximum shear stress τ_m along the radius at a distance roughly one-quarter the diameter of the contact surface (Fig. 2). These stresses have the following ratio:

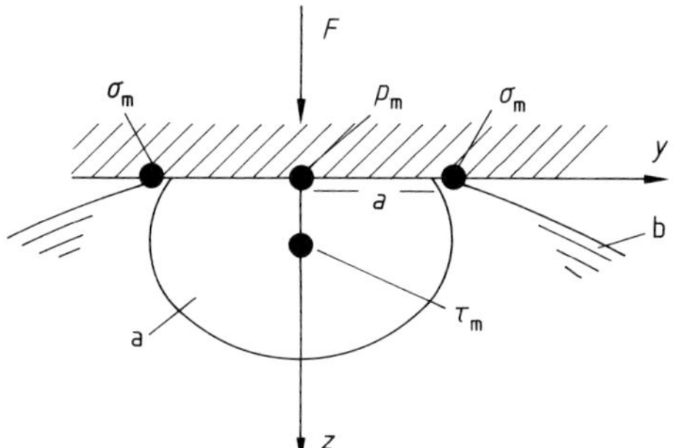

Figure 2. Explanation of the results of the Hertz–Huber theory for the contact region of a spherical particle
a) Region of shear and compressive stress; b) Surface of stressed particle
a = radius of the contact surface; F = applied force; p_m = maximum compressive stress; σ_m = maximum tensile stress; τ_m = maximum shear stress

$$p_m : \tau_m : \sigma_m = 1 : 0.3 : 0.17$$

This ratio is a function of the Poisson number ν, and the above values are for $\nu = 0.25$.

Brittle materials exhibit primarily elastic behavior up to the breaking point. Here, cracking starts in the area of tensile stress at the perimeter of the contact area. During compression, cracks extend between the contact areas (Fig. 3 B). During impacting, cracks diverge within the particle (Fig. 3A) [19].

Despite the existence of triaxial compressive stresses in the contact region, most of the fines are formed here. Since the compressive stresses are not equal, the cracks may trigger fragmentation. Theoretical studies have shown that a compact fracture field, and thus fines, are formed at high compressive stress [19].

By contrast, in plastic behavior, a conical region is pushed into the sphere. Sideways displacement of material produces peripheral tensile stresses and cracks in meridian planes (Fig. 4). Because stresses are determined by the depth of cone penetration, additional shear stress does not significantly affect the fracturing process. Thus, additional frictional stress has no effect, it leads to increased energy losses. The idea that friction is an advantage must be viewed

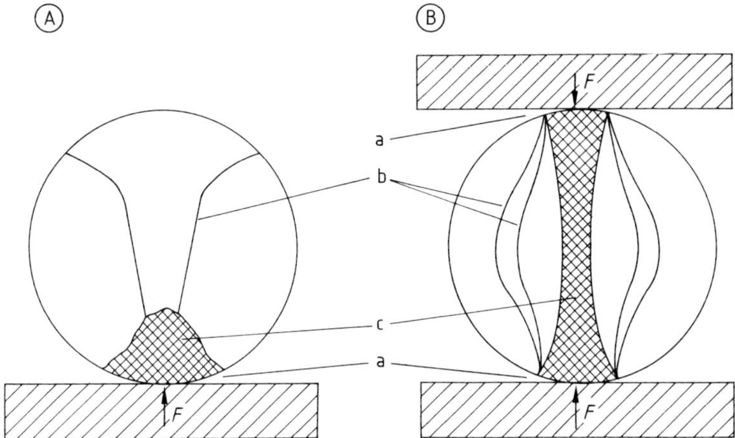

Figure 3. Fracturing in brittle glass spheres subjected to impact (A) and compression (B) loading
a) Contact surface at which stress is applied; b) Cracks; c) Fines

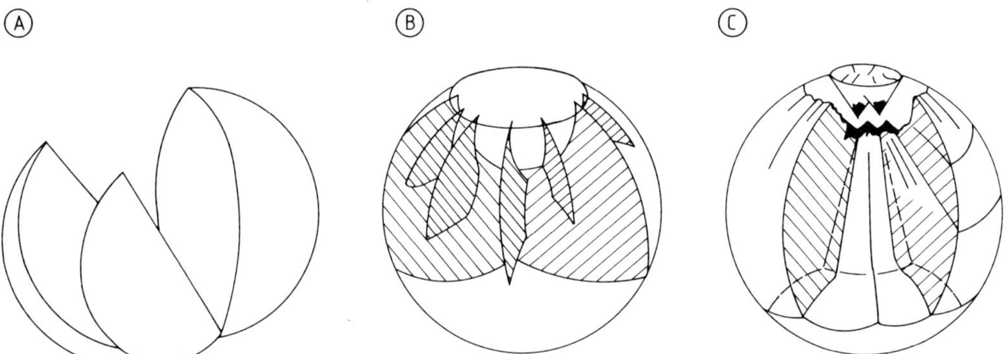

Figure 4. Meridian cracks in inelastically deforming poly(methyl methacrylate) spheres
A) Slow compression at 20 °C; B) Impacting at 90 m/s and 100 °C; C) Impacting at 90 m/s and −196 °C

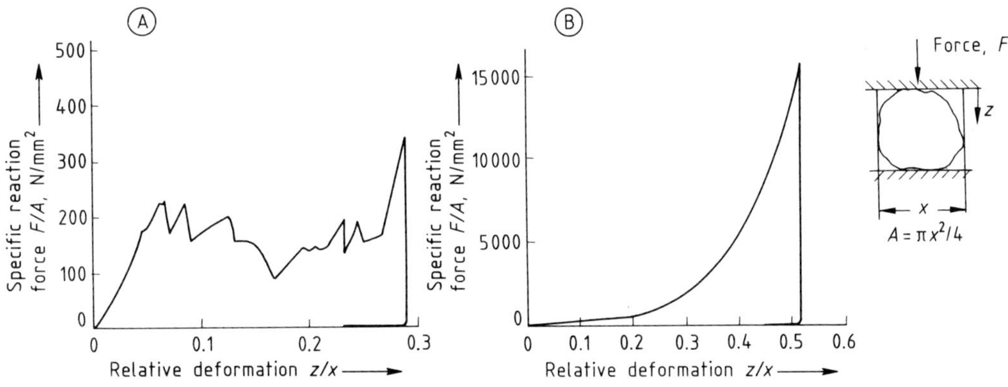

Figure 5. Loading diagram for cement clinker particles with a diameter of 16.1 μm (A) and 2.5 μm (B)

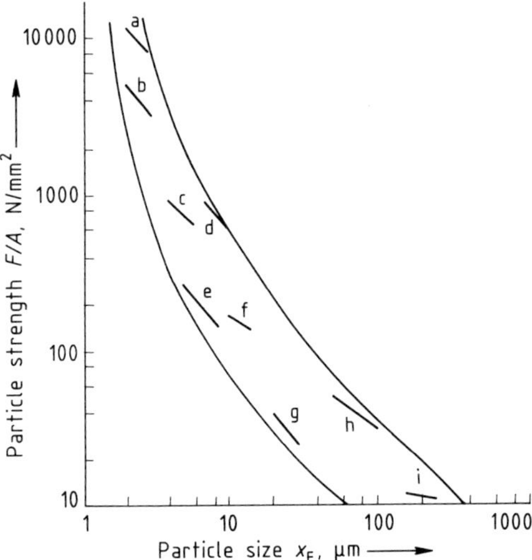

Figure 6. Particle strength of various materials in the transition zone between brittle and plastic deformation
a) Boron carbide; b) Crystalline boron; c) Quartz; d) Cement clinker; e) Limestone; f) Marble; g) Coal; h) Cane sugar; i) Potash salts

with skepticism. However, frictional stresses are useful for the disintegration of agglomerates.

Smaller particles contain only small flaws. Thus, larger stresses are required to reach the yield stress beneath the contact area and initiate fracturing. Deformation behavior, therefore, depends on particle size. For a brittle material, a transition from brittle to plastic behavior is observed in a defined particle-size range. These transitions are evident in loading diagrams. Multiple peaks indicate fracturing and brittle behavior (Fig. 5A), whereas smooth curves indicate plastic deformation (Fig. 5B). Figure 6 shows the transition zones between brittle and plastic behavior for various materials. Corresponding particle strengths are also given and are defined as the breaking force divided by the nominal particle cross section. If the contact surface is assumed to be one-tenth of the nominal cross section, the resulting stresses are the same order of magnitude as the molecular bond strength. Reduction of these materials to sizes below this transition region is very difficult.

All solids display *material flaws* such as crystal dislocations, grain boundaries, voids, and cracks which lead to an increase in stress when a load is applied. Consequently, they are the starting point for inelastic deformation and fracturing. The material flaws within a particle largely determine its reaction and, thus, its strength, breaking energy, and fragment-size distribution. The increased stress also explains why the observed particle strength is far below the theoretical or molecular strength (which is usually one- to two-tenths of Young's modulus). The increased stress at the tip of the cracks can be estimated. According to the notch tension theory, for elastic behavior with a uniaxial tensile stress perpendicular to the crack the following applies (Fig. 7):

$$\sigma_m/\sigma = 1 + 2\sqrt{a/r}$$

where σ is the stress in the unperturbed state, a is half the crack length l, and r is the radius of curvature of the tip of the crack. Substitution of typical values (e.g., $a = 1$ µm, $r = 1$ nm, $\sigma = 100$ N/mm^2) gives very high stresses that exceed the yield stress. This leads to the conclusion that a zone of inelastic deformation, called the fracture zone, forms around the tip of the flaw and later around the tip of the crack.

Fracturing involves *energetic considerations*. Energy is required for the inelastic deformation occurring in the fracture zone, for the electronic processes involved in material separation, and for the creation of new surfaces. The required energy, relative to the fracture surface area A_B ($A_B = B \cdot l$, where B is the thickness of the material, see Fig. 7), is termed the *crack resis-*

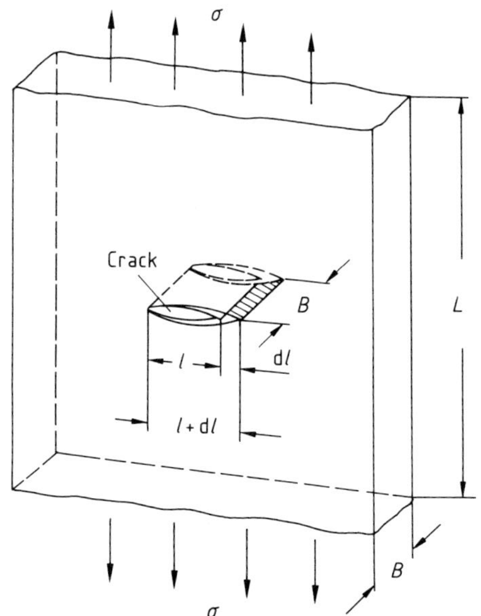

Figure 7. Schematic of a specimen with a crack produced by tensile stress

tance R and, at maximum crack propagation velocity, is a material constant.

The crack generally propagates so quickly that energy cannot be supplied externally; it can only come from the stress field. To determine the energy supply from the stress field, the decrease in elastic energy of the stress field $\mathrm{d}U_{\mathrm{el}}$ caused by a differential increase in crack length $\mathrm{d}l$ is considered

$$G = -\,\mathrm{d}U_{\mathrm{el}}/\mathrm{d}A_{\mathrm{B}}$$

where G is the *energy release rate*. The fracture surface A_{B} corresponds to half the total surface area produced by the fracture. The parameter G is a function of the specimen shape, the position of the crack, the type of loading, and the mechanical properties of the material. Theoretical solutions for calculating G are available for simple cases. For a crack or fracture of length $l = 2a$, which propagates perpendicular to a tensile stress in an infinitely large plate (Fig. 7) at a velocity less than one-tenth the velocity of the longitudinal wave, the following applies:

$$G = \pi a \sigma^2/E$$

where E is Young's modulus. The energy release rate increases with crack length, provided the stress remains constant; this is always the case for relatively large specimens.

The expression $G \geq R$ represents the differential energy balance. The parameter G can be calculated from measured values of a and σ. In addition, the heat value of the fracturing process can be measured and used to determine the inelastic deformation energy. These values agree within a few percent, which implies that the fracturing energy is primarily determined by inelastic deformation. For glass and similar materials, the crack resistance ranges from 1 to 10 J/m^2; for plastics, it is 10–10^3 J/m^2; and for metals, 10^2–10^5 J/m^2. The specific surface energy of the newly created surface is, however, only 0.01–0.5 J/m^2. These figures clearly show that changing the surface energy via the medium surrounding the solid, for example with *grinding aids*, cannot have any effect on the fracturing process. Effective aids for very fine grinding in ball mills improve processing conditions by preventing agglomerate formation and adhesion of the ground material.

A *minimum specimen size* is necessary to ensure that a fracture crosses the entire particle. This can be estimated by using a cylindrical model of cross section A and length L. Fracturing starts at the breaking stress σ_{B}. Thus, the integral energy balance is

$$(\sigma_{\mathrm{B}}^2/2E)\, A \cdot L > R \cdot A$$

$$L \geq 2\,R \cdot E/\sigma_{\mathrm{B}}^2$$

Typical values for the minimum specimen length of glass are 10–100 µm and of steel 2–20 mm. Therefore, in the case of fine particles (<100 µm), the integral energy balance is not met. Furthermore, a single contact stress produces several breaks rather than a single fracture. Therefore, to ensure the disintegration of fine particles, additional energy must be supplied after the initial fracture, or stress must be applied repeatedly.

The *length of the fracture zone* δ can be estimated if it is much smaller than the length of the crack. The following applies [20]:

$$\delta \approx 0.4\,R \cdot E/\sigma_{\mathrm{y}}^2$$

where σ_{y} is the yield stress. Values of δ for glass and brittle minerals are 1–10 nm and for brittle plastics 1–10 µm. The length of the fracture zone determines the minimum separation between fractures and, consequently, the size of the smallest fragments (ca. 5–10 δ).

Inelastic deformation causes an *increase in temperature*. At the maximum crack propagation velocity (1000–2000 m/s for glass, 500–800 m/s for plastics), a sudden temperature change occurs. The maximum temperature increase can be calculated from adiabatic considerations:

$$\Delta T_{\mathrm{m}} = Q/\delta^3\,\varrho\,c$$

where Q is the heat produced during fracturing, ϱ the density, and c the specific heat. Measurements give the following values; 4000 K for quartz, 3000 K for glass, 1300 K for limestone, 200 K for sodium chloride, and 150 K for poly(methyl methacrylate) [21].

Extreme conditions exist in the fracture zone. Structural changes (mechanical activation) can be expected, which may be either favorable or harmful to the product [22]. Different findings show a higher reactivity of the fracture surface (e.g., the amorphous structure of the fracture surfaces of crystalline quartz or sugar) and a decrease in the molecular mass of polymers. The original state of the fracture surface changes immediately after size reduction due to the action of the atmosphere. Microcracks and microstructures disappear. Gas adsorption measurements made immediately after size reduction yield substantially larger surface area values than later measurements. Values increase by the following factors: quartz 1.6, calcite 1.6, calcium fluoride 2.0, lithium fluoride 2.7, glass 2.7, sugar cane 3.1, and sodium chloride 6.4 [23].

1.2. Material Properties

To plan, predict, optimize, and control a process, knowledge of the relevant material properties is required. In the case of size reduction, this poses several difficulties. The material is not a continuous phase but a dispersion composed of a group of particles with various shapes and sizes. The energy applied to the group of particles reaches only a portion of them. The size of this fraction is unknown, and the type of energy transfer can be described only approximately. The reaction of a particle to an applied stress depends on its size and shape, as well as the conditions under which stress is applied. As described in Section 1.1 (Fig. 1), the effect of applied stress is not solely a function of the energy supplied.

The material properties relevant to size reduction can be divided into two groups:

1) Parameters describing the resistance to breakage: particle strength as well as the energy and probability of fracturing, and the breakage fraction
2) Parameters pertaining to the effect of the applied stress: breakage function (size distribution of the resulting fragments), increase in surface area, and energy utilization

The irregular shape of the particles and the effects of particle size cause special difficulties. The above parameters are defined in such a way that they can be measured and are of practical use. They cannot be derived from known material properties such as Young's modulus, tensile strength, yield strength, or hardness.

Material properties that are important during size reduction depend on the nature of the feed in the mill and the loading conditions.

Feed parameters are

1) type, origin, and prior treatment of material;
2) particle size;
3) particle shape; and
4) particle homogeneity.

Loading conditions include

1) type of applied stress as described by RUMPF [24]: stress can be applied between two surfaces (compression and shear stress), to a single surface (impacting), via the surrounding medium, or via nonmechanical energy transfer;
2) particle configuration: single particles or groups of particles;
3) stress intensity quantified by the energy of fracturing relative to volume or mass, the reaction force relative to surface area, and the reduction ratio (see pp. **5**-9–**5**-10);
4) loading velocity: in impacting, the energy supplied relative to the mass is a function of the impact velocity ($W_M = v^2/2$);
5) frequency of stress application: single or multiple applications;
6) shape and hardness of the surface of size-reduction tools: smooth or molded surfaces; and
7) temperature.

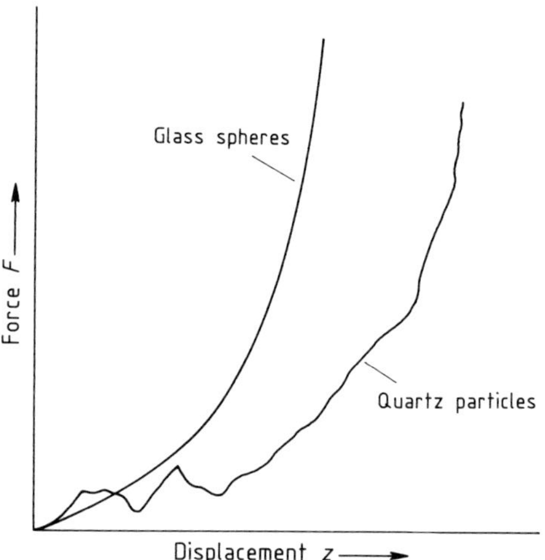

Figure 8. Loading diagram for glass spheres and quartz particles

In the usual strength tests, material breakdown occurs when the applied load reaches a critical value. This value can be used to derive a measurement of strength or energy of fracturing. However, the loading diagram of small, irregularly shaped particles displays numerous peaks indicating fractures (Fig. 8). Therefore, the first large peak must be defined as the breaking point.

Parameters at the Breaking Point. The particle strength and the energy of fracturing are derived from the breaking point. *Particle strength* is the force at the breaking point relative to the nominal cross section $\pi x_V^2/4$ (x_V is the

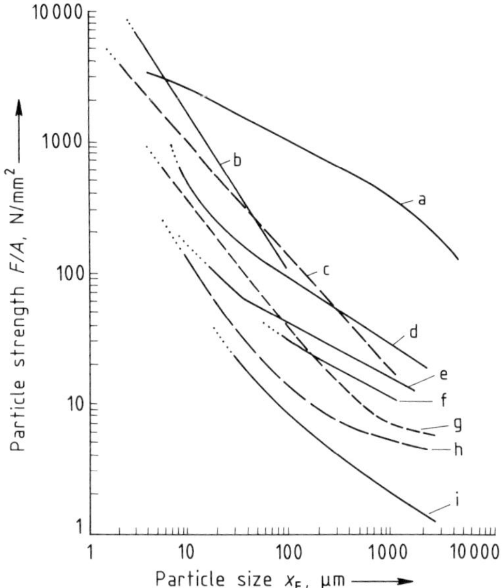

Figure 9. Particle strength of various materials as a function of particle size
a) Glass spheres; b) Boron carbide; c) Crystalline boron; d) Cement clinker; e) Marble; f) Cane sugar; g) Quartz; h) Limestone; i) Coal

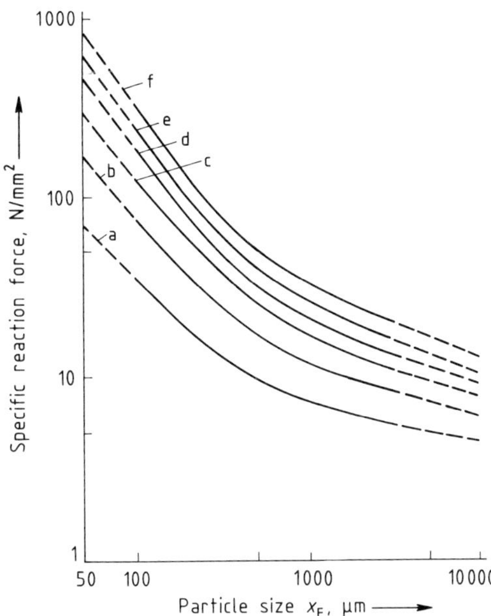

Figure 10. Specific reaction force of quartz particles as a function of particle size and reduction ratio
Reduction ratio, x_F/s (particle size divided by gap width): a) 1.5; b) 2; c) 2.5; d) 3; e) 3.5; f) 4

equivalent diameter of a sphere which has the same volume as the particle in question). The *energy of fracturing* E_B is the energy supplied up to the breaking point, relative to the mass of the particle.

Although particle strength has the same dimensions as stress, it does not represent the stress on the particle because the contact area is unknown. Figure 9 shows particle strengths for various materials [25]. At particle sizes below several millimeters, the strength increases sharply because as the particle size decreases, the material flaws become smaller and the particles more homogeneous. Because the crack resistance remains constant, the differential energy balance necessitates that increasing stress be applied to initiate fracturing (see Section 1.1, p. **5**-7). However this leads to considerable deformation because the compressive and shear stresses in the contact region are much higher than the tensile stresses that initiate fracturing. As a result, the tensile stresses increase only slightly or may even be prevented from increasing.

Parameters at Loads above the Breaking Point. In contrast to the customary strength tests, energy can be supplied to a particle under compression or impacting at levels that exceed the breaking point. This yields a larger number of finer particles. In the case of compression, both the particles and their fragments exert a reaction force against energy transfer, which is related to the nominal cross section.

To distinguish parameters above the breaking point from those related to the breaking point, the terms *specific reaction force* (measured force relative to the nominal cross section) and *specific work of comminution* W_M (work done relative to the mass) are used. The parameters at loads above the breaking point are not resistance parameters but are useful, measurable quantities. For example, the *reduction ratio* x_F/s (feed particle size divided by gap width) provides a useful standard when compression is employed. Reaction force and work of comminution can be measured as functions of the reduction ratio. An example is shown in Figure 10, which again displays the typical increase of reaction force with decreasing particle size. Figure 11 compares the energy of fracturing W_B, measured by means of impacting, with the specific work of comminution at three reduction ratios for limestone (Fig. 11 A) and quartz (Fig. 11 B). To a first approximation, the work of comminution for re-

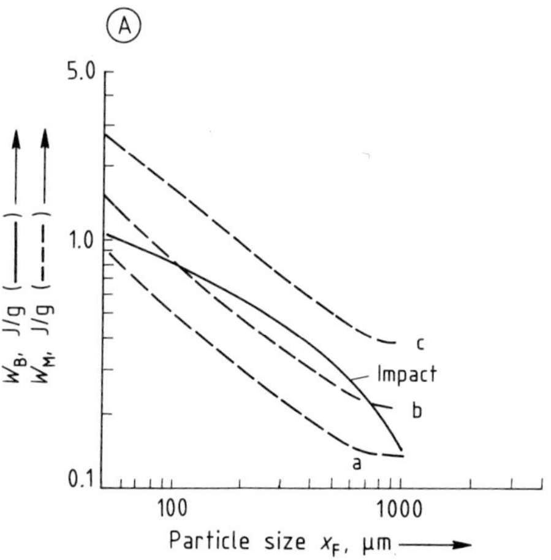

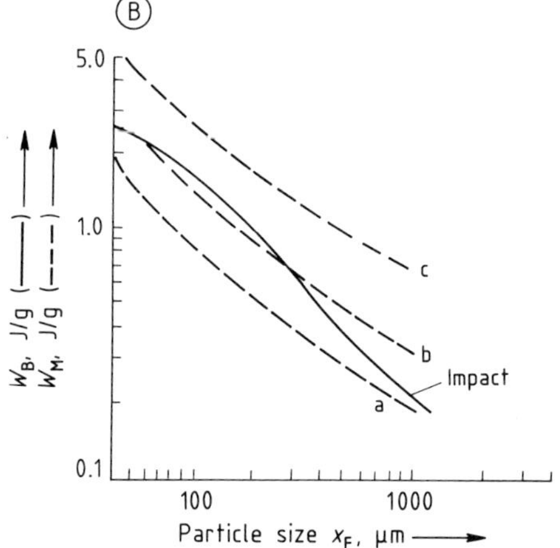

Figure 11. Energy of fracturing W_B and specific work of comminution W_M for quartz (A) and limestone (B)
Reduction ratio, x_F/s: a) 1.2; b) 1.3; c) 1.5

duction ratios of 1.2 – 1.3 is equal to the energy of fracturing. This is probably also valid for other brittle materials.

Probability of Fracturing. The probability of fracturing indicates the fraction of particles broken under a specific load. Determination of breakage by visual inspection is simple in the case of spheres, but more difficult in the case of irregularly shaped particles. Figure 12 shows the probability of fracturing of glass spheres. The increase in particle strength at lower particle sizes displaces the curves to the right, in the direction of higher impact velocities.

Breakage Fraction. The breakage fraction B_f can be considered a modified fracture probability for irregularly shaped particles. It is also a very informative resistance parameter for practical applications. A narrow particle-size fraction between $x_F - (\Delta x/2)$ and $x_F + (\Delta x/2)$ is subjected to stress by a specified method, and the mass of fragments M_p below the lower limit of the size range is determined; x_F is the feed particle size, defined as the mean value of the particle-size range Δx. The value of M_p depends on the particle-size range Δx. Extrapolation of the function M_p/M_0 (M_0 = total particle mass) to $\Delta x = 0$ gives the breakage fraction:

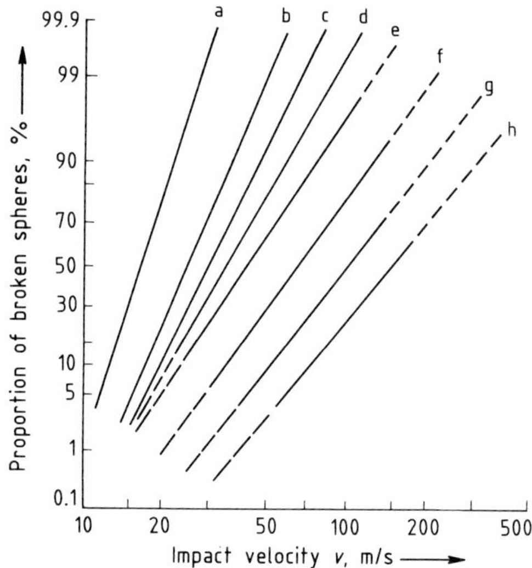

Figure 12. Probability of fracturing for glass spheres
Particle size x_F, mm: a) 8; b) 4; c) 2.3; d) 1.6; e) 1.2; f) 0.5; g) 0.2; h) 0.1

$$B_f = \lim_{\Delta x \to 0} (M_p/M_0) = \lim_{\Delta x \to 0} f(x_F, \Delta x, W_M)$$

The breakage fraction is obtained with sufficient accuracy with the aid of fractions obtained by using standard laboratory screens whose mesh sizes decrease by a factor of $\sqrt{2}$. Theoretical consideration of crack initiation based on (1) the Hertz theory describing stress in spheres and (2) Weibull statistics for the distribution of flaws shows that the dependence of the breakage probability on the specific work requirement obeys a Rosin – Rammler – Sperling – Bennett distribution (→ 2. Particle Size Analysis, p. **2**-6) [26]. Comparison with experimental results confirms that this relationship also holds for a wide range of irregular particles. Empirical studies show that the dependence of the breakage probability on the work requirement approximates to a normal, multiparameter, logarithmic distribution [27].

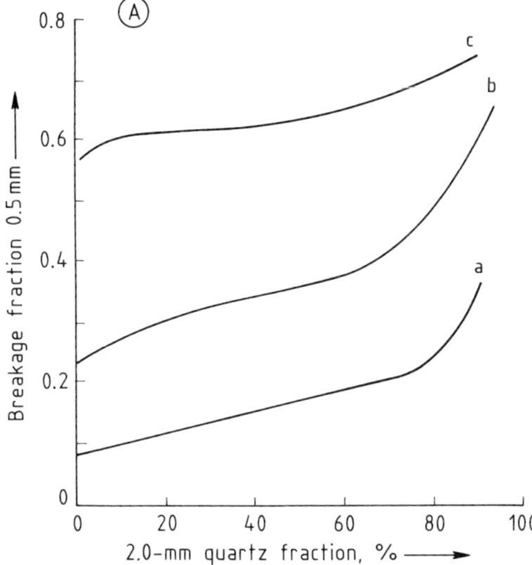

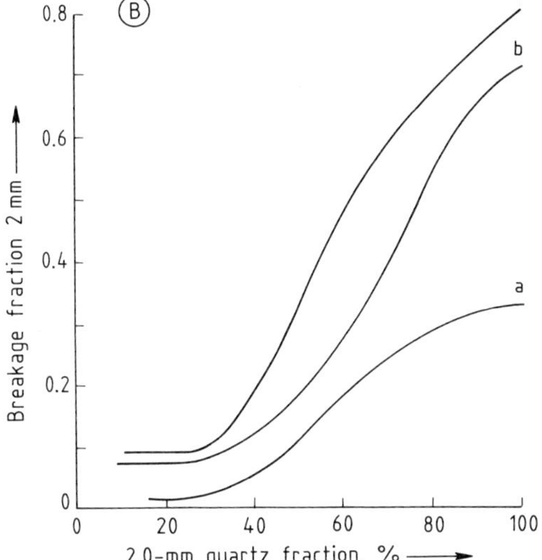

Figure 13. Breakage fractions of 0.5-mm particles (A) and 2.0-mm particles (B) in binary mixtures of 0.5-mm and 2.0-mm quartz fractions
Compression, MPa: a) 14; b) 28; c) 56

The breakage fraction in a stressed bed of particles depends on the particle-size distribution and other factors [28]. Figure 13 shows results obtained with a binary mixture of quartz particles. The addition of coarse particles increases the breakage fraction of fine particles and vice versa. The coarse particles concentrate the force flux on the smaller ones, whereas the fine particles distribute the force flux over the larger surface of the coarse particles.

Resistance to Grinding (Grindability). The specific work of comminution required to achieve a defined degree of fineness in a given process is termed the resistance to grinding. The Bond index is an example of such a parameter (see Section 1.3, p. **5**-14). The applicability of these parameters to other grinding processes is limited.

Breakage Function. The most important result of size reduction is the size distribution of the fragments. The fragment size distribution is referred to as the breakage function when the feed consists of particles of equal size or a narrow size fraction. The breakage function is determined by applying stress to single particles or particle beds; it is also determined by grinding in size-reduction equipment.

To determine the breakage function, stress is applied under predetermined conditions to narrow size fractions, and the resulting particle-size distribution Q is determined. The term $Q(x_F - \Delta x/2)$ gives the mass fraction below the lower limit of the size class Δx. The breakage function B results from the following limiting expression:

$$B(x, x_F) = \lim_{\Delta x \to 0} [Q(x, x_F)/Q(x_F - \Delta x/2, x_F)]$$

The width of a fraction corresponding to a $\sqrt{2}$ relation is sufficiently small. The breakage function depends on the specific work of comminution and other loading conditions:

$$B = B(x, x_F, W_M, \ldots)$$

The breakage functions for impacting and compression are different, as explained in Section 1.1 (p. **5**-4). However, experimental results indicate that they may be the same in the fine particle range, i.e., when $x < 0.1\ x_F$. These fine fragments are formed primarily in the contact region where conditions are the same for both types of applied stress.

When individual particles are compressed in the size range $0.05\ \text{mm} < x_F < 5\ \text{mm}$ and the fragment size x is expressed relative to the initial particle size x_F, the breakage function depends on the reduction ratio (see Fig. 14):

$$B(x, x_F, W_M) = B(x/x_F, x_F/s)$$

The breakage function depends principally on the supplied energy and, thus, on the specific work of comminution. Generally, in a bed of

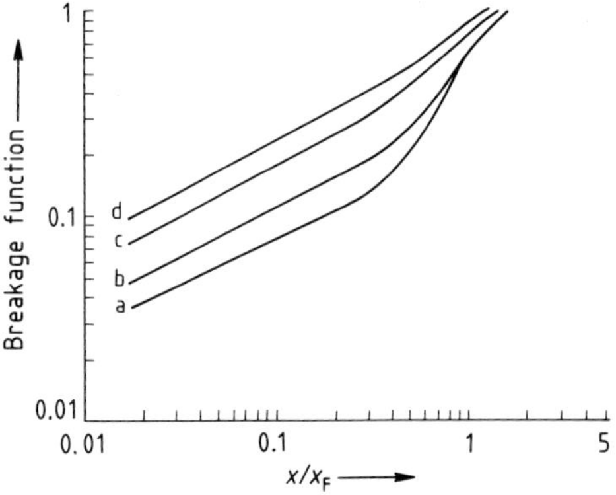

Figure 14. Breakage function of single particles of limestone subjected to compressive stress
Reduction ratio, x_F/s: a) 1.5; b) 2; c) 3; d) 4

particles, only energy up to the energy of fracturing can be transmitted to a particle when an increased load is applied. This is because the fragments can avoid additional stress in the voids as long as the solid fraction remains below 60–65%. In this case, the breakage function is not dependent, or only slightly dependent, on the energy supplied to the particle bed; on the other hand, the breakage fraction increases. Therefore, in particle beds a material-dependent breakage function is always determined, which is referred to as the *natural breakage function*. This does not apply to impact and jet mills, in which stress is applied to individual particles.

Repeated attempts have been made to describe the breakage function mathematically. However, generally valid functions still have to be found. For example, SCHUBERT and coworkers stress that the breakage function can always be represented as the sum of a three-parameteric, logarithmic probability distribution [27].

Energy Utilization. The quotient of the newly created surface area ΔS and the work supplied W is called the energy utilization E_u:

$$E_u = \Delta S/W = S_M/W_M$$

Because the value of the specific surface area S_M depends on the method of measurement, this parameter must be specified. The gas adsorption method (BET method, → 9. Adsorption, **B3**, **9**-14) is not suitable because it also measures internal surface area and surface microstructures. More suitable methods include measuring the permeability of the bulk particles (Blaine method), the photoextinction method, and calculation from the particle-size distribution (→ 2. Particle Size Analysis, p. **2**-7).

Figure 15 illustrates results obtained for the size reduction of individual cement clinker particles. The increase in specific surface area is plotted against the specific work of comminution. Particle size and reduction ratio are variable parameters. The straight line with a gradient of 1 corresponds to constant energy utilization. The measured curves have a smaller gradient, which means that energy utilization decreases as more energy is supplied. This has also been found to be true for size reduction in a particle bed [29].

Because of this phenomenon, size-reduction equipment combined with a downstream classifier must operate at minimal stress levels with multiple applications of stress. However, this means that the amount of final product decreases and the amount that must be recirculated to the

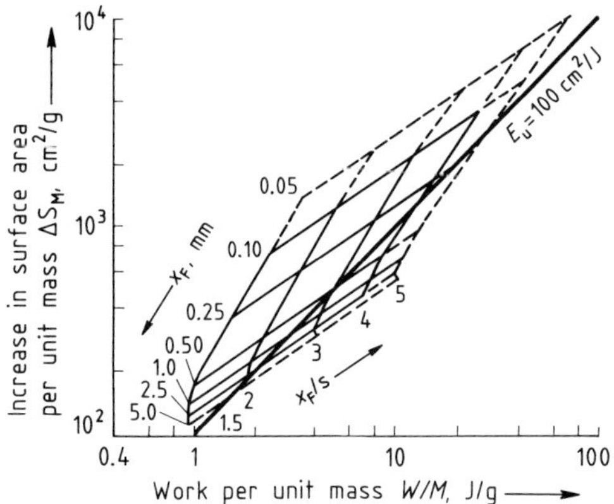

Figure 15. Dependence of surface area on specific work of comminution W_M for size reduction of individual cement clinker particles

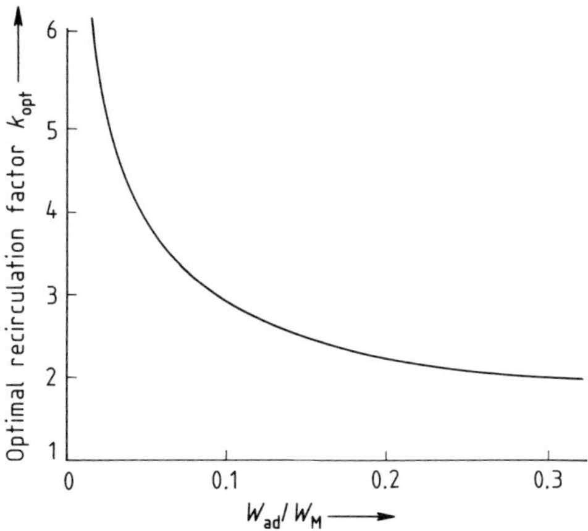

Figure 16. The optimum recirculation factor k_{opt} as a function of the energy required by additional process steps W_{ad}

mill increases. If the total energy requirement (milling, classifying, transporting) is taken into account, an optimal recirculation factor k_{opt} can be obtained (where k_{opt} is the ratio of the mill and product throughputs) [30]. The optimal recirculating level depends on the ratio of the specific energy required for the additional process steps (W_{ad}) to the consumption energy of the mill (W_M) (Fig. 16). Calculations also show that deviations from the optimal recirculating level of ±20% lead to an additional total energy consumption of ca. 2%. At larger deviations, the energy requirements increase significantly.

Figure 15 also shows that energy utilization is higher with smaller particles than with larger ones. This result is at first surprising, because the smaller particles are stronger and require more energy when comminuted in size-reduction equipment. However, this is understandable on a physical basis because the stored stress energy can be utilized more efficiently by smaller particles. Cement clinker particles have an energy utilization of 100–200 cm^2/J when stress is applied to individual particles. In a ball mill, values are considerably lower, 20–40 cm^2/J. The energy utilization of compression and impacting is compared in Figure 17. Compression shows better energy utilization. However, this is true only for brittle materials, not for viscous substances. Viscous materials cannot be broken down by compression; they must be impacted.

If application of stress to individual particles is assumed to be the most energy-efficient method of size reduction, the efficiency of other size-reduction processes can be judged by comparison with results from single particle tests [31]. The efficiency is the energy requirement of the ideal process divided by that of the size-reduction machine under the assumption that the products in both cases have the same particle-size distribution.

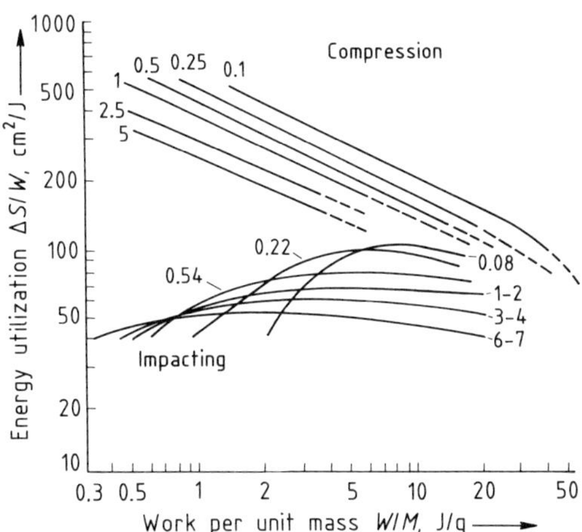

Figure 17. Energy utilization of limestone particles
Values on the curves indicate particle size x_F in millimeters.

The ideal process is defined as a multistage sequence in which stress is applied to single particles, it can be described by a mathematical model [30]. The model requires experimental determination of the breakage function, the breakage fraction, and the energy required to comminute the various particle sizes under different loads. Such data exist only for a few materials. The following standard values for the efficiency of size-reduction equipment are derived from currently available data:

Jaw and roll crushers	0.7–0.9
Impact crushers	0.3–0.4
Roller-ring mills	0.007–0.15
Ball mills	0.05–0.1
Impact mills	0.01–0.1

The poor values for roller-ring and ball mills are due to the fact that stress is applied to a bed of particles. Furthermore, in ball mills the stress is applied stochastically and with a wide variation in intensity. Roller-ring mills require energy for extensive internal circulation of the feed. Impact mills are not very efficient because of the energy needed for ventilation and the limited loading attainable in fine grinding.

1.3. Process Description

An attempt is made to relate the particle size to the energy consumption with three simple relationships, which are referred to as the laws of comminution.

In 1867, RITTINGER postulated that the increase in surface area is proportional to the energy consumption and that the energy utilization (the ratio of these two quantities) remains constant (Rittinger's hypothesis). Based on more recent results with the fracturing process, RUMPF [32] showed that in the case of a geometrically similar distribution of cracks, the product ($W_M \cdot x$) is independent of particle size, and geometrically similar fracture surfaces are formed. The energy utilization is then a constant as postulated by RITTINGER. In 1885, KICK described a model system which assumed that strength is constant and that the propagation of fracture surfaces is geometrically similar. The energy utilization was concluded to be inversely proportional to the particle size x (Kick's law of similarity). Results obtained from single-particle tests confirm this but also show that strength depends on particle size. BOND [33] developed the following relationship:

$$W_M = W_i(10/\sqrt{x_{80,p}} - 10/\sqrt{x_{80,F}})$$

where $x_{80,F}$ and $x_{80,p}$ are the particle sizes of the feed and product in micrometers for 80% cumulative undersize. The parameter W_i is a measure of grindability (Bond index). The Bond index W_i corresponds to the specific work required to grind from a theoretical infinite size to 80% passing 100 µm, i.e., $x_{80} = 100$ µm. Typical values for W_i (in kW · h/t) are: corundum 64; basalt 22; iron ore 16–20; cement clinker 15–18, feldspar, limestone, and quartz 13–15; lead and chrome ores 11–13. Additional values are given in [7], [10]. Grindability is determined in a laboratory mill of specified size, according to a set procedure. The Bond relationship permits a first approximation for the design of ball mills; a number of correction factors give better agreement with actual values [33].

Mathematical modeling allows a better description of the size-reduction process. A sucessful model provides answers to the following questions:

1) What state of dispersity can be attained for known material properties and types of stress?
2) How can experimental results from test apparatus or laboratory-scale mills be scaled up to production plants [34]?
3) From what viewpoint should new mills and plans be designed [35]?
4) What effect do various parameters have on the results of size reduction [36], [37]?
5) How can a plant be controlled?

Extensive development of methods for the mathematical modeling of size-reduction processes began in the late 1960 s. A detailed description is not given here; only the general concept is presented.

The most common method of mathematical modeling, the *population balance model*, describes the change in particle-size distribution with time. To do this, the complete particle-size distribution is divided into fractions (intervals). A mass balance is set up for the individual fractions, and the decrease in mass for each fraction is assumed to be proportional to the mass already present in that fraction. The parameter m_i is the mass fraction of particle-size interval i and w_i is the specific rate of breakage. The term b_{ij} is the breakage distribution parameter of the fragments, i.e., the mass fraction of fragments from fraction j that move into size fraction i. The stepwise size-reduction process is described by a system of coupled first-order differential equations:

$$\mathrm{d}m_i/\mathrm{d}t = -w_i m_i + \sum_{j=1}^{i-1} b_{ij} w_j m_j;\ i = 1, 2, \ldots, n$$

where the class with the coarsest particles is denoted by the index 1. The first term $(-w_i m_i)$ describes the decrease in mass produced by size reduction, and the second describes the increase in mass due to fragments from coarser fractions. If w_i and b_{ij} are constant and $w_i \neq w_j$, this system of equations can be solved explicitly [38]. The problem in mathematical modeling is the determination of the size-reduction coefficients w_j and b_{ij}. For example, dividing the particle-size distribution into ten fractions requires 45 coefficients. Experimental determination is very time-consuming. Usually, the desired parameters are estimated from a small number of measurements (back-calculation method) [39]–[41]. The size-reduction coefficients, especially size-reduction rates, are not constant, e.g., in wet milling in ball mills [42]. On the other hand, the distribution (transfer) coefficients are less strongly dependent on milling conditions.

The performance of continuously operated plants and milling circuits can also be simulated by mathematical modeling. More complex methods are required because transport of particles through the mill must also be considered. The residence-time distribution or the convection–dispersion model is used [39], [43].

2. Size-Reduction Equipment

The first major step in the development of size-reduction equipment was the utilization of steam power to drive crushers and mills. Development proceeded independently for different kinds of materials (e.g., crushers for road and railroad construction, comminution equipment for ore preparation, and mills for processing grain). New types of machines were invented. Subsequent developments were an increase in equipment size and the utilization of new materials of construction. New methods of construction (e.g., welding instead of casting), and new fabrication techniques (e.g., flame cutting) were also introduced. A major advance in improving energy utilization was the introduction of classification and the development of efficient air classifiers capable of performing sharp separations (→ 17. Air Classifying). Recent developments have been largely determined by emphasis on efficient energy utilization. However, this has not led to the development of new types of mills. Instead, existing types of mills have been used with new feed materials and new processing techniques [44]–[49].

Size-reduction equipment cannot be operated without auxiliary devices, such as bunkers, feeders, feed-control and conveying equipment, ventilators, pipes, classifiers, and dust separators. This auxiliary equipment is often responsible for weaknesses and disturbances in size-reduction plants. Additional areas to which attention must be paid include protection against explosions (dust explosion) and, above all, against wear [50].

Size reduction is always associated with wear. In some cases, wear can be minimized, for example, by selecting suitable materials or by modifying the grinding method and the intensity of stress application. However, wear is often intentionally concentrated in one area of the mill, which is then equipped with easily changeable parts. Typical values for wear are 0.5–5.0 g/kW · h for dry grinding and 10–80 g/kW · h for wet grinding of hard materials. (Wear is expressed as the loss in gram per tonne of material passing through the mill relative to the specific work of comminution in kilowatt hours per tonne.)

A comprehensive, detailed description of size-reduction equipment is to be found in [7]. Aspects such as sturdiness, bearings, lubrication, power supply, overload protection, and other design fundamentals are covered. Information on size of construction and operating parameters such as speed of rotation, throughput and power consumption is also given.

A broad selection of different types of equipment is available as a result of historical developments and the wide range of applications based on particle size, throughput, and material properties. No uniform criteria are available for classifying size-reduction equipment. Furthermore, the system used for naming machines is not standardized, so that a particular type of machine may have several names. The classification scheme used in the following sections is based on particle size as well as similarities between process types and machine construction.

A clearly defined boundary between crushers and mills does not exist. The term *crusher* is applied to devices which perform coarse size reduction. The term *mill* describes machines which perform fine size reduction where the particle size of the feed is a few centimeters or less.

2.1. Crushers

Crushers can process particles up to 1.5 m in edge length. They are used in mines and quarries; for ore preparation; and for the production of road construction materials (ballast and fine gravel) and raw materials for cement. Their function is to produce particles ready for transport or to perform a primary reduction step prior to subsequent comminution. A recent application is the recycling of rubble, road surfacing, and metal turnings or scrap.

The use of conveyor belts in mines means that the crusher must be located at the site where mining occurs. This has led to the development of portable or movable crushing installations. The crushers weigh up to 700 t and are equipped with tires, rails, caterpillar treads, or hydraulic advancing mechanisms.

Crushers can be classified into four types: jaw crushers, gyratory (cone) crushers, roll crushers, and impact crushers.

2.1.1. Jaw Crushers

Jaw crushers compress the feed between a stationary and a movable surface.

Double-Toggle Crusher. The double-toggle crusher (Fig. 18) was invented by E. W. Blake in

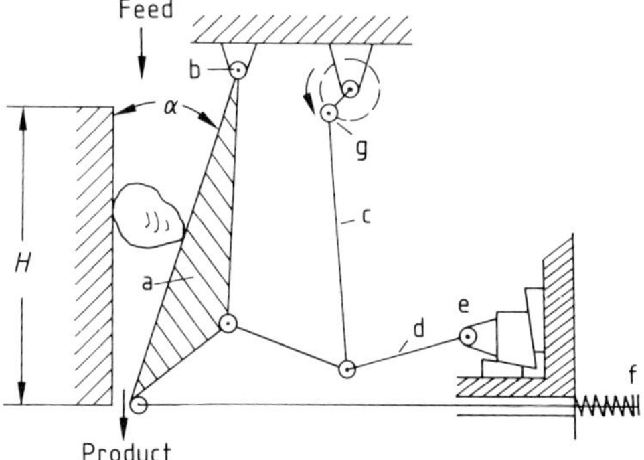

Figure 18. Schematic of a double-toggle jaw crusher
a) Movable jaw; b) Jaw bearing; c) Lifting lever; d) Pressure (toggle) plate; e) Toggle-plate bearing; f) Spring; g) Eccentric

1858 and is also known as the *Blake jaw crusher*. The movable jaw (a) is pivoted at point (b) and stress is applied to the particles in the crusher opening when the jaw moves. In order to smooth out the discontinuous energy requirements, large fly wheels are mounted on the eccentric shaft. Movement of the jaw is brought about by means of a system of bent levers equipped with a lifting lever (c) and pressure (toggle) plates (d) as well as an eccentric drive (g). A return spring (f) is provided to open the crusher discharge opening or to brace the toggle plate. The narrowest gap width can be changed slightly (by a factor of ca. 1.5) by moving the toggle-plate pivot (e).

The most important operating parameter is the crushing or nip angle α. It must be so small (14–22°) that the particles are held in the crusher cavity by friction when the load is applied. The *reduction ratio* (the ratio of feed particle size to product particle size, e.g., $x_{80,\mathrm{F}}:x_{80,\mathrm{P}}$ where x_{80} is the particle size for 80 % cumulative undersize) is determined by setting the angle α and the height of the crusher cavity H. Reduction ratios of up to 8 are achieved by jaw crushers. The smaller the angle α at a constant stroke, the greater is the distance that the particle may fall as the crusher cavity opens. This means that α affects the throughput of the crusher. Consequently, α has a small value in the region of the discharge opening of fine crushers because here the danger of plugging is particularly high.

Since material in the crusher chamber is transported by gravity, an optimal speed of rotation exists in relation to the throughput. At high rotation speeds, the movable jaw returns before the material has moved downward along the entire theoretical path.

Overhead Eccentric Jaw Crusher. In the overhead eccentric jaw crusher (Fig. 19), also called the *single-toggle crusher*, the movable jaw (a) is directly attached to the eccentric pivot (b). At its lower end, the jaw is supported by a pivot pin (c) above a toggle plate (d) which is attached to the pivot (e). Therefore, the motion of the jaw shows a significant tangential component compared to the Blake crusher. Although this is advantageous for moving the feed, it increases wear. Variable positioning of the toggle-plate pivot (e.g., $\beta <$ or $> 90°$, where β is the angle between the jaw and the toggle plate) allows the lift and the motion of the jaw in the exit aperture to be varied and adjusted to the type of crushing required (coarse or fine crushing with specially desired particle shapes). The single-toggle crusher is the simplest and lightest of the jaw crushers, but is suitable only for producing low crushing forces.

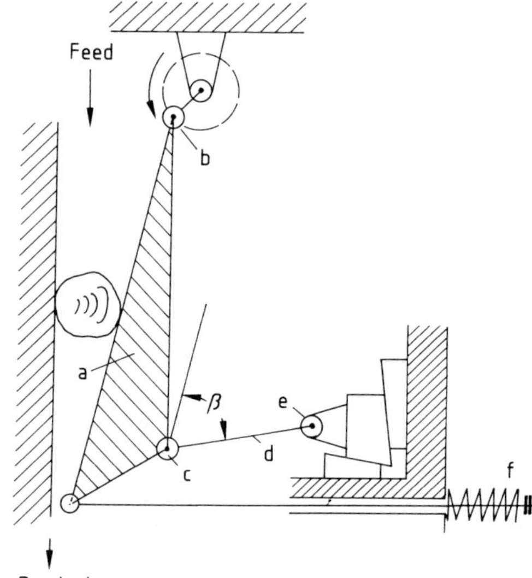

Figure 19. Schematic of an overhead eccentric jaw crusher
a) Movable jaw; b) Eccentric jaw bearing; c) Pivot pin; d) Toggle plate; e) Toggle-plate bearing; f) Spring

In both single- and double-toggle crushers, unbreakable objects may lodge in the crusher chamber. Therefore, they are equipped with overload protection devices which usually consist of a torque overload device built into the fly wheel. Hydropneumatic support of the toggle-plate bearing is sometimes used and has the advantages that the gap width can be easily adjusted or controlled and that the gap can be opened wide.

Impact Jaw Crushers. In an impact jaw crusher, the crusher cavity is inclined (Fig. 20). The

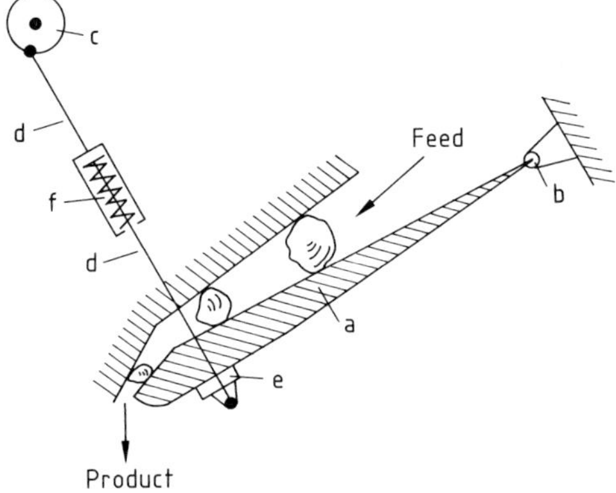

Figure 20. Schematic of an impact jaw crusher
a) Movable jaw; b) Jaw bearing: c) Eccentric; d) Connecting rod; e) Crossbar; f) Spring

jaw (a) below it is attached to the pivot pin (b) and is moved by two laterally mounted, split rods (d), which engage with the crossbar (e). The spring (f) between (c) and (d) serves as an overload protector. The rods are driven by the eccentric bearing (c).

As a result of the larger stroke and higher rotation speed (400 rpm compared to 200–250 rpm for other jaw crushers), stronger impact is achieved. As a result, hard, tough materials, such as ferrochrome or ferromolybdenum, can be processed.

2.1.2. Gyratory (Cone) Crushers

Stress is applied in a gyratory (cone) crusher in the same way as in the jaw crusher, i.e., between a stationary and a movable surface.

Primary Gyratory Crusher. In the primary gyratory crusher, stress is applied to the feed by pressure as the conical head periodically approaches the bowl (a) (Fig. 21). This periodic movement occurs because the shaft (b) of the crushing head (c) follows a conical surface, i.e., gyrates. The shaft is suspended from a bearing (d) and is guided along a circular path by the lower bearing (e). The angle between the cone axis and the vertical is ca. 1–3°. The bearing (e)

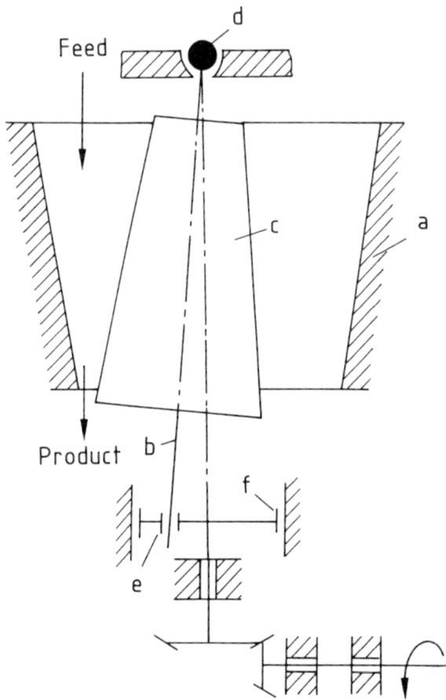

Figure 21. Schematic of a primary gyratory crusher
a) Bowl; b) Shaft; c) Head; d) Upper bearing; e) Lower bearing; f) Eccentric

Table 1. Applications of gyratory crushers [7]

Parameter	Primary gyratory crusher	Shallow cone crusher
Feed particle size x_F, mm	150–1800	25–300
Product particle size x_F, mm	25–250	5–40
Throughput $\dot{m}$, t/h	35–3500	10–600
Specific work of comminution W_M, kW · h/t	0.15–0.5	0.4–2.2
Type of applied stress	pressure	pressure/impact

is set in an eccentric hole in the lower bearing bushing (f). The narrowest gap between the head and the bowl rotates at the same rate as the head shaft; however, the head does not roll along the circumference of the bowl. The gap width can be varied slightly by axial displacement of the head, particularly if the crusher shaft is supported vertically by a hydropneumatic device, as is now common practice. This feature simultaneously serves as overload protection when, for example, a piece of iron enters the crusher. The ring-shaped form of the crushing chamber allows a constant power input and substantially higher throughputs than in jaw crushers.

The primary gyratory crusher is a large, heavy, and thus expensive machine. Consequently, it is only used for special materials and when a certain throughput is required (e.g., as a primary crusher at $>600\ m^3/h$). It is employed for the coarse and intermediate comminution of hard and moderately hard materials. Operating parameters are listed in Table 1.

Symons Cone Crusher. The shallow cone crusher, also called the *Symons cone crusher*, has a shallower crusher cavity than the primary gyratory crusher (Fig. 22). The crushing head (a) is supported from below (b), requiring less machine height. The design of the bowl (d) produces high reduction ratios up to 18. The relatively long, parallel gap preceding the discharge aperture ensures a uniform product particle size and good particle shape. The shallow cone crusher cannot handle such large feed particles as the primary gyratory crusher because the diameter of the discharge aperture is larger than that of the feed aperture. As in the impact jaw crusher, the stroke is larger and the speed of rotation higher than in the gyratory crusher. The stroke h is ca. 2–5 times the gap width s; the speed of

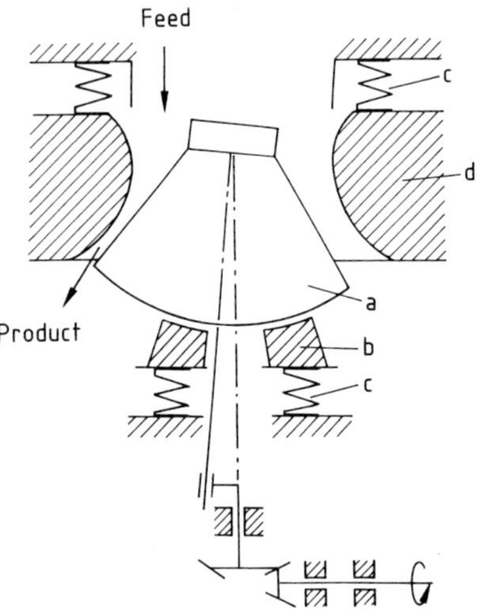

Figure 22. Schematic of a shallow cone crusher
a) Head; b) Bearing; c) Spring; d) Bowl

rotation is 200–300 rpm compared to 100–200 rpm for primary gyratory crushers. These values are favorable for material transport and give a better shaped particle, i.e., more cubic. Gap adjustment and overload protection can be provided by supporting the bowl (d) with springs or by supporting the bearing (b) with springs or a hydraulic device.

Shallow cone crushers are used mainly for the fine crushing of hard and moderately hard materials. Operating parameters are listed in Table 1.

Disk Crusher. The disk crusher (Fig. 23) was developed to overcome the disadvantages of the shallow cone crusher (i.e., plugging and low throughput) and is basically a centrifugal gyratory crusher. The crushing head (a) is extended to form a disk, resulting in a radial gap. The support (b) is driven at an angular velocity ω_1. A gear (c) ensures that the narrowest gap rotates with velocity ω_2. The feed falls onto the center of the rotating disk and is transported through the gap by centrifugal force. The stator disk (d) rotates and is driven either by friction in the gap or by a separate motor. To provide overload protection, the cover and the stator disk are supported with springs; they can both be moved axially to adjust the gap.

2.1.3. Roll Crushers

Double-Roll Crushers. Double-roll crushers consist of two rollers, which rotate toward each other and are separated by an adjustable gap (Fig. 24). One roller (a) is held by stationary bearings, the bearings of the other (b) are held by springs or a hydraulic system and can move along a linear or circular path. The material is fed in from above. The rollers, which may have a shaped surface, draw in the feed and apply stress to it. The crushers are designated as cam, toothed, gear, knob grooved, or helical according to the roll profile. The profile improves the drawing in of the feed and therefore helps to increase the reduction ratio. Rollers with diameters of up to 2 m can handle feed sizes up to 1 m. With soft to moderately hard materials, throughputs up to 3500 t/h may be obtained.

Both rollers usually rotate at the same rate, but they may be set at different speeds in the case

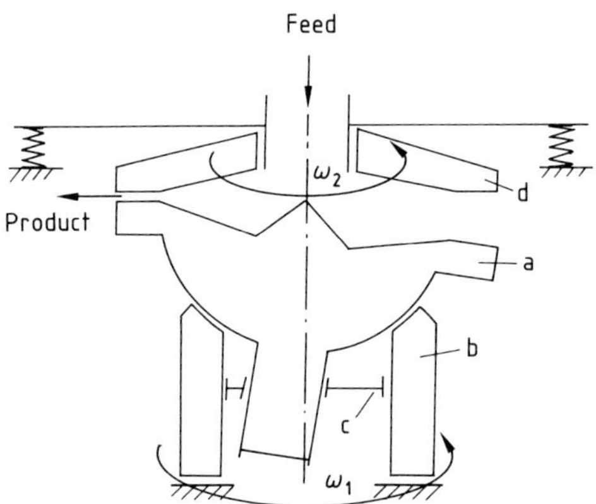

Figure 23. Schematic of a disk crusher
a) Crushing disk; b) Support for crushing disk; c) Gear; d) Stator disk

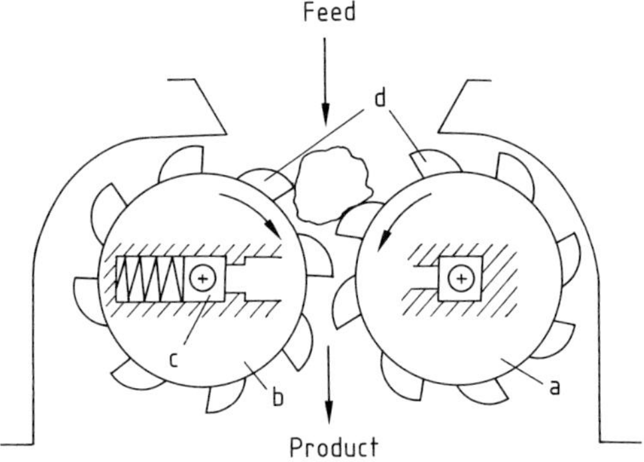

Figure 24. Schematic of a double-roll crusher
a) Roller held by stationary bearing; b) Roller held by movable bearing; c) Movable bearing; d) Tooth

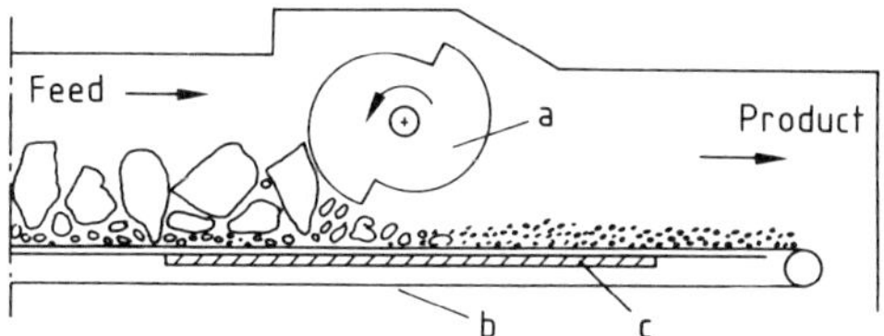

Figure 25. Schematic of a continuous single-roll crusher
a) Crushing roller; b) Mechanical conveyor; c) Support for conveyor

of soft, sticky, wet materials such as coal, lignite, peat, salt, clay, or pumice. This has the advantage of producing an additional shearing or cutting stress. The peripheral velocity of the roller controls the amount of material passing through the gap. For coarse crushing, values range up to 10 m/s.

Single-Roll Crusher. In the single-roll crusher, a shaped crushing roller acts against a crushing plate, which is held either by springs or a hydraulic system. This simple design is suitable for coarse crushing of moderately hard to soft materials.

In the *continuous crusher* (Fig. 25), the toothed roller (a) acts against a conveyor belt (b) rather than a crushing plate. This feature makes the crusher especially suitable as a mobile comminution device for use in mines and quarries. This crusher has long been used underground for coarse crushing of coal and salt. Recently, it has been used for the comminution of limestone and abrasive materials such as rubble.

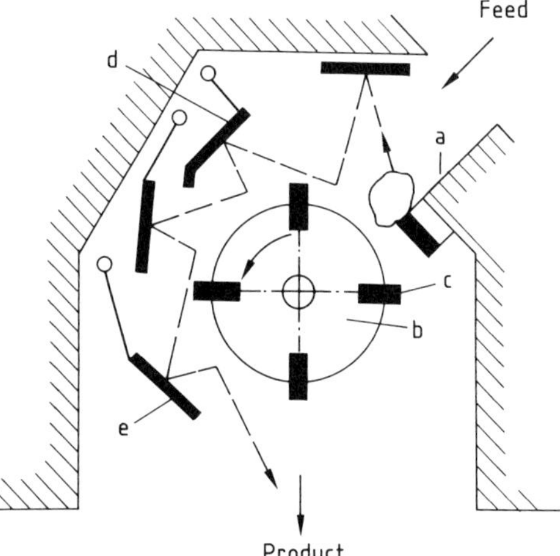

Figure 26. Schematic of an impact crusher
a) Feed chute; b) Rotor; c) Impact bar; d) Impact plate; e) Grinding path

2.1.4. Impact and Hammer Crushers

Impact Crushers. A typical impact crusher is depicted in Figure 26. The feed slides through the feed chute (a) and is hit by the impact bars (c), which are firmly attached to the rotor (b) and occupy its entire width. The particles and fragments are thrown against the impact faces or plates (d), and are then redirected to the impact bar zone. The impact plates are mounted flexibly, so that they can retract when a large piece of material enters the crusher. A grinding path (e) can also be included; it primarily applies additional stress to coarse particles and promotes a better particle shape. The crusher housing is open at the bottom. The reduction ratio and the particle size are determined by the peripheral speed of the rotor and not by the gap between the rotor and the impact path. Reduction ratios of 20 or higher are possible.

Hammer Crushers. In the case of hammer crushers (Fig. 27), impact devices (hammers) (a) are attached to the rotor via pivots so that they are deflected when they hit strong, particularly large particles. In most cases the crushing zone is surrounded by grate bars (b), so that fragments which are larger than the openings of the grating are retained in the crushing zone. Hammer crushers with rotor diameters up to 3 m are available and can crush ca. 1500 t/h of limestone.

Although hammer crushers wear more quickly than impact crushers, they can process moist materials more efficiently. Impact and hammer crushers operate with peripheral velocities between 20 and 50 m/s (maximum velocity 70 m/s). In principle, only soft to moderately

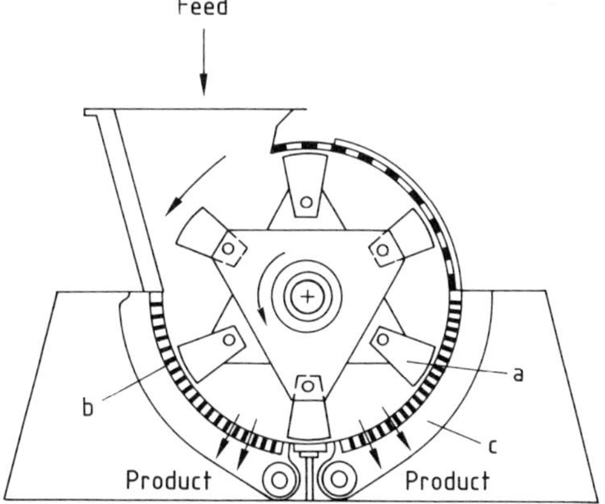

Figure 27. Schematic of a hammer crusher with a grating
a) Pivoted hammer; b) Grate bar; c) Support for grating

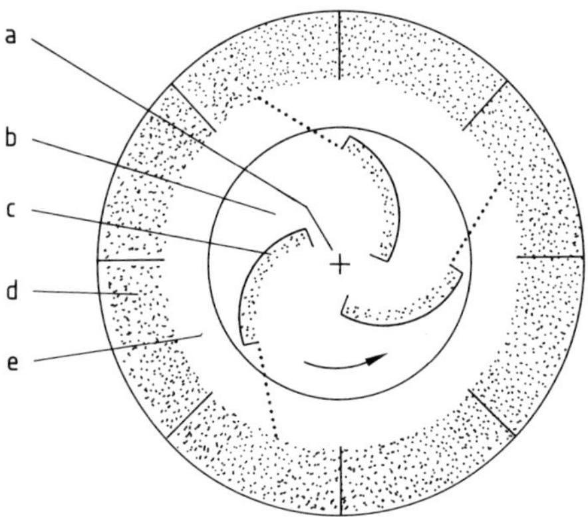

Figure 28. Schematic of a crusher with a vertical shaft viewed from above
a) Central feed inlet; b) Rotating disk; c) Acceleration bars with particle cushions; d) Fixed ring of particles; e) Product discharge

hard materials can be processed because of wear considerations at these high peripheral velocities. However, these crushers are simpler than jaw and cone crushers and units with equivalent throughputs are much smaller in size. Consequently, impact crushers have been developed for the processing of hard rock such as basalt, diabase, and graywacke. The peripheral velocity is then decreased to 25–35 m/s.

The *crusher with a vertical shaft* (Fig. 28) has been specially developed for low-wear operation. The feed is introduced into the center (a) of the rotating disk (b). The particles are accelerated by radial bars (c) and fly off the rim of the disk against a surrounding stator (d) where stress is applied by impact. To reduce wear, the acceleration bars and the stator have recesses in which the particles collect and interparticle contact is thus maximized. However, interparticle contact produces relatively little size reduction. The crusher is often used to improve particle shape. The reduction ratio can be improved by using a metal stator.

Special hammer crushers known as *shredders* are used in the disintegration of scrap metal for the recovery of raw materials. A shredder is depicted in Figure 29 A. Two rollers (a) control feed entry; their speed is variable and the clearance between them can be adjusted according to the feed particle size. Stress is applied to the pieces of scrap on the anvil (b) at the full peripheral velocity. Two types of anvil are shown in Figure 29 B. In the toothed design, correspondingly longer and shorter hammers are used, so that they can mesh with the anvil. The fragments or deformed pieces are thrown against a grating (d). Pieces which are not sufficiently reduced in size can be removed from the shredding zone via a door (e).

2.2. Grinding Media Mills

The size reduction devices discussed in this section consist of a mill vessel containing a moving grinding medium. Balls, rods, short bars (cylpebs), or coarse particles of the feed material itself (autogenous grinding) are used as grinding media. The feed trapped between the media particles is ground by their movement.

In the case of *tumbling mills* (Fig. 30 A), the tubular mill vessel rotates around its own axis. The grinding medium is elevated by the vessel wall, and then rolls, slides, or falls toward the lowest point of the vessel.

The mill vessel in a *vibration mill* moves in a circular path, without rotating around its own

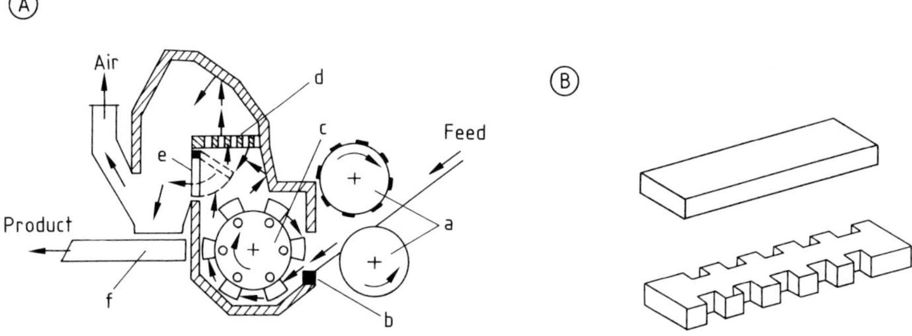

Figure 29. Schematic of a shredder (A) and two anvil designs (B)
a) Feed rollers; b) Anvil; c) Rotor with hammers; d) Grating; e) Hinged door for removing uncrushed material; f) Vibration feeder

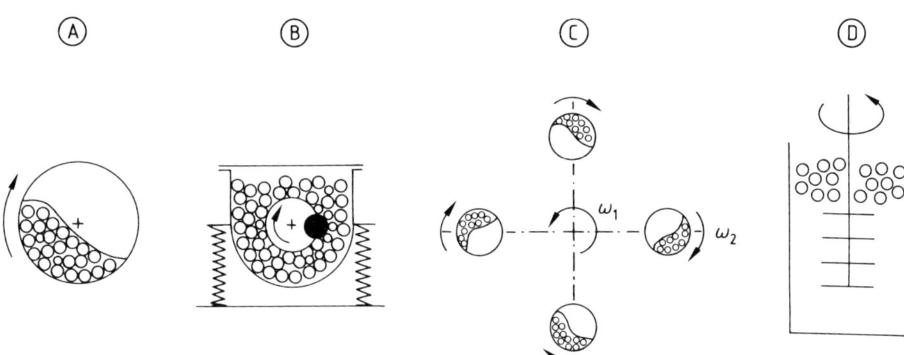

Figure 30. Schematic of mills using grinding media
A) Tumbling mill; B) Vibration mill; C) Planetary mill; D) Agitated ball mill

axis (Fig. 30 B). This sets the grinding medium in motion and produces a grinding action. The pathway of the moving grinding media depends on the diameter of the circular oscillation and the charge of the grinding media.

In the *planetary mill* (Fig. 30 C), the mill vessels rotate with an angular velocity ω_2 on a carrier system which itself rotates with an angular velocity ω_1. Depending on the combination of rotational speeds selected, the type of grinding media motion can be adjusted to correspond to that of either a tumbling mill in a centrifugal force field or to a vibraton mill with a very large diameter of oscillation.

In theory, the *agitated ball mill* (Fig. 30 D) can also be classified in this group, because the agitator moves a grinding medium. However, since this type of mill is primarily operated as a wet mill (the material is fed and ground as a suspension), it is described in Section 2.6.

2.2.1. Tumbling mills

Tumbling mills have a horizontally supported cylindrical (or conical–cylindrical) mill vessel. The mills can be named, for example, according to (1) the shape of the grinding medium or the mill vessel (ball, rod, drum, tube, or cone mills); (2) to its connection with a classifier (open or closed circuit); or (3) according to the method used for product removal (air stream mill). The length to diameter ratio (L/D) varies from ca. 0.25 (autogenous mill) to ca. 6 (continuous mill). Mill sizes range from 300 × 200 mm ($L \times D$) for laboratory use, up to 18 × 6 m. In most cases, the mill is operated continuously in combination with a classifier (closed circuit), which determines the particle-size distribution of the resulting product.

Tumbling mills are suitable for dry or wet grinding. General uses are the fine grinding of hard (to soft) and abrasive materials such as ores, cement clinker, coal, quartz sand, and limestone at high throughputs (up to 500 t/h for closed circuit mills, the product stream then being ca. 200 t/h).

Figure 31 shows a vertical section through a tube mill. The mill vessel is supported at both ends with trunnion bearings (a). The inside of the vessel is lined (b). The lining primarily protects against wear, but it also assists the motion of the grinding medium (better elevation by means of lifter bars) and classifies the grinding balls. The large balls are mainly found near the mill inlet (c) and the smaller ones near the outlet (d) (classification lining). This classification effect is necessary because an increasing degree of fineness along the mill axis requires smaller grinding media; it is often achieved by dividing the mills into chambers (multicompartment mills). The partition walls (e) have perforations which are large enough to allow the feed to pass through but are small enough to retain the grinding medium.

In the absence of an air stream, the particles can only pass through the walls if an axial force is exerted. This is the case in the ascending zone of the solids. The *throughput* $\dot{m}$ through a perforated barrier can be estimated as described in [51]:

$$\dot{m} \approx 0.35\,(1 - 0.6\,d_L/d)\,q_L\,q_A\,\varrho_S\,D^2\,\sqrt{g d_L}$$

where d is the diameter of the ball, d_L the diameter of the perforations, q_L the relative area of the

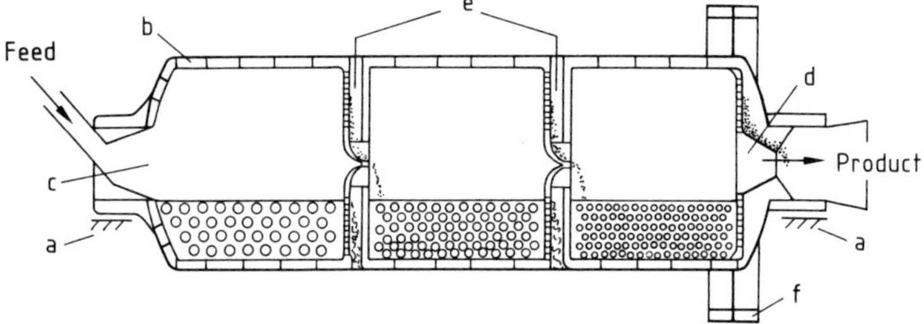

Figure 31. Vertical section through a tube mill
a) Trunnion bearing; b) Liners; c) Tube inlet; d) Tube outlet; e) Intermediate diaphragm with lifter; f) Drive gear

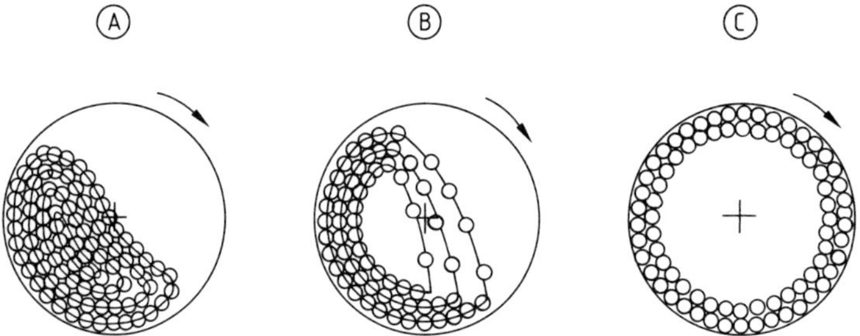

Figure 32. Types of motion in a ball mill
A) Cascading; B) Falling or cataracting; C) Centrifugal

perforations, q_A the relative area of the ascending zone, ϱ_S the density of the solid, D the mill diameter, and g the acceleration due to gravity. Thus, critical factors for the throughput are the relative area of the perforations (as a constructional parameter) and the size of the ascending zone (as an operating parameter).

The *motion of the grinding medium* can be described as cascading, falling (or cataracting), or centrifugal (Fig. 32).

Centrifugal motion (Fig. 32C) begins at the critical rotational speed at which the centrifugal acceleration is equal to the acceleration due to gravity. The critical speed in revolutions per minute (n_c) is commonly defined as

$$n_c = 42.3/\sqrt{D}$$

where D is the mill diameter in meters. Tube mills are operated at speeds below this value. Because a significant amount of slippage may occur between the mill vessel and its contents, the type of motion cannot be deduced directly from the speed of rotation.

The *power consumption* of the mill can be explained by means of the motion of the mill charge. When the mill is stationary, the center of gravity of the charge is adjacent to the mid-point of the mill (Figs. 32A and B). The weight of the charge times the unknown lever arm gives the driving torque from which the following performance formula can be derived:

$$P \sim n/n_c \, M\sqrt{D}$$

where P is the operating power and M the mass of the charge. The mass fraction of the feed is usually less than a fifth of the total mass of the charge. This relationship was described by Blanc and Eckhardt in 1928. The corresponding proportionality constant has values ranging from 6 to 9, if $n/n_c = 0.75$. If the mass is given in tonnes and the diameter in meters, the calculated power is in kilowatts. The mass of the charge can be expressed with the aid of the mill volume. The effect of the length L and diameter D of a mill on power consumption is given by

$$P \sim n/n_c \, LD^{2.5}$$

The parameter P is only proportional to the speed of rotation up to a certain value, i.e., $n/n_c = 0.68-0.75$. Above this range the power consumption decreases sharply due to the centrifugal motion of individual balls as shown in Figure 33 [52].

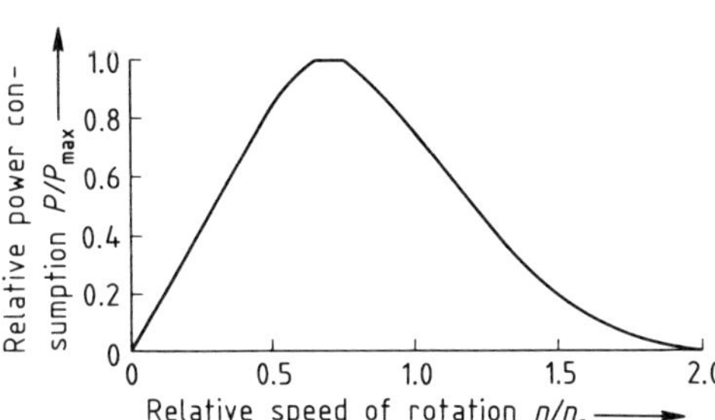

Figure 33. Dependence of power consumption on the relative speed of rotation for dry grinding with lifters

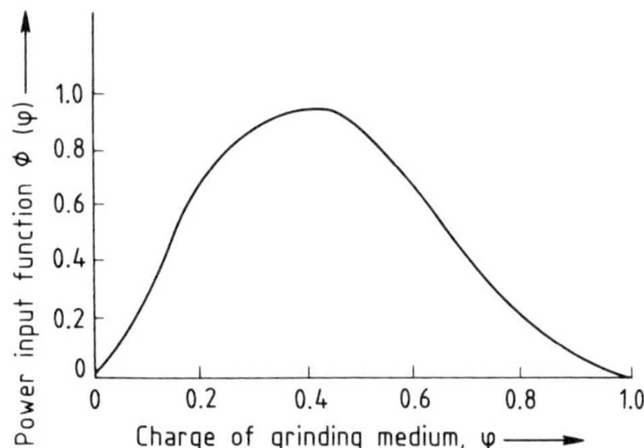

Figure 34. Dependence of power consumption on the charge of grinding medium

The relationship between the charge of grinding medium (expressed as the ratio of the bulk volume of the grinding medium to the mill volume) and power consumption was investigated by ROSE and SULLIVAN [53]. A maximum was found at a loading ratio of 0.4 (Fig. 34). Optimal operating values are in the range 0.27–0.33.

The amount and the motion of the grinding medium determine the power consumption of tube mills. However, the energy must be efficiently transmitted to the particles by selecting a correct size for the grinding medium. For example, the impact energy of a ball must be sufficient to break the largest particle, but must not compact the resulting fines into agglomerates. Examples of empirical formulas for the selection of grinding media size are

$$d = 28 \sqrt[3]{x} \quad \text{or} \quad d = 6 \log (x_P \sqrt{x_F})$$

where x_F and x_P are the feed and product particle sizes in millimeters. To avoid sliding and oscillation of the ball charge, the diameter of the balls should be at least $\leq D/20$ and is usually 20–80 mm.

The literature often states that wet grinding is more energy efficient than dry grinding by as much as 20–30%. However, systematic comparison has shown that under comparable operating conditions a maximum energy saving of 10% can be expected [52]. In wet grinding, wear is up to 10 times higher. For wet grinding the volume fraction of solid c_v should be in the range 0.35–0.4.

Autogenous Mills. Tumbling mills with a small length: diameter ratio of 0.25–0.4 which use the coarse feed particle as grinding bodies (possibly with up to 10% of a steel grinding medium) are termed autogenous mills (Fig. 35). They are used to process ores, refuse, and slag. They can either be operated wet (*Hydrofall mills*) or dry (*Aerofall mills*) in a continuous mode. The

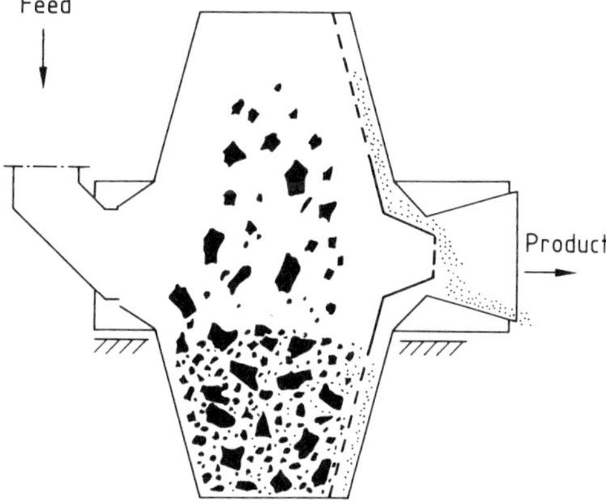

Figure 35. Schematic of an autogenous mill

mill diameters are large compared to tube mills (up to 14 m with a power consumption of ca. 15 MW) and their relative speed of rotation is somewhat higher ($n/n_c = 0.75$–0.8). Lifting and breaking bars increase the elevation of the feed. Addition of steel grinding balls (diameter 75–150 mm) makes operation independent of fluctuations in the size of the feed particles. It also increases impact stress which leads either to a decrease in mill diameter or to an increase in throughput.

2.2.2. Vibration Mills

In a typical vibration mill, the mill vessel is a tube (a) which is filled 65–80% with spheres, rods, or cylpebs (Fig. 36). The tube is held in a frame (b), which is supported by springs (c). It is moved in an almost circular fashion by an unbalance (d).

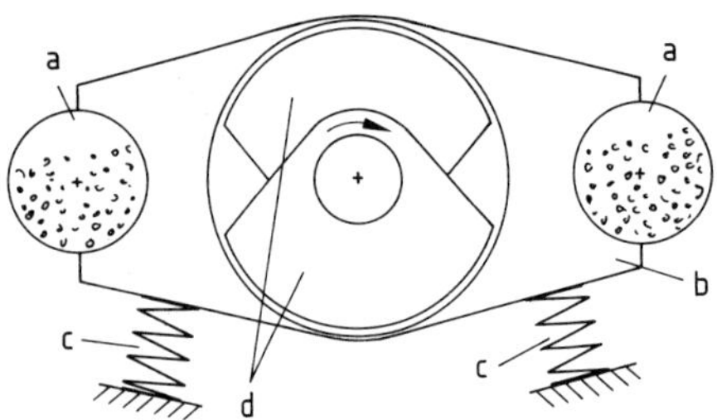

Figure 36. Schematic of a vibration mill
a) Mill tube; b) Frame; c) Spring; d) Unbalance

Vibration mills are much smaller than tube mills (maximum diameter = 0.65 m, length = 4 m) and their throughput is lower (maximum is ca. 15 t/h). Like tube mills, they are suitable for processing hard and abrasive materials. The vibration mill can easily be adjusted to handle different feeds (for example, by changing the oscillation diameter or the residence time). The feed particle size is smaller than that of the tube mill (5–10 mm compared to 25–40 mm). The throughput largely controls the fineness of the product. As a result, the mill is usually operated in a continuous mode without a classifier.

Vibration mills are used to grind fired clay, bauxite, silicon carbide, quartz sand, barium ferrite and even limestone and talc. Wet grinding is possible but is rarely carried out in practice. Special applications include the grinding of dried lignite (because of the low risk of dust explosion); use as a leaching reactor in hydrometallurgy; and activation grinding, (i.e., very frequent, prolonged application of stress to produce crystal modification, see Section 1.1, p. **5**-7).

Under typical operating conditions (oscillation diameter 6–14 mm, speed of rotation 1000–1500 rpm) and with a grinding medium charge > 50 % of the mill volume, each particle of the grinding medium follows an almost circular path. The circle diameter is roughly equal to the oscillation diameter and the angular velocity is equal to that of the mill. In addition, the entire charge rotates slowly at a speed about one hundreth that of the mill and in the opposite direction [54].

Above oscillation diameters of ca. 20 mm (which are not yet used on an industrial scale), a different type of motion takes place. It is characterized by rotation of the center of gravity of the grinding-medium charge at the frequency of the mill and in the same direction. Raasch has derived an expression for the associated power consumption of the mill, based on a force balance [55]:

$$P = c/r_s \, M \, r_s^2 \, \omega^3$$

where r_s is the amplitude of oscillation, M is the mass of the grinding medium, and c is the distance between the center of gravity of the grinding-medium charge and the center of the mill tube. The parameter c is not constant and its value accounts for the variation of the power consumption as a function of the grinding medium charge which is similar to that shown in Figure 34. Even though the type of motion is different at lower oscillation diameters, the equation may also be used in this range. The equation can also be derived from the impact energy of the charge [55]. The author's own studies have shown that the product fineness is primarily determined by the energy per impact.

At high grinding medium charges (above ca. 70 % of the mill volume), the power consumption decreases as the amount of solid increases. This is because the amplitude of oscillation decreases and the added substrate moves the center of gravity of the solid charge toward the center of the tube. With lower grinding medium charges, the diameter of oscillation is scarcely affected and the power consumption increases with the amount of solid.

2.2.3. Planetary Mills

In planetary mills (Fig. 30 D), high stress intensities can be obtained as a result of the greater acceleration. Considerable design problems are associated with this device on a production scale, especially if it is to be operated in a continuous mode. This has hindered the broad application of this system. Although new developments have been made and their associated advantages have been reported, e.g., [56], the only units available commercially are laboratory models. They are used for the batchwise preparation of solids for chemical analysis. The use of grinding media and mill vessels made of semiprecious stone prevents iron contamination.

2.3. Roller and Ring-Roller Mills

In *roller mills* stress is applied to the feed in the gap between two counterrotating rollers (Fig. 37 A). In a *ring-roller mill*, different types of rollers roll over a compressed layer of the feed (particle bed), which is supported on a grinding ring (Fig. 37 B).

Roller mills have two modes of operation: at the start of compression, the feed between the rollers consists of a layer of single particles (Fig. 37 A) or, in the case of small particles, a bed of particles (Fig. 37 C). In the latter case, a higher pressure is applied in the gap and, as a result,

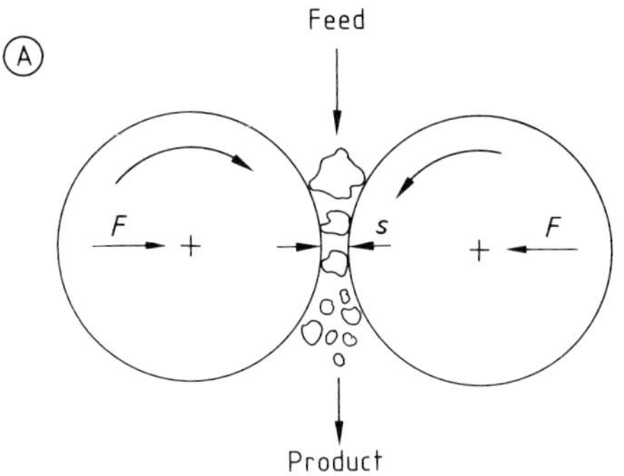

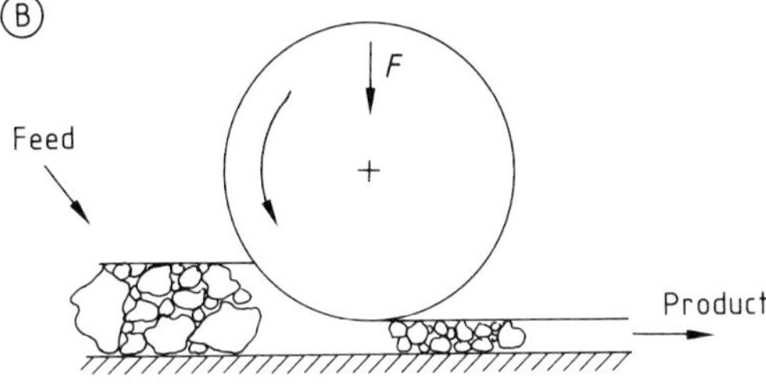

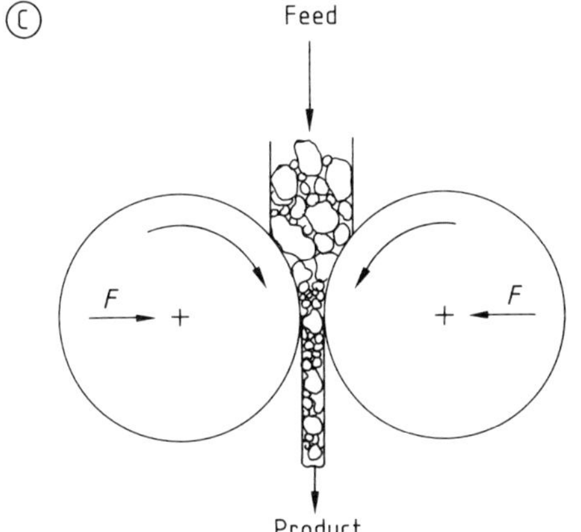

Figure 37. Schematic of a roller mill (A), a ring-roller mill (B), and a high-compression roller mill (C)
F = force; s = gap width

agglomerates are usually formed. The term *high-compression roller mill* has been applied to this mode of operation [48], [49].

2.3.1. Operation Characteristics

Even though ring-roller, roller, and high-compression roller mills are very different in construction, they can be discussed together because of the similarity between the zones in which stress is applied. The conditions in the grinding gap furnish information concerning throughput, power consumption, and the product fineness. Operating characteristics of the three mill types are listed in Table 2. The feed particle size x_F has a maximum value relative to the diameter of the roller D (see Table 2). The *throughput* between the rollers is given by

$$\dot{V} = c_v \, s \, L \, u_p \quad \text{or}$$

$$\dot{V} = c_v s/D \cdot D^2 \, (L/D) \, u \cdot u_p/u$$

where $\dot{V}$ is the volumetric flow rate of the solid, c_v its volume concentration, s the gap width, D the roller diameter, L the length of the roller, u the peripheral speed of the roller, and u_p the velocity of a particle in the gap.

The term $c_v s/D$ is dimensionless and is called the *specific throughput*. It is specific to the mill and the feed material, and is a characteristic of the drawing-in process which is a complex interaction between the feed and the roller. The term $D^2 \, (L/D) \, u$ contains design parameters. The term u_p/u is approximately equal to 1 because, in most cases, slippage between the feed and the roller can be prevented. The equation shows that, without slippage and with constant drawing-in conditions, the throughput increases linearly with the peripheral speed of the roller. However, an upper limit exists, above which the expression is no

Table 2. Comparison of roller, ring-roller mills, and high-compression roller mills

Parameter	Roller mill	High-compression roller mill	Ring-roller mill
Stress application	single particles	particle bed	particle bed
Adjustable parameter	gap width	force	force
Specific grinding force F/DL, MPa*		1–5	0.5–1
Volume fraction of solid in gap	<0.10	0.7–0.9	0.6–0.7
Gap width	>0.5 mm	$D/100$–$D/50$	$D/200$–$D/70$
Maximum feed particle size x_F	$D/20$	$D/40$	$D/40$
Feed material	moderately hard to soft	hard to soft	moderately hard to soft

* D = roller diameter; F = applied force; L = roller length

longer valid. Operating throughputs for roller mills range from 2 to 6 m/s and for high-compression roller mills from 0.5 to 1.0 m/s. In the case of geometrically similar mills (i.e., equal values of L/D and u, and equal drawing-in conditions), the throughput is proportional to the square of the roller diameter. A similarly simple method cannot be used for estimating throughput in ring-roller mills, because the feed is not forced through. An estimate for the throughput of ring-roller mills with disk-shaped grinding rings is cited in the literature as

$$\dot{V} \sim d^{2.65}$$

where d is the diameter of the disk.

The *driving torque* T of a single roller in a roller or ring-roller mill is given by

$$T = F\,D/2 \sin \beta$$

where β is the angle of force action (Fig. 38). The angle β can be determined by measurement of the grinding force F and the torque, values are between 2.5 and 8 °.

Multiplication of the driving torque by the angular velocity gives the *power consumption* P:

$$P = 2 \sin \beta \cdot D^2\, L/D\, u \cdot F/LD$$

The term F/LD is called the *specific grinding force* F_{sp}. It has the dimensions of pressure and is a measure of the maximum pressure in the gap ($p_m \sim F_{sp}$). Measurements give $p_m \approx 45\, F_{sp}$.

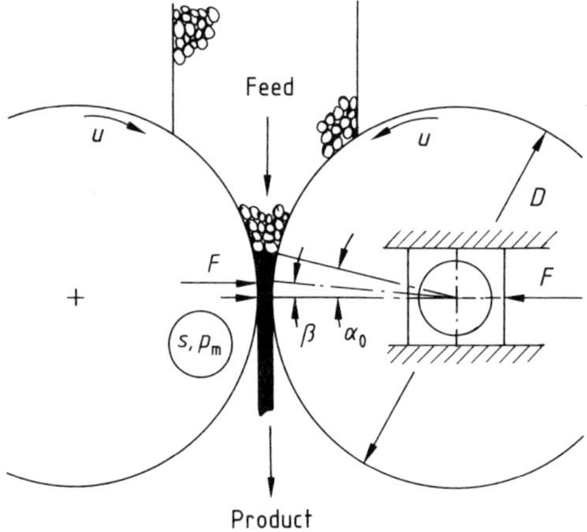

Figure 38. Comminution of material between two rollers
D = roller diameter; F = grinding force; p_m = maximum pressure in gap; s = gap width; u = peripheral velocity; α_0 = compression angle; β = angle of force action

The equation for power consumption is also suitable for scale-up purposes. For geometrically similar mills and equal specific grinding force, they give the result $P \sim D^2$. Ring-roller mills with disk-shaped grinding rings are operated so that the radial acceleration at the rim of the disk remains constant, i.e., $u \sim \sqrt{d}$. Thus, for ring-roller mills $P \sim d^{2.5}$.

Grinding in roller mills approximates very closely the comminution of single particles. As a result the reduction ratio x/s determines the specific energy input (see also Section 1.2, p. **5**-12) and thus the product fineness. In the case of particle-bed comminution, the pressure in the gap is the major parameter affecting product fineness. Use of the above equations and the equation defining the specific work of comminution shows that the reduction ratio is primarily determined by the specific work:

$$W_M = P/V\varrho_s \sim F_{sp} \sim p_m$$

2.3.2. Roller Mills

The design of roller (roll) mills is equivalent to that of roll crushers (Section 2.1.3). The movable bearing is usually supported hydropneumatically; this provides overload protection and allows easy adjustment of the grinding force. The rollers are 0.5 to 1.5 m in diameter and, in contrast to crushers, are smooth or grooved. The reduction ratio lies between 3 and 5. The feed must have a narrow particle-size distribution to ensure that stress is not only applied to the coarsest particles in the mill gap. The size of the coarsest product particles is of the same order of magnitude as the gap width. Due to constructional factors, the gap width has a minimum value (0.5 mm) and ultrafine grinding is therefore not possible.

Roller mills are especially suited for the abrasives industry and the milling of cereals. They are also used for processing clays, especially for the comminution of hard mineral inclusions. Operating parameters are listed in Table 2.

Roller Frame. The roller frame is a roller mill whose rollers operate at different speeds of rotation (friction). The peripheral speed of the faster roller can range up to 8 m/s. The ratio of rotation speeds lies between 1:1.3 and 1:3. The rollers are often shaped (grooved) and their diameters are ca. 0.25 m.

Roller frames are used primarily for milling cereals. A double-roller frame is depicted in Fig-

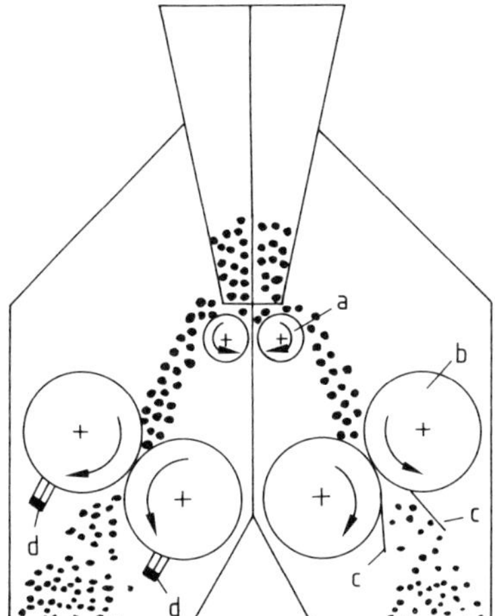

Figure 39. Schematic of a double-roller frame
a) Grooved feed roller; b) Crushing roller; c) Scraper; d) Brush

ure 39. The feed passes over grooved rollers (a) to ensure a uniform feed coating. Scrapers (c) or brushes (d) remove adhering feed particles from the crushing rolls (b). Machines with rollers turning at the same speed are called *crushing roller frames*. They are used to produce flakes from vegetable products (e.g., oat flakes).

Roller frames are also used for the comminution of hard waxes or brittle materials in the chemical industry, because the amount of fines in the product is low. This is especially true at low reduction ratios x_F/s (Section 1.2, p. **5**-12).

High-Compression Roller Mills. The construction of a high-compression roller mill differs from that of a roller mill in that it is stronger because it must exert a higher grinding force. Because agglomerates form as the feed passes through the mill, additional comminution equipment is required. The desired degree of fineness cannot be obtained in a single pass even at very high pressure. Consequently, the mill must work in conjunction with a classifier, the classifier determines the product fineness. Particle-size distributions with upper limits of ca. 0.01 – 1 mm can be achieved.

Mills with diameters D up to 2.8 m have been built, but a diameter of 1 – 1.5 m is most common. The L/D ratio lies between 0.3 and 1.0. Throughputs range from 500 to 550 t/h.

The mills can operate with dry feeds or moist feeds containing up to 15 – 20 vol% water. Materials such as cement clinker, limestone, marble, slag, foundry sand, quartz, silicon carbide, ores, and kimberlite have been processed with high-compression roller mills. The specific grinding force lies between 2 and 5 N/mm^2 depending on the hardness of the material. The energy required for deagglomeration ranges from 0.2 to 0.8 kW · h/t, depending on the material, and corresponds to ca. 2 – 5% of the energy expended in the comminution process. The specific wear is roughly 10 – 20 times smaller than that of a ball mill. This is because a major portion of the applied energy is distributed within the particle bed and only a small amount of material contacts the rollers.

2.3.3. Ring-Roller Mills

Many different mill designs are available for applying stress to bulk solids between rollers and a grinding surface. They differ in the shape of the roller and the grinding ring as well as in the way in which the grinding force is exerted. The ring, pendular roller, bowl, and ring ball mills depicted in Figure 40 operate in closed circuit with an integrated classifier. An air current carries the ground material to the classifier which returns the coarse material to the grinding ring. The classifier determines the final degree of fineness; particle sizes smaller than 30 – 40 µm can be obtained.

Depending on the required fineness and the feed material, the recirculation rates of the milling system may reach very high values. In a plant for milling cement raw material, recirculation rates up to 18 were observed (i.e., the amount of recirculated material was 18 times the product output) [46], [57]. However, usual values range from 3 to 6. A large fraction of the total energy consumption (ca. 40%) can be attributed to recirculation of material. However, the total specific energy requirement of such mills is ca. 25% less than that of tube mills [45].

A milling process with external recirculation is described in [7], [46]. Here, the ground material is transported to a classifier above the mill by means of a bucket conveyor. This significantly lowers the transport energy requirements. The large volumes of air required for material transport are often available as hot gas supplied at a temperature up to 600 °C, drying by as much as 20 wt% is then possible.

Ring Mill. The ring mill depicted in Figure 41 comprises three rollers (a) which support a freely rotating grinding ring (b). One of the three rollers is driven. The grinding ring rotates as a

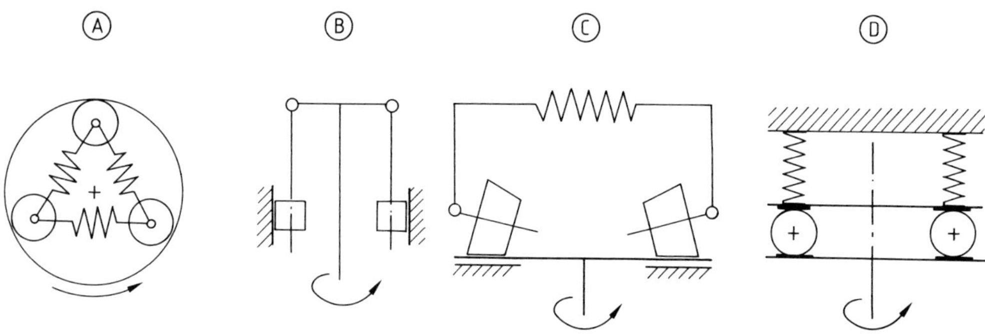

Figure 40. Schematics of the most important industrial ring-roller mills
A) Ring mill; B) Pendular mill; C) Bowl mill; D) Ring ball mill

result of friction and in turn drives the two other rollers. Usually, the rollers are cambered and the grinding ring is correspondingly concave. The springs (c) produce the applied pressure. The feed is fed in from lateral chutes but remains on the grinding ring due to centrifugal force, until it is displaced by additional material. This American design has not been able to penetrate the European market.

Pendular Mill. The pendular mill (Fig. 42) has 2–5 suspended, swinging rollers (a) that are pressed in the direction of the stationary grinding ring (b) by centrifugal force (peripheral speed ca. 7 m/s, maximum diameter of the grinding face 2.5 m). The rollers are driven by rotation of the shaft (c). A stop prevents direct contact between the rollers and the ring. The feed flows in front of the rolls under the action of gravity. In front of each roller is a blade resembling a plowshare, which brings up the feed from below and directs it toward the roller. The product is carried upward by a stream of air to the classifier. These mills are used to grind soft to moderately hard materials, especially sticky, temperature-sensitive, or low-density substances. Examples are lignite, sulfur, barite, ceramic materials, and clay.

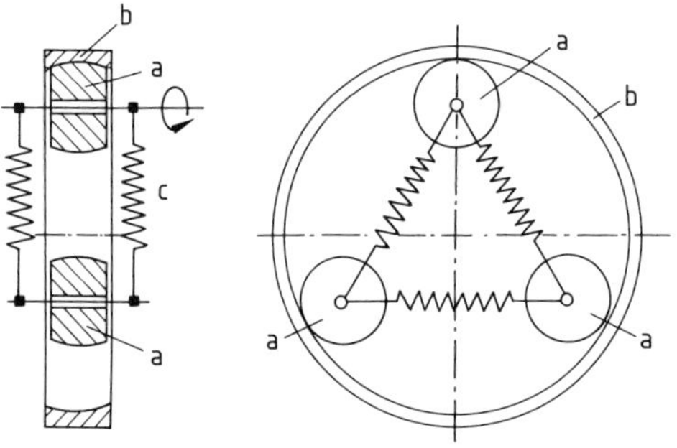

Figure 41. Schematic of a ring mill with three rollers
a) Roller; b) Grinding ring; c) Spring

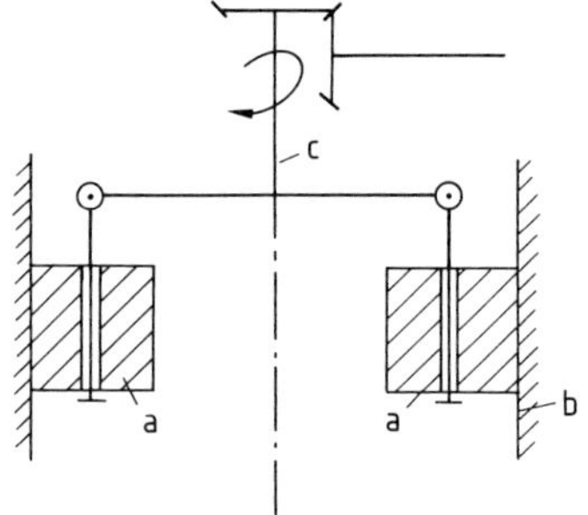

Figure 42. Schematic of a pendular mill
a) Pendular roller; b) Stationary grinding ring; c) Shaft

Ring-Roller Mills with Powered Grinding Rings. The most widely used ring-roller mill designs have powered disk- or pan-shaped grinding rings (Fig. 43). The feed falls onto the rotating grinding ring (a) and is driven radially under the rollers (b) by centrifugal force. After being rolled over repeatedly, the material is taken from the edge of the disk and carried upward to the classifier by a stream of air (c). These designs have found two main applications—the grinding of coal fed into furnaces and the grinding of limestone in the cement industry. In the grinding of coal, large amounts of exhaust gas provide an inert atmosphere and can be used for drying and for transporting the fines to the burner. Maximum throughput is 80 t/h. Mills with disk diameters of up to 6 m and rollers with diameters of ca. 3 m are used for grinding limestone. Because of their lower energy requirements, these mills have largely displaced tube mills. Maximum througput is 400 t/h. Additional uses are

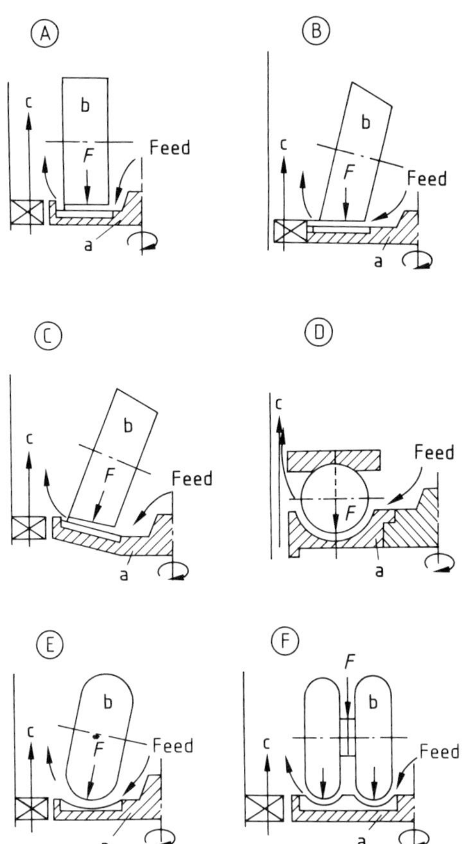

Figure 43. Schematic of ring-roller mills with powered grinding rings
a) Grinding ring; b) Roller; c) Airstream
F = grinding force

the grinding of cement, phosphate, magnesite, burnt lime, and talc.

Edge Runner Mill. The edge runner mill (Fig. 44) is the oldest of the ring-roller mills. It was developed for gold ore preparation in Chile and is therefore also called a *Chilean mill.* It is now used in grinding and mixing plants in the ceramic and refractory ceramic industries as well as in the production of olive oil. Two heavy cylindrical rollers called mullers (a) roll over a pan (b) and may be positioned at different distances from the axis of rotation. Either the muller or the pan can be powered. The design with a powered pan has the advantage that the mullers are individually suspended and therefore the beds can be loaded uniformly with different particle bed depths. In addition, material transport is accomplished by centrifugal force. The weight of the

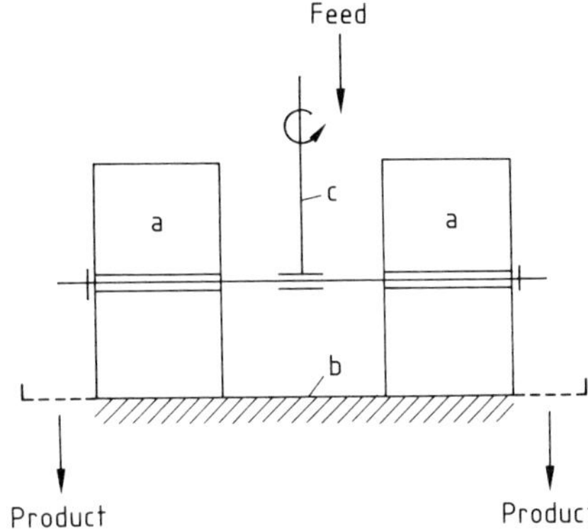

Figure 44. Schematic of an edge runner mill
a) Muller; b) Pan; c) Shaft

muller and, in the case of the driven mullers, the momentum of rotation supply the grinding force.

The peripheral speed of the pan is a function of the radius. In contrast, all points on the surface of the muller have the same peripheral speed. As a result the muller and the pan run at the same speed only at one particular value of the radius. At a larger radius the pan is faster, at a smaller radius the surface of the muller is faster. Thus, the applied pressure is reinforced by friction that promotes the homogenization and mixing of the materials.

Ring-roller mills with conical rollers (Fig. 45) have flat or pan-shaped grinding rings (a). A velocity difference exists between the ring and roller. The grinding force is controlled by a hydraulic device (b) which also permits lifting of the rollers during startup. A stop (c) prevents contact of the rollers with the grinding ring. The rollers can often be swung out (d) in order to change them.

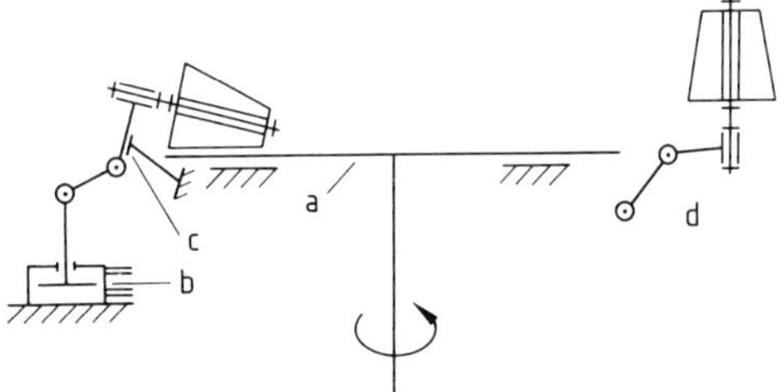

Figure 45. Schematic of a ring-roller mill with conical rollers
a) Grinding ring; b) Hydraulic device for pressure control; c) Stop; d) Roller swung out for replacement

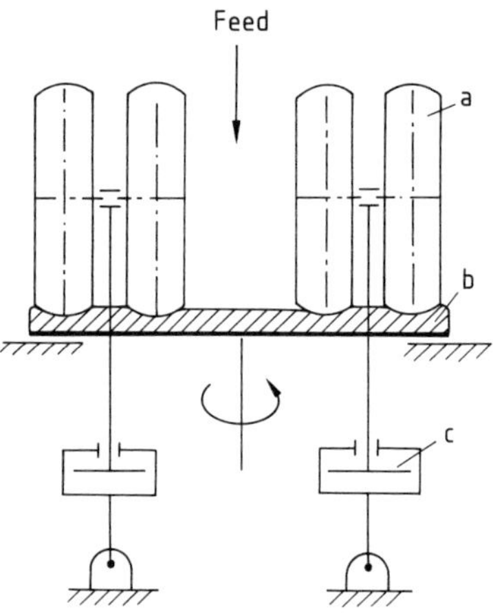

Figure 46. Schematic of a ring-roller mill with cambered rollers
a) Roller; b) Grinding ring; c) Hydraulic device for applying pressure

Ring-roller Mills with Cambered Rollers and Pan-Shaped Grinding Rings. In the mill shown in Figure 46, the rollers (a) are cambered and move in the correspondingly pan-shaped grinding rings (b). The contact pressure is applied by a system of springs or hydropneumatically (c). The mill housing does not have openings for shafts or axles. Consequently, no complicated seals are required because a slight excess pressure exists within the housing; pressure-resistant operation (coal grinding) is possible.

Ring Ball Mill. The ring ball mill is basically a large axial ball bearing (Fig. 47). The lower race (b) is powered. The upper race (c) is stationary and is held against the housing by an array of springs or by a hydraulic system. The balls (a) lay on the lower race next to each other without spacers and can touch each other. Heavy-walled, wear-resistant, cast, hollow spheres are used with diameters up to 1.25 m. Because there are no bearings inside the mill, the gas temperature can be somewhat higher than in other roller mills (up to 800 °C for the entering gas). Openings for shafts or axles in the housing are also unnecessary.

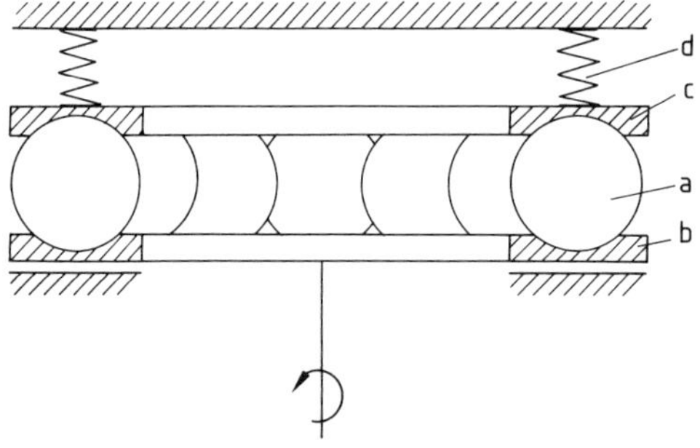

Figure 47. Schematic of a ring ball mill
a) Ball; b) Lower race; c) Upper race; d) Spring system

2.4. Impact and Jet Mills

In impact and jet mills, stress is applied at a single surface; the particles fly against an impact plate or a rapidly moving grinding tool impacts slowly moving particles. Stress can also result from interparticle collisions (see Section 1.2). In jet mills, gas jets are used to accelerate the particles.

Stress application differs from that in previously described mills where stress is primarily applied by means of pressure. The main differences between impacting and compression are

1) stress is applied to individual particles and not to a particle bed,
2) the energy supplied to the individual particle is proportional to the square of its velocity,
3) values of the fracturing energy are limited to ca. 5 J/g (corresponding to 100 m/s), and
4) the velocity at which stress is applied is higher, however this is mainly relevant to materials which exhibit viscous behavior (Section 1.1, p. 5-4).

Very fine particles can be generated under dry conditions with impact and jet mills. However, the final particle fineness (in the range 1 – 10 μm) is limited by two factors. The first factor is that the finest particles flow around the moving impact tools of the mill in the air stream and can no longer be struck by them. The second factor is the slowing down of the particles; for example, a 10-μm particle in stationary air slows down from 100 to 0 m/s in a distance of 10 mm. This means that any particle can lose its kinetic energy over a short distance, so that little size reduction occurs when a collision takes place [3]. Jet mills are somewhat better suited to deal with this problem, because they use interparticle collisions that take place at relatively high particle concentrations.

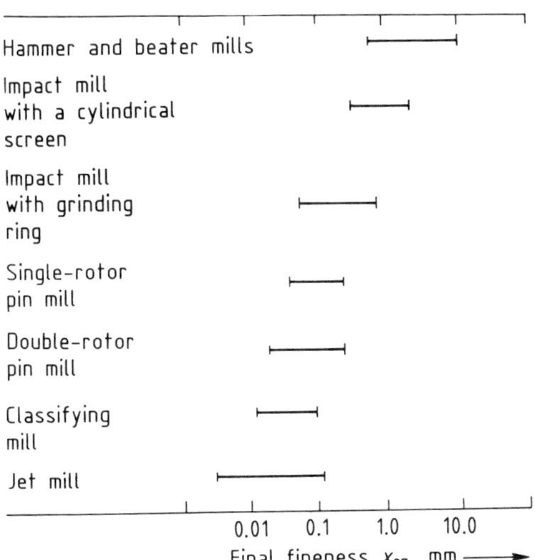

Figure 48. Classification of impact and jet mills according to product fineness

2.4.1. Impact Mills

The range of products processed in impact mills is extraordinarily large. Impact mills are found in all branches of industry which deal with bulk materials or powders. A few examples of the many substrates are cocoa, herb tea, plant seeds, spices, fertilizers, plant protectants, resins, waxes, gypsum, limestone, wood, and lignite.

Several limits are set for the economic industrial use of impact mills. The product must not be abrasive (usually < 4 on the Mohs hardness scale). Hard minor components must be considered, for example the quartz content must not exceed 0.15 %. Brittle materials (i.e., substances which exhibit elastic behavior up to the breaking point, see Section 1.1, p. 5-4) are especially suitable. The material should not be too heat sensitive. However, some heat-sensitive materials can be ground in an impact mill because of the large volume of air used and the relatively low number of collisions. Moisture, or oil or fat content can cause coating or plugging of the mill.

Impact mill designs can be divided into beater mills, hammer mills, and mills with or without peripheral grinding faces. In Figure 48 the mills are classified according to the final product fineness. The fineness is given by x_{97}, the size at which 97 % of the material passes.

In impact mills, the ventilator action of the rotor ensures that a stream of air flows through the mill (provided, of course, that entry of air into the feed inlet is not obstructed or blocked). Volumetric flow rates up to 360 000 m^3/h can be handled by beater mills, which are specially designed for high gas throughputs. However, in conventional impact mills with a maximum diameter of ca. 1.5 m, the flow rates range up to 20 000 m^3/h. Consequently, a large portion of the total energy supplied (ca. 30 – 80 %) must be used for air or gas transport.

Beater Mills. Beater mills (Fig. 49 A) serve two purposes, they operate as blowers and simultaneously as impact mills. They are used for grinding lignite fed into steam boilers. Hot flue gas is used to dry the lignite and provide an inert atmosphere in the mill. Lignite and gas enter axially, pass radially through the blower cascade, and collect in the housing. From there, they flow to a classifier which returns the coarse material to the mill. The mill gas carries the fines to the burner. The largest mills have rotor diameters of ca. 4 m and process ca. 100 t/h.

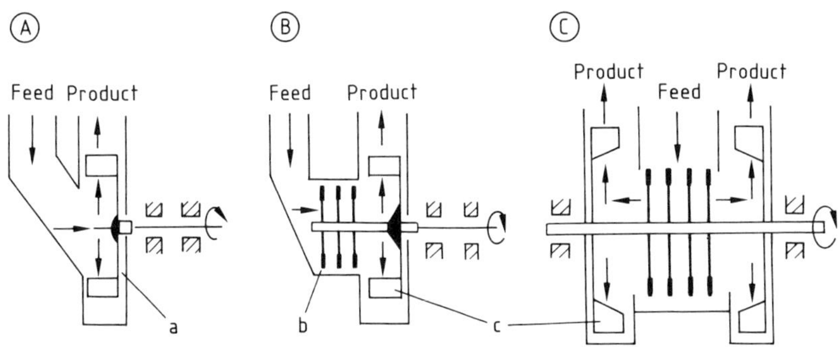

Figure 49. Schematic of beater and blower – beater mills
A) Beater mill with horizontal wheel shaft; B) Blower – beater mill; C) Double flow blower – beater mill
a) Beater wheel; b) Swinging hammer; c) Fan

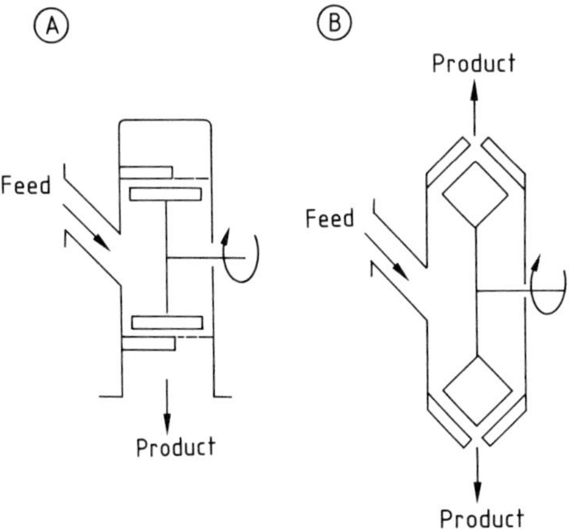

Figure 50. Schematic of impact mills with horizontal shafts
A) Mill with grinding bars and screen; B) Mill with a V-shaped grinding chamber and peripheral product outlets

Blower–Beater Mills. Blower–beater mills (Figs. 49 B and C) are also used for grinding furnace feeds. Separate devices are employed for comminution and gas transport. As in hammer mills, comminution is performed by a rotating array of swinging beaters attached to a shaft.

Hammer Mills. Hammer mills are basically similar to hammer crushers (Fig. 27, p. **5**-19) but their construction is less sturdy. The boundary between hammer mills and hammer crushers is not clearly defined but can be taken to be a diameter of ca. 1.5 m. The hammers are usually simple iron plates. The screen openings surrounding the rotor range between 2 and 10 mm. Applications include rough milling of grain, chipping of wood, and processing of waste such as wood turnings, fruit crates, paper bags and corrugated cardboard.

Universal Mills. The characteristic feature of the mills depicted in Figure 50 is a horizontal axis that is surrounded by a grinding face. The feed is fed in laterally into the center of the rotor and then moves radially outward. The material is crushed by the comminution tools of the rotor, accelerated, and crushed once again at the grinding face. The grinding face holds back the particles until they are sufficiently fine.

The grinding face consists of a perforated or slotted screen, which is usually interrupted at intervals by impact bars. The bars direct the material sliding along the screen back toward the circular impact zone. Recent findings on the operation of screen-impact mills are given in [58].

The side wall which holds the feeder is constructed as a door. This permits easy access to the grinding chamber for cleaning purposes or for changing the grinding tools. These mills are referred to as universal mills because the grinding devices and grinding surfaces can easily be changed to adapt the mill for different feeds. These mills are extraordinarily versatile in their application. Their names correspond to their construction, e.g., cross-flow, turbo, blower, beater, pendular hammer, or disintegrator mill.

Peripheral speeds range from 30 to 120 m/s. With diameters up to 1.5 m they can achieve throughputs up to 10 t/h depending on the product fineness required.

Impact Mills with Vertical Shafts. Impact mills with vertical shafts are depicted in Figure 51. Figure 51 A shows a design with a conical

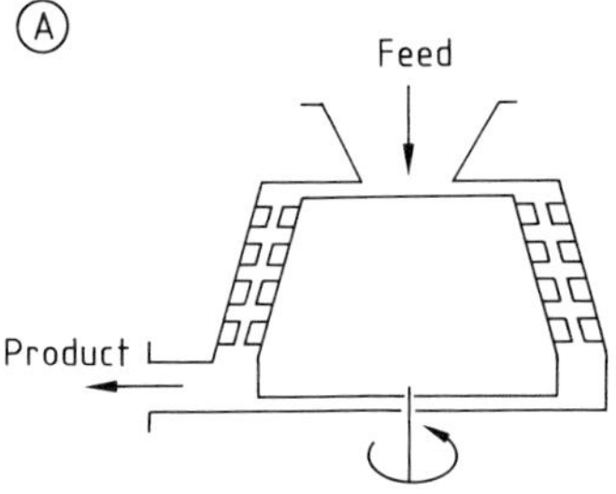

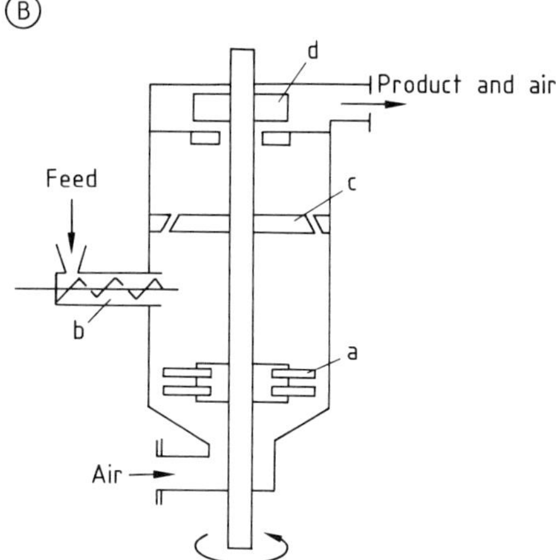

Figure 51. Schematic of impact mills with vertical shafts
A) Impact-bar mill with a conical rotor; B) Impact mill with rejector classifier
a) Impact hammer; b) Screw feeder; c) Classifying whizzer blades; d) Fan

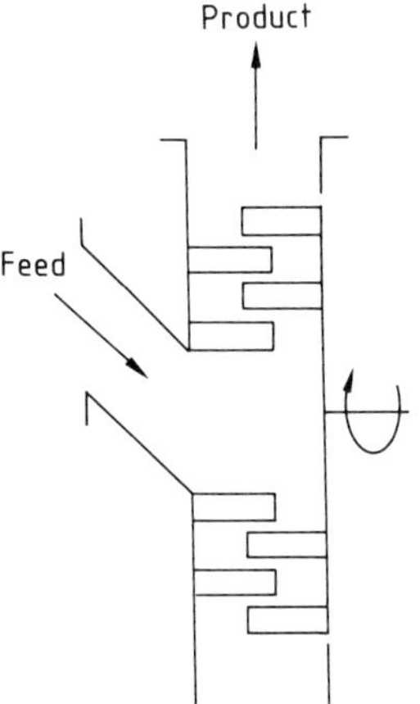

Figure 52. Schematic of a pin mill

rotor and stator. The gap width can be adjusted by axial movement of the stator. The impact bars are arrayed in multiple tiers along the circumference of the rotor. Air and feed flow down through the mill cocurrently. In a version with a cylindrical rotor, flow is upward.

Disk-shaped rotors with peripheral impact tools (hammers) (Fig. 51 B) can, as in roller mills, have an airstream which carries the feed to a classifier. The degree of fineness is determined by the adjustable classifier.

Pin or Pin-Beater Mills. The impact devices used in pin mills (Fig. 52) are a large number of round pins or elongated squares (beaters). The devices are arranged in concentric rings of varying radius. A stator of corresponding design meshes with the rotating rows of pins.

The pin mill differs from the universal mill in that the rotor is not surrounded by a grinding face. Thus, the degree of fineness is determined primarily by the peripheral speed of the rotor.

The number of collisions is relatively low because there is no grinding face (ca. 5–8 collisions for three rows of pins). Consequently, materials which are temperature-sensitive or tend to cake are often processed in these mills.

Designs are available in which two arrays of pins rotate in opposite directions, so that the impact velocity is theoretically doubled (maximum 230 m/s).

2.4.2. Jet Mills

In jet mills the feed particles are accelerated by means of propellant jets (air, gas, or steam) with high velocities (sonic or supersonic speeds, i.e., 500–1200 m/s). Comminution occurs by interparticle collision or by impact with an impact surface (target). These mills are always operated in closed circuit with an air classifier. The particles do not attain the velocity of the gas because they cannot reach the center of the jet. In addition, a single particle collision is insufficient for obtaining the desired final particle size for particles smaller than ca. 20 μm. The specific energy consumption ranges from 300 to 3000 kW · h/t and is mainly determined by the generation of propellant [e.g., 0.6 MPa (6 bar), 3000–12 000 m^3/h depending on design] and inefficient particle acceleration.

The actual milling or comminution volume is small compared to the sections which provide the propellant gas and separate the fine particles (cf. p. **5**-15).

Spiral Jet Mill. In the spiral jet mill (Fig. 53 A), the feed is sucked into the mill by means of an injector (a). The propellant is fed via a circular distributor (b) to individual nozzles (c) distributed on the circumference of the mill. The angles of the nozzles and the gas velocity are adjusted such that the feed rotates around the chamber in the form of a ring without touching the wall. The particles are slowed and accelerated at the margins of the jets and interparticle collisions occur. The gas motion allows size classification as in spiral air classifying. The equilibrium between the centrifugal force acting on a particle and the radial component of the drag of the spiral flow determines the cut point (→ 17. Air Classifying, pp. **17**-2–**17**-4). The fine particles leave the mill through the central opening.

Spiral jet mills can handle explosive powders as well as temperature- and contamination-sensitive materials because they have no moving parts and there is no contact between the feed and the wall. The mills have a maximum diameter of 800 mm, and therefore do not require much space; they also offer the advantage of easy cleaning and sterilization. Throughputs range from 1–500 kg/h. Their main application is the comminution of chemicals and pharmaceuticals.

Oval-Tube Jet Mill. The oval-tube jet mill (Fig. 53 B) is based on the same comminution principle as the spiral jet mill. The particles are classified by a sharp change of direction of the flow stream. Particles that can follow this bend

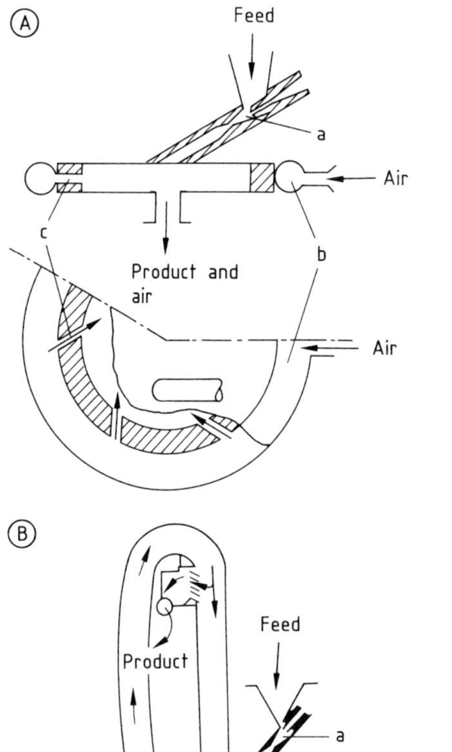

Figure 53. Schematics of the spiral jet mill (A) and the oval-tube jet mill (B)
a) Injector; b) Air-supply pipe; c) Nozzle

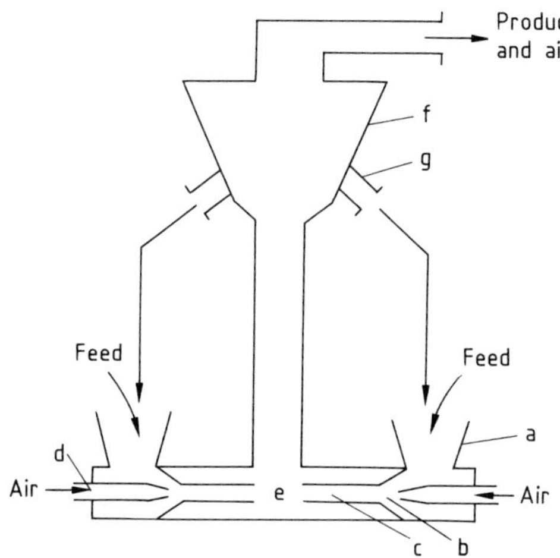

Figure 54. Schematic of an opposed jet mill
a) Feed funnel; b) Injector; c) Acceleration pipe; d) Propellant inlet; e) Grinding chamber; f) Classifier; g) Outlet for recycling coarse fraction

can leave the mill. This mill is used in the preparation of minerals (e.g., talc and mica), because of its sturdy construction.

Opposed Jet Mill. In the opposed jet mill (Fig. 54), the particles are accelerated through two pipes (c) and directed against each other in the central grinding chamber (e). The feed streams may be the coarse material from the classifier and the feed, as shown. Alternatively, the feed may first pass through a classifier; the coarse material is then distributed between the two pipes. Acceleration and comminution occur in a restricted region which can easily be protected against wear. Consequently, abrasive material such as quartz or zircon sand can be ground in this type of mill.

The opposed jet mill differs from the spiral jet and oval-tube mills in that it has somewhat lower energy requirement and achieves a higher degree of fineness. In the case of an independently powered air classifier, better control of coarse particles is possible.

Fluidized-Bed Jet Mill. The fluidized-bed jet mill (Fig. 55) does not require an armored acceleration tube or a grinding chamber, as do the previously described opposed jet mills. The grinding zone (a) consists of a fluidized bed. Jets of air are blown into the bed of particles toward the center of the mill through 3–6 nozzles (e) arranged on the perimeter of the cylindrical mill vessel (b). The jets fluidize the particles. The particles collide with each other at the margins of the jets and in the center. The air carries the ground particles to a classifier (c) in the upper part of the cylinder. The classifier determines the particle size of the product. The mill cannot be used with nonfluidizable particles.

The particles come into contact almost exclusively with other particles, thus this mill is especially suited for abrasive substances (e.g., abrasives, ceramic raw materials, as well as metal carbides, oxides, and nitrides) and materials that must be free of contaminants (e.g., powdered phosphors, dental materials, silicic acid, and silicon).

Target Mills. In particle jets that are directed against each other, the probability that the parti-

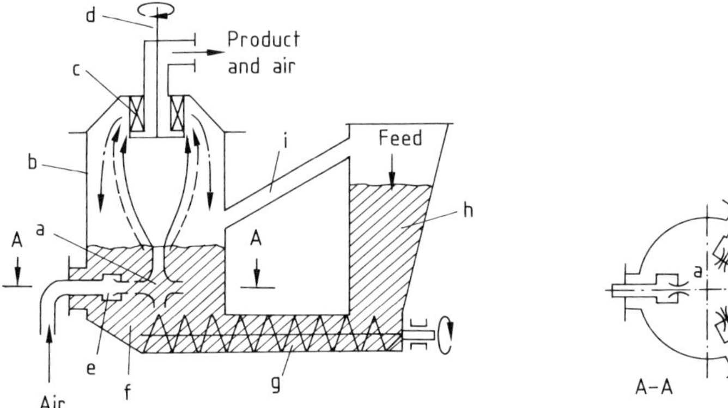

Figure 55. Schematic of a fluidized-bed jet mill
a) Grinding zone; b) Grinding vessel; c) Classifier; d) Drive for classifier; e) Nozzle; f) Fluidized bed; g) Screw feeder; h) Hopper; i) Pressure equalization tube

cles will not collide is relatively high. Consequently, mills have also been developed in which the particles accelerated in the gas jet are directed against a solid surface or target. Wear can be controlled by using suitable construction materials, because the impact area is not very large and the target can be made sufficiently thick.

2.5. Cutting Mills

Paper and wood waste, tough plastic films, rubber, leather, and feeds cannot be comminuted with any of the mills described so far. Equipment with a cutting action is required for these applications as well as for the disintegration of bundles of plant fibers, the production of wood chips, and cellulose pulp. Shredders, chippers, knife-choppers, and, especially for fibers, pulp machines are used.

Shredders. The construction of shredder mills resembles that of impact crushers (see p. 5-19). The rotor and stator are equipped with blades that pass by each other, and are separated by a narrow clearance. The peripheral speed ranges from 5 to 20 m/s. The basket screen determines the final product fineness.

Other designs use disks mounted on two parallel shafts and mesh with each other to give a shearing action (Fig. 56 A). Each disk varies in radius along its circumference. The part with the largest radius forms a cam, which helps to draw in the feed. The cams are set at different positions along the shaft to ensure uniform energy consumption during rotation. In case of plugging or

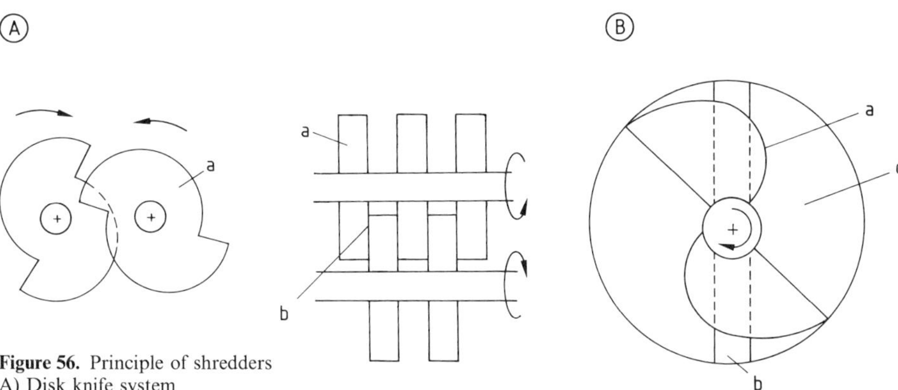

Figure 56. Principle of shredders
A) Disk knife system
a) Cutting disk; b) Cutting edge
B) Top view of a rotating shredder
a) Cutting edge; b) Cutting beam; c) Funnel outlet

overload, the direction of motion reverses automatically.

In another design, which can be used in the outlets of hoppers, two propeller-like blades work against a stationary knife beam (Fig. 56 B).

Pulp Mills. Mechanical separation of fibers is of great importance in the paper and cellulose industries. If possible, only shear stress should be exerted in the direction of the fiber. No tensile stress should be applied since this could lead to tearing of the fiber. The paper pulp is prepared in a pulp machine, in which stress is applied between drums equipped with blades and blades attached to the bottom of a trough. Many different designs of this type of machine are described in [59] (see also → Paper, treated in the A series).

2.6. Wet Grinding

Wet grinding is defined as the comminution of solids suspended in a fluid. The principal reasons for using this technique are

1) the feed is already in the form of a suspension,
2) the product is desired as a suspension,
3) the feed material has to be finely ground and has a tendency to agglomerate, or
4) the material is explosive or toxic.

Wet grinding combines the actual comminution of the solid particles with the breaking up of agglomerates and the dispersion of the individual particles. In some machines comminution is the primary objective, whereas in others deagglomeration is the primary function.

All the previously described comminution devices with the exception of crushers, impact mills, and jet mills are used for wet grinding. They operate primarily with low-to-moderate viscosity suspensions. However, roller and high-compression roller mills can only be employed at moisture levels of 20% or less; although some roller mills are made for wet grinding [60], [61].

The Szego mill, the roller frame, the agitated ball mill, and the disk mill have been specially designed for wet grinding.

Szego Mill. The Szego mill has been developed recently and is basically a centrifugal roller mill (Fig. 57). The roller shafts (c) are held between two disks (a), which rotate inside a cylindrical housing (b). Because the roller shafts can be moved radially, centrifugal force presses the rollers against the stator and they roll over it. A helical channel is cut into the surface of the rollers and, together with gravity, provides for material transport.

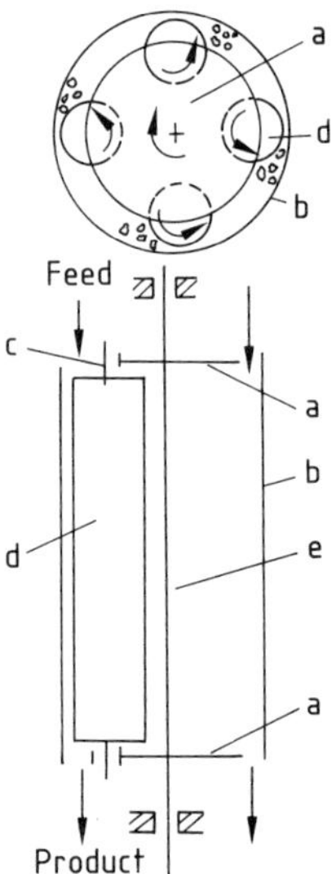

Figure 57. Schematic of the Szego mill
a) Rotating disk; b) Housing; c) Shaft of grinding roller; d) Helical grooved roller; e) Central shaft

A production-scale mill was investigated with coal–oil and coal–water suspensions (60–70 wt% coal) [62]. The mill can also be used for the dry grinding of materials such as mica, talc, wheat flour, cocoa, and soy beans. A similar design is reported in [7] and has been used in the Soviet Union for the wet grinding of ores.

Another horizontal design employs rods that are driven by a thick central shaft. This system is comparable to a needle bearing.

Roller Frames. Roller frames with two or more smoothly polished rollers (Fig. 58) are finding application in the processing of moderately viscous suspensions, for example in the

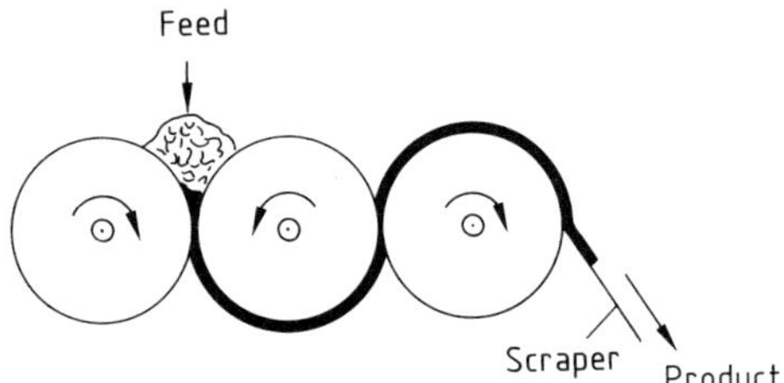

Figure 58. Schematic of a triple-roller frame

printing ink industry or the manufacture of chocolate. The rollers turn at different speeds, each being faster than the one following it. In this way the suspension coating is transferred between the rollers without any difficulty. In some machines, one of the rollers is replaced by a plate, usually called the bar. The gap width is adjusted by changing the grinding force, which is usually applied hydraulically. The smallest gap width is 20 μm. Coarse particles are stressed between the rollers, according to the gap width. Fine particles can also be stressed if particle bridges form at sufficiently high concentrations, otherwise only the shear stress transmitted by the medium is effective. Multiple passes through the gap are usually required to achieve the desired degree of fineness.

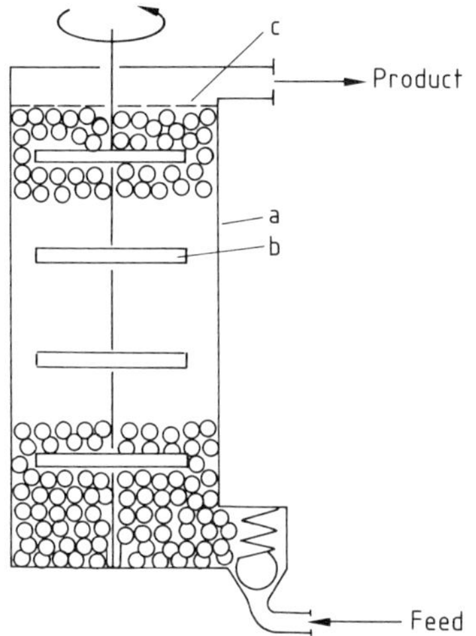

Figure 59. Schematic of an agitated ball mill
a) Cylindrical mill housing; b) Agitator; c) Screen

Agitated (Stirred) Ball Mills. The agitated ball mill (Fig. 59) was developed from the "sand-mill" and has recently gained importance in wet grinding. It consists of a horizontal or vertical cylindrical housing (a), which is filled to ca. 80 % with glass beads, steel balls, or a ceramic grinding medium, with sizes ranging from a few tenths of a millimeter to several millimeters. An agitator shaft runs along the axis of the mill and is equipped with perforated disks, rings, or pins. The agitator (b) rotates with a peripheral speed of 4–20 m/s. The suspension is forced through the mill by a pump. A screen (c) or slot holds back the grinding media at the product discharge. Product fineness of less than 2 μm can be obtained. Examples of applications are the paint and lacquer industry, coatings for the paper industry, ceramic materials, chocolate production, and the disruption of cells in biotechnology. Mill volumes may be up to 1 m^3.

Other designs employ a narrow conical or cylindrical grinding gap with smooth walls (gap width is ca. four times the diameter of the grinding balls). The grinding balls move through the gap together with the suspension and the recycled within the mill by appropriate devices. The narrow gap permits high energy densities to be achieved.

Under certain operating conditions (especially with aqueous suspensions), the specific work measured in a lab-

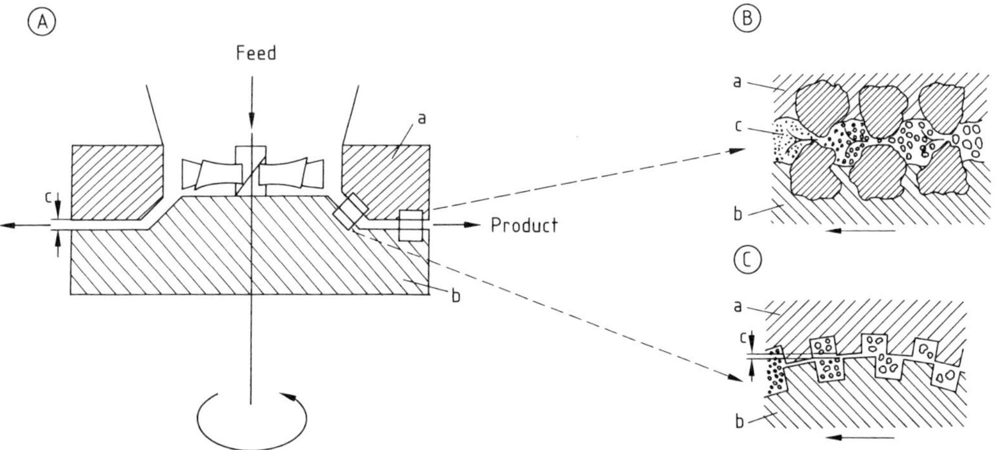

Figure 60. Schematic of a disk mill (A) showing close-ups of the gaps of a corundum disk mill (B) and a toothed colloid mill (C)
a) Stator; b) Rotor; c) Gap width

oratory-scale mill can be used for scale-up purposes. It determines the attainable fineness [63].

Agitator mills are sometimes used for dry grinding. The separation of the grinding medium then takes place outside the mill. The grinding medium is fed in with the feed. A considerable amount of additional equipment is required: conveying devices and intermediate bunkers for the grinding media as well as classifiers and dust separators for the fine material.

Disk Attrition Mills. Disk attrition mills (Fig. 60) are rotor–stator configurations with maximum diameters of ca. 500 mm. Large mills are equipped with a horizontal drive. The profiled or perforated disks can be made of steel, sintered material, or corundum (Fig. 60 B). They rotate with peripheral velocities up to 35 m/s. The gap width can be very finely adjusted (in the micrometer range) and is in the order of 30–500 μm.

Examples of applications are the comminution of pharmaceuticals and chemicals to finenesses in the range of 10–20 μm. The mills are also used in the comminution of leather, hides, bone, cardboard, and wood, as well as for the pulping of foods such as fruit, meat, bacon, and almonds. Throughputs are usually a few hundred kilograms per hour but depend on the application; several tonnes per hour may be reached. A special rotor–stator configuration with a conical clearance, the toothed colloid mill (Fig. 60 C), is used primarily for the homogenization, dispersion, and mixing of creams, salves, toothpaste, baby food, and mayonnaise.

Kneaders are required to disperse pigments and fillers in high-viscosity suspensions. These devices are discussed elswehere (→ 26. Mixing of Highly Viscous Media).

3. References

[1] H. Rumpf, *Staub-Reinhalt. Luft* **27** (1967) 3–13.
[2] R. Polke, *Part. Charact.* **4** (1987) 54–62.
[3] H. Rumpf, K. Schönert in *Ullmann*, 4th ed., **2**, 1–23.
[4] *6th European Symposium "Comminution" Preprints*, Part II, Nürnberger Messe- und Ausstellungsgesellschaft, Nürnberg 1986.
[5] *European Symposium "Particle Technology," Preprints*, Vol. A, Dechema, Frankfurt 1980.
[6] *4th Europäisches Symposium "Zerkleinern," Dechema Monographie*, vol. 79, Part A, Verlag Chemie, Weinheim 1976.
[7] K. Höffl: *Zerkleinerungs- und Klassiermaschinen*, Springer Verlag, Berlin-Heidelberg-New York-Tokyo 1986.
[8] K. Schönert in *Winnacker-Küchler*, **1**, 80–93.
[9] K. Dialer, V. Onken, K. Leschonski: *Grundzüge der Verfahrenstechnik und Reaktionstechnik*, Hanser Verlag, München-Wien 1986.
[10] M. Zoog: *Einführung in die Mechanische Verfahrenstechnik*, Verlag Teubner, Stuttgart 1987.
[11] H. Schubert: *Aufbereitung fester mineralischer Rohstoffe*, vol. 1, VEB Deutscher Verlag für Grundstoffindustrie, Leipzig 1975.
[12] C. L. Prasher: *Crushing and Grinding Process Handbook*, J. Wiley and Sons, New York 1987.
[13] B. Beke: *The Process of Fine Grinding*, M. Nijhoff, W. Junk Publishers, Den Haag–Boston–London 1981.
[14] J. Beddow: *The Production of Metal Powders by Atomisation*, Heyden and Sons, London 1978.
[15] A. J. Lynch: "Mineral Crushing and Grinding Circuits" in *Developments in Mineral Processing*, Elsevier, Amsterdam 1977.
[16] V. C. Marshall: *Comminution, Report of a Working Party* Chap. 2, Inst. of Chemical Engineers, London 1975.
[17] G. C. Lowrison: *Crushing and Grinding*, Butterworth, London 1974.
[18] R. H. Snow, B. H. Kaye, C. E. Capes, G. C. Sresty in R. H. Perry, D. Green (eds.): *Perry's Chemical Engineers Handbook*, McGraw Hill, New York 1984, pp. **8**-7–**8**-60.
[19] H. Horii, S. Nemat-Nasser, *JGR J. Geo-phys. Res.* **90** (1985) 3105–3125.
[20] D. S. Dugdale, *J. Mech. Phys. Solids* **8** (1960) 100.
[21] R. Weichert, K. Schönert, *J. Mech. Phys. Solids* **26** (1978) 151–161.
[22] V. Boldyrev, K. Meyer: *Festkörperchemie, Beiträge aus Forschung und Praxis*, VEB Verlag für Grundstoffindustrie, Leizig 1973.
[23] W. Schwenk: *3rd European Symposium "Zerkleinern,"* Cannes (1971) Dechema-Monogr. vol. 69, Verlag Chemie, Weinheim 1972, pp. 121–138.
[24] H. Rumpf, *Chem. Ing. Tech.* **37** (1965) 187–202.
[25] W. Hess: *Einfluß der Schubbeanspruchung und des Verformungsverhaltens bei der Druckzerkleinerung von Kugeln*, Dissertation Universität Karlsruhe 1980.
[26] R. Weichert, *Int. J. Miner. Process.* **22** (1988) 1–8.
[27] H. Schubert, *Aufbereit. Tech.* **28** (1987) 237–246.
[28] N. Hoffmann et al., *Chem. Ing. Tech.* **48** (1976) 329.
[29] G. Göll, J. Hanisch, *Aufbereit. Tech.* **28** (1987) 582–590.
[30] K. Schönert in [4].
[31] C. J. Stairmand in [6].
[32] H. Rumpf, *Aufbereit. Tech.* **16** (1975) 59–71.
[33] F. C. Bond, *Aufbereit. Tech.* **5** (1964) 211–218.
[34] P. S. B. Stewart, C. J. Restarick, *Proc. Australas. Inst. Min. Metall.* **239** (1971) 81–92.
[35] L. G. Austin, *Ind. Eng. Chem. Process Des. Dev.* **12** (1973) 121–129.
[36] V. K. Gupta, P. C. Kapur, *Powder Technol.* **10** (1974) 217–223.
[37] R. K. Jaspan et al. in [6].
S. G. Malghan, D. W. Fuerstenau in [6].
[38] K. J. Reid, *Chem. Eng. Sci.* **20** (1965) 953–963.
[39] L. G. Austin, P. T. Luckie, *Powder Technol.* **5** (1971/1972) 267–271.
[40] A. L. Mular, *Bulletin Canad. Mining and Metall.* (1970) 821–826.
[41] R. R. Klimpel, L. G. Austin, *Ind. Eng. Chem. Fundam.* **9** (1970) 230–237.
[42] J. A. Herbst, T. S. Mika, *Proc. 11th Intern. Symp. Comp. Applications in the Minerals Industry*, Tucson Arizona (1973) E 78–E 123.
K. Shoji, L. G. Austin, *Powder Technol.* **10** (1974) 29–35.

[43] K. Schönert, *Chem. Ing. Tech.* **6** (1971) 361–367. R. P. Gardner, K. Verghese, *Powder Technol.* **11** (1975) 87–88.
[44] J. Decasper, *ZKG Zement Kalk Gips* **52** (1980) 219–222.
[45] D. Hochdahl, *ZKG Zement Kalk Gips* **54** (1982) 1–10.
[46] F. Feige et al., *ZKG Zement Kalk Gips* **55** (1983) 628–632.
[47] K. Schönert, *ZKG Zement Kalk Gips* **51** (1979) 1–9.
[48] H. Kellerwessel, *Aufbereit. Tech.* **27** (1986) 555–559.
[49] M. V. Seebach, N. Parzelt, *ZKG Zement Kalk Gips* **40** (1987) 337–344.
[50] DIN 50 320. H. Metz (ed.): *Abrasion und Erosion,* Hanser Verlag, München–Wien 1986.
[51] G. Roth: *Transport in Drehrohren und Kugelmühlen,* Dissertation Techn. Universität Karlsruhe 1982.
[52] J. Leluschko: *Untersuchungen zum Einfluß der Flüssigkeit bei der Naßmahlung in Kugelmühlen,* Dissertation Techn. Universität München 1985.
[53] H. E. Rose, R. M. E. Sullivan: *Ball, Tube and Roll Mills,* Constable, London 1958.
[54] K. E. Kurrer, E. Gock, *Chem. Ing. Tech.* **57** (1985) 64–65.
[55] J. Raasch, *Chem. Ing. Tech.* **36** (1964) 125–130.
[56] H. Meiler et al. in [4].
[57] L. T. Schneider et al., *ZKG Zement Kalk Gips* **38** (1985) 705–708.
[58] D. Landwehr, M. H. Pahl, *Aufbereit. Tech.* **28** (1987) 57–65, 188–192.
[59] K. Rosenfeld: *1st European Symposium "Zerkleinern,"* Verlag Chemie–VDI Verlag, Weinheim, Düsseldorf 1963.
[60] T. Loesche, D. Anstey in [4].
[61] D. Schwechten, K. Schönert in [4].
[62] O. Trass et al., *Powder Technol.* **40** (1984) 269–282.
[63] H. Weit, *Chem. Eng. Technol.* **10** (1987) 398–404; *Chem. Ing. Tech.* **6** (1988) 318–319.

6. Spraying and Atomizing of Liquids

PETER WALZEL, Bayer AG, Dormagen, Federal Republic of Germany

In addition to the standard symbols defined in the front matter of this volume (p. IX), the following symbols are used:

A	m^2	cross section area of the nozzle
A_e	m^2	cross section area of emerging jet
a	m/s^2	centrifugal accelleration
C_1, C_2, C_3	-	constants
d	m	drop diameter
d_{32}	m	sauter mean diameter
$d_{v,0.5}$	m	volume median diameter
d_{th}	m	thread or jet diameter
d'	m	nozzle diameter of liquid orifice
d'_C	m	diameter of swirl chamber
d'_g	m	outer diameter of gas annulus, see Figure 13
d'_p	m	equivalent diameter of gas orifice
E	V/m	strength of electrical field
f	1/s	frequency
g	m/s^2	acceleration of gravity
j, k	–	exponents
K	m^2	sheet thickness parameter
l_{th}	m	length of threads
m	–	exponent
$\dot{m}$	kg/s	mass flow rate of liquid
$\dot{m}_g$	kg/s	mass flow rate of gas
n	–	exponent
Δp	$kg s^{-2} m^{-1}$	pressure drop of the nozzle
Q	C	electrical charge
r	m	radius of atomizer
s	m	spacing between threads
U	V	voltage
t	s	time
v	m/s	gas velocity
v_{rel}	m/s	relative gas velocity
$\dot{V}$	m^3/s	volumetric flow rate
w	m/s	liquid velocity
x	m	distance from orifice in jet direction
δ	m	sheet thickness
ε	$C V^{-1} m^{-1}$	dielectrical constant of the fluid
η	$kg m^{-1} s^{-1}$	viscosity of the liquid
η_g	$kg m^{-1} s^{-1}$	viscosity of the gas
θ	–	spray angle
λ	m	wavelength
ϱ	kg/m^3	density of the liquid
ϱ_g	kg/m^3	density of the gas
σ	kg/s^2	surface tension
ω	1/s	angular frequency

a^*	$= a\eta^4/\sigma^3\varrho$	Kapitza number
Bo	$= d'^2 \varrho g/\sigma$	Bond number
d^*_{th}	$= d_{th}\sqrt{r\omega^2\varrho/\sigma}$	dimensionless thread diameter
d^*	$= d\sqrt{r\omega^2\varrho/\sigma}$	dimensionless drop diameter

$\Delta p^* = \Delta p\, d'/\sigma$ Laplace number
$\dot{V}^* = \dot{V}\sqrt{\varrho/(r\sigma)}/2\pi r$ throughput number
$Re_p = d'\sqrt{\Delta p\, \varrho}/\eta$ pressure Reynolds number
$s^* = s\sqrt{\varrho\, a/\sigma}$ dimensionless thread spacing
$We = w^2 d'\, \varrho/\sigma$ Weber number
$We_g = v^2 d'_p\, \varrho_g/\sigma$ gas-Weber number
$We_r = \varrho\, r^3 \omega^2/\sigma$ Weber number for rotary atomizers
$Z = \eta/\sqrt{d'\, \varrho\, \sigma}$ Ohnesorge number of the nozzle
$Z_d = \eta/\sqrt{d\, \varrho\, \sigma}$ Ohnesorge number of drops
$Z_r = \eta/\sqrt{r\, \varrho\, \sigma}$ Ohnesorge number for rotary atomizers

$\alpha = A_e/A$ contraction coefficient
$\varkappa = \delta x/A$ sheet number
$\mu = \dot{V}/(A\sqrt{2\,\Delta p/\varrho})$ discharge coefficient
$\mu_m = \dot{m}/\dot{m}_g$ mass flow ratio of liquid to gas
$\varphi = w/\sqrt{2\,\Delta p/\varrho}$ velocity coefficient

Atomization is the formation of drops from liquids in a gas phase or vacuum. The gas density ϱ_g and liquid density ϱ differ by one or more orders of magnitude. In most cases, the disintegration processes differ fundamentally from the dispersion processes in systems with small density differences.

In chemical engineering, liquid atomizers are employed in four major fields:

1) Atomization of liquid fuels
2) Production of granular products
3) Execution of mass-tranfer operations
4) Coating of surfaces

Depending on the given problem, the desired characteristics of the spray differ. A narrow drop-size distribution is often desired in the production of granular products. A small average drop size is usually necessary in mass-transfer processes. Intensive, turbulent mixing between gas and liquid, as well as a small drop size, is required in combustion and mass-transfer operations. In coating, all of the drops should precipitate on a body to form a uniform film.

Frequently, the atomized products are not pure liquids in the usual sense but rather solutions, melts, or suspensions, thus demanding different atomization techniques than do pure liquids. Trouble is often caused by wear and plugging, which however can be reduced to a minimum by suitable apparatus. With highly abrasive suspensions, easily exchangeable hard-metal or ceramic inserts are used at the points of high flow velocity. To avoid plugging, nozzles that produce small drops in spite of relatively large orifices are used: hollow cone nozzles, air blast nozzles, and rotary cup atomizers are especially advantageous.

Although there is a large variety of atomization apparatus and nozzles on the market, specialized designs are only considered here to a limited extent. Instead, the emphasis is on the fundamentals of drop formation and their significance in practice, which apply to all nozzles. In general, an estimation of the drop size is possible at least for low viscosity liquids and liquids obeying Newtonic flow behavior.

1. Fundamentals of Drop Formation

Drops can be formed generally by dripping, jet disintegration, sheet disintegration, and dispersions of liquids by gases.

1.1. Dripping and Jet Disintegration

In these drop formation ranges, the influence of the ambient gas is subordinate except for very high discharge velocities.

Dripping. The simplest case of drop formation is *dripping* directly from a solid surface. This occurs for low liquid throughputs, and a static equilibrium between the gravitational (or mass forces) and the surface force develops. The drop constricts from the base as soon as the gravitational force exceeds the surface force. Thus, drops of very uniform size are produced. This is often important in dosing, e.g., in pharmaceutical applications.

When dripping occurs from the underside of horizontal plates [1], [2], drops with diameter d arise upon complete wetting, (Fig. 1A),

$$2.9 < d\sqrt{\varrho\, g/\sigma} < 3.3 \quad (1)$$

where the surface tension σ and the density ϱ of the liquid in the gravitational field with acceleration g control the process. Drops can be obtained by dripping from horizontal edges, such as at

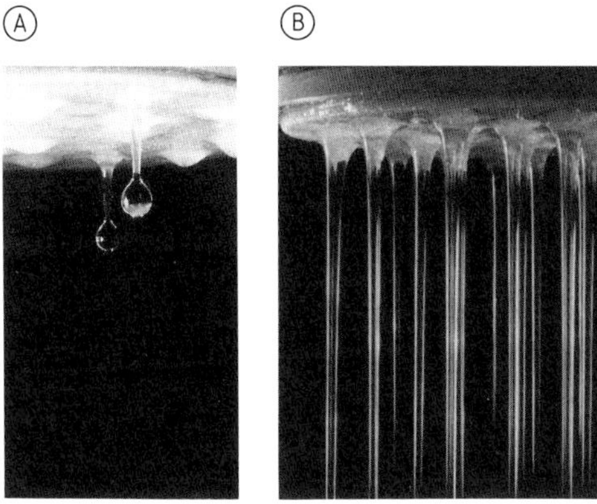

Figure 1 A. Dripping from the underside of a porons penetrated plate by the field of gravity (water)

Figure 1 B. Formation of regular spaced laminar jets on the underside of a porous penetrated plate by the field of gravity (water)

the end of a vertical draining surface [3], with diameter

$$d = 2.7\sqrt{\sigma/\varrho\, g} \tag{2}$$

In these dripping processes the influence of the viscosity η on drop size is small.

In practice, dripping from capillary tubes or tubular nozzles is a frequent case (Fig. 2 A). Here the drop diameter d depends mainly on the capillary tube diameter d' and the liquid–nozzle wetting [4], [5]. Figure 3, Curve a, shows the relationship between d/d' and the Bond number,

$$\sqrt{B_0} = d'\sqrt{\varrho\, g/\sigma} \tag{3}$$

For $\sqrt{B_0} > 6$, this case becomes indistinguishable from a saturated horizontal surface.

Jet Disintegration. With higher liquid throughput, i.e., increased discharge velocity w, liquid jets or threads form and first disintegrate into drops at some distance from the outlet point; this is then called *laminar jet disintegration* (Fig. 2 B). Jet formation occurs only above a certain discharge velocity with the pertinent minimum dynamic pressure $p_\varrho \sim \varrho w^2$. The ratio p_ϱ/p_σ is called the Weber number; when $p_\sigma \sim \sigma/d'$:

$$We = w^2\, d'\,\varrho/\sigma \tag{4}$$

For jet formation [6],

$$We > 8$$

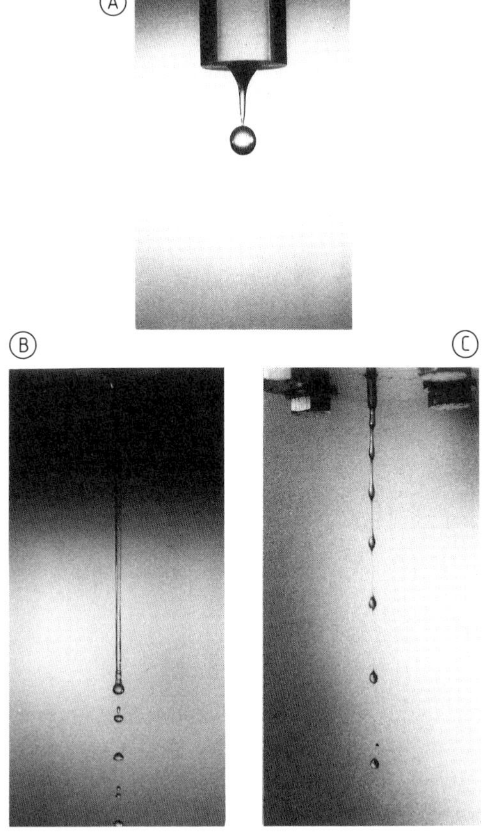

Figure 2 A. Dripping from tubular nozzle at very low flow velocities (water)

Figure 2 B. Laminar jet desintegration from tubular nozzle (water)

Figure 2 C. Laminar pulsed jet desintegration at a tubular nozzle forming droplets of uniform size (viscous oil)

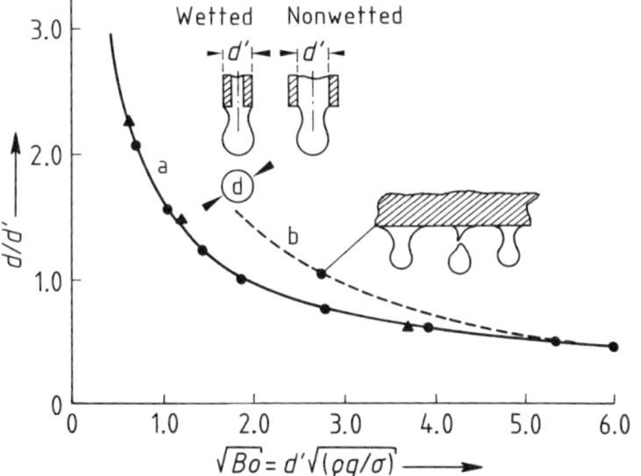

Figure 3. The ratio of drop diameter to capillary diameter (d/d') as a function of the Bond number B_0
a) Small capillary tubes [4], [5]; b) Wetted horizontal surfaces

A bundle of many capillary tubes connected in parallel increases the liquid throughput; in this way, a shower is obtained. If the tubes are too close together, however, jet coalescence takes place [7].

Drops form from liquid threads because the liquid threads are unstable and on their own form centrally symmetrical waves; these grow quickly when the wavelength is $\lambda > \pi d_{th}$ [8]. The jet diameter d_{th} can usually be set equal to the nozzle diameter d'. The fastest growing waves are those with the optimal wavelength [9],

$$\lambda_{opt} = \pi d_{th} \sqrt{2 + 6Z} \tag{5}$$

with $Z = \eta/\sqrt{\varrho \sigma d_{th}}$, where Z is the Ohnesorge number. For low viscosity liquids $\lambda_{opt} = 4.4\, d_{th}$. This produces drops with diameter

$$d = d_{th} \sqrt[6]{44 + 6Z} \tag{6}$$

or in the case of low viscosity $d \approx 1.9\, d_{th}$. The disintegration time for complete drop formation is given by the following equation [9]:

$$t = 12 \sqrt{\frac{\varrho d_{th}^3}{\sigma}} + 36 \frac{\eta d_{th}}{\sigma} \tag{7}$$

Within the laminar jet disintegration range, the jet length before complete disintegration increases with the discharge velocity w: $l_{th} = wt$. In natural jet disintegration all the drops do not have the same size, but rather, although relatively narrow, a drop size distribution arises. This is due to occasionally formed satellite drops between the main drops, to drop coalescence, and to the natural disturbance frequency. Monodisperse drops, i.e., drops of a uniform size, can be produced by oscillatory stimulation, either by jet pulsation or nozzle vibration [10]–[14] (see Fig. 2C). The required stimulation frequency is $f = w/\lambda_{opt}$.

On edges covered with liquid or on the underside of porous horizontal plates with sufficient liquid throughput, dripping points form on their own with regular spacing. The drop size is

$$d = 2.85\, d' \sqrt{\frac{1}{B_0}} = 2.85 \sqrt{\frac{\sigma}{\varrho g}}$$

(Fig. 3, Curve b). The same process occurs on rotating atomizers such as atomizer disk-cups or rotating porous rings [15], [16]. Centrifugal acceleration, $a = r\omega^2$, then replaces the gravitational acceleration g. The liquid threads formed at the dripping points also disintegrate according to the laws of jet disintegration. On edges covered with liquid or on rotating atomizer cups, the average spacing s between the liquid threads is $5.6 < s\sqrt{\varrho a/\sigma} < 7.0$ [3]. On the horizontal underside of saturated porous plates, a regular triangular dripping pattern forms, with a jet spacing of $s^* = s\sqrt{\varrho g/\sigma} = 7.6$ [2] (see Fig. 1 B).

As the liquid throughput is increased, the thread thickness at the "dripping" points increases without a significant increase in drop size. This is due to the greater jet length, which permits pronounced jet contraction. In the laminar jet range, the drop size is ca. $d = 1.1\sqrt{\sigma/\varrho g}$, about 1/3 the drop size in the dripping range. If throughput is increased further, the liquid threads progressively coalesce until finally a liquid sheet is formed on the dripping edges. In the case of plates, with increasing throughput jet coalescence also takes place and finally leads to an irregular pattern of thick jets [2].

In addition to the Weber number and the Ohnesorge number, a further characteristic number takes account of the influence of the acceleration, namely the Kapitza number $a^* = a\eta^4/\sigma^3\varrho$, which describes the flow conditions for *capillary tubes* vertical to the gravitational field or in a centrifugal field. In the case of drop or jet formation due to gravity the Kapitza number is defined as $a^* = g\eta^4/\sigma^3\varrho$. Figure 4 shows a diagram for the various flow conditions. When Z (i.e., liquid properties and nozzle diameter) is given, as the Weber number (i.e., flow velocity w) increases, the sequence of flow conditions is dripping, laminar jet disintegration, sinusodial jet disintegration, and turbulent jet disintegration. If the capillary tubes are large, the liquid column in the capillary becomes unstable and can break: the liquid partially leaks out. This occurs when $d'\sqrt{\varrho g/\sigma} > 5.2$ for $Z < 0.43 \cdot a^{*1/4}$ [17], [18].

Sinusodial jet disintegration is a transitory condition during which the jet is helically deformed [19]. A distribution of drop sizes is produced, but the sizes lie in the range of the jet diameter. With further increasing throughput there is *turbulent flow* in the jet (Fig. 5A, B; 6), which leads to smaller drops. This also occurs at very low ambient pressures, hence the conclusion that the gas pressure has only a subordinate influence. The turbulence in the jet can be significantly pronounced by special flow duct geometries, as will be shown in Section 2.1. At extremely high discharge velocities, on the order of magnitude of the velocity of sound of the gas, the gas influence becomes increasingly noticeable.

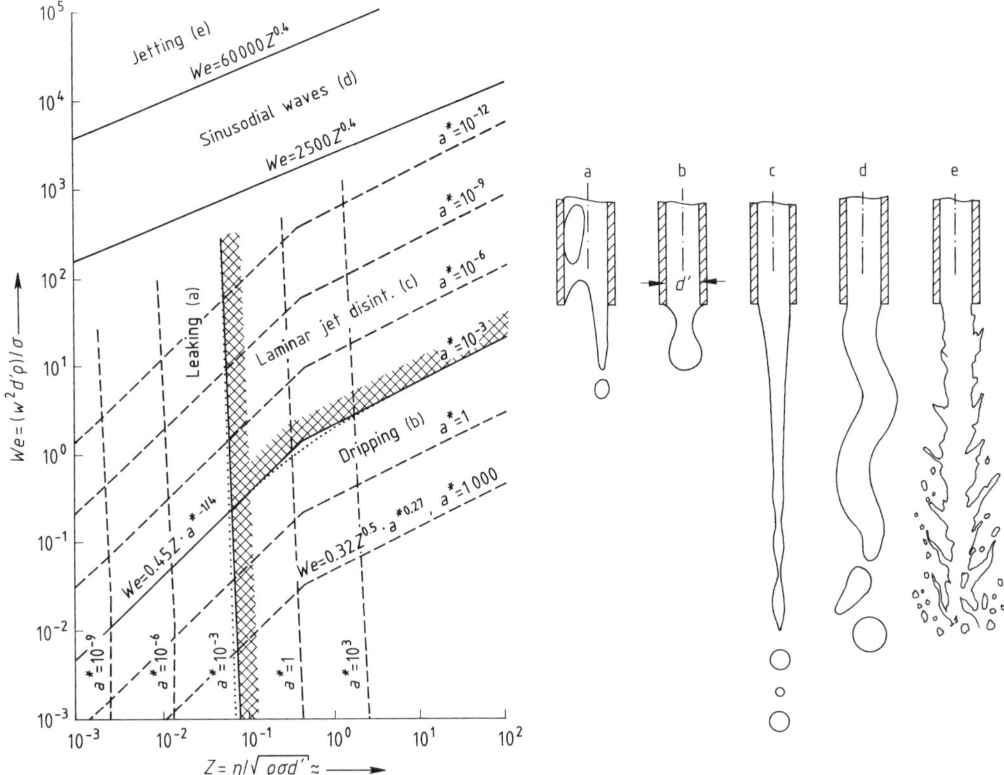

Figure 4. Diagram showing regions of various flow conditions for capillary tubes or nozzles either vertical in the gravitational field or in a centrifugal field with radial tube axis

At low discharge velocities the flow condition also depends on the Kapitza number a^*. As an example, the boundaries for $a^* = 10^{-3}$ are shaded.

$a^* = g\,\eta^4/\sigma^3\,\varrho$ or $a^* = a\eta^4/\sigma^3\varrho$ and $a = r\,\omega^2$

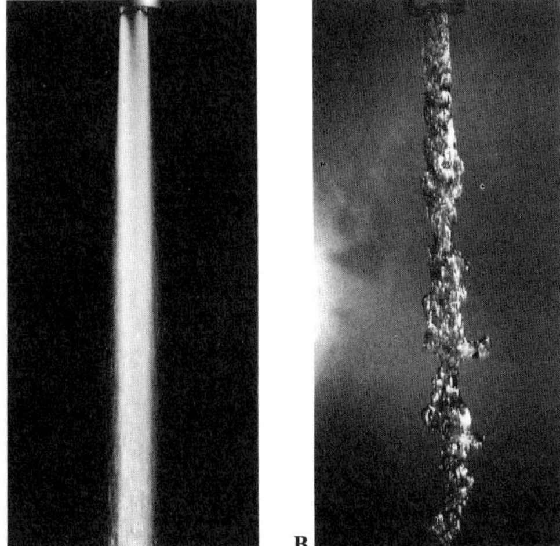

Figure 5 A. Visual appearance of turbulent jet from a tubular nozzle at high exit velocities (water)

Figure 5 B. Short time photograph of a turbulent jet emerging from a tubular nozzle at high exit velocities (water)

1.2. Sheet or Lamella Disintegration

The production of fine drops using capillary tubes has the disadvantage that fine nozzles and high pressures are needed. A better approach is to first produce thin sheets or lamella, which subsequently break up into drops by several breaking mechanisms:

1) Edge contraction
2) Aerodynamic wave formation
3) Turbulent disturbances

Frequently these disintegration mechanisms are superimposed.

In contrast to liquid threads, in which wave formation with $\lambda > \pi d_{th}$ leads to a smaller surface and thus to increasing instability, sheets are stable because each deformation enlarges the surface and must work against a restoring force [20]–[22].

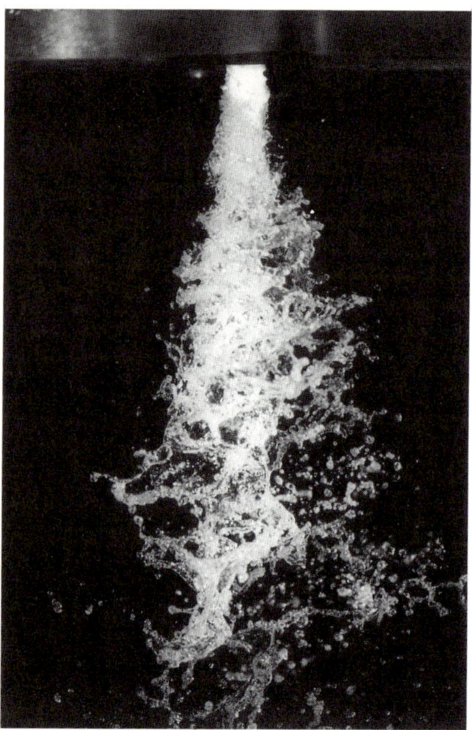

Figure 6. Jet with high degree of turbulence from a knee shaped nozzle duct (water)

Most industrial atomizers form sheets that expand transversely to the main flow direction and thus decrease in thickness δ as the distance x increases. The product δx is generally constant and is designated as the sheet thickness parameter $K = \delta x$ [23], [24]. If K is related to the cross-sectional area of the nozzle outlet (orifice) A, the sheet number $\varkappa = \delta x/A$ is obtained [25]. With circular orifices, $A = d'^2\,\pi/4$ and $\varkappa = 4\,\delta x/\pi d'^2$. For small Weber numbers, $\varkappa\,We = \varkappa\,d'\,\varrho\,w^2/\sigma < 220$, *edge contraction* takes place. The sheet edge contracts in the form of a liquid bulge, which breaks up into relatively coarse drops according to the principles of jet disintegration. At the site of the bulge the propagation velocity w of the sheet is equal to the contraction velocity, or $w^2\,\varrho\,\delta/\sigma = 2$ (see also Fig. 7).

For $\varkappa\,We > 220$, the sheet begins to oscillate due to the influence of the gas. With increasing distance, the quickly growing amplitudes cause the sheet to break up. According to one model, liquid threads are formed by contraction from the broken-up sheet parts and these threads then break up into drops [26], [27].

For low viscosity liquids the drop size is

$$d_{32} = C_1\,(\delta x)^{1/3}\left(\frac{\sigma}{v_{\mathrm{rel}}^2\,\varrho}\right)^{1/3}\left(\frac{\varrho}{\varrho_{\mathrm{g}}}\right)^{1/6} \tag{8}$$

Here, $1.2 < C_1 < 1.7$, v_{rel} is the relative velocity between sheet and gas, ϱ_{g} is the gas density, and d_{32} is the Sauter mean diameter of the drop size distribution. The Sauter mean diameter has the same ratio of surface to drop volume as the whole drop distribution. This *aerodynamic wave formation* takes place for most sheet nozzles and most rotating atomizers when they are operated in the range of sheet formation under conditions of low turbulence. At very high velocities aerodynamic wave formation recedes and *turbulent disintegration* comes up which shows a stronger decrease of drop size with increasing discharge velocity, depending on degree of turbulence.

1.3. Dispersion of Liquids by Gas

If a gas jet strikes a liquid surface with sufficient relative velocity v_{rel}, then the liquid is dispersed. Disintegration of drops in gas flows has been extensively examined [28]–[30]. Dispersion occurs when the dynamic pressure of the gas exceeds the internal pressure in the drops. The ratio of dynamic pressure to surface pressure is described by the gas Weber number, $We_{\mathrm{dg}} = v_{\mathrm{rel}}^2\,d\varrho_{\mathrm{g}}/\sigma$. Blowing excites the drop surface into oscillation, thus promoting disintegration. Increasing the viscosity further damps the oscillation. For $Z_{\mathrm{d}} = \eta/\sqrt{\varrho\,d\,\sigma} > 4$, dispersion of drops in gas flows is no longer possible. In terms of the Weber number, the splitting limit can be approximated by

$$We_{\mathrm{dg}} = 12 + 14\left(\frac{\eta}{\sqrt{d\,\varrho\,\sigma}}\right)^{0.8} \tag{9}$$

At high relative velocities, the drop is blown up to a hemispherical sheet, which subsequently bursts into small drops. The types of splitting arising at various Weber and Ohnesorge numbers are described in more detail in [31].

A liquid film that is blown over by a gas forms waves from whose peaks drops detach. Although this process has been examined repeatedly [32]–[35], a uniform method of calculation has yet to be found. According to KATAOKA and co-workers [36], for drops produced in a two-phase film flow in tubes with a diameter d′,

$$\frac{d_{32}}{d'} = 8 \cdot 10^{-3}(We_{\mathrm{g}} \cdot Z)^{-2/3} \tag{10}$$

where $We_{\mathrm{g}} = d' v_{\mathrm{g}}^2 \varrho_{\mathrm{g}}/\sigma$, $Z = \eta/\sqrt{\varrho\,\sigma\,d'}$, and v_{g} is the gas velocity. Drop formation on pneumatic atomizers is described most often by the relationship

$$\frac{d_{32}}{d'_{\mathrm{p}}} \approx We_{\mathrm{g}}^{j}(1 + C_2\,\mu_{\mathrm{m}})^{k} \tag{11}$$

where d'_{p} is a characteristic nozzle dimension. The principle of the conservation of linear momentum yields $C_2 = 1$ and $k = -j$. However, experimental results deviate from these values. Equation 11 shows that for small loading ratios, $\mu_{\mathrm{m}} = \dot{m}/\dot{m}_{\mathrm{g}} < 0.1$, a minimum drop size is obtained. Up to $\eta \approx 0.5$ Pa s the viscosity of the liquid has no noticeable influence on the drop sizes.

2. Atomizing Apparatus

The basic principles of drop formation can be applied to a great number of technically employed atomizers.

2.1. Single-Substance Pressure Nozzles

Single-substance pressure nozzles are the simplest atomization devices. They convert the liquid pressure Δp into velocity. The geometry of the nozzle determines the conversion efficiency. The discharge velocity is $w = \varphi\sqrt{2\,\Delta p/\varrho}$; φ is the velocity coefficient. The throughput of single-substance pressure nozzles is $\dot{V} = A\alpha\varphi\sqrt{2\,\Delta p/\varrho}$ where $\alpha = A_{\mathrm{e}}/A$ is the contraction number. A_{e} is the cross-sectional area of the emerging jet and A is the nozzle cross-sectional area. The discharge coefficient is $\mu = \alpha\varphi$. The quantities α, φ and μ depend on the nozzle geometry and on the pressure Reynolds number $Re_{\mathrm{p}} = d'\sqrt{\Delta p\,\varrho}/\eta$ respectively. At high pressures or low viscosities, φ and α and therefore μ reach a constant limiting value, which is frequently used for calculation in practice. In practically all single-substance pressure nozzles, variation in throughput leads to a different drop size; hence, throughput variation is often only possible by switching single nozzles on or off.

2.1.1. Sheet Nozzles

For production of extremely small drops, thin sheets are produced and subsequently broken into drops. Especially thin sheets, compared to the orifice, arise from hollow cone nozzles. For more viscous liquids, fan jet nozzles are usually employed because they still ensure a high discharge velocity.

Hollow Cone Nozzles. Hollow cone nozzles have a swirl chamber into which the liquid flows tangentially through one or more inlets (see Fig. 8 A and B). A gas core forms in the middle of the swirl chamber. At the orifice, with diameter d', a hyperbolic sheet forms; with increasing distance it becomes thinner and breaks up into drops by aerodynamic wave formation (Fig. 9). A minimum pressure is required [25], [37]. For liquids with low viscosity this minimum pressure is a result of the inequality

$$\varkappa\,\Delta p^{*}\varphi^{2} \geq 110 \tag{12}$$

with $\Delta p^{*} = \Delta p\,d'/\sigma$. Common values for the ve-

Figure 7. Tulip formation and edge contraction of a lamella formed at an hollow cone nozzle at low exit velocities (water)

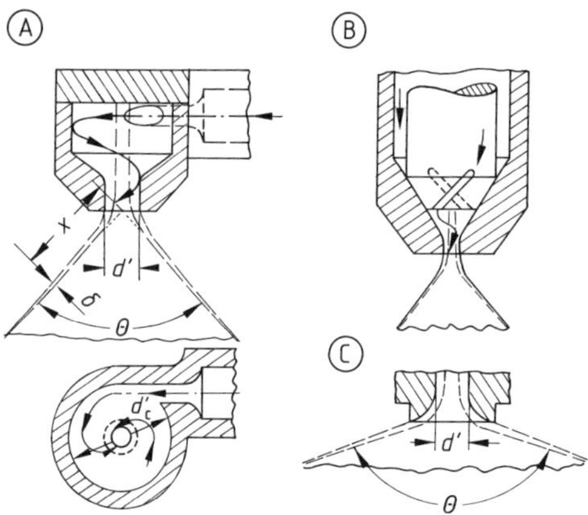

Figure 8. Hollow cone nozzles, also called swirl atomizers A) Nozzle with tangential inlet (side and top views); B) Nozzle with slitted cone insertion; C) Deviation aperture

locity coefficient for hollow cone nozzles operated at high Re_p numbers lie in the range $0.7 < \varphi < 0.8$. The sheet number is obtained from the spray angle θ,

$$\varkappa = \frac{\mu}{2\pi\,\varphi\sin(\theta/2)} \tag{13}$$

with $\mu = 4\,\dot{V}/d'^2\pi\,(2\,\Delta p/\varrho)^{1/2}$. In practice, the dependence of $\dot{V}$ on Δp noted in the nozzle manufacturer's pamphlet is best taken. For low viscosity liquids, sheet number and discharge velocity coefficient can also be calculated from potential theory [38], [39]. Considerable deviations, however, arise even at relatively low viscosities [40]. In this case, the velocity coefficient can then drop to $\varphi = 0.35$ [41].

The spray angle decreases as the viscosity increases. The throughput, however, increases because in the nozzle the annular area occupied by the liquid increases while the gas core area shrinks. The drop size is approximately [25]

$$\frac{d_{32}}{d'} = 1.2\left(\frac{\varkappa}{2\,\varphi^2\,\Delta p^*}\right)^{1/3}\left(\frac{\varrho_g}{\varrho}\right)^{-1/6} \tag{14}$$

In conical spray jets formed at hollow cone nozzles or swirl nozzles as shown in Figure 8 A and 8 B a graduation arises, which leads to retro-transport of small drops in a back-flow zone. The large drops remain on the jet edge (Fig. 9) [42], [43].

The spray angle can be enlarged by a special aperture as in Figure 8 C, where the Coanda Effect is utilized [25].

Figure 9. Aerodynamic wave formation on a lamella formed at an hollow cone, swirle type nozzle resp. at high exit velocities (water)

To adjust the throughput, part of the liquid can be removed from the swirl chamber [44] without significant effect on the drop size.

Fan Jet Nozzles. Sector-shaped sheets can be produced by suitable nozzle geometries. Figure 10 shows two designs for fan jet nozzles. A frequent design consists of a nozzle with a spherical bore slitted by a cross groove (Fig. 10 A). Another construction called tongue nozzle is shown in Figure 10 B. The circular jet is deflected by an inclined plate producing a sector shaped lamella.

The flow through such a nozzle causes streamlines that diverge in the slit direction but converge transversally, forming a sector-shaped sheet fan (Fig. 11). Discharge coefficient, velocity coefficient, and sheet number (μ, φ, $\varkappa$) must be determined by experiment [23], [25]. With low viscosity liquids or for $Re_p > 1000$, fan jet nozzles have much higher $\varkappa$ values, in the range $0.4 < \varkappa < 0.7$, than do hollow cone nozzles. For lower Re_p numbers or highly viscous liquids, however, the velocity coefficient decreases much less than in the case of hollow cone nozzles. The discharge coefficient μ also decreases, which counteracts an increase in sheet thickness. For these reasons, fan jet nozzles are also suitable for highly viscous liquids. Approximated it is true

$$\varkappa = \frac{360\,\mu}{2\pi\theta\,\varphi} \tag{15}$$

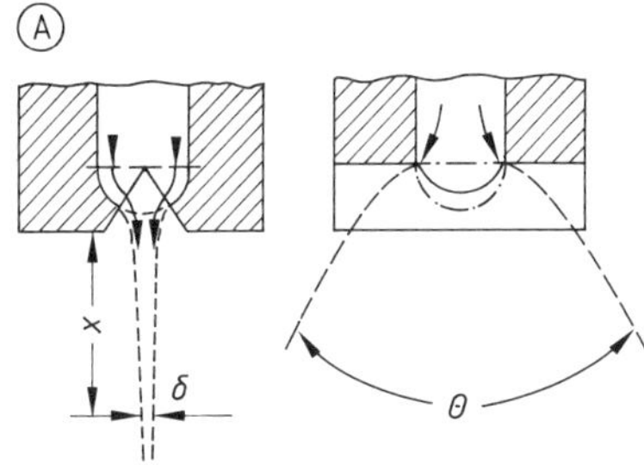

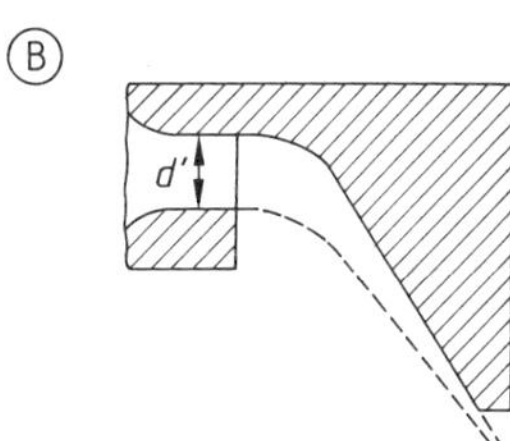

Figure 10. Fan jet nozzles
A) Nozzle with spherical bore and transverse slit; B) Tongue nozzle

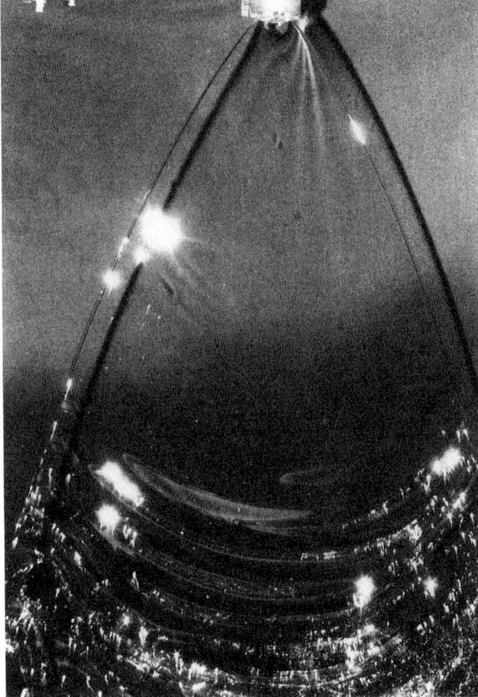

Figure 11. Aerodynamic wave formation of a fluid sheet emerging from a fan jet nozzle (viscous oil)

The projection surface A of the contour in the flow direction should be used for the nozzle cross-sectional area. The equivalent nozzle diameter is then $d' = \sqrt{4\,A/\pi}$. Calculation of the drop size is equal to Equation (14). For low viscosity liquids the velocity coefficient lies in the range $0.9 > \varphi > 0.8$.

Fan jet nozzles are employed for atomizing lacquers and the like because of the still relatively high φ values even at higher viscosities (airless atomization). They are also used for coating of moving sheets and in washing plants, which require high jet impulse. Deceleration of the spray jet by the ambient gas is less than in the case of hollow cone nozzles [45].

Turbulence Nozzles. For uniform spraying of large surfaces, full cone nozzles are frequently employed. Compared with swirl nozzles they produce large drops [46]. By insertion of a swirl body in the hollow cone nozzle a swirl and strong turbulence are simultaneously produced (see Fig. 12 C and D). The swirl also makes large spray angles, up to 120°, possible. Formation of an air core must be prevented because otherwise a hollow jet forms. Other possibilities of turbulence amplifications are the Borda aperture, as in Figure 12 A, or sharp edged deflections in the nozzle duct (see Fig. 12 B [103]). Jet disintegration is caused by the strong internal turbulence. Here, both temporally and spatially irregular sheet structures develop and form drops (see Fig. 6). This process is also effective in vacuum.

Turbulent jets emerging from smooth tube nozzles disintegrate to fine drops only at high discharge velocities. The spray angle is small. According to Glaser [47] the relative drop size from a nozzle with a turbulence-amplification Borda aperture (Fig. 12 A) is

$$\frac{d_{32}}{d'} = 400\,\Delta p^{*\,-3/4} \left(\frac{\varrho}{\varrho_g}\right)^{-1/5} Z^{1/7} \tag{16}$$

in the range $20\,000 < \Delta p^* < 100\,000$ and $2 \times 10^{-3} < Z < 7 \times 10^{-2}$. With extremely high discharge velocities, of the order of the velocity of sound in the gas, fine drops can also be produced by smooth nozzles, e.g., diesel injection [48]–[51].

For a nozzle as in Figure 12 D spraying low viscosity liquids into air in the range $2 \times 10^4 < \Delta p^* < 1.5 \times 10^5$, measurements with a laser diffraction spectrometer showed that

$$\frac{d_{32}}{d'} = C_3\,(\Delta p^*)^n \tag{17}$$

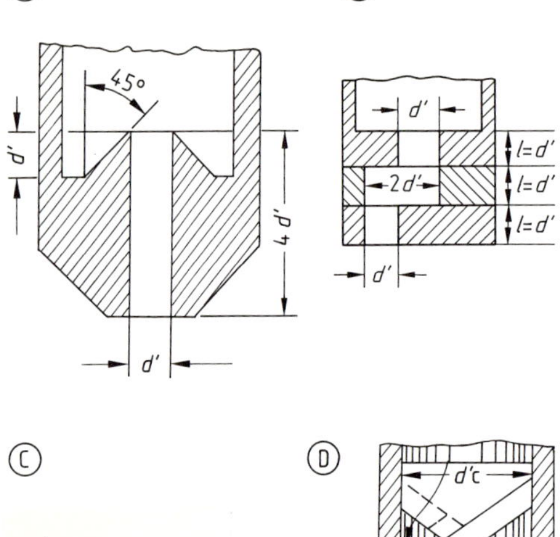

Figure 12. Full cone nozzles (turbulence nozzles)
A) Nozzle with Borda aperture; B) Nozzle with knee shaped, sharp edged duct; C) Nozzle with turbulence-amplifying and swirl producing insertion to avoid the formation of an air core, the insert has a central drill hole; D) Nozzle with sharp edged, swirl and turbulence inducing insert

For the ratio $d'_c/d' = 2.7$, a spray angle $\theta = 110°$ and the parameters $C_3 = 23$, $n = -0.6$, and $\mu = 0.72$ were found.

2.2. Pneumatic Atomizers

By using gases, small drops can be produced even at low pressures because high relative velocities arise. The cross section of the liquid channel plays a subordinate role, so it can be selected large enough to avoid clogging. Most atomizer nozzles with external mixing depend on the principle of *prefilming* [52]. First, the liquid is spread by the gas flow as a thin film on the atomizer surface. A wave flow arises on the film surface and is blown over an edge, where the actual drop formation occurs [53]. The thinner the liquid film on the tearing edge the smaller the drops.

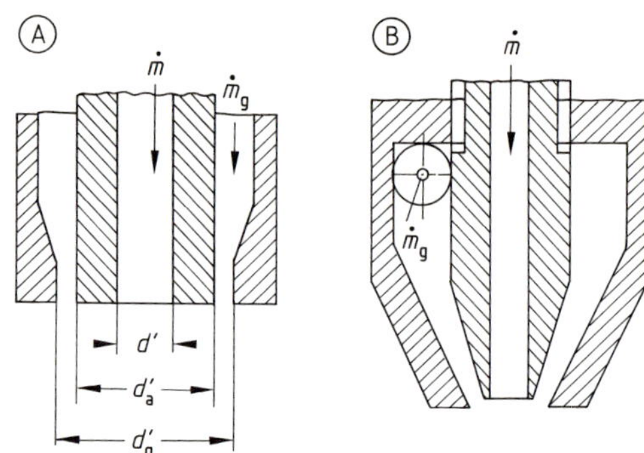

Figure 13. Pneumatic or twin-fluid nozzles with external mixing
A) Nozzle with parallel flow; B) Nozzle with adjustable gas annulus and tangential gas inlet for spray angle $\theta > 30°$

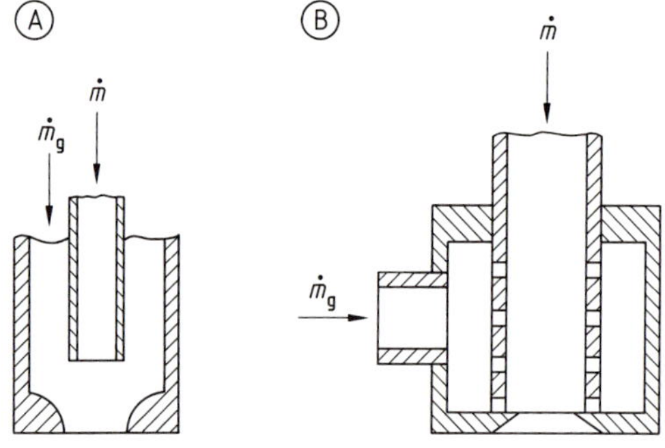

Figure 14. Pneumatic nozzles with internal mixing
A) Nozzle with concentric ducts for gas and liquid;
B) Nozzle with predispersion chamber [59]

The drop sizes are given by the equation

$$\frac{d_{32}}{d'_p} = 8\, We_g^{-1/2} \left(\frac{\varrho_g}{\varrho}\right)^{-1/2} (1 + 13\,\mu_m)^{1/2} \qquad (18)$$

in the range $0 < \eta < 1$ Pa s, $20 < v_g < 500$ m/s, $1 < d'_p < 12$, and $0 < \mu_m < 20$ with $We_g = v_g^2\, \varrho_g\, d'_p / \sigma$ [54].

This equation correlates well over a wide range with the results of the frequently cited Reference [55], but also considers the influence of nozzle geometry. For a comparison see also [56].

The definition of d'_p (see Fig. 13 A) is $d'_p = \sqrt{d_g'^2 - d'^2}$. Normal loading of twin fluid atomizers with external mixing lies in the range $0.1 < \mu_m < 5$. The spray angle lies in the range $20° < \theta < 40°$. To enlarge the spray angle, the air is frequently fed with swirl (Fig. 13B). For atomization of lacquers using twin fluid atomizers, see [57].

While nozzles with external mixing produce suction in the liquid channel, in nozzles with internal mixing (see Fig. 14 A and B) the liquid must be fed or dosed under pressure. The two-phase mixture emerges from the nozzle aperture with the velocity of sound, which is in the range $20 < w < 50$ m/s. Then the gas suddenly expands to ambient pressure, suddenly accelerating and breaking up the large drops that have been carried along (see Section 1.3 and also [58]–[60]).

2.3. Atomization with Propellants

Spray cans contain propellants, formerly halogenated hydrocarbons, now simple hydrocarbons, e.g., butane, together with the substance to be sprayed. When the valve is opened, the liquid components flow through a nozzle, usually a hollow cone nozzle. The propellant vaporizes in the nozzle and promotes disintegration of the emerging sheet by the expanding bubbles of gas [61].

2.4. Rotary Atomizers

Rotary atomizers produce a bell-shaped spray, which, e.g., is utilized in spray drying. A tower with a large diameter is required so that the drops flying furthest do not touch the side of the tower before drying. The drying gas therefore often is supplied through an annular clearance near to the atomizer in order to achieve a more conical spray pattern together with high mass-transfer rates.

The liquid is fed centrally on rotating cups, disks, or plates. Figure 15 shows two types of rotating atomizers. The liquid flows to the edge of the atomizer due to the centrifugal force. The drop formation mechanisms depend on the amount of liquid. The processes correspond to the flowing from edges in the gravitational field (Section 1.1). For calculation, however, instead of the acceleration of gravity g, the centrifugal acceleration $a = r\omega^2$ is used, where r is the atomizer radius and ω the angular frequency of rotation. Dripping is attained at low throughputs. The drop size is calculated with Equation (2). Increasing the throughput causes regular liquid threads to form at equal intervals along the circumference of the atomizer. This occurs [3], [62], [63] when

$$\log(Y_C - Y_{C,S}) = \log\left(\frac{\dot{m}_F}{\dot{m}_S}\right) + \log(X_C - X_{C,R}) \quad (19)$$

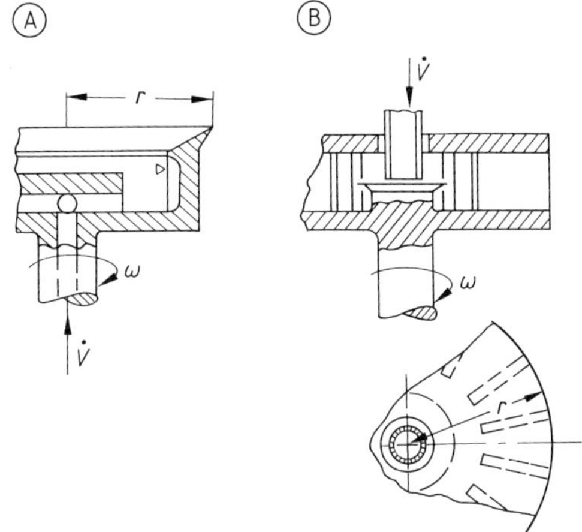

Figure 15. Rotary atomizer
A) Cup atomizer with distribution nuts and liquid feed through a hollow shaft; B) Vaned atomizer disk

where $\dot{V}^* = \dot{V}\sqrt{\varrho/\sigma r}/2\pi r$, $We_r = \varrho r^3\omega^2/\sigma$, and $Z_r = \eta/\sqrt{\varrho\sigma r}$; $\dot{V}$ is the liquid volume flow fed onto the atomizer. The range of thread formation is preferred to other ranges because it leads to a narrow drop size distribution. The thickness of the liquid threads is [62], [64], [65]

$$d_{th}^* = 0.85\, We_r^{1/8}\, Z_r^{1/3}\, \dot{V}^{*1/6} \quad (20)$$

with $d_{th}^* = d_{th}\sqrt{r\,\omega^2\,\varrho/\sigma}$. The drop size arising by jet disintegration is calculated with Equation (6). The result is the diameter d of the main drops. Coalescence allows double or triple drops to form, leading to a multimodal distribution [3]. Moreover, with low viscosity liquids, satellite drops, with $d_s \approx d/4$, are also formed.

Most rotary atomizers, however, are operated in the range of sheet formation due to the higher throughput. The drop size distribution is much broader than in the thread range, $2 < d_{max}/d_{32} < 3$. Furthermore, a considerable amount of fine material is obtained. The transition from thread to sheet formation for $Z < 0.01$ occurs [62], [63] at

$$\dot{V}^* \geq 0.34\, We_r^{-0.3}\, Z_r^{-1/6} \quad (21)$$

For higher viscosities and $Z > 0.01$

$$\dot{V}^* \geq 0.3\, We_r^{-1/3}\, Z_r^{-1/3} \quad (22)$$

The Sauter mean diameter of the drop size distribution can be approximated by [66]:

$$d_{32}^* = 1.44\,(We_r \cdot Z_r\, \dot{V}^*)^{0.20} \quad (23)$$

The relationships in Equations (19)–(23) can be used for vaned disks when in the definition for $\dot{V}^*$—the entire quantity [(circumference + vane height) × vane number]—is inserted instead of the circumference $2\pi r$. Especially on vaned disks the liquid distribution is often imperfect so that local thick liquid strands cause thick drops. Special constructions of rotary atomizers are the rotating porous body [15], [16], the twin roller atomizer [67], [68], and the spin top atomizer with pneumatic drive [69]: Rotary atomizers are also employed in combination with electrostatic charging, among other effects, for lacquer coating of car bodies [70].

2.5. Ultrasonic Atomization

Various principles can be differentiated in ultrasonic atomization [71], namely

1) Capillary waves [72]–[75]
2) Cavitation [74]
3) Stationary ultrasonic waves in a gas [76]
4) Excited jet disintegration (Section 1.1)

The *capillary wave* is the most important; it enables production of drops with extremely low residual impulse. The drop size can be approximated by

$$d_{32} \approx 0.8 \sqrt[3]{\frac{\pi \sigma}{\varrho f^2}} \tag{24}$$

where f is the excitation frequency. With water, when $f = 100$ kHz, a medium drop size of $d_{32} = 25$ μm results. Here the atomization rate of 10 cm^3/cm^2 s (= cm s^{-1}) is obtained at an oscillator amplitude $A = 5$ μm. Capillary wave atomization is only suitable for low viscosity liquids, $\eta < 20$ mPa s, and produces a drop distribution.

2.6. Electrostatic Atomization

In addition to the dispersion of liquids by high potentials or field forces, the charging of drops produced mechanically is also sometimes advantageous [77], [78]. In both cases, charging causes the spray particles to drift apart. In this way, coalescence is avoided. Subsequently, the droplets can be deposited with a high yield on an oppositely charged, or simply a grounded, body. If a drop is charged, it breaks apart when the repulsive force of the charges on the surface exceeds the surface force [79]–[84],

$$d \pi \sigma \leq \frac{Q^2}{8 \pi \varepsilon d^2} \tag{25}$$

where Q is the charge on the drop and ε the dielectric constant of the liquid. The same mechanism is valid for the detachment of drops from highly curved liquid surfaces, e.g., the tip of a capillary tube. Even at relatively low potentials a high field strength arises and produces high charges.

The throughput of electrostatic atomizers is relatively low, $\dot{m}_{max} \approx 1$ kg/h. However at high field strengths, $E \geq 10^{10}$ V/m, drops can be produced in the range $0.1 < d < 5$ μm, the diameter depending on the liquid characteristics. The usual potentials lie in the ranges $5 < U < 100$ kV.

3. Drop Size Measurements and Drop Size Distributions

Drop sizes can be measured by short time photography. A holographic figure, which enables evaluation of the spatial spray and presents no problems with the focus depth, is even better. In this way, in combination with double exposures at defined time intervals, the magnitude and direction of the velocity can also be determined [85]. The laser diffraction method can measure drop sizes in the range $0.5 < d <$ 500 μm quickly and online. The method presents the drop size distribution found in the measuring volume (laser beam).

In addition to the measurement of single drops by the scattered light counting method [86]–[87], recently laser–Doppler anemometry has been employed in combination with the scattered light method for simultaneous velocity and drop size determination [88]–[91]. Mechanical methods [92] are sometimes also employed.

All known distribution functions are used as suitable model functions for the measured drop size distributions: RRSB [93], [94] and log normal distribution [95], [96] for example. The ULLN distribution has the advantage that it permits the definition of a maximum drop diameter [97], [98]. For the significance of drop size distributions in jet scrubbers, see [99].

An approximation of the distribution width in different atomization processes in given in [97], [100]–[102] and Table 1.

Table 1. Distribution widths in atomization processes

Atomization process	Measure of distribution	
	$d_{v,0.5}/d_{32}$	$d_{max}/d_{v,0.5}$
Dripping	1	1.1
Natural laminar jet disintegration	1.05–1.1	1.4
Sheet nozzles	1.1–1.3	1.5–2.5
Turbulence nozzles	1.3–1.5	2.5–3.5
Twin-fluid atomizers	1.1–1.5	2.5–3.5
Rotary atomizer		
threads	1.05–1.1	1.4–2.0
sheets	1.1–1.5	2.5–3.5
Capillary wave atomizer	1.05–1.3	1.5–2.5

The efficiency of the atomization processes and the ratio of surface formation energy to the expended discharge energy is usually low and, depending on the process, lies in the range $0.005 < e < 0.01$. Higher values up to $e \approx 0.2$ can only be reached in the laminar jet region [103].

Special cases are the atomization of metals [104]–[106] and drop impingement [107], [108]. For the application of similarity theory to atomizers see [109]. The drag coefficients of drops in gas flows can be found in [110]–[113].

4. References

General References

R. Clift, J. R. Grace, M. E. Weber: *Bubbles, Drops, and Particles,* Academic Press, New York 1978.

The *Journal of Atomization and Spray Technology* (Elsevier Applied Sience), which began in 1985 with vol. 1, is a good source of information on spraying and atomization of liquids. Other good sources are the *Proceedings of the International Congresses of Liquid Atomization and Spraying Systems* (ICLASS): 1) Tokyo (1978), 2) Madison, Wisc., (1982), and 3) London (1985). The Fourth Congress is to be held in Sendai, Japan (1988).

Specific References

Chapter 1

[1] Y. Tanasawa, S. Toyoda, *Technol. Rep. Tohoku Univ.* **21** (1955) 135.
[2] P. Walzel, U. Klaumünzner, *Ger. Chem. Eng. (Engl. Transl.)* **4** (1981) 154–160.
[3] M. R. Mehrhardt, Dissertation, TU Berlin 1978.
[4] W. D. Harkins, F. E. Brown, *J. Am. Chem. Soc.* **41** (1919) 499–524.
[5] M. Hozawa, T. Tsukada, N. Imaishi, K. Fujinawa, *J. Chem. Eng. Jpn.* **14** (1981) no. 5, 358–364.
[6] M. R. Lindblad, J. M. Schneider, *J. Sci. Instrum.* **42** (1965) 635–658.
[7] P. Walzel, *Chem. Ing. Tech.* **52** (1980) no. 8, 652–654.
[8] Lord Rayleigh, *Proc. London Math. Soc.* **10** (1978) 4/13.
[9] C. Weber, *Z. Angew. Math. Mech.* **11** (1931) 136–154.
[10] R. N. Berglund, Y. H. Lin, *Environ. Sci. Technol.* **7** (1973) no. 2, 147–153.
[11] H. Hausner, P. W. Taubenblat: Proc. Int. Powder Met. Conf. Princeton, N. J., (1976).
[12] P. Walzel, *Chem. Ing. Techn.* 51 (1979) no. 5, MS 692–679.
[13] P. Schümmer, K. Tebel, *Chem. Ing. Tech.* **12** (1981) MS 961/81.
[14] Ch. Müller, F. Widmer, *Swiss. Chem.* **8** (1986) no. 2a, 31–43.
[15] P. Schmidt, *Chem. Ing. Tech.* **39** (1967) nos. 5/6, 375–379.
[16] W. Gösele, *Chem. Ing. Tech.* **40** (1968) nos. 1/2, 37–43.
[17] M. Dosoudil, *Chem. Ing. Tech.* **43** (1971) no. 21, 1172–1176.
[18] P. Walzel, H. Michalski, *VI Verfahrenstechnik* **14** (1980) no. 3, 157–159.
[19] W. v. Ohnesorge, *Z. Angew. Math. Mech.* **16** (1936) 355–358.
[20] W. N. Bond, *Proc. Phys. Soc. London* **47** (1935) no. 4, 549–558.
[21] G. I. Taylor, *Proc. R. Soc. London. Ser. A* **253** (1959) 313–321.
[22] H. B. Squire, *Br. J. Appl. Phys.* **4** (1953) 167–169.
[23] N. Dombrowski, D. Hasson, D. E. Ward, *Chem. Eng. Sci.* **12** (1960) 35–50.
[24] D. Hasson, J. Mizrahi, *Trans. Inst. Chem. Eng.* **39** (1961) 415–422.
[25] P. Walzel, *Ger. Chem. Eng. (Engl. Transl.)* **7** (1984) 1–12.
[26] R. P. Fraser, P. Eisenklam, N. Dombrowski, D. Hasson, *AIChE J.* **8** (1962) 672–680.
[27] N. Dombrowski, W. R. Johns, *Chem. Eng. Sci.* **18** (1963) 203–214.
[28] J. O. Hinze, *AIChE J.* **1** (1955) no. 3, 289–295.
[29] F. C. Haas, *AIChE J.* **10** (1964) no. 6, 920–924.
[30] R. S. Brodkey: *The Phenomena of Fluid Motion*, Addison-Wesley, Reading, Mass., 1967.
[31] S. A. Krzeczkowski, *Int. J. Multiphase Flow* **6** (1980) 227–239.
[32] J. W. Miles, *J. Fluid Mech.* **3** (1957) 185–204.
[33] S. Ostrach, A. Koestel, *AIChE J.* **11** (1965) no. 2, 294–303.
[34] G. F. Hewitt, N. S. Hall-Taylor: *Annular Two Phase Flow*, Pergamon Press, Oxford 1970.
[35] P. Andreussi, J. C. Asali, T. J. Hanratty, *AIChE J.* **31** (1985) 119–126.
[36] J. Kataoka, M. Ishii, M. Kaichiro, *Trans. ASME* **105** (1983) 230–238.

Chapter 2

[37] I. C. P. Huang, *J. Fluid Mech.* **43** (1970) no. 2, 305–319.
[38] G. J. Taylor, *Proc. 7th Int. Congr. Appl. Mech.* **2** (1948) 392–400.
[39] W. C. Nieuwkamp, *Proceedings ICLASS 85*, III C/1 (1985) pp. 1–9.
[40] M. Horvay, Dissertation, TU Karlsruhe 1981.
[41] N. K. Rizk, A. H. Lefebvre, *Atom. Spray Technol.* **2** (1983) no. 3, 219–233.
[42] P. H. Rothe, J. A. Block, *Int. J. Multiphase Flow,* **3** (1977) 263–272.

[43] S. Y. Lee, R. S. Tankin, *Int. J. Heat Mass Transfer* **27** (1984) no. 3, 351–361.
[44] K. Lohoff, *Wärme-Gas, Int.* **31** (1982) nos. 2–3, 87–93.
[45] F. E. J. Briffa, N. Dombrowski, *AIChE. J.* **12** (1966) no. 4, 708–717.
[46] E. Sada, K. Takahashi, K. Morikawa, S. Ito, Can. J. Chem. Eng. **56** (1978) 455–459.
[47] H. W. Glaser, *Brennst. Wärme Kraft* **38** (1986) no. 5, 193–200.
[48] R. P. Grant, *AIChE. J.* **12** (1966) no. 4, 669–678.
[49] R. E. Phinney, *Phys. Fluids* **16** (1973) no. 2, 193–196.
[50] G. T. Sato, H. Tanabe, H. Fujimoto, *Proceedings ICLASS 82* (1982) 229–235.
[51] F. Ruiz, N. Chigier, *Proceedings ICLASS 85*, VI B/3/1.
[52] N. Dombrowski, R. P. Fraser, *Philos. Trans. R. Soc. London, Ser A* **247** (1954) 101–130.
[53] R. P. Fraser, P. Eisenklam, N. Dombrowski, *Br. Chem. Eng.* **2** (1957) 610–613.
[54] J. Y. Deysson, J. Karian, *Proceedings ICLASS 78*, (1978) 243–249.
[55] S. Nukiyama, Y. Tanasawa, *Nippon Kikai Gakkai Ronbunshu* **5** (1939) 1–4.
[56] N. K. Rizk, A. H. Lefebvre, *Trans. ASME* **106** (1984) 634–638.
[57] K. W. Thomer, Dissertation, Stuttgart 1981.
[58] T. Sakai, M. Kuroda, M. Saito, *J. Chem. Eng. Jpn.* **12** (1979) no. 2, 86–90.
[59] J. M. Chawla, *Proceedings ICLASS 85*, LP/1 A/5/1–7.
[60] R. W. Tate, *Proceedings ICLASS 85*, II C/1/1–13.
[61] N. Namiyama, N. Tokuka, P. Eisenklam, G. T. Sato, *Proceedings ICLASS 85*, III B/2/1–9.
[62] J. O. Hinze, H. Milborn, *J. Appl. Mech.* **17** (1950) 145–153.
[63] S. Matsumoto, K. Saito, Y. Takashima, *J. Chem. Eng. Jpn.* **7** (1974) no. 1, 13–19.
[64] S. Matsumoto, Y. Takashima, *Kagaku Kogaku* **33** (1969) 357–362.
[65] S. Matsumoto, D. W. Belcher, E. J. Crosby, *Proceedings ICLASS 85*/I A/1/1–21.
[66] S. J. Friedman, F. A. Gluckert, W. R. Marshall, *Chem. Eng. Prog.* **48** (1952) no. 4, 181–191.
[67] A. R. Singer, A. D. Roche, L. Day, *Powder Metall.* **23** (1980) 673–682.
[68] P. Walzel, *Ger. Chem. Eng.* (*Engl. Transl.*) **5** (1982) 121–129.
[69] T. P. Mitchell, R. L. Stone, *J. Phys. E.* **15** (1982) 565–569.
[70] G. Schäfer, *Ind.-Lackierbetr.* **49** (1981) no. 6, 193–199.
[71] P. Schmidt, *Maschinenmarkt* **91** (1985) no. 72, 1419–1421.
[72] W. Eisenmenger, *Acustica* **9** (1959) 327–340.
[73] U. Reimann, R. Pohlmann, *Forsch. Ingenieurwes.* **42** (1976) no. 1, 1–7.
[74] M. K. Li, H. S. Fogler, *J. Fluid Mech.* **88** (1978) no. 3, 499–528.
[75] L. H. Berger, *Proceedings ICLASS 85*, I A/2/1–13.
[76] H. Haberfelner, *Z. Umwelttechn.* **22** (1978) 10.
[77] K. Heberlein, *Met. Oberfläche* 39 (1985) 355–358.
[78] E. Moser, H. Ganzelmeier, K. Schmidt, *Nachr. Deut. Pflanzenschutz* **33** (1981) no. 10, 145–157.
[79] B. Vonnegut, R. L. Neubauer, *J. Coll. Sci.* **7** (1952) 616–622.
[80] V. G. Drozin, *J. Coll, Sci.* **10** (1955) 158–164.
[81] W. Simm, *Chem. Ing. Tech.* **41** (1969) no. 8, 503–507.
[82] A. Kelly, *J. Appl. Phys.* **47** (1976) no. 12, 5264–5270.
[83] A. Kelly, *J. Appl. Phys.* **49** (1978) no. 5, 2621–2628.
[84] A. G. Bailey, *Atom. Spray. Technol.* **2** (1986) no. 2, 95–134.

Chapter 3

[85] M. Schäfer, *Fortschr. Ber. VDI Z* **3** (1986) 126.
[86] K. Leschonski, *Chem. Ing. Tech.* **50** (1978) no. 3, 194–203.
[87] G. Schuch, H. Umhauer, *VI Verfahrenstechnik* **14** (1980) no. 4, 237–241.
[88] R. Semiat, A. E. Dukler, *AIChE J.* **27** (1981) no. 1, 148–153.
[89] F. Durst, *Trans. ASME* **104** (1982) 284–296.
[90] G. Dibelius, H. Voß, *VDI Ber.* **487** (1983) 131–136.
[91] K. Bauckhage, V. Wassmansdorff, *Chem. Ing. Tech.* **57** (1985) no. 8, MS 1388–1385.
[92] A. Buerkholz, *Chem. Ing. Tech.* **45** (1973) no. 1, 1–7.
[93] E. Rammler, A. Bahr, *Freiberg. Forschungsh. A* **A 590** (1978) 95–109.
[94] J. S. Chin, A. H. Lefebvre, *Proceedings ICLASS 85*, IV A/1/1–12.
[95] S. Matsumoto, Y. Takashima, *J. Chem. Eng. Jpn.* **4** (1971) no. 3, 257–263.
[96] L. P. Bayvel, *Atom. Spray. Technol.* **1** (1985) 3–20.
[97] R. A. Mugele, *AIChE. J.* **6** (1960) no. 1, 3–8.
[98] A. Lekic, R. Bajramovic, J. D. Ford, *Can. J. Chem. Eng.* **54** (1976) no. 5, 399–402.
[99] D. D. Schegk, F. Löffler, *Chem. Ing. Tech.* **59** (1987) no. 4, 319–322.
[100] R. P. Fraser, P. Eisenklam, *Trans. Inst. Chem. Eng.* **34** (1956) 294–319.
[101] R. A. Mugele, H. D. Evans, *Ind. Eng. Chem.* **43** (1951) no. 6, 1317–1324.
[102] D. E. Dietrich, *Proceedings ICLASS 78* (1978) 69/77.
[103] P. Walzel, *Chem. Ing. Tech.* **52** (1980) no. 12, 985, MS 857–880.
[104] H. A. Kuhn, A. Lawley: *Powder Metallurgy Processing*, Academic Press, New York–San Francisco–London 1978.
[105] J. K. Beddow: *The Production of Metal Powders by Atomisation*, Heyden & Son, London–Philadelphia–Rheine 1978.
[106] G. Rai, E. Lavernia, N. J. Grant, *J. Met.* **37** (1985) no. 8, 22–26.
[107] P. Walzel, *Chem. Ing. Tech.* **52** (1980) no. 4, 338–339.
[108] R. Reske, *Fortschr. Ber. VDI Z. Reihe 7* (1987) Nr. 115.
[109] P. Walzel, *Proceedings ICLASS 82* (1982) 187–194.
[110] A. Reinhart, *Chem. Ing. Tech.* **36** (1964) no. 7, 740–746.
[111] R. R. Hughes, E. R. Gilliland, *Chem. Eng. Prog.* **48** (1952) no. 10, 497–504.
[112] K. A. Cliffe, D. A. Lever, *AIChE Symp. Ser.* **80** (1984) 236, 61–66.
[113] J. A. Wesselingh, *Chem. Eng. Process* **21** (1987), 9–14.

7. Size Enlargement

KARL SOMMER, Technische Universität München, Lehrstuhl für Maschinen- und Apparatekunde Weihenstephan, Freising-Weihenstephan, Federal Republic of Germany

In the broadest sense of the term, size enlargement includes all processes in which fine particles, dispersed in either gas or liquid, aggregate to form a coarser product. The collection of particles that results is called an agglomerate or granule. Depending on the process, the size of the agglomerate is between 0.02 and 50 mm. In most cases, the preferred particle shape is spherical. In many processes, the product is a cylindrical section such as a tablet or some other regular geometrical shape. Size enlargement is used in many industries, such as fertilizer production, iron ore, nuclear fuel, pulverized fuel ash, ceramics, light-weight aggregate production, carbon blacks, catalysts, pesticides, and pharmaceutical products.

There are other reasons for the production and use of agglomerates:

1) Plant contamination and nuisance dusts in the workplace are reduced
2) Hygiene is easier to maintain
3) Air quality is easier to achieve, and in some controls can be avoided

4) The danger of dust explosions is reduced
5) With the increase in air and water pollution control regulations there is a need to install dust and sludge removal equipment. As a result, there is a need to process and dispose of the resulting fines. With agglomeration processes, perhaps with the addition of a binder, dust and sludge can be recycled or disposed of without pollution.

There are many problems in the handling of fine particles. The flow properties and consequently dosage control are poor. Size enlargement can eliminate these disadvantages while retaining the desired particulate properties. The chemical industry, with countless intermediate and final products, has a great need for size enlargement technology. Some examples are the agglomeration of simple or complex fertilizers, to avoid segregation; agglomeration of fillers used in plastics, to add exact amounts of iron oxide, zinc oxide, stearate, or silicate to the molten plastic; production of dust-free products, especially those with toxic or corrosive materials. Size enlargement increases the commercial worth because of the improved physical properties. A typical example is detergents. Some materials, such as fertilizers, pigments, pesticides, and instant foods, are sold in agglomerated form, but instantly decompose when added to liquids [1], [2]. Many raw materials are only suitable for processing or end use as coarse particles. These coarse solids have the desired strengths and porosities, for example, catalysts.

Ore beneficiation produces a large number of fine-grained products. Examples are flotation concentrates from selective beneficiation processes, dusty products from roasting, blast furnace by-products, filter cakes, and particulates from dust removal equipment. Ores or ore mixtures must be agglomerated for smelting to obtain optimal flow conditions in the roasting process. The principal area of application in the ceramic industry is in the preparation of pressed articles made of barium titanate and manganese–zinc, nickel–zinc and barium ferrites [3]. These materials are initially fine powders, which must be agglomerated to obtain good dimensional replication in the pressed articles.

In addition, size enlargement can cause a delayed action, which is important for pharmaceutical and agricultural applications. This allows good dosage or improves the appearance of the product.

The product properties desired determine which enlargement process is used. Size enlargement processes are classified by the principal mechanism by which the particles are made to come together [4], [5] (Table 1). The selection of a specific process is only possible if the user clearly defines the properties required of the product. Special attention must be paid to the following points: state of the material, desired agglomerate

Table 1. Agglomeration processes

Growth Agglomeration (Agitation Methods). Fine particles are brought into contact with each other in a flowing system or in air when the concentration is higher. This is usually done in the presence of liquid and binders. The particle size enlargement occurs by coalescence or accretion (snowballing) based on capillary forces. In a few exceptional cases, the major cohesive force is the van der Waals force. Usually the agglomerates are spherical with diameters between 0.5 and 20 mm. Typical equipment types are inclined drums, cones, pans, paddle mixers, and plowshare mixers. The maximum throughput is about 100–200 t/h for iron ore pellets and 50 t/h for fertilizers.

Spray Agglomeration (Spray Methods). This is one of the most commonly used methods in the chemical, pharmaceutical, and food industries. Pumpable suspensions are atomized, and the liquid is evaporated from the droplets by means of hot air, as a preliminary drying step. The first cohesive forces are the capillary forces, which are followed by crystal bridges at the contact points. The agglomerates are 20–500 μm. For chemical applications, throughputs of up to 50 t/h are possible.

Selective Agglomeration (Spherical Agglomeration). The most recent agglomeration process, a second immiscible phase is added to the suspension. This wets the solid phase and binds the particles together by means of capillary forces. As a result, rounded flocs or agglomerates form with diameters up to 5 mm. Selective agglomeration can be achieved for mixtures of solids. It is currently being studied for many substances. In the case of coal suspensions, throughputs up to and exceeding several tonnes can be achieved.

Pressure Agglomeration (Pressure Methods, Compaction). Particles with only slight amounts of moisture are formed in tablets and briquettes in stamp presses, tablet presses, and roller presses. The principal binding force is van der Waals attraction. The agglomerates have uniform shapes and range in size from a few millimeters (pharmaceutical tablets) to decimeter size (fuels). In the case of smooth rollers, the resulting flakes are broken up into the desired size. The throughput for ores is roughly 100 t/h; for chemicals, up to 30 t/h.

Sintering (Thermal Methods). Fine particles are made into a paste by adding moisture and then processed in a horizontal sintering oven into sinter (burnt agglomerates). This is especially common in the mining and preparative industries. The final sintered product usually has an irregular shape and is usually coarser than other agglomerates. The mechanism of binding is the formation of solid bridges at the contact points. Sintering plants have throughputs up to 1000 t/h.

size, size distribution, and agglomerate shape, required strength and porosity, and required throughput.

State of the Material. The starting material may be fluid, pumpable, sprayable, pasty, or dry. The product may be dryable. Although temperature sensitive, it may have to be dried. Frequently the cost of drying determines the economic feasibility of the process [6].

Desired Agglomerate Size, Size Distribution, and Agglomerate Shape. Many processes permit only a narrow range of agglomerate size. Spray agglomeration gives the finest products; briquetting and sintering, the coarsest. In the case of pressure agglomeration with smooth rollers, the resulting flakes must be broken up. The cost of this size reduction and the recycling of the resultant fines must be considered. The shape of the agglomerate may be of critical importance to the process. Some examples are flow through a bed of solid catalyst, in the case of bunkering and for storage.

Required Strength and Porosity of the Agglomerate. Many end uses require a large specific surface area and ready dispersibility (e.g., instant food products). Examples are pigments, agricultural sprays, food, and pharmaceuticals. Other products must be dispersible but with a delay time. Catalyst and iron ore pellets, however, must withstand high stresses without being crushed. The type of binding material (and thus binding force) has a major effect on strength and porosity.

Required Equipment Throughput. Some size enlargement processes, for example, tableting, are not suitable for producing high tonnages.

Other Factors. Amont other factors are compressibility drying behavior (including possible cracking), and high- and low-temperature behavior. For ore pellets, there are properties important in smelting, reducibility, swelling, and shrinking [7].

Once an appropriate process has been selected, laboratory tests must be run to determine the fluid dosage that gives a suitable moisture level for agglomeration. Laboratory tests must be run to determine applicability and dosage of agglomeration aids such as binders and wetting agents (wetting rate). After the laboratory tests, a pilot plant must be set up, and the process checked on a larger scale. At this point the production process, involving recycling, can be tested, allowing prediction of the technical and economic feasibility of the process on an industrial scale [8], [9].

1. Binding Mechanisms of Agglomerates

The attractive forces between particles cause the size enlargement process and are essential to the strength and dispersibility of the agglomerate. Rumpf [10] has given a complete summary of binding mechanisms. Schubert [11] considers the presence or absence of material bridges as the overriding criterion for binding mechanisms (Fig. 1). Material bridges can be further categorized as solid or liquid. Solid bridges include those formed by sintering, chemical reaction at contact points, hardening of binders, and the crystallization of dissolved materials. Fluid bridges are formed by capillary forces between the particle and a fluid or high viscosity binder. In the case where there are no material bridges van der Waals forces are the principal binding force. In general, for similar materials with the same surface ions, electrostatic forces are repulsive.

Mechanical interlocking is rarely a mechanism for agglomeration, for example, the balling up of "burrs."

Attractive forces can be calculated for model systems, such as smooth, fixed, ideal spheres. Even though the values calculated for these models apply only crudely for real systems, they do indicate the effects of important parameters on the agglomeration process. Models are essential for the understanding of the size enlargement process. To calculate the actual attractive forces between real particles is not currently possible because of their irregular shape and usually rough surfaces [11].

1.1. Tensile Strength of Agglomerates Derived from Attractive Forces

Rumpf [12] presents the following equation, based on a statistical, geometric approach:

$$\sigma_z = \frac{1-\varepsilon}{\pi x^2} \cdot k \cdot F \quad (1)$$

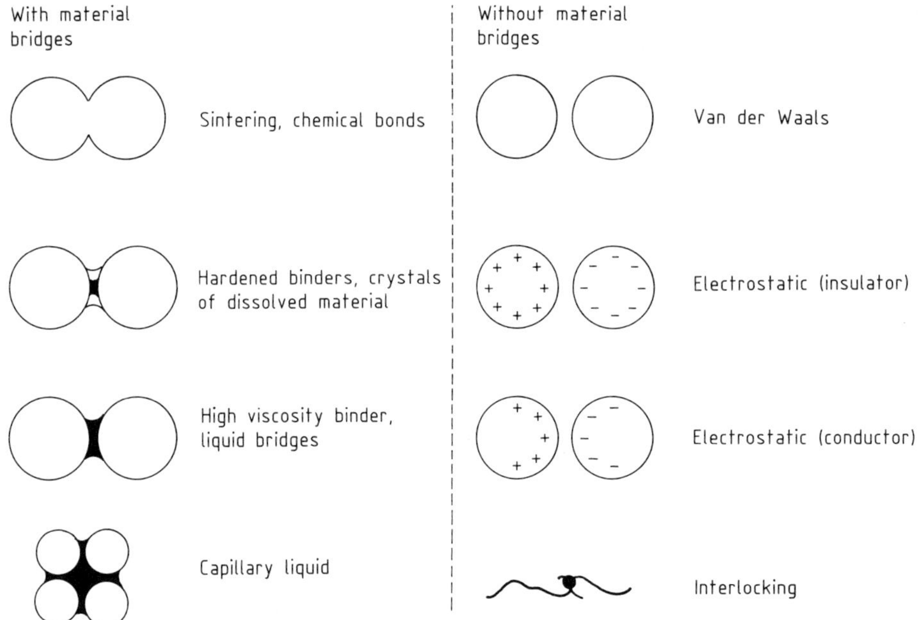

Figure 1. Schematic representation of major binding mechanisms in agglomerates [11]

where σ_z is the tensile strength of randomly packed, equally sized spheres with diameter x; ε is the porosity. F is the attractive force at the point of contact, and k is the coordination number.

RUMPF [13] has derived a more general equation for the nonspherical particle-size distribution using the approximation of SMITH et al. [14] for the coordination number ($\varepsilon \cdot k \approx \pi$)

$$\sigma_z = \frac{1-\varepsilon}{\varepsilon} \cdot \frac{\bar{F}}{x_s^2} \quad (2)$$

where $\bar{F}$ is the average attractive force at the contact points and x_s is the equivalent diameter. Equation (2) can be experimentally confirmed [15] for moist agglomerates with force transmission via fluid bridges. In the case of short range forces, such as van der Waals forces, surface roughness at the contacts causes problems. In most cases of this type, the strength is measurably less than that calculated using Equation (2) [15]. Even though the van der Waals forces can only be calculated for model systems (Section 1.2.2.). Equation (2) does give an indication of the effect of various parameters.

The strength of fluid-saturated agglomerates can be calculated by using the capillary pressure curve [16]. The primary force in fluid-filled capillaries is the capillary pressure p_k. A negative pressure, it is equal to the tensile stress transmitted to the pores. The tensile strength is given by the following equation only for the case when the degree of saturation $S \approx 0.8$ (Section 1.2.3.), and the stress is transmitted to the pores.

$$\sigma_z = S \cdot p_k \quad (3)$$

1.2. Attractive Forces in Size Enlargement in Air

Agglomeration with mixers, drums, and pans; fluidized bed agglomeration; compaction; and tableting. The dominant force acting on particles larger than 1 cm is gravity. As the size decreases,

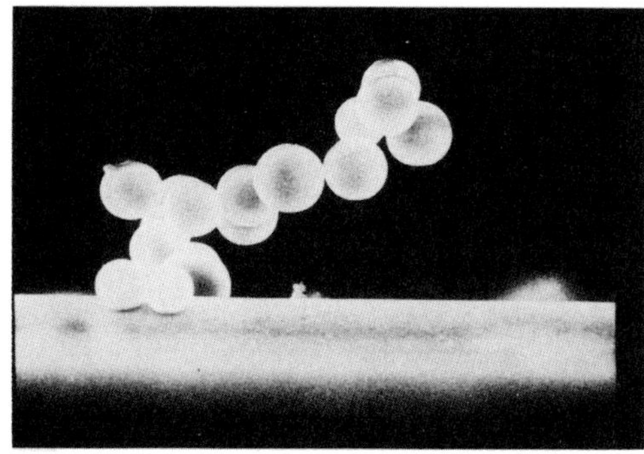

Figure 2. Glass spheres on a glass fiber from Polke [17]

the effect of gravity rapidly decreases—in proportion to the third power of the particle diameter—and attractive forces become more important. For example, for a 1 μm particle the van der Waals forces are larger than the weight by about six orders of magnitude. Figure 2 shows glass spheres about 10 μm in diameter, which adhere to the surface of a glass fiber, solely by van der Waals attraction.

1.2.1. Electrostatic Forces [18]

The attractive force F_{el} between two charged surfaces is derived from the energy of the two electrostatic fields. The force has been calculated by KRUPP [19] as well as MOSER [20] for the case of two electrically conducting spheres (Figs. 3 and 4).

$$F_{el} = \frac{1}{4} \varepsilon \cdot \varepsilon_0 \cdot \pi \frac{U^2 \cdot x}{a + z_{oel}} \tag{4}$$

ε is the relative dielectric constant of the medium. In the case of air ε is 1; x is the diameter of the sphere, and a is the distance between the surfaces, U is the potential difference, z_{oel} is a fitting parameter of the order of magnitude of an atomic dimension, for example, 4×10^{-8} cm.

For nonconductors the attractive forces are noticeably smaller than for conductors. In nonconductors the charge is not concentrated on the surface, but can extend below the surface to a depth exceeding 1 μm. Two spheres with surface charges σ_1 and σ_2 exert an attractive force

$$F_{el} = \frac{\pi}{4 \cdot \varepsilon \cdot \varepsilon_0} \cdot \frac{\sigma_1 \cdot \sigma_2 \cdot x^2}{\left(1 + \frac{a + z_{oel}}{x}\right)^2} \tag{5}$$

In the case of ideal nonconductors in contact ($a = 0$), the quantity z_{oel}/x is small compared to 1 and can be neglected.

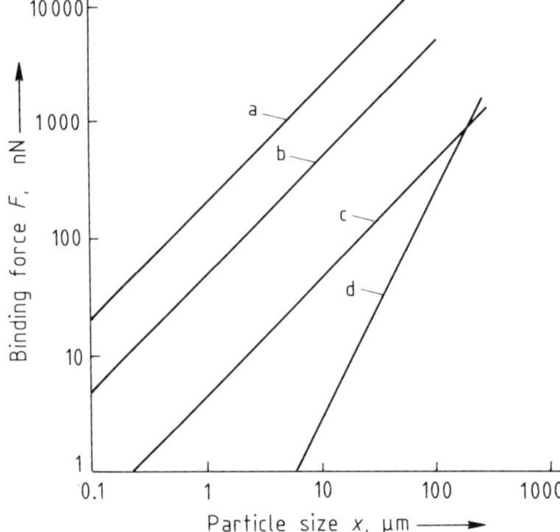

Figure 3. Theoretical binding forces derived from different mechanisms calculated for two identical ideal spheres [18]
Separation $a = 0$, $z_{el} = z_0 = 4 \cdot 10^{-8}$ cm
a) Liquid bridges $V_L/2\ V_S = 10^{-4}$; b) van der Waals $\hbar\omega = 5$ eV; c) Conductor $U = 0.5$ V, $\varepsilon_0 = 8.86 \times 10^{-12}$ As/Vm; d) Insulator $\sigma_1 = \sigma_2 = 100\ e\ \mu m^{-2}$

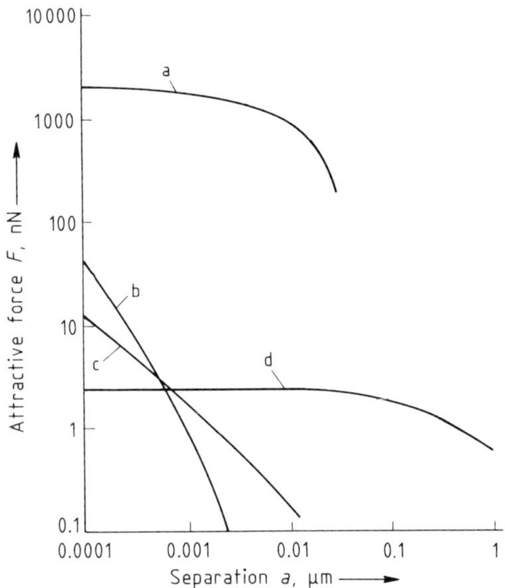

Figure 4. Theoretical attractive forces derived from various mechanisms calculated for two equal-sized spheres as a function of distance between surfaces
Particle size 10 μm
a) Liquid bridges $V_L/2\ V_s = 10^{-4}$, $\gamma = 0.072$ Nm; b) van der Waals $\hbar\omega = 5$ eV; c) Conductor $U = 0.5$ V; $\varepsilon_0 = 8.86 \times 10^{-12}$ As/Vm; d) Insulator $\sigma_1 = \sigma_2 = 100\ e\ \mu m^{-2}$

1.2.2. Van der Waals Forces [21]

There are two principal methods of calculating the van der Waals attraction. In the microscopic theory the attractive energy is calculated as the sum of the interaction energy of two atoms [22]. The assumption is made that the interactive forces are additive and do not influence each other. Another, more physically satisfactory theory starts with the optical and electrical properties of the two interacting particles and calculates the van der Waals force from them [23].

KRUPP [19] gives the attraction for the case of two spheres in which the distance a is less than 500 Å.

$$F = \frac{\hbar\bar{\omega}}{32 \cdot \pi} \cdot \frac{x}{(a + z_0)^2} \quad (6)$$

where $\hbar\bar{\omega}$ is the Lifschitz–van der Waals constant. Depending on the material, the value is 0.2–9 eV and must be determined experimentally. It can be calculated in a few cases [19] (see Figs. 3 and 4).

1.2.3. Capillary Forces

Because of surface tension, there is an attractive force between wetted particles, which can hold agglomerates together. NEWITT and CONWAY-JONES [24] have classified binding mechanisms based on mobile liquids into four types: pendular, funicular, capillary, and droplet (Fig. 5). Liquid bridging is the term used to describe the case in which the fluid droplets are concentrated at the contact points and are separated from each other [25]. Figure 6 shows a liquid bridge between two spherical particles. The simplifying assumption is made that the boundary of the bridge is a circular arc. For the case of wetted particles in contact, PIETSCH [26] gives the following equations:

$$F = \gamma \cdot x \cdot \pi \cdot \sin^2 \beta \cdot \left[1 + \frac{x}{4}\left(\frac{1}{R_1} - \frac{1}{R_2}\right)\right] \quad (7)$$

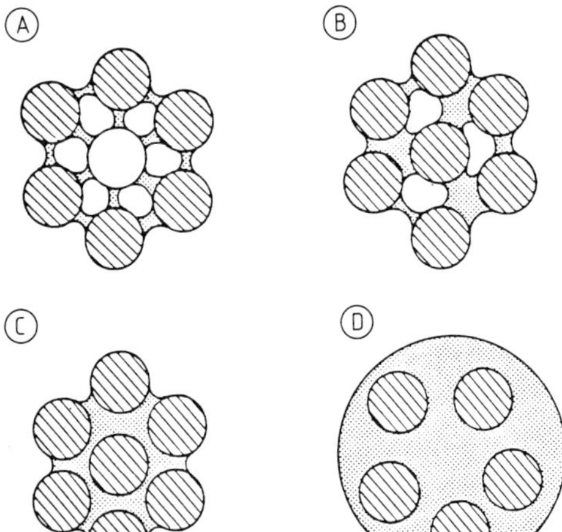

Figure 5. Agglomerate binding mechanisms due to mobile liquids [24]
A) Pendular; B) Funicular; C) Capillary; D) Droplet

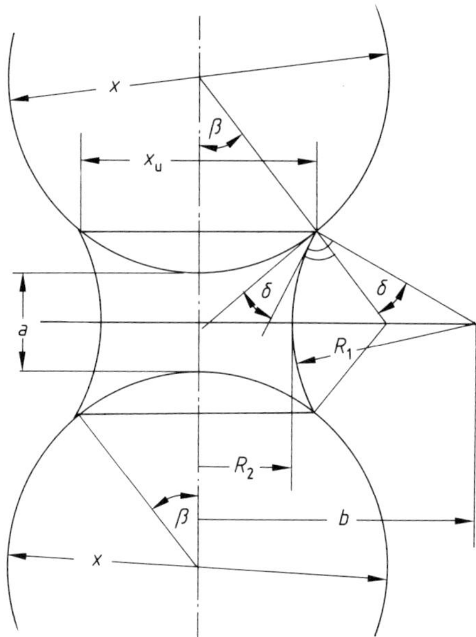

Figure 6. Liquid bridges between two equal-sized spherical particles

$$R_1 = \frac{x(1 - \cos\beta) + a}{2\cos\beta}$$

$$R_2 = \frac{x}{2}\sin\beta + R_1(\sin\beta - 1) \quad (8)$$

where γ is the surface tension, β is one-half the central angle (see Fig. 6), and x is the diameter (see Figs. 3 and 4). SCHUBERT [27], [28] has calculated the appropriate differential equations exactly corresponding to the attractive forces transmitted by rotationally symmetric fluid bridges. The numerical results with the parameters separation distance, bridge volume, and contact angle are presented in chart form [29]. The maximum divergence of these values is less than approximately 20%.

1.2.4. Solid Bridges

In many cases, the attractive forces due to mobile liquids are only an aid in forming an agglomerate. They are crucial to the process of formation, but are not often important for the physical properties of the final product. These properties are produced by binding processes that follow agglomerate formation. In this step, solid bridges are formed by pressure or thermal treatment. The major mechanisms are as follows:

1) Crystallization of dissolved material
2) Hardening of fluid binders

3) Local melting and coalescence processes at the contact points
4) Particle deformation in combination with sintering and chemical reactions

Salt Bridges. If wet agglomerates were only held together by surface forces and the capillary pressure, the cohesive force would disappear when the liquid was evaporated. To give the dry agglomerates a definite amount of strength, dissolved salts are often used. These crystallize out during drying and form salt bridges at the contact points between the particles. The strength of such an agglomerate can be estimated to an order of magnitude if the total volume of the material in the salt bridges V_{salt} is known [31].

$$\sigma_z = \frac{V_{salt}}{V_{solid}} \cdot (1 - \varepsilon) \cdot \sigma_s \tag{9}$$

V_{salt}/V_{solid} is the volume ratio of the crystallized salt to the total volume of the particles (V_{solid}), and σ_s is the tensile strength of the dissolved salt. In this idealized case, there is only a small amount of water, primarily at the contact points (pendular state). Alternatively, the assumption is made that the solubility of the salt is so high that it remains completely dissolved in the water at the contact points during the drying process [32], [33].

Binders. The mechanical properties that can be obtained by using binders, which harden and increase in strength with time, are similar to those obtainable with salt bridging. However, the strength can be better controlled. CAPES [34] has shown that the addition of a binder, for example, corn starch, decreases crust formation during drying, leading to a more even distribution of the solid forming bridges, noticeably increasing the strength (Fig. 7). Spherical agglomerates that contain both, a binder and salt crystals, are more resistant to breakage under pressure than agglomerates that have only one of the two.

Sinter Bridges. Sinter bridges can form in the contact area between particles when the temperature exceeds roughly 60 % of the melting temperature of the material that makes up the particles. The strength of the sinter bridges is of a similar order of magnitude as the strength of the material which makes up the particles. For this reason, the maximum attractive force present

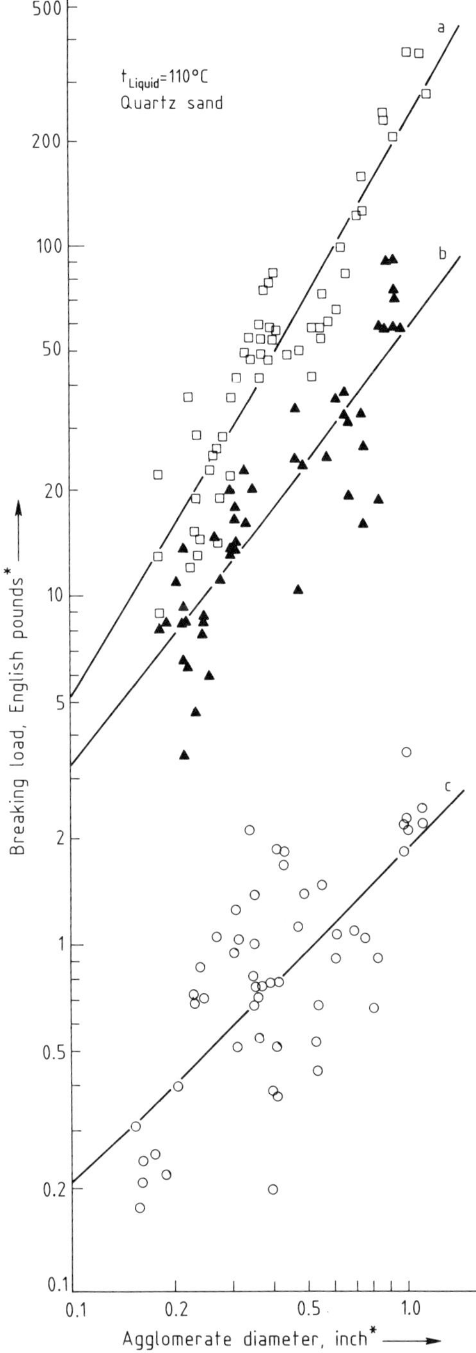

Figure 7. Breaking load of spherical agglomerates with NaCl salt bridges and cornstarch as binder from Capes [34]
* To convert English pound into kilograms multiply by 0.4536; to convert inch into millimeter multiply by 25.4
a) NaCl solution and cornstarch; b) Distilled water and cornstarch; c) NaCl solution

depends primarily on the diameter $2b$ of the neck of the sinter bridge. During the sintering process, the sintered area grows with time t.

Only the initial growth stage of a sinter bridge is of interest for agglomeration. Complete sintering, down to pore-free material, which is important in metallurgy, is not considered here.

Research on sintering began in 1949 with the work of FRENKEL [35] on the sintering of viscous materials. In this case the kinetics are determined solely by the reduction of surface free energy and viscous dissipation energy. At almost the same time, PINES [36] published an article on the sintering of solids, which showed that the actual molecular diffusion mechanism could be derived from a mathematical treatment of the diffusion of vacancies in the crystal lattice. The growth of the sinter neck was explained by the diffusive migration of vacancies from concave points of contact to convex parts of the particle's surface. In a similar way, atoms migrate from the particle surface to the sinter neck. This results in diffusion both in the bulk material (volume diffusion) and on the surface (surface diffusion). There is also gas-phase material transport because of the difference in equilibrium vapor pressures near concave surfaces and convex surfaces [36]–[41], [72]. An exact description of all the physical processes that occur in the various stages of sintering is extremely difficult because of the complex geometry and many factors that affect the surface, such as adsorbed and oxide layers.

1.2.5. Influence of Particle Surface Roughness on Binding

The idealized-spheres model cannot completely describe what actually happens. In actuality, surfaces are rough, and this can have a major effect on the attractive forces. The effect of surface roughness can be visualized by means of the following model. A semicircular peak with radius r sits on the surface of a sphere with diameter x and is in contact with another sphere. As a first approximation, the electrostatic and van der Waals forces of the individual sections can be assumed to be superimposable.

Electrostatic Forces (Fig. 8) (with roughness $r \ll x$).

Electrical Conductor

$$F_{el} \approx \frac{1}{4}\pi \cdot \varepsilon \cdot \varepsilon_0 \, U^2 \left[\frac{x}{r + z_{0el}} + \frac{4 \cdot r}{z_{0el}}\right] \qquad (10)$$

Nonconductor

$$F_{el} \approx \frac{1}{4}\pi \cdot \frac{\sigma^2 \cdot x^2}{\varepsilon \cdot \varepsilon_0} \qquad (11)$$

Van der Waals Forces (see Fig. 8)

$$F = \frac{\hbar\bar{\omega}}{32\pi}\left[\frac{x}{(r + z_0)^2} + \frac{2r}{z_0^2}\right] \qquad (12)$$

Capillary Forces. For capillary forces, the effect of roughness is closely related to the size of the bridges. If the fluid bridge is so small that only the projection is wetted, and not the surface of the rest of the particle, then the appropriate model is the attraction of two contacting spheres with different diameters (x and $2r$) [16, p. 150]. If the fluid bridge is so large that the projection is completely covered, so that it acts merely as a spacer ($a = r$), then the appropriate model is for two large spheres with diameter x [16, p. 139].

For a given particle size, the van der Waals force passes through a minimum, which is roughly 300 times smaller than for an idealized, smooth sphere. A similar result applies for the electrostatic attractive force between conductors. However, the attractive force between insulators is largely independent of roughness. Even though van der Waals forces are an order of magnitude greater than electrostatic forces in the smooth sphere model, electrostatic forces can be dominant when roughness is present. Like for van der Waals forces, the attractive force in the case of small fluid bridges is almost exclusively determined by the size and shape of the rough-

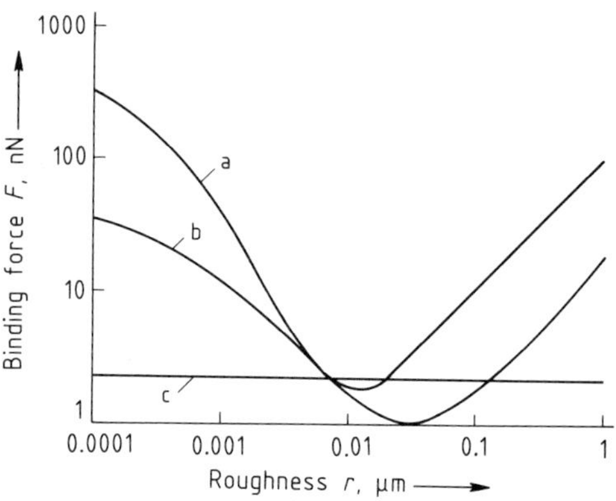

Figure 8. Effect of a hemispherical roughness on the surface of a sphere in the contact region on the binding force; Particle size 10 µm
a) van der Waals $\hbar\omega = 5$ eV; b) Conductor $U = 0.5$ V; c) Insulator $\sigma = 100\,e\,\mu m^{-2}$

ness projections. The fact that these are mostly unknown explains the difficulty in trying to measure the attractive forces between real particles. In comparison to ideal spheres, results in real systems commonly deviate by orders of magnitude.

1.2.6. Strengthening of Binding Forces

Deformation in the contact area, caused by external or internal forces, can affect the binding of real particles, just as roughness can. Strengthening of the binding forces means an increase relative to the value calculated for a model system with no particle deformation [38]. Strengthening of binding forces can also occur directly without any external agent. This may be brought about by the binding forces of the undeformed system alone, or for example, the dead weight of a pile of agglomerates. In addition, external force can cause a rearrangement or further deformation in the contact areas. This is especially the case for tableting and compacting. The mechanisms for strengthening of binding forces are classified as follows [44]:

1) Rearrangement
2) Elastic deformation at the contact point
3) Plastic deformation at the contact point
4) Viscoelastic flattening
5) Sintering

The driving forces (or energies) are:

1) The binding force itself, deformation or rearrangement increasing it relative to the binding force of the undeformed system F_0
2) An external force F_{CO}, the sum of the binding force F_0 and the external force F_{CO} being the total force

$$F_t = F_0 + F_{\mathrm{CO}} \tag{13}$$

3) The interfacial energy

1.3. Attractive Forces of Particles in a Liquid Medium

In principle, the attractive forces between particles in liquids and in gases are based on the same mechanisms. This similarity is especially true in regard to the bonding mechanisms of the material bridges if the changes with time because of solution processes and corrosion in liquids are considered and if the surface tension for particles in gases is replaced for particles in liquids by the interfacial tension when considering capillary forces. However, the interaction of the van der Waals forces and the repulsive electrostatic forces, which are caused by absorbed ions, give rise to anomalies in liquid media.

1.3.1. DLVO theory

For particles suspended in a fluid, the medium acts as a dielectric that has an effect on the dipole–dipole interaction between the particles. As a result the van der Waals attraction is modified. Equivalent to Equation (6), Equation 14 expresses the van der Waals potential V_{vdw} as the product of a constant related to the type of material and a geometric term [22].

$$V_{\mathrm{vdw}} = -\frac{A_{\mathrm{PlP}}}{12} \cdot \frac{x}{a + z_0} \tag{14}$$

An approximate relationship for the Hamaker–van der Waals constant A_{PlP} between the particle P and the liquid l has been given by **Derjagin** and **Landau** [45] and **Vervey** and **Overbeck** [46].

$$A_{\mathrm{PlP}} \approx (\sqrt{A_{\mathrm{PP}}} - \sqrt{A_{\mathrm{ll}}})^2 \tag{15}$$

The resultant Hamaker–van der Waals constant increases as the difference between the Hamaker–van der Waals constants for the particle A_{PP} and the liquid A_{ll} increases. The Lifschitz constants for the usual macroscopic theory for the corresponding dielectric spectra have a similar relationship.

The source of the electrostatic charge in a liquid medium is a layer of tightly adsorbed ions on the particle surface (Stern layer). This layer causes an electrostatic repulsion. Over the tightly adsorbed ionic layer there is a diffuse layer of counter ions (Gouy–Chapman layer) as shown in Figure 9. A characteristic value for the thickness of the double layer is the Debye length l_{D} over which the potential decreases by a factor of $1/e$. The potential V_{el} of the electrostatic repulsion increases with the number of adsorbed ions and the resulting increase of the Stern layer potential V_0. The combination of the van der Waals attraction and the electrostatic repulsion determines the interaction potential as a function of the interparticle distance (Fig. 10). A potential barrier prevents particle agglomeration when the

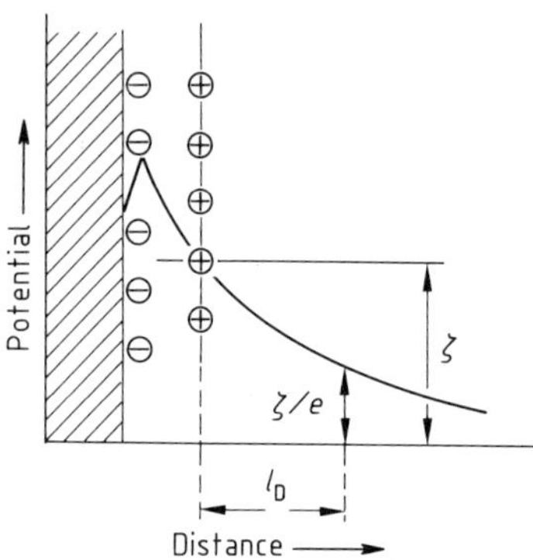

Figure 9. Potential function of the electrical double layer around a particle

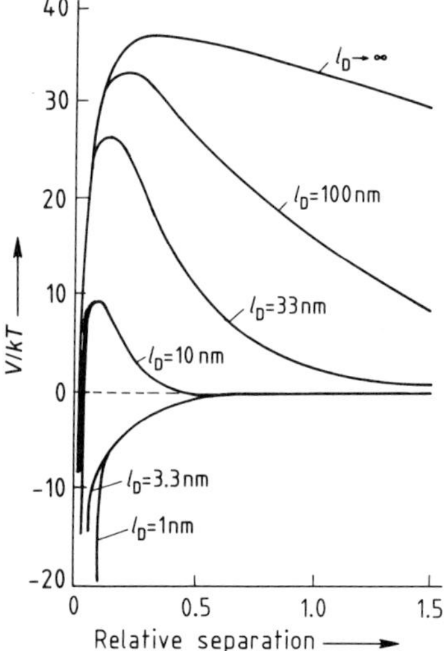

Figure 10. Interactive potential function according to DLVO theory [45], [46]

electrostatic charge is high enough. While the van der Waals forces cannot be varied much in a given system (although one possibility is perhaps a suitable adsorbed layer) the electrostatic repulsion can be greatly decreased, for example, by the addition of electrolytes. In this case the van der Waals forces dominate, and the particles agglomerate.

1.3.2. Flocculation by Polymeric Flocculants

The use of synthetic polymers for the agglomeration in suspensions is steadily gaining in industrial importance as replacements for inorganic electrolytes. Synthetic polyelectrolytes are used in a large number of diverse applications. Polymeric flocculants are classified by their charge. Polymers with a negative charge are termed anionic, those with a positive charge are termed cationic. There is also a group of nonionic water-soluble polymers such as poly(ethylene oxide), poly(vinyl alcohol), and polyacrylamide.

Two mechanisms have been proposed [47] for the action of polymers in promoting inter-particle attraction [47]. The macromolecule also affects the electrostatic repulsion, which in turn allows aggregation of the particles due to van der Waals forces (Section 1.3.1). The polymer must have a charge opposite to that of the surface to be adsorbed in a flat configuration and neutralize the surface charge. If there is an excess of polymer, long range electrostatic attraction of other particles can occur.

The second mechanisms is based on bridging. In this case, only a few segments of the polymer are attached to the surface, while the rest project out into the liquid. The distance they extend must exceed the range of the repulsive forces.

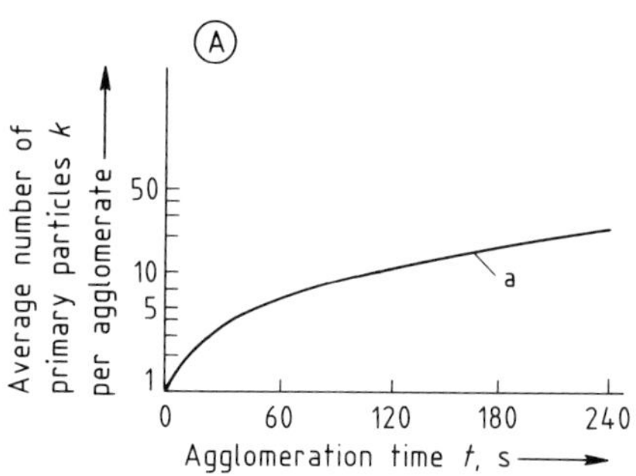

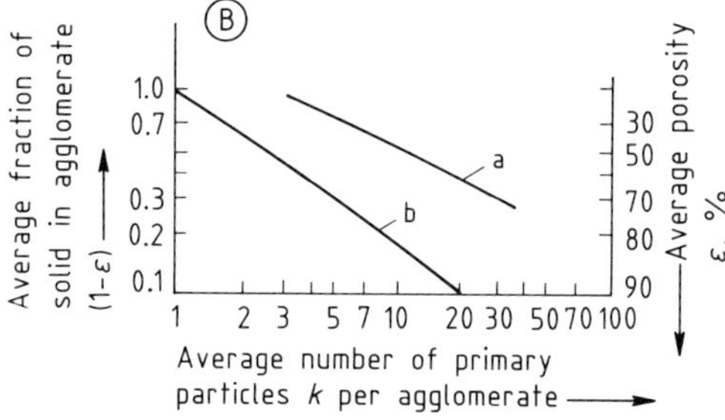

Figure 11. Effect of gravity on polymer induced agglomerates [47]
A) a) Effect of shear force on agglomerates of fine solids; B) a) Quartz; b) 100% cationic polymer

This type of binding can only take place at high solids levels or when the particles are sufficiently close. PUSCH has used the scanning electron microscope to show that polymer bridges of considerable length, as well as flat layers covering the surface and crosslinks can occur on the particle surface, depending on the polymer concentration [47]. The polymer molecules do not appear to act as individuals, but are themselves agglomerated. These aggregates are somewhat flattened and are very elastic. They fill the interparticle regions, forming thin threads, bridges or configurations resembling pendular water bridges. Because of their porosity, the strength of flocs produced with the aid of flocculants is quite low. The shear forces encountered in flow through a pipe, nozzle, or pump can break the aggregates up into the original particles (Fig. 11). However, polymer particles show a considerable tendency to reagglomerate.

2. Characterization of Agglomerates

An understanding of the properties of agglomerates is of importance for the various steps encountered during further processing as well as the application by the end-user:

1) Process optimization
2) Quality control
3) Production of specific properties

2.1. Particle Size, Particle Size Distribution [49]

Agglomerates can be characterized by their size, shape, and size distribution, as is usually done with finely dispersed materials. For most applications the optimal shape is spherical because this is the shape least subject to breakage and abrasion. Cylindrical shapes (pellets) are also common. In the optimal case, the size distribution is narrow and the fines fraction small. Size and size distribution can be determined by a large number of techniques [48]. For practical applications, a quantitative size criterion that is relevant to the end use should be found. Therefore, great care must be taken in selecting a suitable measurement process. Even when a suitable process has been established, serious problems in sampling and sample preparation can be encountered.

Up to the present, mechanical screening is the most common analytical method because it covers a broad size range. In addition to an analytical method, screening may be part of the process. In this case, the product stream is divided into three fractions: fines, product, and oversize. Care must be taken that screening does not result in excessive abrasion or even agglomerate breakage.

Even though agglomerate size is recognized as one of the most important parameters, particle size analysis is seldom used in industry for process control. The principal reason for this is that product application during additional processing as well as by the customer depends on other parameters. Therefore, an application related test is often a better standard of quality [50]. The pigment industry offers one example. The color in the standard formulation is required as a definitive test of the quality of pigment agglomerates.

2.2. Agglomerate Density and Porosity

2.2.1. Definition

Porosity is defined as the ratio of the void volume V_v to the total volume V_{tot}.

$$\varepsilon = V_v/V_{tot} \tag{16}$$

Density is defined as the ratio of the mass m to the volume V of a particular material.

$$\varrho = m/V \tag{17}$$

For porous particles, agglomerates and piles of agglomerates, the difficulty in defining porosity and density lies in the fact that there are various methods for determining volume.

Figure 12 illustrates the different kinds of pores. Pores may be either accessible or inaccessible. Some pores have a constant diameter, while others tend to narrow down and are accessible through narrow capillaries. Other pores are open enough to permit flow through the material. Surface roughness also has to be considered.

2.2.2. Densities

Solid Density ϱ_S. The solid density is defined as the density of the pure solid. This value determines the sedimentation rate of solids without any included void space, i.e., in wetting fluids. The density of such solids is measured with a

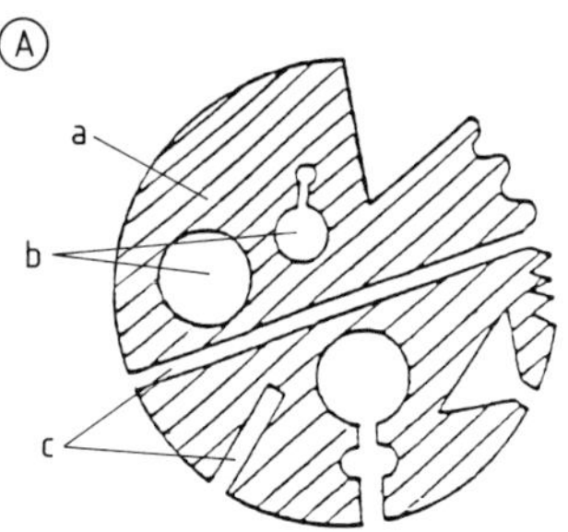

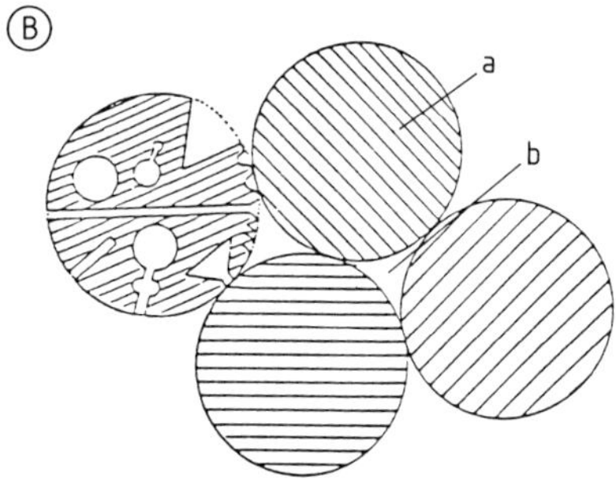

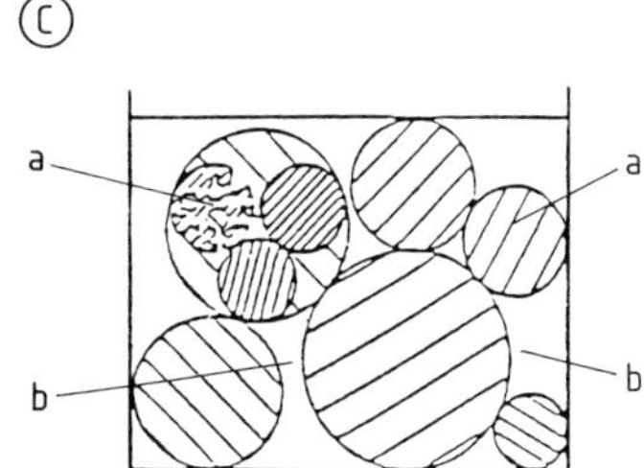

Figure 12. Structure of packed agglomerates composed of individual particles
A) Particle: a) Solid; b) Inaccessible pores; c) Accessible pores
B) Agglomerate: a) Particle; b) Interstitial volume
C) Bulk agglomerates: a) Agglomerate; b) Interagglomerate volume

pycnometer. If the solid may contain inaccessible void space this must be made accessible by grinding.

$$\varrho_S = \frac{m_S}{V_S} \quad (18)$$

Apparent Solid Density ϱ'_S. The apparent solid density is defined for solids that contain inaccessible void space V_{cP}. This value determines the sedimentation rate of such solids in a wetting fluid, under the assumption that the void volume does not become accessible by some dissolution process. The apparent density is determined with a pycnometer directly, i.e., without grinding.

$$\varrho'_S = m_S/(V_S + V_{cP}) \quad (19)$$

Particle Density ϱ_P. Both accessible and inaccessible pores are taken into account in the particle density, which determines the sedimentation rate of solids in nonwetting fluids.

$$\varrho_P = m_S/V_P \quad (20)$$

Agglomerate Density ϱ_A. Agglomerate density controls the sedimentation rate of agglomerates in a nonwetting fluid.

$$\varrho_A = m_S/V_A \quad (21)$$

Bulk Density ϱ_B. A large group of particles or agglomerates is characterized by the bulk density, which is determined by measuring the total mass of bulk material contained in a given volume.

$$\varrho_B = m_S/V_B \quad (22)$$

2.2.3. Porosities

Particle Porosity ε_P. The ratio of the pore volume (accessible V_{oP} and inaccessible V_{cP}) to the total particle volume V_P is the particle porosity.

$$\varepsilon_P = \frac{V_{oP} + V_{cP}}{V_P} \quad (23)$$

If the inaccessible pore volume cannot be measured, the apparent particle porosity is used, instead of the particle porosity.

$$\varepsilon'_P = V_{oP}/V_{cP} \quad (24)$$

Agglomerate Porosity ε_A. The ratio of the volume of the void space between the particles V_{iP} and the volume of the agglomerate V_A is the agglomerate porosity. The agglomerate porosity allows the difference in sedimentation rate of the same particles in wetting and nonwetting fluids to be determined. The rate of sedimentation is always greater in wetting fluids. The agglomerate porosity is the ratio of the particle density to the agglomerate density.

$$\varepsilon_A = V_{iP}/V_A \quad (25)$$

$$\varepsilon'_A = 1 - \varrho_A/\varrho'_S \tag{26}$$

where ε'_A is the apparent agglomerate porosity.

Bulk Porosity ε_B. In a group of agglomerates, there is an interstitial volume V_{iA} between the agglomerates themselves. From this value the bulk porosity ε_B is defined as follows:

$$\varepsilon_B = V_{iA}/V_B \tag{27}$$

The apparent bulk porosity is defined as:

$$\varepsilon'_B = 1 - \varrho_B/\varrho'_s \tag{28}$$

Total Porosity ε_{tot}. The total porosity of the system must be distinguished from the bulk porosity.

$$\varepsilon_{tot} = \frac{V_{oP} + V_{cP} + V_{iP} + V_{iA}}{V_B} \tag{29}$$

If the inaccessible pore volume is not included in the measurement, then the quantity calculated is the apparent total porosity ε'_{tot}.

$$\varepsilon'_{tot} = \frac{V_{oP} + V_{iP} + V_{iA}}{V_B} \tag{30}$$

The different porosity values are related as follows:

$$(1 - \varepsilon_{tot}) = (1 - \varepsilon_P) \cdot (1 - \varepsilon_A) \cdot (1 - \varepsilon_B) \tag{31}$$

$$\varepsilon'_{tot} = 1 - \varrho_B/\varrho'_S \tag{32}$$

In addition, that pore sizes are different must be considered. Figure 13 illustrates the pore radius size distribution function for a bulk agglomerate system. In general, pore sizes are distinguished clearly for three systems: individual particles, agglomerates, and bulk systems. This is the result, in ideal cases, of three different maxima in the distribution function [51]–[55].

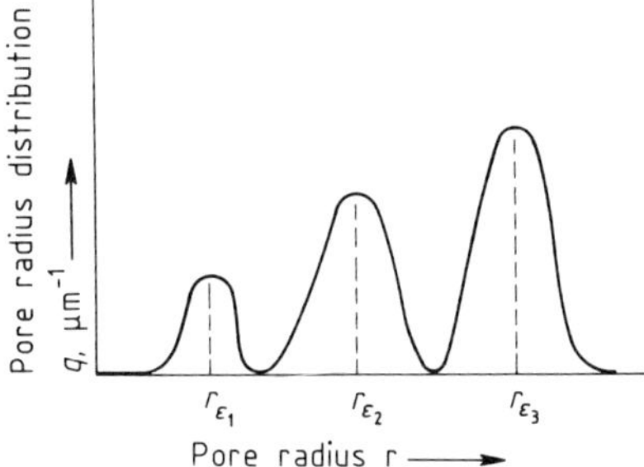

Figure 13. Idealized pore size distribution [51]

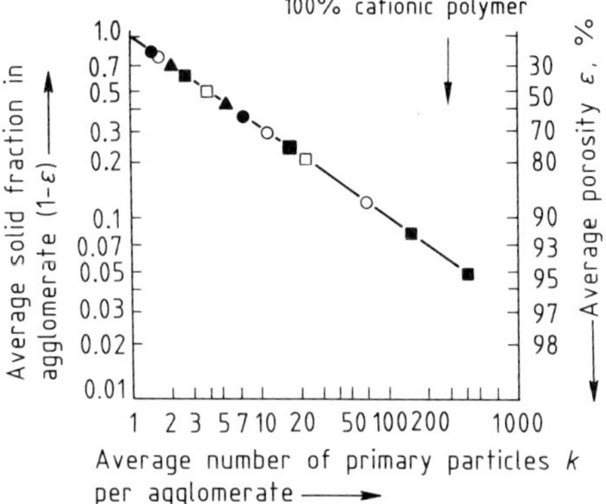

Figure 14. Structure of agglomerates in suspension [58]

Type of solid	x, µm
● Glass spheres	16
▲ Quartz	14.5
■ Coal	8.5
○ Cornstarch	13
□ Cellulose	14

Because of the limited strength of flocs, the determination of the size and porosity of flocs formed by agglomeration in liquids is especially difficult. None of the usual methods can be used in this case. In general, another measurable parameter is used, one which depends on both the size and porosity. At Technische Hochschule Karlsruhe, a method has been devised to determine the average size and porosity of agglomerates in undisturbed suspensions. Simultaneous measurements are made with a photometer and a sedimentation balance [56], [57]. Figure 14 shows the average floc density during flocculation as a function of particle size. The process is characterized by the increase in the number of particles per agglomerate [58].

2.3. Agglomerate Strength

In general, the term strength refers to all of the properties of the agglomerate that act against whatever stresses the agglomerate may encounter: abrasion, fracturing, compression, or impact. The breakability that is desired for the end use of the product determines the upper limit of the strength that the agglomerate may have. At the same time, the agglomerate should have as high a strength as possible to prevent breakage during handling. Because the handling stresses are of the same order of magnitude as the breakability, the permissible stress range is narrow.

The measurement of agglomerate strength is divided into two categories: agglomerates smaller than 5 mm, and agglomerates whose diameter is greater than 5 mm. A number of scientific and industrial tests are used for the larger agglomerates [59]–[61]. Figure 15 illustrates some of the usual methods that are used in the pharmaceutical industry: the diametral compression test, the bending test, the drop test, and the abrasion test. There are numerous types of commercial apparatus that can be used to determine the strength of a pressed product with these tests. The strength values obtained on different machines are not usually comparable with each other. Some reasons for this are differences in geometry and the rate at which the stress is applied, as well as differences in external factors, handling, and calibration.

The strength of smaller agglomerates can be determined by the rate of deagglomeration in a specific shear field [60] or with a dispersion apparatus specific to the industrial application [61]–[65]. This could mean the ratio of optically determined surface areas after various types of dispersion processes, or it could mean the relative values of the desired property, such as the color intensity or average agglomerate size after two dispersants have been used.

Pneumatic transfer over a specific distance under simulated industrial conditions can be used as a measure of agglomerate abrasion.

2.4. Redispersion

When fine particulate properties are required for processing or application, as with pigments, the agglomerates must be easily broken down mechanically. If the agglomerate is dissolved during use, the required redispersibility also depends on the wettability. Good wetting and rapid dissolution are desirable for instant products and pharmaceuticals, while slow dissolution is desired for fertilizers.

Today, a large number of foods, such as coffee, cocoa, and cereal are sold as instant products. Often this property can only be obtained by means of agglomeration. Different manufacturers may define this property in different ways, therefore a method to measure it quantitatively is required. A quantitative method of measuring redispersion rate is necessary for industrial pow-

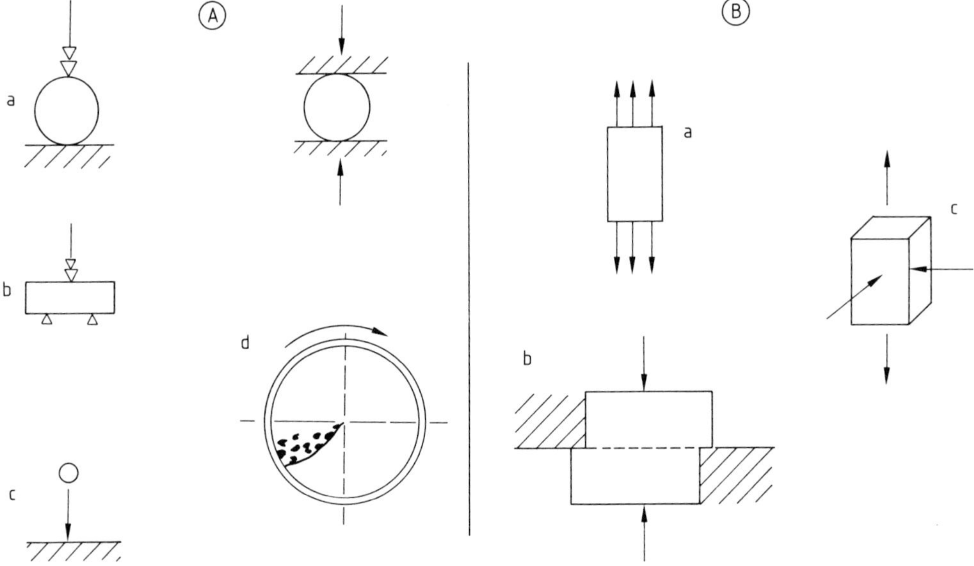

Figure 15. Operation of strength measurement techniques that simulate actual industrial conditions
A) Industrial tests: a) Compressive strength; b) Bending test; c) Drop test; d) Abrasion test
B) Deformation stress: a) Tensile strength; b) Shear strength; c) Shearing strength

ders that are added to and dissolved in liquids. The reconstitution process takes places in the following steps:

1) Fluid soaks into the bulk material and the individual agglomerates
2) Agglomerates sink in the liquid
3) Agglomerates break up
4) Particles dissolve in the liquid

None of these steps can be clearly isolated from the others. They do not happen sequentially, but rather occur simultaneously, affecting each other. Therefore, determination of the individual properties separately is difficult.

Penetration of the Liquid [66]–[70]. The first step in the reconstitution process is the penetration of the bulk material by the liquid. This results from capillary fluid transport through the void spaces in the bulk material. If the contact angle $\delta = 0$, wetting is rapid and complete. For most powdered foods, δ is greater than zero so that liquid penetration of the bulk is slower. Agglomeration of these powdery foods reduces the time required for the liquid to penetrate the bulk [50].

Sinking. If the wetting angle of a powder is 0°, then the powder can sink only if its apparent density is greater than the density of the liquid or the material dissolves or disperses in the liquid. If the powder dissolves, inaccessible void spaces are opened up and the apparent density increases with time. A powder with a wetting angle of greater than 0° requires a higher apparent density to sink than one with an angle of 0°.

Dispersion [71]–[73]. Dispersion means the breakup of the agglomerates into their primary particles. This process should be as rapid as possible. Dispersion greatly increases the solid–liquid contact area, and consequently the rate of dissolution. To obtain rapid agglomerate breakup, the bridges between the particles should be weak. However, weak bridges may allow more abrasion, which would tend to slow down liquid penetration. PFALZER [68] therefore recommends having a large number of weak interparticle bridges. The weak bridges would be easily dissolved, while the large number of bridges would provide the necessary mechanical stability. SCHUBERT [71] seeks an optimum in promoting the desired, antithetical properties of rapid breakup and low abrasion. As strength increases, the liquid penetration time decreases because of the reduced amount of abrasion. However, the time required for the agglomerate to break up increases because of the stronger interparticle bridges. There is a relative minimum in the overall time and a corresponding optimal agglomerate strength. The optimal strength must be determined for various porosity levels. The optimal porosity is the one that gives the shortest overall reconstitution time.

So-called burst agents are added to highly compacted pharmaceutical preparations. These agents are materials that swell to considerable degree when they absorb water. Some typical release agents are starch and modified starches. The action of a burst agent is only apparent when the swollen dimensions of the burst agent exceed the pore size of the material in which it is incorporated. There are no literature references to the application of burst agents in granulated products. Any improvement in dispersibility from such an application cannot be predicted with certainty. The effect of such a burst agent is influenced by its concentration, its swelling properties, the particle size distribution, and the agent's distribution in the instant product.

3. Size Enlargement Processes

3.1. Growth Agglomeration (Pelletization)

Growth agglomeration is defined as the growth of more or less solid agglomerates in either of two environments. The first is a rotating apparatus that produces both a mixing and a rolling motion. The second is a turbulently agitated suspension of particles that generates interparticle collisions. There is a stable accumulation if the attractive forces in the industrial equipment always are greater than the destructive forces present in the system. The binding forces in this type of agglomerate are primarily capillary forces. Binders are often added to the liquid used to form the agglomerates. The strength of the final product is usually a function of the processing that occurs after agglomeration, such as drying and sintering. The most common types of equipment for growth agglomeration are pans, drums, mixers, and fluidized beds (Fig. 16). Growth agglomeration can result from several mechanistic processes: coalescence of fine parti-

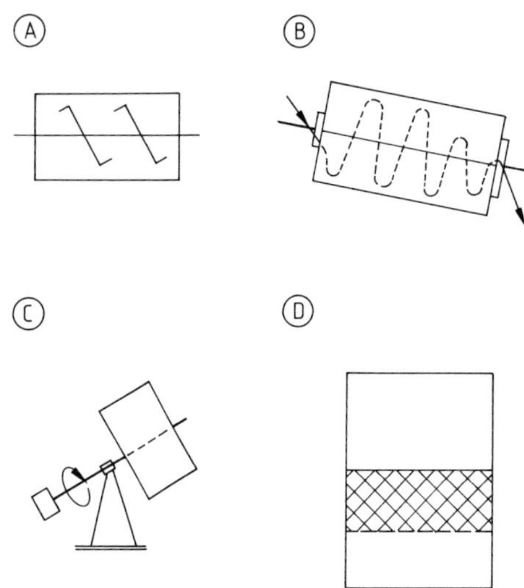

Figure 16. Schematic representation of equipment used for growth agglomeration
A) Mixer; B) Drum granulator; C) Pan granulator; D) Fluidized bed granulator

cles (starting materials and recycled fines), i.e., the nucleation phase, followed by growth through accumulation of further starting materials and the debris from other particles. The course of these processes with time is described by agglomeration kinetics.

3.1.1. Agglomeration Kinetics

CAPES and DANCKWARTS [74] and KAPUR and FÜRSTENAU [75], [76] have developed kinetic models for particle growth. If certain assumptions are made for the probability of agglomeration, changes in particle size distribution for a batchwise operated drum can be predicted theoretically. CAPES and DANCKWARTS use the following assumptions in their model of particle growth:

1) The smallest size fraction in the feed material will be abraded
2) The abraded material from this process will be distributed among the remaining agglomerates according to their size
3) The rate of growth of the diameter of each agglomerate is proportional to the difference between its diameter and the diameter of the smallest agglomerate present in the system.

Then they obtained a dimensionless form of the agglomerate size distribution, which is not explicitly time dependent, and which they called "self-preserving."

KAPUR and FÜRSTENAU determine the change in an agglomerate size distribution as a function of time by means of a coalescence model. This approach is similar to that used by SMOLUCHOWSKI [77] in his droplet coalescence theory. KAPUR and FÜRSTENAU make the following assumptions:

1) Mixing of all the fractions is completely homogeneous
2) Particle size distribution is relatively narrow
3) Coalescence rate is independent of granule size
4) Probability of collision between the granules is independent of their size

This model leads to the same results as CAPES and DANCKWARTS, the so-called "self-preserving" distribution. These kinetic models have the disadvantage that it is difficult or impossible to apply them to continuous systems. Over 90% of growth agglomeration is performed in continuously operated mixers, drums, and pans. The agglomeration model can be modified by dividing the process into two stages. The first is the aggregation stage, in which the particles coalesce. The second stage is called the snowballing stage (Fig. 17), in which agglomeration proceeds by layerwise deposition. By suitable selection of particle size classes, a system of differential equations can be written which describe the change in particle size distribution as a function of time [78].

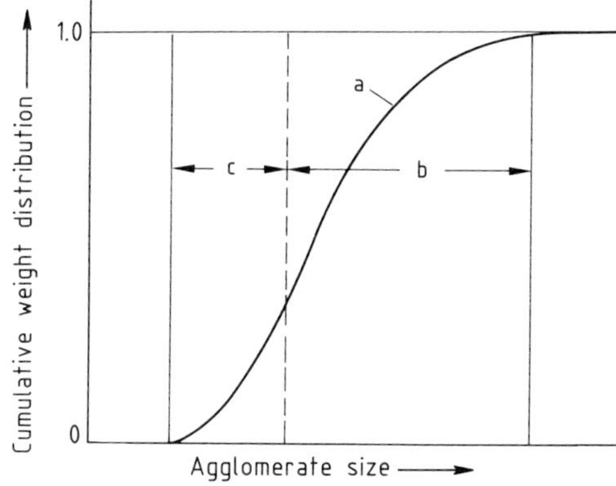

Figure 17. Division of the agglomerate size distribution in the aggregation and snowballing zones
a) Agglomerate size distribution in agglomerator; b) Snowballing zone; c) Aggregation zone

3.1.2. Pan and Drum Agglomeration

[79]–[85]

Growth granulation using a pan or drum granulator converts finely dispersed material to a coarser product. The fine feed (diameter < 1000 μm) enter the disk rotating on an inclined axis or the rotating trommel, along with some added liquid. They are exposed to a rolling action (Fig. 18). Because the rolling action is primarily random, spherical agglomerates form, with diameters up to 50 mm.

In the pan agglomerator, the agglomerator nuclei and small agglomerates move toward the base of the disk. Because of increased friction, due to their largely irregular shape, the particles are transported higher. The larger, more rounded agglomerates roll easily over the smaller ones. Ultimately, at a certain size they are removed. In this way the pan classifies the agglomerates. The overflow product is so uniform in size that subsequent classification is usually unnecessary. Classifying pans producing granules having a narrow size range may have diameters up to 7.5 m and have iron-ore-pellet capacities greater than 100 t/h.

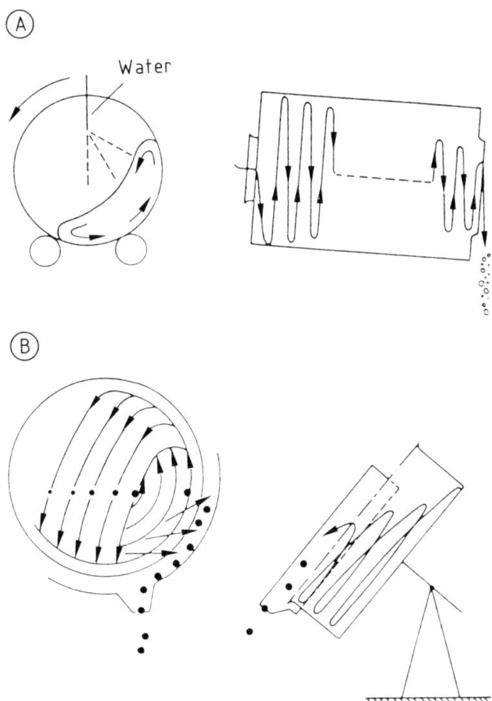

Figure 18. Schematic diagram of A) pan and B) drum agglomeration

In the agglomeration drum, the inclination of the axis is too slight to cause classification. It serves primarily to transport the feed through the cylindrical drum. The agglomerate size distribution is substantially broader than that obtained with a pan. To obtain a specific agglomerate diameter, the desired size agglomerates must be screened out of the exit stream. The oversize material must be reduced in size and fed back into the drum along with the fines to form a closed loop. The principal users of agglomeration drums are the iron and fertilizer industries. Sizes range up to 10 m in length and 3.5 m in diameter.

3.1.3. Mixer and Fluidized Bed Agglomeration

In theory, all solid and solid–liquid mixers are suitable for agglomerate production. In contrast to pan and drum agglomerators, the powder is rolled about in mixers by means of a mechanical impeller or, in the case of a fluidized bed, pneumatically. When a suitable amount of agglomeration liquid is sprayed in, most of the particles form either solid agglomerates or a low-density instant product due to the stresses generated by stirring.

3.1.3.1. Pan and Drum Mixers

Horizontal pan and vertical drum mixers have been constructed with capacities up to 5 m^3. Their first application was in the agglomeration of fertilizers [86]. A typical example is the Eirich countercurrent mixer-granulator. This device has a rotating mixing drum supported by a vertical or slightly tilted axle. A mixing impeller rotates in the opposite direction. When used for fertilizers, throughputs of up to 30 t/h, with residence times of 2–3 min, can be obtained. If the conventional impeller is replaced with a specially designed impeller, intensive mixing action can be obtained, even permitting the agglomeration of molten materials while they are cooling. Drum granulation of a similar type is widely used for pulpy, plastic, or pasty starting materials in the pharmaceutical industry. In some cases, dry material is added to produce small agglomerates as feed for tableting. The agglomerates produced by pan or drum mixers are usually stable, with high strength, after drying.

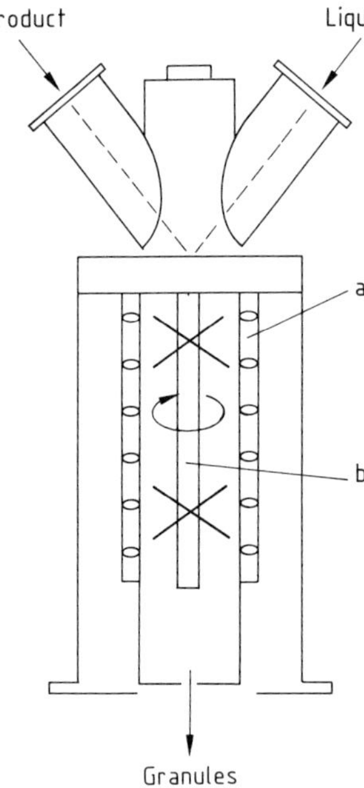

Figure 19. Schugi mixer
a) Roller cage; b) Rotor with blades

3.1.3.2. Horizontal Mixers [87]

The mixing action necessary for agglomeration is provided in the case of the horizontal mixer by a horizontal mixing shaft. This shaft may be equipped with paddles, screw strips, or special mixing blades that resemble plowshares. A typical example is the Lödige plowshare mixer. The rolling action is therefore more gentle than that in a pan or vertical drum mixer. As a result, the agglomerates are usually more porous and have a lower strength. Horizontal mixers can be operated either continuously or batchwise.

3.1.3.3. High-Speed Mixers

High-speed mixers are continuous mixers with impellers operating at relatively high speeds. The finished product has a brief residence time in the unit, only a few seconds. Porous, easily dispersible agglomerates are formed with the aid of sprayed in liquid. The product has a size range of 0.5–2 mm and finds application in the food industry, especially as instant products (Fig. 19).

3.1.3.4. Fluidized-Bed Agglomeration [90]–[92]

In this process, pumpable solutions, suspensions, pastes, or melts are converted into agglomerates mostly in combination with drying (see Section 3.3.1.3.). Fluidized bed drying technology has been known for several decades. Even though agglomerate formation was observed early in this process, the process was not used for agglomeration until 1970. The first use was the production of feed material for tablet production in the pharmaceutical industry [88]. Today the range of application has grown from pharmaceuticals to the chemical, ceramics, and food industries and is steadily growing in importance. In fluidized bed agglomeration, the feed particles are moved about by flowing gas (Fig. 20). Externally heated gas (air) is forced or drawn upward through a porous distributor plate and then through the particle bed. A heterogeneous solid–gas fluidized bed is formed. The solid particles are intensively mixed by the rising gas bubbles [89]. This promotes good heat and material exchange. The agglomeration fluid is sprayed with either single- or two-fluid nozzles, which are located above the bed surface or project just below it. The motion of the bed serves to distribute the

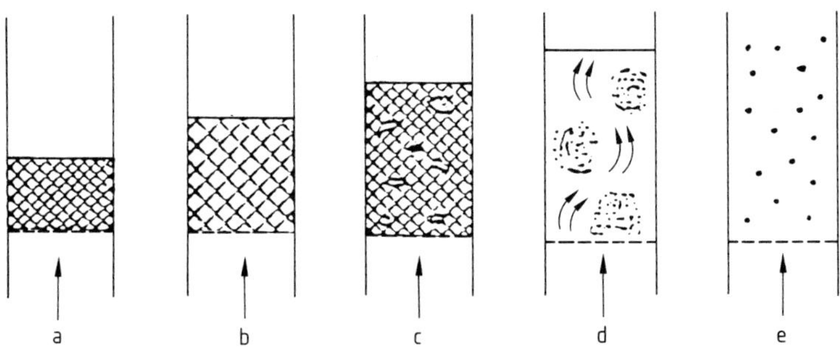

Figure 20. Schematic diagram of fluidized bed formation
a) Solid bed; b) Homogeneous fluidized bed; c) Inhomogeneous fluidized bed; d) Transition zone; e) Pneumatic transport

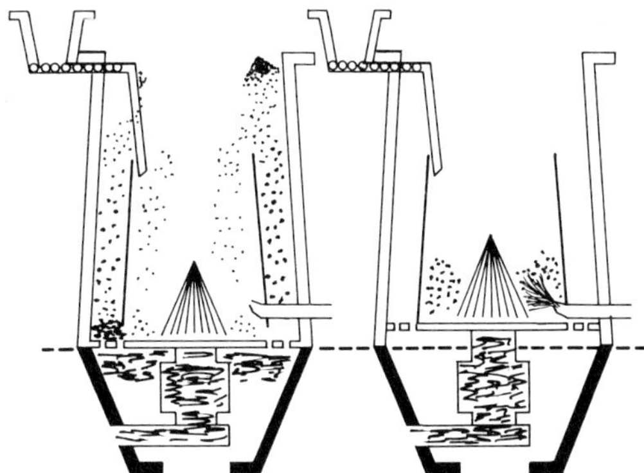

Figure 21. Schematic representation of rotor agglomeration

fluid uniformly among the particles. A portion of the agglomerates in the bed are drawn through an opening beneath the bed surface or through the overflow port. The product, in most cases, is separated into oversize, product, and fines. The fines and crushed oversize are recycled to the fluidized bed. To produce round agglomerates in the fluidized bed, the fixed perforated plates are replaced by rotating disks. The gas flow is either through perforations in the outside of the disk, or through a slot between the rotating disks or a slot between the rotating disk and the stationary housing. This process is called rotor agglomeration (Fig. 21). Many manufacturers offer fluidized bed equipment with a rectalinear cross section. These offer nearly the same possibilities of variation with regard to construction and operation as does equipment with a round cross section.

3.2. Agglomerate Formation from Moist Material

The production of fine, dry powders entails high drying costs, which are greatly increased by the addition of fluid during agglomeration. Economic considerations (energy savings up to 50%) would favor combining the drying and granulation process segments. This would prevent having to granulate a dry powder or enable the granulation to be done before drying. The goals of this process are the following:

1) Processing of pasty products without repulping to reduce drying energy
2) Production of a dust-free agglomerate
3) Production of the desired size of agglomerate having the required product characteristics

One possible way of achieving the first goal lies in combining the granulation process with fluidized bed drying. In conventional fluidized bed drying (see Section 3.3.1.3.), dust from the filter is remixed with the bed material, producing a dusty product. A better method is to mix the dust from the filters into the moist feed so that it is incorporated into the agglomerate (Fig. 22).

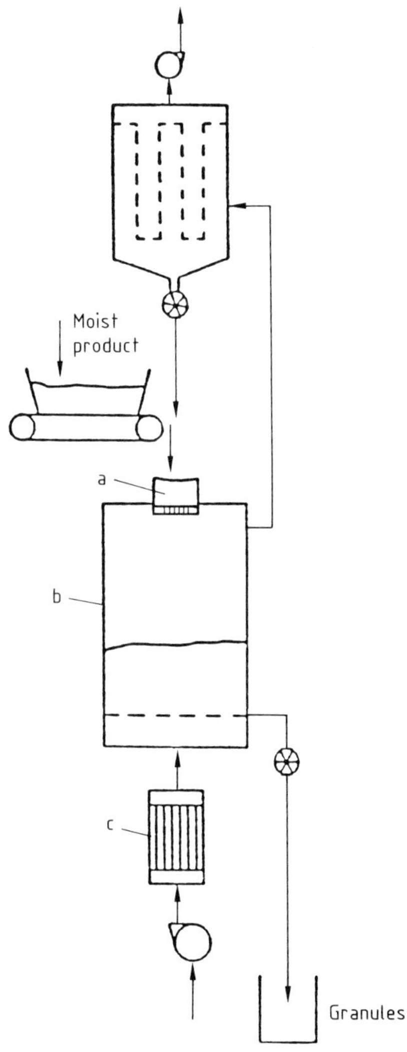

Figure 22. Agglomeration prior to drying
a) Screen granulator; b) Fluidized bed dryer; c) Air heater

3.2.1. Agglomeration Equipment for Plastic Materials

There are a large number of different machines for this purpose. They differ in their construction, the configuration of the kneading surfaces, and their kinematics. There are more than 50 types commercially available (Fig. 23).

Extrusion through Screens and Perforated Plates. Pasty products, when pressure is applied, are forced to flow through the provided openings. Pressure can be applied by the following physical processes:

Internal Friction (*Friction Formation*). The amount of pressure is greatly dependent on the properties of the product. The maximum pressure is ca. 50 kPa, which is relatively low. The product is strongly sheared by the frictional forces.

Volume Reduction (*Shape Formation*). The amount of pressure that can be applied is only dependent on geometric factors. Pressures up to 2.0 MPa can be generated. In this case there is little shearing.

Movement of Excluded Volume (*Shape Formation*). High pressure can be generated in the space allowed. The product must be capable of flowing.

Breaking up of Product Clumps. The following methods are used for pasty products:

Cutting. Thin cakes, produced by rollers, are cut by the action of a knife roller. The method is also applicable to harder clumps.

Scalping, breaking. A rapidly rotating knife roller either peels off or breaks off by inertia, pieces of the product stream. The shape of the pieces cannot be well controlled. The method is not suitable for soft pastes.

Squeezing. The product is squeezed into the grooves of a grooved roller made of an elastic material. The shaped material is released when the grooves are retracted. This method is suitable for soft pastes.

Pellet mills are used to shape pasty materials such as filter cakes and others. The most important types of machines are illustrated in Figure 24. In all cases, the material is forced through openings and in so doing is compressed more or less into granules, also called pellets. In the roll pellet mill, the material is introduced into the space between a perforated roller and a solid one. It is compressed and forced through cylin-

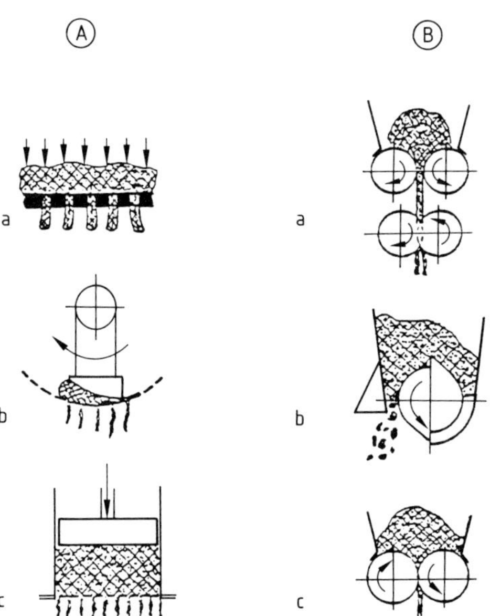

Figure 23. Operation of machines for agglomeration of moist materials
A) Extrusion in pellet mills: a) Application of pressure; b) Frictional press; c) Pressure by volume reduction
B) Breaking of clumps: a) Cutting of sheets; b) Crushing or milling; c) Squeezing into molded products

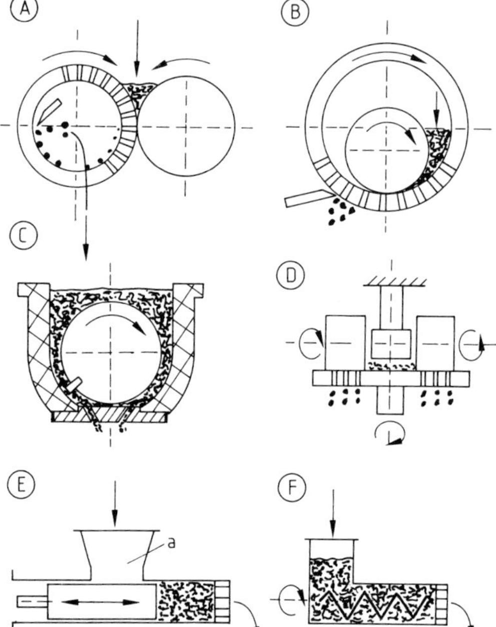

Figure 24. Operation of pellet mills
A) Roll pellet mill; B) Ring collar mill; C) Haas-granulator; D) Disk collar mill; E) Extrusion mill; F) Screw mill

drical openings. Inside the perforated roller, the granules are cut off by a knife and flow outward in an axial direction. A variation of this type of machine has two perforated rollers.

In the *ring collar mill,* the material is introduced inside and forced out through the openings in the die ring by the collar. The granules are then cut off on the outer side by a knife. This type of machine can have various sized collars as well as different numbers of collars. The force exerted in a ring collar mill can be up to 15 t [93]. Through puts up to 15 t/h for feed materials can be achieved with 22 mm openings.

The so-called *Haas granulator* is commonly used before belt drying. In this machine, the in and out motion of a roller equipped with a ridge forces the material through a perforated plate into a collection trough. In the *disk collar mill* the collar as well as the perforated plate move. In the extrusion and *screw mills* the perforated disk is stationary.

The swivel press has a rectilinear body and is open on the top. The bottom is closed off with a removable perforated plate. It is also equipped with a swivel arm. The feed is distributed across the perforated plate by a conveyor. Due to the swivel motion of the displacement arm, a triangular space bounded by the wall, the perforated plate, and the displacement member is enclosed. A squeezing or load stroke forces the material through a perforated plate. Because the plates are removable, various geometries can be selected. If the product is highly resistant to extrusion through the plate, it may be squeezed out sideways. It is drawn in again on the return stroke. After the pressure stroke, the swivel arm quickly moves back. The return angle for the no load and the pressure strokes is adjustable. If there is a foreign body in the paste, the machine stops after the overload alarm sounds. Because of the low height, it can easily be removed by hand. The following parameters are involved in the construction and operation of the machine; geometry of the perforated plate (number of openings and their configuration), number of strokes, extrusion rate, return angle, extrusion pressure, and throughput [93]–[95].

3.3. Drying Processes

In the case of fluid raw materials, agglomeration can take place during drying, i.e., drying and agglomeration are combined into one process step. The available methods are classified as follows:

1) Atomization drying
2) Contact drying
3) Vacuum drying

Atomization drying includes tower spray drying [98], fluidized bed spray drying, and multistage spray drying. Contact drying includes roller and belt drying. Under vacuum drying, only freeze drying and contact vacuum drying are described.

3.3.1. Atomization Drying

Atomization drying processes are important for industry. A number of arrangements are used. The principal methods are classified as follows:

1) Simple spray drying
2) Fluidized bed spray drying
3) Spray drying with an integrated or downstream fluidized bed

The feed to atomization processes are pumpable products, such as solutions, suspensions, dispersions, and pumpable pastes. These are converted into dry powder by spraying them into a hot gas stream.

3.3.1.1. Fundamentals of Atomization Drying

Four different types of equipment are used to achieve atomization (Fig. 25):

1) Ultrasonic atomizer
2) Centrifugal atomizer
3) Two-fluid nozzle
4) One-fluid nozzle (pressure nozzle)

The use of the ultrasonic atomizer in drying is limited to special cases.

In centrifugal atomization, a wet product is accelerated to velocities between 100 and 200 m/s. A thin, fluid film enters the hot gas stream of the dryer through an opening in the atomization disk. It then breaks up into droplets. The resulting particle size is independent of throughput over a wide range. The average particle can be varied by changing the rotational speed.

A narrow particle size distribution can be obtained with a one-fluid nozzle. The average drop size is controlled by the diameter of the nozzle opening, the atomization pressure, the viscosity,

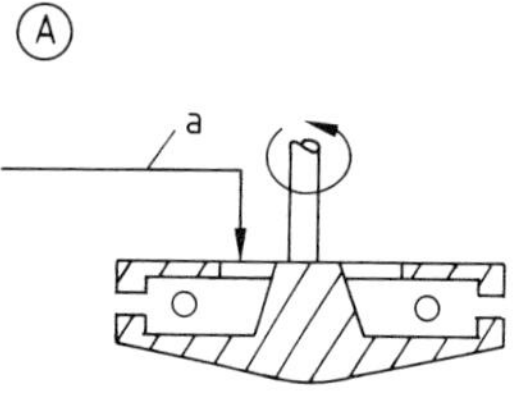

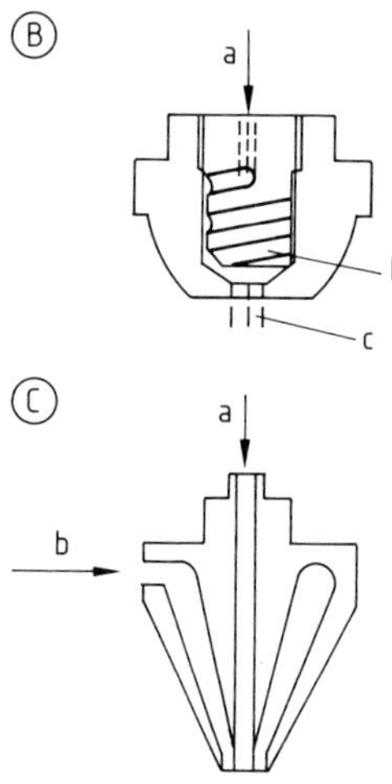

Figure 25. Types of atomizers
A) Centrifugal atomization disk: a) Moist product (not under pressure)
B) One-fluid nozzle: a) Moist product (under pressure up to 10 MPa); b) Mouthpiece; c) Nozzle opening
C) Two-fluid nozzle: a) Moist product (not under pressure); b) Air for atomization

the solids content, and the throughput requirements. One-fluid nozzles tend to clog easily.

In a two-fluid nozzle, the energy required for atomization is provided by the air stream. In most cases the air stream is compressed air, although in special applications steam or inert gas may be used. The wet product is introduced into the two-fluid nozzle under little or no pressure. Energy requirements are higher for the two-fluid nozzle than for the one-fluid type. With two-fluid nozzles, fine grained dry products can be obtained, although the particle size distribution is relatively broad. Compared to one-fluid nozzles, two-fluid nozzles present fewer handling and control problems. Special designs, which lessen the risk of clogging [96], are used for fluidized beds.

The average particle size has a major effect on the instant properties of a product. The drying process is slower for larger particles compared to smaller ones. This has an effect on the design of a drying tower. The longer residence time worsens the solubility properties. Temperature control is important for the properties of a spray-dried product. In the case of sensitive products, such as powdered milk, too high a temperature can damage the product. BRUMMELHUIS [97] has shown the effect of the temperature of the air leaving the dryer on various product properties using as the example the atomization drying of whole milk concentrate. Lower dryer exhaust temperatures improve the instant properties. The apparent density increases. The free fat content and the solubility index decreases. However, there is an increase in the moisture content of the powder, so that a second drying step is required.

3.3.1.2. Fluidized-Bed Spray Drying

If gas flows through a solid, either wet or dry, resting on a porous platform, the piled material begins to loosen. At a certain velocity it begins to whirl about. Vibrators can be used to aid fluidization. The pumpable product is sprayed into the receiver on top of the material already there. Either one- or two-fluid nozzles are installed over the pile or brought into it. In addition, stirrers and disintegrators are used to loosen up the product and to break up the larger clumps.

The separation of solid particles can be done externally by cyclones, filters, or wet scrubbers or internally by an integrated filter. The use of an integrated filter has the advantage that the product falls directly into the receiver. The risk of having moisture condense in the exhaust is minimized because of the short pathway. However, moisture condensing in the filter can cause stickiness and fluctuations in flow rate. Integrated filters are available with automatic shakers or continuously acting blow-off devices.

The units can be operated either in suction or pressure mode. Suction operation is clearly predominant among the major manufacturers because no dust escapes under low-pressure conditions. Also, suction spray towers can be operated without any material escaping. The dry product can be removed by means of belts, rotary bin values, overflow, or exit pipes. The product can subsequently be classified by screening or sintered, and the undersize material can be recycled. If necessary, the oversize material can be broken up and recycled. Product removal can be combined with classification. This is done with an ascending tube, which discharges at the screen. The tube is used as a sorter. By regulating the flow velocity in the tube, the size of the particles that are removed can be controlled.

Plants using only fluidized beds are mainly used to dry solutions or melts of organic acids, organic or inorganic salts, tanning agents, and dyes. Binders can be added to the solution or melt to improve cohesion. Exhaust temperatures are lower than in spray tower drying. The dry product comes in contact with the high-temperature, incoming gas. This is similar to the situation in countercurrent spray tower drying. As a result, however, the same disadvantages are present.

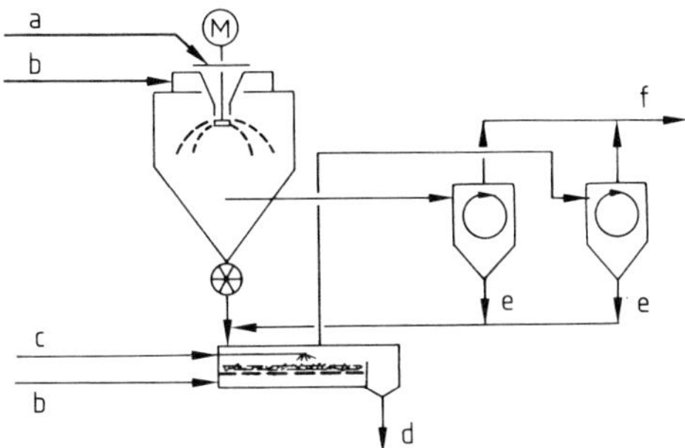

Figure 27. Spray dryer with downstream fluidized bed dryer with injection of agglomeration liquid
a) Moist product; b) Drying gas; c) Agglomeration liquid; d) Dry product, agglomerates; e) Recycling of fines; f) Exhaust gas

3.3.1.3. Combination Processes

During the last ten years, spray towers with downstream or integrated fluidized bed dryers have become more common. This is due to their lower energy usage, the primarily dust-free product, and the reduced thermal stress. The disadvantage of these types of processes is their high capital costs. A spray drying process with a downstream fluidized-bed drying stage (Fig. 26) is called a two-stage drying process. Lower residual product moisture with less harm to product quality as well as more efficient energy utilization can be achieved. The solids leave the drying tower with 5–8% residual moisture. The final drying step takes place under mild conditions with low energy usage. Two-stage drying gives unsatisfactory results when the product with high residual moisture has a tendency to stick or form hard clumps. In these cases it may be possible to dry the product to a lower moisture in the drying tower. An agglomeration fluid is then sprayed in, either directly on the wet product or into the downstream fluidized bed (Fig. 27). As an alternative to having a fluidized bed dryer follow a tower drying stage, one can be integrated into a drying tower. In this case, rather than removing the powder from the spray tower, it is allowed to fall, with its high moisture content, directly into a fluidized bed. In the fluidized bed zone, drying takes place in a cloud of fine particles. This powdering permits the drying of sticky products with a high moisture content without their sticking to the walls (Fig. 28).

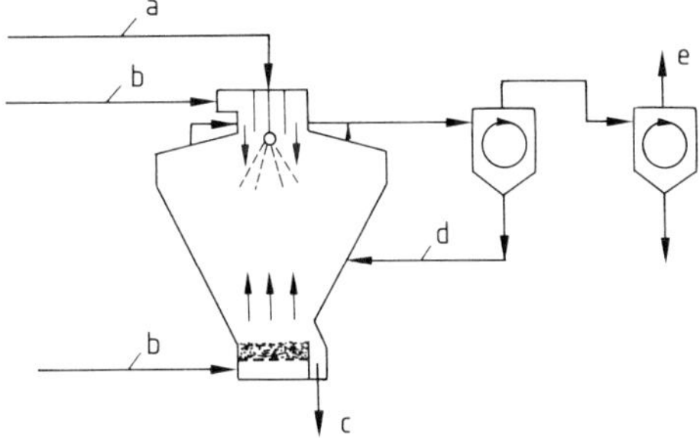

Figure 28. Spray dryer with integrated fluidized bed
a) Moist product; b) Drying gas; c) Dry product, agglomerates; d) Recycling of fines; e) Exhaust gas

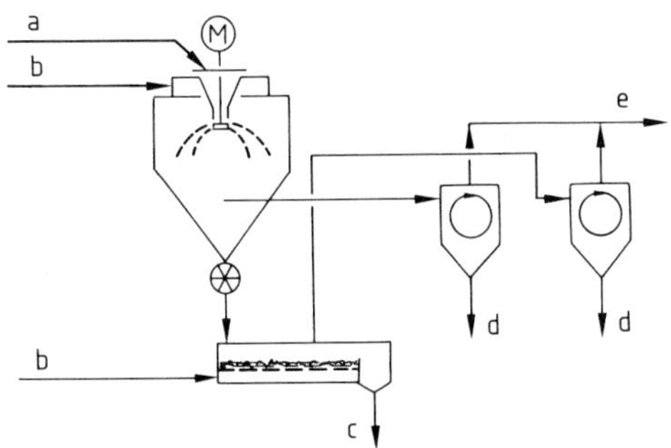

Figure 26. Spray dryers with downstream fluidized bed dryers
a) Moist product; b) Drying gas; c) Dry product, coarse fraction; d) Dry product, fine fraction; e) Exhaust gas

3.3.2. Contact Drying

Contact drying is roller drying (Fig. 29). Pumpable products, including low- and high-viscosity materials as well as pastes, can be dried on a roller [99]. Metal rollers are heated internally

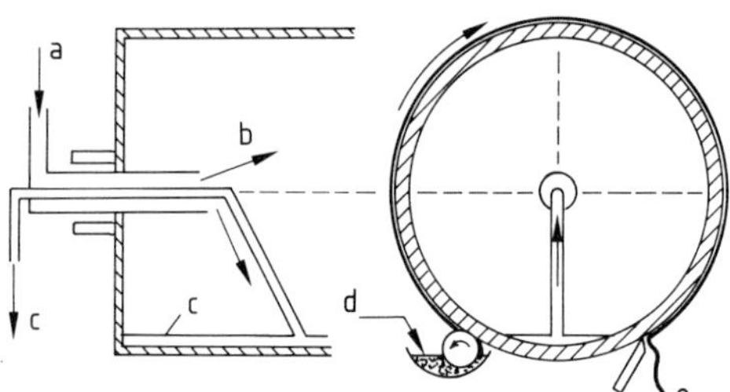

Figure 29. Roller dryer [99]
a) Hot steam; b) Steam; c) Condensate; d) Liquid product; e) Dry product

with steam. The heat used for drying is obtained by condensing the steam. The resulting condensate must, therefore, be removed from the lowest point within the roller. The product to be dried is applied as a thin layer onto the hot roller. This coating process can be done in different ways. The principal methods are classified either as roller application methods or as spray methods.

The roller can dip directly into the material to be dried. This type of equipment is referred to as a sump roller dryer. Temperature and concentration gradients are caused by heat transfer to the fluid. These can be removed by means of an agitator. The disadvantages of this design are the nonuniform coating on the roller and exposing the product to heat, the latter having quite negative effects in the case of heat-sensitive products.

An improvement is achieved by using a so-called coating roller (Fig. 30). In this case, only the coating roller dips into the liquid. It coats a liquid film of uniform thickness on the drying roller. Another variation of these design makes use of an additional roller, called a transfer

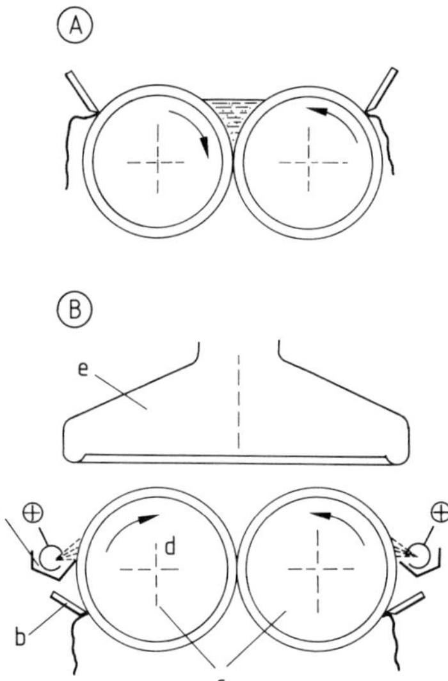

Figure 31. A) Dual roller steam dryer and B) Dual roller dryer with spray coating
a) Distributor trough; b) Scraper; c) Drying rollers; d) Steam space; e) Exhaust hood

roller. This roller may be coated with hard rubber and cooled internally with water. In addition to single roller dryers, there are also heated twin rollers (dual roller dryers) in use (Fig. 31). Examples of this type are the sump dual roller dryer and the dual roller dryer with spray coating. The dry product is scraped off by a knife. It may be in the form of a film, or flocs, flakes, or powder [99]. The products are usually removed with a screw conveyor. If necessary the particle size can be reduced to the desired value by means of a hammer mill or crusher. In designing this type of dryer, the product feed point and the product

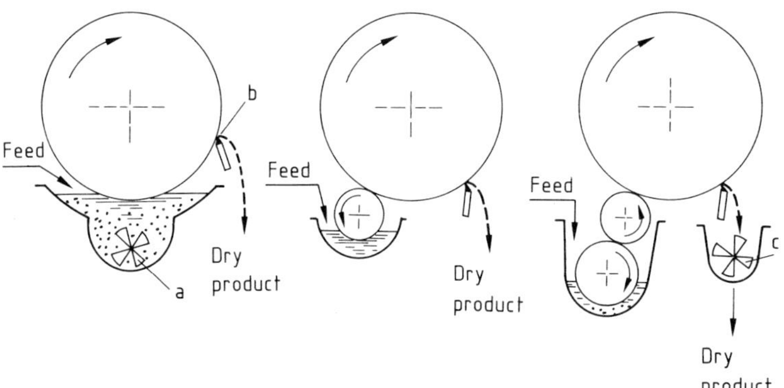

Figure 30. Roller dryers with various feed systems [99]
a) Stirrer; b) Blade; c) Screw conveyer

removal point must be separated. For this reason, only 70–80 % of the circumferance is effectively used for drying.

3.3.3. Vacuum Drying

In the case of temperature-sensitive products, the external pressure must be lowered to avoid drying at a high temperature. This is called vacuum drying. One simple type of vacuum drying is vacuum roller drying. In this method, either one or two rollers are installed in a vacuum housing. The resulting vapor precipitates in a condensor located between the vacuum chamber and the pump. The product is removed by a screw conveyor (Fig. 32).

Freeze drying is also widely used for sensitive products. Freeze drying, also called sublimation drying, is defined as the drying of frozen material. At pressures below 0.6 kPa, ice goes directly into water vapor, without passing through the liquid state. Dissolved material (for example, salts) lower the freezing point. Actual sublimation is only possible at the eutectic temperature (the point of maximum freezing point depression) [99]. KLUGE and HEISS [100] used the nonenzymatic browning reaction to demonstrate qualitatively that freeze drying was the best drying process. The biggest disadvantage of this process is its high capital and operatings costs. Products with good instant properties can be produced by freeze drying because sublimation produces a highly porous solid structure. The product must first be frozen before drying. The

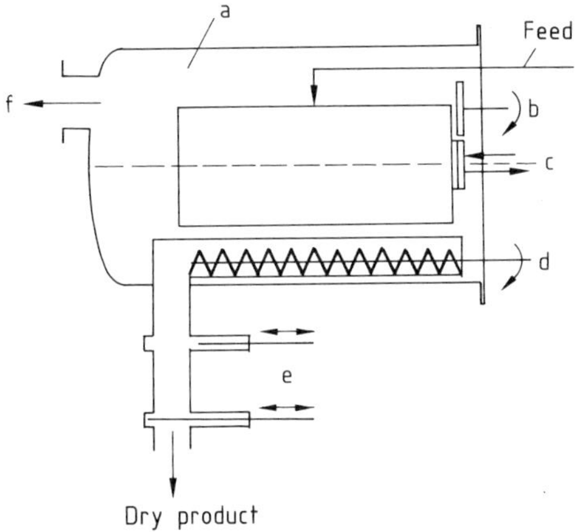

Figure 32. Vacuum roller dryer [99]
a) Vacuum chamber; b) Roller drive; c) Heat; d) Screw conveyer drive; e) Air lock; f) To condenser and vacuum pump

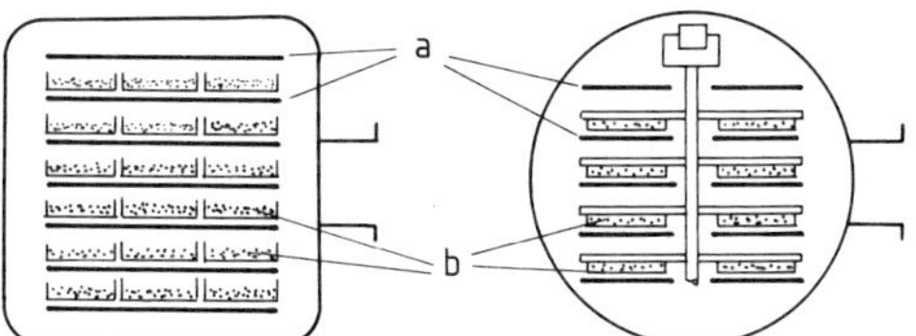

Figure 33. Cross-sectional view of batch freeze dryers
a) Heating surfaces; b) Product trays

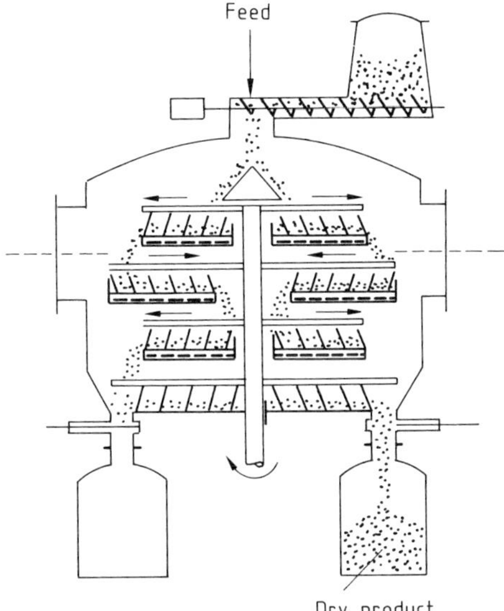

Figure 34. Continuous freeze dryer

rate of freezing has a strong effect on both product quality and the rate of drying in the freeze dryer. Rapid freezing produces small ice crystals, a strong product structure, and fine pores. On the other hand, these factors increase the resistance to the flow of moisture during drying. Freeze drying can be performed batchwise or continuously. Figure 33 shows a cross sectional diagram of a batch freeze dryer. Continuously operating freeze dryers are illustrated in Figure 34. A vibrational freeze dryer and a disk freeze dryer are shown.

3.4. Spherical Agglomeration [101]

Another possible method of forming agglomerates in a fluid medium is the addition of a second immiscible liquid [102]. This binding liquid must be able to wet the particles better than the liquid in which they are suspended. This process, called spherical agglomeration takes place in a

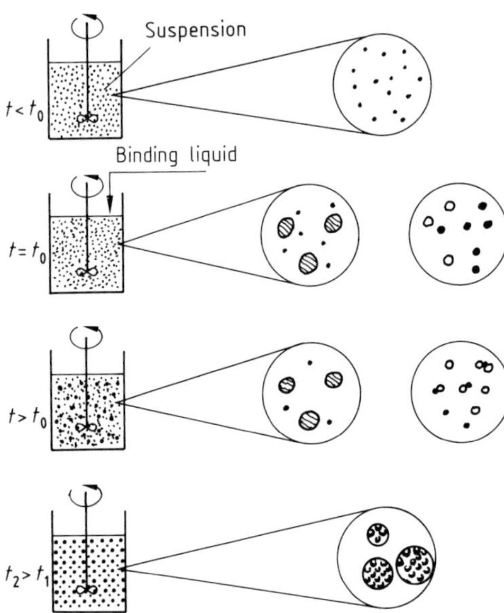

Figure 35. Stages of spherical agglomeration according to Bröckel [103]

three-phase system. The three phases are the suspending liquid, the immiscible binding liquid, and the solid dispersed phase. This phenomenon, which occurs during the progressive addition of the bridging liquid, is illustrated in Figure 35. The finely dispersed droplets of binding liquid encounter the particles in a dynamic system. The particles cling to the droplets and become concentrated in a manner not unlike flotation. This process may take place in a few seconds or last for hours, depending on the wetting properties and composition of the system. BEMER [103], [107] calls this process the flocculation phase. He used 19 μm glass spheres in carbon tetrachloride and a water–glycerin mixture as the binding liquid. Flocs roughly 100–300 μm in size were formed. This was followed by an extended period of apparent stability. This was followed suddenly by a period of rapid agglomerate growth. The agglomerates grew by coalescence until they arrived at a stable state, in which the cohesive and dispersive forces were in equilibrium.

FARNAND [104] has stated three factors that control the spherical agglomeration process, without describing their effect in a quantitative manner. These are, in order of importance:

1) Relative interfacial free energies
2) Ratio of binding liquid to solid
3) Type and intensity of mixing

The effect of interfacial free energy on spherical agglomeration has not been described in the literature. During the wetting stage, flocs, agglomerate nuclei, spherical agglomerates, even pastes can form depending on the composition of the three-phase system and the ratio of binding liquid to solid [104]–[110].

The effect of the wetting properties of the solid and liquid components makes the agglomerates sensitive to composition. For certain combinations of materials, the agglomeration process can be made selective. Therfore in a mixture of solids, particles with a certain composition can be made to form agglomerates, while the remaining material can be separated by screening. By changing the liquids, additional components can be obtained in succession in a way resembling preparation by flotation. For this reason spherical agglomeration is also called selective agglomeration.

Application Cases. A major area of application of spherical agglomeration is the removal of fine coal particles dispersed in water. The fraction of fines produced during coal preparation is increasing. This is due to the increased use of mechanized mining methods and finer milling to produce a cleaner final product. Coal particles in aqueous suspension can be agglomerated with special oils and removed. Inorganic or ash components are left behind.

Olifloc Process. The flowsheet of the olifloc process (Bogenschneider, Ruhrkohle AG, 1976) is shown in Figure 36. The pilot plant was designed for a throughput of ca. 3 t/h. This process is a further development of the Convertol pro-

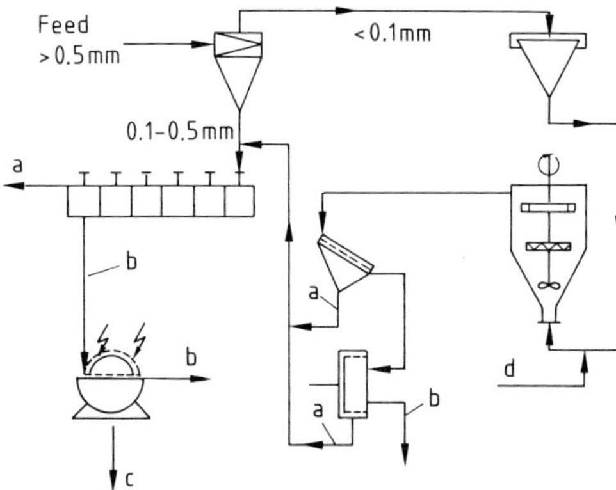

Figure 36. Process flowsheet of the olifloc pilot plant
a) Gangue; b) Coal; c) Water; d) Oil

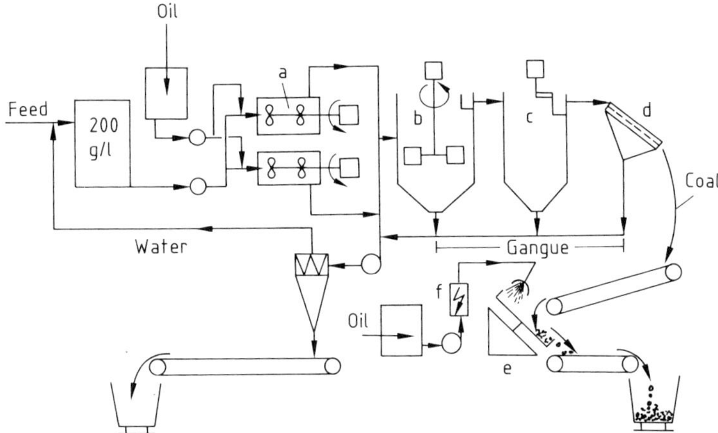

Figure 37. Process flowsheet for the spherical agglomeration process
a) High capacity mixer; b) Agglomeration and settling vessel; c) Stripping vessel; d) Dewatering screen; e) Disk pelletizer; f) Continuous heater

cess developed by the Essen Mining Association in the 1950s.

Operation. The local slurry with a particle size of < 500 µm is classified in a hydrocyclone. The fraction < 100 µm is sent to a clarifier to increase its solids concentration. The underflow with a solids content of 200–400 g/L is removed and mixed with oil, which serves as a binder. The components are intensively mixed and agglomerate in the upper section of the mixer. An inclined dewatering screen is used to separate the agglomerates from the gangue, which remains behind in the water. The residual moisture in the oil–coal agglomerates is lowered to 15% with a scrubbing centrifuge. The centrifugates from cyclone and screen go to flotation to further recover the coal content. BOGENSCHNEIDER states that the residual moisture in a coal preparation plant can be reduced to about 3% by using this process.

Spherical Agglomeration Process. Spherical agglomeration, as illustrated in Figure 37, was developed in the 1960s by the National Research Council of Canada [111].

Operation. Feed slurries are mixed in 2 high capacity mixers (a) with, for example, diesel fuel in order to completely disperse the binder. The agglomerates form in the agglomeration vessel (b). They are partially separated from the gangue by sedimentation in vessel (c). An inclined dewatering screen also serves to separate the agglomerates from the gangue. In this variation of the process, the agglomerates are pelletized with hot heavy fuel oil on a disk pelletizer. The pellets reach a diameter of ca. 6 mm with high strength.

Pelletizer–Separator. The pelletizer–separator (Fig. 38) [111] has a simple design. It generates solid pellets 1–6 mm in diameter by churning with rotating paddles. Nearly all fine coal slurries can be selectively agglomerated by adding 9–15 wt% of heavy oil. The binder is almost completely recovered in the pellets. An industrial plant with a capacity of 10 t/h has been in operation in Japan since 1972.

Operation. A coal–water suspension and heavy fuel oil are fed simultaneously from the left. After mixing in the first third of the vessel,

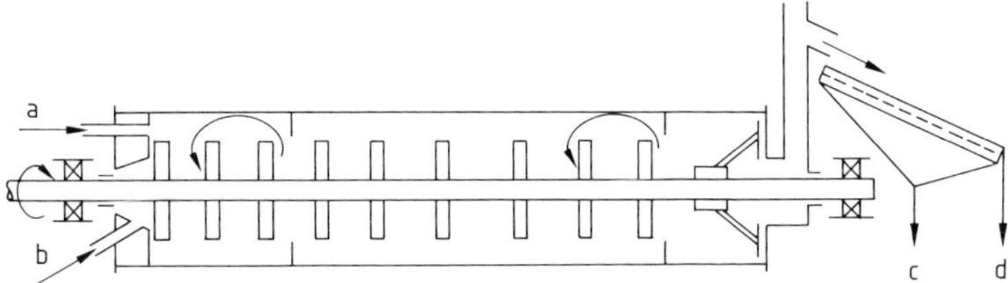

Figure 38. The Shell pelletizer-separator
a) Heavy fuel oil; b) Coal water suspension; c) Gangue; d) Pellets

the coal is selectively agglomerated in the second third. Next it is separated from the gangue on an inclined dewatering screen.

3.5. Pressure Agglomeration

The use of liquids in size enlargement is dependent on the compatibility of the product with the liquid. In some cases, the necessity of adding a drying step increases the cost. Pressure agglomeration is used when it is desirable to proceed without the use of liquids. During compression the number of interparticle contact points is increased in the substrate. Then by plastic deformation the contact points are converted to areas of interfacial attraction. Van der Waals forces or sintering are the principal factors affecting strength in the case of compression. Both increase with increasing surface contact area. Pressed articles with high strength can be obtained with materials that can be deformed plastically. It is difficult to compress elastic particles with unfavorable properties. Compressive properties can be improved by addition of a binder. If a binder is used, the compressive properties depend on what type of binder it is and its binding capacity. The particle-size distribution of the starting material is also a factor. Above all, coal, low-temperature coke and ores are briquetted with binders.

The pressure agglomeration process has a long history in the preparative and coal industries as the concept of briquetting. Coal tar pitch, bitumen, or 50% solids waste sulfite liquors are added to plasticize coal powder. Fine grained ores and other fine grained metal containing materials are mixed with fine limestone, magnesium hydroxide, or clay before being compressed. In addition to coal and ores, large amounts of ceramic precursors, fertilizers, plant protectants, and pharmaceuticals are produced with pressure agglomeration processes.

In *roller compaction,* the powder is introduced in between two smooth or mold rollers and is compressed. The product is either thin sheets (flakes) or molded shapes. In addition, the process is widely used as an intermediate step in agglomeration processes. One example is the pregranulation of feed materials for tableting.

The oldest process is *compression* with *stamp* and *die*. This process is widely used in powder metallurgy and the coal industry, as is roller compaction.

In *isostatic compression,* the powdered starting material is loaded into an elastic membrane and is compressed from all sides by fluid pressure. The resulting pressed object is homogeneous and has little internal stress. Compared to tableting, the pressure is relatively low. Because of its cost, this process is not suitable for high volume applications. It is used primarily in powder metallurgy.

Vibration compaction is especially suited for the production of long pressed articles with uniform properties. It is used in the ceramic industry and, for example, for the production of reactor fuel rods. The substrate is homogeneously compressed by the wall impact of the vibrating container as well as the uniform pressure due to interparticle collisions.

3.5.1. Compression Equations

Figure 39 illustrates the force–displacement diagram for compression of cornstarch, with its elastic and plastic regions. There are over 15 different mathematical models in the literature [113], [114] for compression. These described the volume reduction of a powdery material when a compressive force is applied. Depending on the magnitude of the compressive force or the type of material, most of these models have only a limited validity. The best known compression equations are listed in Table 2.

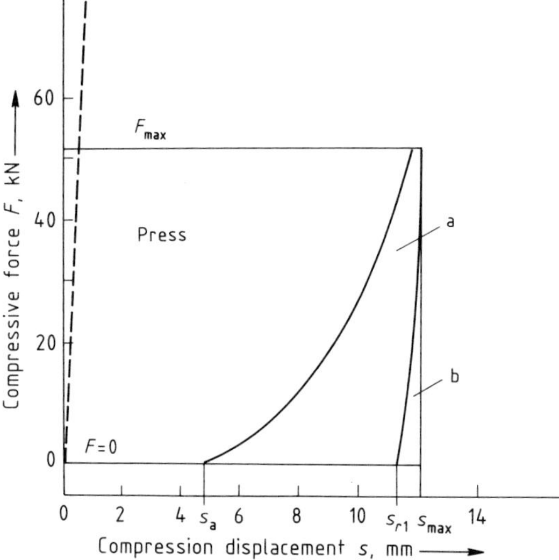

Figure 39. Force–displacement diagram for compression, cornstarch compression rate $v_v = 0.24$ cm/s
a) Compression energy E_a plastic; b) Compression energy E_2 elastic

Table 2. Mathematical models of compression [115]

Athy	$\frac{V-V_\infty}{V_0} = \frac{V_0 - V_\infty}{V_0} e^{-k_1 \sigma_c}$	(a)
Ballhausen	$\ln \frac{V_\infty}{V - V_\infty} = K_2 \sigma_c + K_3$	(b)
Balshin	$\ln \sigma_c = -K_4 \frac{V}{V_\infty} + K_5$	(c)
Cooper	$\frac{V_0 - V}{V - V_\infty} = k_6 e^{-k_7 \sigma_c} + k_8 e^{-k_9 \sigma_c}$	(d)
Führer	$(s - k_{10})(\sigma_c + k_{11}) = k_{12}$	(e)
Gurnham	$\sigma_c = k_{13} e^{k_{14}/V}$	(f)
Heckel	$\ln \frac{V}{V - V_\infty} = k_{15} \sigma_c + \frac{V_0}{V_0 - V_\infty}$	(g)
Jones	$\ln \sigma_c = -k_{16} \left(\frac{V}{V_\infty}\right)^2 + k_{17}$	(h)
Kawakita	$\frac{V_0 - V}{V_0} = \frac{k_{18} k_{19} \sigma_c}{1 + k_{20} \sigma_c} \quad k_{18} = \frac{V_0 - V_\infty}{V_0}$	(i)
Konopicky	$\ln \frac{V}{V_0 - V} = k_{21} \sigma_c + \ln \frac{V_0}{V_0 - V_\infty}$	(j)
Murray	$\ln \frac{V}{V - V_\infty} = k_{22} \left(\frac{V_\infty}{V - V_\infty}\right)^{1/3} + k_{23} \sigma_c$	(k)
Nishishara	$\ln\left(\frac{V_0}{V}\right) = -\left(\frac{\sigma_c}{k_{24}}\right)^{1/k_{25}}$	(l)
Nutting	$\ln\left(\frac{V_0}{V}\right) = k_{26} \sigma_c^{k_{27}}$	(m)
Smith	$\frac{1}{V} - \frac{1}{V_0} = k_{28} \sigma_c^{1/3}$	(n)
Tanimoto	$\frac{V_0 - V}{V} = \frac{k_{29} \sigma_c}{V} + \frac{k_{30} \sigma_c}{\sigma_c + k_{31}}$	(o)
Terzaghi	$\frac{V - V_\infty}{V_\infty} = k_{32} \ln(\sigma_c + k_{33}) - k_{34}(\sigma_c + k_{35}) - k_{36} \sigma_c + k_{37}$	(p)
Tsuwa	$\frac{V_0 - V}{V_0} = \frac{V_0 - V_\infty}{V_0} \frac{1/k_{38} \sigma_c}{1 + 1/k_{39} \sigma_c}$	(q)
Walker	$\frac{V}{V_\infty} = k_{40} - k_{41} \ln \sigma_c$	(r)
Leuenberger	$P = P_{max}(1 - e^{\gamma \cdot \sigma_c \cdot \varrho_r})$	(s)
Ryshkewitch	$F = F_0 \cdot e^{-b \cdot \varepsilon}$	(t)

σ_c Compressive pressure
V Volume $V(\sigma_c)$
V_∞ Minimum possible volume ($\sigma_c \to \infty$)
V_0 Initial volume
k_i Constants specific to the model and material
s Compressive displacement
P Deformational strength of the compressed body
P_{max} Deformational strength of the compressed body when $\varrho_r = 1$
ϱ_r Relative density $(1 - \varepsilon)$
F Strength of compressed body
F_0 Strength of compressed body at porosity = 0
b Constant related to composition

Aggregates of powdery material can be described by their compressibility, that is, the ease of volume reduction under pressure, or by their pressability, that is to say, the ease with which they can be pressed into a molded object with sufficient strength. All of the compression equations presented in Table 2 have the disadvantage that they describe the compressibility of a powder rather than its pressability. For practical applications, the pressability is the more important parameter. For this reason, a mathematical model was developed which linked the two parameters of a powder, the compressibility and the pressability (Table 2, Eq. (s)). Equation (s) is in good agreement [115] with the experimental data. This has been demonstrated for a number of pharmaceutical products and additives. These equations are derived from mathematical models based on binding and nonbinding contact points. In addition, there are empirically derived equations in the literature. Various workers [116] cite the formula found by Ryshkewitch [117] for ceramic materials. This is the relationship (Table 2, Eq. (t)) between the strength and the porosity of a molded object.

3.5.2. Stamp Presses

Closed-form stamp presses, as shown in Figure 40, work in the following way: the material to be compacted is loaded into a mold, which is

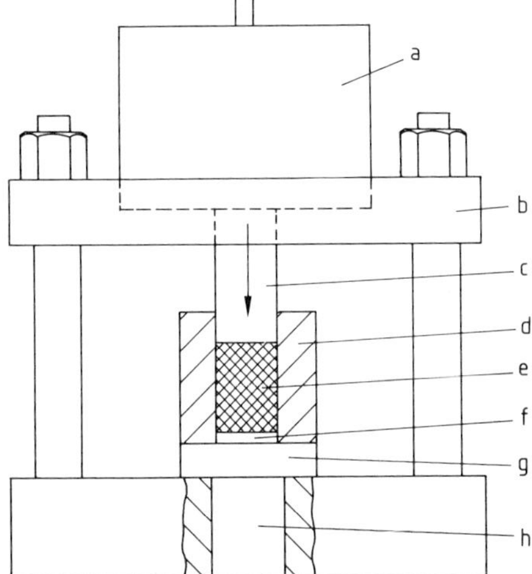

Figure 40. Hydraulic stamp press [118]
a) Hydraulic cylinder; b) Press frame; c) Upper stamp; d) Mold; e) Substrate; f) Closing plate; g) Bottom plate; h) Exit port

closed on all sides except one. A stamp, which closely fits the mold, is applied and presses the material into the desired form of a briquette. The stamp is operated automatically with an electric motor. The pressure can be transmitted hydraulically with oil or water, or by means of a screw spindle. The finished briquette can be forced out of the mold in the original direction or in the opposite direction by a counter stamp. In the preparative industry a typical size for cylindrical briquettes is 12 × 7 cm. Typical values for compressive force are 318 t supplied by an hydraulic oil pump of 56 kW capacity. Throughput is 4–6 t/h. An alternative high-capacity press is an automatic table press (couffinhal press). It is used for producing the briquettes with binders in the construction industry. The press consists of a rotating, horizontal, circular platform in which are set molds. After each pressing step, the platform is rotated to the next mold by means of guide rollers. The pressing action is commonly assisted by vibration. The production capacity for cubic or round briquettes can reach 20 t/h or even higher. The tablet presses of the pharmaceutical and catalyst industries are types of stamp presses, as are some extruders.

3.5.3. Roller Presses

In a roller press, the material is introduced between two oppositely turning rollers and is compressed (Fig. 41). Smooth rollers produce flakes. This is called roller compaction. Specially shaped rollers produce pressed objects. This is termed roller briquetting. Gravity feeding from bins is the preferred method in the beginning stages of roller press compaction. With increasing fineness of the feed, screw feeders are used. The specific operational configuration, such as the number of screw feeders, the size of the feed opening and the screw feeder pressure, is determined experimentally for the most part.

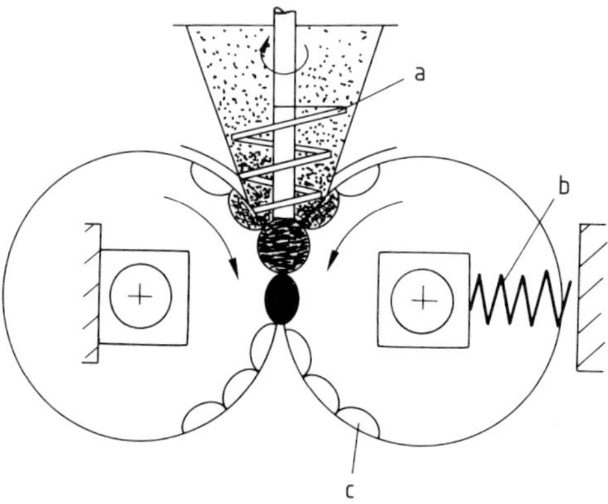

Figure 41. Operation of a roller press with screw feeder
a) Screw feeder; b) Applied pressure; c) Smooth or shaped roller

An essential factor in pressure agglomeration by roller presses is the external compressive forces. These generate a pressure distribution in the space between the rollers, which has a maximum. An average pressure can be calculated from the pressure distribution. Usually the specific compressive force is specified rather than the pressure. This is defined as the total compressive force relative to the width of the roller (linear force). The specific compressive force usually has a range of 10 to 140 kN/cm for rollers 1000 mm in diameter. The specific compressive force required for various materials is shown in Table 3.

The rotational speed of roller presses is limited by behavior of the feed as trapped air is expelled. This can only be determined experimentally by simulation of industrial conditions. A fine grained, loose material has a large void volume. This is diminished drastically during pressure agglomeration. As a result, a correspondingly large volume of air is released. In order to maintain a steady compression process, the air must be able to escape upward from the space between the rollers. This is opposed by the downward flow of the feed. The air removal problem also directly affects the maximum throughput of

Table 3. Specific compressive forces [119]

Material	Specific compressive force (for 10 cm diameter roller), kN/cm
Coal	10–30
Ceramic raw materials	40
Rock salt	60–80
Mixed fertilizer (with urea)	40–60
Mixed fertilizer (without phosphate rock and ground slag)	50–80
Mixed fertilizer (with phosphate rock and ground slag)	≥80
Calcium cyanamide (nitrolime)	60
Potassium chloride (120 °C)	50
Potassium chloride (20 °C)	70
Potassium sulfate (70–100 °C)	70
Scrubber sludge	95
Magnesite (cold and hot)	110–130
Quicklime	130
Spongy iron (cold and hot)	130–140
Ores (cold with binder)	60–80
Ores (hot without binder)	120–140

Table 4. Roller rotational speeds [119]

Material	Maximum peripheral speed, m/s	Briquette size, cm^3	Type of roller surface	Roller diameter, mm	Roller width, mm
Scrubber sludge	0.35	5–10		750	600
Potassium chloride	0.70		molded	1000	1250
LD-process dust (fine)	0.25	20		650	220
Magnesium oxide (natural)	0.40	5–7		650–750	320
Lead–zinc oxide	0.27	100		750	265
Zirconium tetrachloride	0.17		smooth	500	200
Dolomite dust	0.17	6.5		650	250
Quicklime (< 5 mm)	0.50	10		650–1000	250–540

the roller press, which is primarily a function of the circumferential speed of the roller. The problem of air removal is especially acute with smooth rollers. Many variations in shape have been tried in order to find a remedy. The circumferential speeds conforming to the air removal properties of different materials are listed in Table 4.

The specific energy requirement is dependent on the power consumption and the throughput. While the throughput is generally of a pre-supposed size, the power consumption must be determined. It depends on the torque and the rate of rotation. Both of these quantities must be determined experimentally for the material being processed (Table 5).

The type of material feeding system is governed by the properties of the material such as its bulk density and flow properties. It is always desirable to have a continuous and uniform material feed in order to have optimal loading of the press. Either gravity feed equipment (hoppers) or power feeders (screw feeders) can be used. In cases where the material tends to cake or form bridges, the flow of material may be assisted by vibrators installed in the hoppers. If the material has good flow properties, a hopper is sufficient.

Table 5. Specific energy requirements [119]

Material	Specific energy, kW h/t
Scrubber sludge	11–12
Potassium chloride	7–8
Caustic magnesite	18–20
Spongy iron	8–12
Ore	7–9
Quicklime	16–19
LD-process dust (hot)	7–9
Coal (binder)	2–3
Coal (hot)	4–6
Mixed compost (100–120 kN/cm)	12–13
Mixed compost (< 80 kN/cm)	9–11

It is provided with a slide plate in order to regulate the amount being fed. This is done by using the slide plate to change the effective size of the loading area cross section. Screw feeders are used in cases where adequate feeding cannot be obtained by gravity. This may occur if the bulk density is too low or the loading capacity is insufficient. Screw feeders of this type transport the material directly into the loading area. In this manner, they perform a type of pre-compression. The machines must be designed as precisely as possible in order to obtain optimal product properties. For this reason the interrelation between the compressive characteristics of the material (rheology) and the machine parameters, especially the compressive force and the roller diameter, must be known [120]–[127].

3.5.4. Tableting

Tablets originated in the field of pharmacy. From a technological viewpoint, they were defined as an individual dose of a medicinal substance. These were produced in various shapes by compression of powders or granules (Fig. 42).

K. Munzel, I. Buchi, and O.-E. Schultz [128] give the following definition. "Tablets are rigid aggregates or pressed articles (pressed bodies) produced by compression by mechanical means. They are made from dry dispersions such as powders, macroscopic crystals and/or granules and the medicinal material that they contain. As a rule, these pressed articles exhibit radial symmetry or at least have a plane of symmetry."

The origin of the term tablet is the Latin word "tabyletta" (the diminutive form of tabula), which means "little board." This designation is derived from the usually flat shape of early tablets. It is only appropriate today in very rare cases, since most tablets are cylindrical. Even

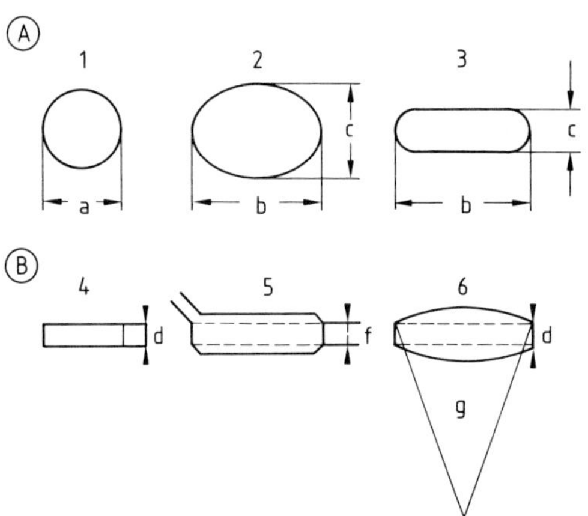

Figure 42. The most important tablet shapes [128]
A) Top view: 1) Round; 2) Oval; 3) Oblong; a) Diameter; b) Length; c) Width
B) Cross section: 4) Biplanar; 5) Biplanar with beveled edges; 6) Domes (biconvex); d) Tablet height; e) Facette; f) Median height; g) Dome radius

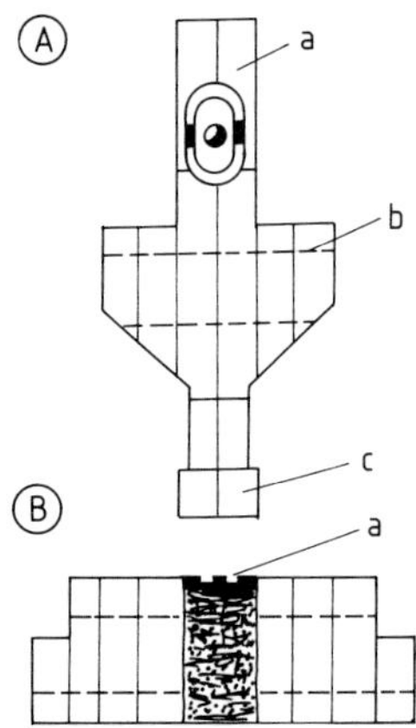

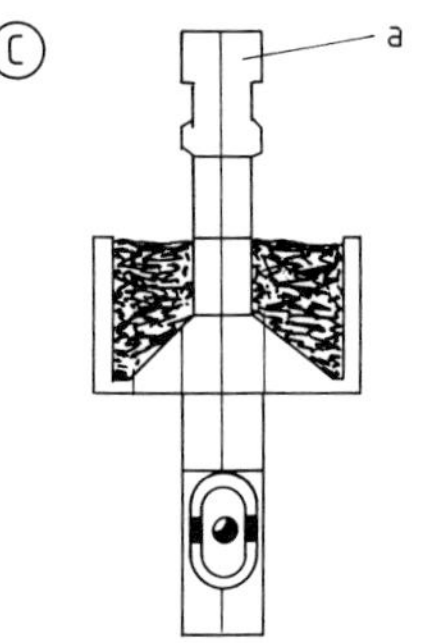

Figure 43. Working components of tableting machines [130]
A) Upper punch: a) Set pin; b) Support plate; c) Punch head
B) Die: a) Die bore
C) Lower punch: a) Punch head

though tableting today is in the domain of pharmaceutical practice, other areas of application are growing. For example, it is used in chemistry for the production of catalysts and in the ceramic industry for the production of reactor packing material. More recently it has been used for making carriers for immobilized enzymes in bioreactors.

The production capacity of tableting machines has been developed further. Centrally rotating machines produce about 700 000 tablets per hour today. For a tablet weight of 500 mg, this gives an impressive hourly production of 350 kg/h.

3.5.4.1. Tableting Equipment

In order to produce a tablet, the material to be used is loaded into a mold (die) and then compressed by two punches (upper and lower punches) as shown in Figure 43. The size and geometric shape of the stamp affects the weight and shape of the tablet as well as its outer surface. As a result, it also affects the properties of the tablet, such as its compressive strength and disintegration time. The care and maintenance of the pressing equipment is of great importance to the success of the tableting process.

3.5.4.2. Tablet Pressing

Today there are two principal types of tablet presses in use; the eccentric press and the rotating press.

Eccentric Press (Fig. 44). The compressive force is transmitted from a cam to the substrate by the upper punch. The lower punch is usually stationary during the compression process. It can, however, be adjusted to change the depth of the loaded material. There can also be a precompression step by means of an upward motion of the lower punch, instead of the unidirectional pressure of the upper punch. After the completion of the pressing process, the tablet is removed by an expulsion stroke of the lower punch (Fig. 45). With their simple design, eccentric presses only have a production rate of 3000 tablets/h, which is rather low. In order to in-

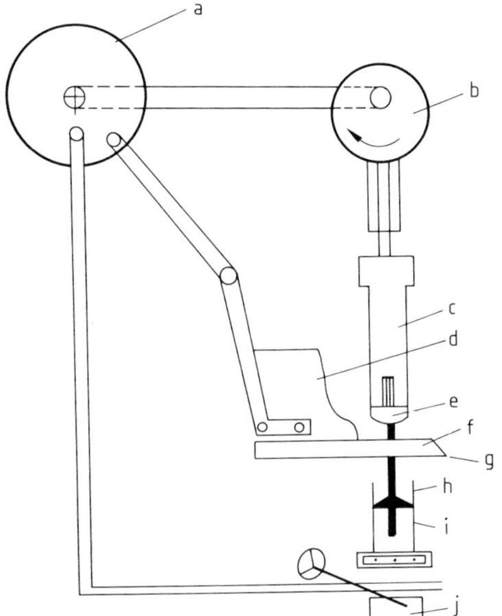

Figure 44. Diagram of an eccentric press [130]
a) Driving wheel; b) Eccentric disk; c) Upper punch support; d) Filling device with hopper; e) Upper punch; f) Die; g) Die plate; h) Lower punch; i) Lower punch support; j) Loading depth adjustment

crease capacity, multiple arrays of eccentric presses are used. They are clearly less expensive than rotating presses and are preferred for the production of smaller batches as well as the production of larger tablets, such as catalyst supports.

Rotating Press. It operates in the same way as a revolver. The compressive force is generated between an upper and a lower pressure roll. In contrast to the usual type of eccentric press, the tablet is compressed from both above and below

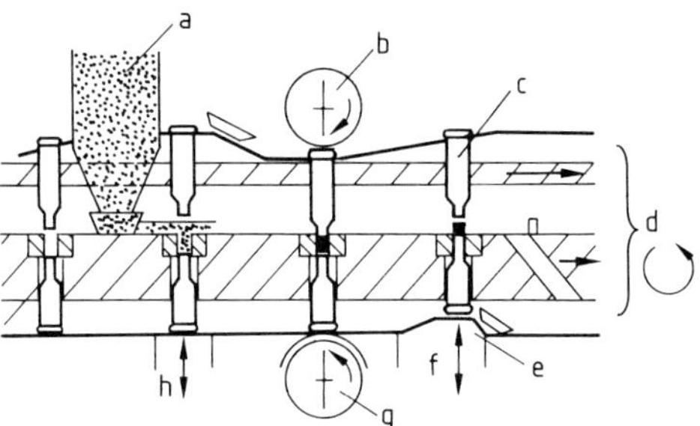

Figure 46. Schematic illustration of the compression process in a rotating tablet press [128]
a) Hopper (stationary); b) Upper pressure roller; c) Upper punch; d) Rotating die platform; e) Lower punch; f) Tablet removal mechanism; g) Lower pressure roller; h) Loading depth adjustment

(Fig. 46). The depth of loading is controlled by a system which uses either a wire strain gauge or a piezoelectric sensor. The pairs of punches along with their dies are mounted in a rotating die platform. Each pair of punches produces one tablet per revolution. The capacities of the rotating presses range from $10-700 \times 10^3$ tablets/h, depending on the number of punch pairs and the rate of rotation. For very high speed presses the persistence of the flow velocity of the material being tableted can present problems. For this reason, pre-granulation to produce a free-flowing feed material is of considerable importance.

3.5.4.3. Tableting Problems

If the tablet press (rotating or eccentric) runs properly without material and the adjustable parameters such as tablet thickness are set correctly, then any perturbations of the process are limited to those related to the properties of the

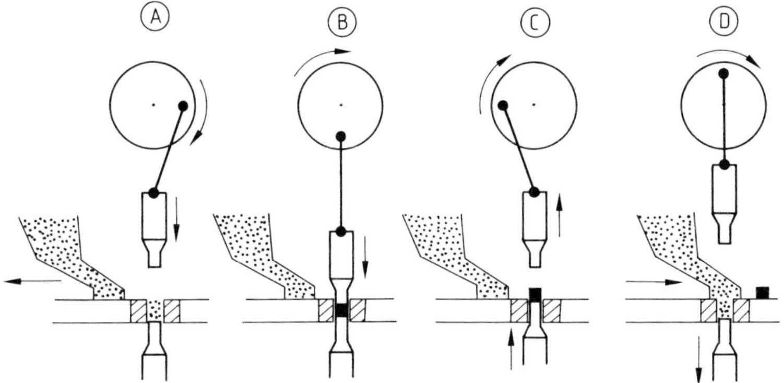

Figure 45. Schematic illustration of the compression process in an eccentric tablet press [128]

substrate. The weight of the product can be faulty. This is exhibited by an intolerable fluctuation in the weight of the tablets. The cause usually is poor flow properties of the substrate. It can form bridges in the hopper which interrupt the flow, and consequently the loading process. In addition, undesirable vibrations can cause segregation that impairs the uniformity of the tablets. Too high a moisture content of the substrate or too little lubrication leads to sticking, crackling, and coating. The resulting inadequate release of the tablet from the surface of the stamp causes surface defects on the tablet (nonuniform color, roughness). Coating occurs when the tablet breaks up into upper or lower fragments which form layers on the surfaces.

The interaction of the particle size distribution, the crystal shape and other physical properties of the substrate is the principal factor affecting the compressive strength, the disintegration time, and the release of the active ingredient.

The properties of the tablet are thus substantially affected by the properties of the fillers. In case of difficulties therefore, there is always the option of changing the formulation (binder, lubricant). If this is not possible, the other process variables such as the pressure and speed and other environmental conditions must be investigated.

3.5.4.4. Effect of Compression Rate and Humidity [133]

For practical application, the most important major parameters affecting tableting and tablet properties are humidity and compression rate. Even under extreme conditions, such as high humidity, high press speed, and difficult substrates, it should be possible to avoid tablet defects. A range of techniques and types of measurements are used to investigate the tableting process and the resulting tablet properties. The aim is to comprehend the microscopic processes that are occurring during tableting and to be able to draw conclusions about tablet properties from them [137]. The example of tableting powdered limestone in moist air (45% relative humidity) shows the effect of the speed of compression on the compressive behavior. The force-displacement curves are shown in Figure 47. Tablets with constant weight and the same diameter were produced. The average values for the three different compression rates v_v which were used were 0.3, 2.5, and 9.0 cm/s. The compressive pressure was held constant at ca. 110 MPa. In tableting, the compression rate is defined as the ratio of the displacement of the punch from the first application of a detectable force to the point of maximum force to the time

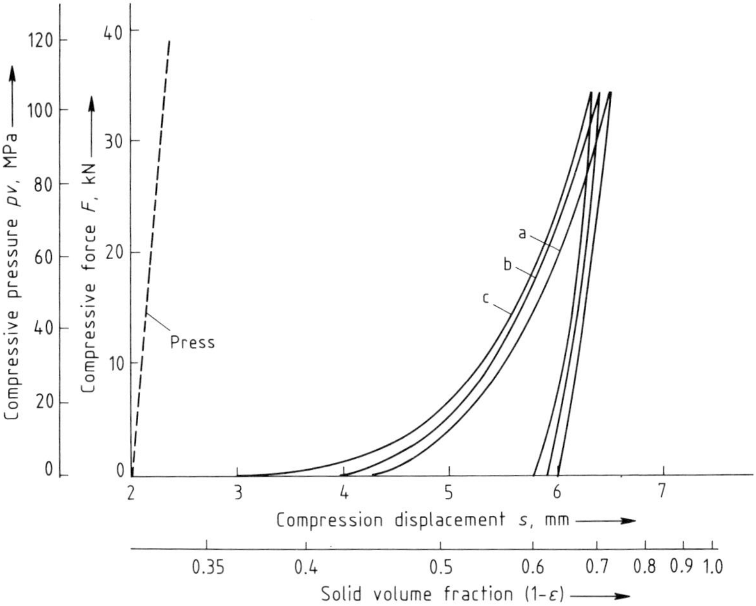

Figure 47. Force–displacement diagrams of powdered limestone (x_{50} = 13.5 μm) for tableting for three compression rates v_v [133]
a) v_v = 0.3; b) v_v = 2.5; c) v_v = 9.0 cm/s. Parameters of tablets: m = 2.25 g; h_{loading} = 10 mm; $\varnothing_{\text{Tablet}}$ = 20 mm

required for this displacement to occur. The compression pressure is plotted as a function of the compressive displacement of the lower stamp or the volume fraction of solid $(1 - \varepsilon)$ in the pressed object. Compression occurs earlier with increasing compression rate and the rate of pressure increase is lower. However, the rates of increase become equal at higher levels of compression. The principal reason for the lower rate of pressure increase at high compression rates is the frictional effect of air rapidly escaping from the powder pressed between the punch and the die. In addition, sliding and rearrangement processes of the particles are hindered by turbulence in the powder due to rapid compression.

The density of tablets made with the same pressure is lower for rapidly made tablets. The elastic recovery when they are released and removed from the die is larger. The ratio of the energy incorporated into the tablet to the energy delivered by the machine is consistent with this. The relative energy incorporation is higher for slow compression compared to rapid compression. The elastic recovery of the tablets after release and removal from the die is limited by the recovery of the elasticly stressed particles. On one hand, consequently, the contact area between the particles is reduced, resulting in a loss of strength. On the other hand, there is a reduction of stress inside the tablet, which has a favorable effect on strength. Strength studies must be done to determine which mechanism predominates.

The example of the effect of compression rate on the shape of the compression diagram illustrates how such a diagram can be used to characterize changes in compression parameters, for the same material under otherwise constant conditions. The first conclusions can be drawn concerning the trends in properties of the resulting tablets by determining the compression energy and the elastic recovery behavior. It should be noted that the compression diagram is only an integrated representation of all the complex processes which occur during tableting. The evidence obtained from the compression diagram must therefore first be checked and proven by the appropriate types of measurements. Figure 48 shows the diametrical compressive strength as a function of compression rate v_v for limestone tablets produced under vacuum (6×10^{-4} MPa) and in moist air (45% and 80% relative humidity). The compressive pressure p_v was 110 MPa. The production and testing conditions for the tablets being studied are abbreviated as illustrated in the following example. The designation air/vacuum tablet means that the tablet was made in air and tested under vacuum.

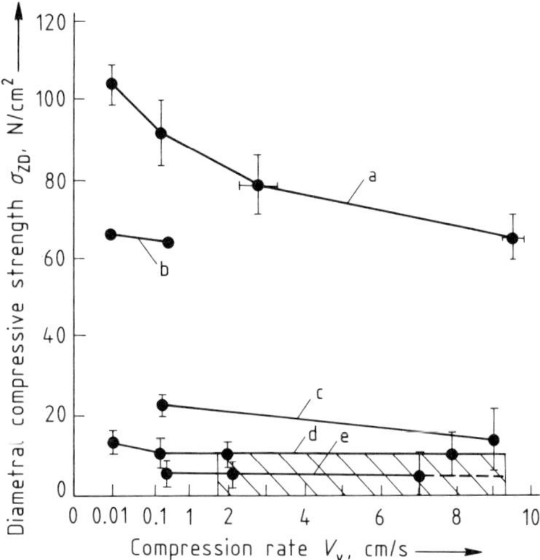

Figure 48. Diametrical compressive strength as a function of compression rate [133]
Limestone $x_D = 13.5$ µm; $m_{Tablet} = 2.25$ g; $\varnothing_{Tablet} = 20$ mm; $h_{Tablet} = 3.6-4.4$ mm; $p_v = 110$ MPa
The rectangular region marked by diagonal lines is the region where 10–50% of the tablets produced are defective.
a) Vacuum/vacuum; b) Air/vacuum; c) Vacuum/air (45% rel. humidity); d) Air/air (45% rel. humidity); e) Air/air (100% rel. humidity)

The following conclusions, which are applicable to actual tableting, can be drawn from this investigation of strength:

1) Substances that exhibit problems when tableted at high humidity and high compression rates should be tableted at low humidities.
2) Certain compression rates may not be exceeded.
3) Strength testing must be done by a test that is designed to match the actual requirements of the tablet as closely as possible. Furthermore, testing must be done in a controlled atmosphere, otherwise the test results cannot be compared with each other.
4) Low humidity is favorable for resistance to stress when the tablets are stored or transported [134]–[139].

4. References

General References

[1] R. H. Perry, C. H. Chilton: *Chemical Engineer's Handbook,* McGraw Hill, New York 1973.
[2] A. W. Adamson: *Physical Chemistry of Solids,* Wiley-Interscience, New York 1967.
[3] B. Ries, *Ind. Anz.* **21** (1961) 1–4.
[4] P. J. Sherrington, R. Oliver: *Granulation,* Heyden & Son, London—Philadelphia—Rheine 1981.
[5] C. E. Capes: *Particle Size Enlargement,* Elsevier, Amsterdam—Oxford—New York 1980.
[6] W. Herrmann, K. Sommer: *Preprints 3rd Int. Symposium Agglomeration,* Nürnberg 1977.
[7] G. Astarita, *Chem. React. Eng. Proc. Eur. Symp.* **5th 1972**, Amsterdam.
[8] W. Herrmann: *Das Verdichten von Pulvern zwischen zwei Walzen,* Verlag Chemie, Weinheim 1973.
[9] K. Sommer, W. Herrmann, *Chem. Ing. Tech.* **50** (1978) 518–524.

Specific References

[10] H. Rumpf, *Chem. Ing. Tech.* **30** (1958) 144–158, 329–336.
[11] H. Schubert, *Chem. Ing. Tech.* **51** (1979) 266–277.
[12] H. Rumpf in W. A. Knepper (ed.): *Agglomeration,* Wiley-Interscience, New York 1962, pp. 379–414.
[13] H. Rumpf, *Chem. Ing. Tech.* **42** (1970) 538–540.
[14] W. O. Smith et al., *Phys. Rev.* **34** (1929) 1271–1274.
[15] H. Schubert, *Powder Technol.* **11** (1975) 107–119.
[16] H. Schubert: *Kapillarität in porösen Feststoffsystemen,* Springer Verlag, Berlin—Heidelberg—New York 1982.
[17] R. Polke: *Zur Haftung zwischen Festkörpern bei erhöhten Temperaturen,* Dissertation, Technische Universität Karlsruhe, 1971.
[18] H. Krupp: *Static Electrification,* The Institute of Physics, London—Bristol 1971.
[19] H. Krupp, *Adv. Colloid Interface Sci.* **1** (1967) 111–234.
[20] St. Moser: *Van der Waals- und elektrostatische Haftkräfte,* Dissertation, Technische Universität Karlsruhe, 1976.
[21] D. Langbein: "Van der Waals Attraction in and between Solids," *Festkörperprobleme* **13** (1973) 85–108.
[22] H. C. Hamaker, *Physika* **4** (1937) 1058 ff.
[23] E. M. Lifschitz, *Sov. Phys. JETP (Engl. Trans.)* **2** (1956) 73 ff.
[24] D. M. Newitt, J. W. Conway-Jones, *Trans. Inst. Chem. Eng.* **36** (1985) 422 ff.
[25] N. G. Stanley-Wood: *Enlargement and Compaction of Particulate Solids,* Butterworths, London 1983.
[26] W. Pietsch, H. Rumpf, *Chem. Ing. Tech.* **39** (1967) 885–893.
[27] H. Schubert, *Chem. Ing. Tech.* **45** (1973) 396–401.
[28] H. Schubert, *Chem. Ing. Tech.* **46** (1974) 333–334.
[29] H. Schubert: *Untersuchungen zur Ermittlung von Kapillardruck und Zugfestigkeit von feuchten Haufwerken aus körnigen Stoffen,* Dissertation, Technische Universität Karlsruhe, 1972.
[30] W. Schütz, H. Schubert, *Chem. Ing. Tech.* **52** (1980) 451–453.
[31] H. Rumpf, *Chem. Ing. Tech.* **30** (1958) 144–158.
[32] W. Pietsch: *Festigkeit und Trocknungsverhalten von Granulaten,* Dissertation, Technische Universität Karlsruhe, 1965.
[33] J. Charè: *Trocknung von Agglomeraten bei Anwesenheit auskristallisierender Stoffe,* Dissertation, Technische Universität Karlsruhe, 1976.
[34] C. E. Capes, *Powder Technol.* **4** (1970–1971) 77–82.
[35] J. Frenkel, *J. Phys.* **9** (1945) 385–391.
[36] B. Pines, *Z. Techn. Fiz.* **16** (1946) no. 6, 137.
[37] G. Kuczynski, *J. Appl. Phys.* **20** (1949) 1160–1163.
[38] H. Rumpf, K. Sommer, K. Steier, *Chem. Ing. Tech.* **48** (1976) 300–307.
[39] F. Steiner et al., *Polym. Prepr. Am. Chem. Soc. Div. Polym. Chem.* **11** (1970) no. 2, 1168–1175.
[40] Wei Hsuin Yang, *J. Appl. Mech.* **33** (1966) 395–401.
[41] H. Hertz: *Gesammelte Werke,* Barth, Leipzig 1895.
[42] B. Dahneke, *J. Colloid Interface Sci.* **40** (1972) no. 1, 1–13.
[43] F. Faulhaber: *Einfluß der Mikroplastizität auf das Bruchverhalten von Quarzkugeln,* Dissertation, Technische Universität Karlsruhe, 1966.
[44] W. Schütz, H. Schubert, *Chem. Ing. Tech.* **48** (1976) MS 375–376.
[45] B. V. Derjagin, L. D. Landau, *Acta Physicochim. USSR* **14** (1941) 633 ff.
[46] E. J. W. Verwey, J. Th. G. Overbeck: *Theory of the Stability of Cyophobic Colloids,* Elsevier, Amsterdam 1948.
[47] W. Pusch: *Agglomeration mit polymeren Flokkungsmitteln,* Dissertation, Technische Universität Karlsruhe, 1982.
[48] K. Leschonski, Kurs "Partikelgrößenanalyse" 1984.
[49] K. Sommer, R. Fath: "Automatische Bildanalyse für die Qualitätskontrolle in der Lebensmittelindustrie," *ZFL* **38** (1987) no. 7, 290–292.
[50] L. Pfalzer, W. Bartusch, R. Heiss, *Gordian* **74** (1974) 80–83, 126–129.
[51] R. Polke, W. Herrmann, K. Sommer, *Chem. Ing. Tech.* **51** (1979) 283–288.
[52] E. A. Flood: *The Solid/Gas Interface,* Marcel Dekker, New York 1967.
[53] A. Baiker, W. Richarz, *Chem. Ing. Tech.* **49** (1977) 399–403.
[54] W. W. Washburn, *Proc. Natl. Acad. Sci. USA* **68** (1971) 115–116.
[55] *Powder Technol.* **9** (1974) no. 4 (special number).
[56] B. Koglin, *Proc. 1st Int. Conf. on Particle Technology,* Chicago, Aug. 1973, pp. 272–278.
[57] B. Koglin, *Powder Technol.* **17** (1977) 219–227.
[58] W. Pusch, B. Koglin, *Chem. Ing. Tech.* **55** (1983) 137–139.
[59] H. Rumpf, W. Herrmann, *Aufbereit. Tech.* **11** (1970) no. 3, 117–127.
[60] H. Reichert: "Vortrag zum Jahrestreffen der Verfahrensingenieure," Stuttgart 1977.
[61] H. Schubert, *Chem. Ing. Tech.* **8** (1970) 541–545.
[62] H. Schubert, *Powder Technol.* **11** (1975) 107–119.
[63] U. Kaluza, *DEFAZET Dtsch. Farben Z.* **29** (1975) 102–116.
[64] R. Polke, R. Rieger, *Chem. Ing. Tech.* **50** (1978) 149–154.
[65] H. Reichert: *Desagglomeration organischer Farbpigmente in Couetteströmungen,* Dissertation, Technische Universität Karlsruhe, 1973.
[66] W. Mohr, *Milchwissenschaft* **15** (1960) 215–222.
[67] J. J. Mol, *Int. Dairy Fed. E-Doc 90,* Paris 1973.

[68] L. Pfalzer, W. Bartusch, R. Heiss, *Chem. Ing. Tech.* **45** (1973) 510–516.
[69] H. Schubert, *VT Verfahrenstechnik* **12** (1978) 296–301.
[70] H. Schubert, B. Bauer, A. Löflath, paper read at the ZDS Instantisieren, Solingen 1986.
[71] H. Schubert, *Chem. Ing. Tech.* **47** (1975) 86–94.
[72] H. Schubert, *ZFL* **36** (1985) no. 3, 149–152.
[73] A. Krefeld, J. M. Van der Vertroog, *Neth. Milk Dairy J.* **17** (1963) 209–232.
[74] C. E. Capes, P. V. Danckwarts, *Trans. Inst. Chem. Eng.* **43** (1965) T116–T130.
[75] P. C. Kapur, D. W. Fürstenau, *Ind. Eng. Chem. Process Des. Dev.* **5** (1966) 5–10.
[76] P. C. Kapur, D. W. Fürstenau, *Ind. Eng. Chem. Process Des. Dev.* **8** (1969) 56–62.
[77] M. Smoluchowski, *Z. Phys. Chem. (Leipzig)* **92** (1918) 129–168.
[78] W. Herrmann, K. Sommer, *Preprints 3rd Int. Symposium Agglomeration,* Nürnberg 1977.
[79] W. Dötsch: *Agglomerationskinetik zur Stimulation von Agglomerationsprozessen im Agglomeriesteller,* Dissertation, Technische Universität München, 1987.
[80] H. Klatt, *Zem. Kalk Gips* **11** (1958) 144–154.
[81] H. B. Ries, *Aufbereit. Tech.* **16** (1975) 639–646.
[82] U. N. Bhrany, R. Manz, *Aufbereit. Tech.* **18** (1977) 641–647, *Zem. Kalk Gips* **23** (1970) 407–412.
[83] M. Papadakis, J. P. Bombled, *Rev. Mat. Construct.* **549** (1961) 289–299.
[84] W. Pietsch, *Chem. Tech. (Leipzig)* **19** (1967) 259–266.
[85] H. E. Rose, R. M. E. Sullivan, *Ball, Rube and Rod Mills,* Constable, London 1957.
[86] J. Nordengren: *Granulation of phosphate fertilizers—theory and practice,* Proc. Fertil. Soc., London 1947.
[87] H. Leuenberger, G. Imanidis, *Chem. Ing. Tech.* **56** (1984) 322–323.
[88] J. R. Farzand, I. E. Puddington, US 3 528 809, 1970.
[89] J. Werther, *Powder Technol.* **15** (1976) 155 ff.
[90] N. A. Shakhova, B. G. Yevdokomov, N. M. Ragozina, *Process Technol. Int.* **17** (1972) 946–947.
[91] P. E. Wurster, *J. Am. Pharm. Assoc.* **48** (1959) 451 ff.
[92] E. P. Nichols, GB 1 039 177, 1966.
[93] H. Schwanghardt, *Aufbereit. Tech.* **12** (1969) 713–722.
[94] H. Schwanghardt, *Aufbereit. Tech.* **12** (1973) 818–830.
[95] W. Herrmann, *Chem. Ing. Tech.* **48** (1976) 333.
[96] R. Herbener, *Chem. Ing. Tech.* **59** (1987) 112–117.
[97] I. A. J. Brümmelkuis, *Milchwissenschaft* **30** (1975) 75–80.
[98] J. Pisecky, V. Westergaard, *Milchwissenschaft* **26** (1971) 557–560.
[99] H. G. Kessler: *Lebensmittelverfahrenstechnik,* Verlag A. Kessler, Freising 1976.
[100] G. Kluge, R. Heiss, *Verfahrenstechnik (Mainz)* **1** (1967) 251–260.
[101] U. Bröckel, Kurs "Agglomeration" Weihenstephan 1987.
[102] N. K. Stuart, World Coal, August 1979.
[103] G. G. Bemer, F. J. Zuiderweg, Int. Symposium *Fineparticles Processing,* Las Vegas 1980.
[104] J. R. Farnand, *J. Sep. Process Technol.* **3** (1982) 1–15.
[105] S. Bosa: Dissertation, University of Technology of the Netherlands, Delft 1983.
[106] K. V. S. Sastry, Agglom. 77, Am. Inst. of Minery Met. and Petroleum Eng. 1977.
[107] G. G. Bemer: Dissertation, University of Technology of the Netherlands, Delft 1983.
[108] Y. Kawashima, C. E. Capes, *Powder Technol.* **10** (1974) 85–92.
[109] C. E. Capes, J. P. Sutherland, *Ind. Eng. Chem. Process Des. Dev.* **6** (1967) 1.
[110] J. R. Farnand, H. M. Smith, I. E. Puddington, *Can. J. Chem. Eng.* **39** (1961) 4.
[111] B. Bogenscheider, H. J. Behrenbeck, K. H. Kubika, *Glückauf* **112** (1976) 12.
[112] K. Sartor, Seminar "Verdichten von Stäuben," Graz 1977.
[113] K. H. Lüdde, K. Kawakita, *Pharmazie* **21** (1968) 393 ff.
[114] K. Kawakita, K. H. Lüdde, *Powder Technol.* **4** (1970/1971) 61 ff.
[115] H. Leuenberger, E. N. Hiestand, H. Sucker, *Chem. Ing. Tech.* **59** (1987) 77–78.
[116] R. Spang: Dissertation no. 5081, ETH Zürich 1973.
[117] E. Ryshkewitch, *J. Am. Ceram. Soc.* **6** (1953) 65 ff.
[118] W. John, *Ullmann,* 4th ed., **14**, 317.
[119] R. Zisselmar, Seminar: "Kompaktieren," TA Wuppertal 1986.
[120] W. Herrmann, *Das Verdichten von Pulvern zwischen zwei Walzen,* Verlag Chemie, Weinheim 1973.
[121] J. R. Johanson, *Trans. ASME* (1965) 12.
[122] J. R. Johanson, *Proc. of 9th Biennal Briquetting Conference,* 1965 (IBA).
[123] J. R. Johanson, *Conference of International Briquetting Association,* 27.–29. Aug. 1969, Sun Valley, Idaho.
[124] A. W. Jenike, *Exp. Station Bulletin,* No. 108, vol. 52, Oct. 1961.
[125] O. Molerus: *Schüttgutmechanik,* Springer Verlag, Heidelberg 1985.
[126] G. Hauser, K. Sommer, CHISA, Prag, Sept. 1984.
[127] K. H. Sartor, H. Schubert, *Chem. Ing. Tech.* **50** (1978) 708–709.
[128] K. H. Bauer, K. H. Frömming, C. Führer: *Pharmazeutische Technologie,* Thieme Verlag, Stuttgart 1986.
[129] K. Münzel, J. Büchi, O. E. Schultz: *Chemisches Praktikum,* Thieme Verlag, Stuttgart 1959, p. 727.
[130] H. Sucker, P. Fuchs, P. Speiser: *Pharmazeutische Technologie,* Thieme Verlag, Stuttgart 1978.
[131] N. G. Stanley-Wood: *Enlargement and Compaction of Particulate Solids,* Butterworths, London 1983.
[132] D. Train, *J. Pharm. Pharmacol.* **8** (1956) 745.
[133] K. Sator, *Chem. Ing. Tech.* **50** (1978) 708–709.
[134] K. H. Lüdde, K. Kawakita, *Pharmazie* **21** (1968) 393 ff.
[135] K. Kawakita, K. H. Lüdde, *Powder Technol.* **4** (1970/1971) 61 ff.
[136] R. Spang: Dissertation no. 5081, ETH Zürich 1973.
[137] E. Ryshkewitch, *J. Am. Ceram. Soc.* **6** (1953) 65 ff.
[138] K.-H. Sartor, H. Schubert: Vortrag auf dem Jahrestreffen der Verfahrensingenieure, Stuttgart 1977.
[139] W. Herrmann, *Chem. Ing. Tech.* **43** (1971) 418–425.

8. Solids Handling

Roy D. Marcus, Cargo Carriers Ltd., Isando, Republic of South Africa
F. Rizk, BASF Aktiengesellschaft, Ludwigshafen, Federal Republic of Germany
Stephen J. Meijers, University of Pretoria, Pretoria, Republic of South Africa

In addition to the standard symbols defined in the front matter of this volume, the following symbols are used:

$A = \pi D^2/4$	cross-sectional area of pipe or product, m^2
A^*	cross-sectional area of a particle normal to flow, m^2
ΔA	effective cross-sectional area
c	solids velocity, m/s
C_d	delivered concentration
C_{vd}	concentration delivered by volume
$C_D = \frac{2F_D}{\varrho w^2 A^*}$	drag coefficient
D	pipe inner diameter, mm
D_{min}	minimum cone diameter, m
d	particle diameter, mm
D_s	shaft diameter, mm
d_v	volume equivalent particle diameter, mm
f	friction factor
f_c	unconfined yield stress, N/m^2
$f_{c, crit}$	critical unconfined yield stress, N/m^2
F_D	drag resistance, N
$\dot{G}$	solid mass flow rate, kg/h, t/h
K	constant
k	effective pipe wall roughness, mm
L, ΔL	length, length of element, m
L_v	vertical length, m
$\dot{m}$	mass flow rate, kg/h
$M_I = (\sigma_1 + \sigma_2)/2$	
$M_{II} = f_c/2$	
n	flow index for power law fluid or number of revolutions per minute
p	pressure, kPa
Δp_L, Δp_z, Δp_{tr}	pressure change due to air, solids, transportation, respectively, kPa
Δp	pressure difference, kPa
$Re = vD/\nu$	Reynolds number related to pipe diameter

$Re_p = vd/\nu$	Reynolds number related to particle diameter
Re^*	generalized Reynolds number
Re_{crit}	critical Reynolds number
s	pitch of screw, m, or slip or velocity ratio
S_p	surface area of particle, m^2
U	shape factor
$V_p = \pi d^3/6$	volume of particle, m^3
$\dot{V}_p = \dot{G}/\varrho^*$	solids volume flow rate, m^3/s
v	flight or superficial velocity, m/s
$\bar{v}$	average fluid velocity, m/s
v_m	velocity of mixture, m/s
v_D	deposition velocity, m/s
v_{min}	minimum conveying velocity, m/s
v_s	gas conveying velocity at point of particle settling velocity, m/s
$w = (v-c)$	relative velocity, slip velocity, m/s
β	trough angle, degrees
γ	rate of shear
δ	angle of surcharge, degrees
$\varepsilon = 1 - \frac{\varrho^*}{\varrho_p}$	voidage
ε_H	head loss gradient
η	dynamic viscosity, Ns/m^2
η_a	apparent viscosity, Ns/m^2
η_e	effective viscosity, Ns/m^2
η_p	plastic viscosity, Ns/m^2
θ	cone angle, degrees
ν	kinematic viscosity, m^2/s
ϱ	air (gas) density, kg/m^3
$\varrho^* = \varrho_p(1-\varepsilon)$	conveying bulk density, kg/m^3
ϱ_f	fluid density, kg/m^3
$\varrho = \varrho_p - \varrho_f$	density difference between particle and fluid, kg/m^3
ϱ_B	bulk density, kg/m^3
ϱ_p	particle density, kg/m^3
σ	normal stress, N/m^2
$\bar{\sigma}_1$	bridging stress, N/m^2
σ_1	principal or major consolidating stress, N/m^2
σ_2	minor consolidating stress
τ	shear stress, N/m^2
τ_o	shear stress at pipe wall, N/m^2
τ_y	yield shear stress for Bingham plastic, N/m^2
φ_e	effective angle of internal friction, degrees
φ_w	angle of wall friction, degrees
φ_i	kinematic angle of friction, angle of internal friction, degrees
φ	filling efficiency
$\psi = \pi d_v^2/S_p$	sphericity

Handling bulk materials plays an important role in the chemical industry. The effective operation of most modern chemical manufacturing plants depends on the ability to move raw materials, semiprocessed and fully-processed products, both within the manufacturing environment and ultimately to the end user.

Bulk materials handling comprises a wide range of techniques incorporating a number of handling modes. An attempt is made here to highlight some of these modes by providing some basis on which both the method of operation and the basic conveying requirements and parameters are given. In the limited space available, it is not possible to do justice to any one system or to provide full details on all aspects of the various techniques. The reader is advised to obtain more extensive information if a full system design is required; adequate references are listed to facilitate further reading.

1. System Selection

Selection of a specific handling system for bulk materials requires a knowledge of the physical and chemical properties of the material to be handled. Frequently, both the material characteristics and the prevailing plant conditions play an important role in determining the flow behavior of the product, thereby influencing the type of equipment to be selected.

It is thus essential to have prior knowledge of the product flow behavior before any reliable system selection can be attempted. As particle characterization techniques are improved, so the selection of suitable equipment to handle the material is simplified. Nevertheless, there are still various conveying techniques which require pilot plant testing in order to obtain the necessary flow parameters for reliable system selection and design. It is frequently essential that a representative sample (of the product to be conveyed) be tested under realistically simulated conditions.

Two of the most salient features for selecting a suitable handling system are a knowledge of the *distance* and the *throughput* required of the system. When specifying distance, details of the *route profile* are important: elevation, horizontal and vertical conditions to be traversed, head room, and route flexibility are critical issues for effective system selection. The handling system will often incorporate storage facilities, which also have to be carefully designed to meet a number of requirements.

In terms of solid mass flow rate, it must be clearly identified whether the system can operate in a batch mode or if continuous flow is mandatory, together with an estimation of peak loading. Future expansion must also be taken into account; this could include, for example, increasing the length or the flow rate.

1.1. Selection of a Transportation System

Essentially two categories of bulk materials handling systems exist: indoor (or in-plant)

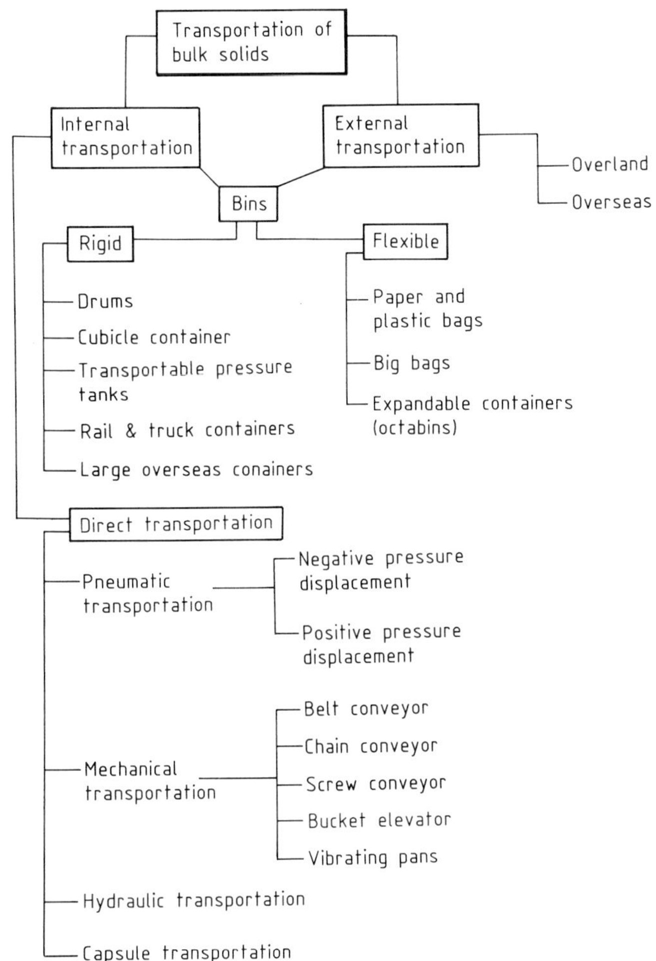

Figure 1. Categorization of the transportation system

transportation and long distance external transportation. The type of equipment selected can sometimes be suitable for both types of operation, but internal transportation systems are usually associated with specialized equipment which is normally designed to meet fairly stringent process requirements. The categorization of the transportation system is illustrated by Figure 1. Within the overall classification of bins there is a number of techniques used to transport, weigh, load, and unload these systems (Fig. 2).

The ultimate selection of suitable hardware depends on the conveying rate. A basis for selection has been made for so-called continuous conveying systems (Fig. 3). As an example, the appropriate conveying system for a given flow rate as a function of the conveying length can be selected, using Figure 3, as follows:

1) For a solid mass flow rate $\dot{G} = 300$ kg/h and conveying lengths $L = 2.5$, 25, and 250 m, the following results are found:

   $L = 2.5$ m: systems (b), (d), and (f) may be used;
   $L = 25$ m: system (a), (b), and (d) are applicable;
   $L = 250$ m: only system (a) can be considered.

2) For a solid mass flow rate $\dot{G} = 10\,000$ kg/h and a length $L = 200$ m, Figure 3 shows that systems (a), (b), and (c), i.e., a pneumatic conveyor, air-activated gravity conveyor, or a belt or chain conveyor would be acceptable. The final selection, however, depends on the route which the conveyor is to traverse. If a requirement to raise the material vertically for, say, $L_v = 35$ m, is then included in the route, systems (b) and (d) must be combined with system (e).

The final selection will also depend on laboratory tests which could indicate that the prod-

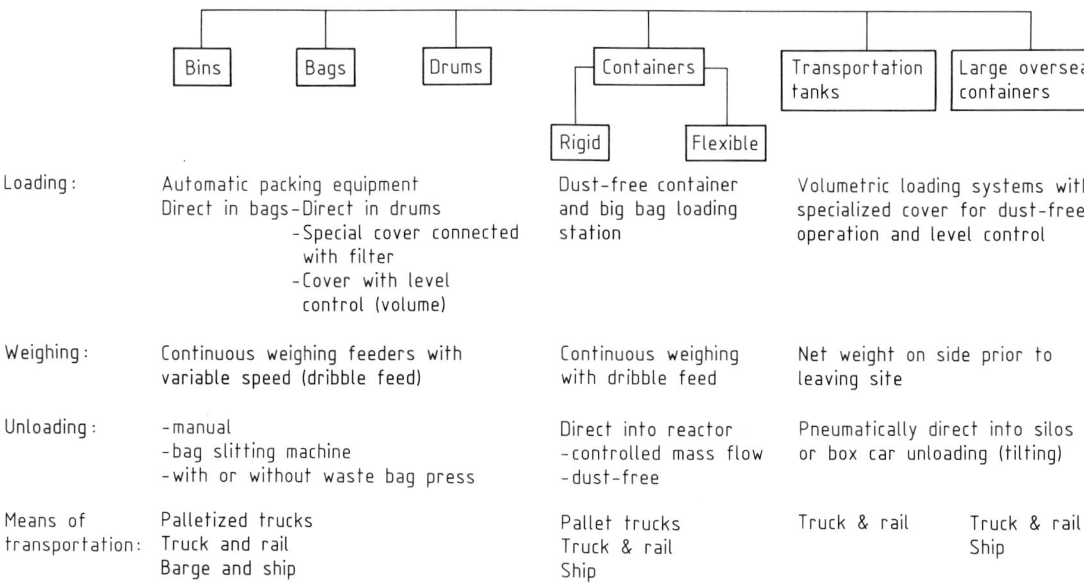

Figure 2. Loading, unloading, and transportation techniques

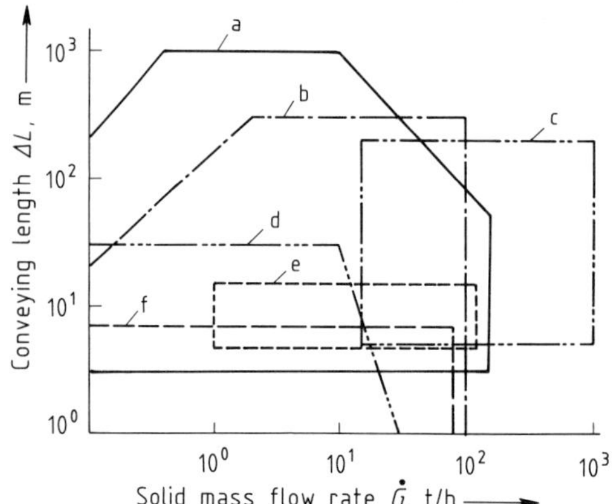

Figure 3. Transportation system selection chart
a) Pneumatic conveyors; b) Air-activated gravity conveyors; c) Belt and chain conveyors; d) Vibrating pans; e) Elevators; f) Screw conveyors and vibrating conduits

uct might not be ideal for, say, a pneumatic conveying system.

1.2. Flow Properties of Bulk Solids

The flow characteristics of bulk solid materials depend on their physical and chemical properties, on the pressure to which the solids are subjected, and the duration of the pressure. When fine particles are pressurized, the particles are drawn closer together, resulting in deformation (elastic/plastic) with a possible simultaneous reduction in porosity and voidage; these effects can drastically change the flow properties of the solid.

The main characterizing factors are the angle of friction and the so-called flow function; both can de determined by means of a shear cell analysis, which is discussed later (Section 2.5.).

Product Grouping. In the selection of conveying systems, the flowability of the product yields two main categories of bulk solids:

Group I. This group includes free-flowing materials, i.e., noncohesive products: those that do not undergo any plastic deformation when subjected to high pressures. When the load is removed, the particles return to their original condition in terms of both shape and flow characterisitics.

Group II. All products which undergo plastic deformation when subjected to external pressure: cohesive products. The degree of deformation is strongly influenced by both temperature and moisture. When the load is removed the particles do not regain their original shape, thereby yielding poor flow conditions.

Fluidization Characteristics. The ability of the material to fluidize and whether the product has an affinity to trap air or gas is of major importance when designing and selecting handling and storage systems. Note that in various cases noncohesive products flow extremely well and tend to flood when withdrawn from bins.

Angle of Friction. There are two angles of friction, the internal angle φ_i and the hopper wall-product angle φ_w. The internal angle of friction yields the friction between the particles being sheared. The angle of friction between the hopper wall and the product is critical in determining whether the product will flow or not.

Flow Function. For both short and long residence times in a silo, the flow function must be taken into account. Additional complications occur when the product is stored at elevated temperature or when humidity could influence the moisture content of the product.

1.3. Properties of Bulk Solids

An attempt will be made here to highlight some of the more important physical properties of bulk solids which influence flowability.

Voidage. All bulk solids, irrespective of their shape, are packed together in a particular volume, leaving areas of free space. The percentage of the total volume not occupied by the particles is referred to as voidage ε. The actual volume of solids is then $(1-\varepsilon)$. The magnitude of the voidage varies from bulk solid to bulk solid and depends considerably on particle shape. For spheres of uniform size, the voidage $\varepsilon = 0.48$.

Bulk Density. An important parameter required for the design of both storage and conveying systems is the bulk density ϱ_B, defined as the ratio of the mass of the product to the total volume occupied by the solid.

There are two types of bulk density: loose bulk density and packed bulk density. Each is a function of the measuring technique and will yield results which depend on the ultimate process to which the material is being subjected. In the former the measurement is made when the product is poured into a container of known volume. In packed bulk density the product is tamped in the container, thereby reducing the voidage.

Particle Density. The actual particle density ϱ_p is defined as the ratio of the mass of the individual particle to its volume. Numerous techniques are available for measuring particle density. In the case of fine particles these techniques are relatively complex.

In terms of the definitions of voidage and bulk density ϱ_B above, particle density can also be defined as:

$$\varrho_B = \varrho_p(1-\varepsilon)$$

Particle Size and Shape. The design of a number of conveying systems relies on knowledge of the particle size and in many cases the so-called particle size distribution. For a number of products the determination of particle size does not present a problem: e.g., for pellets, spheres, and cubes there is obviously no doubt about the particle size.

Difficulties arise when trying to determine the particle size of irregularly-shaped solids, and it is usual to determine the size distribution. Several techniques are used to determine particle size distribution (→ 2. Particle Size Analysis).

The data obtained from sizing are then plotted on a cumulative graph in which the cumulative percentage mass passing is plotted as a function of the particle size (Fig. 4). From the cumulative graph it is possible to find the 50% passing value, which is then taken as the average diameter of the particles.

Particle shape determination is more complex and in many cases knowledge of the shape is an important consideration. Two materials with identical size distribution, density, and chemical composition may behave completely differently as a result of a variation in particle shape [1].

Although much research is being conducted on particle shape determination [2], [3] the most common shape factor currently used is the

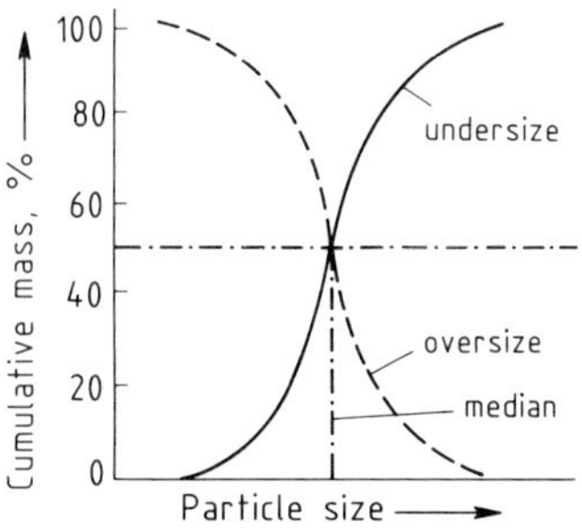

Figure 4. Typical particle size analysis

sphericity factor:

$$\psi = \frac{\text{Surface area of a sphere with same volume as particle}}{\text{surface area of particle}} = \frac{\pi (6\, V_p/\pi)^{2/3}}{S_p}$$

where V_p and S_p are the particle volume and surface area, respectively. From the above definition, $\psi = 1$ for a spherical particle.

Cohesion and Agglomeration between Particles. Cohesion between particles is of fundamental importance when determining the flow properties of the bulk solid, which can be described as either free flowing or cohesive. Cohesion can be defined as the resistance of a bulk solid to shear at zero compressive normal stress [4].

A number of mechanisms promote cohesive forces in bulk solids. In particular, van der Waal's forces, flow turbulence [5], static electrical effects, moisture content, and particle interactions are all important contributors.

Cohesive forces are the most important, but the tendency of particles to agglomerate can also present the designer with major problems: e.g., if a system is designed to meet a particular particle size and due to agglomeration the resulting particle size is significantly larger.

The cohesion of a material can be obtained from shear cell analysis (Section 2.5.).

2. Flow of Solids from Bins and Silos

For the successful operation of any materials handling system the flow of solids from bins and silos must be controlled. This section highlights some of the important features relating to the suitable design of such storage facilities.

2.1. Common Flow Problems

Two common flow problems are associated with the incorrect design of a silo or hopper: a) bridging; b) rat holing.

Bridging results in formation of an arch or a bridge across the opening of a hopper. The strength of the bridge is such that it supports the total load of solids above it. Bridging occurs when the outlet of the silo is too small on the basis of a calculated minimum diameter D_{min} required to effect uninterrupted solids flow, although the cone angle θ might be of the same magnitude as that obtained from silo design theory. Bridging can also occur when θ is too large.

Rat-holing is characterized by the formation of a channel through the solid mass and is the result of incorrect selection of the cone angle. Rat-holing will occur even though the hopper opening conforms to the correct specifications.

The various bridging and rat-holing phenomena are shown in Figure 5.

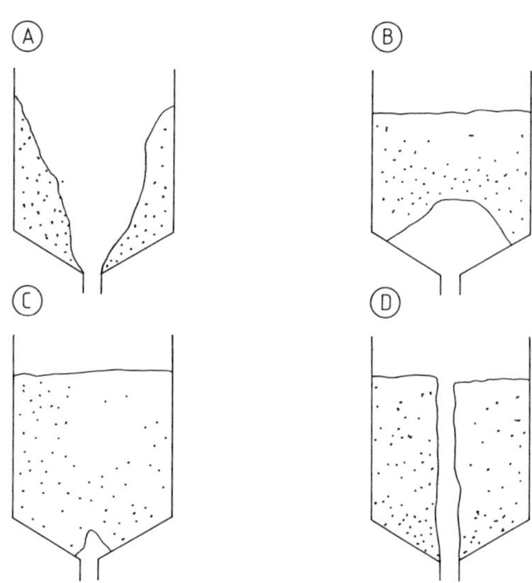

Figure 5. Typical flow problems from silos
A) Clinging; B) Arching; C) Bridging; D) Rat-holing

2.2. Flow Characteristics

The flow of a product from a storage bin conforms essentially to two types of flow: (A) mass flow or (B) funnel flow, the characteristics of which are depicted in Figure 6.

Mass flow from a bin implies that the total volume of the stored solids is in motion; uniform and steady state flow can be attained. A mass flow bin is devoid of channeling, surging, flooding, hang-ups, or bridging. The flow is also independent of the head of the stored solids and there is minimum consolidation of product, no dead space, and minimal degradation or spoilage. The flow from a mass flow bin is aptly described as first in-first out.

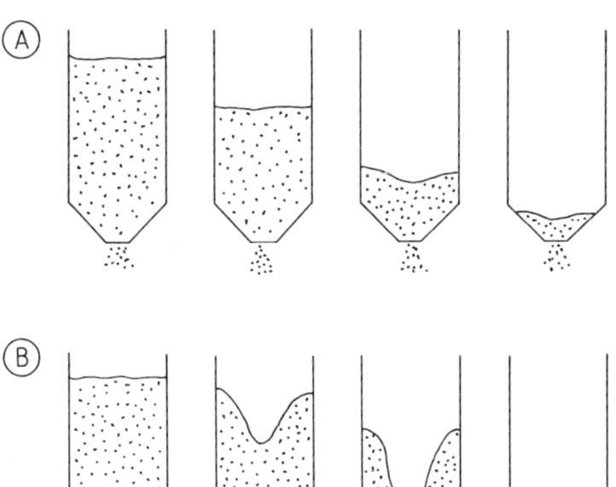

Figure 6. Flow of a solid from a silo
A) Mass flow; B) Funnel flow

Funnel flow is best understood by considering the flow from a bin with a flat bottom or gentle slope: flow occurs in a channel in the center of the bin and is always from the top of the stored solids to the center. During the flow of the solids through the channel, a dead zone occurs in all other parts of the bin. Funnel flow leads to first in-last out and has disadvantages such as flooding, material degradation, and consolidation.

The ultimate selection between mass flow and funnel flow must, in addition to taking into consideration the product characteristics and the process, recognize that a mass flow bin is generally more costly and will occupy considerably more head room.

Factors Influencing Flow. Three essential factors must be considered when designing a storage hopper or bin:

1) *Geometric form of the hopper:* the elements which must be considered include a) cone angle; b) size of outlet; c) shape (circular or rectangular); d) hopper construction material.
2) *Product characteristics:* a) particle size and shape; b) particle size distribution; c) particle density and bulk density; d) cohesiveness of the product; d) fluidizability; f) floodability; g) deaeration characteristics.
3) *Additional factors:* a) influence of humidity; b) temperature of product and process; c) storage time; d) ambient conditions.

2.3. Stress Distribution in a Silo

In order to appreciate the various forces imposed by the stored solids on the wall of the silo, the stress distribution for a typical conical bottom silo will be discussed using Figure 7. The stress distribution for a silo carrying a liquid is shown as the hydrostatic pressure. Clearly, the stress distribution is uniform and a function of the liquid level in the silo.

Substituting the liquid with a granular product results in a stress distribution known as the principal stress or major consolidating stress curve. At the highest lcvel in the silo the stress distribution for the solid is similar to that of the liquid, but the curve for the former deviates with increasing depth because of the friction between the particles and the wall of the silo.

Above a certain head of solid the stress distribution remains constant until the transition zone, in which the silo changes shape from a cylinder to a cone. Beyond the transition the stress in the cone decreases linearly towards the outlet. Theoretically, the stress reaches zero at the peak of the cone.

An important feature of the storage of products in a silo is the advent of consolidation. The stress arising from consolidation is referred to as the unconfined yield stress f_c (see Fig. 7).

A critical component in the design of a silo is knowledge of the minimum opening to ensure free flow of the solid. The determination of this minimum diameter depends on the so-called critical unconfined yield stress $f_{c,crit}$, which occurs at

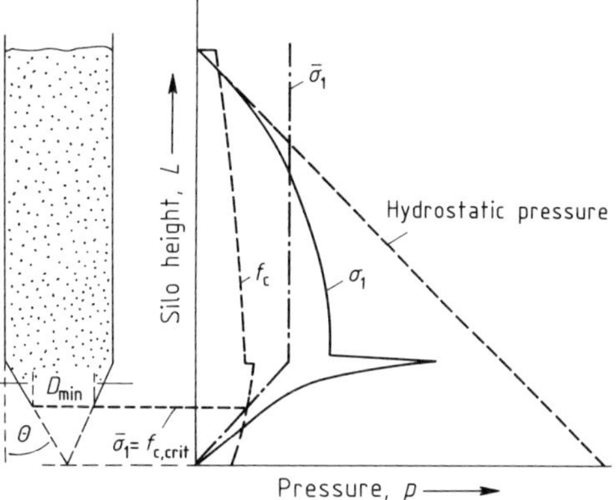

Figure 7. Stress distribution in a silo
a) Principal stress, σ_1; b) Unconfined yield stress, f_c; c) Bridging stress, $\bar{\sigma}_1$; d) Critical unconfined yield stress $f_{c,crit}$; θ = cone angle; D_{min} = min. cone diameter

the intersection of the so-called bridging stress $\bar{\sigma}_1$ and the unconfined yield stress.

2.4. Measurement of Flow Characteristics

The final design procedures for silos and hoppers depend on the acquisition of certain fundamental flow properties of the material. Although there are several techniques available to provide design information, perhaps the most common and successful technique is the Jenike shear cell which enables evaluation of the following [6]:

1) The effective angle of internal friction, φ_e
2) The kinematic angle of friction, φ_i
3) The major consolidating stress, σ_1
4) The unconfined yield stress, f_c
5) The bulk density of the product as a function of bulk solids head.

Other techniques include the ring shear cell developed by Walker, the powder test of Hosakawa, and the torsional shear cell of Peschl [7].

2.5. Shear Cell Analysis

Once an accepted testing procedure has been performed on the bulk solids using a shear cell, the observations obtained are plotted on a shear stress versus a normal stress diagram (Fig. 8).

From the shear cell analysis it is possible to draw three yield loci, obtained by a Mohr semicircle technique as shown in Figure 8. When the large Mohr semicircle is drawn tangential to the obtained yield locus and passed through the initial consolidating load, the major consolidating stress σ_1 is obtained. The small Mohr semicircle passes through the origin and the tangent to the yield locus curve yields the unconfined stress f_c. In addition, from tests carried out using the plate

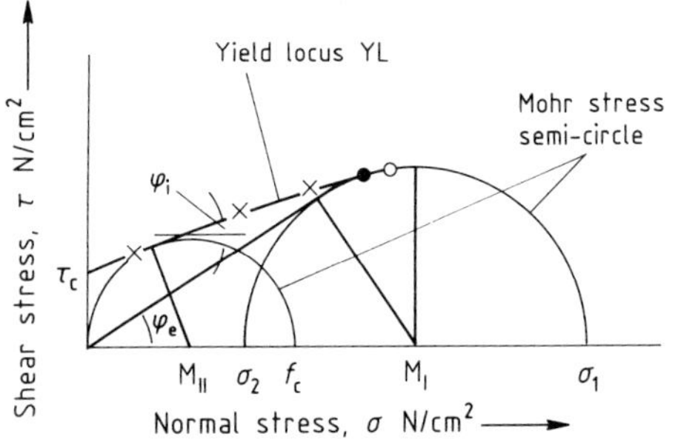

Figure 8. Typical shear-stress diagram

of the material to be used for the construction of the silo, the shear cell analysis permits the construction of another diagram from which the friction factor between the bulk solid and the silo construction material is obtained. Once the various flow charcteristics have been obtained the exact design of a silo, in which all the major dimensions are specified, can be carried out.

3. Mechanical Conveyors

Mechanical conveying techniques are the most widely used form of materials handling in the chemical industry. Although a large number of devices are available, most systems conform to the basic elements in which either belts, chains, or moving flights are used to move material.

Mechanical conveyors have distinct advantages in terms of the ability to effect accurate control in the monitoring of material from one process to another. Further, by careful selection of construction materials, it is possible to design systems to meet a wide variety of operating conditions including corrosive environments and extremes of temperature.

3.1. Belt Conveyors

Belt conveyors are the most widely used and versatile mode of mechanical conveying systems. They are able to handle larger tonnages over longer distances and at a lower cost than most forms of conveying system. A belt conveyor is an endless length of flexible material stretched between two drums and supported at intervals. The first belt conveyor recorded was that used for the transport of wheat almost 200 years ago [8, p. 127].

The belt conveyor system consists of several main components:

1) *Belt.* The belt consists of a carcass and a cover; numerous forms of construction are used.
2) *Idlers.* Troughing idlers enable the belt load to be distributed, and the belt volume and mass to be optimized. Idlers are strategically located along the length of the conveying system to ensure maximum belt support whilst minimizing the number of idlers.
3) *Drive Unit.* The belt is driven by one drum at the turnabout side of the conveyor. The drive may be positioned at either the feed end or the head end.

4) *Loading and discharging.* The belt is usually loaded by a gravity feeder such as a hopper, a preceding conveyor, or a vibratory feeder. The art of loading is to provide a steady flow of material onto the belt. Discharge of the bulk material is usually by gravity dumping, but the trajectory of the material must be taken into account.
5) *Belt cleaners.* This accessory to the belt conveyor is probably the most important. There are several types of belt cleaners designed to minimize the amount of carryover, increasing the life of the conveying system.

Belt Speed. The capacity of a belt conveying system is clearly a function of belt speed. Considerable effort is being made to increase belt speeds and many novel design features are currently being introduced in order to facilitate higher speeds. The average belt speed in practice ranges from 1 m/s to 5 m/s, although some belts operate at speeds of up to 10 m/s [8, p. 141]. The belt speed is not only a function of the belt width, but also of the type of material to be conveyed (see Fig. 9).

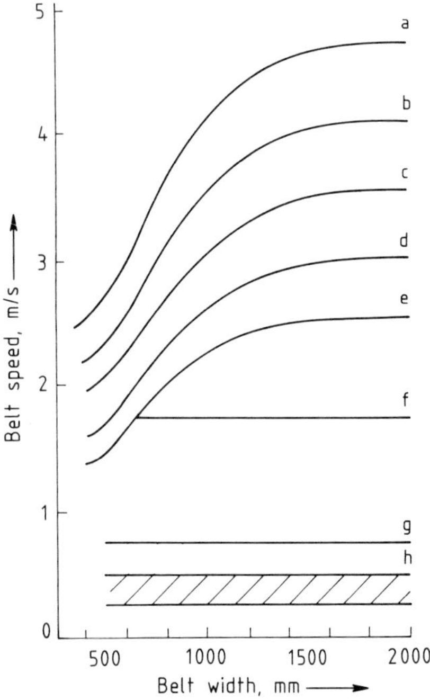

Figure 9. Conveyor belt speed as a function of belt width
Conveyed material
a) Fine: free-flowing, nonabrasive; b) Fine: mildly abrasive or lumpy; c) Granular: abrasive or lumpy, mildly abrasive; d) Granular: very abrasive or lumpy, moderately abrasive; e) Lumpy and very abrasive; f) Belts used with belt-propelled trippers; g) Belts used with ploughs; h) For picking and feeding belts

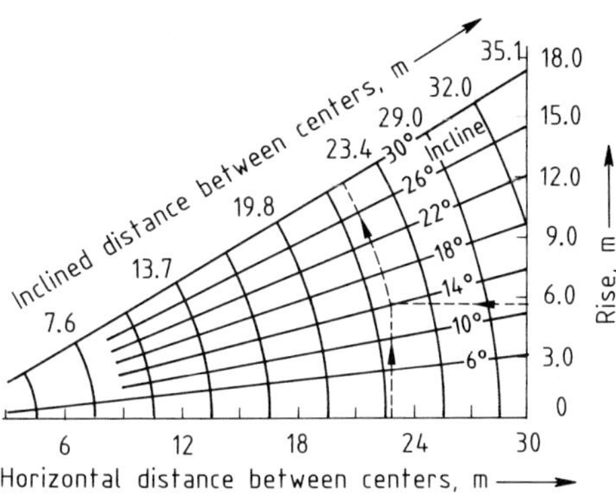

Figure 10. Inclined length of a belt conveyor

Angle and Length of Incline. The angle of the incline and the length of the inclined portion of the conveyor may be obtained when the horizontal length and the rise of the inclined portion is known (see Fig. 10).

The use of Figure 10 is best illustrated with an example. Consider a belt conveyor with a horizontal distance between centers of 46 m and a vertical rise of 11.6 m. To use the chart for this example, divide the figures by two giving 23 m and 5.8 m, respectively. The intersection of a vertical line from the 23 m mark and a horizontal line from the 5.8 m mark occurs at a point corresponding to approximately the 14° line to a radius of 23 m. Multiplying this by 2 results in an incline length of 46 m for the conveyor. The angle is 14° for this set of conditions.

The belt width and speed are dependent on a number of factors including the capacity, angle of inclination, belt tension, average particle size, and configuration. A number of combinations are usually considered before a final decision is made on the design.

Idler Selection. A typical chart for a three idler system is given for rollers of identical length in Figure 11; the optimum trough angle β for a maximum shape factor U is indicated. Only open belt conveyors, i.e., without any capsulation, have been considered in this text. Conditions do not, however, vary for closed or pipe conveyors.

3.2. Oscillating Conveyors

The vibratory conveyor accelerates the particle in both the horizontal and vertical directions. This mode of transportation must not be con

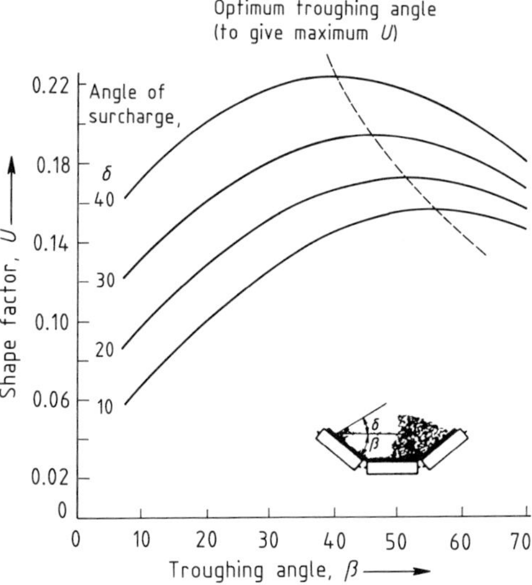

Figure 11. Shape factor for a three idler system

fused with that of vibratory and reciprocating feeders. The major difference is the range of frequency and amplitude under which they operate, as can be seen below.

Device	Frequency, Hz	Amplitude, mm
Vibratory feeder	13–60	12–1
Vibratory conveyor	3–17	50–5
Reciprocating Conveyor	1–3	300–50

These conveyors are ideally suited to the handling of granular, free-flowing materials and hot, fine, dusty, lumpy, stringy, or other materials which are awkward to handle. The basic operation of the vibratory conveyor is shown in Figure 12 [9]. The relationship between the frequency and the amplitude of the oscillation is vitally important when tuning and selecting a vibratory conveyor.

Some typical arrangements for several products using a vibratory conveyor are given in Table 1. This table should be used as a guide only if applied to the products mentioned. Four main types of vibratory conveyor are available:

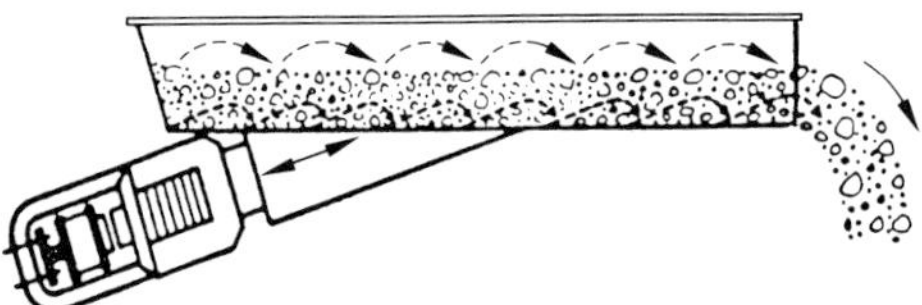

Figure 12. Arrangement of a vibratory conveyor

Table 1. Typical data for vibratory conveyors [10]

Material	Approximate size, mm	Average bed depth, mm	Average transport velocity, m/s
Alumina	0.15	75	0.15
Bagasse	0.25–5	150	0.4
Carbon black	1.5*	75	0.18
Cement clinker	6–10	125	0.36
Cereal	6–10	150	0.36
Coal	18–26	125	0.3
Crumb rubber	6	100	0.3
Detergent powder	0.15	75	0.25
Glass cullet	3–12	100	0.3
Gravel	6–10	125	0.33
Limestone	10.3	100	0.36
Milk powder	0.075	35	0.13
Plastic pellets	3–6	100	0.36
Sand-damp	0.8	100	0.4–0.45
Sand-dry	0.8	75	0.25–0.3
Salt (table)	0.4–0.8	50	0.3
Steel shot	1.5–3	50	0.36
Steel turnings	6–12	100	0.28
Sugar (granulated)	0.5–0.8	60	0.25
Tobacco	Cut	250	0.36
Wood chips	10	250	0.4

Positive Mechanical Drive. The low frequency-high amplitude feature is predominant in this mode of mechanical drive. These features enable high loads to be transported over long distances. The drive mechanism is coupled to the trough to prevent any restrictions to movement. Because there is no restrictive movement, large vibratory forces can be transmitted to surrounding structures, requiring that the mechanical drive unit be substantially secured. Conveying speeds vary between 0.2 and 0.8 m/s.

Eccentric Mass Mechanical Drive. Comparatively high frequencies and variable amplitudes are predominant in eccentric mass conveyors. The amplitudes can vary from 1 to 10 mm depending on the load. This layout consists of a rotating single eccentric load or, more often, two masses of equal configuration placed 180° apart on a rotating shaft. The twin load can be easily adjusted to produce the desired motion.

Electromagnetic Drive. This technique relies on a number of electromagnets to provide the required motion. By applying impulses at the correct frequency to the electromagnets it is possible to establish the conveying motion. The amplitudes are generally small compared with other forms of drives; the conveying velocity is up to 0.3 m/s.

Hydraulic Drive. The advantage of using hydraulic drives (in some cases pneumatic) is that they are not susceptible to explosions, enabling vibratory conveyors to be used in areas of high fire or explosion risk. These conveyors are capable of heavier duty cycles than electromagnetic conveyors.

One of the biggest advantages of vibratory conveyors is that they can convey products such as sugar and milk powder under conditions that almost eliminate degradation. They are also well suited to abrasive products and are amenable to fine tuning and control. With the incorporation of sensing devices and a feedback loop it is possible to use these conveyors for very accurate dosing in many chemical operations.

3.3. Screw Conveyors

The screw conveyor moves mass products at a controllable rate. The screw conveyor, commonly known as the Archimedes screw, was first used 2000 years ago as a water pump. The largest recorded screw conveyor in operation today is at a clarification plant delivering polluted water at a rate of 20 000 m^3/h.

The screw conveyor consists primarily of a rotating helicord placed in a stationary trough. Material is propagated by the action of the rotating screw. Inlet and outlet ports control the loading and delivery of the product. The helicord flight conveyor screw is made from a helix, formed from a flat bar. This helix is then mounted on a pipe or round bar. A typical screw conveyor arrangement, showing the more important parameters of such a conveyor, is illustrated in Figure 13.

There is a diversified number of uses for screw conveyors, such as controlled heating or cooling, mixing, and blending.

The mass flow rate can be calculated using the following expression:

$$\begin{aligned}\dot{G} &= \dot{V}_p \varrho_B \\ &= \Delta A\, s\, n\, \varrho_B\, 60\, \varphi \\ &= \Delta A\, c\, \varrho_B\, \varphi\end{aligned}$$

where $\dot{G}$ = solids mass flow rate, $\dot{V}_p$ = bulk solids volumetric flow rate, ϱ_B = bulk density, ΔA = cross-sectional area available for solid flow, s = pitch, n = revolutions per minute, φ = filling efficiency, c = solids velocity = $sn/60$. The slip factor is assumed to be unity.

The filling efficiency is defined as the ratio between the volume of product in the screw conveyor at any one time to the volume which is available to convey product. The filling efficiencies are illustrated schematically in Figure 14.

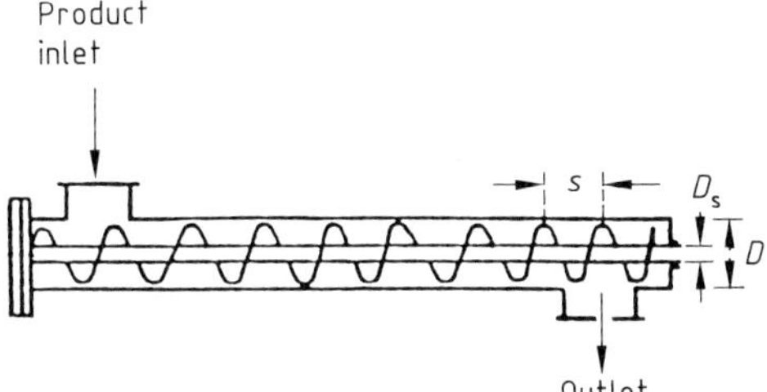

Figure 13. Arrangement of a screw conveyor

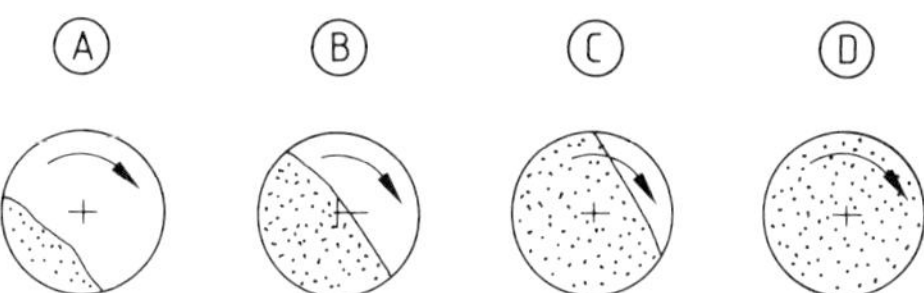

Figure 14. Screw conveyor filling efficiency
A) 25%; B) 50%; C) 75%; D) 100%

3.4. Bucket Elevators

Bucket elevators are the most commonly used device in the continuous unloading of ships, tankers, and other large storage depots. They have a distinct advantage over other methods of high payload conveyors in that they are completely maneuverable, simple, efficient, and reliable. The continuous elevator system consists of a digging element, a horizontal transfer element, and a junction element.

The design and selection of bucket elevators depends predominantly on the capacity requirements and characteristics of the material. There are two distinct categories of bulk elevators: the continuous discharge and the centrifugal discharge (spaced buckets) type [11].

The centrifugal type bucket elevator is used for free-flowing or granular products. The name arises from the discharge of the product which occurs at the head of the elevator. As the bucket passes over the head the product is discharged by centrifugal forces. The buckets usually have a 60° vertical/lip angle. The speed of the carrying chain must be high enough to create sufficient centrifugal force; speeds of 1.2 to 2.0 m/s are normal.

Table 2. Size and power requirements for a centrifugal bucket elevator

Pitch, mm	Width, mm	Velocity, m/s	Bucket volume, dm^3	Solid mass, flow t/h	Relative power consumption, kW/m	Additional power consumption, kW
250	200	0.3	6.3	5.5	0.25	0.37
250	400	0.3	12.5	11.0	0.30	0.37
320	300	0.3	18.0	12.0	0.32	0.37
320	500	0.3	30.0	20.0	0.38	0.37
400	400	0.3	38.0	20.5	0.18	0.74
400	600	0.3	57.0	31.0	0.21	0.74
400	800	0.3	76.0	41.0	0.27	0.74
400	1000	0.3	95.0	51.0	0.32	0.74
500	600	0.3	93.0	40.0	0.29	0.74
500	800	0.3	121.0	52.0	0.32	0.74
500	1000	0.3	151.0	65.0	0.41	0.74

Table 3. Size and power requirements for a continuous bucket elevator

Bucket volume, dm^3	Solids mass flow, t/h	Velocity c, m/s	Related power consumption, kW/h	Additional power, kW
400	45.0	0.3	0.32	0.74
600	70.0	0.3	0.38	0.74
400	61.0	0.3	0.35	1.10
600	94.0	0.3	0.47	1.10
800	124.0	0.3	0.54	1.10
1000	155.0	0.3	0.69	1.10

Bulk density taken as 0.5 kg/cm^3, filling efficiency 0.8 and 1.0 for spaced and continuous bucket elevators, respectively.

Materials that range from light to heavy and from fine to large lumps are usually handled by a continuous bucket elevator, where the buckets are spaced close together. These elevators can handle high loading factors and continuous loading. Normal speeds for continuous elevators range between 1.0 and 1.3 m/s.

Tables 2 and 3 give data for the sizing and power requirements of centrifugal and continuous bucket elevators, as obtained from DIN 22 201 and DIN 22 203, respectively.

A typical arrangement of bucket elevators can be seen in Figure 15.

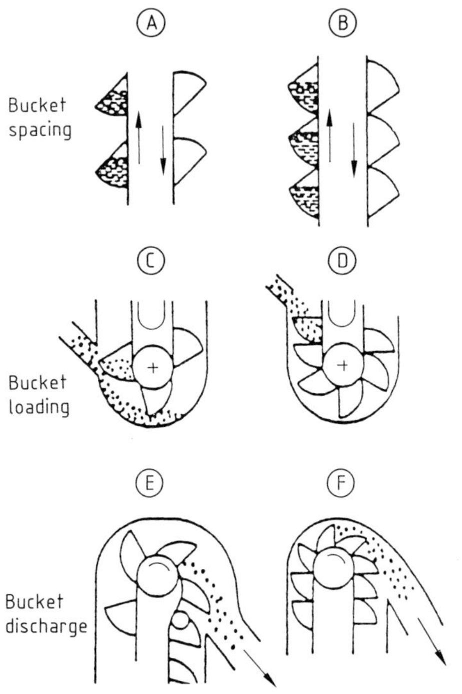

Figure 15. Arrangement of bucket elevators
A) Spaced bucket; B) Continuous bucket; C) Sloping and direct loading; D) Direct loading; E) Gravity discharge; F) Centrifugal discharge

3.5. En Masse Conveyors

The en masse, or continuous flow, conveyor was first used in England in the 1920 s. Transport of the bulk product relies on the friction between the product and the chain or flight and the friction between the particles themselves. If the friction is said to be 100 % then there would be zero slip between the product and the conveyor and hence ideal efficiency would result [9].

The product is introduced to the system to move with the flights at a steady speed. Since the movement is relative to the flight or chain there is a minimum amount of frictional movement between the particles and hence degradation is minimized. En masse conveyors can operate in the vertical, horizontal, and inclined modes. The most common applications are, however, horizontal conveyance of products. Distances in excess of 100 m are not uncommon, and the conveyors may reach capacities of 1000 t/h.

The flight profile depends on the characteristics of the material to be conveyed. Flat flights are most commonly used in the horizontal direction, whilst skeletal flights are used in the vertical direction (Fig. 16) [10, p. 322]. The material from which the flights are made plays an important role in the design of the system.

The flights are housed in a trough or casing which consists of a steel box with or without a removable lid. The product introduction points are at the top surface of the box or trough and the discharge point is normally at the bottom. The volumetric conveying rate of an en masse conveyor is dependent upon the velocity of the chain, the cross-sectional area of the mass being conveyed, and the slip between the flight and the material. The slip, or the ratio of product velocity to chain velocity, is the efficiency at which the product is conveyed.

The volumetric flow rate is then:

$$\dot{V}_p = A\, v\, s$$

and the mass flow rate is simply:

$$\dot{m} = \dot{V}_p \varrho_B$$

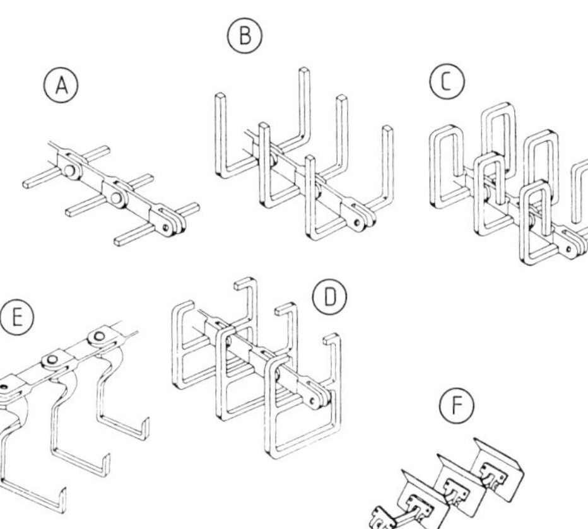

Figure 16. Flight arrangements for en masse conveyors [10]
A) Flat flights; B), C), D) Skeleton flights for vertical conveying; E) Suspended flights; F) Solid peaked flights

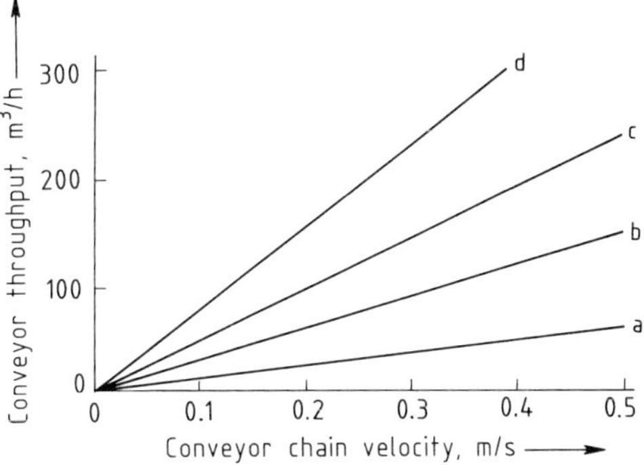

Figure 17. En masse conveyor throughput
Conveyor width, mm: a) 190; b) 290; c) 390; d) 490

where A = cross sectional area of product, m^2
v = flight velocity, m/s
s = slip or velocity ratio
ϱ_B = bulk density, kg/m^3

For horizontal conveyors with a product of large particle size, the slip s can usually be taken as unity, whereas when dealing with vertical and incline systems the slip is between 0.6 and 0.85 [10, p. 324].

The velocity of the conveyor depends on the nature of the material. Products which are easily fluidized have flight velocities in the region of 0.25 m/s while larger products can have flight velocities of up to 0.5 m/s.

Typical capacities for an en masse conveyor are shown in Figure 17 [12].

3.6. Shuttle Belt Conveyors

A shuttle belt conveyor is a simple horizontal track-mounted belt conveyor which receives material from some type of fixed conveying configuration and distributes the material along the length of a bin or a stock pile. Shuttle belt conveyors are often used for lumpy, sticky or highly abrasive materials that would be undesirable in, for example, tripper chutes. They require little head room and can be designed in a variety of belt widths and lengths [8, p. 87].

3.7. Apron Feeders

Apron feeders are essentially an endless apron of overlapping pans supported by suitable intermediate supports. The pans are attached to two or more strands of chain and the apron acts as the conveying medium. These feeders can be used to handle practically any

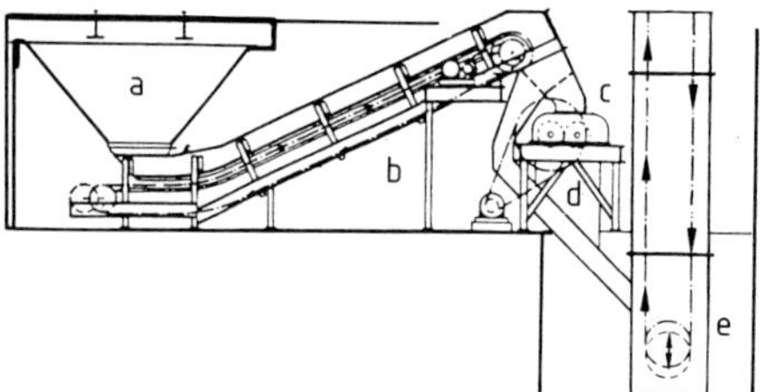

Figure 18. General layout of an apron feeder
a) Hopper; b) Apron feeder; c) Crusher; d) Bypass chute; e) Elevator

loose bulk material such as ores, sand, gravel, coal, stone, or industrial refuse. They are suitable for sharp, heavy, or hostile materials which are incompatible with other types of material handling systems. There are four basic types of apron feeders:

1) Feeders which are suitable for handling both light and heavy materials where maximum lump size is limited and impact and service are not severe.
2) Feeders which are designed for heavier service than type 1 apron feeders; they can handle larger lump sizes and withstand higher impacts. These feeders are compact and economical, requiring minimum headroom.
3) Feeders which are designed for medium duty but are more flexible as to length and application than the self-contained types 1 and 2. They are mostly used for light materials, such as coal.
4) Heavier duty feeders which are extremely ruggedly built to withstand severe impact and abrasion. They can handle materials containing a high percentage of large lumps, such as ore and rock.

The general layout of an apron feeder is shown in Figure 18 [8, p. 89].

3.8. Rotary Table Feeders

The rotary table feeder consists of a power-driven circular plate rotating directly below a bin opening. The volume of the material delivered to the feeder is regulated by a type of feed collar located directly above the table. These feeders are normally used with vertical bins for the handling of materials which tend to arch, such as damp sand or wood chips. The general layout of the rotary table feeder is shown in Figure 19 [8, p. 118].

Numerous other types of feeder are available and a brief description of the more common types is given below .

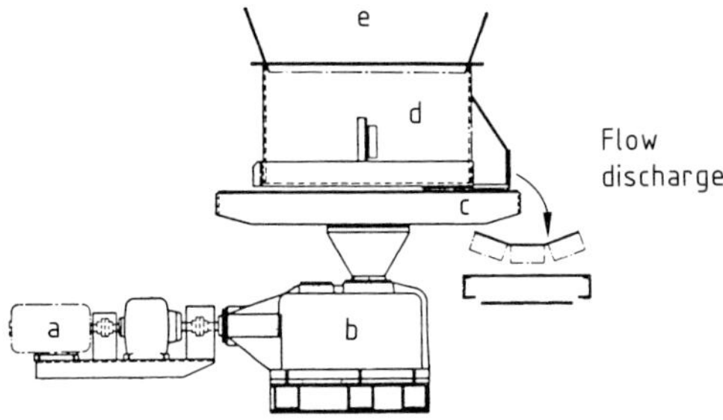

Figure 19. Layout of rotary table feeder
a) Motor; b) Spiral bevel gear drive; c) Revolving table; d) Collar; e) Hopper

Belt feeders consist of a short length of belt operating over idlers. They can handle nearly all types of materials and deliver the product at a uniform rate evenly distributed over the belt.

Reciprocating feeders consist of reciprocating driven plates which operate under a head of material. The motion delivers a forward pulsating action to the material, thereby promoting flow. Non-sticky materials and large particle size products can be handled.

Bar flight feeders are simply bars or flights attached to two strands of chain which slide along the flat surface of troughs. They are used for free-flowing non-abrasive materials.

Rotary vane feeders are one of the most common types of feeding devices for the controlled feeding of a product from a storage vessel. They consist of radial vanes rotating in a close-fitting housing and are generally used for the fine free-flowing materials which are not very abrasive.

4. Hydraulic Conveying

Hydraulic conveying technology has made significant advances since its appearance in the 1600s. Today it provides the end user with a technique to move product reliably in the form of a slurry. Although there has been significant progress in the development of theoretical predictions over the past twenty years, there are still uncertainties in the prediction of pipeline characteristics. The information presented in this text can by no means be used as accurate design guidelines but will provide the user with an opportunity for sizing a system [13].

4.1. Pump Selection

There are two main categories of pumps: reciprocating and centrifugal. Reciprocating pumps can be divided into two sub-categories: plunger configuration and piston configuration. Pump selection depends on three fac-

Table 4. Performance capabilities of various pump types

Type	Max.[a] pressure, MPa	Max.[b] flowrate, m^3/h	Mechanical efficiency, %	Max. particle size; mm	Comments
Plunger	24–28	12 540	85–90	2.4	For abrasive slurries[c]
Piston	17–21	36 800	85–90	2.4	
Centrifugal		681 400	40–75	150	Series installation, low capital cost

[a] Maximum pressure refers to commercially viable slurry pipeline volumes
[b] Positive displacement pumps are only capable of these maximum flow rates at pressures considerably lower than the maximum pressure.
[c] Defined as having a Miller No. > 50.

tors: the pressure required, the flow rate required, and the nature of the slurry, taking particle characterization into account. The discharge pressure determines whether a centrifugal or a positive displacement pump should be used.

Table 4 outlines the characteristics of various pumps [14].

4.1.1. Centrifugal Pumps

There are two basic configurations for the centrifugal pump pipeline: low pressure pumps spaced equally along the pipeline, and high pressure pumps stationed at only a few positions along the pipeline. Some of the advantages of centrifugal pumps over positive displacement pumps include:

1) Large flow rates and large particle sizes (up to 150 mm)
2) Ability to take high discharge pressures
3) Relatively smooth pressure characteristics
4) Long life, especially with abrasive particles
5) The capital costs of multiple staged of centrifugal pumps can be significantly lower than those of positive displacement pumps for high flow rates

High pressure centrifugal pumps were used predominantly for the transportation of tailings. These pumps are, however, being used increasingly to transport coarse mineral concentrates and coal [15].

For a homogeneous slurry the transport velocity is usually chosen to promote turbulent flow. The system can be operated, however, with a completely nonsettling homogeneous slurry where laminar flow exists, but the transition area must be avoided due to inherent instability.

Heterogeneous slurries are operated above the particle settling velocities. The approximate settling velocity can be obtained from numerous equations, the better being that of Cave's modification of Durand's equation [15].

Some high pressure centrifugal pipeline installations are given in Table 5.

4.1.2. Reciprocating Pumps

Plunger pumps are used for high pressure conveying due to the relatively small plunger diameters. Their use for abrasive materials can only be affected if a suitable plunger flushing system is installed. The flushing system keeps the conveyed material from contact with the surface of the plunger and hence minimizes wear. Plunger pumps operate between 80 and 120 strokes per minute, depending on the particle characteristics. Plunger pumps are relatively inexpensive but require more maintenance than piston pumps.

Piston diaphragm pumps are used for the conveyance of very abrasive materials. The product to be conveyed does not come into direct contact with the moving parts of the pump, from which it is separated by a diaphragm that reduces rapid wear propagation. The normal life expectancy of a diaphragm can be up to 6000 h [16].

4.2. Classification of Slurries

There are two categories of slurries, namely *settling* and *nonsettling*. The characteristics of solid-liquid flow are not the same as those of Newtonian fluids for the following reasons:

The effects of the particles on the mixture are combined with the liquid properties, and the

Table 5. Some high pressure centrifugal pipeline installations [15]

Owner (Location)	Slurry type	Solids SG*	Particle size distribution, % ± (mm)	Water, wt%	Capacity, t/h	Pipe diameter, mm	Pipe length, km	Number of pumps in series	Displacement pressure, kPa
BHP Minerals (Australia)	Coal Tailing	1.8	50–0.075	52	81	150	5.6	4	4135
BHP Minerals (Australia)	Sand Tailing	3.9	100–1	50	200	–	–	5	2700
Alcoa Australia (Australia)	Alumina (sand) Tailings	2.7	50–0.28	55	127	150	3.3	6	3550
N.Z. Steel Mining (New Zealand)	Iron Sands	4.9	100–0.3	50	1000	325	3.6	6	3450
Consolidated Coal (U.S.A.)	ROM Coal	–	100–100	–	425	300	–	7	3500
Consolidated Coal (U.S.A.)	ROM Coal	–	100–100	–	425	300	–	7	3500
Hanna-National Steel (U.S.A.)	Taconite Tailings	3.0	100–3.5	30–40	1500	590	9.7	7	2400
Hanna Butter (U.S.A.)	Taconite Tailings	3.0	100–3.5	35–40	850	340	7.3	6	3880
Nchanga Consolidated Copper Mine (Zambia)	Copper Tailings	2.7	100–0.85	40	1300	600	8.2	4	2545
Folldal-Werk (Norway)	Pyrites Tailings	4.9		33	20	90	1.0	3	2661
Funtana Raminosa (Italy)	Copper Tailings	2.7		50	37	100	1.55	3	2954

* Solids specific gravity

flow characteristics will be determined by the particular conditions under which the slurry finds itself [13].

4.2.1. Settling Slurries

This category of slurry can be further divided into four flow regimes:

1) Homogeneous flow
2) Heterogeneous flow
3) Saltation
4) Moving bed

An *homogeneous slurry* has a uniform size distribution or concentration of solid particles across the section of pipe through which it flows [13]. If the ratio of concentration of particles at any plane y to the concentration at some arbitrary reference plane at a height a above the bottom of the pipe is greater than 0.8, then the slurry is homogeneous.

In *heterogeneous slurries* the concentration of particles increases towards the bottom of the pipe. A critical deposition velocity exists, at which point the particles are just held in suspension. At lower velocities the particles fall out of suspension and are deposited on the bottom of the pipe.

Saltation flow occurs when the particles have an energy lower than the critical minimum and have fallen out of suspension. The particles move along the bottom of the pipeline in consecutive bounces.

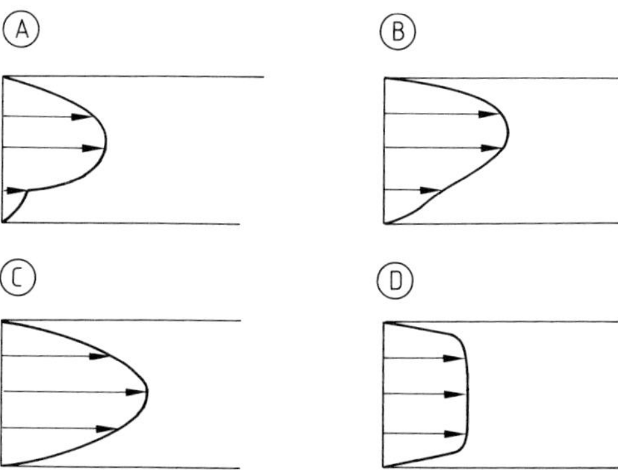

Figure 20. Flow regimes [17]
A) Sliding bed; B) Saltation; C) Suspended heterogeneous; D) Homogeneous

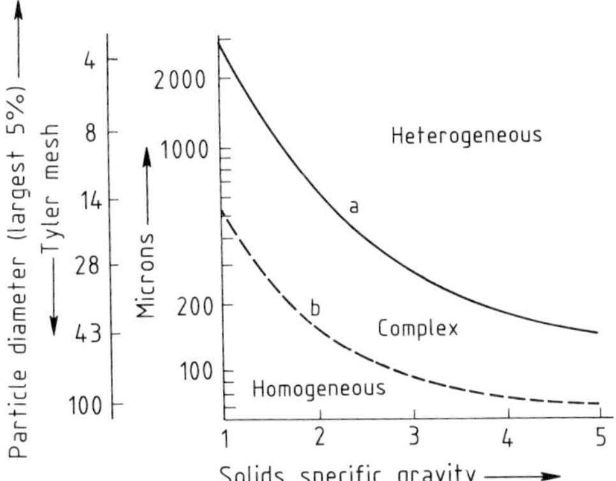

Figure 21. Slurry flow regimes (velocity = 1.2–2.1 m/s) as a function of solids size and specific gravity [12]
a) Based on thick slurries with fine (0.325 mesh) vehicle; b) Based on thin slurries with graded particle size

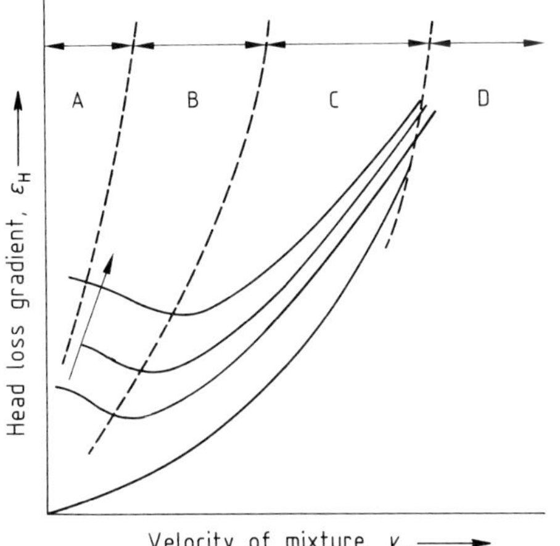

Figure 22. Head loss gradient versus mixture velocity [18], for given values of *d*, *D*, and specific gravity
A) Stationary bed; B) Fully moving bed, C_d increasing; C) Heterogeneous flow, critical deposition velocity; D) Homogeneous flow

The *moving bed regime* occurs when the particles are found on the bottom of the pipe and move by sliding along the bottom of the pipe.

The various flow regimes are shown schematically in Figure 20. The flow regime may be classified by analyzing the particle diameter and density (Fig. 21).

The influence of the velocity on the flow regime is shown in Figure 22. Note that most practical slurries have both homogeneous and heterogeneous flow patterns, due to the range of particle size in any given slurry.

4.2.2. Nonsettling Slurries

Nonsettling slurries comprise *Newtonian* and *non-Newtonian* fluids.

Newtonian Fluids. The viscosity of a Newtonian fluid is independent of the flow conditions. Newtonian fluids are not common, however, in the hydraulic transport of solids.

Non-Newtonian Fluids. This flow pattern is characterized by a continually changing viscosity and can be divided into two categories:

Time-independent fluids require two parameters to define them, unlike Newtonian fluids which require only one. Those parameters needed for classification of flow are known as the rheology of the fluid [13].

The shear stress τ is a function of the applied rate of the shearing strain for those fluids in laminar flow, and is defined as:

$$\tau = K \cdot \frac{\mathrm{d}u}{\mathrm{d}y}$$

where K is a constant and $\mathrm{d}u/\mathrm{d}y$ is the rate of change of velocity u relative to the distance y from the pipe wall.

The time-independent fluids can be divided into several categories (Fig. 23) [13].

The viscosity used to describe the stress-strain relationship of the various time-independent fluids depends on the applied rate of shearing strain and is illustrated in Figure 24. There is still, however, a great deal of controversy in acquiring the viscosity from analytical curves.

Time-dependent fluids. There are two categories of time-dependent fluids, thixotropic and

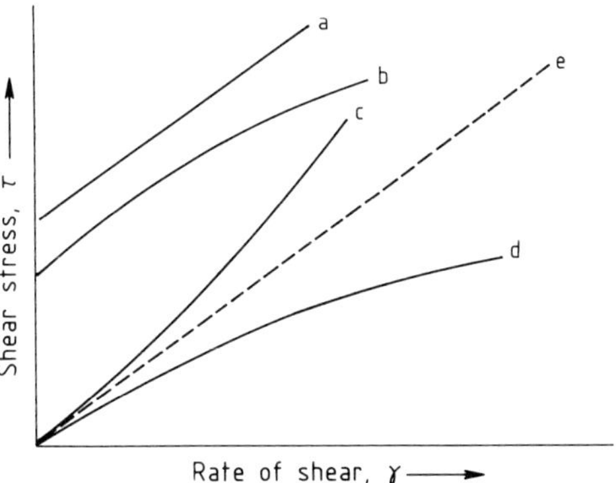

Figure 23. Shear stress–shear strain curves for typical time-independent non-Newtonian fluids
a) Bingham plastic; b) Yield pseudo-plastic; c) Dilatant; d) Pseudo-plastic; e) Newtonian fluid, for comparison

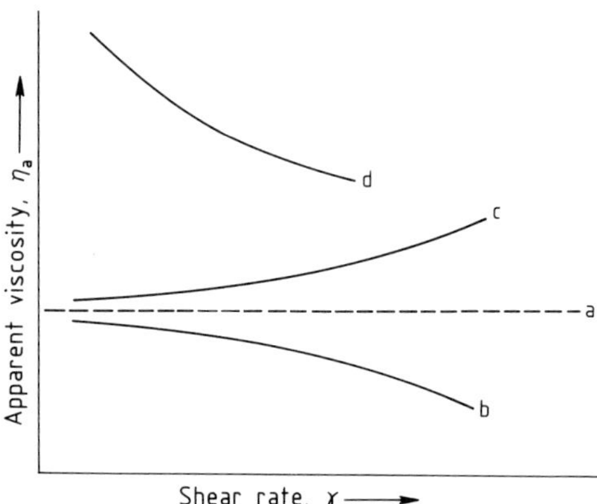

Figure 24. Apparent viscosity versus shear rate for time-independent fluids
a) Newtonian; b) Pseudo-plastic; c) Dilatant; d) Bingham

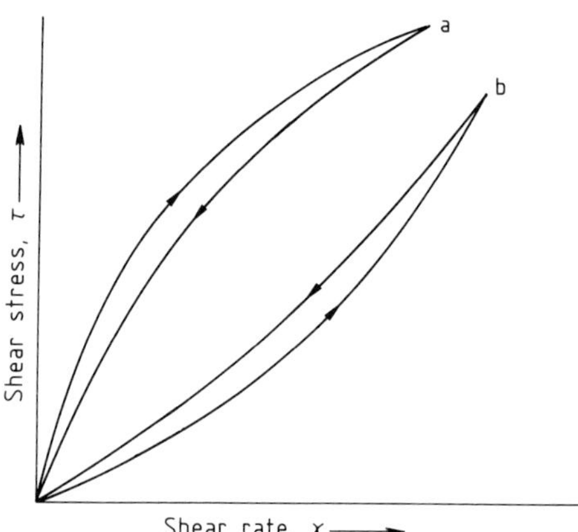

Figure 25. Shear diagram for time-dependent fluids
a) Thioxotropic; b) Rheopetic

rheopetic fluids. These fluids are indicated in Figure 25 in the stress-strain curves.

4.3. Principles of Design Calculations

The complete area of design principles cannot be treated in this text; some of the more important parameters only will be discussed.

The drag resistance F_D of a solid particle in a fluid medium is proportional to the fluid density, the cross-sectional area of the particle and the square of the relative flow velocity. The drag coefficient C_D is a factor of proportionality and is a function of the Reynolds number [19].

$$F_D = C_D(Re_p)\frac{\varrho_f}{2}(v-c)^2\frac{\pi d^2}{4}$$

where $(v - c)$ is the velocity difference between the liquid and the product, ϱ_f is the fluid density, and d is the particle diameter.

For a circular pipe, with a laminar fluid flowing through it, of diamter D flow index n, and effective wall roughness k, the average volocity $\bar{v}$ is:

$$\bar{v} = \frac{n}{(3n+1)}\frac{D}{2}\left(\frac{1}{k}\frac{D\Delta p}{4L}\right)^{\frac{1}{n}}$$

from which the pressure drop Δp per length L is expressed as:

$$\frac{\Delta p}{L} = \frac{4k}{D}\left(\frac{3n+1}{n}\cdot\frac{2\bar{v}}{D}\right)^n \qquad (1)$$

where k is the effective pipe wall roughness.

Substituting for the shear stress at the pipe wall

$$\tau_o = \frac{D\Delta p}{4L}$$

gives:

$$\tau_o = k\left(\frac{3n+1}{4n}\right)^n\left(\frac{8\bar{v}}{D}\right)^n$$

Manipulating this equation for a non-Newtonian fluid by defining the effective viscosity as the wall shear stress divided by the average shear rate at the boundary, it can be seen that for a power law fluid the effective viscosity η_e is [20]:

$$\eta_e = k\left(\frac{3n+1}{4n}\right)^n\left(\frac{8\bar{v}}{D}\right)^{n-1}$$

The Reynolds number (as a function of the effective viscosity) at which the transition from laminar to turbulent flow occurs is a function of the flow index n [20]. Figure 26 shows the approximate relationship between the critical Reynolds number and the flow index n [21]. It is then possible to obtain the velocity at which the transition from laminar to turbulent flow occurs for a non-Newtonian fluid.

If the flow occurs, however, in the turbulent region, the calculations are not so simple. Dodge and Metzner [22] suggested that the generalized Reynolds number, defined as follows, be used for pressure drop calculations:

$$Re^* = \frac{\varrho_f D^n \bar{v}^{2-n}}{\eta}$$

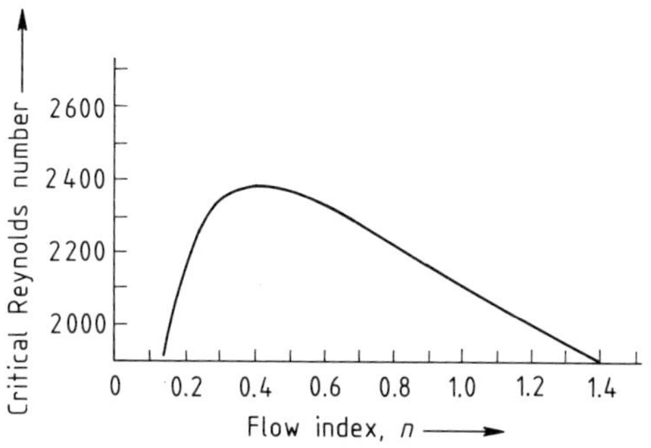

Figure 26. Relationship between critical Reynolds number and flow index n for power law fluids [21]

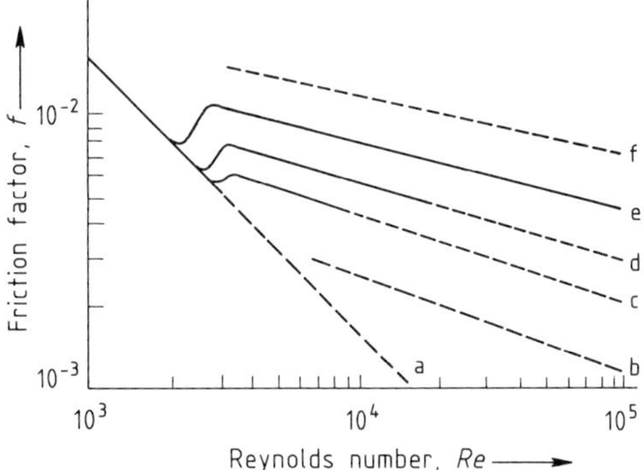

Figure 27. Friction factor design chart for power law fluids [21] —— Experimental region; ---- Extrapolated region
a) $n = 0$; b) $n = 0.2$; c) $n = 0.4$; d) $n = 0.6$; e) $n = 1.0$; f) $n = 2.0$

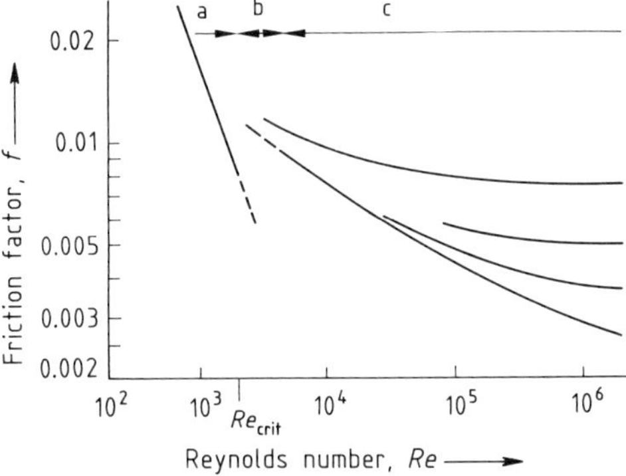

Figure 28. Moody chart for the determination of friction factors for the flow of fluids in pipes of circular cross-section
a) Laminar flow, $f = 16/Re$; b) Critical flow; c) Turbulent flow

Using the above Reynolds number and the design chart (Fig. 27) [23], the pipe friction factor f can be calculated and the pressure drop obtained by the normal method.

For laminar flow the pressure drop is simply calculated using Equation (1) above.

The pressure drop calculation for a Bingham plastic is given by the equation for laminar flow:

$$\frac{\Delta p}{L} = \frac{32\eta_p \bar{v}}{D^2} + \frac{16\tau_y}{3D}$$

For turbulent flow a convenient method is to obtain the friction factor from the common Moody diagram (Figure 28) and to use the plastic viscosity for the Reynolds number.

The most predictable method of designing an installation is clearly to use theoretical correlations backed up by extensive practical loop tests.

5. Pneumatic Conveying

The pneumatic transport of bulk solids represents one of the most widely used material handling techniques in the chemical industry. In addition to providing a relatively compact (high density) handling technique, pneumatic conveying offers the following *advantages*:

1) *Flexibility*. The system can follow any route; a change in flow direction is achieved by simple addition of a pipe bend; it is relatively easy to extend pipeline length.
2) *Pollution-free transport*. Because solids are contained in a closed conduit, toxic materials can be transported safely and the product is also protected from environmental contamination.
3) *Reduction in manpower and minimal maintenance*. Pneumatic conveying systems have virtually no moving parts, require minimal maintenance, and have relatively modest manpower requirements.
4) *Control*. With the latest feeding techniques, pneumatic conveying systems can be easily integrated into modern chemical plants employing sophisticated process control.
5) *Range of products transported*. Pneumatic conveying techniques have been succesfully applied to a wide range of products from large lumps of ore to ultrafine submicron powders. Modern handling techniques have facilitated the transportation of flakes, powders, and granules.

Although the technique provides the user with a wide range of attractive features, there are also certain *disadvantages*:

1) *High power* consumption, when compared with other mechanical handling systems.
2) *Limited distance*. Pneumatic handling techniques are generally limited to the indoor transportation of materials, although there are systems operating which move products over a maximum distance of 4000 m.
3) *Limited throughput*. Pneumatic conveying techniques are best suited to handling products at rates of hundreds, rather than thousands, of tonnes per hour.

5.1. Classification of Pneumatic Conveying Systems

There is much confusion about the classification of pneumatic conveying systems. Scientific classification techniques have been confused by commercial vendors highlighting certain flow features of their own systems; it is not always clear in exactly which flow regime such systems operate.

5.1.1. State Diagram

In order to eliminate any confusion pneumatic conveying systems are now classified according to the so-called state diagram (Fig. 29), constructed by plotting the pressure gradient $\Delta p/L$ for a gas-solid suspension against the superficial air velocity v.

A state diagram can only be obtained by conducting tests in a test loop and monitoring the pressure loss over a certain length of pipe in the so-called established flow regime (after the solids have been accelerated). The superficial air velocities are obtained by using standard air flow measuring techniques.

The flow conditions for a horizontal pipe are depicted in Figure 29 A. The following considerations are important:

Dilute Phase and Dense Phase Flow: The state diagram distinguishes clearly between these two flow regimes. Dilute phase flow is characterized by fully suspended flow in which all particles are moved along the pipe supported by the gas flow (Fig. 29 A, a).

The transition between dilute and dense phase flow is known as the saltation point; at velocities below the saltation velocity the solids will settle out of suspension. Dense phase flow is thus characterized by the solids moving as a bed along the bottom of the conveying pipe. A number of flow profiles can occur in the dense phase mode, including plug flow, dune flow, and slug flow. These flow profiles are generally a function of the solids being conveyed.

Between dilute phase flow and the so-called steady state dense phase flow there is a region known as the unsteady state. This flow region is normally characterized by a surging type flow pattern in which there is a continual buildup and breakdown of dunes. The flow in this region is normally associated with pressure surging. A state diagram with five flow regimes is shown in Figure 30 A. For each regime, a typical flow pattern appears in Figure 30 B.

The state diagram (Fig. 29) also illustrates that the pressure gradient at the saltation point is at a minimum. In dilute phase flow it is desirable to operate as close to the saltation velocity as possible; in general, dense phase flow occurs at a much higher pressure gradient.

Two important observations must be noted: an increasing solids mass flow rate results in an increase in the pressure gradient; the locus drawn through the saltation points for each

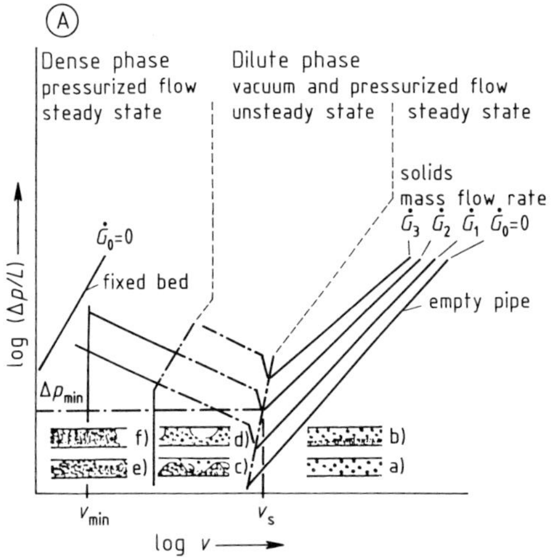

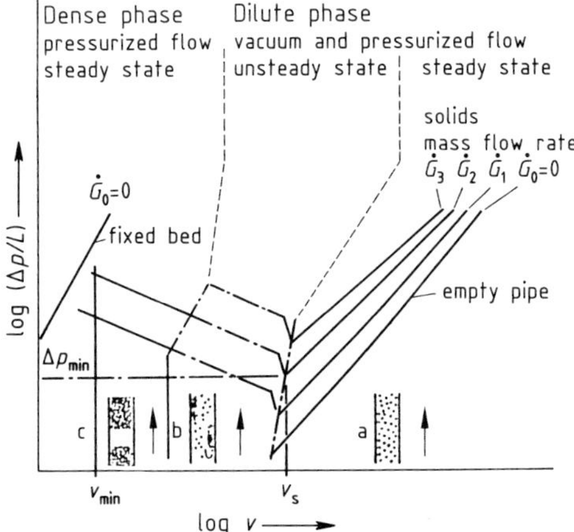

Figure 29. State diagram for A) Horizontal conveying;
a) Fully dispersed; b) Layer dispersed; c) Dune flow; d), e) Plug flow; f) Full bore plug flow
B) Vertical conveying; a) Uniformly dispersed; b) Non uniformly dispersed; c) Full bore plug flow;
Solids Mass flow rates $\dot{G}_3 > \dot{G}_2 > \dot{G}_1$

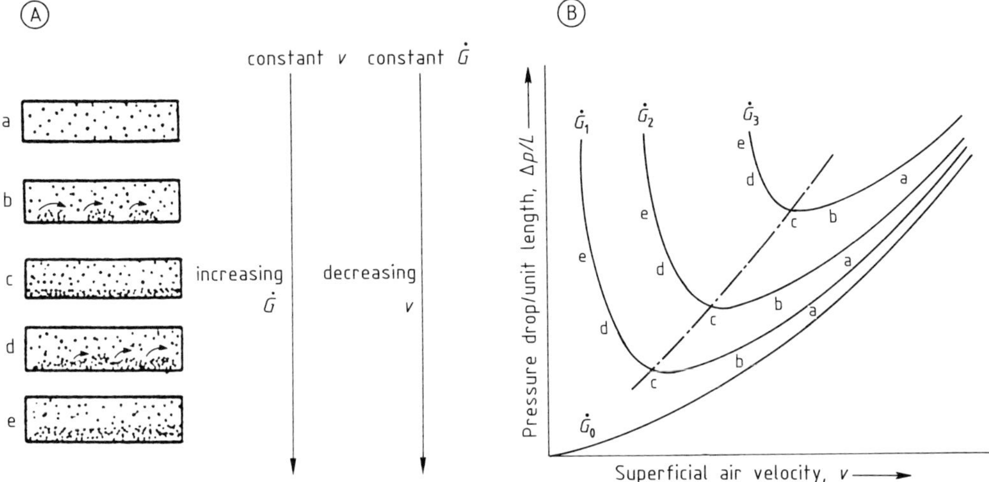

Figure 30. State diagram with corresponding flow regimes, $\dot{G}_3 > \dot{G}_2 > \dot{G}_1$

solids flow rate increases with increasing air velocity (see Fig. 30 B). This observation is critical in terms of uprating a system, when it is important to ensure that the air compressor is capable of providing the additional pressure requirements and that there is also a corresponding increase in air flow rate. By not increasing the air flow rate it is quite conceivable that the system could move from dilute phase flow at the lowest solids mass flow rate to dense phase flow at the higher solids mass flow rate. Many problems could then arise, particularly if the system has not been designed to withstand a higher operating pressure.

In general, dense phase flow (which employs far less conveying air) is more energy efficient than dilute phase systems. The added pressure requirements, however, dictate the need for more expensive equipment. Dense phase conveying systems employ high pressure feeders which require high pressure compressors. Although the current trend is towards dense phase conveying, the most popular pneumatic conveying systems still operate in the dilute phase mode.

Because dilute phase systems employ more conveying air and as the requirements are for all the solids to move in a suspended flow mode, dilute phase systems operate at conveying velocities which are substantially greater than those employed in the dense phase systems. Typically, for, say, ordinary portland cement moving as a dilute phase, flow air velocities of the order of 25 m/s are employed. In dense phase flow air velocities of the order of 10 m/s are normal practice. Dilute phase systems are, due to the higher velocities, limited to conveying products in which product degradation is unimportant. The high velocities cause very high solids degradation because of particle collisions and, more importantly, particle-pipe wall interaction.

With dense phase conveying there is a number of techniques used to promote the so-called plug phase conveying. External mechanical devices are placed along the pipe to promote the movement of solids as discrete plugs. Plug phase or low velocity conveying has been developed to ensure minimal product degradation. The ability of a product to move as a discrete plug depends on the physical and chemical characteristics of the product itself. Vendors will often market a dense phase system for a product which either has a tendency to self plug generation or for products whose characteristics are not amenable to plug formation at all.

5.1.2. Vertical and Horizontal Flow

The state diagram for vertical flow (Fig. 29 B) has the same overall characteristics as that for horizontal flow (Fig. 29 A). There are, however, some important differences between horizontal and vertical flow. On average the saltation point in vertical flow occurs at a much lower velocity, and the overall pressure loss in a vertical pipe is substantially lower than that in a horizontal pipe.

These differences can be ascribed to particle-wall interaction. In the case of horizontal flow the conveying air must have sufficient energy to overcome the tendency of the particles (which have a density some 1000 times greater than that of the conveying air) to drag along the bottom of the pipe. Designers are thus cautioned not to base their design criteria on a vertical pipeline if it is intended to move the product through both vertical and horizontal pipes.

5.2. Pressure Drop Components

The components of the pressure drop required for the transportation of a gas-solid suspension are depicted in Figure 31. For simplicity the pressure drop may be broken into components:

$$\Delta p_{\mathrm{tr}} = \Delta p_{\mathrm{L}} + \Delta p_{\mathrm{Z}}$$

where Δp_{tr} is the total pressure drop required to effect the transportation of solids, Δp_{L} is the pressure drop due to air alone, and Δp_{Z} is the additional pressure drop due to the presence of the solid.

The greater contribution to the transportation pressure drop is because of the air alone and because the solids contribute a relatively minor portion of this pressure drop (see Fig. 31).

5.3. Modes of Pneumatic Conveying and Product Characteristics

Although the state diagram provides a useful description of dilute and dense phase conveying

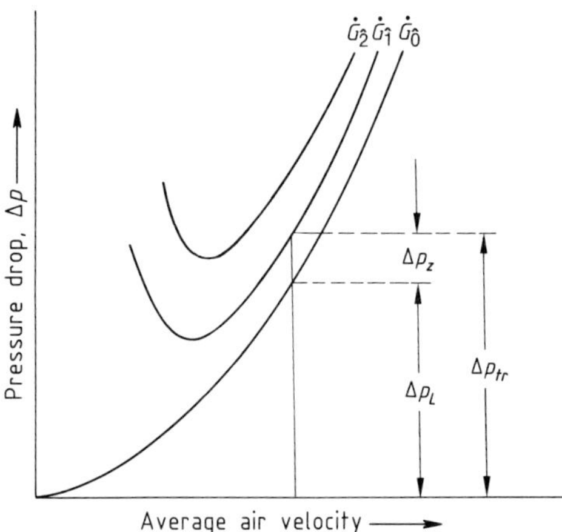

Figure 31. Components of pressure drop required to transport product, constant solids mass flow rates $\dot{G}_1$, $\dot{G}_2$

modes, there is still the question of product characteristics and whether a product is amenable to either dilute or dense phase conveying or both.

Products can be categorized into four regimes: The so-called Group A materials (e.g., cement, lime, fly ash) are considered to be good candidates for dense phase conveying. Group B materials (e.g., sugar, sodium sulphate) are more granular in nature and can be conveyed in the dense phase mode; special attention must be given to the method of feeding the product into the pipe line. (Without controlled feeding, these products tend to flood the pipeline and the system invariably blocks almost instantaneously.) Group C materials are those ultrafine submicron powders (e.g., talc powder) that, because of their particle size, tend to build up on the conveying pipe wall leading eventually to complete pipe blockage. These products, although amenable to dense phase conveying, must be transported in a flexible rubber pipe; the continuous movements of the pipe wall minimize any buildup. The Group D materials are those products (e.g., coal, mine ore) with large lump sizes. Because of their size (lumps up to 80 mm have been conveyed) it is suggested that the only reliable conveying mode is to transport the products as a fully suspended flow in the dilute phase mode.

5.4. Pneumatic Conveying Systems

Certain basic components are required for a pneumatic conveying system. The solid product, at atmospheric pressure, must be fed into a flowing gas stream, conveyed over a prescribed route, and on reaching its destination, is separated from the conveying gas. A basic pneumatic conveying system consists of the following components (Fig. 32):

1) *Prime mover.* Prime movers vary from low pressure fans to blowers and positive displacement compressors or vacuum pumps, and are selected according to system requirements.
2) *Feeding system.* Perhaps the most important item in any pneumatic conveying system. The feeder introduces the solid into a positive or negative pressure environment, simultaneously providing an airtight seal. Feeders vary in construction and operation and are selected according to the nature of the product, the process, and the system operating pressure.

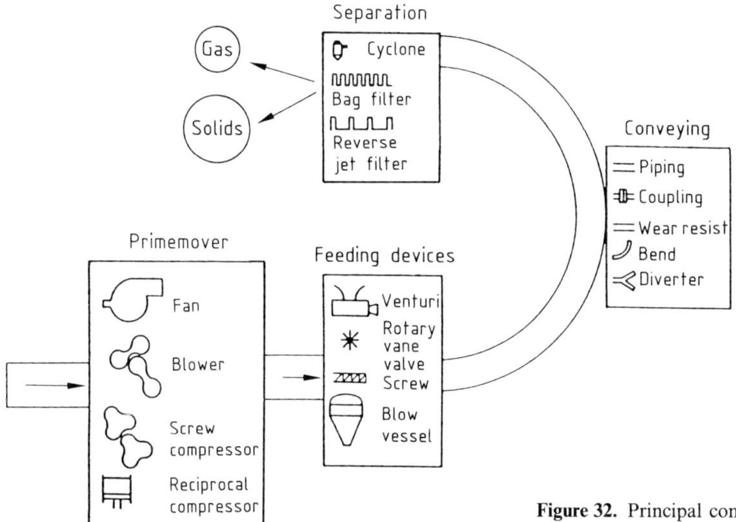

Figure 32. Principal components of a pneumatic conveying system

3) *The conveying pipe.* With more sophisticated feeding techniques it is now possible to convey highly abrasive materials and the selection of the conveying pipe is rather critical. A large number of different piping materials is available and piping materials should be selected on the basis of both pressure and the particular requirements of the product. In certain cases, pipe walls are specially prepared to prevent the formation of so-called angle hairs in the conveying of, say, plastic pellets.
4) *The gas-solid separation system.* In order to separate the solid from the conveying gas stream, it is necessary to select a suitable separation system which is also dependent upon system requirements. Gas-solid separation systems vary from cyclone separators to highly sophisticated reverse jet bag filters. The separation efficiency is determined by the process and the possible pollution problems arising from inadequate separation. Separating devices are selected on the basis of separation efficiency and are designed to ensure complete separation of the solid from the conveying stream. Although the gas-solid separation system is located at the end of the process, it is a vital component to the effectiveness of a pneumatic conveying system.

There are four basic configurations in pneumatic conveying systems, each being arranged to meet a particular industrial application.

5.4.1. Positive Pressure System

A positive pressure system facilitates the feeding of the product into a positive pressure air stream. These systems are normally used where a product is conveyed from a single pick-up point to multiple delivery points. Such a system is illustrated in Figure 33. Alternatively, there are also positive pressure systems that pick up product from several feeding points.

5.4.2. Negative Pressure System

Negative pressure or vacuum systems are well suited to applications where product has to be picked up from several feed points and delivered to a single receiving point. These systems are limited in capacity and distance because of the

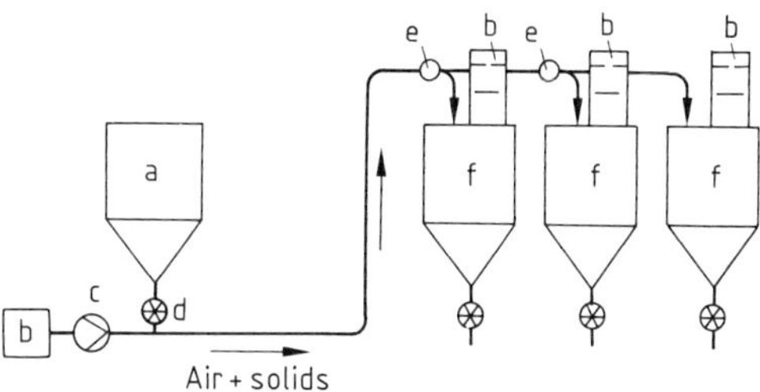

Figure 33. Typical positive pressure pneumatic system
a) Storage hopper; b) Filter; c) Air supply; d) Feeder; e) Diverter valves; f) Discharge hoppers

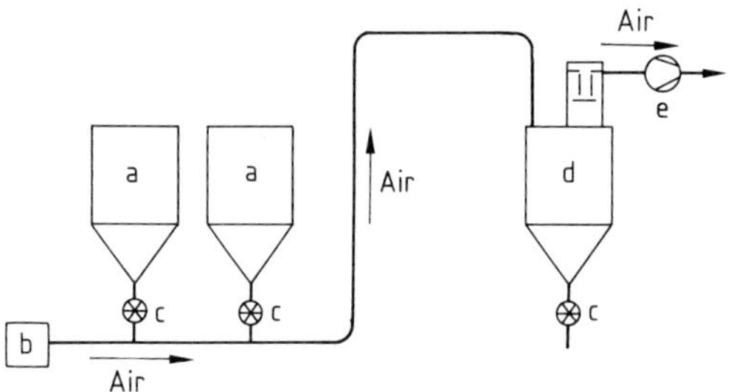

Figure 34. Typical negative pressure pneumatic system
a) Storage hoppers; b) Filter; c) Feeder; d) Discharge hopper; e) Exhauster

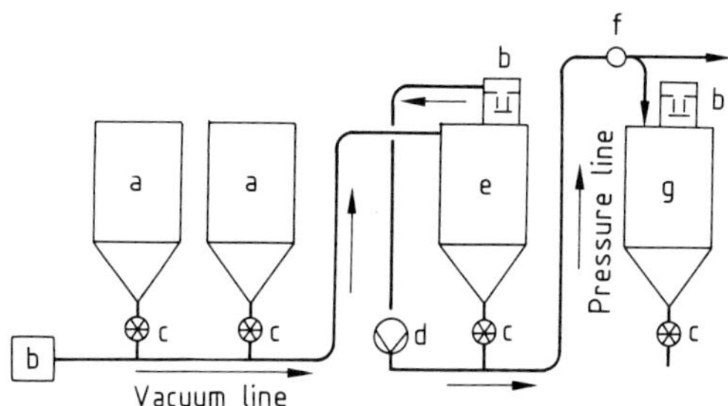

Figure 35. Combined positive and negative pressure pneumatic system
a) Storage hoppers; b) Filter; c) Feeder; d) Exhauster/blower; e) Intermediate hopper; f) Diverter to alternative discharges; g) Discharge hopper

vacuum which can be generated. A schematic of a negative pressure system is shown in Figure 34.

Vacuum systems are extensively used for the movement of toxic products, where any hole in the pipeline would cause pollution because of product spillage. Vacuum systems are also extensively used to handle product from open stockpiles, rail and road vehicles, and are often used as primary feeder for positive pressure systems.

5.4.3. Combined Positive and Negative Pressure Systems

The advantages of the negative and positive pressure systems can be consolidated into one system (Fig. 35). The hardware is arranged to facilitate the use of a single blower/compressor in which both the intake and outlet are used to provide the negative and positive pressure air flow, respectively.

6. References

[1] N. Clark, T. E. Durney, T. P. Meloy, "Recent Advances in Particle Shape Characterisation," *J. Powder Bulk Solids Technol.* **8** (1984) 21–24.

[2] H. H. Hausner, "Powder Characteristics and their Effect on Powder Processing," *Powder Technol.* **30** (1981) 3–8.

[3] T. P. Meloy, "A Hypothesis for Morphological Characterisation of Particle Shape and Physiochemical Properties," *Powder Technol.* **16** (1977) 233–253.

[4] C. R. Woodcock, J. S. Mason: *Bulk Solids Handling*, Chapman and Hall, London 1987, p. 30.

[5] R. G. Boothroyd: *Flowing Gas-Solid Suspension*, Chapman and Hall, London 1971, p. 34.

[6] P. C. Arnold, A. G. McLean, A. W. Roberts: *Bulk Solids: Storage, Flow and Handling*, Tunra Bulk Solids Handling Research Associates, 1981.

[7] R. L. Carr, "Evaluating Flow Properteis of Solids," *Chem. Eng. NY*, **72** (1965) 163–168.

[8] LINK BELT – Materials Handling and Processing Equipment, 1958.

[9] H. A. Bolz, G. E. Hagemann: *Materials Handling Handbook*, Ronald Press, New York 1958, p. 24.24.

[10] C. R. Woodcock, J. S. Mason: *Bulk Solids Handling*, Chapman and Hall, London 1987, p. 276–369.

[11] T. Kobayashi, "The Mitsubishi Continuous Ship Unloader," *Bulk Solids Handling*, **5** (1985) no. 3, 63, Trans. Tech. Publications.

[12] G. C. King, "The Application and Design of Enmasse Conveyors," Proc. Solidex 80 conf, Harrogate, UK, March/April 1980.

[13] C. G. Verkerk, *An Investigation into the Design of Slurry Pipelines with Reference to Low and High Concentration Slurries*, MSc Dissertation, University of the Witwatersrand, Johannesburg, S. Africa, July 1982, pp. 22–60.

[14] E. J. Wasp, J. P. Kenny, R. L. Gandhi, "Solid-liquid flow Slurry Pipeline Transportation," Trans. Tech. Publications, 1977.

[15] C. I. Walker, "High Pressure Centrifugal Pumps and their Pipeline Applications," *Bulk Solids Handling*, **3** (1983) no. 4, 131, Trans. Tech. Publications.

[16] G. Wallrafen, "Piston Pumps for the Hydraulic Transport of Solids," *Bulk Solids Handling*, **3** (1983) no. 1, 149, Trans. Tech. Publications.

[17] E. Anderson, S. Bouchardy, G. Capitaine, J. Davies, M. Lavitte, J. C. Leyssieux, R. Meslah: *Evolution of Pipeline Transport Techniques* **4**, Pipelines Geneva Research Centre, March 1977.

[18] P. J. Baker, B. E. A. Jacobs: *A Guide to Pipeline Slurry Systems*, BHRA Fluid Engineering, Cranfield, Bedford, England 1979.

[19] M. Weber, "Principles of Hydraulic and Pneumatic Conveying in Pipes," *Bulk Solids Handling*, **1** (1981) no. 1, 2, Trans. Tech. Publications.

[20] C. R. Woodcock, J. S. Mason: *Bulk Solids Handling*, Chapman Hall, London 1987, p. 142–145.

[21] E. J. Wasp. J. P. Kenny, R. L. Gandhi: *Solid-Liquid Flow Slurry Pipeline Transportation*, Gulf Publishing Co, Houston, TX, 1979.

[22] D. W. Dodge, A. B. Metzner, "Turbulent Flow of non-Newtonian System," *AIChE J.*, **5** (1959) 1–9.

[23] C. G. Verkerk, "Some Practical Aspects of Correlating Empirical Equations to Experimental Data in Slurry Pipeline Transport," *Bulk Solids Handling*, **5** (1985) no. 4, 21, Trans. Tech. Publications.

9. Solid–Liquid Separation, Introduction

Christian Alt, München, Federal Republic of Germany

1. General Considerations

A wide variety of equipment and machinery is available for the separation of solids and liquids. Counting the many possible optional variations, a potential user can choose from among more than 100 separation devices. Since some of the equipment was developed to solve specific problems, its use is confined mainly to those or to other closely related tasks. However, most equipment is universally applicable and lends itself to comparisons.

The equipment considered in this article is used principally in the following separation processes:

1) screening (wet screening or screen filtration)
2) sedimentation
3) filtration
4) expression, and
5) electrokinetic or magnetic separation.

In this classification, equipment for elutriation occupies a special place. Although it includes separation of the two phases of a solid–liquid mixture, its major task is selective separation within the solid phase (→ 16. Elutriation, p. **16**-1).

Each of the listed separation processes can be further categorized by additional characteristics. Figure 1 shows a classification according to the forces that are necessarily active in every separation process. Accordingly, separation processes are basically differentiated by the motion of the phases to be separated, and the equipment can be classified into *sedimentation apparatus* and *filters*. Nevertheless, this classification is not always unambiguous, since a large number of devices comprise both mechanisms.

Separators are used for coarsely dispersed solid–liquid mixture systems with particle sizes of several millimeters down to finely dispersed systems of molecular dimensions. In addition, highly concentrated slurries are separated as well as dilute suspensions. For the evaluation of separation equipment such inherent parameters as different geometries of the particles, interfacial effects, and electrostatic interactions are considered first; they can be influenced by pretreatment. Therefore, a number of important factors must be observed.

A systematic collection of factors to be considered may be helpful for selecting separation equipment and avoiding basic mistakes. Some criteria for the use of liquid separation processes are listed below; see also [1]:

1) Material Factors

1.1) Feed

Solids: concentration
density
particle size
particle size distribution
particle shape
particle structure (aggregates, flocks, and crystals)
stability and hardness
solubility
corrosiveness
chemical composition

Liquids: chemical composition
density
viscosity
pH value
volatility, vapor pressure

Impurities
Foaming tendency
Temperature
Toxicity

Transport mechanism		Free motion of dispersed particles	Free motion of liquid
Designation		Sedimentation Flotation	Filtration
Driving forces	Gravity	Thickener/Clarifier	Screen drainage Dewatering screens Hydrostatic filter Microstrainer
	Pressure gradient		Cake filter Pressure filter Vacuum filter Capillary filter Expression device Deep bed filter Cross-flow filter Strainer
	Centrifugal field	Static wall Hydrocyclone Rotating wall Solid bowl centrifuge Rim centrifuge Scroll discharge centrifuge Disk-type centrifuge	Centrifugal filter Peeler centrifuge Continuous solids discharge
	Electric field	Electrophoresis	Electro-osmosis
	Magnetic field	Low-field magnetic separator High-gradient magnetic separator	

Figure 1. Classification of equipment for solid–liquid separation

1.2) Wash Liquor
Material properties; see "liquids"

1.3) Equipment
Corrosion protection
Abrasion protection
Load capacity

2) Performance Requirements
Amount of feed
Residual moisture
Effluent clarity
Wash liquor concentration
Residual mother liquor
Solids transportability

3) Operating Conditions
Continuous versus discontinuous
Throughput
Feed variation (concentration and amount)
Residual moisture
Solids breakage
Washing (separate collection of effluents)
Foaming tendency
Temperature range
Auxiliary processes: flocculation
precoating
filtration aids
foam suppression
Preceding and subsequent process steps

4) Equipment Conditions
Energy requirement
Auxiliaries' requirement
Space requirement
Pollution (e.g., gases or noise)
Corrosion tendency
Abrasion tendency
Personnel requirements
Automatic control
Durability
Spare parts
Accessories
Special adaptation
Apparatus combination
Availability

This list does not claim to be complete for extreme cases. However, very often only a fraction of the factors mentioned above must be observed.

The properties of materials to be separated sometimes vary widely in similar processes and fluctuate frequently in the course of production, which further complicates comparisons and evaluations made for the purpose of selecting the optimal separation equipment.

Nevertheless, some generally valid relations narrow the comparison of different separation processes. A separation process utilizing gravity usually requires less energy than most other processes. Thus, for example, a sedimentation tank can be operated with only a few joules per cubic meter of liquid to be separated, while filters and centrifuges rarely require less than 10 000 J per cubic meter of liquid to be separated. Both machines require less than half the energy needed to operate a drier, although the latter is capable of achieving a lower residual moisture content. Thus, energy consumption and residual moisture content determine to some extent the type of separation process to be used. Another factor to be considered is the intensity of operation of the separation equipment; thus, the energy requirement for the same apparatus can be reduced considerably if, for example, a higher residual moisture content of the solids can be tolerated.

The particle size of the dispersed solids also shows a relationship between material parameters and the equipment that can be used. Figure 2 surveys proven application ranges of typical separators. The plot along the ordinate is roughly proportional to the optimum feed concentration for the various separator devices; in Figure 2 the ordinate extends to a solids concentration of about 40 wt%. However, in individual cases a lower or even a higher feed concentration may be preferable.

A comparison of separation processes should also include *investment costs*. In addition to the purchase price of each piece of equipment and of the necessary auxiliaries, one must consider the costs of installation, erection, and connections as well as safety, both of the operating personnel and of the environment. The investment cost divided by the throughput always decreases as the capacity of the equipment increases. As an example, the net purchase price of filters is

$$I_A = k_1 A^{-k_2} \tag{1}$$

where I_A is the purchase price per square meter of filter area, k_1 and k_2 are constants that depend

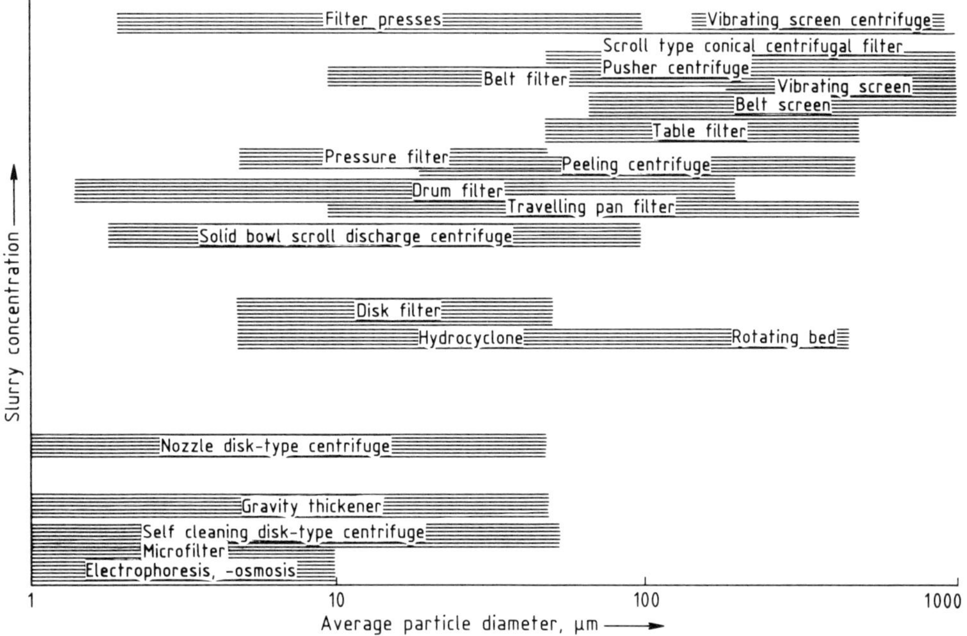

Figure 2. Ranges of application of solid–liquid separation equipment as a function of particle size and slurry concentration

on the type of equipment and the material of construction, and A is the filter area.

Only an expert is able to compare easily and effectively separation processes and associated equipment, considering all nuances of the factors and criteria mentioned above. If, in addition, economic factors are to be taken into account, it is almost impossible to select the optimum separation equipment without expert knowledge and years of experience. Therefore, it is, frequently impossible to avoid the type of laboratory studies that until recently were practiced routinely. Certain types of testing equipment are useful, in such cases, e.g., long-tube apparatus, hand filter plates or compression filter cells, and bottle centrifuges. Frequently, pilot-plant or small-scale production testing cannot be avoided. The corresponding production technique can then be simulated, if necessary.

Even the relatively favorable energy consumption in the solid-liquid separation, mentioned at the outset, should be compared in relation to the subsequent process steps. Initially, the energy consumption of the equipment under consideration certainly forms the basis of planning studies, especially for new plants and within the context of cost saving measures. However, it should be kept in mind that the situation can be altered by overriding factors. For example, dilutions should be avoided in a production process wherever possible, since the solids-related separation costs are increased by reductions in feed concentration. Nevertheless, dilution of a suspension prior to separation may be useful; it can, for example, advance flocculation. In certain filtration processes a more compact filter cake can result from a dilute suspension than from a concentrated one.

2. Separation Processes

Each separation process exhibits its own specific behavior based on physical and structural situations which characterize the ranges of application from the outset [2].

2.1. Sedimentation

Gravity Sedimentation. Sedimentation in the gravitational field should be considered in principle only for highly dilute dispersions with relatively finely divided solids particles. It is, indeed, the simplest separation process and entails the lowest cost. Even broadly dispersed and extremely structured, flaky solids particles can be separated at relatively high throughputs with very few problems, if suitably long sedimentation times are provided.

Thickeners and clarifiers are, as a rule, not operated in a free-standing, single-step mode, but often as preliminary stages if both of the phases to be separated are to be obtained with high efficiency. The liquid contents (residual moistures) of the sludges are comparatively high. Nevertheless, the degree of solids separation is good and may be as high as 90 to 100%. The degree of liquids separation, on the other hand, is lower.

At throughputs of up to $1\ m^3\ m^{-2}\ d^{-1}$ the specific performances are relatively modest and require suitably large units at higher throughputs. Concrete basins are cheap, but require a lot of space. Because of their low operating costs, simplicity of automation devices, and the small number of operators required, thickeners are used widely in the separation and wastewater treatment plants.

Centrifugal Sedimentation. Sedimentation in a centrifugal force field clearly requires more complicated equipment and more careful assessment than gravity sedimentation. For example, with solid bowl centrifuges and disk-type centrifuges operating at an acceleration of 1000 g (g: acceleration due to gravity), high sedimentation rates can be achieved, even with fine-grained particles, which permit high throughputs per machine unit. With scroll discharge of solids, in continuous decanter centrifuges, about 200 m^3/h of wastewater slurry can be dewatered in large modern units. With the aid of flocculation, the sedimentation rate can be increased further by about a factor of 10. The flocculating agent, which in this case remains in the separated solids, may be undesirable at times; for that reason, separation processes free of flocculating agents, for example, in disk-type centrifuges, are then preferred. Also suitable, although at reduced throughput, are special modifications of solid bowl scroll discharge centrifuges. For further information, see → 11. Centrifuges and Hydrocyclones, p. **11**-8. Ternary systems—two liquid phases and one solid phase—can be successfully and completely separated both in differently modified solid bowl scroll discharge centrifuges and in disk-type separators with peripheral solids discharge through jets or annular slots.

A comparison of sedimentation equipment should usually also include the wear that is expected. Centrifuges are considerably more subject to wear than gravity basins; therefore, a number of devices are available to protect against wear, especially for the highly stressed conveyors in continuous decanter centrifuges.

Hydrocyclones. Hydrocyclones are simple devices and require very little expenditure. To be sure, they do not provide sharp phase separations; for this reason, they are used more as prestages for concentrating dilute suspensions ahead of a filter and for similar purposes (→ 11. Centrifuges and Hydrocyclones, p. 11-19.

A key disadvantage of most sedimentation equipment is that sludges cannot be washed. If the washing of solids is strictly required, solid bowl scroll discharge centrifuges and disk-type centrifuges are applicable, though only to a limited extent.

2.2. Filtration

A very large number of different filters are available on the market. They can be used for coarse-particle suspensions of up to a several millimeters particle size and even with a broad particle size distribution on the one hand, and for colloidal suspensions on the other hand. The very fine-particle suspensions have only recently become amenable to mechanical separation because membranes of various configurations have become available. In addition, filters can be used for almost any solids concentration.

In contrast to sedimentation equipment, filters are independent of phase density. Even slurries with only minute differences in density can be filtered without problems.

Commensurate with their importance, filters are also available for extremely high performance. Filtration areas can be up to 1000 m^2 in size and permit throughputs of several hundred cubic meters per hour.

Particle Size and Solids Concentration. *Coarse-grained suspensions* are reserved for vacuum filters or occasionally for hydrostatic filters. Both devices are energetically more favorable and simpler than pressure filters. For *slurries containing fine particles* deep-bed filters and dynamic (cross flow) filters are available in addition to prevalent pressure filters. The *feed concentration* largely determines the preferred type of filter. Pressure filters generally require higher solids concentrations than the others which are selected more on the basis of what is expected subsequently of the separated phases. For example, it is difficult to further concentrate solids retained in deep-bed filters; in dynamic filters such concentration is routine.

When highly concentrated slurries contain solids that are finely divided or otherwise difficult to separate, such as slimy particles, they can also be separated by precoat filtration, preferably on drum filters, centrifugal discharge plate or candle filters; this involves the use of filter aids. However, relatively dry solids obtained in this way contain the filter aids, except for those shaved off the drum filter.

Coarse-grained and simultaneously fast settling solids are less suitable for drum and disk filters since the solids form a sediment in the tanks. Likewise, slurries with relatively large fractions of fine particles are more difficult to handle in these filters since the finest particles, which build up the filter cake first, may frequently blind the filter medium.

Residual Moisture. Filters in comparison to thickeners clearly yield lower residual moisture. However, minor differences in individual filters should be noted. Thus, lower residual moisture is obtained in pressure filters, filter presses, and especially in press filters. In this connection, dynamic filters give above-average residual moisture. Deep-bed filters generally are not intended for the production of dry solids with low residual moisture, but are preferred to clarify liquids.

Particularly low residual moisture is obtained in centrifugal filters, since centrifugal forces act on the retained liquid more efficiently than do the other liquid displacement mechanisms usually employed in filters.

Bleeding. If everything else is equal, bleeding of solids to the filtrate is higher in centrifuges than in filters. Therefore, slurries containing a relatively wide particle size distribution are preferably conveyed to filters.

Solids Discharge. Typical cake filters occasionally have problems with removing the filter cake from the filter medium. For example, if the filter cake clings strongly to the filter medium, whether because the solids have adhesive surface properties or because some of the solids are entrapped at higher rate within the interstices of the filter medium, the filter cake no longer drops off the filter medium. Only drum and disk filters with scrapers and the like permit safe removal; with pan, table, leaf, and candle filters, as well as with filter presses, cake removal can present difficulties.

Washing of the Filter Cake and Filter Medium. Some separations also require complete removal of the liquid that remains in the filter cake after dewatering, for example, when a corrosive mother liquor or solvent is involved. For this purpose, filter cakes can be washed. Good possibilities for washing exist in belt, plate, and table filters, because the washing liquid does not remove the filter cake from the filter medium. Newer spraying devices make it possible to wash the filter medium in pressure leaf filters and filter presses as well; for the latter, high pressure spraying devices with up to 10 MPa water pressure are on the market.

Accessibility of Filter Area. It is often desirable that the filter area be accessible to operating personnel for instant intervention in case of a breakdown. Vacuum filters offer the best conditions for access. In line with the general trend toward process automation, however, this need may recede more and more into the background.

Cake Compression. Compressible filter cakes are less suited to pressure filtration than to vacuum filtration, since the relatively high pressure drops which exist in pressure filters favor consolidation of the filter cake. The same is true for centrifuges. However, by combining clarifiers and filters, as in solid bowl centrifuges in which the lip ring is perforated for filtration, compression can become almost insignificant.

Durability. Some filters are distinguished by long durability. This includes particularly filter presses. However, drum and disk filters also have a relatively long durability. The reason for this lies in the design of these filters which minimizes component parts exposed to abrasion and wear. Centrifuges, in contrast, are more subject to wear, especially in those component parts which come into contact with solids at high velocities.

Economy. Some filters have relatively favorable economics such as nutsche filters and candle filters. Centrifugal-discharge filters and filter centrifuges are considerably more expensive. Considering their specific throughputs, however, it is largely found that they also increase with construction cost. Many cost advantages are balanced in this way. However, minimizing costs is always an important development objective; it results in improved and novel filters. An example is seen in the electrokinetic filter, much discussed recently.

Throughput. The throughputs attainable per filter surface area for equivalent products are generally greater in centrifuges than in the conventional filters. For the same reason, slurries which are difficult to filter, can be more efficiently treated in centrifuges (Fig. 3) [3]. Accordingly, filter cakes with small cake formation times also require shorter cycles in centrifuges.

Energy Consumption. The energy consumption per mass rate of liquid or solid separated is only slightly higher in centrifuges than in filters, given the same conditions. Only extremely simple designs such as nutsche filters behave differently.

Occasionally, however, filters consume more energy, for example, if greater demands are made with respect to the degree of washing rate or residual moisture, which frequently requires additional equipment. As a rule, filters and filter centrifuges, including the necessary auxiliary equipment, consume between 1 and 10 kW h per ton of dry solids. The lower values apply to

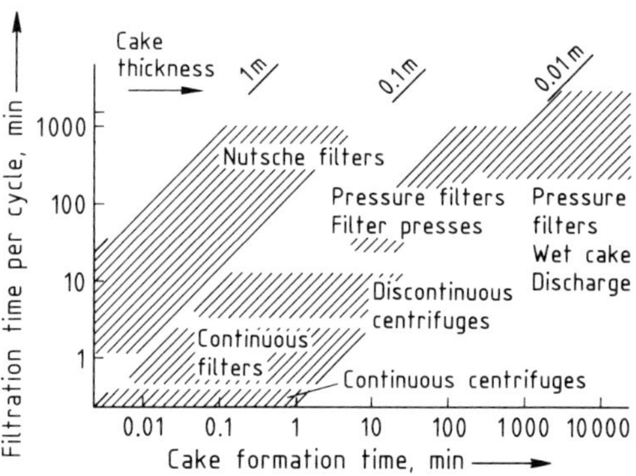

Figure 3. Typical ranges of application of cake filtration equipment
The cake formation time is the time required for the buildup of the first centimeter of filter cake.

coarse-grain and crystalline products. The higher values must be expected for fine-grain and especially for pasty materials. The observed behavior of filters often varies widely with a corresponding broad scatter in the separation efficiencies.

Quantitative Comparison. Because the effectiveness of a separation is strongly related to the product, it is difficult to find quantitative comparisons of different separators. A causative factor is the impossibility of making precise studies on identical products in industrial-scale operations.

Some examples, however, have recently been published [4]. In the separation of a pigment that was difficult to filter, one comparison was made between a filter press and a centrifugal discharge plate filter. Higher residual moistures resulted in the latter, especially in opposite membrane filter presses. Of course, to achieve equal elutriation the centrifugal discharge plate filter requires two to three times the amount of wash water. The availability was equal in both cases. Cleanup of the centrifugal discharge plate filter following possible product changes is disproportionately more expensive.

Another comparison between a pusher centrifuge and a belt filter, using a crystalline product, indicated a better durability for the filter. Bleeding of solids is also lower in the filter than in the centrifuge which is provided with relatively large wide slots for economic reasons. Total energy consumption, however, was greater for the filter. Total operating costs of the filter are lower because of the high wear of the centrifuge.

3. References

[1] D. A. Dahlstrom: "Selection of Solid-Liquid Separation Equipment", in H. S. Muralidhara (ed.): *Advances in Solid-Liquid Separation,* Battelle Press, Columbus, OH, 1986.

[2] H. F. Trawinski: "Current Liquid-Solid Separation Technology", *Filtr. Sep.* **17** (1980) no. 4, 326–335.

[3] W. Gösele: "Filterapparate—eine Übersicht", *Aufbereit. Techn.* **18** (1977) no. 5, 210—214.

[4] W. Gösele: "Vergleich von ausgeführten Anlagen zur Fest-Flüssig-Trennung", *Preprints GVC-Dezember-Tagung 1987,* Verein Deutscher Ingenieure, Düsseldorf 1987.

10. Filtration

Christian Alt, München, Federal Republic of Germany

In addition to the standard symbols defined in the front matter of this volume, the following symbols are used:

a_V	specific surface
A	cross-sectional area of the bed
d	differential
E	efficiency
Eu	Euler number
f	friction factor
H	bed thickness
K	Kozeny number
L	quantity of liquid
M	mass
$\dot{M}$	mean output of a filter
Pe	Peclet number
Re	Reynolds number
S	quantity of solids, mean saturation
St	Stokes number
u	superficial velocity
U	average consolidation ratio
v	fluid velocity
$\dot{V}$	volumetric flow rate of the fluid
w	mass of dry cake solids per unit volume filtrate
z	cake formation rate
α	flow resistance, cake resistance
β	bulking factor, resistance of filter medium
ε	porosity
λ	filtration coefficient
σ	loading
Φ	shape factor, dewatering parameter

Subscripts

c cake
F fluid
I inlet
L liquid
LF filtrate
O outlet
p particle
S solid

1. Terminology

Filtration is the separation of solid particles or liquid ones (droplets) from liquids and gases with the help of a *filter medium*, also called a septum, which is essentially permeable to only one phase of the mixture being separated. In earlier times, this process was nearly always carried out with felts, and the word "filter" has a common derivation with "felt."

Often, however, purification of a liquid or gas is called filtration even when separation does not involve a semipermeable medium (as in electrokinetic filtration). This article deals only with the filtration of solid–liquid mixtures (suspensions, slurries, sludges); for the treatment of gases by filtration, see → 13. Dust Separation.

The liquid more or less thoroughly separated from the solids is called the *filtrate, effluent, permeate* or, more rarely, *clean water*. As in other separation processes, the separation of phases is never complete; liquid exits with the solids and/or solids with the filtrate. The liquid content of the separated solids is called *cake* or *residual moisture*; a filtration operation is also characterized by the *solids content in the filtrate* (clarity).

Cake Filtration and Clarification. Filtration is termed *solids-recovery* or *surface filtration* on the one hand, or *clarification* or *polishing filtration* on the other, depending on the purpose of the operation. In cake filtration the majority of the solid particles is always larger than the openings in the filter medium and the cake; therefore, the solids are retained as a cake of increasing thickness on the surface of the medium. They are usually the product to be recovered. Because a filter cake is formed, this operation is also called *cake filtration*.

In clarification, also referred to as *filter-medium, bed* or *deep-bed filtration*, the solids are retained in the interior of the filter medium. Here the effluent is the product. An exception is clarification performed in specialized cake filtration (precoat and strainer filtration). Because of the limited capacity of the filter bed to trap solids within its body, only dilute slurries with fine-grained solids should be treated by deep-bed filtration.

Cake filtration is especially well-suited to feed with higher solids content containing large solid particles. Sharp distinctions are not possible. On economic grounds, however, bed filtration should not generally be used for influents containing more than 0.1 % solids. Either by design or because of process characteristics, it often happens that both cake and deep-bed filtration are active in the same filtration process. For example, some of the solid particles in cake filtration initially penetrate into the openings of the filter medium and are attached there, while other, finer particles exit with the filtrate. As the pores and channels in the filter medium become narrower and narrower, subsequent solid particles cannot get into it.

Screening and Cross-Flow Filtration. A filtration process in which the collected solids are continuously removed, mostly in tangential flow to the filter medium, is called *screen drainage, membrane*, or *cross-flow filtration*; as in dry screening, it is always necessary to avoid leaving a residue. Screen drainage used to be confined to the dewatering of harder solids with a size range from about 5 mm down to about 0.03 mm in a microstrainer. *Strainers* are filters used to filter liquids containing small amounts of particles.

The development of filter media with extremely fine pores (membranes), down to the microscopic range (i.e., from a few micrometers to molecular sizes), has allowed new applications of cross-flow filtration. In *membrane filtration, ultrafiltration*, and *reverse osmosis*, steps are generally taken to prevent significant accumulation of solids that would hinder flow of the liquid.

Especially in the range of molecular-sized pores, higher molecular mass compounds are retained on the membranes from their solutions. Such membranes even make it possible to filter colloidal systems. Osmotic processes also take place in solutions, so the operation is also referred to as *reverse osmosis*. Because of its importance, it is discussed in a separate article (→ 11). Biochemical Separations, **B3**.

In contrast to cake filtration, the retained particles in this type of filtration are continuously moved; the operation—especially on membrane filters—is therefore called *dynamic filtration*, sometimes also *delayed cake filtration*. The teminology described so far is shown in Figure 1.

Driving Force in Filtration. In the simplest case, the liquid is allowed to flow through the filter medium by gravity. Such operations of *gravity* or *hydrostatic filtration* are often referred to as simply dewatering or drainage. More often, however, a pressure drop is imposed on the separating system. In *pressure filtration*, the influent is under pressure; in *suction* or *vacuum filtration*, a vacuum is applied to the effluent. Centrifugal forces are used to drive filtration in *centrifugal filters*.

Expression. Finally, it is possible to start with a compact disperse system and, by mechanical pressure, separate from it part of the liquid dis-

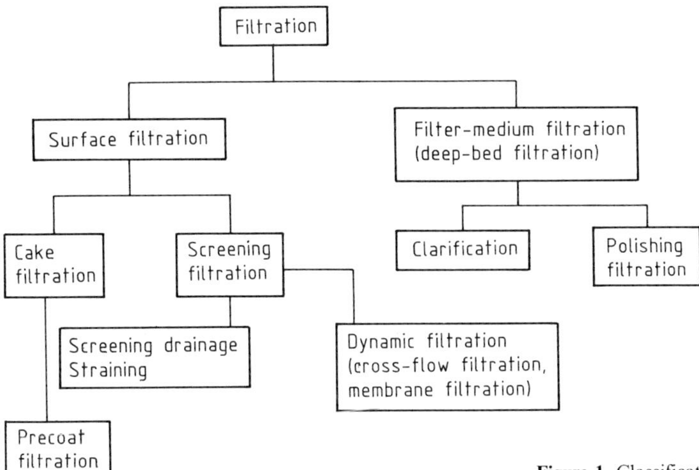

Figure 1. Classification of filtration

tributed through the solids. This operation, long used in the production of fruit juices and essential oils, removes moisture from bulky, deformable filter cakes. Expression in presses is described by laws similar to those that govern other filtration processes. Reviews on the subject of filtration in general are given in [2], [5]–[8], [10], and [12].

2. Basic Theory of Filtration

Though filtration has been practiced for a very long time and the technology has reached a comparatively high level, its theory has been developed only in recent decades. The lateness of this development may be due to the extremely complicated laws governing the hydrodynamics of filtration and the mechanism of separation. Filtration parameters and material characteristics cause great difficulties in practice, especially since—as in no other unit operation—they change from case to case.

As a result, it is often thought that the complicated physics of actual filtration processes makes the efficient assessment of filters for practical use impossible. Current filtration theory, however, gives reliable formulas with which filter performance can be predicted and laboratory tests can be scaled up to full scale. Where the physical description of filtration does not yield satisfactory results, mathematical modeling offers an alternative solution.

Only the theory of fluid flow in filtration is treated in this section. The theory of cake and deep-bed filtration and the separation efficiency will follow.

Fluid Flow in Filtration. Filtration is essentially a flow of fluid through a porous medium, commonly a granular bed. The filtrate moves in capillary fashion through the bed. If the bed is assumed to be characterized by a uniform voidage, the general formula for the superficial fluid velocity u is [1]

$$\frac{\dot{V}}{A} = u = 2\left[\frac{f \cdot H}{d_{\mathrm{p}} \cdot \Delta p}\left(\frac{1}{2}g\right)\right]^{\frac{1}{2}} \tag{1}$$

where , $\dot{V}$ is the volumetric flow rate of the fluid, A is the cross-sectional area of the bed, f is the friction factor, H is the bed thickness, d_{p} is the effective particle diameter of the solids, and Δp is the dynamic pressure drop across the bed [5]. In granular beds, d_{p} is a function of the porosity and of the specific surface a_V, that is, the total surface area of solid particles per unit volume of particles:

$$d_{\mathrm{p}} = 6/(\Phi\, a_V) \tag{2}$$

where Φ is a shape factor for nonspherical particles.

The friction factor f depends on the Reynolds number Re, which is defined in a modified form for flow through porous media as

$$Re = \frac{d_{\mathrm{p}} \cdot \varrho_{\mathrm{L}} \cdot u}{\eta} \tag{3}$$

where ϱ_L is the density of the liquid and η is its dynamic viscosity. The superficial velocity u is the product of the fluid velocity v through the capillaries of the bed and the porosity ε of the bed (ε = volume of capillaries/total volume of system).

If the capillaries or pores are taken as having the shape of circular tubes, the fluid velocity in laminar flow is given by Poiseuille's law

$$\bar{v} = \frac{r^2 \Delta p}{8 \eta H} \tag{4}$$

where $\bar{v}$ is the average flow velocity in the tubes and r is the tube radius. It should be emphasized here that the flow velocity $\bar{v}$ (in other words, the flow rate of the fluid per unit cross-sectional area of the tubes) is directly proportional to the pressure drop Δp. Since most practical filters operate in the laminar range, a linear correlation should be expected between the effluent flow rate and the pressure drop.

Because the capillaries in porous media may be more complicated in shape than a circular tube, the hydraulic radius is introduced. This quantity is defined as the ratio of the flow cross section of the capillaries to the wetted perimeter, or equivalently as the porosity of the system divided by the wetted area. Flow in the capillaries is described by the Kozeny equation [13]

$$u = \frac{\varepsilon^3}{K \cdot \eta (1-\varepsilon)^2 \cdot a_V^2} \cdot \frac{\Delta p}{H} \tag{5}$$

where K is the Kozeny constant, which has the value 2 for circular tubes. It is often assumed that K is a material constant and that its value lies between 4 and 5 for granular beds, as found in many experiments. In fact, K depends on the porosity itself, and for granular beds with porosities between 0.32 and 0.47 it has values from 3.5–5.5 [2]. In porous media consisting of fibers, where the porosity is much higher, the Kozeny constant is much greater than 5.

Sometimes it is useful to replace the specific surface area a_V in Equation (5) with the particle diameter. This is justified only in the case of monodisperse beds. From Equation (2), the modified Kozeny equation (with K taken equal to 25/6) becomes [1]

$$u = \frac{\varepsilon^3 d_p^2}{150 (1-\varepsilon)^2} \cdot \frac{\Delta p}{H} \tag{6}$$

Equation (6) is often preferred in filtration when the cake is made up of particles of a tighter size range.

For granular beds the Kozeny constant K also depends on the grain shape. It is therefore not surprising that the Kozeny constant varies between 1 and 10. Thus, the product of the Kozeny constant and the porosity in Equations (5) and (6) should be considered as a mathematical model for the flow function in porous systems. Modified models have been proposed and compared with experimental data [14], [15]. According to those, the Kozeny porosity function agrees widely in porosity about 0.4.

With the Kozeny model, Equation (6), the friction factor in Equation (1) becomes

$$f = \frac{(1-\varepsilon)^2}{\varepsilon^3} \frac{75}{Re} \tag{7}$$

The experimental determination of the porosity and the Kozeny constant often requires a great many measurements. Dimensionless groups [5] can be used to generalize the experimental data. Fluid flow through porous media is described by

$$Re \cdot Eu = 32 \frac{H}{d_0} \tag{8}$$

where the Reynolds number Re and the Euler number $Eu = \Delta p/(\varrho v^2)$ are referred to flow through cylindrical capillaries with diameter d_0. This is a modified form of Poiseuille's law. For any porous medium, Equation (8) can be rewritten with modified groups Re' and Eu', which are referred to the superficial velocity u and the particle diameter d_p:

$$Re' \cdot Eu' = k' \frac{H}{d_p} \tag{9}$$

where k' depends on the shape of the grains and the porosity of the medium. The factor k' is an important material property and is commonly found by laboratory tests. With the Kozeny model, Equation (6), the factor k′ becomes

$$k' = 36 \frac{(1-\varepsilon)^2}{\varepsilon^3} K \tag{10}$$

The use of dimensionless groups obviously presupposes that the geometries of all particles in the medium are known. In practice, this can be more complex than measuring the filtrate flow through media.

DARCY introduced the permeability coefficient, the rate of fluid flow through a unit volume of a granular bed under a unit differential pressure, with no allowance for viscosity [16]. In a modified form now widely used, the permeability coefficient is replaced by its reciprocal, the flow resistance α, so that

$$u = \frac{1}{\alpha \eta} \cdot \frac{\Delta p}{H} \qquad (11)$$

Equation (11) is the modified Darcy equation, on which filtration theory is most often based. The flow resistance α is related to the Kozeny model as

$$\alpha = \frac{K(1-\varepsilon)^2 \cdot a_V^2}{\varepsilon^3} \qquad (12)$$

All the correlations mentioned before are applicable to a medium provided the porosity is rigorously constant during service. This may be true of some few practical processes, for example where the porous medium consists of fairly coarse material such as salts, sand, or rigid plastics. In most cases, the voids in the medium (especially in cake filtration) become smaller through consolidation of the bed. The theory of filtration with a compressible cake is discussed in the section on cake filtration (p. 10-5).

Another effect is the decrease in pore volume due to solids deposited during filtration. Some particles seal the smaller pores, and the medium may become blocked. The blocking varies with the mechanism of particle deposition [9], [12].

Complete blocking occurs when all particles down to the last caused pore clogging, i.e., all particles in the influent are retained in the pores of the filter medium. The *standard blocking* is characterized by a continuous decrease in pore volume. Here the influent particles are deposited on the walls of the pores. The rate of fluid flow through the medium decreases with the cross-sectional area of the pores. The decrease in pore volume is assumed to equal the decrease in cross section. Finally, intermediate situations are observed, in which one solids deposition mechanism or the other dominates.

Blocking filtration at constant pressure can be described by

$$\frac{d^2 t}{dV^2} \sim \left(\frac{dt}{dV}\right)^n \qquad (13)$$

where t is the filtration time and V is the effluent volume. The superscript n varies with the type of blocking: $n = 1$ for intermediate blocking, 3/2 for standard blocking (deep-bed filtration), and 2 for complete blocking. When blocking filtration further depends on time, the superscript becomes greater than 2 [12]. For cake filtration, $n = 0$, so that Equation (13) becomes

$$\frac{dt}{dV} \sim V \qquad (14)$$

3. Cake Filtration

3.1. Description of Process

When the slurry is fed to a clean filter medium, the residue held up by the medium forms an increasing layer of solids, the filter cake, which then takes over the separating function of the filter medium. As the cake thickness increases, so does the resistance to flow; at a given filtration pressure, the quantity of effluent per unit time declines. Figure 2 shows filtrate volume versus time in cake filtration, for two different filtration pressures p_1 and p_2. In order to keep effluent flow high, the feed would have to be cut off as quickly as possible. A number of these relatively thin filter cakes will take several times longer to be removed than will one thicker cake. The filtration time before cake removal (service time) depends on the quantity and concentration of the

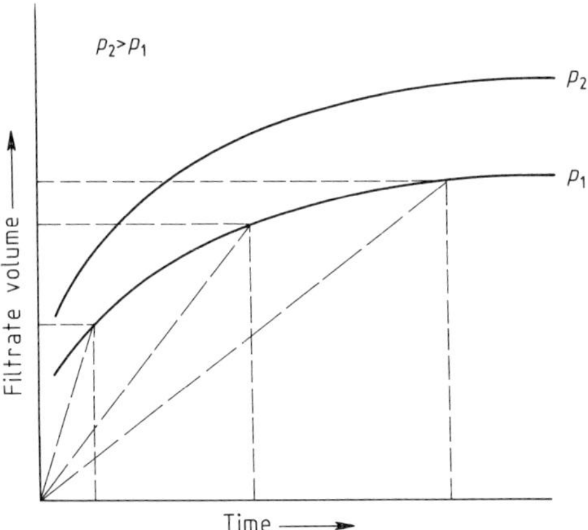

Figure 2. Filtrate volume vs. filtration time

influent and the properties of the filter cake. If the cake is thick, mechanical and hydrodynamic loads often compress it until the specific flow resistance increases and the service time is impaired. The entrapment of fine particles within the filter medium and within the existing cake also reduces the effluent rate and shortens the service time.

Often, however, the decrease of the pore and channel volumes in the filter medium and cake due to penetration of fine-grained fractions is welcome, for example if a slurry of particles of a relatively broad size spectrum is to be filtered at a low pressure drop with a filter medium that is widely spaced and has a correspondingly low flow resistance. But the advantage of low cost for establishing the filtration pressure must then be balanced against the necessity of accepting solids in the filtrate at the start of filtration. The dirty or raw filtrate can be recycled to the product stream.

The appearance of solids in the effluent depends heavily on the condition of the filter medium; a new filter cloth is, as a rule, more permeable than one that has been washed several times but still contains some included solid particles.

After the filter cake has been removed and the filter cloth washed if necessary, filtration can be resumed. The filter cake is sometimes washed as well (see Section 3.7, p. **10**-16) or treated in some other way, for example dewatered and dried. Like cake filtration, these are also batch operations. In continuous systems such as endless belts, disks, pans, or drums, they can be made quasi-continuous with relatively little expenditure (see Chapter 8, p. **10**-25). On the other hand, the batch process may have considerable advantages, since filter performance can be improved by, for example, the application of a precoat on the filter medium (see Chapter 11, p. **10**-58).

Many slurries containing coarse solid fractions or materials of high specific gravity show a marked tendency to settling. With horizontal filter surfaces, this tendency can be utilized to form a coarse-grained bottom cake layer with good permeability; for example, initial feed can be drawn from the deeper portion of a slurry tank.

The steps in cake filtration can be summarized in the following filtration cycle:

1) Filtration proper, which ends after formation of a filter cake from a few millimeters to several centimeters in thickness.
2) Dewatering of the filter cake by displacement of residual moisture with air, gas, steam, or immiscible liquids.
3) Washing of the filter cake. This step may replace dewatering or may follow it. Preferably done with mother liquors or solvents, it has the purpose of removing adhering liquid.
4) Another dewatering step to reduce the amount of wash liquor remaining in the voids of the filter cake.
5) Expression of the residual moisture, an operation that has gained in importance with compressible filter cakes.
6) Removal of the filter cake by dropping, scraping, backblowing, or some other method.
7) Cleaning of the filter medium to remove trapped particles by washing or rinsing, at times with water at pressures of up to 10 MPa.

Depending on quality requirements and economics, a filtration cycle includes all or only some of these steps. Further operations may be considered, for example when the solid is preflocculated in order to improve its filtration properties, or when a precoat is deposited on the filter medium so that even slurries containing very fine particles can be filtered, or when the filter cake must be lifted pneumatically for removal. These special practices are discussed elsewhere in this article (p. **10**-25, **10**-58).

Multiple washing steps may be advantageous on occasion; they can be run in countercurrent to concentrate the wash liquor.

The several stages in a filtration cycle determine the specific output of a filter. Cake removal is important in setting the cycle length; it is rather variable, depending on the process and the filter design. For example, cake removal from a pressure vessel, even by ingenious methods, requires more time than cake removal from an open, unpressurized device. As the time needed for dewatering, washing and removal of the filter cake grows longer, an effort is generally made to extend the service time as well.

The mean output of a filter is calculated as

$$\dot{M}_s = \frac{M_s}{t_c} \tag{15}$$

where M_S is the mass of dry solids collected in the length t_c of one filtration cycle. For the cycle described above,

$$t_c = t_f + t_{d_1} + t_w + t_{d_2} + t_r \tag{16}$$

where t_f is the filtration time, t_{d_1} is the time for the first dewatering stage, t_w is the washing time, t_{d_2} is the time for the second dewatering stage, and t_r is the time for removal of the filter cake.

Calculations on this basis are often simplified by stating the times as fractions of the cycle length or filtration time. In rotary vacuum filters (see p. **10**-37), which have a relatively short service time because the filter cake is not very thick (usually 6–10 mm), cake removal accounts for some 25% of the cycle time.

If the filtrate, as a mother liquor, is valuable along with the filter cake, liquid produced during dewatering is collected separately, as is the wash liquor from the next operation.

3.2. Categories of Cake Filtration

The driving force in cake filtration can be pressure (applied slurry pressure or hydrostatic pressure) or suction (vacuum applied to the filtrate). In special cases, capillary pressures and electrostatic forces can also be used.

The most important parameters are the differential pressure across the filtration operation, the viscosity of the liquid, the permeability of the medium, and the filter loading (feed rate). At low cake resistance, satisfactory output can generally be achieved at relatively low filtration pressure produced by vacuum pumps. Higher pressure is recommended chiefly for relatively finely-woven filter cloths and high cake resistances.

Comprehensive and universally valid rules for selecting optimal filtration conditions cannot be given. In general, pressure filtration is employed with slurries that form fine-grained, slimy, compact cakes. Vacuum filtration is suitable for slurries of moderate to good filterability, which form loose cakes, and for cases where filter aids are used (see p. **10**-58).

Pressure Filtration. In pressure filtration, the slurry is at the desired pressure and fed to the filter by appropriate slurry pumps (centrifugal, Moyno single-rotor screw, and diaphragm pumps). The filter is an enclosed, pressure-tight device. A medium such as a pressurized gas (air) is occasionally used to set up the desired filtration pressure. Differences in process conditions have led to two designs with corresponding modes of operation.

In the first type, filter elements with filter media are assembled into *compact systems* in pressure-tight casings. After the filtration step is completed and it is time to initiate dewatering, the remaining slurry is drained from the vessel, the pressure is let down, and the casing is opened so that the filter cake can be removed. Special apparatus or practices such as back-blowing are often necessary for cake removal. Washing before cake removal is possible only after all the influent has been drained from the pressure casing. In general, therefore, this filter design is economically suitable only for operations with few or long filtration cycles, and thus chiefly for dilute slurries.

The second design type has *pressure chambers* that are opened after filtration (or after washing and dewatering, if practiced) for cake discharge. In a continuous version, the cake is conveyed through sealing devices into unpressurized zones. The filter cake can be washed and dewatered more efficiently and economically. Filters of this type do require somewhat more expensive equipment and practices. Above all, however, the cake discharge is much simpler than in the first type. Slurries containing finer solids at higher concentration are preferred.

Pressure filters operate at up to 3 MPa. The generally higher equipment and operating costs limit applications to those cases where vacuum filters meet with difficulties.

The relatively high pressure drop nearly always results in compaction of the filter cake, whereas cake compression is hardly a factor in vacuum filters operating at a maximum pressure difference of 50 kPa. Compaction of the cake does reduce the cake moisture. But the residual moisture after dewatering depends also on the fineness of the filter medium. A widely spaced filter cloth will, to be sure, give a lower residual moisture content, but plugging of the cloth is more of a problem in high-pressure filtration. In most cases, the optimal ratio of pressure to filter cloth fineness can be established only by testing or on the basis of experience. Blocking of the filter cloth can be diminished by bringing the filtration pressure up to the design value slowly; this mode of operation also pertains when the filter is fed directly from a centrifugal pump with no intermediate pressure-compensating vessel.

Vacuum Filtration. In vacuum filtration suction is applied to the filtrate with liquid-piston type rotary blowers; in smaller plants, water-jet pumps (aspirators) are used. Because of the usually slight pressure drop, up to a maximum of 50 kPa, the term vacuum filtration is not wholly correct, and the operation is sometimes called *suction filtration*. The feed to the filter surface, as well as washing and dewatering, can be set up in an open and readily accessible manner, so that filtration and service is easier to carry out. A special advantage arises from the possibility of using relatively simple means to remove the filter cake from the cloth. As a disadvantage, the pressure drop between the influent and the filtrate-collecting space is limited to less than 100 kPa and is still further restricted by increased evaporation of the filtrate.

Because of the simplicity of the process and of the design, vacuum filters have found wide use, from the vacuum nutsches commonly used in laboratories up to drum and disk filters, which also lend themselves to continuous filtration. The relatively low filtration pressure can result in low capacities per filter area, particularly when fine suspensions are filtered, even if cake compaction is largely eliminated.

Hydrostatic Filtration. In gravity or hydrostatic filtration, the cake plus filter medium must offer relatively little resistance to flow if outputs interesting for industrial use are to be achieved. These conditions can be expected only with suspensions containing coarse particles and with a relatively widely spaced filter medium. The process entails the lowest equipment cost. Sedimentation is always present in hydrostatic filtration. It does not cause major problems because the suspension is always fed on top of the filter medium, anyway.

3.3. Cake Formation and Structure

Economical cake filtration requires that the cake has an appropriate permeability from the start to the end of filtration. Above all, the solids retained at the start of the process must not plug the filter cloth; solids with a high proportion of very fine particles show a particular tendency to plug the medium.

The resistance α of the filter cake (Eq. 11), must not exceed some value to be established. The usual values of α_S are between 10^8 and 10^{14} m/kg, as Table 1 shows. For comparison, flow rates for geometrically well-defined granular (particulate) beds are given at the end of the table; these values give some idea of appropriate particle sizes of the bed.

Table 1. Filter cake resistance of various filtration media

Material	Resistance $\alpha_S \cdot 10^{-10}$, m/kg	Pressure, MPa
Aluminum hydroxide	2200	0.17
Calcite	5–10	0.01–7.0
Clay-filter aid (50 : 50)	0.1–0.4	0.01
Coal sludge	0.4–0.8	
Iron hydroxide	1500	0.17
Iron oxide	85	
Kaolinite	40–87	0.01–7.0
Kaolinite korean	230	
Kieselgur	12	0.17
Magnesium hydroxdye	1.4–3800	
Solca-floc	0.006–0.1	0.01–7.0
Talc	5–35	0.01–7.0
Titanium dioxide	18	0.01–7.0
Waste water sludge	$(1-3) \cdot 10^4$	
Packages		
Average particle size 10 μm	0.065	
1 μm	6.5	
($\varrho = 2650$ kg/m^3; $\varepsilon = 0.4$)		

In order to simplify the theoretical derivations, the flow resistance per unit mass α_S (in m/kg) is often used in the literature:

$$\alpha_S = \frac{\alpha}{\varrho_s(1-\varepsilon_c)} \tag{17}$$

where ϱ_S is the density of the solids and ε_c is the porosity of the dry cake.

Higher resistances call for special practices such as the use of filter aids, p. **10**-58, and flocculation (→ Flocculants, **A11**, p. 251) or the use of a different separation process (→ 9. Solid – Liquid Separation, Introduction).

Because the permeability decreases with the size of the solid particles (Eqs. 2 and 6) the lower limit of applicability of cake filtration corresponds to mean particle sizes in the range of a few μm. Low contents of finer particles cause relatively little trouble, even if the filterability declines markedly. Rules for the expected cake structure and permeability are always of restricted validity. The extreme variability of the retained solids as to both cake structure and cake stability make it even harder to predict filterability.

Factors with significant influence on cake structure include not only the particle size but also the particle-size distribution, the slurry concentration, the feed rate, and the direction of flow against the filter surface. Nor is the effect of the particle-size spectrum clearly recognizable: The specific surface in the Kozeny flow formula (Eq. 5) also depends on the size distribution, as the conversion formulas (Eqs. 2 and 4) show. Cake formation is strongly influenced by the content of fines in the solids. Fine particles incident on the surface of the filter cake are often entrained by the liquid and deposited inside the cake. This fact also explains the observation that cake resistance increases with decreasing slurry concentration. Further movements of the smallest fractions during filtration are discussed in the next paragraph on cake compression.

Such structural variabilities naturally diminish cake porosity and are more pronounced, the less stable the cake is. *Cake stability* depends, in particular, on the nature of the particles. Crystals and other brittle particles usually form relatively stiff cakes, whose permeability is not much affected by frictional drag of the liquid. This family also includes beds of angular particles, which reinforce one another in such a way that motion inside the bed is virtually impossible. Such cakes have a relatively high initial permeability, but the particles can become interlocked and form relatively dense cakes, especially at low influent concentrations.

Another factor is the surface structure of the particles. For given size relationships, smooth surfaces generally result in denser cakes than rough, uneven surfaces. Finally, the sphericity of the particles also affects cake formation. Rod-shaped particles form relatively loose cakes of high porosity, while flake-shaped particles behave in the opposite way.

Soft, deformable particles, such as those of organic substances or metal hydroxides, exhibit different properties. As do aggregates of finer primary particles, they allow *deformation* and *cake compaction* (compression). The stresses in the cake structure which result from the transmission of hydrodynamical and mechanical forces via the particle contact points, are not involved at pressures below a certain minimum. Cake compression in pressure filtration usually occurs only above a filtration pressure of 0.1 MPa. The reduction in cake permeability associated with this effect may decrease the filter capacity so much that it is desirable to stop the filtration operation after relatively short time. Compression is also promoted when pores and interstices produced at the start of filtration by bridging become filled by the deposition of finer particles and collapse in the course of filtration. This process, referred to as *consolidation*, is accompanied by deformation and disintegration in the case of aggregates and bulky flocs. Further compression may result from expression or filtration in a press, leading to substantial losses of moisture.

The compressibility of a filter cake is expressed as the ratio of the volume fraction of solids (solidosity) at the moment of cake formation (between about 0.02 and 0.1 for bulky flocs) to the volume fraction of solids in the compacted cake (about 0.65 as a rule). Irregularly-shaped particles give higher compressibilities than spherical ones, which can hardly be compressed further at grain sizes over 20 μm.

Filter aids affect filterability mainly in the case of these soft, often slimy and adhesive solids, which otherwise do not form permeable cakes.

Filtration characteristics are predicted on the basis of either laboratory studies or experiments in small pilot units. In rare cases, there is no substitute for in-plant trials with industrial filters. Results obtained on test filters in the laboratory (see p. **10**-23) can usually be scaled up only on the basis of empirical findings. Pilot experiments offer more confidence. However, sometimes several types of filters must be tested in order to find the optimal conditions, with the result that the trials quickly become unaffordable. For this reason, it may be useful to supplement field trials with information from laboratory tests, theoretical calculations, and pilot filtration experiments.

3.4. Theory of Cake Filtration

The theory of cake filtration is based on the classical laws of fluid flow through porous media (p. **10**-3). For most cases, the following assumptions are made:

1) Filtration is steady-state, that is, the volumetric flow rates of the volumes and masses are independent of time.
2) The slurry composition does not vary.
3) All particles are retained on the filter medium or deposited solids. That is, no particles get into the interstices in the cake, so that its permeability is independent of time.
4) Fluid flow is laminar and one-dimensional.
5) The characteristics of the filter medium do not change during the filtration period.
6) The effect of gravity can be neglected.

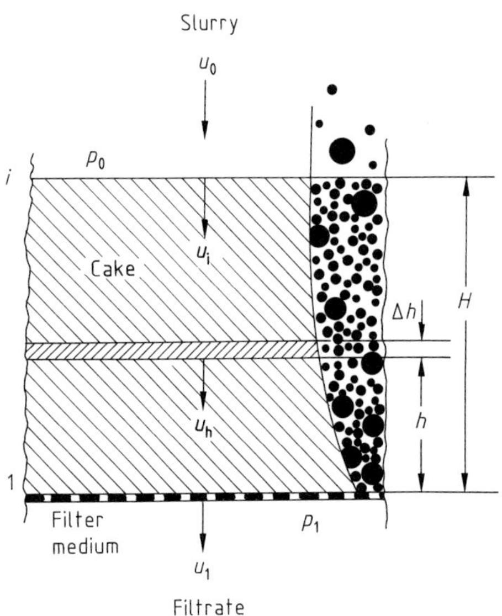

Figure 3. Schematic of cake filtration

Because of differences in the flow conditions in cake filtration, a variety of differential equations are obtained when the internal flow mechanism is rigorously analyzed.

These, have to do mostly with the case of cake filtration with short times and higher slurry concentrations, which occasionally comes up in vacuum filters (drum and disk filters).

In cake filtration theory, the equations of fluid flow through porous media are supplemented by the fact that the influent contains solids (Fig. 3). It is then assumed that the solids deposited on the filter medium form a layer whose structure is consistent and unchanging throughout the filtration process. Further, the filter medium is supposed to have no effect on cake formation.

Fluid flow through porous media is commonly described by Darcy's law (p. **10**-5), which takes on a more general form for capillary flow:

$$u_h = \frac{k_h}{\eta} \cdot \frac{\Delta p}{\Delta h} \tag{18}$$

where

$$u_h = \frac{dV_L}{A\,dt} \tag{19}$$

is the volume flow rate of liquid with V_L being the liquid volume and t the time through a unit cross-sectional area A of the deposited solids plus liquid; k_h is the permeability of the porous medium; η is the viscosity of the liquid; Δp is the pressure drop; and Δh is the thickness of the cake. Since in any filtration the pressure drop dominates the effect of gravity, neglecting gravity does not bring about an incorrect result.

With the specific volume of filtrate (volume per unit cross-sectional area) V_{LF} the mass balance for filter cake formation, including the liquid remaining in the interstices in the cake, yields

$$\frac{\Delta h}{V_{LF}} = \frac{1}{\dfrac{c_{vc}}{c_{v0}} - 1} = \frac{1}{\dfrac{1-\bar{\varepsilon}_c}{c_{v0}} - 1} = \frac{\varrho_L}{\varrho_S}\frac{c_{m0}}{c_{vc} - c_0} = K_m \tag{20}$$

where $\bar{\varepsilon}_c$ is the average porosity of the dry cake, c_{vc} and c_{v0} are the volume fractions of solids in the cake and slurry respectively, c_{m0} is the mass fraction of solids in the slurry, ϱ_S and ϱ_L are the densities of the dry solids and the liquid, respectively, and K_m is a cake formation factor. Average values are used in this equation because the porosity in a compressed cake varies locally.

Substituting for Δh in Equation (18) and using the specific filtrate volume V_{LF} of Equation (19) gives

$$u_h = \frac{dV_{LF}}{dt} = \frac{k_h}{\eta} \frac{\Delta p}{V_{LF} \cdot K_m} \tag{21}$$

It is common in filtration theory to replace the permeability with the average flow resistance $\bar{\alpha}$, by analogy with the average cake porosity; then $\bar{\alpha}$ is equated with $1/\bar{k}_h$.

Then, the first form of the basic formula for cake filtration is

$$\frac{dt}{dV_{LF}} = \frac{\eta\,\bar{\alpha}\,K_m}{\Delta p} \cdot V_{LF} \tag{22}$$

Integration of Equation (22) yields

$$t = \frac{\eta\,\alpha\,K_m}{2\,\Delta p} V_{LF}^2 \tag{23}$$

Equation (23) describes the *filtration parabola*, which is often used in practice to verify the performance of a filter.

Interpretation of Equation (21) shows that the instantaneous specific volumetric flow rate of filtrate u_h (the superficial velocity of filtrate) is inversely proportional to the volume of filtrate V_{LF}. Consequently, the maximum quantity of filtrate is obtained at the beginning of the filtration period.

The filtrate leaving the cake has to pass the filter medium, whose resistance adds to that of the cake. If it is much less than the cake resistance, the filtration parabola, Equation (23), is preferably used in practice, since it obviates the determination of the usually unknown filter medium permeability. The filtration parabola gives satisfactory values, especially when the cake has a tendency to bridge. For example, unused filter cloths of needled felt may have resistances ranging from $(2-7) \cdot 10^7\ \mathrm{m}^{-1}$ (corresponding to a mass per area of 300–600 g/m²). The cake resistance is commonly several orders of magnitude greater (Table 1). When this is not the case, for example with comparatively thin cakes or extremely tight filter media, a term for the pressure drop over the medium is added into Equation (22):

$$\frac{\mathrm{d}t}{\mathrm{d}V_{\mathrm{LF}}} = \frac{\eta\,(\bar{\alpha} K_{\mathrm{m}} V_{\mathrm{LF}} + \beta)}{\Delta p} \tag{24}$$

where β is the specific resistance of the filter medium.

Integration yields

$$t = \frac{\eta \bar{\alpha} K_{\mathrm{m}}}{2\Delta p} V_{\mathrm{LF}}^2 + \frac{\eta \cdot \beta}{\Delta p} V_{\mathrm{LF}} \tag{25}$$

In linearized form—long preferred in practice because it makes the evaluation of experimental data easier— the formula becomes

$$\frac{t}{V_{\mathrm{LF}}} = \frac{\eta \bar{\alpha} K_{\mathrm{m}}}{2\Delta p} V_{\mathrm{LF}} + \frac{\eta \cdot \beta}{\Delta p} \tag{26}$$

Figure 4 is a plot of this relation.

Cake resistance $\bar{\alpha}$ and medium resistance β must be used with caution. As shown in Section 3.3 the two are not really considered constant factors.

Hence three categories of cake filtration are available for practical use:

1) *Constant-pressure filtration*, in which the pressure p is kept at the set level by a pump or compressed air. As the cake forms, the effective pressure drop across the cake increases from zero at the start of the filtration period, and the filtrate rate decreases.
2) *Constant-rate filtration*, in which the derivative $\mathrm{d}V_{\mathrm{LF}}/\mathrm{d}t$ is held constant. A pump with constant volume discharge is needed for this process.
3) *Variable-pressure filtration*, in which the effective filtration pressure depends on the characteristics of the (usually centrifugal) pump.

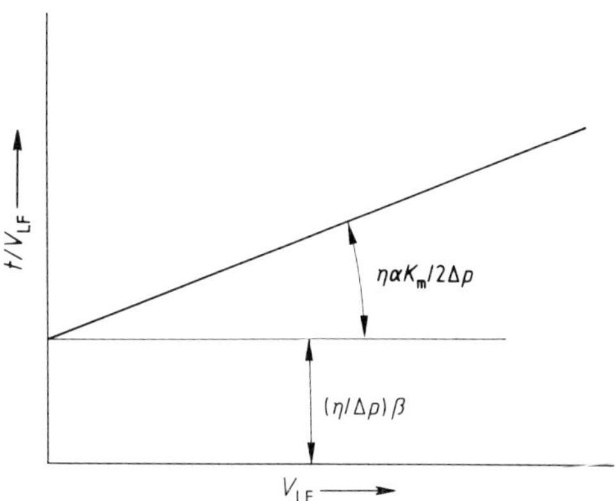

Figure 4. Plot for constant pressure filtration

Constant-pressure filtration is described by the "filtration parabolas," (Eqs. 23 and 25.). These are probably the most appropriate equations in cake filtration practice and may be permissible when cake compression is negligible. Filtration of crystalline products on vacuum filters may fall into this category. Constant-rate filtration can be regarded as a special mode of pressure filtration. Variable-pressure filtration occurs often in pressure filtration with centrifugal pumps. In many cases this regime goes over to constant pressure after some initial period.

Departures from the parabolic curve are easily recognized if the filtrate rate ($\eta\,\alpha\,K_{\mathrm{m}}\,V_{\mathrm{LF}}^2/t$) is plotted against the hydraulic pressure across the cake Δp (curves b and c in Figure 5). The reason for such departures is the variation of the cake resistance during the filtration period.

Filtration with Compressible Cakes. Cake filtration with a compressible cake involves the consolidation and compression of the deposited solids.

Initially, a cake is formed with the deposited particles being agglomerated and bridging into a more-or-less porous layer of comparatively great instability. As filtrate flows through the channels in the cake in the direction of the pressure gradient, friction (drag) at the particle surfaces leads to migration of the particles. The drag is transmitted through points of contact between the

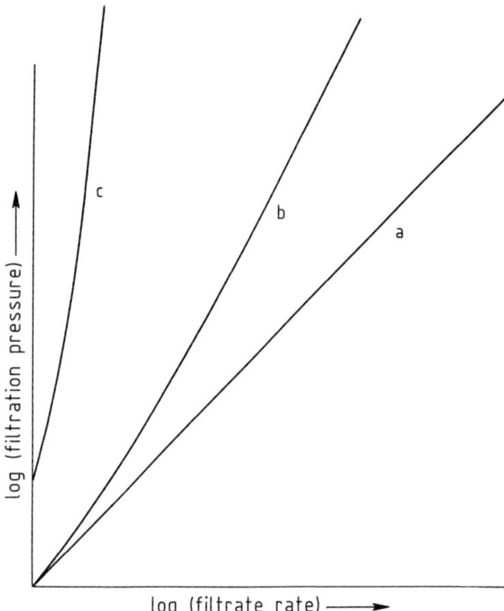

Figure 5. Filtrate rates of compressible cakes vs. hydraulic pressure across the cake [4]
a) Incompressible cake; b) Moderately compressible cake; c) Highly flocculated material

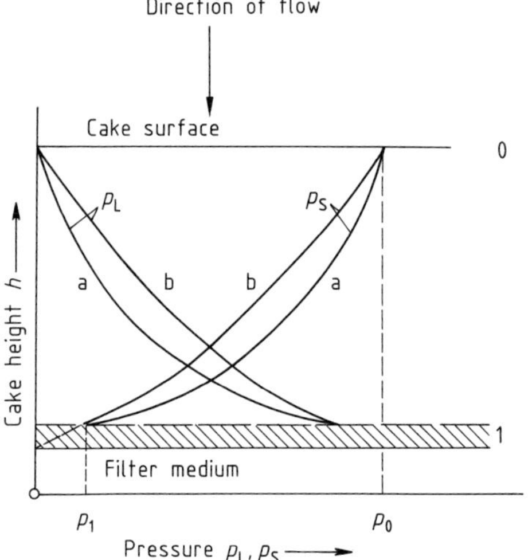

Figure 6. Relationsship between hydraulic pressure p_L and the effective pressure of solids p_S and the cake height. Curve (a) represents a more compressible cake than curve (b)

particles supported on the filter medium. More details are given in Section 3.3.

The force balance is described as

$$p_L + p_S = 0 \tag{27}$$

where p_L is the hydraulic pressure and p_S is the effective pressure over a cross section of the cake. The relationship between hydraulic pressure and the effective pressure and the cake thickness is illustrated in Figure 6.

A rigorous analysis of compressible cake filtration was developed by TILLER and SHIRATO [17], who allowed for variation of the cake structure. The liquid component of the slurry flows through the cake at a velocity $v = \dot{V}_L / A_p$, where A_p is the cross-sectional area of the pores. The quantity of this liquid at the time of cake formation is divided into filtrate flowing through the rest of the cake and residual moisture remaining in its interstices. When the cake is consolidated or compressed by drag, the volume of the interstices is reduced, so that the liquid in them is expressed and added to the effluent.

To eliminate the migration of solids within the cake, the calculated material balance over a cross section was modified by including the mass or volume of deposited solids per unit area.

If the hydraulic pressure does not decrease linearly through a compressible filter cake, as shown in Figure 6, the pressure difference Δp in Equation (22) must be replaced by the derivative $\mathrm{d}p_L/\mathrm{d}h$, where h is the cake coordinate (Fig. 3). With the differential $\mathrm{d}m_S$ of the solids mass per unit area or $\mathrm{d}V_S$ of the solids volume per unit area, the formula becomes

$$\begin{aligned} \mathrm{d}m_S &= \varrho_S (1 - \varepsilon_c)\, \mathrm{d}h \\ \mathrm{d}V_S &= (1 - \varepsilon_c)\, \mathrm{d}h \end{aligned} \tag{28}$$

where ε_c is the porosity of the dry cake. The hydraulic pressure p_L and the pressure acting on the solids p_S vary with the cake thickness, so that with Equation (27) the following is obtained:

$$-\frac{\mathrm{d}p_S}{\mathrm{d}V_S} = \frac{\mathrm{d}p_L}{\mathrm{d}V_S} = \eta \alpha' u_h \tag{29}$$

where

$$\alpha' = \alpha_S \cdot \varrho_S \text{ (see Eq. 17)} \tag{30}$$

If p_1 is the pressure upstream of the filter medium and downstream of the filter cake, the pressure dependence of the cake resistance is described by

$$\eta u \int_0^{V_S} \mathrm{d}V_S = -\int_0^{p-p_1} \frac{\mathrm{d}p_S}{\alpha'} \tag{31}$$

The calculation can be simplified by using the average resistance

$$\frac{1}{\bar{\alpha}'} = \frac{1}{p - p_1} \int_0^{p-p_1} \frac{\mathrm{d}p_S}{\alpha'} \quad (32)$$

In general, the average cake resistance should be found by regression analysis or plotting the observed values from experimental or equipment tests, as shown in Figure 4.

With Equation (32) and with Equation (29) for flow through the medium, the following is obtained for the effluent rate downstream of the cake:

$$u_1 = \frac{\Delta p}{\eta (\bar{\alpha}' V_S + \beta)} \quad (33)$$

where

$$V_S \bar{\alpha}' = V_{LF} K_m \bar{\alpha} \quad (34)$$

The solution of Equation (33) is restricted by the variation of the two factors, the cake resistance and the pressure drop during the filtration period. TILLER [4] offers a theoretical solution based on the empirical function

$$\alpha' = \alpha'_0 \left(1 + \frac{p_S}{p_a}\right)^n \quad (35)$$

where α'_0, p_a and n are constants. Figure 5 is a plot for several types of compressible cakes.

For practical use, either a more accurate determination of the filtration process may be dispensed with, or the $\bar{\alpha}$ values may be found under test conditions similar to those of the operation. It turns out that the rigorous application of the theory can be skipped particularly for vacuum filtration when errors of a few percent relative to the desired throughput are acceptable.

A simplified solution is based on a constant average cake resistance and thus a constant cake porosity. The specific cake formation can be written

$$V_S = V_{LF} W \quad (36)$$

where

$$W = \frac{1 - \varepsilon_c}{\frac{1 - \varepsilon_c}{c_v} - 1} \quad (37)$$

Using the easily measured maximum residual moisture content w_0 and the slurry concentration c, Equation (37) becomes

$$W = \frac{c}{\varrho_S - \frac{c(1 + w_0)}{w_0}} \quad (38)$$

The substitution into Equation (33) leads to

$$\bar{\alpha}' W \frac{V_{LF}^2}{2} + \beta V_{LF} = \frac{\Delta p}{\eta} t \quad (39)$$

The cake thickness is often of interest not only in the assessment of filter capacity, but also in filter selection. The cake thickness is calculated with Equation (39), with the effluent volume per unit area being substituted from Equation (20):

$$h = -\frac{\beta}{\bar{\alpha}'(1 - \varepsilon_c)} + \left[\left(\frac{\beta}{\bar{\alpha}'(1 - \varepsilon_c)}\right)^2 + \frac{2 \Delta p \, W t}{\bar{\alpha}'(1 - \varepsilon_c)^2 \eta}\right]^{0.5} \quad (40)$$

Apparently, the cake thickness increases as the square root of time during filtration. Thus the rate of cake formation decreases with $1/t^{0.5}$.

On the other hand, the two typical relations for cake formation strongly depend on the cake structure.

This fact is the basis of the *standard cake formation time* (SCFT), introduced by PURCHAS as a general term characterizing the filtration of a given suspension [6]. A similar characterization has been in practical use for some time.

An alternative way of characterizing filtration is offered by the cake formation rate proposed by DAHLSTROM [4]. The rate is described by a modified filtration equation with no allowance for the resistance of the filter medium (Eq. 23):

$$z = K \left(\frac{w \Delta p}{\alpha \eta t_f}\right)^r \quad (41)$$

where z is the cake formation rate (i.e., the mass of dry solids per unit area and unit time), w is the mass of dry cake solids per unit volume of filtrate, t_f the filtration time, and $r = 0.5-0.65$.

3.5. Applicability of Analytical Results

In the assessment and evaluation of cake filtration, it is scarcely practical to use all the results from Section 3.4. As a rule, the treatment must be restricted to the most important of these.

Cake filtration is, first of all, a separation process for suspensions of relatively high concentration, with a solids content of at least 3–5 wt %. For economic reasons, a filtrate volume rate of 200 L $m^{-2}h^{-1}$ is required for vacuum filters and at least 40 L $m^{-2}h^{-1}$ for pressure filters. Applications include raw materials recovery, the production of all kinds of chemicals, and to an increasing degree the treatment and disposal of sewage sludges.

The upstream and downstream operations are also important for filtration. These are determined by the properties of the influent and the requirements imposed on separation quality. For example, if the feed has too low a concentration, it can be prethickened in a settler or hydrocyclone. An unfavorable particle size or size distribution can often be remedied by changing the process parameters in upstream operations such as crystallization or precipitation. In the case of extremely fine solids, it is possible to obtain aggregates and flocs that allow the use of cake filtration by addition of flocculants. Care must be taken, however, that the particles produced have the requisite properties for cake formation (see Section 3.3).

If the influent is at a high temperature, it should be kept as close as possible to this temperature upstream of the filter. Since the viscosity of the liquid phase generally decreases with rising temperature, filter performance can sometimes be improved considerably in this way.

3.6. Cake Moisture Content

The cake moisture content is almost always a factor of major importance. For example, satisfactory pelleting of iron concentrates depends on the moisture not exceeding a certain limit. What is more, if materials must be conveyed, corresponding costs increase with the moisture content. Finally, many processes require dry materials for chemical processing and for packaging.

In view of these facts, most filtrations are followed by operations to reduce the moisture content. These are commonly referred to as dewatering, even if the liquid to be removed (the moisture) is not water. Mechanical dewatering processes are preferred, because they cost much less than thermal drying. On the other hand, mechanical dewatering does not completely separate the two phases (→ 4. Drying of Solid Materials).

The first factor determining the residual moisture is the pore volume of the cake. If the feed rate is assumed to be at the greatest possible filter capacity, the voids and pores in the filter cake will be filled with liquid immediately after filtration is complete. The saturation moisture referred to the moist cake is

$$(w_f)_{max} = \frac{\varrho_L \varepsilon_c}{\varrho_S - \varepsilon_c (\varrho_S - \varrho_L)} \tag{42}$$

The degree of saturation is given by

$$S = \frac{w_f}{(w_f)_{max}} \tag{43}$$

where w_f is the residual moisture per unit of moist cake mass.

The residual moisture can be reduced by simply draining off the liquid by gravity, blowing with gases (usually air), and expression.

Each of these dewatering methods is associated with an equilibrium degree of saturation, which does not decline even after a long time.

The liquid contained in the cake is made up of three components:

1) *Adhesive liquid* is bound to the surfaces of the particles by adsorption and adhesion. The quantity depends only on the wettability of the solid by the liquid. This moisture fills only that part of the bed that is not affected by the capillary liquid; it is usually not crucial for the residual moisture level.
2) *Interstitial liquid* is held near the contacts between particles; it is subject to the capillary laws. A model analysis shows that the quantity of interstitial liquid is directly proportional to the surface tension of the liquid and the acceleration due to gravity. For example, adding a surfactant to lower the surface tension by 50% decreases interstitial and adhesive liquids by 15%. On drainage, the interstitial liquid remains only in the part of the bed not filled with capillary liquid, that is, the part of the cake oriented toward the influent.
3) *Capillary liquid* fills the pores and capillaries up to a certain height in the filter-cake layer. By Laplace's or Young's equation, the capillary elevation h_c is inversely proportional to the capillary radius and is thus greater at smaller radii. In beds, however, an equivalent radius can be determined only to a rough ap-

proximation. It is therefore desirable to describe the elevation as a function of particle size:

$$h_c = \text{const.} \frac{\sigma \cdot \cos \Theta}{\varrho_S g} f(d) \qquad (44)$$

where σ is the surface tension (N/m), Θ is the angle of contact, ϱ_S is the density of the solids (kg/m^3), g is the acceleration due to gravity (m/s), and d is the particle diameter (m). The function f is determined either in an elevation test or from the flow through the bed, Equation (6).

The total residual moisture of a filter cake dewatered by draining is

$$w_L = (w_{La} + w_{Li}) \frac{H - h_c}{H} + \frac{h_c}{H} w_{Lc} \qquad (45)$$

where w_{La} is the mass fraction of adhesive liquid, w_{Li} that of interstitial liquid, w_{Lc} that of capillary liquid, and H is the thickness of the cake.

Dewatering by Displacement. The cake moisture can be displaced out of the voids and pores in the filter cake by a gas flowing through the cake under pressure. Displacement is, on the one hand, a purely mechanical process; on the other, it is based on the greater affinity of gases for solids. Air displacement is by far the most common method, since it is easiest to carry out, especially in vacuum filters. The inlet gas pressure must exceed the capillary pressure. The content of capillary liquid, which is in general fairly substantial, is displaced in this way. The degree to which interstitial liquid is also removed depends on the rate of flow.

The cake moisture will not decrease to a uniform level over the height of the filter cake. If suction is applied in the direction of gravity, the remaining moisture increases somewhat toward the bottom of the cake.

The decrease in saturation with time is described by

$$\frac{S - S_\infty}{1 - S_\infty} = \left[1 + \frac{2(n-1)}{1 - S_\infty} K\right]^{\frac{1}{1-n}} \qquad (46)$$

where S is the mean saturation of the filter cake at time t [18]; K is a dimensionless constant taking care of the permeability, the gas pressure drop, the capillary pressure, the time, the liquid viscosity and the cake thickness; and n is a superscript. With limestone having a mean particle size of 13 μm, a porosity of 0.41, and a permeability of $1.2 \cdot 10^{-9}$ m^2, experiments at a differential pressure of 300 kPa yield $n = 3$.

The mean saturation can be calculated with the capillary flow formula extended to allow for film flow [19].

The evaluation of the equilibrium saturation appearing in these equations often entails some expense. This can be avoided if the saturation is approximately calculated as

$$S = (a \cdot \Phi + 1)^{-b} \qquad (47)$$

where a and b are experimentally determined constants and Φ is the dewatering parameter

$$\Phi = \frac{\Delta p\, t}{\eta\,(\alpha H + \beta)\, H} \frac{p_I}{p_O} \left(1 - \frac{p_c}{\Delta p}\right) \qquad (48)$$

where Δp is the gas pressure drop, t is the dewatering time, η is the effluent viscosity, α is the cake resistance, H is the thickness of the cake, β is the resistance of the filter medium, p_I and p_O are the inlet and outlet pressures of the gas, and p_c is the capillary pressure of the filter cake [20].

The equilibrium saturation S_∞ can also be calculated from an experimentally determined diagram [21].

In general, filter cakes deposited on a coarse filter medium can be dewatered better than cakes on a finer medium.

Improvements in dewatering and a reduction in residual moisture are achieved with *hyperbaric filters*; these are ordinary vacuum filters (which allow moderate levels of residual moisture because of the low pressure drop available) placed in pressure casings (see p. **10**-42).

Expression. The expression of the filter cake involves a mechanical reduction in the void volume produced during filtration. Smaller solids particles are preferentially squeezed into the voids; this process often entails deformation of the solids.

This method of dewatering originated in the technique of juice and oil pressing from fruits, which has long been known. In filtration, it is used above all with compressible filter cakes, such as occur especially in sewage treatment, where relatively bulky, floccular sludges are produced. According to rough calculations for sewage sludges the specific energy requirement, per cubic meter of water recovered, is at least 10 times as great in drying as in filtration with expression. There are other advantages with regard to the reduction in sludge volume. If the residual moisture is reduced just from 70 vol% to 69 vol % (which is relatively easy to achieve by expression of sewage sludge), the sludge volume decreases by more than 35%. Another point is the gain in sludge strength associated with dewatering, which is often the key factor in making sludge disposal possible.

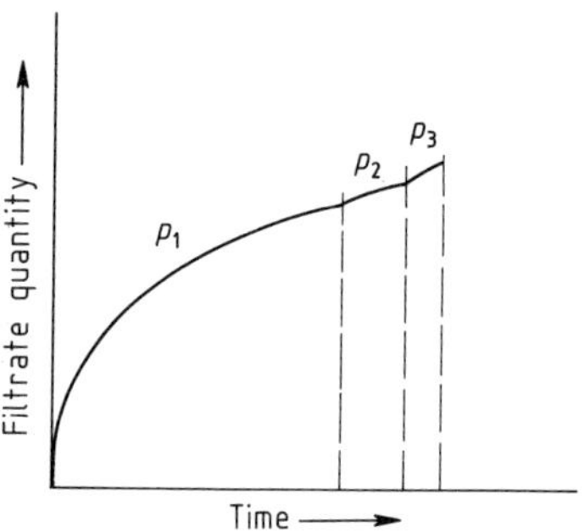

Figure 7. Typical expression curves at constant pressures $p_1 < p_2 < p_3$

Expression equipment is usually incorporated into filters; for description, see Sections on membrane filter presses and pressure belt filters, p. **10**-29, 40, 42. Available forces per unit length vary considerably, from 200 N/cm in continuous drum pressure belt filters to 2000 N/cm in twin-belt presses. The membranes used in filter presses are at a pressure of up to 1.6 MPa. In twin-belt pressure filters, expression is combined with shearing of the filter cake, because simultaneous shearing promotes the compaction of the filter cake.

Because of the decline in porosity and the resistance to flow, the quantity of filtrate per unit time at constant pressure decreases toward zero, as Figure 7 shows. The curve is plotted for a pressure increasing in steps and corresponds roughly to expression in twin-belt pressure filters (see p. **10**-29).

The calculation of solid–liquid separation in expression is based on principles from soil mechanics (Terzaghi 1925) and fruit-juice recovery (Körmendy 1965). Expression can be mathematically simulated with the so-called Terzaghi deformation model, a piston-and-spring system with liquid displacement, and with a sinusoidal pressure variation in the filter cake. The simulation gives the following equation for U_c, the ratio of liquid collected to liquid initially present (average consolidation ratio [22]):

$$U_c = 1 - \exp\left(\frac{\pi^2 T_c}{4}\right) \tag{49}$$

where

$$T_c = \frac{C_e \Theta_c}{w_0^2} \tag{50}$$

and, further, C_e is a constant (modified consolidation coefficient) for the compression properties, Θ_c is the consolidation time, and w_0 is the total solids volume in the cake per unit cross-sectional area. Experimental studies have given satisfactory agreement.

Theoretical calculations confirm the observation that increases in pressure give clear improvements in dewatering, provided enough time is available for expression.

3.7. Cake Washing

The purpose of cake washing is to remove the liquid left behind after dewatering when this liquid is regarded as a contaminant or must be recovered as a valuable component. Washing practice makes use of two processes, relatively old in principle: (1) rinse (flow of wash water through the cake) with an adequate amount of liquid (displacement) and (2) reslurrying of the cake with wash liquor outside the filter. The second process is economic only when purity requirements are stringent.

The consumption of wash liquor (which must usually be treated after use) is minimal in *countercurrent washing* (Fig. 8). This method can be carried out in only a few filters.

In *wash penetration*, wash water is passed through the cake still on the filter medium, in almost the same way as the slurry was fed. The wash liquor follows a special path in filter presses (see p. 49). In washing, the wash liquor more-or-less completely displaces mother liquor present in the voids, pores, interstices, and on the particle surface in the cake. Good washing also depends on uniform distribution of the wash liquor. Often, however, the cake has cracks due to dewatering, and these together with structural anisotropies have a detrimental effect on washing. This is especially true with mother liquors that are more viscous than the wash liquor. In such

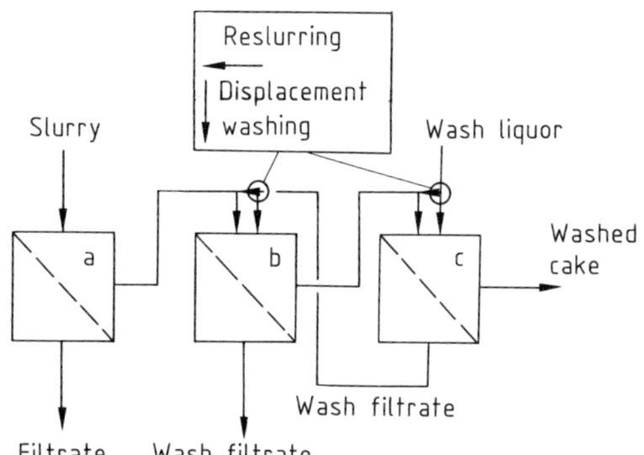

Figure 8. Countercurrent washing
a); b); and c): filters or filter stages

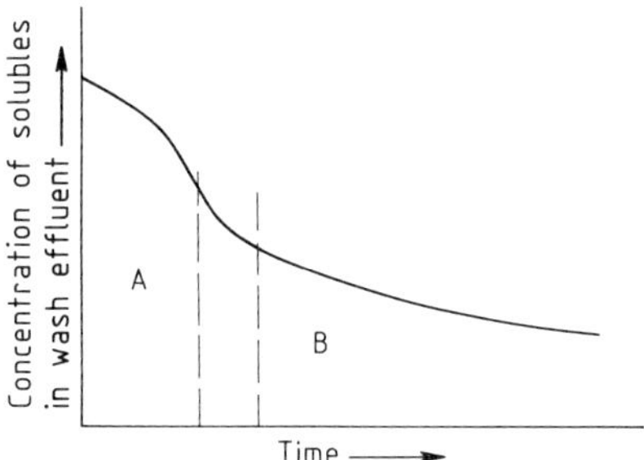

Figure 9. Concentration in wash effluent vs. wash time
A) Displacement region B) Diffusion region

cases it may be desirable to smooth or pre-compact the filter cake, for example with an elastic membrane in a press. Another possibility is to lower the viscosity of the mother liquor by heating. Special liquor-distribution practices, such as washing belts on drum filters (see p. 37) and washing grooves in belt filters (see p. 27), are also suitable ways to improve the effect of washing.

The mother liquor still adhering after displacement washing is taken up into the wash liquor by *diffusion*. There is an intermediate region in which both displacement and diffusion are effective.

Washing performance depends on local flow velocities, material characteristics, wash-liquor pressure, and the diffusion coefficient. Therefore, it is impossible to say in general whether it is better to use high wash-liquor rates and short washing times or low liquor rates and longer times.

An exact analysis of washing shows that many factors influence the process. If the analysis is restricted to the most significant, these can be reduced to three: αH^2 for displacement washing is the resistance to flow times the square of the cake thickness; δ_v is the volume ratio of wash liquor to cake; and $\alpha\ D_w t$ is the diffusion constant for washing times the flow resistance and the time [23].

Figure 9 shows the time dependence of the mother-liquor concentration in the wash effluent. The effective ranges of the several washing mechanisms can be seen in the figure. The displacement region is characterized by $\alpha\ H^2$ and the diffusion region by $\alpha\ D_w t$. It is impossible to state general values for the correlation between these parameters, from which the washing process can be determined ahead of time, because of the large number of controlling factors. Instead, experimental or plant results are combined with empirical relations, from which parameters such as the quantity of wash liquor per unit of residual mother liquor or the volume of wash liquor per unit filter-cake area can be obtained. When experimental values are extended to full-scale washing processes, however, a lower washing efficiency should usually be assumed.

4. Deep-Bed Filtration

In deep-bed filtration (as opposed to cake and cross-flow filtration), the solid particles are separated mainly by deposition within the pores of the filter medium (sand grains, fibers, etc.); pores and holes in the medium that are smaller than the particle diameter play a small part in retaining the solids. They lead to straining and complete blockage, see p. **10**-5. Achieving the highest possible separation efficiency requires that there be ample opportunities for particles to be deposited, therefore, most depth filters consist of relatively thick layers of filter medium.

Increase of the deposits weakens the filtering action, and the filter medium must be cleaned. Thus deep-bed filtration, like cake filtration, is essentially a batch process. Here, however, the filter cycle is limited to a filtration period and a cleaning period. As a rule, deep-bed filtration should be used not to recover solids from a suspension such as cake filtration, but instead to produce a very clean effluent.

The particles to be removed in deep-bed filtration must be so small that they can penetrate into the pores and capillaries of the filter medium. A special area of application is thus the filtration of slurries with much finer solids than those met with in cake filtration. Often the two processes are not sharply separated, for example when the slurry has a broad particle-size spectrum; in such cases, a cake can build up and act as a depth filter as in precoat filtration (see Section 11).

Another restriction, arising chiefly from economics, is that the influent concentration must be fairly low. High solids contents in the feed would quickly block the short pores and voids in the filter medium, creating large pressure drops and necessitating frequent cleaning periods, and thus greatly lowering the filter throughput. In addition, more highly concentrated influents have a stronger tendency to form bridges during filtration; such bridges are very undesirable in clarifying filtration, since they allow only the topmost layers of the filter medium to become loaded with particles, while the downstream layers contribute but little to filtration. The solids content of the influent should not be much higher than 0.1 g/dm^3 in ordinary deep-bed filtration. Very dilute slurries, with solids down to the finest particle sizes, are good to very good candidates for this process. A lower limit is set by the separation mechanism when there are no opportunities for solids deposition.

In certain cases, such as the food industry, feeds with solids concentrations higher than 0.1 g/dm^3 or with fibrous components (arising from extracts of plant and animal origin) must be filtered. So that this can be done without great difficulty, auxiliary materials (precoats) can be applied to the filter medium, as in cake filtration. Precoats have the function of pre-collecting all those solids fractions that would quickly block the pores in the medium.

Deep-bed filtration is used extensively by nature itself, in the clarifying flow of groundwater through beds of gravel and sand. This process is even today imitated by a number of practical techniques for the clarification of liquids, so that it has become customary to speak generally of "sand filters" or "gravel filters" in connection with these deep-bed filtration operations.

In most cases, the sizes of individual bed particles (elements) in clarification vary between 0.3 and 5 mm. The elements should be as uniform as possible within the bed, so that ordinary backwashing at increased flow velocity—which occasionally fluidizes the bed—will not lead to classification of the particles.

The depth of the filter bed is usually 0.5–3 m. The deeper the bed, the longer it can operate between backwashings. In practice, the bed depth is limited on economic grounds by the increasing pressure drop, which requires appropriate pumps and so forth. With rigid, porous filter beds (sheets), thicknesses are in the range of a few millimeters, since these beds generally feature smaller pores and thus have greater pressure drops.

Mechanisms of Deep-Bed Filtration. The separation mechanisms in clarifying filtration involve, on the one hand, the transport of particles to the surfaces (contact surfaces) of the filter medium and, on the other, the presence of adequately great surface forces insuring the adhesion of the particles. Transport takes place by interception, mass forces, diffusion, and electrostatic forces (see also [3]).

Mass forces are gravity and inertia. The flow velocity is calculated with Stokes' law. By virtue of inertial forces, solid particles persist in moving in their initial direction when the flow is deflected at filter-bed particles. The efficiency of collection E_T by inertia is given by

$$E_T = f\left(St, Re, \frac{\varrho_S}{\varrho_L}\right) \quad (51)$$

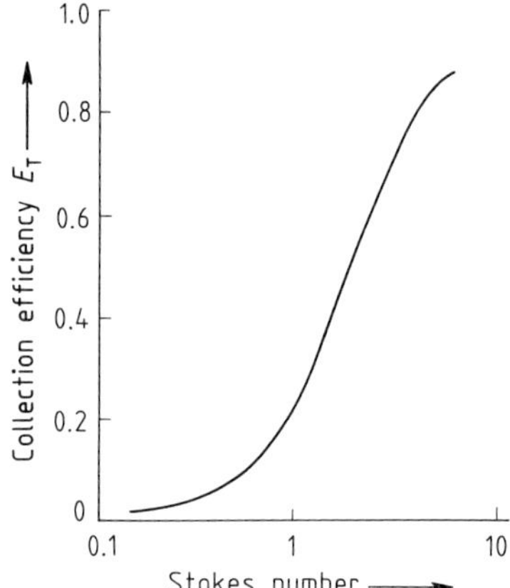

Figure 10. Collection efficiency as a function of Stokes number (innertia parameter)

where

$$St = \frac{v \cdot d_p^2 \varrho_S}{18 \cdot \eta d_A} \quad (52)$$

is the Stokes number, *Re* is the Reynolds number, ϱ_S is the density of the solids, ϱ_L is the density of the liquid, v is the liquid velocity, d_p is the solid particle size, d_A is the size of the bed elements, and η is the viscosity of the liquid. The Reynolds number used here is referred to the dimensions of the bed elements. The function f is plotted versus the Stokes number (inertia parameter) in Figure 10.

Solids can also be collected by a blocking effect (interception), which has to do with the touching of the solids particles in flow around the bed elements. This effect, *I*, is characterized by the dimensionless ratio

$$I = d_p/d_A \quad (53)$$

Particles smaller than 1 μm are further subject to Brownian motion, which can lead to collision with the elements in the bed. This diffusional process is described by a modified Peclet number

$$Pe = \frac{3\, d_p \cdot d_A \cdot v}{kT} \quad (54)$$

where k is Boltzmann's constant and T is the temperature. Collection by diffusion is a function of Pe^n.

Finally, hydrodynamic effects contribute to the collisional motions. Such effects arise when irregularly-shaped particles, rotating in the flow, move transverse to the flow direction; this is a well-known hydrodynamic effect similar to airfoil action. Electrostatic forces, which result from charges on the surfaces, are not very important in ordinary clarifying filters, but in electrokinetic filters (see p. **10**-54) with strong electric fields they can cause collisions with the bed elements and thus effect separation.

The collision probability alone is not a measure of the separating effect. The particles must also adhere, at least until filtration is completed and cleaning begins. Adhesion involves interactions at the surfaces of the contacting solids; these interactions are due to surface phenomena such as adsorption, electrical potentials, electrical double layers, and van der Waals forces.

Satisfactory separation performance can be achieved at filter loadings between 5 and 15 $m^3m^{-2}h^{-1}$ (m/h) where the filtration period, for example in water treatment, lasts no longer than 24 h. The optimal loading and service time are usually determined by experiment.

Sometimes, especially with slurries that contain extremely fine solids (smaller than 1 µm) such as viruses, "slow" filters are operated at only 0.1–0.2 m/h. Such filters are often used in biological processes. Biological sludges are produced on the filter [7]; they aid in the collection of solids and are mechanically removed from time to time.

Depth filters are usually cleaned by backwashing at high flow velocities, around 36 m/h. Relatively high velocities are needed to remove the collected particles by drag and impact. Air blowing along with the backwash liquor can also be useful.

Depth filters in which the filter medium is wound into a tubular sheet, a *cartridge*, have found wide use for the clarification of dilute slurries containing particles smaller than 20 µm, see Section 8.3 on candle filters.

The *filtration cycle*, that is, the sequence of all steps including solids removal, is quite variable in deep-bed filtration, because of the wide variation in slurry composition. Highly dilute feeds (less than 100 mg/L), especially, require fairly long filtration times, and no general guidelines can be stated without an exact knowledge of the solids properties. In sand filters, backwashing times are from 3–8 min, and the amount of wash water is roughly 1–5 % of the slurry throughput.

Design of Depth Filters. Calculations of particle motion and solids collection are mathematically very demanding, and even then cover only special cases based on simplified model conditions. In reality, the conditions are extremely complicated, and at present the outlook is poor for extending model treatments to practical situations.

The practical calculation of collection efficiency is based on a semiempirical model incorporating transport and diffusion. The solids concentration c in the slurry decreases as

$$\frac{\partial c}{\partial h} = \lambda \cdot c \tag{55}$$

where h is the thickness of the filter bed and λ is a filtration coefficient taking care of collection by the bed as a whole. This coefficient depends on a number of other factors, but in somewhat simplified form it can be thought of as a function of the collection efficiency λ_0 at the start of filtration ($t = 0$) and the local bed loading (i.e., the volume of deposited solids). IVES gives the filtration coefficient by the generally valid formula [3],

$$\lambda = \lambda_0 \left[1 + \frac{B\beta\sigma(h,t)}{\varepsilon_0}\right]^y \cdot \left[1 - \frac{\beta\sigma(h,t)}{\varepsilon_0}\right]^z \left[1 - \frac{\beta\sigma(h,t)}{\beta\sigma_u}\right]^x \tag{56}$$

where B is a packing constant, β is a bulking factor, $\beta = 1/(1-\varepsilon_S)$, ε_S is the porosity of the deposited solids, σ_u is the maximum load, $\sigma(h,t)$ is the bed loading as a function of position and time, and x, y and z are superscripts to be determined by experiment.

The first term in Equation (56) represents the geometric change in the bed elements due to solids deposition; the second, the decrease in internal surface area; and the third, the flow through the bed at maximum loading.

With the solids balance for a filter bed

$$-\frac{\partial c}{\partial h} = \frac{A}{\dot{V}}\frac{\partial \sigma}{\partial t} = \frac{1}{u}\frac{\partial \sigma}{\partial t} \tag{57}$$

(where A is the cross-sectional area of the bed, $\dot{V}$ is the flow rate, u is the superficial liquid velocity, and σ is the loading), the collection efficiency from Equation (55) and the filtration coefficient from Equation (56), the decline in slurry concentration and the increase in loading can be calculated. A closed solution is not possible, but at $x = y = 0$ and $z = 1$ in Equation (56), a linear relation between the filtration coefficient λ and the loading σ is obtained. This assumption forms the basis for the *bed depth service time* (BDST), another method of determining the efficiency of a clarifying filter proposed by BAUMANN [3]. Then

$$\frac{c(h,t)}{c_0} = \frac{\exp(\lambda_0 \beta u c_0 t/\varepsilon_0)}{\exp(\lambda_0 h) + \exp(\lambda_0 \beta u c_0 t/\varepsilon_0) - 1} \quad (58)$$

where ε_0 is the porosity of the filter bed at the start of filtration ($t = 0$).

The loading is given by

$$\frac{\sigma(h,t)}{\varepsilon_0} = \frac{\exp(\lambda_0 \beta u c_0 t/\varepsilon_0) - 1}{\exp(\lambda_0 h) + \exp(\lambda_0 \beta u c_0 t/\varepsilon_0)} \quad (59)$$

Since all the quantities here are known or can be determined with ease, Equations (58) and (59) are used as approximate design equations for depth filters.

Pressure Drop Across the Bed. The pressure drop and thus the energy consumption as an important factor in filter operation, are governed by the laws of flow through porous media, (Eqs. 6 and following). However, the porosity of the filter bed varies as solids are deposited in it during filtration. In order to take account of this fact in the calculation, a relation, similar to Equation (55), between the specific surface before and during filtration can be used; evaluation of such an expression for practical use is relatively expensive. For this reason, simplified models with empirically determined parameters have been developed, for example by IVES [3]

$$p(h,t) = p(h,0) + k_h \beta u c_0 t \quad (60)$$

where k_h is a constant.

Another way of handling the decline in flow at constant filter pressures due to decrease of the voids is based on Equation (13).

Application. The principal uses of deep-bed filtration are sanitary and cooling water treatment processes, in which feed concentrations range up to about 50 ppm, and the purification of all kinds of beverages, liquid fuels, and pharmaceuticals.

With slurries that contain very fine solids, it is often useful to add flocculating agents to make the individual particles combine into larger forms. This technique usually results in a large gain in collection efficiency (→ Flocculants, **A11**, p. 251).

When small-pored filter beds are recommended for feeds containing small particles, especially those encountered in the food, pharmaceutical, cosmetics and other industries, pressure filtration with depth filters should be preferred. In order to prevent solids appearing in the effluent, the pressure should be held as low as possible and should be slowly and steadily increased only as the effluent flow declines.

5. Screen Filtration

Like cake filtration screen filtration (including membrane filtration and straining) uses a thin filter medium on whose surface the solid particles are retained. In order to keep the screen or membrane efficiency up, provision must be made for the immediate removal of retained particles. In screen filtration, it is only the openings in the filter medium, not some filter cake with its highly variable structure, that effect separation. Screen filtration can therefore be used with confidence only to separate particles all of which are larger than the holes in the filter medium (i.e., the screen openings or the pores in the membrane).

There are two forms of screen filtration:

1) Relatively coarse solid particles can be removed mostly in tangential flow to the screen, usually metal fabrics or perforated sheets, which may be vibrated if necessary. Motion of the solids is caused by gravity (tilted screens) or mechanical means. The liquid (effluent) flow rate thus depends exclusively on the resistance of the screen, which is comparatively low so that the irrelevant differential pressures can be of considerable economic value when flow rates are very high, as in the mining and metallurgical industries.
2) Suspensions containing very small particles, down to the order of collodial size, would form virtually impermeable cakes in cake filtration. On the other hand, filter media with pore diameters down to some 0.002 µm

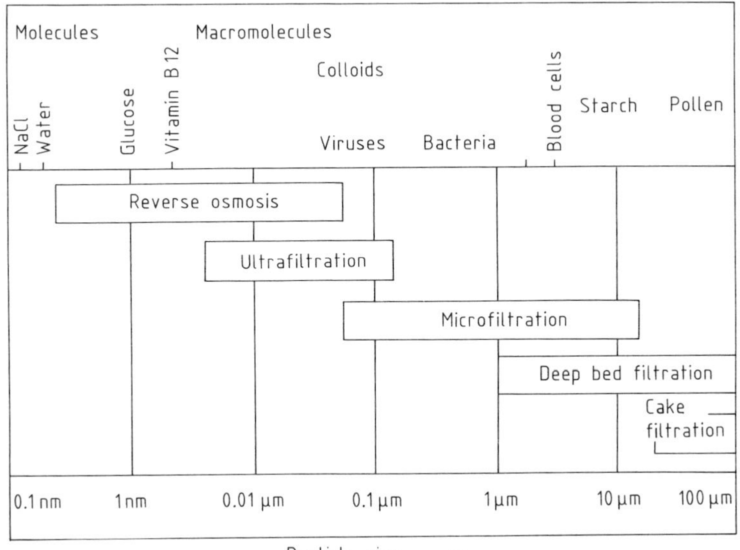

Figure 11. Range of application of microfiltration

(2 nm), the so-called *membranes*, are available (see page **10**-58). With the dynamic filtration technique such slurries can be filtered using these membranes. This technique opens an area of solid–liquid separation, the *microfiltration*. Figure 11 shows approximately in what ranges microfiltration and ultrafiltration for macromolecular solutions are employed. Ultrafiltration is dealt with elsewhere (→ Membranes and Membrane Separations, treated in the A Series).

Dynamic or membrane filtration is also called "cross-flow filtration" because shear forces of the tangential flow of the suspension are used to prevent cake formation (delayed-cake filtration). The filtrate flows crosswise. Retention of a thin cake layer with a relatively high resistance, between 10^{15} and 10^{17} m^{-2} on the filter medium cannot be avoided. This polarized cake layer strongly influences filter performance.

The shearing needed to prevent cake formation can be increased in the following ways :

increased flow velocity;
internals (smaller clearances, vanes to induce turbulence);
Taylor vortices;
pulsations.

In dynamic filters with mechanical shearing by rotating elements, increasing the rotation speed also helps prevent cake formation.

There are at present only a few design models for cross-flow filtration. Important factors include particle motions in liquid flows, the conditions for particle adhesion to the membrane surface, and resistances to flow. A recent model provides some results implying that the effluent rate is proportional to the flow velocity through the filter, not—as would be expected—to the driving pressure drop [24].

Experimental results and some plant experience relate to specific devices (see pressure plate filters, Section 8.10, and tubular filters, Section 8.11). They cannot be extended to arbitrary systems.

Filter elements of fine-pored membranes are often assembled into modules. A module consists of tube bundles containing tubes 2–5.5 mm in diameter, star-shaped membranes, plates, spiral cells with rectangular channels, or coils. An example is described on p. **10**-54. Satisfactory results have been achieved in the filtration of slurries with low concentrations of colloidal solids and similar materials (pigments, microorganisms).

Membrane filters are most commonly operated as pressure filters (0.2–1 MPa). Flow velocities are 2–4 m/s, less frequently 6–10 m/s. Success in this operation strongly depends on the quality of the filter medium, especially the pore-size distribution in the membranes, which are usually made of sintered metals, ceramics or powdered synthetic materials. Membranes with round openings have particularly narrow pore-size distributions.

6. Separation Efficiency

The separation of the "solid" and "liquid" components of a suspension by filtration can never be complete for both components. If an efficiency figure E_T is desired so that the operation or separating equipment can be assessed, there are several possibilities. If the definition

$$E_T = \frac{\text{quantity collected}}{\text{quantity inlet}}$$

is adopted, then in general the efficiency must be calculated for both components.

If S is the quantity of solids, L is the quantity of liquid, and E_S and E_L are the separation efficiencies with respect to the components, then an overall separation efficiency E_T can be derived from the material balance

$$E_T = E_S - E_L \tag{61}$$

If this formula gives a negative value, the terms should be interchanged.

Often, especially in the mineral industry, the efficiency is referred to a partition ratio a, which relates the total inlet rate to one of the exit rates (filtrate and sludge). If, for example, the inlet quantity is given by $S_0 + L_0$ and the outlet by $S_1 + L_1$, then

$$a = (S_1 + L_1)/(S_0 + L_0) \tag{62}$$

The overall separation efficiency is stated in terms of the fractions $m_0 = S_0/(S_0 + L_0)$ and $m_1 = S_1/(S_1 + L_1)$:

$$E_T = a\,\frac{m_{S1} - m_{S0}}{m_{S0}\,(1 - m_{S0})} \tag{63}$$

where m_{S1} is the solids fraction in the effluent and m_{S0} is the solids fraction in the influent. In terms of the concentrations c (c in kg/m^3)

$$E_T = \frac{(c_0 - c_2)\,(c_1 - c_0)}{(c_1 - c_2)\,c_0\,(1 - c_0/\varrho_S)} \tag{64}$$

or

$$E_T = \frac{2\,(c_2 - c_0)\,(c_0 - c_1)}{c_0\,(c_2 - c_0) + (c_0 - c_1)\,(\varrho_S - c_0)} \tag{65}$$

where c_0 is the solids concentration in the influent, c_1 is that in the effluent and c_2 that in the sludge (filter cake), and ϱ_S is the density of the solids.

An overall separation efficiency defined in this way makes it possible to evaluate the component separation efficiencies. Without such a figure, the separation process might be characterized in an incomplete way. If, for example, it is assumed that the component S can be completely separated in the separating equipment, then, since $S_1 = S_0$, the separation efficiency for S is $E_S = 1$, i.e., 100 %. In the same apparatus, however, the separation with respect to component L might be only 70 %. Then E_L is 0.3, since the separation efficiency must always relate to just one outlet flux. For this case, the overall separation efficiency is $E_T = 0.7$. It follows that partial separation efficiencies are meaningful only if the separation of just one component is of interest. The overall figure gives no information about the residual quantity of the other component; this is often unsatisfactory. In the above example, the 23 % cake moisture content may be quite unacceptable even if all the solids S are separated.

Each separation efficiency can also be defined in relation to the other outlet flux; for example, $E_S = 1 - (S_2/S_0)$. This does not fundamentally change the picture except that the numerical values are complementary.

For economic reasons, it is sometimes necessary to dispense with complete separation of the components. This is the case especially when the suspended solid matter has a broad particle-size distribution extending into the very fine range. An appropriate filter is then selected under the criterion that the coarse fraction must be completely retained while only a much smaller proportion of the finer material is separated. If such a filtration process is to be correctly assessed in terms of the degree of separation, a fractional separation efficiency must be stated; this is described, for example, with the Tromp curve (→ 2. Particle Size Analysis and Characterization of a Separation Process, p. **2**-11).

A separation efficiency oriented to the material fluxes is frequently not adequate for an overall assessment. As in some other operations such as drying, distillation, and extraction, the costs incurred (especially the energy consumption) rise with increasing separation. The relations between the degree of separation and the energy costs must be taken into account when different equipment types are compared, and even when operating data for one and the same filter are compared.

The separation efficiency depends heavily not only on the operating conditions, but also on the properties of the mixtures being treated. Important factors of this type include

1) the nature of the solids,
2) the particle-size distribution (which does figure, in part, in the fractional separation efficiency),
3) the slurry concentration,
4) the physical and chemical properties of the filtrate,
5) the quantity of filtrate, and
6) the process temperature.

7. Experimental Determination of Filtration Properties, Small-Scale Tests

The use of proven laboratory testing methods is recommended for the complete determination of filtration behavior under practical conditions. Such experiments not only help in finding the material values needed for theoretically predicting expected throughputs, but also provide a basis for describing the behavior of the components before, during, and after filtration.

A filtration test by itself cannot, however, yield an overview of the factors that control filtration. Other factors that must be known include the particle size, size distribution and grain shape of the solids, the viscosity of the effluent, and the pH of the liquid. A lack of knowledge on these points will make equipment design and selection more difficult; more important, however, it means that the systematic data needed for calculations are not available.

Cake Filtration. The *bench leaf test apparatus* (Fig. 12) is a suitable apparatus for cake filtration tests. It is intended chiefly for studying vacuum filtration, but the device can be regarded as a portion of a drum filter, with which the entire filtration cycle, including cake removal, can be implemented. The measured throughputs and collection efficiency provide a basis for subsequent assessment.

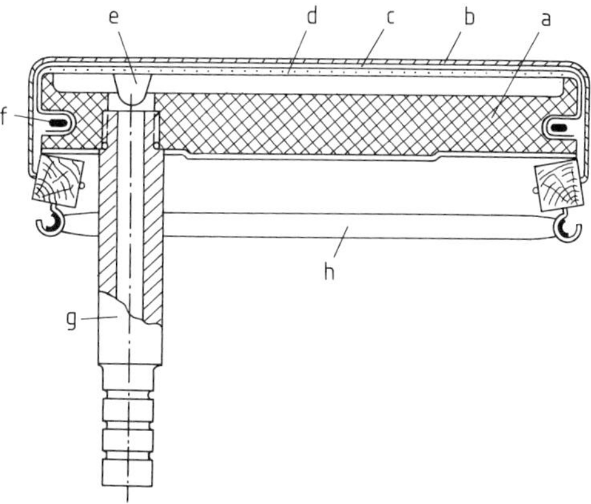

Figure 12. Bench leaf test apparatus
a) Rectangular leaf (total length 10–15 cm; b) String covering for cake discharge; c) Filter medium; d) Drainage grid; e) Filtrate collecting channel; f) Clamp; g) Filtrate pipe; h) Stretch tighter ring

The test leaf consists of a hard rubber plate with a cellular pattern of grooves feeding into one transverse groove, from which filtrate can be withdrawn through a tube. The filtration area is from 100–200 cm^2, and the plate can be round or rectangular. A belt holds the filter medium (cloth) on a support. With a rectangular leaf, a string arrangement can be provided for cake removal. A round leaf has the advantage that an adjustable ring is placed around it; this is useful in a trial with a filter aid. The test leaf can be employed in any position, so as to reproduce the special conditions in practical filtration equipment.

The leaf test yields a filtration time for a given cake volume, which is a measure of filterability under given conditions. A more extensive analysis, for example with the aim of calculating the filtration resistances, is not usually very meaningful and can easily lead to erroneous conclusions. The leaf test can, however, be useful in determining the efficiency of a filter medium, the amount of solids passing it, the adhesion of the cake to the filter cloth, and several other qualities of the cake, such as surface structure, cracking and homogeneity. Sizable problems occur when the results from this test are extended directly to filtration equipment in which the service conditions are much different from those in the test leaf, for example in the tanks of drum or disk

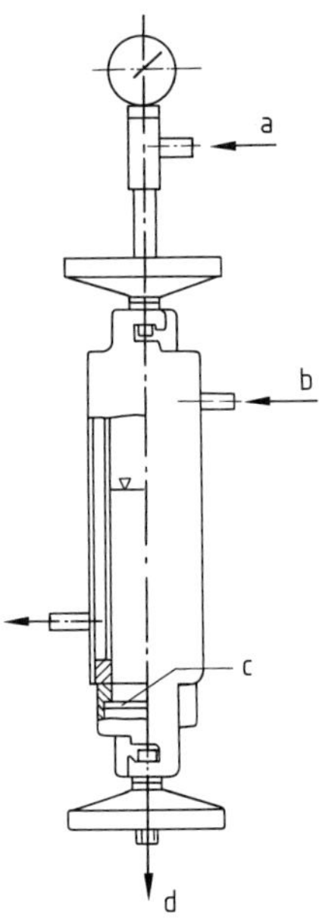

Figure 13. Bench pressure filtration apparatus (BHS)
a) Compressed air; b) Heating medium; c) Filter medium; d) Filtrate

filters (see Sections 8.5 and 8.6). The only help here comes from long experience.

Nor do common tests with laboratory nutsches (Büchner funnels) provide generally usable results; only ample experience can yield information about the performance of a given filter.

These two test methods have remained current simply because they are easy to perform. They make it possible to plan series of pilot filter tests in such a way that the optimization of filter performance requires a minimum of effort.

Somewhat more comprehensive results come from testing with the laboratory pressure filter (Fig. 13), which consists of a pressure vessel some 16 cm long with a filtration area of 20 cm^2 (corresponding to a diameter of 5.5 cm). A pressure of up to 0.4 MPa can be applied with compressed air or other gas. Suspension (250 cm^3) is charged into the filter, and filtration is begun after a few seconds. The filtrate is caught in a graduated cylinder and measured as a function of time. For dewatering, the compressed air is applied to the filter for about another minute after filtration. It is important to hold all the conditions constant (filter cloth quality, prehistory of suspension, test temperature, time variation of pressure, cake weight).

Other data, especially for determining the cake compressibility, can be obtained with a "C – P" (compression–permeability) cell (Fig. 14) [25]. The device consists of a cylindrical cell with a porous bottom and a porous piston. Mechanical loads are placed on the piston and the filter cake is compressed. According to extensive investigations there is no crucial difference between this mechanical com-

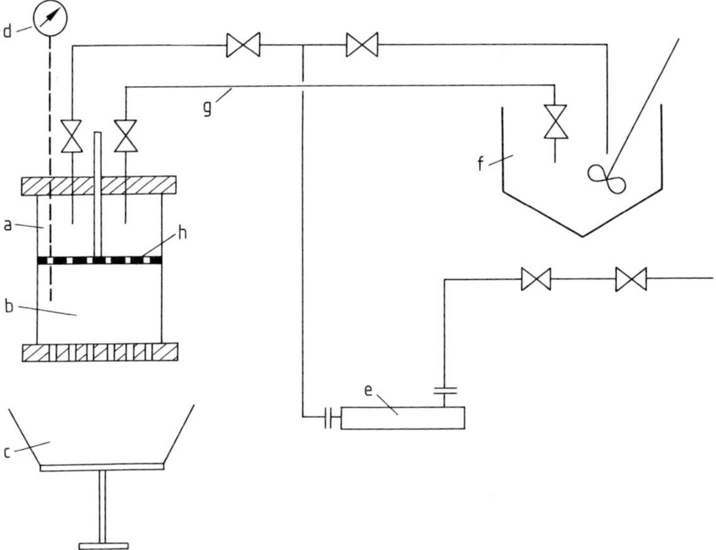

Figure 14. Compression–permeability cell
a) Filter cell; b) Filter cake; c) Filtrate receiver; d) Pressure recorder; e) Pump; f) Slurry tank; g) Recirculation; h) Movable plate

pression and the hydraulic compression that takes place in filtration equipment, as far as the effect on filtrate flow is concerned. The quantity of slurry in the cell is controlled with the overflow.

This apparatus makes it possible to determine the needed material values and service conditions. What is more, it is suitable for determining the properties of filter media and filter aids and also for testing various operating modes (which may be influenced by the type of pump). The interpretation must involve the complete filter equation and, in some cases, may entail high expenditure. This device does, however, offer more adaptability to a range of filtration tasks than would a study in actual plant equipment.

The time dependence of filtrate volume as described by Equation (25) is retained for the interpretation of test results. After just a few measurements, the quantities of importance for the filter equation can be found in regression analysis and often in a diagram (Fig. 4). This method of interpretation holds for any type of filtration.

The test results can, of course, be interpreted by suitable computer programs, with a resulting gain in accuracy and versatility of the conclusions drawn.

8. Filtration Equipment

Filters can be classified in accordance with several criteria:

Solids retention:	surface, deep-bed
Filtration technique:	cake, cross-flow, screening
Driving force:	pressure, vacuum, expression, magnetic or electric field, capillary
Operation:	discontinuous, continuous, quasicontinuous, automatic self-controlling, programmed, precoating
Filter element:	1) bag, belt, candle/cartridge, disk, drum, leaf/plate, nutsche, press plate, tube 2) horizontal, vertical, single-element, multi-element, stationary, rotating 3) open, closed
Solids discharge:	scraper removal, vibration, centrifugation, back blow, tipping, manual
Application [7]:	rapid settling systems, moderate and slow settling systems, clarification of liquids, pastes, pulps, sludges, and nonfluid systems.

Because of numerous overlaps with respect to process engineering and design, any systematic arrangement entails a loss of clarity. In the descriptions that follow, the design element is given top priority.

The principles of operation and the design of filter equipment have changed little for many years. Only with recent efforts to improve the economics of processes have some crucial design changes and additions taken place. The most important of these involve

1) automation of operating modes and cycles
2) increases in specific throughput capacities and separation efficiencies, especially aimed at reducing residual moisture
3) optimization in the context of the overall process
4) improvement of operational reliability
5) increased concern for environmental protection and worker safety
6) adaptation to new, more difficult separation tasks
7) development of new separation methods for existing filtration problems

For example, classical filter presses have been replaced by completely automatic, unattended equipment. Mechanical dewatering devices have been installed on belt and drum filters. Classical open nutsches have been modified into "process filters," which combine several downstream operations (e.g., washing, dewatering, and drying) in a single apparatus. For difficultly filterable suspensions, dynamic filters with membranes have been designed; together with electrokinetic filters, still in development, these open up new possibilities for mechanical liquid–solid separations.

Important advances in the performance of existing filters have come about through better understanding of the theoretical relationships in all areas, such as filtration, washing, and dewatering and expression. In particular, the automation and optimization of processes, based on knowledge about the interactions of the many process parameters, are currently being pursued in extensive programs.

Figure 15. Pocket-type bag filter (reproduced with permission of Scheibler)

8.1. Bag Filters

The filter element in this type of filter is a bag made of the filter medium. A typical bag has an opening at one end. The bags are mounted in various kinds of containers; they allow an uncomplicated design, are easy to service, and are relatively versatile in view of the large selection of filter media available. For example, solids in the range of 80–800 μm are successfully collected with bags made of fabrics woven from monofilament yarns; solids between 1 and 200 μm are separated with needled felt bags or by precoat filtration (in which deep-bed filtration is active at the same time).

As batch filters, which are opened by hand, bag filters are employed only where relatively small charges with extremely low solids contents are to be clarified. They therefore find use chiefly in clarifying and polishing filtration (sterilization), where the solids to be removed are impurities or undesirable byproducts. Applications include chemical and pharmaceutical products, paints and varnishes, juices, edible oils, waxes and resins.

Bags open on one side are charged from the inside, so that the solids are collected as a filter cake in the bag. Service times range from hours to days. In pressure bag filters, where the bags are inserted in supporting cages, pressures between 0.5 and 1 MPa are common; in special designs, the pressures are up to 2.5 MPa. Capacities are in the range of $1\ m^3 m^{-2} min^{-1}$ at 1.6 MPa. Filter units with several bags in housings are available at total capacities of up to $2.4\ m^3/min$ ($4.8\ m^3/min$ maximum) and down to quite small values.

Pocket-shaped filter bags are charged from outside; they are slipped over collapsible filter inserts attached to the filtrate pipe (Fig. 15). The open corner of the filter bag is sealed against the filtrate pipe. In order to increase the filtration area, the bags are made up to three times wider than the inserts, so that folds are produced when they are slipped on; in this way, the space requirement per square meter of filtration area is greatly reduced. Filter units up to 250 m^2 are available. After filtration, the bags can be cleaned by spraying and blowing with compressed air.

8.2. Belt Filters

A belt filter resembles an ordinary belt conveyor driven by one of the guide rollers. Slurry is fed onto the filter belt from above and the filtrate drains or is pulled by suction from the bottom of the belt. Belt filters are operated as simple gravity, vacuum and pressure filters. Whether operation is continuous or intermittent (semicontinuous), slurry is fed onto one end of the belt. The filtrate drains into vacuum tanks, which can be

Table 2. Filterability of cakes

Filtration characteristics	Group of suspension, symbol				
	Fast (F)	Medium (M)	Slow (S)	Dilute (D)	Very dilute (VD)
Initial cake formation rate, min/cm	0.005–0.1	0.1–1	1–10	10–100	no cake
Slurry concentration, %	> 20	10 to 20	1 to 10	< 5	< 0.1
Settling rate	rapid	medium	slow	slow	
Leaf test rate, $kg/m^2 h$	> 2500	250 to 2500	25 to 250	< 25	no test
Filtrate rate, $m^3/m^2 h$	> 10	0.5 to 10	0.0025 to 0.05	0.025 to 0.05	0.025 to 0.05
Typical slurry	crystalline solids	salts	pigments	wastewater	water

Ⓐ

Ⓑ

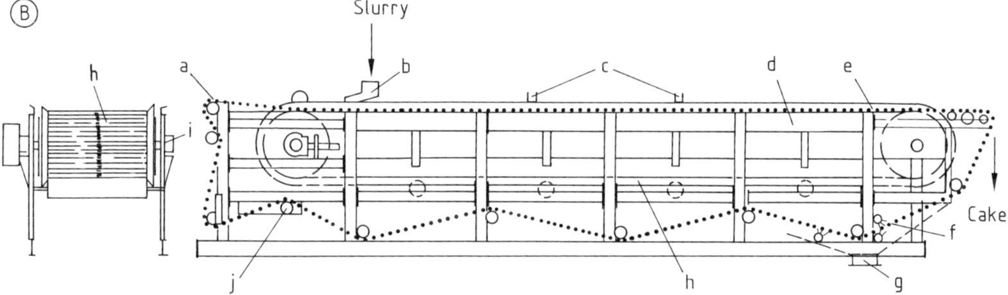

Figure 16. A), B) Vacuum belt filter (reproduced with permission of Dorr-Oliver)
a) Filter cloth take-up assembly; b) Feed box; c) Cake wash; d) Vacuum pan; e) Filter cloth; f) Filter cloth wash; g) Drip pan drain; h) Drainage belt; i) Tensioning device; j) Filter cloth aligning; k) Drainage belt

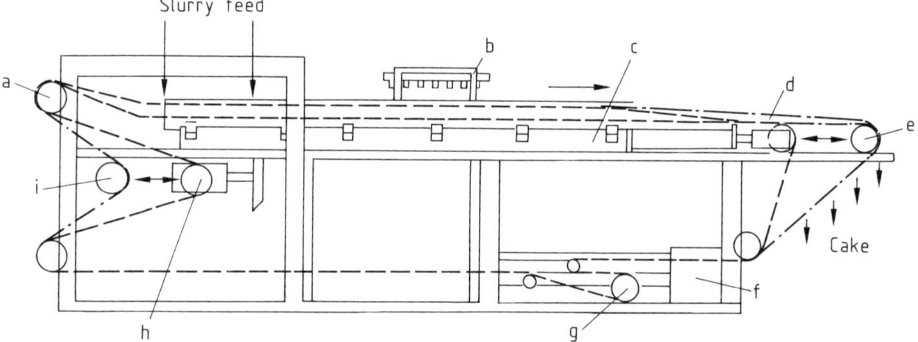

Figure 17. Jerking-type vacuum belt filter (BHS)
a) Filter cloth take-up assembly; b) Cake wash; c) Vacuum trays; d) Filter cloth; e) Movable discharge roller; f) Cloth wash assembly; g) Blocking roller; h) Tensioning assembly; i) Floating roller

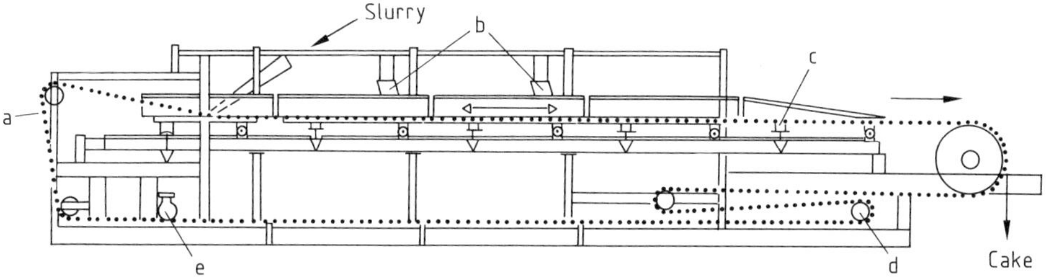

Figure 18. Reciprocating vacuum tray-type belt filter (Pannevis)
a) Filter cloth; b) Cake wash; c) Reciprocating vacuum trays filtrate receiver; d) Cloth tensioning device; e) Filter cloth wash

stationary or can move intermittently with the belt. The cake can be washed and dewatered in downstream zones. If necessary, the filter belt is washed during recycling. In another design, the belt is made up of a number of individual cells, which bear the filter cake. In this way, a more careful separation of the filtration stages is achieved.

In general, belt filters are applied for readily filterable suspensions that contain coarser, and therefore easily sedimented, solids (groups F and M in Table 2), where the filtrate and the wash liquor are to be collected separately and the solids must have gentle handling. The maximum filtration area is 120 m^2. Fluctuations in product call for control of the belt speed or slurry rate.

For especially sensitive filtrations, belt filters are used with filter media that remove the liquid from the cake by capillary action alone. The filter cloth runs over felt or similar filter media.

Vacuum Belt Filters. A continuous vacuum belt filter consists of the moving filter cloth supported on a profiled elastomer transport belt (Fig. 16). To provide better sealing for the vacuum space, sliding belts can be placed under the transport belt, or a specially shaped belt can travel along with the filter cloth. The difficulties met with in ensuring a good vacuum with moving seal systems are overcome when the filter belt or filtrate box move intermittently. Figure 17 shows an example. When the belt is stationary, slurry is fed to the filter and at the same time the filter cake is washed, dewatered and, if necessary, pressed. In the intermittent movement of the belt, the discharge roller shifts the belt, while the retaining roller holds the belt in place. The movement is compensated by movement of the compensating roller in the opposite direction. The vacuum is let down when the belt is moving.

The continuous vacuum belt filter of Figure 18 has vacuum tanks of pan-shaped sections each 1.4 m long. Each section is divided into two parts with flexible connections to the vacuum system (vacuum pumps). In this way the individual filtration stages are divided into arbitrary lengths as needed. The vacuum tanks are designed with grids that allow the filtrate to drain freely. During filtration, the tanks move along with the belt at the same speed, as the vacuum builds up. At the end of the co-moving path, the vacuum is broken and the tank vented, while the filter cloth moves on. Slurry and wash liquors used are fed continuously.

Vacuum belt filters are usually installed in open frames, or in closed housings if no vapor is to be released into the environment. Good filtration capacities and washing effects can be achieved only with uniform loading of the filter medium or filter cake. For this reason, the feed and wash liquor are distributed over the width of the filter belt by flat nozzles or washing grooves. Downstream dewatering of the filter cake is done by double-belt expression (see twin-belt press, p. **10**-29) or a plate press.

Because of the versatility of vacuum belt filters, they have found use in nearly every field of liquid–solid separations where slurries are relatively easy to separate.

Even difficultly filterable slurries with comparatively finely dispersed solids can be separated on belt filters if suitable belt materials are employed and the solids do not immediately block the medium. In such cases, pressure is most commonly used instead of vacuum (see pressure belt filters).

Belt filters for gravity and suction filtration offer low equipment costs. With a suitable filter medium (nonwovens are generally used), they can successfully clean up slurries that are not too highly concentrated (solids down to 50 mg/m^3) and of the filterability group F in Table 2, p. **10**-26. Feeds include machine-tool cooling and cutting lubricants (emulsions), electrolysis cell slimes, beer (for removal of turbidity and yeast), wastewaters, and chemical solutions.

A *gravity or hydrostatic filter* with a continuous belt (Fig. 19) consists of a tank, in which a moving support belt made of coarse woven fabric or a flexible belt forms a depression. The filter belt proper rests on this support belt. Operation is intermittent: slurry is fed in until the increasing resistance to flow causes the liquid above the filter belt to reach a predetermined depth. The hydrostatic head promotes the filtration rate. The belt is then advanced, the filter cake carried out of the tank, and new filter medium transported into it. The filtrate flow rate is $0.2\ m^3\,m^{-2}\,min^{-1}$.

There are two designs of *suction belt filters*, also called *flat-bed filters*. One, used chiefly for wastewater treatment, consists of a receiving tank for slurry, with a drag chain conveyor. The perforated or slotted floor of the tank holds the filtrate receiver, which is connected to the pump. The filter medium is guided between the scrapers and the tank floor. After a filter cake has built up to a predetermined thickness, the increasing

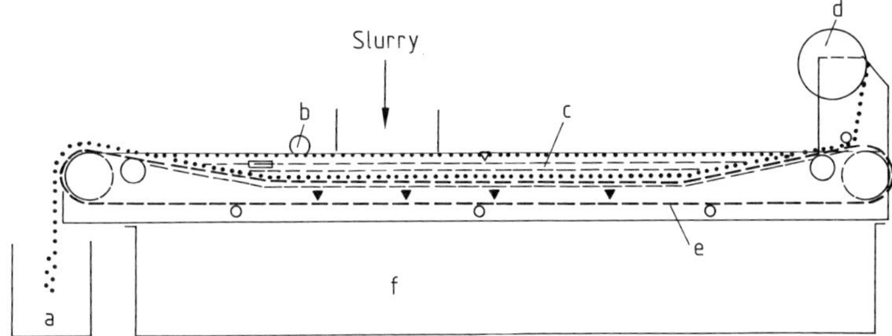

Figure 19. Hydrostatic belt filter
a) Sludge bin ; b) Float control; c) Filter trough; d) Filter medium; e) Carrier belt; f) Filtrate tank

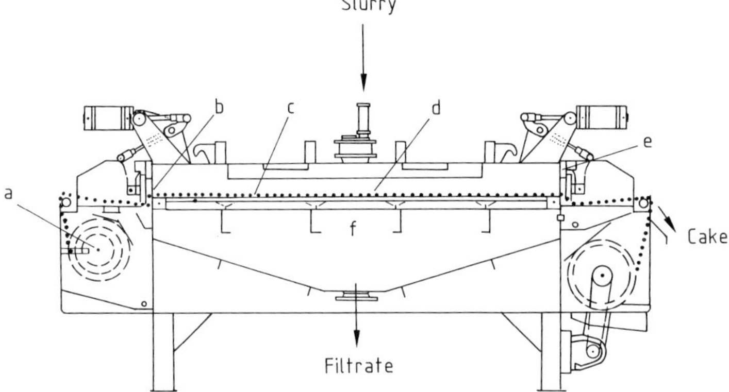

Figure 20. Pressure belt filter (Sack) (typical flat-bed filter)
a) Feed roll; b) Entry sealing flap; c) Filter medium; d) Top shell; e) Outlet sealing flap; f) Bottom shell

pressure drop causes the conveyor to begin moving. It advances some 20–50 cm along with the filter cloth. These filters are made with filtration areas up to 21 m^2.

The other design consists of a vacuum tank with lid; the paper filter tape is fed between the tank and the lid. This type of filter also operates intermittently. At a filter area of 1 m^2 a throughput of 0.1–0.5 m^3/h can be achieved, depending on the amount of solids in the suspension and the presence of filter aids.

Pressure Belt Filters. A pressure belt filter of the flat bed type consists of a filtrate receiver and a pressure chamber (Fig. 20). The filter medium is guided between the two chambers, which are situated in a housing. Operation is necessarily intermittent, since the pressure chamber, which is provided with slotted gates, is pressurized during filtration. In smaller units, the use of slotted gates is replaced by pivoting up the pressurized section. Pressure belt filters are also made with expression devices (up to some 2 MPa).

A *twin-belt pressure filter* is a belt filter with a second, co-moving belt, which exerts an additional mechanical compressive load on the filter cake with pressure rollers. Such a filter can be employed provided the cake is deformable. This condition holds especially for bulky flocculated

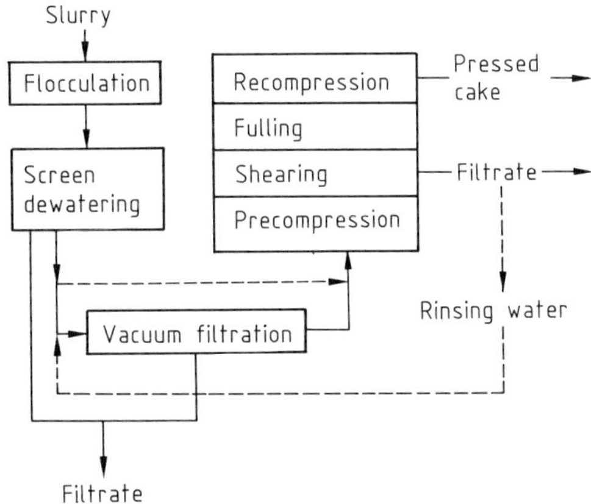

Figure 21. Schematic processing in twin-belt pressure filters

slimes such as occur with filter cakes made up of relatively fine solid particles (mainly of organic origin) having a broad size distribution. Twin-belt pressure filters have proved themselves for the *dewatering of sewage sludges* that have been treated with polyelectrolytes (based on poly(acrylic acid) and polyacrylamide) to produce relatively large, stable flocs.

This type of filter has made it possible to reduce the relatively high 70 vol% residual moisture in municipal sewage sludges by as much as 10 vol%, corresponding to a decrease of 35.7% in the moisture content per cubic meter of sludge. Further applications have also been tested, and this class of filter has been adopted for some dewatering jobs, such as slimes from coal washing, large-scale animal husbandry, cellulose and pulp production, electrolysis, and off-gas treatment. An average dewatering capacity for designs now current is 6 m^3 of municipal sewage sludge per meter of belt width and hour. With sludges that are readily dewaterable, the influent rate can be increased to 15 m^3/h.

Machines with multistage dewatering are used in nearly all these sample applications. Figure 21 shows a simplified diagram. The flocculat-

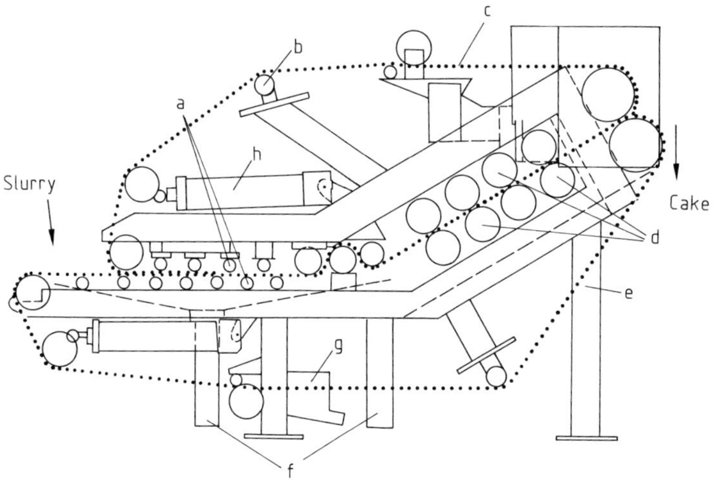

Figure 22. Twin-belt pressure filter (Andritz)
a) Supporting and dewatering rolls; b) Belt tracking devices; c) Top filter belt; d) Compression rolls; e) Bottom filter belt; f) Filtrate drains; g) Belt cleaning device; h) Tensioning devices

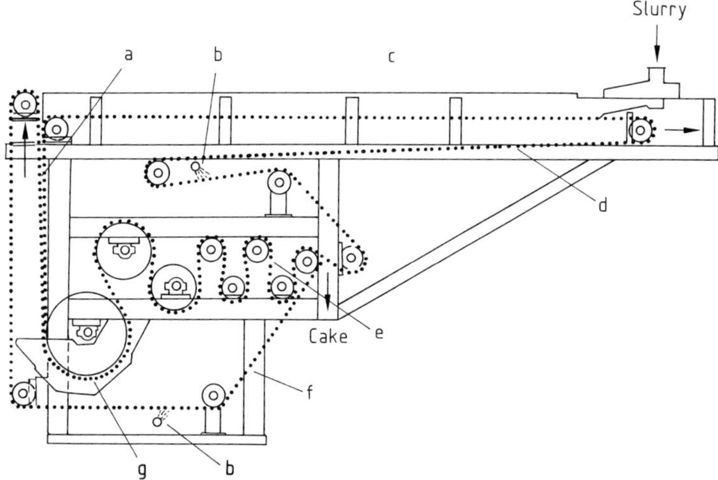

Figure 23. Twin-belt pressure filter "Winkel press" (Bellmer)
a) Variable tapered press zone; b) Belt wash; c) Gravity filtration zone; d) Top filter belt; e) S-Press and shear zone cake discharge; f) Bottom filter belt; g) Press egoutteur

ed sludge is put through preliminary dewatering in a simple screen drainage (rotating or flat screens), since part of the liquid bound up in the sludge is released in the flocculation step. Vacuum filtration usually follows before the filter cake reaches the expression zone.

A special feature of twin-belt pressure filters is that the belts change direction several times, on rollers that shear and press the cake, thus promoting its compaction. Usually the filtrate produced in the several stages, which is not entirely clear, is recycled to the first stage, where the solids still dispersed in the filtrate are re-flocculated. The filtrate resulting here is largely free of solids and can be passed on to, for example, biological treatment.

Quite a variety of arrangements of the individual process stages and design features has come into being. The performance of these machines thus varies widely, being influenced as well by the properties of the influent sludges.

Figure 22 shows a sample embodiment. This machine, mounted in a frame, has the gravity dewatering stage followed by an expression zone. The pressing section is modular to allow displacement, so that a range of requirements on cake compression properties can be satisfied. The version shown is intended for slurries of average compressive strength.

Top gravity filtration and dewatering are featured in the filter of Figure 23. Here however, the partly dewatered filter cake goes to a vertical pre-compression stage in which the filter belts approach each other in wedge fashion; this operation is aided by sedimentation of the solids. Again, subsequent compression takes place in a system involving bends of various radii.

In the modification of Figure 24 A, the sludge enters the expression zones from below, through a straining zone (gravity dewatering). Passing over rolls that vary in diameter, the cake is sheared and compressed (up to somewhat over 100 kPa). The modification of Figure 24 B applies a further increased compressive load with a third moving flat belt.

Figure 24 C shows a filter in which an overflow (for gentle handling of the flocculated solids) directly connects the flocculation vessel to the gravity filtration and dewatering stages; the filter cake passes downward through the succes-

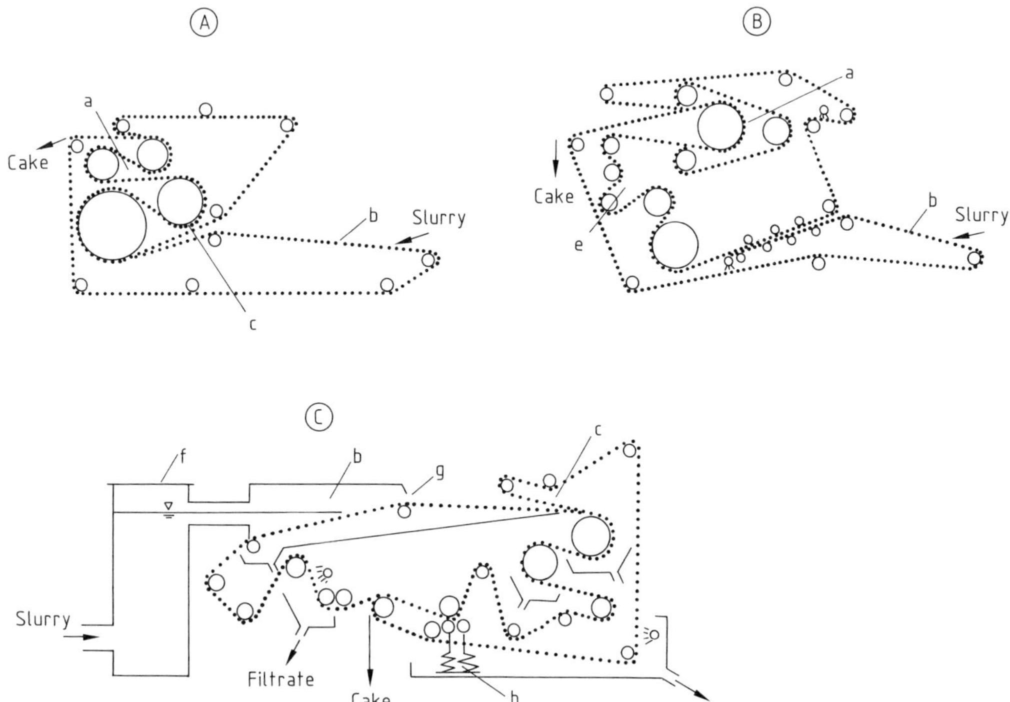

Figure 24. Modifications of twin-belt pressure filters (A, B, and C)
a) Press zone; b) Gravity strainer zone; c) Pre-compression zone; d) Press belt high compression; e) Low compression zone; f) Flocculation; g) Cake break-up; h) Line compression

sive expressing zones. The last stage is supplemented by a device that exerts a controlled linear pressure.

Cake compression in twin-belt pressure filters is rather modest in comparison with the compressions achieved in membrane filter presses (p. **10**-52). The twin-belt devices have the advantage of continuous operation and relatively low cost. A recent development demonstrates the possibility of reaching pressures in twin-belt expressing filters that are similar to those in membrane filters (up to 2 MPa). A filtration zone with filter belts converging in wedge fashion is followed by compaction as the cake turns around a drum pressurized by hydraulically actuated elements.

Experience and theoretical analyses (see p. **10**-15) lead to the conclusion that optimal performance is attained in twin-belt pressure filters when the cake is relatively thin, the filtrate drains on both sides, the expressing times are long (especially for readily compressible cakes), and the expressing force is as great as possible.

8.3. Candle Filters

Candle filters (cartridge filters) consist of tubular elements, mostly covered by sleeves of a filter medium. They are widely used in the filtration of weakly to extremely weakly concentrated slurries that contain solids more or less as impurities at concentrations down to the ppm range. This class includes all filters with cylindrical elements mounted, singly or multiply, more or less vertically in tanks (Fig. 25 A).

All candle filters have the fundamental advantages that the filter elements are relatively simple in construction and cheap to fabricate, and that the devices are adaptable to requirements imposed on separation efficiency and throughput capacity. The elements are mounted either suspended from an overhead tube sheet or upright on a plate (except where each element has its own housing). There are advantages to the suspended design, but great care is required with elements that break easily.

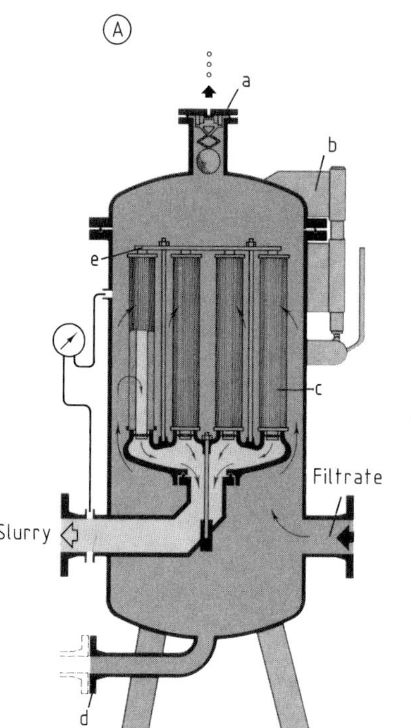

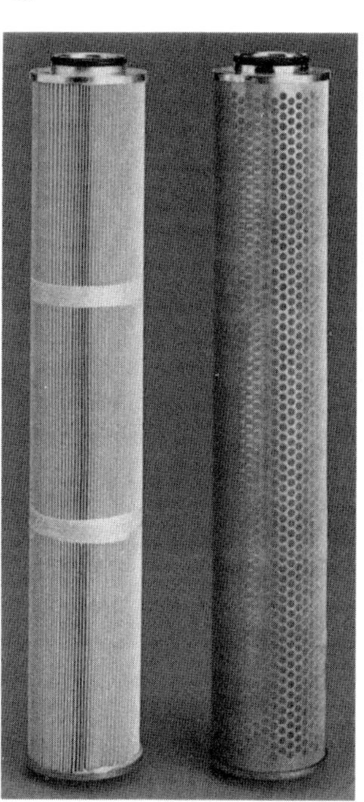

Figure 25. A) Sectional view of candle filters and B) tubular filter elements (reproduced with permission of Faudi Feinbau GmbH, Oberursel)
a) Deaeration; b) Opening device; c) Filter candle d) Discharge; e) Tensioning plate

Most candle filters are operated as pressure filters. The elements can be cleaned by backwashing, or they may be replaceable. Some multiple-element filters offer filtration areas of far more than 100 m^2. As typical clarifying filters, these devices have a broad field of application (groups D and VD in Table 2, p. **10**-26). The filtration surface can take a wide variety of shapes, depending on the job being done.

Sheet Candles. A simple candle of this type consists of wire mesh or perforated screen made of AISI 316 stainless steel, phosphor bronze or manmade fibers supported on a tubular metal core (Fig. 25 B). Slotted screens and wire-wound filters can also be employed. Such elements are used mainly to collect weakly-concentrated solids from liquids such as oils and soap and dyestuff solutions; mesh openings range from 0.01 mm to 0.5 mm (down to 2 μm in special meshes). Backwashing is desirable for wire-mesh elements.

Disposable sleeves of paper or nonwoven synthetics are used to retain finer solids. In part, these filter media simultaneously act as bed filters, the solids being deposited in the medium itself. These simple filter elements offer comparatively high flow rates. Available filter units cover the range from a few liters per minute to several hundred cubic meters per hour. Enclosed in pressure-tight vessels, they can be operated at absolute pressures of up to 10 MPa. The pressure drop across such a filter is around 10 kPa.

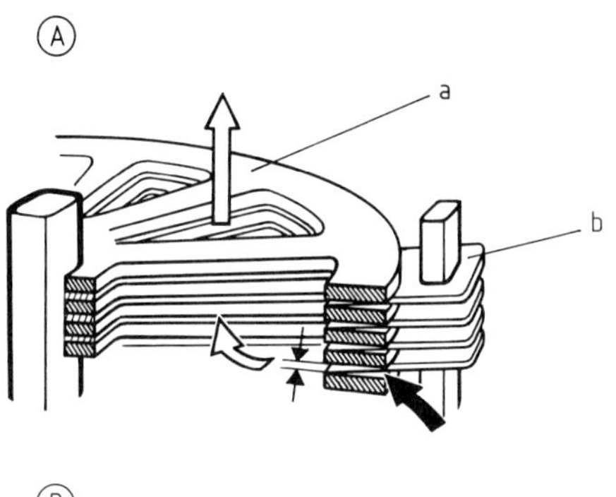

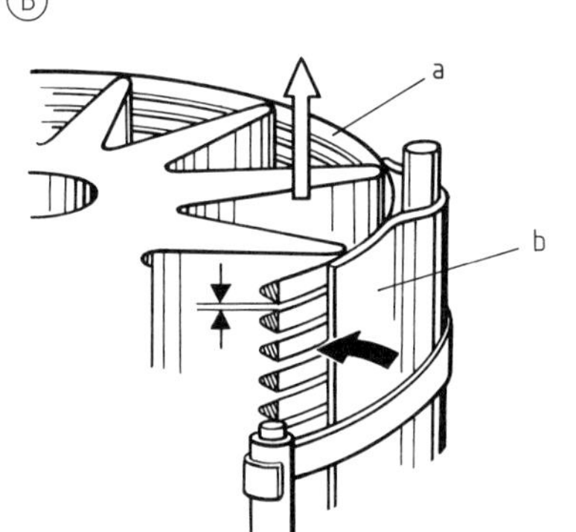

Figure 26. Gap-type candle filter (Mann & Hummel)
A: a) Steel disk package (turned for solids discharge); b) Fixed gap cleaner
B: a) Steel wire spiral (turned for solids discharge); b) Fixed scraper

Gap-Type Candle Filters. Gap-type candle filters can be made of disks, profiled wire, or tubes (Fig. 26). Most are held together by straps. In one design, the gap-forming elements are slipped or welded onto a tube; filters of this type are used as well strainers. A special drinking water filter element is the gravel-filled prepack cartridge.

A continuously operating gap-type cartridge filter has several filter elements, charged on the inside, arranged on a rotating mount that brings them one after another to a stationary backwash device.

Cartridge Filters. Cartridges are designed for deep-bed filtration and are thus used mainly for fine filtration of dilute slurries, concentrations of less than 0.01 % and particle sizes less than 20 μm. They usually comprise replaceable cartridges made of a variety of materials such as sintered metals (AISI 316 stainless steel), ceramics (also usable at high temperatures), plastics, cellulose or glass fiber (resistant to most liquids). Wound cartridges of cotton, cellulose or manmade polymer fibers, often combined with nonwovens, are distinguished by relatively large spaces (honeycomb) for retaining solids. The simple construction and low material costs open up technically and economically interesting possibilities.

Large-Area Filters. Large-area filter elements are made by folding the filter medium along the direction of the cartridge (Fig. 27) or by using pouch and disk constructions. The filter medium is again mounted on a center tube. Elements can also be built up in several layers. Folded-paper cartridges of this type offer 170 m^2 of filtration area per unit. Pouch- and disk-type filters have far less area per unit (up to 0.5 m^2) but can be cleaned and are therefore reusable. Up to 3.45 m^2 of filtration area per unit is available in pouch filters. A disk filter consists of disks of pulp, metal screen, or membrane filter medium

Figure 27. Candle-type strainer filters (reproduced with permission of Mann & Hummel)
A) Radial fold filter; B) Disk filter; C) Multitube filter

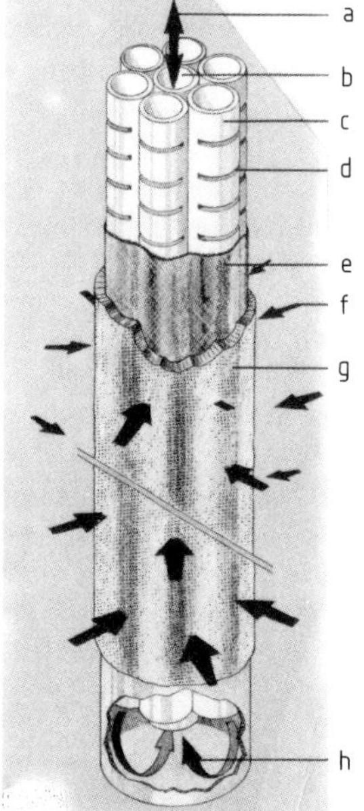

Figure 28. Candle-bag filter (reproduced with permission of Dr. Müller, Männedorf)
a) Filtrate/blowback-air; b) Central tube; c) Concentric tubes; d) Horizontal slots for filtrate entry–blowback; e) Filter medium; f) Slurry; g) Filter cake; h) Filtrate coming down concentric tubes and rising in central tube

slipped on a central support, alternating with support disks, which have slots to admit the slurry into the elements.

Other Designs. A number of other cartridge designs have been devised for special tasks. For clarifying difficultly filterable liquids, such as some pharmaceutical and biological liquids (blood, among others), elements made of fibrous bases or materials coated with activated carbon give satisfactory performance in the micrometer range (microfiltration).

A recently developed element is assembled from seven individual tubes having slots for the passage of filtrate (Fig. 28). A tube bundle is covered with a suitable high-pressure fabric, which fits tightly around the curved surfaces during filtration. During the backwash cycle, the fabric is lifted off the tubes; this movement causes the cake residue to be thrown off. The filter medium has a tubular weave, that is, it forms a seamless tube that can withstand several bars of backwash pressure. Therefore, its cleaning does not cause any concern. This large-area filter element is stable and can be used especially well as a backwash filter with a cake thickness of between 3 and 5 mm. With up to 200 m^2 of filtration area, it is suitable for thickening and clarifying polymer suspensions and various kinds of solutions, as well as cane syrups.

Design and Selection. Candle filters are generally designed to the flow capacity, filtration pressure and separating power stated by the manufacturer. The selection of the filter medium calls for some experience.

It can often be advantageous to use several units with small filtration areas in place of one larger-area unit. With several units in parallel,

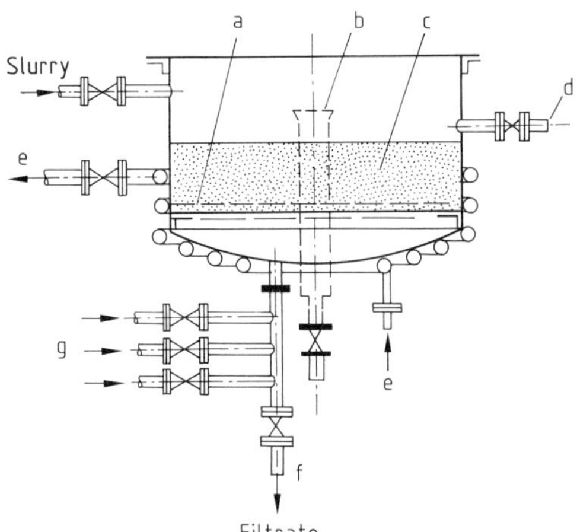

Figure 29. Open "low" sand filter
a) Screen; b) Effluent; c) Sand bed; d) Wash water; e) Heating or cooling system; f) Filtrate; g) Cleaning

for example, one element at a time can be cut out of the line for washing while the remaining ones stay in the filtration stage.

8.4. Deep-Bed Filters

This section deals only with depth filters having beds of sand, gravel or similar loose materials.

A depth filter is usually a simple container with a porous bottom, on which a bed of filter medium between 0.5 and 1.5 m deep is placed. If the bed material is gravel or sand, such a filter is also called a sand filter. Figure 29 shows a simple form of a gravity sand filter. The filter bed (c) is supported by the screen (a), and the filtrate is discharged through pipe (e). Water, steam or air can be admitted through pipes (g) for cleaning. Wash water overflows at (b), if at all. When agitation is used in cleaning, the wash liquor is discharged at (f). The cooling or heating coils (e) make it possible to temper the filter. Depth filters are also operated as pressure filters, either in order to increase the throughput or when they are installed in pressurized lines. To increase the filtration area of pressure sand filters, several beds can be arranged one above another.

Most filters of this type are used in the treatment of drinking water and process water. Basin-type sand filters have been introduced where quantities are very large, since these fixed structures offer relatively large areas and depths.

Flow in a depth filter is usually from top to bottom. The flow is reversed in order to clean the bed. The basic operating modes differ little even though a variety of techniques are possible, especially for cleaning.

With finer-grained bed materials used to collect finely dispersed solids, the topmost layer of the bed tends to become plugged, so that cleaning may become necessary before the entire bed is loaded with solids. In such cases, layers differing in particle size can be used; a new cleaning method is needed, however, to avoid mixing the fractions. In such multilayer filters, the liquid can be introduced at different points. This approach gives a 70% increase in flow velocity, substantially longer service time, and enhanced sludge capacity.

When a depth filter is employed to clarify juices, solutions and similar valuable liquids, enclosed tanks are used instead of open vessels, and higher pressures can also be applied. Pressure sand filters usually permit two to three times as great a pressure drop as gravity filters, and so finer-grained bed materials can be used as well. With a few exceptions, bed depths in such filters are much less than the depths usually found in water-treatment plants. In industry, depth filters with beds of special materials such as activated carbon or ion-exchange resins are often installed so that dissolved substances can be removed simultaneously.

In a *continuous process* (Fig. 30), the influent is initially in an unpacked internal space, whence it enters the bed through conical internals. The clarified liquid (filtrate) drains from the bed into an external space. The filter medium is hydraulically removed at the bottom and cleaned in a

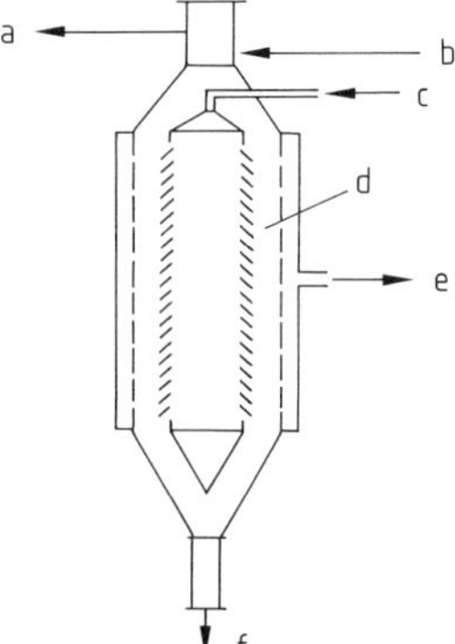

Figure 30. Continuous sand filter
a) Sludge; b) Sand and sludge; c) Slurry; d) Sand bed; e) Filtrate; f) Sand and sludge

Figure 31. Vacuum disk filter (reproduced with permission of Krupp)

regeneration tank at the top. The clean sand is returned to the filter vessel.

For smaller units (up to roughly 600 m^3/h), a backwash-cleaning filter has been developed for largely unattended operation. When solids deposited in the filter bed cause the differential pressure to increase, the liquid level in the tank will also rise until its hydrostatic pressure initiates backwashing. Because the cross-sectional area of the backwash pipe is larger than that of the influent pipe, the flow velocity can be roughly four times as great (44 m/h).

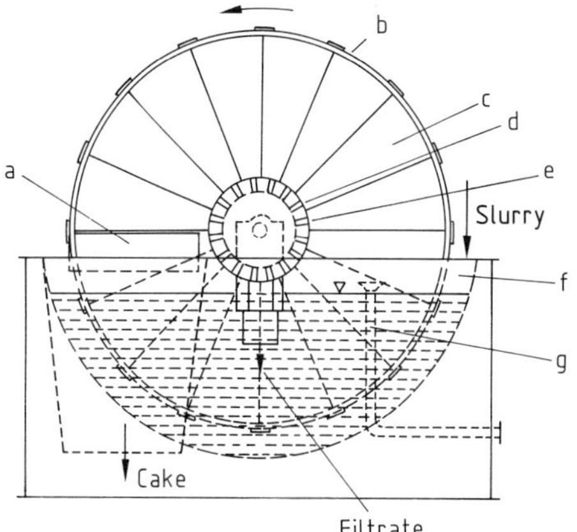

Figure 32. Vacuum disk filter
a) Scraper; b) Filter disk; c) Trapezoidal sectors; d) Outlet nippels ; e) Automatic valve; f) Filter tank; g) Overflow

8.5. Disk Filters

The element in a disk filter is a rotating vertical disk (Fig. 31); one or more such disks are arranged on a shaft, with roughly half of each disk submerged in a slurry tank (Fig. 32). The device allows continuous filtration. The disks are made up of 8–10 individual sectors, each of which is an independent filter. The interior space of a sector is connected to the filtrate discharge by a rotary valve, so that a cake can form on the outside when the sector is submerged in the tank. After the submersion part of the cycle, air is drawn or blown through the sector in order to effect dewatering; finally, the cake is removed by simultaneous application of a scraper and an air blast, or by spraying.

The disks are 1.2–3.6 m in diameter; in a special design, disks up to 5 m in diameter were used. Multi-disk filters have filtration areas of up to 280 m^2. In devices with relatively large diameters, hydrostatic effects may cause nonuniform cake thickness.

When several disks are mounted in a single tank, the tank is fitted with an agitator (usually a paddle mixer) or an air turbulizer in order to prevent sedimentation of the solids, which would lead to uneven cakes and poor dewatering. If not with an agitator, the multi-disk filter is built with a divided tank. The slurry level in the tank dictates the disk rotation speed.

The disk filter is commonly operated as a vacuum filter. The tank is open and readily accessible. If vapors generated at elevated temperatures must be kept from escaping into the environment, the filter can be mounted in a pressure-tight housing and operated under pressure; for hyperbaric operation, see also p. **10**-15).

In principle, the filter cake in a vacuum disk filter can be washed, but it is certainly less effective than in, e.g., belt filters. The basic advantage of the continuous vacuum filter is that it requires little space per unit of filtration area. Another advantage is that the filter elements are readily accessible. In some cases, this type of filter has the drawback that cracks develop in the cake and cannot be smoothed.

Applications center on high-tonnage products where large filtration areas are needed, e.g., iron ore slimes, flotation concentrates, white water from papermaking machines, blast-furnace dust from scrubbers, and aluminum hydrate (groups M and S, Table 2, p. **10**-26).

8.6. Drum Filters

Wrapping the filtration surface around a cylinder yields significant advantages that have made drum filters one of the most widely used types. Their advantages are continuous operation, easy charging, a choice of cake-removal methods, low sensitivity to suspension variations, good possibilities for washing in some cases, satisfactory residual moisture, relatively simple designs, reliability, and economy.

In one revolution, the drum passes through the processing stages, thus allowing continuous movement of filter cake and filtrate. Filters can be operated as gravity, vacuum or pressure filters.

Gravity Drum Filters (Revolving Screens). Applications of gravity drum filters are limited to simple screening, where cake formation is not crucial to the filtration operation, that is, where the cake does not greatly increase the filtration resistance. Thus the nature of the solids governs the application: gravity drum filters are suitable for filtering coarse-grained or fibrous mixtures, for example, treating surface waters, papermaking wastewaters, and the like. They can be charged on the inside or the outside. When the filter is internally charged, the untreated liquid flows through a distributing channel to the inside of the drum. As the drum turns, the solids collected by gravity move upward, separate and are removed, for example with a screw conveyor.

The drum is covered with a screen fabric. With microscreens, it is possible to collect solids that are somewhat more difficult to filter, such as occur in sewage treatment.

A filter charged from outside is used in the treatment of recycle waters and wastewaters in paper and pulp manufacturing. The drum is almost completely submerged in the slurry. While the solids are held up on the screen surface (an easily removable cake forms in the case of fibrous solids), the virtually fiber-free filtrate runs into the lowermost part of the drum. For a drum roughly 2 m in diameter turning at a peripheral speed of 3–15 m/min, throughputs of 0.5–5.0 m^3/min per meter of drum width are possible. Activated sludge flocs are collected with a drum 3 m in diameter and 3 m long, lined with a microscreen having a mesh opening of 10 μm; the throughput is 630 m^3/h.

Vacuum Drum Filters. In order to increase throughputs and extend the range of applications, the inside or outside of the drum is enclosed in a housing connected to a vacuum pump. In the simplest case, the drum has heads on either end, so that the entire inside space is under a uniform vacuum. Roughly 30% of the drum is submerged in a tank (Fig. 33), which fits

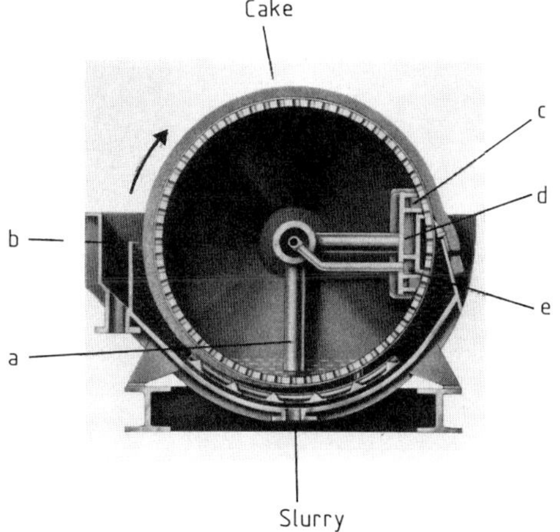

Figure 33. Sectional view of single-cell vacuum drum filter (KHD Humboldt Wedag)
a) Suction pipe; b) Overflow; c) Reverse blow; d) Cake discharge shoe; e) Sealing press liquid

fairly closely around it. Slurry in the tank is held at a constant level by means of an overflow. The feed is distributed by a stirrer, whose oscillating movement along the drum is intended to prevent partial settling of the suspended solids. In the submerged part of the drum, the vacuum pulls liquid through the filter medium while a cake builds up on the outside. The cake exposed on the top part of the drum is dewatered by air pulled through it. Shortly before the drum re-enters the tank, the cake is removed by an air blast from inside the drum, falling through a chute into a collector.

Single-cell vacuum drum filters are built with filtration areas of up to 40 m^2. They are suitable for all slurries that can be held suspended in the tank, that is, slurries that do not contain readily sedimenting solids (group S in Table 2, p. **10**-26), and where the cake need not be washed. They feature a high working vacuum, since there are no high losses; good dewatering, since relatively long blowing times are possible; and simple construction.

Drum filters generally have continuously variable drives so that they can accommodate a range of product properties. The filter medium (a finely-woven or fine-pored fabric) is supported on a widely spaced wire mesh.

Because of its simple design, the single-cell drum filter is well-suited to precoat filtration. In a special version where the common path of filtrate and air is short, it is preferred in the pro-

Ⓐ

Ⓑ

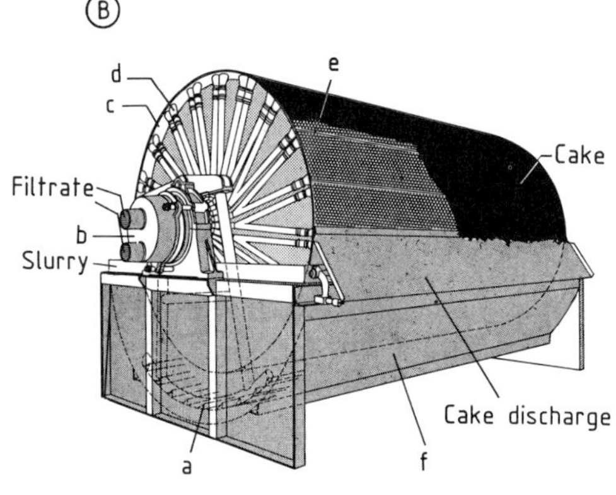

Figure 34.
A) Vacuum drum filter (reproduced with permission of Krupp)
B) Vacuum drum filter (Dorr-Oliver)
a) Agitator; b) Valve; c) Drum; d) Filter pipes; e) Division strips; f) Tank

cessing of cloudy wine and beverages. Throughputs varies from 0.2–1.2 $m^3 m^{-2} h^{-1}$, depending on the kind of beverage. In order to produce a uniform precoat, which may be 10–12 cm thick (this layer is continuously scraped off during filtration), the concentration of precoat material (diatomaceous earth, Chapter 11) is low, the depth of submersion is slight, and the drum is rotated rapidly. If the drum shell is divided into sections 50–60 mm wide and a tank is added to receive the wash liquor, cake washing is also possible in single-cell drum filters.

A special variety of this filter is the *suction roll* used in papermaking. It has a suction box, which restricts the vacuum to a relatively narrow zone at the top (Voith, Heidenheim). Suction rolls make it possible to increase the speed of the papermaking machine from 250 m/min with couch rolls to over 1000 m/min.

Multicompartment Drum Filters. Washing, as well as the separate recovery of filtrates from the several process steps, are made easier in multicompartment drum filters. The filtration surface in such a device is subdivided into flat, closed cells, which have their filtrate sides connected through separate pipes to vacuum pump(s) (Fig. 34). The compartments usually have perforated inserts bearing, for example, drainage channels on the cake side and filtrate collection channels on the vacuum side. The compartment insert bears a fabric, which supports the actual filter medium. While this fabric is installed separately for each compartment, the filter cloth itself covers as much of the drum periphery as possible.

The filtrate pipes rotate with the drum and have a sliding seal connection (rotary valve) to the fixed lines leading to the supply tanks and pumps. The valve also serves as control means, allotting the desired process stages to the several sectors of the drum. The control arrangement makes it possible to set up any desired filtration cycle; this is an undisputed advantage of the compartment filter.

The slurry feed is on the ascending side of the drum. An overflow pipe provides level control and thus maintains steady filtration conditions; the pipe can be adjustable. A stirrer in the bottom of the tank has the function of preventing the solids from settling in order to keep from upsetting the filtration. It is not always desirable to start filtration immediately after the drum is submerged, since with sedimenting feeds this would mean picking up the fine fraction first and thus impairing the permeability of the filter medium. The cake emerging from the tank should first be dewatered and then washed.

If cracks appear in the cake on dewatering, they will interfere with uniform washing. The cake surface can then be smoothed with pressure rolls. Washing should be spread over a large portion of the drum circumference, and so with large quantities of wash liquor there is a danger that the cake will slip off. To prevent this, washing belts are placed around the drum. Like the expressing belts described further on, these surround the drum and have the dual function of closing up cracks and effecting good distribution of the wash liquor. The wash liquor is usually delivered to the cake through drip tubes, channels or nozzles.

A variety of devices and techniques are used to remove various cakes from the filter medium. *Scraper discharge* is the earliest and simplest method (Fig. 34 B); it is best suited to rigid, nondeformable cakes at least 4 mm thick. Lifting of the cake can be reinforced by an air blast supplied below the filter medium through the filtrate pipes.

In *string and chain discharge*, parallel strings or chains are passed over the drum in the circumferential direction. After the last dewatering step, these pass over a deflecting roll with a relatively small diameter, and the cake then drops off. Chain discharge is suitable for relatively thick cakes such as are produced in the filtration of coarsely crystalline materials. "Felted" cakes, on the other hand, call for string discharge.

Roller Discharge. Thin filter cakes, down to 1 mm and thinner, can be taken up by rollers (Fig. 35A). Contact between roller and drum should not be too firm. The roller turns at the same speed or slightly (3–5 %) faster. The cake is removed from the roller by a scraper, comb or other means; part is left on the roller to promote separation from the drum.

Belt discharge involves leading the filter cloth with the cake over small rollers (Fig. 35 B). This method can be used for continuous filtration of slurries containing fine solids that would generally block the filter cloth too quickly. Spraying the cloth from both sides makes it possible to handle finely-dispersed suspensions without precoating.

When filtrates are volatile or tend to produce vapors, or when an improvement in filter capacity is desired (see p. **10**-43), it may be useful to enclose the filter and pressurize the housing slightly.

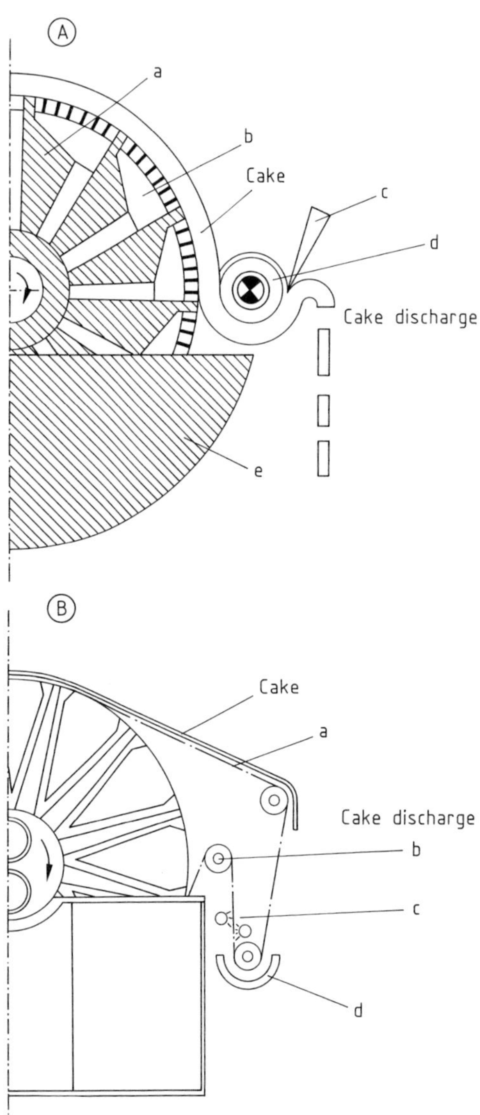

Figure 35. Cake discharge assemblies in vacuum filters
A) Roller discharge; a) Filter drum; b) Filtrate pipe; c) Scraper; d) Roller; e) Slurry tank;
B) Belt discharge; a) Filter cloth; b) Tensioning device; c) Cloth wash; d) Wash liquid receiver

Expression devices have recently been added to drum filters (Fig. 36). Besides lowering the cake moisture by 20–200 %, compression also promotes cake removal. The rubber expressing belt is held against the drum with several pneumatically adjustable rollers; forces per unit length of up to some 250 N/cm are used. The belt path is such that the cake gradually enters the region of maximum pressure.

For the filtration of finely dispersed feeds, the drum filter can also be operated as a precoat filter. Before filtration proper, a 5–6 cm layer of filter aid is applied from a slurry. It may be advantageous to smooth the precoat layer. During filtration, the filter aid must remain adhering to the filter cloth over the entire periphery of the

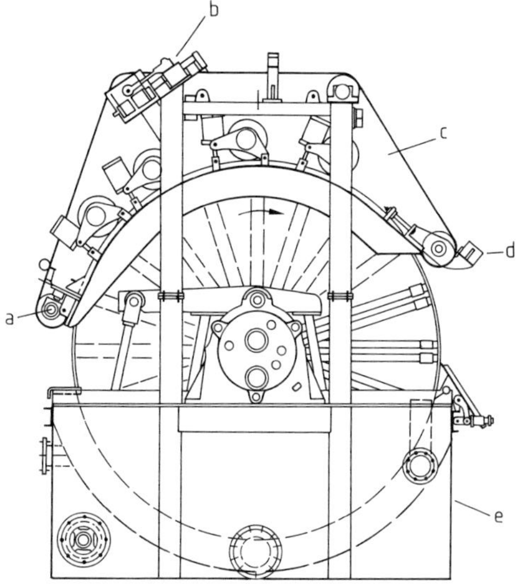

Figure 36. Combined vacuum drum pressure belt filter (Dorr–Oliver)
a) Press belt drive; b) Filter cloth aligning assembly; c) Press rolls; d) Press belt tensioning device; e) Vacuum drum filter

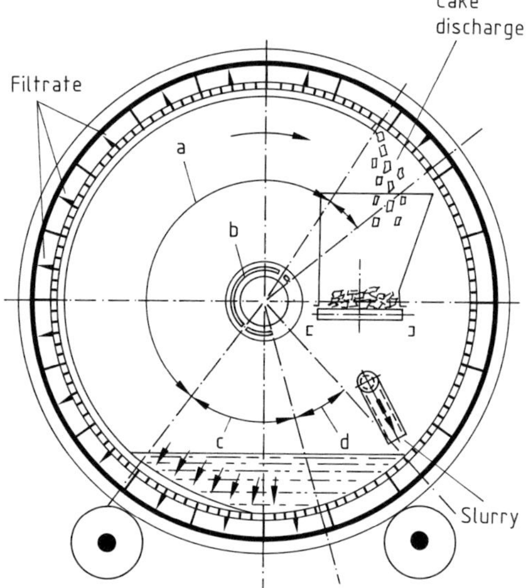

Figure 37. Inner face vacuum filter
a) Dewatering zone; b) Rotary valve; c) Settling zone; d) Suction zone

drum, so as to maintain the suction in all the compartments of the drum. No air blast is used to help in cake removal. During filtration, the scraper is advanced 0.05–3 mm per drum revolution until a permanent precoat thickness of a few mm is reached.

If the suspension is one that settles rapidly (group F of Table 2, p. **10**-26), it can be supplied to the tank from a box located at the top of the drum (*top-feed filter*). This arrangement creates sealing problems with respect to the drum and the cake. In general, the lower tank is either eliminated or filled with wash liquor. The *inner-face vacuum drum filter* (Fig. 37) is also suitable for this type of feed; sedimentation and filtration in such a device are in the same direction, so that no stirring is needed. The filter medium is on the inside of the drum shell, while the compartments are on the outside. The drum is open at one end to allow slurry feeding and cake removal; the closed end bears the filtrate pipes and rotary valve. By virtue of gravity, the coarse fraction of solids is deposited first on the filter cloth, forming a more permeable ground layer, which separates more easily from the cloth.

Vacuum multi-compartment filters are made in standard series (about 30 sizes) with filtration areas up to 100 m^2 and drum diameters up to 4.2 m. The smaller sizes, roughly 0.25 m^2, do good service as laboratory filters. The filters are irreplaceable for many continuous filtration jobs, being distinguished by reliability, low operating costs, versatility and adaptability.

Rotation speeds depend on filter size and desired cake moisture content and range from 0.1–3 revolutions per minute.

Sizing of Vacuum Drum Filters. From the filtration equation (Eq. 23), using Equation (42) and

$$t = \frac{60}{a\,n} \tag{66}$$

where a is the submergence ratio (ratio of submerged filtration area to total filtration area), t is the filtration time, and $60/n$ is the time for one rotation, one obtains

$$(M_S)_A = \frac{465\,\varrho_S}{\dfrac{100-c}{c} - \dfrac{\varrho_S\,w_f}{\varrho_S + w_f(\varrho_S - \varrho_L)}} \sqrt{\frac{\Delta p\,a\,n}{\bar{\alpha}\,\eta\,K_m}} \tag{67}$$

Here $(M_S)_A$ is the quantity of solids per unit time and unit filtration area; ϱ_S and ϱ_L are the densities of the solids and liquid, kg/m^3; c is the influent concentration, vol%; w_f is the residual cake moisture (weight basis); Δp is the pressure drop across the cake, N/m^2; $\bar{\alpha}$ is the average filtration resistance, m^{-2}; η is the viscosity of the liquid, N s/m^2; and K_m is a cake formation factor from Equation (20). The resistance of the cloth has been neglected.

The equation gives a parabolic increase in cake production with increasing rotation speed; in other words, the maximum solids output corresponds to the highest possible rotation speed. On the other hand, the cake thickness decreases exponentially with increasing throughput. Since this thickness is limited by the cake discharge device, the maximum solids production is also limited. The limit further depends on the material values contained in Equation (67).

Table 3 presents some examples illustrating the performance range of the vacuum drum filter.

Pressure Drum Filters. To realize the advantages of continuous drum operation in pressure filtration, with its higher throughputs, and to make the filtration of volatile liquids more economical, FEST developed a drum filter with a pressure housing around the drum (Fig. 38). The drum turns at a continuously variable speed. The annular space between drum and housing is sealed with stuffing boxes and partitioned into four or five chambers. The drum shell consists of compartments (Fig. 39), which are connected to

Table 3. Performance of vacuum drum filters (Dorr–Oliver).

Material	Feed concentration, % solids	Filtrate rate, kg m^{-2}h^{-1}	Cake moisture, %	Discharge
Calcium carbonate	20	34–50	40–50	scraper
Coal (flotation concentrate)	25	360–500	25	scraper
Coal (refuse)	30	120–145	25–30	scraper
Blast furnace flue dust	45	45	25–30	scraper
Iron oxide (pigment red)	13	17	30	roll
Corn starch	12–28	130–165	30–40	belt
Wine lees	low	82–820 (L/m^{2}h)	–	precoat

Figure 38 Pressure drum filter, 2.1 m^2 filtration area (reproduced with permission of BHS)

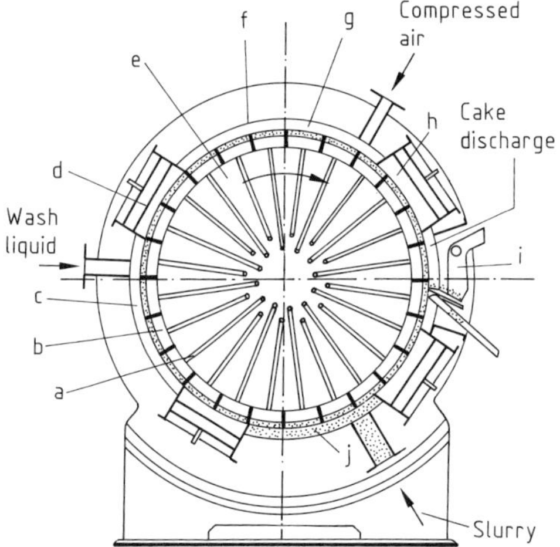

Figure 39. Pressure drum filter – (FEST, BHS)
a) Filtrate pipes; b) Drum cell; c) Wash chamber; d) Partitioning element pressed by compressed air; e) Filter drum; f) Pressure tight housing; g) Dewatering chamber; h) Partitioning element; i) Scraper; j) Filtration chamber

Table 4. Throughputs of pressure drum filters, $kg\,m^{-2}h^{-1}$, based on dry cake (BHS, Sonthofen, Federal Republic of Germany)

	Throughput	Cake moisture, %
Copper sulfate	525	24
Sodium hydrosulfite	700	20
Zinc stearate	45–160	60–70
PVC (suspension)	420	51
Fuller's earth	96	30
Titanium dioxide	150	45

a rotary valve by filtrate pipes. Slurry under pressure is continuously pumped into the filtration space. The cake builds up in the filter compartments; as the drum rotates, the cake is mechanically compacted by passing through a wedge, then moves on to the washing chamber. If no provision is made for multistage washing, the next step is dewatering by blowing in a separate chamber, which is followed by the cake removal section. In the unpressurized region, the cake is removed by back-blowing and pivoting scrapers. Washing of the filter cloth can also take place in this zone.

The advantages are: continuous operation, relatively high throughputs, filtrations up to near the boiling point of the liquid, no vapor losses, very good washing, countercurrent washing

Figure 40. Pressurized drum filter–Hyperbar filter

when multistage arrangements are used, low cake moisture, no cracking before washing, and mechanical compaction. The filter is naturally more expensive than a conventional rotary drum filter, so that additional filtration costs are calculated as some 50%. The application of this type of filter is thus somewhat limited; it finds use mainly in the chemical industry. Table 4 gives some throughput figures, which show substantial gains in some cases. Because cleaning is difficult, it is recommended that the product not be changed.

Increased capacity sometimes results from the use of pressure drum filters created by housing conventional vacuum drum filters in pressure tanks (Fig. 40). These devices have long been known and have recently come to be called *hyperbaric filters*. Many such designs provide better dewatering than the straight vacuum filter.

8.7. Leaf and Plate Filters

The element in a leaf or plate filter is essentially a leaf-shaped structure covered with filter medium on one or both sides. Slurry is supplied to the leaf from the outside, and the filtrate can drain through the void inside, while the cake collects on the outside. Filter elements are usually assembled into units of varying size. They can be operated as open-tank devices, immersed in the feed, with the filtrate being withdrawn by suction (open-tank leaf filter, Moore filter). Because the cake often drops spontaneously into the feed tank, these devices are also used as filter thickeners. Alternatively, the filter elements are enclosed in pressure vessels (which may be heatable) and operated as pressure filters, especially for influents of relatively low concentration (clarifying filtration). They are also well-suited to precoat filtration.

The shape and arrangement of the filter leaves or plates are adapted to the properties of the cake and the filter operating requirements. If the cake is easily detached from the filter medium, a vertical leaf configuration can be chosen, so that the cake falls off the medium by gravity (removal can be promoted by an air blast or vibrating). Horizontal leaves are suitable for poorly-adhering filter cakes.

Possible element shapes are innumerable, from simple perforated and mesh constructions to multi-layer designs with internal stiffening. The design of Figure 41 is typical for bed elements which are preferred because of their pressure resistance. Each element consists of five fabric plies with graded mesh openings. The core ply, which allows the filtrate to drain, bears a supporting fabric on either side, with the filter cloths proper on the outside.

In another type, the effective filtration area in each element is made up of a base with perforat-

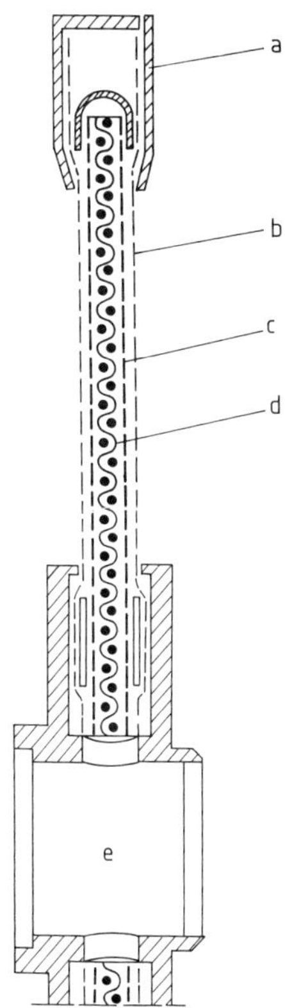

Figure 41. Wire filter leaf
a) Binding; b) Filter medium; c) Metal mesh; d) Wire mesh support (Chamber screen); e) Filtrate manifold

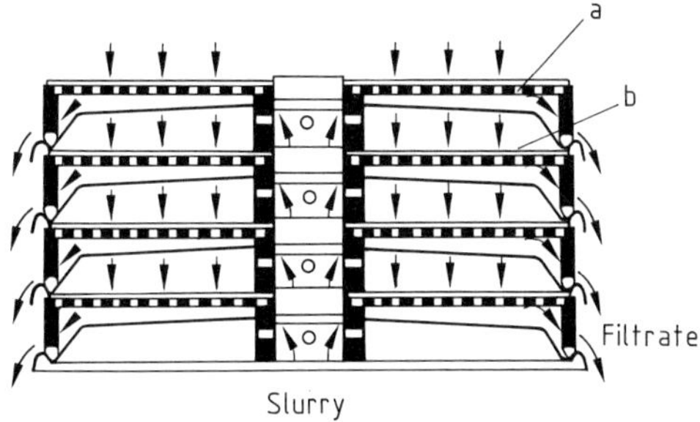

Figure 42. Horizontal plate filter (Tank sheet filter, Seitz/Schenk)
a) Perforated plate; b) Filter sheet

Figure 43. Horizontal-tank pressure leaf filter (Niagara)

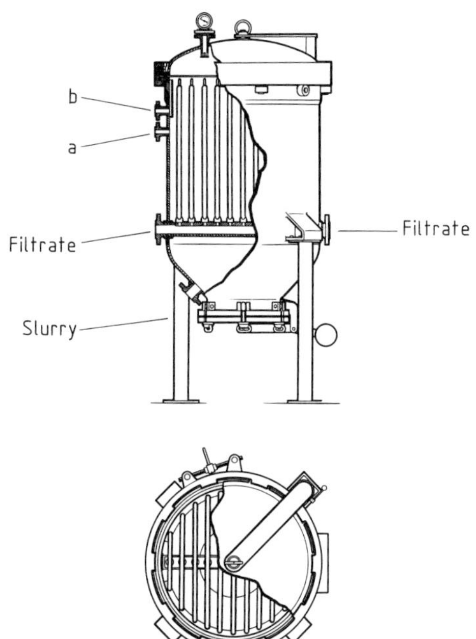

Figure 44. Vertical pressure leaf filter (Loeffler)
a) Valve security; b) Overflow

ed plates, which is inserted in a massive frame to withstand greater mechanical loads. The filter cloth or a wire mesh is placed over the base.

For clarifying filtration with filter beds, woven fabrics, filter paper or membranes, the filter media are placed right on the horizontal plate structure. Sealing is accomplished by pressing the filter elements together with a spindle (Fig. 42). The elements are cleaned by backwashing.

Leaf filters are made in a smooth design with up to 60 m^2 of filtration area. In large plant units, the leaf assemblies can be moved mechanically or hydraulically (Fig. 43), or the housings are designed for easy cake loosening and removal. Because the pressure on the cake is relieved when the tank is opened, a vacuum is often pulled on the filtrate pipes at this time so that the cake will not drop off the leaves until the desired moment.

Pressure leaf filter with relatively high througputs call for special designs. In the Kelly filter (Dorr–Oliver), vertical filter leaves were arranged axially in the cylidrical pressure housing, so that the leaf size decreases to either side. The leaves are vertical in smaller units or horizontal in larger ones (up to 140 m^2).

The cylindrical housing of the Sweetland filter (Dorr–Oliver) is divided; the lower half can be pivoted to the side to allow cake removal by hosing off.

Where process conditions do not allow the slurry heel (slurry remaining in the pressure housing after filtration is complete) to be recycled to the filter feed or otherwise utilized, some designs have one or more filter leaves in the bottom of the pressure space to filter this residual

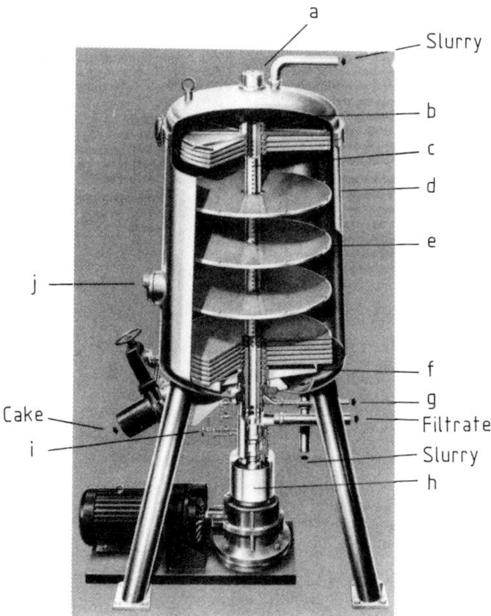

Figure 45. Centrifugal cake discharge pressure plate filter (Schenk)
a) Upper bearing housing; b) Compression flange; c) Hollow filter shaft; d) Filter vessel; e) Filter plate; f) Scavenge plate; g) Scavenge filtrate; h) Drive shaft with bearing housing; i) Gland irrigation; j) Sightglass

feed. These leaves do not function during the main part of the cycle. The heel is usually driven through those filter leaves by pressurized gases (compressed air).

Many designs address the problem of cake removal. Modern filtration systems are generally provided only with completely automatic filters; with vertical leaves as shown in Figure 44, automatic removal is often a simple step.

In horizontal systems, flowable and non-flowable cakes must be distinguished. Flowable cakes are the rule in precoat filtration for clarification; here the cake is commonly discarded as waste. The round filter leaves or plates, with slurry feed from the top, are arranged on a hollow shaft in a pressure vessel and positioned a few centimeters apart with spacing disks (Fig. 45). Cake thickness is controlled through the filtration pressure or with special sensors. Cakes in these devices are nearly always so loose that they

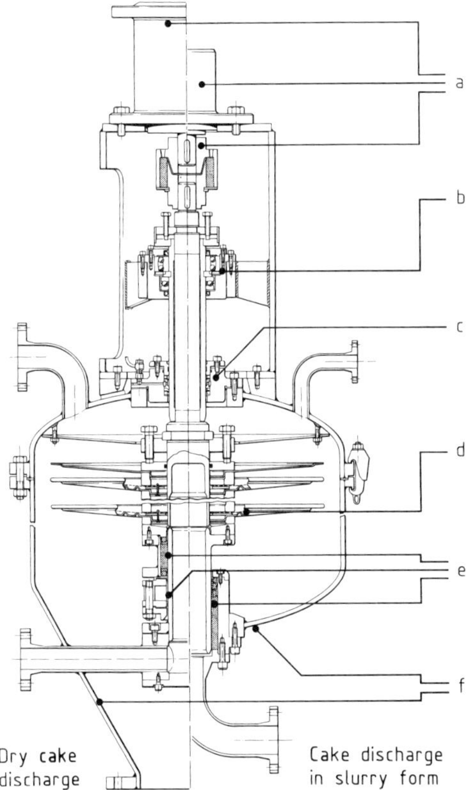

Figure 46. Centrifugal-discharge pressure plate filter (Stawag)
a) Filter drive; b) Upper bearing; c) Upper seal; d) Filter plates; e) Lower bearing and seal assembly; f) Filter vessel

can be discharged by briefly spinning the stack of plates; separation can be helped by backwashing if necessary. Cake removal, required when leaves cease to function, is effected by vertical vibration in addition to rotation. Filters of this type are also called *centrifugal-discharge filters*. They are often classified as plate filters.

To empty the filter completely before removing the cake, most filters of this group have leaves at the bottom of the stack to allow separate filtration of the slurry heel.

For relatively dry filter cakes, a conical discharge is preferred (Fig. 46, left). The stack of plates in such a device must be driven from the top of the tank, which, in addition facilitates the sealing of the driving shaft.

Non-flowable cakes are generally obtained in cake filtration where filter aids must be avoided. Help in these instances comes only from mechanical cleaning devices. Washing in these pressure filters usually involves some difficulty, because the slurry heel must first be removed and the housing filled with wash liquor. Special devices have been developed to reduce the consumption of wash liquor.

In the pressure leaf filters described above, the feed is pumped into the housing, and its distribution over the leaves is naturally somewhat random. The slurry can also be supplied to the leaves through a hollow center shaft, with the filtrate draining into the housing (Seitz, Bad Kreuznach). In this way, uniform loading of the filter elements is ensured.

Range of Application. The applications of pressure leaf filters are extraordinarily broad, and extension to special filtration problems is relatively easy. As cake filters, these devices are employed for filtering all kinds of chemical products, such as glycerol, caustic soda solution, phosphoric acid, or pharmaceuticals. Run times between cleaning periods are several hours, and throughputs of $1-4\ m^3 m^{-2} h^{-1}$ (up to $6\ m^3 m^{-2} h^{-1}$ in extreme cases) have been achieved. Pressure leaf filters are applied in group S and D of Table 2 p. **10**-26.

When used as precoat filters, these units are especially well-suited for the collection of relatively fine-grained solids from low-concentration feeds (beverages, cane syrup, or water). Average throughputs are $0.3-1\ m^3 m^{-2} h^{-1}$. Run times are strongly dependent on the level of impurities to be removed; in mixed and thick juices in sugar refining, runs of 7–16 h (depending on, e.g., the

A

B

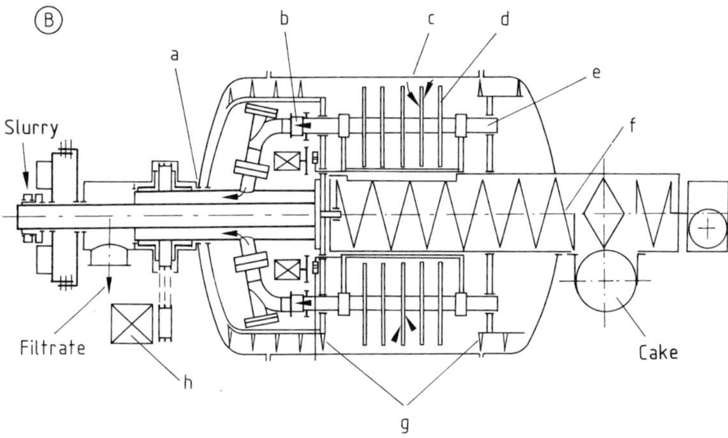

Figure 47. Continuous pressure leaf filter (A and B) (Amafilter)
a) Drive shaft; b) Pressure tank box with gear box; c) Filter vessel; d) Filter leaf; e) Rotating axis; f) Screw conveyer; g) Mixing blades; h) Drive

quantity of diatomaceous earth used as filter aid) are achieved.

Centrifugal-discharge filters are available with filtration areas of 1–60 m^2 (up to 150 m^2 from one manufacturer); the leaves are 420–1500 mm in diameter.

A novel, *completely automatic pressure leaf filter* (Fig. 47) is intended for especially high throughputs, such as are needed in coal cleaning. In a horizontal pressure vessel, round two-sided filter leaves are mounted vertically on six shafts. The elements rotate during filtration. At the same time, the stacks of filter leaves revolve about the center axis of the vessel, so that one stack after another is situated in the slurry sump at the bottom of the tank. A uniform cake forms on the filter leaves connected to the high shafts with unpressurized filtrate drainage. After dewatering by air blowing, the cake in the top part of the vessel is blown off with compressed air and

Figure 48. Pressure Nutsche filter–Process unit (reproduced with permission of Schenk)

discharged through a gate. With coal, concentrates with 14% cake moisture have been produced in a filter rated at 625 kg $m^{-2} h^{-1}$.

8.8. Nutsche (Pan) Filters

The simplest type of nutsche filter is an open tankor box with a porous bottom, into which the slurry is charged or flows continuously. Filters of this kind are common everywhere, especially in laboratories. They do not impose stringent requirements on suspension properties, and they can function even without difficulty with highly-concentrated or rapidly-settling slurries. As a rule, they are operated manually. For poorly filterable feeds, suction can be applied to the filtrate side.

Nutsches are widely applied in the chemical industry because all the steps in filtration, including washing, dewatering, and drying if practiced (with hot air), can be performed in a clean, distinct manner. There has accordingly been no lack of efforts to automate the operation of nutsches or make it continuous.

In order to increase filtrate flow rates, an enclosed nutsche can also be pressurized with air or inert gas. Pressure nutsches are especially good with volatile solvents. They can be drained mechanically through a retractable bottom. Several process steps can be carried out with no change of apparatus when nutsches in process service are fitted with a variety of accessories (e.g., agitator, mixer, heating and cooling devices); in this way operations such as mixing, stirring, reacting, filtration, extraction and drying can be performed in sequence. Low equipment costs, non-polluting operation, and optimal techniques make these devices suitable for small product batches. Nutsches can be placed on pivots and repositioned for the various process steps (Fig. 48).

Nutsches are good candidates for filtration with relatively thick cakes. To keep cracks from appearing, some filters have smoothing attachments, which can also provide advantageous mixing of wash liquor and cake and can assist in cake discharge.

Tanks with internals such as agitators and solids-discharge devices (plough) are more versatile nutsches. These accessories can homogenize the filter cake and render it especially washable (Fig. 49).

In some advanced developments, intended for comparatively high throughputs, a number of individual nutsches circulate on either a belt or a carousel. (A system of the first type is usually classed as a belt filter.) This kind of nutsche has proved useful especially in phosphoric acid manufacturing, where it is employed to remove gypsum in the sulfuric acid digestion of phosphate rocks. Careful countercurrent washing is practiced in order to attain a high degree of separation. Equipment sizes range up to a filtration area of some 200 m^2.

Automated nutsches share the advantages stated earlier: They are insensitive to slurry properties and thus suitable for groups F, M, and S in Table 2 (p. **10**-26), and they provide a clear separation of the process steps and thus insure a uniform, reliable filtrate composition.

For the filtration of slimy feeds, it is advantageous to keep the filter cake in motion, for example by mechanical vibration or an appropriate slurry flow, so that the cake will remain permeable for a longer time.

8.9. Table Filters

A table or bowl filter (Fig. 50) consists of flat filter elements, usually forming compartments,

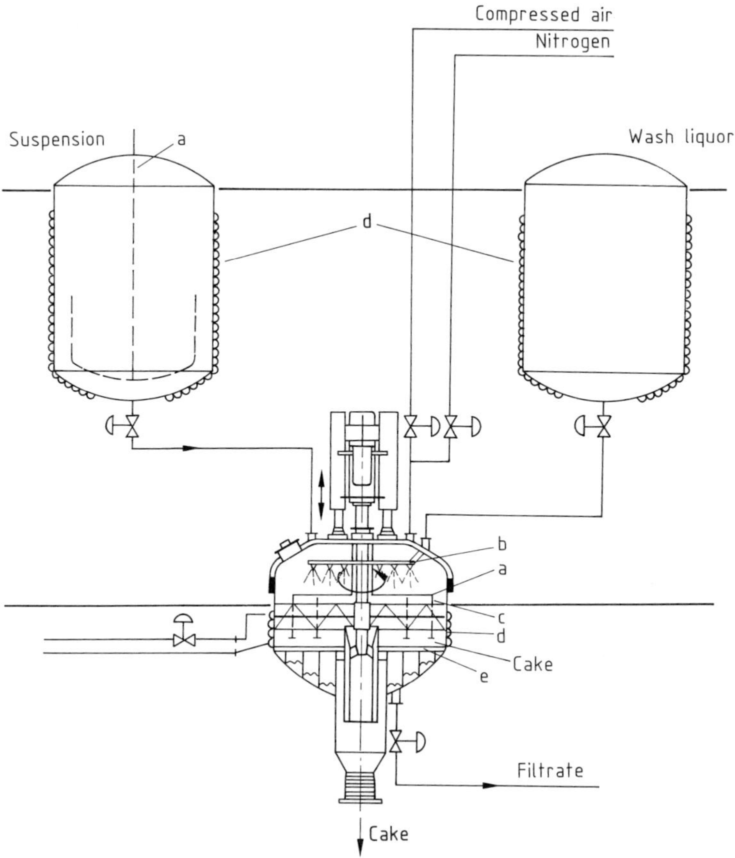

Figure 49. Pressure Nutsche filter
a) Agitator; b) Wash device; c) Screw conveyer; d) Heating pipes; e) Filter plate

which successively pass through the process steps as the table rotates. A rotary valve pressed against the bottom of the table connects it to the vacuum pump, filtrate tanks, and compressed-air supply (for lifting the cake). As usual, the filter elements are lined with suitable filter media. To prevent slurry overflow, the table has an attached or co-moving vertical rim (table). Slurry is fed to the filtration surface by a groove distributor. The horizontal arrangement makes it possible to filter rapidly-settling suspensions (group F in Table 2, p. **10**-26) and allows the formation of relatively thick cakes (up to 30 cm), which improves the economics of filtering coarse-grained slurries. After further dewatering, the cake is commonly removed from the table by a screw solids discharge.

To ensure uniform washing, the cake is often smoothed by a bar immediately downstream of the feed position. Table filtration is especially suitable for countercurrent washing. Load ratings are up to 10 000 $kg\,m^{-2}\,h^{-1}$ or higher, and filters are built with diameters up to some 7 m, corresponding to a filtration area of 40 m^2. Some models are gas- and pressure-tight.

A special table filter, which avoids problems and complications in filter-cake removal and rim sealing, is provided with a co-moving rubber rim. At the cake-removal position, the rim is pulled away from the table by rollers [6]. Filters are available with filtration areas of up to 250 m^2. The centrifugal-discharge filters described on p. **10**-45 are often classified as table filters.

8.10. Pressure Plate Filters

The filter press is the simplest of all pressure filters; it allows relatively easy cake removal after the individual filtration steps. It consists of a

Figure 50. Horizontal rotary table filter (reproduced with permission of KHD Humboldt Wedag AG)

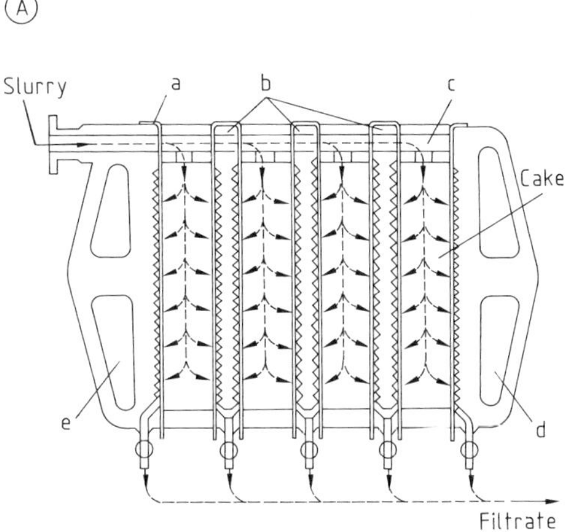

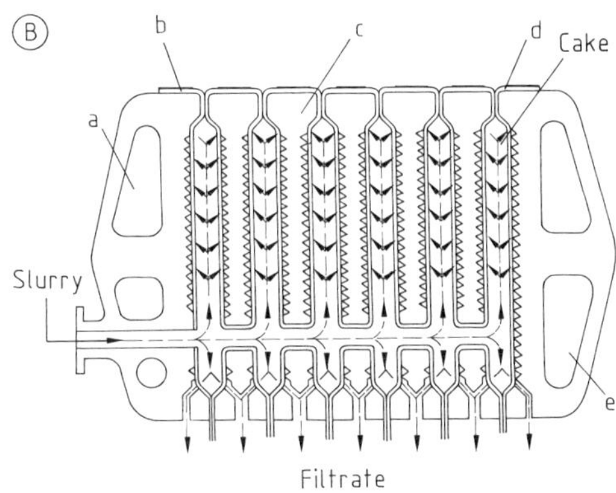

Figure 51. A) Plate-and-frame filter press
a) Filter cloth; b) Filter plates; c) Hollow frames; d) Movable end half plate; e) Fixed end half plate
B) Recessed-plate filter press
a) Fixed end half plate; b) Filter cloth fastening; c) Recessed filter plates; d) Filter cloth; e) Movable end half plate

number of flush plates, up to 100 or more, and frames whose width (between 25 and 150 mm) can be matched to the service conditions. The plates and frames form rather flat pressure filtration spaces with a relatively large filtration area (*plate-and-frame filter press*, Fig. 51A). During filtration, the plates and frames must be squeezed by mechanical or hydraulic means in order to seal the filtration spaces. The faces of the plates are studded or grooved to permit filtrate drainage.

The plates can also be made with a raised edge (*recessed-plate filter press*, Fig. 51B). There are several essential differences between the two types: In the plate-and-frame press, slurry enters through channels in the corners of the plates and frames. The filter cloths can therefore be shaped as flat strips and pulled over the plates; this arrangement is advantageous when the cloths are changed. In recessed-plate presses, slurry is introduced into a center orifice of the plates. The filter cloths must be made in two parts and inserted in the slurry feed channels; this arrangement involves much more effort. Further, it is harder to vary the cake thickness in recessed-plate devices than in plate-and-frame presses, where all that is necessary is to replace the frames. The recessed-plate press has, however, the advantage that only about half of the filter elements need be moved to open the press; the equipment also costs much less for a given filtration area. Because of the deflection of the filter cloths at the edge of the recessed plate, thinner cakes (up to about 35 mm) are generally produced than in plate-and-frame devices. With either type of press, it must be noted that capillary flow takes place despite the relatively high density of the filter cloths, and this flow—at least in filtration with solvents— causes the undesirable escape of filtrate through the seals between plates.

The filtrate is either discharged separately from each plate through cocks (open discharge) or collected in a channel (closed discharge), which is preferred when toxic or volatile materials are handled.

At cake washing time, valves are changed over and special lines carry wash water only to every other plate, so that the liquor can flow transversely through the cake to the opposite plate. Occasionally the wash water is also delivered through the slurry channel; this arrangement is, of course, not as effective. After washing, the cake can be dewatered by passing steam or (possibly heated) air through it.

For cake discharge, the press is opened and the plates are moved, one by one, through a predetermined distance. A variety of mechanical devices are now available for this purpose. If the cake is to drop out by itself, it should not adhere too tightly to the filter cloth and should not be too densely filtered. Cake separation can be assisted by rapping, pulling out the filter cloth, or other means. An automatic filter press for high throughputs has more than 100 chambers with a filtration area of 1000 m^2 and more.

Filter presses are generally insensitive to slurry filtration properties, with perhaps one exception: In large presses with long slurry channels, varying sedimentation may cause a slurry of group S (see Table 2, p. **10**-26) to fill the filtration spaces unevenly. Under some conditions, an undesired classification occurs, with a detrimental effect on washing, and may even lead to pressure differences between chambers, causing the plates to deform or break. In such cases, it is desirable to provide a hole for pressure equalization or a supporting button.

In addition to filter presses with vertical plates, some designs also feature horizontal plates.

Filtration Cycle. Cake filtration in filter presses involves two indistinctly separated stages. At the start of filtration there is no cake, and the hydraulic resistance is therefore very low, so that the filtration pressure cannot rise very high at this point. The process is similar to constant-volume filtration (see p. **10**-11). The pressure increases as the cake thickness grows, until the rated pressure (of the pump) is reached. Constant-pressure filtration then begins, with a declining quantity of filtrate. The exact sequence of filtration in filter presses is still unclear and cannot be calculated, because not enough research has been done, but a rough estimate suggests that the highest throughput is achieved if

$$t_f = t_k Z \tag{68}$$

where t_f is the filtration time, t_k is the time for filling and emptying per chamber, and Z is the number of chambers.

On the basis of a cost calculation, the maximum number of chambers is

$$Z = \left[\frac{t_f (C_M - C_p)}{t_k C_p}\right]^{\frac{1}{2}} \tag{69}$$

where C_M is the cost of mechanical equipment (fixed and movable end plates, adjusting device) and C_p is the cost per plate (chamber). For conventional designs, Z is found to be around 60.

To determine t_f, the total cycle time including washing and other steps, experimental data must be used. Unfortunately, experience has shown that measurements in laboratory-scale filter presses cannot generally be extended to arbitrary sizes, since there may be discrepancies in the filling of the filtration chambers and thus in the pressure relationships.

Filter presses are commonly charged at pressures of up to 2 MPa, more in special cases. Centrifugal or diaphragm pumps are preferred for the cake filtration of highly-concentrated slurries. The possibility of employing relatively high filtration pressures, along with the mechanical advantages in charging and discharging, lend filter presses some advantages over other types of pressure filter, even though these devices are always operated batchwise:

1) Insensitivity to feed fluctuations
2) Relatively high throughputs, even with difficultly filterable slurries, so that far smaller amounts of filter aids are needed
3) Low cake moisture
4) Denser cakes that are easier to handle
5) Clear filtrate
6) Good separation of different filtrates (wash liquor, etc.)
7) Arbitrary filtration steps (washing, steam or air blowing, etc.)
8) Low maintenance costs, few spare parts, low depreciation
9) Simple operation
10) Comparable equipment costs per square meter of filtration area

The drawbacks are obvious: Batch operation; personnel requirements in cases where cake

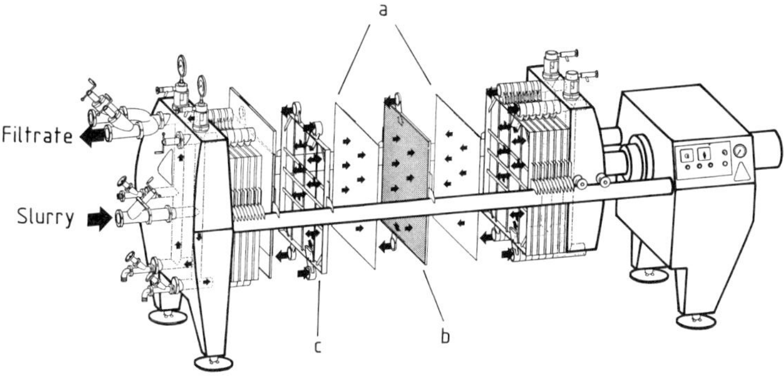

Figure 52. Sheet filter press (reproduced with permission of Schenk)
a) Filter sheets; b) Plate; c) Frame

does not drop off by itself; relatively long cleaning times.

Applications. Even though they are batch devices, the good washing and low cake moisture available in filter presses have kept them in wide use. Contributing factors, along with comparatively long service times and good reliability, include mechanical plate movement, used almost exclusively for the opening and discharge of large units; automatic filter-cloth cleaning with high-pressure water sprays; and the availability of special plate designs. Plates up to 2.6 m on a side give far more than 1000 m^2 of filtration area in one press. The design of the plates and the use of polypropylene as the main plate material have remedied many of the shortcomings of filter presses, such as plate fractures, corrosion, and plate mass.

Filter presses can be employed almost anywhere, even if the mode of operation and the equipment must be adapted to the feeds in question. Filter presses are used successfully in either cake or clarifying filtration, depending on the concentration and properties of the solids; in the latter case, filter aids are usually added, or filter beds are employed instead of cloth media.

In the production of chemicals (e.g., dyes), ceramics and raw materials, and in wastewater treatment, nuclear technology and other fields, cake filtration is applied chiefly to groups M and S of Table 2, p. **10**-26. Slurry feed rates in these fields depend on the widely varying slurry composition and range from 0.2–1 $m^3 m^{-2} h^{-1}$ on the average.

Clarification is carried out in *sheet filter presses*, which resemble plate-and-frame filter presses in design (Fig. 52). Because of the very much lower solids concentration, service times are much longer (several hours). For this reason, automation is commonly dispensed with. Filter

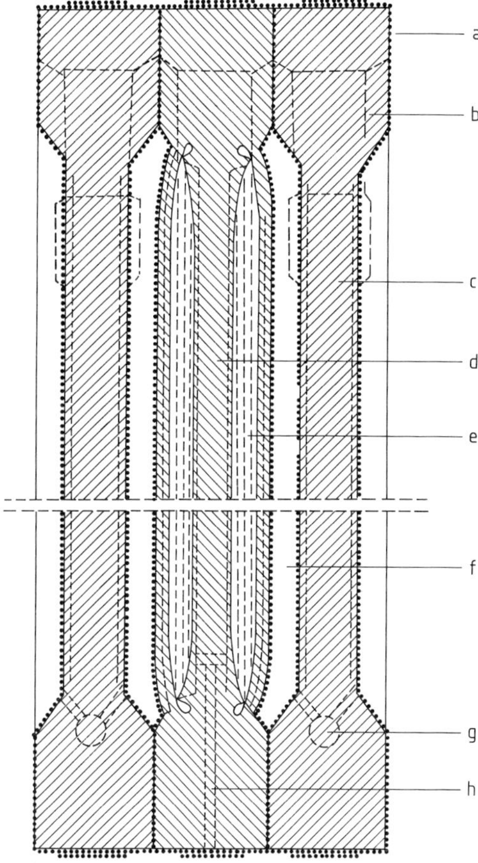

Figure 53. Membrane filter press plate assembly
a) Filter cloth; b) Slurry corner feed; c) Recessed plate; d) Membrane recessed plate; e) Press medium; f) Cake; g) Filtrate; h) Air or water inlet (Pressuring membranes)

sheets include ordinary precoats and special papers (see p. 10-57).

The most important applications of clarifying filtration include beer, wort, wine, fruit juices, pharmaceutical liquids, or water sterilization (group D of Table 2). Here too, filter loading varies widely, depending both on the properties of the solids being removed and their concentration in the feed and on the nature of the liquid. They range from 0.05 $m^3 m^{-2} h^{-1}$ for viscous varnishes up to 10 $m^3 m^{-2} h^{-1}$ for dilute suspensions, such as beverages.

Membrane Filter Presses. In a membrane press, the filter plate is coated with an elastic material (rubber). After filtration, the membranes are pressurized so that they mechanically compress the filter cake. In this way the cake moisture is reduced by ca. 1 % – 20 %, depending on the cake compressibility. Figure 53 shows an example of a recessed-plate membrane press. A similar expedient is also possible in a plate-and-frame press. The pressing either shortens filtration time or the filtration pressure can be made smaller, which is an eminent practical advantage.

Membrane filter presses can be operated at up to 2 MPa; they produce a uniformly dewatered cake.

Plate Press with Belt Discharge (Automatic Press). This type of filter press, originally developed in Russia, combines positive automatic cake discharge with cake compression. The horizontal filter plates are stacked and, with the filter frames, combined into a three-chamber system: filtrate chamber, cake chamber, and high-pressure water space. The water space is separated from the cake by a membrane (Fig. 54). For cake

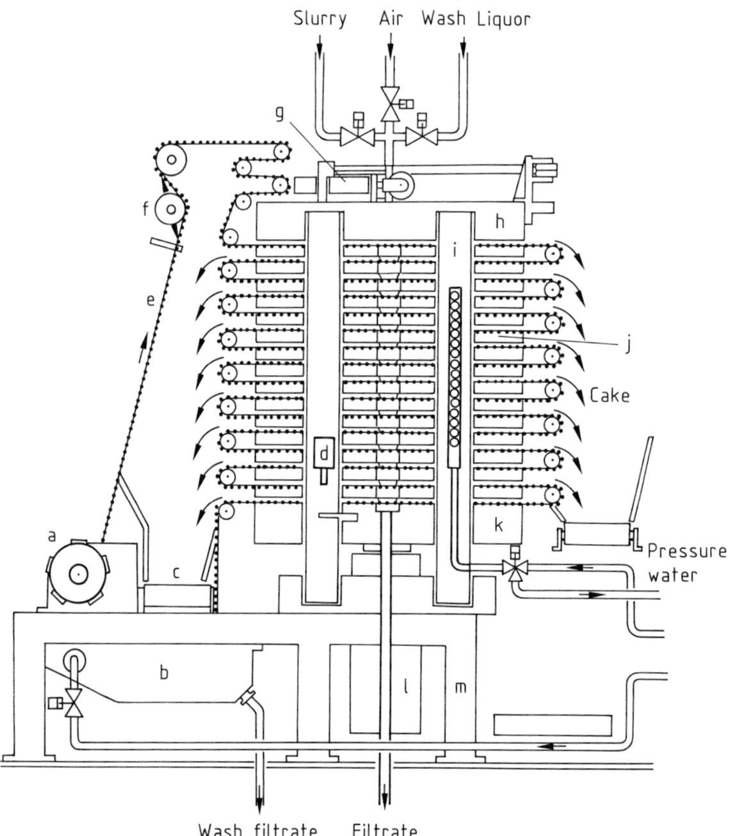

Figure 54. Automatic filter press (Hoesch)
a) Cloth drive; b) Cloth wash; c) Cake removal; d) Pressure control; e) Filter cloth; f) Filter cloth aligning assembly; g) Tensioning device; h) Yoke plate; i) Yoke anchor; j) Filter plate with inflatable membrane; k) Lifting table; l) Hydraulic closure device; m) Filter preframe

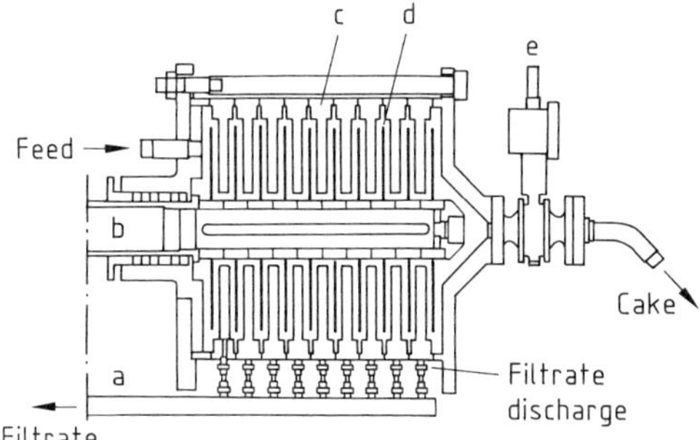

Figure 55. The dynamic filter press
a) Filtrate discharge pipe; b) Main shaft; c) Filter plate; d) Agitating disk; e) Cake discharge valve

discharge, a cloth belt on the frame is advanced until the cake from each plate is completely stripped off as the belt is deflected into the next lower plate.

Filter Press with Stationary and Rotating Plates. This filter contains alternating stationary and rotating plates on a shaft (Fig. 55). A velocity difference is produced in the cake between the plates, leading to breakdown of the cake. A thin layer of filter cake, whose thickness depends on the filter medium employed, is left on the filtration surface. In the extreme case where cake formation is totally prevented, this is pure cross-flow filtration (see. p. **10**-2). The intermediate states are referred to as dynamic filtration.

If the filter cake is thixotropic, the shearing makes it flowable, so that it can be conveyed by pressure through the system of plates. According to studies of some dye suspensions, this filter can collect the same quantity of solids as a conventional filter press with roughly 20 times as large a filtration area.

8.11. Tubular Filters

Tubes as filter elements are similar in structure to cartridges. This section deals only with tubular filters whose dimensions do not usually correspond to the concept of a filter cartridge (diameter either too large or too small). This group includes, on the one hand, tubular pressure filters and, on the other, tubular membranes used in cross-flow filtration.

Tube Filters. The tubular pressure filter—the *tube press*—developed just a few years ago in

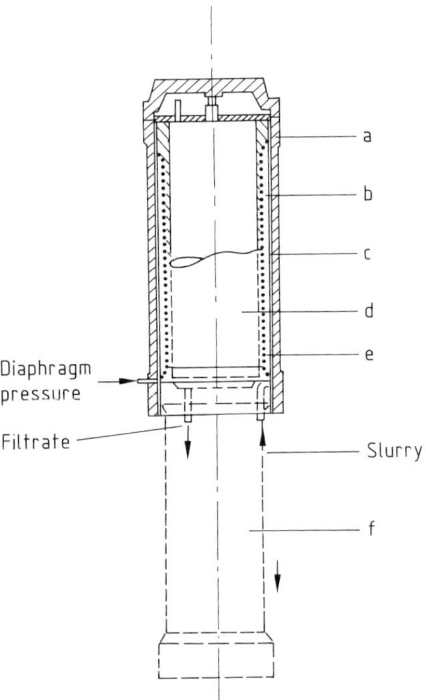

Figure 56. Tube press (Rittershaus & Blecher)
a) Tubular shell; b) Filter chamber; c) Inflatable sleeve; d) Filter tube; e) Filter cloth; f) Lowered filter tube for cake discharge

England, consists of a pressure tube with a filter tube inserted in it. After cake filtration in the annular space, a membrane placed inside the pressure tube compresses the cake (Fig. 56). A pressure of up to 10 MPa can be applied. Next (or immediately after filtration), the cake can be washed. For cake discharge, the filter tube is swung downward and the cake is loosened from the filtration surface with compressed air. This filter has the advantage of the cylindrical filtration chamber, which is economical for relatively high pressures.

In another type of tubular pressure filter, filtration and compaction are similar but the tube is positioned horizontally, so that the cake can be scraped out of the filter tube.

A third type of tubular pressure filter has a cylindrical rotor inside the filter tube. Slurry is continuously compacted in the annular space. The solids are retained in the annular space on the filtration surfaces of the tube and the rotor, where they are subjected to shear. In this way, the cake is caused to move downward to the discharge. As in the filter press with rotating plates (see p. **10**-53), the filter cake must remain flow-

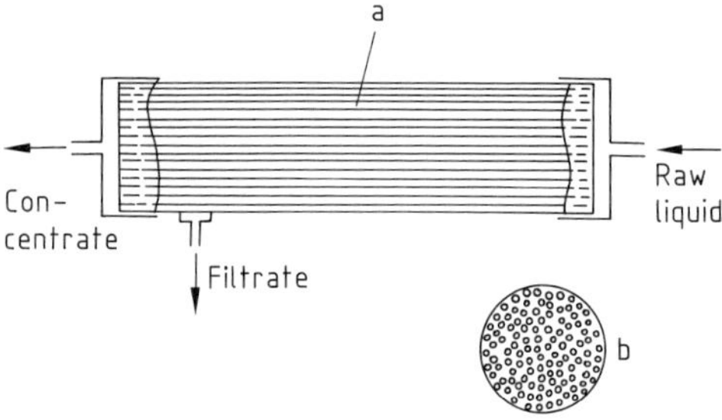

Figure 57. Schematic of tubular membrane filter (Berghof)
a) Tube bundle; b) Cross section

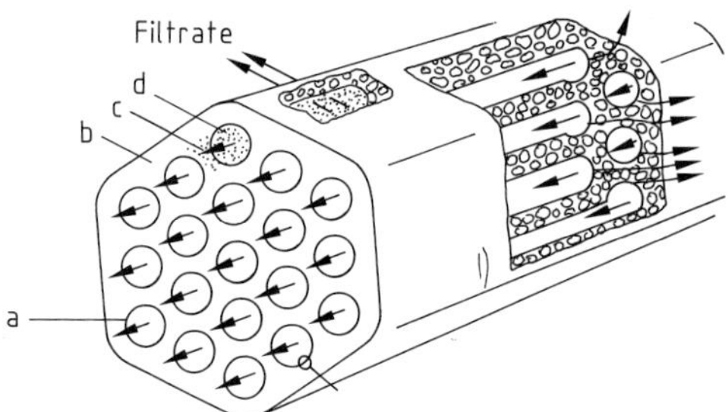

Figure 58. Ceramic multi-channel membrane filter element (Schenk)
a) Membrane; b) Support; c) Retentate; d) Channel

able after substantial dewatering; that is, it must be thixotropic. In one or more washing zones, wash liquor can be fed radially through the filter tube. Several filter sizes with filtration areas from 0.25–1.6 m^2 are available.

Cross-Flow Tubular Filters. The filter elements in the tubular cross-flow filter are small tubes or hoses, a few millimeters in diameter, which are bundled into systems of tubes carrying parallel flows (Fig. 57). The filtrate passes the system through walls of the tubes whereas the suspension flows tangentially to them dragging the retained particles, see p. **10**-21. The tubes can be made of various porous materials, preferably polymers or ceramics, and have quite narrow pore-size ranges, from a few micrometers down to hundredths of a micrometer. They are, above all, microfilters for the concentration of dilute suspensions and for clarifying and sterilizing filtration. Such filter systems with the finest membrane pore sizes are also suitable for ultrafiltration and reverse osmosis; multilayer or asymmetric membranes are used in them (→ Membranes and Membrane Separations, treated in the A Series).

Throughput rates range up to 3000 L m^{-2} h^{-1}, depending on pore size and filtration pressure. The possible applications vary with capacity. The principal users are chemical, pharmaceutical and foodstuffs producers and, increasingly, bioengineering concerns.

Figure 58 shows an example of a novel membrane filter made of a ceramic material. The rod-shaped body, made of high-mechanical-strength aluminum oxide, has 19 parallel tubular channels 4 or 6 mm in diameter, which give a total filtration area of several square meters. The ceramic base material allows use at high and changing temperatures and the presence of acids and alkalies. The pores in the body have diameters between 15 and 20 μm, in the corresponding membranes between 0.2 and 5 μm for microfiltration.

8.12. Special Filter Types

Coarsely-dispersed systems with relatively high solids contents are dewatered with conventional screens like the vibrating screens used for classification. One such filter is the vertical sieve containig a *bent screen* plate, which is employed to separate coarse and fibrous solids at lower concentrations from process waters and wastewaters with screen filtration. The sieve, which is equipped with an adjustable screen slope, is charged from the top, so that the slurry flows downward by gravity. The screen, with slots varying from 0.125–2.5 mm wide, retains the solids, which drop downward. The simple, inexpensive design has made this device a widely-used one in food preparation, the textile and manmade fiber industry, paper and pulp manufacturing, and the chemical industry.

Several new filtration mechanisms are being explored with an eye to filtration problems that have not yet been solved. These include, in particular, the dewatering of suspensions having relatively high concentrations of extremely fine solids (below 10 μm down to colloidal sizes). The conventional separation techniques—flocculation and microfiltration—generally yield filter cakes or concentrates that require considerable energy for complete phase separation (e.g., by drying). Research on ionized phases has shown that the electrokinetic effects of *electro-osmosis* for liquid transport and *electrophoresis* for solids transport along an electric field allow separation, even of difficultly filterable suspensions. Both these effects have been known for a long time.

A filter based on the above processes consists of separating chambers with anodes, where the ionized solids collect, and cathodes made as filter media, where the electrolytic filtrate is discharged [4]. Several separating chambers form filter systems similar to filter presses.

Difficultly filterable suspensions are also filtered by *high-gradient magnetic separation* (HGMS) (→ 19. Magnetic Separation). This type of filter employs a steel wool medium. In a very strong magnetic field, high field gradients are generated at the individual filaments. These gradients attract fine magnetic particles. The HGMS filter thus resembles a depth filter. Successes have been achieved in the purification of kaolin and wastewaters.

9. Filter Selection

Because certain qualities, largely related to design, characterize the ranges of application of filters, some guidelines can be set forth. These together with a few other points can provide a basis for filter selection. In general, a good selection takes account of factors such as slurry properties, throughput, or cake formation rate (see Table 2). Other points to be considered are the density and viscosity of the liquid, the size, size distribution and shape of the solids particles, the sensitivity of the product to impurities, corrosion problems, the sensitivity of the equipment to sedimentation and to product fluctuations, the cost of operation, and so on. Table 5 can be used in the selection of special vacuum filter types under a number of criteria. The tabulated data should, however, be regarded only as rough guidelines. Filter types are shown in Figure 59 together with the particle sizes for which they can be used.

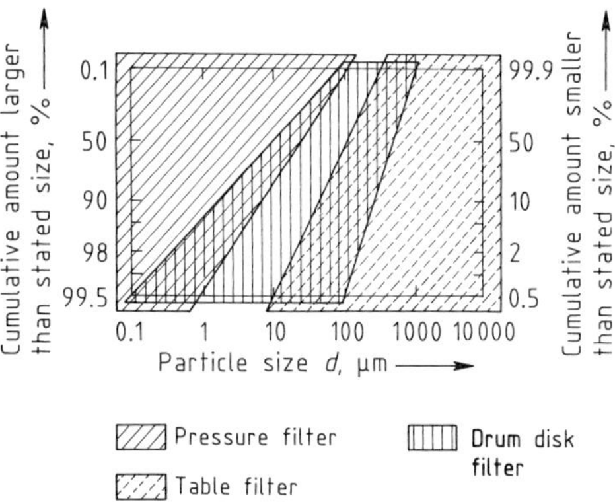

Figure 59. Filter selection as a function of solids particle sizes

The higher cost of pressure filtration in comparison with vacuum filtration is usually not justified except when the cake has a resistance of $3.5 \cdot 10^{11}$ m/kg. Flocculated solids, which should not lose their structure since it is advantageous for filtration, require careful handling, especially in feeding to the filter. Because most such solids are somewhat compressible, not too much pressure should be applied, and so vacuum filtration is commonly preferred for these materials.

A safe selection without trials or appropriate experience is possible only in the rarest instances.

Little general information and few computational formulas can be given in the field of filtration, which is one of the oldest operations in process engineering. Regrettably, filtration (like other essential processes involving the handling

Table 5. Applicability of filters

Filter type	Suspension of fine particles	Operation			
		Pressure	Washing	Compression	Continuous (c) automatic (a)
Belt			×		c
Belt with compression	×		×	×	c + a
Nutsche	×	×	×	×	a
Tilting pan			×		c
Leaf/plate	×	×			a
Drum	×				c
Pressure drum	×	×	×		c
Membrane filter press	×	×	×	×	a
Twin-belt pressure filters	×	×	×	×	c
Tube press	×	×		×	a

of bulk solids, dumped and packed beds, etc.) must deal with the effects of material properties that can be determined only with difficulty or cannot be measured in a predictable way, and these effects are so large and pervasive that there is no comprehensive, unified way of allowing for them. There is no doubt, however, that selection criteria exist and that they imply at least certain filter categories, so that in many cases the number of open options is not very large.

10. Filter Media

The outcome of filtration, that is, the degree to which the filter separates the two phases, is governed above all by the filter medium. Where the available filter media are not suited to the properties of the phase (solid) being separated, filter aids can close the gap.

The filter medium should be selected to provide the desired separating effect and to fit the special properties of the suspension (such as the particle size of the solids and the viscosity of the liquid). The selection of a medium should also take account of service conditions such as cleaning, washing, reaction to back-blowing, and chemical and mechanical stability, which are no less important than the slurry properties. In cake filtration, cake removal must also be considered.

In line with the multiple purposes of filtration, media must perform the following tasks:

1) Optimal retention of all solid particles outside the filter medium
2) Retention of the solid particles, mainly inside the filter medium, by virtue of effects generally lumped together as "entrapment," which essentially have to do with adhesion forces and electrical forces.

Often these functions cannot be unambiguously separated; this is particularly the case with slurries having a very broad particle-size spectrum. In such cases, some concession nearly always has to be made in terms of separation efficiency, or further operating practices, such as careful cleaning or frequent replacement of the filter medium, have to be introduced.

Filter media can be characterized in terms of the following overall criteria:

1) Cut size, that is, the particle size that just passes through the filter medium.
2) Permeability; high permeability means low pressure drop.
3) Chemical stability with respect to the filtrate.
4) Blocking tendency, particularly for fabrics in cake filtration.
5) Mechanical strength in relation to loads imposed in back-blowing or the movement of filter cloths.
6) Smooth surface to promote cake removal.

As is appropriate after many years' experience in filtration, a great many types of filter media have been introduced in practice. New materials and processing methods continue to offer a wide range of improvements. The diversity of filter-media properties has led to a number of classifications in the literature. The following seems to be an advantageous one:

Perforated plates, screens, slotted screens
Fabrics
Beds, felts and other nonwoven materials
Packed beds and precoat layers
Porous materials
Membranes

Within these groups, there are differences in properties, especially with regard to chemical stability (including aging). However, often physical and biological requirements are also imposed, such as high-temperature stability, water uptake, swelling capacity, weathering behavior (stability to light), and resistance to insects (moths) and bacteria (rotting).

Aside from material properties, the overall filtration qualities can be represented only in an approximate way, because they depend heavily on the properties of the suspension being filtered. Even the often-used permeability and/or mean pore size figures should be considered—at best—rough reference values for the behavior during filtration. A reliable assessment is usually possible only if practical tests or, within limits, laboratory trials (e.g., with the test leaf filter, p. **10**-23) are performed.

Perforated Plates, Screens, Slotted Screens. Aside from screens used in screen filtration, filter media in this group serve mainly to support filter cloths or papers, which as a rule have finer pores.

With perforated plates, the largest possible free surface area is usually desired so that the pressure differences to be overcome in filtration will be small. Free areas range from some 20–30% for ordinary plates with circular holes to over 90% for high-quality screens with square openings in the millimeter range. The free area drops to less than 10% for very fine sieves with openings a few micrometers across.

Slotted screens with very small openings, used chiefly as strainers and candle filters (see p. **10**-32), are made by stacking plates, rings or other elements around a permeable cylinder. The fine gaps between the elements allow only the liquid to pass. Strainers can be fabricated more simply by winding profiled (wedge) wires or strings on a core. These units have little free area, roughly 10%. As a result, they are economic only when solids concentrations are very low.

Fabrics. By far the most widely used filter media are woven fabrics. Manmade fibers dominate, but fibers of glass, minerals, and metals are also used to an increasing extent, especially in view of their good chemical stabilities.

The advantages of multifilament-yarn fabrics include the depth action of the filter medium, which permits slight penetration of solids into the medium, thus effecting some reduction in pore size and also promoting bridge formation on the medium. The ease of working with fabrics is an incomparable advantage, of course.

The fabric weave is important for filtration proper as well as service properties. Another essential factor is the structure of the fiber, which varies widely in natural and synthetic materials.

Where particularly high strengths are needed, as in twin-belt pressure filters, monofilament-yarn fabrics are preferred. High precision in fabrication allows mesh openings in the micrometer range, especially with metal fibers.

Within some limits it is possible to start with typical fabric data, such as weight per unit area, numbers of warp and filling yarns, fiber diameter, fabric thickness and fiber density, and calculate the pore size and hydraulic resistance. The resistance calculation is based on the pore radii of the fabric and the yarn, and Poiseuille's law, Equation (4), is used as a first approximation.

The permeability of a clean, unused fabric can be calculated (with about the same reliability) from Kozeny's equation (6); the starting data are the densities and porosities of the yarn and the fabric. But empirical relations should give better agreement with experimental values.

However, it is vital to know how the fabric influences cake formation and the narrowing and possible blocking of the pores. The resulting change in hydraulic properties of the filter medium is described by Equation (13). If blocking does not occur, n will be 0 and the equation corresponds to the ordinary filter equation. If there is a tendency to partial blocking, n can equal 1.5.

Nonwovens, Papers, and Felts. These filter media consist of relatively short fibers (staple fibers a few centimeters long) formed into a random layer and then compacted in some way. *Nonwoven media* are usually fabricated in fairly thick layers and therefore function as depth filters (→ Nonwoven Fabrics, treated in the A Series).

Papers made of cellulose fibers are available in a bewildering range of grades for use in laboratories, small plants, and daily life (→ Paper and Pulp, treated in the A Series). Heavier papers and paperboards (2–6 mm thick) can also be fabricated from mixed materials, such as cellulose plus diatomaceous earth or activated carbon. These filter materials meet a wide range of requirements and are employed in large quantities for the clarification of beer, wine, or juices, and in the dyes and coatings, food, and cosmetics industries. By virtue of the raw materials and fabrication processes, one-time use of the filter medium is often economic; this feature is especially desirable in sterile filtration.

Conventional *felts* of relatively short natural and manmade fibers are usually not adequate to the filtration and service loads in plant equipment. Felts that incorporate woven-fabric reinforcement are often preferred for this reason. Some loss in terms of the basic advantage of felt must be accepted when these composite materials are used. As a rule, needled felts have lower resistances than woven fabrics. They have now become the most commonly employed non-wovens (→ Felts, **A10,** p. 297).

Packed Beds and Precoat Layers. Loose beds of granular materials can also be used in clarifying and some cake filtrations. For economic reasons, cheap natural products such as sand, gravel, or coal are used, but it is desirable to remove the excessively fine-grained fractions from them by washing. Common particle sizes are 3–50 mm for gravel, 0.35–0.45 mm for fine sand, 0.45–0.55 mm for medium sands, and over 0.55 mm for coarse sands. Other candidate materials include minerals (quartz, limestone), coke, anthracite, slag, wood pulp, glass beads, wadding, or fibers.

The hydraulic properties are calculated with the relations developed in the theory section, in particular Darcy's law, Equation (11), or—if the appropriate material data are known—the Kozeny equation (5).

Rigid Porous Materials. Porous materials, which are fundamentally similar in structure, are suitable for any kind of filtration. Their mechanical stability allows their use as filter media without supporting structures (up to pressures of 2 MPa and more). They are preferably employed as filter cartridges or in the form of sheets and, recently, honeycomb and other three-dimensional shapes. Because they are generally too valuable to use just once, and because they can withstand vigorous backwashing without special preparation, solids should be collected mainly by surface filtration. Deep-bed filtration is recommended only when the particles are very small in comparison with the pores and are collected by an adsorptive mechanism. In the intermediate region, solid particles block the pores too quickly, and backwashing may not adequately remove them. Materials for porous filters include minerals, diatomaceous earth, porcelain, clay, coal, graphite, corundum, rubber, plastics, and metals, including highly resistant nickel – chromium – molybdenum alloys. The most important criterion for selection is chemical and thermal stability.

Porous materials made from equal-sized particles are distinguished by uniform porosity. In order to reduce the blocking tendency, layers with differing particle sizes can be superimposed with the fine-pored layer toward the filter cake; thus only the topmost layer need be cleaned, and the filter offers better service qualities.

Porcelain filters are available with pore radii from 0.08 to 6 μm and porosities of 30–70%. Sintered metal filters range from 1 to over 100 μm in pore size and from 30 to 60% in porosity. Plastic filter media with about the same pore spectra can easily be fitted to the equipment. Their relatively poor temperature stability, however, restricts their use. Besides poly(vinyl chloride) and polyethylene, porous masses can also be fabricated from poly(tetrafluoroethylene), with its excellent stability.

Membranes. At present a great number of distinct membranes are on the market, not just for ultrafiltration and reverse osmosis, but also for microfiltration (→ Membranes and Membrane Separation, treated in the A Series). They are made of plastics and are suitable for clarifying as well as pure surface filtration . If very small pores are needed, membranes are assembled from two layers, a coarse-pored supporting layer (with, e.g., 15–20 μm pore openings) topped with a filter layer (0.2 μm pores).

All membranes have a relatively narrow range of pore sizes, so that they can also be used for separation jobs.

A special form of membrane filtration employs hollow fibers made of, e.g., polyamides, with diameters of 50 μm and wall thicknesses of 13 μm. The solution being separated is introduced into the hollow fibers under an appropriate pressure. Pure liquid escapes through the wall of the fiber, while the dissolved material remains behind and can easily be carried out of the fiber.

11. Filter Aids

Some suspensions contain solids that quickly fill or block the openings in the filter medium. This group includes suspensions substantially made up of relatively fine, soft (especially gelatinous), compressible solids. Such feeds can still be filtered well if the blocking of the filter medium is prevented by the inclusion of inert, readily filterable granular materials. These filter aids form a layer permeable to the filtrate, and at the same time they carry out the functions of a loose filter cake. The particles being retained are deposited on the filter aid. This process must still leave sufficiently free openings through which the filtrate can pass. The main candidates for filter aids are thus highly porous materials. Diatomaceous earths are very economical examples; their fineness is characterized by 10% > 40 μm at the high end and between 1 and 7 μm at the low end. Also used are perlites and, for special jobs, Fuller's earths, powdered glass, coal preparations, cellulose fibers, wood pulp, paper stock, bagasse (a sugar cane residue), talc, or plastics, although these products are less effective than diatomaceous earths and perlites. The requirements for a good filter aid are uniform quality, correct particle size and shape, and the highest possible wet volume (specific volume in the wet condition, usually greater than the wet volume per mass of filter aid).

In diatomaceous earths, particles consisting of needle-shaped diatoms have proved advantageous; disk-shaped particles filter poorly and produce too dense a cake. Unsuitable constituents such as clay, or sand must be removed from the raw materials, because they hinder filtration by blocking the pores.

A synthetic filter aid is manufactured from volcanic rocks. The raw material, bound with some water, is ground, then suddenly heated to a high temperature; the grains inflate to form spheres. Regrinding yields half-shell-shaped fragments, whose large surface area makes them suitable as a filter aid once the fines are screened out. Their action in filtration is somewhat worse than that of diatomaceous earth, with a porosity 20–40 % lower.

In *precoating*, a layer of filter aid several millimeters thick is deposited on the filter medium before filtration proper. After the filtration is completed, the filter aid is often discarded along with the collected solids.

Several steps take advantage of the effect of the loosened filter cake. First, a precoat some 1–2 mm thick, containing 200–800 g/m^2 of filter aid, is filtered onto the medium (base layer) so that a clear filtrate can be obtained right from the start of filtration. Filter aid is also added to the slurry continuously during filtration. In batch operation, this is done by stirring in the slurry tank; in continuous filtration, dry or wet filter aid can be metered into the slurry delivery line. No generally valid rules exist for the proper ratio of filter aid to solids concentration. A variety of empirical values are available for some processes and solids types. If there is too little filter aid, the filtration goes badly; if too much, the filtrate flow rate drops off too sharply.

The main applications for filtration with filter aids are the precoat technique for poorly filterable slimes of chemical and mineral products and the clarifying filtration of beverages (beer, wine, juices) containing soft, gelatinous impurities, as well as the clarification of gelatins, cane juice, or edible oils.

12. References

General References

[1] R. B. Bird, W. Stewart, E. N. Lightfoot: *Transport Phenomena*, J. Wiley & Sons, New York 1960.

[2] R. E. Coulson, J. F. Richardson: *Chemical Engineering Practice*, vol. 2, Pergamon Press, Oxford 1955.

[3] K. J. Ives: *Scientific Basis of Filtration*, NATO ASI Series E No. 2, Noordhoff, Leiden 1975.

[4] H. S. Muralidhara: *Advances in Solid–Liquid Separation*, Battelle Press, Columbus 1986.

[5] R. H. Perry, C. H. Chilton: *Chemical Engineer's Handbook*, 6th ed., McGraw-Hill, New York 1984.

[6] D. B. Purchas: *Solid–Liquid Equipment Scale-up*, Uplands Press, Croydon 1977.

[7] D. B. Purchas: *Solid/Liquid Separation Technology*, Uplands Press, Croydon 1981.

[8] P. Rivet: *Guide de la Separation Liquide-Solide*, ID-EXPO, Cachan 1981.

[9] A. Rushton: *Mathematical Models and Design Methods in Solid-Liquid Separation*, NATO ASI Series E No. 88, Nijhoff, Dordrecht 1985

[10] L. Svarovsky: *Solid–Liquid Separation*, 2nd ed., Butterworths, Sevenoaks 1981.

[11] *Lehrbuch der chemischen Verfahrenstechnik*, 2nd ed., VEB Deutscher Verlag für Grundstoffindustrie, Leipzig 1969.

[12] *Verfahrenstechnischen Berechnungsmethoden*; „Mechanisches Trennen in fluider Phase", Part 3; VCH Verlagsgesellschaft, Weinheim 1985.

Specific References

[13] J. Kozeny: *Über die kapillare Leitung des Wassers im Boden*, Ber. Math.-naturwissensch. Abt. Akad. Wien 1927, pp. 271–306.

[14] P. C. Carman: "Fluid flow through granular beds," *Trans. Inst. Chem. Eng.* **15** (1937), 150–166.

[15] C. Alt: "Stand der Filtrationstheorie und seine Bedeutung für die praktische Anwendung", *Aufbereit. Tech.* **21** (1980) no. 4, 177–183.

[16] H. P. G. D'Arcy: *Les fontaines publiques de la ville de Dijon*, Dalmont, Paris 1956.

[17] F. M. Tiller, M. Shirato: "The role of porosity, Part 6: New definition of filter resistance", *AIChE J.* **10** (1964), 61–67.

[18] H. Schubert: *Kapillarität in porösen Feststoffsystemen*, Springer Verlag, Heidelberg 1982.

[19] M. Shirato et al.: "Gravitational drainage of granular beds", *Kagaku Kogaku Ronbunshu* **5** (1979) no. 6, 630–636.

[20] O. Molerus et al.: "Durch Verdrängung mit Luftströmung erzielbare Restfeuchten bei der Vakuumfiltration feinkörniger Feststoffe", *Verfahrenstechnik (Mainz)* **15** (1981) no. 11, 805–810.

[21] D. Redeker, K.-H. Steiner, U. Esser: "Das mechanische Entfeuchten von Filterkuchen", *Preprint Symp. Filtertechn. Wiesbaden*, GVC Düsseldorf 1983, pp. 32–64 (Nachtrag).

[22] M. Shirato, T. Murase, M. Iwata: "Theoretical and Experimental Studies in Expression", *Mem. Fac. Eng. Nagoya Univ.* **38** (1986) no. 1, 42–85.

[23] W. Bender: "Das Auswaschen von Filterkuchen", *Preprints Symp. Filtertechnik Wiesbaden*, GVC Düsseldorf 1983, pp. 3–31.

[24] S. Ripperger: „Dynamische Filtrationsverfahren – Apparative Ausführungen, Anwendungen, Entwicklungen", *Preprints GVC-Dezembertagung*, VDI Düsseldorf 1987.

[25] F. M. Tiller: "Laboratory Analysis for the Solution of Industrial Filtration Problems", *Filtr. Sep.* **7** (1970) 430–433.

11. Centrifuges and Hydrocyclones

HELMUT TRAWINSKI, Amberger Kaolinwerke GmbH, Hirschau/Oberpfalz, Federal Republic of Germany

Symbols

In addition to the standard symbols defined in the front matter of this volume, the following symbols are used:

A clarifying area, m^2
A_s material cross section, m^2
b centrifugal acceleration, m/s^2
$\bar{d}$ mean diameter, m
d_p diameter of solid particle, m
d_T cut size, m
f force, N
i index for inner zone
L length of centrifuge rotor, m
l length of hydrocyclone cone, m
N number of disks (in radial section) in centrifugal separator
n rotational speed, min^{-1}
o index for outer zone
Δp pressure drop, $N\,m^{-2}$
Q volume flow rate, m^3/s
q_F clarifying area load, $m^3/m^2\,s \equiv m/s$
R index for rotor
u, u_0 settling velocity in centrifugal or gravitational field, respectively, m/h
v circumferential velocity, m/s
z acceleration factor b/g
γ relative density of a wall (metal), kg/m^3
δ wall thickness, m
ε relative void volume
η dynamic viscosity, $kg\,m^{-1}\,s^{-1}$
λ slenderness ratio L/d
χ normalization factor
ϱ_L density of flow medium, kg/m^3
ϱ_s density of suspension, kg/m^3
ϱ_p density of solid, kg/m^3
Σ equivalent clarifying area, m^2
σ tensile stress in material , N/m^2
ω angular velocity, s^{-1}

Filtration and sedimentation can be effectively accelerated by the use of centrifugal force. In solid–liquid phase separations, a clarified liquid phase is obtained, with extensive thickening or dewatering of the granulated solid phase. Liquid–liquid phases, i.e., mixtures of immiscible liquids, can also be separated in a centrifugal field. With liquid–liquid–solid multiple-phase systems (simultaneous separation and sedimentation), it is possible to obtain separation into two phases (normally a clarified light liquid and a heavy liquid that takes up solid impurities) or three phases (two liquids and a thickened solid slurry).

1. Fundamentals

The centrifuge owes its effect to centrifugal acceleration and its associated mass forces, which cause separation of the components of a mixture according to their mass. If the centrifuge drum or rotor is a solid bowl, i.e., imperforate, the liquid phase cannot escape. With mixtures of liquids, this means that stratification (*separation*) occurs in the radial direction. With suspensions, it results in relative movement of the granulated solids radially outwards (*sedimentation*). With perforate centrifuge drums, i.e., those lined with fine-mesh screens of metal or plastic or with cloth, the liquid phase escapes and the granulated solids are deposited on the screen or cloth, forming a cake (*filtration*). There is normally no attempt at obtaining a screen classification of the solid collective, as the cake forms so rapidly that this must always be obscure.

Filtration [23], [24]. Static filtration depends on the effect of external forces on the liquid phase, whereas the mass forces in the centrifugal field act from within. Formally, however, these forces can also be described as pressure drops.

The following formula describes the pressure p on the bottom of a container filled to a height h:

$$p = g \varrho_L h \qquad (1)$$

In the centrifugal force field, *centrifugal acceleration*

$$b = r\omega^2 = \frac{1}{2}(d_o + d_i)\frac{\pi^2 n^2}{1800} \qquad (2)$$

replaces the acceleration due to gravity g and the radial distance between the ring dam and the rotor wall

$$r_o - r_i = \tfrac{1}{2}(d_o - d_i) \qquad (3)$$

replaces filling height h. In Equation (2), $\frac{1}{2}(d_o + d_i)$ is the mean diameter of the free space between rotor wall and ring dam, also designated $\bar{d}$. The factor 1800 in the denominator of Equation (2) allows for conversion of the units of time from $\min^{-2}$ (ω^2) to s^{-2} (b).

The pressure on the bottom of a filter centrifuge can therefore be described using the following formula [24]:

$$p = \varrho_L (d_o - d_i)(d_o + d_i)\frac{\pi^2 n^2}{7200} \qquad (4)$$

Filtration speed could be calculated using this pressure according to filtration theory (→ 10. Filtration), with filter resistance being defined according to Darcy's law and depending on the bulk properties of the solid collective (particle-size distribution, relative void volume, cross flow, shape of particles, cake compressibility, etc.). However, it is usually determined empirically. A device suitable for this purpose (the long-arm bucket centrifuge) has been developed by BENDER [25].

Sedimentation. Stokes' law can be used to describe sedimentation in the centrifugal field, because solids separable in this way are usually sufficiently fine grained and the flow around the particles during sedimentation has a low Reynolds number, i.e., is laminar. The Stokes settling velocity can be defined as follows:

$$u = \frac{d_p^2(\varrho_p - \varrho_L)}{18\eta} b f(\varepsilon) \qquad (5)$$

where again Equation (2) is used for centrifugal acceleration b. If sedimentation is hindered by the concentration of solids in the suspension the value is lower, but this is allowed for by the value $f(\varepsilon)$, a semiempirical function of the relative void volume ε. Corresponding relations have now been established by RUMPF and GUPTE [26] for the related problems of percolating flow, expanding on the ideas of KOZENY [27], STEINOUR [28], and KRIEGEL [29].

For static tests in a glass cylinder, the acceleration due to gravity g replaces the centrifugal acceleration b in Equation (5). Settling velocities measured in the glass cylinder can then be converted into settling velocities in the centrifugal field by multiplying by the factor b/g, designated as acceleration factor z:

$$z = \frac{b}{g} = \frac{\pi^2 \bar{d} n^2}{g\ 1800} \approx \frac{\bar{d} n^2}{1800} \qquad (6)$$

$$u = u_0 z \qquad (7)$$

(The last term in Eq. (6) is a convenient approximation for z that is frequently employed in the industry, although it is incorrect with respect to dimensions.) In overflow centrifuges, continuous flow through the rotor is superimposed on sedimentation. With a volume flow rate Q (in cm^3/s or m^3/h) and an effective cylindrical clarifying area (in cm^2 or m^2)

$$A = \bar{d}\pi L \qquad (8)$$

the *clarifying area load* is [24]

$$q_F = \frac{Q}{A} = \frac{Q}{\pi \bar{d} L} \qquad (9)$$

in cm/s or m/h. All those particles whose settling velocity u in the radial direction is the same or higher than the clarifying area load have a chance of being separated. The obtainable *cut size* d_T is calculated by setting q_F and u equal and eliminating the obstruction factor $f(\varepsilon)$

$$u = \frac{g(\varrho_p - \varrho_L)}{18\eta} z d_p^2 \qquad (10)$$

so that

$$d_T = \sqrt{\frac{18\eta}{g(\varrho_p - \varrho_L)} \frac{q_F}{z}} \qquad (11\,a)$$

or

$$d_T = \sqrt{\frac{18\eta}{g(\varrho_p - \varrho_L)} \frac{Q}{\pi z \bar{d} L}} \qquad (11\,b)$$

is obtained [30]. The *volume flow rate* Q of a

decanting centrifuge follows from Equation (9) if clarifying area load q_F is equated with settling velocity u (cf. Eq. 7):

$$Q = A\,u = A\,z\,u_0 \tag{12}$$

The product of rotor area A and acceleration factor z is the *equivalent clarifying area* [24], [31]:

$$\Sigma = A\,z \tag{13}$$

The throughput

$$Q = u_0\,\Sigma \tag{14}$$

is obtained from the settling velocity u_0, measured statically, and the inner surface of the rotor A multiplied by the acceleration factor z.

The separation performance of a centrifuge can be improved by increasing the equivalent clarifying area Σ, either by increasing the flow rate Q, for a given separating efficiency, or by reducing the cut size d_T, because in Equation (11) Σ occurs as the product $\pi \bar{d} L z$ in the denominator of the radicand. Reduction of d_T can be achieved by enlarging the surface area A, increasing the rotor length L, or increasing the mean rotor diameter $\bar{d}$ and therefore the overall rotor diameter d. It is also possible to raise the acceleration factor z by increasing the rotational speed n (or again enlarging the rotor diameter d). There is no point, however, in raising the Σ value beyond the absorption capacity of the feed and discharge devices of the centrifuge [32].

Unfortunately, rotor diameter d and rotational speed n cannot be increased simultaneously without limit, as they are restricted by the strength of the rotor material: d and n are coupled with each other so that their product, dn, the circumferential speed of the rotor,

$$u_R = \frac{\pi}{60}\,d\,n \tag{15}$$

may not exceed a certain critical value. This can be explained as follows [24]: if a rotor has diameter d, length L, and wall thickness δ, if the relative density of the material is γ, and if the rotational speed n results in an acceleration factor $z = dn^2/1800$ as in Equation (6), then the area of the double longitudinal seam $A_s = 2\,\delta L$ is loaded with a force

$$f = d\,L\,\delta\,\gamma\,z \tag{16a}$$

i.e.,

$$f = \frac{\gamma\,\delta}{1800}\,L\,d^2\,n^2 \tag{16b}$$

The tensile stress is thus

$$\sigma = \frac{f}{A_s} = \frac{\gamma}{3600}\,d^2\,n^2 \tag{17}$$

or with Equation (15)

$$\sigma = \frac{\gamma}{\pi^2}\,u_R^2 \tag{18}$$

Combining Equations (13), (8), and (6) gives the equivalent clarifying area, which determines the separation performance:

$$\Sigma = \bar{d}\,\pi\,L\,\frac{\bar{d}\,n^2}{1800} = \frac{\pi}{1800}\,L\,(\bar{d}\,n)^2 \tag{19}$$

and after introduction of Equations (15) and (18)

$$\Sigma = \frac{2\,L}{\pi}\,u_R^2 = \frac{2\,\pi\,\sigma}{\gamma}\,L \tag{20}$$

For this model the mean diameter $\bar{d}$ has been replaced by rotor diameter d. The corresponding proportionality factor is ignored in Equation (20).

The equivalent clarifying area Σ, and thus the separation performance, is therefore proportional to the breaking length of the rotor material σ/γ and to the length of the rotor L. Rotor diameter d does not enter into the relation: increasing the diameter d reduces the rotational speed n in proportion to $1/d$ because of Equation (17). The acceleration factor z (Eq. (6)) is reduced by the factor $1/d$ and the diameter d is compensated in the product $\Sigma = \pi L d z$ (Eq. (13)).

This realization led to the development of slim solid-bowl centrifuges, such as the tube centrifuge and the cylindrical decanter.

However, large Σ values are only useful for increasing the clarifying effect. Cylindrical decanters are therefore used basically as clarifying decanters. They have a relatively small diameter (250–400 mm), high rotational speed (3000–4000 rpm), and slenderness ratios of $\lambda = 3-4$ (i.e., rotor lengths of 800–1600 mm). When decanters (solid-bowl scroll discharge centrifuges) are used for slurry dewatering, the important factor is not the clarifying effect but the high slurry throughput. These centrifuges therefore have rotors with a large diameter and correspondingly smaller slenderness ratios ($\lambda = 1-2$).

The concept of the equivalent clarifying area Σ [24], [31], so often used elsewhere, cannot be applied in this case. It is only applicable to processes of near-complete clarification with low quantities of solids. GÖSELE has reviewed model equations for decanters where the amount of slurry is higher [33], [34].

When clarifying decanters are used for accelerating sedimentation with added flocculants (polyacrylamides), a large Σ value is not necessary. The rotational speed can be lower (1000–1500 rpm), i.e., rotors with a larger diameter (500–800 mm) can be used. Such decanters play an important role in wastewater engineering.

Another possibility for increasing the equivalent clarifying area is to install *disk stacks*. These subdivide the separating compartment into a number of sloping ring-shaped zones where the sedimentation distance is much shorter. If N is the number of disks cut by a radius (e.g., $N = 50-100$ in large sludge separators, although the total number of disks may be up to twice that), then the equivalent clarifying area is

$$\Sigma = \bar{d} \pi L z N \qquad (21)$$

where $\bar{d}$ is the mean diameter of the disk stack and L its length.

The following is an exact formula [24], [31], [35]:

$$\Sigma = \frac{\pi}{1800} L \, (d n)^2 \, N \times \left\{ 1 - \frac{2h}{d} + \frac{4}{3} \left(\frac{h}{d} \right)^2 \right\} \qquad (22)$$

where d is the outer and $d - 2h$ the inner diameter of the disk stack and h therefore its radial height. In the case of disk centrifuges, i.e., separators and slurry nozzle centrifuges, this Σ value is particularly useful as an objective figure for indicating their comparative efficiency.

Table 1 shows the equivalent clarifying areas Σ for the most important types of solid-bowl centrifuge. The Σ values shown for disk-type centrifuges, especially nozzle disk centrifuges, contain the amplifying effect of the disks. Dimensions and operational characteristics were selected on the basis of centrifuges available at the time.

The maximum permissible throughput load, or *absorption capacity*, has nothing to do with the separating effectiveness. It is normally only an indication of the conveying capacity of the machine's scoop devices.

Hydrocyclone. In the hydrocyclone, the liquid medium rotates not because it is in a rotating

Table 1. Characteristic values of solid-bowl centrifuges (decanting centrifuges)

Type of centrifuge	Rotor		$\lambda = L/d$	n,	u_R,	z	N*	A,	$\Sigma = A \cdot N \cdot z$*
	d, mm	L, mm		rpm	m/s			m²	m²
Overflow basket centrifuge (1)	630 1 600	400 800	0.64 0.5	1 900 750	62.7 62.8	1 265 500		0.79 4.02	1000 2010
Tube centrifuge (2)	45 105	250 750	5.56 7.14	45 000 17 000	106 93.5	50 600 16 850		0.035 0.247	1770 4160
Classifying decanter (3 a)	355 900	585 1800	1.65 2.0	3 250 1 800	60.4 84.8	1 875 1 540		0.40 4.3	750 ** 6620
Dewatering decanter (3 b)	400 1 400	650 2500	1.63 1.78	2 500 850	52.4 62.3	1 280 520		0.82 11.0	1 050 ** 5 720
High speed clarifying decanter (3 c)	350 800	950 2800	2.71 3.5	4 750 2 200	87.0 92.2	4 210 2 085		0.94 6.5	3 960 13 550
Flocculation clarifying decanter (3 d)	350 900	600 1800	1.71 2.0	2 500 1 100	45.8 51.8	1 130 575		0.53 4.3	600 2 470
Nozzle disk centrifuge (4)	250/110 800/350	(120) (350)		9 000 3 600	118 150	11 250/5 000 5 750/2 520	30 80	0.042 0.385	6 300 77 600

* $N = 1$ where there are no disks

** Recent results [34] suggest that in the case of higher solids content in the feed slurry transport becomes a bottleneck. The Σ value is then no longer suitable for a layout of decanters.

body, but because it is fed under pressure, i.e., a circumferential speed u is generated. Instead of using Equation (2), the centrifugal acceleration b is formulated as follows:

$$b = \frac{u^2}{r} = \frac{2\,u^2}{d} \tag{23}$$

in which the relationship between u and the available pressure drop Δp is given by the formula

$$u = \sqrt{\frac{2\,\Delta p}{\varrho_s}} \tag{24}$$

so that $2\Delta p/\varrho_s$ can replace u^2. Thus the acceleration, according to Equation (23), is

$$b = \frac{4\,\Delta p}{\varrho_s\,d} \quad \text{or} \quad z = \frac{4\,\Delta p}{\varrho_s\,g\,d} \tag{25}$$

If this is inserted into Equation (11 a), the following expression is obtained for the cut size:

$$d_T = \sqrt{\frac{18\,\eta}{(\varrho_p - \varrho_L)}\,\frac{q_F\,d\,\varrho_s}{4\,\Delta p}} \tag{26}$$

The semiempirical formula for the volume flow rate capacity Q of a hydrocyclone [36] is

$$Q = \chi\,d^2\sqrt{\frac{\Delta p}{\varrho_s}} \tag{27}$$

If the following equation is used for a cylindrical mean clarifying area A

$$A = \pi\,d\,l \tag{28}$$

i.e., for the clarifying area load

$$q_F = \frac{Q}{\pi\,d\,l} = \frac{\chi\,d}{\pi\,l}\sqrt{\frac{\Delta p}{\varrho_s}} \tag{29}$$

and introduced into Equation (26), then the following equation is obtained for the cut size [37]:

$$d_T = d\left[\frac{\chi}{4\,\pi\,l}\,\frac{18\,\eta}{(\varrho_p - \varrho_L)}\right]^{1/2}\left[\frac{\varrho_s}{\Delta p}\right]^{1/4} \tag{30}$$

This highly simplified representation of the very complex relationships gives only a first approximation for judging the cut size in the sense of a model law. The dependence on the fourth root of the pressure drop proves in practice to be even less, because of the increasing proportion of friction losses with higher pressure. For geometrically similar cyclones ($l \sim d$), the dependence of the cut size on the root of the nominal diameter d also applies only over a narrow range of sizes. For the essential operational parameters of the cyclone Equations (27) and (30) show a tendency toward dependence on the characteristics of the suspension, the geometry, and the pressure. The estimation of cut size valid for conical hydrocyclones (Eq. 30) must be extended for the fully cylindrical cyclone introduced during the last decade [38], [39]. As this can be over-throttled to a high degree, without blocking the underflow, mass recovery can be considerably reduced. As a result, the cut size d_T (Eq. 30) shifts by a factor of up to 10 (and possibly more) in the direction of coarse. This means that the range of application of the hydrocyclone now exceeds the lower limit for industrial fine screening (→ 15. Screening).

2. Centrifuges

The separations possible with centrifuges give rise to an abundance of applications and to a great number of centrifuge designs. Centrifuges can be divided into centrifugal filters, sedimentation centrifuges, and centrifugal separators, according to the method of liquid–liquid separation. Centrifugal filters (also known as filter or screen centrifuges) have a perforate bowl, whereas sedimentation centrifuges and separators (disk or chamber centrifuges) have a solid-wall bowl [40]–[42]. In a systematic sense, hydrocyclones could be considered as solid-bowl centrifuges.

For further information, especially concerning the design of individual centrifuges and hydrocyclones, the reader is directed to the trade fair reports that appear regularly.

Schematic sectional diagrams of the most important types for screen and filter centrifuges are shown in Figure 1, p. **11**-6, and for decanting centrifuges and separators in Figure 2 [43]. Further details, particularly for special types, are shown in subsequent illustrations. The three-column centrifuge (Figure 1 E) is also built with automatically controlled scraping devices and is then called a vertical scraper-type centrifuge. The pusher centrifuge (Figure 1 C) has three rotor stages. It can be supplied with one stage, or with two to four stages moving against each other. The most common version has two stages.

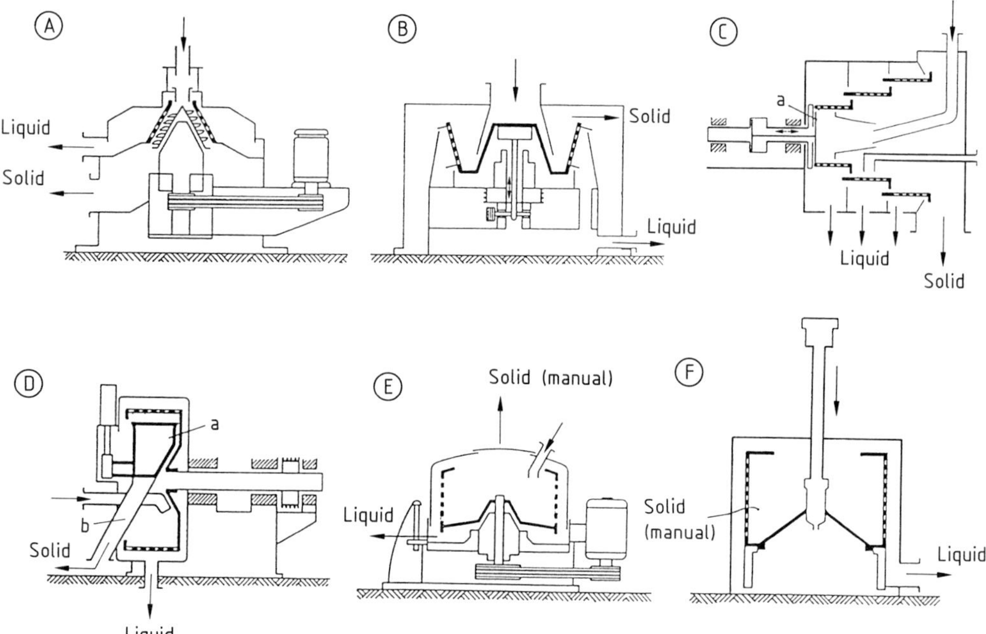

Figure 1. Filter and screen centrifuges
A) Screen centrifuge with scroll discharge
B) Vibrating screen centrifuge
C) Pusher centrifuge: a) Pusher plate
D) Scraper-type centrifuge: a) Scraping knife; b) Chute
E) Three-column centrifuge
F) Top-suspended basket centrifuge

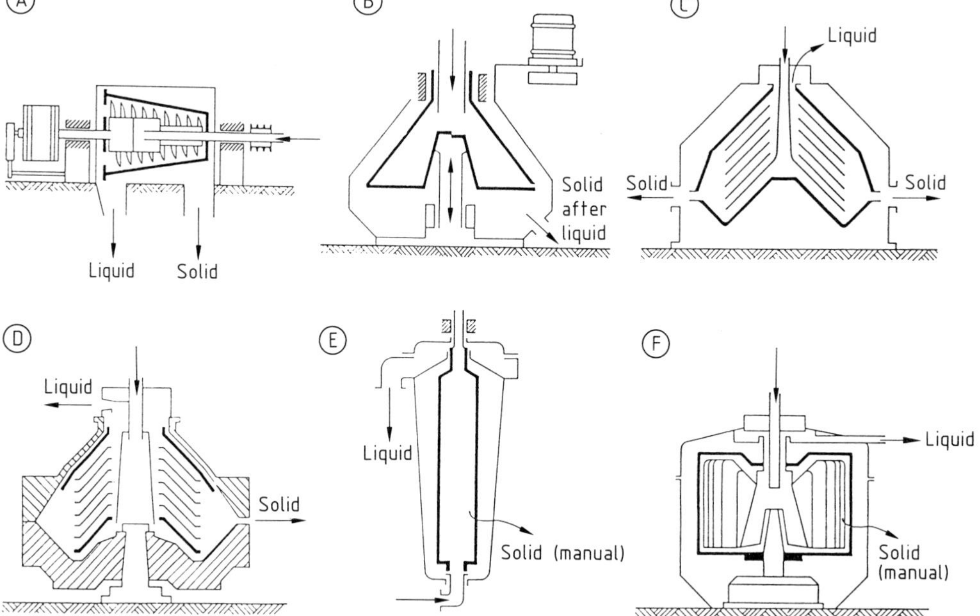

Figure 2. Sedimentation and decanting centrifuges
A) Classifying decanter
B) Granulate centrifuge
C) Nozzle disk centrifuge
D) Circular-slot disk centrifuge
E) Tube centrifuge
F) Circular compartment centrifuge

The decanter shown in Figure 2 A, p. **11**-6, is most often manufactured with a cylindro-conical rotor. In cylindrical decanters the cylindrical section is the longest. The helical conveyors for the slurry are designed so that in the conical section of the rotor the sides also stand perpendicular to the cone wall. The granulate centrifuge in Figure 2 B has now faded in significance in favor of a filter centrifuge with a terraced rotor.

The correlation between rotor diameter, rotational speed, circumferential speed, and acceleration factor of the various types of centrifuge can be shown clearly in graphs [43]. Figures 3–5 show the currently used ranges of industrial centrifuges: Figure 3 for continuous and automatic screen and filter centrifuges, Figure 4 for batch basket and for continuous and automatic sugar centrifuges, and Figure 5 for solid-bowl centrifuges. More detailed explanations of Figure 5 are given on p. **11**-16 in connection with sludge separators.

In Figures 3–5, the abscissa gives the rotor diameter d and the ordinate gives the acceleration factor z. The corresponding rotational speed n can also be read at the upper edge of the graphs

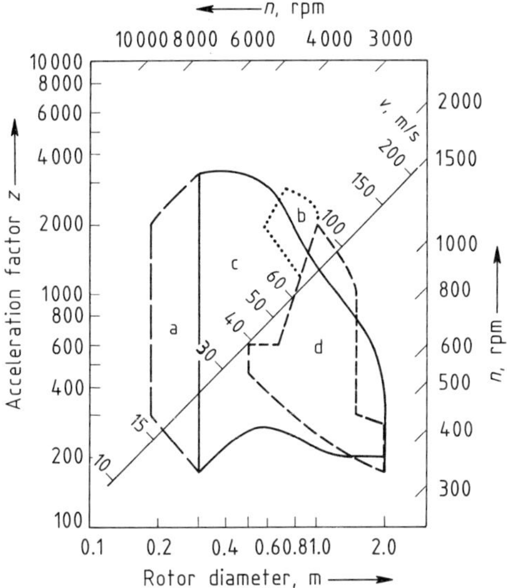

Figure 4. Operational ranges for pendulum and top-suspended centrifuges (for explanation, see Fig. 3)
a) Basket with elastic axis; b) Thin-layer flow; c) Three-column or pendulum; d) Top-suspended basket

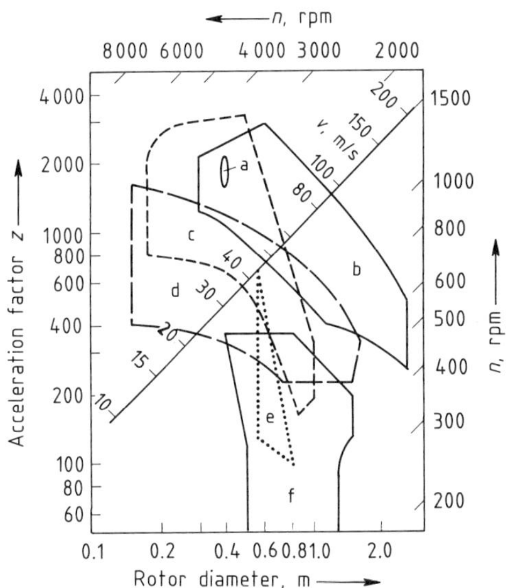

Figure 3. Operational ranges for screen, scraper, and pusher centrifuges
The vertical lines perpendicular to the rotor-diameter axis and the rotational speed (n) lines from top right to bottom left intersect at the operating point. The ordinate of this intersection gives the acceleration factor z. Secondary parameter lines from top left to bottom right give the circumferential velocity v.
a) Conversion filter; b) Scraper type; c) Screen with scroll; discharge d) Pusher; e) Oscillating screen; f) Pendulum

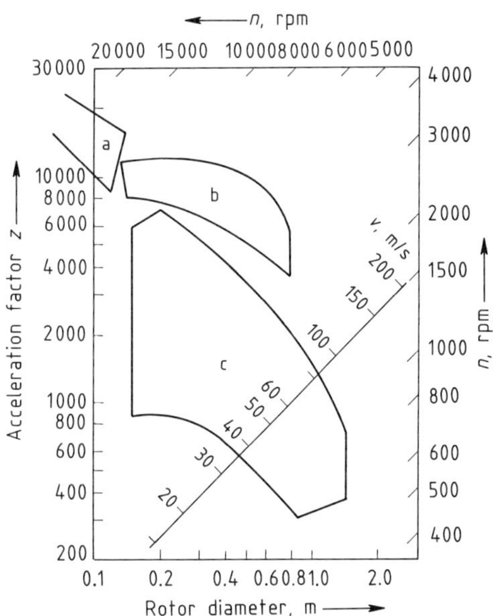

Figure 5. Operational ranges for tube centrifuges, liquid–liquid centrifugal separators, and decanters (for explanation, see Fig. 3)
a) Tube; b) Liquid–liquid centrifugal separators; c) Decanters

Table 2. Phase separations with centrifuges

Centrifuge type	Separations
Pusher centrifuge	LS
Scraper-type	LS, LLS*
Pendulum	LS
Decanting	LS, LLS
Nozzle	LS, LLS
Self-discharging	LS, LLS
Solids-collecting	LS, LLS, LL

LS liquid–solid; LLS liquid–liquid–solid;
LL liquid–liquid; LLS* liquid–liquid–solid possible but not usual

for the rising system lines, and the relevant circumferential speeds v (related to the maximum inner rotor diameter) on the diagonal scale for the falling system lines. The acceleration factor z can be computed from d and n using Equation (6) and the circumferential speed u_R from Equation (15).

The fields of application for the various types of centrifuge depend partly on their design with regard to the method of solids removal, i.e., cake in the case of filter centrifuges or sedimented sludge in the case of solid-bowl centrifuges, and partly—with regard to the attainable cut size—on the screen mesh or clarifying area load (Table 2). Table 3 shows the particle-size and solids content distributions for 7 types of centrifuge, published by Alfa-Laval in 1972 [44].

2.1. Filter and Screen Centrifuges

The design of filter and screen centrifuges varies according to the method of cake removal. Periodic removal may be carried out manually, with automatic scraping devices, or the solids may be continuously removed by a helical conveyor [40], [41]. The filtration rate of filter and screen centrifuges depends on the drainage properties of the solid cake [45]. As with static filters, permeability depends on particle structure and cake thickness. The solidified, stationary cake in batch centrifuges is less permeable than one that moves, as in continuous centrifuging. Under otherwise identical operating conditions, pusher centrifuges therefore have a higher filter area than scraper-type centrifuges, but are inferior to continuous screen centrifuges. The constant turning over of the cake in multiple-stage pusher centrifuges, screen centrifuges with scroll discharge, and vibrating screen centrifuges makes the cake looser and more permeable. More fine particles pass into the filter medium this way, however, leading to increased effluent losses [46]. Another disadvantage is crystal breakage caused by the cake being conveyed at unreduced speed. Each type of centrifugal filter has its own special applications, in accord with its particular operational characteristics. Where there are very fine particles, or sensitive crystal products, and where loss-free operation is required, batch basket centrifuges are preferred. In addition to top-suspended centrifuges, these include three-column carried centrifuges and, for smaller diameters, basket centrifuges with elastic axis (see p. **11**-11). For automatic systems, basket centrifuges are built with scraping devices, i.e., scraping spoons or knives moving to and fro, covering the whole width of the basket (see p. **11**-10). Quasi-continu-

Table 3. Selection of centrifuges by granulometry (A) and solids content of the feed (B)

Ⓐ Particle size, µm	0.1	1	10	100	1000	10 000	100 000
Pusher centrifuge							
Scraper-type							
Pendulum							
Decanting							
Nozzle							
Self-discharging							
Solids-collecting							

Ⓑ Solids, %	0	10	20	30	40	50	60	70	80	90	100
Pusher centrifuge											
Scraper-type											
Pendulum											
Decanting											
Nozzle											
Self-discharging											
Solids-collecting											

ous pusher centrifuges form a link to the continuous screen centrifuges (see pp. **11**-8, **11**-9).

Continuous Screen Centrifuges. The basic type of continuous screen centrifuge is the *screen centrifuge with scroll discharge* [47] (Fig. 1 A, p. **11**-6). It has a particularly high capacity and is used for dewatering crystalline products that can tolerate rough handling.

The main areas of application are the dewatering of fine coal, residual salts and fertilizers (e.g., ammonium sulfate), other inorganic salts (e.g., Glauber salt, soda, borax), plastic granules and pearl polymers, as well as organic crystalline products (e.g., oxalates, naphthalene). Applications in the food processing industry involve milk sugar (from whey), raw sugar, starch, protein, and similar products; in the pharmaceutical industry, gland extracts. Further examples are cellulose and textile fibers, nitrocellulose, and iron sulfate from pickling baths.

A helical screw conveyor driven by a differential gear (cyclo gear) is not always necessary for transporting the cake. Light flowing cake moves as a stream of slurry over the conical screen basket without assistance. In the steep cone centrifuge or *thin-layer centrifuge* with no screw conveyor, the angle of friction and the cone angle balance each other. This type is mainly used for raw sugar.

The cake often has to be retarded to allow it to dwell long enough on the perforated basket to achieve sufficient dewatering (e.g., pearl polymers). In this case, it is not necessary to drive the screw conveyor at differential speeds. The *guiding channel centrifuge* has a retarding screw conveyor rotating at synchronous speed. The screw conveyor can be adjusted by ring-shaped cone elements with paddles attached, which can be turned to various angles of inclination against the tangent [48].

The *gliding ridge bar centrifuge* possesses a knife-like transport scraper, driven at differential speeds, whereas the *removing paddles centrifuge* has louver-shaped scraper systems. These special screen centrifuges occasionally compete with the screen centrifuge with scroll discharge.

A further possibility for transporting the cake consists of neutralizing the angle of friction through vibratory movements of the conical basket. In *vibrating screen centrifuges*, an axial vibration is superimposed orthogonally on the rotary movement of the basket. Figure 1 B, p. **11**-6, shows the vertical version and Figure 6 shows a cut-away of the horizontal type [49]. These centrifuges operate at low speeds (500–1000 rpm) and are used for dewatering coarse crystalline

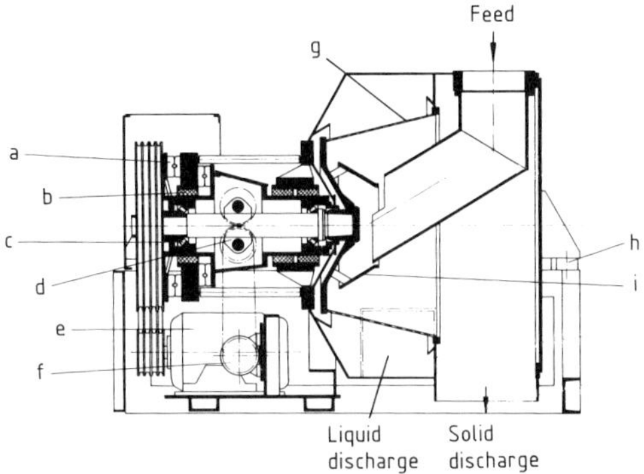

Figure 6. Horizontal vibrating screen centrifuges (Siebtechnik, Mühlheim/Ruhr)
a) Resilient pillow; b) Rubber ring bumper; c) Main bearings; d) Unbalance masses; e) Motor; f) Unbalance drive; g) Perforated drum; h) Rubber bumper; i) Feed distributor

products, fine coal, and silica sands (the last at only 250 rpm) [50], [51]. This applies in the same measure to the *reeling screen centrifuge* (manufactured by Krauss-Maffei, München), in which the perforated basket oscillates and tumbles (eccentric bearing) [52]–[54].

Automatic Filter Centrifuges. The areas of application for *pusher centrifuges* and screen centrifuges overlap. Pusher centrifuges filter short batches, until the cake has become sufficiently stiff [55], [56]. A pusher ring plate intermittently moves the cake over the screen or filter area in the axial direction up to the edge, over which it is discharged. Quasi-continuous operation is achieved by the rapid succession of small batches. The material is handled more gently and the screen mesh is smaller; however, the cost is higher due to the lower specific output. The two-stage design has become the one used most (Fig. 1 C) [40], [41], [52]. Here the rim of the first stage is used as the pusher plate for the second stage. The pusher plate and the second stage move axially. Pusher centrifuges are used mainly in the potash and salt industries and for other fertilizers. They also play an important role in other branches of the chemical industry. The pusher plate is usually powered hydraulically. Pusher frequencies are around 60–100 min^{-1}. The capacities of the biggest centrifuges are ca. 60–80 t/h. Pusher centrifuges are also effective for washing the product [57].

The diameter ranges in vibrating screen centrifuges, screen centrifuges with scroll discharge, and pusher centrifuges are ca. 400–1500 mm,

200–500 mm (chemical industry) or 650–1000 mm (ore dressing), and 300–1400 mm, respectively. The corresponding rotational speeds are 1300–500, 5000–3000 or 1000–600, and 3000–800 rpm; the acceleration factors 400–200, 3000–2500 or 350–200, and 1500–500; and the circumferential speeds 35, 60 or 30, and 50 m/s.

Scraper-type centrifuges with horizontal axis are built as large-volume centrifuges with diameters up to 2 m. Historically, some designs even have a double rotor (i.e., two rotor halves with a common back, with feed and scraper devices in both sides) to obtain maximum cake output. The design with a scraping knife and a chute is shown in Figure 1 D [40], [41]. Knives that turn tangentially are also used, as well as spoons that may be moved axially and radially at the same time. This also applies to vertical three-column scraper-type centrifuges, which are built with diameters in the narrow range 1000–1600 mm, whereas horizontal scraper centrifuges have diameters varying from 700 to 2000 mm. With rotational speeds of 1300–750 rpm (vertical) and 2000–750 rpm (horizontal), the circumferential speeds are 68–63 or 78–73 m/s, and the acceleration factors 900–500 or 1500–600. The number of batches can vary between 5 and 20 per hour, the higher figures being achieved where there is no washing. When necessary, scraper-type centrifuges can also be supplied vapor and pressure tight (0.3–0.5 MPa gauge). Increased output in scraper centrifuges is achieved by mounting a ring-shaped filtrate overflow pan (cup) outside the filter cloth on the rotor. This is subdivided by a baffle plate into two ring-shaped sections with different levels, the outer level being regulated by a scraping tube. The difference in level (ca. 40 mm) results in suction, or a kind of siphon effect, on the outside of the filter cloth, hence the name *siphon* (Fig. 7). In this way the filtrate output (and thus the total capacity) can be nearly doubled and the residual moisture of the cake significantly reduced [58].

Typical applications for scraper-type centrifuges are inorganic and organic crystalline products such as rock salt, potash, chlorates, perborates, aluminum hydroxide, cryolite, iron oxides, titanium dioxides, paraffin, *p*-xylene, urea, and adipic acid. They are also used for polyolefins, silicone and melamine resins, aspirin, insecticides, etc., and products such as fruit marc, fish meal, protein, starch, and antibiotics.

Among scraper centrifuges in the sugar industry (so-called A quality), the *top-suspended centrifuge* is most generally used. It operates with a circumferential speed of nearly 100 m/s. For example, a centrifuge with a diameter of 1250 mm has a rotational speed of 1500 rpm and can attain an acceleration factor of 1550.

In baskets with a height of almost 1100 mm, batches of 1000 kg can be processed; with a cycle time of 3–4 min, this gives an output capacity of 15–20 t/h. Unloading takes

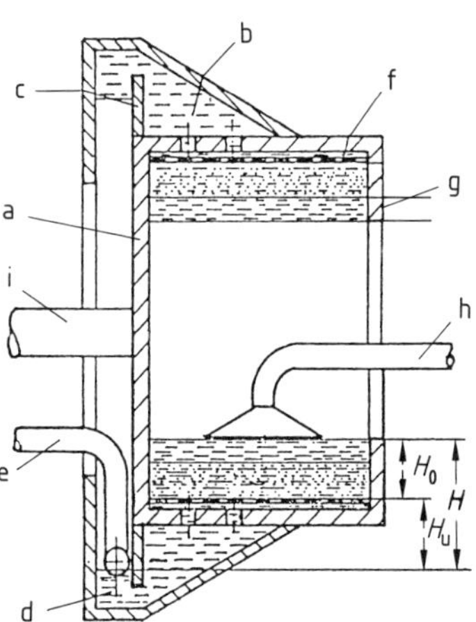

Figure 7. Schematic diagram of a horizontal scraper basket centrifuge with siphon effect
a) Centrifuge rotor; b) Filtrate compartment; c) Siphon disk; d) Ring cup; e) Peeling tube f) Filter medium; g) Dipping ring; h) Feed; i) Axis
H = Height of liquid; H_0 = Height of liquid above the filter medium; H_u = Height of liquid below the filter medium (Krauss-Maffei, München)

Figure 8. Top-suspended basket centrifuge for white sugar, the largest centrifuge of its type in the world (Braunschweigische Maschinenbauanstalt, Braunschweig)

almost 20 s, at a reduced speed of 50 rpm. A braking time of nearly one minute is needed to reduce the speed from 1500 to 50 rpm, using motors that then work as generators. Filling only takes about 10 s at speeds of 200–250 rpm. Therefore, 30–40 % of the operating time per batch is used for centrifuging and washing at full speed. The oversize multiple-speed motor, with its intensive cooling system, gives these sugar centrifuges a characteristic appearance. Figure 8 shows one of the largest sugar centrifuges of this type. Another application of top-suspended centrifuges is the recovery of sea salt.

Basket Centrifuges. Another type of batch-operated centrifugal filter is the basket centrifuge, still very significant in terms of numbers. Small basket centrifuges with an elastic axis are used domestically as juice extractors and washing units, and in industry mainly for removing excess oil from chips and hardware [40]. Three-column centrifuges (Fig. 1 E, p. **11**-6) play an important role in the chemical industry, especially dyes, textiles, wax, resins and pharmaceuticals, fruit juice and cereal processing, tanneries, pickling plants, and chip deoiling [40], [42]. These centrifuges may have top or bottom discharge. Batch operation is also common. The lifting bottom centrifuge is an automatic version. In addition to electric motors, explosion-proof hydraulic drives are also used. Automatic unloading with pneumatic scrapers (suction filter centrifuge) is a variant of this type [59]. Figure 9 shows the principal of operation. A further modification is the vertical centrifuge with automatic

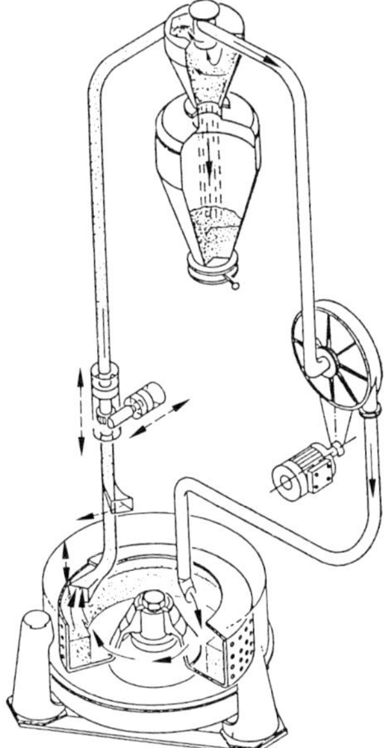

Figure 9. Pneumatic peeling system for top discharge pendulum centrifuge (Alfa-Laval, Hamburg-Bergedorf)

pusher bottom for chip deoiling, which operates at constant speed with about 300 g acceleration and a pusher frequency of 10–20 min^{-1}. The applications of the manual unload pendular centrifuge (Fig. 1 F) [40], [41] are similar to those of the three-column centrifuge. The practice of discharge bags for vertical basket centrifuges led to the development of the horizontal automatic reversible filter bag. This *conversion filter bag centrifuge* with a horizontal shaft (manufacturer: Ernst Heinkel Maschinenbau, Karlsruhe) has now proved useful in solving the most difficult filtering problems, e.g., the dewatering of hydrated iron hydroxides. The effective regeneration of the filter medium by back-rinsing, which otherwise may easily become blocked, is of great advantage. With a rotor diameter of 350 mm and at 2400 rpm (circumferential speed 44 m/s), an acceleration of 1000 g can be attained. The slenderness ratio is extremely low, $\lambda = 0.5$. For a filter area of 0.2 m^2, the slurry holding volume is about 12 L. With 40–60 automatic batches per hour, 0.5–0.7 m^3/h slurry volume or about 200–300 kg/h absolutely dry solid is produced. The filtrate output is about 10 m^3/h.

2.2. Decanting and Sedimentation Centrifuges (Solid-Bowl Centrifuges)

Sedimentation in the centrifugal field is employed for both clarifying liquids, with simultaneous thickening of the separated solids, and for stream classifying (separation of the suspended solid granules according to particle size into fine and coarse fractions, with simultaneous thickening of the coarse fraction and dilution of the fine fraction). The bowl wall is not perforated. Such machines are called *solid-bowl centrifuges* or *hydrocyclones,* depending on whether the bowl wall and the suspension rotate together or only the suspension rotates.

There are different types of solid-bowl centrifuges, according to whether operation is batch, automatic, or continuous. As long as only two liquid phases are to be separated, continuous operation is no problem (liquid–liquid centrifugal separators). However, if one of the two components, or a third component, is a suspended solid, then special discharge devices (nozzles or helical screw conveyors) are required. Batch operation with automatic slurry removal is possible through the use of circular slots that open intermittently or by means of scraping devices. Batch manual unloading is also practiced; for this, the machine must be standing still. The common feature of these centrifuges is that they are operated as overflow centrifuges, i.e., with constant overflow of the clear liquid or the fine suspension, and simultaneous constant feed of the raw suspension, at least during the batch cycle.

Overflow Centrifuges. Solid-bowl overflow centrifuges can also be called decanting centrifuges. The special continuous design with a helical conveyor for discharging the slurry, i.e., the solid-bowl scroll discharge centrifuge, is known quite simply as a *decanter*. It is the most important type of centrifuge in mineral dressing, but also plays an important role in chemical engineering because of its continuous mode of operation. Sedimentation takes place in the decanter bowl; the slurry that collects on the bottom, e.g., the rotor inner surface, is transported continuously by a differential-speed screw conveyor [32], [60]. Decanters are used for clarifying suspensions, for thickening slurries, and for stream classification of dispersed, fine collectives. They can handle both thin and concentrated suspensions economically [40], [41], [61].

The rotor diameter *d*, the slenderness ratio $\lambda = L/d$, and the rotational speed *n* vary in the different types of decanters, depending on the application (cf. p. **11**-12, **11**-13). Operational characteristics of four decanter designs are listed in Table 1, p. **11**-7, (nos. 3a–3d), each with two typical rotor diameters that easily fit into the operational field for decanters, shown in Figure 5, p. **11**-7. The classifying decanter with conical rotor (Fig. 2A, p. **11**-6) operates with medium rotational or circumferential speeds (60–70 m/s, as shown under item no. 3a in Table 1 for diameters 355 and 900 mm. The acceleration factors are between 2000 and 1500. As equivalent clarifying areas in the sense of Equations (13) to (20), relatively low values, between 750 and 6500 m^2, are obtained. Typical applications are the classification of kaolin within the range 2–5 µm, pigments (e.g., titanium dioxide, lithopone) in the 2 µm range, and abrasives (hematite, corundum) in the range 5–10 µm. In wastewater engineering, cylindrical decanters with high Σ values are used, and the *cocurrent principle* has been found to be valuable. In classifying decanters the slurry and the supernatant liquid phase move countercurrently, whereas in these machines they move in the same direction (Fig. 10).

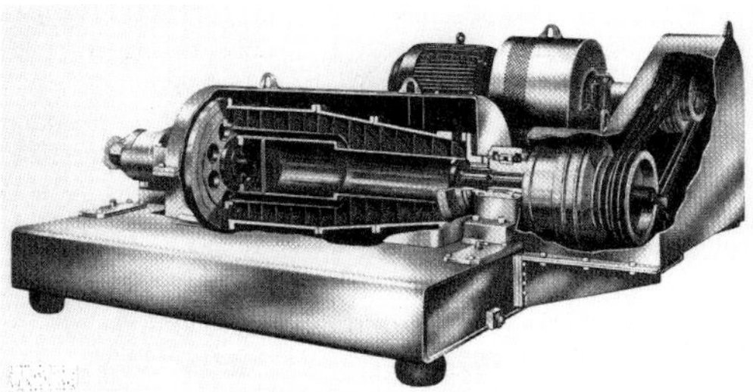

Figure 10. Cocurrent decanter (KHD Humboldt-Wedag, Köln)

Decanters that are used for dewatering crystalline products have still lower speeds [13]: about 50 m/s circumferential speed. The equivalent clarifying areas for this group (no. 3b in Table 1) are also small, and can only be increased above 2000 m^2 by building rotors with very large diameters (1.4 m) (manufacturer: Bird Machine Corp., United States). These rotors, in groups 3a and 3b in Table 1, have low slenderness ratios, mostly below 1.8.

Applications of dewatering decanters include, e.g., fine coal, potash, rock salt, phosphate rock, soda, borax, etc., as well as *p*-xylene, caprolactam, polymers, etc.

High-speed clarifying decanters (no. 3c in Table 1) are used for clarifying without addition of flocculation agents. These machines can attain a circumferential speed of 90 m/s. The value for the equivalent clarifying area is very high, ca. 4000–13 000 m^2, and the slenderness ratio is large, ca. 2.5–3.5 [24], [62]. Applications are limited to products where abrasion remains tolerable despite an acceleration factor of 3000 or even above 4000, e.g., certain wastewater slurries, hydroxides, and fine dust (cocurrent principle).

If flocculants are added to the suspension, the circumferential speed of overflow centrifuges is reduced to about 45–50 m/s to avoid breaking the flocculus. The decanters in Group 3d in Table 1 belong to this type. The acceleration factors are then mostly below 1000, the slenderness ratios of the rotors are 1.7–2, and the equivalent clarifying areas are within the range 500–2500 m^2.

This design could prove successful in wastewater engineering. Decanters using a flocculant, which is usually added only within the rotating bowl, have also been found useful for clarifying coal wash water and tailings water and for thickening flotation concentrates and paper wastes.

Further applications for clarifying decanters, with or without the addition of flocculants, include milk of lime, caustic potash lye, caustic soda lye, catalyst recovery, tar clarification, drilling mud, fish press liquor (fish meal production), fruit and citrus juices, vegetable oils, lactose from whey, lard and tallow, chocolate, wool grease, woodpulp, nitrocellulose, rubber solution, and gelatin. Some special designs are vapor or pressure tight up to about 0.6 MPa.

Combination of both screw-conveyor centrifuge types, i.e., the decanter and the screen centrifuge with scroll discharge, led to the development of the *screen decanter*. Entirely different applications, such as the dewatering of potash and rock salt [30] or separating lactose from whey, have become much more efficient through the use of this type of centrifuge.

The helical screw conveyors of decanter and screen centrifuges have a differential speed (higher or lower) relative to the rotor (gear ratio about 1:40 to 1:120) obtained through a planetary or cyclo gear. They are subject to shear stress caused by the slurry or cake sliding along the sides. This causes *wear*, which is particularly strong in the conical section of the rotor of decanters, as there the slurry is transported against the centrifugal force. For a long time these conveyors have been protected by hardfacing, e.g., with Hastelloy. During the last ten years, the practice of cladding the sides of the conveyor with a hard material, e.g., Al_2O_3 ceramics or tungsten carbide, has been developed. The inner wall of the whole rotor is sometimes also lined in the same way. RECORDS has reviewed this topic [62].

Decanting centrifuges without a screw conveyor require nozzles if they are to operate with continous slurry discharge. There are also decanters that combine a screw conveyor and nozzles as a slurry-conveying aggregate. The nozzle principle has become generally accepted, especially in *nozzle disk centrifuges* (Fig. 2 C, p. **11**-6). In order to be able to operate the nozzles continuously there must be an adequate amount of slurry. The thickened slurries have a large angle of repose and to prevent them from collecting in the annular slurry space there must be many (8–12) nozzles at the periphery of the bowl [61]. The load for each nozzle is then only a small portion of the slurry mass and the small nozzles, which tend to wear quickly, need to be only a few millimeters in diameter. Reducing the number of nozzles by adding dispersants (pyrophosphates) inside the rotor to make the slurry flow better rarely succeeds. However, slurry recycling, returning a partial stream of slurry to the feed inlet [61], has proved of value. The most effective method is slurry recycling directly to the slurry chamber, for example, to the catalyst recovery unit in polyolefin production (manufacturer: Dorr-Oliver) [63], [64]. The recycle factors in this method reach values up to 20. A retaining dam within the slurry recycle chamber has also proved valuable in maintaining a constant slurry level inside the rotor (Gravitrol principle for purifying diesel oil on ships) (manufacturer: Sharples Centrifuges, Camberley) [65].

Nozzle disk centrifuges are used for thickening kaolin and bentonite suspensions, purifying diesel and residual oils, phosphoric acid, caustic soda solution, tar, benzene, beer, wine, animal and vegetable oils, juices, antibiotics, etc., and for separating hydroxides, catalysts, activated carbon, etc.

Nozzles for slurry discharge can also be operated intermittently, if they can be covered by valves. This makes the stream of slurry greater during the opening time, so the nozzle diameters can be larger. The application of valve–nozzle centrifuges with disk inserts in rendering plants has been reviewed [65].

For intermittent automatic slurry discharge, closable circular slots are generally used. Disk centrifuges have been pioneers in the form of *self-cleaning disk centrifuges*, which are now being used again as liquid–liquid centrifugal separators [40], [66]–[68]. A schematic diagram of this type is shown in Figure 2 D. Control devices with a time switch have long been in use as accessory units. Devices for automatic level control are also on the market, where either a photoelectric turbidimeter or a slurry level indicator triggers the control pulses [67].

To a large extent, applications of the circular-slot disk centrifuge overlap those of the nozzle disk centrifuge, the former being preferred where the amount of slurry is small (Table 3, **11**-8). Further applications include hydrated iron hydroxides, lacquers, meat extract, sugarcane juice, molasses, alcoholic liquors, essential oils, vinegar, shellac, and ammonia solution.

For the separation of flocculated substances, there are circular-slot centrifuges without disks, e.g., the *double cone centrifuge*, which is particularly useful for wastewater slurries [30]. It is simply a sedimentation overflow centrifuge. Slurry is discharged by intermittent opening of the circular slot, one cone half being axially displaced for a short time. The pulse for this is given by the slurry level indicator. On the other hand, in centrifuges used for dewatering granules the circular slots serve as drainage slots and are never completely closed. These machines represent a compromise between the solid-bowl centrifuge and the centrifugal filter. A schematic diagram of the original design of the granule centrifuge is shown in Figure 2 B. The reflecting annular pads centrifuge, now used for dewatering granules, has a terraced rotor [69].

The *free jet centrifuge*, used for purifying motor oil, also possesses filter elements. The solid-bowl section is designed as a reaction turbine. The rotor is driven by the oil itself, which is fed under pressure. This small-scale centrifuge must be emptied periodically by hand [41].

Very high acceleration factors are achieved by *tube centrifuges*, solid-bowl centrifuges built without any device for discharging the slurry [40], [41]. They have been used with great success for fine purification, i.e., removal of the slightest turbidity. Economical operation depends on a very low concentration of solids, as the (compact) cake has to be removed manually, and the batch cycles should therefore be as long as possible. Tube centrifuges (or super centrifuges) have a slim rotor ($\lambda = 5-7$) and diameter 45 (laboratory model), 75, 105, or 145 mm. An acceleration factor of about 13 000–17 000 is attainable at a rotational speed of 18 000–14 000 rpm (laboratory model 50 000 g at 45 000 rpm). This corresponds to a circumferential speed of 80 to 110 m/s. The equivalent clarifying area ranges from 1500 to 4000 m^2. With a dirt-holding space of two to ten liters, the capacity is ca. 200–2000 L/h. Some characteristics of industrial and laboratory tube centrifuges are given in Table 1, no. 2, p. **11**-4.

Tube centrifuges are used for purifying mineral and vegetable oils, in soap manufacture, in the lacquer and varnish industry, for the separation of viruses, in the production of chewing gum, and in the preparation of vaccines and blood serum. There are special designs for specific applications, e.g., with cooling or heating coils, hermetically sealed, hygienic, etc.

The inner clarifying area of a tube centrifuge can be increased by nesting of concentric cylinders, as in the *circular compartment centrifuge* (multichamber centrifuge) (Fig. 2 F) [40], [41]. It is used for fine clarifying (polishing filtration), for example, for beer, wine, juices, oils, paints and varnishes, and chemical solutions. The rotational speed is 5000–8000 rpm, with 500–250 mm diameter, corresponding to 7000–9000 g acceleration, and is thus considerably lower than the speed of tube centrifuges. The rotors are also comparatively short ($\lambda = 0.7$). Nevertheless, the number of chambers (up to 5) and the larger diameter result in higher Σ values (about 4000–12 000 m^2) than in tube centrifuges.

Decanting Basket Centrifuges. Large-volume decanting basket centrifuges have lower Σ values than tube centrifuges. They correspond to the types discussed under filter and screen centrifuges, but with solid-wall rotors instead of perforate bowls: specific examples include the *basket centrifuge with elastic axis*, the *three-column centrifuge*, the *top-suspended centrifuge*, the horizontal and vertical *scraper-type centrifuges*, and the *solid-bowl top-suspended centrifuge* with automatic knife discharge. The equivalent clarifying area of these types of decanting centrifuge is 1000–2000 m^2 (see no. 1 in Table 1). They have large dirt-holding spaces and even without automatic knife discharge are superior to the two manually unloaded centrifuges. They are suitable for the most diverse clarifying problems, but not for extremely fine grains: for these the tube or multichamber centrifuge is necessary.

Decanting basket centrifuges are used for purifying juice, must, wine, and beer, for vegetable oils, rubber solution, lacquers, varnishes, printing colors and ink, and occasionally the recovery of lyes and catalysts.

2.3. Separators

The most frequently used centrifuge types, in terms of numbers, are liquid–liquid centrifugal separators, especially milk separators, from which all others have been developed. The complete range of separators is given in Figure 5, p. **11**-7 [70]. Inside rotor diameters vary from 250 to 800 mm, with rotational speeds between 9000 and 3000 rpm. Circumferential speeds are 120–160 m/s, and acceleration factors are 6000–12 000. (The effect of the number of disks in the separator and the length of the rotor in the decanter are not included in Fig. 5. Actual Σ values therefore have a considerably broader spread than the acceleration factors, see Table 1, p. **11**-4.) The separator and decanter designs given in Table 1 are distributed in Figure 5 over the field of operational characteristics as follows: in the upper right section of area c (maximum circumferential speeds) are high-speed clarifying decanters (no. 3c in Table 1), towards the middle are classifying decanters (3 a), not far from the lower left limit (minimum circumferential speeds) are flocculation clarifying decanters

(3 d), and between 3 a and 3 d are crystal dewatering decanters (3 b). Overflow basket centrifuges (decanting basket centrifuges) are in the lower right section (low rotational speeds). Industrial tube centrifuges (Fig. 5, area a), with a diameter of 75–145 mm, fall on the left section whereas laboratory sizes would appear outside the diagram. The superiority of the tube centrifuge over disk separators (Fig. 3, area b) is only apparent. When the number of disks is included, their positions with regard to the Σ values are reversed.

Disk Separators. The classic centrifugal separator is the two-phase disk centrifuge separator. One of the largest machines has the following dimensions (see also Table 1, no. 4): inside rotor diameter 700 mm; mean disk diameter 460 mm; length of the disk stack 380 mm. Thus the area of the mean effective cylinder is 0.55 m^2.

Of about 160 disks, 100 are amplifying (i.e., lie on a radial section), giving 55 m^2 effective disk surface. With a rotational speed of 4400 rpm, a maximum centrifugal acceleration of 7500 g (with 160 m/s circumferential speed) can be developed. At the mean disk diameter, the acceleration is 4900 g, so that an equivalent clarifying area of 270 000 m^2 results. This is more than two magnitudes greater than that of manually operated agricultural milk centrifuges. A large range of intermediate sizes and designs exists for diverse applications in industry and food processing. In batch operation, the solids are removed manually after stopping the rotor and siphoning off the supernatant liquid. In automatic operation, the sedimented slurry is unloaded periodically by a scraper or by the opening of circular slots. Until the maximum possible cake thickness has been reached, the carrier liquid drains away continuously at a constant throughput rate (overflow centrifuge) and a corresponding amount of suspension is continuously loaded. With continuous slurry discharge devices (nozzles or helical conveyors) in overflow centrifuges, the whole separation process is continuous. In the liquid–liquid centrifugal separator, both liquid phases are discharged over separate overflow weirs or scoop devices: lighter liquids inside, heavier outside.

These separators have large disk stacks and the remaining space is small; they are used to separate two pure liquids, e.g., to remove water from fatty acids, mineral oils, tar, and hydrocarbons. However, there is usually some separation of contaminating solids, i.e., three-phase separa-

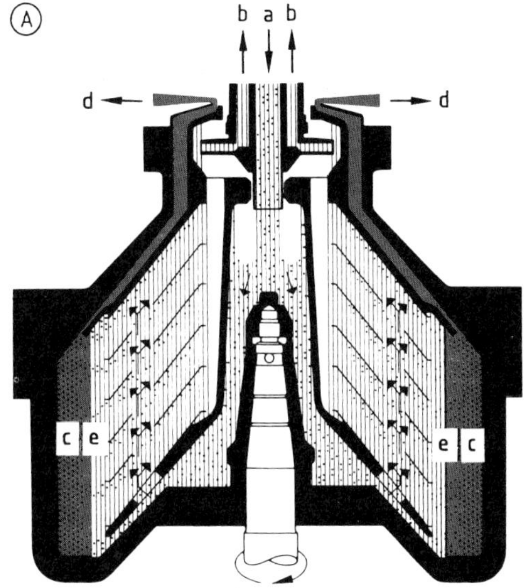

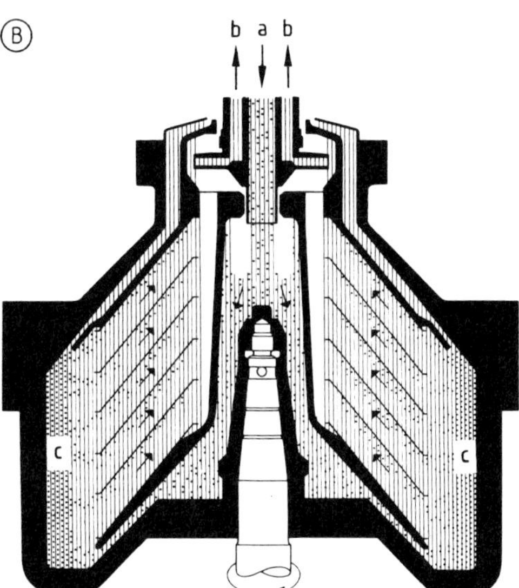

Figure 11. Rotors of a two-phase separator (Alfa-Laval, Industrie-Technik)
A) Purifier, used to purify oil that contains significant quantities of water
B) Clarifier, used to purify oil that contains sludge but little water
a) Inlet; b) Outlet for purified oil; c) Sludge; d) Outlet for water; e) Boundary between oil and water

tion. Disk separators are therefore built with a larger dirt-holding space (*purifier*), as shown on the left of Figure 11. The rotor shown on the right (*clarifier*) belongs to the group of pure clarifying centrifuges (only one liquid phase). This

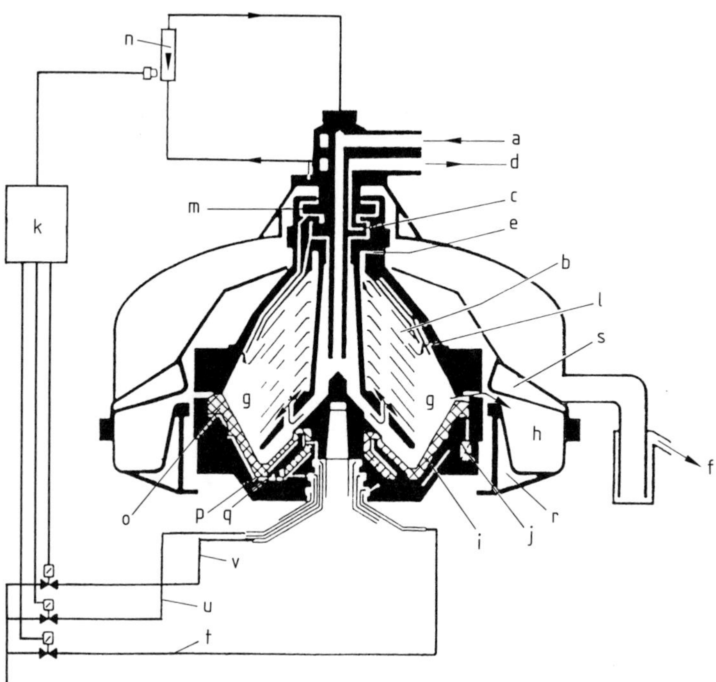

Figure 12. Schematic drawing of a self-cleaning rotor SA 160-33-177
a) Feed; b) Disk; c) Grip peeler; d) Heavy phase discharge; e) Outlet for light phase; f) Light phase discharge; g) Solids compartment; h) Solids discharge; i) Piston slide valve; j) Annular valve; k) Control board; l) Disk for sensor liquid; m) Control peeler; n) Flow meter with initiator; o) Outer closing compartment; p) Inner closing compartment; q) Forced opening compartment; r) Cooling compartment; s) Sound insulation; t) Fluid for opening; u) Fluid for closing; v) Fluid for opening under force; w) Sensor liquid discharge (Westfalia Separator, Oelde)

increases the range of applications to include the regeneration of used oils and solvents, the purification of Buna milk, lacquers and varnishes, vegetable oils, animal fats, citrus juices, essential oils, soap stock, beeswax, and wine. In some of these cases there is no second liquid phase. When contamination is only slight, wide-chamber disk centrifuges are also used as clarifiers or sedimentation centrifuges. Disk separators of the most diverse designs are being increasingly used in biotechnology, pharmaceuticals (e.g., antibiotics), and genetic engineering [71].

When the two liquids to be separated contain a considerable amount of dirt, thus constituting a genuine case of three-phase separation, the disk centrifuges are fitted with nozzles or circular slots. These self-opening or *self-cleaning separators* have a second liquid outlet. A cross section of a self-cleaning three-phase separator is shown in Figure 12 [40], [61], [72]. This type of centrifuge has now become the standard design for the purification of heavy oil and diesel oil, where both water and dirt taken up by the oil during storage and transportation must be removed. Self-cleaning separators are particularly valuable for shipping. The many possible applications have already been discussed in connection with two-phase disk circular slot centrifuges. There is a similar correlation between the applications of disk nozzle centrifuges and three-phase nozzle separators.

Disk separators are available in a large range of sizes, with or without slurry-removing devices, with all three variants (three-phase liquid–liquid–solid centrifugal separators, two-phase liquid–liquid separators, and two-phase liquid–solid centrifuges) being distributed over the whole range. In Table 1 (no. 4) the typical data for two nozzle-type separators are listed. The larger one, with a nozzle ring diameter of ca. 800 mm, represents the maximum size currently available (Fig. 13). Its equivalent clarifying area, however, is only a third of that of the biggest two-phase separator, being about 80 000 m^2. One reason for the lower clarifying area is that the dirt separator operates at a lower speed, but

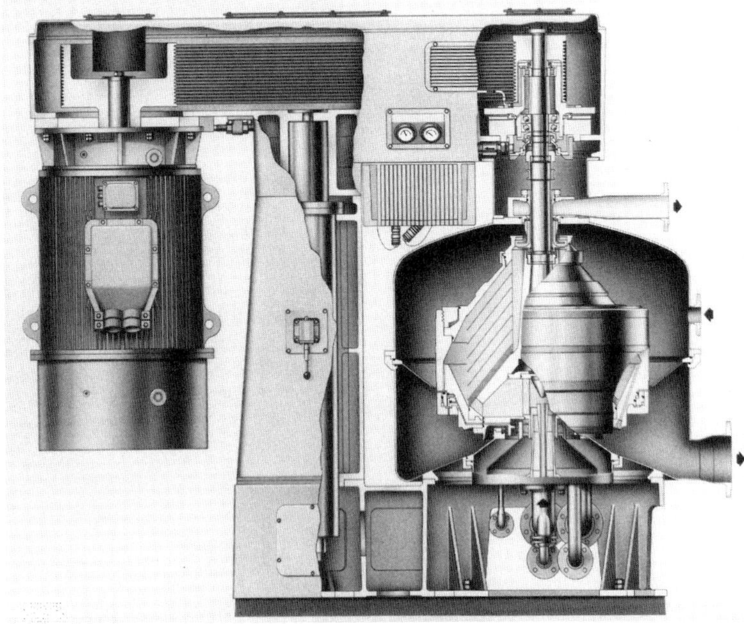

Figure 13. Top-suspended nozzle-type separator type HDA (Westfalia Separator, Oelde)

the main factor is the geometry: the disk stack must be considerably smaller causing the effective mean cylinder area and the acceleration factor on the radius to drop sharply.

Tube Separators. Tube centrifuges can also be operated as separators [40], [42], or, to be more exact, as two-phase liquid–liquid separators and as three-phase separators. The latter, however, like clarifying centrifuges, must be cleaned periodically by hand. These machines have lost their original importance for purifying diesel oil on board ships to the self-opening disk separators. Multiphase tube separators can only compete to a limited degree with automatic disk separators, because manual cleaning is time consuming. In the field of pure two-phase liquid–liquid separation, however, tube centrifuges are still used, e.g., in soap making (soap stock), refining of vegetable oils (miscella), and blood serum preparation.

Centrifugal Extractors. The separation of liquids also plays a major role in the case of centrifugal extractors [40], [42], [62]. In liquid–liquid extraction (solvent extraction), substances dissolved in a carrier liquid are extracted by adding a solvent in which the substances are more soluble. Provided the carrier liquid and the solvent have different densities, they can then be separated in centrifugal separators, which can be used in combination with mixers or emulsifiers. In single-stage extractors, the success of the extraction depends entirely on how closely the theoretically optimum dispersion figures are approached.

The contact between the extracting and the extracted liquid can be increased by using baffles, for instance to divide up the inside of the rotor spirally, like a rolled-up sandwich (manufacturer: Podbielniak, Chicago, Ill.), or guide screws can be built into a multichamber rotor (with 15–25 chambers). Radial mixing between the concentric tube zones is occasionally achieved by removable perforated disk systems (shashlik design) [9]. Disks are often used to bring the liquids flowing countercurrently into contact with each other [40].

After extraction, the liquids must be separated as completely as possible, which is achieved by using high acceleration factors. Mixing and separation stages can be combined quite elegantly in one machine.

Higher degrees of extraction can be achieved with *multi-stage extraction:* the two liquid components flow countercurrently. The rotors and the mixers are mounted in separate but concentric shafts and flow is forced through appropriate channels. These machines were built with two or three stages on the basis of the disk separator [40].

When operating with a higher number of stages and very high acceleration factors it is possible to do without the packing effect of the inserts or disks and yet still achieve good results. The separating elements of such extractors are like those of basket or tube centrifuges. In both cases, both liquids flow countercurrently. The two scraper disks (or grip devices) built into each stage transport both the carrier liquid and the solvent on to the next stage (or the outlet) and at the same time the liquid mixture formed in each stage is homogenized. The extraction equilibrium is immediately verified, resulting in a high degree of efficiency. In large spiral or circular compartment extractors and in multi-stage extractors, this is equivalent to an ideal plate number of up to 20.

Standard sizes of this type have 48″ or 1220-mm rotor diameter and a rotational speed of max. 1800 rpm. At a circumferential speed of 115 m/s, an acceleration of 2200 *g* is attained. With 1 m^3 rotor holding capacity and a volume throughput of 130 m^3/h, (both phases together), the residence time is ca. 30 s. The biggest types, with a rotor diameter of 72″ or 1830 mm and a rotor length of 64″ or 1620 mm, can even reach absorption capacities of 300 m^3/h. Such a machine weighs 4.5 t.

Applications of centrifugal solvent extraction include the benzene-lye process for dephenolating wastewater, uranium leaching, the production of pyridine, and the production of antibiotics. The same machines are also used with modifications to carry out chemical reactions, such as the treatment of acids for refining mineral oils and edible oils, bleaching, ion exchange, neutralization, and production of hydrogen peroxide. Washing processes (selective purification, polymer purification) are also possible.

Batch Centrifuges for Laboratory and Industry. Laboratory centrifuges, from the ordinary bottle centrifuge up to ultracentrifuges, have always been used in chemical, food-processing, medical, and pharmaceutical laboratories. They are also used in biochemistry and genetic engineering to produce special high-grade products in relatively small batches. The boundary between laboratory and industrial centrifuges is not fixed.

There are two basic types of rotor in laboratory centrifuges: the swinging-bucket type and the fixed-angle type, in which small glass tubes inclined in a fixed position are used as sample containers. There are also special capillary rotors, which are used to separate erythrocytes from blood (in serum preparation). The hematocrit centrifuge belongs to this category. Rotors that hold solid particles are also used, e.g., for drying semiconductor elements after rinsing. Many analytical centrifuges can be sterilized and may also be fitted with a cooling device, in order to counter the rise in temperature that occurs during rotation over a long period. High-speed centrifuges, especially ultracentrifuges, also have a safety device against excessive speed. They are constructed in such a way that sedimentation can be measured optically while the machine is running. Modern laboratory centrifuges are equipped with sophisticated electronic systems, including digital display of operational data. In biochemistry, layering of solvents or carrier liquids of different densities is used in the investigation of material such as bacteria, viruses, enzymes, proteins, and RNA.

Two basic types of laboratory centrifuges should be distinguished: analytical centrifuges, which have rotors with holding capacities of 1 – 10 L and which operate with accelerations between 5000 *g* and 100 000 *g*, and ultracentrifuges, which have rotors with holding capacities from 50 to 500 mL and accelerations between 300 000 *g* and 600 000 *g*.

The capacity of rotors naturally depends on the use for which they are designed. Many small tubes provide a greater volume than a smaller number of large containers.

Suppliers of analytical centrifuges with large capacity and low acceleration values include Beckman Instruments, München; Christ Leonberg, Leonberg; Heraeus-Christ, Osterode; Hettich-Zentrifugen Zentrifugen- und Apparatebau, Tuttlingen; International Equipment Company, a Division of Damon Corp. (IEC), Needham Heights, United States; Kontron Analytik, Eching; Kubota Medical Appliance Supply Corp., Tokyo, Japan; Runne, and VEB-MLW Zentrifugenbau GDR. Of these, the following also supply ultracentrifuges: Beckman, IEC, Kontron, VEB-MLW.

Within each range, the centrifuge types from the various manufacturers are quite similar to each other. Typical operational data for selected sizes are given in Table 4.

The criterion used for classifying a centrifuge as a high-speed centrifuge is the same as that for industrial centrifuges, i.e., the circumferential

Table 4. Typical operational data for laboratory centrifuges

Rotor holding capacity, mL	Rotor diameter, max., mm	Speed, max., rpm	Acceleration at largest radius, g*	Circumferential velocity at largest radius, m/s
9 000	530	4 800	6 800	130
6 000	575	4 000	5 100	120
3 000	390	4 800	5 000	100
3 000	330	9 000	15 000	155
3 000	300	14 000	33 000	220
3 000	210	21 000	51 000	230
1 500	225	20 000	50 000	235
1 650	175	32 000	100 000	293
460	260	50 000	360 000	680
310	185	70 000	500 000	680
300	220	65 000	510 000	750
110	170	80 000	600 000	710
25	245	55 000	410 000	705

* $g = 9.81\ m/s^2$

speed. For *slow* analytical centrifuges this is 100–150 m/s (5000 g–15 000 g); for *faster* centrifuges, it is 200–300 m/s (30 000 g–100 000 g); and for *ultracentrifuges* it is ca. 700 m/s (400 000 g–600 000 g). These machines are imposing, not only because of their acceleration values of 600 000 g, which almost reach the value of the classic Svedberg ultracentrifuge ($10^6 \times g$) but also because of the enormous circumferential speeds of 750 m/s (2700 km/h, faster than a standard jet aircraft). At 80 000 rpm, balancing, exactly even loading, and precise bearing arrangement are of the utmost importance.

The fastest separators operate with a circumferential speed of 130 m/s (e.g., 750 mm diameter, 3300 rpm, 4500 g). However, the rotating mass is greater by a factor of 100. If the masses are related to the same relative radius (e.g., 2/3 of the maximum radius), then the stored energies are comparable: $100 \times 130^2/1 \times 750^2 = 3$ [66], [72].

3. Hydrocyclones

The hydrocyclone can be classified as a mechanical separation device in which sedimentation takes place in a centrifugal field. It is rather like a tube centrifuge with a nonrotating body. If the rotation of the basket in a centrifuge is regarded purely as a feature of design, and not as an essential functional factor, then the hydrocyclone can be regarded as a solid-wall overflow centrifuge with automatic slurry discharge.

The hydrocyclone is an important operational unit for separation and process technology. Often scarcely noticed, it is a link in almost every flowsheet involving liquid-phase processes. Its operation presents no special problems, its function is uncomplicated, and no special training is required for its maintenance. The simple construction of the hydrocyclone, and in particular the lack of rotating parts, means it can be made of a variety of materials and in different shapes, an advantage in combating wear and corrosion [121], [122]. The investment and operating costs are relatively low, due to the high-volume throughput and moderate expenditure of energy. In addition, space requirements and the cost of the foundations are low. There are problems, however, arising from the lack of any useful theoretical basis for flow in the three dimensional cyclone under continuous operation with a heterogeneous medium, and the lack of model laws going beyond a wide range of sizes. Prediction of results is thus extremely unreliable. Laboratory tests on a semicommercial scale are indispensable, and in general must be carried out with the original size of cyclone.

To plan a hydrocyclone system requires a great deal of practical experience, especially in evaluating tests from the engineering department and in scale-up. The simpler a device is in design and function, the more knowledge and understanding required.

Rotational motion in the hydrocyclone is produced by the suspension entering tangentially under pressure [123]–[126] (Fig. 14). Flow is at first down the inner wall of the cylindrical section and the cone, as far as the stagnation point near the cone apex. There, because the opening is small, the downward primary vortex is forced to turn upwards again, thus forming the secondary vortex. This revolves tightly around the air core near the axis and finally leaves the cyclone body through the upper axial nozzle, or vortex finder. In a cross section of the body, both spirals rotate in the same direction; in longitudinal section, however, they move in opposite directions.

Phase separation of the suspension takes place during this double vortex flow. The suspended particles, at least the coarse fractions, move outwards to the wall of the cyclone, where they flow downwards as a screw-shaped slurry stream. This leaves as the underflow stream through the apex opening, or underflow nozzle.

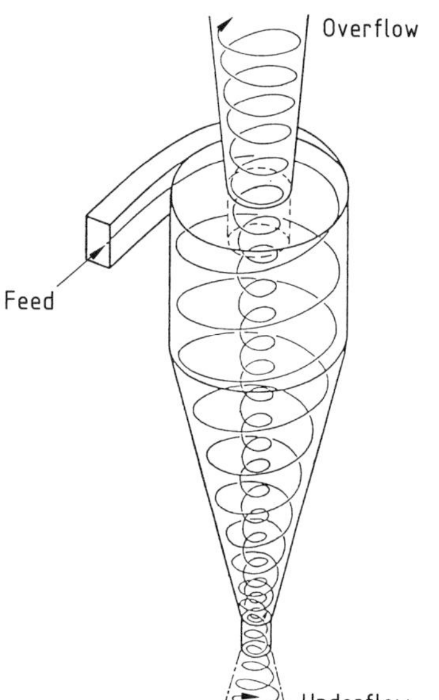

Figure 14. Flow pattern of the hydrocyclone

The nozzle is closed hydraulically by the stream of slurry, thus preventing the main stream of suspension, now thinned or clarified, from entering. The main stream is then forced to move through the inner vortex and leave the cyclone via the vortex finder, which acts as overflow outlet (and therefore is also known as the overflow nozzle). Equation (27) shows that for high volume flow rates, cyclones with a large cross section or large nozzle opening are required. However, cyclones should not be designed only according to the size of the flow to be handled. With regard to separation cut, Equation (30) must be considered. Small cut sizes can only be obtained with large pressure drops, involving high expenditure of energy, which is uneconomical. A better way is to use a small cyclone diameter d, with the length l as long as possible. Large capacities can be obtained by using many small cyclones in parallel. For coarse separations, cyclones with a large diameter are used [124], [127] whereas for fine separations groups of small cyclones are preferred (e.g., annular distributors or distributors with planetary cyclone groups) [132], [133].

Multihydrocyclones were also built in multistage blocks [128]–[130]. They are no longer in use, except for the separation of starch, because blockage of the underflow nozzles in single cyclone units within the blocks cannot be checked.

Open annular distributors are also used in ore dressing for large hydrocyclones (e.g., 500 mm diameter) (Fig. 15).

The hydrocyclone is a good stream classifier in the 5–100 µm particle-size range. At the same time it is useful as a thickener for the separated coarse fraction (underflow). It is less suited as a clarifier, because its vortex flow causes high shear forces, which counteract flocculation. In comparison to an overloaded settling tank operated as a hydroseparator, the hydrocyclone is the better classifier because of these shear forces [131]–[133]. Thickening, therefore, is necessarily tied up with classification. This is ocassionally considered as an undesired loss of fines in the overflow, but frequently it is used for desliming the coarse fraction [134], [135]. Applications are listed below [136], [137]:

- Solid–liquid phase separation
  - Thickening
  - Partial clarification
    - Circuit water clarification (in bypass stream)
    - Preclarification (followed by conventional clarification)
- Treatment of quasi-stable mixed phases
  - Purification of fine suspensions
  - Purification of emulsions
- Stream classification
  - Desliming (removal of fine particles)
    - Underflow recovery
  - Degritting
    - Overflow recovery
    - Countercurrent grinding
    - Formation of precoat layers
  - Refining (recovery of very fine products)
- Stream sorting
  - According to equal settling
  - According to particle shape
  - Selective classification (antiparallel particle distribution)
  - Heavy media process (sink-float sorting)

Hydrocyclones have limited success in clarification and are mostly used as preliminary stages to relieve other, more sophisticated and costly machines: prethickening, or the removal of large amounts of liquid, increases the capacities of filters and centrifuges. Preclarification, i.e., previous separation of most of the solid material, helps to relieve any slurry conveying elements in thickeners, centrifuges, etc.; alternatively, it may protect the machines (centrifuges, small-scale cyclones, etc.) against heavy wear. Hydrocyclones are therefore frequently used in conjunction with other mechanical separation devices [134], [137].

In clarifying and thickening, suspensions from some previous process are treated, but in

Figure 15. Hydrocyclone cluster (spider)
Annular distributor for six hydrocyclones of 500-mm diameter (4 mounted) made of vulkollan (designed by Amberger Kaolin-Werke, Hirschau/Oberpfalz

stream classification the particle collectives are only converted to the suspended condition by dispersion in a carrier liquid. In addition to desliming, i.e., the removal of dust from the coarse fraction, the opposite, degritting, is often also required. This feature of the hydrocyclone makes it useful for grinding circuits. To prevent dead grinding (production of too fine a product) the unfinished mill product is degritted in a cyclone stage, while the recovered coarse particles are recycled to the mill [126], [132], [134], [138], [139]. In the case of degritting, the undesired coarse particles are removed from a fine fraction. This process is also known as refining, especially when applied for kaolin benefication [132], [134], [140]. In desliming the product is represented by the cyclone underflow, whereas in degritting it is the overflow. There are many other applications for wet classification of fines with hydrocyclones. For example, the dewatering of very fine slurries on vacuum horizontal filters or on vibratory screens can be simplified by first separating a coarse fraction with hydrocyclones; this is then applied as a precoat layer to the filter or screen surface and the remaining fraction, after thickening in a fine cyclone, is then filtered through this layer [132]. A hydrocyclone can also be used to recover the effluent, especially from vibratory screens. Simple recycling to a prethickening cyclone is generally not successful, as fines collect in this circuit. A second cyclone whose underflow is filtered through the cake from the preliminary cyclone, as a kind of precoat layer, is a preferred arrangement. In this way, screen machines are used for dewatering fine fractions without any loss, even when the particles are finer than the screen mesh.

Thickening cyclones with a small diameter are also used in heavy media systems to bring the quasi-stable slurry back to the required density before recycling to the beginning of the process. These units are called densifier cyclones [39].

Hydrocyclones can also be used for sorting. Enrichment as a form of sorting occurs in normal classification by sedimentation, in gravity equipment and in the centrifugal field. This happens, for instance, when the components have different weights. On the basis of the Stokes settling velocity (Eq. (5)), coarse, light particles sink faster than finer, heavy particles. However, if the components have different particle-size spectra (antiparallel particle distribution), then sedimentation has a selective effect even where there is no difference in density. This selective classification is important in the refining of kaolin [132], [140].

Finally, particle shape can also have a sorting effect. Flat-bladed particles appear as faulty size particles in finer classes, which can lead, e.g., to mica enrichment in fine tailings slurries. Successful sorting is possible using the *heavy media process*. Tailings and concentrate can be separated by the sink and float method, by introducing them into a quasi-stable suspension. The slurry density must lie between that of the two components. Hydrocyclones are particularly well suited for sink and float processes with particles of 1–8 mm [39], [125], [127], [136], [141]–[147]. In particular, the required slurry densities can be lower than in the gravity field, because of the radial concentration gradient that builds up in the hydrocyclone. This means that a greater slurry density is built up in the zone near the wall (where sorting takes place) than corresponds to the mean slurry consistency [141]. Cyclone sorting of finer fractions (< 1 mm particle size) can in some cases be carried out in water because a stationary heavy medium is formed in the zone near the wall by the material to be separated (water-only cyclone) [148], [149].

The cone is not essential in hydrocyclones. In the last 15 years, completely *cyclindrical hydrocyclones* have come into general use, allowing a coarser classification than was previously possible. In the cylindrical cyclone, a bed of slurry is allowed to accumulate above the underflow nozzle. Reduction of the rotational speed at the lower cyclone end due to friction on the bottom surface (the *tea-cup effect*) sets up a convective current in the rotating bed (in the vertical cutting plane); consequently radial inward transport takes place on the bottom. Figure 16 shows this process schematically [39]. RIETEMA and VAN DIJN [150] have described this condition as a fluidized layer. The height of the accumulated bed can be adjusted by throttling the underflow nozzle. If the hydrocyclone is over-throttled, mass recovery is lowered and the cut size is raised. This cut size, according to Equation (30), is determined by a length L, defined here as the free inner length between the lower edge of the upper vortex finder and the surface of the bed. VISMAN [145], [151] uses the expression "vortex finder clearance length." In the cylindrical cyclone, the cut size can be regulated from outside by throttling the underflow nozzle, which controls the bed height and thus the effective length L. Figure 17 shows two parallel cylindrical hydrocyclones (type CBC) in operation. Variations of the bed level in the CBC, as shown in Figure 16 B, cause the VCL to change, and thereby the cut size. Cylindrical hydrocyclones are also useful for heavy media sorting. One design with tangential discharge of the bottom slurry is supplied under the name Triflow by Sula Maschinenfabrik [146]. The CBC cyclone can also be used for sorting in water, with the accumulated stationary bed of slurry functioning as the heavy medium. This is known as the autogenous heavy medium effect. For difficult sorting this allows a great reduction in the requirement for a circulating dense medium (magnetite or ferrosilicon).

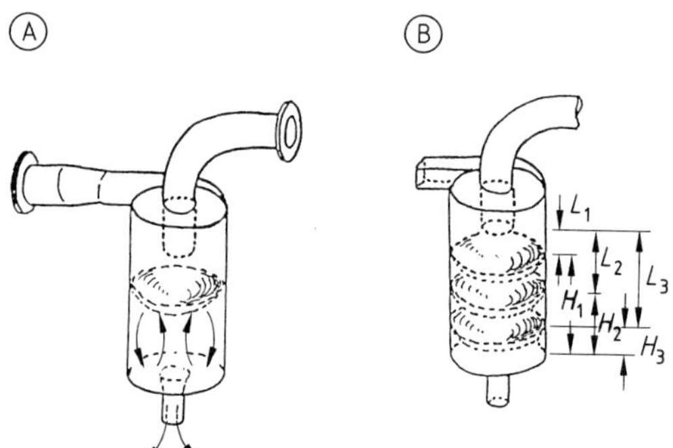

Figure 16. Functional diagrams of the CBC-hydrocyclone
A) Rotating stationary bed with convective slurry flow
B) Varying bed levels H_{1-3} and corresponding vortex finder clearance lengths, L_{1-3}

Figure 17. Two 500-mm CBC hydrocyclones made of urethane rubber (vulkollan)
Separation at 250 μm for pre-enrichment of nonferrous heavy-metal ores
CBC = circulating bed classifier

Wear is also a factor in the operating costs of a hydrocyclone, the whole body of which is exposed to heavy wear [121]. There is an optimum differential speed between inner flow and wall, leading to high shear forces. Nevertheless, there are differences of degree in the zones subject to wear. The underflow nozzle is most vulnerable, followed by the lower section of the cone, the feed nozzle, and the cover plate. The overflow nozzle and the cylindrical section show the least wear [122]. With regard to wear, however, the hydrocyclone has a marked advantage over centrifuges: it has no rotating parts and after repair does not need to be balanced. It can be lined, e.g., with rubber, or it can be manufactured wholly from self-supporting elastomers (e.g., polyurethane). Simultaneous erosion and corrosion are also easier to combat in the hydrocyclone. A further advantage is the possibility of manufacturing hydrocyclones from cast or machined metals, elastomers, resins, or oxide ceramics (e.g., Al_2O_3).

Some recent uses for hydrocyclones are the regeneration of drilling mud for tunneling and the recycling of plastics from waste [152]–[156].

Conical Screen. Hydrocyclones can be designed with a perforated wall. The suspension is fed tangentially through nozzles so that it flows spirally downwards over the screen. These hydrocyclones can be employed for the initial dewatering of suspensions with medium-sized particles, before centrifuges, filters, or other dewatering equipment. They can also be used for washing.

4. References

General References for Chaps. 1–2

[1] H. H. Hülsen, Neue Bauweisen von Filtern und Schleudern, *Verfahrenstechnik* (Berlin) **1939** no. 8, 187–195.

[2] W. Buddeberg, Probleme der Trenntechnik und ihre Lösungen unter besonderer Berücksichtigung der Zentrifugen, *Chem. Techn.* **10** (1958) 241–345.

[3] S. Kiesskalt, Die Reibung des Gutes in den Trommeln stetig arbeitender Zentrifugen, *Chem. Ing. Tech.* **34** (1962) 10–12.

[4] F. W. Schneider, Fest-Flüssigkeits-Trennung mit Zentrifugen, *Maschinenmarkt* **73** (1967) 1733–1737.

[5] H. H. Hülsen, Entwässerung von Massengütern in kontinuierlichen Zentrifugen, *Aufbereit. Tech.* **10** (1969) 354–360.

[6] C. M. Ambler, How to Select the Optimum Centrifuge, *Chem. Eng.* (N.Y.), **76** (1969) 96–103.

[7] H. H. Hülsen, Kontinuierliche Zentrifugen zur Trennung fest-flüssig in der chemischen Industrie, *Verfahrenstechnik (Mainz)* **6** (1972) 7–15.

[8] R. J. Woolcock, The Adaption of the Centrifuge to the Processing of Industrial Effluents, *Filtr. Sep.* **12** (1975) 174–180.

[9] K. Theiler, Klären und Trennen flüssiger Medien mit Tellerzentrifugen und Dekantern, *Maschinenmarkt* **83** (1987) nos. 35 and 44.

[10] W. Stahl, "Alternative Auslegung von Dekantierzentrifugen," *Info Krauss-Maffei* 1971.

[11] W. Stahl, H. Topp, Der Einsatz einer Mess-Rechenanlage zur Ermittlung der Optimaleinstellung von Dekantierzentrifugen, *DECHEMA Monogr.* **67** (1971).

[12] W. Gösele, Strömungen in Überlaufzentrifugen und ihr Einfluß auf die Sedimentation, *Chem. Ing. Tech.* **46** (1974) 67.

[13] W. Stahl, "Über die Entfeuchtung des Feststoffes in Dekantierzentrifugen," *Info Kraus-Maffei* 1974.

[14] W. Stahl, Über die Waschung des Feststoffes in Dekantierzentrifugen, *Chem. Ing. Tech.* **47** (1975) 853.

[15] T. Faust, W. Gösele, Untersuchung zur Klärwirkung von Dekantierzentrifugen, *Chem. Ing. Tech.* **57** (1985) 698–699.

[16] O. Knott, Systematik in der Konstruktion von Maschinen und Apparaten für die chemische Industrie, *Konstruktion* **19** (1967) no. 6.

[17] W. Buddeberg, Moderne Zentrifugen in der Zukkerindustrie, *Z. Zuckerind.* **3** (1953) 271.

[18] A. Schulz-Walz, Verfahrenstechnische Auslegung von Axial- und Drehschwingzentrifugen, *Aufbereit. Tech.* **15** (1974) 635–639.

[19] J. Nemeth, R. Horányi, Eine neue Leistungsbestimmungs-Methode für vertikale Rohrzentrifugen, *DECHEMA Monogr.* **66** (1970).

[20] H. Balster, Fortschritte in der Entwässerungstechnik durch neuartige Zentrifugierverfahren, *Aufbereit. Tech.* **21** (1980) 217–218.

[21] H. Hemfort, Diversifikation von Separatoren mit Düsen-, selbstreinigenden und Vollmantel-Trommeln, *Chem. Ing. Tech.* **51** (1979) 479–484.

[22] K. H. Brunner, H. Scherfler, M. Stölting, Erfahrungen mit Gegenstrom-Extraktions-Dekantern, *Verfahrenstech.* **15** (1981) 619–622.

Specific References

[23] G. Hultsch, H. Wilkesmann, "Filtering Centrifuges," in D.B. Purchas, *Solid–Liquid Separation*, Upland Press, Croydon 1977, pp. 493–559.

[24] H. Trawinski, Die äquivalente Klärfläche von Zentrifugen, *Chem. Ztg.* **83** (1959) 606–612.

[25] W. Bender, *Zur Berechnung des Durchsatzes von Schälschleudern,* Dissertation, Technische Universität Karlsruhe 1971.

[26] H. Rumpf, A. Gupte, Einflüsse der Parosität und Korngrößenverteilung im Widerstandsgesetz der Porenströmung, *Chem. Ing. Tech.* **43** (1971) 367–375.

[27] J. Kozeny, Content: Settling of particle bulks, *Sitzungsb. Ber. Akad. Wiss. Wien* **136** (1927) II a, 271.

[28] H. H. Steinour, Rate of sedimentation of non-flocculated suspensions of uniform spheres, *Ind. Eng. Chem.* **36** (1944) 618–624.

[29] E. Kriegel, Kornbewegung bei der Sedimentation, *Chem. Ing. Tech.* **38** (1966) 321–330.

[30] H. Trawinski, Mechanische Trennverfahren für Suspensionen und Schlämme, *Aufbereit. Tech.* **7** (1966) 709–717.

[31] C. M. Ambler, The Evaluation of Centrifuge Performance, *Chem. Eng. Progr.* **48** (1952) 150–158.

[32] H. H. Hülsen, Strömungs- und Bewegungsvorgänge in Dekanterzentriges, *Chem. Ing. Tech.* **41** (1969) 375–381.

[33] W. Gösele, Auslegungsversuche und Maßstabsübertragung bei Dekantern, *Chem. Ing. Tech.* **52** (1980) 178–179.

[34] C. Alt, W. Gösele, Einsatzkriterien für Dekanter, *Chem. Ing. Tech.* **54** (1982) 425–430.

[35] M. M. Hebb, F. H. Smith, "Centrifuges," *Kirk-Othmer*, 1st ed., **3**, pp. 501–521.

[36] H. Trawinski, Näherungsansätze zur Berechnung wichtiger Betriebsdaten für Hydrozyklone und Zentrifugen, *Chem. Ing. Techn.* **30** (1958) 85–95.

[37] G. M. Kosoj, Der Einfluß der Konstruktionsparameter des Hydrozyklons das Feld der Flüssigkeitsgeschwind., *Obogashch. Rud* (Leningrad) **13** (1968) 48–53.

[38] H. Trawinski, Neuentwicklungen auf dem Hydrozyklon-Gebiet, *Chem. Ing. Tech.* **50** (1978) 789–792.

[39] H. Trawinski, Zum Stand der Hydrozyklon-Technologie, *Verfahrenstech.* **12** (1978) 710–716.

[40] H. Trawinski, Zentrifugen, Trenngeräte mit höchster Abscheidungswirkung, *Chem. Ing. Tech.* **26** (1954) 189–201.

[41] F. Damrat, Trennung heterogener Stoffe durch Zentrifugieren, *Konstruktion* **12** (1960) 447–456.

[42] W. J. Sokolov, *Moderne Industriezentrifugen,* 2nd ed., Maschinostrojenije, Moskwa 1967, and VEB Verlag Technik, Berlin 1971.

[43] H. Trawinski, *Ullmann, 4th ed.,* **2** pp. 204–224.

[44] J. C. Malterra, La force centrifuge outil de separation en genie chemique, *Inf. Chim.,* n'109 special, Juin 1972, pp. 239–249.

[45] W. Batel, Menge und Verhalten der Zwichenraumflüssigkeit in körnigen Stoffen, *Chem. Ing. Tech.* **33** (1961) 541–547.

[46] W. Heckmann, Die physikalischen Grundlagen der mechanischen Feststoff-Flüssigkeits-Trennung und ihr Einfluß auf den Bau von Entwässerungsvorrichtungen, *Glueckauf* **96** (1960) 1281–1287.

[47] H. H. Hülsen, Kontinuierliche Zentrifugen zur Trennung fest-flüssig in der chemischen Industrie, *Verfahrenstechnik (Mainz)* **6** (1972) no. 1.

[48] H. Quetsch, Neue Siebzentrifuge für Fest-Flüssig-Trennung, *Chem. Ing. Tech.* **42** (1970) no. 4, 183–184.

[49] E. V. Szantho, H. Mathiak, Entwicklung der Schwingsiebschleuder und Einsatzmöglichkeiten zum Entwässern von aufbereitenden Produkten, *VDI Z.* **108** (1966) 493–499.

[50] O. Molerus, K. Brunner, Verfahrenstechnische Auslegung von Schwingsiebschleudern, *Chem. Ing. Tech.* **48** (1976) no. 2.

[51] G. Fengler, Siebschnecken-Zentrifugen, *Chem.-Anlagen + Verfahren,* **1984** nos. 10 and 11.

[52] H. Trawinski, Der neueste Entwicklungsstand auf dem Gebiet der Zentrifugen- und Hydrozyklontechnik, *Aufbereit. Tech.* **8** (1967) 470–483.

[53] G. Hultsch, Einige Kriterien von Massengutschleudern und Beschreibung einer neuen Siebzentrifuge, *Aufbereit. Tech.* **7** (1966) 720–723.
[54] H. Wilkesmann, Kinematik und verfahrenstechnische Optimierung einer Massengut-Zentrifuge mit Präzessionsaustrag (Taumelzentrifuge), *Aufbereit. Tech.* **19** (1978) 587–592.
[55] E. Rüegg, Kontinuierliche Zentrifugen für die Zukkerindustrie *Zucker* **5** (1952) 548–551.
[56] W. Schnurrenberger, 50 Jahre Schubzentrifugen, System Escher Wyss, *Techn. Rundsch. Sulzer* **67** (1985) no. 2.
[57] H. Zürrer, Waschprozeß in mehrstufigen Schubzentrifugen, *Escher Wyss Mitt.* (1973) no. 2.
[58] G. Hultsch, J. Wangermann, Eine neuartige Zentrifuge zur Filtration und gleichzeitigen Abtrennung von Zwickelkapillarflüssigkeit, *Aufbereit. Tech.* **9** (1968) 487–490.
[59] H. J. Titus, Schälpneumatik-Stromtrocknung mit Lösungsmittel-Rückgewinnung, *Chem. Ing. Tech.* **42** (1970) no. 6.
[60] H. Trawinski, Kapazität, Trenneffect und Dimensionierung von Vollmantelschleudern, *Chem. Ing. Techn.* **31** (1959) 661–666.
[61] H. Hemfort, Separatoren zur Trennung und Klärung von Flüssigkeitsgemischen, *Verfahrenstechnik (Mainz)* **4** (1970) 167–170.
[62] F. A. Records, "Sedementation Centrifuges," in D. B. Purchas: *Solid–Liquid Separation*, Upland Press, Croydon 1977, pp. 199–240.
[63] J. A. Weedman, W. E. Payne, O. W. Johnson, Development of Pressurized Continuous Centrifuge, *Chem. Eng Prog.* **55** (1959) 49–53.
[64] G. A. Frampton, Evaluating the Performance of Industrial Centrifuges, *Chem. Process. Eng.* (London) **44** (1963) no. 8, 229–243.
[65] F. Hess, Zentrifugen in der Verfahrenstechnik, *Maschinenmarkt* **67** (1961) 17–21.
[66] H. Hemfort, Die Konstrukion selbstreinigender Separatoren, *Chem.-Ztg.* **90** (1966) 247–267.
[67] H. Hemfort, Steuerungssysteme selbstreinigender Separatoren, *Chem. Ing. Tech.* **47** (1975) 14–20.
[68] T. Hermes, Automatisierung von Trennverfahren mit lichtelektrischen Zellen und Radioisotopen, *Chem. Ing. Tech.* **30** (1958) 213–219.
[69] G. Hultsch, R. Lidl, Neue Wege der Granulatentwässerung beim Konfektionieren von Kunststoffen, *Kunststoffe* **58** (1968) 334–337.
[70] H. Hemfort, Über Zentrifugalseparation und den Vergleich von Separatoren, *Motortech. Z.* **21** (1960) no. 3.
[71] K. H. Brunner, Separatoreneinsatz in der Biotechnologie, *Chem. Ing. Tech.* **59** (1983) no. 4.
[72] H. Hemfort, W. Kohlstette, Entwicklungstendenzen im Zentrifugalseparatorenbau, *Chem. Ind.* (Düsseldorf) **27** (1985) no. 6.

General References for Chap. 3

[73] H. Trawinski, *Ullmann's*, 4th ed., **2**, pp. 204–224.
[74] L. Svarovsky, Solid–Liquid Separation, Butterworths, London 1977, pp. 101–123.
[75] D. Purchas, "Cyclones (wet)" in D. Purchas: *Solid/Liquid Separation Technology*, Upland Press Ltd. Croydon 1981, pp. 249–276.
[76] G. Tarjan, On the theory and use of hydrocyclones, *Acta Techn. Acad. Sci. Hung.* **7** (1953) 389–441.
[77] G. Tarjan, Contribution to the Kinematics of hydrocyclones, *Acta Chim. Acad. Sci. Hung.* **28** (1957) 349–382.
[78] D. Bradley, A Theoretical Study of the Hydraulic Cyclone, *The Ind. Chemist.* Sept 1958, p. 473–480.
[79] K. Rietema, Performance and design of hydrocyclones, *Ing. Science* 1961, p. 298–309.
[80] K. Rietema, The Mechanisme of the Separation of Finely Dispersed Solids in Cyclones in "Cyclones in Industry," p. 46, *Elsevier*, Amsterdam 1961.
[81] K. Rietema, Liquid solid separation in a cyclone. The effect of turbulence on separation, *Inst. Chem. Eng.*, London 1962 paper C44.
[82] F. J. Fontein, J. G. van Kooy, H. A. Leniger, The influence of some variables upon hydrocyclone performane, *Br. Chem. Eng.* **7** (1962) no. 6, p. 410–21.
[83] E. O. Lilge, Hydrocyclone Fundamentals, *Trans. Inst. Min. Met.* **71** (1962) 285–331.
[84] G. Tarjan, Über besondere Kraftwirkungen in Hydrozyklonen, *Freiberg Forschungsh.* **A326** (1964) 105–123.
[85] V. V. Klajacin, Kornscheide und Durchsatz geometr. ähnlicher Hydrozyklone, *Izv. cysa vcabn. saved. Goryj z. sverdl* **7** (1964) 142–148.
[86] H. Schubert, W. Schneider, Zur Frage der Berechnung von Hydrozyklonen, *Freiberg Forschungsh.* **A354** (1965) 67–88.
[87] H. Schubert, T. Neese, The role of turbulence in wet classification, *Inst. Min. Met.* 1973, p. 213–239.
[88] H. Trawinski, Trennwirkungsgrad und toter Fluß des Hydrozyklons, *Keram. Z.* **26** (1974) 21–24.
[89] A. J. Lynch, T. C. Rao, Modelling and scale-up of hydrocyclone classifiers, *Proc. 11th Congr. Inst. Min.* 1975.
[90] B. Müller, T. Nesse, H. Schubert, Berechnung von Hydrozyklonen nach dem Turbulenzmodell, *Freiberg Forschungsh.* **A544** (1975) 31–43.
[91] L. R. Plitt, A Mathematical Model of the Hydrocyclone classifier, *CIM Bull.*, Dec. 1976, p. 114–123.
[92] H. Schubert, T. Neesse, A hydrocyclone separation model in consider of the turbulent multi-phase flow, *Int. Conf. Hydrocyclones*, Cambridge BHRA 1980, paper 3, p. 23–36.
[93] P. I. Pilov, Tubulent transport of solid particles in hydrocyclones, *Symp. Theory and Ind. Applications of Hydrocyclones*, Gorki 1981, p. 49–53.
[94] M. Bohnet, Optimalauslegung von Hydrozyklonen, *Chem. Techn.* **34** (1982) 564–568.
[95] H. Trawinski, Der Einfluß von Sedimentations-Behinderung und totem Fluß auf die Korngröße, *Aufbereit Tech.* **24** (1983) 527–534.
[96] M. I. G. Bloor, D. B. Ingham, A theoretical Investigation of the Fluid Mechanics of the Hydrocyclone, *Filtr. Sep.* **21** (1984) no. 4, pp. 266–269.
[97] T. Neesse, W. Dallmann, D. Espig, Effect of Turbulence on the Efficiency of Separation in Hydrocyclones at high Feed Solid Concentrations, *2nd, Intern. Converence on Hydrocyclones* 1984, paper **B3**, p. 51–66.
[98] D. F. Kelsall, A study of the motion of solid particles in a hydraulic cyclone, *Trans. Inst. Chem. Eng.* **30** (1952) 87–101.
[99] D. F. Kelsall, A further study of the hydraulic cyclone, *Chem. Ing. Sci.* 1953, p. 254–272.
[100] D. Bradley, D. I. Pulling, Flow patterns in the hydraulic cyclone and their interpretation of terms of performance, *Trans. Inst. Chem. Eng.* **37** (1959) 34–45.

[101] D. A. Dahlstrom, Cyclone operation factors and capacities on coal refuse slurries, *Min. Eng.* (1949), p. 331–344.
[102] I. I. Moder, D. A. Dahlstrom, Fine-size, close-specific-gravity solid separation with the liquid-solid cyclone, *Chem. Eng. Prog.* **48** (1953) 75–88.
[103] F. I. Fontein, in *Cyclones in Industry,* p. 118, Elsevier, London 1961.
[104] P. H. Fahlström, Studies of the hydrocyclone as a classifier, *IMPC* Cannes 1963.
[105] I. L. Bouso Acagonés, Aplication de hidrociclones, *rocas y minerales* **15** (1986) 48–65.
[106] M. G. Driessen, *Trans. Am. Inst. Min. Met. Eng.* (1948), p. 177–240.
[107] G. Göll, H. I. Schulze, Beitrag zur Hydrozyklonklassierung von Kreidesuspensionen, *Neue Bergbau T.* **13** (1983) 698–701.
[108] R. Hochscheid, The horizontal cyclone in closed circuit grinding, *SME-AIME meeting* 1985, 85-85, 1–10.
[109] K. D. Partz, Dekanter in H. Uetz *Abrasion und Erosion,* p. 588–596, Carl Hanser Verlag München 1986.
[110] E. Hoffmann, Bauart und Anwendung nass arbeitender Zentrifugalabschneider in der Kohlenaufbereitung, *Glückauf* **89** (1953) 105–120.
[111] K. Krebs, New cyclone design (double cyclone unit), *Eng. Min. J.* **155** (1954) 35–37.
[112] D. F. Kelsall, I. A. Holmes, Improvement of classification efficiency in hydraulic cyclones by water injection, Proc. 5th *Min. Congr.* IMM London, 1960. paper 9, 159–170.
[113] F. Molyneux, Electrostatic Cyclone Separator, *Chem. Process Eng. At. World.* **44** (1963) 517–519.
[114] I. Visman, Bulk processing of fine materials by compound water cyclones, *Can. Inst. Min. Met.* **69** (1966) 85–98.
[115] D. F. Kelsall, The double entry cyclone, Declone, *Symp. Int. Feder. Autom. Control* Sydney Aug. 1973.
[116] I. Novacek, Aufbereitung von Magnetiterzen in einem Hydrozyklon unter Nutzung eines Magnetsystems, *Freiberg Forschungsh.* **A544** (1975) p. 141–157.
[117] L. Simek, Eine neue Hydrocyclonebauart, *VT Verfahrenstech.* **11** (1977) 18–24.
[118] W. Dallmann, T. Nesse, G. Thomas, Hochleistungshydrozyklone aus dem FIA Freiburg, *Freiberg Forschungsh.* 1983, p. 366–370.
[119] A. Jowett, C. J. Restarick, Sizing of fine coal with a compound cyclindercyclone system, *IMM Sympos. Coal. Min. Siz.,* Aug. 1984.
[120] H. Eronen, A new high-performance hydrocyclone (Twin Vortex), *Finn. Trade Rev.* Juli 1986.

Specific References

[121] M. Clement, Verschleißerscheinungen an Trübepumpen und Hydrozyklonen in Erzaufbereitungsanlagen, *Erzmetall* **15** (1962) 53–57.
[122] H. Trawinski, Chapter 3.3 Hydrozyklone in H. Uetz *Abrasion und Erosion,* pp. 573–588, Carl Hanser Verlag München 1986.
[123] H. Trawinski, Der Hydrozyklon als Hilfsgerät der Grundstoffveredelung, *Chem. Ing. Tech.* **25** (1953) 331–341.
[124] D. Bradley: *The Hydrocyclone,* Pergamon Press Ltd, Oxford 1965.
[125] H. Schubert, *Aufbereitung fester mineralischer Rohstoffe*, vol. I, Grundstoffverlag, Leipzig, pp. 360 ff.
[126] A. J. Lynch, *Mineral Crushing and Grinding Circuits,* Elsevier, Amsterdam 1977, Chap. 5.3 and 6.1.
[127] G. Tarjan, *Mineral Processing,* Vol. 1, Akadémiai Kiadó Budapest 1981, pp. 531–563.
[128] C. Krijgsman, Versuchs- und Betriebsergebnisse mit Hydrozyklonen, *Chem. Ing. Tech.* **23** (1951) 540–542.
[129] F. I. Fontein, Stand der Entwicklung und Anwendung von Hydrozyklonen, *Chem. Ing. Tech.* **27** (1955) 190–192.
[130] S. Bednarski, Hydrozyklone in der Stärkeindustrie, Teil II: Konstruktion der (Multi-) Hydrozyklone, *Die Stärke* **15** (1963) 171–181.
[131] N. Yoshioka, Y. Hotta, Liquid cyclone as a hydraulic classifier, *Chem. Eng.* (Tokyo), 1955, p. 632–640.
[132] H. Trawinski, Die Betriebspraxis des Hydrozyklons, *Ton-Ind. Ztg.* 1984, 9, 10 and 11.
[133] H. Trawinski, Practical Hydrocyclone Operation, *Filtr. Sep.* **22** (1985) no. 1.
[134] H. Trawinski, Theory, applications, and practical operation of hydrocyclones, *Eng. Min. J.* **177** (1976) 115–127.
[135] D. A. Dahlstrom, High-efficiency desliming by use of hydraulic water editions to the liquid-solid cyclone, *Trans. Am. Inst. Mi. Engrs.* **4** (1952) 788–793.
[136] H. Trawinski, "Hydrocyclones" in D.B. Purchas: *Solid–Liquid Separation,* Upland Press, Croydon 1977, pp. 241–287.
[137] H. Trawinski, Practical aspects of the design and industrial applications of the hydrocyclone, *Filtr. Sep.* **6** (1969) 361–367, 651–657
[138] H. Trawinski, Nassklassieren von feinkörnigem Gut, besonders in Mahlkreisläufen, *Techn. Mitt.* Essen **59** (1966) 249–257.
[139] S. K. De Kok, Symposium on recent developments in the use of hydrocyclones in mill operation—a review, *J. Chem. Soc. S. Afr.* (1956), p. 287.
[140] K. Baumann, Der Hydrozyklon als Klassiergerät für Kreide-, Kaolin, und Tontrüben, *Silikattechnik* **6** (1955) 247–251.
[141] A. Hundertmark, Untersuchungen über die Feststoffverteilung im Hydrozyklon mit Hilfe radioakt. Durchstrahlung, *Erzmetall* **18** (1964) 403–412.
[142] H. Schubert, Die Sortierung nach der Dichte in Zentrifugal-Kraftfeldern, *Bergakademie* **17** (1965) 293–297.
[143] M. G. Driessen, A new process in the washing of coal, *Univ. Mines.* 8th series (1939) 177–193.
[144] F. I. Fontein, New coal washing systems, recently developed in the Netherlands, *Geol Mijnbouw* **15** (1953) p. 414–424.
[145] I. Visman, The compound water cyclone, *Coal Min. Process* **2** 1964.
[146] G. Ferrara, H. J. Ruff, Dynamic dense medium separation process, Tri-Flow separator, *Erzmetall* **35** (1982) 294–299.
[147] P. Commack, Der Larcodems—ein neuer Schwertrübescheider für Rohkohle, *Aufbereit. Tech.* **28** (1987) 427–434.
[148] E. J. O'Brien, Water-only cyclones, their function and performance, *Coal Age,* **81** (1976) no. 1, p. 110–114.
[149] H. H. Dreissen, A. T. Basten, Reclaiming products from shredded junked cars by the water-only cyclone and heavy medium cyclone processes, *Proc. 5th Min. Waste Utililisation Sympos.* **378,** Chicago 1976.

[150] G. van Duijn, K. Rietema, M. G. Atjak, A hydrocyclone for the recovery of minerals from sand, *Europ. Sympos. Particle Technol.* Amsterdam, Juni 1980.
[151] H. Trawinski, Sind die Anwendungsgrenzen des Hydrozyklons konstruktiv zu erweitern? *Masch. Markt* **87** (1981) p. 684–687.
Der Hydrozyklon hat seine Anwendungsgrenzen noch nicht erreicht, *Masch. Markt* **87** (1981), p. 1068–1072.
[152] H. Trawinski, Technologie für die Regenerierung der Bentonit-Spülung, die der gleichzeitigen Stützung der Ortsbrust beim Schildvortrieb dient, *TIZ* **109** (1985) no. 2.
[153] A. Bahr, V. Voigt, Sortiertechnologie bei Kunststoffabfällen, *Ind. Anz.* **99**(1977) 2021–2025.
[154] A. Bahr, Zur Sortierung reiner und NE-Metalle enthaltender Kunststoffabfälle, *Erzmetall* **33** (1979) 324–330.
[155] G. Roperts, Nassmechan. Aufbereitung von Kunststoffabfällen, *Chem. Anl. + Verfahren* (1984), p. 17–22.
[156] G. Roperts, Wirtschaftliche Kunststoffrückgewinnung aus kommunalen und industriellen Abfällen, *Aufbereit. Tech.* **27** (1986) 489–494.

12. Sedimentation

YORAM ZIMMELS, Department of Civil Engineering, Technion IIT, Haifa, Israel

In addition to the standard symbols defined in the front matter of this volume, the following symbols are used:

a_f	acceleration of fluid
a_i	constant ($i = 1, 2, \ldots$)
a_1, a_2, a_3	constants
$a(d)$	size-dependent constant
A, A_1	constants
A_d	constant dependent on ratio of viscosities
A_s	area under a schlieren peak; also used to denote area across which the dispersion flows in a settling pool
b	length of shorter wall of inclined channel
b_j	constants ($j = 1, 2, \ldots$)
b^*	dimensionless length
$b(d)$	size-dependent constant
B_d, B_v	functions of volume fraction and viscosity
B_p	pool width
B_1	constant
C	concentration of particles
C_a	concentration at a reference level y_a or at the center of a conduit
C_D, $C_{D\varphi}$, C_{D0}	drag coefficient: general, generalized concentration dependent, and at infinite dilution, respectively
C_h, C_{ht}	concentration in the vehicle and in the heterogeneous part of the suspension, respectively
C_m	concentration at $r = r_m$

C_0, $C(t)$ initial and time-dependent concentration, respectively
C_1, C_2 concentrations at regions 1 and 2 in differential sedimentation system
d, d_i particle size: general and characterizing the ith size fraction, respectively
d_L limiting size for capture in a settling pool
d_{max} limiting size of particles for which $G(d > d_{max}) = 1$
d_{50} equiprobable size of particles, defined by $d_{50} = d[G(d) = 0.5]$
D Stokes–Einstein diffusion coefficient; also used to denote pipe diameter
D' diffusion coefficient
D_a, D_T diameter of impeller and tank, respectively
D_c diameter or characteristic size of container
D_e eddy diffusion coefficient
D_m fluid momentum transfer coefficient
D_s mass-transfer coefficient
D'_{sed} diffusion coefficient obtained from measurement in an ultracentrifuge
$D_\varphi(r,\eta)$, $D_d(r,\eta)$ position- and η-dependent diffusion coefficient in the fluid and the dispersion frame of reference, respectively
$\overline{D_\varphi(r)}$ expectation of $D_\varphi(r,\eta)$ with respect to η
E electric field strength
f friction factor
f_a activity coefficient of diffusing particles
f_D drag coefficient of a particle
f_1, f_s fluid and slurry friction factors, respectively
f_p driving force density
$f(\xi)$ distribution function of ξ
$f(\varphi)$, $f_D(C)$ hydrodynamic hindrance factors
$f_1(\varphi)$, $f_2(\varepsilon)$, $f_3(\varepsilon)$ functions of φ and ε
F_D hydrodynamic drag force
$F_{D\varphi}$ generalized concentration-dependent hydrodynamic drag force
F_G, F_ω, F_q, F_M, F_E gravitational, centrifugal, electrical, magnetic, and electrical polarization forces, respectively
F_i ith force acting on a particle
F_R reaction force due to accelerated fluid
F_γ force associated with γ
F_μ diffusion driving force
Fr Froude number
Fr_L modified Froude number
g gravitational acceleration
g' effective acceleration
$G(d)$ grade efficiency
h height of sedimentation column in nonsteady sedimentation models
h_c channel height in a disk centrifuge
h_s height of a schlieren peak
h_1 thickness of one layer of a sedimentation column
H magnetic field strength
H_a height of impeller above tank floor
H_j height
H_j^* dimensionless height
H_p pool height
i integer
j, j' integers denoting current and start-off layer numbers
J, J_{max} particle flux (and flux densitiy) and maximum particle flux in the dispersion frame of reference, respectively
J_D, J_F particle flux density due to diffusion and sedimentation, respectively
k number of time increments elapsed
k^* surface effect retardation constant
k_B Boltzmann constant
K, K_1, K_2, K_3, K_w constants
K_a design parameter of agitation tanks
K_{ac} agitation concentration factor
K_f Karman constant
l_m eddy mixing length
L length
L_p, L_e length of straight pipe and equivalent pipe length of fittings, respectively; is also used as length of inclined channel
L_t total effective pipe length
m, m_f mass and mass of fluid, respectively
M_i ith component of magnetization
M_p, M_f, M_φ magnetization of particles, fluid, and dispersion, respectively
M_r molecular mass
n exponent; also an integer
n_i number of particles in the ith fraction per unit volume of dispersion
$\tilde{N}$ Avogadro number
N number of particles per unit volume of dispersion; also used to denote number of sedimentation zones in an inclined channel
P pressure; also used to denote electrical polarization of particles (i.e., dipole moment per unit volume) and power

P_{EP}, $P_{E\varphi}$ electrical polarization (i.e., dipole moment per unit volume) of particles and dispersion, respectively
P_0 power number
ΔP_{exp}, ΔP_{cal} measured and calculated pressure drop, respectively
ΔP_l, ΔP_t, ΔP_i pressure drop across a bed of particles in the laminar, turbulent, and intermediate flow regimes, respectively
q_v, q_s, $q_{v\varphi}$, $q_{s\varphi}$ charge per unit volume and charge per unit area of particles and dispersion (subscript φ)
$q(t)$ volume flow rate of clarified fluid per channel per unit length
Q flow rate in a pool; also used to denote volumetric flow rate in a centrifuge
Q_p particle flux in a centrifuge
$Q(t)$ volumetric rate of fluid clarification in an inclined channel
r radial distance from axis of rotation; also used to denote position vector and radial coordinate
$\bar{r}$ radial distance to a point in the boundary which moves at the same velocity as particles in the dispersion ahead
r_m, r_p radial distance of meniscus and of a point in the region of concentration plateau, respectively
R gas constant
Re, Re_φ Reynolds number and generalized concentration-dependent particle Reynolds number
Re_a Reynolds number of an agitator
S cross-sectional area of particle perpendicular to direction of flow
S_g, $S_{g(20, W)}$ sedimentation coefficient and its value at standard state, respectively
S_{gD} differential sedimentation coefficient
S_p shape factor
$S(i, j', k)$ array of sedimentation distances (at time t)
$S(t)$ time-dependent distance
$S_j^*(t)$ dimensionless instantaneous volumetric rate of formation of suspension devoid of particles faster than (or equal to) particles of the jth fraction
t time
t_H, t_R time for sedimentation to the pool floor and for crossing the pool, respectively
T absolute temperature
u_f flow velocity of a carrier fluid
U flow velocity of fluid in a conduit; also used as velocity of particles
U_c superficial velocity; also used to denote critical deposition velocity of particles on a pipe wall
U_d, U_f velocity of particles and fluid, respectively, in the dispersion frame of reference
U_{di}, $U_{\varphi i}$ velocity of ith particles in the dispersion and the fluid frame of reference, respectively
U_{exp} measured velocity of fluidizing gas
U_g free sedimentation velocity of d_{50} particles
U_h transition velocity from homogenous to heterogenous flow
U_j velocity of a particle
U_R crossflow velocity of fluid in a settling pool
U_x, U_y x and y components of particle velocity in the dispersion frame of reference
U_φ velocity of particles in the fluid frame of reference
$U_{\varphi a}$ velocity of aggregates in the fluid frame of reference
U_0 $= U_\varphi$ at $\varphi = 0$
U_i' energy per diffusing particle
U'_{M1}, U'_{M2}, U'_{H1}, U'_{H2} magnetic energies
U^* friction velocity
U_{dj}^* dimensionless velocity
$U_d(t)$, $U_d(\infty)$ time-dependent and terminal settling velocity in the dispersion frame of reference
v velocity of slurry; also used to denote particle specific volume
V, V_p general volume and volume of a particle, respectively
V_b volume of dispersion in centrifuge bowl
V_v volume per unit length of inclined channel sedimentation vessel
V_T volume of tank
V_φ volume of dispersion
w weight fraction of solids
w_s weight of dispersed phase
x_s function of U_d/U_c
y height
y_a reference level above conduit bottom
y_m distance from bottom wall of conduit to a point in the dispersion

Greek Symbols

α	size parameter that determines the four size fractions in the computer model
α_s	safety factor
α_φ	concentration-dependent factor
$\alpha_1, \alpha_2, \alpha_3, \alpha_4$	variables in the equation of motion
β_s	constant
β_φ	concentration-dependent factor
$\beta(t)$	time-dependent force transmission parameter
γ	interfacial tension; also used to denote U_d/U_c
$\gamma_p, \gamma_f, \gamma_\varphi$	physical parameters (size, density, etc.) of particles, fluid, and dispersion, respectively
Γ	adsorption density
δ	$= \varrho_p - \varrho_\varphi$; also used to denote thickness of viscous sublayer
ε	void fraction
ε_m	effective magnetic porosity
ζ	parameter
η	parameter denoting size, density, etc.
η_p	pool efficiency factor
θ	angle of inclination
λ	resistance to flow of a carrier fluid in a pipe
μ	chemical potential
μ_d	viscosity of fluid of the dispersed phase
μ_f	viscosity of pure fluid
$\mu_\varphi, \mu'_\varphi$	concentration-dependent extended viscosities
$\mu_{\varphi d}$	apparent "Taylor" viscosity of a continuous fluid containing dispersed fluid particles
μ_0	magnetic permeability of vacuum
μ'_f	effective viscosity of a continuous fluid surrounding a single spherical fluid particle
μ_d^*	extended viscosity inclusive of interfacial effects
ν_f	kinematic viscosity
$\xi, \bar{\xi}$	distribution variable characterizing particles (i.e., size, density, etc.) and its expectation, respectively
ϱ	density
$\varrho_p, \varrho_{pi}, \varrho_f$	density of particle, *i*th particle, and pure fluid, respectively
ϱ_φ	density of dispersion
σ_k	surface parameter of viscosity
Σ, Σ_φ	general and extended concentration-dependent index of centrifuge size
τ_w	shear stress at wall
τ_φ	factor accounting for degree of particle fixation in space
φ	total volume fraction occupied by particles in dispersion
φ_a	volume fraction occupied by aggregates in dispersion
$\varphi_i, \varphi_i(d_i)$	volume fraction occupied by *i*th particle fraction and by *i*th size fraction (d_i) in dispersion, respectively
φ_{max}, φ_m	volume fraction at maximum particle and aggregate flux, respectively
φ_s	volume fraction occupied by particles in a single aggregate
φ'	probability density function
$\varphi(i, j)$	time-dependent volume fraction occupied by *i*th particles in the *j*th layer
$\varphi(i, j')$	invariable volume fraction of *i*th particles that start sedimentation at the *j*th layer
$\varphi(j)$	$= \sum_i \varphi(i, j)$
$\varphi_s(i, j)$	volume fraction occupied by *i*th particles in the sediment
Φ	cumulative size distribution
χ_p	susceptibility of particles
ω	angular velocity
ω_1, ω_φ	effective, concentration-dependent, angular velocities

Sedimentation is the motion of particulates in a fluid which is generated by a force field. This motion often results in increased concentration of the particulates, either in the form of a concentrated dispersion or as a sediment.

1. Scope

Sedimentation involves the motion of particulates under the action of driving forces that are generated by force fields, such as gravitational, centrifugal (→ 11. Centrifugation and Hydrocyclone Separation), magnetic (→ 19. Magnetic Separation), and electrical (→ 16. Electrostatic Separation) fields. The use of a particular field depends on the process required and its technical and economical feasibility. To date, sedimentation techniques can handle particulates with a variety of properties (e.g., size and density). This is because the force fields that are available today can produce force densities equivalent to 300 000 g and higher. Such high force densities are readily produced by commercial ultracentrifuges and, in some cases, also by electromagnetic fields. Thus large particulates, as well as those having dimensions comparable to molecular sizes, can be handled by conventional sedimentation techniques. This capacity is applied in chemical, biochemical (→ 11. Biochemical Separations **B3**), mineral, and environmental processing technologies. It is also used as a standard means of biomedical testing and sample preparation. A world without sedimentation phenomena, whether natural or man-made, would be incomprehensible. In some way, sedimentation exists around us every day in the form of natural (i.e., rain, snow, dust, sedimentation of silt in rivers) or artificial (i.e., processing and pollution products) precipitation. Living in a world where gravity prevails means that sedimentation exists, to different degrees of significance, in every system in which one or more phases are dispersed in a fluid.

2. Theory

Sedimentation can be part of a process in which particulates dispersed in a fluid are involved, or else it can be used to separate them from the dispersion in the form of a concentrated sludge or sediment. Analysis of sedimentation requires a distinction between steady-state, continuous and nonsteady, time-dependent systems, in conjunction with either mono- or polydisperse dispersions. Depending on the frame of reference, time-dependent sedimentation may be part of a flow system running continuously at steady state. In this article, the theory of steady sedimentation of mono- and polydisperse mixtures is reviewed first; it is then extended to include nonsteady conditions (for a theoretical treatment of the motion of particles in fluids, see → 16. Elutriation).

2.1. Steady Sedimentation of Nondiffusing Particulates

2.1.1. Monodisperse Particulates

The term *monodisperse particulates* implies that all dispersed particulates are identical. The following material considers the steady-state "terminal velocity" of monodisperse spherical particulates uniformly dispersed in a Newtonian fluid [1]. Let U_φ denote the velocity of the particulates in the fluid frame of reference. Thus, U_φ is the velocity of particulates relative to the fluid being displaced by their motion; it can be evaluated in the creeping (i.e., laminar) flow regime where $Re < 1$ ($Re = 0.2$ defines the upper limit of the creeping flow regime) by Equation (1), and in intermediate and turbulent flow where $Re > 1$, by Equation (2). Equation (2) also provides a good approximation for creeping flow, but in this regime Equation (1) is preferable. The flow regimes that apply to Equations (1) and (2) pertain to flow around the particles and not to the whole dispersion. Thus, turbulent flow means the formation of eddies due to unstable flow around the particles and not due to other sources, such as agitation or turbulent flow in a conduit.

$$U_\varphi = \frac{(\varrho_p - \varrho_f)\, g'\, d^2}{18\,\mu_f} f(\varphi),$$

$$f(\varphi) = \frac{(1-\varphi)}{(1+\varphi^{1/3})\exp\left(\frac{5}{3}\cdot\frac{\varphi}{1-\varphi}\right)} \tag{1}$$

$$U_\varphi = \left(\frac{-a_2 + \sqrt{a_2^2 + 4a_1 a_3}^{\,1/2}}{2a_1}\right)^2 \tag{2}$$

$$a_1 = 0.63\,(d\varrho_f)^{1/2}$$

$$a_2 = 4.80\,\mu_f^{1/2}\exp\left(\frac{5}{6}\cdot\frac{\varphi}{1-\varphi}\right)$$

$$a_3 = 4.0\,(\varrho_p - \varrho_f)\,(1-\varphi)\,g'\,d^2/\,[3\,(1+\varphi^{1/3})]$$

where φ is the volume fraction occupied by the particulates in the dispersion; d is particle diameter (i.e., a measure of particle size); ϱ_p and ϱ_f are densities of particulates and fluid, respectively; μ_f is the viscosity of the pure fluid (i.e., at $\varphi = 0$); and g' is the effective acceleration defined by Equations (3)–(5):

$$g' = f_p/\delta \tag{3}$$

$$\delta = \varrho_p - \varrho_\varphi \tag{4}$$

$$\varrho_\varphi = \varphi\,\varrho_p + (1-\varphi)\,\varrho_f \tag{5}$$

where f_p is the driving force density (i.e., force per unit volume) due to which steady motion occurs. In *gravitational* and *centrifugal fields*, $g' = g$ and $g' = \omega^2 r$, respectively, where g is acceleration due to gravity, ω is angular velocity, and r is

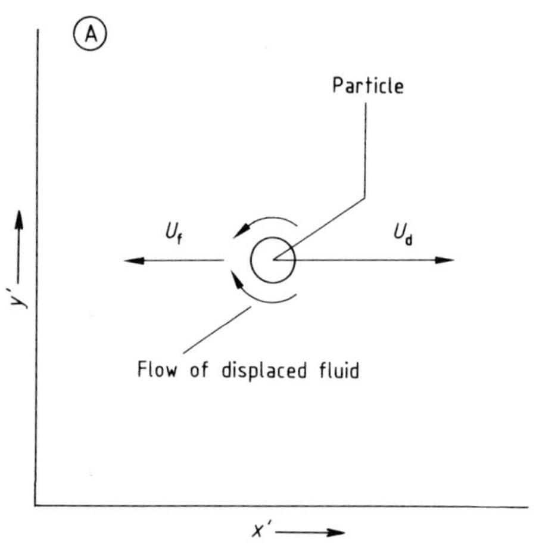

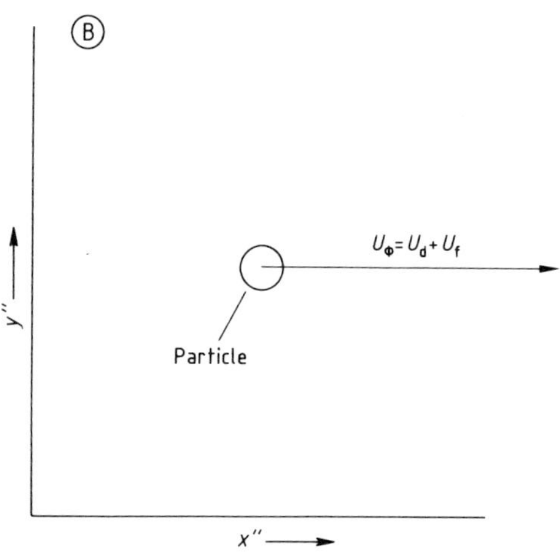

Figure 1. Definition of U_φ and U_d in the fluid and dispersion frame of references
A) Dispersion frame of reference x', y';
B) Fluid frame of reference x'', y''

radius of rotation. In *magnetic fields*, g' can be evaluated by Equation (6):

$$g' = \mu_0 \nabla [(M_p - M_\varphi) \cdot H]/(\varrho_p - \varrho_\varphi) \quad (6)$$

where M_p and M_φ represent the magnetization (i.e., magnetic moment per unit volume) of particulates and of the dispersion, respectively, and H is magnetic field strength. The electrical counterpart of Equation (6) is obtained by replacing $\mu_0 (M_p - M_\varphi)$ by $P_{Ep} - P_{E\varphi}$ and H by E, where P_{Ep} and $P_{E\varphi}$ denote polarization of particulates and dispersion, respectively, and E is electric field strength. Let U_d be the velocity of particulates relative to the dispersion frame of reference. Then, if the dispersion is at rest relative to its container, U_d is also the velocity in the container frame of reference. Figure 1 shows the definition of U_φ and U_d. The velocity U_f of the displaced fluid in the dispersion frame of reference is given by

$$U_f = U_\varphi \varphi \quad (7)$$

Hence,

$$U_d = U_\varphi - U_f = U_\varphi (1 - \varphi) \quad (8)$$

Note that Equation (8) applies when by continuity the displaced fluid flows counter to the particulates so as to maintain volumetric balance (→ 16. Elutriation). Other forms of $f(\varphi)$ are known [2]–[4], particularly for sedimentation involving low Reynolds numbers [2]. The correlation given by Equation (9) is attributed to RICHARDSON and ZAKI [5]:

$$f(\varphi) = (1 - \varphi)^{n-1} \quad (9)$$

The results of GARSIDE and AL-DIBOUNI [6] suggest a value of $n = 5.1$. Dimensional analysis indicates that n should depend solely on wall effects for both creeping and turbulent flow, and on wall effects and Reynolds number Re evaluated at $\varphi = 0$ in the intermediate flow regime [5]. The results [7] corresponding to flow ranges of Re excluding wall effects, are given in the following material:

Range of Re	n
$Re < 0.2$	4.65
$0.2 < Re < 1.0$	$4.4\ Re^{-0.03}$
$1.0 < Re < 500$	$4.4\ Re^{-0.1}$
$Re > 500$	2.4

Note that n is highest in laminar flow and drops as turbulence increases. For more information on generalized $f(\varphi)$ functions, see Section 2.6. For dilute dispersions [2] ($\varphi \to 0$), $f(\varphi)$ given by Equation (1) reduces to

$$f(\varphi) = (1 - \varphi^{1/3})/(1 - \varphi) \quad (10)$$

whereas Equation (9) for $n = 5.1$ gives

$$f(\varphi) = (1 - 5.1\ \varphi)/(1 - \varphi) \quad (11)$$

Figure 2 shows a plot of $(1 - \varphi) f(\varphi)$ vs. φ for four correlations, two of which are given by Equations (1) and (9) [2]. Equation (1) [3] (curve b) predicts a steeper decrease in $f(\varphi)$ for $\varphi < 0.2$ than Equation (9) does; the reverse is true for $\varphi > 0.2$.

The discrepancy between the four correlations is clear. Motion of particulates at very low Reynolds numbers may involve factors other

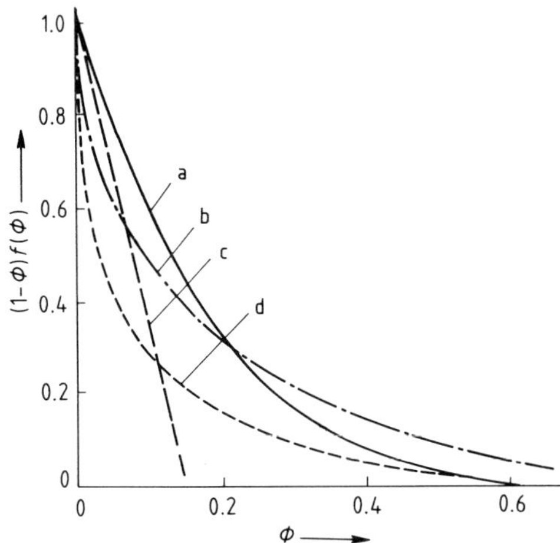

Figure 2. A plot of $(1-\varphi)\cdot f(\varphi)$ vs. φ
a) Equation (9); b) Equation (1); c) Batchelor's theory of dilute suspension [8]; d) Model of simple cubic arrays (Sangani and Acrivos [9])

than driving and hydrodynamic drag forces. In such cases, the discrepancy between theory and experimental results may not be due entirely to the inadequacy of the theory. In this article, $f(\varphi)$ is defined according to Equation (1) because of the vast amounts of experimental data that support it [3].

Sedimentation volume flux density J is defined as the volume of particulates crossing a unit area per unit time. The flux density is given in the dispersion frame of reference by

$$J = U_d\,\varphi = U_\varphi(1-\varphi)\,\varphi \tag{12}$$

In the laminar flow regime,

$$J = \frac{(\varrho_p - \varrho_f)\,g'\,d^2}{18\,\mu_f}(1-\varphi)\,\varphi\cdot f(\varphi), \quad Re < 1 \tag{13}$$

and in the intermediate and turbulent regimes,

$$J = \left(\frac{-a_2 + \sqrt{a_2^2 + 4a_1 a_3^{1/2}}}{2a_1}\right)^2 \cdot(1-\varphi)\,\varphi, \quad Re \geq 1 \tag{14}$$

Figure 3 shows a plot of J vs. φ calculated for sedimentation of particles ($\varrho_p = 3000$ kg/m^3, different sizes) in water [1], [10]. The dark circles are results obtained by using the generalized correlation of Barnea and Mizrahi [3] (→ 16. Elutriation, p. **16**-4), and the open circles were obtained by using Equation (14). The two sets of calculated results are in good agreement; hence, Equation (14) can be used safely. The volume fraction φ_{max} that maximizes the flux density varies with particle size. Calculated values of φ_{max} vs. d are plotted in Figure 4. In the laminar flow regime, $\varphi_{max} = 0.1792$ and is independent of particle size. At this volume fraction, $J_{max} = 0.2994\,\varphi_{max}\,U_{d0}$, where U_{d0} is the free sedimentation velocity (i.e., at $\varphi \to 0$). Therefore, J_{max} is 29.94% of the flux limit (in the given fluid and at φ_{max}), which is expected if the hindrance due to φ could have been eliminated so that free sedimentation prevails. Increase of particle size or transfer of motion into the turbulent regime results in an increase of φ_{max} (see Fig. 4):

$$\log \varphi_{max} = A_1 \log d + B_1 \tag{15}$$

where for settling in dry air (18 °C), $A_1 = 0.1320$, $B_1 = -0.9505$ and for settling in water (25 °C), $A_1 = 0.1309$, $B_1 = -0.9946$.

These data may be used for process evaluation regarding the capacity of phases involving sedimentation. The following points are suggested for consideration in evaluation of sedimentation processes:

1) For most practical purposes, $0.1792 < \varphi_{max} < 0.40$.
2) The flux density of finer and less dense particles is sensitive to changes in φ and φ_{max} is independent of particle size in the laminar flow regime.
3) The flux density of larger and more dense particles is less sensitive to variation of φ in the vicinity of φ_{max}. Dispersions containing larger particles may be processed at higher levels of φ.
4) The gain in particle flux density, obtained due to increase in particle size, is largest in the vicinity of $Re = 1$.

For further details, see [1].

The evaluation of flux densities is important in mass balance calculations. If the incoming flux exceeds the limiting outgoing flux in a sedimentation zone, then undesirable accumulation and surging are expected, which can produce instability and process shutdown.

2.1.2. Polydisperse Particulates

The term *polydisperse particulates* means that the dispersed particulates are not identical and may differ in properties such as size, density,

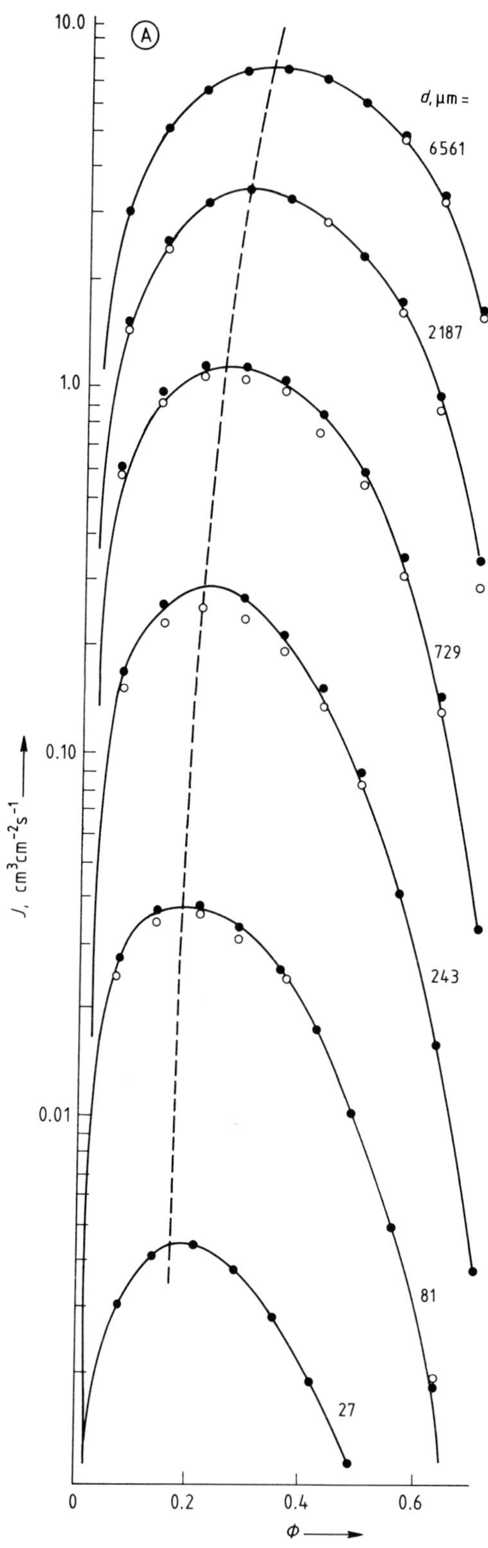

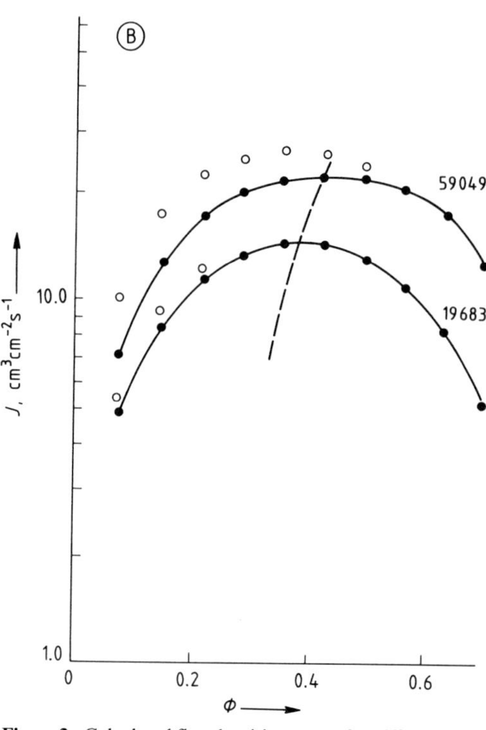

Figure 3. Calculated flux densities J vs. φ for different particle sizes, $\varrho_p = 3000$ kg/m^3, sedimentation in water [1], [10]
-●- General correlation; -○- Dallavalle

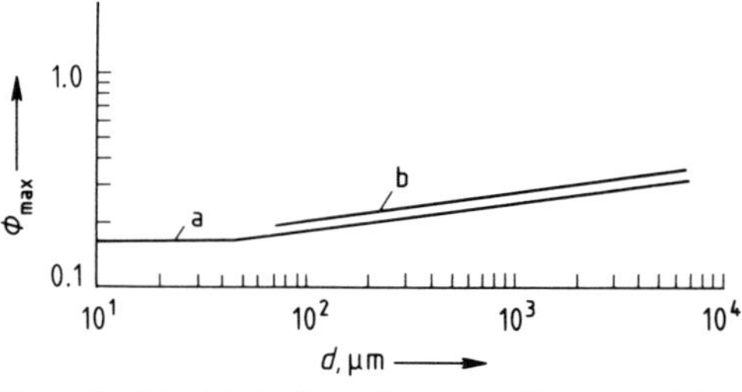

Figure 4. Calculated values of φ_{max} vs. d (data from Fig. 3 [1], [10])
a) Water; b) Air

shape, polarizability, and charge. Alternatively, it implies that properties of the dispersed particulates can be described by distribution functions.

Distribution Functions. (For a review of useful distribution functions, see → 2. Particle-Size Analysis and Characterisation of a Classification Process, p. **2**-6 and → 16. Elutriation, p. **16**-9). Because particle size is a dominant factor in many sedimentation systems, and in powder technology generally, a wealth of size distribution functions is available from the literature. If a known distribution function cannot be made to fit particle-size data, then the data must be divided into appropriate size fractions small enough to be assumed uniform. In general, information related to the distribution of properties that are significant in the sedimentation of polydisperse particle mixtures, must be obtained and processed to a usable form prior to implementation of sedimentation theories or to evaluation of velocity and flux.

Sedimentation Velocities and Flux Densities. *Discrete Distributions.* If the total volume fraction occupied by the particulates in a dispersion is expressed in terms of uniform fractions as

$$\varphi = \sum_{i=1}^{n} \sum_{j=1}^{m} \ldots \varphi_{ij} \ldots (d, \varrho, \ldots) \quad (16)$$

where the dots denote the inclusion of summation and the indexing of variables other than d and ϱ. Thus, if three variables characterize the polydisperse mixture, then three summation symbols, three indexes, and three independent variables will be used in Equation (16). For the sake of clarity, only one independent variable (i.e., size) is considered here, for which

$$\varphi = \sum_{i=1}^{n} \varphi_i(d_i) \quad (17)$$

Equations (1) and (2) can also be used to determine the sedimentation velocities of fractions that are part of a polydisperse mixture, with φ given by Equation (17). This involves the assumption that hydrodynamic hindrance to motion of the particulates is (to a good approximation) solely dependent on the total volume fraction occupied by the whole distribution, i.e., that it is independent of the distributions characterizing the dispersion. Thus although $U_{\varphi i}$, pertaining to the ith fraction, is assumed to be independent of effects of the distributions, U_f and hence also $U_{\mathrm{d}i}$ are directly affected by them. The effect of interparticle collision is neglected; otherwise, $U_{\varphi i}$ must be corrected by a collision factor $\alpha(\varphi)$ which is assumed here at a value of unity.

$$U_\mathrm{f} = \sum_{i=1}^{n} U_{\varphi i}\, \varphi_\mathrm{i} \quad (18)$$

$$U_{\mathrm{d}i} = U_{\varphi i} - U_\mathrm{f} = U_{\varphi i} - \sum_{i=1}^{n} U_{\varphi \mathrm{i}}\, \varphi_\mathrm{i} \quad (19)$$

Equations (18) and (19) reflect the effect of counterflowing fluid set in motion by the whole polydisperse mixture. These currents have the strongest effect on slower particles. Observe that in Equations (18) and (19), $U_{\varphi i}$ represents vector quantities; hence, the terms $U_{\varphi i}\, \varphi_i$ should be added accordingly. Equation (19) facilitates an extended definition of "equal sedimentation" (or equal settling) particles:

$$U_{\mathrm{d}i} = U_{\mathrm{d}j}, \quad i \neq j \quad (20)$$

Thus different particles satisfy this definition under different conditions, i.e., at different values of φ and different distribution functions.

Continuous Distributions. If the distribution variable is denoted by ξ and its probability density function by $f(\xi)$, the differential volume fraction $\mathrm{d}\varphi$, U_f, and $U_\mathrm{d}(\xi)$ are given by

$$\mathrm{d}\varphi = \varphi f(\xi)\,\mathrm{d}\xi, \quad \int_{-\infty}^{\infty} f(\xi)\,\mathrm{d}\xi = 1 \quad (21)$$

$$U_\mathrm{f} = \varphi \int_{-\infty}^{\infty} U_\varphi(\xi) \cdot f(\xi)\,\mathrm{d}\xi \quad (22)$$

$$U_\mathrm{d}(\xi) = U_\varphi(\xi) - \varphi \int_{-\infty}^{\infty} U_\varphi(\xi) \cdot f(\xi)\,\mathrm{d}\xi \quad (23)$$

The total flux density in the dispersion frame of reference is then

$$J = \int_{-\infty}^{\infty} U_\mathrm{d}(\xi)\,\mathrm{d}\varphi = \varphi \int_{-\infty}^{\infty} [U_\varphi(\xi) - \varphi \int_{-\infty}^{\infty} U_\varphi(\xi) f(\xi)\,\mathrm{d}\xi] f(\xi)\,\mathrm{d}\xi \quad (24)$$

and the flux density of the $\xi_1 < \xi < \xi_2$ fraction is given by

$$J(\xi_2) - J(\xi_1) = \varphi \int_{\xi_1}^{\xi_2} [U_\varphi(\xi) - \varphi \int_{-\infty}^{\infty} U_\varphi(\xi) f(\xi)\,\mathrm{d}\xi] f(\xi)\,\mathrm{d}\xi \quad (25)$$

where $J(\xi_i)$ denotes cumulative flux density, i.e., for $-\infty < \xi \leq \xi_i$, $i = 1, 2$. Letting $\xi_1 \to -\infty$ gives $j(\xi_2)$, i.e., (without the subscript) the cumulative flux density distribution $J(\xi)$. Equations (19) and (25) show that the velocity and flux density of a given fraction depend on φ as well as on $f(\xi)$. This can significantly affect the velocity and flux of slower fractions, even to the extent of changing their direction of motion. Figure 5 [11] shows a plot of $U_\mathrm{d}(\xi)$ vs. ξ at different levels of φ, which were calculated by assuming a normal size distribution. The distribution was truncated on the left at $\xi = 0$; average particle size was $\bar{\xi} = 10^{-4}$ m, and the standard deviation was

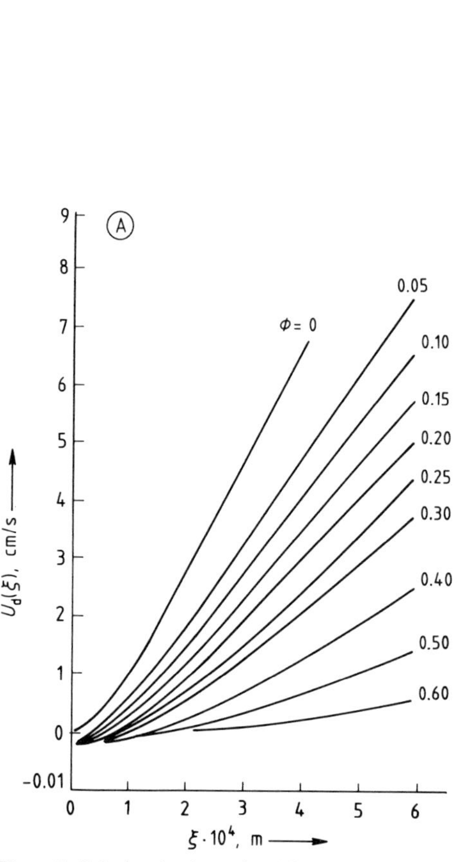

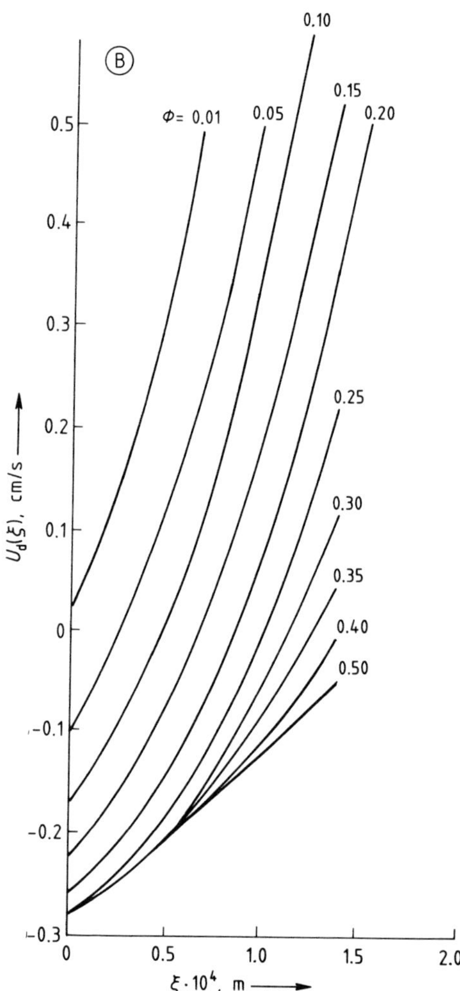

Figure 5. Calculated values of $U_d(\xi)$ vs. ξ
A) Truncated normal distribution $\bar{\xi} = 10^{-4}$ m, $\sigma = 3\,\bar{\xi}$ [11];
B) Detail of A at the finer size end

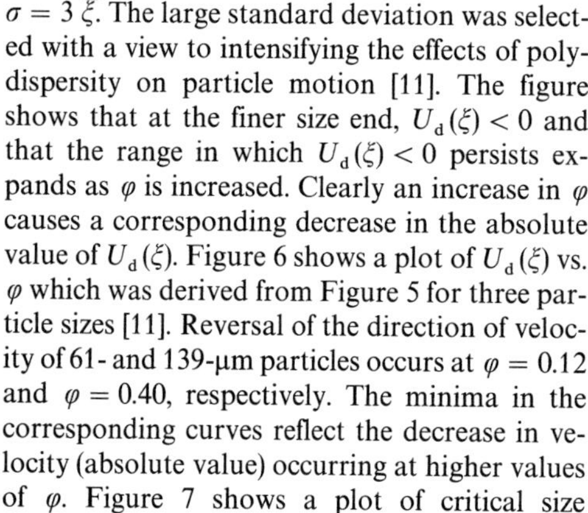

$\sigma = 3\,\bar{\xi}$. The large standard deviation was selected with a view to intensifying the effects of polydispersity on particle motion [11]. The figure shows that at the finer size end, $U_d(\xi) < 0$ and that the range in which $U_d(\xi) < 0$ persists expands as φ is increased. Clearly an increase in φ causes a corresponding decrease in the absolute value of $U_d(\xi)$. Figure 6 shows a plot of $U_d(\xi)$ vs. φ which was derived from Figure 5 for three particle sizes [11]. Reversal of the direction of velocity of 61- and 139-μm particles occurs at $\varphi = 0.12$ and $\varphi = 0.40$, respectively. The minima in the corresponding curves reflect the decrease in velocity (absolute value) occurring at higher values of φ. Figure 7 shows a plot of critical size $\xi\,[U_d(\xi) = 0]$ vs. φ at two levels of standard deviation of the normal distribution. Note that the critical size is defined as one at which the particle is stationary in the dispersion frame of reference. The value $\xi\,[U_d(\xi) = 0]$ increases with φ and also with an increase in σ. For further details, see [1] and [11].

2.1.3. Aggregated Dispersions

For details of the theory of sedimentation of aggregated dispersions [12], see → 16. Elutriation, p. **16**-6. The hydrodynamic hindrance function $f(\varphi)$ (Eq. 1) is given by Equation (26) for

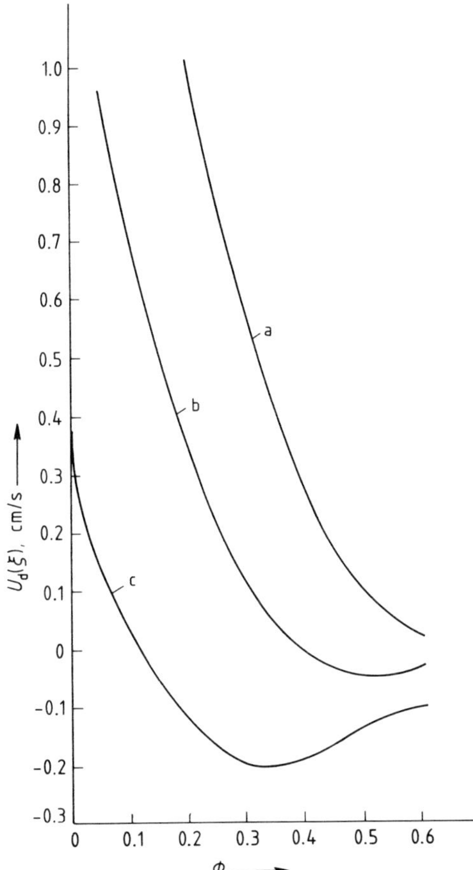

Figure 6. Calculated values of $U_d(\xi)$ vs. φ for three particle sizes [11] (data derived from Fig. 5 [11])
a) $d = 2.17 \times 10^{-4}$ m; b) $d = 1.39 \times 10^{-4}$ m; c) $d = 0.61 \times 10^{-4}$ m

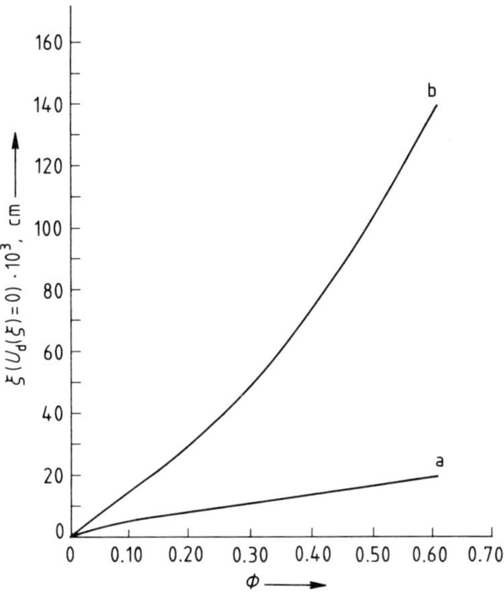

Figure 7. Critical values of particle size $\xi(U_d(\xi) = 0)$ vs. φ Normal size distribution, $\bar{\xi} = 10^{-4}$ m, at two levels of σ [11]:
a) $\sigma = 3\,\bar{\xi}$; b) $\sigma = 9\,\bar{\xi}$

sedimentation of monodisperse aggregates:

$$f(\varphi_a) = \frac{\varphi_s - \varphi}{[1 + (\varphi/\varphi_s)^{1/3}] \exp\left(\frac{5}{3} \cdot \frac{\varphi}{\varphi_s - \varphi}\right)},$$

$$\varphi_s > \varphi \qquad (26)$$

where φ_s is the volume fraction occupied by particulates within an aggregate and φ_a is the volume fraction occupied by the aggregates in the dispersion. The meaning of φ is unchanged, i.e., the volume fraction occupied by particles in the dispersion irrespective of their aggregation.

In the turbulent flow regime where Equation (2) is applied, a_1 remains unchanged, and a_2 and a_3 are given by

$$a_2 = 4.80\,\mu_f^{1/2} \exp\left(\frac{5}{6} \cdot \frac{\varphi}{\varphi_s - \varphi}\right)$$

$$a_3 = 4.0\,(\varrho_p - \varrho_f)\,(\varphi_s - \varphi) \cdot g' d^2 / \{3[1 + (\varphi/\varphi_s)^{1/3}]\} \qquad (27)$$

For polydisperse aggregates, the same calculation procedure suggested for unaggregated particles (see Eqs. 19–25) can be applied, with Equations (26) and (27) replacing their counterparts in Equations (1) and (2). The following calculated results show the effects of aggregation on sedimentation velocities and flux densities in the dispersion frame of reference. For monodisperse aggregates, $d = 183$ µm (where d is the diameter of the aggregates), $\varrho_p = 2500$ kg/m^3, $\varrho_f = 1000$ kg/m^3, and $\mu = 0.001$ kg m^{-1} s^{-1} were used in the calculations.

Figure 8 shows a plot of U_{da} vs. φ at three levels of φ_s; U_{da} is the velocity of aggregates in the dispersion frame of reference. The strong effect of φ_s is clear, and the combined effect of φ and φ_s intensifies at lower values of φ_s. The flux density of aggregated particles is given by

$$J = U_{\varphi a}\,(\varphi_s - \varphi)\,\varphi/\varphi_s \qquad (28)$$

where $U_{\varphi a}$ (similar to U_φ) is the sedimentation velocity in the fluid frame of reference. Figure 9

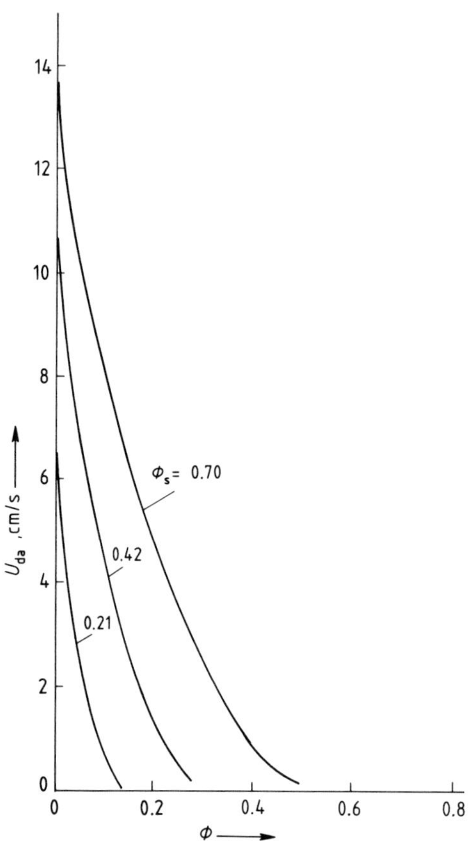

Figure 8. Calculated velocities of monodisperse aggregates vs. φ for three levels of φ_s, $d = 183$ µm [12]

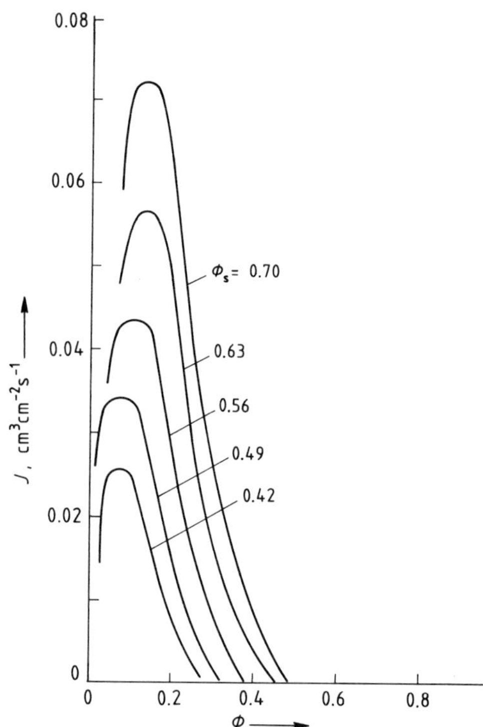

Figure 9. Calculated values of J vs. φ for different levels of φ_s, $d = 183$ µm [12]

shows a plot of J vs. φ calculated for different values of φ_s by using the data given above. Low φ_s values lower the flux densities considerably. Figure 10 shows plots of maximum flux density J_{max} vs. φ_s calculated for two aggregate sizes. The strong effect of φ_s is clear to the extent of possibly overriding the effect of aggregate size. Equation (29) relates J_{max} to φ_s and d (183 µm $< d < 0.01$ m):

$$\ln J_{max} = a(d) \ln \varphi_s + \ln b(d),\ 0 < \varphi_s < 0.7 \quad (29)$$

Values of $a(d)$ and $b(d)$ vs. d are given in Table 1. Figure 11 shows a plot of volume fraction φ_m that yields J_{max} vs. φ_s for $0 < d < 0.01$ m and $0 < \varphi_s < 0.7$. Aggregate density is plotted on the horizontal axis. Equation (30) correlates the results well [$\varphi_m = (0.0391 \ln d + 0.340)\, \varphi_s$, 183 µm $< d < 0.01$ m]:

$$\begin{aligned}\varphi_m &= [0.0391 \ln(0.0183) + 0.340]\, \varphi_s \\ &= 0.1836\, \varphi_s,\ d < 183\ \mu m\end{aligned} \quad (30)$$

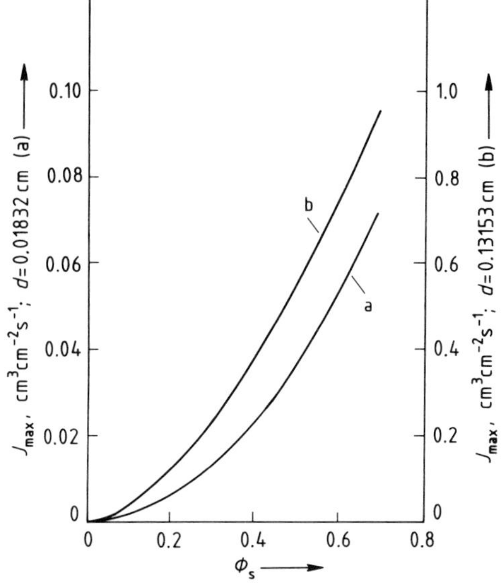

Figure 10. Calculated values of J_{max} vs. φ_s for two aggregate sizes [12]

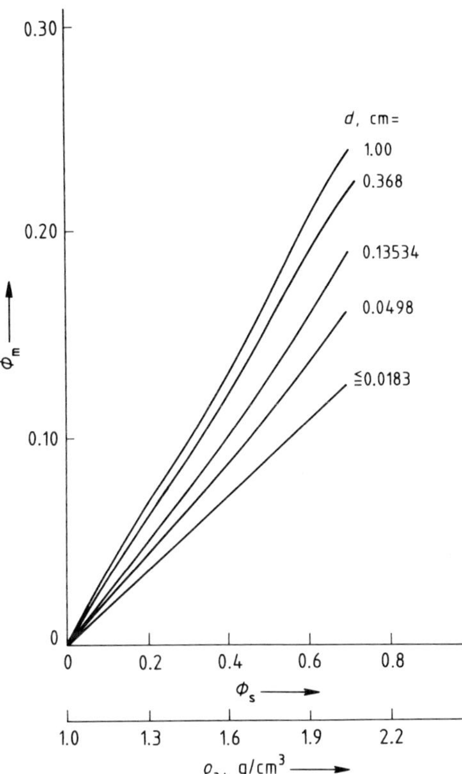

Figure 11. A plot of φ_m vs. φ_s for different aggregate sizes with $d \leq 0.01$ m [12]

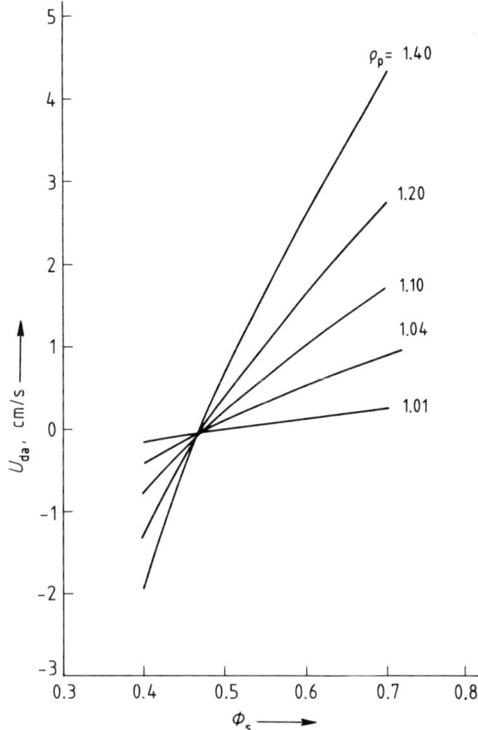

Figure 12. A plot of calculated U_{da} vs. φ_s for different values of ϱ_p, 0.01 m aggregates, $\varphi = 0.36$, distribution of packing density φ_s given below [12]

Table 1. Values of a and b obtained from linear regression of Equation (29) for different aggregate sizes

d, cm	a (d)	b (d)	Correlation factor
0.01832	2.0018	0.1471	> 0.9999
0.1353	1.7205	1.8013	> 0.9999
0.3679	1.6217	4.2131	> 0.9999
1.0000	1.5638	8.3352	> 0.9999

Equation (30) shows that for a high packing density ($\varphi_s = 0.6$) of $d < 183$ μm aggregates in the creeping flow regime, $\varphi_m = 0.11$ is recommended. This value is substantially lower than the value of 0.1792 that maximizes the flux of nonaggregated particles in the same flow regime (see Section 2.1.1).

Aggregation results in size enlargement, but at the same time the density of the aggregate may be lower than that of the unaggregated particles. This is particularly significant for particles that are "near-density" in the sense that their density is close to that of the fluid. Near-density particles can be found in biological, agricultural, and chemical systems, e.g., when the water content of particles (which are dispersed in water) is high. In near-density polydisperse aggregate systems, the effects of φ_s, φ, and the distribution on sedimentation are expected to be pronounced.

Figure 12 shows a plot of U_{da} vs. φ_s for different values of ϱ_p, where the diameter of the aggregates is 0.01 m, $\varphi = 0.36$, and the distribution of aggregate packing density is given below. The dominant effect of φ_s (at this high value of φ) in determining the magnitude and direction of U_{da} is demonstrated clearly for aggregates as large as 0.01 m. Figure 13 shows a plot of "partition values" $\varphi_s(U_{da} = 0)$ vs. d, calculated for $\varrho_p - \varrho_f = 400$ kg/m^3 at $\varphi = 0.36$. The discrete distribution of ϱ_s characterized by higher packing density of aggregates is given in the following material:

φ_s	0.400	0.475	0.550	0.625	0.700
$\psi(\varphi_s)$*	0.05	0.10	0.15	0.30	0.40

* The parameter $\psi(\varphi_s)$ denotes the distribution of φ_s.

The smallest aggregates cannot undergo sedimentation at $\varphi_s = 0.54$, and decreasing φ_s to 0.46

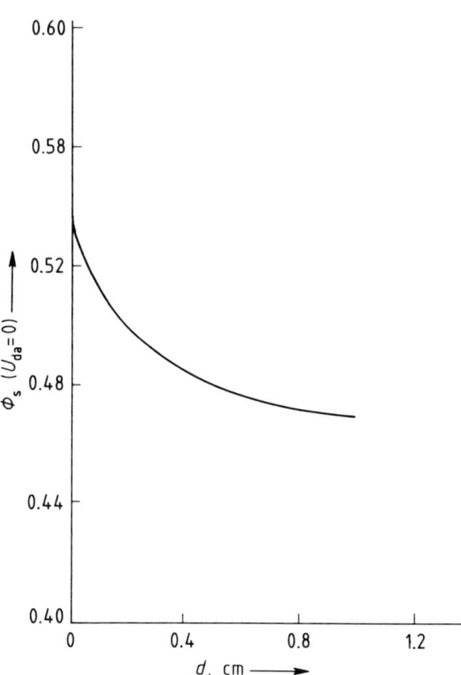

Figure 13. A plot of calculated partition values $\varphi_s(U_{da} = 0)$ vs. d; $\varphi = 0.36$, $\varrho_p - \varrho_f = 400\ kg/m^3$ [12]

is sufficient to prevent sedimentation of aggregates as large as 0.010 m. In summary: if higher sedimentation flux is required, then higher packing density of aggregates, with the smallest scatter about the packing density mean, should be used. The formation of large aggregates may be insufficient to produce a desirable sedimentation flux if the combination of φ_s, size, density, and the value of φ is unfavorable. For additional information, see [1].

2.2. Time Dependence of Steady Sedimentation Systems

Steady sedimentation systems are supposed to be time independent. This must be true at least in one frame of reference. However, in other frames of reference the system may no longer be time independent. Figure 14 shows schematically a dispersion flowing steadily in a pipe. A force field applied across the flow produces crossflow sedimentation. For an observer in the pipe frame of reference, the flow system, excluding the rising sediment, is time independent. This frame of reference is an Eulerian representation of flow. For an observer moving with the stream (i.e., in the fluid frame of reference), particle sedimentation appears time dependent, sedimentation being a batchwise rather than a continuous process (see Fig. 14 B). A frame of reference moving with the fluid is a Lagrangian representation of flow. If the dispersion is monodisperse, then in the Lagrangian representation, the time-dependent sedimentation involves steady sedimentation. This does not hold if polydisperse mixtures are involved, where in this case nonsteady sedimentation is observed. Thus, analysis of sedimentation systems that are steady in an Eulerian representation is likely to require the use of a Lagrangian representation to uncover their nonsteady features. This applies to analysis of the multiphase flow of slurries and dispersions, reactors, contacting devices, separation processes involving particulates, and sedimentation phenomena from flowing streams in nature. Thus the ability to analyze batch nonsteady sedimentation provides the means for analysis of sedimentation in steady flow systems. In the following, nonsteady sedimentation is considered.

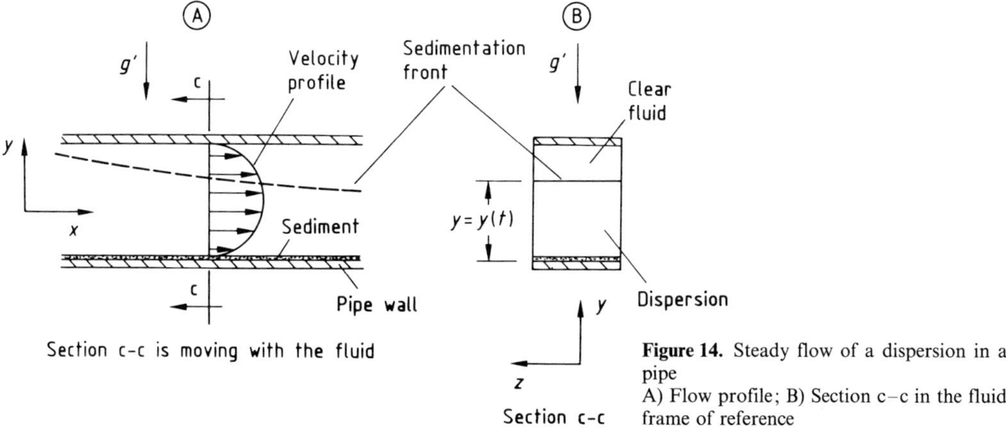

Figure 14. Steady flow of a dispersion in a pipe
A) Flow profile; B) Section c–c in the fluid frame of reference

2.3. Nonsteady Sedimentation

2.3.1. System Characterization

Continuity must be satisfied by all parts of a distribution of a polydisperse mixture undergoing sedimentation. Thus in the conduit frame of reference and with respect to the ith fraction,

$$\nabla[(U_{di} + v)\varphi_i] + \frac{\partial \varphi_i}{\partial t} = 0 \tag{31}$$

where v is the velocity of the dispersion relative to the conduit. If the distribution is continuous, then from Equations (21) and (23),

$$\nabla[(U_d + v)\varphi f(\xi)] + \frac{\partial[\varphi f(\xi)]}{\partial t} = 0 \tag{32}$$

$$\nabla[(U_\varphi(\xi) - \varphi \int_{-\infty}^{\infty} U_\varphi(\xi)\cdot f(\xi)\,\mathrm{d}\xi + v)\,\varphi f(\xi)] + \frac{\partial[\varphi f(\xi)]}{\partial t} = 0 \tag{33}$$

where as before ξ is a distribution parameter such as size, density, etc.

The justification for using U_φ in nonsteady sedimentation depends on the relaxation time involved in attaining this velocity. If the relaxation time for attaining, for example, 99 % of $U_\varphi(\xi)$ is significantly shorter than that associated with changes in $U_\varphi(\xi)$, then the assumption that terminal velocity conditions apply for the time- or position-dependent $U_\varphi(\xi)$ is justified. In other words, if the hydrodynamic response to changes in U_φ is significantly faster than the time rate of these changes, then using terminal sedimentation velocities values for the time- and position-dependent particle velocities provides a good approximation and is justified. Hence, Equations (1) and (2) can be used for evaluation of $U_\varphi(\xi)$ in nonsteady sedimentation. If the relaxation time associated with hydrodynamic response to changes in $U_\varphi(\xi)$ is not short enough, then the use of Equations (1) and (2) may introduce an unacceptable error. Sedimentation of particles in liquids generally involves short relaxation times, whereas this is not always true for sedimentation in gases.

Simulation of nonsteady sedimentation of polydisperse mixtures requires solution of Equation (31) or (33) for each fraction ξ at every point in the dispersion at a given time. The number of simultaneous equations is $n_1 \times n_2 \times n_3$, where n_1, n_2, and n_3 denote the number of fractions, the number of segments (i.e., volume elements) into which the dispersion is divided, and the number of time increments, respectively. In the following material, an outline and sample outputs of a computer program using an algorithm of direct tracing of each sedimenting fraction are described briefly [13].

2.3.2. Computer Program for Simulation of Nonsteady One-Dimensional Sedimentation of Polydisperse Mixtures

The computer model incorporates a principle of tracing positions and concentrations, along the sedimentation trajectory, of all size fractions, given initial conditions and elapsed time from start of sedimentation [13]. The fluid is divided into m thin layers (of thickness h_1) which are set perpendicular to the direction of sedimentation (see Fig. 15 B). Each layer is assumed uniform with respect to composition and concentration, both being time dependent. The sedimentation system is defined with respect to two arrays: the (i, j') "primed array" and the (i, j) unprimed array (see Fig. 15 C). The (i, j') primed array is invariable and characterizes the volume fraction $\varphi(i, j')$ of the ith fraction originating at the jth layer at the onset of sedimen-

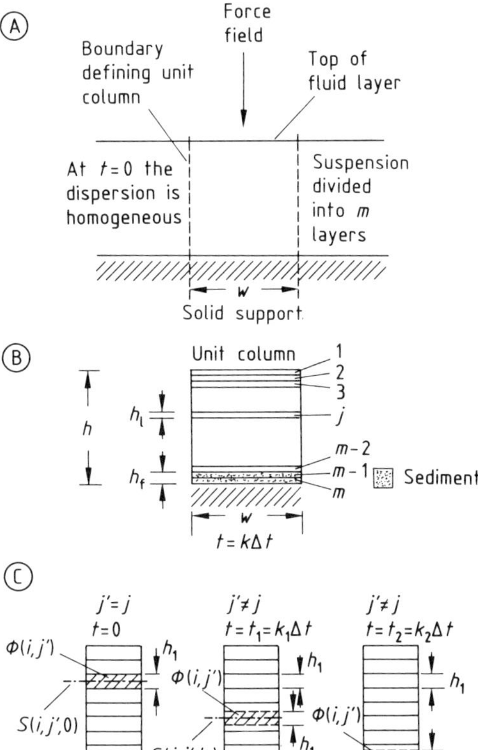

Figure 15. Characterization of the sedimentation system A), B) Definition of a unit sedimentation column; C) Definition of sedimentation variables

tation ($t = 0$). The (i, j) unprimed array characterizes the time-variable volume fraction $\varphi(i, j)$ of the ith size fraction in the jth layer at time t. The fraction $\varphi(i, j')$ occupies a space of one layer and moves through the dispersion as one invariable unit (see Fig. 15 C). The distance traveled by $\varphi(i, j')$ is denoted by $S(i, j', k)$, where k denotes the number of time increments elapsed. The fraction $\varphi(i, j)$ consists of all contributions from $\varphi(i, j')$ fractions present in the jth layer at time t. Note that a contribution of $\varphi(i, j')$ means that a nonvanishing part of it entered and remained in the jth layer at time t. The total layer concentration at time t is given by

$$\varphi(j) = \sum_{i=1}^{n} \varphi(i, j)$$

where summation is over all size fractions.

The tracing algorithm is rather involved; therefore, only a sample of outputs is presented here. The following data apply. One-dimensional sedimentation takes place in water at room temperature, with the number of layers $m = 20$; at the onset of sedimentation, the dispersion is uniform consisting of four $i = 1,2,3$, and 4 size fractions $d_i = 20 \times 9^{(i-1)/4}$ µm (giving a size range from 20 to 103.9 µm), $\varphi(i, j') = 0.1\, i$, at $t = 0$, $\varphi = 0.3$. The field is gravitational, acting perpendicular to the m layers. Packing density of the sediment is $\varphi = 0.6$. Height of the sedimentation column h is either 0.05 or 0.00625 m. High φ and skewed size distribution were selected so that the effects of concentration and distribution on sedimentation patterns would be pronounced. The computer graphical outputs provided the following information:

1) particle trajectories $S(i, j', k)$ and residence time in dispersion,
2) profiles of total concentration $\varphi(j)$ as a function of time and position,
3) profiles of specific concentrations $\varphi(i, j)$ as a function of time and position, and
4) structure and composition of sediment as a function of time and position.

2.3.3. Graphical Output

Trajectory and Composition of Dispersion and Sediment. Figure 16 shows plots of sedimentation trajectories $S(i, j', k)$ vs. time for the four size fractions positioned at the tenth layer ($j' = 10$) at $t = 0$, the column height being 0.05 m. The largest particles $d_4 = 103.9$ µm follow practically a steady sedimentation, reaching the sediment in less than 2 s. The smallest particles $d_1 = 20$ µm follow an inverted trajectory, first counterflowing the large particles and then changing direction toward the sediment. This fraction required 25 s for velocity inversion to occur, 57 s to return to the tenth layer, and 65 s to reach the sediment. The $d_2 = 34.6$ µm fraction also follows a smaller inverted trajectory. The curves disclose abrupt changes of slope. In the ith ($i = 1, 2, 3, 4$) curve, $5 - i$ such changes occur, including the final slope change due to transfer into the sediment. Thus the smallest fraction $i = 1$ shows four abrupt slope changes, the first, second, and third being due to disappearance of d_4 (at 10 s), d_3 (at 25 s), and d_2 (at 55 s) fractions from the layer in which the d_1 fraction moves. The increase in slope after its abrupt change indicates a shock, or discontinuity, occurring due to an abrupt fall in φ.

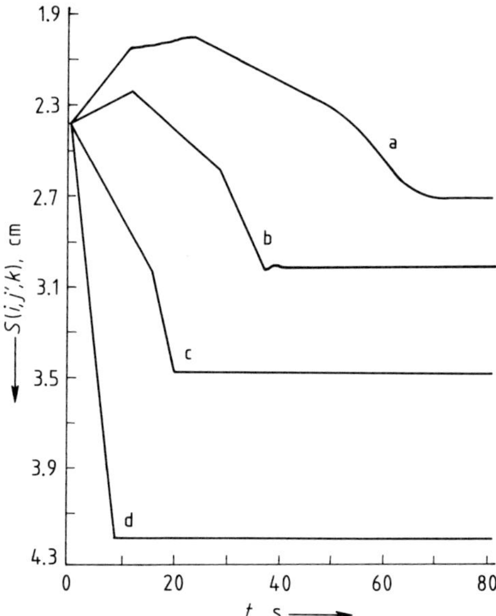

Figure 16. Trajectories of the four $\varphi(i, j')$ fractions, a) $i = 1$; b) $i = 2$; c) $i = 3$; d) $i = 4$, starting sedimentation at the tenth layer ($j' = 10$), $\varphi = 0.30$, $h = 0.05$ m

Thus the full sedimentation record of the four size fractions is available from Figure 16, i.e., starting position, trajectory, and position in the sediment, which shows a clear segregation effect.

Figure 17 is a plot of $\varphi(j)$ vs. t in the tenth layer $j = 10$, for four levels of initial (i.e., at $t = 0$) concentration: 0.05, 0.10, 0.20, and 0.30. A stepwise decrease of $\varphi(j)$ can be observed for the four levels of φ. Furthermore, in a given step, $\varphi(j)$ exhibits a train of concentration waves. Each step corresponds to partial or full exit of one size fraction from the tenth layer, the largest first and the smallest last. The concentration waves are an integrated effect of the contribution from each size fraction, with their amplitudes diminishing after a decrease in φ. The train of concentration waves intensifies with increasing layer depth, i.e., larger j. An example is shown in Figure 18 where $\varphi(j)$ vs. t is plotted for different layers j and $\varphi = 0.3$. In the first layer $j = 1$, the amplitude of the concentration waves is small, whereas in the tenth layer $j = 10$, large waves exist. Furthermore, in the upper layers ($j = 1$ and 2), sedimentation steps [wherein $\varphi(j)$ fluctuates] are small, whereas in lower ones ($j = 5$ and 10), they expand considerably. A plot of $\varphi(i, j)$ vs. t for the four size fractions $i = 1, 2, 3$, and 4, in the fifth $j = 5$ layer, given $\varphi = 0.3$, is shown in Figure 19. Observe that at $t = 0$, $\varphi(i, j) = 0.03 \times i$; for example, the initial volume fraction of the $d_1 = 20$ µm, $i = 1$ fraction in the fifth layer was 0.03. Time-variable composition occurs due to accumulation and size segregation effects, prior to depletion of the layer from particles. For example, after 20 s the layer consists of $\varphi(1, 5) = 0.04$, $\varphi(2, 5) = 0.13$, $\varphi(3, 5) = \varphi(4,5) = 0$ fractions, indicating depletion of the third and fourth fractions, with simultaneous upgrading of the second and first fractions. Concentration waves of individual size fractions pass the fifth layer, increasing in relative intensity for smaller particle sizes. The maximum amplitude of these waves is of the order of twice the initial concentration. Superposition

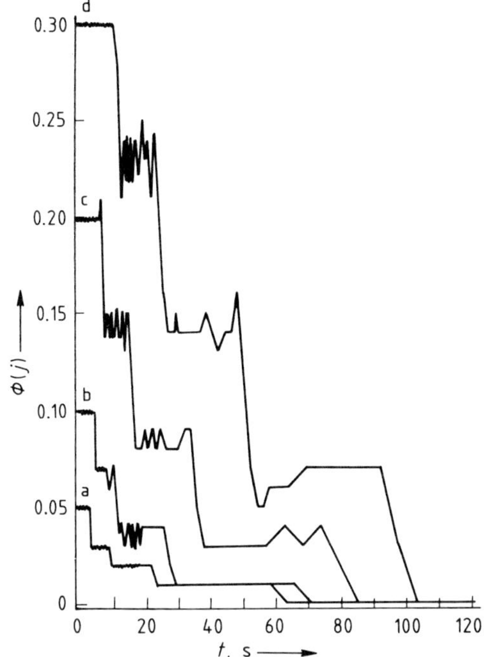

Figure 17. A plot of $\varphi(j)$ vs. t for $j = 10$ and four levels of φ:
a) $\varphi = 0.05$; b) $\varphi = 0.1$; c) $\varphi = 0.2$; d) $\varphi = 0.3$; $h = 0.05$ m

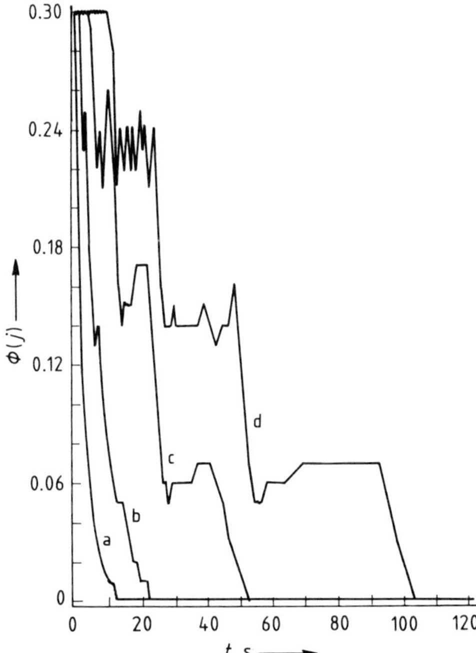

Figure 18. A plot of $\varphi(j)$ vs. t for different layers j:
a) $j = 1$; b) $j = 2$; c) $j = 3$; d) $j = 4$, $\varphi = 0.30$; $h = 0.05$ m

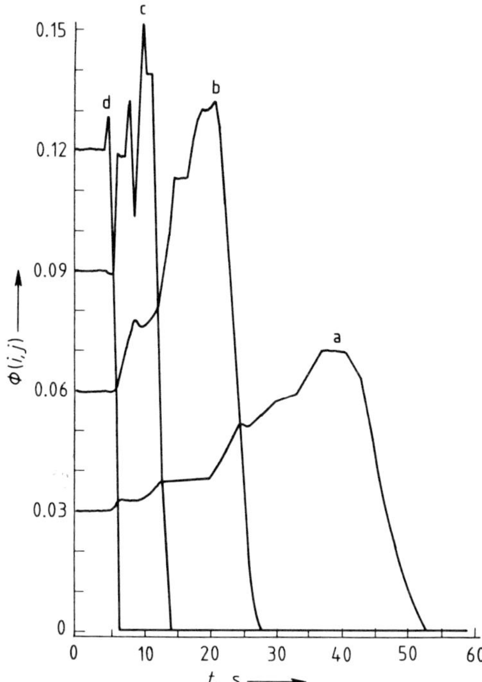

Figure 19. A plot of $\varphi(i, j)$ vs. t in the $j = 5$ layer for:
a) $i = 1$; b) $i = 2$; c) $i = 3$; d) $i = 4$, $\varphi = 0.30$, $h = 0.05$ m, $\alpha = 9.0$

of the individual concentration waves yields total concentration waves of the type shown in Figure 18. The individual as well as total concentration waves expand as they move down the layers. This has been verified by outputs not shown here. The effect of the range of size distribution in the polydisperse mixture was tested by using a size parameter α, with the size range being defined by $d_i = 20\alpha^{(i-1)/4}$ μm for $\alpha = 1.0$, 2.5, and 9.0. Figure 20 shows the variation of particle trajectories $S(i, j', k)$ of the $i = 1$ fraction vs. time, for the three values of α tested. At $t = 0$, particles are in the tenth layer and $\varphi = 0.3$. The $\alpha = 1.0$ curve corresponding to monodisperse 20 μm particles is a straight line as expected. Slope transitions, but no inversions, can be observed in the $\alpha = 2.5$ curve. Increasing α to 9.0 (size range from 20 to 103.9 μm) produces an inverted trajectory with four slope transitions (as also shown by the highest curve in Fig. 16). Increasing α from $\alpha = 1$ decreases the overall sedimentation velocity initially, and then the reverse occurs. Thus the sedimentation of the 20-μm particles is slower at $\alpha = 2.5$ than at $\alpha = 9.0$. The computer program also provided information regarding composition of the sediment and effects of perturbation of concentration in a given position on the stability of the sedimentation system.

Figure 21 shows the volume fraction of the 20-μm fraction $i = 1$ in the sediment $\varphi_s(i, j)$ vs. the layer number j for different elapsed times of sedimentation, $\alpha = 9$, $h = 0.625$ cm, $\varphi = 0.30$. Accumulation of this fraction around the $j = 11$ layer (and other fractions around other layers, see Fig. 22) increases with time, indicating a size segregation effect. Figure 22 shows a plot of $\varphi_s(i, j)$ vs. t in the $j = 14$ layer; the rest of the data remain unchanged. The

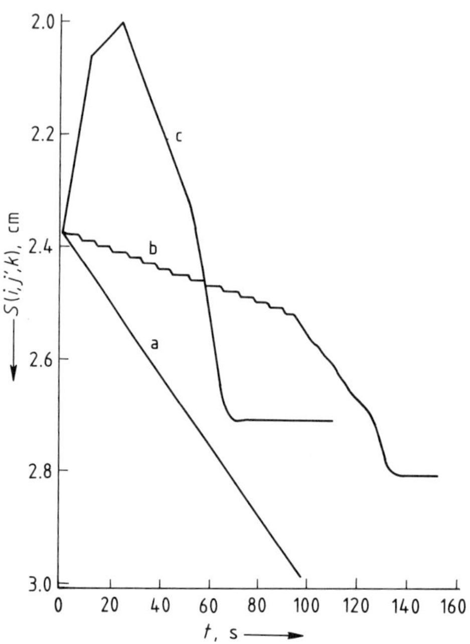

Figure 20. A plot of $S(i, j', k)$ vs. t of the 20 μm particles ($i = 1$), originating at the $j = 10$ layer, $\varphi = 0.30$, $h = 5.0$ cm, for three levels of the distribution parameters:
a) $\alpha = 1$; b) $\alpha = 2.5$; c) $\alpha = 9$

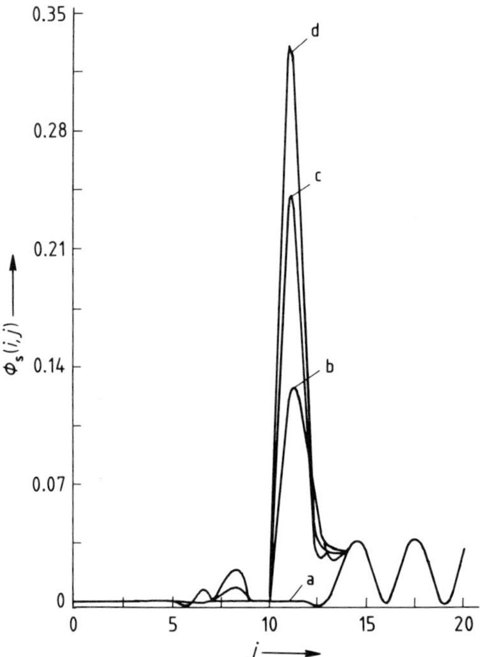

Figure 21. A plot of $\varphi_s(i, j)$ vs. j of the 20 μm particles, at different elapsed times from onset of sedimentation:
a) $t = 3.8$ s; b) $t = 10.1$ s; c) $t = 12.2$ s; d) $t = 15$ s, $\varphi = 0.30$, $h = 0.625$ cm, $\alpha = 9.0$

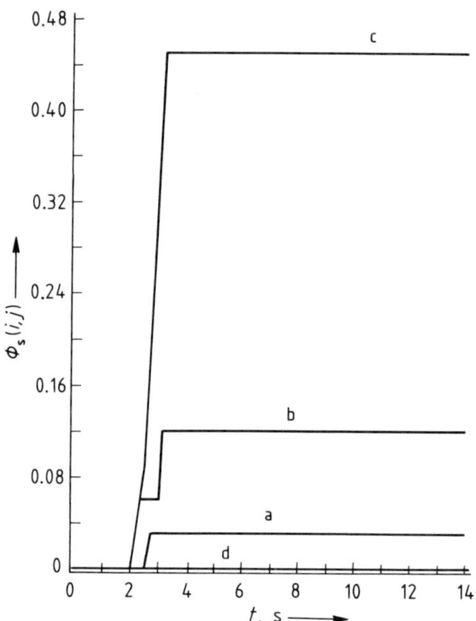

Figure 22. A plot of $\varphi_s(i, j)$ vs. t for:
a) $i = 1$; b) $i = 2$; c) $i = 3$; d) $i = 4$, $\varphi = 0.30$, $h = 0.625$ cm, $\alpha = 9.0$

figure shows that after 3 s the sediment in the 14th layer consists of 75% $i = 3$ particles, 20% $i = 2$ particles, 5% $i = 1$ particles, and no $i = 4$ particles. Compositions of other sediment layers can be obtained in a similar manner.

Effect of Distribution and Total Volume Fraction on Sedimentation Behavior. The effect of size range on sedimentation trajectories of particles in water is illustrated in Figure 20. High φ dispersions of polydisperse mixtures in air are characteristic of fluidization and sedimentation systems in which volume fractions occupied by particles range typically from 0.4 to 0.6. High φ dispersions are affected mainly by deviation of size distribution from uniformity, being prone to instabilities caused by perturbation of local concentrations. In the following examples, size distributions that satisfy $d = 60 \times \alpha^{(i-1)/4}$ μm and $i = 1, 2, 3$, and 4 are used, the fluid being air $\mu_f = 182 \times 10^{-7}$ kg m^{-1} s^{-1}, $\varrho_f = 1000$ kg/m^3, $\varrho_p = 3000$ kg/m^3. Figures 23 and 24 contain plots of $\varphi(j)$ vs. t for the $j = 9$, 10, and 11 layers, and $\alpha = 1.05$ and 3.0, respectively. At $t = 0$, $\varphi = 0.40$.

Figure 23 shows that a narrow size distribution (i.e., low α) and high value of φ generate strong fluctuations of $\varphi(j)$ in the $j = 9$ and 10 layers, and particle accumulation in the $j = 11$ layer. At $t = 0.4$ s, $\varphi(9)$ increases above 0.50, and at $t = 0.8$ s, the eleventh layer becomes practically blocked with a dense phase of particles $\varphi(11) \geq 0.6$. After 0.9 s, the ninth layer is almost depleted of particles, the tenth layer remains close to $\varphi(10) \simeq 0.40$, and the eleventh layer is highly packed with particles, i.e., $\varphi(11) \geq 0.6$. Fluidization of such polydisperse particle system would be inherently unstable, since no single sedimentation velocity can be balanced by the drag of the fluidizing gas. Higher packing density (i.e., $\varphi(j)$) in a given layer decreases sedimentation

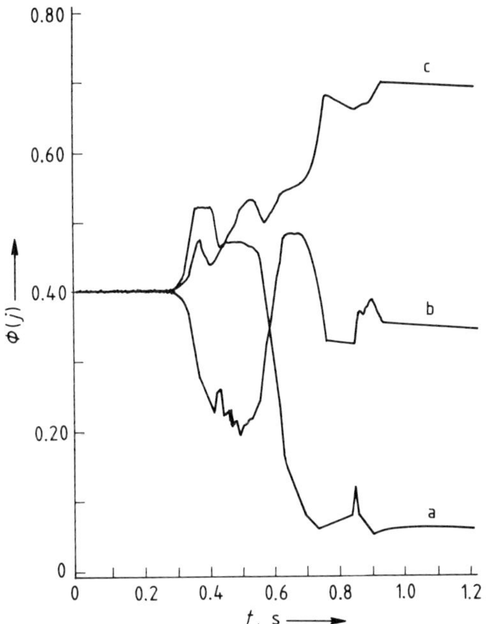

Figure 23. A plot of $\varphi(j)$ vs. t in the a) $j = 9$; b) $j = 10$; c) $j = 11$ layers, $d_i = 60 \times \alpha^{(i-1)/4}$, $\alpha = 1.05$, $\varphi = 0.40$, $h = 5.0$ cm, sedimentation in air

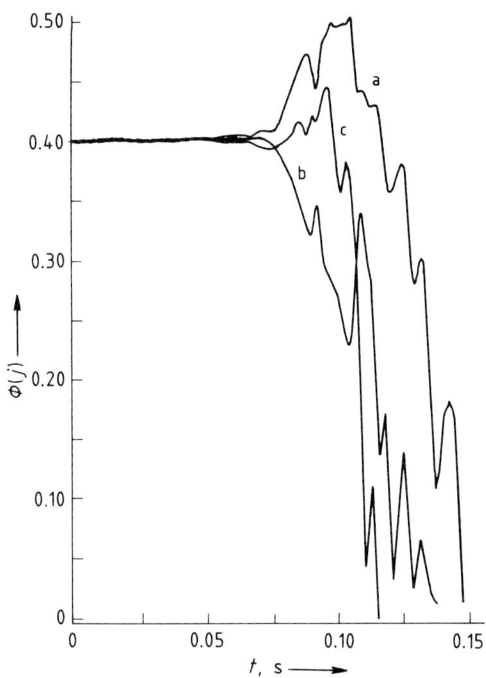

Figure 24. A plot of $\varphi(j)$ vs. t, data as in Figure 23 except for $\alpha = 3.0$

velocity there. Hence, if the velocity of the fluidizing gas is unchanged, material in this layer will be lifted and probably separated from lower, less dense layers. This can be related to the phenomenon of slugging which is well documented in fluidization engineering [14], [15].

Figure 24 shows similar results of $\varphi(j)$ vs. t but for a wider size distribution (i.e., $\alpha = 3.0$). Here the largest increase [to $\varphi(j) \simeq 0.50$] in concentration occurs in the $j = 9$ layer after 0.10 s of sedimentation. This rise may be sufficient to effect instability, but the overall effect has diminished compared to $\alpha = 1.05$ (Fig. 23). Further increase of α to 9.0 (not shown here) eliminates the surging of concentration. This indicates that strong sedimentation instability is expected in dispersions that are characterized by high initial particle concentration and narrow size distribution. Thus the computer model for nonsteady sedimentation of polydisperse mixtures predicts that for high concentration and narrow distribution, instability is an inherent property of the sedimentation system.

Figures 25, 26, and 23 show for $\alpha = 1.05$ the effect of increasing φ on the formation of surging concentrations in the $j = 9$, 10, and 11 layers. The rest of the data remain unchanged. If at $t = 0$, $\varphi = 0.20$, the maximum surge (occurring in all three layers) in $\varphi(j)$ is 10% above its initial value (see Fig. 25). The surge intensifies at $\varphi = 0.30$, reaching in the $j = 11$ layer a 20% increase in concentration over the initial value of 0.30 (see Fig. 26). Further increase of φ (for the same $\alpha = 1.05$ size distribution) to 0.4 produces the effect already shown in Figure 23, where the $j = 11$ layer accumulates particles to the extent of becoming a sort of plug capable of obstructing further sedimentation. The

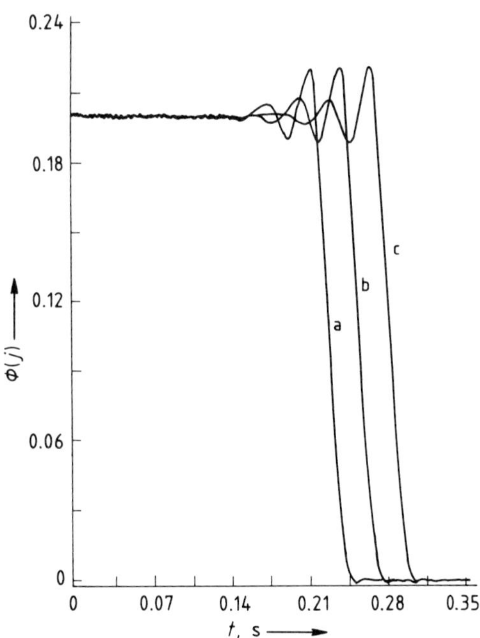

Figure 25. A plot of $\varphi(j)$ vs. t, data as in Figure 23 except for $\varphi = 0.20$

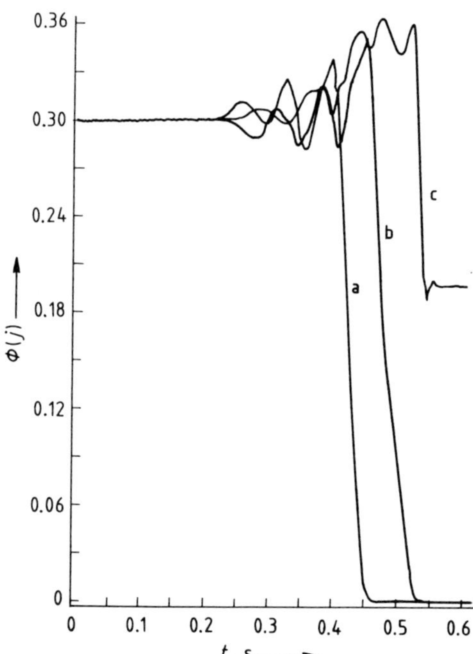

Figure 26. A plot of $\varphi(j)$ vs. t, data as in Figure 23 except for $\varphi = 0.30$

above results agree with the strong dependence of velocity and particle flux on φ; see Equations (1), (2), (13), (14), and (24), and Figures 3 and 6, in particular, for sedimentation in the laminar flow regime. Self- or *internal perturbation* occurs in polydisperse mixtures, which as shown here can produce substantial instability due to surges in φ. Such surges can develop into a self-accelerating process in which further increase in φ results from the incoming flux being increasingly higher than the one capable of leaving. Clearly, perturbations induced by external sources (i.e., *external perturbations*) are expected to produce similar concentration-dependent instabilities. Further computer outputs (not shown here) verify that the sedimentation system becomes more sensitive to external perturbation as the particle concentration (prior to the perturbation) increases. The complexity of nonsteady batch sedimentation has been known for a long time [16]. Figure 27 [17] shows three categories of batch sedimentation defined for uniform size and density (A), narrow size distribution (B), and wide size distribution (C). Figure 28 describes [2] the formation of discontinuities of composition, or "shock," in batch sedimentation of a polydisperse particle system consisting of three size fractions. The existence of such discontinuities is predicted by the computer model which, as shown above, also provides in-depth information about the internal structure and dynamics of nonsteady sedimentation as a function of position and time.

2.3.4. Transient Sedimentation

Definition. Transient sedimentation means nonsteady sedimentation of particles under the action of constant driving forces. If the driving forces vary along the sedimentation trajectory, then transient sedimentation is associated with velocity changes at every point along the trajectory. When the relaxation time associated with transient motion is sufficiently small relative to that associated with velocity changes, the effect of transient sedimentation may be neglected. This is assumed in earlier sections (Section 2.3.3) of this article.

Equation of Motion. Newton's law of motion reads

$$\sum_j \mathrm{d}(\mathrm{m}U)_j/\mathrm{d}t = \sum_i F_i \qquad (34)$$

where the summation is over all forces F_i acting on accelerated masses m_j whose velocity is U_j at time t. All masses directly or indirectly accelerated by the driving forces are included in the momentum balance equation. The forces acting on a particle can be divided into body and surface forces, as well as driving and retarding forces. If the motion of particles is coupled with that of the fluid, then the accelerated mass of the fluid must be also accounted for in Equation (34). This results in a so-called added or virtual mass [18]–[22], which accounts for the fact that part of the driving force must be expected to accelerate the fluid displaced by motion of the particles. The significance of this added mass becomes apparent when the density of the fluid is comparable to, or higher than, that of the particles, an extreme example being dispersion of bubbles in a liquid. The average velocity of displaced fluid counterflowing particles in the dispersion frame of reference is given (see Eqs. 7 and 8) by

$$U_f = U_\mathrm{d}\,\varphi/(1-\varphi) \qquad (35)$$

In a given volume V of dispersion, the mass of flowing fluid is $m_\mathrm{f} = \varrho_\mathrm{f} V(1-\varphi)$; hence the average fluid momentum in V is $m_\mathrm{f} U_\mathrm{f} = \varrho_\mathrm{f} V \varphi U_\mathrm{d}$. Thus the fluid momentum per particle is $\varrho_\mathrm{f} V_\mathrm{p} U_\mathrm{d}$, where V_p is particle volume and the number of particles in V is $V\varphi/V_\mathrm{p}$. Hence, the equation of motion of monodisperse particles in a fluid is given by

$$\mathrm{d}[V_\mathrm{p}(\varrho_\mathrm{p}+\varrho_\mathrm{f})U_\mathrm{d}]/\mathrm{d}t = \sum_i F_i \qquad (36)$$

If V_p, ϱ_f, and ϱ_p are constants, then Equation (36) reduces to

$$V_\mathrm{p}\varrho_\mathrm{p}(1+\varrho_\mathrm{f}/\varrho_\mathrm{p})\mathrm{d}U_\mathrm{d}/\mathrm{d}t = \sum_i F_i \qquad (37)$$

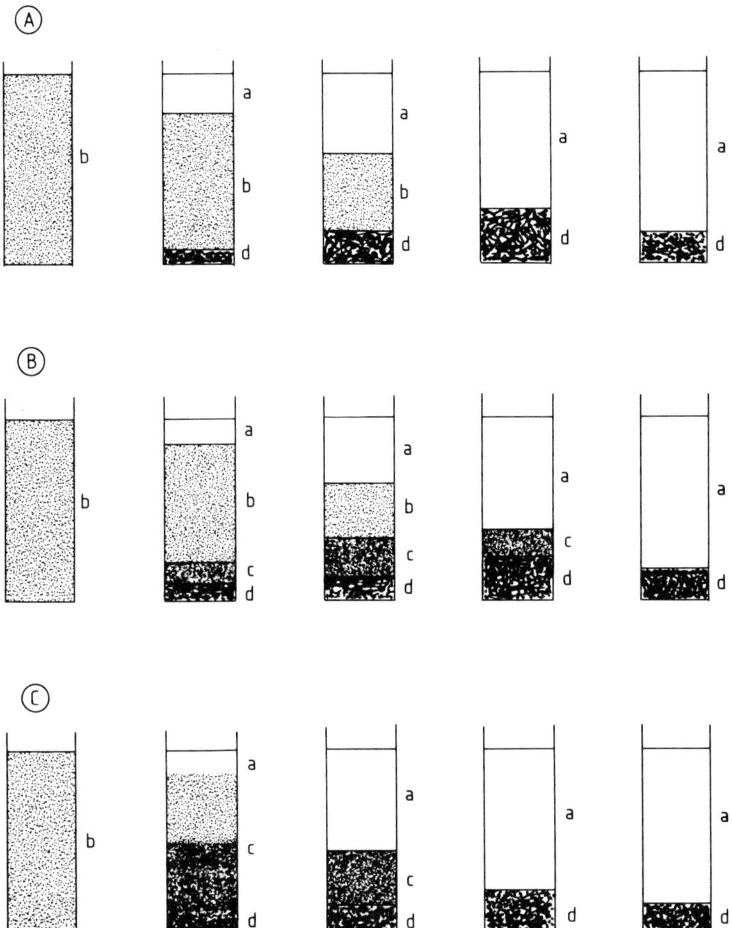

Figure 27. Effect of particle size distribution on profile of batch sedimentation [17]
A) Uniform size and density; B) Narrow size distribution; C) Wide size distribution
a) Clear liquid; b) Dispersion; c) Polydisperse mixture; d) Sediment

Equation (37) applies to a sedimentation system at constant total volume of incompressible particles and fluid. If sedimentation involves variable density of either particles or fluid, or variable volume of particles, then Equation (36) must be applied. Changes in particle density and size can occur due to aggregation, flocculation, and crystallization (or dissolution) processes. Properties of the fluid can be changed by thermodynamic means such as temperature gradients. For nonsteady sedimentation of polydisperse mixtures, Equation (36) must be applied to each fraction of the distribution. Thus for a distribution consisting of n size fractions, n equations are required to describe the system.

Driving Forces. At steady state, the local properties of a dispersion consisting of two phases can be expressed as a weighted average of properties of the individual phases:

$$\gamma_{\varphi} = \varphi \gamma_{p} + (1 - \varphi) \gamma_{f} \qquad (38)$$

where γ_{φ}, γ_{p}, and γ_{f} are the values of this property in a given small volume of the dispersion for the dispersion, for particles, and for fluid, respectively. Equation (5) is a special case of Equation (38), i.e., for $\gamma = \varrho$. The postulated steady state implies that the forces exerted on the particles are transmitted fully to the fluid, thus generating pressure gradients having the same effect as Archimedes

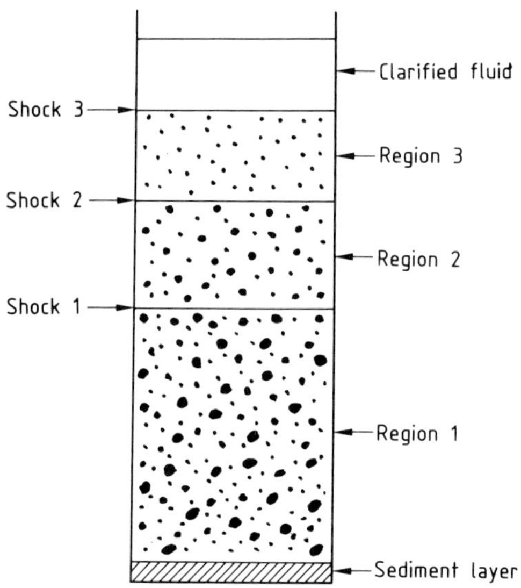

Figure 28. Formation of distinct regions that are separated by composition discontinuities ("shocks") in batch sedimentation of a mixture of three different particle fractions. Region 1 contains all three fractions, region 2 is devoid of the fastest settling particles, and region 3 contains only the slowest settling particles [2]

levitational force. If the degree of force transmission $\beta(t)$ is time dependent, then γ_φ is given by

$$\gamma_\varphi = \beta(t)\,\varphi\gamma_p + (1 - \beta(t)\,\varphi)\,\gamma_f,\ 0 \leq \beta(t) \leq 1 \qquad (39)$$

$$\beta(t) = F_D/F_\gamma = F_D/(\varrho_p V_p g'),\ 0 \leq \beta(t) \leq 1 \qquad (40)$$

where F_D and F_γ are the drag force and the force associated with the property γ, respectively, and g' (see Eqs. 1–3) is acceleration due to F_γ. Thus at the onset of sedimentation when no drag exists $\beta(t) = 0$, whereas at terminal velocity conditions $\beta(t) = 1$. The following categories of driving forces may be used in sedimentation systems.

Gravitational Force:

$$F_G = (\varrho_p - \varrho_\varphi)\,V_p g \qquad (41)$$

At the onset of sedimentation of monodisperse particles $F_G = (\varrho_p - \varrho_f)\,V_p g$, whereas at steady-state terminal velocity conditions $F_G = (\varrho_p - \varrho_f)(1 - \varphi)\,V_p g$. Thus full transmission of the particle weight to the fluid results in a decrease of F_G by a factor of $1 - \varphi$.

Centrifugal Force:

$$F_\omega = (\varrho_p - \varrho_\varphi)\,V_p \omega^2 r \qquad (42)$$

where r is the radius of rotation and ω is the angular velocity.

Electrical Charge Force:

$$F_q = \int_{V_p} E(q_v - q_{v\varphi})\,\mathrm{d}V_p + \int_{S_p} E(q_s - q_{s\varphi})\,\mathrm{d}S_p \qquad (43)$$

where q_v and q_s are charge densities per unit volume and unit area of the particles. For monodisperse particle mixtures at $\beta(t) = 1$,

$$q_{v\varphi} = \varphi q_{vp} + (1 - \varphi)\,q_{vf},$$

$$q_{s\varphi} = \varphi q_{sp} + (1 - \varphi)\,q_{sf}$$

where subindex f denotes fluid.

Forces due to Magnetic Polarization:

$$F_M = \mu_0 V_p \nabla[(M_p - M_\varphi)\cdot H] \qquad (44)$$

For monodisperse particle mixtures and $\beta(t) = 1$, $M_\varphi = \varphi M_p + (1 - \varphi)\,M_f$; hence, $F_M = \mu_0 V_p \nabla[(1 - \varphi)(M_p - M_f)\cdot H]$.

Forces due to Electrical Polarization:

$$F_E = V_p \nabla[(P_{Ep} - P_{E\varphi})\cdot E] \qquad (45)$$

For monodisperse particle mixtures and $\beta(t) = 1$, $P_{E\varphi} = \varphi P_{Ep} + (1 - \varphi)\,P_{Ef}$; hence, $F_E = V_p \nabla[(1 - \varphi)(P_{Ep} - P_{Ef})\cdot E]$.

Other forces such as those arising from interaction of moving charges (i.e., currents) and magnetic fields may also be relevant.

Note that for polydisperse mixtures, Equation (38) can be applied, with γ_p being replaced by its average value $\bar{\gamma}_p = \frac{1}{\varphi}\int_0^\varphi \gamma_p\,\mathrm{d}\varphi$. For example, in Equation (41), $\varrho_{pi} - \varrho_\varphi = \varrho_{pi} - [\varphi\bar{\varrho}_p + (1 - \varphi)\varrho_f]$ should be used for the ith fraction.

Retarding Forces. Equations (46) and (47) give the *hydrodynamic drag force* exerted on a spherical particle in the laminar $Re < 1$ and general flow regimes, respectively.

$$F_D = -3\pi d\mu'_\varphi U_\varphi,\quad \mu'_\varphi = \mu_\varphi(1 + \varphi^{1/3}),$$

$$\mu_\varphi = \mu_f \exp\left(\frac{5}{3}\cdot\frac{\varphi}{1 - \varphi}\right) \qquad (46)$$

$$F_D = -\frac{\pi}{8}[0.63(U_\varphi d\varrho_f)^{1/2} + 4.80\,\mu_\varphi^{1/2}]^2 \cdot d(1 + \varphi^{1/3})\,U_\varphi \qquad (47)$$

In Equations (46) and (47), F_D is the drag force existing in a fully developed flow around the particle, which corresponds to terminal velocity (here U_φ) flow conditions. Under transient flow conditions (such as those existing in turbulent flow) where flow around the particles is not steady, the history of flow may be significant. This is accounted for by an additional term from BASSET [23], [24]. The effects of Basset force and virtual mass on the amplitude of particle oscillations in an oscillating flow were reported in [25].

Force due to Wall Hindrance:

$$F_W = -f(d/D_c) \tag{48}$$

where f denotes a function and D_c is the diameter or characteristic distance between the walls. The particle wall hindrance effect is accounted for in F_D by the factor $1 + \varphi^{1/3}$ [3].

Reaction due to Coupled Acceleration of Displaced Fluid:

$$F_R = \int_{m_f} a_f \, dm_f \tag{49}$$

where a_f is the acceleration of the differential fluid mass dm_f. In polydisperse mixtures, the motion of faster particles is expected to displace not only fluid but also slower particles. Hence, the coupled acceleration of displaced particles must be taken into account. For nonspherical particles, appropriate shape factors should be applied to d, where d is the diameter of an equivalent sphere [26], [27].

Solution of Equation of Motion of Monodisperse Particles in a Constant Force Field. If the driving forces are expressed as in Equation (3) by an equivalent acceleration g' and retarding forces are exclusively hydrodynamic, then Equation (37) takes the following form:

$$V_p \varrho_p (1 + \varrho_f/\varrho_p)(1 - \varphi) dU_\varphi/dt = (\varrho_p - \varrho_f)(1 - \varphi) V_p g' + F_D \tag{50}$$

where $U_d = (1 - \varphi) U_\varphi$ (see Equation 7) and $\varrho_p - \varrho_\varphi = (\varrho_p - \varrho_f)(1 - \varphi)$ if $\beta(t) = 1$. Solution of Equation (50) in the laminar flow regime, where F_D is given by Equation (46) gives

$$U_d(t) = (1 - \varphi) U_\varphi [1 - \exp(-At)] \tag{51}$$

$$A = \frac{18 \mu'_\varphi}{(\varrho_p + \varrho_f) d^2 (1 - \varphi)}, \quad Re < 1$$

where U_φ is given by Equation (1) and μ'_φ is defined in Equation (46).

The distance $S(t)$ traveled at t is then [10]

$$S(t) = U_\varphi (1 - \varphi) \{t - A^{-1}[1 - \exp(-At)]\}, \quad Re < 1 \tag{52}$$

In the general flow regime (i.e., for $Re > 0$), the equation of motion is given by

$$dU_\varphi/dt = \alpha_1 + \alpha_2 (\alpha_3 U_\varphi^{1/2} + \alpha_4)^2 U_\varphi \tag{53}$$

$$\alpha_1 = (\varrho_p - \varrho_f)(\varrho_p + \varrho_f)^{-1} g'$$

$$\alpha_2 = (3/4)(1 + \varphi^{1/3}) [(\varrho_p + \varrho_f) d^2 (1 - \varphi)]^{-1}$$

$$\alpha_3 = 0.63 (d\varrho_f)^{1/2}$$

$$\alpha_4 = 4.80 \mu_\varphi^{1/2}$$

The solution of Equation (53) is given in [10].

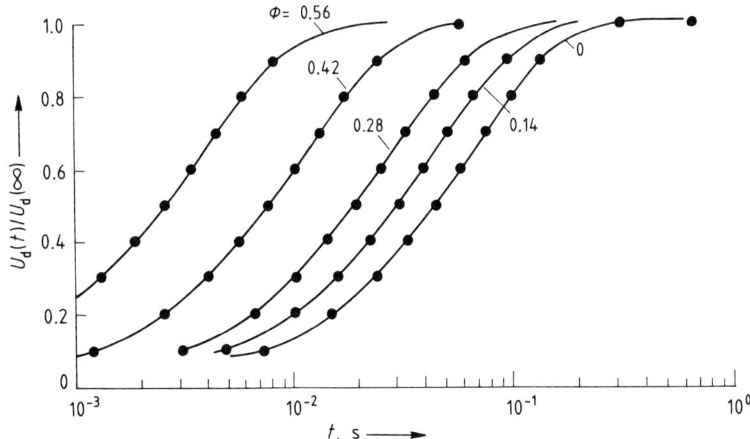

Figure 29. A calculated plot of $U_d(t)/(U_d(\infty)$ vs. t for different values of φ, gravitational settling in air (18 °C), $d = 243$ µm, $\varrho_p = 3000$ kg/m^3

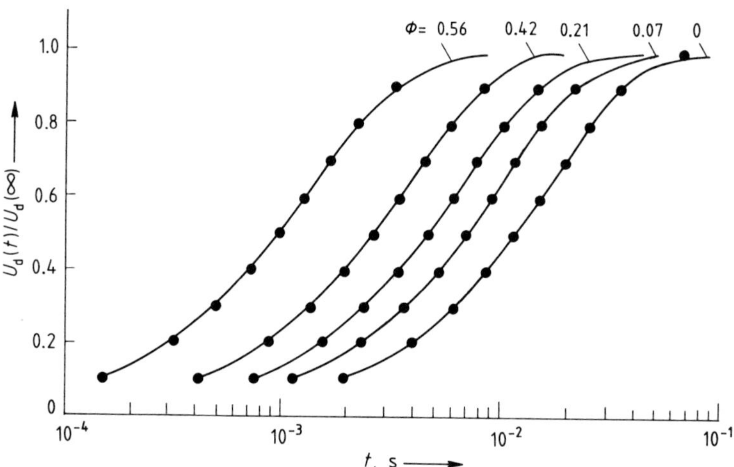

Figure 30. A calculated plot of $U_d(t)/U_d(\infty)$ vs. t for different values of φ, gravitational settling in water (25 °C) $d = 243$ µm, $\varrho_p = 3000$ kg/m^3

Figures 29 and 30 show a plot of $U_d(t)/U_d(\infty)$ vs. t for $d = 243$ µm, $\varrho_p = 3000$ kg/m^3 particles settling in a gravitational field in air (18 °C) and water (25 °C), respectively. Clearly, transient sedimentation can be significant in air for it takes (at $\varphi = 0$) the 243-µm particles nearly 1 s to almost reach their terminal velocity $U_d(\infty)$. During this time, the particles covered approximately 4 m. Increased φ decreases relaxation time of the transient sedimentation. Increasing φ from 0.0 to 0.56, a value in the range of φ used in fluidized beds, decreases relaxation time by an order of magnitude. Changing the fluid from air to water also results in a decrease of relaxation time by more than an order of magnitude. In this example, 0.03 s is required for the 243-µm particles to attain nearly their terminal velocity, i.e., $U_d(t)/U_d(\infty) > 0.99$. The inverse of the slope of the $\varphi = 0$ curve (Fig. 30) at $U_d(t)/U_d(\infty) = 0.5$ is 0.0168 s. If the time Δt required for a driving force to double the velocity of a particle [e.g., from $0.4\,U_d(\infty)$ to $0.8\,U_d(\infty)$] satisfies $\Delta t \gg 0.0168$ s, then neglecting the effect of transient sedimentation is justified, and terminal velocity conditions can be assumed along the sedimentation path. Alternatively, if the time t required for a particle to attain a given velocity U_d in a variable force field satisfies $t \gg t_0$, where t_0 is the time required to attain U_d during transient sedimentation in a constant force field, then the effect of transient sedimentation can be neglected, and the assumption that $\beta(t) = 1$ is justified.

2.3.5. Sedimentation in Turbulent Suspension

Hitherto, the flow regime has been postulated to be solely dependent on the motion of particles relative to the fluid. Thus the nature of flow around the particles determines whether it is laminar or turbulent. However, the suspension may be set in motion by external means such as agitators and pumps, in which case the flow is different from that induced by particle motion. In the extreme, for a completely turbulent suspension, the flow in the vicinity of the particles may be totally independent of their motion. Examples are turbulent agitation so as to maintain a uniform particle suspension, turbulent slurry transportation in pipes, and turbulent pneumatic conveying. In such cases it can be assumed that the fully turbulent drag coefficient applies; hence, the drag force is given by

$$F_D = -\frac{\pi}{8} f_D \varrho_f U_\varphi^2 d^2 (1 + \varphi^{1/3}) \tag{54}$$

where f_D is the drag coefficient, having a value of 0.40 at high values of Re for the particle. If a time-averaged steady state is assumed and $V_p = \pi d^3/6$ Equation (50) gives

$$U_\varphi = \left[\frac{4}{3} \cdot \frac{g'd}{f_D} \cdot \frac{(\varrho_p - \varrho_f)}{\varrho_f} \cdot \frac{(1 - \varphi)}{(1 + \varphi^{1/3})}\right]^{1/2} \tag{55}$$

where turbulence is postulated to eliminate only the effect of viscosity, not the hydrostatic (see

Eq. 41) and the interparticle wall hindrance (see Eq. 48) effects.

Because f_D is taken at its lowest value, the highest values of U_φ are obtained for a given driving force field. In this sense, Equation (55) provides a conservative evaluation of U_φ.

2.4. Sedimentation of Diffusing Particulates

2.4.1. Monodisperse System

The diffusion equation in the dispersion frame of reference can be derived from continuity as follows:

$$\partial C/\partial t + \nabla \cdot J = 0 \quad (56)$$

$$J = J_0 + J_F \quad (57)$$

$$J_0 = -D' \nabla C \quad (58)$$

$$J_F = U_d C \quad (59)$$

where C is concentration, D' is the diffusion coefficient, and J_D and J_F are flux densities due to diffusion and sedimentation drift in the force field, respectively; C is related to φ by

$$\varphi = Cv,\ v = \tilde{N} V_p \quad (60)$$

where N and v are the Avogadro number and the partial molar volume, respectively. Combining Equation (56)–(59) with Equations (7) and (60) gives

$$\partial C/\partial t - \nabla \cdot (D' \nabla C) + \nabla \cdot [U_\varphi (1 - C \tilde{N} V_p) C] = 0 \quad (61)$$

where U_φ is given by Equation (1). Equation (61) can be solved provided that information regarding D' is available. In the following, some properties of D' are outlined which can help to theoretically predict its value in diffusion sedimentation systems. The driving force F_μ per diffusing particle is often derived by using the gradient of the chemical potential μ [28]:

$$F_\mu = -\frac{1}{N} \nabla \mu \quad (62)$$

where μ is the chemical potential of 1 mol of particles. The hindrance force due to diffusion, as to other driving forces, in the laminar flow regime is given by Equation (46). Particles are assumed to be large, compared to fluid molecules, so that Stokes drag law can be applied. The chemical potential may be expressed as [29]

$$\mu = RT \left(1 + \sum_i U'_i / k_B T\right) \ln C + RT \ln f_a \quad (63)$$

where R and T are the gas constant and the absolute temperature, respectively; $k_B = R/\tilde{N}$ is the Boltzmann constant; f_a is the activity coefficient; and U'_i are energies (per particle), such as interfacial and electromagnetic energies, which may be associated with the particles. Combining Equations (62) and (63) gives

$$F_\mu = -k_B T \left(1 + \sum_i U'_i / k_B T\right) \nabla \ln C - k_B T \nabla \ln f_a \quad (64)$$

Hence for steady diffusion where $F_\mu + F_D = 0$ is satisfied U_φ is given by

$$U_\varphi = \frac{-k_B T \left(1 + \sum_i U'_i / k_B T\right) \nabla \ln C - k_B T \nabla \ln f_a}{3 \pi d \mu'_\varphi} \quad (65)$$

Note that the effect of hindrance due to acceleration of fluid displaced by the diffusing particles has been neglected. For steady diffusion and vanishing total flux, by using Equations (61) and (65),

$$D' = D\left(1 + \sum_i U'_i / k_B T + \nabla \ln f_a / \nabla \ln C\right) f_D(C) \quad (66)$$

where D is the well-known [28], [30] Stokes–Einstein diffusion coefficient

$$D = k_B T / (3 \pi \mu_f d) \quad (67)$$

and $f_D(C)$ is the hydrodynamic hindrance factor

$$f_D(C) = \frac{1 - C\tilde{N} V_p}{\left[1 + (C\tilde{N} V_p)^{1/3}\right] \exp\left(\frac{5}{3} \cdot \frac{C\tilde{N} V_p}{1 - C\tilde{N} V_p}\right)} \quad (68)$$

Note that $1/(3 \pi d \mu'_\varphi)$, which appears in Equation (65), is a generalization of the well-known particle mobility, i.e., drift velocity per unit driving force in the fluid frame of reference. In the dispersion frame of reference, this mobility is $(1 - C \tilde{N} V_p)/(3 \pi d \mu'_\varphi)$. In centrifugation technology, particle mobilities are also referred to as sedimentation coefficients, which are given in terms of velocity per unit acceleration g' [31]. In this case, the mobility S_g in the dispersion frame of reference is given by

$$S_g = V_p (\varrho_p - \varrho_\varphi)(1 - C \tilde{N} V_p)/(3 \pi d \mu'_\varphi)$$

In centrifugal fields where the driving force is radial, the diffusion equation takes the following form

$$\partial C/\partial t = D'\left(\frac{\partial^2 C}{\partial r^2} + \frac{1}{r}\cdot\frac{\partial C}{\partial r}\right) - S_g\omega^2\left(r\frac{\partial C}{\partial r} + 2C\right) + \frac{\partial C}{\partial r}\left(\frac{\partial D'}{\partial C}\cdot\frac{\partial C}{\partial r} - C\omega^2 r\frac{\partial S_g}{\partial C}\right) \quad (69)$$

and at steady state,

$$-D'\nabla C + U_\varphi(1 - C\tilde{N}V_p)C = \text{const.} \quad (70)$$

Equation (70) can be solved, given the boundary conditions, for the steady diffusion and sedimentation profile of particle concentration.

2.4.2. Polydisperse System

Let φ' denote the probability density function of a random variable η (e.g., size, density, etc.) that characterizes the particle population:

$$\varphi' = \frac{1}{\varphi}\cdot\frac{\partial\varphi}{\partial\eta} \quad (71)$$

$$\int_{-\infty}^{\infty} \varphi'\,\mathrm{d}\eta = 1 \quad (72)$$

$$\mathrm{d}\varphi = \varphi\varphi'\,\mathrm{d}\eta \quad (73)$$

Let φ' be a function of position r and $D_\varphi(r, \eta)$ be the diffusion coefficient of sedimenting particles in the fluid frame of reference. The expectation $\overline{D_\varphi}(r)$ (i.e., weighted average of the population) of $D_\varphi(r, \eta)$ at position r is given by [32]

$$\overline{D_\varphi}(r) = \int_{-\infty}^{\infty} D_\varphi(r, \eta)\,\varphi'\,\mathrm{d}\eta \quad (74)$$

The diffusion coefficient $D_d(r, \eta)$ of the fraction η at position r in the dispersion frame of reference is given by [29]

$$D_d(r, \eta) = D_\varphi(r, \eta) - \varphi\overline{D_\varphi}(r) \quad (75)$$

Equation (75) provides the transformation from the frame of reference of the fluid (which is displaced by the diffusing particles) to that of the dispersion. Equation (75) shows that at a given position in the diffusion system, the diffusion coefficient of particles pertaining to the η fraction depends on the expected diffusion coefficient of the whole distribution (in the fluid frame of reference) and on φ. Only in the fluid frame of reference can the diffusion of a particular fraction be assumed to be independent of the diffusion of the rest of the distribution. Furthermore, Equation (75) shows that reverse diffusion can occur provided $D_\varphi(r, \eta) - \varphi\overline{D_\varphi}(r) < 0$ is satisfied. Thus, faster diffusing particles can affect slower ones to the extent of reversing their local diffusion coefficient and also their diffusion flux. If reverse diffusion occurs, then the sedimentation and diffusion system will be divided into two counterdiffusing groups. Analysis of such systems involves a model of nonsteady diffusion that can be solved numerically. This, however, is outside the scope of this article.

2.5. Sedimentation of Fluid Particles

Fluid particulates differ from solid ones in view of their deformability and their capacity to maintain internal flow. Properties that may affect sedimentation of fluid particulates are listed below:

1) The stability of the fluid–fluid interface depends on flow conditions; slip velocity can occur.
2) Fluid particulates may be deformed; larger drops are more susceptible to deforming forces.
3) Fluid particulates can undergo coalescence or disintegration.
4) The viscosity of the dispersed fluid phase is finite; no limitation is imposed on its value relative to that of the continuous fluid phase.
5) Interfacial effects may be significant, for example, due to adsorption of surface-active agents or contaminants.

Barnea and Mizrahi extended their generalized approach toward the fluid dynamics of solid particulate systems [3] to dispersed liquids in fluid systems [33], [34]. If the assumptions made regarding the general properties of solid dispersions remain valid also for liquid in fluid dispersions, then the theory available for solids can be used for liquid dispersions once an appropriate modified viscosity is known. This modified viscosity must account for the finite viscosity of the dispersed phase and the hindrance effect of neighboring drops [35]. Equation (76) gives the effective viscosity μ_f' of a continuous fluid surrounding a single spherical drop, i.e., at $\varphi \to 0$ [35], [36]:

$$\mu_f' = \mu_f \frac{2/3 + \mu_d^*/\mu_f}{1 + \mu_d^*/\mu_f} \quad (76)$$

where

$$\mu_d^* = \mu_d + \sigma_k, \quad \sigma_k = -\frac{1}{3} k^* \partial\gamma/\partial\Gamma,$$

μ_d is the viscosity of the dispersed phase, k^* is the surface effect retardation constant, and γ and Γ are the interfacial tension and the adsorption density of the dissolved material, respectively. Equation (76) shows that $\mu_f' \to \mu_f$ if $\mu_d \to \infty$, whereas $\mu_f' \to 2\,\mu_f/3$ for $\mu_d \to 0$. For dilute dispersions, the variation of the apparent viscosity of the continuous fluid $\mu_{\varphi d}$ with φ is given in [37].

$$\mu_{\varphi d} = \mu_f (1 + 2.5\,\varphi A_d),$$
$$A_d = (\mu_f + 2.5\,\mu_d)/(2.5\,\mu_f + 2.5\,\mu_d) \quad (77)$$

Equation (78) is an extension of Equation (77) due to LEVITON and LEIGHTON [38]:

$$\mu_{\varphi d} = \mu_f \exp[2.5\,A_d(\varphi + \varphi^{5/3} + \varphi^{11/3})] \quad (78)$$

Further extension of $\mu_{\varphi d}$ can be obtained by use of the generalized drag coefficient $C_{D\varphi}$ and Reynolds number Re_φ for monodisperse dispersions [3], [33], [34] (see Section 2.6). The drag coefficient $C_{D\varphi}$, which depends on hydrostatic effects and on the distribution of particles in the dispersion, is unaffected by changing the solid into fluid particulates. The finite internal viscosity of each particle and the effect of this finite viscosity on hydrodynamic interactions between neighboring particles (Taylor correction) affect Re_φ. The parameter $\mu_{\varphi d}$, given by Equation (79), accounts for these effects:

$$\mu_{\varphi d} = \mu_f B_d B_v,\ B_d = \exp\left(\frac{5\,\varphi A_d}{3(1-\varphi)}\right),$$
$$B_v = \frac{\frac{3}{2} B_d + \mu_d^*/\mu_f}{B_d + \mu_d^*/\mu_f} \quad (79)$$

where the Taylor correction is accounted for by B_d and the effect of finite viscosity of the dispersed phase on that of the continuous phase by B_v, in accordance with Equation (76). Substitution of $\mu_{\varphi d}$ for μ_φ in Equations (1) and (2) facilitates evaluation of steady sedimentation velocities of fluid particulates that are dispersed in a continuous fluid.

2.6. Generalized Correlations for Hindered Sedimentation Fluidization and Flow through Fixed Beds

Hindered sedimentation fluidization and flow through fixed beds pertain to the class of physical phenomena in which fluid is forced to flow between particles. This flow can be modeled as being inside a bundle of pipes that are formed between particles (when they are part of a fixed bed) or around individual particles with no reference to specific path of flow. FOSCOLO et al. [39] proposed a model for a fluidized bed that "enables its steady state particulate expansion to be predicted as a function of superficial velocity (U_c) from initial (packed bed) conditions to the final fully expanded (single suspended particle) state." Their results, which can also be applied to hindered sedimentation, are given by Equation (80) for laminar and turbulent flow, and Equation (81) for intermediate flow:

$$U_c/U_o = \left[\frac{\varepsilon^4}{\alpha(1-\varepsilon)+\varepsilon^3}\right]^m, \quad \varepsilon = 1-\varphi \quad (80)$$

For laminar flow, $Re < 0.2$, $\alpha = 4.00$, and $m = 1.0$. For turbulent flow, $Re > 500$, $\alpha = 3.55$, and $m = 0.5$.

$$U_c/U_o = \frac{[0.0777\,Re(1 - 0.0194\,Re)\varepsilon^{4.8} + 1]^{0.5}}{0.0388\,Re}, \quad (81)$$

$$0.2 \le Re \le 500$$

where U_c is superficial velocity, related to U_φ by

$$U_c = U_\varphi \varepsilon = U_d,\ U_o = U_\varphi(\varphi = 0), \quad \text{and} \quad Re = \frac{\varrho_f U_o d}{\mu_f}$$

Equations (80) and (81) were derived by using an extended tortuosity function and the model of flow through a bundle of pipes in accordance with the Blake–Kozeny flow model. Because $\varepsilon = 1 - \varphi$, these equations provide another set of generalized $f(\varphi)$ functions such as that given by Equation (1). Pressure drop per unit length is given by

$$\Delta P_l/L = 17.3(\mu_f U_c/d^2)(1-\varepsilon)\varepsilon^{-4.8}, \quad Re < 0.2 \quad (82)$$
$$\Delta P_t/L = 0.336(\varrho_f U_c^2/d)(1-\varepsilon)\varepsilon^{-4.8}, \quad Re > 500 \quad (83)$$
$$\Delta P_i/L = (\Delta P_l + \Delta P_t)/L, \quad 0.2 \le Re \le 500 \quad (84)$$

The model of flow around individual particles was used to derive a generalized correlation for hindered sedimentation fluidization and flow through fixed beds [40]. Figure 31 shows the drag coefficient $C_{D\varphi}$ vs. the Reynolds number Re_φ [41]. Equations (85) and (86) provide definitions of $C_{D\varphi}$ and Re_φ:

$$Re_\varphi = U_\varphi \varrho_f d/\mu_\varphi, \quad \mu_\varphi = \mu_f \exp\left(\frac{5}{3}\cdot\frac{\varphi}{1-\varphi}\right) \quad (85)$$

$$C_{D\varphi} = 2F_{D\varphi}/[\varrho_f S U_\varphi^2(1 + \beta_\varphi \varphi^{1/3})], \quad S = \pi d^2/4 \quad (86)$$

The subscript φ denotes that the variable is given for a dispersion at φ. For fixed beds $\beta_\varphi = 2.6$, whereas for hindered sedimentation $\beta_\varphi = 1$. To evaluate U_φ and $F_{D\varphi}$, a correlation must be made between $C_{D\varphi}$ and Re_φ, and β_φ must be specified. To this end, the Dallavalle correlation [40–42] can be used:

$$C_{D\varphi} = (0.63 + 4.8\,Re_\varphi^{-1/2})^2 \quad (87)$$

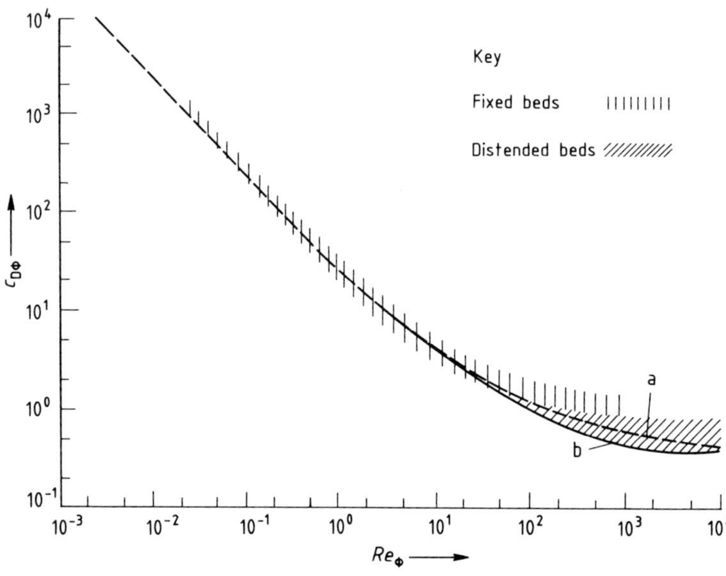

Figure 31. A plot of $C_{D\varphi}$ vs. Re_φ for calculated and experimental data (compiled from [41])
a) Equation 87; b) Lapple–Shepherd curve for single sphere

A plot of Equation (87) is shown in Figure 31. Equation (87) agrees with experimental data for fixed beds in the laminar flow regime, but it involves a substantial discrepancy in the turbulent $Re > 500$ regime. For hindered sedimentation, Equation (87) also fits experimental data in the turbulent regime and for the full range of practical values of φ. Equation (88) is an extension of Equation (87) [40]:

$$C_{D\varphi} = (0.63 + 4.8\, Re^{-1/2} + \alpha_\varphi \tau_\varphi\, Re_\varphi^{1/8})^2 \quad (88)$$

$$\alpha_\varphi = (\varphi/0.6)\,[1/8 + \sum_{i=2}^{n} a_i (\varphi/0.6)^i], \quad i = 2, 3, \ldots, n \quad (89)$$

where α_i represents constants by which $\alpha_\varphi = \alpha(\varphi)$ can be determined and τ_φ relates to the degree of constraint of particle motion. For particles that are fully constrained, i.e., fixed in space as in fixed beds, $\tau_\varphi = 1$; for hindered sedimentation where no constraints other than hydrodynamic drag exist, $\tau_\varphi = 0$. If partial fixation of particles prevails, then $0 \le \tau_\varphi \le 1$. Using only the first term of α_φ provides (Eq. 90) a good first-order fit between Equation (88) and experimental data (Fig. 31) for both fixed and hindered sedimentation curves:

$$C_{D\varphi} = (0.63 + 4.8\, Re_\varphi^{-1/2} + \tau_\varphi (\varphi/4.8)\, Re_\varphi^{1/8})^2 \quad (90)$$

At $\varphi = 0$ Equation (90) reduces to Equation (87), and at $\varphi = 0.6$ corresponds well with literature data for fixed beds $\tau_\varphi = 1$ and for hindered sedimentation $\tau_\varphi = 0$. Equation (91) gives $\beta_\varphi = \beta(\varphi)$ in the form of an infinite series:

$$\beta_\varphi = 1 + \tau_\varphi (\varphi/0.6)\,[1.6 + \sum_{j=2}^{m} b_j (\varphi/0.6)^j], \quad j = 2, 3, \ldots, m \quad (91)$$

The first constant of β_φ satisfies $\beta_\varphi = 2.6$ at $\varphi = 0.6$, $\tau_\varphi = 1$, $b_j = 0$ for $j > 2$, as required. Taking only the first term as a first-order approximation yields

$$\beta_\varphi = 1 + 8\, \tau_\varphi\, \varphi/3 \quad (92)$$

and the corresponding extended drag coefficient is then

$$C_{D\varphi} = C_{D0}/[1 + \varphi^{1/3} + (8\,\tau_\varphi/3)\, \varphi^{4/3}] \quad (93)$$

where $C_{D0} = C_{D\varphi}(\varphi = 0)$ (see Equation 86).

Note that in the laminar flow regime Equation (94) is preferable:

$$C_{D\varphi} = 24/Re_\varphi \quad (94)$$

The pressure drop per unit length across the particle assembly (e.g., as a fixed bed or as a suspension) is the sum of individual (i.e., per particle) drag forces:

$$\Delta P/L = N F_{D\varphi}, \quad N = \varphi/\left(\frac{\pi d^3}{6}\right) \quad (95)$$

Equations (96) and (97) give the results for the laminar and intermediate or turbulent flow regimes, respectively:

$$\Delta P/L = \frac{18\, \mu_f U_c\, \varphi \left[(1 + \beta_\varphi \varphi^{1/3}) \exp\left(\frac{5}{3} \cdot \frac{\varphi}{1 - \varphi}\right)\right]}{d^2 (1 - \varphi)} \quad (96)$$

$$\Delta P/L = \frac{(\frac{3}{4})(0.63 + 4.8\, Re_\varphi^{-1.2} + \tau_\varphi \varphi\, Re_\varphi^{1/8}/4.8)^2 \varrho_f U_c^2\, \varphi (1 + \beta_\varphi \varphi^{1/3})}{d (1 - \varphi)^2} \quad (97)$$

Note that Equation (97) also provides a good approximation in the laminar flow regime; hence, if Re_φ is not known, it can be used as a first trial. In cases involving polydisperse mixtures and consisting of i_n size fractions, Equation (95) should be replaced by Equation (98), with due changes in Equations (96) and (97).

$$\Delta P/L = \sum_{i=1}^{i_n} n_i F_{D\varphi i}, \quad n_i = \varphi_i/\left(\frac{\pi d_i^3}{6}\right), \quad N = \sum_{i=1}^{i_n} n_i \quad (98)$$

Extending the above model, in which the contribution to pressure drop due to the drag exerted on individual particles has been taken into account, to polydisperse mixtures is straightforward. This applies to fixed, distended, and stable fluidized beds, and to hindered sedimentation in the full range of flow regimes and volume fractions occupied by the particles. Moreover this generalized model does not involve any assumptions as to the appropriate averaging of particle size, which is necessary when the model of flow through a bundle of pipes is used.

2.7. Enhancement of Sedimentation

Sedimentation of small and near-density particles can be very slow; hence, a way is needed to enhance the throughput of devices involving sedimentation of such particles. *Near density* means that the density of particles is close to that of the fluid. The following options for enhancement of sedimentation rate should be considered:

1) size enlargement by processes such as flocculation (→ Flocculants; **A11**, p. 251), coagulation, precipitation, and crystallization (→ Crystallization);
2) increase of driving forces, i.e., of the effective acceleration g'; and
3) decrease of sedimentation retention time by decreasing the sedimentation path and increasing the area available for sedimentation and for deposition of sediment.

Size enlargement is a well-known method in solid–liquid separation and water treatment operations [17], [42], [43]. Size enlargement (or the reverse, i.e., dispersion) can occur spontaneously through a change in the properties of the suspension (i.e., pH and ionic strength) and through addition of flocculants (or dispersants) that bind particles (or disintegrate aggregates) into floccules.

Increase of driving force can be achieved by applying force fields (of higher intensity) to which the particles are or can be made susceptible. *Centrifugal fields* are often employed when separation of slowly settling particles by density or size differences is desirable. *Sedimentation of polarizable particles* toward polarized elements occurs in magnetic (→ Magnetic Separation) and electrical separation (→ Electrostatic Separation) and in filtration (→ Filtration). *Attachment of polarizable particles to nonpolarizable ones* renders the aggregate susceptible to sedimentation in the relevant polarizing field.

A *decrease of sedimentation retention time* can be achieved by increasing the volumetric density of collecting elements toward which the particles undergo sedimentation. This involves a corresponding decrease in the working volume available for holding and flow of the dispersion. Enhanced sedimentation in inclined channels is a good example of this approach (see Section 2.8). Enhanced sedimentation occurs also in filtration processes where particles are partially or fully carried by the filtering fluid toward the filtration cake on which they are deposited. In this case, drag forces that drive particles can reinforce or counteract other sedimentation forces, depending on the geometry of the system. In electromagnetic separation and filtration (→ 19. Magnetic Separation, p. **19**-6) of polarizable particles, a matrix of highly polarizable elements is distributed in the separation chamber, which is placed in a high-intensity field, thus decreasing the sedimentation path of particles and increasing the area available for their collection [44].

2.8. Sedimentation in Inclined Channels

Enhanced sedimentation of blood corpuscles in inclined tubes was observed by Boycott in 1920; since then, this phenomenon of enhanced sedimentation in inclined channels has been a subject of long-standing research [45]. Reviews are given in [3] and [46]. Enhanced sedimentation in inclined channels is used commercially in *lamella thickeners* [17], [43], *sedimentation centrifuges* [47], [48], and *sedimentation cones* [17] (→ Wastewater Treatment).

Figure 32 shows five regions characterizing the flow field in an inclined channel sedimentation system [2]. Particles reach the final sediment by first forming a deposit on the inclined channel floor, which slides at a faster rate toward the channel bottom. In this way, the higher retention time associated with sedimentation in suspension is cut down in favor of the lower retention time associated with the sliding deposit. The inclined channel system incorporates the principle of increased volume density of collection elements for the sedimenting particles or else increased surface area for their deposition. The displaced fluid flows toward the top of the channel. A thin clear layer is then formed above the inclined sedimentation front, and a thicker clear layer develops above the shorter horizontal sedimentation front at the top (see Fig. 32). Ponder [49] and Nakamura and Kuroda [50] proposed the so-called PNK theory for the clarification rate in inclined

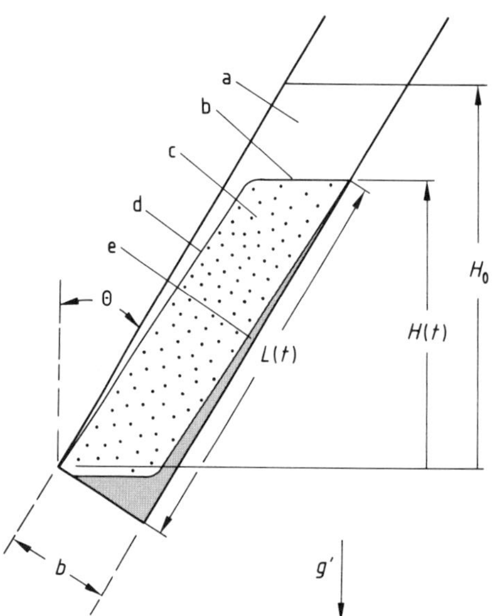

Figure 32. Formation of flow regions during sedimentation of monodisperse particles in an inclined channel [2]
a) Clear fluid; b) Horizontal interface between fluid and suspension; c) Suspension; d) Clear fluid above inclined sedimentation front; e) Sediment

channels. This theory states that the rate of production of clarified fluid is given by the product of the settling velocity of particles and the projection of the channel area available for particle deposition on a plane set perpendicular to the settling direction. For the parallel plate geometry shown in Figure 32, this theory predicts

$$g(t) = U_d(b \cos \theta + L \sin \Theta) \tag{99}$$

where $g(t)$ is the volume rate of clarified fluid per channel per unit length L_p perpendicular to the plane of the drawing, U_d is given by Equation (7) for monodisperse particles, θ is the angle of inclination from the vertical axis, and b and L are the dimensions of the shorter and the larger lower walls of the channels, respectively (see Fig. 32). Let the volume per unit length L_p of the sedimentation vessel be V_v, and assume that it is large enough to contain a number of channels for which Equation (99) applies. From this, the steady-state rate of clarification for such volume (fully packed with inclined channels) is

$$Q(t) = V_v U_d(\cos \theta / L + \sin \theta / b) \tag{100}$$

Equation (100) shows that $Q(t)$ is inversely related to L as well as to b. Hence, given L, the use of

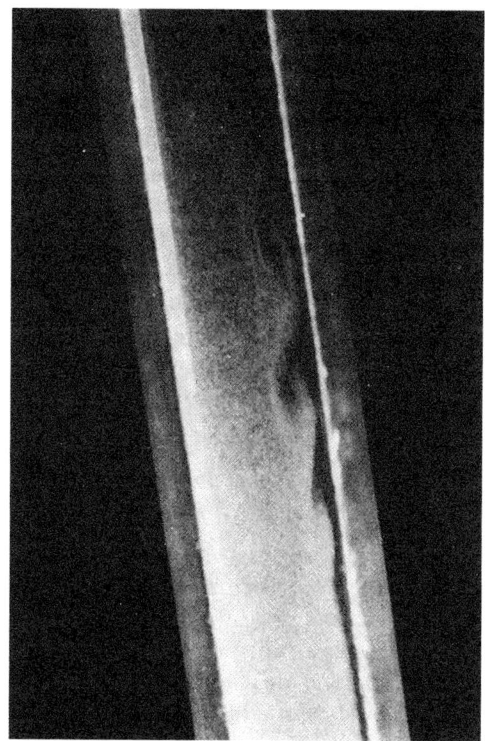

Figure 33. Formation of unstable waves during sedimentation in an inclined channel [2]

narrower channels, with due regard to design constraints, facilitates a corresponding increase in $Q(t)$. For analysis of sedimentation in inclined channels, see [51]–[56]. Equations (99) and (100) are valid for laminar flow. If instabilities followed by deformation of sedimentation fronts occur, then the efficiency of the process can be curtailed significantly. Figure 33 shows an example of the formation of unstable waves in the sedimentation fronts during sedimentation of 130-μm glass beads, $\varrho_p = 2440$ kg/m^3, which were suspended in a synthetic Newtonian lubricant, $\varrho_f = 1050$ kg/m^3, $\mu_f = 0.01$ kg m^{-1} s^{-1}. For further information regarding linear stability analysis of sedimentation in inclined channels, see [2] and [57]–[59]. The sedimentation of polydisperse suspensions in vessels having inclined walls is considered in [60]. Figure 34 shows the different sedimentation regions that develop during sedimentation of a polydisperse particle mixture containing N discrete fractions. Equation (101) gives the dimensionless instantaneous volumetric rate $S_j^*(t)$ at which a suspension devoid of particles faster than or equal to the jth fraction is

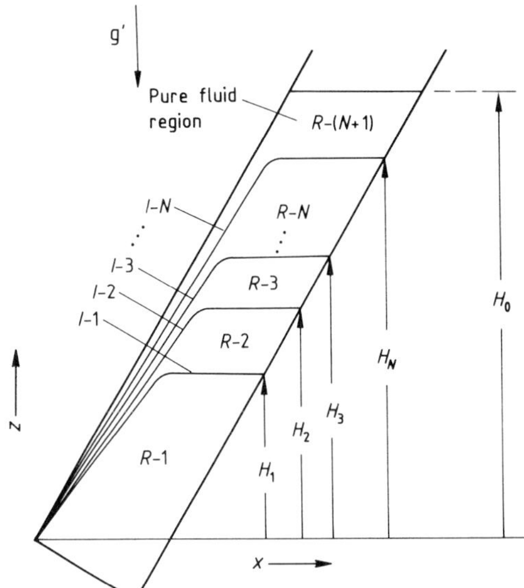

Figure 34. Formation of $N+1$ flow regions during sedimentation of polydisperse particles (comprising N fractions) in an inclined channel [60]. Region j, $1 \leq j \leq N$ contains particles pertaining to fraction i, $i+1, \ldots, N$, $i=j$. Fraction $j = N+1$ denotes clear fluid. Larger i denotes slower settling particles

formed [60]:

$$S_j^*(t) = U_{dj}^* \left[1 + \frac{H_j^*}{b^*} \sin\theta \right] \frac{b^*}{\cos\theta} \qquad (101)$$

where all starred quantities have been rendered dimensionless by using the initial height of the suspension H_0 as the characteristic length and the velocity (at $\varphi = 0$) of the fastest settling particle as the characteristic velocity. For steady-state continuous settling, letting $j = N$ in Equation (101) gives the dimensionless rate of formation of clear fluid. This can be used for design purposes or for evaluation of the maximum capacities that can be expected from existing equipment. For batch sedimentation [60],

$$\mathrm{d}H_j/\mathrm{d}t = -U_{dj} \left[1 + \frac{H_j}{b} \right] \sin\theta \qquad (102)$$

where at $t = 0$, $H_j = 1$; $j = 1, 2, \ldots, N$. The rate of descent of the region containing all particles ($j = 1$) is fastest, with the one corresponding to $j = N$ being the slowest. Equations (101) and (102) can also be applied to continuous distributions by letting $N \to \infty$.

Sedimentation phenomena of polydisperse mixtures in inclined channels are generally considered by assuming low volume fractions of solids. Furthermore, the analysis is concerned mainly with discontinuities between regions and not with processes occurring inside these sedimentation regions. As with nonsteady sedimentation, the formation of concentration waves can also be expected in sedimentation in inclined channels. Such waves can provide further hindrance to sedimentation and cause instabilities (see Section 2.3.3).

2.9. Principle of Hydrodynamic Particle Competition

The principle of hydrodynamic particle competition (HPC) states that hydrodynamic hindrance to motion of particles in the direction of a driving force is enhanced due to the competition of neighboring particles for available flow space. The probability of particles, entering a flow space is determined not only by their own response to the driving force but also by their relative response compared to other particles. Thus according to this principle, less responsive particles may be driven out of a flow zone by more responsive ones, notwithstanding their being driven by a positive force toward this zone. The term "response" is used in the most general way; for example, larger particles are more responsive to body forces. In the analysis of particle motion (e.g., in separation processes), assessing whether a certain part of a distribution can reach a given position in space in a given time may be of interest. In this context, the expected effect of changing operating parameters on the probability of separation must be determined. This probability is affected by "side variables," which do not constitute the basis for separation but inherently enter the force terms. The relative HPC effect of such variables is expected to be more pronounced due to the following:

1) increase of φ;
2) shift of size distribution toward the finer size end;
3) decrease of difference between particles and fluid with respect to the variable that constitutes the basis for particle motion;
4) increase in standard deviation of variables that constitute the basis for particle motion, of side variables, and of particle size; and
5) increased skewness of distribution variables, specifically those directly affecting particle motion.

3. Processes and Equipment Involving Sedimentation

3.1. Settling Pools

The settling pool can be used as an idealized model for crossflow sedimentation (see Section 2.2). Figure 35 shows an ideal settling pool in which plug flow is assumed to prevail. The time t_H required for a particle introduced at the feed port to reach the pool floor is related to the pool height H_p by

$$H_p = \int_0^{t_H} U_d \mathrm{d}t = \int_0^{t_H} (U_\varphi - U_f) \mathrm{d}t \tag{103}$$

If φ is low throughout the sedimentation path, then $U_d = U_\varphi =$ const. can be assumed; hence

$$t_H = H_p / U_d \tag{104}$$

For particles that travel simultaneously the full length L and depth H_p of the pool, $t_H = t_R$, with the residence time t_R related to the crossflow velocity U_R by

$$t_R = L / U_R \tag{105}$$

$$U_R = Q / (B_p H_p) \tag{106}$$

where Q is the volumetric flow rate across the pool and B_p is the pool width. Thus the limiting particle size reaching the pool floor within the distance L can be determined from

$$U_\varphi = \frac{Q}{A_s}, \quad A_s = BL \tag{107}$$

where A_s is the pool area across which sedimentation occurs. Equation (107) shows that in ideal pools, the ability to capture a particle is independent of pool depth. This property is also known as the area principle. Real pools and systems that are simulated by sedimentation pools deviate from ideality due to the effects of increased particle concentration, particle distribution, flow profiles other than plug flow, and inhomogeneous force fields. The actual surface area required for a given limiting particle size is then

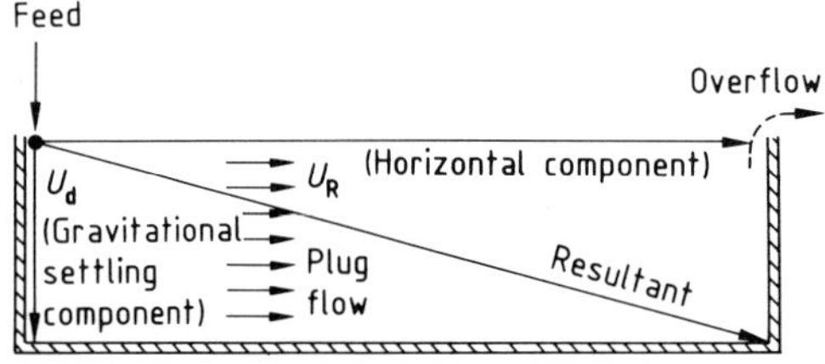

Figure 35. Plug flow and sedimentation of particles in an ideal settling pool [17]

$$A_s = \frac{Q}{\eta_p U_\varphi} \tag{108}$$

where η_p denotes the efficiency, i.e., the ratio of theoretical to required surface area. The customary assumption is that $0.5 \leq \eta_p \leq 1.0$, but lower and in theory even vanishing values of η_p can be expected in pools and crossflow systems that involve sedimentation of polydisperse mixtures at elevated concentration. For fixed values of A_s and η_p, Q/U_φ is constant. Hence, increasing the concentration of the suspension, when the limiting size d_L is fixed, decreases U_φ and also Q. All particles having size larger than d_L are expected to settle in the pool. Thus the captured fraction is given by

$$\Phi(d \geq d_L) = 1 - \Phi(d \leq d_L) = \int_0^{d_L} \varphi' \mathrm{d}d_c,$$

and the total mass of solids capture is then

$$\begin{aligned} &\varrho_p \varphi [1 - \Phi(d_L)] Q / A_s \\ &= \eta_p \varrho_p \varphi [1 - \Phi(d_L)] U_d(d_L) \\ &= \eta_p \varrho_p [1 - \Phi(d_L)] \varphi (1 - \varphi) U_\varphi(d_L) \end{aligned}$$

For a given d_L, the maximum mass that can be captured is obtained by maximizing $\varphi(1 - \varphi) U_\varphi(d_L)$, which gives $\varphi_{max} = 0.1792$ in the laminar regime. However, if d_L is not fixed, then $\Phi(d_L) \varphi (1 - \varphi) U_\varphi(d_L)$ must be maximized instead.

3.2. Suspension of Particles in Fluids

3.2.1. Suspension of Particles in Mixed Containers

Particles must often be suspended in fluids in which sedimentation must be prevented at various degrees. Uniform suspensions of particles in fluid streams can be achieved by *fluidization*, *intense agitation*, and *turbulent flow*. If particles are suspended by turbulent agitation, then the process can be expressed in terms of the *eddy diffusion coefficient* D_e:

$$D_e \cdot \frac{\mathrm{d}C}{\mathrm{d}y} = -C U_y \tag{109}$$

where C is particle concentration at height y and $U_y = U_d$. Equation (109) is incomplete in the sense that it relates to a mass-transfer process that is restricted exclusively to the direction of sedimentation. The following equation also accounts for mass transfer across the direction of sedimentation:

$$\frac{\partial}{\partial Y}\left(D_e \cdot \frac{\partial C}{\partial Y}\right) + D_e \cdot \frac{\partial^2 C}{\partial x} = U_x \cdot \frac{\partial C}{\partial X} - U_y \cdot \frac{\partial C}{\partial Y} \quad (110)$$

where U_x is the time-averaged component of particles velocity in a direction perpendicular to U_y (i.e., horizontal if U_y is vertical). Increased intensity of agitation decreases concentration gradients. The power required for complete dispersion in a tank is given as follows [17], [61]:

$$P/V_T = 0.092\, g U_\varphi \frac{D_T}{D_a}\left(\frac{\varphi}{1-\varphi}\right)^{1/2} \cdot \exp\left(5.3 \cdot \frac{H_a}{D_T}\right)(\varrho_p - \varrho_f) \quad (111)$$

where P, V_T, D_a, D_T, and H_a are the power, volume of slurry in the tank, agitator diameter, tank diameter, and distance of agitator impeller from the bottom of the tank, respectively. For turbulent agitation, U_φ is given by Equation (55). Equation (111) shows that the power density is directly related to the sedimentation velocity at $\varphi \to 0$ and to $[\varphi/(1-\varphi)]^{1/2}$. This equation provides conservative estimates of power consumption because in many cases, a completely uniform dispersion is not required. A less conservative estimation of power requirements is given by the *scale of agitation* approach [62], which classifies the degree of suspension into ten levels. In level 1, particles are partially suspended, being in contact with the bottom of the agitator; in level 3, all particles are suspended; and in level 10, uniform suspension prevails. In level 5, the lower part of the agitated volume (i.e., closer to the impeller) is practically uniform, whereas the upper part contains concentration gradients. Figure 36 shows the observed levels of agitation vs. K_a, where K_a (Eq. 112) relates to design parameters of the tank and properties of the slurry:

$$K_a = 154\, K_{ac}\, \omega^{3.75} D_a^{2.81} / U_0 \quad (112)$$

where ω is expressed in revolutions per minute and K_{ac} is the agitation concentration factor that is plotted vs. concentration in Figure 37. Figure 38 shows a plot of $P_0/(Fr)^n$ vs. Re_a for six-blade turbine generator data [63], [64], where Re_a is the Reynolds number of the agitator,

$$Re_a = D_a^2\, \omega \varrho_f / \mu_f \quad (113)$$

P_0 is the power number,

$$P_0 = \frac{P}{\varrho_f\, \omega^3 D_a^5} \quad (114)$$

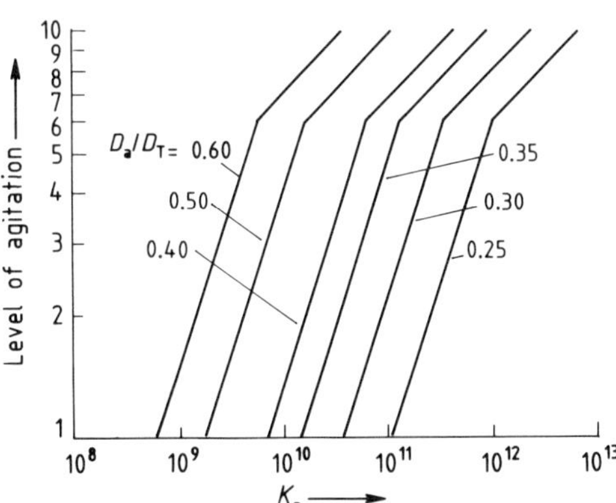

Figure 36. A plot of level agitation vs. K_a for different values of D_a/D_T [17] after [62]

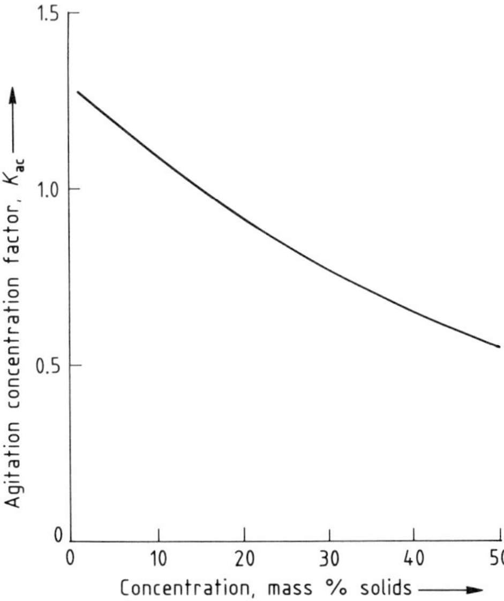

Figure 37. A plot of agitation concentration factor K_{ac} vs. concentration [17] after [62]

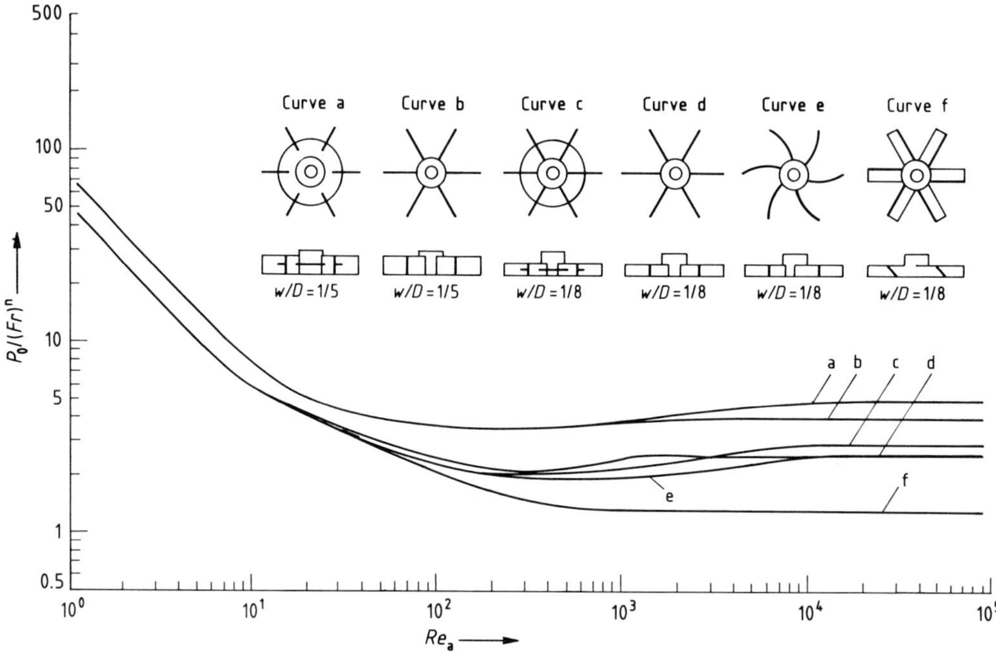

Figure 38. Turbine power correlation [63], [64]: $P_0/(Fr)^n$ vs. Re_a

and Fr is Froude number, i.e., the ratio of inertial to gravitational force:

$$Fr = \omega^2 D_a/g \quad (115)$$

In baffled tanks, $Fr = 1.0$; otherwise, an appropriate value of n must be selected [17].

3.2.2. Suspension of Particles in Conduits

Horizontal Open Channels. Suspension and sedimentation of particles in conduits occur in flows of slurries and dispersions in channels or pipes [65]. As with agitators and mixers, suspension of particles against their sedimentation is achieved by turbulent flow. Equation (109) was used by O'Brien [66] in the early 1930s to describe the mechanism of particle suspension in turbulent transportation of particles in conduits. The eddy diffusion coefficient D_e given in Equation (109) is also a mass-transfer coefficient. This mass-transfer coefficient D_s can be related to the eddy viscosity or fluid momentum transfer coefficient D_m as follows:

$$D_s = \beta_s D_m = \beta_s k_f U^* y(1 - y/y_m) \quad (116)$$

$$U^* = \sqrt{\tau_w/\varrho_f} = U/\sqrt{f/2} \quad (117)$$

$$k_f = l_m/y \quad (118)$$

where β_s is a constant believed by some researchers to be unity; l_m is the eddy mixing length, which has the same role as the mean free path in the kinetic theory of gases; y and y_m are distances from the conduit bottom wall to a given point and to the point of maximum flow velocity, respectively; U^* is the frictional velocity defined by Prandtl; τ_w is shear stress at the wall; U is mean flow velocity in the conduit; f is the friction factor of the flow system; and k_f is the Karman constant rated at 0.40 for pure fluids but lower for slurries due to suppression of the intensity of turbulence [67]. Substituting Equation (116) in Equation (109) and integrating give

$$\ln(C/C_a) = \frac{U_d}{\beta_s k_f U^*} \ln(y'/y'_a) \quad (119)$$

$$y' = y_m/y - 1, \quad y'_a = y_m/y_a - 1$$

where C and C_a are particle concentrations at a level y and at a reference level y_a above the conduit bottom, respectively.

Equation (119) shows that C/C_a depends on y and on the ratio of the sedimentation velocity at y to the group $\beta_s k_f U^*$, which is a measure of the intensity of turbulence. Good agreement between experimental data for open channels and Equation (119) has been achieved [67].

Closed Channels and Pipes. A correlation for slurry flow in pipes was developed by an engineering team of the Consolidated Coal Company in the United States [68]. For the design of an Ohio coal pipeline, they collected data from a 30.5-cm pipeline and arrived at the following correlation:

$$\ln(C/C_a) = -4.145\, U_d/(\beta_s k_f U^*) \qquad (120)$$

where C_a is the concentration at the pipe axis. A plot of calculated vs. observed values of C/C_a for two (1.19–0.59 mm, i.e., 14/28 mesh; 0.595–0.297 mm, i.e., 28/48 mesh) size fractions at two positions $L_p = 0$ and $L_p = 160$ km (L_p is distance along the pipe) is shown in Figure 39 [65], [69]. The reliability of Equation (120), within a 95% confidence level, in predicting the distribution of relative concentration across the flow profile is demonstrated clearly. Often the definition of the state of flow depends on setting accepted values for C/C_a where, for example, C is taken at the topmost part of the pipe. Thus letting $C/C_a \geq 0.8$ across the full flow profile can be considered a definition of homogeneous flow in which vertical concentration gradients can be neglected. Because C/C_a is a power function of U_d/U^*, the effect of sedimentation velocity and hence also of particle size can be critical. This is seen in Figure 40 [65], [70], where C/C_a vs. particle size d is plotted for a 30.5-cm pipe carrying a 50 wt% coal slurry at a flow velocity $U = 1.82$ m/s (C is concentration at the top of the pipe). By using the $C/C_a \geq 0.8$ criterion, the slurry can be considered homogeneously flowing if $d < 700$ µm. Increasing d by an order of magnitude under the same operating conditions results in practically complete sedimentation. Flow velocities and concentration of solids have a strong effect on C/C_a, whereas variation of pipe diameter has little effect [70]. Increasing flow velocities and using higher slurry concentration provide enhanced conditions for operating at larger C/C_A. This is due to a decrease in U_d (following the increase in C) and an increase in $\beta_s k_f U^*$ (following the increase in flow velocities).

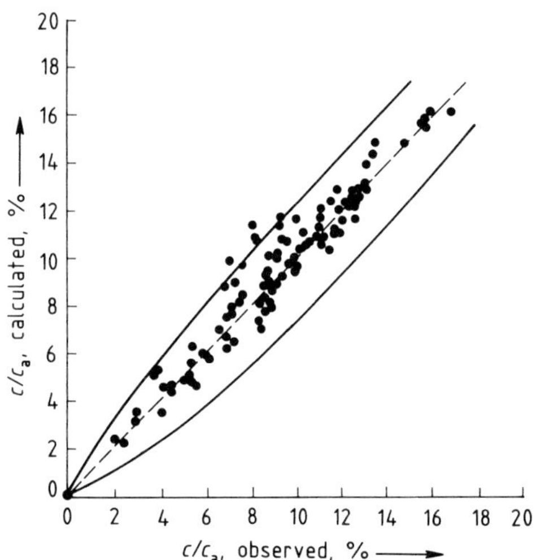

Figure 39. A plot of calculated vs. observed values of C/C_a, for 1.19–0.59 mm and 0.595–0.297 mm size fractions of coal [65], [69]
——— = 95% confidence limit for probe size analyses

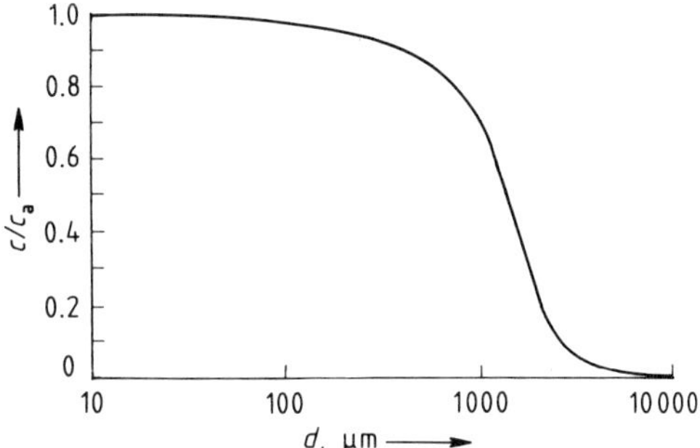

Figure 40. A plot of C/C_a vs. particle size d, 30.5 cm pipe inner diameter, flow velocity 1.8 m/s, 50 wt% coal

3.3. Fluidization and Sedimentation

In *fluidization*, particulates are suspended in a fluid stream so that their time-averaged velocity in the dispersion frame of reference vanishes. However, in the fluid frame of reference, particles undergo a time-averaged steady sedimentation. In stable fluidization, no displaced fluid exists because the fluid provides the driving force for suspension of the particles, whereas in sedimentation (which takes place in a container at fixed volume), fluid is displaced due to particle motion. Thus in fluidization of monodisperse particles, $U_d = 0$, $U_\varphi = U_f$, and the right-hand side of Equation (8) does not hold. Therefore, in cases of particulate fluidization Equations (1) and (2) can be used to evaluate gas velocity directly or, alternatively, to evaluate sedimentation velocity in the fluid frame of reference. Note that particulate fluidization means stable fluidization of randomly and uniformly dispersed particles. However, if particles attain ordered structures, then their sedimentation velocities can be expected to deviate significantly from those predicted for random structures. Thus, different structures formed by particles in sedimentation and fluidization systems, at the same levels of φ, are expected to yield different sedimentation velocities. In lateral segregation in sedimentation of bidisperse suspensions, formation of particle streams in a suspension containing a mixture of dense and neutrally buoyant particles has been observed [71]. A remarkably enhanced sedimentation rate was also noted for the denser particles. Similar phenomena have been observed in a bidisperse mixture containing particles more and less dense than the fluid [72]; the fingering effect also occurs in the sedimentation of bidisperse mixtures containing particles that are all more dense than the fluid [73]. The settling rate of glass spheres in carbon tetrachloride can be increased up to sixfold by adding poly(vinyl chloride) (PVC) particles to the dispersion [74].

Notwithstanding the fact that these phenomena are of great interest from the point of view of sedimentation and fluid mechanics, they seem to have relatively little importance in enhancing overall sedimentation rate [2]. The effect of structuring the particle assembly on sedimentation and fluidization velocities, has been demonstrated in a magnetically stabilized fluidized-bed system [75]. Figure 41 depicts schematically an ordinary (Fig. 41 A) and a magnetically stabilized fluidized bed (Fig. 41 B) [76]. The bed contains magnetizable particles, and the field is generated by a coil wound around the fluidization vessel. Low-intensity uniform magnetic fields are sufficient to stabilize a bed of ferromagnetic particles such as magnetite (Fe_3O_4) [76]. In the absence of a magnetic field or for sufficiently high fluid velocities, the bed becomes unstable and functions as an ordinary unstable bed. Figure 42 shows different structures that can be formed in the fixed-, fluidized-, and intermediate-bed states [75]. The structures are characterized by the degree of alignment of particles with the field H. Figure 43 shows a typical ΔP vs. U_c plot of fluidization and defluidization curves at different field strengths, where ΔP and U_c are the pressure drop across the bed and the superficial velocity of the gas, respectively. The structural effect of particle alignment with the field is clearly seen in defluidization path III, $H > 0$. Along this path, ΔP is significantly lower compared to that required to fluidize the bed, despite the gas velocity being higher between points C and D than the minimum fluidization velocity U_{mf}. The field produced by the coils was $H = 2.395 \times 10^5$ A/m. Figure 44 is a plot of U_{exp} vs. U_φ at different currents in the coils, where U_{exp} is the measured sedimentation velocity in the fluid frame of reference and U_φ is the calculated (i.e., theoretical) value. The bed consists of 197–297-µm polyester-encapsulated magnetite particles, defined as magnetic sand, $\varrho_p = 2670$ kg/m^3, having a magnetic permeability of 1.64 µm and a shape factor $S_p = 0.85$. The volume fraction occupied by particles φ varied in the range of 0.45–0.50. The transition zone for the 10–25-A current levels (see Fig. 44) denotes transition from a stable to a bubbling bed. Initial bubbling observed in the vicinity of $U_\varphi = 0.3$ m/s indicates that ordered structures are turning into random ones. Only at

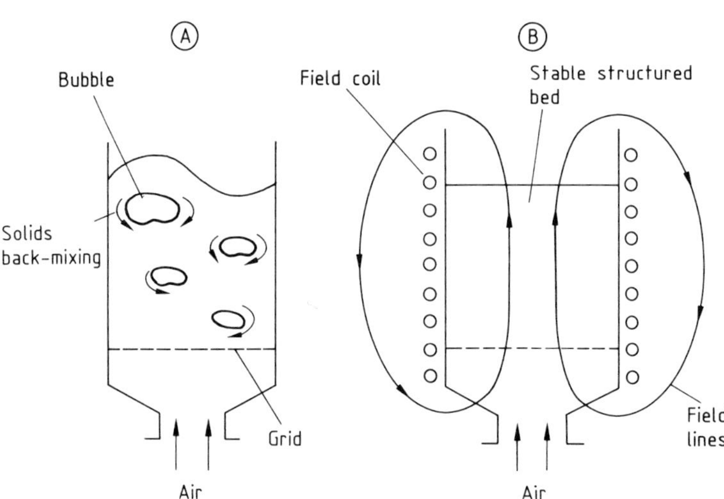

Figure 41. A) Ordinary and B) magnetically stabilized fluidized beds [76]

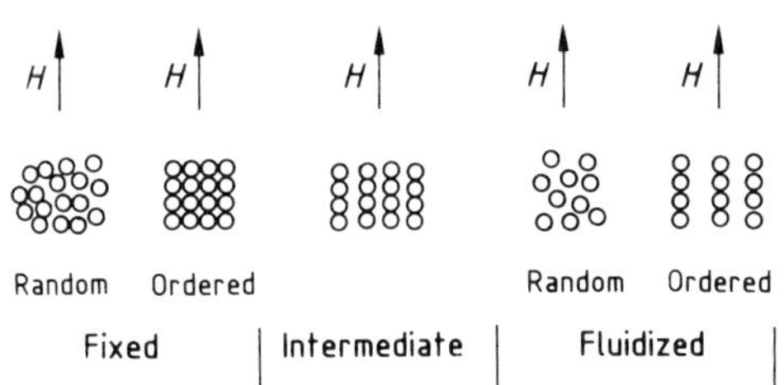

Figure 42. Particle structures in fixed, intermediate and fully fluidized magnetizable beds [75]

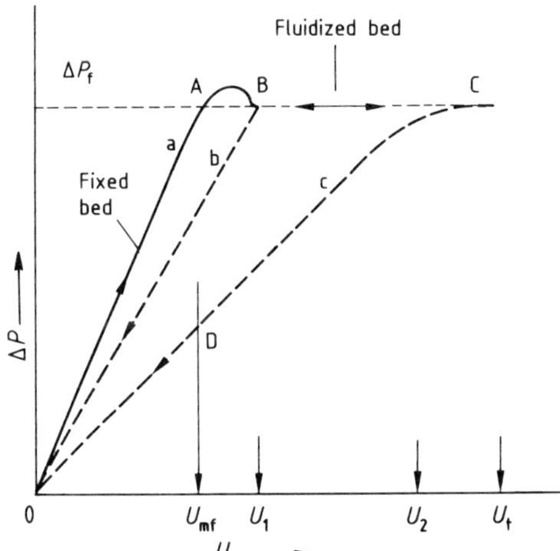

Figure 43. A schematic plot of ΔP vs. U_c showing fluidization and defluidization curves at different field intensities [75]
a) Fluidization curve; b) Defluidization curve, $H = 0$; c) Defluidization curve, $H > 0$

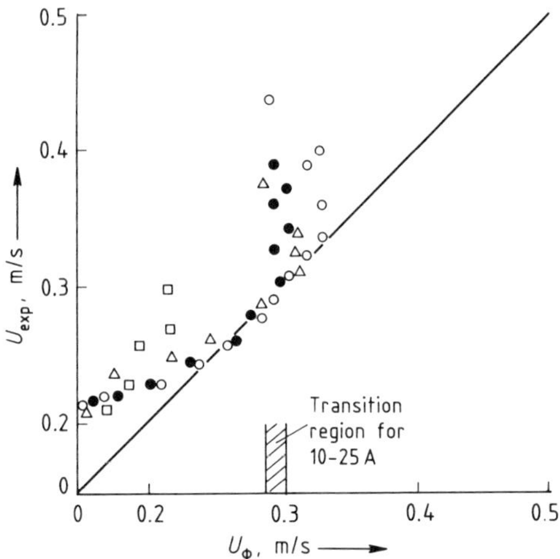

Figure 44. A plot of measured sedimentation velocity U_{exp} vs. calculated velocity U_φ, at different current levels [75]: □ 5 A; △ 10 A; ● 15–20 A; ○ 25 A

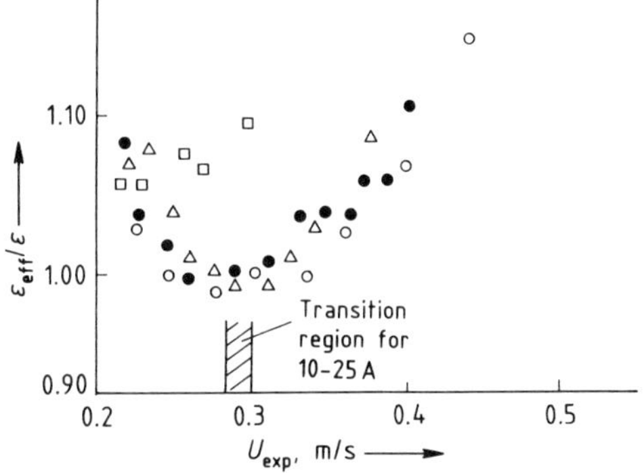

Figure 45. A plot of $\varepsilon_{eff}/\varepsilon$ vs. U_{exp} [75], symbols as in Figure 44

this transition zone is $U_{exp} = U_\varphi$ expected to be satisfied. In the region prior to the transition zone, $U_{exp} > U_\varphi$ reflects the reduced resistance to the flow of the aligned particle structure. In the posttransition zone, $U_{exp} > U_\varphi$ persists due to bypassing of gas in the form of bubbles. The results in Figure 44 conform with this. At $U_\varphi \simeq 0.3$, $U_{exp} = U_\varphi$ for the 15–25-A current levels, and at $U_\varphi = 0.15$ m/s, U_{exp} is 46 % higher than U_φ. Figure 45 shows the *structural effect* in terms of calculated effective porosity $\varepsilon_{eff} = \varepsilon + \varepsilon_m$, where ε is the actual bed porosity and ε_m is the magnetic structural porosity, $\varepsilon_m \leq 0.1\ \varepsilon$, at the transition velocity $\varepsilon_m = 0$. The structural effects persist along defluidization paths (see Fig. 43). Figure 46 shows plots of ΔP_{exp} vs. ΔP_{cal} for fluidization (A) and defluidization (B) paths; ΔP_{exp} and ΔP_{cal} are pressure drops across the bed obtained experimentally and calculated by the Blake–Kozeny correlation, respectively. A good fit between ΔP_{exp} and ΔP_{cal} was obtained in the fluidization path; deviations increasing with current level were observed in the defluidization path. The lower pressure drops observed indicate that the increased alignment of particles persists throughout the transition from fluidized- to fixed-bed states. In summary, sedimentation of polarizable particles in a polarizing field (set parallel to the sedimentation velocity) can be enhanced by increasing the alignment of particles with the field lines.

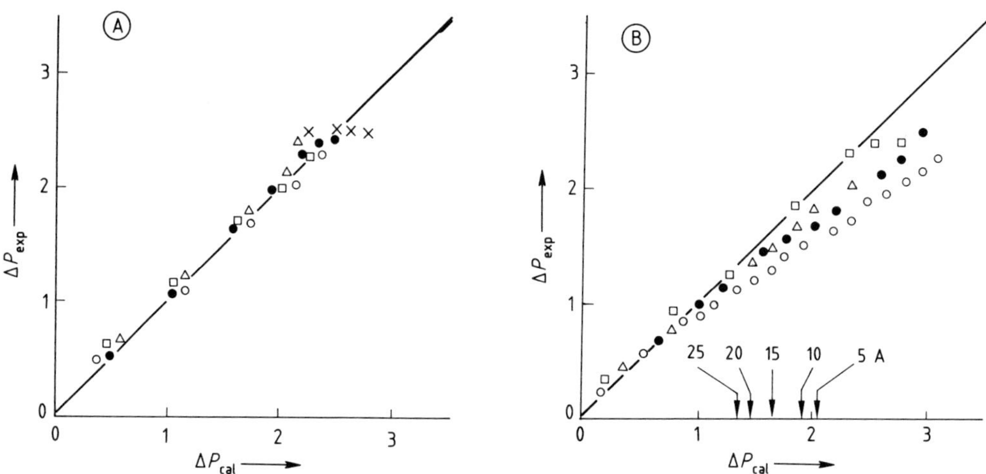

Figure 46. A plot of ΔP_{exp} vs. ΔP_{cal} for different current levels
A) Fluidization path; B) Defluidization path
× Fluidization, other symbols as in Figure 44

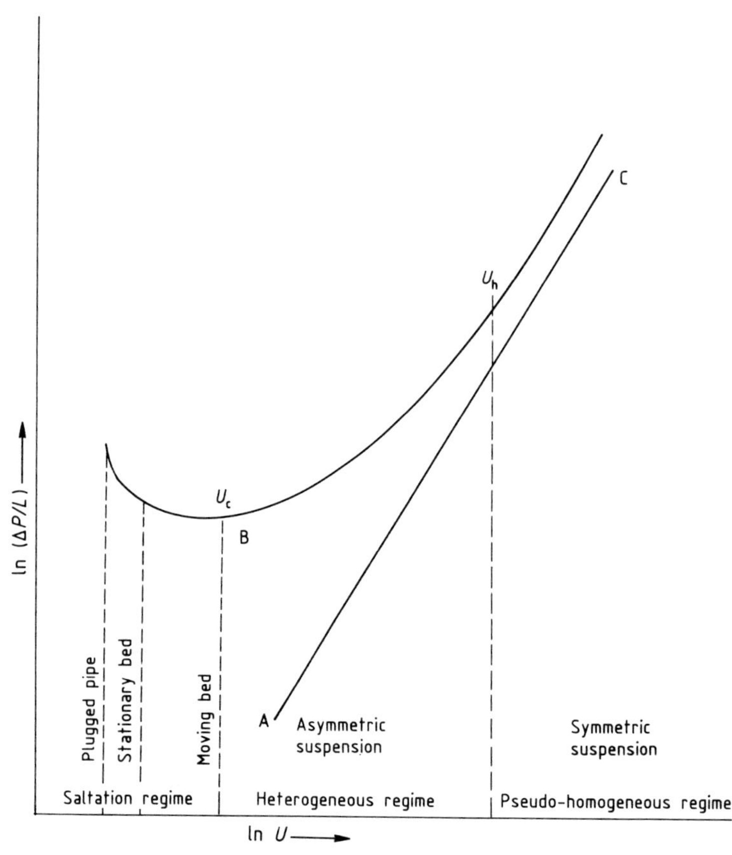

Figure 47. A schematic plot of ln ($\Delta P/L$) vs. ln U for slurry flow in horizontal pipes

3.4. Slurry Transportation in Horizontal Pipes

Slurry transportation in pipes must provide the necessary conditions to maintain a predesigned proportion of the solids, which are being carried by the fluid, in suspension. Prevention of undesirable sedimentation across the flow profile can be achieved at the expense of increasing pressure gradients along the pipe. Thus a definite relation must exist between pressure gradients associated with the axial flow and concentration gradients (formed by inertial effects, i.e., sedimentation) across it. Slurry flows in pipes are commonly characterized by the so-called *critical velocity*. This velocity marks the transition from a state in which particles are deposited on the pipe wall to a state in which they are in full inhomogeneous suspension in the stream. Figure 47 is an idealized plot of $\ln(\Delta P/L)$ vs. $\ln U$ for the flow of a heterogeneous slurry, where $\Delta P/L$ is the pressure drop per unit pipe length due to frictional losses and U is the mean flow velocity. At high velocities, the slurry tends to behave as a uniform fluid, such as that represented by line AC. The critical or deposition velocity U_c corresponds to the minimum (point B) in the heterogeneous slurry curve. Above U_c the slope is positive in accordance with the effect of increasing flow velocity on increasing friction losses, whereas below U_c the slope is negative due to the existence of particle deposits that narrow the space available for flow. The critical velocity is a strictly turbulent phenomenon; the transport of slurries in pipelines must be carried out above the critical velocity, otherwise excessive sedimentation is likely to cause instabilities, erosion, and finally plugging of the pipe. In the transportation of polydisperse slurries, excessive segregation of large from small fractions and the ensuing action of the pipe as a classifier are undesirable and should be avoided. In Section 3.2, $C/C_a \geq 0.8$ was selected as a criterion for homogeneity of the flowing suspension. Similarly, heterogeneous flow can be defined by $C/C_a \leq 0.1$ (see Fig. 40). The condition $C/C_a \leq 0.1$ applies also to the moving bed and saltation regime in the $U < U_c$ range (see Fig. 47). Note that *saltation* implies a process whereby particles advance along the pipe by being alternately lifted from, and deposited on, a bed that may be either moving or stationary. Other criteria that depend on sedimentation velocity have been suggested [77], [78] in the form of $U_d/U^* > 0.2$ and $U_d/U^* > 0.13$, respectively. Using Equation (120) with $\beta_s = 1.0$ and $k_f = 0.4$, gives $C/C_a < 0.13$ and $C/C_a < 0.27$, respectively [65]. Figure 48 shows a different approach for determination of slurry flow regimes [79]. Data are given for 1.2–2.1-m/s flow velocities as a function of particle size and density. This figure summarizes data collected from commercial pipelines. It shows that homogeneous flow of dilute slurries is expected provided that particle size is smaller than 70 µm, whereas a twofold increase in this limiting size is expected for dense slurries. In a more conservative estimate, a limiting size of 200 µm, above which heterogeneous flow prevails, has been suggested [65]. Figure 48 shows that for dilute slurries (given the particle density), the heterogeneous flow regime is extended in terms of a two- to fourfold increase in particle size. This is expected in view of the increased sedimentation rates prevailing in low φ slurries (see Eqs. 1 and

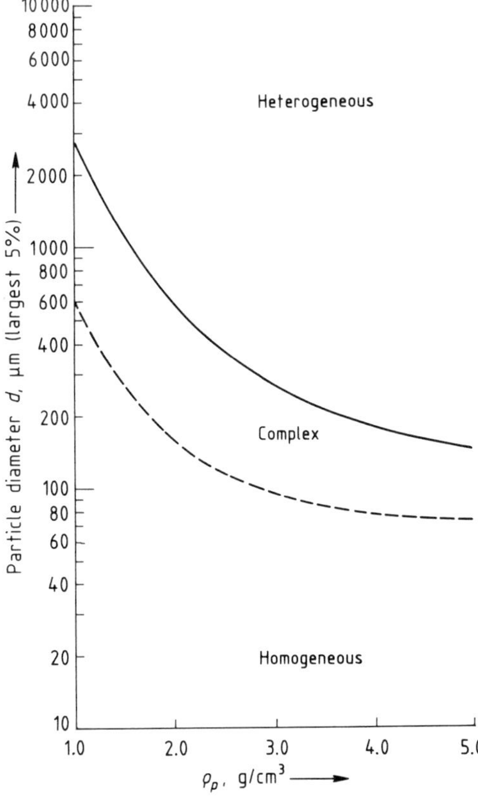

Figure 48. Slurry flow regimes as a function of particle diameter d and particle density ϱ_p, velocity = 1.2–2.1 m/s [78], [79]
——— Based on thick slurries with fine (< 44 µm) vehicle; ---- Based on thin slurries or slurries with graded particle size

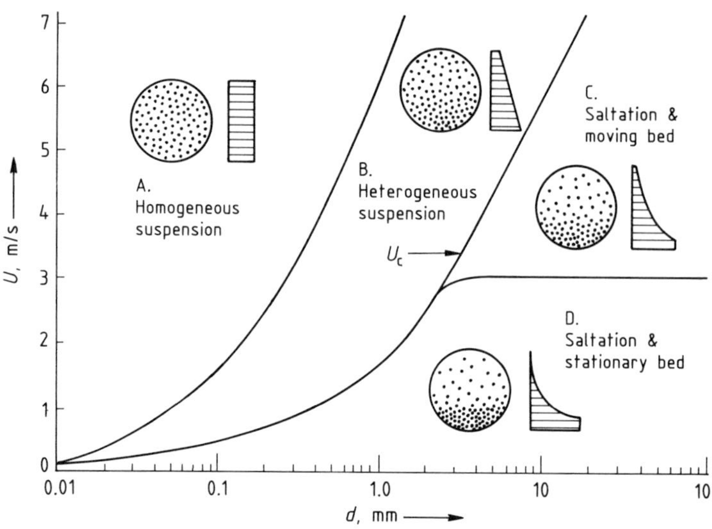

Figure 49. Flow regimes in a slurry pipeline defined in a plot of flow velocity U vs. particle size d [17], [80]

2). The effect of particle size and flow rate, drawn for a specific set of ϱ_p, φ, and D data, on the formation of different types of slurry flows is shown in Figure 49 [17], [80]. The curve of critical velocity is marked by U_c. Homogeneous and heterogeneous suspensions characterize the regions marked A and B, respectively. Saltation and moving bed, and saltation and stationary bed prevail in the regions marked C and D, respectively. Schematic particle concentration profiles are also shown in Figure 49. Many correlations have been produced over the years for estimation of critical velocity. These correlations show either explicitly or implicitly the dependence of U_c on the sedimentation velocities of the transported particles. In formulating these correlations, use is often made of free sedimentation velocities. This does not conform to the substantial dependence of sedimentation velocity on φ. Hence although experimental data can be made to fit empirical correlation, imposing $\varphi = 0$ on sedimentation velocity does not seem warranted. The necessary correction factors that make the correlations work may not disclose the true mechanisms that control the flow of slurries and the critical velocity; more work is needed to resolve this discrepancy. In 1942, WILSON contended that U_c is proportional to the one-third power of U_d (evaluated at $\varphi = 0$) and suggested the following correlation

$$U_c = \frac{[K_w(\varrho_p/\varrho_f)\,\varphi g D U_d(\varphi = 0)]^{1/3}}{1 + \varphi(\varrho_p/\varrho_f - 1)} \tag{121}$$

where D is pipe diameter and K_w is a constant [81]. The pioneering work of DURAND and coworkers [82], [83] provided a widely recognized empirical correlation. This correlation depends on a modified slurry Froude number Fr_L:

$$U_c = Fr_L[2\,gD(\varrho_p/\varrho_f - 1)]^{1/2} \tag{122}$$

Figure 50 shows a plot of Fr_L vs. particle size d for different levels of φ: Fr_L is most sensitive to d for particles smaller than 0.5 mm; for particles larger than 2 mm, Fr_L is practically independent of particle size. It has a value of 1.34, regardless of the system properties, with U_c being a function of $D^{1/2}$ and of ϱ_p/ϱ_f. These data [82], [83] were obtained using coal, sand, and gravel slurries con-

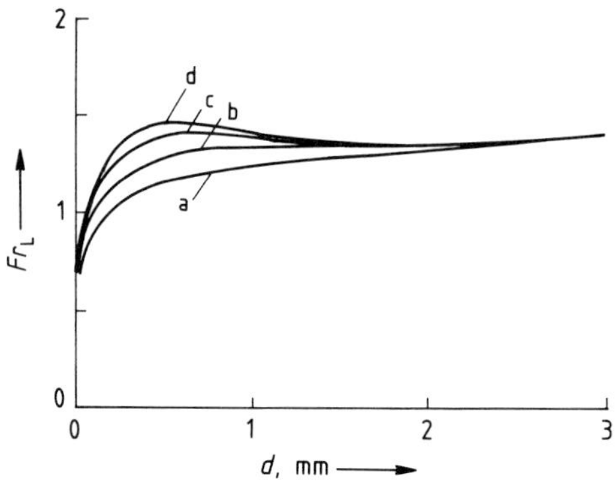

Figure 50. Variation of the modified slurry Froude number Fr_L as a function of particle diameter d [65], [82]
a) $\varphi = 2\%$; b) $\varphi = 5\%$; c) $\varphi = 10\%$; d) $\varphi = 15\%$

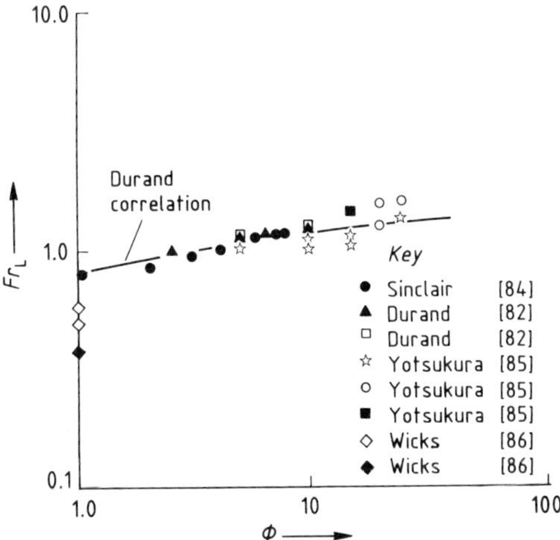

Figure 51. A plot of Fr_L vs. φ, literature data, and the DURAND Correlation [65], [70]

taining 0.2–25-mm particles and flowing in pipes of 3.8–5.8-cm diameter. A plot of Fr_L vs. φ (Fig. 51) discloses a weak dependence of Fr_L and hence of U_c on φ [65], [70]. The plot shows the excellent fit provided by the Durand correlation to the experimental results of several researchers [82]–[86]. Results similar to the ones plotted in Figure 50 were reported for transportation of tailings [17], [87]. The significance of sedimentation velocities in affecting U_c is also reflected in the pressure drops necessary to maintain a stable flow. The pressure drop ΔP in a pipe due to frictional losses is given by

$$\Delta P = f_s (L_t/D)\, 2\, \varrho_f U^2 \qquad (123)$$

where the slurry friction factor in heterogeneous flow is given by

$$f_s = f_l \left\{ 1 + K\varphi \left[\frac{gD}{U^2} (\varrho_p/\varrho_f - 1)/\sqrt{C_D} \right]^{3/2} \right\} \qquad (124)$$

where f_s and f_l are friction factors for the slurry and the pure fluid, respectively; K is a constant ranging from 80 to 150 (for $U > U_c$, $K = 82$ [17], [65]); C_D is the particle drag coefficient, evaluated at $\varphi = 0$; L_t is the total effective length of the pipe:

$$L_t = L_p + L_e \qquad (125)$$

where L_p and L_e are the length of straight sections of the pipe and the equivalent length of other components along the flow path (e.g., fittings, constrictions, valves, elbows, lees), respectively. If U is proportional to U_d^n, $n \leq 1$, then ΔP (via U^2) is proportional to U_d^{2n}. Equation (124) can be arranged as follows:

$$\frac{f_s - f_l}{f_l} \cdot \frac{1}{\varphi} = 82 \left[\left(\frac{U^2}{gD} \right) \cdot \frac{\varrho_f}{\varrho_p - \varrho_f} \sqrt{C_D} \right]^{-3/2},$$

$$U > U_c \qquad (126)$$

A plot of

$$\frac{f_s - f_l}{f_l} \cdot \frac{1}{\varphi} \text{ vs. } \frac{U^2}{gD} \cdot \frac{\varrho_f}{\varrho_p - \varrho_f} \sqrt{C_D}$$

is shown in Figure 52 [65]. Lower flow velocities, when D, ϱ_p, ϱ_f, C_D, and φ are fixed, give a larger relative increase of f_s (up to two orders of magnitude), whereas at high velocities this increase may be as small as 10%. This increased friction factor (i.e., over pressure) results from concentration gradients across the pipe which arise due to sedimentation effects. NEWITT et al. studied the flow of different slurries containing polydisperse particle-size mixtures [80]. Table 2 gives details of materials used in these studies, which produced the flow regimes depicted in Figure 49. The range of densities of solids tested was 1180–4600 kg/m³, the size range being 20–5970 µm. They proposed a direct relation between U_c and sedimentation velocity:

$$U_c = 17\, U_\varphi (\varphi = 0) \qquad (127)$$

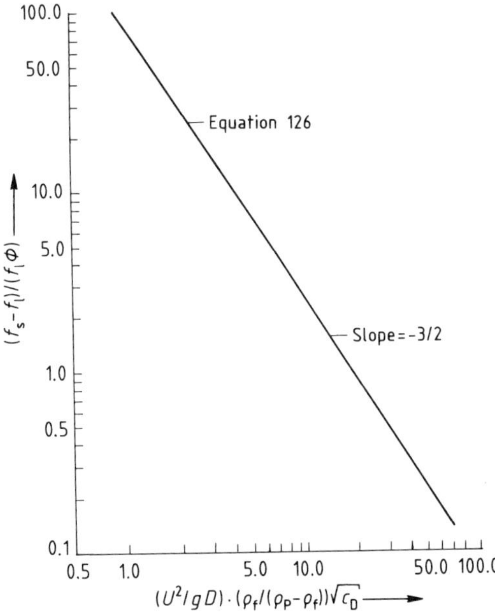

Figure 52. A plot of $(f_s - f_l)/(f_l\varphi)$ vs. $\frac{U^2}{gD} \frac{\varrho_f}{\varrho_p - \varrho_f} \sqrt{C_D}$ [65]

Table 2. Properties of materials used by Newitt et al [80]

Material	Density ϱ_p, kg/m^3	Terminal velocity U_d, m/s	Mean particle size d, cm
Perspex	1180	0.024	0.0584
	1180	0.036	0.1092
	1180	0.067	0.1981
Coal (anthracite)	1400	0.068	0.1321
	1400	0.102	0.2286
	1400	0.148	0.5969
Sand	2640	—	0.0020
	2640	0.001	0.0097
	2640	0.028	0.0208
	2640	0.113	0.0762
	2560	0.079	0.0483
Gravel	2600	0.192	0.1981
	2600	0.314	0.3124
	2550	0.326	0.3810
Manganese dioxide	4100	0.174	0.0965
	4100	0.253	0.1575
Zircon sand	4600	0.025	0.0109

Combining Equations (123) and (127) and letting $U = (1 + \alpha_s) U_c$, where $\alpha_s > 0$ denotes a safety factor, give

$$\Delta P = 578 f_s \left(\frac{L_t}{D}\right) \varrho_f [(1 + \alpha_s) \cdot U_\varphi(\varphi = 0)]^2 \quad (128)$$

Because for turbulent flow $U_\varphi \sim d^{1/2}$, according to the Newitt model $\Delta P \sim d$. A linear dependence of pressure drop on particle size has direct implications on pipeline economics. The cost of size reduction prior to transportation should be evaluated against expected savings in energy consumed by the pumps. The longer the pipe, the more favorable size reduction may be. Constraints on size reduction due to specification of end users and increased cost of dewatering should also be considered. Optimization involves determination of the pipe diameter that yields a minimum annual cost, where the annual cost consists of operating and investment costs. Increasing pipe diameter requires larger investment at lower operating cost; hence an optimum can be found. An illustration of single-phase pipe flow optimization is given in [88]. A nearly linear dependence of U_c on U_φ has also been reported [89].

$$U_c = (g D)^{0.5} \left[\frac{0.094\, \varphi (\varrho_p/\varrho_f - 1)}{\lambda} \cdot \frac{\varrho_f U_\varphi^3 (\varphi = 0)}{g \mu_f}\right]^{0.375} \quad (129)$$

where λ is resistance to the flow of carrier fluid through the pipe:

$$\lambda = 0.3164\, Re^{-0.25} \quad (130)$$

Equation (129) shows that $U_c \sim U_\varphi^{1.125}$, where U_φ is evaluated at $\varphi = 0$.

Published correlations diverge widely in predictions of critical deposition velocity U_c [90]. The value of U_c has been postulated as

$$U_c = f(d, D, \varphi, (\varrho_p - \varrho_f) g, \varrho_f, \mu_f)$$

or, in dimensionless form,

$$U_c = f\left[\left(\frac{d}{D}\right), \frac{D \varrho_f \sqrt{g d (\varrho_p/\varrho_f - 1)}}{\mu_f}, \varphi\right] \quad (131)$$

The result of this analysis is

$$U_c = \left[\frac{15}{4} \cdot \frac{\varphi}{(1 - \varphi)^{1 - 2n}} \cdot C_D U_0^2 \left(\frac{D}{d}\right) \left(\frac{D \varrho_f}{\mu_f}\right)^{1/8} \cdot \frac{1}{x_s}\right]^{8/15} \quad (132)$$

where $U_0 = U_\varphi(\varphi = 0)$, n is the exponent in $U_d = U_0 (1 - \varphi)^n$ which depends on the particle Reynolds number, $2.3 \leq n \leq 4.6$ [91], and x_s is a function of U_d/U_c.

$$x_s = \frac{2}{\sqrt{\pi}} \left[\frac{2}{\sqrt{\pi}} \gamma \exp\left(-\frac{4\gamma^2}{\pi}\right) + \int_\gamma^\infty \exp\left(-\frac{4\gamma^2}{\pi}\right) \mathrm{d}\gamma\right] \quad (133)$$

where $\gamma = U_d/U_c$.

Figure 53 shows a plot of x_s vs. U_d/U_c calculated from Equation (133). The decrease in x_s is

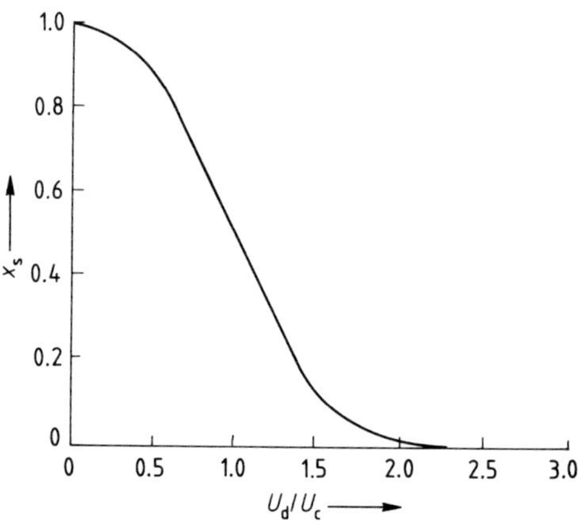

Figure 53. A plot of x_s vs. U_d/U_c [90], [91]

nearly linear in the 0.5–1.5 range of U_d/U_c. A simplified version of Equation (132) is given by

$$U_c = \left[5\,\varphi (1-\varphi)^{2n-1} \left(\frac{D}{d}\right) \frac{D \varrho_f \sqrt{g d (\varrho_p/\varrho_f - 1)}}{\mu_f} \cdot \frac{1}{x_s} \right]^{8/15} \sqrt{g d (\varrho_p/\varrho_f - 1)} \quad (134)$$

The following is a correlation obtained from regression analysis of published data [91], [90]:

$$U_c/\sqrt{g d (\varrho_p/\varrho_f - 1)} = 1.85\, \varphi^{0.1536} \cdot (1-\varphi)^{0.3564} \left(\frac{D}{d}\right)^{0.378} Re^{0.09}\, x_s^{0.3} \quad (135)$$

Equations (134) and (135) provide the fewest deviations from experimental data compared to other known correlations [91]. Transition velocity from heterogeneous to homogeneous flow (U_h in Fig. 47) depends on sedimentation velocities (see Section 3.2 and Eq. 120). Other correlations that do not make use of an arbitrary ratio of concentrations are listed below. These correlations show a weaker dependence on sedimentation velocity compared to those concerned with critical deposition velocity. Equation (136) was derived by Govier and Charles [92], [93]:

$$U_h = 11.9 (U_0 D)^{1/2} d^{-1/4}, \quad U_0 = U_\varphi(\varphi = 0) \quad (136)$$

whereas according to Newitt [92], [80], U_h is given by

$$U_h = 38.7 (g D U_0)^{1/3} \quad (137)$$

The plot of U_0/U^* vs. dU^*/ν_f in a generalized phase diagram for evaluation of transport of suspensions in pipes is shown in Figure 54; ν_f is the kinematic viscosity $\nu_f = \mu_f/\varrho_f$ [65], [94]. The solid line indicates initial particle movement from a state of rest on the pipe wall. Regions of formation of transverse and longitudinal waves, and of heterogeneous and homogeneous flow, are also indicated (δ denotes the thickness of the viscous sublayer and is given by $\delta = 11.6\, \nu_f/U^*$ [65].

This plot shows that the effect of U_0/U^* depends on d/δ, being largest for $d/\delta < 1$ and smallest in the $d/\delta > 6$ range. Particle Reynolds numbers are marked on the solid line indicating the Stokes, intermediate, and turbulent flow regimes associated with particle motion.

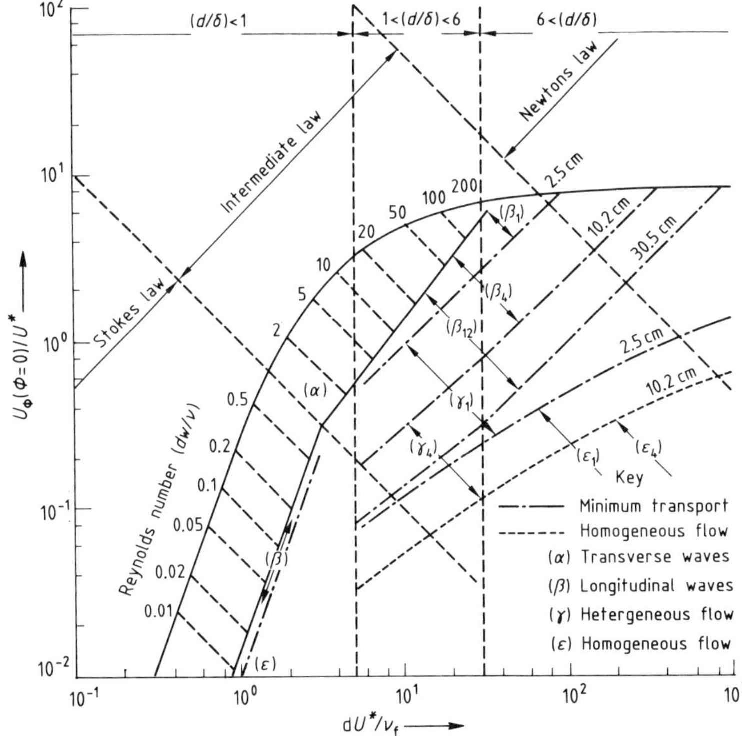

Figure 54. A generalized phase diagram for transport of suspensions [65], [94] Range of β, γ, and ε phases apply to density ratio $(\varrho_p - \varrho_f)/\varrho_f = 1.65$

Centrifugal forces affect sedimentation of particles suspended in slurries that flow in bent pipes. Examples are flow in elbows and in helical pipes. If the resulting value of g' exceeds g, then prevention of deposition or transition from homogeneous to heterogeneous flow must occur at the expense of increasing the frictional velocity U^*.

In the transportation of *polydisperse slurries*, the effect of attendant distributions must be accounted for. WASP et al. used the concept of a two-phase vehicle [65], [69], [95]. According to this, the slurry is divided into homogeneous and heterogeneous parts. The concentration and the size distribution of particles occurring at the top of the pipe are assumed to exist throughout. This concentration is considered part of the homogeneous or "vehicle" portion of the flow. The remainder of the polydisperse mixture is considered as a heterogeneous suspension transported by this vehicle. Analysis of flow requires the determination of particle split between the vehicle and the heterogeneous portion of the slurry. This split depends on U^*, which cannot be determined until the friction factor is found; hence an iterative solution is called for. Applying Equation (120), with $\beta_s = 1$, $k_f = 0.4$, to the ith fraction gives

$$C_i = C_{ai} \exp(-10.362\, U_{0i}/U^*) \qquad (138)$$

in which C_i is added to the vehicle and the balance to the heterogeneous part of the suspension. Thus $C_h = \sum_{i=1}^{n} C_i$ and $C_{ht} = \sum_{i=1}^{n} C_{ai} - C_i$, where C_h and C_{ht} are total concentrations in the vehicle and in the heterogeneous part of the suspension, respectively. Thus in each iteration, U_{0i} for all particle fractions and U^* must be determined, and the contribution of each heterogeneous fraction must be accounted for, the total frictional loss being due to both the vehicle and the heterogeneous part of the slurry [65], [92]. Again, evaluation of free sedimentation velocity (i.e., U_φ at $\varphi = 0$) does not conform to the physics of the flow and sedimentation systems. The successfulness of using U_0 instead of U_φ at $\varphi > 0$, particularly for dense slurries, seems to be due primarily to the adjustment of empirical constants.

Furthermore, slurry flow and particle sedimentation in dense slurries are likely to involve non-Newtonian rheology. Details of slurry pipeline hydraulics, treatment of different slurry rheologies and description of computational models are given in [96].

3.5. Sedimentation in Centrifugal Fields

The use of centrifugal fields to enhance sedimentation of fine and slow-settling particles is well known. Sedimentation of diffusing particles in centrifugal fields is currently used for the purposes of separation, concentration, pellet formation, sample preparation, and as an analytical means for particle characterization. *Ultracentrifugation* generally implies sedimentation of diffusing particles in centrifugal fields inclusive of those that provide the highest (i.e., ultra) g forces known to date in commercial centrifuges. Because this technique incorporates many aspects of sedimentation, it is introduced first.

3.5.1. Ultracentrifugation

Ultracentrifugation is used extensively in biochemistry (→ 11. Biochemical Separations, **B3**, p. **11**-9), where analysis of molecules, macromolecules, cells, and other microscopic matter of organic origin is often required [31], [97]. Particulates in this category are generally dispersed in a carrier fluid, the dispersions thus formed being uniform under ordinary conditions. Analysis of the contents of such dispersions often requires redistribution of the dispersed entities so that their characteristic signals can be picked up. This can be achieved during or after sedimentation in ultracentrifugal fields. The ultracentrifuge generally consists of a driving mechanism, control unit, centrifugation chamber, and rotor that holds the sample containers. Different types of rotors are available, and the purpose of the runs dictates rotor selection.

Figure 55 shows the LKB Ultrospin preparative ultracentrifuge, and typical material bands produced in sample tubes that were run in different rotors. This high-speed programmable unit is available in three models that can be operated at maximum speeds of 55 000, 70 000, and 83 000 rpm, which are equivalent to ratings of 393 600 g, 504 000 g, and 615 700 g as maximum values of g', respectively.

Figures 56–58 show typical *fixed-angle*, *swing-out*, and *vertical rotors*, respectively. One of the many available models of each type of rotor is shown at the side of each figure. The tilted tubes in the *fixed-angle rotor* provide conditions for sedimentation of diffusing particles that can be considered as inclined channels (see Section 2.8); observe the inclined bands in the

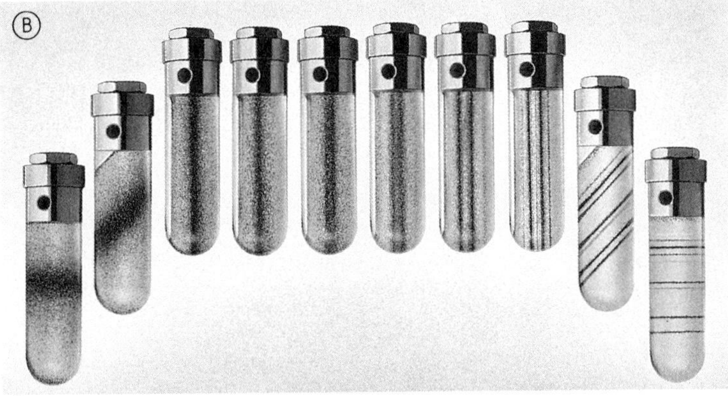

Figure 55. A) The LKB Ultrospin preparative ultracentrifuge; B) Material patterns in sample tubes spun with different types of rotors

second and ninth tubes of Figure 55. *Swing-out rotors* provide ordinary conditions for batch sedimentation; note the bands of particles in the first and tenth tubes in Figure 55. *Vertical rotors* produce sedimentation across the tube, see bands in the third to eighth tubes in Figure 55. Standard ultracentrifugation equipment, such as the Beckman series, is not designed to provide g' levels exceeding 300 000 g. The Beckman *elutriator rotor* incorporates counterflow of the carrier fluid to separate slower settling from fast-settling particles. Figure 59 shows the elutriator rotor and the mechanism of elutriation. The suspension is introduced via a narrow inlet port into a tapered cell (stage 1), where it is subjected to centrifugal force and to a counteracting hydrodynamic drag of the carrier fluid (stage 2). Larger and heavier particles undergo faster sedimentation against the fluid currents, whereas slower ones are washed away and thus separated (stage 3). This high-recovery method is used for separation of whole cells and large subcellular particles. Ultracentrifugal analysis can provide information on diffusion coefficient D and sedimentation coefficient S_g which were defined in conjunction with Equation (69). These parameters can then be used to evaluate molecular mass and to predict sedimentation under different operating conditions. Measurement of D and S_g can be done by either the sedimentation velocity or the sedimen-

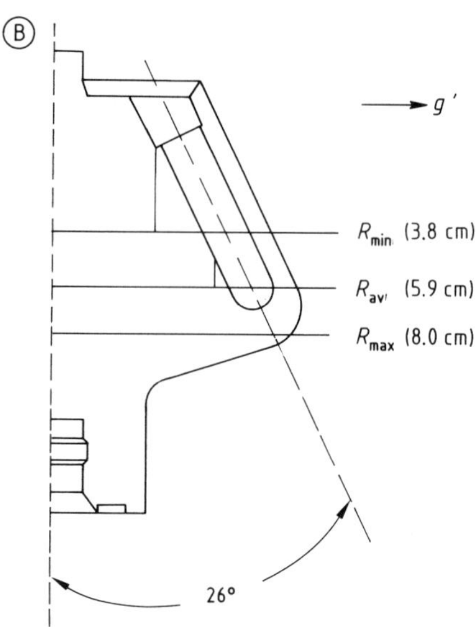

Figure 56. Fixed-angle rotors
A) Apparatus; B) Cross-section

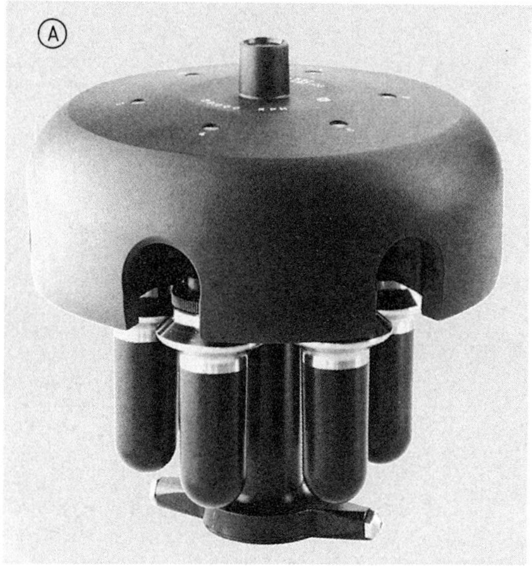

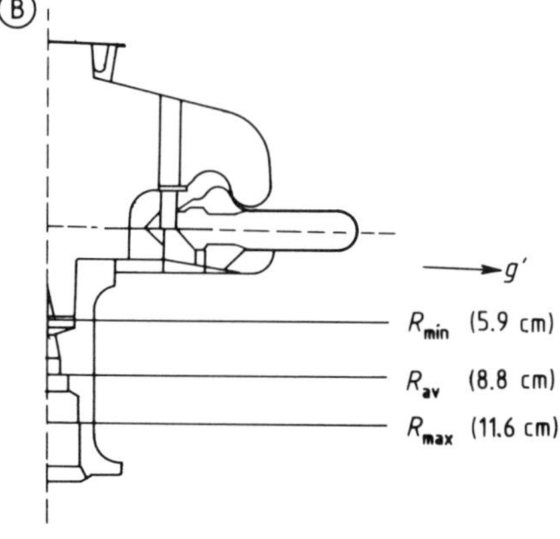

Figure 57. Swing-out rotors
A) Apparatus; B) Cross-section

tation equilibrium method. In the sedimentation velocity method which involves relatively high rotor speeds, a moving boundary that separates particle-devoid fluid from the dispersion is driven by centrifugal force toward the cell boundaries. If $\varrho_p - \varrho_\varphi > 0$, motion is away from the axis of rotation and toward the cell bottom. Figure 60 shows a plot of concentration and concentration gradient profiles vs. distance from the axis of rotation [31]. As time proceeds, the boundary of fluid devoid of particles moves to the right and the variance of concentration gradients (i.e., their spread) increases due to diffusion.

Such boundary motion is generated in sedimentation velocity experiments. In sedimentation equilibrium, sufficient time is allowed for the system to reach steady, zero-flux conditions in the sedimentation cell. Relatively low rotor speeds are used to bring the dispersion to a state where sedimentation drift is balanced by coun-

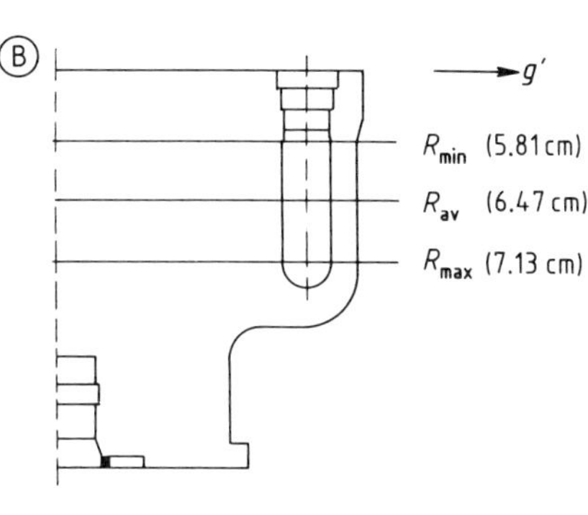

Figure 58. Vertical rotors
A) Apparatus; B) Cross-section

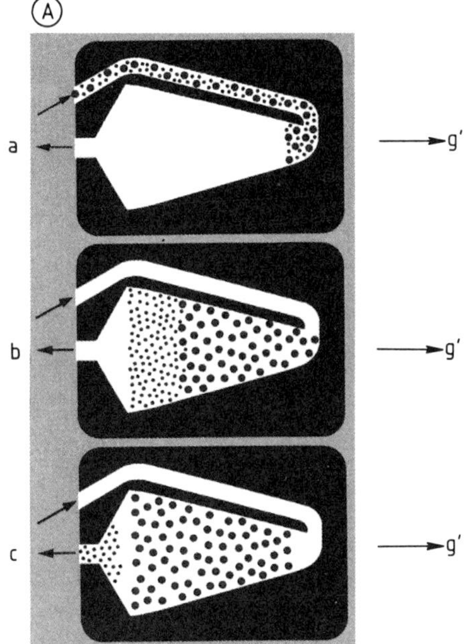

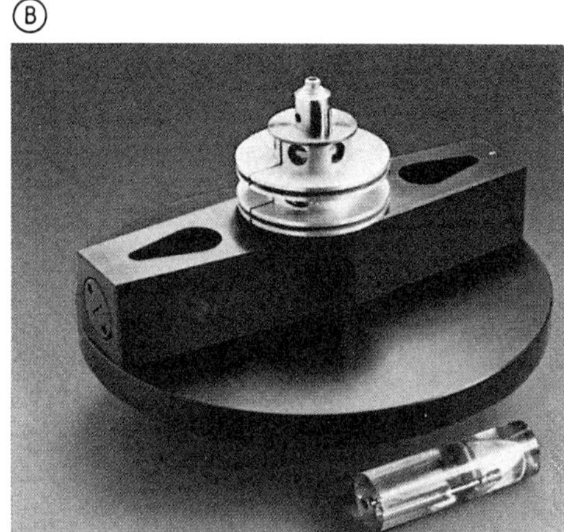

Figure 59. The Beckman elutriator rotor
A) Mechanism of elutriation (schematic):
a) Sample suspended in medium enters chamber; b) Sedimentation tendency of particles balanced by counterflow; c) Flow increased. Slow sedimenting particles elutriated from chamber;
B) Rotor

terdiffusion throughout the cell. Figure 61 shows schematically the variation of concentration and concentration gradient profiles vs. distance from the axis of rotation occurring during a typical sedimentation equilibrium experiment. The curves are numbered in order of increasing time. A typical equilibrium profile is characterized by equilibrium concentration having approximate values of $C_0/2$, C_0, and $2\ C_0$ at the cell meniscus, midpoint, and bottom (6.0, 6.6, and 7.2 cm, respectively, in Fig. 61).

Sedimentation Coefficient. The sedimentation coefficient is defined as sedimentation velocity per unit acceleration g'. In the fluid frame of reference, the sedimentation coefficient $S_{g\varphi}$ of

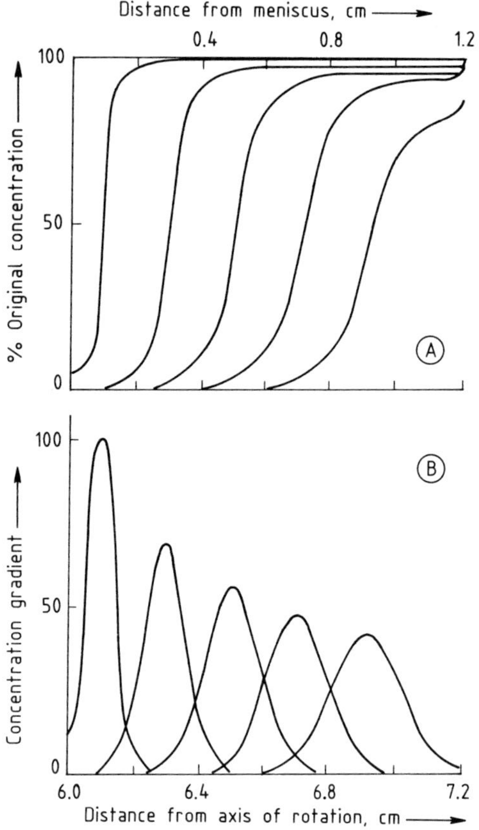

Figure 60. Schematic plots of: A) concentration, and B) concentration gradient vs. distance from axis of rotation at different elapsed times from the start of a sedimentation velocity experiment [31]

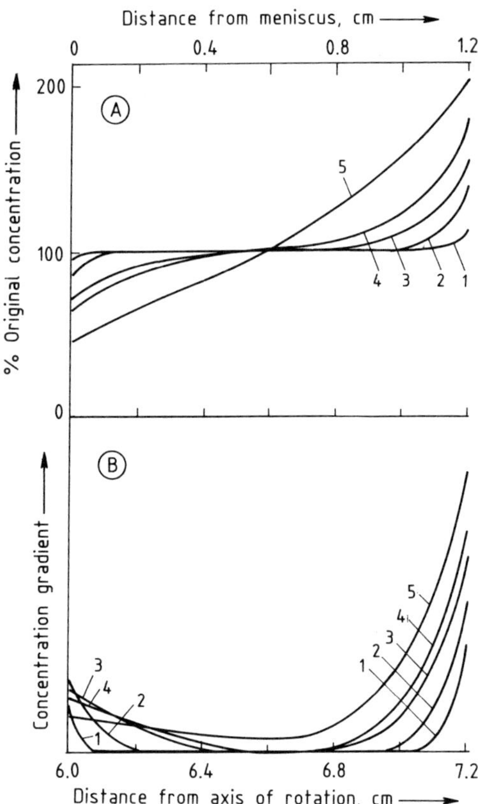

Figure 61. Schematic plots of: A) concentration, and B) concentration gradient vs. distance from axis of rotation, at different elapsed times from the start of a sedimentation equilibrium experiment

monodisperse particles is then given by

$$S_{g\varphi} = \frac{U_\varphi}{g'} = \frac{U_\varphi}{\omega^2 r} \tag{139}$$

whereas in the dispersion frame of reference,

$$S_g = \frac{U_d}{\omega^2 r} = \frac{U_\varphi(1-\varphi)}{\omega^2 r} = \frac{U_\varphi(1-C\tilde{N}V_p)}{\omega^2 r} \tag{140}$$

The sedimentation coefficient (in the dispersion frame of reference) of the ith particle fraction pertaining to a polydisperse mixture depends on coefficients of the rest of the distribution as follows:

$$S_{gi} = S_{g\varphi i} - \sum_i S_{g\varphi i}\,\varphi_i \tag{141}$$

Because $U_d = dr/dt$,

$$S_g = \frac{dr/dt}{\omega^2 r} = \frac{1}{\omega^2}\cdot\frac{d\ln r}{dt} \tag{142}$$

The dimension of S_g is seconds.

In ultracentrifugation, one unit of sedimentation coefficient is called one Svedberg. One Svedberg equals 10^{-13} s. The rate of change of S_g with C is given by

$$\partial S_g/\partial C = \frac{1}{\omega^2 r}\left[\frac{\partial U_\varphi}{\partial C}(1-C\tilde{N}V_p) - U_\varphi \tilde{N} V_p\right] \tag{143}$$

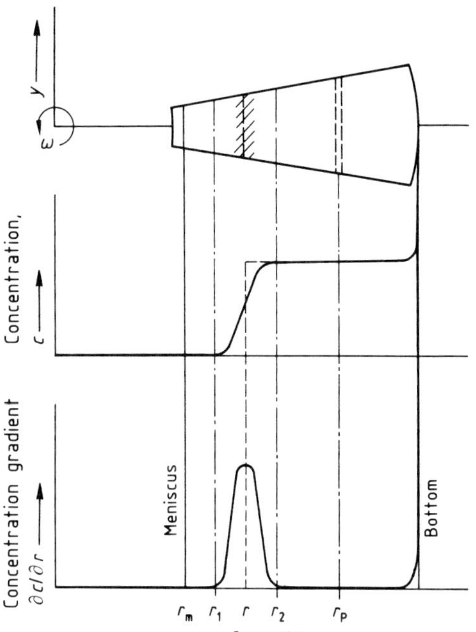

Figure 62. Concentration and concentration gradient profiles during sedimentation in a sector cell [31], [99]

in which Equation (140) has been used. Combining Equations (69) and (143) gives

$$\partial C/\partial t = D'\left(\frac{\partial^2 C}{\partial r^2} + \frac{1}{r}\cdot\frac{\partial C}{\partial r}\right) - S_g\omega^2\left(r\cdot\frac{\partial C}{\partial r} + 2C\right) + \frac{\partial C}{\partial r}\left\{\frac{\partial D'}{\partial C}\cdot\frac{\partial C}{\partial r} - \left[\frac{\partial U_\varphi}{\partial C}\cdot C(1 - C\tilde{N}V_p) - U_\varphi C\tilde{N}V_p\right]\right\} \tag{144}$$

The relationship $\frac{\partial U_\varphi}{\partial C} = U_0\cdot\frac{\partial f(\varphi)}{\partial C}$ can be obtained from Equation (1), and if the dependence of D' on C is known (or use is made of Equation 66), then Equation (144) can be solved without the need to impose the constraint that D' and S_g be constants. In dilute dispersions, provided that concentration and concentration gradients are low, the assumption that D' and S_g are constants seems to be justified. In the range $r < r_1$ and $r > r_2$, $\partial C/\partial r$ and thus $\partial^2 C/\partial r^2$ practically vanish (see Fig. 62). Hence, Equation (143) reduces to

$$\partial C/\partial t = -2\omega^2 S_g C$$

which upon integration gives

$$C(t) = C_0 \exp\left(-2\omega^2 \int_0^t S_g\, dt\right) \tag{145}$$

$C(t)$ and C_0 are concentrations at time t and $t = 0$, respectively. If S_g is constant, then

$$C(t) = C_0 \exp(-2\omega^2 S_g t) \tag{146}$$

In the sedimentation of nondiffusing monodisperse particles, the transition zone would be infinitely sharp, i.e., $r_2 - r_1 \to 0$ in Figure 62. Under these conditions, the velocity of the resultant moving boundary would be the same as the sedimentation velocity of particles ahead of the boundary within the dispersion. In sedimentation of diffusing particles, the boundary tends to spread out as time elapses (see Fig. 60). In this case, the peak of the concentreation gradient profile does not move at the same velocity as particles within the dispersion. The point $\bar{r}$ that does satisfy this condition (i.e., having the same velocity as particles that move within the suspension) is given by the square root of the normalized second moment of the concentration gradient curve [31], [98]:

$$\bar{r}^2 = \int_{r_m}^{r_p} r^2 (\partial C/\partial r)\, dr/C_p\,, \quad C_p = \int_{r_m}^{r_p} (\partial C/\partial r)\, dr \tag{147}$$

where r_m and r_p are the distances of the meniscus and of a point in the dispersion at which $\partial C/\partial r = 0$ and $C = C_p$, respectively. In deriving Equation (147), a meniscus devoid of particles is assumed. The actual difference between $\bar{r}$ and the location of the peak may be small. In such cases, use of the latter can be justified. As the transition zone, i.e., the peak in Figure 62 (see also Fig. 60), moves away from the meniscus, interparticle spacings become larger due to the increased intensity of the centrifugal field. This phenomenon is known as radial dilution, the end result being time-decreasing concentration. Using Equation (142) for $r = \bar{r}$ in the plateau region (see Fig. 62) gives $\partial C/\partial t = -\frac{2C}{r}\cdot\frac{\partial \bar{r}}{\partial t}$, which upon integration yields the radial dilution equation:

$$C(t)/C_0 = (r_m/\bar{r})^2 \tag{148}$$

Thus the advance of the concentration transition zone (i.e., the peak) from the meniscus at r_m to a new position $\bar{r}$ is associated with a decrease in concentration of the dispersion from C_0 to $C(t)$. In a polydisperse mixture that forms a series of moving boundaries [99],

$$C_0 = \sum_{i=1}^{n} (\bar{r}/r_m)_i^2 C_i(t) \quad (149)$$

where i denotes the ith component of a mixture comprising n components. If the dependence of S_g on concentration is significant, then radial dilution must also be accounted for. The suggestion has been made that the standard sedimentation coefficient $S_{g(20,w)}$ be defined with respect to sedimentation in pure water at 20 °C; $S_{g(20,w)}$ is related to S_g by [100]

$$S_{g(20,w)} = S_g \cdot \frac{\mu_f}{\mu_{f(20,w)}} \cdot \frac{(1 - v\varrho_\varphi)_{20,w}}{1 - v\varrho_\varphi} \quad (150)$$

where v is the partial specific volume, which can be considered as the volume occupied by 1 g of the dispersed phase (e.g., solute) in the dispersion [97].

$$v = (dV_\varphi/dW_s)_{V_\phi \to \infty} \quad (151)$$

where V_φ is the volume of the dispersion and W_s is the weight of the dispersed phase. Note that $v = 1/\varrho_p$, where ϱ_p is the density of the dispersed phase in the dispersion. Different values of v and ϱ_p may apply to materials in pure and in dispersed (or dissolved) states.

Equation (150) does not account for the full effect of concentration and is valid if $\varphi \to 0$, i.e., $C \to 0$. In view of Equations (1) and (140), $S_{g(20,w)}$ can be related to S_g as follows:

$$S_{g(20,w)} = S_g \frac{[U_\varphi(1-\varphi)]_{20,w}}{U_\varphi(1-\varphi)} = S_g \frac{(U_0)_{20,w}}{U_0} \cdot \frac{[f(\varphi)(1-\varphi)]_{20,w}}{f(\varphi)(1-\varphi)} \quad (152)$$

Equation (152) shows that in transforming S_g to $S_{g(20,w)}$, the combined effect of U_0 and of $f(\varphi)(1-\varphi)$ must be accounted for. In the moving-boundary method (Eq. 142), ln $\bar{r}$ is plotted vs. t and the slope yields $\omega^2 S_g$. If S_g is constant, a straight line is obtained. Figure 63 shows a plot of log r vs. t for sucrose, which indicates a constant S_g over the range of distance measured [101]. In the *transport method*, S_g is determined by measuring the total amount of solute transported across an arbitrary surface in the sedimentation cell. This method is useful for measuring S_g of small molecules. Material balance in the sector $r_m \le r \le r_p$ (see Fig. 62) gives [31]

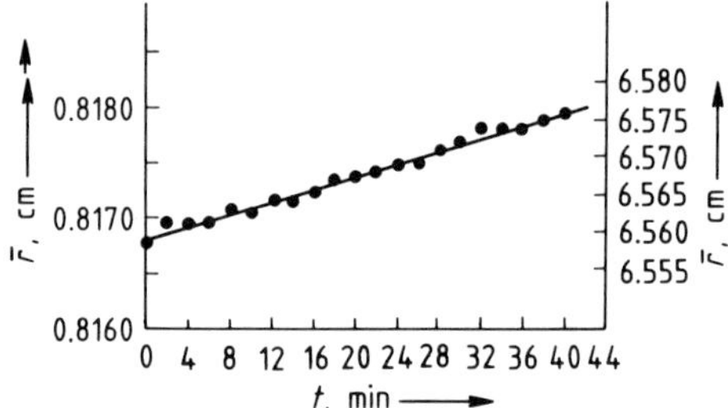

Figure 63. A plot of log $\bar{r}$ and $\bar{r}$ vs. t for sedimentation of sucrose [31], [101]

$$\tfrac{1}{2}\beta_c C_0 (r_p^2 - r_m^2) - \int_{r_m}^{r_p} \beta_c C r \, dr - \int_0^t \beta_c r_p C \omega^2 r_p S_g \, dt = 0 \quad (153)$$

where β_c is a constant depending on cell geometry. The first, second, and third terms represent the amount of dispersed material at $t = 0$ and $t > 0$, and the amount accumulated from $t = 0$ to t, respectively. Combining Equations (146) and (153), if S_g is constant, and integrating give

$$C_0 (r_p^2 - r_m^2) - 2\int_{r_m}^{r_p} Cr \, dr = r_p^2 C_0 \cdot [1 - \exp(-2\,\omega^2 S_g t)] \quad (154a)$$

Rearranging Equation (154 a) yields

$$S_g = -\frac{1}{2\,\omega^2 t} \cdot \ln\left[2\int_{r_m}^{r_p} Cr\, dr/(Cr_p^2 C_0) + r_m^2/r_p^2\right] \quad (154b)$$

If t is given, and $C = C(r)$ is known in the $r_m \le r \le r_p$ range, then S_g can be determined directly from Equation (154). Equation (154) is valid, irrespective of the formation or absence of a clear boundary between solution and supernatant, if a plateau region exists at $r = r_p$. Thus it is useful in the ultracentrifugation of small molecules that do not form well-developed moving boundaries. Application of Equation (154) requires direct measurements of C. This can be done by using optical absorption or analytical chemistry techniques. BALDWIN [102] transformed Equation (154 b) so that methods in

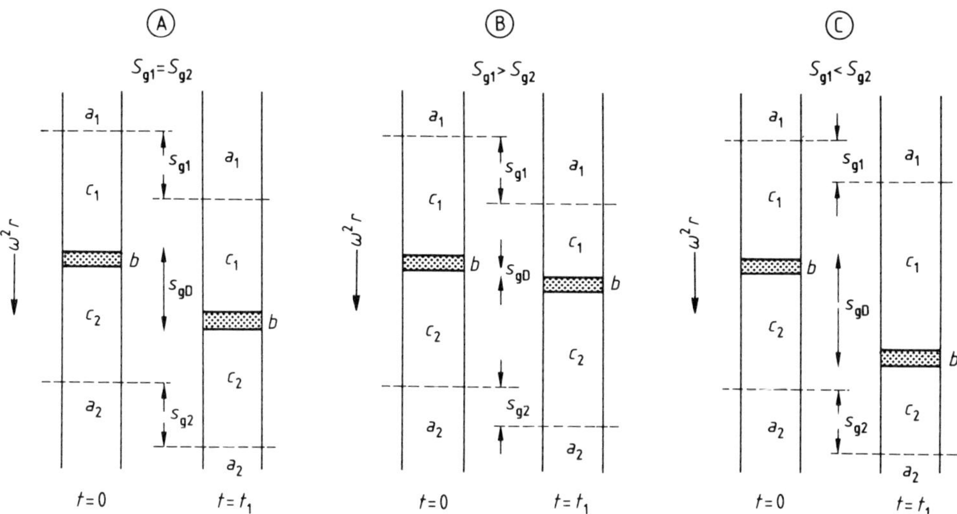

Figure 64. Schematic setup for measurement of differential sedimentation rates [31]

which $\partial C/\partial r$ is measured (i.e., *schlieren optics*) can be applied [31], [97]. The result is as follows:

$$S_g = -\frac{1}{2\,\omega^2 t}\ln\left\{1+\left[r_m^2\int_{r_m}^{r_p}(\partial C/\partial r)\,dr - \int_{r_m}^{r_p} r^2(\partial C/\partial r)\,dr\right]/(r_m^2 C_0)\right\} \quad (155)$$

A plot of the negative of the logarithmic expression vs. t for data corresponding to Equation (154) or (155) is expected to give a straight line, whose slope is $2\,\omega^2 S_g$ [103], [104].

Differential Sedimentation Rates of Monodisperse Particles. Differential sedimentation rates can be used to determine the dependence of S_g on concentration. Differential sedimentation rates are defined with respect to movement of a boundary formed between solutions (or, generally, dispersions) that are identical except for having different concentrations. Figure 64 shows schematically a system for measuring differential sedimentation rates; a_1 and a_2 mark imaginary moving planes of "zero flux." These planes move at the same velocity as local particles; hence no particles cross them, and particle flux (in the plane frame of reference) vanishes. A boundary is shown between the upper (concentration C_1) and lower (concentration C_2) sections. Because no flux enters or leaves the control zone between a_1 and a_2, the associated amount of solute (e.g., particles) is independent of time. The volume of the control zone is nevertheless time dependent. Differential sedimentation rates in the C_1 and C_2 sections give rise to motion of the boundary in the control zone. Let the variation of $\omega^2 r$ across the control zone be negligible and assume that S_g is solely a function of C such that $\partial S_g/\partial C < 0$. If $C_1 = C_2$, then (see Fig. 64 A) the relative position of the boundary b (which here is also imaginary) with respect to the control zone is time independent. If $C_1 < C_2$ (Fig. 64 B), then $S_{g1} > S_{g2}$ and the solute enters at a faster rate than it leaves the boundary b. Accumulation of solute in b occurs until the concentration rises to C_2. The net result is displacement of the boundary in the frame of reference of the control zone toward a_1. If $C_1 > C_2$, then $S_{g1} < S_{g2}$ and particles in the C_2 zone will move faster than the boundary. An unstable low-density zone will be formed below the boundary, which is then expected to be deformed and probably eliminated by convection currents. Stable motion of the boundary toward a_2 (see Fig. 64 C) can occur if $\partial S_g/\partial C > 0$ and $C_2 > C_1$ are satisfied. From continuity (see Eq. 31), because the system is one dimensional and U_d is a sole function of C,

$$\mathrm{d}C/\mathrm{d}t + C\,\mathrm{d}U_d/\mathrm{d}r + U_d\,\mathrm{d}C/\mathrm{d}r = \mathrm{d}C/\mathrm{d}t + C(\mathrm{d}U_d/\mathrm{d}C)(\mathrm{d}C/\mathrm{d}r) + U_d(\mathrm{d}C/\mathrm{d}r) \quad (156)$$

Multiplying Equation (156) by $(\mathrm{d}r/\mathrm{d}C)/(\omega^2 r)$ gives

$$S_{gD} + S_g + C\,\mathrm{d}S_g/\mathrm{d}C = 0 \quad (157)$$

where $S_{gD} = (dr/dt)/(\omega^2 r)$ is the differential sedimentation coefficient of the boundary that moves at velocity dr/dt. For $dS_g/dC < 0$ and $C_1 < C_2$, the boundary will move toward a_1 at a slower rate than the sedimentation rate of particles in zone C_1.

Diffusion Coefficient. The diffusion equation, for the static (i.e., $U_d = 0$) case and constant diffusion coefficient, is obtained readily from Equation (61):

$$\partial C/\partial t = D' \nabla^2 C \tag{158}$$

Equations (159) and (161) present two integrated forms of Equation (158) for a one-dimensional (x-oriented) diffusion,

$$dC/dx = (C_0/\sqrt{4\pi D' t}) \exp(-\zeta^2) \tag{159}$$

$$\zeta = x/\sqrt{4D't} \tag{160}$$

$$C = \frac{C_0}{2}\left[1 - \frac{2}{\pi} \cdot \int_0^y \exp(-\zeta^2)\, d\zeta\right] \tag{161}$$

Figure 65 shows a plot of time-dependent concentration (Eq. 161) and concentration gradient (Eq. 159) profiles around a static and initially sharp boundary between solution and solvent. Various methods for measuring D' have been described in detail [105]. Equation (159) is suitable for use when the experimental system provides direct values of dC/dx, whereas Equation (161) requires direct determination of C. Because Schlieren optics is the standard optical system in ultracentrifuges [97] and this method provides direct data on concentration gradients, Equation (159) seems to be a convenient choice.

At the center of the boundary, $x = 0$ and the value of dC/dx is at a maximum.

$$(dC/dx)_0 = C_0/\sqrt{4\pi D' t}\,, \quad x = 0 \tag{162}$$

Schlieren optics provides peaks that are directly related to the dC/dx peaks shown in Figure 65. The area under the schlieren peak is directly proportional to C_0, which is given by the area under the peak in Figure 65. If h_s and A_s denote the height and the area of the schlieren peak, then

$$D' = \frac{(A_s/h_s)^2}{4\pi t} \tag{163}$$

A plot of $(A_s/h_s)^2$ vs. $4\pi t$ gives D' as the slope of a straight line. Note that Equation (159) can be

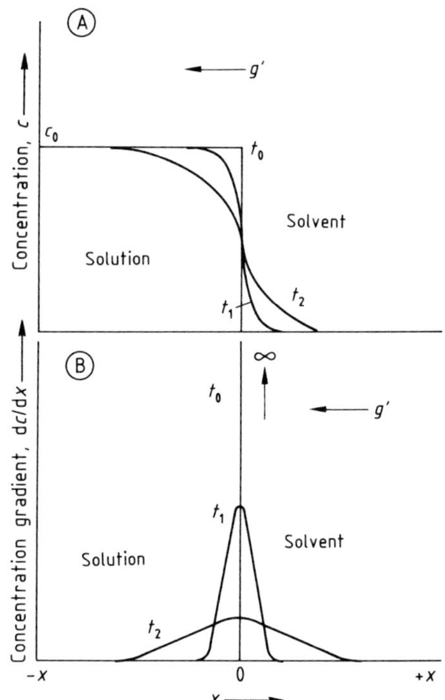

Figure 65. A plot of: A) Equation (161), and B) Equation (159). Area under the time dependent curve in B) is constant [97]

obtained as a solution for diffusion of a concentration pulse [28]. In diffusion that follows a pulse, the variable dC/dx in Equation (159) (and Fig. 65) is equivalent to the concentration. Alternatively, the pulse effect is to transform ordinary concentration gradients that evolve from a boundary into concentration profiles and vice versa. Determination of D' by ultracentrifugation can be done by starting with Equation (144) and applying simplifying assumptions such as constant D'.

The fact that the boundary is moving and its concentration gradient profile may deviate from a Gaussian curve introduces factors that must be corrected. This is the reason for the superiority of the static over the dynamic method. At low speed the effect of S_g on the shape of the diffusing boundary may be neglected, but at higher speed [31], [97], [106]

$$D' = D'_{sed}(1 - \omega^2 S_g t) \tag{164}$$

where D'_{sed} denotes the sedimentation coefficient derived from experimental ultracentrifugal data. Allowance for a variation of S_g with concentration must be made, and an advanced version for determination of D' has been suggested [107], [108]. The main reason for presenting the methods of evaluating D' in an ultracentrifuge is the desirability of obtaining S_g and D' from the same equipment. Otherwise the standard static methods for determination of D' are preferable.

Determination of Molecular Mass. Ultracentrifugation techniques provide the means for evaluating the molecular masses of materials which range from as low as 300 up to 50×10^6 or higher. The following three methods and their attendant equations can be used to determine molecular mass [31], [97]:

1. Sedimentation–diffusion:

$$M_r = \Omega S_{g0}/D\,, \quad S_{g0} = S_g(C \to 0) \tag{165}$$

2. Equilibrium:

$$M_r = 2(\Omega/\omega^2)\,\mathrm{d}\ln C/\mathrm{d}r^2 \tag{166}$$

3. Approach to equilibrium:

$$M_r = \Omega(\mathrm{d}C/\mathrm{d}r)_m/(\omega^2 C_m r_m) \tag{167}$$

where

$$\Omega = RT/(1 - v\varrho_\varphi) \tag{168}$$

Equations (165)–(167) are valid at infinite dilution. Note that in Equation (165)—the Svedberg equation—D (not D') is given by Equation (67), hence the requirement of infinite dilution. Furthermore, at infinite dilution $\varrho_\varphi \to \varrho_f$, where ϱ_f is the density of the carrier fluid which may be a solution containing dissolved species other than the one being tested. In the following, an extended Svedberg equation with regard to the hydrodynamic effect of concentration is derived by using the microscopic particle drift model [31]. The drift of molecules in centrifugal fields is assumed to involve full transmission of stresses from the sedimenting molecules to the carrier fluid. Under these conditions $F + F_D = 0$ prevails along the sedimentation path (see Section 2.3.4). Hence, using Equations (42) and (46) for monodisperse molecules gives

$$\varrho_p V_p(1-\varphi)(1-\varrho_f/\varrho_p)\omega^2 r = 3\pi d\mu'_\varphi U_\varphi \tag{169}$$

where μ'_φ is given by Equation (46). Multiplying both sides of Equation (169) by $\tilde{N}$; substituting the molecular mass M_r for $\tilde{N}\varrho_p V_p$, $S_g/(1-\varphi)$ (Equation 140) for $U_\varphi/(\omega^2 r)$, RT/D for $3\pi\tilde{N}d\mu_f$, v for $1/\varrho_p$, and $C\tilde{N}V_p$ for φ; and rearranging give

$$\begin{aligned} M_r &= \frac{RTS_g}{D(1-v\varrho_f)} \cdot \frac{1}{(1-C\tilde{N}V_p)f_D(C)} \\ &= \frac{\Omega(S_g/D)}{(1-C\tilde{N}V_p)f_D(C)} \end{aligned} \tag{170}$$

where $f_D(C)$ is given by Equation (68) and S_g is the sedimentation coefficient at concentration C.

Combining Equations (66) and (170) gives

$$M_r = \frac{RTS_g}{D'(1-v\varrho_f)} \cdot \frac{(1 + \sum_i U'_i/k_B T + \nabla\ln f_a/\nabla\ln C)}{(1-C\tilde{N}V_p)} \tag{171}$$

Note that D' is the diffusion coefficient at concentration C. For dilute solutions of the sedimenting molecules, $(1-C\tilde{N}V_p)f_D(C)$ can be approximated by Equation (10) or (11). Equation (170) reduces to the Svedberg equation for $c \to 0$. Use of the diffusion coefficient as in Equation (171) calls for further considerations. If $C \to 0$, then D' is equal to D only if $\sum_i U'_i/k_B T = 0$. For example, electromagnetic and interfacial energies may be associated with large sedimenting molecules or colloids, in which case the above condition does not hold. Consider now the Svedberg equation for polydisperse molecules or colloidal mixtures. In this case, combining Equations (42) and (46) for molecules pertaining to the ith fraction and using

$$U_{\varphi i} = U_{di} + U_f = U_{di} + \sum_{j=1}^{n} U_{dj}\varphi_j/(1-\varphi)$$

give

$$\begin{aligned} &V_p(\varrho_{pi} - \varrho_\varphi)\omega^2 r \\ &\quad = 3\pi d_i \mu'_\varphi \left[U_{di} + \sum_{j=1}^{n} U_{dj}\varphi_j/(1-\varphi)\right] \end{aligned} \tag{172}$$

$$\varrho_\varphi = \sum_{j=1}^{n} \varrho_{pj}\varphi_j + \varrho_f(1-\varphi) \tag{173}$$

By using the same procedure as above and recalling that

$$S_{gi} = U_{di}/\omega^2 r\,,$$

$$M_{ri} = \frac{RT\left[S_{gi} + \sum_{j=1}^{n} S_{gj}\varphi_j/(1-\varphi)\right]}{D(1-v_i\varrho_\varphi)} \cdot \frac{(1-\varphi)}{f_D(C)} \tag{174}$$

where

$$\varphi_j = \tilde{N} C_j V_{pj} \quad \text{and} \quad \varphi = \sum_{j=1}^{n} \varphi_j$$

Note that Equations (170) and (174) are extensions of the Svedberg equation accounting only for the hydrodynamic effect of concentration.

If intermolecular interactions other than hydrodynamic exist, then further corrections are needed. For determination of molecular mass, see [31] and [97].

3.5.2. Magnetic and Magnetocentrifugal Sedimentation

Magnetic forces can be made comparable to diffusion and centrifugal forces that act on particles down to the colloidal size range [29], [109]. Magnetic forces that act on polarizable particles can arise due to magnetic field gradients, as well as to the formation of concentration gradients in the dispersion of such particles (which are placed in an externally uniform field). Nonuniform fields generate forces that directly affect the sedimentation coefficient in the same way the centrifugal field does. This force per particle can be expressed by

$$F_H = \mu_0 V_p \cdot \sum_{i=1}^{3} M_i \nabla H_{ei} \tag{175}$$

where i denotes the coordinate system used (for example, $i = 1, 2, 3$ can represent the cartesian x, y, z system) and subscript e means that the field H is external. High g' fields (see Eq. 6) can be generated by using high-intensity, high-gradient magnetic fields (→ 19. Magnetic Separation, p. **19**-6). The formation of nonuniform concentration profiles in an externally uniform (or nonuniform) magnetic field generates a magnetic diffusion effect that can be expressed in terms of the energy U_i' (see Eq. 66). In externally uniform magnetic fields [29],

$$U_{M1}' = \mu_0 V_p H \cdot M \tag{176}$$

$$U_{M2}' = \mu_0 V_p C \cdot \sum_{i=1}^{3} H_i \partial M_i / \partial C \tag{177}$$

where subscript M denotes that magnetization via concentration is the thermodynamic unconstrained variable, and U_{M1}' and U_{M2}' are the magnetic energies of particles and those due to particle-particle interaction, respectively.

In an externally nonuniform field,

$$U_{H1}' = \mu_0 v V_p C M^2 \tag{178}$$

$$U_{H2}' = \mu_0 v V_p C^2 \cdot \sum_{i=1}^{3} M_i \partial M_i / \partial C \tag{179}$$

where subscript H denotes that the field strength is the thermodynamic unconstrained variable. In many cases, $U_{M2}' \ll U_{M1}'$ and $U_{H2}' \ll U_{H1}'$ can be assumed. The significance of U_{H1}' (being directly proportional to CM^2) increases with concentration and particle magnetization, i.e., for highly permeable particles in concentrated dispersions; U_{M1}' becomes dominant at high field strengths. The value of U_{M1}' can be made comparable to $k_B T$ for a wide range of particle magnetization and size. For example, at 298 K, 10-nm paramagnetic particles having susceptibility as low as $\chi_p = 10^{-6}$ require fields of 30 T to this end, whereas only 1 T is sufficient if particle size is 0.1 μm. If particles are made of ferri- or ferromagnetic material, then the field can be substantially reduced. Thus for a 1-nm magnetite particle, the above condition is fulfilled at a field intensity of 2 T, the magnetite magnetization being 0.56 T. Note that in ferrofluids, particle size is generally an order of magnitude larger. Hence for ferromagnetic particles in the colloidal range and larger, U_M' and U_H' can be made to exceed $k_B T$, thus providing the means for magnetically controlled diffusion. For example, imposing a sufficiently high magnetic field on a mixture of colloids having differential magnetizations and subjecting it to a centrifugal field can be used for subsequent separation. This subject is yet to be developed.

Magnetocentrifugal separation (*MCS*) incorporates the combined action of centrifugal and nonuniform magnetic fields [109]. Particles undergo sedimentation in a centrifugal field and then are subjected to a high-gradient magnetic field along the sedimentation path. High gradients can be produced by placing a ferromagnetic matrix in an external field. Such a matrix can be an array of ferromagnetic wires, spheres, etc. Three alternative MCS systems using a one-dimensional array of wires with different orientations toward the centrifugal field are shown in Figure 66. Note that the high-gradient zones are generated in the vicinity of the wires and that typical wire diameters range between 10 and 1000 μm depending on particle size. Figure 67 shows schematic trajectories of two particles of different sizes and the same susceptibility. In Fig-

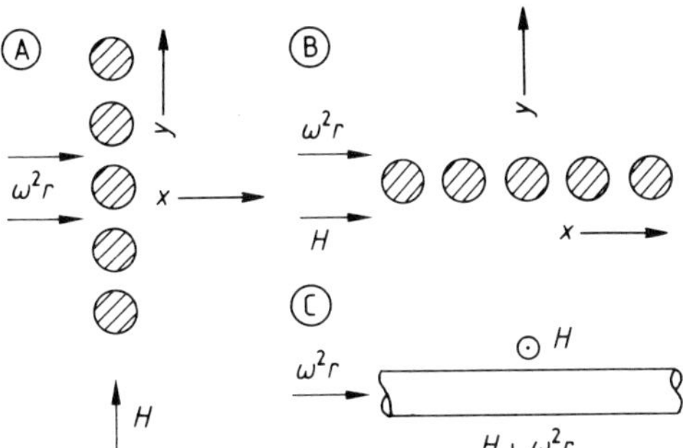

Figure 66. Three alternative magnetocentrifugal separation systems using arrays of ferromagnetic wires set at different orientations towards the centrifugal field [109]

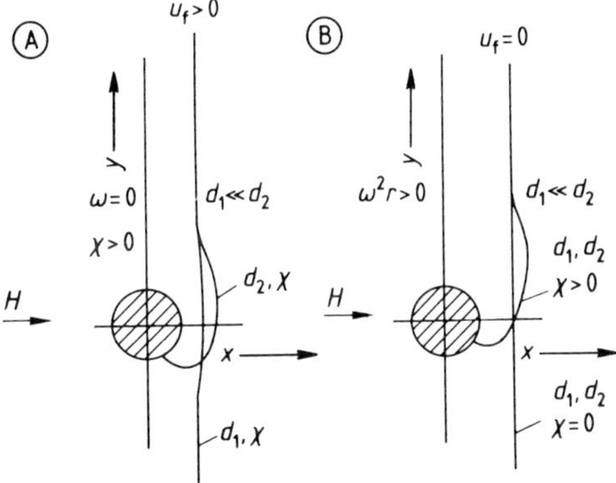

Figure 67. Trajectories of two particles of different sizes and the same susceptibility [109]
A) HGMS system: $U_f > 0$, $\omega^2 r = 0$; B) MCS system: $u_f = 0$, $\omega^2 r > 0$

ure 67 A, $\omega = 0$ and $u_f > 0$; in Figure 67 B, $\omega > 0$ and $u_f = 0$. In the first case, which is typical of high-gradient magnetic separation (HGMS), particle trajectory depends on particle size, whereas in the second case (MCS) it does not. Note that the wire is perpendicular to the plane of the drawing. Thus in MCS, if particle concentration is low, trajectories depend solely on magnetization. In magnetocentrifugal separation, particles of higher magnetization can be captured and separated from those that flow past the magnetic barrier. Figure 68 is a schematic of a MCS cell. The "magnetic" fraction is filtered out from the "nonmagnetic" fraction that flows on toward the cell bottom. The MCS technique can also be used for density separation in magnetizable fluids [109].

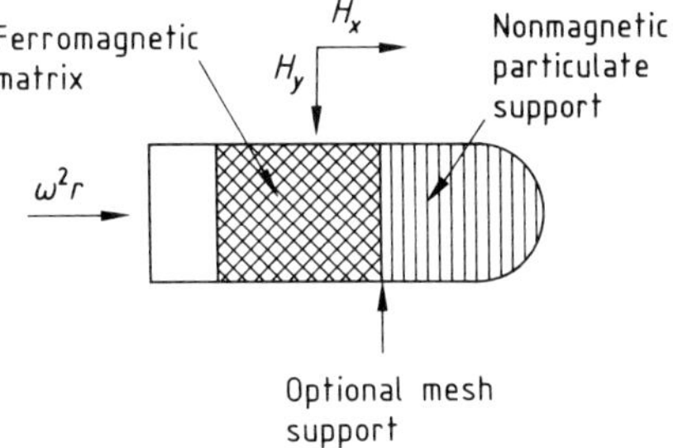

Figure 68. MCS separation cell [109]
Ferromagnetic matrix is on top of a nonmagnetic support

3.5.3. Centrifugal Sedimentation in Flowing Dispersions

Trajectory, Grade Efficiency, and Cut Size. In Section 3.5.1. and 3.5.2., ultracentrifugation and magnetocentrifugal separation involving sedimentation of particles in a constrained fluid, are treated. An exception is the elutriator rotor for separation of whole cells. This section deals with cases involving the combined effect of fluid flow and centrifugal field on particle sedimentation. Figure 69 shows schematically a simple tubular centrifuge. Feed is introduced axially, and clarified fluid overflows the exit port. Although the flow of the dispersion is nearly steady in the bowl frame of reference, for an observer in the disper-

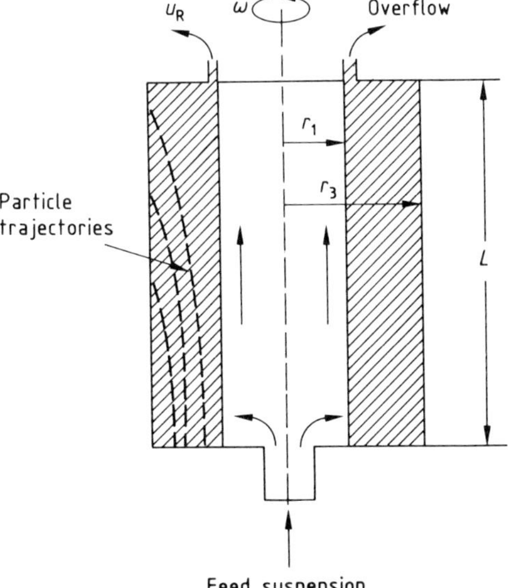

Figure 69. Schematic diagram of a simple tubular centrifuge [110]

sion frame of reference particles undergo non-steady sedimentation. Description of the sedimentation of a polydisperse mixture consisting of n fractions requires the simultaneous solution of a set of n pairs of the following equations:

$$\mathrm{d}r_i/\mathrm{d}t = U_{\mathrm{d}i}(r)\,, \quad \mathrm{d}z/\mathrm{d}t = v(r_i, z)\,, \quad i = 1, 2, \ldots, n \qquad (180)$$

where v denotes the axial fluid velocity. The use of the computer model presented in Section 2.4 is another alternative for a numerical solution. Such detailed solutions are, however, outside the scope of this chapter. Instead a set of simplifying assumptions that facilitate acceptable approximation of centrifuge performance is applied. Monodisperse particles, fine enough so that creeping flow sedimentation prevails, are assumed, and the theoretical background reviewed in [110], [111] is followed. With $g' = \omega^2 r$ and U_φ given by Equation (1), the radial sedimentation velocity is given by

$$U_\mathrm{d}(r) = \mathrm{d}r/\mathrm{d}t = U_\varphi(1 - \varphi) \qquad (181)$$

If L/r_3 is large enough, dispersion entry and fluid discharge effects on axial flow can be neglected. In this case, the axial flow velocity is solely a function of r, which is given by [112]

$$\mathrm{d}z/\mathrm{d}t = \frac{Q}{\pi(r_3^2 - r_1^2)} K_1 \left(r_1^2 \ln\frac{r}{2} + \frac{r_3^2 - r_1^2}{2} \right) \qquad (182)$$

where K_1 is a bowl constant

$$K_1 = \frac{r_3^2 - r_1^2}{\frac{3}{4} r_1^2 + \frac{1}{4} r_3^4 - r_1^2 r_3^3 - r_1^4 \ln(r_1/r_3)}$$

and Q is the volumetric flow rate of the liquid. Particle trajectories can be determined by dividing Equation (181) by Equation (182) followed by integration. Considerable simplification is achieved by using an axial plug flow model, i.e.,

$$\mathrm{d}z/\mathrm{d}t = \frac{Q}{\pi(r_3^2 - r_1^2)} \qquad (183)$$

Substituting Equation (1), with $g' = \omega^2 r$, in Equation (181) and integrating the ratio of Equations (181) and (183) between the limits $r = r$ at $z = 0$ and $r = r_3$ at $z = L$ give

$$\ln(r_3/r) = \frac{K_2 d^2 L \pi (r_3^2 - r_1^2)}{Q} \qquad (184)$$

$$K_2 = \frac{(\varrho_\mathrm{p} - \varrho_\mathrm{f})\omega^2}{18\,\mu_\mathrm{f}} (1 - \varphi) f(\varphi) \qquad (185)$$

The limit of separation is

$$d_{\max} = \left[\frac{Q \ln(r_3/r_1)}{K_2 \pi L (r_3^2 - r_1^2)} \right]^{1/2} \qquad (186)$$

where all particles $d < d_{\max}$ are expected to escape sedimentation if they start at $r = r_1$, $z = 0$. For a given size d, all particles in the annulus between r and r_3 will reach the bowl wall, where r is determined from Equation (184).

$$r = r_3 \exp[-\pi K_2 L (r_3^2 - r_1^2)\, d^2/Q] \qquad (187)$$

For a given size fraction, the ratio of particles deposited on the bowl wall, (i.e., between $z = 0$ and $z = L$) to the feed is defined as the grade efficiency $G(d)$ of this size (For a definition of grade efficiency, → 2. Particle-Size Analysis and Characterization of a Classification Process, p. **2**-10). If the dispersion feed is uniform, $G(d)$ is given by

$$G(d) = (r_3^2 - r^2)/(r_3^2 - r_1^2) = \frac{r_3^2}{r_3^2 - r_1^2} \cdot [1 - \exp(-2\,K_2 K_3 d^2)] \qquad (188)$$

where r is given by Equation (187) and K_3 is a constant equal to the residence time of the slurry in the bowl because

$$K_3 = \frac{\pi(r_3^2 - r_1^2)L}{Q} = \frac{V_\mathrm{b}}{Q} \qquad (189)$$

in which V_b denotes the volume of dispersion in the bowl. The equiprobable cut size d_{50} is defined as the size that has an equal probability of escape or deposition on the bowl wall, i.e., $d_{50} = d[G(d) = 0.5]$. Hence, when $G(d) = 0.5$, Equation (188) gives

$$d_{50}^2 = \frac{1}{2\,K_2 K_3} \ln \frac{2\,r_3^2}{r_1^2 + r_3^2} \qquad (190)$$

The value of d_{50} is often used as a criterion related to the efficiency of separation (→ 2. Particle-Size Analysis and Characterization of a Classificatin Process, p. **2**-10). The effect of particle concentration on $d_{\max}$, r, $G(d)$, and d_{50} occurs via the constant K_2 (Eq. 185). Increase of φ increases $d_{\max}$, r, and d_{50}, and decreases $G(d)$.

The Sigma Concept [111]. A simplified relation of efficiency of the centrifuge in terms of d_{50}, total volumetric flow rate Q, and index of machine size Σ has been developed [113]. This relation involves the sigma concept for which a free sedimentation reference velocity (at $g' = g$) of d_{50} particles is defined by

$$U_g = \frac{(\varrho_p - \varrho_f)\, g\, d_{50}^2}{18\,\mu_f}$$
$$= d_{50}^2 K_2 (g/\omega^2)/[(1-\varphi) f(\varphi)] \quad (191)$$

Combining Equations (189)–(191) gives

$$Q = 2\,U_g \Sigma \quad (192)$$

$$\Sigma = \left(\frac{\omega_1^2}{g}\right) \pi L (r_3^2 - r_1^2)/\ln \frac{2\,r_3^2}{r_1^2 + r_3^2} \quad (193)$$

where ω_1 is defined as an effective angular speed

$$\omega_1^2 = \omega^2 (1-\varphi) f(\varphi) \quad (194)$$

In dilute dispersions, $\omega_1 \to \omega$ and Σ is a constant of the centrifuge. However as Equations (193) and (194) show, it is also generally a function of slurry concentration. Thus Equation (192) facilitates scaleup (in the form of Eq. 195) of a given type of centrifuge as well as "scaleup" in the sense of increasing concentration.

$$Q_1/\Sigma_1 = Q_2/\Sigma_2, \text{ for a constant } d_{50} \quad (195)$$

The parameter Σ has the dimension of area and can be viewed as representing half the area of an ideal settling pool that is required to handle gravitational sedimentation of the same particles. This parameter relates to the grade but not to the throughput of particles, and to this end the following extension is proposed. The particle flux Q_p is related to the fluid flux by $Q_p = \varphi Q$; hence, by using Equation (192),

$$Q_p = 2\,U_g \Sigma_\varphi$$

$$\Sigma_\varphi = \left(\frac{\omega_\varphi^2}{g}\right) \pi L (r_3^2 - r_1^2)/\ln \frac{2\,r_3^2}{r_1^2 + r_3^2} \quad (196)$$

$$\omega_\varphi^2 = \omega^2 \varphi (1-\varphi) f(\varphi) = \varphi \omega_1^2 \quad (197)$$

Equations (196) and (197) show that from the point of view of flux, $\varphi \to 0$ gives a vanishing effective angular speed; hence given ω, an infinitely large sedimentation area [i.e., $(r_3^2 - r_1^2) \to \infty$] must be provided if Σ_φ is to remain finite. On the other extreme, high values of φ call for increased ω so that the d_{50} specifications can be met, but this incurs increased operating and possibly investment costs. Hence the optimum is expected at intermediate values of φ. The estimation of centrifuge capacity, by using classical analysis of gravitational thickening and sedimentation flux curves, has been described [114].

Centrifugal Sedimentation in Rotating Inclined Channels [111]. The enhanced sedimentation of particles in inclined channels in a constant force field is described in Section 2.8. In this section, the position-dependent centrifugal field is considered. The operation of disk centrifuges involves sedimentation in inclined channels, or chambers, that are formed between disks. The inclined disks are mounted on a rotating shaft, as shown in Figure 70. An exact analysis of flow and sedimentation in each channel requires data about the flow profile of the fluid and the solution of nonsteady sedimentation equations. Instead, a simplified plug flow model of flow in the channels is assumed. The flow velocity along the

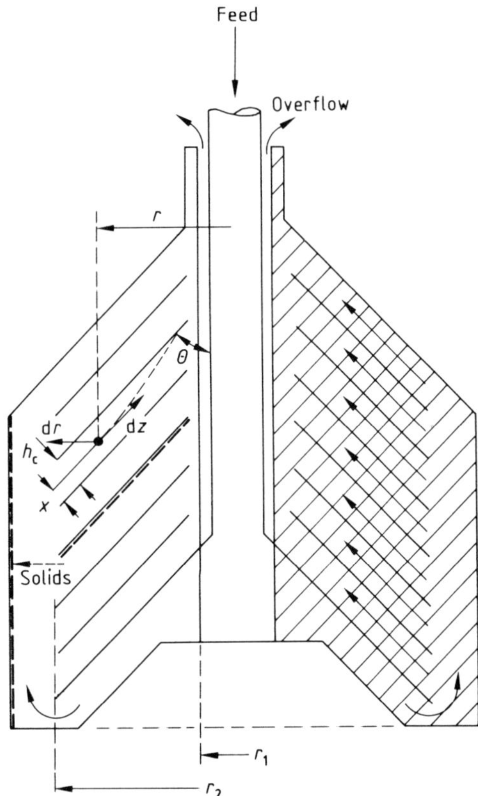

Figure 70. Structure of sedimentation and flow zones in a disk centrifuge, schematic [110]

channel, in the z direction (see Fig. 70), is related to the total flow rate Q by

$$\mathrm{d}z/\mathrm{d}t = (Q/n)/(2\,\pi\, r\, h_c) \tag{198}$$

where h_c is the spacing between adjacent disks. The sedimentation velocity toward the upper disk downward facing wall is related to the radial velocity by

$$\mathrm{d}x/\mathrm{d}t = (\mathrm{d}r/\mathrm{d}t)\cos\theta \tag{199}$$

where θ is the angle between the disk and the shaft. Combining Equations (181), (198), and (199) gives

$$\mathrm{d}x/\mathrm{d}z = K_2 d^2 r^2 \frac{2\pi n h_c}{Q}\cos\theta \tag{200}$$

$$\mathrm{d}x/\mathrm{d}r = -K_2 d^2 r^2 \frac{2\pi n h_c}{Q}\cot\theta \tag{201}$$

In deriving Equation (201), use was made of $\mathrm{d}z = -\mathrm{d}r/\sin\theta$. Integration of Equation (201) between $x = x_1$ at $r = r_2$ and $x = h_c$ at $r = r_1$ gives

$$h_c - x_1 = K_2 d^2 \frac{2\pi n h_c}{Q}(r_2^3 - r_1^3)\cot\theta \tag{202}$$

Equation (202) determines the limiting size given the entry point $x = x_1$ and $r = r_2$—or, alternatively, the limiting entry point given the size d—for which the sedimentation trajectory still crosses the upper wall of the channel (i.e., at $x = h_c$ and $r = r_1$). The size limit $d = d_{max}$ of separation is readily obtained by letting $x = x_1 = 0$, $r = r_2$ at the entry point. Under specific operating conditions, $d > d_{max}$ particles would not report to the overflow. The grade efficiency of particles of size d is given by

$$G(d) = \frac{h_c - x_1}{h_c} \tag{203}$$

where x_1 satisfies Equation (202). Because d_{max} is obtained by substitution of $x_1 = 0$ in Equation (202),

$$(h_c - x_1)/h_c = d^2/d_{max}^2$$

hence,

$$G(d) = d^2/d_{max}^2 \text{ for } d \le d_{max} \tag{204}$$

and

$$G(d) = 1 \quad \text{for } d > d_{max}$$

Letting $G(d) = 0.5$ and using Equation (204) give

$$d_{50} = d_{max}/\sqrt{2} \tag{205}$$

The Σ factor for a disk centrifuge is obtained by combining Equations (191), (192), (202) for $x_1 = 0$ and $d = d_{max}$, and (205).

$$\Sigma = \frac{2}{3}\cdot\frac{\omega_1^2}{g}\pi n(r_2^3 - r_1^3)\cot\theta \tag{206}$$

Replacing ω_1 (Eq. 194) by ω_φ (Eq. 197) in Equation (205) gives Σ_φ instead of Σ.

Centrifugal Sedimentation in Rotating Fluid Streams. Centrifugal sedimentation in rotating fluid streams is well known and is used for solid–liquid (→ 11. Centrifugation and Hydrocyclone Separation), solid–solid (→ 17. Air Classifying, → 16. Elutriation), liquid–gas, and solid–gas (→ 13. Dust Separation) separation. The fluid stream is forced to rotate in the separation zone as it flows through, thus producing centrifugal forces due to its own flow. Such flow ordinarily occurs through devices having fixed curved walls. This technique is used in various dry and wet separators, classifiers, and elutriators (→ 17. Air Classifying; → 16. Elutriation; → 11. Centrifugation and Hydrocyclone Separation). Analysis of sedimentation in rotating streams requires data from the flow field and the solution of n simultaneous nonsteady sedimentation equations, where n denotes the number of sedimentating particle fractions. Computer models of the type described in Section 2.3 can also be used to this end. The present state of the art provides mainly empirical correlations for prediction of the performance of such devices with respect to the products of feed containing relatively concentrated polydisperse mixtures.

3.5.4. Solid–Liquid Separation by Sedimentation

Sedimentation methods are versatile in solid–liquid separation processes. Solid–liquid separation often involves sedimentation, filtration (→ 10. Filtration), or both. Large-scale sedimentation operations have had long-standing success in water treatment (→ Wastewater Treatment, treated in the B-series) and reclamation. Different literature sources provide outlines for the selection of equipment for solid–liquid separation. Sedimentation and filtration are de-

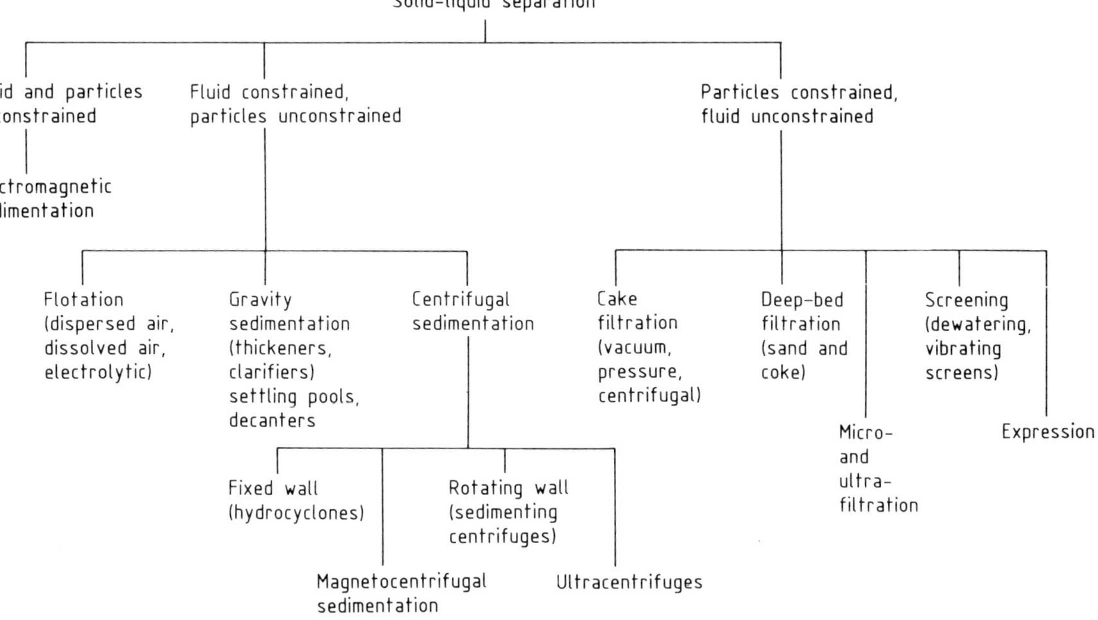

Figure 71. Classification of solid – liquid separation techniques

scribed under the heading of mechanical separation, along with practical detailed information about related equipment, in [115]. A systematic approach to the solution of solid – liquid separation problems is contained in [116]. A comparison of sedimentation and filtration for solid – liquid separation and time scales for sedimentation in different groups of equipment are provided in [117]. These time scales, given for different sludge concentrations, provide initial elimination and selection options for sedimentation equipment, which must then be processed further by experts in the field. Figure 71 shows an extended version of the diagram provided by SVAROVSKY [111] in which the solid – liquid separation is classified according to whether or not the fluid and particles are constrained. Centrifugal sedimentation is introduced in Section 3.5.1 – 3.5.3, and thickeners and clarifiers are considered under Wastewater Treatment, treated in the B series. Flotation (→ 23. Flotation) can be viewed as reverse sedimentation or "buoyant sedimentation," in which particle-laden bubbles float or undergo negative sedimentation. In electromagnetic sedimentation, fluid and particles are generally unconstrained, with magnetocentrifugal sedimentation being an exception. Sedimentation techniques in many ways complement filtration and vice versa; in solid – liquid separation processes, sedimentation usually precedes filtration as the first step. This is due to the fact that often none of these techniques provides a completely viable economic solution if used as the sole separation step.

4. References

[1] Y. Zimmels: "Accelerated and Steady Particle Flows in Newtonian Fluids," *Encyclopedia of Fluid Mechanics,* vol. 5, Chap. 3, Gulf Publishing Co., Houston 1986, pp. 93 – 153.
[2] R. H. Davis, A. Acrivos, *Ann. Rev. Fluid Mech.* **17** (1985) 91 – 118.
[3] E. Barnea, J. Mizrahi, *Chem. Eng. J. (Lausanne)* **5** (1973) 171 – 189.
[4] P. U. Foscolo, G. L. Gibilaro, S. P. Waldram, *Chem. Eng. Sci.* **38** (1983) no. 8, 1251 – 1260.
[5] J. F. Richardson, W. N. Zaki, *Trans. Inst. Chem. Eng.* **32** (1954) 35 – 53.
[6] J. Garside, M. R. Al-Dibouni, *Ind. Eng. Chem. Process Des. Dev.* **16** (1977) 206 – 214.
[7] J. M. Coulson, J. F. Richardson: *Chemical Engineering,* 3rd ed., vol. 2, Pergamon Press, Oxford 1978.
[8] G. K. Batchelor, *J. Fluid Mech.* **52** (1972) 245 – 268.
[9] A. S. Sangani, A. Acrivos, *Int. J. Multiphase Flow* **8** (1982) 343 – 360.
[10] Y. Zimmels, *Powder Technol.* **34** (1983) 191 – 202.
[11] Y. Zimmels, *AIChE J.* **29** (1983) no. 4, 669 – 676.
[12] Y. Zimmels, *J. Appl. Phys.* **56** (1984) no. 7, 2133 – 2141.
[13] Y. Zimmels, *Powder Tech.* **56** (1988) no. 4, 227 – 250.

[14] J. F. Davidson, D. Harrison: *Fluidization*, Academic Press, New York 1971.

[15] D. Kuni, O. Levenspiel: *Fluidization Engineering*, J. Wiley & Sons, New York 1969.

[16] S. Mirza, J. R. Richardson, *Chem. Eng. Sci.* **34** (1979) 447–454.

[17] E. G. Kelly, D. J. Spotiswood: *Introduction to Mineral processing*, J. Wiley & Sons, New York 1982.

[18] H. Lamb: *Hydrodynamics*, Dover, New York 1945.

[19] L. M. Milne Thompson: *Theoretical Hydrodynamics*, 5th ed. Macmillan Publ. Comp, London 1974.

[20] V. L. Streeter: *Fluid Dynamics*, McGraw-Hill, New York 1948.

[21] Chia-Shun Yih: *Fluid Mechanics*, McGraw-Hill, New York 1969.

[22] G. B. Wallis: *One Dimensional two phase flow*, McGraw-Hill, New York 1969.

[23] A. B. Basset: *Hydrodynamics*, vol. 2, Dover, New York 1961.

[24] G. Rudinger: "Fundamentals of Gas Particle Flow," *Handbook of Powder Technology*, Elsevier, Amsterdam 1980.

[25] A. T. Hjelmfelt, L. F. Mockros, *Appl. Sci. Res.* **16** (1966) 149–161.

[26] T. Allen: *Particle Size Measurements*, 3rd ed. Chapman and Hall, London 1981.

[27] B. H. Kaye: *Direct Characterization of Fine Particles*, J. Wiley & Sons, New York 1981.

[28] E. L. Cussler: *Diffusion Mass Transfer in Fluid Systems*, Cambridge University Press, Cambridge 1984.

[29] Y. Zimmels: "Diffusion of Polarizable Colloids in Electromagnetic Fields" *J. Colloid Interface Sci.*, in press.

[30] R. B. Bird: *Theory of Diffusion, Advances in Chemical Engineering*, vol. 1, Academic Press, New York 1956, pp. 155–239.

[31] H. K. Schachman: *Ultracentrifugation in Biochemistry*, Academic Press, New York 1959.

[32] P. L. Meyer: *Introductory probability and Statistical Applications*, 2nd ed., Addison-Wesley, Reading, Mass. 1970.

[33] E. Barnea, J. Mizrahi, *Can. J. Chem. Eng.* **53** (1975) 461–467.

[34] E. Barnea, J. Mizrahi, *Ind. Eng. Chem. Fundam.* **15** (1976) 120–125.

[35] V. G. Levich: *Physicochemical Hydrodynamics*, Prentice Hall, Englewood Cliffs, N.J. 1962.

[36] J. Newman, *J. Chem. Eng. Sci.* **22** (1967) 83.

[37] G. I. Taylor, *Proc. R. Soc. London Ser. A* **138** (1932) 41.

[38] A. Leviton, A. Leighton, *J. Phys. Chem.* **40** (1936) 71.

[39] P. U. Foscolo, L. G. Gibilaro, S. P. Waldram, *Chem. Eng. Sci.* **38** (1983) no. 8, 1251–1260.

[40] Y. Zimmels: "A Generalized Approach to Flow through Fixed Beds Fluidization and Hindered Sedimentation," *Chem. Eng. Commun.*, **67** (1988) 19–42.

[41] E. Barnea, R. L. Mednick, *Chem. Eng. J.* (*Lausanne*) **15** (1978) 215–227.

[42] J. M. Dallavalle: *Micromeritics*, 2nd ed., Pitman, (London) 1948.

[43] N. L. Weiss (ed.): *SME Mineral Processing Handbook*, Society of Mining Engineers, American Institute of Mining, Metallurgical and Petroleum Engineers Inc., New York 1985.

[44] R. Gerber, R. Briss: *High Gradient Magnetic Separation*, Research Studies Press, Chichester 1983.

[45] A. E. Boycott, *Nature* (*London*) **104** (1920) 532.

[46] W. D. Hill: *Boundary-Enhanced Sedimentation due to Setting Convection*, Ph. D. Thesis, Carengie-Mellon Univ. Pittsburgh, PA 1974.

[47] R. H. Perry, C. H. Chilton (eds.): *Chemical Engineers Handbook*, 6th ed., McGraw-Hill, New York, 1984.

[48] L. Svarvosky (ed.): *Solid-Liquid Separation*, Chap. 7, Butterworths, London 1977.

[49] E. Ponder, *Q. J. Exp. Physiol.* **15** (1925) 235–252.

[50] N. Nakamura, K. Kuroda, *Keijo J. Med.* **8** (1937) 256–296.

[51] W. D. Hill, R. R. Rothfus, K. Li, *Int. J. Multiphase Flow* **3** (1977) 561–583.

[52] I. Robinstein, *Int. J. Multiphase Flow* **6** (1980) 473–490.

[53] R. F. Probstein, R. Yung, R. Hicks in M. P. Freeman, J. A. Fitzpatrick (eds.): *Physical Separation, "A model of lamella settlers,"* Engineering Foundation, New York, 1981, pp. 53–92.

[54] W. F. Leung, R. F. Probstein, *Ind. Eng. Chem. Process Des. Dev.* **22** (1983) 58–67.

[55] A. Acrivos, E. Herbolzheimer, *J. Fluid Mech.* **92** (1979) 435–457.

[56] E. Herbolzheimer, A. Acrivos, *J. Fluid Mech.* **108** (1981) 485–499.

[57] E. Herbolzheimer, *Phys. Fluids* **26** (1983) 2043–2054.

[58] R. H. Davis, E. Herbolzheimer, A. Acrivos, *Phys. Fluids* **26** (1983) 2055–2064.

[59] W. F. Leung, *Ind. Eng. Chem. Process Des. Dev.* **22** (1983) 68–73.

[60] R. H. Davis, E. Herbolzheimer, A. Acrivos, *Int. J. Multiphase Flow* **8** (1982) no. 6, 571–585.

[61] J. Weisman, L. Efferding, *AIChE J.* **6** (1960) 419–426.

[62] L. E. Gates, J. R. Morton, P. L. Pondy, *Chem. Eng.* (*N.Y.*) **83** (1976) 144–150.

[63] F. Aerstin, G. Street: *Applied Chemical Process Design*, Plenum Press, New York 1978.

[64] R. L. Bates, P. L. Pondy, R. R. Corpstein, *Ind. Eng. Chem. Process Des. Dev.* **2** (1963) no. 4, 310–314.

[65] E. J. Wasp, J. P. Kenny, R. L. Gandhi: *Solid–Liquid Flow Slurry Pipeline Transportation*, 1st ed., 1978, Trans. Tech. publications, Clausthal, Federal Republic of Germany, Gulf Publishing Co., Houston 1979.

[66] M. P. O'Brien, *Trans. Am. Geophys. Union* **14** (1933) 487–491.

[67] V. A. Vanoni, *Trans. ASCE* **111** (1946) 67–133.

[68] H. M. Ismail, *Trans. ASCE* **117** (1952) 409–447.

[69] E. J. Wasp et al.: "Cross Country Coal Pipeline Hydraulics," Pipe Line News (July 1963) pp. 20–30.

[70] E. J. Wasp et al.: "Deposition Velocities, Transition Velocities and Spatial Distribution of Solids in Slurry Pipelines," *1st. Int. Conf. on Hydr. Transport of Solids in Pipes*, BHRA Fluid Engn., Cranfield U.K., paper H4, Sept. 1970.

[71] R. L. Whitmore, *Br. J. Appl. Phys.* **6** (1955) 239–245.

[72] R. H. Weiland, R. R. McPherson, *Ind. Eng. Chem. Fundam.* **18** (1979) 45–49.

[73] R. H. Weiland, Y. P. Fessas, B. V. Ramarao, *J. Fluid Mech.* **142** (1984) 383–389.

[74] Y. P. Fessas, R. H. Weiland, *Int. J. Multiphase Flow.* **110** (1984).

[75] W. Resnick, Y. Zimmels, D. Boadi: *Magnetic Structural Effects on Flow through Beds of Magnetizable Particles*, In MT10, Boston, MA, Oct. 1987.
[76] R. E. Rosensweig: *Ferrohydrodynamics*, Cambridge University Press, Cambridge 1985.
[77] D. G. Thomas, *AIChE J.* **8** (1962) 373–378.
[78] M. E. Charles, G. S. Stevens: "The Pipeline Flow of Slurries-Transition Velocities," *2nd Int. Conf. on Hyd. Transport of Solids in Pipes*, BHRA Fluid Eng. Cranfield, U.K., paper E3, Sept. 1972.
[79] T. C. Aude et al., *Chem. Eng.* (June 28, 1971) 74.
[80] D. M. Newitt et al., *Trans. IChemE* **33** (1955) 93–102.
[81] W. E. Wilson, *Trans. ASCE* **107** (1942) 1576–1594.
[82] R. Durand: "Basic Relationships of the Transportation of Solids in Pipes," *Experimental Research, Proceedings*, International Hydraulics Convention, Minnesota 1953, pp. 89–103.
[83] R. Durand, F. Condolios: "Hydraulic Transport of Coal and Solid Materials in Pipes," *Proceedings of Colloquium on Hydraulic Transport of Coal* (Nov. 5–6, 1952). National Coal Board London England 1953, pp. 39–52.
[84] C. G. Sinclair: "The Limit-Deposit Velocity of Heterogeneous Suspensions," *Interactions between Fluid and Particles*, Inc. Chem. Eng. London 1962.
[85] N. Yotsukura: *Some Effects of Bentonite Suspensions on Sand Transport in Smooth 4-inch Pipe*, Ph. D. Thesis, Colorado State University, 1961.
[86] M. Wicks: "Transportation of Solids at low Concentration in Horizontal Pipes," *ASCE Int. Symposium on Solid–Liquid Flow in Pipes*, Univ. of Pennsylvania, March 1968.
[87] R. E. McElvain, I. Cave: "Transportation of Tailings," *World Mining Tailings Symposium*, 1972.
[88] M. S. Peters, K. D. Timmerhaus: *Plant Design and Economics for Chemical Engineers*, 3rd ed., McGraw-Hill, New York 1980, pp. 377–383.
[89] H. Braur, E. Kriegel: "Hydraulic Conveying of Solids in Horizontal Pipes," *Bänder Bleche Rohre* **6** (1965) no. 6, 315–324; also in Proc. Hydrotransp. **2** (1972) E1–1.
[90] A. R. Oroskar, R. M. Turian, *AIChE J.* **26** (1980) no. 4, 550–558.
[91] G. A. Wani in N. P. Cheremisinoff (ed.): "Critical Velocity in Multisize Particle Transport Through Pipes," *Encyclopedia of Fluid Mechanics*, Chap. 4, vol. 5, Gulf Publ. Co., Houston 1986, pp. 155–187.
[92] R. Darby in N. P. Cheremisinoff (ed.): "Hydrodynamics of Slurries and Suspensions," *Encyclopedia of Fluid Mechanics*, Chap. 2, vol. 5, Gulf Publ. Co., Houston 1985, pp. 49–91.
[93] G. W. Govier, M. W. Charles, *Eng. J.* **44** (1951) 50.
[94] D. G. Thomas, *AIChE J.* **10** (1964) 303–308.
[95] E. J. Wasp et al.: "Hetero-Homogeneous Solid-Liquid Flow in the Turbulent Regime," *ASCE Int. Symposium on Solid Liquid Flow*, Univ. of Pennsylvania, March 1968.
[96] R. W. Hanks in N. P. Cheremisinoff (ed.): "Principles of Slurry Pipeline Hydraulics," *Encyclopedia of Fluid Mechanics*, Chap. 6, vol. 5, Gulf. Publ. Co., Houston 1986, 213–276.
[97] T. J. Bowen: *An Introduction to Ultracentrifugation*, Wiley-Interscience, London 1970.
[98] R. J. Goldberg, *J. Phys. Chem.* **57** (1953) 194.
[99] R. Trautman, V. N. Schumaker, *J. Chem. Phys.* **22** (1954) 551.
[100] T. Svedberg, K. O. Pedersen: *The Ultracentrifuge*, Oxford Univ. Press, London, Johnson Reprint Corporation, 1940.
[101] E. G. Pickels, W. F. Harrington, H. K. Schachman, *Proc. Natl. Acad. Sci. U.S.* **38** (1952) 943.
[102] R. L. Baldwin, *Biochem. J.* **55** (1953) 644.
[103] R. V. Webber, *J. Am. Chem. Soc.* **78** (1956) 536.
[104] R. Trautman, *J. Phys. Chem.* **60** (1956) 1211.
[105] L. J. Gosting, *Adv. Protein Chem.* **11** (1956) 429.
[106] O. Lamm, *Z. Phys. Chem. Ser. A* **143** (1929) 177.
[107] H. Fujita, *J. Chem. Phys.* **24** (1956) 1084.
[108] K. Kawahara, *Biochemistry* **8** (1969) 2551.
[109] Y. Zimmels, *J. Appl. Phys.* **60** (1986) no. 6, 2162–2174.
[110] L. Svarovsky in L. Svarovsky (ed.): "Separation by Centrifugal Separation," *Solid Liquid Separation*, Chemical Engineering series, Butterworths, London 1977, pp. 125–147.
[111] L. Svarovsky, in L. Ricci (ed.): "Advances in Solid–Liquid Separation II: Sedimentation Centrifugation and Flotation," *Separation Techniques 2: Gas/Liquid/Solid Systems*, Chemical Engineering, McGraw-Hill, New York 1980, pp. 19–31.
[112] H. K. Schachman, *J. Colloid Phys. Chem.* **52** (1948) 1034–1045.
[113] C. M. Ambler, *Chem. Eng. (N.Y.)* **48** (1952) 150–158.
[114] P. A. Vesilind in L. Ricci (ed.): "Estimating Centrifuge Capacities," *Separation Techniques 2: Gas/Liquid/Solid Systems*, Chemical Engineering, McGraw-Hill, New York 1980, pp. 169–172.
[115] *Solid/Liquid Separation Equipment Scale-Up*, edited by D. B. Purchas, Uplands press, 1977.
[116] F. M. Tiller, "How to select Solid–Liquid Separation Equipment," in *Separation Techniques 2: Gas/Liquid/Solid Systems*, Chemical Engineering, McGraw-Hill, New York 1980, pp. 42–45.
[117] H. G. W. Pierson in L. Svarovsky (ed.): "The Selection of Solid–Liquid Separation Equipment," *Solid–Liquid Separation*, Chemical Engineering Series, Butterworths, London 1977, pp. 313–326.

13. Dust Separation

FRIEDRICH LÖFFLER, Universität Karlsruhe (TH), Karlsruhe, Federal Republic of Germany

The following symbols are used in this article:

a	plate spacing
A	surface area
c	particle concentration
c_m	limiting dust load
C_D	drag coefficient
d_G	diameter of granules
d_l	droplet diameter
D_F	fiber diameter
e	diameter of flow tube
E	overall collection efficiency
f	circular cross-sectional area
f	factor
f'	fitting parameter
F	filtration area
h	adhering fraction
H	height
K_1, K_2	empirical parameters
K_1	residual resistance of filter medium after cleaning
K_2	specific resistance of filter cake
L	collection length
L	specific scrubbing liquor ratio
m	specific cleaned volume
m	particle mass, filter mass per unit filter area
$\dot{M}$	mass flow rate

Δp	pressure drop
P	penetration
$q(x)$	particle-size distribution
r	radius
R	resultant of external forces
Re	Reynolds number
t	time
T	absolute temperature
$T(x)$	fractional collection efficiency
U_0	gas velocity, superficial velocity
v	velocity
v	air to cloth ratio
v_{rel}	relative velocity between the droplet and the gas
$\dot{V}$	volume flow rate
w	particle velocity
$w(x)$	migration velocity
W	mass of dust deposited per unit filter area
x	particle size (diameter)
x_{cr}	critical particle diameter
x_t	cut size
Z	thickness
ε	porosity
η	collision efficiency
η_D	diffusion effect
η	transport parameter
μ	dynamic viscosity
μ_m	maximum dust load
ξ	pressure-drop coefficient
ϱ	gas density
ϱ_F	fiber density
φ	collection efficiency of single fiber or granule
ψ	intertial (Stokes) parameter

Subscripts

c	collected
e	entry
f	fine
F	fiber
g	gas
i	internal radius of exit duct
l	scrubbing liquor
p	particle

1. Introduction

Whether desired or not, many industrial processes inevitably generate particles entrained by a carrier gas. These particles may be solid (often called "dust") or liquid ("droplets" or "mist"). As a rule, the particles must be separated from the transporting gas, either because they are the desired product or because they are an undesirable or even harmful emission.

Product recovery by means of dust (gas–solid) separation has been practiced long and intensively in process engineering. The importance of emission control has grown in recent years, and standards have become more stringent. Rapid increases in industrial production, on the one hand, and a better knowledge of potential hazards and harmful effects, on the other, have led to more stringent regulations for the control of air pollution. One example is the emission limits established in "TA Luft 1986" [5], an administrative regulation issued by the Federal Republic of Germany, which tolerates a maximum emission of 50 mg/m^3 of inert dust only. In many cases, which are defined in these regulations, the specified emission value is much lower than this. For dust that is harmful to health, the limits are geared to the potential hazard and run from 5 mg/m^3 (e.g., antimony, lead), through 1 mg/m^3 (e.g., arsenic, nickel) and 0.2 mg/m^3 (e.g., cadmium, mercury), to 0.1 mg/m^3 (carcinogens).

Because dust concentration upstream of collection equipment is frequently in the range of 1–10 g/m^3, collection efficiencies greater than 99.99 % are often necessary to meet the above limits. Such values can be achieved only through the most painstaking selection and design and, in many cases, only by combining several types of equipment. One difficulty is that the size distribution of the particles to be collected generally extends over a wide range, from < 0.1 µm to > 100 µm; the harmful components (e.g., heavy metals), which demand especially efficient collection, often fall in the fine range below 1 µm.

Dust collection methods used in practice can be classified as

1) inertial and centrifugal separation (cyclones),
2) wet collection (scrubbers),
3) filtration (fabric filters and granular-bed separators), and
4) electrostatic collection (precipitators).

All of these collectors separate particles from gas by a common principle: forces acting on the particles cause them to enter regions from which the gas cannot transport them away. These regions may be the inner wall of a cyclone, the droplet surface in a scrubber, the fiber or grain surface in a filter, or the collecting electrode in a precipitator.

An elementary theory of gas–solid separation thus consists of calculating particle trajectories by solving the equation of motion [6, p. 22]:

$$m \cdot \mathrm{d}w/\mathrm{d}t = R \tag{1}$$

where m is the particle mass, w is the particle velocity, t is time, and R is the resultant of the external forces acting on the particle (such as gravitational, hydrodynamic and electrical forces).

As a rule, Equation (1) is solved numerically. A problem arises because the flow field of the gas must be entered into the equation and this field is often not known precisely. Furthermore, a random component of motion may also be involved, allowance for which can be difficult. In practice, therefore, approximations are used or a purely empirical approach is taken, in which design is based on experimental findings for a given type of equipment. The drawback of the empirical method is that the extent to which experience can be extrapolated to new situations is unclear.

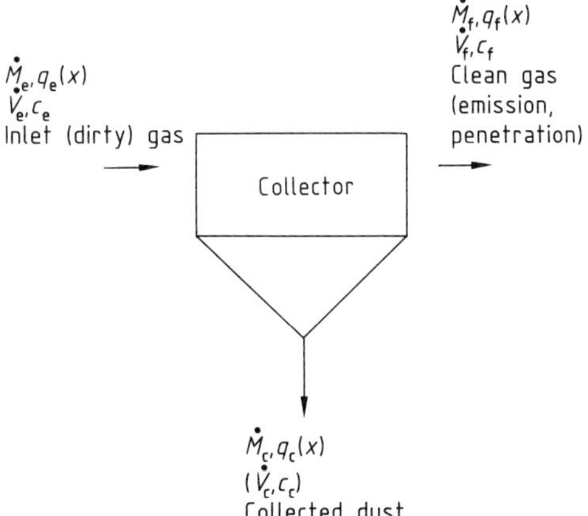

Figure 1. Notation used in dust collectors
$\dot{M}$ = mass flow rate of particles; $\dot{V}$ = volume flow rate of gas; c = particle concentration; $q(x)$ = particle-size distribution (density distribution)
Subscripts: c = collected (coarse); e = entry (inlet); f = fine (penetrating)

2. Evaluation of Dust Collection Equipment

To evaluate dust collection equipment, both the energy consumption (or pressure drop Δp) and the collection efficiency must be determined. The latter can be found by integral techniques (for the totality of particles) or differential methods (for each size fraction).

A general plan for characterizing a separation can be found in DIN 66 142. Here, only the case is considered where the solid material entering along with the crude gas is assumed to be separated into two components. The nomenclature has been adapted to common usage in separation technology; important parameters are depicted and defined in Figure 1.

The *overall collection efficiency* (integral mass balance) E, also known as the total or cumulative collection efficiency, is given by

$$E = \frac{\dot{M}_c}{\dot{M}_f} \tag{2a}$$

or

$$E = 1 - \frac{\dot{M}_t}{\dot{M}_e} \tag{2b}$$

In view of the importance of emissions, especially for collection efficiencies close to 100%, the *penetration* P is more informative:

$$P = 1 - E \tag{3}$$

or

$$P = \frac{\dot{M}_f}{\dot{M}_e} \tag{4}$$

As a rule, dust loads upstream and downstream of the collection equipment are measured experimentally and E is determined as follows:

$$E = \frac{\dot{V}c_e - \dot{V}_f c_f}{\dot{V}_e c_e} \tag{5}$$

with $\dot{V}_e = \dot{V}_f$, then

$$E = 1 - \frac{c_f}{c_e} \tag{6}$$

or

$$P = \frac{c_f}{c_e} \tag{7}$$

Note that, if a collection device is rated in terms of overall collection efficiency, this quantity depends not only on the equipment design and operating conditions, but also on the type of dust and, especially, the particle-size distribution. The stated overall collection efficiency or overall penetration thus always refers to a particular case; this severely restricts its applicability to other situations.

A more useful and informative measure is the *fractional collection efficiency,* also called the "grade efficiency," "separation efficiency," or "separation function." The fractional collection efficiency $T(x)$ measures collection as a function of particle size, x; it is defined as

$$T(\mathrm{x}) = 1 - \frac{\mathrm{d}\dot{M}_c(x)}{\mathrm{d}\dot{M}_e(x)} = 1 - \frac{\mathrm{d}c_f(x)}{\mathrm{d}c_e(x)} \quad (8)$$

Hence,

$$T(x) = 1 - \frac{Pq_f(x)}{q_e(x)} \quad (9\,a)$$

or

$$T(x) = \frac{Eq_c(x)}{q_e(x)} \quad (9\,b)$$

Procedures for measuring $T(x)$ are based on Equations (9 a) and (9 b); Equation (9 a) is employed most commonly in separation technology.

Typical fractional collection curves are presented in subsequent chapters on individual types of collection equipment. The *cut size* x_t is the particle size at which a given fractional collection efficiency is achieved. A value of 50 % is often used (x_{50}) but other cut sizes (e.g., x_{90}, x_{99}) may be chosen depending on technical requirements and specific applications. Cut size and collection (separation) efficiencies are discussed in detail elsewhere in this volume (→ 2. Particle Size Analysis).

One advantage of the fractional collection efficiency is that, being a ratio, it is independent of the quantity (e.g., number or mass) used to express the amount of particulate material in the particle-size ranges upstream and downstream from the collection equipment. The fractional collection efficiency is a property of the apparatus and can thus be used to estimate the expected overall collection efficiency for any given particle size distribution; this is important in the selection and sizing of dust collectors:

$$E = 1 - P = \int_{x_{\min}}^{x_{\max}} T(x) \cdot q_e(x)\,\mathrm{d}x \quad (10)$$

The overall collection efficiency calculated by means of Equation (10) is, of course, expressed in the same quantity as the particle-size distribution of the inlet gas $q_e(x)$; it is usually a mass fraction but can also be a number fraction.

3. Centrifugal Collectors (Cyclones)

3.1 General

The simplest way to separate particles from gases is to allow them to settle out in so-called gravity settlers into zones of low gas velocity [7]. Because this method is technically feasible only for coarse particles that are significantly larger than 100 µm, gravity settlers now have negligible importance and are not discussed here.

Even before the end of the 19th century, collection by means of centrifugal forces was known to be much more efficient than simple gravity separation. This led to the development of centrifugal collectors; the most commonly used form employs flow reversal and is called a gas cyclone.

Because of their simple design, reliable operation, small space requirement, and low cost, cyclones are used widely in many branches of industry. They are frequently employed in material recovery from gas recycle systems, pneumatic conveying, and other areas.

Cyclones can be operated at pressures of 0.001 – 10 MPa and temperatures exceeding 1000 °C. They currently provide the only method of particle collection at high temperature for use on an industrial scale. Growing interest in dust removal from hot gases has given a new impetus to research and development work on cyclones.

The advantages of cyclones over other collectors are offset by the fact that they are less efficient collectors for particles below ca. 1 – 5 µm in size. Therefore, in view of the increasingly stringent requirements on collecting capacity, the use of cyclones is restricted; they are often used as first-stage collectors (precollectors, see Section 7.3).

3.2. Mode of Operation and Basic Designs

Figure 2 shows the most common type of cyclone with flow reversal. The gas path is designed to impart a twist to the particle-laden gas entering the rotationally symmetrical apparatus.

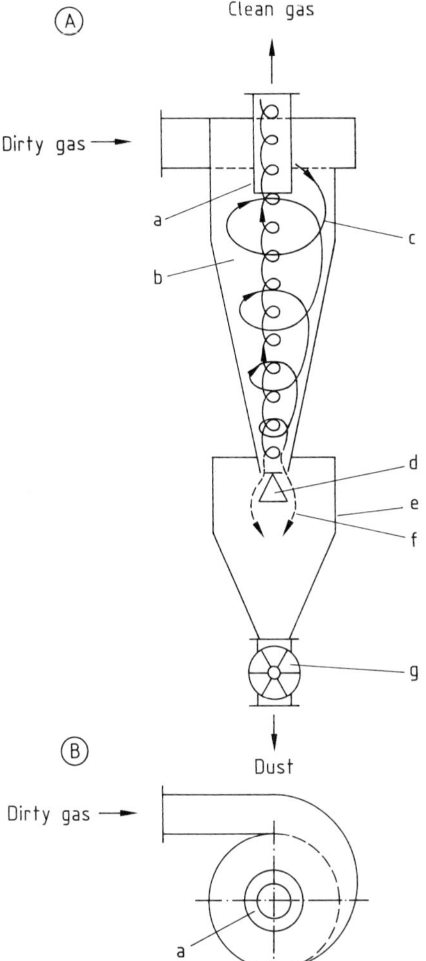

Figure 2. Schematic of a reverse-flow cyclone
A) Vertical section; B) Top view
a) Exit duct; b) Separating chamber; c) Gas flow pattern; d) Conical shield; e) Dust hopper; f) Dust; g) Lock

A turbulent, three-dimensional, rotational flow is produced. In the separating chamber spiral, downward flow takes place at larger radii. Flow is then reversed, and the gas spirals upward at smaller radii, and reaches the exit duct. The particles entrained in the revolving gas stream experience centrifugal forces hundreds to thousands of times greater than the force of gravity. Thus, the larger particles migrate outward to the cyclone wall, where they collect as strands; they are carried downward along the tapered body and out of the separating chamber by boundary-layer flow. The conical shield (d), often built into the bottom of the tapered body, prevents already collected material from being reentrained. The particles that are not collected are carried inward by the gas flow and removed through the exit duct.

Cyclones vary in the form of the swirl-producing inlet (tangential, axial with swirl vanes); the shape of the separating chamber (cylindrical, conical); the design of the exit duct; and their dimensions. Diameters range from 5 m to 2 cm (the latter for measurement purposes); flow rates vary from 1 to 100 000 m^3/h. Reviews of cyclone construction are given in [3], [8], [9].

3.3. Design Calculations

Important parameters in cyclone design and operation are the fractional collection efficiency $T(x)$ (defined in Eq. 8, p. **13**-4) and the energy consumption, which is measured in terms of the pressure drop Δp.

3.3.1. Collection Efficiency

In theory, particle trajectories inside the cyclone can be calculated to determine the fractional collection efficiency; such calculations have been attempted [10]. This method provides insight into processes occurring in the cyclone which improve the understanding of particle separation. However, because the flow field is complicated and turbulence must be considered, this procedure is quite complex and not yet adequately proven. Thus far, it has been primarily of scientific interest and has not been used practically.

Models for Low Dust Loads. Several models have been devised for the approximate calculation of collection efficiency at low dust loads [12]–[16]. They describe the flow field in more or less simplified terms. The flow field is shown schematically in Figure 3. An especially important parameter is the tangential velocity component v_φ, which initially increases with decreasing radius and can be approximated in this region by

$$v_\varphi \cdot r^n = \text{constant} \qquad (11)$$

Here r is the radius and n is an empirical exponent that takes into account the loss of angular momentum due to friction. In a loss-free vortex, n would be unity; in real gas flow, it lies between 0.5 and 0.8. The value of n depends on other factors, such as wall roughness, particulate load, and cyclone geometry [11].

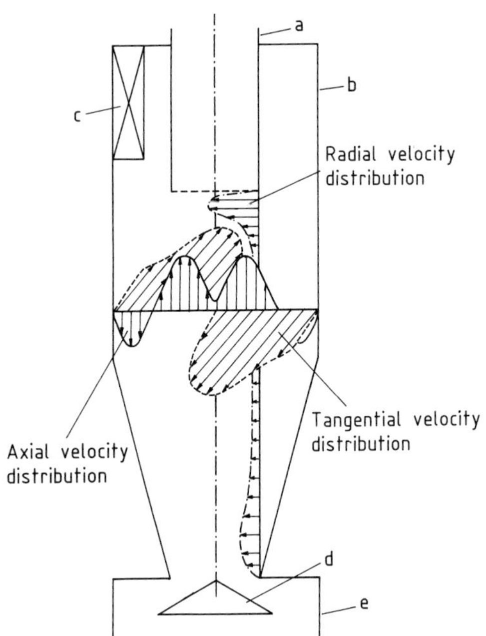

Figure 3. Flow field in a gas cyclone
a) Exit duct; b) Separating chamber; c) Gas inlet; d) Conical shield; e) Dust hopper

Because of a strong "vortex" in the cyclone, the tangential velocity depends mainly on the radius and not on the height. In the core of the cyclone, the tangential component decreases rapidly and approaches zero at the axis.

Design models are based either on a residence-time concept [12] or a separation-zone concept [13], [14]. Recent theoretical work combines both concepts [15], [16].

The separation-zone model proposed by BARTH [13] and improved by MUSCHELKNAUTZ [14] is now employed widely because of its simplicity and ease of use. However, this approach does not give the complete fractional efficiency curve; only a *critical particle diameter* x_{cr} can be calculated [17]:

$$x_{cr} = \left[\frac{18\, \mu_f \cdot v_{r,i} \cdot r_i}{(\varrho_p - \varrho_f) \cdot v_{\varphi,i}} \right]^{\frac{1}{2}} \tag{12}$$

where μ_f is the dynamic viscosity of the gas, ϱ_p is the particle density, ϱ_f is the gas density, r_i is the internal radius of the exit duct, $v_{r,i}$ is the radial velocity at radius r_i, and $v_{\varphi,i}$ is the tangential velocity at radius r_i.

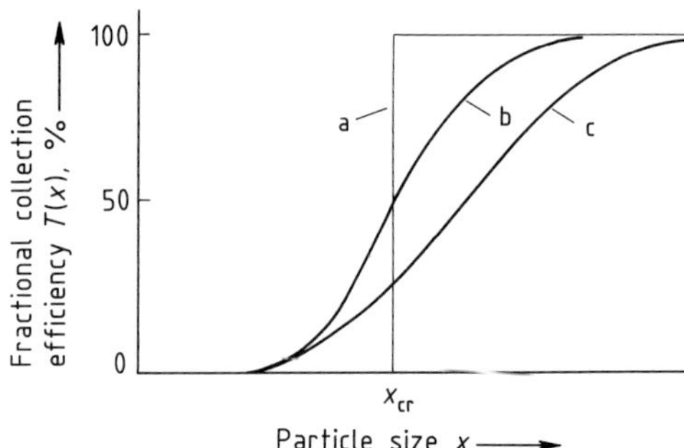

Figure 4. Fractional efficiency curves for cyclones
a) Theoretical cut diameter [13]; b) Assumed effective curve [13]; c) Experimental curve [14]

According to this concept, all particles whose diameter x is less than x_{cr} pass through the device; all particles with $x > x_{cr}$ are collected.

In theory, the fractional efficiency curve jumps from 0 to 1 at x_{cr} (curve a, Fig. 4). BARTH assumed that x_{cr} corresponds to the median diameter x_{50} of the separation function $T(x)$, i.e., the 50% value on the effective S-shaped fractional efficiency curve (curve b, Fig. 4). According to experimental results, the shape of $T(x)$ depends on the cyclone geometry (curve c, Fig. 4). Therefore, the suggestion has been made that the actual cut diameter be estimated as $x_{50} \approx 1.3\, x_{cr}$ [14].

Although using Equation (12) to predict the critical diameter involves some uncertainty and the complete fractional efficiency curve cannot be derived, experience shows that this formula has an important advantage: experimental values of x_{50} can be converted easily and reliably to different conditions, i.e., the change in the cut diameter produced by a change in cyclone geometry, gas velocity, or gas temperature (viscosity) can be estimated.

In a more recent approach [16], particle motion is treated as a process in which diffusional particle transport is superimposed on a predetermined motion. Particle flux calculations are based on a modified flow field [11]. The advantage of this model is that the complete fractional efficiency curve can be obtained and the entire cyclone geometry allowed for. The computing time needed for numerical calculation is short, and processing can be done with an ordinary personal computer. Comparison has shown good agreement with measured values [16].

Models for High Dust Loads. The computational techniques described up to this point have all presumed that the particle concentration entering the cyclone is low and, therefore, has a negligible effect on flow conditions in the separating chamber. This is the case up to a particle load of ca. 10 g/m^3. In practice, however, cyclones often operate at higher concentration. Experiment has shown that the collection efficiency of a cyclone improves with increasing particulate load in the inlet gas, even though the tangential velocity (which is responsible for collection in the vortex) and the centrifugal acceleration decrease simultaneously.

At higher particle concentration, additional collection mechanisms contribute to particle separation [16], and these must also be taken into account in design.

By analogy with pneumatic conveying, MUSCHELKNAUTZ [14] reasons that a fluid stream can transport only a limited particle mass flow. If a limiting dust load c_m is exceeded, the excess particle mass is accelerated to the wall immediately after entering the cyclone and moves down into the dust trap in the form of a strand. Collection due to a dust load in excess of c_m is assumed to be selective only at low total loads and increasingly nonselective (i.e., independent of particle size) at higher loads. If c_m is exceeded and the particles thrown to the wall are collected in the inlet, the remaining particles can be collected selectively in the separating chamber. Thus, particle collection at high concentration can be considered as two consecutive processes. Equations for calculating the overall collection efficiency of a cyclone are given in [17].

Along with particle collection due to excess loading, agglomeration of small particles with large ones can be assumed to occur at high particle concentration; this also improves overall collection efficiency. A model that takes account of agglomeration is described in [16]. Larger particles migrate to the wall of the cyclone at a higher radial velocity than the small particles. This difference in settling velocity leads to collisions between particles of different sizes and hence to agglomeration. Separation of the small particles adhering to the larger ones is thus improved.

3.3.2. Pressure Drop

The total pressure drop Δp of a cyclone consists of pressure losses in the inlet, separating chamber, and exit duct; it is usually calculated as follows:

$$\Delta \varrho = \xi \frac{\varrho_f}{2} v_i^2 \tag{13a}$$

where

$$\xi = \xi_{inlet} + \xi_{separator} + \xi_{exit} \tag{13b}$$

Here ξ is the pressure-drop coefficient, ϱ_f is the density of the gas, and v_i is the axial velocity in the exit duct.

The calculation of ξ is described in [17]. Experience has shown that the pressure drop in the exit duct accounts for as much as 90 % of the total cyclone pressure drop. Usually, ξ varies from 5 to 40, depending on cyclone geometry; values of 10–20 are common. Typical values for Δp are in the range of 500–2000 Pa.

The collection efficiency and the pressure drop depend on cyclone geometry and volume flow rate. These quantities can be combined and used to find the "optimum" cyclone with either a minimum pressure drop [18] or minimum equipment volume [14], [19] for a given cut diameter. Calculations of this kind are not implemented, however, because they yield unrealistically long cyclone designs, especially for fine particles. Empirical shapes are, therefore, generally used.

3.4. Operating Characteristics

The collection performance of a cyclone depends chiefly on its geometry, gas throughput, inlet gas concentration, and properties of the material to be separated. The change in cut size under different operating conditions can be estimated with Equation (12). For example, an increase in gas viscosity (e.g., when the temperature is raised) decreases performance. If the cyclone geometry is altered (e.g., exit-duct diameter or inlet cross section is diminished), the tangential velocity component $v_{\varphi,i}$ at the exit-duct radius increases so that a greater centrifugal force acts on the particles. As a result, the cut size x_{50} decreases and particle collection improves.

Collection efficiency is also improved if the volume flow rate through the cyclone increases. Figure 5 shows some fractional efficiency curves for an experimental cyclone in which the inlet velocity v_e, and thus the volume flow rate $\dot{V}$ are varied. The cut size decreases with increasing $\dot{V}$. The following relation between cut size and gas throughput can be derived from Equation (12):

$$x_{50} \sim \frac{1}{\sqrt{\dot{V}}} \tag{14}$$

When the tangential velocity in the cyclone increases, however, the pressure drop Δp also increases. To achieve good collection performance and acceptable energy consumption when gases with high flow rates have to be cleaned, several small cyclones connected in parallel are preferable to a large cyclone because the reduction in cyclone diameter shifts the cut size toward smaller particle sizes. Figure 6 presents typical fractional efficiency curves for three cyclones with different diameters but the same pressure drop (the gas flow rate $\dot{V}$ decreases at the smaller diameters). Cleaning a gas with a given flow rate, therefore, requires more cyclones as the cyclone diameter decreases. If geometrically similar cyclones are assumed and Δp is constant, the result of Equations (12) and (13) is that the cut size for n cyclones in parallel decreases in proportion to $(1/n)^{1/4}$.

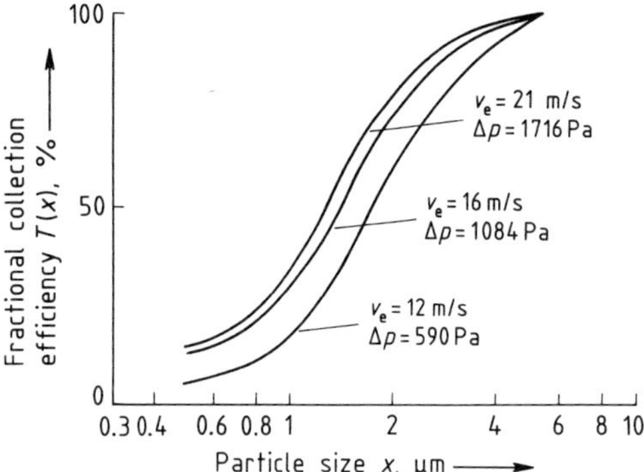

Figure 5. Fractional efficiency curves for a cyclone at various inlet velocities v_e and pressure drops Δp
Particle concentration = 0.5 g/m^3; external radius of cyclone = 95 mm; radius of exit duct = 37 mm; cyclone height = 612 mm; cross-sectional area of inlet duct = 0.003 m^2

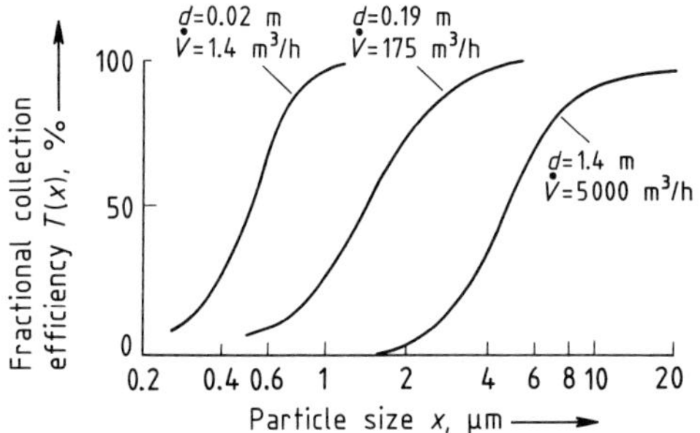

Figure 6. Fractional efficiency curves for cyclones of varying diameter d at a constant pressure drop
$\dot{V}$ = volume gas flow rate

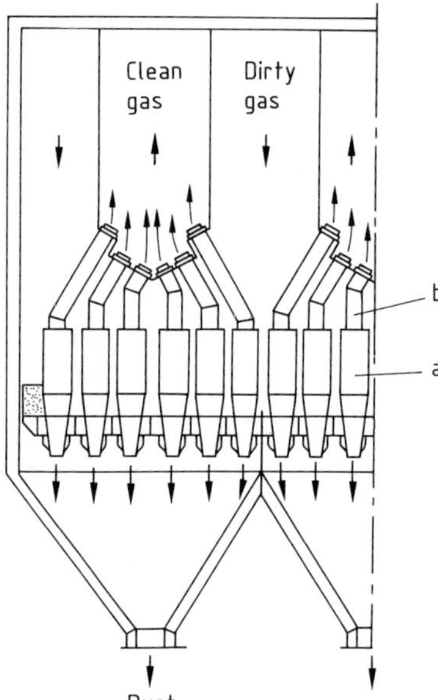

Figure 7. Multicyclone (Lurgi, Multiklon) with bent clean-gas pipes
a) Separating cell; b) Clean-gas pipe

Systems composed of many small cyclones are often called *multicell collectors* or *multicyclones*. An example is shown in Figure 7.

Cyclones with diameters of a few centimeters and particle cut sizes less than 1 µm are used mainly for testing purposes rather than dust collection.

Substantial differences are often found between calculated and measured collection efficiencies. This discrepancy occurs because the cyclone design is unsuitable for the internal flow pattern and, thus, unfavorable for collection.

An especially important point is the correct design of the dust removal system. Long vertical tubes have proved valuable, as have conical shields which must be mounted below the cyclone cone. The cyclone should not be too long, and the ratio of the dust discharge opening to the exit-duct radius should be greater than one; this prevents the inner vortex core from being deflected toward the wall and reentraining already collected particles.

In multicell collectors, the cells should be loaded uniformly; short-circuit flow at the dust removal ends of the cells leads to poorer separation and must be prevented.

Finally, no welded joints or other flow-obstructing ridges should protrude into the cyclone. Such irregularities can yield secondary flow patterns that reduce collection performance. Detailed information on cyclone design can be found in [20].

4. Scrubbers

4.1. General [21]

In wet collectors (scrubbers), both particulate and gaseous components can be separated from the carrier gas. Only particle collection is discussed below; for gas separation see →8. Absorption.

Even very fine particles (≤ 0.1 μm) can be collected in scrubbers. The particles are brought into contact with a scrubbing liquor, generally water; they bind to the liquor and are removed from the gas stream in bound form as a slurry. Thus two operations must take place in a scrubber: (1) incorporation (binding) of particles into the scrubbing liquor and (2) separation of the dust-laden liquor. In most scrubbers, especially those with the best collection efficiency for fine particles, the scrubbing liquor is in droplet form. The first operation can, therefore, be regarded as the agglomeration of (small) solid particles and (larger) liquid particles. These larger agglomerates are readily collected from the gas stream. However, care must be taken to ensure the efficient separation of dust-laden liquid droplets, especially in high-energy scrubbers which use relatively fine droplets.

Scrubbers are relatively easy to adapt to various operating conditions. They are employed mainly when products are to be recovered from a wet phase or when the danger of dust explosion or other hazards forbids dry collection. They have proved suitable chiefly for moderately large-volume gas flow rates (often $< 30\,000$ m^3/h).

The solid materials collected in scrubbers are transferred from a gas into a liquid, usually water; therefore, their use entails either wastewater treatment (which may also be dictated by the presence of absorbed gases such as hydrogenchlorate or fluorine) or recycling of the scrubbing liquor in the plant. In either case, water consumption has to be minimized by setting up a closed liquor cycle and optimizing collection conditions. Finally, the danger of corrosion or icing should be mentioned.

4.2. Mode of Operation and Basic Designs

Many scrubber designs are available; examples can be found in [1] and [22]. However,

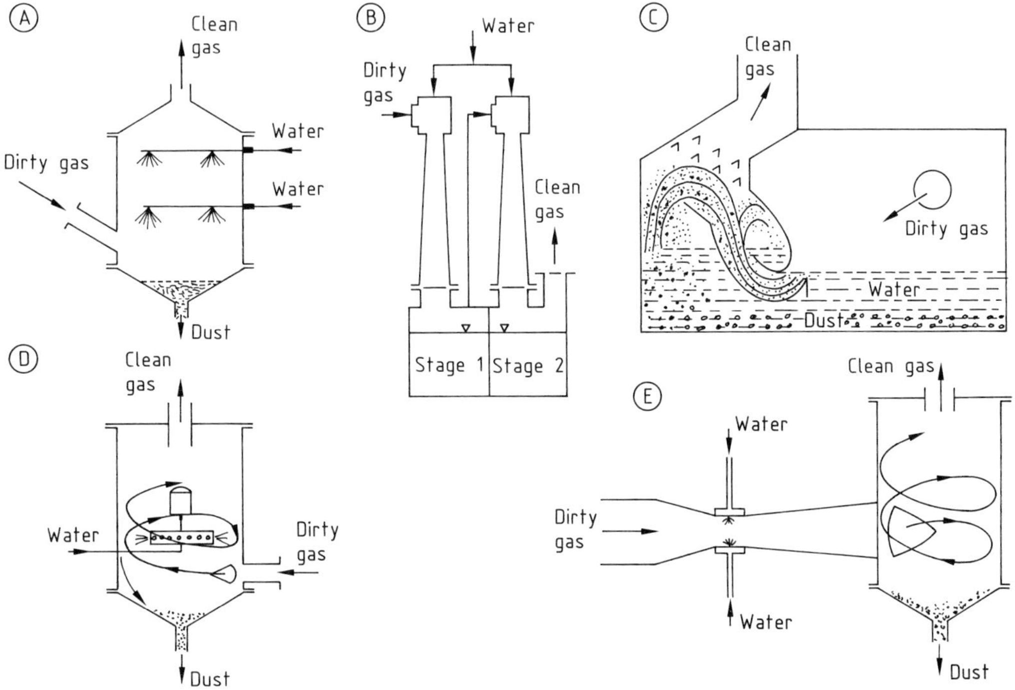

Figure 8. Basic types of scrubbers
A) Spray tower; B) Ejector venturi scrubber; C) Self-induced spray scrubber; D) Rotational scrubber; E) Venturi scrubber

most designs fall into five basic types, shown in Figure 8.

The *spray tower* (Figure 8 A) is the oldest type of scrubber. The dirty gas flows upward through a large tube at a relatively low speed of ca. 1 m/s. The washing fluid is dispersed countercurrent to the gas by spray nozzles that are mounted at different levels. The dust-laden liquid leaves the bottom of the scrubber. Packings, e.g., glass beads, are sometimes used to improve collection efficiency (packed-bed scrubbers).

The *ejector venturi scrubber* (Fig. 8 B) is basically a water-jet pump. Water is sprayed into the scrubber at a speed of ca. 25–35 m/s and provides a draft for moving the gas which often attains a flow rate of 10–20 m/s. Two successive stages are often used because the collection efficiency of ejector venturi scrubbers is not much higher than that of spray towers.

Self-induced spray scrubbers (Fig. 8 C) are available in many designs. In contrast to other scrubbers, the dust-laden gas is fed through the wash fluid. The wash fluid is thus atomized, entrained, and mixed with the gas. This effect is reinforced by a redirecting vane. Entrained, dust-laden fluid collects inside the scrubber in the collection tank. The resulting slurry must be removed at intervals. Water consumption is limited to that required to replace losses associated with the gas or slurry removal.

In *rotational scrubbers* (Fig. 8 D), the washing fluid is dispersed radially in the gas by rotating spray nozzles. The gas spirals upward and the dirty water is removed from the bottom of the scrubber.

The *venturi scrubbers* (Fig. 8 E) have the highest collection efficiency. The gas is accelerated to up to 100 m/s in the venturi throat (orifice). The washing fluid enters the middle of the throat just in front of its narrowest point or is fed radially into the throat. The liquid is dispersed by the inflowing gas. The dust-laden water droplets are separated from the gas in a cyclone.

Typical operating data are listed in Table 1. Increased collection (i.e., reduction in critical particle diameter) is associated with greater relative velocity and higher energy consumption. With venturi scrubbers, for example, a cut size (50 % value on the fractional efficiency curve) of about 0.1 μm can be attained, but the pressure loss increases to 20 kPa.

Scrubbers operate by transporting dust particles to the scrubbing liquor and binding them to it. The wettability of the particles does not play a crucial role in binding; wettable particles penetrate the interior of the droplet, while nonwettable particles adhere to the surface [1]. The transport step is decisive in separation. Diffusion and condensation may certainly take place, but in general, inertial forces appear to dominate. Furthermore, because the scrubbing liquor is assumed to be mostly in droplet form, the transport process can be described in clear terms.

The gas flow lines diverge as they approach the liquor droplet (Fig. 9). By virtue of their inertia, the particles follow the flow lines only partially, if at all; instead, they fly toward the droplet surface. Particles strike the droplet if

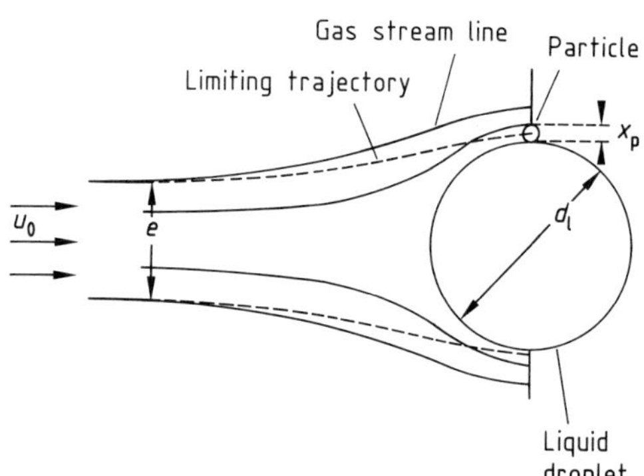

Figure 9. Trajectories and flow lines for flow past a spherical droplet
U_0 = gas velocity; e = diameter of flow tube; d_l = droplet diameter; x_p = particle diameter

Table 1. Operating data for scrubbers

Parameter	Packed-bed scrubber	Ejector venturi scrubber	Self-induced spray scrubber	Mechanical scrubber	Venturi scrubber
Cut size, μm*	0.7–1.5	0.8–0.9	0.6–0.9	0.1–0.5	0.05–0.2
Relative velocity, m/s	1	10–25	8–20	25–70	40–150
Pressure drop, kPa	0.2–2.5		1.5–2.8	0.4–1.0	3–20
Water consumption, L/m^3	0.05–5	5–20**		1–3**	0.5–5
Energy consumption, kW · h/1000 m^3	0.2–1.5	1.2–3	1–2	2–6	1.5–6

* Particle density ϱ_p = 2.42 g/cm^3. ** For each stage.

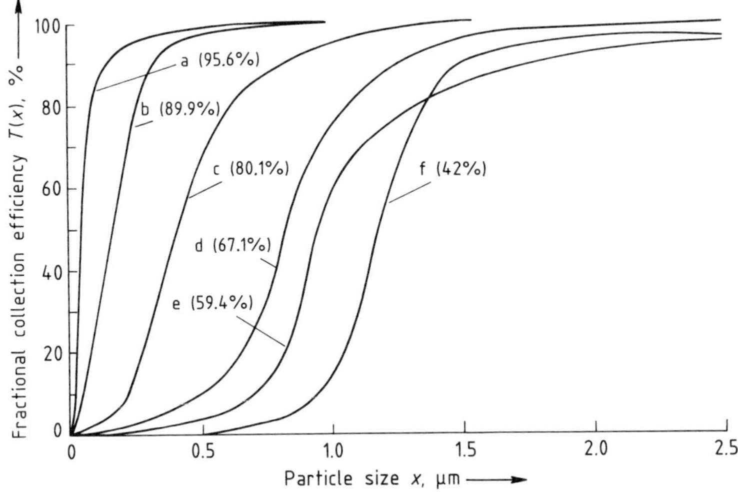

Figure 10. Fractional collection efficiencies for various scrubbers [24]
Measurements were made using sillitin, mean particle size = 1.5 µm, particle density = 2.42 g/cm^3; percentages indicate the overall collection efficiency for sillitin.
a) Venturi scrubber, Δp = 10 kPa; b) Two-stage rotational scrubber; c) Single-stage rotational scrubber; d) Self-induced spray scrubber; e) Single-stage ejector venturi scrubber; f) Packed-bed spray scrubber

their initial position in the undisturbed flow upstream of the droplet is inside a flow tube with diameter e. Particles located outside the tube bypass the droplet.

A *limiting trajectory* can thus be defined as the line separating particles that collide with the droplet from those that pass by it. The *collision efficiency* η is defined as

$$\eta = \left(\frac{e}{d_1}\right)^2 \tag{15}$$

where d_1 is the droplet diameter. Observations indicate that practically all particles striking the droplet adhere to it [23]; therefore, the collision efficiency η corresponds to the single-droplet collection efficiency. Collection in the scrubber is then a result of the overall action of individual droplets. Figure 10 gives fractional efficiency curves and overall collection efficiencies for the scrubbers shown in Figure 8. Measurements were performed with sillitin, a very fine dust composed of silicon dioxide [24].

4.3. Design Calculations

4.3.1. Collection Efficiency

Development and sizing of scrubbers are often purely empirical because the processes that occur in these complicated pieces of equipment are complex and difficult to simulate with theoretical models [22]. However, Barth [25] and Calvert [26] have proposed models that are quite helpful because (1) they allow clear identification of important parameters and trends; (2) they have been at least partly tested by experiment [6], [27]; and (3) they contain geometrical dimensions that are important for size calculations.

In these models, the scrubbing liquor is assumed to be in droplet form, and inertial forces are considered crucial in the transport of dust particles to the droplet (cf. Section 4.2). The calculation consists of three steps:

1) calculation of single-droplet collection,
2) calculation of the gas volume cleaned by the single droplet, and
3) calculation of the change in dust concentration due to the action of all droplets.

The model depicted in Figure 9 is the basis for calculating the single-droplet collection efficiency; the limiting trajectory is found by solving the equations of motion for dust particles numerically. Important parameters are the Reynolds number and the Stokes parameter.

The Reynolds number Re for flow around the droplet is given by

$$Re = \frac{v_{\mathrm{rel}} \cdot d_1 \cdot \varrho}{\mu} \tag{16}$$

where ϱ is the gas density, μ is the viscosity of the

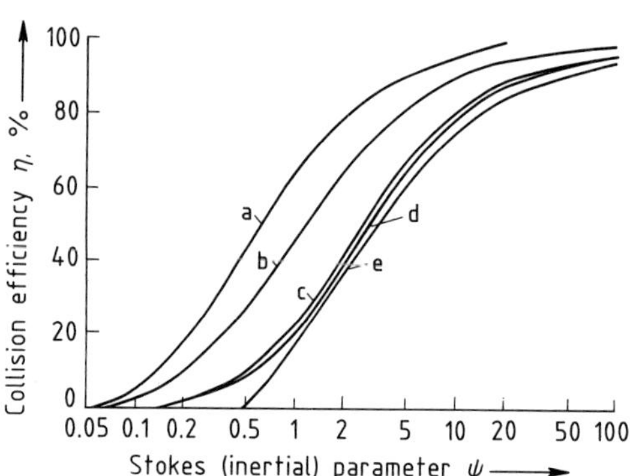

Figure 11. Calculated collision frequencies η vs. Stokes (intertial) parameter ψ for various Reynolds numbers Re
Values of Re: a) $\gg 1$, potential flow; b) 60, 80; c) 40; d) 10, 20; e) < 1, viscous flow

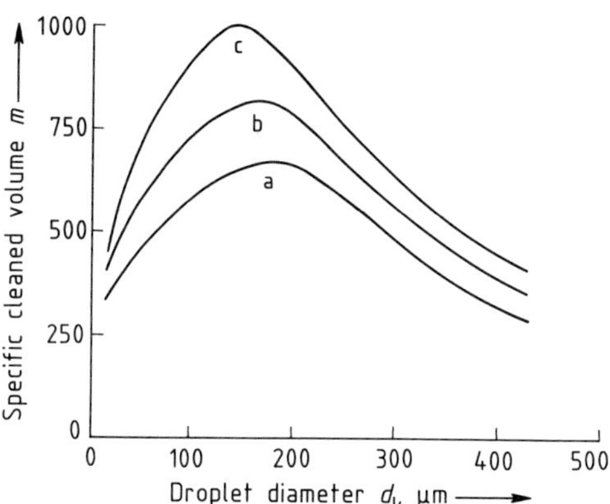

Figure 12. Specific cleaned volume vs. droplet diameter d_l for different particle sizes
Particle size x_p, µm: a) 2.0; b) 2.5; c) 3.2
Velocity of the entering droplets $w_e = 7$ m/s; gas velocity $U = 7$ m/s

gas, and v_{rel} is the relative velocity between the droplet and the gas.

The inertial parameter (Stokes parameter) ψ describes the dynamics of the particles:

$$\psi = \frac{\varrho_p \cdot U_0 \cdot x_p^2}{18\mu \cdot d_l} \tag{17}$$

where ϱ_p is particle density and x_p is particle size.

The numerical results presented in Figure 11 have been confirmed experimentally. To a good approximation [27], they can be calculated as

$$\eta = \left(\frac{\psi}{\psi + a}\right)^b \tag{18}$$

where η is the collision efficiency (Eq. 5) and a and b are constants that depend on the Reynolds number. For most operating conditions, the collection process takes place primarily in the range $Re < 40$. Here the results can be approximated well with $a = 0.65$ and $b = 3.7$ [27].

As the droplets move through the apparatus, they clean a certain volume of gas. The ratio of the cleaned volume to the droplet volume, the *specific cleaned volume m,* is the most important parameter for sizing a scrubber:

$$m = \frac{3}{2d_l}\int_{t_1}^{t_2} \eta(t)\, v_{rel}(t)\, dt \tag{19}$$

where t is time.

The value of m is found by (numerically) solving the equation of motion (Eq. 1) for the droplet. In this way, the relative path length between the droplet and the gas in the time difference $t_2 - t_1$ (residence time) is determined.

The results depend on scrubber geometry, operating conditions (gas velocity and direction), and particle properties (size, density, initial velocity). Equation (19) thus provides an important and useful basis for sizing scrubbers. The drawback of this procedure is that the calculated results cannot be generalized but are valid only for the selected boundary conditions. Furthermore, numerical calculation requires a certain amount of computational effort, but this should not be a problem in view of the present state of computer technology.

An optimal droplet size usually exists for the specific cleaned volume, as can be seen in Figure 12. Under other operating conditions, this optimum can also occur at droplet sizes much smaller than 100 µm [6].

In the third step, the total effect of all the droplets in changing the dust concentration is calculated. The result is the fractional or overall collection efficiency.

The particle concentration c in the scrubber is expressed by

$$c = c_e \cdot \exp(-L \cdot m) \tag{20}$$

where c_e is the particle concentration in the gas inlet to the scrubber; m is the specific cleaned volume; and L is the specific scrubbing-liquor ratio ($L = \dot{V}_l/\dot{V}_g$, where $\dot{V}_l$ is the volume flow rate of scrubbing liquor and $\dot{V}_g$ is the volume flow rate of gas). Typical L values are

$(0.5-5)\times10^{-3}$, i.e., 0.5–5 L of liquor per cubic meter of gas.

To find the fractional separation efficiency $T(x_p)$, m is calculated as a function of particle size x_p:

$$T(x_p) = 1 - \exp[-Lm(x_p)] \qquad (21)$$

The overall collection efficiency is finally obtained by integrating the product of the fractional efficiency and the particle density (cf. Eq. 10, p. **13**-4). The complete droplet-size distribution must be used for these calculations. If a mean, often arbitrarily selected value is used, considerable errors can occur [28].

The model proposed by CALVERT [26] utilizes the same basic approach as that of BARTH. The effort required for the numerical computations is shortened substantially by fitting the equations to experimental values, so that the equation of motion for the droplets need not be solved. However, this procedure entails some loss of general applicability and information content.

For the venturi scrubber, the following equation is given:

$$T(x_p) = 1 - \exp\left[-\frac{2\cdot\dot{V}_1\cdot d_1\cdot\varrho_1\cdot v}{55\cdot\dot{V}_g\cdot\mu}F(\psi,f')\right] \qquad (22)$$

where ψ is the inertial parameter (Eq. 17, p. **13**-12) and f' is an empirical fitting parameter, often in the range of 0.25–0.5.

The following approximation is given for the function $F(\psi, f')$ in the range $1 \leq \psi \leq 4$:

$$F(\psi, f') \approx 0.312\cdot\psi\cdot f'^2 \qquad (23)$$

Equation (22) can also be used, along with a pressure-drop equation, for optimization [6]. In general, model calculations indicate that higher relative velocities and increased addition of scrubbing liquor lead to better collection efficiency; this is also observed experimentally (Table 1, p. **13**-10).

4.3.2. Pressure Drop

Calculation of the pressure drop is of interest chiefly for venturi scrubbers. Empirical formulas and model calculations are available for this purpose.

The *empirical formulas* are based on the procedure for evaluating flow in pipes:

$$\Delta p = \xi\frac{\varrho}{2}v_K^2 \qquad (24)$$

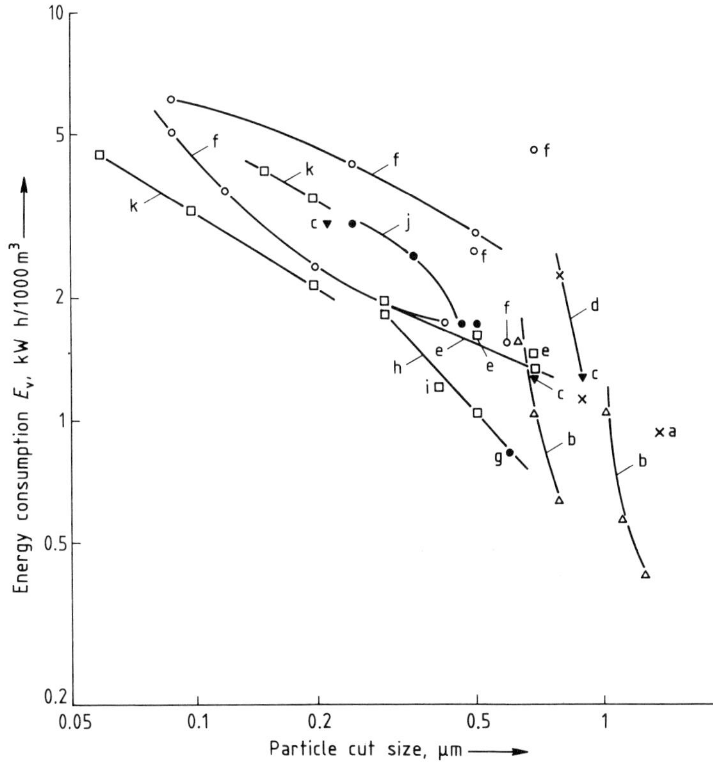

Figure 13. Energy consumption of scrubbers as a function of cut size
a) Wet cyclone; b) Spray tower; c) Self-induced spray scrubber; d) Ejector venturi scrubber; e) Low-pressure venturi scrubber; f) Rotational scrubber; g) Impinger; h) Low-pressure venturi–impinger scrubber; i) Wet cyclonette; j) Scrubber based on different principles; k) High-pressure venturi scrubber

where v_K is the gas velocity in the venturi throat and ξ is the pressure-drop coefficient:

$$\xi = K_1 + K_2 \frac{\dot{V}_1}{\dot{V}_g} 10^3 \qquad (25)$$

where K_1 and K_2 are empirical constants that depend on the equipment and thus cannot be generalized; typical values are $0 \leq K_1 \leq 0.5$ and $0 < K_2 \leq 1.7$.

In *model calculations,* the pressure-drop contributions of the gas and droplets are added. The droplet contribution is taken into account by solving the equation of droplet motion and by introducing the laws of two-phase flow [29]. The resulting formulas can be solved only by numerical techniques; they are in good agreement with experiment. No further details are discussed here; however, the extent of agreement between calculation and experiment depends on whether the scrubbing liquor is uniformly distributed over the total flow cross section or whether a significant fraction flows down the walls of the scrubber as a film.

Experiments have repeatedly shown that energy consumption grows as the particle cut size decreases, i.e., as collection efficiency increases. Empirical equations have also been derived on the basis of this observation [30]. Such equations are not, however, well suited to the sizing or optimization of scrubbers, because the several independent variables (equipment dimensions, dust properties, operating conditions) are not explicitly stated. The correlation between energy consumption and cut size holds only for a given size and type of equipment. Considerable discrepancies can occur between different types of scrubbers (Fig. 13) [24].

4.4. Droplet Separators [6], [30]

Efficient dust collection in a scrubber requires separation of the dust-laden droplets in a separator that is either integrated into the scrubber or connected downstream. Because droplet sizes generally range from 1 μm to several hundred micrometers, commercial separators utilize inertial forces to separate droplets from the gas stream. Layers of wire mesh or fabric, stacks of plates, or centrifugal devices are employed.

Lamellar (plate) separators are often used because of their simple construction and low pressure drop. These devices consist of parallel channels with multiple deflections and liquid-retaining grooves (Fig. 14). The droplet-laden gas stream undergoes many changes of direction in the channels. By virtue of their inertia, droplets collide with the wall to create a film which is "peeled" off in the retaining grooves. To prevent product caking, the plates can also be irrigated; a second downstream stage is then installed to trap the droplets entrained at the leading edges of the plates in the first stage. The

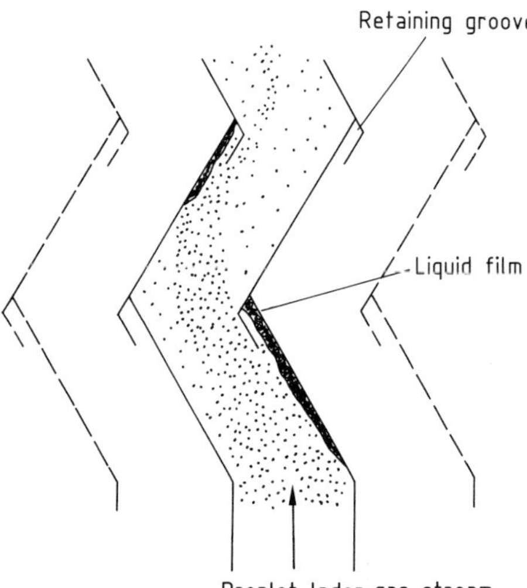

Figure 14. Droplet collection by stacks of plates

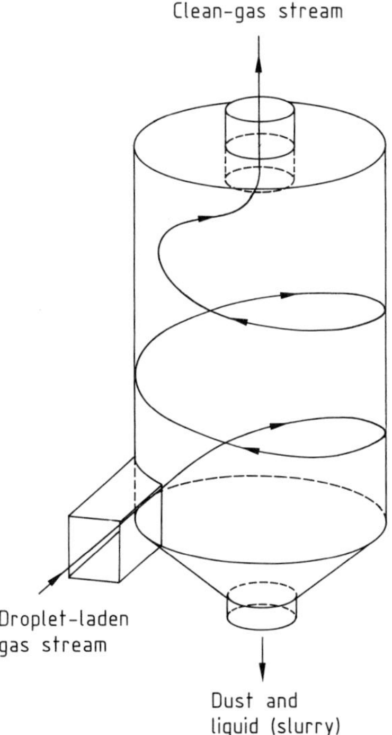

Figure 15. Centrifugal droplet separator

plates can have various profiles. Typical channel widths are 20–30 mm; gas velocities are 5–10 m/s for horizontal flow and 2–3 m/s for vertical flow; pressure drops are a few hundred pascals. The usual cut diameters are ca. 5–10 μm.

Centrifugal separators (cyclones) are often employed if high separation efficiencies are required, especially downstream of venturi scrubbers. Figure 15 shows a simple form.

Like all inertial collectors, lamellar and centrifugal separators become more efficient as flow velocity increases. However, an upper velocity limit must not be exceeded; otherwise, collected liquid is reentrained (cyclones), or the formation of secondary droplets due to impact on the liquid film becomes significant (lamellar separators). These limits must be determined empirically for each type of equipment and application.

5. Fibrous Filters

5.1. General

Fibrous filters constitute an important aid in modern particle collection technology, especially when highly efficient collection in the finest particle-size range is required. Many versions are available. Of all collectors, these have the widest spectrum of applications and thus a large share of the market.

With regard to application, design, and mode of operation, fibrous filters for dust collection can be classified as deep-bed filters or surface filters.

Deep-bed filters are used where dust loads are low, on the order of several milligrams per cubic meter of gas. Typical applications include air conditioning and ventilation, laboratory exhaust systems, clean rooms, and respiratory protection. The filter consists of a relatively loose fiber mat with a pore fraction over 90 % (often > 99 %). As the gas flows through the fibrous layer, particle collection takes place in the interior of the layer where the dust accumulates (deep-bed filtration). After becoming saturated with dust, these filters are usually discarded; some can be cleaned by washing or blowing. Typical gas velocities are 0.1–3 m/s.

Surface filters are used for high dust loads, on the order of several grams to hundreds of grams per cubic meter; this is the typical range for industrial dust removal and air pollution control. The fibrous layers used to be mainly woven fabrics; today, nonwovens such as mats or felts, sometimes with surface coatings, are preferred. The pore-volume fraction in these media is 70–90 %. Collection occurs mainly at the filter surface, in the resulting dust layer (filter cake). The dust bed is the true filter medium and is highly efficient: emission values of a few milligrams per cubic meter can be attained. Because the pressure drop increases as the cake grows, these filters must be cleaned periodically. Collected dust, which may be the product, is recovered. Depending on operating conditions, cleaning can take place at intervals of a few minutes to several hours. Typical gas velocities are 0.5–5 cm/s.

Recently, combinations of surface and deep-bed filters have been discussed with a view to satisfying extreme emission limits (e.g., 0.1 mg/m^3).

Because filter design and functioning as well as system design differ greatly for deep-bed and surface filters, the two groups are discussed separately.

5.2. Deep-Bed Filters

5.2.1. Mode of Operation and Basic Designs

The fibrous layers in filter-medium systems consist mainly of synthetic fibers (glass or plastic). Table 2 summarizes typical geometrical data. The spacing between the fibers is large compared with the size of the particles to be collected. This implies that the filters do not function as a sieve; the purely geometrical blocking action is not crucial for particle collection. Various mechanisms (diffusion, inertia, electrostatic attraction) cause the particles to be transported to the fibers, where they must then adhere.

Filter mats are installed in a variety of ways in accordance with plant requirements and available space. Figure 16 shows the most important types. In the simplest case (A), the flat filter mat is inserted perpendicular to the direction of flow. To increase filtration area (lower flow veloci-

Table 2. Geometrical data for deep-bed

Parameter	Coarse filter	Submicron filter
Fiber diameter D_F, μm	50–100	1–5
Mat thickness Z, cm	1–3	0.1–0.3
Fiber volume fraction $(1-\varepsilon)$, %	$\lesssim 1$	< 5–10
Mean fiber spacing	9 D_F	3 D_F

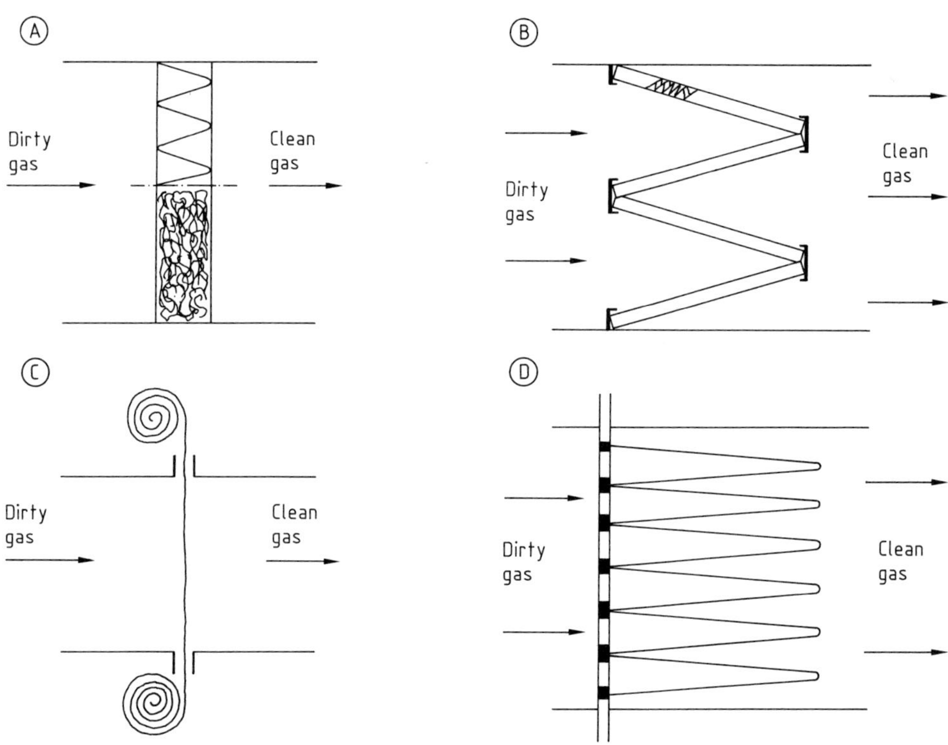

Figure 16. Designs of deep-bed filters
A) Flat filter; B) Stationary zigzag filter; C) Roll filter; D) Pocket filter

ty → lower pressure drop → longer lifetime for a given channel cross section), the filter is often folded and mounted in a frame. A further gain in filtration area is obtained by a zigzag placement of frames with folded filter mats (B).

Because the filters become plugged with collected dust over time, they must be replaced at certain intervals (usually several months to years). The method used for filter mounting is therefore very important. The mount must be tight but easy to handle. Filter replacement can be effected, for example, with a roll filter arrangement (C). This rather costly design has been increasingly superseded by stationary zigzag filters (B) or by pocket filters (D). This avoids the danger of releasing collected dust when the filter strip is advanced.

The main application for deep-bed filters is inlet-air cleaning, i.e., for low dust loads. However, since the early 1980s high-efficiency filters have been used in exhaust-gas systems where particle loads are higher but the clean gas must have especially low particle concentration, e.g., for toxic or harmful dusts composed of heavy metals or radioactive substances. Two-stage submicron filters are then used (Fig. 17). The first stage (a) is periodically cleaned by reverse blowing with a reciprocating nozzle carriage (b). The second stage (c) is replaced after it is saturated with dust; because of the low dust concentration, replacement intervals are sufficiently long. The usual cleaning procedure for such a system involves shutting off the gas stream, so at least two units must be connected in parallel. Development work aimed at finding application limits, optimal designs, materials, and operating conditions is already under way. Problems must be expected for very fine, highly adhesive dust, or if the folds in the first filter stage are too narrow. The collection efficiency for such a system may be very good; an emission value of 0.05 mg/m^3 has been reported with a dirty-gas dust load of 10 g/m^3 [31].

The range of collection efficiencies for deep-bed filters varies widely and depends on the design and operating conditions. It extends from a single prefilter for relatively coarse dust to submicron particulate filters for clean rooms, where

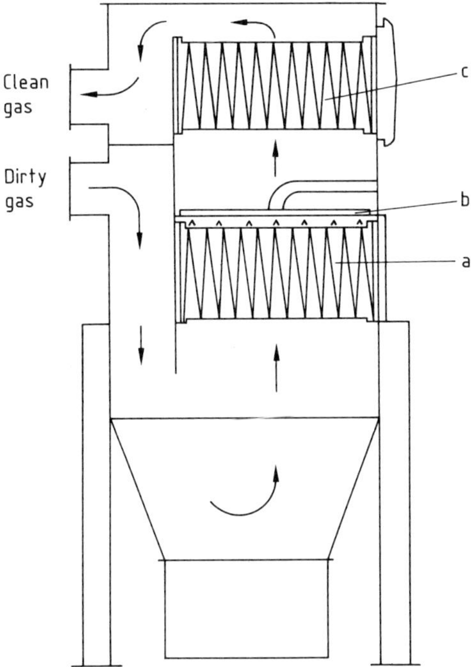

Figure 17. Exhaust-gas cleaning system employing submicron filters
a) First filter stage; b) Nozzle carriage for cleaning filter; c) Second filter stage

the DIN standard requires a collection efficiency of at least 99.97% for 0.3-μm particles [32].

With regard to temperature, deep-bed filters are used mainly at < 100 °C. In paint shops, temperature resistance up to 120 °C is required. Some designs can also be used up to 300 °C. These filters generally employ glass fibers in metal cartridges with appropriately temperature-resistant sealing materials.

Another application of deep-bed filters is the collection of liquid droplets, especially oil or acid mist. Here, the collected liquid flows continuously downward out of the filter, so that the pressure drop is stable. A variant of this use is the collection of soluble dusts with simultaneous solvent feed. Again, the collected material flows continuously out of the filter layer.

5.2.2. Design Calculations

The models for calculation presented in this section refer to unloaded filters, i.e., to the initial conditions. With regard to collection efficiency, this is usually the critical state because collection generally improves with increasing dust load (electret filters are an exception, see p. **13**-18).

The methods proposed in the literature for calculating filtration behavior over time are not discussed because they are very complex and of limited practical use.

5.2.2.1. Collection Efficiency

The fractional collection efficiency $T(x)$ of a fibrous layer is given by

$$T(x) = 1 - \exp[-f \cdot \varphi(x)] \qquad (26)$$

For fibers of circular cross section,

$$f = \frac{4}{\pi} \cdot \frac{1-\varepsilon}{\varepsilon} \cdot \frac{Z}{D_F} \qquad (27\text{a})$$

$$f = \frac{4}{\pi} \cdot \frac{1}{\varepsilon} \cdot \frac{M}{D_F \cdot \varrho_F} \qquad (27\text{b})$$

where ε is the porosity, Z is mat thickness, D_F is fiber diameter, M is filter mass per unit filter area, and ϱ_F is the fiber density.

The geometric factor f represents the ratio of the projected fiber area to the filter face area. For prefilters $f \approx 3-10$; for high-efficiency submicron filters, $f \approx 100-300$.

The factor $\varphi(x)$ is the collection efficiency of a single fiber inside the mat. Because collection in a medium (depth) filter includes transport and adhesion of the particles to the fiber,

$$\varphi = \eta \cdot h \qquad (28)$$

where η is the collision efficiency and h is the adhering fraction. This factor is included because at the usual filtration velocities and especially for particle sizes > 1 μm, not all the particles hitting the fiber surface for the first time adhere [33], [34].

Since the 1930s, research has been concerned with determining η. The importance of the adhering fraction h was not recognized until the 1960s [6], [33], [34], but research on this factor has been increasing.

Particles can be transported to the fiber surface in three ways: Brownian diffusion, inertial forces, and electrostatic interaction (Fig. 18).

1) *Brownian Diffusion.* The random motion of particles about their mean trajectory [Fig. 18 (a)] is due to the thermal motion of gas molecules (Brownian motion); it can be calculated with Fick's diffusion equations. For engineering purposes, this effect is important chiefly for particles

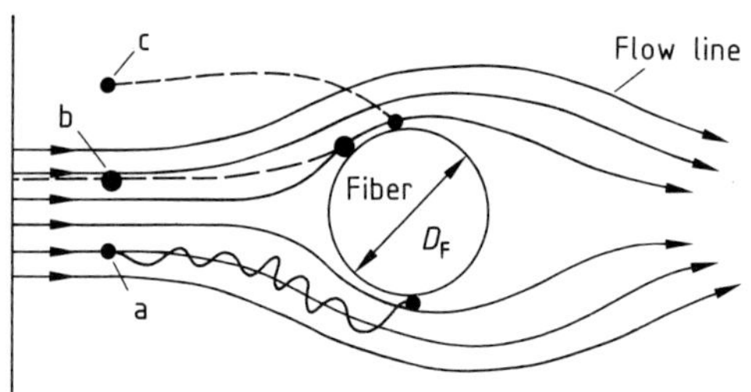

Figure 18. Transport mechanisms for particle collection on fibers
a) Brownian diffusion; b) Inertial forces; c) Electrostatic interaction

smaller than ca. 0.5 μm and velocities lower than 10 cm/s.

This diffusion effect η_D is governed by the following relation:

$$\eta_D \sim \left(U_0 \cdot D_F \cdot x \cdot \frac{\mu}{T} \right)^{-2/3} \tag{29}$$

where U_0 is the velocity of the incident gas stream, μ is the viscosity of the gas, and T is the absolute temperature.

2) *Inertial forces* [Fig. 18, (b)] cause a predetermined movement of particles toward the fiber. These processes can be described by the equation of motion (Eq. 1).

The decisive parameter for the inertial effect is the inertial (Stokes) parameter

$$\psi = \frac{\varrho_p x^2 \cdot U_0}{18 \, \mu \, D_F} \tag{30}$$

Inertial effects are important mainly for particles larger than 0.5–1 μm. As ψ increases, the collision efficiency number rises rapidly. The Reynolds number for flow past the fiber is also important because the collision efficiency number increases with the Reynolds number.

3) *Electrostatic interactions* [Fig. 18 (c)] can also be described by the equation of motion.

Figure 19 presents measurements of the fractional collection efficiency of a fibrous layer [35]. Characteristic influences of the diffusion and inertial effects can be seen at particle diameters < 0.5 and > 0.5 μm, respectively. In the transition region (ca. 0.1–0.5 μm), both effects are very slight. This minimum in the fractional efficiency curve is typical; it represents the critical particle-size range for the filter. For this reason, filter tests are often performed with particle sizes ca. 0.3 μm.

One way of overcoming the efficiency minimum is to utilize electrostatic interactions [Fig. 18, (c)]. Electrostatic forces act over long ranges and can thus attract particles from the flow even if they are far from the fiber.

Calculation of electrostatic effects is very complicated because many configurations of charges and fields are possible. "Electret" filters have attracted much interest in recent years.

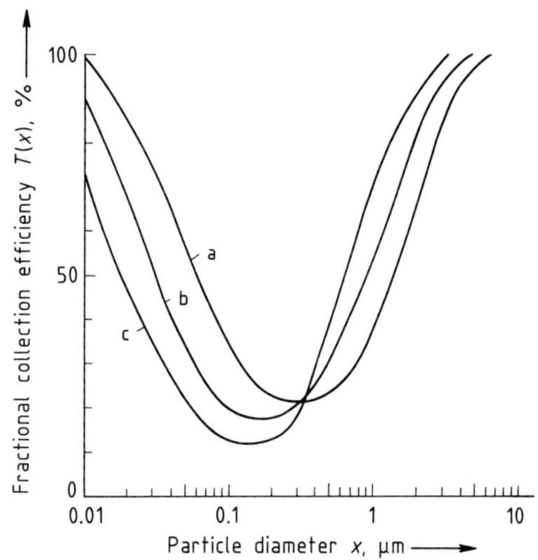

Figure 19. Fractional efficiency curves for a glass-fiber filter at various superficial velocities
Superficial velocity U_0, m/s: a) 0.055; b) 0.2; c) 0.5
Values for $x > 0.5$ μm were determined with a quartz aerosol and for $x < 0.2$ μm with a sodium chloride aerosol

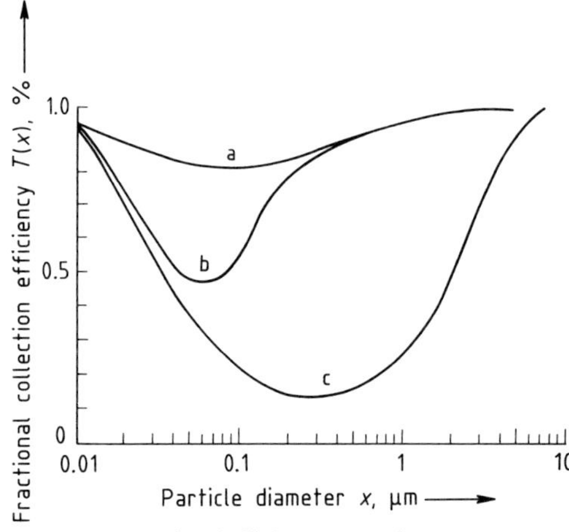

Figure 20. Fractional efficiency curves for an electret filter spun from a polymer solution
a) Fibers and particles charged; b) Fibers charged, particles uncharged; c) Fibers and particles uncharged
Superficial velocity = 0.1 m/s

These devices consist of fibers on which a charge is imposed during manufacture [36]. The charge is localized not at the fiber surface but in its interior. Such fibers greatly improve collection, even of uncharged particles (Fig. 20). If the particles carry only a small charge (e.g., are in equilibrium with the ion concentration in air), even the size range below 0.1 μm is collected quite efficiently [Fig. 20, curve (a)]. The advantage of the electret filter is that, with comparatively slight pressure drops, the collection efficiency is much higher than that of conventional filters, especially in the critical transition region. Its drawback is that its behavior deteriorates with timc bccause, as more particles are deposited on the fibers, the electrostatic effects weaken (depending on particle properties). Further experimental and theoretical results can be found in [35].

To obtain the single-fiber collection efficiency φ from the transport parameter η, information is needed about the adhering fraction h (Eq. 28); this is important not only for single fibers but also for fibrous beds. Extensive studies have shown that the adhering fraction generally depends on flow velocity, fiber geometry, particle geometry (size and shape), and surface properties of the fiber and particle [34]. Bouncing of particles from the fibers can begin at velocities lower than 10 cm/s. As the velocity and particle size increase, the adhering fraction usually decreases. Therefore, adhesion in the particle-size range > 1 μm should always be given special attention.

The fiber configuration in the filter layer is another aspect that has been neglected until recently. Theoretical calculations usually assume a homogeneous distribution of fibers; this is certainly not the case, and uneven distribution strongly affects the collection efficiency. As the inhomogeneity increases, collection becomes poorer. However, quantitative study and description of this factor are limited to the region of inertial effects, i.e., particle sizes > 1 μm [37].

5.2.2.2. Pressure Drop

The pressure drop generated in flow through a filter medium can be described by the "drag" model. The pressure drop is attributed to the drag involved in flow past the fibers; in view of the low volume fraction of fibers (< 5%), this is certainly realistic in physical terms.

Accordingly,

$$\frac{\Delta p}{Z} = \frac{2}{\pi} \cdot \frac{\varrho U_0^2}{D_F} \cdot \frac{(1-\varepsilon)}{\varepsilon^2} \cdot C_D(Re) \tag{31}$$

where Z is the thickness of the fiber layer, ε is the porosity of the fiber layer, and U_0 is the incident flow rate. The drag coefficient C_D depends on the Reynolds number for flow past the fibers:

$$Re < 1: \quad C_D = \frac{8\pi}{Re(2 - \ln Re)} \tag{32 a}$$

$$1 \leq Re \leq 50: \quad C_D \approx \frac{10}{Re} + 1.5 \tag{32 b}$$

where

$$Re = \frac{U_0 \cdot D_F \cdot \varrho}{\mu \cdot \varepsilon} \tag{32 c}$$

Essential assumptions for derivation of these equations are that the fibers are circular–cylindrical and homogeneously distributed and that the incident flow is perpendicular to the fiber axes. Furthermore, the equations hold only for the initial filtration state, i.e., with no dust on the filter. Within these constraints, agreement with experiment is good. For practical purposes, however, time-dependent behavior is often more important, i.e., the dependence of pressure drop on the quantity of dust trapped. However, because no reliable way of calculating this relationship is known, the time dependence must be determined by experiment. Optimization can be based on the pressure-drop equation (Eq. 31) and the efficiency equation (Eq. 26) [6] and can give useful information on the design of medium filters.

5.3. Surface Filters

Surface filters are used widely because of their excellent collection performance, even for the finest particles; their adaptability to the most diverse conditions; and significant advances in the development of new filter media (e.g., improved chemical resistance, heat resistance, and cleaning properties). In the Federal Republic of Germany, they hold a 50% share of the dust collector market. Cake filters are available with filtration areas from a few to several hundred thousand square meters. For a properly designed, well-maintained filter, emission values can be held to a few milligrams per cubic meter (in the best cases less than 1 mg/m^3). The allowable temperature range extends to ca. 250 °C for

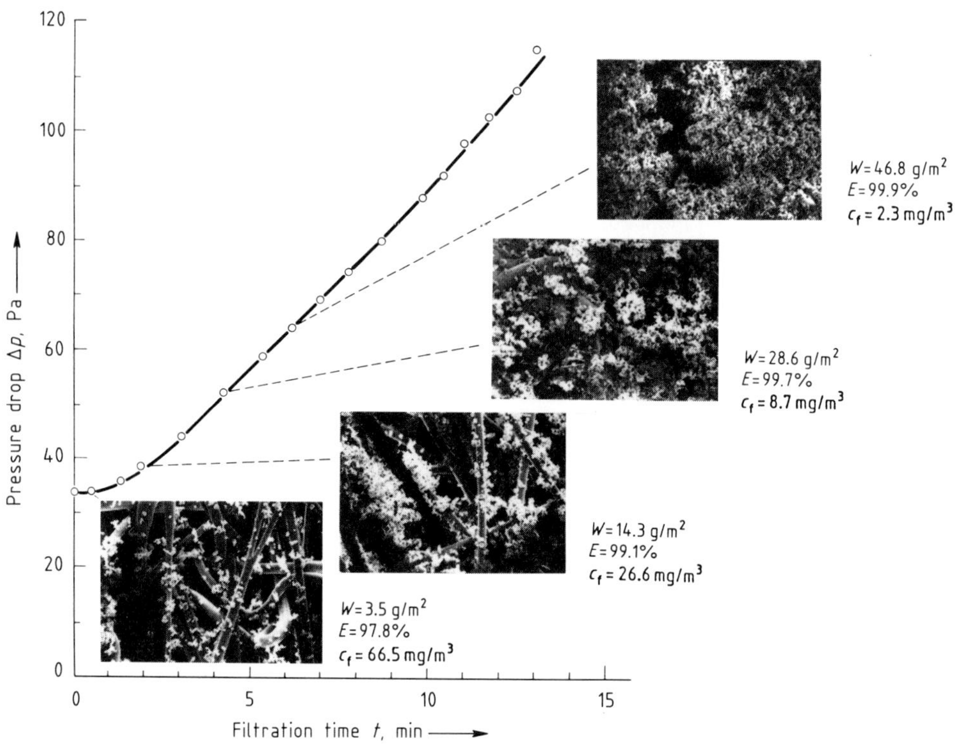

Figure 21. Time dependence of filter-cake formation, pressure drop, and dust collection on a needled felt
W = amount of dust deposited; E = total collection efficiency; c_f = dust content of clean gas; dust content of dirty gas = 3 g/m^3; gas velocity = 150 m/h

glass-fiber filters. Special metal fibers permit operation up to ca. 600 °C; sintered silicon carbide filters can function to 1000 °C, but they cost substantially more. Filtration rates are typically 40–180 $m^3 h^{-1} m^{-2}$ and reach 300 $m^3 h^{-1} m^{-2}$ in exceptional cases; typical pressure drops are 1–3 kPa.

5.3.1. Mode of Operation

The principal surface-filter media in current use are nonwovens and (needled) felts that are composed of randomly arranged glass, plastic, or ceramic fibers. Woven materials, chiefly constructed from glass-fiber yarns or metal wires, are also employed.

Only at the beginning of filtration do these media trap particles in the interior of the fibrous layer. As more dust is deposited, it bridges the fibers and forms a dust layer (filter cake) on the surface.

Figure 21 shows a typical example of filtration behavior for a needled felt, in which the initial emission value was 66.5 mg/m^3 and decreased to 2.3 mg/m^3 after 46.8 g/m^2 of dust had been deposited.

The fractional efficiency curves in Figure 22 clearly show the improvement in collection performance as more dust is collected. The new, clean filter gives a typical depth-filter curve (a); after 5 min, the efficiency is over 99 % throughout the particle-size range covered (f). Since the cake-formation phase lasts only a short time for the dust loads usually encountered in practice, the interception effect inside the cake is dominant. This process is called surface filtration.

As the amount of dust deposited in and on the filter grows, the pressure drop also increases. After a certain time, or when a predetermined pressure-drop value is reached, the dust must be removed (cleaned off) to regenerate the filter. Filtration periods may be 10 min or less when the amount of incident dust is large (i.e., high concentration and high velocity); if the dust load is small, filtration can last several hours. As a rule, filters should be regenerated only when absolute-

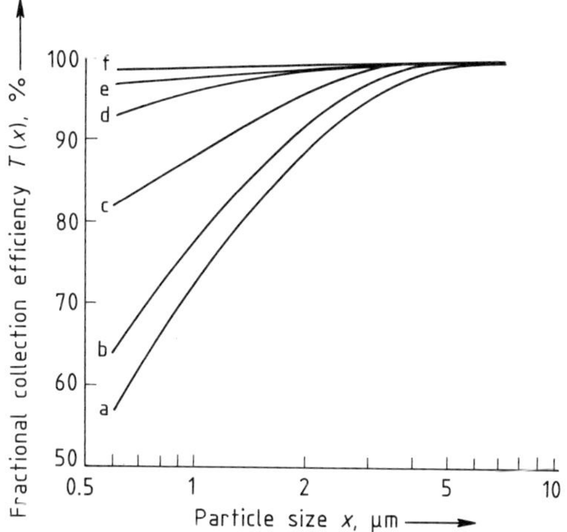

Figure 22. Variation in fractional collection efficiency due to dust deposition on a polyester needled felt
Amount of dust deposited, g/m^2: a) 4, Δp = 33 Pa; b) 14; c) 29; d) 47; e) 75; f) 95, Δp = 110 Pa
Dust content of dirty gas = 3 g/m^3; gas velocity = 150 m/h

ly necessary (cf. Section 5.3.2); this saves operating costs and reduces emissions.

In newly developed commercial filter media, surface coating or treatment is used to attempt collection on the surface right from the start and to prevent particles from penetrating into the medium. This approach yields high initial collection efficiencies, avoids the danger of plugging, and facilitates removal of the filter cake.

5.3.2. Basic Designs

The three most important basic surface-filter designs are round bag, envelope, and cartridge (Fig. 23). A filter unit contains the requisite number of filter elements connected in parallel. Designs differ primarily in the method of cleaning and sometimes in the way the dirty gas is introduced [39], [40].

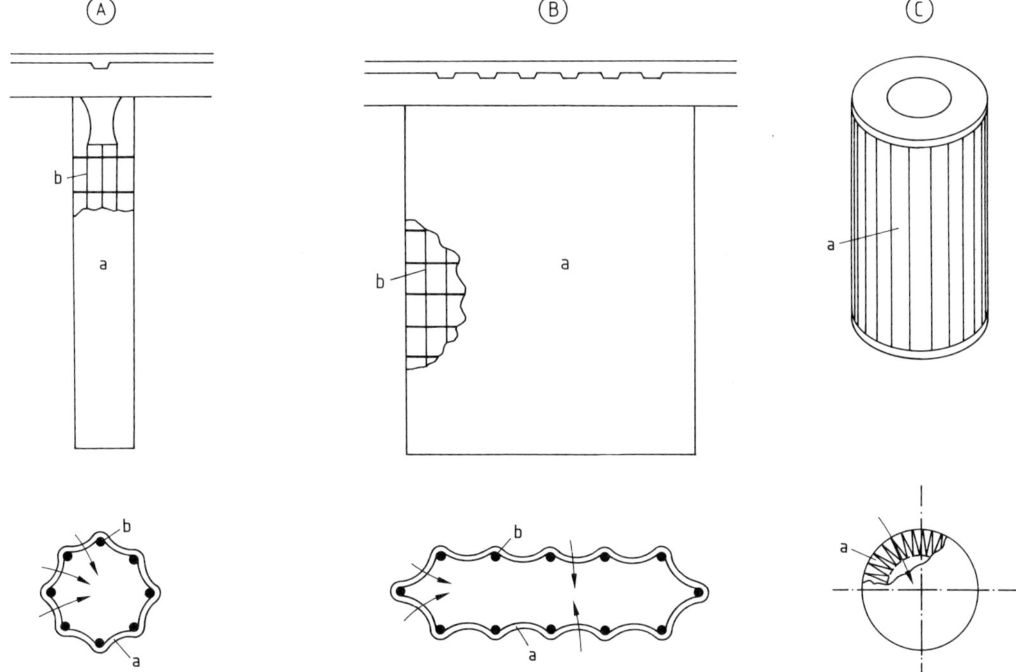

Figure 23. Basic filter designs
A) Round bag filter; B) Envelope filter; C) Cartridge filter a) Filter medium; b) Supporting cage

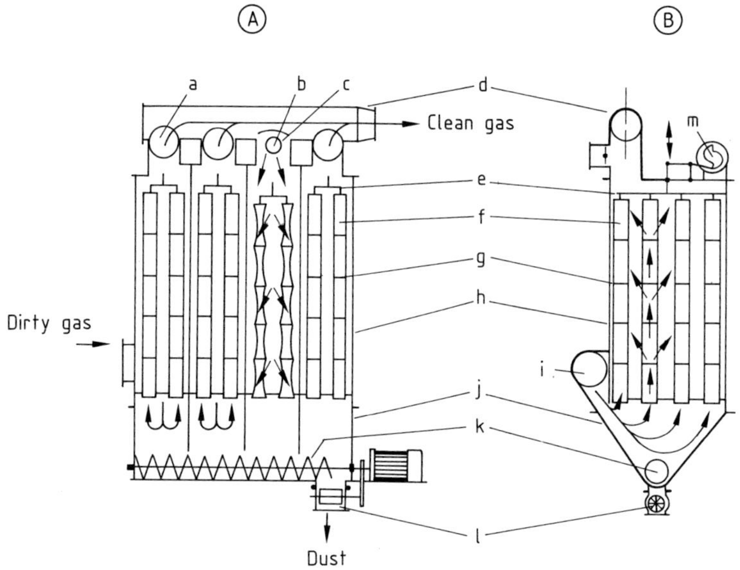

Figure 24. Multichamber bag filtration plant with shaker cleaning
The two vertical sections (A) and (B) are at right angles to each other a) Valve open for filtration; b) Valve closed for cleaning; c) Purge air inlet; d) Clean-gas duct; e) Bag support; f) Filter bag; g) Stiffening ring; h) Filter housing; i) Dirty-gas duct; j) Dust hopper; k) Screw conveyor; l) Air lock; m) Shaker mechanism

Round (tubular) bag filters are used most widely. If gas flow during filtration is inward, as shown in Figure 23 A, a retainer (supporting cage) is required. Often, however, flow is outward and, in this case, only a few stiffening rings need be incorporated in the bag. Typical bags are 100–300 mm in diameter and 1.5–10 m long. The filtration area per bag can thus be as much as 10 m^2.

The classical design, introduced in 1887 by the Beth company, is a *multichamber filter with shaker cleaning* (Fig. 24). Dirty gas is fed into the bags from below; it then flows through the filter medium (f) and is led to a clean-gas duct (d) at the top of the device. Particle collection takes place on the inside of the bags. For cleaning, the dirty-gas supply is cut off from one chamber at a time, and the dust is loosened by shaking or rapping the bag support. The dust removed from the bags falls into a dust hopper; purge air drawn into the bags from outside aids this process. Because the bags being cleaned are cut off from the gas flow, several filter chambers must always be used in parallel.

Large multichamber filtration plants (e.g., for cleaning flue gas from furnaces or use in the cement and metallurgical industries) are often designed as *baghouses* (Fig. 25). The bags are usually large (300 mm in diameter, 10 m long). The dirty gas flows from the inside to the outside of the bags, which are cleaned by low-pressure reverse flushing (Fig. 25 B). The reversal of flow for cleaning causes the bags to collapse and the dust to drop out. In connection with the novel surface-coated filters mentioned in Section 5.3.1, this design offers a particularly economical approach.

Pulse-jet filters (bag filters with reverse-pulse cleaning) became more widely used as progress was made in needle felts. This design dominates in small and intermediate-size units. The gas always flows from outside to inside (Fig. 26), so the bags must be pulled over supporting cages. For cleaning, a jet of compressed air is admitted to the bag through its open top end. The pressure needed in the air tank depends on the design (0.3–0.7 MPa). The compressed-air jet suddenly raises the pressure in the bag, which is rapidly inflated and gas flow is reversed. The dust is thus dislodged from the outside of the bag; the process lasts ca. 0.1–0.3 ms. Because single bags or single rows of bags must be removed from the filtration process for only a short time, the filter does not have to be divided into chambers. Cleaning is done on-line; however, off-line cleaning, in which the dirty-gas flow is cut off in a multichamber filter, may be preferable in extreme cases, e.g., for very fine dusts. Further de-

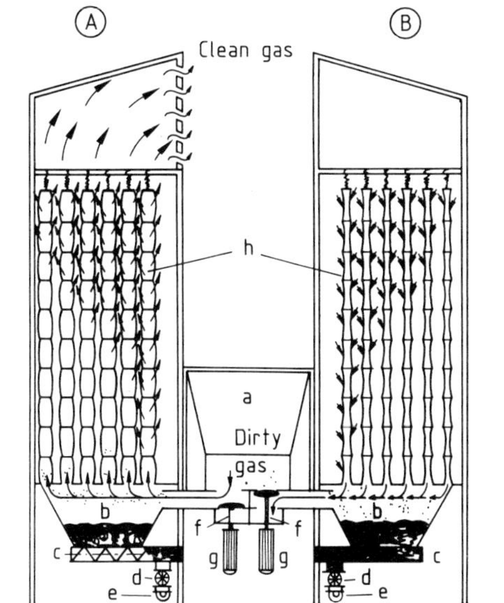

Figure 25. Baghouse with low-pressure reverse flushing
A) Operating filter chamber; B) Filter chamber being cleaned
a) Dirty-gas pressure duct; b) Dust hopper; c) Hopper screw conveyor; d) Air lock; e) Screw conveyor; f) Flushing gas duct; g) Cleaning valve; h) Filter bag

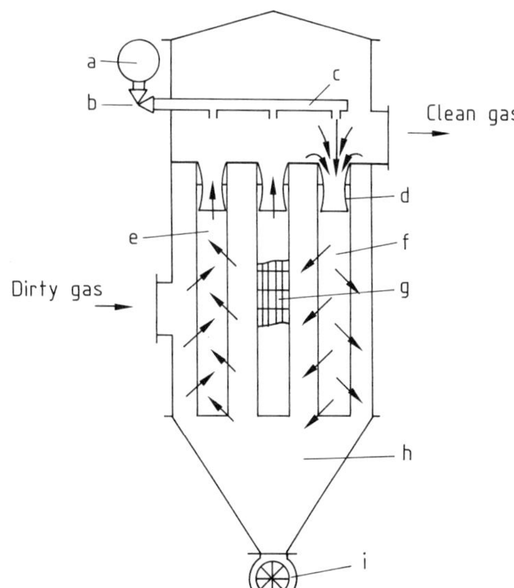

Figure 26. Round bag filter with reverse-pulse cleaning
a) Compressed-air tank; b) Right-angle diaphragm valve; c) Compressed-air duct with nozzles; d) Venturi nozzle; e) Filter bag in operation; f) Filter bag being cleaned; g) Retainer; h) Dust hopper; i) Air lock

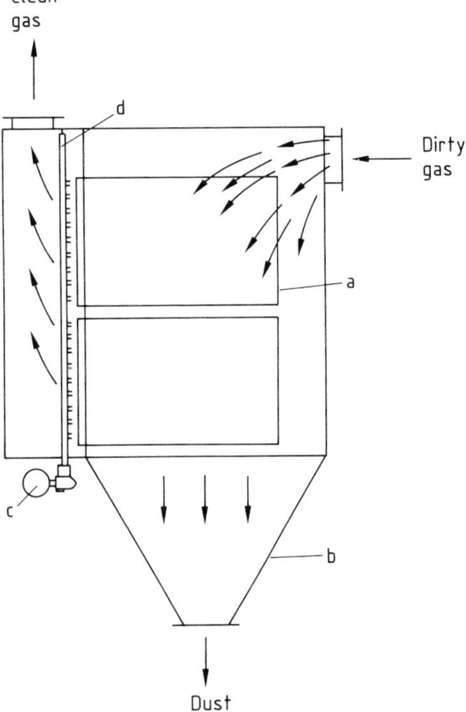

Figure 27. Schematic of an envelope filter unit with reverse-pulse cleaning
a) Envelope filter; b) Dust hopper; c) Compressed-air tank; d) Compressed-air duct with nozzles

tails on the design and properties of pulse-jet filters can be found in [6] and [39].

Envelope filters (Fig. 23 B) are not as widely employed as round bag filters. They are used mainly to remove dust from small amounts of gas, e.g., in silo venting, and as bunker top filters [39].

Flow is always from outside to inside. The filter medium is stretched over a flat, rectangular, supporting cage; dust is collected on the outer surface of the medium. Envelope filters are usually smaller than round bags; filtration areas are often 0.1 – 1.5 m^2. An envelope filtration unit is shown in Figure 27. The usual cleaning method is intermediate-pressure reverse flushing or reverse pulsing. As a rule, multichamber design is not necessary.

Cartridge filters are a new development in surface filters. The filter medium is folded in star fashion (Fig. 23 C) to pack more filtration area into a unit volume. With a standard cartridge

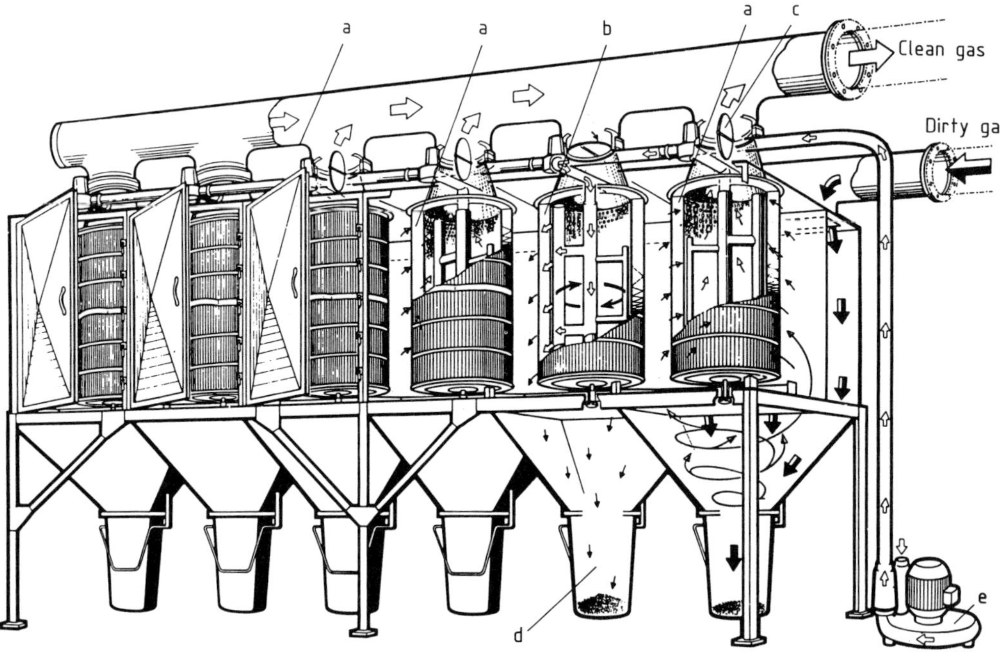

Figure 28. Multichamber filtration unit (Mann & Hummel)
a) Filter in operation (valve open, dust collects on the outside of the filter); b) Filter cleaning (valve closed, cleaning air is pumped outward through the filter and dust falls into the hopper); c) Valve; d) Dust hopper; e) Pump for cleaning air

height [38] of 606 mm, the filtration area is between 5 and 20 m^2, depending on the depth and angle of the folds. Gas flow is always from outside to inside. A multichamber cartridge filter unit is shown in Figure 28.

The filter cake on the outside of the filter is removed either by cutting off the dirty-gas stream and flushing at low pressure with circulating nozzles (multichamber design) or by applying a reverse pulse (on-line). Cartridge filters are very compact devices that offer a relatively large filtration area in the enclosed volume. As a result, filtration velocities are possible; this is an advantage with respect to emission levels and costs. Until now, however, cartridge filters have been useful only for easily removable dust; their development continues.

5.3.3. Operating Characteristics

Particle collection in a surface filter occurs mainly in the dust layer formed on the surface (Section 5.3.1); efficiencies greater than 99.9 % are possible. Therefore, this highly efficient cake must be retained on the surface as long as possible.

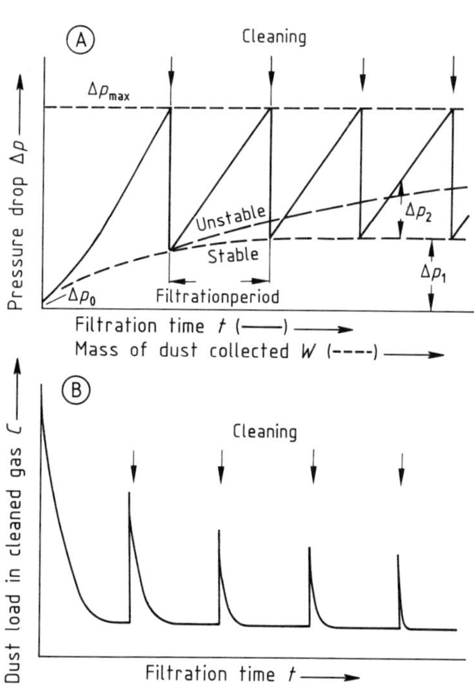

Figure 29. Pressure drop (A) and dust content (B) in clean gas vs. filtration time at constant filtration velocity
The values Δp_1 and Δp_2 denote the pressure-drop contributions of the filtration medium and filter cake, respectively

Figure 29 A shows that the pressure drop Δp increases as more dust is deposited, and periodic cleaning is necessary. After cleaning, the pressure drop falls to a value Δp_1 that depends on the flow resistance of the filter medium (including residual dust not removed); Δp_1 is higher than the initial pressure drop Δp_0 of the new, clean filter. For stable filter operation, Δp_1 must reach an approximately constant value after an induction period. The filter medium, filtration velocity, and the method and intensity of cleaning determine whether stable filtration is attained for given gas and dust properties.

During and immediately after cleaning, the particle content of the clean gas increases (Fig. 29 B). This occurs partly because the filter cake, which acts as a layer of filter aid, has been removed and must be rebuilt. Especially with reverse-pulse cleaning, some of the dust passes through the filter medium because after the pressure pulse inside the bag has decayed, the filter medium falls back against the retainer and knocks part of the dust through the filter (seepage effect).

Many laboratory and plant studies have shown that the peak concentrations after cleaning may be crucial for the emission value [6]. The maintenance of low emission levels thus demands that cleaning be done as seldom and as carefully as possible. Accordingly, filtration velocities should not be too high, and pressure drop—rather than time—should be used as a criterion for cleaning.

5.3.4. Design Calculations

The main purpose of surface-filter design is to determine the required filtration area F, which is given by

$$F = \frac{\dot{V}}{v} \tag{33}$$

where $\dot{V}$ is the volume flow rate of dirty gas and v is the filtration velocity; v, the ratio of flow rate to filtration area, is often called the *air to cloth ratio* and is expressed in $m^3 h^{-1} m^{-2}$. A proper selection of filter load determines the pressure drop and, ultimately, the emission value.

The filtration velocity is selected on the basis of empirical values because of a lack of generally valid theoretical models. The VDI 3677 standard gives guideline values for a filter with shaker cleaning that collects a wide variety of dusts. A base load value is also employed, which is modified for actual conditions by means of various correction factors [39].

$$v = v_0 \cdot c_1 \cdot c_2 \ldots c_n \tag{34}$$

where v is the effective filtration velocity, v_0 is the base value, and $c_1 - c_n$ are correction factors. This method is used chiefly for filters with reverse-pulse cleaning. The correction factors always refer to a defined table of base values. They differ from one author to another, being purely empirical and specific in nature [39].

In describing the time dependence of the pressure drop (Fig. 28), Δp is assumed to be the sum of the contributions of the medium (Δp_1) and the filter cake (Δp_2):

$$\Delta p = \Delta p_1 + \Delta p_2 \tag{35}$$

Because flow through the filter takes place at low Reynolds numbers (< 1), Darcy's law is also assumed to hold:

$$\Delta p = K_1 \mu \cdot v + K_2 \mu \cdot W \cdot v \tag{36}$$

where K_1 is the residual resistance of the filter medium after cleaning, K_2 is the specific resistance of the filter cake, W is the mass of dust collected per unit filter area, and μ is the viscosity of the gas. Since W is proportional to filtration time, the second term in Equation (36) increases with time and describes the time dependence of Δp.

Important parameters in Equation (36) are the resistances K_1 and K_2; as a rule, they are not constants but functions of several variables. Prediction of these parameters is still problematic because the functions are not known well enough [6], [39]. Even when K_1 and K_2 can be determined by experiment, great care must be taken to ensure that conditions in the experimental device and the full-scale plant are comparable.

6. Granular-Bed Filters

[1], [4], [6], [31], [41]

6.1. General

In granular-bed filters, the dust-laden gas stream flows through the bed and is thereby cleaned. The granular bed functions as a filter medium and can be considered a depth or surface filter. Under some conditions, however, dust bridges and then a filter cake are formed, so that surface filtration can occur.

Besides pure particle collection, the removal of gaseous components in granular-bed filters ("dry sorption" of gases such as sulfur dioxide, hydrogen chloride, and hydrogen fluoride) has recently attracted increasing interest. The granules employed for these applications must be made of appropriate sorbents (often calcium compounds). As a result of more stringent demands for air pollution control, the combined collection of dust and noxious gases may become still more important for small and medium-sized installations. However, only particle collection is discussed here.

The applications of granular-bed filters are determined by the material properties of the granules making up the bed. Examples of such materials are gravel, sand, ceramics, and activated carbon.

Granular-bed filters are used chiefly for the removal of hot, chemically aggressive, abrasive, or adhesive dust or when danger of fire exists. Main applications are in the lime and cement, rock and soil, metallurgical, chemical, and nuclear industries.

The service temperature range is limited by the materials used for plant construction rather than by the filter granules; in some cases, the behavior of the collected particles is also important. Operation up to 450 °C presents no problem; with special designs, 800 °C is attainable.

6.2. Mode of Operation

Granular beds can be either fixed or moving. Table 3 presents typical operating data.

Fixed-bed granular filters are operated batchwise; i.e., when dust deposition has raised the pressure drop to a predetermined limit, the dirty-gas stream is cut off and the bed is regenerated in one of two ways. Either the dust is removed internally by reverse flushing and agitation or stirring, or else it is removed externally by screening or other means. These techniques always require several chambers in parallel, which are regenerated in rotation.

In *moving-bed granular filters*, the granules move (slowly) through the filter chamber by gravity, either steadily or at intervals. Fresh granules are added continuously from above. Regeneration takes place in an external loop. This filtration method operates without interruption of the dirty-gas stream. Multichamber design for regeneration is thus unnecessary.

Dust collection occurs mainly in the interior of the bed. Particles are transported to the filter granules by the same mechanisms as in fibrous depth filters: diffusion, inertial forces, and (less often) electrostatic attraction (cf. Section 5.2.2). For successful collection, the particles must adhere to the granules. Sieve effects become significant only after the deposition of sufficient dust and the formation of bridges or a cake.

Once a critical load or a critical pressure drop has been reached, dust can penetrate to the clean-gas side of the bed. With fixed beds, dust bridges can collapse; with moving granular beds,

Table 3. Typical data for granular-bed filters

Parameter	Value
Granule diameter, mm	0.5–5
Bed heights, cm	5–20
Porosity, %	40–50
Incident flow rate, m/s	0.5–2.5
Pressure drop (no dust), Pa	500–1500

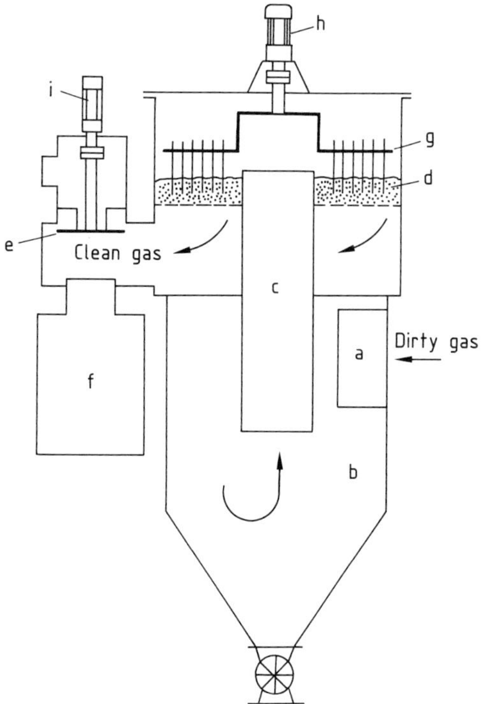

Figure 30. "Drallschichtfilter" in operation (Lurgi)
a) Tangential dirty-gas inlet; b) Precollector; c) Exit duct; d) Granular bed; e) Changeover valve; f) Clean-gas duct; g) Rake; h) Drive; i) Valve actuator motor

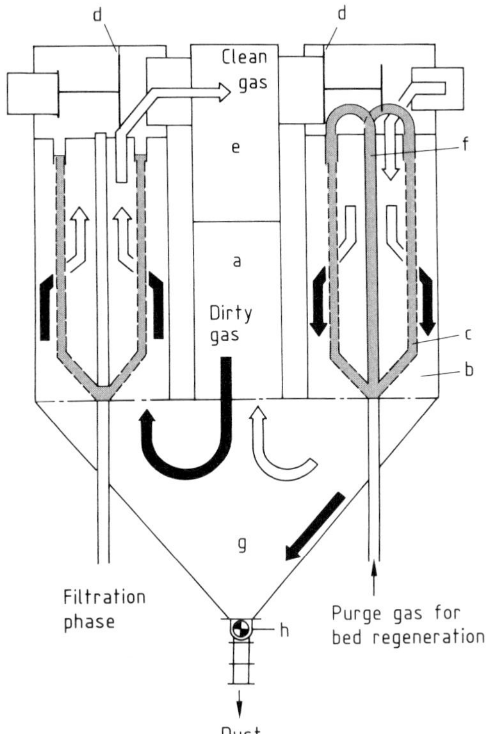

Figure 31. Granulate-tube filter (Lurgi)
a) Dirty-gas duct; b) Filter space; c) Perforated granulate tubes; d) Shutoff valve; e) Clean-gas duct; f) Riser pipe; g) Hopper; h) Screw conveyor

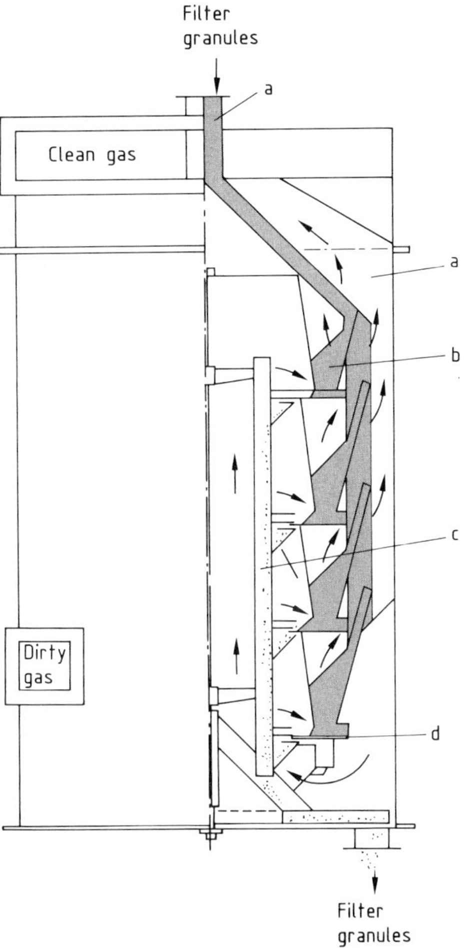

Figure 32. Vertical section of a countercurrent filter (Rüskamp)
a) Clean-gas plenum; b) Granular bed; c) Downpipe; d) Scraper
Arrows indicate direction of gas flow

a danger exists that friction between granules will loosen collected dust particles, which will be carried out of the filter (reentrained).

6.3. Basic Designs

Fixed-Bed Filters. An example of a fixed-bed filter is the so-called *Drallschicht* (*DS*) *filter* shown in Figure 30 [42]. A cyclone-like precollector (b) removes coarse particles. Collection takes place as the gas flows downward through the bed (d). For regeneration, the dirty-gas stream is interrupted by adjusting the appropriate shutoff valve (e), causing clean gas to flow back through the bed while the bed is raked (g). Removed dust is collected partly in the hopper of the precollector and partly in another filter connected in parallel (multichamber design).

In the *granulate-tube filter* (Fig. 31), the granular filter material is contained in perforated double-wall pipes (c); the gas flows horizontally through the pipes. For regeneration (right-hand side of Fig. 31), the dirty-gas stream is again cut off. Purge gas sets the granules in motion and conveys them in a closed loop; at the same time, clean gas flows backward (horizontally) through the pipes and carries the loosened dust to the hopper (g).

Moving-Bed Filters. An example of a moving-bed bilter is the *countercurrent granular-bed filter* (Fig. 32). Gas and filter material move countercurrently. The dust content of the dirty gas is first lowered in a cyclone-like precollector. The gas then flows radially into the filter bed where it is redirected and slowed. The gas then flows upward through the fresh filter material (b) into the clean-gas plenum (a). To increase filtration area, the filter material is distributed in several conical containers stacked one above another; gas flows through these beds in parallel.

Dust-laden filter material is removed at intervals by a scraper (d), collected in the tube (c), regenerated externally in vibrating sieves, and returned to the top of the collector. This design and mode of operation make a multichamber device unnecessary.

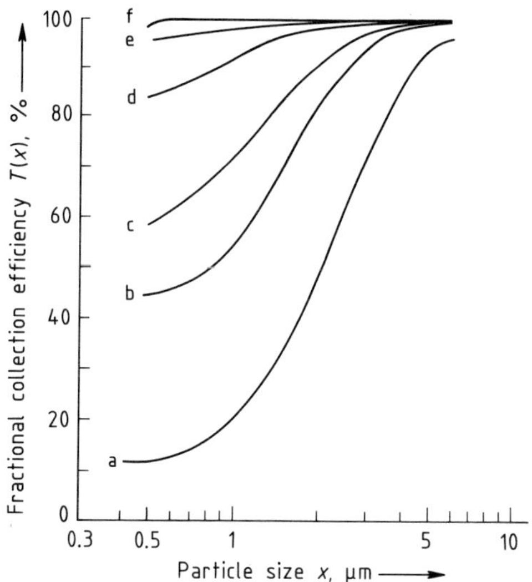

Figure 33. Fractional efficiency curves for a limestone bed (height 5 mm, granule diameter 0.28 mm) with various dust loads of quartz F 500 at a gas velocity of 0.055 m/s
Mass of dust W collected per unit filter area, g/m^2 (pressure drop Δp, Pa): a) 0 (108, initial trial); b) 6.98 (117); c) 11.97 (141); d) 17.95 (244); e) 25.93 (452); f) 50.85 (1110)

6.4. Design Calculations

The design of granular-bed dust filters is mainly empirical (as in collection generally), and development of theories is not yet complete. However, existing formulas still yield valuable information about important variables and trends and, therefore, are discussed briefly. A detailed discussion can be found in [6].

6.4.1. Collection Efficiency

The initial fractional collection efficiency (no dust load in the bed) is described by

$$T(x) = 1 - \exp\left[-1.5\frac{(1-\varepsilon)}{\varepsilon}\cdot\frac{H}{d_G}\cdot\varphi(x)\right] \quad (37)$$

where ε is the porosity of the bed, H is its height, d_G is the diameter of the filter granules, and $\varphi(x)$ is the single-granule collection efficiency.

The dependences of $T(x)$ on filter geometry (ε, H, d_G) have been confirmed experimentally [41], [43]. As the dust load in the bed increases, fractional efficiency increases, at least up to the critical load. The fractional efficiency curves in Figure 33 were measured with relatively small filter granules ($d_G = 0.28$ mm) and at low filtration velocities [44]. Under these conditions, the shift from collection in the bed to collection in the dust layer (i.e., to surface filtration) is relatively quick. As a result, fractional efficiencies are very high and only slightly dependent on particle size.

The difficulty in calculating fractional efficiency with Equation (37) is determination of the single-grain collection efficiency $\varphi(x)$. Transport mechanisms are the same as in fibrous-bed filters (diffusion, electrostatic attraction, inertia; see Section 5.2.2), but defining the flow field near a filter granule (which is assumed to be spherical) is more difficult.

Because the filter granules are close together, the flow fields derived for isolated spheres are too inexact, and correction factors are introduced to allow for the packing density [41]. At higher velocities, even flow fields modified in this way do not closely approximate reality.

The experimental fractional efficiency curves in Figure 34 essentially confirm theoretically predicted effects and trends. Below about 0.5 µm

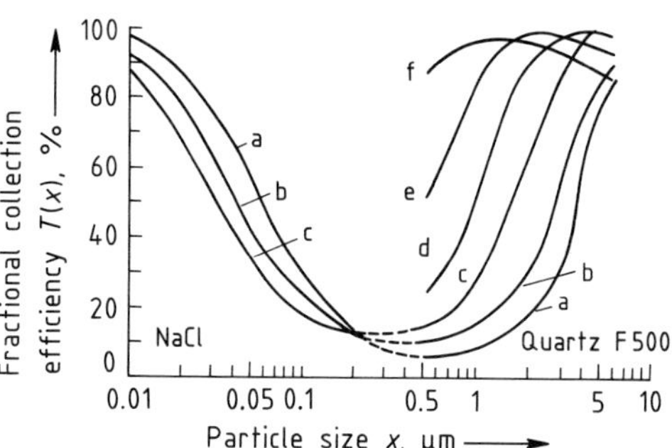

Figure 34. Fractional efficiency curves for a bed of glass beads at various filtration velocities
Bed height $H = 20$ mm; bead diameter $d_G = 0.58$ mm
Filtration velocity U_0, m/s: a) 0.05; b) 0.10; c) 0.25; d) 0.5; e) 1.0; f) 1.5

and 0.5 m/s (sodium chloride particles), diffusion effects are dominant. In this region, collection efficiency increases with decreasing particle size x, decreasing velocity U_0, and decreasing filter-granule size d_G. Inertial forces control collection to the right of the minimum (Quartz F 500), but here the curves pass through a maximum. The maximum shifts toward smaller particle sizes as the velocity increases, due to inadequate adhesion of particles when they collide with the filter granules. This phenomenon still occurs in much thicker beds and becomes even more evident if Equation (37) is used to calculate the single-grain efficiency from measured fractional efficiencies [43]. These tests, performed on beds up to 9.7 cm thick and composed of 4-mm spheres, show that the dust particles bounce off at velocities as low as 2 m/s.

6.4.2. Pressure Drop

The pressure drop across the granular bed in its initial unloaded state can be calculated fairly accurately. Ergun's equation is used primarily because it holds for all Reynolds numbers and has been tested for granular beds.

$$\Delta p = 150 \frac{(1-\varepsilon)^2}{\varepsilon^3} \cdot \frac{\mu \cdot U_0}{d_G^2} H + 1.75 \frac{(1-\varepsilon)}{\varepsilon^3} \cdot \frac{\varrho \cdot U_0^2}{d_G} H \quad (38)$$

where U_0 is the superficial velocity and μ is the gas viscosity. The marked dependence on porosity ε should be noted. Inaccuracy in the determination of ε strongly affects Δp.

The pressure drop increases with time due to incorporation of dust into the bed. This process has been described by a few equations which apply only to special cases and are not generally valid. Accordingly, the time-dependent behavior of the pressure drop must be determined experimentally.

7. Electrical Precipitators

7.1. General

An especially efficient way of separating particles from gases is based on the forces exerted on charged particles in an electric field. This principle is important when inertial forces are no longer effective, i.e., for fine particles with a diameter of ca. < 1 μm. The benefits of such processes have been known and patented for over a hundred years. Large-scale implementation, however, did not start until early in the 20th century, when sufficiently powerful high-voltage equipment was developed. The developmental history, principles, and current status of this field have been described in detail [23], [45]–[48].

Electrical collectors are often simply called *electrofilters;* the term *electrostatic precipitator* has been adopted even though the processes are by no means electrostatic and sizable electric currents are involved.

Because the required capital investment is rather large, the main application of these precipitators is in dust removal from large gas streams (up to several million cubic meters per hour). These includes flue gases from power plants and refuse incinerators (furnaces), cement plants (rotary kilns and mill drying), the iron and steel industry (blast furnaces and converters), foundries, nonferrous metal refineries (furnaces), expanded-clay aggregate plants, and chemical plants. Collection efficiencies over 99.9 % can be obtained. Emission levels significantly lower than 50 mg/m³ can be achieved if the dust properties are favorable. Pressure drops are generally very slight (< 500 Pa).

7.2. Mode of Operation

The collection principle used in electrical precipitators is depicted in Figure 35 and involves three successive steps:

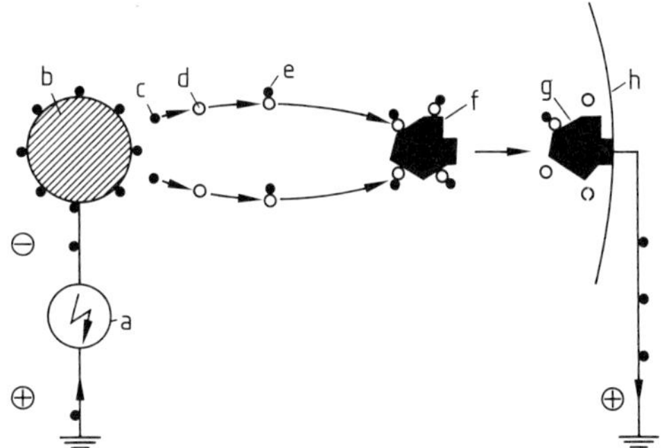

Figure 35. Schematic diagram of charging and collection process
a) High-voltage power supply; b) Emission electrode; c) Electron; d) Neutral molecule; e) Ionized molecule; f) Charged dust particle; g) Collected dust particle; h) Collecting electrode

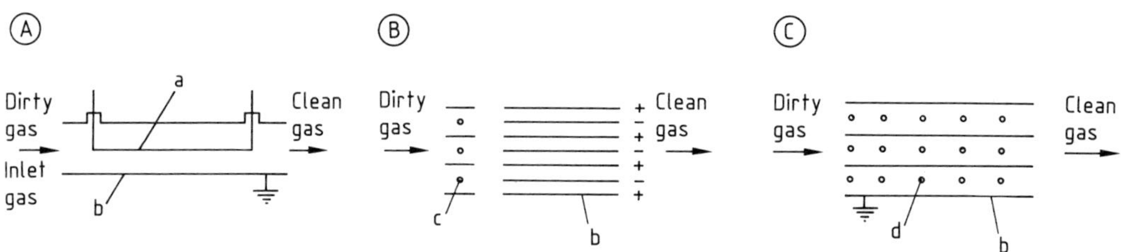

Figure 36. Schematic arrangements of emission and collecting electrodes in electrical precipitators
A) Wire-in-tube precipitator; B) Two-stage parallel-plate precipitator; C) Single-stage parallel-plate precipitator
a) Emission electrode; b) Collecting electrode; c) Positive emission electrode; d) Negative emission electrode

1) charging of particles,
2) transport of particles to the collecting electrode, and
3) removal of particles gathered on the collecting electrode.

Most particles carry an electrostatic charge, but it is usually too small for the collection process. Additional charge must, therefore, be supplied. This is accomplished by gas ions that are generated in a corona discharge. Figure 36 shows the configurations used for charging and collection: wire-in-tube (A) and parallel-plate (B and C) precipitators.

The parallel-plate types are by far more widely used and may be either two-stage or single-stage devices. In the two-stage unit (Fig. 36 B), the charging and collection regions are separate. These systems operate advantageously with a positive emission electrode and are preferably used in air-conditioning applications because a positive corona does not form ozone. However, the currents which can be obtained are not as high as with a negative corona. Higher currents are necessary for high collection efficiency and for operation with high dust concentration. Therefore, for dust removal from industrial exhaust gases, the single-stage negative-corona design is preferred (Fig. 36 C).

Typical voltages in industrial filters are 20–70 kV, with channel widths of 200–600 mm.

Electric charges released at the emission electrode attach themselves to the particles that are to be collected. The charged particles in the electric field between the emission and collecting electrodes are transported to the collecting electrode and retained. Liquid particles flow downward as a film, while solid particles form a dust layer that must be removed from time to time by rapping or flushing with a liquid. The loosened dust falls into a hopper beneath the collecting electrode. Dust removal is done on-line, i.e., the gas stream is not cut off. Consequently, previously collected dust is reentrained and transported further. The collector length is therefore subdivided into several (three to five) successive zones, which are cleaned at different times in order to minimize dust penetration.

7.2.1. Charging of Particles

Production of Charge Carriers. The ions needed for particle charging are produced by corona discharge at the emission electrodes, which are made of thin wires or tapes with barbs. Triggering the corona discharge requires a minimum voltage, the *corona inception voltage,* which depends on the electrode geometry as well as the gas species and its properties; a typical value is in the range 10–30 kV.

If the applied voltage is raised, the current increases until electrical breakdown (sparking) occurs at the *breakdown voltage* (Fig. 37). The breakdown voltage depends on electrode geometry and gas properties. To obtain optimal collection efficiencies, high field strengths and high corona currents are required. The d.c. voltage is, therefore, adjusted to establish a condition as close as possible to electrical breakdown. A recent trend involves use of pulsed voltage (pulse energization), in which the voltage peaks exceed the breakdown voltage [49].

To allow precipitator operation, the difference between the corona inception voltage and the breakdown voltage must not be too small. As the gas temperature increases, however, this difference becomes smaller and smaller, so that stable operation at atmospheric pressure is difficult above ca. 400 °C. However, the difference becomes larger with increasing pressure; this point is especially important in cleaning gases from high-pressure fluidized-bed combustors: at pressures > 1 MPa, operation is possible even at 800 °C.

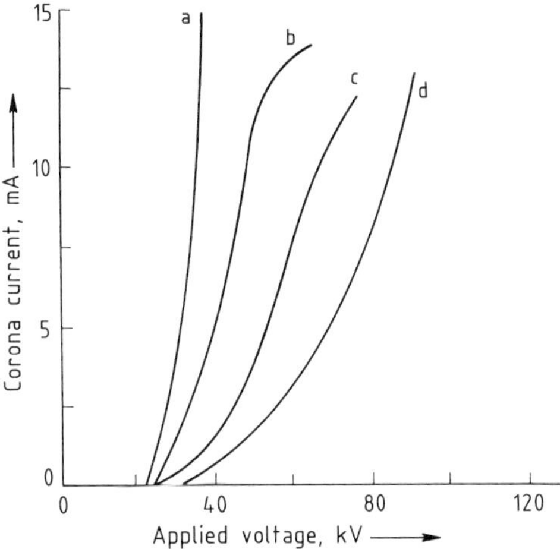

Figure 37. Current–voltage curve for a negative corona in nitrogen, nitrogen–oxygen mixtures, and air (electrical tube precipitator)
a) Nitrogen 100 %; b) Nitrogen 98 %, oxygen 2 %; c) Nitrogen 95 %, oxygen 5 %; d) Air

Attachment of Charge. Depending on particle size, the attachment of gas ions to particles occurs by one of two mechanisms: field charging or diffusion charging.

Field charging dominates in the particle-size range above ca. 1 μm. The ions move toward the collecting electrodes along the electric field lines. As they move, the ions collide with particles and become attached to them. The charging process is time-dependent; 80 % of the maximum charge is acquired after about 0.1 s. The maximum charge is proportional to the surface area of the particle, i.e., to x^2. For a 2-μm particle, this amounts to some 250 elementary charges.

Diffusion charging results from random thermal motion of the ions (Brownian motion) and is important chiefly for particles smaller than 1 μm. This process is slower than field charging: acquisition of 80 % of the maximum charge takes approximately 1 s. In this range, the maximum charge is proportional to the particle size x.

7.2.2. Effect of Dust Resistivity

Charged particles are transported to the collecting electrode and must adhere to it. The electrical properties of the particles, particularly resistance, are important for the behavior of both the particle on impact and the resulting dust layer.

The favorable range of resistivity for collection is $10^4 - 10^{11}$ Ω cm. Particles with a lower resistance ($< 10^4$ Ω cm) give up their charge quickly when they collide with the collecting electrode; they change their polarity and are repelled from the electrode back into the gas stream. If the resistance is too high ($> 10^{11}$ Ω cm), the particles form a layer that continues to accumulate more charge, thus leading to a weakening of the electric field and "reemission." Reemission results from discharge in the porous dust layer.

Dust resistivity depends on the geometry of the dust layer, particle characteristics, and gas properties (especially the dew point, Fig. 38); thus it should be measured under conditions as close as possible to those used in practice. The resistivity is made up of two contributions, surface resistivity and volume resistivity, which both depend on temperature but in opposite ways. The temperature–resistance curves therefore have a maximum (Fig. 37).

The resistivity of fly-ash particles is affected by their sulfur content. Some coals are so low in sulfur that the resulting fly-ash dust has a high resistance, which makes successful precipitator operation difficult. Consequently, electrical precipitation has been abandoned in power plants fired with these coals and bag filters have been installed. Problems caused by high dust resistivity can be overcome through gas conditioning, e.g., by spraying water into the gas or adding sulfur trioxide in low concentrations. A sulfur trioxide level of 10–15 ppm has proved most suitable; it is completely adsorbed by the fly ash so that the sulfur oxide emission level does not increase [50]. Another way of promoting the collection of high-resistance dusts is the pulsed-voltage technique discussed in Section 7.2.1.

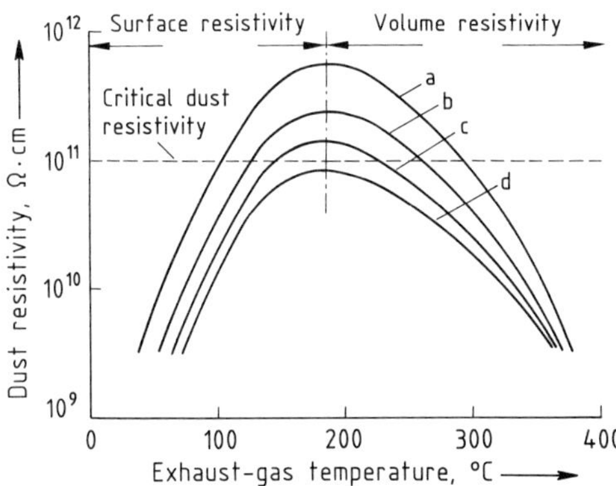

Figure 38. Electrical resistivity of exhaust gas as a function of temperature
Dew point of gas, °C: a) 20; b) 40; c) 50; d) 60

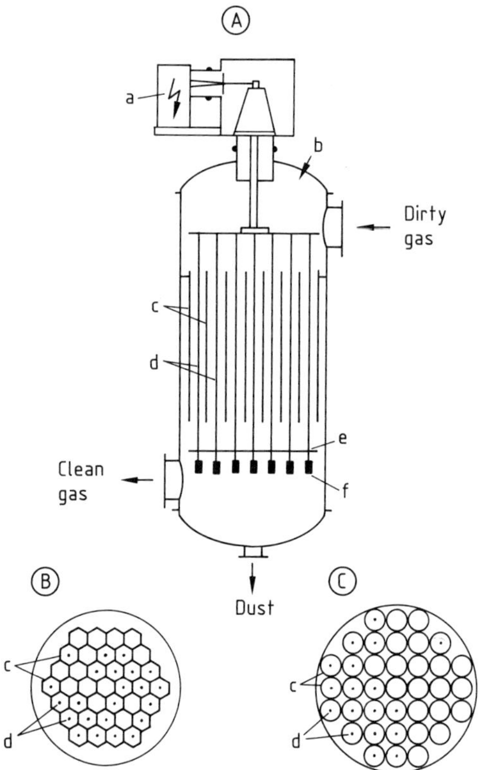

Figure 39. Tube precipitator consisting of individual collectors in parallel
A) Vertical section; B) Tubes with a hexagonal cross section (honeycomb precipitator); C) Tubes with a circular cross section
a) High-voltage power supply; b) Flushing; c) Tube walls (collectors); d) Emission electrodes; e) Electrode spacer; f) Weight

7.3. Basic Designs

Wire-in-tube precipitators (Fig. 39) are employed for small quantities of gas and especially for the collection of liquids (e.g., tar mists and acid mists). The emission electrodes (wires) are stretched along the central axis of parallel tubes. The inside walls of the tubes serve as collecting electrodes. If the tubes are set up vertically, the collected liquid runs down the walls as a film. If solid particles are collected, they are frequently rinsed out by flushing the tubes with water. The tubes are usually 0.2–0.3 m in diameter and 2–5 m long. They may have a circular (Fig. 39 B) or a hexagonal cross section (Fig. 39 C).

Parallel-plate precipitators contain many plane or profiled plates that are suspended at uniform intervals (200–600 mm) as collecting electrodes (Fig. 40). Gas flows horizontally in the channels between plates. Emission electrodes hang in the midplane between neighboring plates. Particle removal may be either dry or wet.

In *dry electrostatic precipitators,* the dust layer collected on the plates is removed periodically by rapping the plates with hammers; the dust falls into the hopper beneath the plates. The total plate length is divided into several (three to five) zones for two purposes: (1) dust reentrainment is reduced because the zones can be cleaned at different times; and (2) separate zone-by-zone power controls allow the voltage to be adjusted in accordance with dust concentration, which declines exponentially along the precipitator.

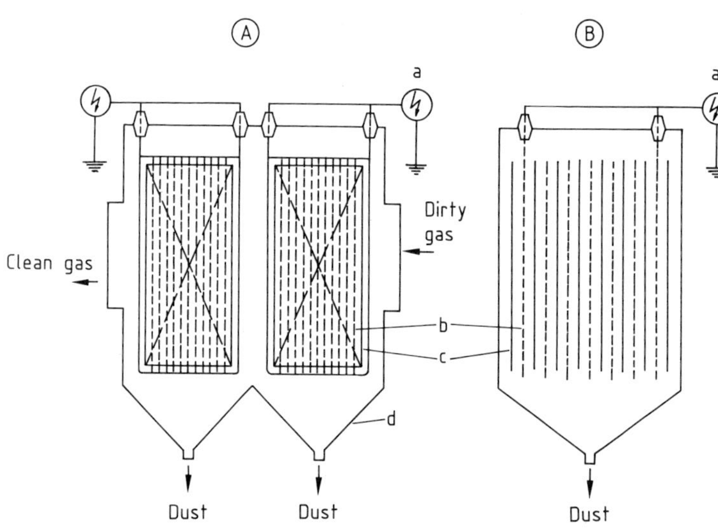

Figure 40. Schematic of a two-zone parallel-plate electrical precipitator
Vertical sections A and B are shown at right-angles to each other
a) High-voltage power supply;
b) Emission electrode;
c) Collecting electrode;
d) Hopper

In *wet electrostatic precipitators,* the collecting electrodes are cleaned by flushing with water. The dust flows down off the plates with the water; reentrainment of the dust is thus effectively prevented.

If the dirty gas has a very high dust content, the corona or the electric field strength is weakened, i.e., collection performance deteriorates. Precollectors are then used to lower dust concentration. However, these should not alter the particle-size distribution because dust removal from the plates is better with a wider distribution than with fine particles only. Simple deflecting vanes have proved suitable for this purpose [51].

If dust concentration in the clean gas must be particularly low, combinations of several types of dust collectors are employed: flue gases from blast furnaces can be cleaned up to 5–10 mg/m^3 by a series consisting of a cyclone, a scrubber, and a wet electrostatic precipitator.

7.4. Design Calculations

Calculations for precipitator design are concerned solely with collection efficiency; the very small pressure drop is not important.

In 1922, DEUTSCH derived the precipitator equation that bears his name. The equation has been challenged, but it is still used as the basis for sizing precipitators and for evaluating and interpreting measurements. If turbulent flow in the precipitator is assumed (i.e., the distributions of gas velocity and particle concentration transverse to the direction of flow are uniform) and if reentrainment and reemission are neglected, the Deutsch equation for fractional efficiency is

$$T(x) = 1 - \exp\left[-\frac{2L}{a \cdot v}\, w(x)\right] \qquad (39\,\mathrm{a})$$

where L is the collection length, a is the plate spacing (parallel plate precipitators) or the diameter of the tube (tube precipitators), v is the gas velocity, and $w(x)$ is the so-called migration velocity. For a parallel-plate device, Equation (39 a) can also be written as

$$T(x) = 1 - \exp\left[-\frac{A}{\dot{V}}\, w(x)\right] \qquad (39\,\mathrm{b})$$

where A is the surface area of the collecting electrodes and $\dot{V}$ is the volume gas flow rate. The ratio $A/\dot{V}$ is also called the specific collecting area.

The migration velocity $w(x)$ depends on the particle size and charge, the electric field, and the gas viscosity. It can be derived theoretically by solving the equation of particle motion [6], [46]. In practice, migration velocity is calculated from the measured fractional efficiencies, by means of Equation (39); it often lies in the range of 10–30 cm/s. The parameter $w(x)$ is basically a transport coefficient that characterizes particle transport to the collecting electrode; it is thus vital for sizing electrical precipitators. Deutsch's assumption that $w(x)$ is constant along the precipitator has not been confirmed experimentally [52]. The mean value of w (the effective migration velocity), recalculated with Equation (39) from measured fractional efficiencies, increases with gas velocity and channel width [53]. The physical significance of these phenomena and their incorporation in the design equation are not yet completely understood because particle behavior inside the precipitator is still not clear. For this reason, the search for an optimal channel width has led to differing results in the range of 40–60 cm.

Nevertheless, Deutsch's equation continues to be of central importance in design. A variety of correction factors in the exponent of Equation (39) have been suggested [54]. One problem in interpreting experimental studies is that measurements usually yield overall collection efficiencies, not fractional efficiency curves. The resulting w values thus represent integral mean values, which depend on particle-size distribution. As the distribution in the dirty gas becomes broader, the overall efficiency for a given precipitator declines. As one of the few fractional measurements shows, the shape of the fractional efficiency curve is rather complicated but also informative (Fig. 41) [55]. The ascending part of the curve, to the left of the minimum at ca. 0.4 µm, can be accounted for by the dependence on flow resistance of particle motion relative to the gas (Cunningham correction). The decline of the curve past ca. 3 µm has been explained [40]

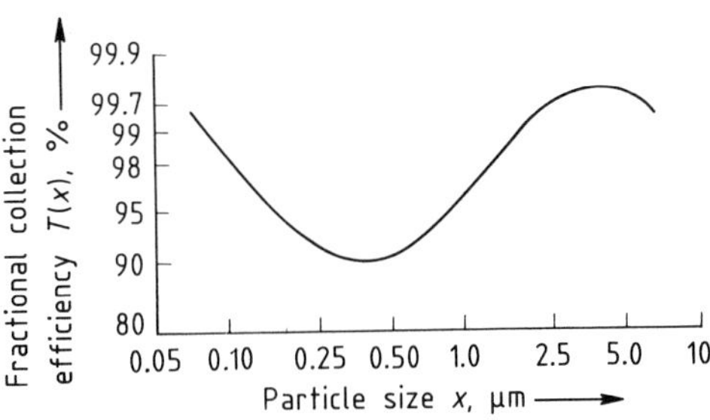

Figure 41. Experimentally determined fractional efficiency curve for an electrical precipitator (specific collecting area = 70 $m^2\,s\,m^{-3}$) [55]

by the typical behavior of incompletely combusted coal particles, which are repelled from the collecting electrode because of their low resistivity (cf. Section 7.2.2). Improved particle measuring instruments have been developed recently; their use in determining fractional efficiency curves for electrostatic precipitators should contribute to the understanding of this process.

8. References

General References

[1] E. Weber, W. Brocke: *Apparate und Verfahren der industriellen Gasreinigung,* vol. 1, "Feststoffabscheidung," R. Oldenbourg Verlag, München 1973.
[2] A. Ogawa: *Separation of Particles from Air and Gases,* vol. II, CRC Press, Boca Raton, Florida, 1984.
[3] W. Strauß: *Industrial Gas Cleaning,* 2nd ed., Pergamon Press, Oxford 1975.
[4] W. Licht: *Air Pollution Control Engineering,* 2nd ed., Marcel Dekker, New York 1988.

Specific References

[5] Technische Anleitung zur Reinhaltung der Luft (TA Luft) 1986, Allgemeine Verwaltungsschrift zum Bundesimmissionsschutzgesetz.
[6] F. Löffler: *Staubabscheiden,* Thieme Verlag, Stuttgart 1988.
[7] R. H. Perry, O. Green (eds.): *Perry's Chemical Engineers' Handbook,* 6th ed., McGraw-Hill, New York 1984, pp. 20-82, 83.
[8] VDI-Richtlinie 3676, Massenkraftabscheider, VDI-Verlag, Düsseldorf 1980.
[9] M. Bohnet, *Chem. Ing.-Tech.* **54** (1982) 621-630.
[10] F. Boysan, W. H. Ayers, J. Swithenbank, *Trans. Inst. Chem. Eng.* **60** (1982) 222-232.
[11] P. Meißner, F. Löffler, VDI Ber. **294** (1978) 69-75.
[12] D. Leith, W. Licht, *AIChE Symp. Ser.* **68** (1972) 196-206.
[13] W. Barth, *Brennst. Wärme Kraft* **8** (1956) 1-9.
[14] E. Muschelknautz, Chem. Ing. Tech. **44** (1972) 63-71.
[15] P. W. Dietz, *Powder Technol.* **31** (1982) 221-226.
[16] H. Mothes, F. Löffler, *Chem. Eng. Process* **18** (1984) 323-331.
[17] E. Muschelknautz, *Verfahrenstechnik (Mainz)* **6** (1972) 5. Beilage Hochschulkurs II.
[18] H. Rumpf, K. Borho, H. Reichert, *Chem. Ing. Tech.* **40** (1968) 1072-1082.
[19] M. Bohnet, *Chem. Ing. Tech.* **54** (1982) 621-630.
[20] W. Krambrock, Chem. Ing. Tech. **51** (1979) 493-496.
[21] S. Calvert, *Chem. Eng.* **54** (1977) 29.
[22] VDI-Richtlinie 3679, Naßarbeitende Abscheider, VDI-Verlag, Düsseldorf 1980.
[23] F. Löffler, G. Schuch, *Filtr. Sep.* **18** (1981) 70-74.
[24] K. Holzer, *Staub Reinh. Luft* **34** (1974) 361-365.
[25] W. Barth, *Staub Reinh. Luft* **19** (1959) 175-180.
[26] S. Calvert, *J. A. P. C. A.* **24** (1974) 929-934.
[27] G. Schuch, F. Löffler, *Verfahrenstechnik Mainz* **12** (1978) 302-306.
[28] C.-D. Schegk, F. Löffler, *Chem. Ing. Tech.* **59** (1987) 319-322.
[29] S. Ripperger, G. Dau, *VT Verfahrenstechnik* **14** (1980) 164-168.
[30] K. T. Semran, *Chem. Eng.* **26** (1977) 87-91.
[31] L. Gail, *Staub Reinhalt. Luft* **37** (1977) 175-178.
[32] DIN 24 184, Typprüfung von Schwebstoffiltern.
[33] C. N. Davies: *Air Filtration,* Academic Press, London 1973.
[34] R. Hiller, *Fortschr. Ber. VDI Z. Reihe 3,* **61** (1981).
[35] Hp. Baumgartner, Dissertation Universität Karlsruhe, 1986.
[36] Hp. Baumgartner, F. Löffler, H. Umhauer, *JEEE Trans. Electr. Ins.* **21** (1986) no. 3, 477-486.
[37] H. Jodeit, *Fortschr. Ber. VDI Z. Reihe 3,* **108** (1985).
[38] DIN 71 459, Filtereinsätze für Trockenluftfilter, 1973.
[39] F. Löffler, H. Dietrich, W. Flatt: *Dust Collection with Bag Filters and Envelope Filters,* Vieweg & Sohn, Braunschweig - Wiesbaden 1988.
[40] VDI-Richtlinie 3677, Filternde Abscheider, VDI-Verlag, Düsseldorf 1980.
[41] G. J. Tardos, N. Abuaf, C. Gutfinger, *J. Air Pollut. Control Assoc.* **28** (1978) 354-363.
[42] G. L. Kinchett, *Filtr. Sep.* **22** (1985) 378-380.
[43] S. L. Goren, T. D'Ottavio, *Aerosol Sci. Technol.* **2** (1983) 91-108.
[44] W. Peukert, F. Löffler, Proceedings of International Conference on Advanced Coal Power Plant Technology, Düsseldorf, 2-4 Dec. 1987.
[45] H. J. White: *Industrial Electrostatic Precipitation,* Addison-Wesley, Publ. Co., Pergamon Press, Oxford 1963.
[46] M. Robinson in E. Strauß (ed.): *Air Pollution Control I,* "Electrostatic Precipitation," J. Wiley & Sons, New York 1971.
[47] J. Böhm: *Electrostatic Precipitation,* Elsevier, Amsterdam 1982.
[48] VDI-Richtlinie 3678, Elektrische Abscheider, VDI-Verlag, Düsseldorf 1980.
[49] K. Darby, *2nd Int. Conf. Electr. Precipitation,* Kyoto 1984, pp. 575-584.
[50] G. Mayer-Schwinning, R. Rennhack, *Chem. Ing. Tech.* **52** (1980) 375-383.
[51] K. Arras, *Zem. Kalk Gips,* **29** (1976) 241-247.
[52] K. Kinkelin, *Fortschr. Ber. VDI Z. Reihe 15,* **24** (1982).
[53] H. Wiggers, *Staub Reinhalt. Luft* **42** (1982) 292-294.
[54] G. Cooperman, *Atmos. Environ.* **14** (1984) 277-285.
[55] S. Maartmann, *Staub Reinhalt. Luft* **34** (1974) 353-355.

14. Solid–Solid Separation, Introduction

Henry E. Cohen, Imperial College of Science and Technology, Royal School of Mines, London, United Kingdom

The need to separate solids occurs in many different industrial activities and for a variety of materials. Examples include the separation of food grains such as nuts, beans, peas, or rice from contaminants or waste such as shells, husks, stones, or vermin. Other examples are the removal of print from recycled paper, the separation of granulated plastics and copper in recycling electric cables, or the recovery of precious-metal contacts from scrapped electronic equipment.

By far the largest application of solid–solid separation, in both tonnage of throughput and value of products, is the recovery and purification of ores for the production of metals, industrial minerals, and fuels. Physical methods for concentrating and separating desired metals or minerals (values) from unwanted ore constituents (waste or tailings) are collectively known as *mineral processing*. "Ore" is defined as any kind of rock or mineral deposit from which marketable products can be extracted profitably. Hence, solid–solid separation employs scientific principles to devise engineering solutions based on property differences, for separating wanted from unwanted constituents of ores within given economic constraints.

Differences in physical and chemical properties of values and tailings are displayed and enhanced as much as possible by appropriate preparatory treatment. Adequate physical liberation of constituents is essential for effective separation. The product quality (grade) and the recovery of values depend on the degree of liberation and the enhancement of differential properties achieved during preparation. In most cases, a trade-off is made between product grade and percentage recovery of values, because an increase in one generally entails a decline in the other.

Broadly, all ores can be divided into two categories:

1) Ores that contain high-priced values (e.g., gold or other precious metals, nickel, tin, diamonds, emeralds, etc.) in low concentrations. A typical gold ore contains a few grams of gold per ton of rock.
2) Ores that contain low-priced values (e.g., iron, manganese, chromium, gypsum, china clay, etc.) in relatively high concentrations. A typical iron ore contains 60 % iron, or nearly two-thirds of a ton of iron per ton of rock.

These differences in ore composition influence the selection, application, and performance of methods of solid–solid separation. The highest possible percentage recovery (better than 90 %) of a small content of high-priced values is essential (e.g., with gold ores) to cover the cost of treatment. With low-priced values (e.g., in iron ores), modest recoveries (60 – 70 %) can be tolerated as long as the product grade is high enough to qualify for the best market prices. For china clay, practically 100 % purity (i.e., 100 % grade) is necessary for its acceptance as a paper-coating material.

1. Preparatory Processes

Essential preparatory processes include *comminution* for solid rocks (or other heterogeneous solids), *dry* or *wet scrubbing* for unconsolidated materials such as sand or clay deposits, and *size separation processes* for populations of particles.

Although preparation processes necessarily precede solid–solid separation, complex alternating sequences of preparation and separation are commonly applied. This sequence, also called stage processing, is advantageous if the values show some inhomogeneous liberation behavior or if different values in the ore require separate recovery after dissimilar preparation.

For example, the tin mineral cassiterite (SnO_2) may occur in granite in natural crystal sizes from 10 mm down to 0.05 mm. During comminution to liberate the cassiterite from waste minerals such as silicates, iron oxides, etc., breakage to < 5 mm may produce substantial proportions of coarse, liberated cassiterite particles. Having to break such particles further during comminution to liberate smaller crystals of cassiterite would be wasteful in terms of processing cost and possible loss of values. Hence, good industrial practice means interrupting comminution and first separating a coarse concentrate of liberated cassiterite. Comminution followed by separation of a product (or of a waste reject) can be repeated several times at progressively smaller size ranges.

Other complex sequences of preparation and separation may be necessary for ores that contain several different value minerals. Another common reason for using secondary grinding and size separation near the end of a process sequence is the persistence of locked middlings.

1.1. Comminution

In industrial practice, comminution—also called size reduction—is usually divided into crushing and grinding (→ 5. Size Reduction). *Crushing* employs processes of compression or impact shattering and is applied to coarse particles, from meters down to centimeters in size. Power consumption is relatively low, up to ca. 3 kW · h per ton of throughput. *Grinding* employs processes of abrasion, with or without impact, and is applied to fine particles, from centimeters down to submicrometers in size. Power consumption is relatively high, commonly in the range of 10–20 kW · h and up to 200 kW · h per ton for very fine products. Grinding is usually the most energy-intensive process in any sequence of treatment and can account for more than 70 % of the total energy consumed.

This provides another reason for stage processing. If waste material can be liberated from values at relatively coarse particle sizes, separating and discarding such waste ahead of the expensive grinding stage may be economically advantageous.

For example, in the treatment of iron ores in Minnesota, some 40 % of the feed tonnage could be removed as coarse waste before grinding. Magnetic separators were used to produce a clean nonmagnetic waste and a magnetic "middling" which passed to the grinding stage. About 10 % of the total iron in the feed was lost with the coarse waste, but this loss was far outweighed by the savings in grinding costs.

Liberation is the most important purpose of comminution and scrubbing processes. Another purpose is size preparation, either to provide better handling characteristics or to meet some specific size requirements imposed by a subsequent method of separation. For example, the recovery in froth flotation usually declines considerably for particles coarser than 0.2 mm or finer than 0.02 mm. If solid–solid separation by flotation is appropriate (e.g., for sulfide minerals), preparation should yield a maximum proportion of the ore in that size range, irrespective of liberation sizes.

1.2. Size Separation

Size separation includes important aspects of preparation requirements. For purposes of handling, separation, and subsequent use or disposal of products, ores are generally divided into three size groups: oversize, product size (one or several), and undersize. *Oversize material* is usually unwanted and insufficiently liberated. It may be subjected to further comminution or discarded if it is only a small fraction of the product. *Product sizes* are those most suitable for a separation process; they may also be defined according to size specifications imposed by users. For example, commercial specifications for iron ores usually include upper and lower size limits. The term *undersize* refers to particles that are too small for a given process of separation or for commercial acceptance of a product. Particle-size analysis is treated elsewhere in this volume.

All three terms—oversize, product size, and undersize—are relative and can cover widely different size ranges for different materials and purposes. For example, in treating gold ores, an upper product size limit of about 0.1 mm is usually necessary to ensure adequate liberation. Particles coarser than 0.1 mm are oversize and require further grinding. In contrast, when jigs are used for gravity concentration of tin ores, particles smaller than around 0.2 mm are undersize. The jig would be inefficient in treating such small particles which must be treated separately by other means.

The two dominant industrial processes for performing size separation are *screening* and *classification*. They can be used for wet particles carried in water (or other fluids) or for dry particle populations that are free-flowing.

1.2.1. Screening

In screening (or sieving), each individual in a particulate material is presented to apertures which may be circular, square, or rectangular and which act as two-dimensional gauges (→ 15. Screening). Ability to pass through depends on size but is also affected by particle shape which may be platy, nearly equidimensional, needle-like, or irregular. Thus, with increasing aspect ratios, the orientation of a particle relative to the aperture becomes an important factor which blurs the definition of particle size. For example, consider a square screen aperture of 10 × 10 mm. With favorable presentation, this aperture would permit passage of a 9-mm square disk, a 9-mm cube, or a rectangular block 9 mm square and 90 mm long. Screening assigns particles roughly into size groups, each group having passed through a larger aperture and having failed to pass through the next smaller aperture used in a given process.

The difficulty of passage increases as particle dimensions approach aperture size, and this can lead to blockage. Therefore, good industrial practice involves avoiding aperture sizes that are near the dominant particle size in a given product.

Screening is used generally for coarse particles, from meters down to about 1 mm and even smaller sizes (e.g., for china clay), occasionally. Under laboratory conditions, woven wire sieves are used down to 0.025 mm. Fragile electroformed perforated sheets and molecular sieves can be employed down to submicrometer sizes, but then throughput is very slow.

1.2.2. Classification

Classification is commonly employed in industry for size separation of particles below about 1 mm. The process relies on different terminal velocities (different settling rates) for particles of different sizes moving through fluids (usually water; → 16. Elutriation) or through gases (usually air; → 17. Air Classification). Particles accelerating under the influence of gravity (or some other force) encounter increasing resistance from the enveloping medium. Terminal velocities are reached when the forces of acceleration and resistance are equal. Larger particles then move away from smaller ones, and separate size groups can be collected.

However, differences in specific gravity must yield different terminal velocities. Hence, each classified size group in heterogeneous populations of particles includes larger particles of lower specific gravity and smaller particles of higher specific gravity. This distinguishes a classified size fraction from a screened product, which contains the same size range for all particles, irrespective of specific gravities. Classified feed is preferred for some separation processes, e.g., gravity separation on shaking tables.

Classification used for the recirculation of oversize material in grinding circuits can cause undesirable overgrinding of value minerals. Values having higher specific gravities than gangue minerals may undergo selective recirculation, with further grinding, even though they are fine enough for processing.

2. Physical Separation Processes

The physical processes of solid–solid separation may be grouped broadly as

1) radiation sorting,
2) gravity separation,
3) magnetic separation,
4) electrostatic separation,
5) separation based on surface phenomena.

Another important group of processes is based on chemical reactivity, e.g., acid leaching, solvent extraction, or other methods of selective dissolution that entail the partial or total destruction of ore constituents. These processes are termed chemical separation or hydrometallurgy and are beyond the scope of physical separation methods, which normally produce concentrates of unaltered minerals (→ 6. Liquid–Liquid Extraction, **B3**; → 7. Liquid–Solid Extraction, **B3**).

2.1. Radiation Sorting

Many types of radiation can be used for solid–solid separation, including visible, ultraviolet, and infrared light; X rays, gamma rays, and electron beams; ultrasound; and a range of radio frequencies. The methods of separation rely on differential effects of transmission, absorption, reflection, emission, or modulation. For example, reflection of visible light can be used to separate particles by color. Similarly, transmission of light can be used to separate opaque from transparent species. Use can be made of radiation emitted selectively by particles in mixed populations; e.g., the tungsten mineral scheelite fluoresces bright blue and the mineral calcite fluoresces bright pink under ultraviolet illumination. The use of emissions from radioactive substances is another example.

The necessary process mechanisms include a sequence of presentation, inspection, and sorting (→ 18. Mineral Sorting). Each particle, in free flight or traveling on a conveyor, must be presented individually to one or several radiation sources. The particle is inspected, and the resultant information is analyzed to trigger an accept–reject signal which operates a sorting mechanism (e.g., a moving splitter, an air or water jet, or a trap). The size range of particles suitable for such sorting extends from about 150 mm down to about 2 mm, but individual feeds must be sized relatively closely. Throughput rates are proportional to particle size, and separation is most efficient if the ejected material is a minority constituent.

2.2. Gravity Separation

Gravity separation is divided into two groups of processes:

1) sink–float separation and
2) differential acceleration.

Sink–float separation, also called *dense medium separation* or *heavy medium separation,* employs heavy liquids or suspensions, which have effective densities intermediate between the specific gravities of species in a mixed population (→ 22. Dense Medium Separation). For example, quartz (*d* 2.65) can be separated from hematite (*d* 5.1) by using a heavy liquid such as tetrabromoethane which has a specific gravity of 2.95. If a mixture of quartz and hematite is immersed in tetrabromoethane, the quartz floats on the liquid and the hematite sinks to the bottom. Similarly, the heavy liquid can be a suspension of finely pulverized heavy minerals in water. Effective specific densities up to about 3.5 can be achieved with suspensions of barites, galena, magnetite, or ferrosilicon. For the separation of coal from shale, quartz sand suspensions with effective densities in the range of 1.4–1.6 are used. Heavy suspensions are much more economical for industrial use than organic heavy liquids, such as tetrabromoethane, which are employed mainly in laboratory work.

The second major group of gravity separation processes relies on *gravitational acceleration* to exploit differences in specific gravity; a difference of not less than 1 is normally required for efficient separation. Gravitational acceleration causes particles of higher specific gravity to advance faster from rest than particles of lower specific gravity, irrespective of their sizes. Path lengths must be short so that the particles cannot reach terminal velocities (i.e., zero acceleration). Thus, gravity separation, which employs differential acceleration over short path lengths, differs from size classification, which depends on differential terminal velocities (→ 21. Gravity Concentration).

Various separator designs employ flowing films or oscillatory motion to produce repeated acceleration over short path lengths. The flow velocities and amplitudes of oscillation must match the dynamic characteristics of the particles; hence, size-classified feeds are usually necessary. Throughput capacities range from hundreds of tons per hour (e.g., jigs for coarse feed) to less than 0.5 t/h (e.g., shaking tables for slimes).

2.3. Magnetic Separation

Most natural minerals are more or less feebly paramagnetic and a few are very weakly diamagnetic. Magnetic separation processes can separate species of relatively higher magnetic susceptibility from others of lower susceptibility. The strength of the available magnetic force limits the applicability of a separator, and the difference in susceptibility determines the possible efficiency of a separation. Magnetic separators employ various methods of presenting feed to convergent magnetic fields which provide directional tractive forces. Counterforces, such as fluid drag, gravitational acceleration, and others, are arranged singly or jointly to act in a divergent direction from the magnetic forces. The balance of forces is arranged so that the more susceptible particles (the "magnetics") follow the magnetic force and the less susceptible particles (the "non-magnetics") follow the counterforces (→ 19. Magnetic Separation).

Magnetic separation can be carried out dry or in fluid suspension. The magnetic force experienced by a particle is mass related; hence, no theoretical upper or lower size limit exists. However, problems of physical handling usually impose an upper limit of about 150 mm. Electrostatic charges in dry feed and fluid drag in wet feed generally impose lower size limits around 0.01–0.05 mm. Throughput rates depend on particle size, but for different types of separators they can range from hundreds of tons to a few kilograms per hour.

2.4. Electrostatic Separation

Electrostatic separation is based on differences in surface electrical properties, particularly surface conductivity, and the associated ability to acquire, retain, or lose charges. Dry particles can acquire charges by induction, convection, or conduction. If different species in a mixture are charged selectively, they can be separated by application of external electrical forces (→ 20. Electrostatic Separation).

For example, a special case of conduction charging arises through spontaneous migration of electrons in a pulverized heterogeneous population of particles. Electrons will pass from a poorer to a better conductor; thus, the latter acquires a negative charge and the former is left with a net positive charge. If this population passes between a pair of electrode plates, one positively and one negatively charged, each particle will experience repulsion from the plate of like charge and attraction from the plate of opposite charge. The geometry, level of charge, and speed of passage can be arranged so that the particles separate into two streams.

In another example, a mixed particle population of conductors and nonconductors can be charged by means of a corona discharge from a wire electrode. The nonconductors retain the charge, but the conductors lose their charge during passage over a rotating drum. A second charged electrode can then "pin" the charged particles to the drum and "lift" the discharged particles. A typical application is the separation of zircon (a nonconductor) from rutile (a conductor) in the treatment of beach sand.

Various separator designs utilize different combinations of charging and discharging to achieve discrete products. A restricted feed size range is usually important for good separation, but several repeated treatments are normally needed to attain acceptable clean products. Throughput rates range up to several tons per hour.

2.5. Separation Based on Surface Phenomena

Surface chemical properties of solids can be manipulated selectively in various ways to generate differential behavior of the different constituents in a mixed population. For example, traces of reagents can be selectively adsorbed on species, rendering them either hydrophobic or hydrophilic. The hydrophobic particles can be made to coagulate and to settle from a suspension. This is known as *selective flocculation* (→ Flocculants, **A11**, p. 254; see also → 12. Sedimentation).

More important, hydrophobic particles can be attached to air bubbles which are made to stream through a pulp suspension. The air bubbles transport the hydrophobic particles to the surface of the pulp, forming a scum or froth which can be collected as a concentrate. This process, known as *froth flotation,* is one of the most important and widespread methods of solid–solid separation in the mineral industry (→ 23. Flotation). A variety of reagents have been developed for specifically enhancing the selectivity of flotation. The use of reagents is usually combined with stringent control of pH and of dissolved ions in the water.

With different reagent regimes, froth flotation has such diverse applications as the recovery of sulfides of copper, lead, zinc, nickel, molybdenum, antimony, and other nonferrous metals; oxides of iron, titanium, tungsten, tin, etc.; calcium fluoride, phosphates, various carbonates, silicates, and many other mineral compounds. Flotation is an extremely versatile process, limited only by the development of selective reagents for specific applications. It operates most efficiently on particles in the range of 0.2–0.02 mm, but coarser and finer sizes can be treated successfully under favorable conditions.

The throughput capacity of individual flotation machines varies, according to unit size, from kilograms to tons per hour. Multiple units in very large flotation plants, achieve throughputs in the range of 30 000–50 000 t/d. The process route is usually arranged so that a first flotation, the rougher stage, is followed by one or more cleaner stages for the float product. Similarly, one or more scavenger stages for the sink residue can rescue additional float material. These stages may be interspersed with regrinding of some products to improve liberation.

Another application of surface chemistry, *oil agglomeration,* employs various reagents for specific conditioning of mineral species so that they are selectively taken into an organic fluid that is immiscible with water.

3. Process Design Economics

The cost of processing must be less than the resultant added value of the product. Processing costs are minimized when the simplest possible process route is used to attain the best possible recovery, at the lowest product grade acceptable to the customer. Variations in the three basic cost components—capital, energy, and labor—can influence process choices in relation to specific local conditions, but the resultant costs for any separation route consist of only two elements:

1) processing costs, which are definable as capital and operating charges, and
2) loss of values due to inevitable process inefficiencies. The recoverable values are always lower than 100% of the values contained in the feed.

Both cost elements are related to the balance between grade and recovery, as well as to the complexity of the processing route.

For high-priced values which are minority constituents of the feed, high recovery is more important than grade; hence, the process design should aim at the earliest possible removal of concentrates. This points to treatment in stages so that values can be extracted as soon as they are liberated in any significant quantity. Repeated further preparation of the remaining bulk of the material and separation of values should continue until the residual values cannot justify the cost of further processing.

Low-priced values which are major constituents of a feed must be worked up to high-grade specifications, but high recovery can be sacrificed. To keep processing costs as low as possible, the process design should aim at early gangue removal, especially to avoid fine grinding. This requires stage separation of liberated gangue and retention of impure values for further processing to attain the desired grade.

Detailed factors influencing the selection of appropriate processes of solid–solid separation are discussed in the other articles to which reference is made here.

4. References

[1] N. L. Weiss (ed.): *SME Mineral Processing Handbook,* Society of Mining Engineers, AIMMPF, New York 1985, pp. 7–5 to 7–29.
[2] B. A. Wills: *Mineral Processing Handbook,* 3rd ed., Pergamon Press, Oxford 1985, pp. 522–530.
[3] G. Tarjan: *Mineral Processing,* Akademiai Klado, Budapest 1981, pp. 419–423.
[4] P. Somasundranan (ed.): *Advances in Mineral Processing,* Society of Mining Engineers, Boulder, Col., 1986.
[5] L. White (ed.): *E & MJ Second Operating Handbook of Mineral Processing,* McGraw-Hill, New York 1980.

15. Screening

Paul Schmidt, Universität Essen, Federal Republic of Germany

In addition to the standard symbols defined in the front matter of this volume, the following symbols are used:

a	acceleration, fraction of fines in the feed
A	area
c	concentration, constant, correction factor
d	wire gauge
e	amplitude
E	modulus of elasticity
f	frequency, fines in fine fraction, function
F	force
Fr	Froude number
g	gravitational acceleration ($g = 9.81\ m/s^2$), fines in coarse fraction
h	height
H	depth of bed
k	sieving rate constant
K	acceleration number, throw number
l	length
L	screen length
m	mass
$\dot{m}$	mass flow, specific capacity
n	number, rpm
p	pressure
P	probability, power
q	particle-size distribution
Q	cumulative particle-size distribution
R	radius
t	time, side of triangle
T	grade efficiency
u	velocity of fluid flow
v	velocity
V	volume
$\dot{V}$	volume flow
w	mesh size
x	particle size
α	casting angle
β	slope
δ	angle of screening
Δ	difference
η	efficiency
ϑ	angle of tangential inclination
κ	sharpness of cut
μ	coefficient of friction
π	dimensionless number
ϱ	density, angle of friction, angle of radial inclination
σ	normal stress
τ	shear stress
φ	open screen area
ψ	form of the mesh
ω	angular velocity

Indices

a	side of rectangle
A	feed
b	side of rectangle
B	width
c	critical
C	near-mesh
F	fine, fluid
G	coarse
h	horizontal, transport
H	bond, adhesive
H	depth of bed
K	plugging, jamming
L	length
m	mean
m	mass
M	medium
n	normal
p	particle
R	radius of sieve bend
S	screen, segregation
T	separation, cut size
v	vertical
V	volume
w	mesh

1. General

Sieving or screening is a method of separating particulate materials. The separation is according to size and is referred to as classification. Separation results from the repeated, random comparison of each individual particle with the characteristic size w of the screen opening, as shown in Figure 1. Particles may be larger or smaller than the mesh size. So-called near-mesh particles, however, present problems. Both the feed to the screen and the resulting fractions are described by the particle-size distribution $q(x)$ or the cumulative particle-size distribution $Q(x)$. The precision of separation is shown by the grade efficiency curve $T(x)$, which gives the sharpness of cut (for further information, → 2. Particle Size Analysis and Characterization of a Classification Process p. **2**-11). In some cases, screening can be used to separate according to composition or particle shape. This is referred to as sorting.

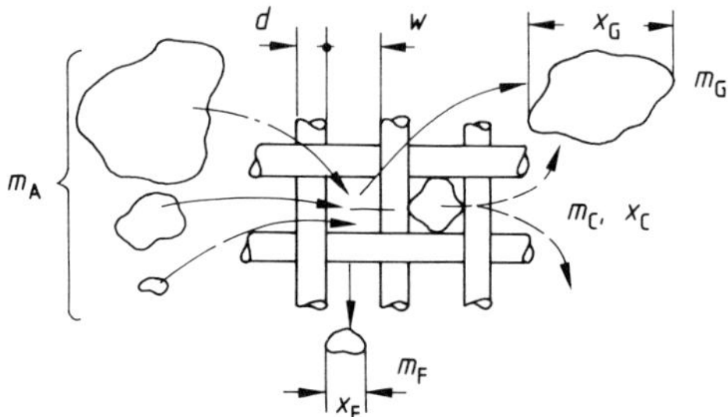

Figure 1. Screening as a comparison between particles in the feed material m_A with size x and the mesh with opening w
Partition into oversize, coarse, or nonpassing (m_G) and fines or passing (m_F); near-mesh particles (m_C) with $x \approx w$ cause blocking of the screen

1.1. Applications of Screening

Almost every process in which particulate materials are handled employs some screening method to obtain a product of desired quality. Coarse screening, with mesh size $w \gtrsim 100$ mm, is used to process earth and rock as well as ore and coal. Intermediate screening, with $100 > w > 10$ mm, is crucial in the preparation of building materials such as gravel, sand, and crushed stone. Fine screening, with $10 > w > 1$ mm, is employed in all industry—agriculture, food processing, particle-board manufacture, etc. Here and in ultrafine screening, where $1 > w > 0.025$ mm, the methods and machines have been developed extensively. Some examples are the screening of plastic powders, fillers, pigments, pharmaceuticals, flour, etc. Especially narrow size fractions are required in screening abrasives.

Screening assists both size reduction and agglomeration processes. Screening machines are responsible for economically removing particles of the proper size from the process stream. Screening machines are also used for solid–liquid separation as dewatering screens and screening centrifugals. Even the generation of fluidized beds is sometimes performed by vibratory screens. Worldwide, about 5×10^9 t of product are screened per year. About 100 companies produce screening machines, and about 50 manufacture wire cloth.

1.2. Fundamentals of Screening

In most cases, screening is a continuous process. Generally, screening is done batchwise only for analytical purposes, where it is referred to as sieving. As shown in Figure 2, a feed stream $\dot{m}_A$

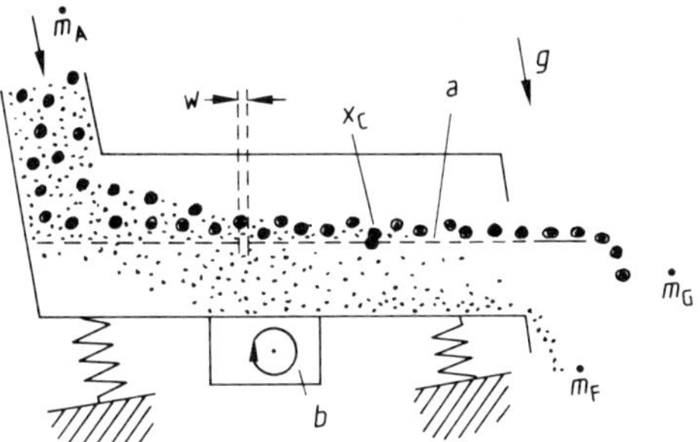

Figure 2. Operation of screen classification
The vibrator (b) causes a mechanical fluidization of the bed, which results in a stratification, and moves the feed material m_A across the screen (a)
Separation into nonpassing material m_G and passing material m_F occurs by diffusion and under the influence of gravity g

is transported across a screen and separated into the passing fraction or fines m_F and the nonpassing fraction or oversize m_G. The coarse fraction consists of oversize grains as well as misplaced particles, i.e., fines $g \cdot m_G$. Theoretically, if the screen is intact, the passing fraction should not contain any misplaced particles. However, because mesh size can vary by the manufacturing tolerance $\pm \Delta w$ from the nominal mesh size w, some misplaced particles can be found in the fine fractions also. The mass of fines is $f \cdot m_F$.

The efficiency of screening η (η = fines passing/fines not passing) can be expressed mathematically

$$\eta = (a - g)(f - a)/(1 - a)(f - g)a \quad (1)$$

where a is the percentage of fines in the feed; g, in the coarse fraction; and f, in the fine fraction. If the fine fraction contains no misplaced grains, Equation (1) is simplified to

$$\eta = (a - g)/a(1 - g) \quad (2)$$

The fraction of fines in the feed as well as in the coarse and fine fractions, is determined by analysis. Unfortunately, these analyses are not absolute. They are performed mainly on laboratory sieves, whose efficiency is often of the same order of magnitude as production screens.

Another important parameter for a screen is its specific capacity, also called the screen surface rating $\dot{m}$. This is the ratio of mass flow to screen surface area for a particular separation; it is measured as kilograms per square meter per second or tons per square meter per hour. The mass flow usually refers to the feed stream. The specific screen capacity is a function of the cut size, the particle-size distribution, the type of feed particle, and the desired sharpness of cut. The specific screen capacity determines the screen size required for a screening machine.

From a practical viewpoint, dry screening can be broken down into three categories, as shown in Figure 3:

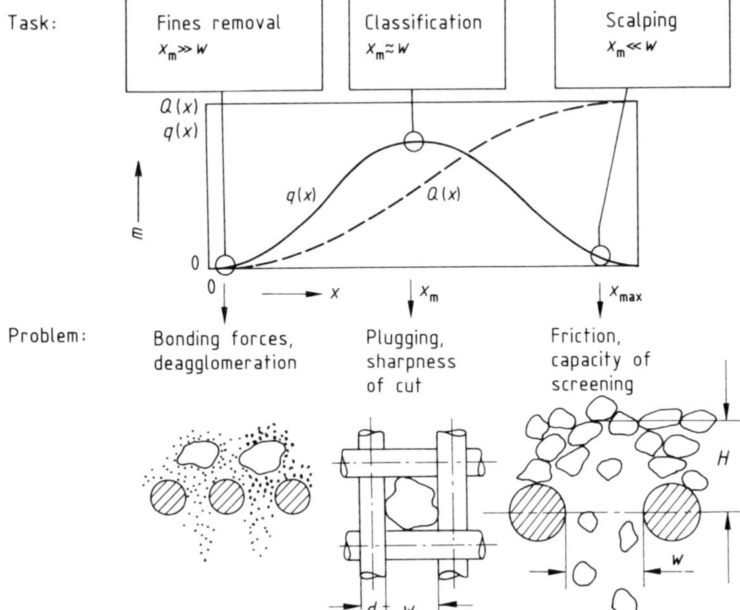

Figure 3. Principal practical applications of screening, which depend on the relationship between mean particle size x_m and mesh size w

1) *Scalping* or *trash removal* is the process in which relatively small amounts of foreign matter, lumps, and coarse impurities are removed from the product. The particles are predominantly smaller than the mesh size $x \ll w$. The main problem in this case is to obtain a high specific screen capacity $\dot{m}$. The process involves a *flowing* of the product through the mesh.
2) *Classification* is the separation of material into fine and coarse fractions. The problem here is to achieve a high sharpness of cut and avoid blinding of the screen by jammed particles. The latter occurs because many particles are close to the mesh size, $x \approx w$ (so-called near-mesh material). The physics of the material motion is *mixing,* with *diffusion* of the particles through the bed.
3) *Fines removal* or *dedusting* is the removal of a small amount of dust that clings to the particles, $x \ll w$. The problem here is deagglomerating and overcoming adhesive forces. Fines removal, normally, requires a long screening time. Theoretically, this is *segregation.*

2. Theory of Screening

Screening is very easy to do. Its use is shown in illustrations from ancient times and from the Middle Ages. However, a precise analytical treatment is extremely difficult, because many parameters influence the screening process. Even today, when screening machines are designed for new products, extensive experimentation must be carried out. The results can be extrapolated to some extent for similar tasks.

The physics of screening is mainly a mechanical fluidization of the product in order to overcome Coulomb friction. Six different processes of screening can be distinguished:

1) free flowing, as applied in scalping;
2) segregation, as in fines removal;
3) mixing or diffusion, as in classification;
4) superposition of mixing and segregation;
5) oblique impact between particles and the screen; and
6) fluid transport through the screen.

The first three processes are shown in Figure 4. (For a definition of k and ω see Section 2.1.)

All these screening processes are dominated by near-mesh size particles, which give rise to plugging or jamming. In addition, cementing of the screen can be caused by fine and adhesive particles. Although the analytical treatment of screening has progressed steadily [1]–[4], experience is still very important in practical applications [5].

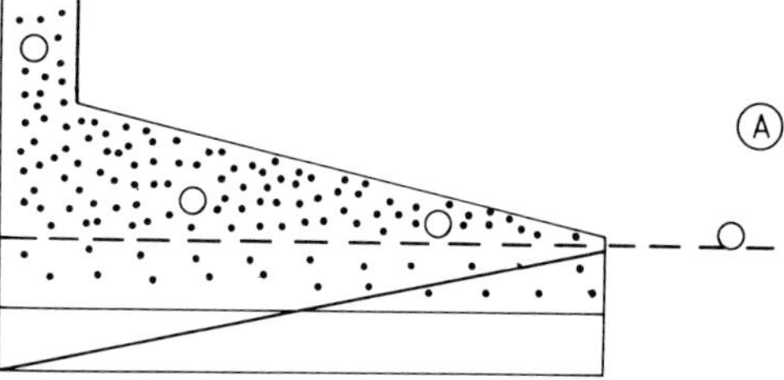

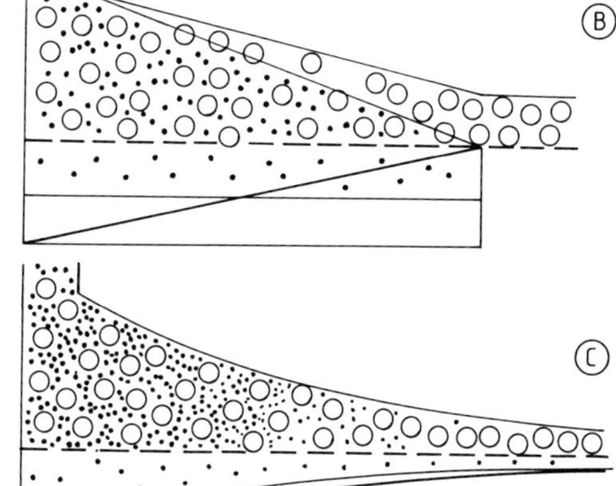

Figure 4. Processes in separation on vibrating screens
A) Scalping, $x \gtrsim 0.1$ mm, $w/x \gtrsim 5$; B) Segregation, fines removal, $K = e\omega^2/g \gtrsim 1$; C) Mixing, classification, $K \gtrsim 3$, $e \gtrsim 3\,w$

2.1. Free-Flow Screening, Scalping

Free-flow screening is analogous in operation to an hourglass. In scalping, it is frequently the dominant component process. The prerequisites are a free-flowing feed, a feed layer of sufficient thickness $H/w \gtrsim 2$, a ratio of mesh size to particle size of $w/x \gtrsim 4$, and only a small amount of coarse material. As shown in Figure 5, the feed forms an arch over the mesh, through which the particles fall. The configuration of the arch is similar for each feed–mesh combination [6]; thus

$$H/w = \text{constant} \tag{3}$$

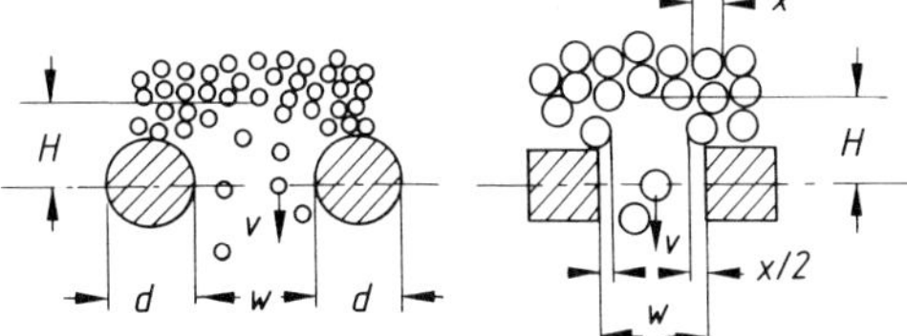

Figure 5. Arching over the screen mesh; H/w is constant according to similarity

The screen capacity is controlled by the downward velocity through the plane of the screen. By using the formula for the velocity of a falling object $v = \sqrt{2\,gH}$ and Equation (3), the proportionality

$$v \sim c\sqrt{gw} \tag{4}$$

is obtained. The throughput per opening is

$$\dot{m}_w = \varrho\, w^2\, v \tag{5}$$

From Equations (4) and (5),

$$\dot{m}_w = c\varrho\sqrt{g}\, w^{5/2} \tag{6}$$

Particles on the edge of the mesh hinder falling within a distance one-half of their diameter. This leads to a reduction in the effective mesh size w to $(w - x)$ (see Fig. 5). With this correction, the specific screen capacity, in kilograms per square meter per second, is finally obtained:

$$\dot{m} = c(w + x)^{-2}\varrho\sqrt{g}\,(w - x)^{5/2} \tag{7}$$

The correction factor c is about 0.2–0.5. For free flow, the important relationship that results is

$$\dot{m} \sim \sqrt{g}\sqrt{w} \tag{8}$$

Often in screening, the fines do not run through the mesh voluntarily, they must be set in motion. In practice, this is achieved by mechanical vibration (see Section 3.1).

The most important parameters influencing mechanical screening are (1) cut size x_T, x_{50}; (2) particle-size distribution $q(x)$, $Q(x)$; (3) efficiency η; (4) sharpness of cut $\kappa_{25/75}$; (5) mesh size w; (6) wire gauge d; (7) form of the mesh ψ; (8) stroke = 2 · amplitude e; (9) frequency or angular velocity ω; (10) gravitational acceleration g; (11) depth of bed H; (12) coefficient of friction μ; (13) adhesion or bonding stress σ_H; (14) slope (down- or uphill) β; (15) casting angle α; and (16) flow velocity v.

To make things simpler, the number of parameters can be reduced to nine or even to six. By applying dimensional analysis, the whole problem can be reduced further to six or three dimensionless numbers.

$$\pi_1 = x/w \quad (0.1\text{–}10) \tag{9}$$

$$\pi_2 = d/w \quad (0.5\text{–}1.0 \approx \text{const.}) \tag{10}$$

$$\pi_3 = e/w \quad (\text{vibration: } 1\text{–}100;\ \text{sifting: } 5\text{–}1000) \tag{11}$$

$$\pi_4 = K = e\omega^2/g \quad (\text{vibration: } 3\text{–}5;\ \text{sifting: } 1) \tag{12}$$

$$\pi_5 = \dot{m}^2/\varrho^2\varphi^2 g x \quad (10^{-5}\text{–}1.0) \tag{13}$$

$$\pi_6 = H/e \quad (10\text{–}0.1) \tag{14}$$

The relationship of these numbers must be determined experimentally and depicted graphically. Exponential products can also be developed especially for computer analysis. Figure 6 shows a typical result from experiments with vibrating screens, as well as the relationship between π_1, π_4, and π_5.

Adhesive Forces between Particles. In screening coarse- and intermediate-size feeds, the particles are affected only by inertial and frictional forces. For very fine particles, however, fluid flow and adhesive forces are dominant. If the former are ignored, liquid bridges between wet particles, van der Waals forces, and electrostatic attraction must be considered. The exact

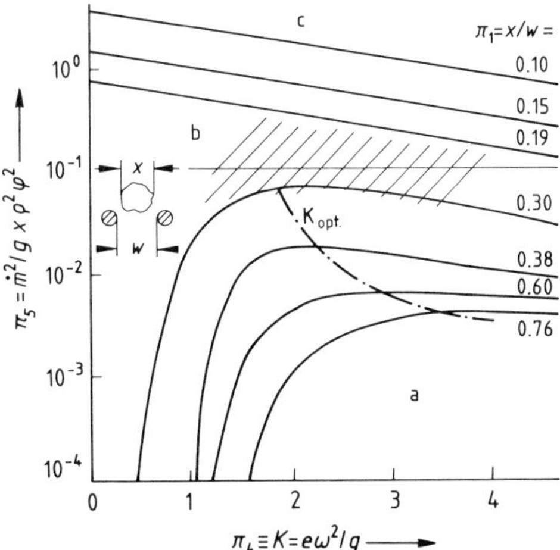

Figure 6. Operating conditions for scalping on vibrating screens, presented as a relationship of dimensionless numbers

a) Vibratory screening; b) Transition range; c) Free-flow screening

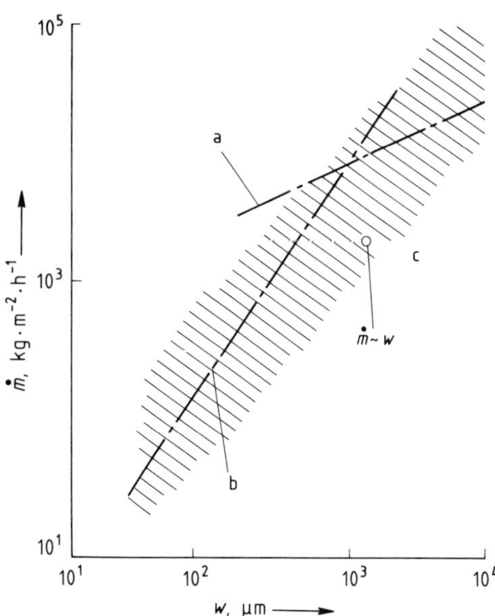

Figure 7. Relation between screening capacity $\dot{m}$ and mesh size w
a) Coarse material, $\dot{m} \sim \sqrt{w}$; b) Fine products with interparticle bonding forces, $\dot{m} \sim w^{3/2}$; c) Practical relation for intermediate material, $\dot{m} \sim w$

determination of interparticle forces or forces between particles and screen is complicated. For practical application, a rough estimation that the attractive or adhesive force between particles is proportional to the particle size is sufficient:

$$F_H \sim x \quad (15)$$

These forces hinder particle motion and, therefore, reduce screen capacity. With the assumption that hindrance of flow is roughly proportional to the adhesive force, the following relationship is derived by using Equation (8); it applies to scalping for a very fine feed:

$$\dot{m} \sim c\varrho \sqrt{g} \sqrt{wx} \quad (16)$$

If $x \approx w$, then for very fine feed the relationship between specific screen capacity and separation size or mesh size is

$$\dot{m} \sim c\varrho \sqrt{g} w^{3/2} \quad (17)$$

with the constant $c = 2.8\ \text{m}^{-1}$. Figure 7 shows that this model accurately predicts the specific screen capacities observed in operation. The commonly given formula $\dot{m} \sim w$ is an approximation for the range $0.1 < w < 10$ mm.

2.2. Segregation, Fines Removal

If the fines flow not only through the mesh, but also through the space between the coarse particles, the screening process is referred to as segregation. This is the case when $x_G/x_F \gtrsim 7$ and $w/x_F \gtrsim 2$. The first requirement makes the flow of fines through the bed possible; the second prevents accumulation of fines at the screen surface. For many nonflowing materials, Coulomb friction and adhesion as well as other surface forces can be overcome by vibrating the screen. Depending on the feed, a high efficiency can be achieved by pure segregation screening. In this case, the yield of fines is constant over the length of the screen and, in theory, the required screen length can be calculated from the velocity of segregation v_S, the transport velocity v_h, and the bed depth H:

$$L = v_h H/v_S \quad (18)$$

Segregation also plays an important role in mixing or elutriation screening, which is discussed in Sections 2.4 and 2.5. It occurs when the feed has a continuous particle-size distribution and vibration is used to loosen the particles by directing them upward. The bed must not be more than about ten particles thick.

2.3. Probability Approach to Screening

Another model of the screening process is derived from the concept that particle–mesh interaction dominates the entire process [7], [8]. The probability P_S that a fine particle x going over the screen will pass through an opening of width w depends on the dimensionless parameters x/w and d/w, as well as the geometric configuration. In the case of square mesh screens, the probability is

$$P_S = (w - x)^2/(w + d)^2 \quad (19)$$

The reciprocal $1/P$ of the probability gives the number n of vibrations necessary for fine particles to pass through the screen. Figure 8 shows the relationship for square mesh screens.

The passage of fine particles through a screen is impeded not only by the wire mesh but also by coarse particles and their own volume fraction V_F. The latter decreases steadily along the length of the screen. For particles that are not directly on the screen surface, the probability P_H decreases with increasing thickness H of the bed, as follows:

$$P_H \approx w/H \quad (20)$$

This fact explains the effectiveness of so-called thin-layer screening.

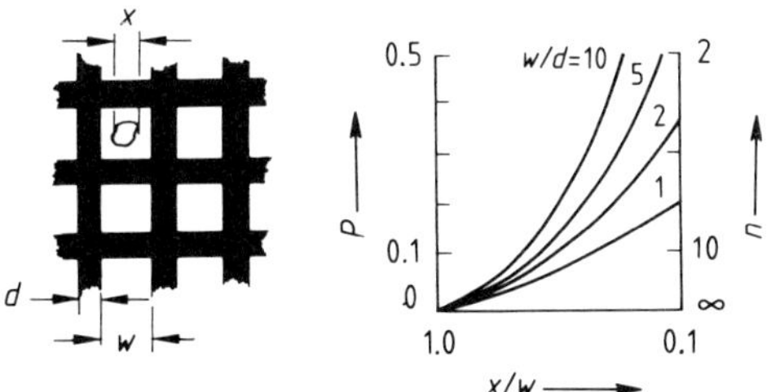

Figure 8. Probability P of a fine particle x passing through a square mesh w with wire diameter d in one vibration or the number n of vibrations needed to make $P = 1$

For square mesh screens, the total probability of fines passing is, therefore,

$$P_{tot} = P_S P_V P_H = [(w - x)^2/(w + d)^2]\, w\, V_F/H \quad (21)$$

If, for example, in fines removal, $w = 200$ µm, $x = 100$ µm, $d = 200$ µm, $V_F = 5\%$, and $H = 1$ cm, then the resulting probability is 0.0000625. The number of vibrations required to yield a separation probability of 1 is 16000. On a screen vibrating at 50 Hz, this corresponds to a residence time $t = n/f$ of 320 s or 5.3 min.

Because of the number of factors that influence screening, applied statistics must be used to study this problem. An example of the use of regressional analysis is given in [9].

2.4. Mixing and Diffusion, Classification

Most products have a continuous particle-size distribution. For this reason, effective segregation cannot take place in the bed. Therefore, the feed is mixed by vibration or rotation in order to move the fines against the screen. However, fines also flow back toward the surface of the bed.

The screening process can be described analytically with the aid of the mixing model. It is identical with a first-order reaction. Therefore, many authors have suggested that the screening process be described as follows:

$$dm_F/dt = -kAm_F/m \quad (22)$$

In sieving, the passage of fines after time t is proportional to the content m_F of fines in the feed at time t and also to the area A of the screen and the sieving rate constant k; it is inversely proportional to the mass m of feed on the sieve at time t.

If the sieving rate constants k_i, in kilograms per square meter per second, of the individual fractions and the composition of the feed are known, then the sieving or screening capacity and precision can be calculated. The sieving rate constant k_i is specific for the fraction i of a certain material, the screen, and its motion.

Sieving rate constants k_i can be determined by sieving a feed for time t or over a length l, analyzing the fines, and comparing them with the feed by use of Equation (22). Integration of this equation gives

$$\ln(m_F/m_{F_0}) = -kAt/m \quad (23)$$

The total feed m at the start has a content m_{F_0} of fines. After sieving for time t, the fines remaining on the sieve are given by

$$m_F = m_{F_0} \cdot \exp[-kAt/m] \quad (24)$$

From this, the sieving constant

$$k = \ln(m_F/m_{F_0})\, m/At \quad (25)$$

or the sieving time

$$t = \ln(m_F/m_{F_0})\, m/Ak \quad (26)$$

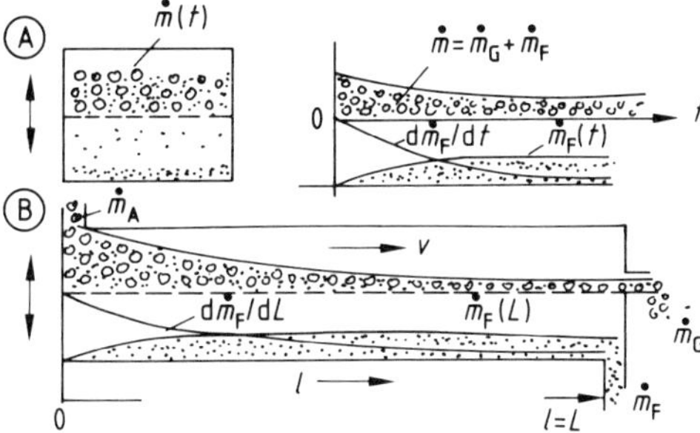

Figure 9. Progress of oversize material and fines in mixing the feed
A) For sieving; B) For screening

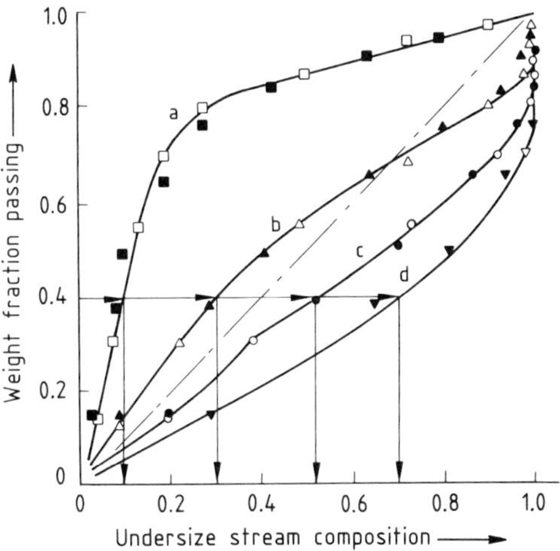

Figure 10. Sieving and screening as a first-order reaction
Undersize stream composition vs. weight fraction passing through [2]
Example: for 40% weight fraction passing through, 70% of the fraction $x/w = 0.3$, but only 10% of the fraction $x/w = 0.92$, have passed through
a) $x/w = 0.92$, $k = 0.2$ kg m^{-2} s^{-1}; b) $x/w = 0.77$, $k = 1.1$ kg m^{-2} s^{-1}; c) $x/w = 0.65$, $k = 2.1$ kg m^{-2} s^{-1}; d) $x/w = 0.3$, $k = 2.8$ kg m^{-2} s^{-1}

can be calculated. With this model, the efficiency can also be calculated. Equation (2) is defined as $\eta = (m_{F_0} - m_F)/m_{F_0}$. Taking m_F from Equation (24) gives the efficiency

$$\eta = 1 - \exp[-kAt/m] \quad (27)$$

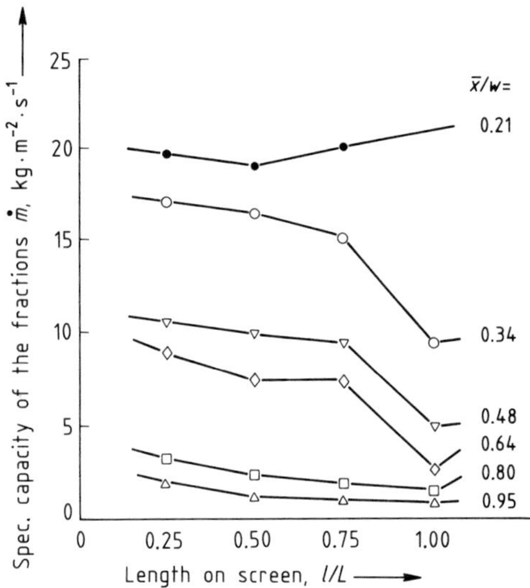

Figure 11. Screening capacity of the fractions $\bar{x}/w$ along the screen length l/L [2]
$w = 3.5$ mm; $K = 4.9$

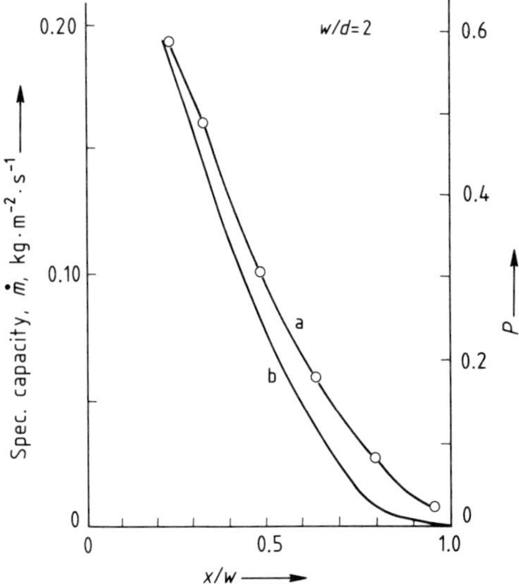

Figure 12. Comparison of experimental values of screening capacity with the probability approach [2]
a) Experimental value; b) Probability approach,
$P = \left(\frac{w - x}{w + d}\right)^2$

Equations (23)–(27) can also be applied to the continuous screening process because the relation $v = l/t$ is generally known. Therefore, in the equations mentioned above l/v must be used instead of t (see Fig. 9).

Figure 10 shows the weight fractions passing through, as a function of the cumulative distribution of the fines, in weight percent. The sieving rate constants k_i are independent of sieving time and length. A decrease can be observed at the end of sieving and for near-mesh-size material only; see Figure 11.

Note that the k values and the calculated probability according to Equations (19) and (21) are closely related (cf. Figs. 8 and 12). The probability approach theoretically also leads to an equation for the first-order reaction [10]. Equation (22) describes the free flow of fines in scalping as well, because for $m = m_F$,

$$dm_F/dt = -kA \quad (28)$$
$$m_F = -kAt \quad (29)$$

2.5. Superposition of Mixing and Segregation

Some screening processes fit between mixing and segregation. The theoretical treatment involves a superposition of these processes and leads to a differential equation, known as the Fokker–Planck–Kolmogorov equation. Solutions are given in [11] and [12]. Further treatment of this type is outside the scope of this article.

2.6. Particle–Mesh Interaction

Screening is influenced primarily by the relation of particle size to mesh size x/w. This is confirmed by Equation (19). Distinction must be made between near-mesh-size particles $x < 0.9\,w$, which hinder screening, and plugging particles $w < x \lesssim 1.2\,w$, which prevent screening. Figure 13 illustrates the way to calculate the largest particle plugging the mesh,

$$x_K = [(d + w)/\cos(\varrho/2)] - d \quad (30)$$

In reality, the largest plugging particle is often larger because of the tolerance Δw of the meshes (3–50 %), the tolerance of the coefficient of friction μ, and the particle form.

Although near-mesh-size particles strongly influence screening capacity, defining these particles is difficult. According to [2], they constitute the particle fraction in the sieving rate curve above the diagonal in Figure 10. Here, the fraction nearest mesh size is always critical; the next one, only to some extent. Near-mesh-size charac-

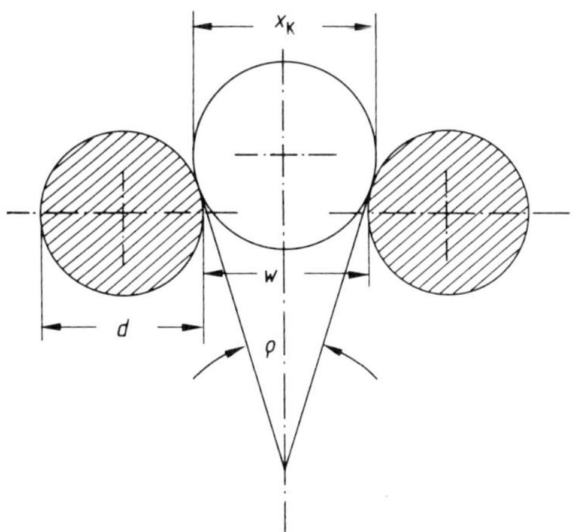

Figure 13. Particle diameter x_K for plugging is a function of the geometry of the mesh w, d, and the angle of friction ϱ

teristics depend also on the screen type, feed, and screen motion.

Oblique Flow to the Mesh. Blinding the screen can be avoided if the feed particles do not enter the mesh perpendicularly. The extreme is parallel flow of the material over the screen as on sifters (see Section 3.2). Another possibility is a sharp inclination of the decks, as in the Mogensen Sizer (see Section 3.4). The result is little blinding of the screen, higher capacity but also smaller cut size, and less precise screening.

For sifters, the cut-size particle x_T shows the relationship

$$x_T/w = f\,[\dot{V}/v_{\text{rel}};\ H/w;\ q(x)] \tag{31}$$

where $\dot{V}/v_{\text{rel}}$ is the specific volume flow. A usual relation is $x_T \approx 0.8\ w$. Multideck screening with oblique particle flow will be treated in Section 3.4.

2.7. Fluid-Flow Screening

In hydraulic or pneumatic transport of feed on the screen and through the mesh, velocities from 1–10 m/s are possible. This results in a very high capacity independent of particle or mesh size. For the probability of particle passage through the mesh, Equation (19) is valid. If the direction of flow is perpendicular to the screen, problems arise with regard to the impact of the particles on the mesh, extremely strong blinding, and transport of oversize particles along the screen. Because of this, oblique flow is preferred. For wet screening, relationships of cut size to other parameters have been determined (see Section 3.6.2).

3. Screening Machines

The function of the screening machine is to move the feed, especially the coarse fraction, across the screen so that the fines encounter the mesh and pass through. The feed is fluidized mechanically, thus overcoming Coulomb friction. Near-mesh particles are brought into repeated contact with the screen in order to increase the probability of their passing through.

In the simplest case, inclined and possibly vibrating gratings, so-called *grizzlies*, can perform coarse screening. A grid of rotating gratings can keep the openings clearer, as shown in Figure 14. Figure 15 surveys the types of screens manufactured.

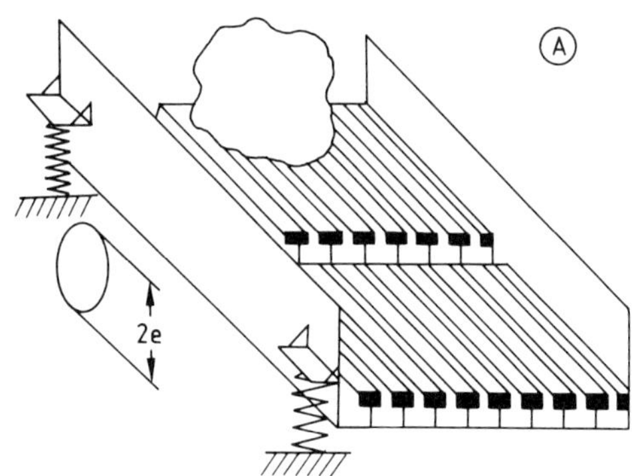

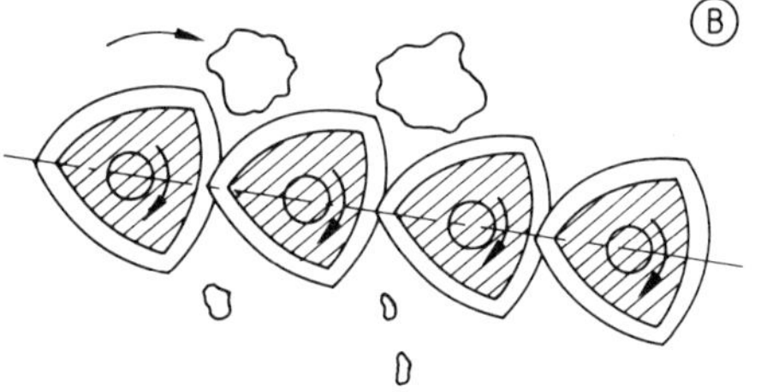

Figure 14. Inclined and possibly vibrating grating
A) Coarse screening (grizzly); B) Grid of power-driven rotating gratings (roll grizzly)

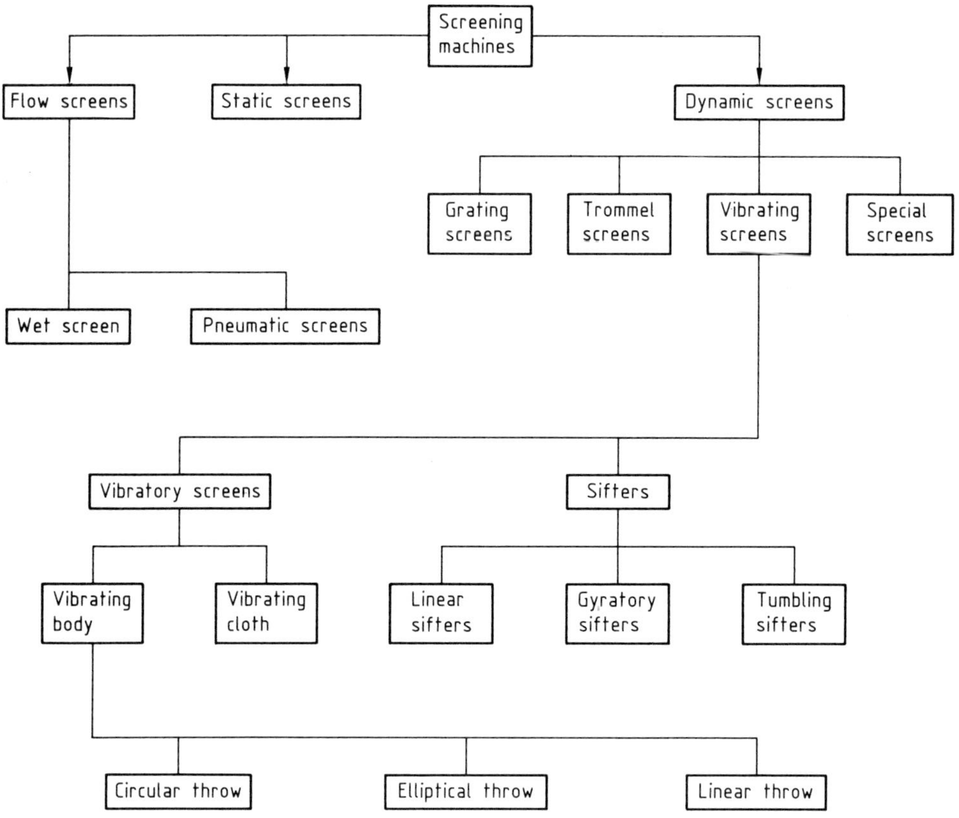

Figure 15. Survey of screening machines

3.1. Vibrating Screens

The vibrating screen is the classical screening machine. It operates on the principal of vibrational feed. The design is based on the ballistics of individual particles. These machines are classified by type of motion—linear, circular, or elliptical. The motion is described by two parameters, the amplitude e and the frequency f. The amplitude is half of the stroke. Angular velocity ω is related to frequency f by the relation $\omega = 2\,\pi f$, and to the number of revolutions n by $\omega = 2\,\pi n/60$. Linear or elliptical oscillation also includes the casting angle α. For the motion of the feed, the acceleration due to gravity g and the slope β relative to the horizontal must be considered as well. The process of vibrational screening is characterized by a dimensionless number, the throw number

$$K_v = a_n/g_n = e\omega^2 \sin(\alpha + \beta)/g \cos\beta \qquad (32)$$

where a is vibrational acceleration. The machine itself is characterized by another dimensionless parameter, the acceleration number K

$$K = e\,\omega^2/g \qquad (33)$$

Figure 16 illustrates the relationship described above. The trajectories of the feed are shown in Figure 17. At a value of $K_v = 1$, the feed begins to jump. At $K_v = 3.3$, statistical resonance occurs; i.e., the time of flight is equal to the period of vibration. This is the starting point for adjustment of screening machines. At K_v between 3.3 and 6.6, the trajectories are irregular. At $K_v = 6.6$, the trajectories are twice as long as at $K_v = 3.3$. At $K_v = 1$, loosening and deagglomeration of the feed begin; however, mechanical stress begins too. The maximum velocity of feed for a horizontal screen occurs at $K_v = 3.3$ with a casting angle of 25–30 °.

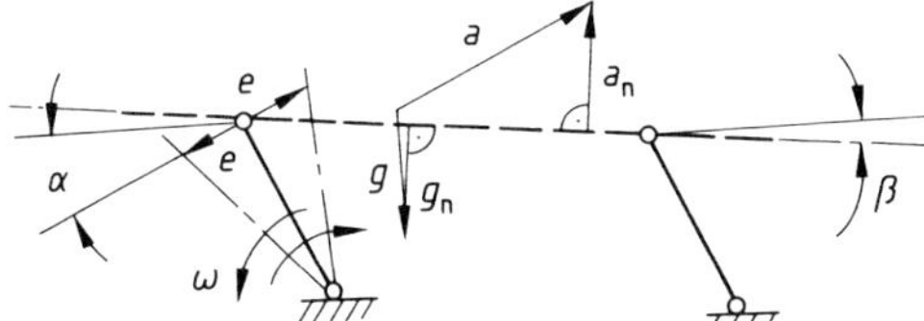

Figure 16. Derivation of the vibrational characteristic number, the throw number K_v for linear stroke
$a = e\omega^2$; $a_n = a \sin(\alpha + \beta)$; $g_n = g \cdot \cos \beta$

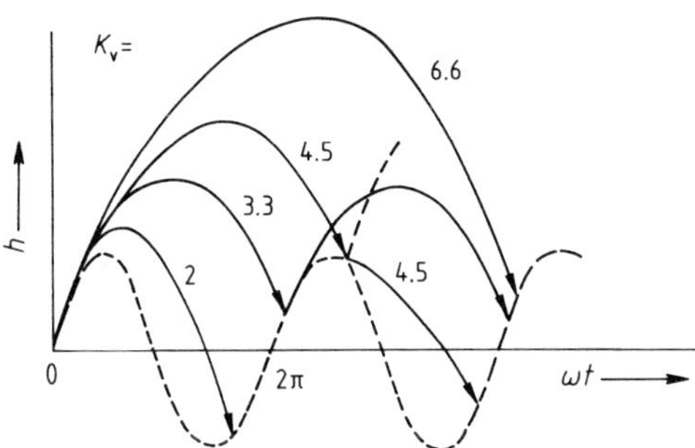

Figure 17. Idealized trajectories of a particle on the surface of a vibrating screen
Height h vs. $\omega \cdot t$ and throw number K_v; $K_v = 3.3$ first statistical resonance

For horizontal screens with circular throw, the casting angle is relatively steep:

$$\alpha = \arcsin(g \cos \beta / e\omega^2) \quad (34)$$

Optimal settings for screening with vibratory screens are $K_v < 3$ for light, polishing screening; $K_v = 3-3.3$ for accurate separation when $\alpha > 45°$; and $K_v = 3.3-6.6$ for difficult feed. Screens with circular throw should be operated at a downward slope of $\beta = 10-15°$. The material bed must be mixed well. Therefore, the bed depth should be $H \lesssim 10\,w$ and the amplitude $e \sim w$. Figure 18 gives an example of a vibrating screen with circular throw.

For fine classification on vibrating screens, the mesh can be driven directly as shown in Figure 19. Magnetic vibrators are attached at various points or in lines on the cloth. Oscillating at 50–100 Hz, they can provide the desired K_v value up to 15. Because the screen is clamped at the rim, the casting or throw angle is approximately 90°. For this reason, the screen is sharply inclined at an angle of 30–45° in order to move the feed alongside the screen. A thick, compact layer of feed is not possible. Actually, a thin layer moves rapidly across the screen. To have a high screening capacity, the screen must be very wide. In this case, problems arise in distributing the feed evenly.

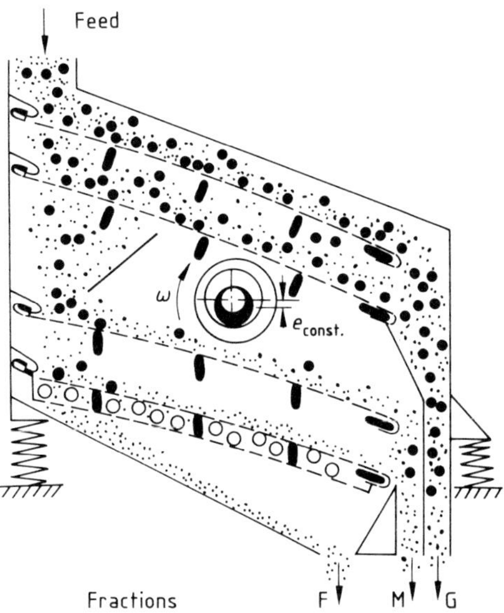

Figure 18. Circular stroke unit for fine screening
Partial parallel configuration furnishes three fractions G, M, and F
Ball deck and balls are used on finest screen
Screens are very taut, $e_{const.} > e$

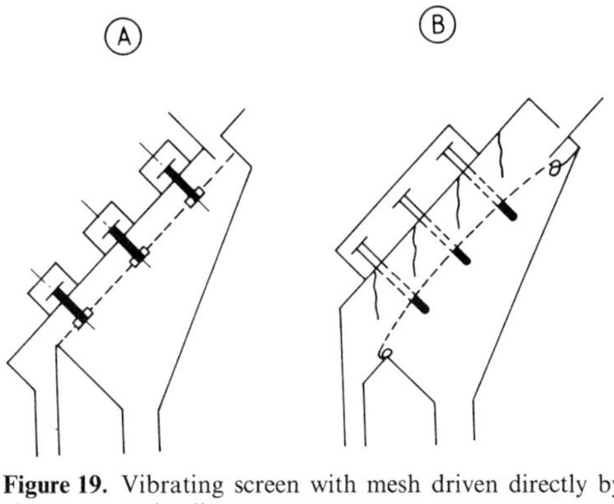

Figure 19. Vibrating screen with mesh driven directly by electromagnetic vibrators
A) Multipoint drive for particles $x \leq 80$ μm; B) Linear drive with screen tensioned lengthwise
Rubber guards limit short-circuiting of the feed

3.2. Sifters

Sifting shows reciprocating or circular motion in the plane of the screen. In the latter case, it is referred to as gyratory screening. Reciprocating motion is possible in the direction of feed flow or perpendicular to it. Gyrating sifters are becoming increasingly more common. They are used mainly for fine and ultrafine screening.

The sifting process is the straining out, through the mesh, of the fines segregated on the screen. The direction of flow for the fines is not perpendicular to the mesh; rather, it is oblique. As a result, the separation size is definitely smaller than the mesh size, $x_T \approx 0.8\, w$. Because of this, few particles are jammed in the mesh. In comparison to vibrating screens, these devices show, for many products, minor blinding of the screen. One screen can also be covered with another that has a coarser mesh, $w \approx 1.2\, x_T$. This kinetic narrowing of the mesh avoids blinding, as mentioned above.

The flow toward the mesh results in uniform separation conditions only if it occurs under constant flow velocity and if the vertical pressure remains constant. This is the case for gyratory, but not for linear, oscillation. In the latter, the thickness of the bed is crucial. As the content of fines decreases, the bed becomes thinner. However, it should not become so thin that the damping effect is lost and the particles jump around individually. In the feed bed itself, strong shear forces overcome friction and enable gravitational segregation to occur as the feed is deagglomerated. On the other hand, this results in high wear for sensitive materials, such as instant products (e.g., instant coffee). The shear stress between the bed and wire cloth is

$$\tau = \varrho_s \cdot g \cdot H \cdot \mu_s \tag{35}$$

The relative velocity is

$$v_{\text{rel}} = \sqrt{e^2 \omega^2 - (\mu_s^2 g^2/\omega^2)} \tag{36}$$

and the angular velocity is

$$\omega^2 = K_h \cdot g/e \tag{37}$$

The specific power input, in watts per square meter, is

$$P = \tau v_{\text{rel}} \tag{38}$$

This power applied to one layer of particles gives the power specific to volume

$$P/V = \tau v_{\text{rel}}/\bar{x} \tag{39}$$

Usually, a specific power of about 10 kW/m^3 exists in the boundary layer. This is even enough to accomplish grinding.

Figure 20 B illustrates the dependence of pressure, horizontal velocity, and shear stress on the thickness of the feed bed, as well as on the resulting flow toward the screen. Figure 20 C shows how the bed begins to slide over the mesh, after overcoming frictional forces, at values of $K_h \approx 1$. This causes a sudden increase in screen capacity $\dot{m}$. As the K_h value or the relative velocity v_{rel} increases further, the capacity slowly decreases because the direction of flow toward the mesh becomes more oblique. For this reason sifters and similar types of screens rarely operate above $K_h = 1$. An increase does not give any higher throughput. Figure 20 C shows that the relative velocity v_{rel} depends on the amplitude or the radius of gyratory motion for constant K_h.

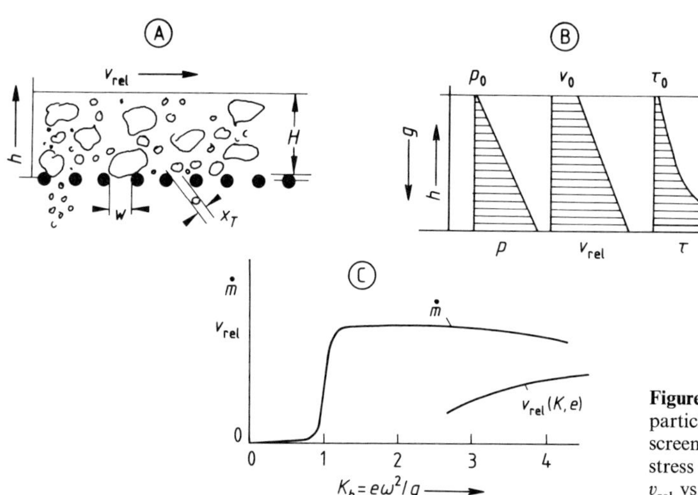

Figure 20. A) Process of sifting, near-mesh particle size $x_T < w$; B) Pressure p on the screen surface, relative velocity v_{rel}, shear stress τ; C) Capacity $\dot{m}$ and relative velocity v_{rel} vs. horizontal throw number K_h

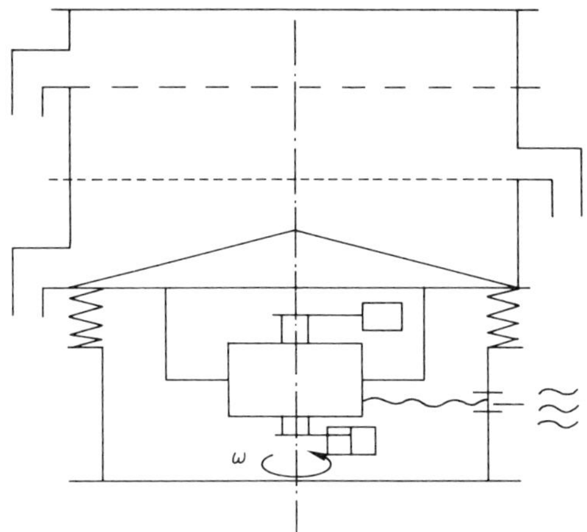

Figure 21. Tumbling sifter driven by a vibrating motor with a vertical shaft and two unbalanced weights
The twisting of the unbalanced weights against one another changes the movement of the feed across the screen: radial toward the outside, tangential to radial toward the inside

Another important advantage of the sifter is the horizontal orientation of flat and elongated particles on the screen, such as wood chips, leaves, and straw. Wood chips, especially, do not get stuck in the mesh. Sifters are the only type of screen suitable for screening such products. However, this positive effect is disturbed at the rim. Each stroke produces an impact on the bed, which results in a mixing action. Thus, blinding of the screen is again possible. Screening of critical products should involve a free overflow for the oversize material, with no rim.

Certain particles, such as metal powder or glass beads, often give rise to blinding when the sifter is started and stopped, because the relative velocity is not sufficient to cause oblique flow toward the mesh.

The gyratory sifter can be improved by controlling the residence time of the feed on the deck, instead of relying only on diffusion. This concept led to the development of the tumbling screen [13]. An example is shown in Figure 21 with simple construction operated by an eccentric drive. According to the relation $K_h = (e\omega^2/g) \approx 1$, selection of a high frequency should be advantageous and favorable to the design. However, for a frequency of 1500 rpm, an amplitude e of only 0.4 mm results. In this case, no flow toward the mesh occurs; rather, the particles dance around on it. For actual screening, a high amplitude and, consequently, a lower frequency must be chosen. This consideration leads

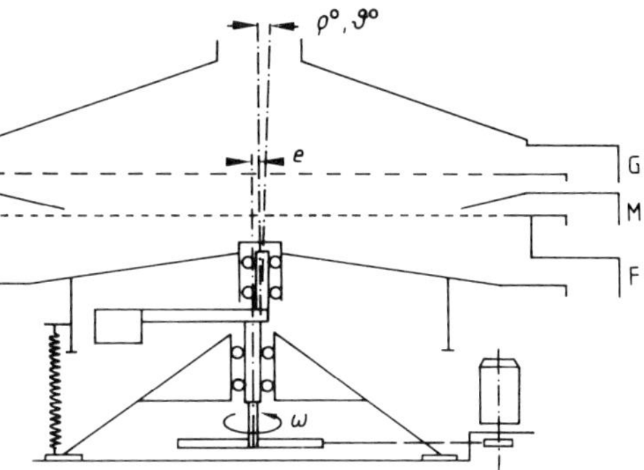

Figure 22. Tumbling sifter driven by a crankshaft with inclined pivot; ϱ^0 angle of radial inclination and ϑ^0 angle of tangential inclination

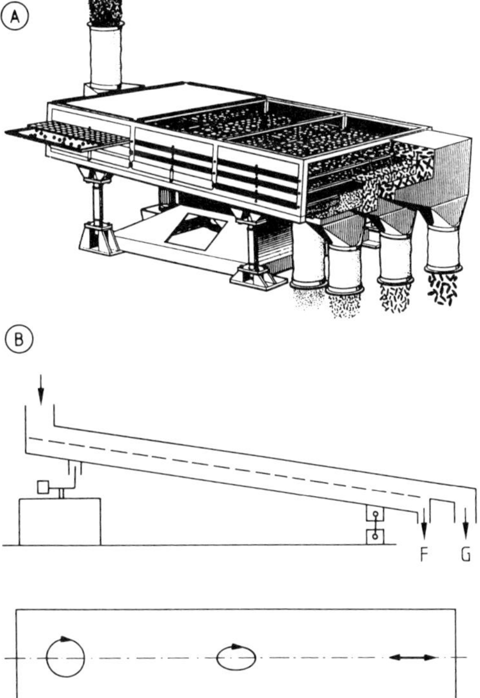

Figure 23. A) Inclined rectangular sifter driven by a crankshaft with vertical acceleration $a_v < 1\,x\,g$; B) Inclined sifter with circular, elliptical, and linear motion along the screen

to a tumbling screen with exact construction requirements, driven by an inclined crankshaft (Fig. 22). Optimal screening conditions can be obtained by adjusting the eccentricity, the radial and tangential inclination of the trunnion, and the rate of rotation. Good results can be obtained for most products.

Combinations of linear and circular motion are also used. Fibrous products, such as wood chips and asbestos, can be better classified at high capacities on a sifter than on a vibrating screen. These sifters are either inclined or driven by an inclined crankshaft (Fig. 23).

3.3. Trommel Screens

Trommel or revolving screens mix the feed thoroughly, without pushing it against the screen by centrifugal force. For this reason, the critical angular velocity ω_c should not be exceeded. It depends on the radius R of the trommel and its value is given by the formula

$$\omega_c = \sqrt{g/R} \quad (40)$$

Most of these devices are run at $\omega \approx 0.6\,\omega_c$. They are filled to less than 50% of capacity. In the case of short trommels, axial transport of the feed is performed by diffusion or mixing. Longer trommels are inclined or have spiral baffles which allow a definite residence time for the feed. Figure 24 shows that only a very small part of the screen surface is involved in the screening process.

Trommel screens have been used primarily for washing and simultaneously classifying gravel. They are also used as special screens for waste and scrap in recycling plants. One disadvantage of revolving screens is that to produce more fractions, the mesh size must increase along the screen, for example, screen surfaces in series. Parallel operation is also possible; this leads to concentric arrays of trommels with practical disadvantages in their construction.

Trommel screens offer the theoretical possibility of achieving very high screen capacities in the ultrafine range, when used as centrifugal screens. If $\dot{m} \sim \sqrt{a}$ and $a = R\omega^2$, the screen capacity can be increased by a factor of ten without any difficulty.

3.4. Special Screens

For very fine particles, blinding of the screen is, by far, the most important factor impairing screen capacity. Most special screening machines are designed to keep the mesh open by special techniques.

The *centrifugal* or *cylinder screen* (see Fig. 25) is similar to the trommel screen. However, in this case, the trommel does not move; vanes turning inside push the feed through the mesh instead. The screen cylinder is not under great tension and the mesh is cleaned because of permanent deformation. These screens have been used successfully for scalping in the range of 100–500 µm for nonabrasive products such as plastic powders.

Stretching wave screening machines, which provide for alternate stretching and sagging of the decks, are able to screen sticky products successfully. The mesh size is limited to above 2 mm because of the difficulty in fabricating more finely perforated sheets from rubber or polyurethane. Figure 26 shows the operation of one type of stretching wave screening machine. Also illustrated is the feasibility of screening wet feed, depending on the particle size and water content. At a critical moisture content of 20–30 vol%, separation can be performed with these screens. By stretching the screen about 10%, a high acceleration is achieved, with K_v values of approximately 100 [14].

The *Mogensen Sizer* (Fig. 27) employs a screenlike process. Generally, five sieve plates are stacked over one another. From top to bottom the mesh size decreases in a ratio 10:10:8:6:4, respectively. The slope of the sieve decks increases by 7° from plate to plate. In

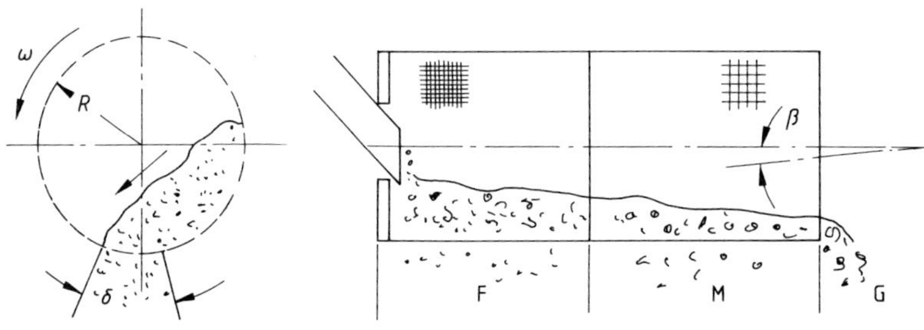

Figure 24. Trommel screen; δ region of active screening Series arrangement of screen surfaces, $\beta = 0-10°$, $\omega \approx 0.5\sqrt{g/R}$, $n \approx 0.5 - 0.741\sqrt{x}$

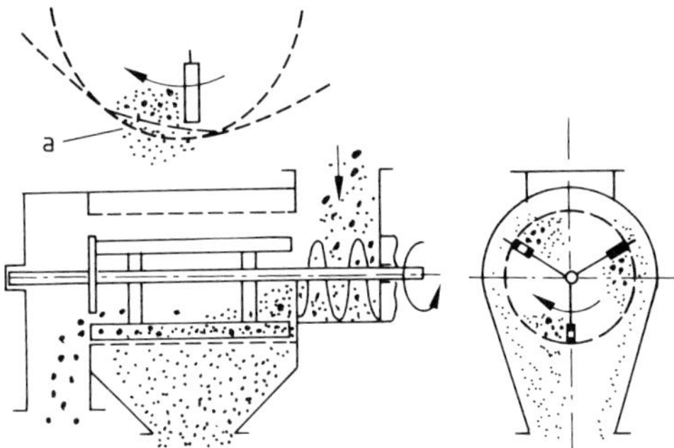

Figure 25. Centrifugal, cylindrical, or whirlpool screening machine
The mesh is bent at (a) by the vane and is cleaned by permanent deformation of the wire cloth

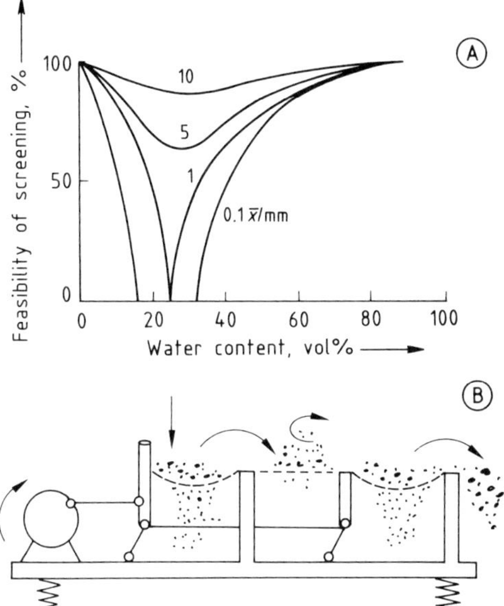

Figure 26. A) Feasibility of screening wet vs. dry products on conventional devices (feasibility plotted as function of average particle size $\bar{x}$ and water content); B) Principle of the stretching wave screening machine for screening wet products

screening dry sand with a medium particle size $\bar{x} = 2.5$ mm, for example, a sharpness of cut $\kappa_{25/75} = 0.6$ can be achieved. The capacity is $\dot{m} = 7.4\ \mathrm{kg\,m^{-2}\,s^{-1}}$, and the percentage of fines in the oversize $g = 5\%$. This sizer is not sensitive to changes in capacity [15].

Because there is no material bed on the screens and no near-mesh-size particles, blinding does not occur. The vibration applied shows K values of 3–6 at high frequency and low amplitudes. It serves only to fluidize the material.

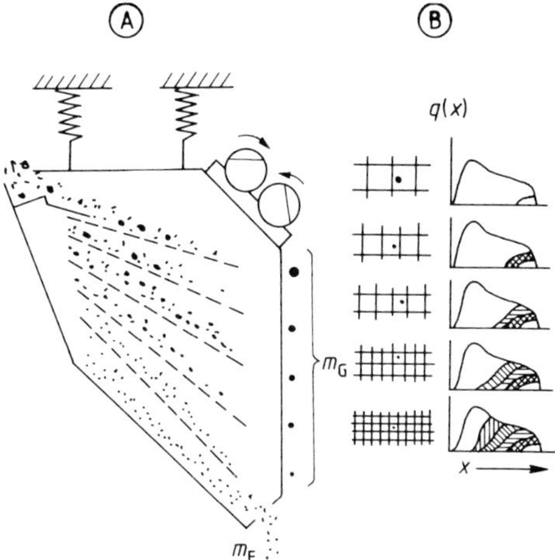

Figure 27. Mogensen Sizer
A) Operation; B) Composition of nonpassing fraction

3.5. Sizing of Screen Surfaces

The design of the required screen surface for a particular feed is the central problem of screening technology. The objective is to minimize the surface area by achieving the highest screen capacity. On the other hand, this depends on many parameters:

1) Product: composition, size, shape, particle size, distribution, sharpness of cut
2) Screen mesh size and type of screen: vibrating, sifting, rolling, or other kinds
3) Screening parameters: amplitude, frequency, casting angle inclination, bed thickness

Today, no generally applicable method exists for calculating the specific screen capacity and, consequently, the required screen surface.

For the special case of scalping, Equation (7) is used for coarse products and Equation (17) for fine ones. In segregation screening or fines removal, the area of the screen depends strongly on the adhesive forces in the feed. Thus, no calculation is possible for the screening area. In mixing–screening, the area of the screen can be calculated by estimation of the sieving rate constant k experimentally. Then by using Equation (25) or (26), and the efficiency desired in Equation (27), the screening area can be determined.

Figure 28 serves as a guide for carrying out the necessary screening tests for a feed whose behavior is not known. It shows average values

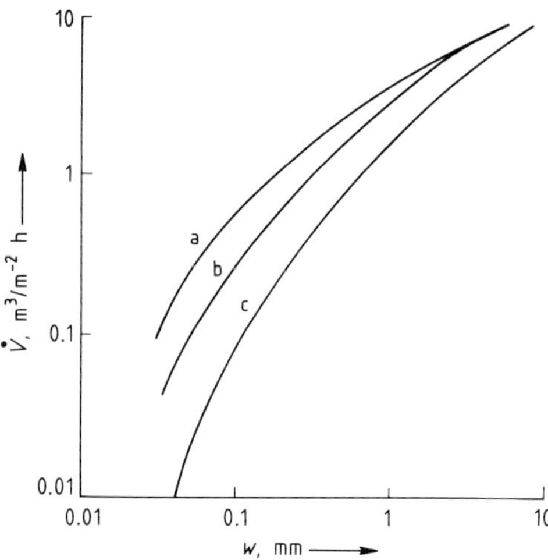

Figure 28. Criteria for calculating the screen surface for common products; specific volume flow $\dot{V}$ as a function of mesh size w and particle density
a) $\varrho = 10^4$ kg/m³; b) $\varrho = 10^3$ kg/m³; c) $\varrho = 10^2$ kg/m³

for the specific volume flow $\dot{V}$ for compact particles as a function of mesh size w.

Figure 29 shows qualitatively the yield of fines and the composition of the passing material over the length of a rectangular screen. As a consequence, the screen capacity can frequently be increased by placing a somewhat coarser screen over the first quarter, with w approximately $1.2\,x_T$. The sharpness of cut decreases only slightly.

3.6. Flow Screens

The capacity of a mechanical screen reaches zero when the attractive forces become greater than the inertial forces that move the particles. Attractive or adhesive forces are roughly proportional to particle size x, and inertial forces are proportional to x^3. Consequently, the two functions intersect. If, as shown empirically, damp sand grains 1 mm in size just cling to each other under normal gravitational forces, then the limit of mechanical screening for this feed occurs at a vibrational acceleration $a = 10 \times g$. The same situation applies to van der Waals forces and, roughly, to electrostatic forces. Only fluid forces are left to overcome adhesive forces. These are provided mainly by air or water. A rough estimation indicates that an airstream with a velocity of 50 m/s can overcome the adhesive forces of particles down to 15 µm in size, as shown in Figure 30.

3.6.1. Pneumatic Screens

By using fluid flow, flow screens can be constructed in which the feed is transported through the mesh either hydraulically or pneumatically. In this case, the screen capacity is very high and is independent of mesh size. In practice, this type of screen has been used only for scalping inert materials such as plastic powder, as shown in Figure 31. These screens are prone to blinding, which occurs both by jamming of the near-mesh

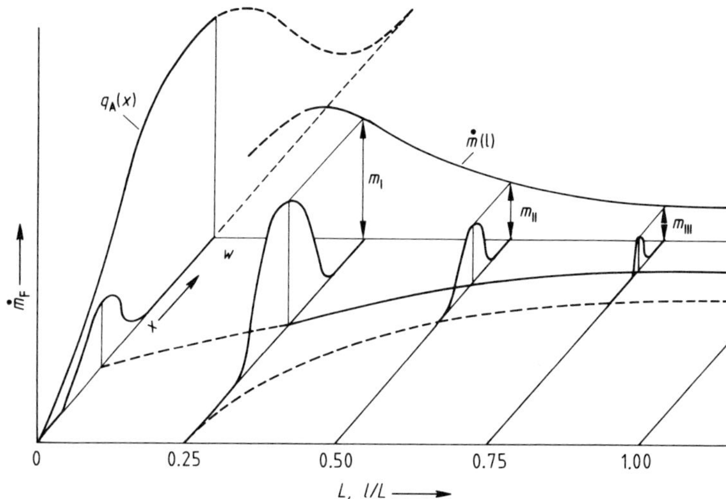

Figure 29. Amount and composition of the passing fraction along the length of the screen

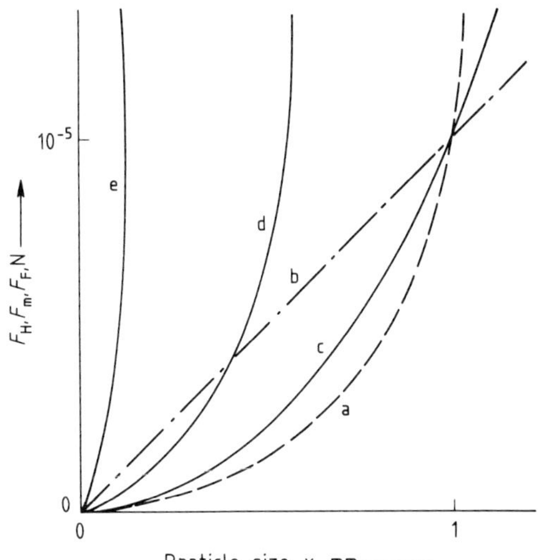

Figure 30. Inertial force F_m, adhesive forces F_H, and fluid forces F_F (F_F for air), respectively, as functions of particle size x

a) F_m, $a = 1 \times g$; b) F_H; c) F_F, $v = 6.3$ m/s; d) F_F, $v = 10$ m/s; e) F_F, $v = 50$ m/s

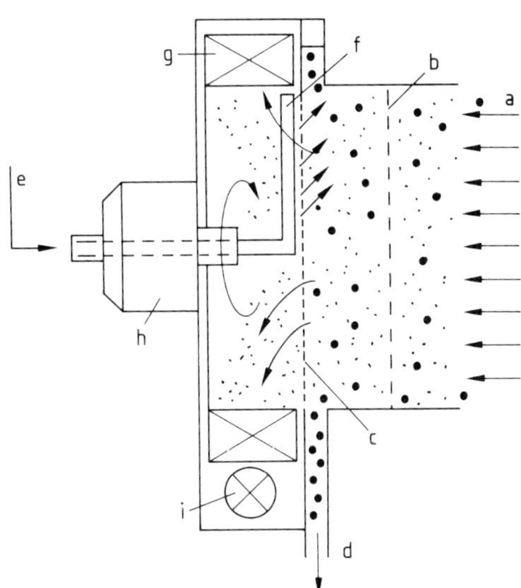

Figure 31. Principle of pneumatic screening

a) Two-phase flow composed of air and solids; b) Coarse protective screen; c) Screen; d) Nonpassing fraction; e) Compressed-air connection; f) Rotating compressed-air duct used to clean the screen and to remove nonpassing material; g) Blower; h) Drive motor; i) Exit for fines and air

particles and by cementing due to fine dust. Nozzles that blow compressed air against the direction of flow through the screen are required to clean the screen [16].

In many cases, for fine screening, supplementation of mechanical screening by blowing air through the screen is sufficient. For vibrating screens, especially directly driven ones, a pressure differential of ≈ 10 Pa is sufficient to measurably increase screen capacity and sharpness of cut. For sifters, a rotating nozzle arm blows air from below. In this case, supplementary traps for fines, such as cyclones or filters, are used. As a result, the machine evolves into an array of devices, as shown in Figure 32.

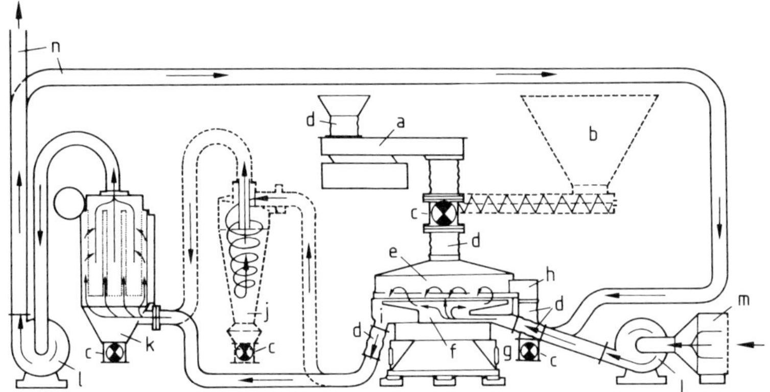

Figure 32. Combined mechanical and pneumatic screen array

a) Dispensing chute; b) Screw conveyor; c) Bucket-wheel valve; d) Flexible connector; e) Tumbling sifter; f) Air duct register; g) Air inlet; h) Coarse outlet; i) Fines exhaust; j) Cyclone separator; k) Filter; l) Blower; m) Air intake filter; n) Circulate–ventilate valve

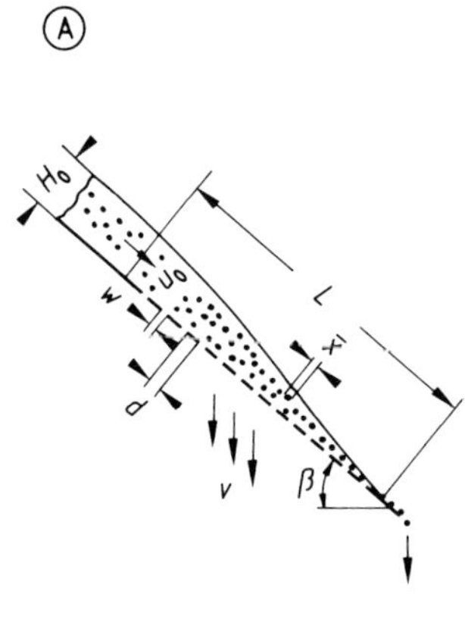

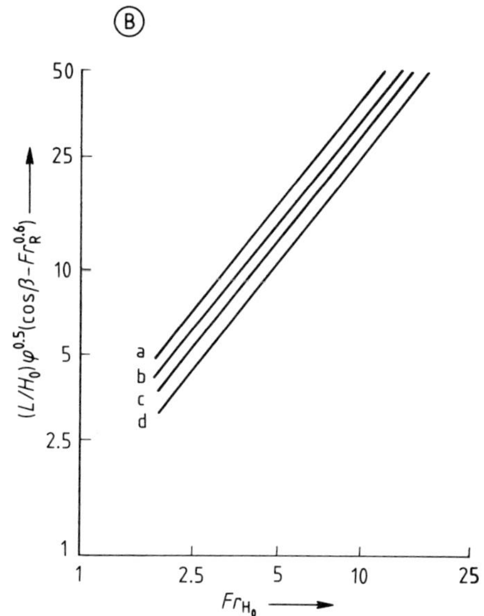

Figure 33. A) Minimum screen length L in wet screening, $\varrho_p/\varrho_F = 2.6$, $Fr = u_0/\sqrt{g R}$ (sieve bend); B) Relation between dimensionless numbers for flat screens and sieve bends
a) $c_V = 0.2$; b) $c_V = 0.1$; c) $c_V = 0.05$; d) $c_V = 0$; $Fr = u_0/\sqrt{g H_0}$, $\bar{x}/w = \text{const.} = 0.24$

3.6.2. Wet Screens

Wet screening may mean underwater vibrational screening. For example, it can be used to improve the classification of a product obtained in water. In principal, it does not differ from the usual vibrational screening. In most cases, however, a suspension is passed over a screen surface. The objective is as much dewatering as it is classification or straining. The fundamental basis of this process is described by the relationship between characteristic dimensionless numbers, e.g., the Froude number Fr. Figure 33 shows the screen length necessary to remove free water from a suspension, as a function of the Froude number $Fr = u_0/\sqrt{g H_0}$ at various solids concentrations c_V. The influence of the particle-size–mesh ratio and the solids concentration on the length of the screen is shown in Figure 34. In conclusion, Figure 35 shows the cut point and sharpness of cut that can be obtained on inclined

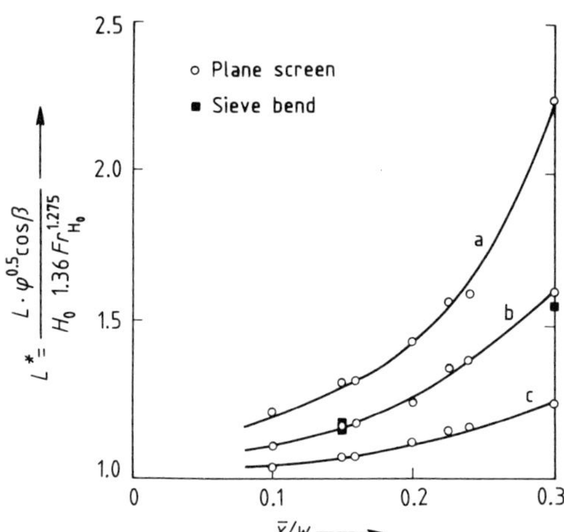

Figure 34. Minimum screen length L as a function of the ratio of average particle size to mesh size $\bar{x}/w$ as well as the solids concentration c_V
Values shown for square mesh flat screens and sieve bends
a) $c_V = 0.2$; b) $c_V = 0.1$; c) $c_V = 0.05$

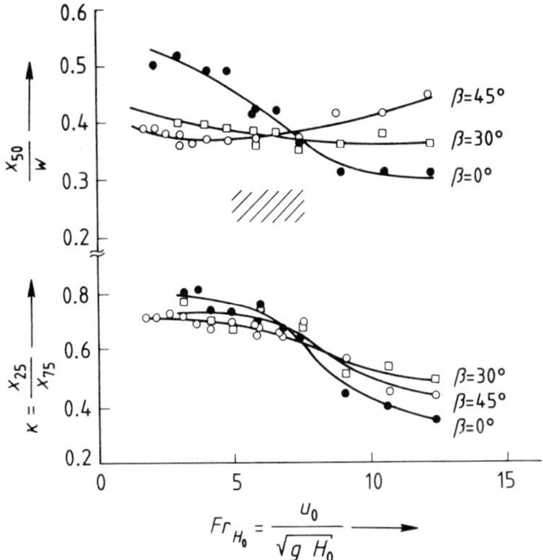

Figure 35. Specific cut point x_{50}/w and sharpness of cut $\kappa_{25/75}$ for wet screening on inclined screens vs. *Fr* number c_V = const. = 0.1, hatched area shows the optimal Fr_{H_0}-range

wet screens as a function of *Fr*. Fundamentally, no difference exists between flat screens and sieve bends [17].

4. Screen Arrays

When more fractions of a material are wanted, use of a screening machine with more screen decks does not always help. In this case, the machines can be arrayed as shown in Figure 36. A combination with decreasing mesh size is generally used. However, increasing mesh size and various combinations are also possible. When a high yield of a particular fraction is expected, screen surfaces can be arranged in parallel. In this case, splitting the feed into equal streams is often a problem.

5. Screen Surfaces

The most important element in a classification device is the screen surface. Figure 37 gives an overview of the important types. Most often, screens made of square mesh from wire with uniform gauge *d* are used. *Mesh sizes* are standardized in ISO R 497, and the *diameters of metal wire* in ISO 4782. In the most general case, screens can have rectangular openings with different diameter wire. The free screen surface φ, the most important parameter, is given by

$$\varphi = w_a\, w_b / (w_a + d_a)\,(w_b + d_b) \tag{41}$$

Twill weaving is also used for fine screens; in this case, the wire skips two or more meshes. The finest industrially produced screens have an opening of about 25 µm. Such screens are made of steel, as well as spring steel and stainless steel. More and more woven cloth made of synthetic fibers, mostly filaments, is used.

The following ISO standards for *sieve plates* are used:

ISO 2194 wire screens and plate screens for industrial purposes

ISO 4783 industrial wire screens and wire cloth
1) general
2) preferred combinations for wire cloth
3) preferred combinations for precrimped and metal wire screens

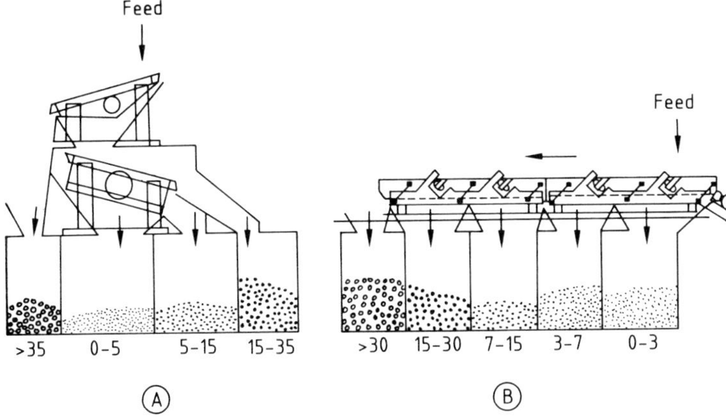

Figure 36. Screen arrays
A) Screening with decreasing mesh size and parallel configuration; B) Screening with increasing mesh size and series arrangement of screen surfaces

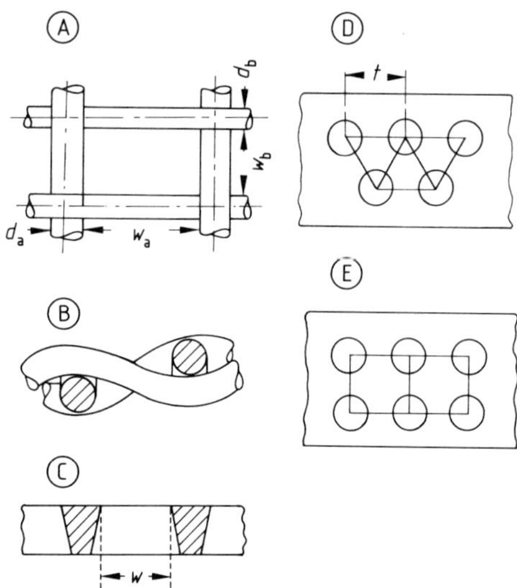

Figure 37. Screens
A) Woven rectangular mesh with opening sizes w_a and w_b, wire diameter d_a and d_b; B) Cross section of woven screen; C) Cross section of perforated plate with wider opening on the bottom; D) Perforated plate with triangular spacing; E) Perforated plate with square spacing

Technical requirements for *wire cloth* are standardized in ISO-DIS 9044. Special standards exist for testing screens (→ 2. Particle Size Analysis and Characterization of a Classification Process **2**-23).

The standard for wire cloth and sieve plates in the United States is given in ASTM E 437-85 standard specifications for industrial wire cloth and screens (square opening series).

The following standards are used in the Federal Republic of Germany:

DIN 4185/1	screens; terms and symbols for woven screens
DIN 4185/2	screens; terms and symbols for screens; perforated plates
DIN 4186	screens; round metal wire
DIN 4189	screens; steel, stainless steel, and nonferrous wire cloth, dimensions
DIN 4195	screens; woven screens from silk or synthetic fibers, dimensions
DIN 4197	screens; woven screens from synthetic filaments, dimensions

The tolerances for woven screens are relatively large; see Table 1, for example.

Perforated screens are generally stamped from sheet metal. The holes are round, square, or rectangular to slotlike. Special types also exist. Plates with fine and precise openings are made by electroplating. The precision of perforated plates, especially those made by electroplating, is

Table 1. Tolerances for woven wire cloth

Mesh diameter	DIN 4188 analytical sieves		DIN 4189 wire cloth
w, mm	Δw, %	Arithmetic mean	Δw, %
0.032–0.056	50	8	50
0.1			30
1.0	14	3	25
10.0	7	3	

much higher than that of woven screens. However, the free screen area is small. Standards for perforated plate screens in the United States are

ASTM E 674-80	standard specification for industrial perforated plate screens (square opening series)
ASTM W 454-80	standard specification for industrial perforated plate screens (square opening series)

and in the Federal Republic of Germany,

DIN 4185 T2	terms and symbols for screens; perforated plates
DIN 24041-043	round-hole, square-hole, and slotted perforated plates

Analytical screens and screens made by electroplating have their own standards (→ 2. Particle Size Analysis and Characterization of a Classification Process).

Ready-to-use screens made of polyurethane have become increasingly important. These are the so-called system screens. They are available with slot openings down to $w > 100$ μm. The advantage of this type of screen is its resistance to abrasion and the fact that it can be changed quickly. Because of the flexibility of the material, it is easy to clean.

Perforated plastic sheets are used on screening devices, which bend or stretch the mesh. Coarse mesh plastic screens with steel reinforcement are also available. Gratings and bars are used for very coarse screening.

5.1. Cleaning of Screens

Near-mesh-size material not only hinders screening, it even stops it (Figs. 11 and 13; Eqs. (7) and (30)). On rigid screens, such as perforated sheets with a mesh size less than 1 mm, screening rapidly comes to a standstill because of jamming.

The jamming force between glass beads and plastic wire cloth with $d = w$ is

$$F_K = 0.01\ E w^2 \qquad (42)$$

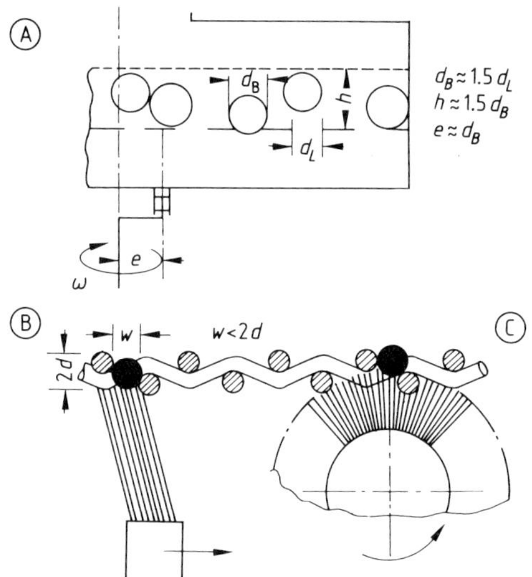

Figure 38. Screen cleaning
A) Rubber-ball, $d_B \approx 1.5\,d_L$, $h \approx 1.5\,d_B$, $e \approx d_B$; B) Flat brush; C) Roll brush

where E is the modulus of elasticity of the wire cloth. To overcome such forces, acceleration of the order of $10^6 \times g$ would be necessary; this cannot be realized. Reliance on the flexibility and deformation of the screen cloth is much easier. Thus about 50 % of the meshes can be kept clean because of the equilibrium between blinding and cleaning [18].

The simplest method of keeping the mesh free is to have or deliberately add coarse material. Only about 10–30 % of coarse material, where $x \approx 1.5-2\,w$, is needed to increase the throughput of near-mesh particles ($0.9 < x/w < 1.1$) by a factor of about 50. This leads to inverted screening with increasing mesh size.

Additional means of cleaning screens are the impact of bouncing rubber balls and the use of brushes, as shown in Figure 38. Cylindrical brushes are better than flat ones because they can reach more deeply into the mesh of fine screens.

The latest way to prevent blinding is the application of high-energy ultrasonics. This is effective in screening ceramic and metal powders with mesh $\lesssim 500$ µm [19]. Using rectangular instead of square meshes is effective too. The ratio of the opening should be $w_a/w_b = 1.5$ or 2.5.

5.2. Electrostatics in Screening

The motion of particles in the bed results in shifting and separating of electrostatic charges. At the points of contact, opposite charges remain. Because of low conductivity, an electrostatic potential arises, resulting in an adhesive or attractive stress in the bed. Compared with van der Waals forces, electrostatic forces have a much greater range. Thus screening is hindered strongly or even prevented completely.

Experiments with plastic powder have shown that screening is possible for particles $\gtrsim 50$ µm by applying air jets or ionizing the surrounding air, e.g., by corona discharge. Screening of particles $x \lesssim 50$ µm can be improved only by raising the electrical conductivity or the humidity of the air. Up to 90 % R.H., no negative influence on flow behavior has been observed [20].

6. Safety Requirements

In the design of a screening machine, the danger of dust explosions must be considered; some necessary precautions are reducing mechanical friction, avoiding generation of sparks, grounding to prevent the buildup of static charges, and blanketing the space around the screen with nitrogen or argon, as is done in the screening of magnesium powder. The screen can also be placed inside a pressure vessel. Finally, an automatic extinguishing system can be installed which sprays an extinguishing fluid when triggered by a sensor.

7. References

[1] N. Standish, *Powder Technol.* **41** (1985) 57–67.
[2] N. Standish, *Proc. Australas. Inst. Min. Metall.* **290** (Mai 1985) 59–66.
[3] M. L. Jansen, J. R. Glastonbury, *Powder Technol.* **1** (1967/68) 334–343.
[4] J. Wessel, *Aufbereit. Tech.* **8** (1967) 167–180.
[5] S. E. Gluck: "Vibratory Screens," *Chem. Eng. (N.Y.)* **72** (1965) Feb. 15, 151–168, March 15, 179–184 and Oct. 25, 121–146.
[6] K. Wieghardt, *Ing. Arch.* **20** (1952) 109–115.
[7] K. Höffl: *Zerkleinerungs- und Klassiermaschinen,* Springer Verlag, Berlin–Heidelberg–New York–Tokyo 1986.
[8] H. E. Rose: "The Derivation of an Equation for the Performance of a Screen," *4th European Symposium on Comminution,* Nürnberg 1975.
[9] N. Standish, *Powder Technol.* **48** (1986) 161–172.
[10] J. E. English, *Filtr. Sep.* **11** (1974) 195–203.
[11] R. B. Hudson, *Powder Technol.* **2** (1968/69) 229–240.
[12] M. A. G. Vorstman, M. Tels, Particle Technology Workshop, Techn. Univ. Eindhoven, 1981.

[13] T. Kanzleiter: "Die Taumelsiebmaschine, Maschinendynamik und Guttransport," *Aufbereit. Tech.* **12** (1971) 119–127, 380–390.
[14] H. Schmidt, *Aufbereit. Tech.* **18** (1977) 327–332.
[15] U. Schmidt, J. Wessel, *Aufbereit. Tech.* **17** (1976) 467–470.
[16] P. Schmidt, *Aufbereit. Tech.* **26** (1985) 393–398.
[17] A. Wignjosaputro, *Fortschr. Ber. VDI Z.*, Reihe 3, **106** (1985).
[18] F. J. Wesselbaum, *Siebmaschinen mit verformenden Sieben,* Dissertation Univ. Essen, 1982.
[19] Russel Finex Ltd., EP-A 0233066 A 2, 1987 (J. Monteith).
[20] H. J. Smigerski, *Chem.-Ing.-Tech.* **54** (1982) MS 983/82.

16. Elutriation

Air Classifying (Air Elutriation) is a separate keyword.

YORAM ZIMMELS, Technion IIT, Haifa, Israel

In addition to the standard symbols defined in the front matter of this volume, the following symbols are used

Symbol	Name
A	area
a_1, a_2, a_3	constants used for evaluation of U_φ
B	normalization factor of the beta distribution
b', b_1, b_2	constants of distribution functions
C	constant
$C_D, C_{D\varphi}, C_{D0}$	drag coefficient: general, extended to include the effect of particle concentration (i.e., φ), and at $\varphi = 0$, respectively
D	normalized sum of squares
d, d_i, d_j	particle size: general of ith particle, and of jth particle, respectively
d_s	Stokes diameter
d_0, d_m	average particle size and maximum size that exist in the beta distribution
E_c, E_f	total "coarse" and "fine" efficiencies
F_D	hydrodynamic drag force
f_p	force per unit volume
$f(\varphi)$	correlation function for dependence of U_p on φ
$f_1(\varphi)$	function of φ used to evaluate dependence of flux on φ
$G_c(x_0), G_f(x_0)$	"coarse" and "fine" cumulative grade efficiencies
g	gravitational acceleration
g'	apparent acceleration of particle set in motion by force field
$g_c(x), g_f(x)$	"coarse" and "fine" specific grade efficiencies
h	depth
J, J_i, J_m	volume flux: total value, value of ith size fraction, and total maximum value
J_u, J_d	"upward" and "downward" volume flux of particles
k	integer or a constant
L	distance along pipe axis; pool length
m	mass fraction; also used as an integer
m_e	value of mass fraction m that maximizes $f_1[\varphi(m)]$
N	size of sample
N_A	Avogadro constant
n	exponent; also used as integer
n_i	number of observed measurements that fall in the ith interval of a sample; hydration number
P_i	probability evaluated from a known distribution

P_0, P_L pressure at reference position and at distance L downstream, respectively

Q_f, Q_u, Q_d flow rate: total, upstream, and downstream, respectively

$q(x)$, $q'(x)$ particle flux densities

$q_u(x)$, $q_d(x)$ upstream and downstream particle flux densities

R Rosin–Rammler cumulative distribution function; also inner radius of a pipe

r radial distance from axis of rotation; axis of a pipe; or parameter of distribution functions

r' $= t - r$, parameter of the beta distribution, or radial distance

r_e elutriation radius

Re, Re_φ Reynolds number and extended definition in which the effect of particle content (i.e., φ) is accounted for

Re_0 Reynolds number at $\varphi = 0$

$\tilde{S}$ cumulative sharpness index

S_1 cross section of particle taken perpendicular to direction of flow

s parameter, also specific sharpness index

t parameter of a distribution

t' parameter of a distribution

U_d average velocity of dispersion relative to the container

U_{da}, U_{dm} average and maximum flow velocities in a pipe

U_f average velocity of fluid which is displaced by motion of particles relative to the dispersion frame of reference

U_φ, U_p average velocity of particles, relative to the frame of reference of the fluid in which they are dispersed and relative to the frame of reference of the dispersion, respectively

$U_{\varphi h}$ velocity of hydrated particles with respect to the free-flowing fluid

$U_{\varphi i}$, U_{pi} average velocity of ith particles relative to the fluid and to the dispersion frame of reference, respectively

U_0 velocity of a particle at infinite dilution, i.e., $\varphi \to 0$

U'_p velocity of particle relative to the container

U'_{pi} velocity of ith particles with respect to the container frame of reference

$U_p(x_0)$ velocity of particles having size x_0 relative to the dispersion

$U'_{p_a}(i, j)$ velocity of particles in i, jth aggregate relative to the container frame of reference

$U_\varphi(x)$, $U_\varphi(x_0)$ velocities of particles having size x and x_0 relative to the fluid, respectively

$U_{\varphi_a}(i, j)$ velocity of i, jth aggregate relative to the fluid at φ_a

V_a, V_s, V_f volume of a single aggregate, of particles, and of fluid in the aggregate, respectively

V_h, V_{h0} volume of hydrated and unhydrated particles, respectively

v molar volume

W width

W_0, W_c, W_f mass of feed, coarse, and fine products respectively

w constant

x, x_0 independent variable of distributions; also a parameter

x_m, x_s variable denoting large and small particle size, respectively

X_φ independent variable in the Barnea–Mizrahi correlation, $(Re_\varphi^2\, C_{D\varphi})^{1/3}$

x_{25}, x_{75}, x_{10}, x_{90} particle sizes used to evaluate the sharpness index; subindex denotes the corresponding value of the cumulative distribution function

x_{50+w}, x_{50-w} particle size used to evaluate the specific index of sharpness, $0 < w < 50$

x'_{50+w}, x'_{50-w} particle size used to evaluate cumulative index of sharpness, $0 < w < 50$

y $= \ln d$

Y_φ dependent variable in the Barnea–Mizrahi correlation, $(Re_\varphi / C_{D\varphi})^{1/3}$

y_0 $= \ln d_0$, the expectation of the log of normal size distribution

z variable of the standard normal distribution

Greek Symbols

α parameter in distributions; also used to denote an angle

β parameter (see Harris equation); also used to denote an angle

$\Gamma(r)$ gamma function

γ parameter (see Harris equation)

δ, δ_a particle–dispersion and aggregate–dispersion density difference, respectively

$\delta_a(i, j)$ i, jth aggregate–dispersion density difference

δ_+, δ_- variables having a value of 0 or 1 that are used to define up- and downstream flux densities

ε porosity

μ_f, μ_φ, μ_{φ_a} viscosity of fluid, of dispersion at φ, and of dispersion containing aggregates at φ_a, respectively

ξ_i variable denoting φ_i/φ; also a coordinate system

ϱ_f, ϱ_p, ϱ_φ density of fluid, particles, and dispersion, respectively

σ_y standard deviation of the logarithmic normal size distribution

φ, φ_i volume fraction occupied by particles: total volume and volume of ith particle-size fraction, respectively

φ_a total volume fraction occupied by monodisperse aggregates in dispersion

φ_f volume fraction occupied by fluid

φ_h volume fraction of hydrated particles

φ_s, φ_{sm} volume fraction occupied by particles within an aggregate and the maximum value of φ_s dictated by packing density considerations

φ_{sj} volume fraction occupied by particles in jth aggregate fraction

$\varphi(i, j)$ volume fraction occupied (in dispersion) by particles pertaining to the i, jth aggregate

$\varphi_h(i, j)$ volume fraction occupied (in dispersion) by hydrated particles

$\varphi_s(i, j)$ volume fraction occupied by particles in the i, jth aggregate

$\chi^2_{n-1,\,1-\alpha}$ parameter of the chi-square distribution with $n - 1$ degrees of freedom and level of confidence $1 - \alpha$

ψ_{LN}	probability density function of the logarithmic normal distribution
$\psi(x)$, $\psi(z)$	probability density function of x and z, respectively
$\psi_0(x)$, $\psi_c(x)$, $\psi_f(x)$	distribution functions characterizing the feed, coarse, and fine products, respectively
$\Psi(x)$, $\Psi(z)$	cumulative distribution functions
$\Psi_c'(x)$, $\Psi_c(x)$	Harris cumulative distribution functions
$\Psi_{GM}(x)$, $\Psi_{GGS}(x)$, $\Psi_s(x)$	Gaudin–Meloy, Gates–Gaudin–Schumann and S-shaped cumulative distribution functions
$\psi_{RR}(x)$	Rosin–Rammler distribution function
ω	angular velocity

Elutriation is a process whereby a polydisperse mixture of particles can be separated into distinct fractions by means of the combined action of a fluid stream and one or more external force fields. The drag force exerted by the fluid on particles that are smaller and less susceptible to the field deflects them (i.e., elutriates them) from trajectories of other, larger, more susceptible particles which tend to follow more closely trajectories dictated by the external force field. The term *elutriation* is sometimes used when the fluid stream is a liquid (wet classifying); with gas streams the operation is called *air classifying* (→17. Air Classifying).

1. Scope

1.1. Historical Aspects

Elutriation and air classifying have traditionally been recognized as reverse sedimentation in which particles move against a current of fluid (usually water or air) that subsequently, by one or more process steps, effects separation and grading of the particle mixture. Elutriation has mostly been applied to fine particles in dilute dispersions that follow the Stokes sedimentation law. In this context, flows in elutriation systems were limited to the laminar regime. The force fields were traditionally of the acceleration type, i.e., gravitational or centrifugal, with the fluid stream originating from sources external to the dispersion.

1.2. Extended Scope

The extended scope that is used here under the title elutriation is in accordance with the definition given above [1] and includes the following elements:

1.2.1. Type of Force Field

No limitation is put on either the type of force field or its distribution in space. Forces may originate from potential and vector (i.e., kinetic) fields that are external to the system, e.g., from gravitational, centrifugal, electromagnetic, and hydrodynamic fields. Forces may also originate internally due to thermal and chemical potential fields, as in diffusion systems.

1.2.2. Particle-Size Distribution

Elutriation has been used traditionally for treatment of fine-particle distributions. Here the finer size fraction is separated from the rest of the distribution by the fluid stream. Elutriation has been applied generally to particles larger than a few micrometers. Under the extended definition used in this article, and in view of the mechanism of elutriation (see Section 1.2.3), submicron particles, colloids, and the large molecules encountered in diffusion are also included.

1.2.3. Mechanism of Elutriation

Elutriation can arise due to internally or externally driven flows:

1) internally counterflowing fluid which is displaced by motion of the particle mixture, or
2) counter- and cross-flowing fluid which is supplied from sources external to the dispersion.

The first mechanism of elutriation is inherent in cases where motion of the dispersed particles is set by external force fields other than hydrodynamic. Here, slower parts of the distribution are partially or fully elutriated counter to other parts that contain faster particles. This mechanism depends on the properties of the dispersion and can be manipulated only by changing them. Its importance increases for dispersions with a finer particle-size distribution.

The second mechanism is better known, but its applicability is limited by particle size. In contrast to internal flows, the external flow can be controlled readily to achieve desirable separation.

1.2.4. Range of Concentration

Elutriation is usually carried out in dilute dispersions. This requires the use of a large ratio of fluid volume to dispersed-phase volume. In this article, no limits are set on the concentration of the dispersion, with the understanding that extremes should be avoided for either economical or technological reasons. When the concentration increases, the effect of displaced fluid on elutriation is enhanced, as is the saving of fluid per unit dispersed phase treated. On the other hand, interparticle hindrance may adversely affect elutriation efficiency, so that an optimum concentration can be expected.

1.2.5. Elutriation in the Context of Physical Separation Processes

Elutriation can have a partial or full role in the separation of particles in flow fields. Thus, classification of particles can be achieved by an elutriation process in a separation zone, followed by auxiliary methods for dispersing and removing the particle stream. Internal flows result in the segregation of particles which, combined with external forces, can be used for density and electromagnetic separation.

2. Theory

2.1. Hindered Sedimentation of Monodisperse Particles

Two kinds of particle velocity can be distinguished: the average steady-state velocity of particles relative to the fluid that has been displaced by their motion, U_φ, and the average steady-state velocity of particles relative to the frame of reference of the whole dispersion, U_p.

In this case, φ is the total volume fraction occupied by particles in the dispersion. The fluid drag on particles depends on U_φ, whereas particle

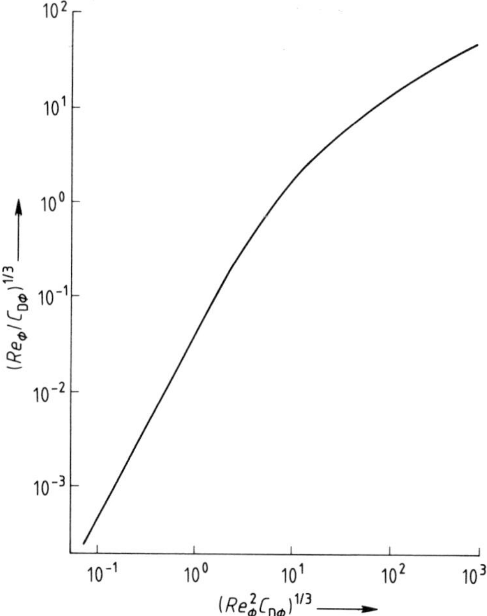

Figure 1. Barnea–Mizrahi correlation for hindered sedimentation of monodisperse spherical particles, $(Re_\varphi/C_{D\varphi})^{1/3}$ vs. $(Re_\varphi^2 C_{D\varphi})^{1/3}$

flux depends on U_p. The velocity of the displaced fluid relative to the dispersion frame of reference is referred to as U_f.

$$U_f = U_\varphi \varphi \tag{1}$$

$$U_p = U_\varphi - U_f = U_\varphi(1 - \varphi) \tag{2}$$

$$\varphi + \varphi_f = 1 \tag{3}$$

where φ_f is the volume fraction occupied by the fluid; φ_f is equivalent to the porosity ε. As shown below, available correlations facilitate the evaluation first of U_φ and then of U_p.

Figure 1 shows the generalized correlation of Barnea and Mizrahi [1] for hindered sedimentation of monodisperse spherical particles in Newtonian fluids. This plot obviates the need to resort to the well-known family of concentration-dependent plots $(Re/C_D)^{1/3}$ vs. $(Re^2 C_D)^{1/3}$, which are traditionally used for calculation of velocity vs. diameter of particles. In the Barnea and Mizrahi plot, the Reynolds number Re is replaced by Re_φ and the drag coefficient C_D by $C_{D\varphi}$; Re_φ and $C_{D\varphi}$ are given by Equations (4) and (6), respectively.

$$Re_\varphi = \frac{U_\varphi d \varrho_f}{\mu_\varphi} = Re_0 \frac{U_\varphi/U_0}{\exp\left(\frac{5}{3} \cdot \frac{\varphi}{1-\varphi}\right)} \tag{4}$$

where d is the particle size, ϱ_f is the density of the fluid and subindex 0 denotes here $\varphi = 0$ (i.e., $U_0 = U_\varphi (\varphi = 0)$) and the extended viscosity μ_φ is given by

$$\mu_\varphi = \mu_f \exp\left(\frac{5}{3} \cdot \frac{\varphi}{1-\varphi}\right) \tag{5}$$

with μ_f being the viscosity of the pure fluid.

$$C_{D\varphi} = \frac{F_D}{\frac{1}{2} \varrho_f S_1 U_\varphi^2 (1 + \varphi^{1/3})} = C_{D0} \left(\frac{U_0}{U_\varphi}\right)^2 \cdot \frac{(1-\varphi)}{(1+\varphi^{1/3})} \tag{6}$$

where F_D is drag force, ϱ_f is fluid density, S_1 is the cross section of the particle perpendicular to the direction of flow and $(1 + \varphi)^{1/3}$ is the particle-interaction equivalent of the wall-hindrance effect. The Barnea–Mizrahi correlation applies to a wide range of φ including the high concentrations encountered in distended beds, and also to the full range of flow regime encountered practically in particle sedimentation and fluidization. This correlation can be described by Equation (7 a) for the creeping-flow regime, and Equation (7 b) for the intermediate- and turbulent-flow regimes, respectively [2].

$$C_{D\varphi} = 24/Re_\varphi \tag{7 a}$$

$$(\log Y_\varphi + 2.563)^2 + (\log X_\varphi - 4.267)^2 = 18.775 \tag{7 b}$$

$$Y_\varphi = (Re_\varphi / C_{D\varphi})^{1/3}, \quad X_\varphi = (Re_\varphi^2 C_{D\varphi})^{1/3}$$

Note that Equation (7 a) corresponds to the straight line, in the creeping-flow regime, in the log–log plot of Y_φ vs. X_φ, see Figure 1.

With the results shown in Figure 1 and Equations (7 a) and (7 b), U_φ can be evaluated directly by Equation (8) for the creeping- and by Equation (9) for the general-flow regimes [2], [3]:

$$U_\varphi = \frac{(\varrho_p - \varrho_f) g' d^2}{18 \mu_f} f(\varphi),$$

$$f(\varphi) = \frac{(1-\varphi)}{(1+\varphi^{1/3}) \exp\left(\frac{5}{3} \cdot \frac{\varphi}{1-\varphi}\right)} \tag{8}$$

$$U_\varphi = \left(\frac{-a_2 + \sqrt{a_2^2 + 4 a_1 a_3^{1/2}}}{2 a_1}\right)^2 \tag{9}$$

$$a_1 = 0.63 (d \varrho_f)^{1/2}$$

$$a_2 = 4.80 \mu_f^{1/2} \exp\left(\frac{5}{6} \cdot \frac{\varphi}{1-\varphi}\right)$$

$$a_3 = 4.0 (\varrho_p - \varrho_f)(1 - \varphi) g' d^2 / [3 (1 + \varphi^{1/3})]$$

where d is the particle diameter, ϱ_p the density of the particles, and g' the apparent acceleration of the particle in the driving force field:

$$g' = f_p / \delta \tag{10}$$

where f_p is the force per unit volume of particle and δ is given by

$$\delta = \varrho_p - \varrho_\varphi \tag{11}$$

$$\varrho_\varphi = \varphi \varrho_p + (1 - \varphi) \varrho_f \tag{12}$$

For gravitational and centrifugal fields, $g' = g$ and $g' = \omega^2 r$, respectively, where g is gravitational acceleration, ω is angular velocity, and r is distance from the axis of rotation.

2.2. Hindered Sedimentation of Polydisperse Particles [2]–[4]

Elutriation is generally applied to polydisperse mixtures. In a mixture consisting of n size fractions,

$$\varphi = \sum_{i=1}^{n} \varphi_i \tag{13}$$

where φ_i is the volume fraction occupied by the ith size fraction of particles having size d_i. Equation (2) applies to each size fraction:

$$U_{pi} = U_{\varphi i} - U_f \tag{14}$$

where U_f is given by:

$$U_f = \sum_{i=1}^{n} U_{\varphi i} \varphi_i \tag{15}$$

If an external source of fluid sets the whole dispersion in motion and the dispersion gains the velocity U_d relative to the dispersion container, then the velocity of particles with respect to the container, U'_{pi}, becomes

$$U'_{pi} = U_{\varphi i} - U_f - U_d \tag{16}$$

Particles will be elutriated (i.e., carried with the stream) if $U'_{pi} < 0$; otherwise they will counterflow the stream and perhaps leave it because of external force fields if $U'_{pi} > 0$. They will remain in the elutriation zone if $U'_{pi} = 0$. Thus the cut size (for definition of cut size →2. Particle Size Analysis and Characterization of a Classification Process, p. 2-10) is determined by

$$U_d = U_{\varphi i} - U_f = U_{\varphi i} - \sum_{i=1}^{n} U_{\varphi i}\, \varphi_i \tag{17}$$

At nonvanishing values of φ, the particle velocity $U_{\varphi i}$ and the particle volume fraction φ_i are functions of the size distribution and of the total volume fraction occupied by particles in the dispersion; the same is true for the cut point of the elutriation process.

Particles that are elutriated by internal fluid current (i.e., no external fluid supply is required to reverse their motion against the driving force field) are defined in this article as self-elutriated particles. Alternatively, "self-elutriation" implies that the system elutriates part of its distribution by internal hydrodynamic drag arising from the remainder of the distribution. The cut point under conditions of self-elutriation is determined simply by letting $U_d = 0$; then

$$U_{\varphi i} = U_f \tag{18}$$

2.3. Effect of Aggregation [2], [3]

2.3.1. Monodisperse Aggregates

Aggregation involves the formation of rigid structures of particles that adhere to each other. Aggregates generally consist of particles, entrained fluid, and a binding material.

The aggregation of particles to form a new particulate (i.e., the aggregate) has the following effects:

1) Increase in the size of the new particulate, compared to the size of the particles from which it was formed
2) Decrease of the average density (magnetic susceptibility, etc.), due to entrainment of fluid, compared to nonaggregated particles
3) Increase of the volume fraction occupied by the particulates
4) Change in shape of the dispersed particulates
5) Change of distribution characteristics of the dispersed particulates
6) Increased probability of interparticle interaction
7) Increased probability of variation in distribution during motion, particularly where shear forces are involved

Aggregation can occur spontaneously, for example, due to existing electrostatic or van der Waals interactions, or as a result of deliberate changes (such as introduction of additives) in the composition of the dispersion. Its effect is important in fine-particle systems, where air elutriation (→17. Air Classifying) cannot be applied because of the tendency of the particles to aggregate electrostatically. Wet elutriation facilitates the use of dispersants or flocculants (→Flocculation, treated in the A Series) to prevent or enhance aggregation. The use of such agents becomes critically important in emulsions and colloidal systems. Although aggregation is generally believed to enhance sedimentation (see item 1 above), situations may exist in which the counteracting factors (see items 2–6) prevail. Let the volume V_a of a single aggregate contain particles and entrained fluid having volumes V_s, and V_f, respectively

$$V_a = V_s + V_f \tag{19}$$

$$V_s = \varphi_s V_a, \quad V_f = (1 - \varphi_s)\, V_a \tag{20}$$

where φ_s is the volume fraction occupied by particles within the aggregate. Let φ_a be the total volume fraction occupied by monodisperse aggregates in the suspension:

$$\varphi_a = \varphi/\varphi_s, \quad \varphi_s \le \varphi_{sm}, \quad \varphi_a \le \varphi_{sm} \tag{21}$$

where φ_{sm} is the maximum volume fraction that can be occupied by spherical particles in the aggregate, which here is assumed to be determined by the maximum packing density of spherical particles. Thus $\varphi \le \varphi_{sm}^2$, and the upper boundary of φ which is 0.6–0.7 in a nonaggregated dispersion decreases to 0.36–0.49 in monodisperse spherically aggregated systems. The aggregate–suspension density difference δ_a and the apparent suspension viscosity μ_{φ_a} are given by

$$\delta_a = (\varphi_s - \varphi)(\varrho_p - \varrho_f) \tag{22}$$

$$\mu_{\varphi_a} = \mu_f \exp\left(\frac{5}{3} \cdot \frac{\varphi}{\varphi_s - \varphi}\right) \tag{23}$$

Equations (22) and (23) show that the density difference (see Eq. 11) decreases and viscosity increases (see Eq. 5) compared to nonaggregated systems at the same φ.

Equations (8) and (9) can be used to evaluate $U_{\varphi a}$ of aggregates, where

$$f(\varphi_a) = \frac{\varphi_s - \varphi}{[1 + (\varphi/\varphi_s)^{1/3}] \exp\left(\frac{5}{3} \cdot \frac{\varphi}{\varphi_s - \varphi}\right)} \tag{24}$$

a_1 remains the same; a_2 and a_3 are given by

$$a_2 = 4.80\,\mu_f^{1/2}\exp\left(\frac{5}{6}\cdot\frac{\varphi}{\varphi_s-\varphi}\right) \tag{25}$$

$$a_3 = 4.0(\varrho_p - \varrho_f)(\varphi_s - \varphi)\,g'\,d^2/\{3[1+(\varphi/\varphi_s)^{1/3}]\}$$

2.3.2. Polydisperse Aggregates

In this section, aggregates are described which vary in size and packing density, i.e., in φ_s. If the aggregate mixture consists of n size fractions and m φ_s fractions, and each fraction denoted (i, j) is characterized by its size d_i and by φ_{sj},

$$\delta_a(i,j) = [\varphi_s(i,j) - \varphi](\varrho_p - \varrho_f) \tag{26}$$

$$\varphi_a = \sum_{i=1}^{n}\sum_{j=1}^{m}\varphi(i,j)/\varphi_s(i,j), \quad \sum_{i=1}^{n}\sum_{j=1}^{m}\varphi(i,j) = \varphi \tag{27}$$

Equation (28) is the counterpart of Equation (16) for polydisperse aggregates:

$$U'_{pa}(i,j) = U_{\varphi a}(i,j) - \sum_{i=1}^{n}\sum_{j=1}^{m} U_{\varphi a}(i,j)\cdot\varphi(i,j)/\varphi_s(i,j) - U_d \tag{28}$$

where $U_{\varphi a}(i,j)$ is calculated from Equation (8) or (9) in conjunction with Equations (24) and (25).

2.4. Composite Particulates

In contrast to aggregates, a composite particulate is defined here as a multicomponent entity subject to the condition that each component is continuous. This excludes the existence of separate particles or the formation of discontinuous islands which characterize aggregated systems. A simple illustration is the example of coated particles in which the core particle and its coating consist of continuous phases. Coated particles are important in solutions and colloidal systems, for example, hydrated species in aqueous solutions. Here, the volume V_h of the hydrated particle is given by [5]

$$V_h = V_{h0} + n\,v/N_A \tag{29}$$

where n, v, and N_A are the hydration number, the molar volume of the hydrating fluid, and the Avogadro constant, respectively. The new volume fraction φ_h of the hydrated particles becomes

$$\varphi_h = \varphi(1 + (n\,v/N_A)/V_{h0}) \tag{30}$$

To evaluate $U_{\varphi h}$ in monodisperse systems, φ is replaced by φ_h in Equations (8) and (9). In polydisperse hydrated particles which may vary in their initial size and also in aggregation number, $\varphi_h(i,j)$ should be defined and used as indicated in Section 2.3.

This procedure applies also to collodial systems such as ferrofluids [6], i.e., magnetic fluids consisting of colloidal magnetic particles, in which ferromagnetic particles (size 5–10 nm) are prevented from aggregation by the steric stabilization that is imparted to them by a dispersant coating. Elutriation of larger composite particles can be applied to the separation of catalysts deposited on substrates and to coated magnetic and pharmaceutical particles.

2.5. Particle Flux

The volume flux J_i of the ith size fraction being elutriated in a fluid of velocity U_d is given by

$$J_i = U'_{pi}\,\varphi_i = (U_{\varphi i} - U_f - U_d)\,\varphi_i \tag{31}$$

Let the fraction j be characteristic of the cut size. This means that particles of diameter d_j are neither lifted, nor do they sink, in the elutriation stream. Applying Equation (17) gives

$$U_d = U_{\varphi j} - U_f \tag{32}$$

hence

$$J_i = (U_{\varphi i} - U_{\varphi j})\,\varphi_i \tag{33}$$

$d_i < d_{i+1}$ for $i = 1, 2, \ldots, n-1$, the "upward" J_u and "downward" J_d total fluxes are given by Equations (34) and (35), respectively.

$$J_u = \sum_{i=1}^{j}(U_{\varphi i} - U_{\varphi j})\,\varphi_i \tag{34}$$

$$J_d = \sum_{i=j+1}^{n}(U_{\varphi i} - U_{\varphi j})\,\varphi_i \tag{35}$$

Note that "upward" means that particles are carried by the elutriation stream and "downward" means the reverse. The total flux J is given by

$$J = J_u + J_d \tag{36}$$

The flux J is a property of the particle distribution and does not depend on the velocity of the elutriation stream. Next, the conditions for maximum flux are determined. If φ_i is expressed as a fraction ξ_i of φ, then

$$\varphi_i = \xi_i \, \varphi \tag{37}$$

The following equation can be used to evaluate J in the creeping-flow regime:

$$J = \frac{(\varrho_p - \varrho_f)\, g'}{18\, \mu_f} f_1(\varphi) \sum_{i=1}^{n} |d_i^2 - d_j^2| \, \xi_i \tag{38}$$

$$f_1(\varphi) = \varphi(1-\varphi) \Big/ \left[(1+\varphi^{1/3}) \exp\left(\frac{5}{3} \cdot \frac{\varphi}{1-\varphi}\right)\right]$$

In deriving Equation (38), use was made of Equations (8), (36), and (37). At a given particle-size distribution, J is solely a function of φ. Figure 2 shows a plot of $f_1(\varphi)$ vs. φ; $f_1(\varphi)$ has a maximum of 66.8744×10^{-3} at $\varphi = 0.218$. This value of φ is higher than the one ($\varphi = 0.1792$) that maximizes the sedimentation flux of monodisperse particles in the creeping-flow regime [2]. The volume or mass flux can be expressed as a function of the total mass fraction m of particles in the dispersion:

$$m = \frac{\varphi\, \varrho_p}{\varphi\, \varrho_p + (1-\varphi)\, \varrho_f} \tag{39}$$

Rearranging Equation (39) gives

$$\varphi = \frac{m}{\varrho_p/\varrho_f - m(\varrho_p/\varrho_f - 1)}$$

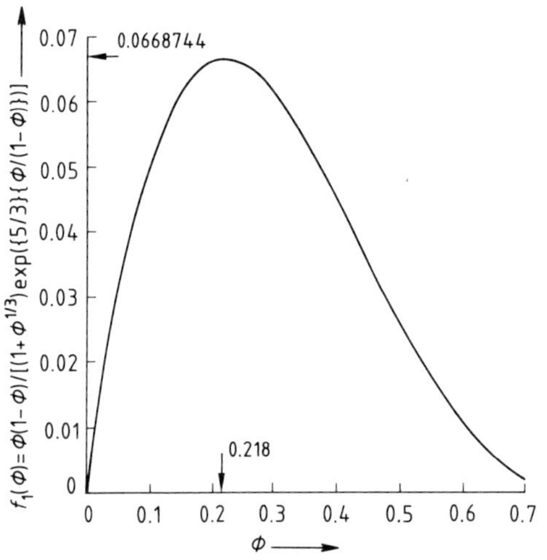

Figure 2. Plot of the flux volume concentration function $f_1(\varphi)$ vs. φ

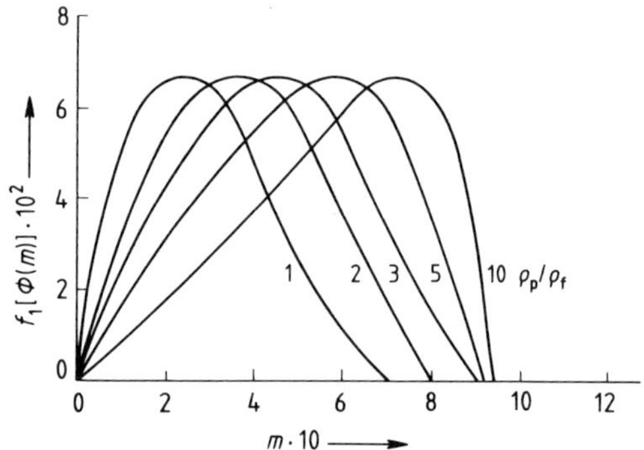

Figure 3. Plot of the flux mass concentration function $f_1[\varphi(m)]$ vs. m

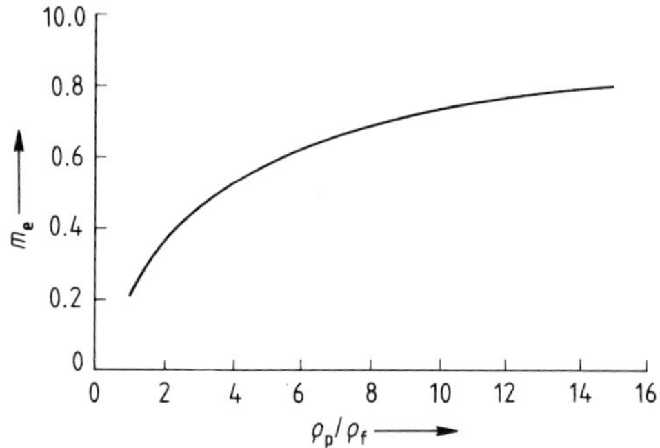

Figure 4. Plot of m values that maximize the flux m_e vs. ϱ_p/ϱ_f

Figure 3 shows a plot of $f_1[\varphi(m)]$ vs. m for different values of ϱ_p/ϱ_f. The curve $\varrho_p/\varrho_f = 1$ is identical to that of Figure 2. Figure 4 shows a plot of m_e vs. ϱ_p/ϱ_f, which was derived from Figure 3; m_e is the value of m that maximizes $f_1[\varphi(m)]$ (i.e., the flux at a given value of ϱ_p/ϱ_f). As expected, increasing ϱ_p/ϱ_f shifts the optimum mass fraction toward higher values. For example, $\varphi = 0.218$ at $\varrho_p/\varrho_f = 3$, $m_e = 0.455$; thus, the total maximum flux becomes

$$J_m = 66.8744 \times 10^{-3} \frac{(\varrho_p - \varrho_f)\, g'}{18\, \mu_f} \sum_{i=1}^{n} |d_i^2 - d_j^2| \, \xi_i \tag{40}$$

Elutriation is often performed in dilute suspensions. This is believed to increase the sharpness of separation. However, the costs in terms of the low throughput due to the low flux can be substantial. An example is the elutriation of particles with $\varrho_p = 3000\ \mathrm{kg/m^3}$ in water with $\varrho_f = 1000\ \mathrm{kg/m^3}$. If elutriation is performed at $m = 0.05$, then the flux is only 19.56% of the maximum, i.e., at $m_e = 0.455$. In other words, five times the amount of water is required compared to the demand had the elutriation been performed at the optimum point. This means larger investment in equipment and higher operating cost.

In summary, unless technical problems occur in working at the recommended values of φ or m, operating closer to the optimal values of these process variables should be considered.

2.6. Other $f(\varphi)$ Correlations for the Creeping-Flow Regime

Equation (41) gives the empirical correlation [7], [8] published by Richardson and Zaki:

$$f(\varphi) = (1 - \varphi)^n \qquad (41)$$

This correlation generally applies to the evaluation of U_p of monodisperse particles, which is given in the form

$$U_p = U_{p0} f(\varphi)$$

According to Equation (2), $n - 1$ should be used instead of n when applying $f(\varphi)$ to U_φ. Thus, in creeping-flow elutriation where $n = 5.1$ fits the data of several investigations [9],

$$f(\varphi) = (1 - \varphi)^{4.1} \qquad (42)$$

The discrepancy between the two correlations is apparent at lower values of φ, with the one given by Equation (8) falling off faster at $\varphi < 0.2$; the reverse is true at $\varphi > 0.2$ [7]. A summary of available models for $f(\varphi)$ is given in [1] and [7]. Here, the model which led to Equation (8) is preferred, in view of the large body of experimental data that support it.

2.7. Particle-Size Distribution

Equations (17) and (18), which give the cut size, and Equations (34)–(40), which give particle flux, show that the outcome of elutriation depends on particle-size distribution. Thus, information regarding size distribution is necessary for analysis of elutriation systems. In the following, some useful size distributions are summarized [10]–[14] (→2. Particle Size Analysis and Characterization of a Classification Process, p. **2**-6).

Logarithmic Normal Distribution. The logarithmic normal distribution in a standard form is given by

$$\psi(z) = \frac{1}{\sqrt{2\pi}} e^{-z^2/2}, \quad z = \frac{y - y_0}{\sigma_y} \qquad (43)$$

$$y = \ln d, \quad y_0 = \ln d_0, \quad d \geq 0$$

where $\ln d_0$ is the expectation, (i.e., weighted mean) of the random variable $y = \ln d$; σ_y is the standard deviation given in units of $\ln d$. The cumulative logarithmic normal distribution is

$$\Psi(Z \leq z) = \int_{-\infty}^{z} \psi_{LN}(z)\,dz \qquad (44)$$

where LN indicates logarithmic normal; $\Psi(Z \leq z)$ is available in standard tables [11], [15].

Thus, being given a cut size d of particles that have a logarithmic normal size distribution facilitates evaluation of the fraction expected to be elutriated. For example, if $d_0 = 50 \times 10^{-6}$ m, $y_0 = \ln d_0 = -9.9035$, and $\sigma_y = 1$, then for a cut size of 40×10^{-6} m, $y = \ln(40 \times 10^{-6}) = -10.1266$; $z = -0.2231$. Using tables of standard normal distribution gives $\Psi(Z \leq z) = 0.4117$; i.e., 41.17% of this distribution will be elutriated. The logarithmic normal distribution is useful for thc approximate description of skewed distributions.

Rosin–Rammler Distribution. The following equation gives the Rosin–Rammler probability density distribution function $\psi_{RR}(x)$:

$$\psi_{RR}(x) = 100\, nb' x^{n-1} \exp(-b' x^n), \quad x \geq 0 \qquad (45)$$

The cumulative function R is given by

$$R = \int_{-\infty}^{x} \psi_{RR}(x)\,dx = 100 \exp(-b' x^n),\ x \geq 0 \qquad (46a)$$

$$\log\log(100/R) = n \log x + \text{constant},\ x \geq 0 \qquad (46b)$$

where n and b' are constants, and b' is a measure of the range of particle size: high values of b' indicate a narrow distribution and vice versa; n is characteristic of the material under consideration. The size for which $R = 36.8\,\%$ is used to characterize the degree of size reduction of the material. This distribution applies to broken, crushed, or ground material, such as broken coal and fan spray droplets [16].

Gates–Gaudin–Schumann Distribution [17], [18]. The cumulative Gates–Gaudin–Schumann distribution is given by

$$\Psi_{GGS}(x) = (b_1 x)^n, \quad x \geq 0 \qquad (47)$$

where b_1 and n are constants.

Gaudin–Meloy Distribution [19]. The Gaudin–Meloy cumulative distribution is given by

$$\Psi_{GM}(x) = 1 - (1 - b_2 x)^n, \quad x \geq 0 \qquad (48)$$

where b_2 and n are constants.

S-Shaped Cumulative Distributions [20]. Equation (49) can be used to fit S-shaped cumulative data

$$\Psi_s(x) = \frac{e^{\alpha x} - 1}{e^{\alpha x} + e^{\alpha} - 2}, \quad x > 0 \qquad (49)$$

The probability density function corresponding to $\Psi_s(x)$ is given by

$$\psi_s(x) = \frac{\alpha\, e^{\alpha x}(e^{\alpha} - 1)}{(e^{\alpha x} + e^{\alpha} - 2)^2} \qquad (50)$$

Increase of α increases the steepness of $\Psi_s(x)$ and, hence, decreases the standard deviation of $\psi_s(x)$.

Gamma Distribution [11], [12]. The probability density function of the gamma distribution is given by

$$\psi(x) = \frac{\alpha}{\Gamma(r)} (\alpha x)^{r-1} e^{-\alpha x}, \quad x > 0, r > 0 \tag{51}$$

where α and r are constants. If r is a positive integer, then

$$\Psi(x) = 1 - \sum_{k=0}^{r-1} e^{-\alpha x} (\alpha x)^k / k! \tag{52}$$

and $\Gamma(r)$ is the gamma function

$$\Gamma(r) = \int_0^\infty x^{r-1} e^{-x} \mathrm{d}x, \quad r > 0 \tag{53}$$

If r is a positive integer, $\Gamma(r) = (r-1)!$. The gamma distribution presents a family of curves that includes the exponential distribution ($r = 1$) and the chi-square distribution ($r = n/2$ and $\alpha = 1/2$).

Beta Distribution. The probability density function of the beta distribution is given by

$$\psi(x) = \frac{1}{B} x^{r-1} (1-x)^{t-r-1}, \quad 0 \le x \le 1, \quad r > 0, \quad t - r > 0 \tag{54}$$

$$B = \Gamma(r)\Gamma(t-r)/\Gamma(t)$$

and if r and $t - r$ are integers,

$$B = (r-1)!(t-r-1)!/(t-1)!$$

The value of the beta distribution lies in the variety of shapes that it can be made to fit by varying the parameters r and t. For example, the beta distribution is skewed right if $r < t/2$ and left if $r > t/2$. It is bell-shaped if $r > 1$ and $t > r + 1$. Interchange of parameters $r' = t - r$ and $t' = t$ yields mirror images. Because $0 \le x \le 1$, normalized data must be used for particle size, such as $x = d/d_m$, where d_m is the maximum size that occurs in the distribution.

Harris Equation [21], [22]. HARRIS suggested that most of the well-known particle-size distributions are special cases of Equation (55):

$$\psi(x) = c x^{\alpha} \left[1 - \left(\frac{x}{x_0} \right)^{\beta} \right]^{x_0^\beta \gamma} \tag{55}$$

where c, α, β, γ, and x_0 are parameters.

For example, if $\alpha = 0$, $\beta = 2$, and $x_0 \to \infty$, the normal distribution is obtained, whereas $\alpha = n - 1$, $\beta = n$, and $x_0 \to \infty$ yields the Rosin–Rammler distribution. If $\alpha = s - 1$, $x_0^\beta \gamma = r - 1$, and $\beta = s$, a three-parameter equation [23] results which, on integration and rearrangement, gives

$$\Psi'_c(x) = 1 - \Psi_c(x) = \left[1 - \left(\frac{x}{x_0} \right)^{s} \right]^{r} \tag{56}$$

where $\Psi'_c(x)$ is the cumulative oversize distribution, x_0 is the maximum size in the sample, s and r are parameters concerned with the shape of the log–ln plot in the fine and coarse ends of the distribution. A convenient form of Equation (56) is given by

$$\log\{\ln[\Psi'_c(x)]^{-1}\} = \log r + \log\{\ln[1 - (x/x_0)^s]^{-1}\} \tag{57}$$

Equations (56) and (57) are flexible and can be fitted to most unimodal distributions. Equation (57) has been used to evaluate results from scanning centrifugal sedimentation instruments for particle-size analysis.

Fitting Known Distributions to Data Obtained from Samples. Procedures to test whether a known distribution (such as the ones listed above) can be taken to describe a given population, by using data obtained from samples of this population, are outlined in standard statistics texts [10], [11]. This procedure involves evaluating the sum

$$D = \sum_{i=1}^{n} \frac{(n_i - NP_i)^2}{NP_i}$$

and testing if

$$D < \chi^2_{n-1,\,1-\alpha}$$

where n_i is the number of observed measurements that fall in a given interval, $N = \sum_{i=1}^{n} n_i$, and P_i is the probability evaluated from the known distribution that the result of the measurement will fall in the ith interval; $\chi^2_{n-1,\,1-\alpha}$ is the chi-square variable having $n - 1$ degrees of freedom and a level of confidence $1 - \alpha$. Usually, $1 - \alpha$ is higher than 90 %.

If $D < \chi^2_{n-1,\,1-\alpha}$, the sample is assumed to be part of the tested known distribution at level of confidence $1 - \alpha$. The known distribution that provides the lowest value of D is preferable.

The wealth of information concerning distributions, especially particle-size distributions, provides sound mathematical description of particulate populations and, hence, prediction of their motion in elutriation system.

2.8. Performance Criteria

Figure 5 shows a single elutriation step that involves feed, as well as coarser and finer product streams. Let $\psi(x)$ and $\Psi(x)$ denote probability density and cumulative distribution functions, and subindexes 0, c, and f the feed, coarser, and finer product streams, respectively. Equations (58) and (59) give total and fractional mass balances, in terms of mass W and probability density functions $\psi(x)$, respectively

$$W_0 = W_c + W_f \tag{58}$$

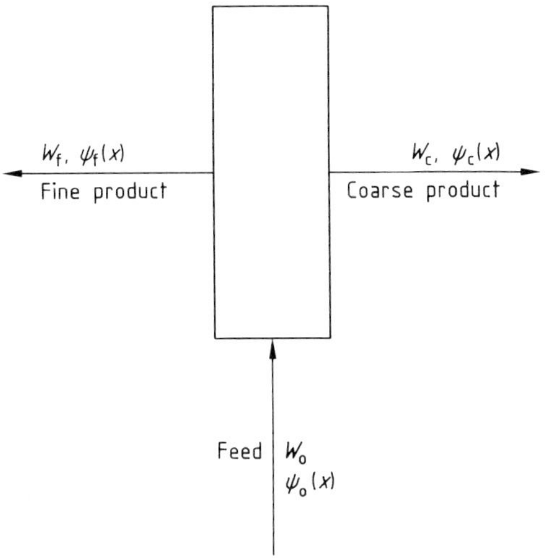

Figure 5. Mass balance around a single elutriation step

$$W_0\,\psi_0(x) = W_c\,\psi_c(x) + W_f\,\psi_f(x) \tag{59}$$

The total "coarse" and "fine" efficiencies are given by

$$E_c = W_c/W_0, \quad E_f = W_f/W_0 = 1 - E_c \tag{60}$$

The "coarse" and "fine" specific grade efficiencies, $g_c(x)$ and $g_f(x)$, are given by

$$\begin{aligned} g_c(x) &= (W_c/W_0)\;\psi_c(x)/\psi_0(x) \\ &= E_c\,\psi_c(x)/\psi_0(x) \end{aligned} \tag{61}$$

$$\begin{aligned} g_f(x) &= (W_f/W_0)\;\psi_f(x)/\psi_0(x) \\ &= E_f\,\psi_f(x)/\psi_0(x) \end{aligned} \tag{62}$$

Dividing Equation (59) by $W_0\,\psi_0(x)$ and comparing with Equations (61) and (62) give

$$g_c(x) = 1 - g_f(x) \tag{63}$$

The grade efficiencies of Equations (61) and (62) are evaluated at x and thus refer to a specific size. The grade efficiencies of sizes coarser or finer than x_0 may be of interest. In this case, the cumulative grade efficiency $G(x)$ is defined as follows:

$$G_c(x_0) = E_c\frac{\Psi_c(x > x_0)}{\Psi_0(x > x_0)} = E_c\frac{1 - \Psi_c(x_0)}{1 - \Psi_0(x_0)} \tag{64}$$

$$G_f(x_0) = E_f\frac{\Psi_f(x_0)}{\Psi_0(x_0)} \tag{65}$$

Note that by definition $\Psi(x_0) = \Psi(x < x_0)$, hence, $\Psi(x > x_0) = 1 - \Psi(x_0)$. Figure 6 shows a plot [14] of $\Psi_c(x)$ vs. $\Psi_0(x)$ over a plot of $\Psi_f(x)$ vs. $\Psi_0(x)$ which was developed for calculation of $g_c(x)$ and $g_f(x)$ [24]; because $\psi(x) = \mathrm{d}\Psi(x)\,\mathrm{d}x$, Equation (61) can be written as

$$g_c(x) = E_c\,\mathrm{d}\Psi_c(x)/\mathrm{d}\Psi_0(x) = E_c(\tan\alpha)_x$$

The largest particles of the distribution (20 µm in Fig. 6) are expected to be completely transferred to the coarser fraction. Hence, for large x, (e.g., x_m) $g_c(x) = 1$ and $(\tan\alpha)_{x_m} = 1/E_c$. Thus the tangent to the $\Psi_c(x)$ vs. $\Psi_0(x)$ curve at $\Psi_c(x) = 1$ intersects the $\Psi_0(x)$ axis at a point $\Psi_0(x) = 1 - E_c$. Now $g_c(x)$ can be evaluated by using $(\tan\alpha)_x$ for different values of x (in Fig. 6 the tangent to the curve at 5 µm is shown). Consider the lower plot of Figure 6: the smallest particles are expected to be completely transferred to the finer

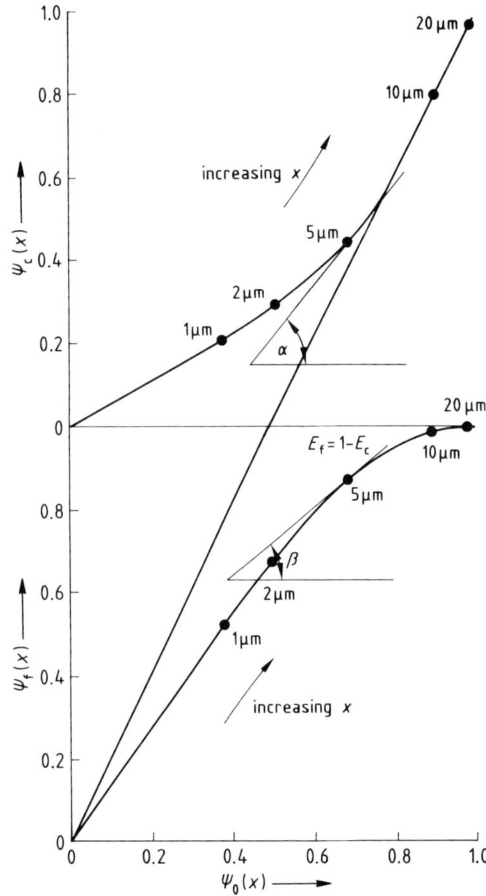

Figure 6. Plot of cumulative distribution
Oversize or coarse fraction $\Psi_c(x)$ appears directly above undersize or fine fraction $\Psi_f(x)$; both are plotted against size of feed $\Psi_0(x)$

fraction; hence, for sufficiently small x (e.g., x_s), $g_f(x) = 1$ and $(\tan\beta)_{x_s} = 1/E_f = 1/(1 - E_c)$. Now $g_f(x)$ can be evaluated by using $(\tan\beta)_x$ for different values of x. Distribution curves of feed, fines, and coarse fractions are evaluated elsewhere (→2. Particle Size Analysis and Characterization of a Classification Process, p. 2-4). For more details and numerical examples see [14] and [25].

2.9. Sharpness Index

Elutriation performance can also be characterized by the specific sharpness index s.

$$s = \frac{x_{(50+w)}}{x_{(50-w)}} \qquad 0 < w < 50 \tag{66}$$

where x_{50+w} and x_{50-w} are particle sizes defined by $g(x_{(50+w)}) = 50 + w$ and $g(x_{(50-w)}) = 50 - w$, respectively, and w is an arbitrary constant usually with a value $w = 25$ or $w = 40$.

The cumulative sharpness index S is defined by

$$S = \frac{x'_{(50+w)}}{x'_{(50-w)}} \qquad 0 < w < 50 \tag{67}$$

where $x'_{(50+w)}$ and $x'_{(50-w)}$ are defined by

$$G(x'_{(50+w)}) = 50 + w,\; G(x'_{(50-w)}) = 50 - w$$

The sharpness of cut obtained by the elutriator increases as either s or S approach 1, which is their lower bound that characterizes an ideal separation system.

2.10. Effect of Flow Profiles

2.10.1. Counterflow System

Figure 7 shows uniform and parabolic flow profiles in a tube. The uniform profile corresponds to nonviscous (i.e., potential) flow; the parabolic profile to laminar viscous flow in a circular tube. A flatter profile is expected to provide higher grade efficiencies and a sharpness of cut index closer to 1. Curved flow profiles introduce a position-dependent flow velocity $U_d = U_d(\xi_1, \xi_2, \xi_3)$, where ξ_i $(i = 1, 2, \text{and } 3)$ is a coordinate system used to describe a steady flow. Different types of flow can be used to effect elutriation. For example, axial flow in circular or rectangular conduits, flow in channels, layer or film flow, and flow around objects such as matrices used to capture particles by electromagnetic fields. In the following material, laminar flow in a circular tube and film flow are considered.

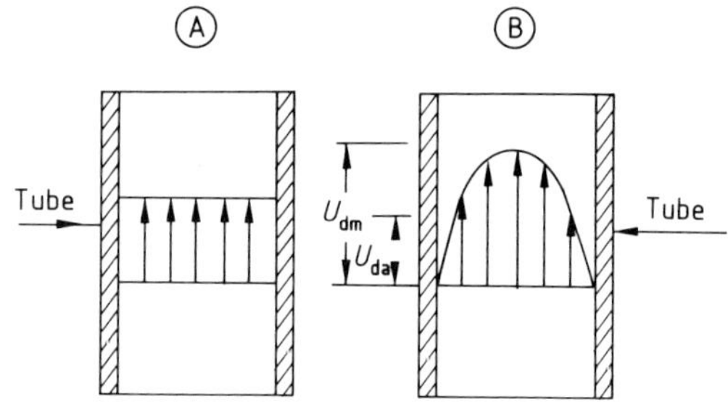

Figure 7. Flow profiles in a tube
A) Uniform velocity profile; B) Parabolic velocity profile

Steady Laminar Incompressible Flow in a Circular Tube. The flow profile is given by the Poiseuille equation [26]:

$$U_d = \frac{(P_0 - P_L)R^2}{4\mu_f L}\left[1 - \left(\frac{r}{R}\right)^2\right] \tag{68}$$

where R is the radius of the tube; r is radial distance from the tube axis; L is distance along the axis; and P_0 and P_L are the pressures at a reference position and at a distance L downstream from this position, respectively.

The flow rate Q and average velocity U_{da} are given by

$$Q_f = \frac{\pi(P_0 - P_L)R^4}{8\mu_f L} \tag{69}$$

$$U_{da} = \frac{(P_0 - P_L)R^2}{8\mu_f L} \tag{70}$$

The maximum velocity U_{dm} occurs at $r = 0$; hence,

$$U_{dm} = \frac{(P_0 - P_L)R^2}{4\mu_f L} = 2U_{da} \tag{71}$$

By using Equation (68),

$$r^2/R^2 = 1 - U_d/(2U_{da}) \tag{72}$$

Combining Equations (72) and (17) gives the cut size d_i (via $U_{\varphi i}$) obtained at r with the average velocity U_{da}:

$$r_e^2/R^2 = 1 - (U_{\varphi i} - \sum_{i=1}^{n} U_{\varphi i}\,\varphi_i)/(2U_{da}) \tag{73}$$

Thus, if U_{da} is large enough, an elutriation radius r_e exists such that when $r < r_e$, particles smaller than d_i (with settling velocities lower than U_d) will flow upstream. Outside r_e (i.e., for $r > r_e$), particles having sizes smaller than d_i will still settle against the elutriation stream. Equation (73) also shows that if

$$U_d = U_{\varphi i} - \sum_{i=1}^{n} U_{\varphi i}\, \varphi_i = 0$$

then $r = R$ and no loss of $d < d_i$ particles is expected. If U_{da} is given, the fractional loss (i.e., relative to the whole distribution) of particles smaller than d_i which arc not elutriated, is expected to maximize at infinite dilution (i.e., at $\varphi \to 0$). In this case, for any cutsize d_i, $U_d = U_{\varphi i}$ must be satisfied.

In the following material, grade efficiencies in elutriators characterized by nonuniform flow profiles are evaluated. Let the elutriator feed be characterized by a continuous distribution $\psi_0(x)$. The particle size is denoted by x, as before. The volume fraction occupied by particles having a size between x and $x + dx$ is $\varphi\psi_0(x)\,dx$, and the corresponding flux density $q'(x)$ is given by

$$q'(x) = U_p'\, \varphi\psi_0(x)\, dx \tag{74}$$

where U_p' is a function of x and of position across the flow stream.

$$U_p' = U_d - U_\varphi + \int_0^\varphi U_\varphi\, d\varphi \tag{75}$$

The steady flux $q(x)$ of particles of size x consists of upstream, $q_u(x)$, and downstream, $q_d(x)$, fluxes.

$$q(x) = q_u(x) + q_d(x) \tag{76}$$

The upward and downward flux is given by the integral over the area A:

$$q_u(x) = \int_0^A \delta_+\, \psi_0(x)\, dx\, U_p' \cdot dA, \qquad \delta_+ = \frac{1}{2}\left(1 + \frac{U_p'}{|U_p'|}\right) \tag{77}$$

$$q_d(x) = \int_0^A \delta_-\, \psi_0(x)\, dx\, U_p' \cdot dA, \qquad \delta_- = \frac{1}{2}\left(1 - \frac{U_p'}{|U_p|}\right) \tag{78}$$

The expression $\psi_0(x)\,dx$ is a constant and can be taken outside the integrals of Equations (77) and (78). If $U_p' > 0$ is associated with dA, then $\delta_+ = 1$ and $\delta_- = 0$; hence, for the given size x, dA is characterized by $q_u(x) > 0$ and $q_d(x) = 0$. If $U_p' < 0$, then $\delta_+ = 0$ and $\delta_- = 1$, and $q_u(x) = 0$ and $q_d(x) > 0$ characterize dA, as required. Dividing Equations (77) and (78) by Equation (76), with upstream and downstream fractions being the finer (f) and coarser (c) fractions respectively, gives

$$g_f(x) = \frac{\int_0^A \delta_+ U_p' \cdot dA}{\int_0^A \delta U_p' \cdot dA} \tag{79}$$

$$g_c(x) = \frac{\int_0^A \delta_- U_p' \cdot dA}{\int_0^A \delta U_p' \cdot dA} = 1 - g_f(x) \tag{80}$$

where $\delta = \delta_+ - \delta_-$.

The respective cumulative grade efficiencies are given by

$$G_f(x_0) = \frac{\int_0^{x_0}\int_0^A \delta_+ U_p' \cdot dA\, \psi_0(x)\, dx}{\int_0^{x_0}\int_0^A \delta U_p' \cdot dA\, \psi_0(x)\, dx} \tag{81}$$

$$G_c(x_0) = \frac{\int_{x_0}^{\infty}\int_0^A \delta_- U_p' \cdot dA\, \psi_0(x)\, dx}{\int_{x_0}^{\infty}\int_0^A \delta U_p' \cdot dA\, \psi_0(x)\, dx} \tag{82}$$

where $G_f(x_0) = G_f(x \le x_0)$ and $G_c(x_0) = G_c(x > x_0)$.

The total upstream (Q_u) and downstream (Q_d) particle fluxes are obtained by integrating Equations (77) and (78) over the complete size range.

$$Q_u = \int_0^\infty \int_0^A \delta_+ U_p' \cdot dA\, \psi_0(x)\, dx \tag{83}$$

$$Q_d = \int_0^\infty \int_0^A \delta_- U_p' \cdot dA\, \psi_0(x)\, dx \tag{84}$$

Dividing Equation (77) by Equation (83) and Equation (78) by Equation (84) gives $d\Psi_f(x)$ and $d\Psi_c(x)$; division by dx yields $\psi_f(x)$ and $\psi_c(x)$, respectively.

$$\psi_{\mathrm{f}}(x)=\frac{\psi_0(x)\left(\int_0^A \delta_+ U_{\mathrm{p}}'\cdot \mathrm{d}A\right)}{\int_0^\infty\int_0^A \delta_+ U_{\mathrm{p}}'\cdot \mathrm{d}A\,\psi_0(x)\,\mathrm{d}x} \quad (85)$$

$$\psi_{\mathrm{c}}(x)=\frac{\psi_0(x)\left(\int_0^A \delta_- U_{\mathrm{p}}'\cdot \mathrm{d}A\right)}{\int_0^\infty\int_0^A \delta_- U_{\mathrm{p}}'\cdot \mathrm{d}A\,\psi_0(x)\,\mathrm{d}x} \quad (86)$$

Exampe of Evaluation of Grade Efficiencies. Let the size distribution of an elutriator feed and the expected cut size be $\psi_0(x)$ and x_0, respectively. The elutriation is performed with a uniform dilute dispersion in a vertical circular tube for which Equations (68)–(73) apply. The average elutriation flow velocity U_{da} is set so as to satisfy Equation (87):

$$U_{\mathrm{da}}=kU_{\mathrm{p}}(x_0)=k[U_\varphi(x_0)-U_{\mathrm{f}}],\; 0<k<\infty \quad (87)$$

where k is an adjustable constant. For example, if particles of size x_0 are to have a settling velocity equal to U_{da}, then $k=1$. However, if particles ten times larger than x_0 must have this velocity, then $k=\sqrt{10}$. In a dilute dispersion and under the assumption that $U_{\mathrm{f}}=\int_0^\varphi U_\varphi d_\varphi \ll U_\varphi(x_0)$, the approximation $U_{\mathrm{da}}=kU_\varphi(x_0)$ is justified. Combining Equations (68), (70), (75), and (87) (where U_{f} is neglected) gives

$$U_{\mathrm{p}}'=2kU_\varphi(x_0)(1-(r/R)^2)-U_\varphi(x) \quad (88)$$

By means of this equation, r' is now defined so that the following two ranges are formed: $r<r'$ where $U_{\mathrm{p}}'>0$, $\delta_+=1$, $\delta_-=0$; and $r>r'$ where $U_{\mathrm{p}}'<0$, $\delta_+=0$, $\delta_-=1$.

$$r'=R\left[1-\frac{1}{2k}U_\varphi(x)/U_\varphi(x_0)\right]^{1/2} \quad (89)$$

Substitution of Equation (88) in Equation (79), with $\mathrm{d}A=2\pi r\mathrm{d}r$ gives

$$g_{\mathrm{f}}(x)=\frac{\delta_+\int_0^{r'}\{2kU_\varphi(x_0)[1-(r/R)^2]-U_\varphi(x)\}\,2\pi r\mathrm{d}r}{\delta_+\int_0^{r'}\{2kU_\varphi(x_0)[1-(r/R)^2]-U_\varphi(x)\}\,2\pi r\mathrm{d}r-\delta_-\int_{r'}^{R}\{2kU_\varphi(x_0)[1-(r/R)^2]-U_\varphi(x)\}\,2\pi r\,\mathrm{d}r}$$

Integration over r and rearrangement give

$$g_{\mathrm{f}}(x)=\frac{\delta_+[2kU_\varphi(x_0)-U_\varphi(x)]\pi r'^2-\pi kU_\varphi(x_0)r'^4/R^2}{\delta_+\{[2kU_\varphi(x_0)-U_\varphi(x)]\pi r'^2-\pi kU_\varphi(x_0)r'^4/R^2\}-\delta_-\{[2kU_\varphi(x_0)-U_\varphi(x)]\pi(R^2-r'^2)-\pi kU_\varphi(x_0)\,(R^4-r'^4)/R^2\}} \quad (90)$$

Consider a simple case where $\psi_0(x)$ exists only if x satisfies creeping-flow sedimentation (i.e., Eq. 8 applies) otherwise $\psi_0(x)$ vanishes. Applying Equation (8) for $\varphi=0$, $f(\varphi)=1$ [hence, $U_\varphi(x)/U_\varphi(x_0)=(x/x_0)^2$] and rearranging yield

$$g_{\mathrm{f}}(x)=\frac{\delta_+[2k-(x/x_0)^2-k(r'/R)^2]}{\delta_+[2k-(x/x_0)^2-k(r'/R)^2]-\delta_-[(2k-(x/x_0)^2)((r'/R)^2-1)-k((r'/R)^{-2}-(r'/R)^2)]} \quad (91)$$

$$g_{\mathrm{c}}(x)=1-g_{\mathrm{f}}(x),\quad r'=R\left[1-\frac{1}{2k}\left(\frac{x}{x_0}\right)^2\right]^{1/2}$$

Equation (91) shows that $g_{\mathrm{f}}(x)$ and $g_{\mathrm{c}}(x)$ are functions of x/x_0 and of k. If $x/x_0<\sqrt{2k}$, then $g_{\mathrm{f}}(x)>0$ and $g_{\mathrm{c}}(x)<1$.

However, if $x/x_0\geq\sqrt{2k}$, then $U_{\mathrm{p}}'\leq 0$, $\delta_+=0$, $\delta_-=1$, and $g_{\mathrm{f}}(x)=0$, $g_{\mathrm{c}}(x)=1$, as expected. Thus the size range that can be split in two product fractions by elutriation is $0\leq x\leq x_0\sqrt{2k}$. Outside this range, no splitting exists. Assume that $x<x_0\sqrt{2k}$; hence, in Equation (91), $\delta_+=\delta_-=1$, and $g_{\mathrm{f}}(x)$ can be reduced further to

$$g_{\mathrm{f}}(x)=\frac{\left[k-\frac{1}{2}\left(\frac{x}{x_0}\right)^2\right]^2}{\left[k-\frac{1}{2}\left(\frac{x}{x_0}\right)^2\right]^2+\frac{1}{4}\left(\frac{x}{x_0}\right)^4} \quad (92)$$

Figure 8 shows a plot of $g_{\mathrm{f}}(x)$ vs. $(x/x_0)k^{-1/2}$. By using this figure, the following typical index of sharpness is obtained: $x_{25}/x_{75}=1.29$, $x_{10}/x_{90}=1.76$. Figure 8 can be used in the following way. Suppose that the average particle size of a particle distribution is $x_0=30\times10^{-6}$ m. If $k=1$, then for the size fraction $x=x_0$, $(x/x_0)k^{-1/2}=1$; hence, from Figure 8, $g_{\mathrm{f}}(x_0)=0.5$. If $g_{\mathrm{f}}(x_0)\geq 0.9$, then from Figure 8, $k^{-1/2}\leq 0.7$ and $k\geq 2.041$. If particles twice as large as x_0 (i.e., $x/x_0=2$) are to satisfy $g(x)\geq 0.9$, then $0.7\,k^{1/2}\geq 2$, and $k\geq 8.163$. By using the following data, $\varrho_{\mathrm{p}}=2700$ kg/m^3, $\varrho_{\mathrm{f}}=1000$ kg/m^3, $\mu_{\mathrm{f}}=0.001$ kg m^{-1} s^{-1}, and $g=9.81$ m/s^2, then

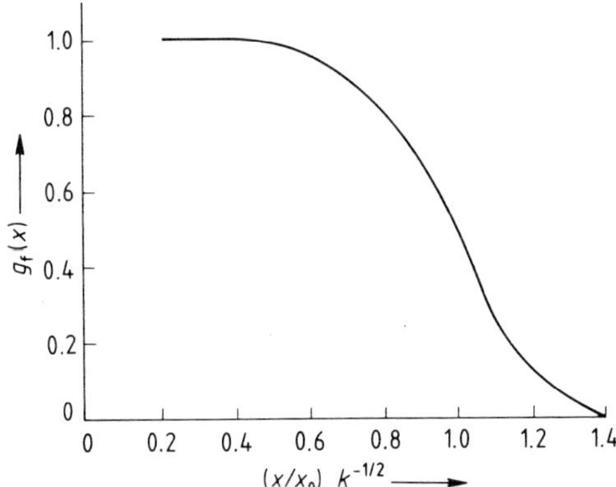

Figure 8. Plot of $g_f(x)$ vs. $(x/x_0)k^{-1/2}$ calculated from Equation (92)

$U_{da} = k U_\varphi(x_0) \geq 2.27 \times 10^{-4}$ m/s is obtained. In dilute dispersions, grade efficiencies are practically independent of particle-size distribution. This is not true for the size distribution of elutriation products (see Eqs. 85 and 86).

In a creeping-flow regime

$$r'^2 = R^2\left[1 - \frac{1}{2k}\left(\frac{x}{x_0}\right)^2\right],$$

thus using Equations (85) and (88), and canceling common terms in the numerator and denominator yields

$$\psi_f(x) = \frac{\psi_0(x) \int\limits_0^{r'} 2k x_0^2 [1 - (r'/R)^2 - x^2]\, 2\pi r\, dr}{\int\limits_0^{\sqrt{2k}\, x_0} \int\limits_0^{r'} 2k x_0^2 [1 - (r^2/R)^2 - x^2]\, 2\pi r\, dr\, \psi_0(x)\, dx} \quad (93)$$

where $x \leq \sqrt{2k}\, x_0$

Consider a simple case where $\psi_0(x)$ is uniform in a given size range $a < x < b$; hence, $\psi_0(x) = 1/(b - a)$. If, for example, $b = 40 \times 10^{-6}$ m and $a = 20 \times 10^{-6}$ m, then $\psi_0(x) = 10^6/20$. In this case, by integrating Equation (93) and performing some simple algebraic manipulations,

$$\psi_f(x) = \frac{\left[1 - \frac{1}{2k}\left(\frac{x}{x_0}\right)^2\right]^2}{(8/15)\sqrt{2k} \cdot x_0}, \quad 0 \leq x \leq \sqrt{2k} \cdot x_0 \quad (94)$$

Figure 9 shows a plot of $\psi_f(x)$ vs. x for $\sqrt{2k} \cdot x_0 = 1$, 2, 4, and 10. An increase in $\sqrt{2k} \cdot x_0$ increases the spread of $\psi_f(x)$ vs. x. Thus, to minimize the spread of $\psi_f(x)$ obtained from a uniform $\psi_0(x)$ in a circular-tube elutriator, $\sqrt{2k} \cdot x_0$ must be small, i.e., if the finer end of the distribution should be elutriated. By using a similar procedure, $\psi_c(x)$ can be evaluated.

Note that the tendency of particles to drift toward the elutriator center, due to velocity differentials across their diameter, is still not accounted for [14]. Thus, the results should be considered as applying to an idealized system in which this tendency is not important.

The effect of particle concentration on grade efficiencies and the expected size distributions in the elutriation products is not discussed in the above example; however, it can be deduced readily by following the same calculation procedures.

2.10.2. Cross-Flow System

Figure 10 shows schematically a two-dimensional steady cross-flow elutriation system. Here, a layer of fluid flows on a solid support set perpendicular to the force field. At position A, the

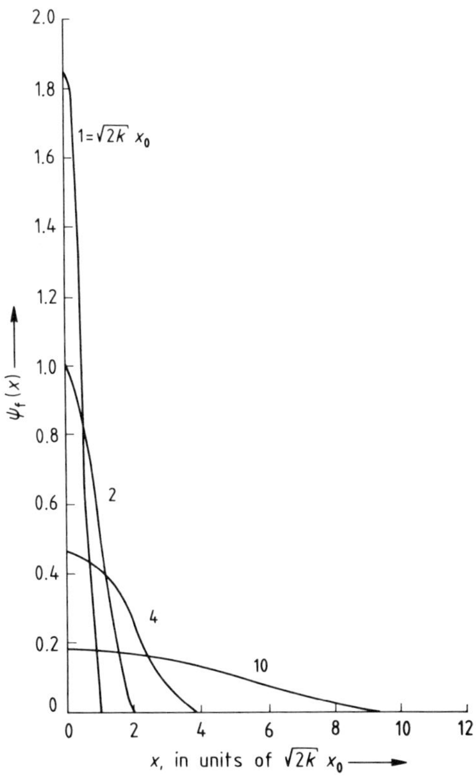

Figure 9. Plot of $\psi_f(x)$ vs. x calculated from Equation (94) for $\sqrt{2k} \cdot x_0 = 1$, 2, 4, and 10

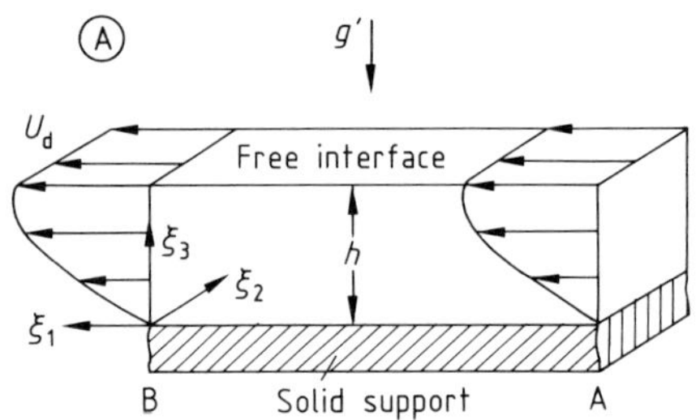

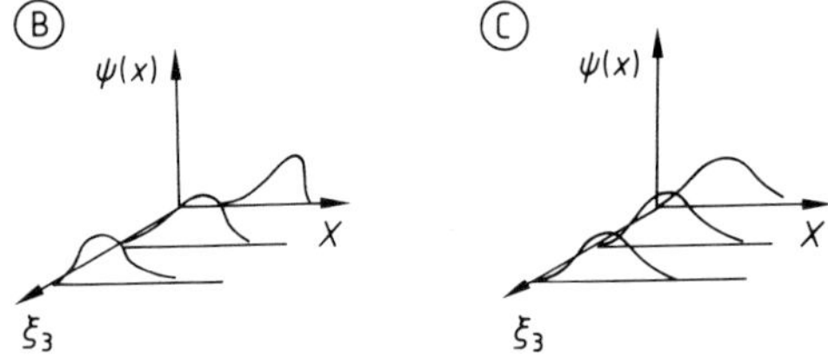

Figure 10. A) Cross-flow elutriation system
The dispersion flows in the form of a layer on a flat solid support. The force field producing accelerations g' is perpendicular to the flow. At position A, $\psi(x)$ is uniform across h; at position B, it is not.
B) Variable $\psi(x)$ and C) Uniform $\psi(x)$ at different position ξ_3

dispersion is uniform and $\psi(x)$ is independent of the vertical dimension ξ_3; at position B, $\psi(x) = \psi(x, \xi_3)$ due to differential sedimentation of particles toward the solid wall. In this system, particles of different sizes are partially segregated because of different sedimentation velocities. This effect is intensified by internal elutriation, in which displaced fluid sweeps slower particles against the driving force. Evaluation of $\psi(x, \xi_3)$ can be made by applying a nonsteady sedimentation model [2] for polydisperse particle mixtures. At a given position, the system shown in Figure 10 appears steady to an observer at rest relative to the solid support, i.e., to the $\xi_i (i = 1, 2, 3)$ frame of reference. However, if the observer moves with the fluid, then he observes a nonsteady system characterized by a time-dependent particle-size distibution which is equivalent to being position-dependent in the ξ_i frame of reference. Let the stream at position B be divided at $\xi_3 = c$ into fine (f) and coarse (c) substreams. Grade efficiency and size distribution for the fine stream are given by Equations (95) and (96), respectively.

$$g_f(x) = \frac{\int_0^c (U_d \cdot dA)\, \psi(x, \xi_3)\, dx}{\int_0^c (U_d \cdot dA)\, \psi(x, \xi_3)\, dx + \int_c^h (U_d \cdot dA)\, \psi(x, \xi_3)\, dx} \tag{95}$$

$$\psi_f(x) = \frac{\int_0^c \psi(x, \xi_3)\, U_d \cdot dA}{\int_0^\infty \int_0^c \psi(x, \xi_3)\, (U_d \cdot dA)\, dx} \tag{96}$$

$$U_d = \frac{3}{2} U_{da} \left[1 - \left(\frac{\xi_3}{h}\right)^2\right], \quad U_{da} = \frac{2}{3} U_{dm}$$

Consider thc simple case of a settling pool, shown in Figure 11. The steady fluid flow and settling velocity of particles are assumed to be uniform. The time required for a particle to settle to the bottom of the pool, i.e., from the surface to depth h, is

$$t_h = h/U_0 \tag{97}$$

while the time required to travel the pool length L (i.e., to be elutriated) horizontally is

$$t_L = L/U_{da} \tag{98}$$

Let Q be the volumetric flow rate of the suspension through the pool. Hence

$$U_{da} = Q/(B\,h) \tag{99}$$

where B is the pool dimension perpendicular to h and L. Setting $t_h = t_L$ means that the particle arrives simultaneously at the bottom and at the end

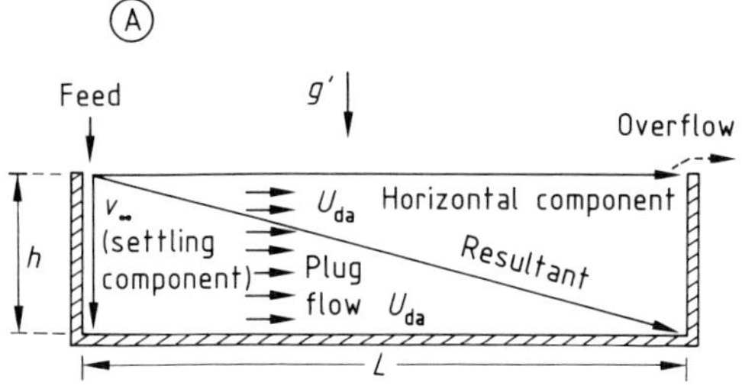

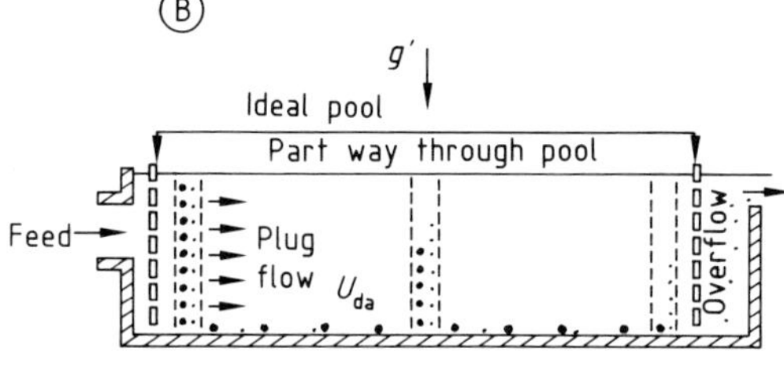

Figure 11. Cross-flow elutriator settling pools
A) Ideal cross-flow and settling velocities; B) Feed containing polydisperse particle mixture

of the pool. For a particle that satisfies this condition, by Equations (97) and (99),

$$U_{\mathrm{p}} = \frac{Q}{BL} \tag{100}$$

Equation (100) shows that U_{p} is independent of the pool depth and is a function only of Q and the pool area. This is the "area principle" of cross-flow elutriation settling pools, which remains valid also for nonuniform flows across the pool such as the one depicted in Figure 10 and described by Equation (96).

If the feed consists of a polydisperse particle mixture as shown in Figure 11, then settling produces a spectrum of particle sediment along the flow path. Furthermore, the pool flow concept in conjunction with dispersion effects [27]–[30] has been used to analyze the performance of classifiers. This is in accordance with the role of cross-flow separation in many modern elutriator classifiers and separators.

3. Elutriation Systems

3.1. Categories of Elutriation Systems

Elutriation systems can be classified according to the following categories of flow mechanism:

1) Steady counterflow
2) Steady transverse or cross flow
3) Nonsteady flow

Elutriators are also classified as vertical or horizontal and gas (e.g., air; →17. Air Classifying) or liquid (e.g., water) operated systems. A *single-stage elutriator* produces coarse and fine products characterized by cumulative oversize and undersize distribution functions. A *multistage elutriation system* can be used for grading into specific, relatively narrow size fractions. If, for example, steady counterflow (of fluid rising against settling particles) is used, then a battery of elutriators connected in series and increasing in diameter will produce progressively finer size fractions. Elutriation systems have been traditionally used for preparation of well-defined size fractions on both a laboratory and an industrial scale. Industrial elutriators are generally used as classifiers, laboratory devices for particle-size analysis and sorting. We shall also use the term "classifiers", along with a clear reference to the elutriation mechanism, when it applies.

3.2. Counterflow Gravity Elutriators

Counterflow elutriators are used for powders in the size range 10–200 μm, with densities greater than 2000 kg/m^3. This category includes the Schöne, Andrews, and Blythe elutriators. The Schöne [14], [31]–[33] and Andrews [14], [34] elutriators are early models of laboratory-size equipment [14], [31]. The Blythe elutriator [35] (Fig. 12) is designed to provide narrow size fractions. Each elutriating chamber, along with its exit tube and sump, is a link in a siphon chain. On-line sampling of each sump facilitates detection of the progress of elutriation. Increasing the diameter of consecutive chambers by a factor of 2 produces size grades that decrease by a factor of $\sqrt{2}$. The Blythe elutriator can be used to fractionate particles as small as 2–3 μm, a process that may take several days. Fractionation of particles as fine as a few micrometers in size requires elimination of convection currents that may interfere with the elutriation process. Other elutriators have been described in [14] and [36]. In the following material, several industrial devices are discussed.

Figure 13 shows the *Dorr–Oliver Monosizer* [37], [38] designed for sharp splitting of a feed

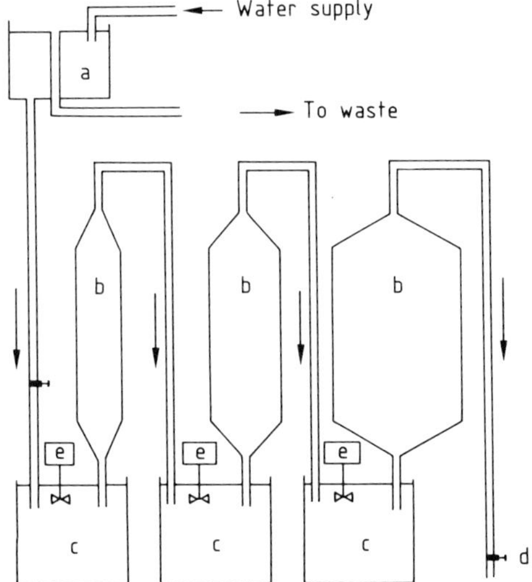

Figure 12. Three-stage Blythe elutriator
a) Constant head apparatus; b) Elutriation chamber; c) Sump; d) Screw clip; e) Stirrer

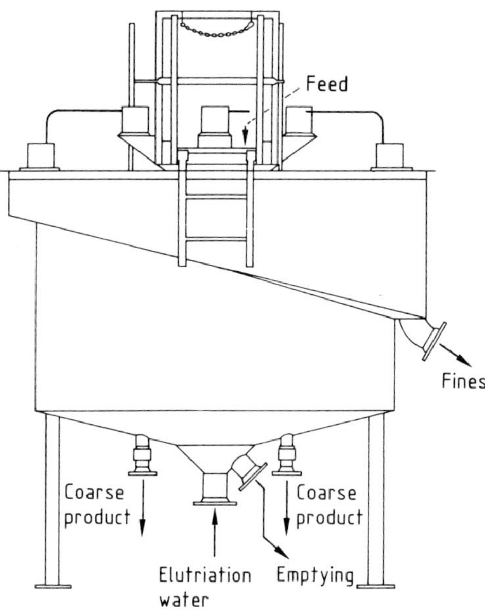

Figure 13. The Dorr–Oliver Monosizer

into two fractions, for cleaning fine sand, of organic contaminants, e.g. coal, (sorting) and generally for desliming. The feed concentration should be about a 1:1 ratio of sand to water. Coarser particles settle to the bottom and form a uniform fluidized layer over the distribution pipes (teeter bed). This layer, which acts as a filter, enhances the sharpness of classification. Coarse product is discharged at the bottom, and fine product at the overflow launder. Capacities range up to 100 t/h for tank diameters up to 3 m.

The *Deister Superscalper* [39], which operates like the Monosizer, is shown in Figure 14. It can handle 35 to 50 t/h of coal feed < 25.4 mm, < 12.7 mm, or < 6.35 mm in diameter, containing 40 % or more rejects. The overflow, stripped of the coarser size fractions and rejects, can normally be processed further by cyclones and shaking tables.

Figure 15 shows the *Krebs C–H Whirlsizer* [40]–[42], a counterflow elutriator incorporating a swirl that is superimposed on the rising water stream. The swirl action, in conjunction with internal design, obviates the need for the teeter bed sizing used in the Monosizer. The Whirlsizer is stable under variable feed conditions, and features compactness and high separation rate per unit sizer area. The separation size range is from 37 to 2000 µm. Capacities per unit are ca. 100 t/h, but units up to 500 t/h have also been designed.

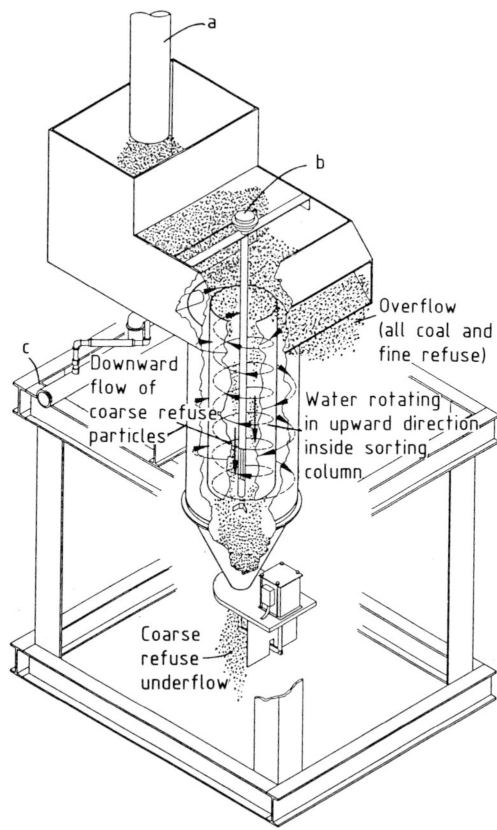

Figure 14. The Deister Superscalper
a) Feed inlet, b) Automatic level-sensing device; c) Water inlet

Typical performance data are shown in Figure 15 B [41]. The curves are demarcated by size, above which recovery to the underflow is practically complete. The steep curves indicate a high sharpness index.

Figure 16 shows the *Humphreys HydroSpec* [43]. This elutriator–classifier is a commercial version of the Lewis hydrosizer [44]. The unit (Fig. 16 A) consists of an elongated box whose height is approximately three to four times greater than the base width *W*. Slurry is introduced via a feed inlet at the bottom of an inclined baffle (a) which generates a higher "density zone" (b) of circulating oversize particles. Coarse solids discharge over the top lip of the baffle into the settling chamber as they are displaced by new feed. Fines are mixed thoroughly in the density zone and pass uniformly across the interface of the density zone into the upper compartment where elutriation in a dilute slurry oc-

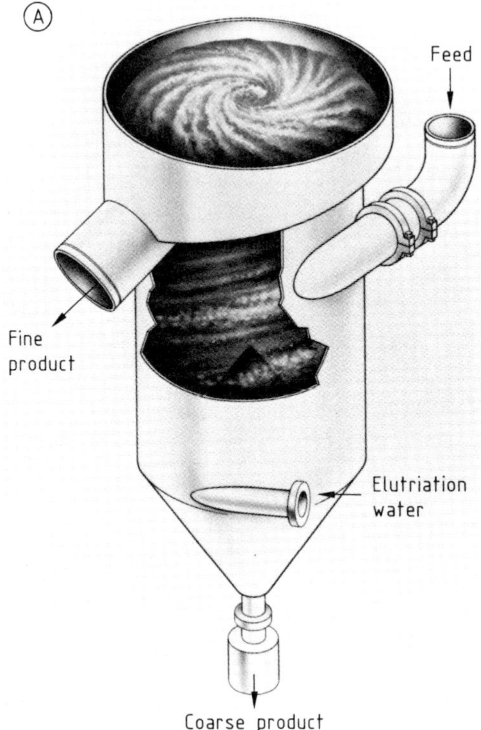

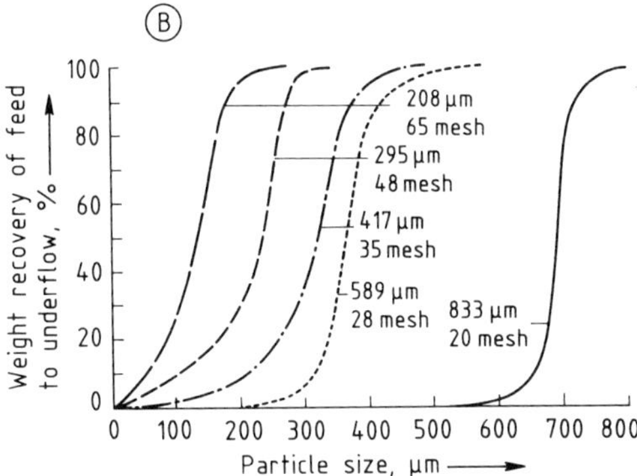

Figure 15. The Krebs C-H Whirlsizer
A) Principle of operation; B) Performance data

curs. Figure 16 B shows typical performance curves that indicate the superiority of the HydroSpec over a typical screw classifier (chart 1) and the near screen efficiency of the HydroSpec for 297-µm (50-mesh) undersize product (chart 2). The HydroSpec is designed to replace screens for sizing particles finer than 1.41 mm. The unit provides two-stage washing of coarse fractions that are discharged at solid contents up to 75 %. By using this unit, circulating loads of rod mill discharge material (in closed-circuit grinding) were reduced from 400 to 100 % (Fig. 16 B, chart 2). A typical unit (1.22 m × 1.22 m × 3.05 m) can classify approximately 40 t/h of sand with a diameter of ≤ 420 µm.

Details of a *single-cone elutriator* that incorporates top feed and wash water, as well as an internal cone surface for material distribution against rising water currents [27], [45], are shown in Figure 17. The feed descends through a central annular pipe (which encloses the down-coming wash-water pipe) and spreads on the cone surface, with coarse particles descending against the rising current that sweeps the fines toward their exit pipe.

Figure 18 shows a *multicone zigzag elutriator* [45]. The alternate cone surfaces and the enclosing guide walls form a three-dimensional zigzag pattern. Top feed and bottom wash water produce results similar to the single-cone elutriator.

The *Dorr–Oliver Hydrosizer* [46] and the *Deister Constriction Plate Classifier* [47] (Fig. 19) are two examples of *multifraction hydrosizers.* Water flows across a row of compartments characterized by diminishing velocity of counterflowing water. The Deister elutriation classifier consists of a row of square cells. Each cell contains a pressure chamber at the bottom, a sorting column separated from the pressure chamber by a perforated constriction plate, and a launder which flares out above the sorting column. Secondary classification occurs at the vortex fitting through which coarse fractions leave the column via a spigot pipe.

A high ratio of solids to water feed, with solids content not lower than 30 % and preferably 50–60 %, is recommended. The constriction plates enhance formation of a teeter bed of coarse particles. This produces a desliming effect similar to the one obtained by the Dorr–Oliver Monosizer described above. The processing capacity for average sand (< 6.35 mm in diameter) is generally rated at 2–6 t/h per cell. The number of cells (regularly 2 to 14) can be varied as required. Generally, makeup water does not exceed 4000 L per ton of feed.

Figure 20 shows the *Dorr–Oliver Bowl Desiltor* [48]. The elutriation bowl is connected to a rake classifier (c) that removes the dewatered coarse fraction. Fines are elutriated by water flowing upward and into the peripheral overflow launder. Settled grit is moved by rakes from the bowl into an inclined channel and then exits

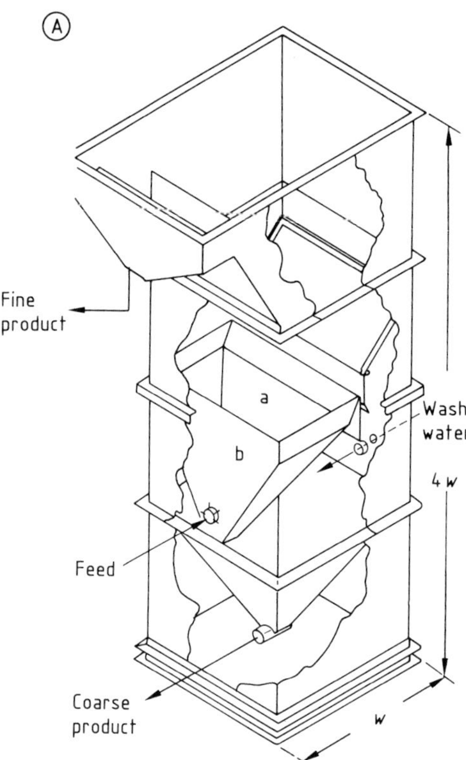

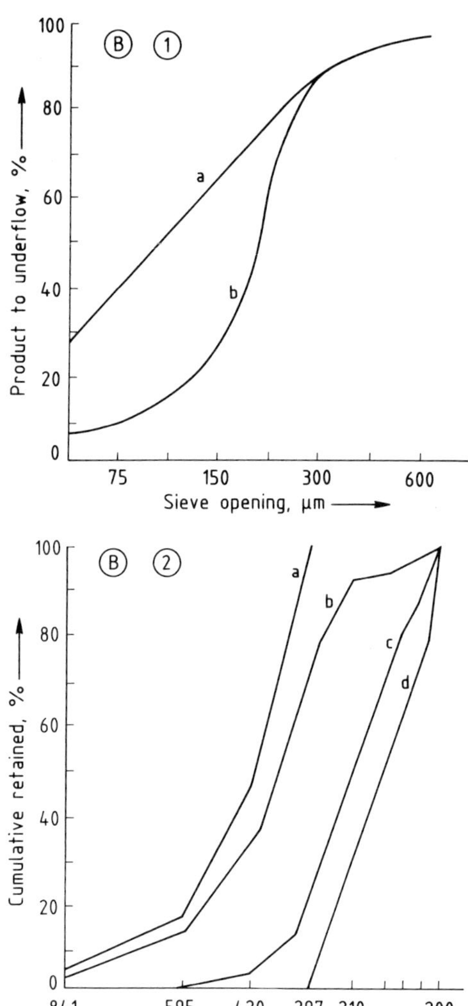

Figure 16. The Humphreys HydroSpec
A) Construction details of one unit: a) Inclined baffle; b) Density zone
B) Performance and comparative data:
1) a) Typical screw classifier, 400% circulating load; b) HydroSpec, 100% circulating load;
2) a) Screen oversize product (> 297 µm); b) HydroSpec, Oversize product; c) HydroSpec undersize product; d) Screen undersize product (< 297 µm)

above the water level from its upper end. Feed concentrations up to 65% solids in the size range < 12 mm, at capacities ranging from 5 to 120 t/h, can be handled. Normal solids content in the overflow is 1–15%, with the underflow generally being 70–85%. The Bowl Desiltor is used mainly in the recovery of relatively fine sand that overflows washing units in the steel (blast furnace scrubber water), metallurgical, concrete, foundry, glass, and limestone-processing industries. It also serves as a scalper used ahead of thickeners and clarifiers, without the need for a higher water head or the pumping capacities associated with other systems such as cyclones. The typical overall power required for the rake drive motors ranges from 2.6 to 5.2 kW.

3.3. Centrifugal Elutriators

In centrifugal elutriators, gravity is replaced by centrifugal force which enhances sedimentation of fine particles, thereby increasing flow capacity so as to satisfy process requirements. Because of design and technological constraints, the radial length of the sorting or elutriation chamber is relatively small; hence, actual classification takes place over a short (radial) distance.

The *hydrocyclone* is a well-known liquid-operated centrifugal elutriator. The conventional hydrocyclone is discussed in detail elsewhere (→11. Centrifugation and Hydrocyclone Separation). The hydrocyclone incorporates complex elutriation mechanisms which range from the

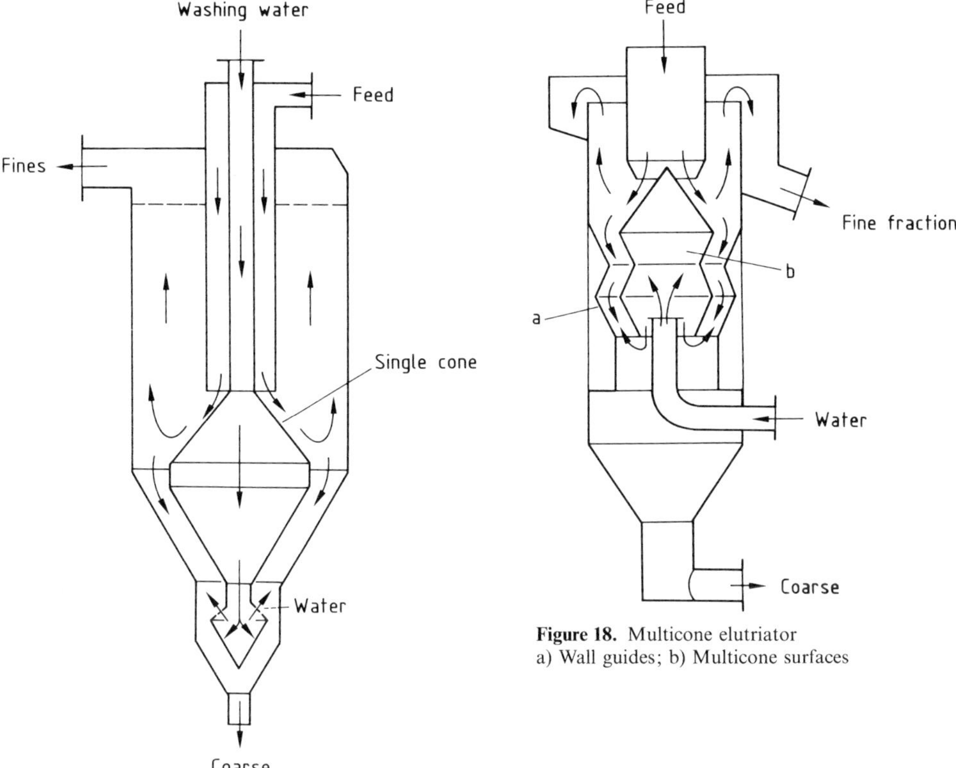

Figure 17. Single-cone elutriator [27], [42]

Figure 18. Multicone elutriator
a) Wall guides; b) Multicone surfaces

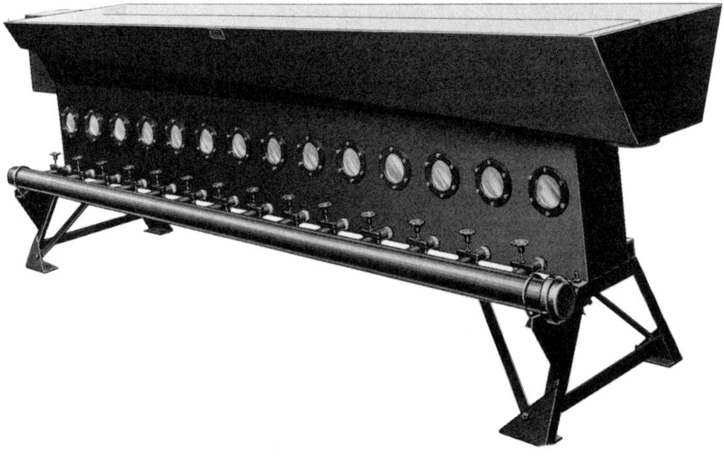

Figure 19. Multifraction hydrosizer: Deister Constriction Plate Classifier

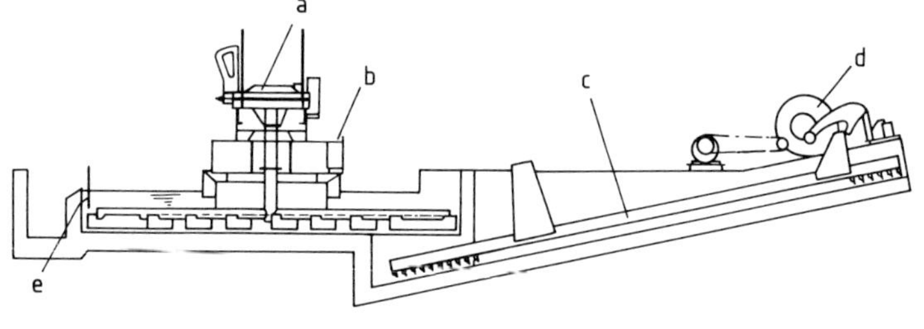

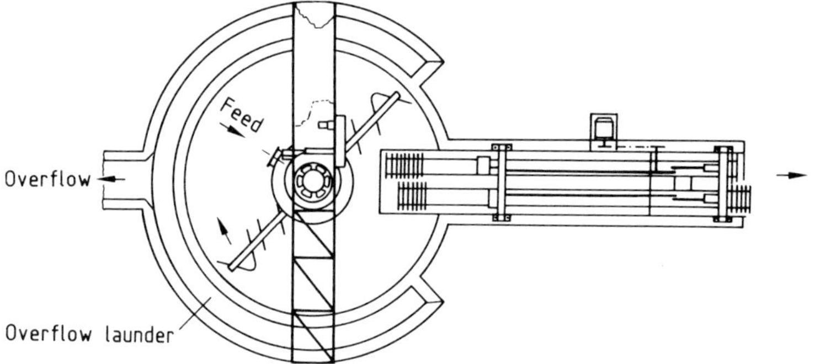

Figure 20. The Dorr Bowl Desiltor
a) Drive head; b) Overflow alarm; c) Rake classifier; d) Rake drive; e) Adjustable overflow weir

one described in Figures 10 and 11 to radial and axial countercurrent flow in streams that flow against the centrifugal and gravitational fields. The analysis of elutriation in hydrocyclones is outside the scope of this article. One drawback of hydrocyclones is that the short residence time of the particles may lead to incomplete sedimentation or elutriation. Alternatively, the internal circulation that characterizes novel air classifiers (→17. Air Classifying) does not exist in conventional cyclones. To overcome this problem, an *inverted cyclone*, shown in Figure 21 A [49], was developed. Here, particles that have reached the walls, either because of closeness to them or because of sedimentation in the outer vortex, climb up toward the closed cylindrical collection box and are then recirculated into the conical section. This circulation facilitates repeated elutriation and classification which enhance the release of trapped fine particles (in the stream or in aggregates) that finally exit through the vortex finder.

Figure 21 C shows a diagram of the *Warman Cyclosizer* [27], [31], [50]; this consists of five inverted hydrocyclones connected in series, which are fed by a pump, and a sample container. The cyclones decrease progressively in size, no. 1 being the largest and no. 5 the smallest. The coarse product of each cyclone can be collected by opening a discharge valve at its otherwise closed collection box. An example of five size fractions, obtained for size analysis of ground silica sand after standard 30-min elutriation, is shown in Figure 21 B. Limiting sizes of quartz particles with ϱ_p of 2700 kg/m^3 for the cyclones are 44, 35, 23, 15, and 9 µm for cyclones no. 1, 2, 3, 4, and 5, respectively [31], [50]. Changing of ϱ_p to ϱ_p' varies the limiting size by a factor of $[(\varrho_p - \varrho_\varphi)/(\varrho_p' - \varrho_\varphi)]^{1/2}$. The Warman Cyclosizer is a laboratory apparatus capable of handling 100 g of < 149-µm (100-mesh) or < 74-µm (200-mesh) material in 10–30 min, depending on the precision required.

Figure 22 A is a schematic diagram of a *counterflow liquid centrifugal elutriator* [31], [51]. An elutriation chamber (a) is engraved in the body of the rotor (b). The elutriation liquid and feed are introduced into this chamber via a concentric inlet (c) liquid channel (d), and suspension feed channels (e) and (f). Liquid counterflowing the centrifugal force drags undersize particles to the fines drain (g), whereas oversize particles are centrifuged toward the oversize drain. Figure 22 B shows performance data in which the cumulative weight percent of particles $\Psi(d_s)$, characterized

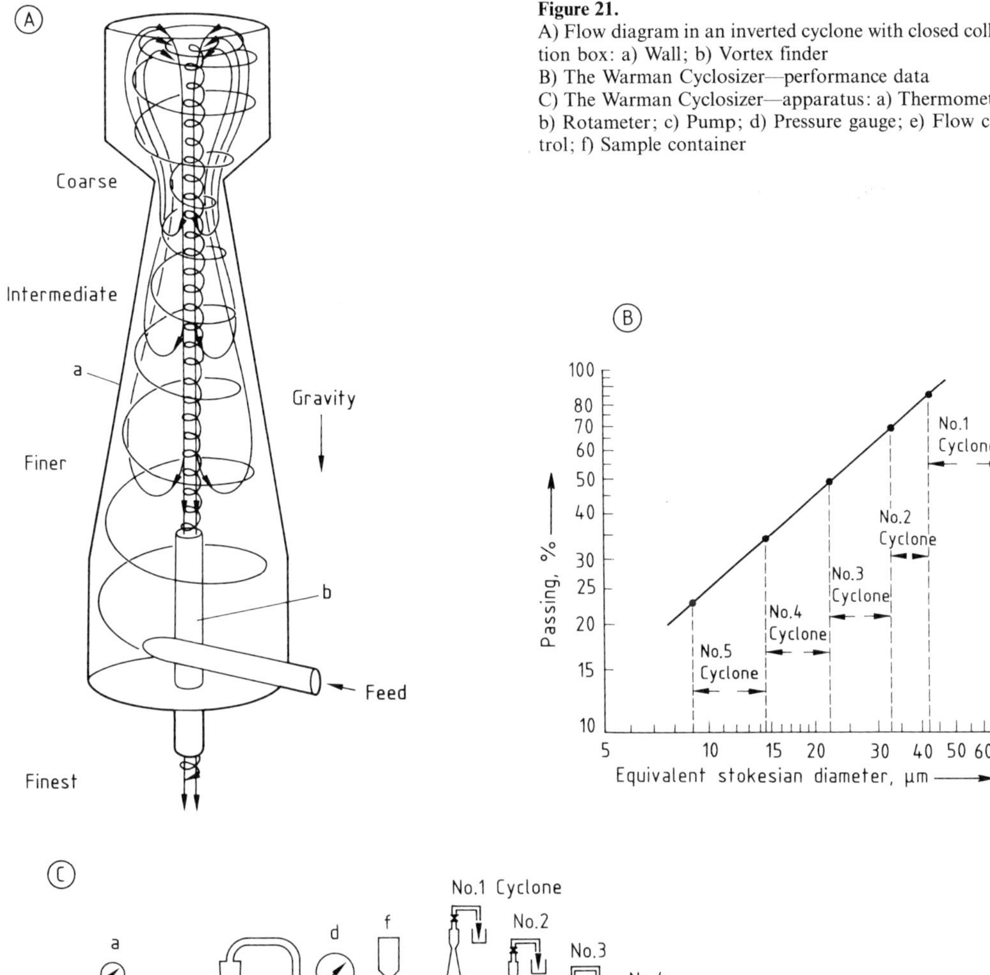

Figure 21.
A) Flow diagram in an inverted cyclone with closed collection box: a) Wall; b) Vortex finder
B) The Warman Cyclosizer—performance data
C) The Warman Cyclosizer—apparatus: a) Thermometer; b) Rotameter; c) Pump; d) Pressure gauge; e) Flow control; f) Sample container

by size $d \leq d_s$, has been plotted as a function of the Stokes diameter d_s for the oversize (a) and undersize (b) fractions. The undersize fraction is sharp, with 90% of particles being in the 1- to 3-µm range and the rest extending up to 5 µm. Loss of fines into the oversize fraction is probably related to the nonuniform flow profile of the fluid in the elutriation chamber.

An improved elutriation chamber is shown in Figure 23 A; the chamber walls are shaped so as to maintain a constant ratio of drag to centrifugal force. The grade efficiency $g_f(x)$ vs, particle size x (of glass spheres at 2.6-µm cut size) is shown in Figure 23 B and gives an index of separation sharpness $x_{25}/x_{75} = 0.61$.

3.4. Axial Cross-Flow Elutriators

Axial cross-flow elutriators are designed to deposit particles axially, i.e., along the flow axis

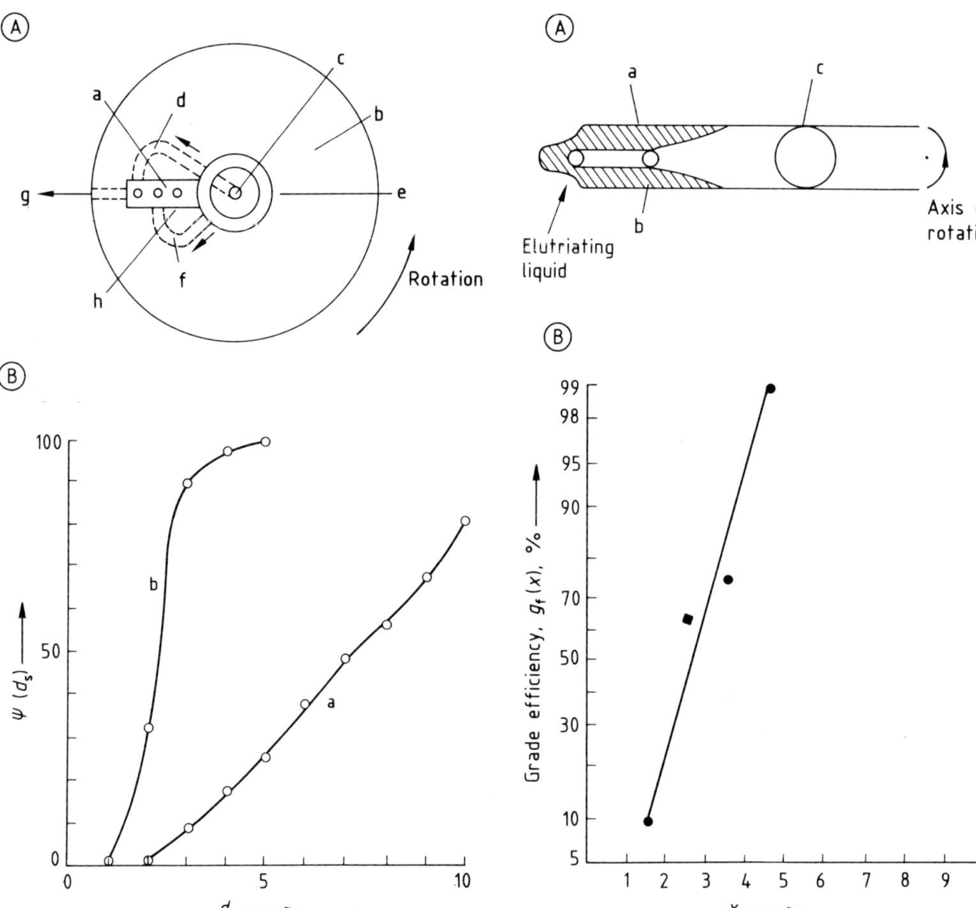

Figure 22. Counterflow liquid centrifugal elutriator
A) Schematic diagram: a) Elutriation chamber; b) Rotor; c) Elutriating liquid inlet; d) Liquid channel; e) f) Suspension feed channels; g) Oversize drain; h) Fines drain
B) Performance data $\psi(d_s)$ vs. d_s: a) Oversize fraction; b) Undersize fraction

Figure 23.
A) An improved elutriation chamber of the counterflow liquid centrifugal elutriator: a) Elutriation chamber with shaped walls; b) Suspension feed hole; c) Fines overflow
B) Grade efficiencies $g_f(x)$ vs. particle size x of glass spheres at 2.6 μm cut size.

of the main fluid stream. The theory of selected cross-flow systems is discussed in Section 2.10.2. Cross-flow elutriation occurs in systems having non-colinear components of flow and of the force field acting on the particles. A simple example of gravitational cross-flow elutriators is a settling pool, where particles are elutriated horizontally as they settle toward the bottom of the pool.

Axial cross-flow elutriation is not limited to separation according to size, it can also be used for fractionation of particles according to density and magnetic susceptibility. Figure 24 shows a *magnetogravimetric horizontal elutriator* [52]. In this system, the channel formed between two identical poles of an electromagnet is filled with magnetizable fluid. Particles elutriated axially experience a decreasing levitational force because of the axial tapering of the poles; thus, they settle at different points along the system axis. Figure 25 A shows typical results of separation tests with galena (ϱ_p 7500 kg/m^3) and copper (ϱ_p 8900 kg/m^3) at different levels of applied electromagnetic current. The axial distribution of these heavy particles is clear. For further details, see [52]–[54].

3.5. Electromagnetic Elutriators

Electromagnetic elutriators (generally referred to as magnetic and electrostatic separators, →19. Magnetic Separation; →20. Electrostatic Separation) are devices in which the force field is generated by an electromagnetic field.

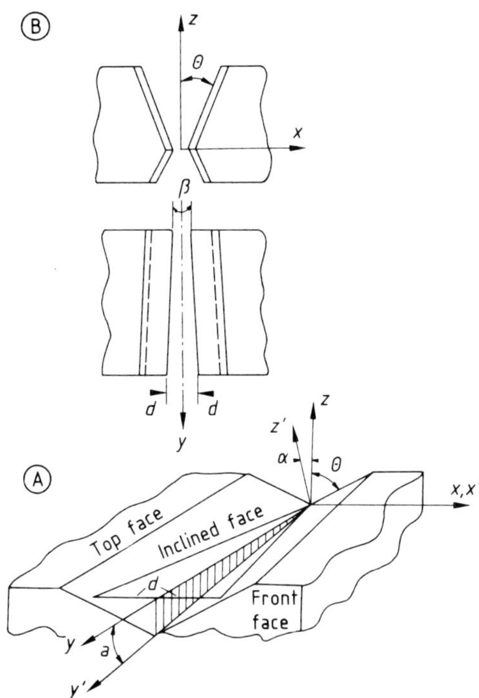

Figure 24. Magnetogravimetric horizontal elutriator
A) Wedge-type profiles, perspective view; B) Axially tapering wedge-type profile, side and top view

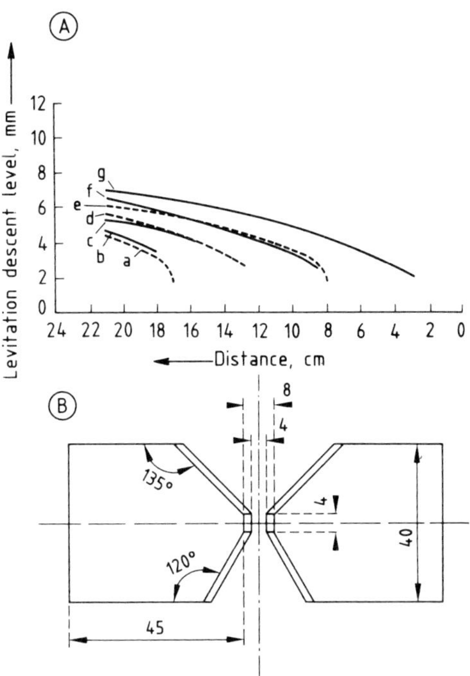

Figure 25.
A) Results of separation tests with galena (G ——) and copper (C·····) particles at different current levels; a) C (1.00 A); b) G (0.90 A); c) G (1.00 A); d) C (1.25 A); e) C (1.50 A); f) G (1.25 A); g) G (1.50 A)
B) Details of pole profiles used for the separation

In these devices, a strong diverging field is generated around matrix elements with two- or three-dimensional geometry (e.g., rods or spheres), which may be energized directly or indirectly (i.e., by a uniform field). Particles elutriated by a fluid stream passing in the vicinity of these elements [55] will experience a force which may or may not result in their removal from the stream, depending on flow, electromagnetic force, and residence time. This subject is outside the scope of this article.

4. References

[1] E. Barnea, J. Mizrahi, *Chem. Eng. J. (Lausanne)* **5** (1973) 171–189.

[2] Y. Zimmels: "Accelerated and Steady Particle Flows in Newtonian Fluids," *Encyclopedia of Fluid Mechanics*, Gulf Pub., Houston, 1986, pp. 93–153.

[3] Y. Zimmels, *J. Appl. Phys.* **56** (1984) no. 7, 2133–2141.

[4] Y. Zimmels, *AIChE J.* **29** (1983) no. 4, 669–676.

[5] E. L. Cussler: *Diffusion, Mass Transfer in Fluid Systems*, Cambridge University Press, Cambridge 1984.

[6] R. E. Rosensweig: *Ferrohydrodynamics*, Cambridge University Press, Cambridge 1985.

[7] R. H. Davis, A. Acrivos, *Ann. Rev. Fluid Mech.* **17** (1985) 91–118.

[8] J. F. Richardson, W. N. Zaki, *Trans. Inst. Chem. Eng.* **32** (1954) 35–53.

[9] J. Garside, M. R. Al-Dibouni, *Ind. Eng. Chem. Process Des. Dev.* **16** (1977) 206–214.

[10] H. Hald: *Statistical Theory with Engineering Applications*, Wiley International Edition, 1952.

[11] P. L. Meyer: *Introductory Probability and Statistical Applications*, Addison Wesley, 2nd ed., Reading, Massachusettes 1970.

[12] J. R. Benjamin, C. A. Cornell: *Probability, Statistics and Decision for Civil Engineers*, McGraw–Hill, New York 1970.

[13] J. K. Beddow: *Particulate Science and Technology*, Chemical Publishing Co., New York 1980.

[14] T. Allen: *Particle Size Measurement*, 3rd ed., Capman and Hall, London, 1981.

[15] B. W. Lindgren: *Statistical Theory, Macmillan Publ. Co.*, New York 1960.

[16] R. P. Fraser, P. Eisenklam, *Trans. Inst. Chem. Eng.* **34** (1956) 304.

[17] A. O. Gates, *Trans. Amer. Inst. Mining Metall. Pet. Eng.* **52** (1915) 875–909.

[18] R. Schumann, *Trans. AIME, Tech. Pub.* (1940) 1189.

[19] A. M. Gaudin, T. P. Meloy, *Trans. Am. Inst. Min. Metall. Pet. Eng.* **223** (1962) 43–50.

[20] A. J. Lynch: *Mineral Crushing and Grinding Circuits*, Elsevier, Amsterdam 1977, p. 116.

[21] C. C. Harris, *Trans. SME* **241** (1968) 343–358.

[22] L. Svarovsky in L. Svarovsky (ed.): *Solid Liquid Separation*, "Characterization of Particles Suspended in Liquids," Butterworths, London 1977, pp. 5–29.
[23] C. C. Harris, *Trans. SME* **244** (1969) no. 6, 187–190.
[24] K. R. Gibson, *Powder Technol.* **18** (1977) no. 2, 169–170.
[25] L. Svarovsky in L. Svarovsky (ed.): *Efficiency of Separation of Particles from Fluids, in Solid Liquid Separation,* Butterworths, London 1977, p. 31–57.
[26] R. B. Bird, W. E. Stewart, E. N. Lightfoot: *Transport Phenomena*, J. Wiley & Sons, New York 1960.
[27] E. G. Kelly, D. J. Spottiswood: *Introduction to Mineral Processing*, J. Wiley & Sons, New York 1982.
[28] W. E. Dobbins, *Trans. ASCE* **109** (1944) 629–656.
[29] T. R. Camp, *Trans. ASCE* **111** (1946) paper 2285, 895–958.
[30] H. Schubert, T. Neesse, *10th Int. Miner. Process. Congr. London*, 1973, pp. 213–239. IMM (1974).
[31] B. H. Kaye: *Direct Characterization of Fine Particles*, Wiley-Interscience, J. Wiley & Sons, New York 1981.
[32] C. J. Stairmand: "Some Practical Aspects of Particle Size Analysis in Industry," *Symposium on Particle Size Analysis* (1947). *Suppl. Transact. Inst. Chem. Eng.*, **25** (1947).
[33] E. Schöne: *Über Schlammanalyse und einen neuen Schlammapparat*, Berlin 1867 (Ref. [2]).
[34] L. Andrews, *Proc. Inst. Eng. Inspection* **25** (1927–8) also *Min. Mag.* (1982), p. 301.
[35] H. N. Blythe, E. J. Pryor, A. Eldridge (1953), *Symp. Recent Developments in Mineral Dressing, Inst. Min. Metall.*, London, 23–5 Sept., 1952, p. 11.
[36] A. H. M. Andreasen, *Ber. Dtsch. Keram. Ges.* **11** (1930) 675.
[37] Dorr–Oliver news release no. 002/72, Sand Beneficiation, Dorr–Oliver information 13/78 d.
[38] W. Schwalbach: "The Processing and Decarbonization of Fine Sand," *Aufbereit. Tech.* **12** (1971) no. 3, 137–139.
[39] Deister Superscalper, The Deister Concentrator Company Inc., Bulletin no. 220-B, 3/82.
[40] P. Hedley, *Pit Quarry,* October 1984.
[41] Krebs Engineering, Leaflets and News Bulletins.
[42] M. Rukavina, *Rock Prod.,* Feb. 1987.
[43] Coors, Humphreys Processing Equipment, Bulletin no. 42 (10/86) LB-1638.
[44] R. M. Lewis: *Hydrosizing of Industrial Minerals*, a publication of Humphreys Mineral Industries Inc.
[45] *Ullmann*, 4th ed., **2,** p. 70.
[46] Dorr–Oliver Technologie Transfer T4, 8/81.
[47] The Deister Concentrator Company Inc., Bulletin no. 200-B, 4/78.
[48] Dorr–Oliver Europe publication, 05-IE-0884.
[49] D. F. Kelsall, J. C. H. McAdam: "Design and Operating Characteristics of Hydraulic Cyclone Elutriator," *Trans. Inst. Chem. Eng.* **41** (1963).
[50] Warman International Ltd., leaflet: Cyclosizer for Sub-Sieve Sizing.
[51] F. J. Colon, J. W. Van Heuven, H. M. Van der Laan: "Centrifugal Elutriation of Particles in Liquid Suspension," in *Proc. Conf. Fine Particle Characterization*, Bradford, England, 1970, British Society for Analytical Chemistry, London, 1971.
[52] Y. Zimmels, I. J. Lin, *IEEE Trans. Magn.* **18** (1982) no. 3, 921.
[53] Y. Zimmels, I. J. Lin, *IEEE Trans. Magn.* **19** (1983) no. 1, 27.
[54] Y. Zimmels, I. J. Lin, *Chem. Eng. Commun.* **21** (1983) 1–3.
[55] Y. Zimmels, *IEEE Trans. Magn.* **20** (1984) no. 4, 597.

17. Air Classifying

Roland Nied, Herbert Horlamus, Alpine Aktiengesellschaft, Augsburg, Federal Republic of Germany

In addition to the symbols defined in the front matter of this volume: (p. VIII) the following symbols are used:

c_W	drag coefficient
F_M	mass force
F_W	drag force
g	acceleration due to gravity
$\dot{m}$	mass flow rate
P	passage
r	radius
Re	Reynolds number
v	velocity
$\dot{V}$	volume flow rate
W	mass fraction
W_c	mass fraction of coarse material
W_f	mass fraction of fines
x	particle diameter
x_{f97}	reference cut size
x_t	cut size (general)
x_{50}	Tromp's cut size
η	dynamic viscosity
κ	sharpness of cut
μ	load
ξ	efficiency
ϱ	density
φ	coefficient

Indices

A	gas
c	coarse material
F	feed material
f	fines
S	solid
v_r	radial component of velocity
v_p	peripheral component of velocity

1. Introduction

Air classification is a process for the dry separation of a disperse phase according to particle size, particle shape or density, or more precisely, settling rate. It is the analogue of hydraulic classification in wet-processing technology (→ 16. Elutriation). The bulk material to be separated is introduced into a gas flow, and the particles are subjected primarily to two forces: (1) the resistance or drag force caused by the gas flowing around the particle, and (2) the mass force. The mass forces acting on the particle can be field forces (e.g., gravity) or inertial forces (e.g., centrifugal force). Depending on magnitude and direction of these forces, particles move along different trajectories and can thus be collected separately after leaving the classification zone.

Air classifiers are used partly as sorters but primarily as classifiers. In sorting, a heterogeneous bulk is separated into components of uniform material. The particles are separated according to density or particle shape. Examples of density sorting are the fractionation of household garbage, the separation of comminuted cable scraps into copper and plastic fractions, etc. Separation according to particle shape is possible for mica (plates)–quartz or mica–feldspar (cubic form) mixtures. In most cases, however, a homogeneous bulk is present, and the air separator operates as a classifier; i.e., it separates according to particle size. In this case, beginning with commercial or natural raw material, the following processes take place:

1) A product of the desired fineness is removed.
2) A relatively coarse-grained product is dedusted.
3) A small amount of oversize material is separated.
4) Fractions with a definite upper and lower limit are produced.
5) Particle size is analyzed.

Air classification is used predominantly in the fine and very fine range, i.e., for effective separation of particles from 300 µm down to 1µm. It dovetails with screen classification, which operates more economically on coarse particles. Although air classifiers can also be used to separate particles in the millimeter-size range, in principle, their use then is primarily for sorting, not for classification.

2. Forces on the Individual Particle

An exact calculation of the forces acting on particles, and thus a calculation of trajectories and, ultimately, of the cut size, is not possible in most cases. The flow prevailing in the separation zone of an air classifier is too unsteady and inhomogeneous. The effects of feed (concentration, particle-size distribution), friction, and secondary flows preclude accurate description of temporal and local velocity patterns in the total classification zone. Theoretical considerations must, therefore, be based on simplified model flows. Nevertheless, even with this approach, some characteristic dependences and properties of air classifiers can be derived.

The *drag force* F_W, i.e., the force that a flow exerts on a particle moving at a relative velocity v_{rel} with respect to it, is calculated according to Newton's law as

$$F_W = c_W \frac{\pi x^2}{4} \frac{\varrho_A}{2} \qquad (1)$$

where x is the particle diameter, ϱ_A is the density of the gas, and c_W is the drag coefficient.

The drag coefficient c_W is known to depend on the Reynolds number ($Re = v_{rel} \cdot x \cdot \varrho_A / \eta_A$, where η_A is the dynamic viscosity).

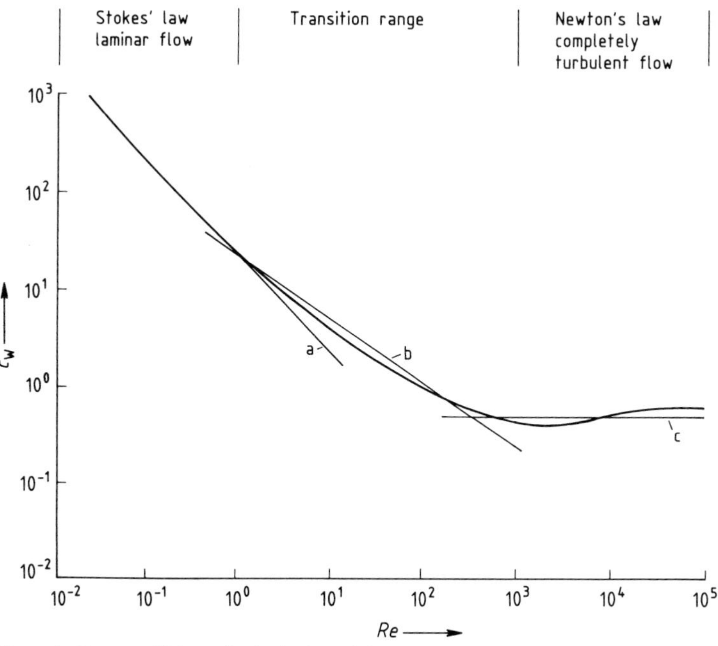

Figure 1. Drag coefficient of spherical particles
a) $c_W = 24/Re$; b) $c_W \approx 24/Re^{2/3}$; c) $c_w \approx 0.45$

Table 1. Reynolds numbers for different types of air classifiers

Classifier type	Pressure in classification space, kPa	v_{rel}, m/s	x_t, µm	*Re*
Rising-pipe analytical classifier	100	0.1	35	0.2
Classifier with high-speed rotor	80	3.0	2	0.4
Impeller classifier	90	6.5	10	0.4
Spiral classifier	95	6.5	25	10
Channel-wheel classifier	95	20	50	60
Recirculating-air classifier	100	10	100	60
Rising-pipe classifier	100	5.0	1000	300

In the particle technique, Reynolds numbers between 10^{-2} and 10^5 occur. A generally valid relationship $c_W = f(Re)$ cannot be given for this large range, and it is therefore subdivided into three parts that characterize the different types of flow around a body (Fig. 1). For practical calculations, which are frequently restricted to only part of a range, more or less accurate approximations are used.

Air classification takes place predominantly in the transition range. Only analytical classifications in the rising-pipe classifier (see Section 4.2.1) and very fine classifications with effective cut sizes x_t of a few micrometers advance into the Stokes range. Table 1 lists some Reynolds numbers for different classifiers under typical operating conditions.

The forces acting on a particle are considered first for the case of ideal flow in a rising pipe (Fig. 2). For the cut-size particle, which theoretically remains in the classification zone for an infinitely long time, the mass force F_M and drag force F_W are in equilibrium. The mass force F_M in this case is the force of gravity and is calculated as

$$F_M = \frac{\pi}{6} \varrho_S x^3 g \qquad (2)$$

where ϱ_S is the particle density and g is the acceleration due to gravity.

By combining Equations (1) and (2) with $v_{rel} = v$, the equilibrium particle size or cut size x_t can be derived:

$$x_t = \frac{6}{8} c_W \frac{\varrho_A}{\varrho_S} \frac{v^2}{g} \qquad (3)$$

For laminar flow around the particles, $c_W = 24/Re$ (Fig. 1). Equation (3) thus becomes

$$x_T = \sqrt{\frac{18 v \eta}{\varrho_S g}} \qquad (4)$$

For the transition range, according to the approximate equation in Figure 1: $c_W = 24/Re^{2/3}$,

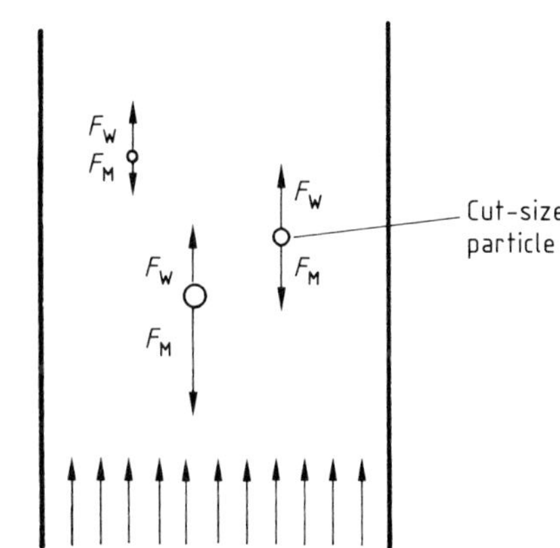

Figure 2. Forces on particles in the rising-pipe classifier

the effective cut size is then calculated as

$$x_t = 5.66\, g^{-0.6} \varrho_S^{-0.6} \varrho_A^{0.2} \eta^{0.4} v^{0.8} \qquad (5)$$

In the following material, the relationship of forces in an ideal spiral flow is investigated (Fig. 3). Spiral flow or at least similar forms of flow are currently found in the separation zones of most industrial air classifiers. Air flows along the trajectories indicated by broken lines; its velocity along the radius r is v, with the peripheral component v_p and the radial component v_r. For a particle of diameter x, if the mass force F_M and the drag force F_W are in equilibrium at the radius r, the particle will describe a circular path. Theoretically, this particle will remain in the classification zone for an infinitely long time. Larger particles are thrown outward, while smaller particles are carried along by the air to the central outlet.

In this case, F_M is the centrifugal force and has a value

$$F_M = \varrho_S \frac{x^3 \pi}{6} \frac{v_p^2}{r} \qquad (6)$$

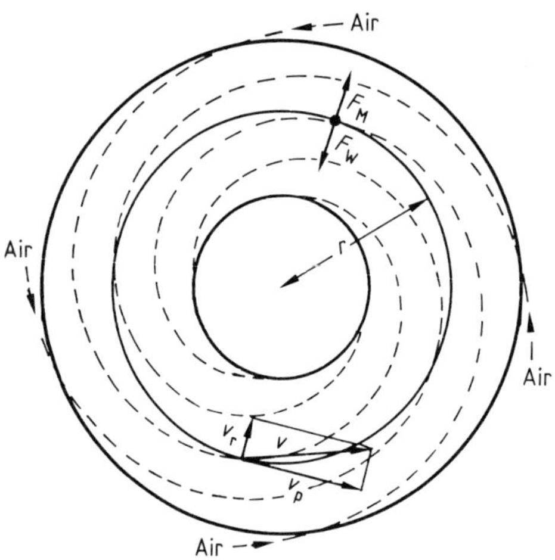

Figure 3. Equilibrium of forces on a particle in spiral flow

From Equations (1) and (6) and $v_{rel} = v_r$, the following value is obtained for the equilibrium particle size x_t:

$$x_t = \frac{6}{8} c_W \frac{\varrho_A}{\varrho_S} r \frac{v_r^2}{v_p^2} \tag{7}$$

Table 1 shows that calculation can be restricted primarily to the transition range; thus Equation (7) can be rearranged to

$$x_t = 5.66 \varrho_S^{-0.6} \varrho_A^{0.2} \eta^{0.4} r^{0.6} v_r^{0.8} v_p^{-1.2} \tag{8}$$

The cut size or the cut-size range to be expected can be estimated roughly by means of these equations when ϱ_S, ϱ_A, and η are constant or given. The cut point is normally adjusted by means of v or by the ratio v_r/v_p. However, the equilibrium radius r, the so-called separation radius, is also used to adjust some spiral classifiers. The separation radius depends on the geometry of the classification zone and, of course, on the absolute size of the classifier. (For a comprehensive description of cut size, density distribution curves, and grade efficiency, see → 2. Particle-Size Analysis and Characterization of a Classification Process)

3. Classifier Performance

The performance of a classifier is characterized by throughput, cut-point range, sharpness of cut, and energy requirements.

3.1. Throughput

The term throughput refers to the mass flow rate of the feed, fines, or coarse material, depending on which stream is considered the product. The capacity of a classifier is generally characterized by the quantity of fines that can be obtained. However, the quantity of fines is not a fixed value, because the output of fines depends on the particle-size distribution of the feed and on the set cut point. The finer the setting of the classifier, the lower its throughput will be. The throughput also has a decisive effect on the sharpness of cut (see Section 3.3). The throughput of both fines and feed material can vary greatly for an identical classifier setting, depending on the objective, e.g., clean coarse material (high sharpness of cut) or "milking" of fines (extraction of a small fraction of fines so that the distribution in the coarse material still corresponds roughly to the distribution in the feed material).

Throughput is dependent on the air flow rate and, thus, primarily on the size of the classifier. To eliminate the effect of size in a more general consideration, the load μ is used. This is defined as a ratio of flow rates: the mass flow rate of either the feed $\dot{m}_F$ or the fines $\dot{m}_f$ is related to the air flow rate $\dot{V}$:

$$\mu_F = \frac{\dot{m}_F}{\dot{V}} \quad \text{or} \quad \mu_f = \frac{\dot{m}_f}{\dot{V}} \tag{9}$$

Most classifiers operate in the range of μ_F from 0.1 (for the finest separations) to 1.0 kg/m³ (for coarser separations).

3.2. Cut-Point Range

The cut point can be varied within a smaller or larger range for all classifiers. The size and absolute location of this range characterize the performance: the largest range of adjustment and smallest possible cut point yield optimal efficiency.

Cut-point values must always be provided, along with the definition on which they were based (see → 2. Particle-Size Analysis and Characterization of a Classification Process). The specific gravity of the material used must also be indicated; the most common test material is limestone with a density ϱ of 2.65 kg/dm³. According to Equations (4), (5), and (8), products with a higher specific gravity produce finer effective cut sizes for the same classifier setting; those with a lower specific gravity result in more coarse effective cut sizes. The particle-size distribution of the feed material and the classifier size also affect the cut-point range (see Sections 5.1 and 5.2).

3.3. Sharpness of Cut

The different classifications of the sharpness of cut are discussed in detail in → 2. Particle-Size Analysis and Characterization of a Classification Process. The sharpness of cut depends primarily on the throughput or on the load μ. It usually remains constant up to a certain throughput and decreases continuously above this.

The properties of the feed material, especially the particle-size distribution and agglomeration tendency, are also important. If a product with a steep distribution curve is fed to the classifier, then for an identical load, a significantly larger number of particles of near-cut size will be in the classification zone than with a flatter distribution of the feed material. The interaction between particles, and thus the danger of misclassified particles, increases and the sharpness of cut decreases.

The agglomeration tendency depends partly on particle size (smaller particles have greater adhesion relative to their mass), and partly on material properties such as conductivity, surface roughness, water or oil content, etc. The stronger the adhesion between individual particles, the more difficult it is to carry out a complete dispersion. However, complete dispersion is a basic requirement for a classification with a sharp cut.

3.4. Energy Requirement

Agglomeration is counteracted by the disintegrating power of the classifier, which is based on high local air velocity as well as frequent and intense particle impact. For very fine classifications, the disintegrating power of the classifier is so high in some cases that larger particles of the material are considerably comminuted.

The high air velocities and the acceleration of particles to high impact velocities consume energy. The specific energy requirement, i.e., the energy neccessary to produce a defined quantity of end product with a specified fineness, is therefore an important parameter in comparing different classifiers. For a correct comparison, the following requirements must be met:

1) Both classifiers must receive identical feed materials.
2) Both classifiers must produce identical fines.
3) Both classifiers must give the same mass fraction of fines W_f.

In this case, the sharpness of cut will be identical.

The driving power of the fan required to produce classifying air is calculated on the basis of the pressure drop in the classifier or the system and on the quantity of air put through. In some cases, the driving power for moving classifier parts must also be considered. The total power, divided by the mass flow rate of the product, yields the specific energy requirement.

4. Description of Individual Classifier Types

4.1. General

Three standard types of air classifier, as well as a variety of combinations, are currently used.

Through-Air Classifier. Material is fed into the classifier with little or no additional air. The classifying air enters without any material and leaves it loaded with fines. It then passes through a dust separator into the open air. An advantage of this system is that the coarse material is very clean because the coarse fraction is rinsed with pure air. One disadvantage is that a relatively large, effective, and therefore expensive, dust separator (filter) must be used.

Closed-Circuit Classifier. Feed is added as already described. Classifying air returns to the classifier after separation of the dust. This allows low-cost blanketing with nitrogen, if required, and a simple dust separator, usually a cyclone, is sufficient. The cyclone can be located outside the classifier (external air circuit), or it can be structurally combined with the classifier and fan (internal circuit). Often, a small amount of air must be removed from the circuit and passed through a filter (leakage-air, caused by rising-air etc.). Dedusting as a whole is, nevertheless, quite inexpensive.

A disadvantage of this method is that the fine dust remaining in the classifying air contaminates the coarse material and can form a deposit in the classifier. Furthermore, distinct heating of the classifying air occurs over the course of time.

Airstream Classifier. The bulk material, suspended in the classifying air, is fed into the classifier. Airstream classifiers are thus used advantageously in grinding–classifying circuits. Material extracted from the mill is fed pneumatically to the classifier, which makes inexpensive plant design possible. A disadvantage is the low purity of the coarse material, which is rinsed only with air loaded with feed material.

Many types of air classifiers are available on the market: some are named according to the mode of operation; others, according to characteristic machine parts or equipment. Therefore, a breakdown into different categories is difficult. In addition, some units have two classification zones (a fines zone and a coarse-material zone) which operate differently, thereby making systematic treatment even more difficult. For example, some air classifiers are distinguished by type of air supply, in others distinction is made between centrifugal and gravity classifiers; finally, division into crosscurrent and countercurrent classifiers is also widespread.

In a *crosscurrent classifier*, the direction of movement of the coarse material (i.e., the direction of mass forces), is at a right angle to the main gas flow or the direction of the drag force. The crosscurrent classifier is also aptly called a fanning-out classifier because separation of a product into any desired number of fractions is possible. In a *countercurrent classifier,* the direction of mass force is counter to the main gas flow. This type of classifier, also called an *accumulating* or *equilibrium classifier,* yields only two fractions.

Accumulating classifiers generally result in sharper separation because the fines and coarse material leave the classification zone in opposite directions, whereas in a *fanning-out classifier,* even the smallest deviation from the limiting trajectory leads to an incorrect assignment of material. Additionally, in an accumulating classifier, cut-size particles remain in the classifier for a longer time so that any incorrect separation is likely to be corrected.

In recent years, a new concept in air classification has been developed: the *elbow classifier*. This terminology is used when the particles of the fines undergo a large change in direction (a deflection of ca. 180°) on their way through the classification zone.

The *recirculating-air classifier* is a type that does not fit well into the scheme developed above. This instrument represents a combination of several classification principles.

In this article, air classifiers are considered as follows:

- Countercurrent classifiers (Section 4.2)
  - Countercurrent gravity classifiers (Section 4.2.1)
  - Countercurrent centrifugal classifiers (Section 4.2.2)
- Crosscurrent classifiers (Section 4.3)
- Recirculating-air classifiers (Section 4.4)

All the different classifiers available on the market cannot possibly be described because the number of manufacturers is too large. The selection made shows the variety of types developed this far.

Capacity figures are largely omitted because, in many cases, they are not generally valid; i.e., they are not applicable to other products (see also Chaps. 3 and 5). In order to find the suitable classifier for a certain problem, tests must always be conducted with the original material.

Where data on the cut-point range are given, these refer to Tromp's cut point x_{50}, unless otherwise explicitly noted. Furthermore, these data always refer to a product with a density ϱ of 2.65 kg/dm³.

4.2. Countercurrent Classifiers

4.2.1. Countercurrent Gravity Classifiers

The *rising-pipe classifier* is the simplest form of countercurrent gravity classifier. Figure 4 is a schematic drawing of a model that is, in part, still

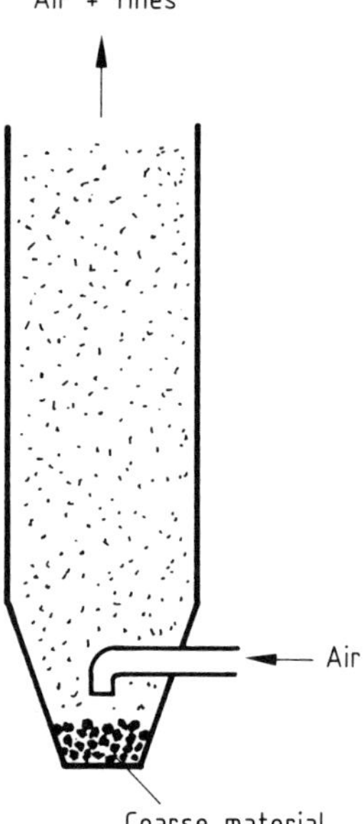

Figure 4. Diagram of a rising-pipe classifier for analytical purposes

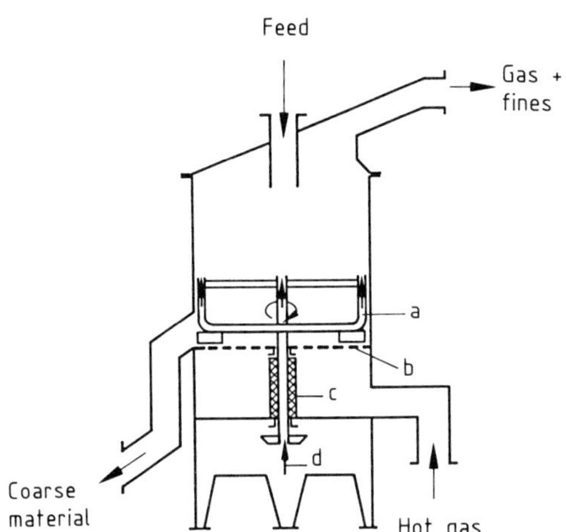

Figure 5. Combination rising-pipe classifier and suspension drier (Keller)
a) Stirring device; b) Screen plate; c) Insulation; d) Cooling air for bearing

used for particle-size analysis [1]. Air enters through a nozzle and stirs up the particles to be investigated. Separation into fines and coarse material takes place in the cylindrical pipe. All particles whose settling rate is lower than the upward air flow rate leave the pipe at its upper end, along with the air. The coarse material remains at the bottom of the pipe and can be removed at the end of the classification process.

An unequal velocity distribution over the cross section and deposition of material on the walls can impair classification. Material deposition can be largely prevented by the installation of knockers. The cut-point range used for analytical classification is located between ca. 10 and 100 µm. The sharpness of cut (κ) in this unit, which is operated in a batchwise manner, depends primarily on the time; for an appropriately long analysis period, κ values of 0.85–0.9 are achieved.

Figure 5 shows a continuously operating gravity classifier in which the drying of moist material can be carried out simultaneously [2]. The material is fed to the center of a screen plate. The hot gas flowing through the screen from below carries the fines or dried product with it toward the top; the coarse material is moved by means of a stirring device to the edge of the screen and discharged. This classifier is used for separating and drying of sawdust, wood chips, splintered wood, and pulpwood.

The *zigzag classifier* is a special variation of the gravity classifier (Fig. 6) [3]. A characteristic

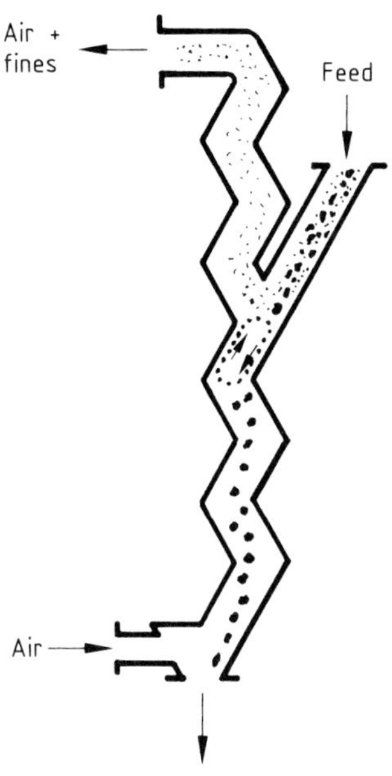

Figure 6. Zigzag classifier (Alpine)

form of movement develops in the zigzag-shaped channel, in which predominantly near cut-size particles form a vortex cylinder in each zigzag section. These particles therefore pass through several classification stages, so that misdirected particles can be brought to the correct path. Accordingly, the sharpness of cut increases exponentially with the number of sections. Large classifiers are obtained by combining many classifying pipes.

In the *elbow classifier* (Fig. 7) [4], feed material is fed pneumatically to the classifier. Before the material reaches the actual classification zone, it is preaccelerated in an annulus. The air flowing through the inlet nozzle slows down the smaller particles and carries them to the outlet. In this process, the fines are deflected by 180° as described in Section 4.1. The classifying air cannot carry along larger particles, which drop down into the coarse-material container. The classifier is used to separate powdery and fibrous components of plastic granulate.

Except for analytical purposes, countercurrent gravity classifiers are used primarily in the separation of coarse bulk material (0.3–10 mm). Even here, however, they are used less for actual classifying purposes (i.e., separation according to particle size) than for separation according to specific gravity or particle shape, as already men-

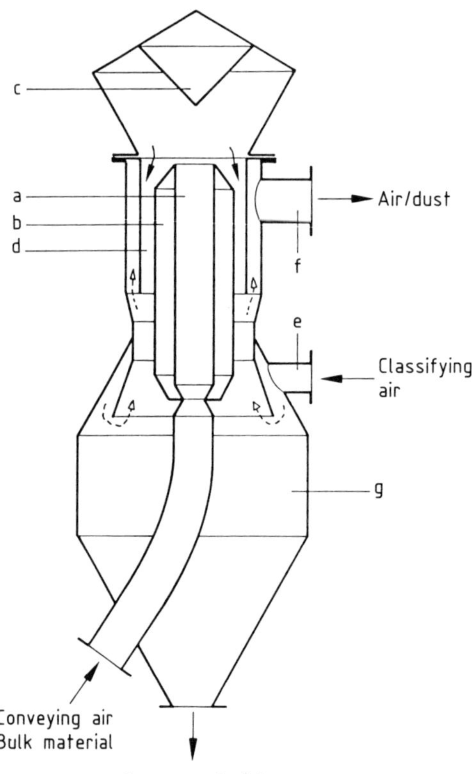

Figure 7. Elbow classifier (Waeschle)
a) Tube; b) Displacement body; c) Deflector cone; d) Annulus; e) Inlet for classifying air; f) Outlet; g) Container for coarse material

tioned. Although separation cuts in the fine range (10–100 µm; see rising-pipe analytical classifier in Table 1) are possible in principle, they are not used as operating classifiers in this case. The cross-sectional areas, and thus space requirements, would become unacceptably large as a result of the low air velocities.

4.2.2. Countercurrent Centrifugal Classifiers

Spiral Classifier. Spiral classification was first used and investigated systematically in the 1930s [5]. Most spiral classifiers have a flat cylindrical separation chamber in which a spiral flow, moving from the outside toward the inside, is produced. Adjustable guide vanes are located on the periphery. Through the setting angle of these vanes, the parameters of the spiral (i.e., the ratio of peripheral to radial velocity) are adjusted and the cut point is thus established. The cut-point range of spiral classifiers is located between ca. 5 and 80 µm. The fines leave the classification space with the air through a central opening. The different possibilities for input of feed material and discharge of coarse material are illustrated by the following examples.

In the *Mikroplex spiral air classifier* (Fig. 8), feed material is introduced through a tangential chute on the periphery of the classification zone. Coarse material flows in a kind of skein over the guide vanes (a) and is freed of adhering fines by

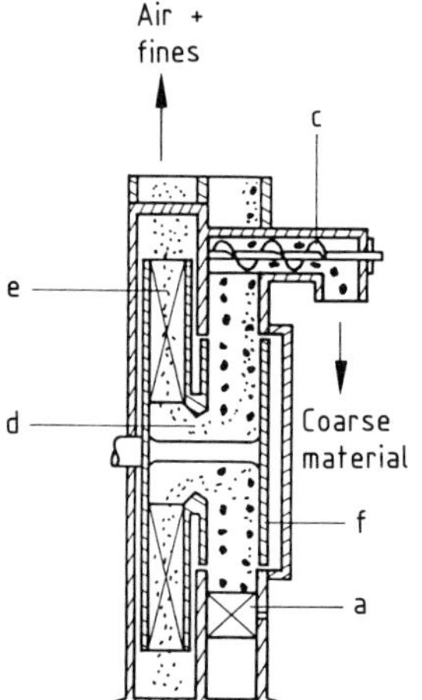

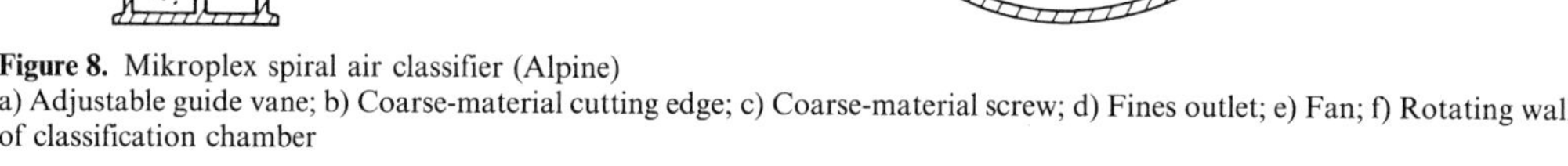

Figure 8. Mikroplex spiral air classifier (Alpine)
a) Adjustable guide vane; b) Coarse-material cutting edge; c) Coarse-material screw; d) Fines outlet; e) Fan; f) Rotating wall of classification chamber

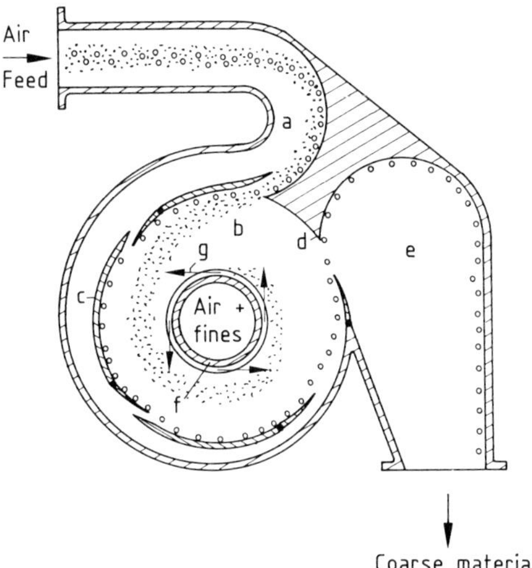

Figure 9. Stratoplex spiral air classifier (Alpine)
a) Intake bend; b) Classification zone; c) Guide vane; d) Coarse-material cutting edge; e) Coarse-material chamber; f) Fines outlet; g) Accelerating air

the inflowing air. Finally, it is peeled off by a cutting edge (b) also located on the periphery of the classification zone and is removed by a screw (c). The fan (e) required for production of the classifying air is integrated into this classifier. A special feature of the Mikroplex classifier is its rotating side walls (f), which largely eliminate the boundary layer close to the wall. This layer, which is characteristic of static classifiers, results in friction and in the transport of coarse particles to the fines outlet (d); this can lead, among other things, to so-called grit.

Figure 9 shows the *Stratoplex spiral air classifier*. This is an airstream classifier in which the feed is dispersed in the classifying air. The coarse material is again peeled off by a cutting edge (d) on the periphery and drops into a coarse-material chamber (e). The ratio between depth and diameter of the classification space, which is of the order of 1:4 to 1:6, for most spiral classifiers was increased to 1:2 for the Stratoplex classifier in order to increase throughput. The air loaded with fines is, therefore, exhausted on both sides (f). To prevent the transport of coarse material in the boundary layer close to the wall, accelerating air (g) is blown in around the central pipes. For products with a tendency toward melt deposition, cooling of equipment parts vital to the operation, such as guide vanes (c) and the coarse-material cutting edge (d), is possible.

The *Statopol* is another airstream classifier (Fig. 10) [6]. The mixture of air and material is fed in over the guide-vane ring on the entire periphery. This classifier has a cyclone-like classification cone instead of a flat-cylindrical separation chamber. Coarse material falls down into this cone; air and fines leave the classification zone through a central pipe.

In the *DS classifier* (Fig. 11) [7], feed material is also introduced already dispersed in air. However, only part of this air is used for classifying. Above the classification zone is an inlet chamber, in which the material is separated from a major part of the transport air and then passed through an annulus into the actual classification zone. The classification zone, in turn, is surrounded by a guide-vane ring, through which the air flows in. Coarse material drops out through an annulus in the lower wall of the classification chamber.

Figure 12 shows another static spiral classifier, the *EC Classifier*, which is used predominantly in cement classification [8], [9]. The simple machine design makes operation with circulating air possible.

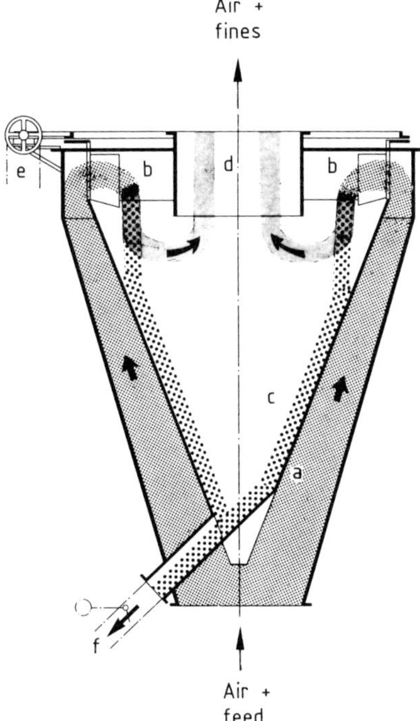

Figure 10. Statopol spiral classifier (Krupp–Polysius)
a) Outer cone; b) Guide-vane ring; c) Grit cone; d) Dip pipe; e) Adjusting device for guide-vane ring; f) Grit outlet

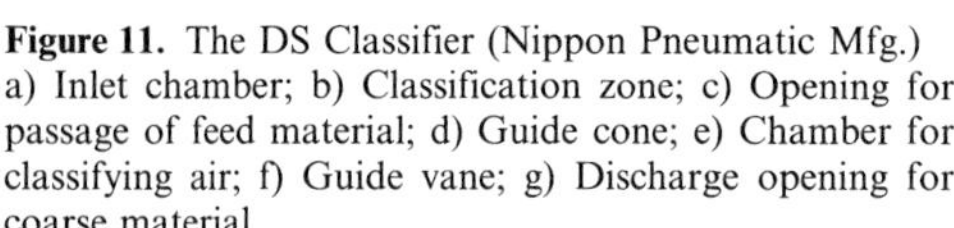

Figure 11. The DS Classifier (Nippon Pneumatic Mfg.)
a) Inlet chamber; b) Classification zone; c) Opening for passage of feed material; d) Guide cone; e) Chamber for classifying air; f) Guide vane; g) Discharge opening for coarse material

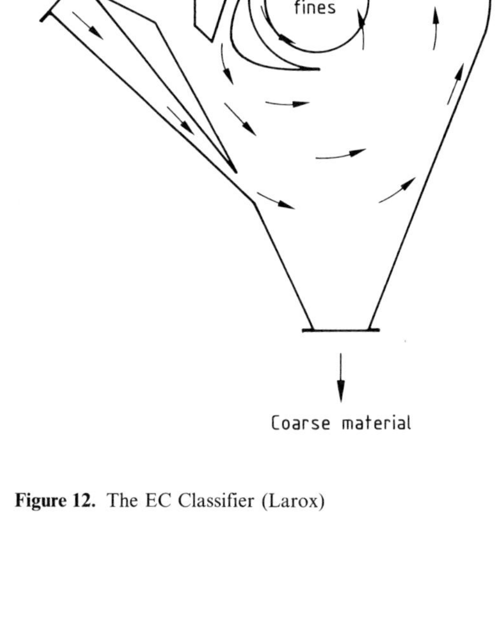

Figure 12. The EC Classifier (Larox)

Deflector Wheel Classifier. The distinguishing feature of a deflector wheel classifier is a rotor equipped with deflecting blades, with air flow from the outside to the inside of this rotor. However, deflection at the blades is limited, and classification takes place in the interior of the rotor in a spiral or spiral-helical flow. The cut point is set by the air flow rate and the rotor speed.

In conventional spiral air classifiers, the flow adjusts itself more or less freely ("free vortex") and, accordingly, can be influenced by several variables such as the mass flow rate of the feed. Contrary to this, significantly more stable classification conditions can be achieved with deflector wheel classifiers because of the bladed rotors ("forced vortex").

The blades, usually made of flattened irons, are attached axially at the rotor periphery; occasionally cylindrical rods are used instead. In most deflector wheel classifiers, the height of the rotor is approximately equal to its diameter. For reasons of strength, the speed of such rotors is limited. At maximum peripheral speed of ca. 65 m/s, a finest cut point of ca. $x_{50} = 3-6$ µm can be achieved, depending on classifier size. If an even finer classification is desired, flatter high-speed rotors with peripheral speed between 80 and 120 m/s must be used.

To perform classifications with a sharp cut, deflector wheel classifiers are generally combined with special classification zones for the coarse fraction. This type of classifier is frequently used in classifier mills.

The *Micron Super Separator* (Fig. 13) [10] has a conical rotor whose blades are set at an angle to the periphery. Feed material, dispersed in the primary air, is introduced into the classifier. Air and fines are sucked through the rotor and pass into the filter system located downstream. Air flows in through a ring of guide vanes located on the outer periphery of the classification zone, at the level of the impeller. Ag-

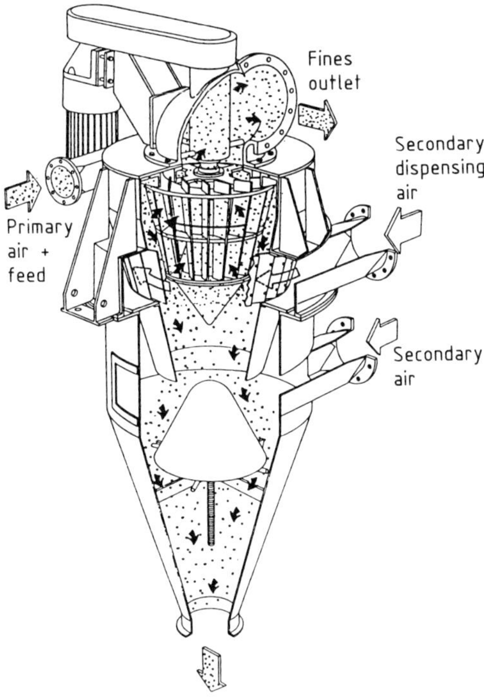

Figure 13. Micron Super Separator (Hosokawa–Nauta)

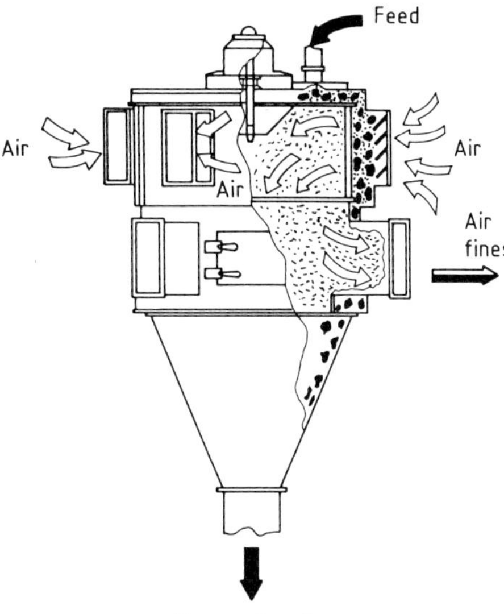

Figure 14. The SD Classifier (Sturtevant)

glomerates rejected by the rotor are dispersed and again conveyed to the classifying wheel. Additional secondary air enters through the funnel-shaped coarse-material outlet.

In the *SD Classifier* (Fig. 14), the rotor is equipped with cylindrical rods. The classification zone for coarse material consists of several conical deflector plates, which are superposed at the periphery of the classifier. Tangentially supplied air flows through these deflector plates. The classifier is loaded by means of gravity pipes [11], [12].

The *Turboplex high-efficiency air classifier* (Fig. 15), in contrast to other classifiers, contains a rotor with a horizontal axis. Problems of wear or deposit in the fines outlet can thus be controlled easily. The classification zone for coarse material consists of a ring of guide vanes located at the lower end of the cylindrical–conical classifier housing. Figure 16 shows a multiwheel arrangement with which very fine separation can be achieved even with high throughput. The product can be fed to the machine either through gravity chutes or dispersed in a partial airstream.

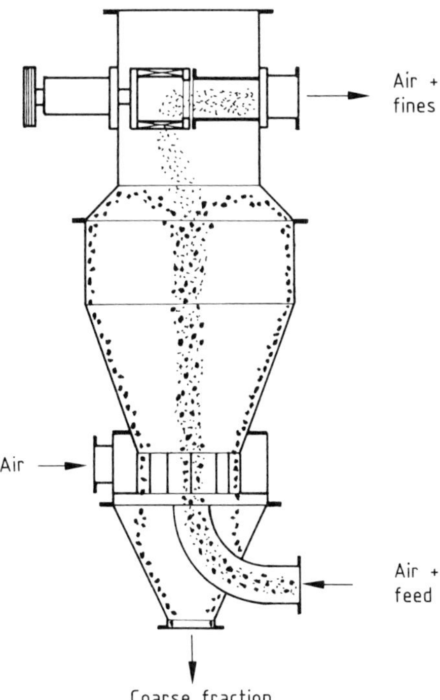

Figure 15. Turboplex high-efficiency air classifier (Alpine)

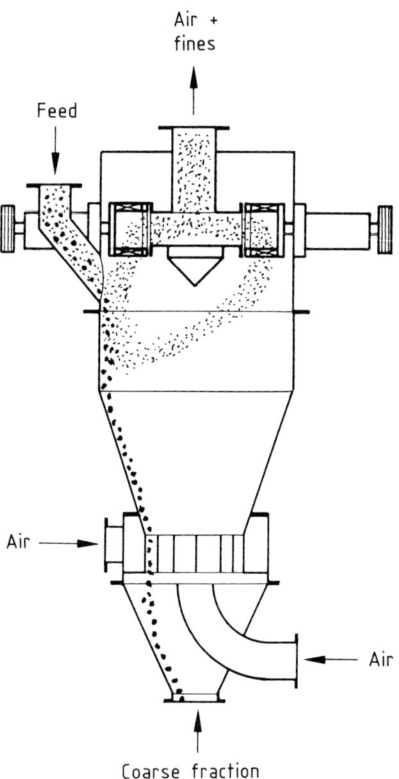

Figure 16. Turboplex multiwheel classifier (Alpine)

Air classifiers with *high-speed rotors* use the kinetic energy of these rotors directly to disperse the feed material. However, this dispersing action based on shock and impact can lead to considerable comminution of larger particles and increased wear. Classifiers of this type are manufactured by Donaldson (Fig. 17) [13],

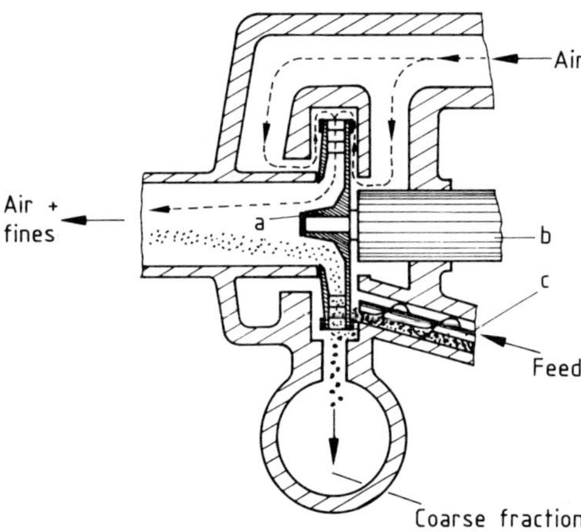

Figure 18. The MZR channel-wheel classifier (Alpine)
a) Rotor with zigzag channels; b) Bearing; c) Feeding screw

Alpine AG (Fig. 18), Nishin Engineering Co. (Fig. 19) [14], and others.

Channel-Wheel Classifier (*Elbow Classifier*). The principle of countercurrent elbow classification was described in [15] and [16]. In a centrifugal field, this led to the so-called channel-wheel classifier, whose design and mode of operation are shown in Figure 20 [6]. The channel wheel is charged from above with the material to be classified. In the individual channels, particles are accelerated in the direction of the periphery and are thrown into the inward flowing classifying air. In this process, the fines are deflected (as described in Section 4.2.1) and exhausted with the air through openings in the upper part of the channel wheel. The adjustment of cut point is carried out by varying air quantity and speed of the channel wheel, as in deflector-wheel classifiers. The cut-point range is 10 – 150 μm.

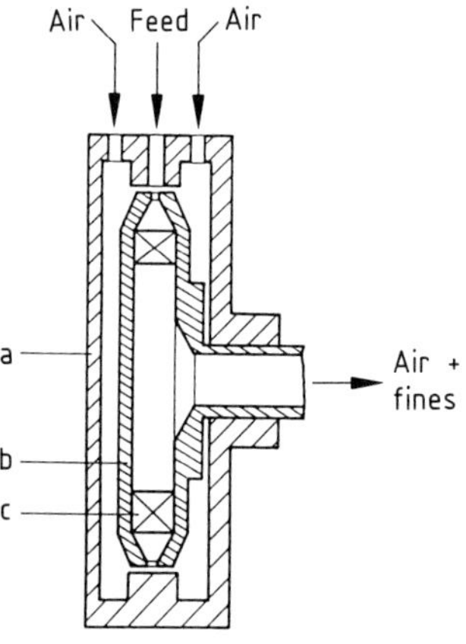

Figure 17. Centrifugal classifier (Donaldson)
a) Housing; b) Classifying wheel; c) Ribs

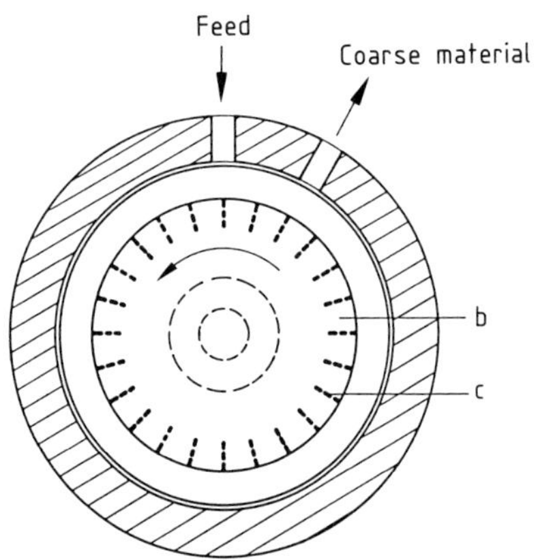

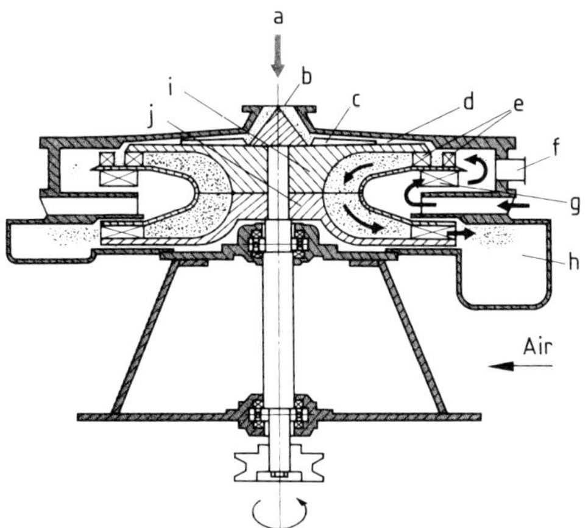

Figure 19. Turbo Classifier (Nishin Engineering)
a) Feed; b) Product inlet; c) Dispersing vane; d) Dispersing disk; e) Classifying vane; f) Coarse-material outlet; g) Auxiliary vane; h) Screw housing (air/fines outlet); i) Classifying rotor; j) Balancing rotor

4.3. Crosscurrent Classifiers

During the 1960s, the principle of crosscurrent classification was developed [17], [18]. By means of special dosing and dispersing devices, particles were fired at a right angle into a flow channel. The particles fanned out in the airstream move along almost fixed trajectories (Fig. 21). In a case of application, the fractions of fanned-out particles are measured quantitatively by photometric means. This unit serves as a fully automatic *particle-size analyzer* [19].

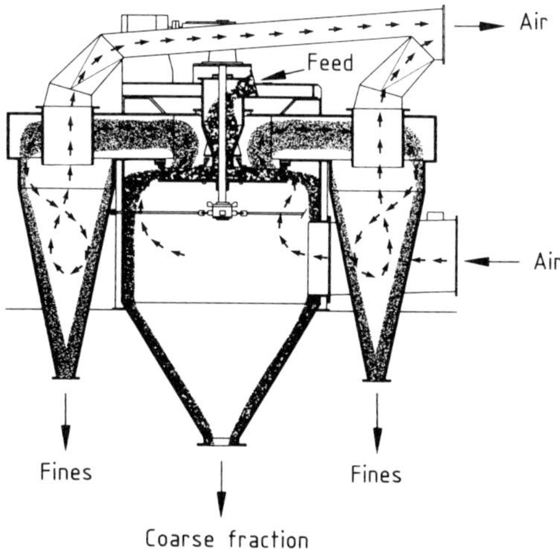

Figure 20. Caropol channel-wheel classifier (Krupp–Polysius)

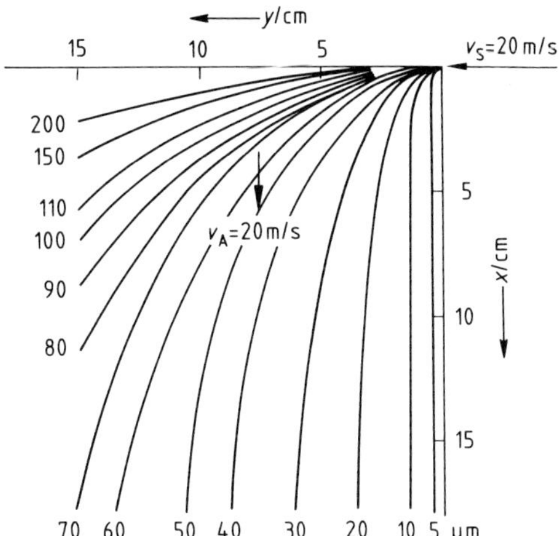

Figure 21. Trajectories of the crosscurrent jet arrangement

A *crosscurrent elbow classifier* has been developed for large-scale industrial purposes (Fig. 22) [20]. The fanned-out stream of material is divided into several partial streams of corresponding particle-size classes by means of cutting edges. The finest particles again undergo a large deflection during this process.

4.4. Recirculating-Air Classifiers

Recirculating-air (or whizzer-type) classifiers are the oldest industrial air classifiers. The first one was built in 1885 by Pfeiffer in Kaisers-

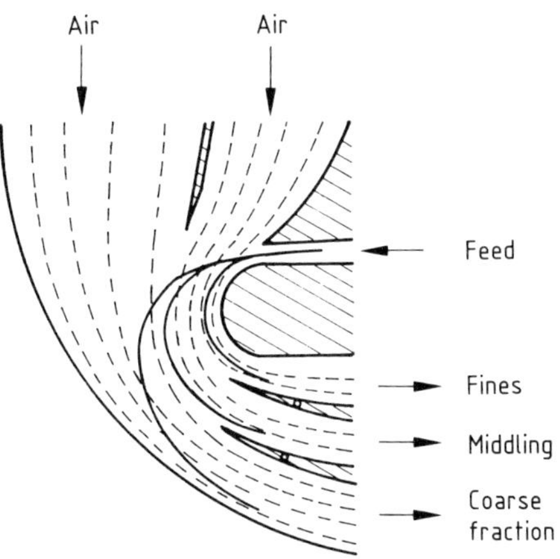

Figure 22. Crosscurrent elbow classifier

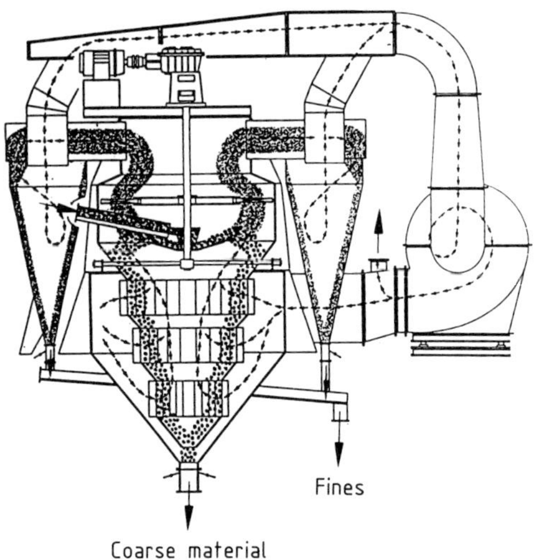

Figure 23. Zyklopol recirculating air classifier/cyclone air classifier (Krupp–Polysius)
External air circuit

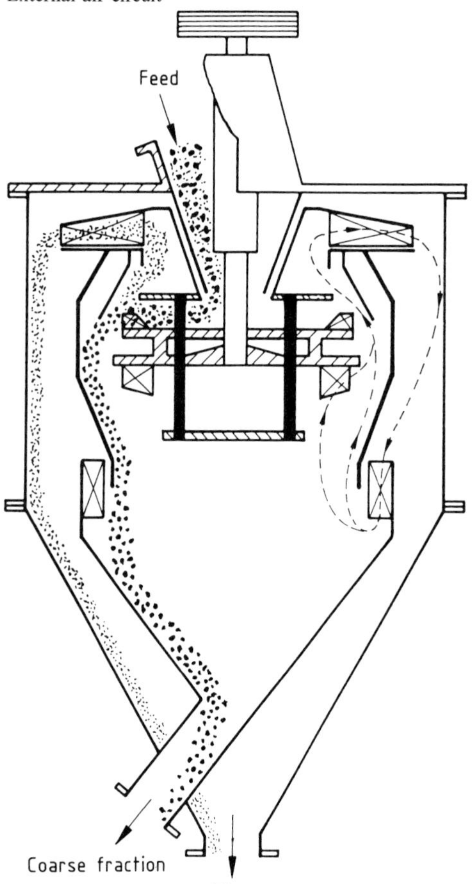

Figure 24. Ventoplex recirculating air classifier (Alpine)
Internal air circuit, main fan and dispersing disk are located on the same shaft

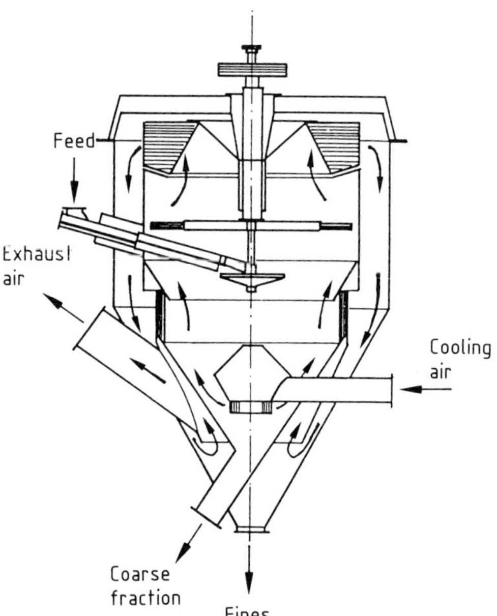

Figure 25. Heyd dispersing air classifier (Pfeiffer)—internal air circuit with supply of cooling air
Adjustable fan, dispersing disk, and main fan are on two separate shafts

lautern (Germany) according to a patent by MOODIE and MUMFORD. This type is still the most frequently used classifier, especially in the production of cement, gypsum, mineral fillers, etc.

The mode of operation is as follows: The material is sent by a dispersing disk into a rotating air flow moving in an upward direction (cross-current classification). The coarse material sliding down the wall of the classification chamber is reclassified as it passes over one or more louvers. Classification of the fines takes place in the conical screw-shaped flow, whose peripheral velocity, and therefore cut point, are adjusted by means of the speed or position of a rotor system. Classifying air may be produced by an external fan. With this *external air circuit,* the fines are separated in externally located cyclones (Fig. 23). If the fan is integrated into the head of the classifier, it is referred to as an *internal air circuit* (see Section 4.1). Here, the fines are separated from the air in a cyclone which envelops the classification space. In some designs, the fan, rotor, and dispersing disk are located on the same shaft (Fig. 24). However, two-shaft designs (i.e., a separate drive for the fan and rotor–dispersing disk) are used more often (Fig. 25).

For the classifier shown in Figure 25, the possibility of fresh-air input also exists. To cool the product, a departure from operation exclusively with circulating air is made [21]. The exhaust air must be dedusted in a filter connected on the downstream side.

Recirculating-air classifiers have a separation range of ca. 30–200 μm. Because these machines operate with relatively low air speed, they have a satisfactory wear behavior. On the other hand, because of the lack of disintegrating power, their sharpness of cut is not very good.

5. Operating Properties

5.1. Effect of Feed

Classifiers can be more or less sensitive to changes in the particle-size distribution of the feed material and in the feed rate. If they are loaded with finer feed material with the same setting (speed, air flow rate, load, etc.), then air classifiers of almost all types produce a finer product or a finer cut point. Figure 26 shows this behavior for a deflector wheel classifier. In this figure, the reference cut point x_{f97} and Tromp's cut point x_{50} are plotted against the parameter φ, where φ is the ratio of the radial velocity v_r (which is proportional to air quantity) to the peripheral velocity v_p (which is proportional to rotor speed). The curves show the effect of two different feed materials A and B, in this case limestone powders produced by an impact grinding mill, whose particle-size distributions are shown in Figure 27.

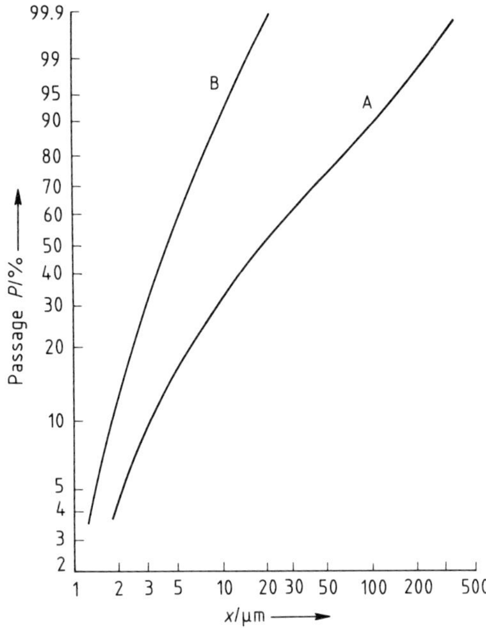

Figure 27. Cumulative passage curves of impact-ground lime stone powders

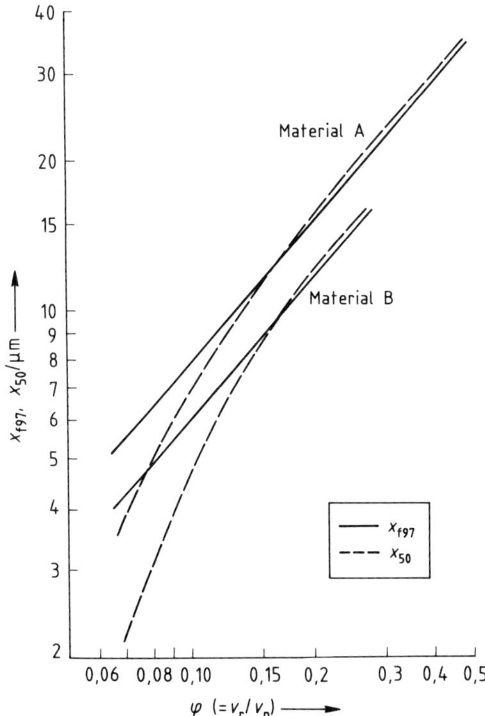

Figure 26. Effect of feed-material distribution on the cut point

Individual classifier types exhibit different behavior when the feed rate is increased. In this connection, the terms *hard* and *soft characteristics* are used. The term "hard"designates a classifier that maintains a constant particle-size distribution of the fines with increasing overload, e.g., deflector wheel classifiers, zigzag classifiers, and recirculating-air classifiers. This behavior is illustrated in Figure 28 for a deflector wheel classifier. The fines even become slightly finer with increasing feed rate, and the coarse material becomes increasingly less pure, as manifest by a distinct decrease in the sharpness of cut.

Spiral classifiers, on the other hand (free vortex flow), have soft characteristics. When the classification zone is overloaded, they show a shift of the cut point, and thus of the fines distri-

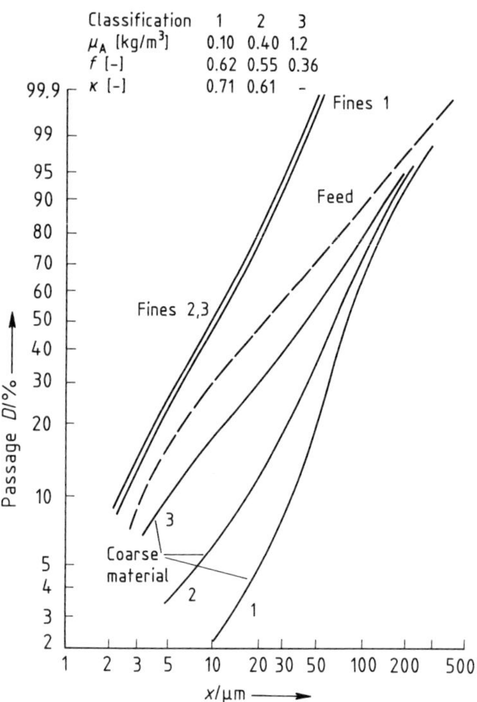

Figure 28. Behavior of a deflector wheel classifier under overload

bution, toward the coarser range. The sharpness of cut also decreases continuously (Fig. 29) until the classification finally breaks down.

In closed grinding–classifying systems, this soft behavior is sometimes desirable. In the case of overloading, the classifier is meant to increase production of coarser fines; otherwise, the circuit will fill up continuously.

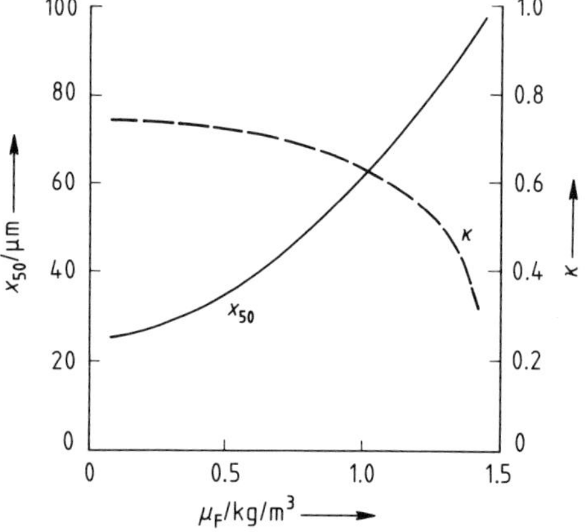

Figure 29. Behavior of a spiral air classifier under overload

5.2. Scaleup

When the size of a classifier is increased, the change in throughput may be accompanied by a change in the sharpness of cut and cut point. A shift in the cut-point range occurs primarily with countercurrent classifiers in a centrifugal field and also with recirculating-air classifiers (i.e., machines in whose classification zones more or less ideal spiral flows predominate). The dependence of cut point x_t on $r^{0.6}$, a characteristic parameter of the classification zone, is shown in Equation (8) (see Chap. 2). This proportionality has been confirmed by investigations on different countercurrent centrifugal classifiers.

Smaller classifiers, therefore, achieve a finer cut point under identical classification conditions. The same cut point can usually be achieved by appropriate setting of larger machines, e.g., by increasing peripheral velocity. However, because the peripheral and radial velocities of a classifier are located within quite specific limits for various reasons (pressure drop, material strength, etc.), very fine classification carried out on a laboratory machine, for example, cannot be duplicated on a large-scale classifier.

5.3. Special Requirements

In the selection of a classifier, performance characteristics (Chap. 3) should be the deciding factor. However, other characteristics, such as operating behavior ("soft" or "hard" classification characteristics), are also involved. Often specific material properties or product requirements must be considered, e.g., deposition behavior, wear, product purity, explosion hazard, etc. In many cases, these requirements can be met only by means of special fittings or a special design.

Deposition Behavior. Very fine dusts have a tendency to agglomerate and form deposits on the walls of classification chambers and on parts of the equipment that are important for the operation. The result can be a shift of the cut point and, in the worst case, breakdown of the classification.

In the matter of material deposition, large classifiers have an advantage over smaller machines. Because of the larger dimensions (cross-sectional area), the deposit layer can grow to a certain thickness without significantly interfering with classification. On the other hand, the layer of deposit then becomes so heavy that it may fall off, be blown away, or be thrown off. High air speeds or elastic, "fluttering" linings can be used to address this problem. With products tending toward melt deposits, lowering the air speed, cooling the classifying air, or cooling critical equipment parts is recommended.

Wear. All materials with a Mohs hardness greater than 4 exert a wearing action on the classifier material. Damage to the classifier related to this becomes a problem primarily in machines with high-speed parts; the wear on classifier elements important to the operation can shift the cut point.

As a countermeasure, easily interchangeable parts and replaceable linings are used. No universally usable wear-resistant material exists as yet; depending on the bulk material and location in the classifier, soft or hard steel, hard metal, hard ceramic, soft rubber, or Vulkollan may be suitable.

Because wear increases exponentially with air velocity it can be minimized by throttling the air quantity. However, this measure decreases the sharpness of cut and the throughput.

Product Purity. In addition to the abrasion produced by wear, a product can also be contaminated by corrosion products and lubricants. Grease may pass easily from the bearings into the classification chamber especially where shafts penetrate. Conversely, the product may get into the bearings and destroy them prematurely. Pneumatic sealing has proved to be a suitable remedial measure. When frequent feed change or a sterile mode of operation is necessary, adequate purification possibilities must be available.

Explosion Hazard. To carry out successful air classification, very fine dusts must be dispersed in large quantities of air. This involves the danger of dust explosions, even with normally harmless products.

Possible protective measures against dust explosions are:

1) avoidance of sparks and ignition sources (e.g., no high-speed machine parts),
2) inert gas blanket, or
3) shock-resistant design.

6. References

[1] O. Lauer: *Feinheitsmessungen an technischen Stäuben,* Alpine AG, Augsburg 1962.
[2] Keller GmbH, DE 1 059 977, 1959 (H. Ide).
[3] F. Kaiser, *Chem. Ing. Tech.* **35** (1963) 273–282.
[4] Waeschle Maschinenfabrik, DE 3 203 209, 1982 (W. Krambrock, H. Hoppe).
[5] H. Rumpf, Dissertation, Technische Hochschule, Karlsruhe 1939.
[6] Krupp-Polysius AG, Beckum (brochure).
[7] Nippon Manufacturing Co., DE 2 949 618, 1979 (N. Nakayama, K. Yonezawa).
[8] R. T. Hukki, *Zem. Kalk Gips Ed. B* (1977) no. 5, 199–205.
[9] Larox News 2/1985, Larox Oy, Lappeenranta, Finland.
[10] Hosokawa-Nauta Europe B.V., Haarlem, The Netherlands (brochure).
[11] J. V. Klumpar, N. N. Zoubov, *World Cem.* **16** (1985) Oct., 302–305.
[12] Sturtevant Inc., Boston, MA, USA (brochure).
[13] Donaldson Co., DE 1 657 122, 1968 (C. E. Lapple).
[14] Nishin Eng. Co., Tokyo, Japan (brochure).
[15] J. Wessel, U. Schmidt, *Aufbereit. Tech.* **12** (1971) no. 7, 402–406.
[16] J. Wessel, *Aufbereit. Tech.* **20** (1979) no. 9, 475–478.
[17] H. Rumpf, K. Leschonski, *Chem. Ing. Tech.* **39** (1967) no. 21, 1231–1241.
[18] K. Leschonski, *Chem. Ing. Tech.* **49** (1977) no. 9, 708–719.
[19] K. L. Metzger, K. Leschonski in Proc. 1. Europ. Symp. "Partikelmeßtechnik," *DECHEMA Monogr.* **79** (1976) no. 1589/1615, Part B, 77–94.
[20] L. C. Rumpf, DE 2 538 190, 1975 (H. Rumpf, K. Leschonski, K. Maly).
[21] E. W. Hanke, *Zem. Kalk Gips* **28** (1975) no. 9, 105–114.
[22] Alpine AG, unpublished results.

18. Mineral Sorting

HENRY E. COHEN, Royal School of Mines, London, United Kingdom

Mineral sorting is man's oldest method of solid–solid separation for minerals. Based upon visual recognition, it was used by the earliest miners to separate desired metals (e.g., gold) or ores (e.g., iron ore) from associated unwanted rocks. To recover obvious values or to remove visible waste before more costly processing (e.g., fine grinding), hand sorting has persisted where inexpensive labor is available. For example, in 1952 all diamonds from the alluvial deposits worked by Consolidated Diamond Mines in South-West Africa (Namibia) were removed by hand at various stages in the concentration process [6]. Today, this has been superseded by machine sorting. However, final grading according to quality is still carried out by hand sorting because the discrimination of the trained eye cannot be matched by any available machine.

Of the 47 major gold–uranium mines in the Witwatersrand area of South Africa, 17 use hand sorting, 1 employs radiometric sorting, and 6 employ both hand and radiometric sorting (a total of 51 %). (The rest use no sorting, but go directly to fine grinding for cyanide extraction of gold.) Manual sorting is confined to material > 50 mm, but photometric and radiation sorting handle feeds down to 25 mm. The main purpose is a substantial rejection of waste (up to 50 %) to reduce the cost of fine grinding. Additional benefits include size reduction of the chemical treatment plant and an increase in overall throughput capacity. In one uranium mine, radiometric sorting reduces the feed tonnage from 3200 t/d of run-of-mine ore to 2200 t/d of feed to the leach plant. This 31 % reduction is accompanied by a 40 % increase in feed grade. In most gold–uranium mines, the rejection of ca. 35 % of the feed mass is accompanied by ca. 95 % recovery of the values. Making low-grade deposits exploitable or reducing the need for selective mining is another economic benefit of preconcentration by ore sorting.

1. Principles of Sorting

Even primitive methods of hand sorting consisted of five consecutive process steps: *preparation, presentation, inspection, data analysis, and separation.* Clean ore must be prepared in relatively narrow size ranges. Generally, particle sizes suitable for sorting most minerals range from about 6 to 180 mm, but treatment has to be carried out on relatively narrow size ranges, usually not exceeding size ratios of about 2:1. Sorting tends to be most effective at the largest particle size permitted by liberation requirements (→ 5. Size Reduction).

Each particle must be examined; hence, rates of throughput are clearly related to particle size. In principle, after preparation, either particles are aligned for individual inspection in single files along one or more paths (trajectories), or they are presented as a wide stream which is large relative to particle dimensions.

Modern sorting processes are fully automated and use sophisticated systems for scanning, recognition, and analysis of information. A variety of sensors have been developed to detect photometric, radiometric, magnetic susceptibility, electrical conductivity [7], and other differences among minerals. Transmission, reflection, and emission of radiation (including infrared, visible, or ultraviolet light; ultrasound; radio frequencies; X rays; gamma rays; etc.) are used or have been tested. Examples of minerals and metals treated by sorting include asbestos, copper, diamonds, fluorspar, gold, limestone, magnesite, scheelite, silver, uranium, and wolframite.

A scanning or inspection system derives information which is analyzed with respect to selected parameters. The location of each particle is defined, and each is either accepted or rejected. The sorting action is applied preferably to the minority fraction, by use of either an air jet or some mechanical deflector.

The operational sequence in ore sorting is shown in Figure 1, with solid arrows indicating the movement of ore. Broken lines indicate the flow of information to the data-analysis unit and the control signals that activate separation.

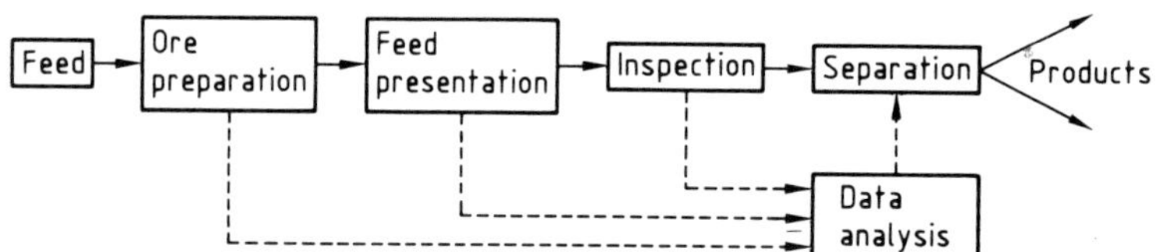

Figure 1. Operational sequence in ore sorting

1.1. Ore Preparation

Liberation is an essential condition for successful sorting. The structural and textural characteristics of an ore determine the particle-size limits for adequate liberation of sufficiently barren gangue which can be rejected without undue loss of values (→ 5. Size Reduction). Mass rejection of 25–30% is usually a minimum requirement for sorting to be economically justifiable.

Normal preparation sequences include coarse crushing to, for instance, particles smaller than 180 mm, followed by washing to remove dust if surface characteristics are important for sensing. For example, washing is unnecessary for sorting by radiometry or magnetic susceptibility but essential for sorting by color, reflectance, or fluorescent emission. The feed must be screened into narrow size groups for separate sorting. Both presentation and inspection are very size dependent. Typical size groups are particles > 100 mm; particles between 100 and 50 mm; particles between 50 and 25 mm; particles between 25 and 12.5 mm; and particles smaller than 12.5 mm. However, specific size limits may have to be chosen in relation to the individual characteristics of an ore.

Surface conditioning can be applied at this stage. For example, an aluminium hydroxide flocculant can be used which adheres selectively to asbestos and can be made fluorescent with a dye [8].

1.2. Feed Presentation

Feed presentation generally delivers the sized material from a surge bin via vibrating or belt feeders to chutes which channel the particles to a conveyor. Assistance may be provided by various types of accelerating rolls or belts to separate the particles. Different manufacturers adopt varying combinations of feeders, chutes, and belts to achieve closely spaced streams of discrete particles. Spaces between particles are kept as small as possible in order to maximize the throughput rate. Typical feed presentation for a radiometric sorter is shown in Figure 2.

Fast movement is important to attain high throughput rates, and linear speeds in the range of 3–5 m/s are common. Feed rates per channel vary from about 60 t/h for coarse material to about 6 t/h for fine sizes. A sectional overall view of a radiometric sorter installation is shown in Figure 3.

Some sorters are designed to inspect and sort particles in flight along a trajectory, rather than during transportation on a conveyor. For this purpose, particles must be aligned in a close-spaced single-file stream that follows an accurate trajectory. The in-line feeder consists basically of two closely spaced counterrotating rolls, whose axes are almost parallel and inclined to the horizontal. This arrangement simulates a V-shaped inclined chute but achieves variable particle speeds by altering the speed of the rolls. Typically, the rolls are 760 mm in diameter and 4.3 m long. The sharp cusplike space between rolls

Figure 2. Feed belt of a 5 channel sorter

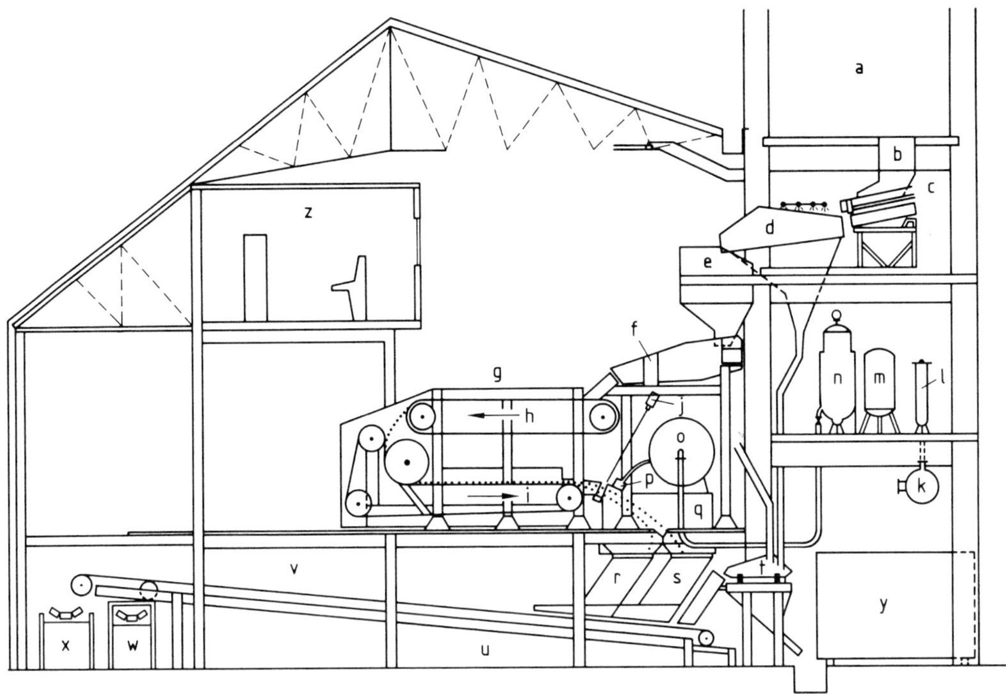

Figure 3. Section of a radiometric sorter plant. a) Main bin; b) Discharge chute; c) Vibrating feeder; d) Washing screen; e) Surge bin; f) Machine feeder; g) Radiometric sorting machine; h) Feeder belt (1.5 m/s); i) Main belt (5 m/s); j) Camera; k) Air manifold; l) Air filter; m) Oil absorber; n) Dryers; o) Air receiver; p) Ejector valve; q) Division plate; r) Accept; s) Reject; t) Screen; u) Accept and reject conveyors next to each other; v) Reef picking area; w) Accept to plant; x) Reject to dump; y) Compressors; z) Control room

breaks up groups of particles and aligns the particle lengths parallel to the rolls. The resulting in-line stream exposes maximum surface areas for inspection [8].

1.3. Inspection and Analysis

The inspection of particles must provide all the information on which decisions to accept or reject are based. This includes the following:

1) Quality—a diagnostic reaction to an applied radiation or signal, or detection of an inherent property such as radioactive emission. Quantitative measurement may be needed, but detection above a set threshold may suffice.
2) Size—a measure of particle size is necessary for quantifying the quality measurement.
3) Speed and position—the speed of the feed stream must be verified and the precise location of a particle in the stream must be defined to operate the ejector mechanism at the correct time lag after inspection.

These various inspection functions usually require complex arrays of sensing heads which repeatedly examine the stream of particles at millisecond intervals. Their signals are integrated and analyzed by the sorting microprocessor in relation to stored memory data. The microprocessor makes corrections (e.g., according to size), calculates the grade of particles, and compares the grade with preset values. A precisely timed decision signal is sent to the appropriate ejector if a particle is assessed as unacceptable.

Higher costs of more refined sensing and microprocessor systems should be balanced against possible gains in resolution: for example, in signal-to-noise ratio, surface detail, or spectral band-width. Resolving power imposes important limitations on the particle size that can be treated, the quality of separation, and the throughput rate.

A separate microprocessor for machine control usually interacts with the sorting microprocessor to ensure correct interlocking of machine parts and sequencing of process events. It also

Table 1. Open circuit operation (three months) at Vaal Reefs Gold Mine*

Material	Total production, t	Split, wt %	Gold grade, g/t	Gold distribution, %	Uranium grade, kg/t	Uranium distribution, %
Feed	8136	100	2.40**	100	0.125**	100
Accept	4358	53.6	4.25	94.7	0.220	93.9
Reject	3778	46.4	0.28	5.3	0.017	6.1

* Total sorting time 260 h. ** Calculated feed grade.

identifies malfunctions (e.g., of sensors, cameras, ejectors, timers, or power supplies) and automatically initiates emergency action such as stopping the feed and indicating the nature of a fault.

1.4. Separation

A highly controlled air jet for ejecting particles from a stream moving in a free flight trajectory is the most widely used method of separation. Both mechanical and hydraulic devices have been described, but they cannot match the speed of air-ejector valves. A blast of compressed air at about 500–700 kPa is triggered at a precise moment to deflect a particle from its trajectory into a separate receiver. The force of the airstream used to eject 30-mm particles is 6 kg/cm^2 [9]. The ejector valves have minimum open times of 3–4 ms and should generate close to "square-wave" impulses with the shortest possible rise and decay times. A typical sorter, for instance for 15–30-mm particles may have a presentation width of 800 mm and employ 80 valves, or one valve every 10 mm. Thus, the ejector resolution is 10 mm across and 12–16 mm in the direction of travel of the particles. Larger particles (up to 180 mm) require longer blasts, of ca. 7–8 ms with 20-mm width and at least 28-mm length in the direction of travel. Each valve can fire up to 120 times per second, with considerable noise emission. Soundproofing and remote operator controls are necessary.

2. Performance

Each sorting application has specific constraints that arise partly from the nature and particle size of the ore and partly from limitations of the equipment. For example, incomplete liberation always enforces an arbitrary grade–recovery trade-off. The contained values of an ore and the price of the end products limit acceptable process costs.

Table 2. Scalping of coarse waste from gold ore

	Mass, %	Gold content, g/t	Gold distribution, %
Feed	100	6.61	100
Product	63.5	10.02	96.3
Reject	36.5	0.67	3.7

Table 3. Separation of magnesite from silicate gangue

	Mass fraction, %	Magnesite content, %	Distribution, %
Feed	100	30	100
Product	32	91	97
Reject	68	2	3

For example, radiometric sorting for scalping coarse waste from a gold–uranium ore at Vaal Reefs Mine in South Africa is shown in Table 1 [10].

Machine availability is quoted as 97.2%. In this example, whether or not the gold loss of 5.3% is acceptable, in return for removing 46.4% of the feed mass prior to fine grinding, remains a purely financial decision.

Another typical performance in scalping coarse waste (25 mm) by radiometric sorting from a gold ore is shown in Table 2.

In separating magnesite from silicate gangue (particle size 40–70 mm) by optical sorting based on color difference at a throughput rate of about 60 t/h, the performance is shown in Table 3.

The products from all three examples proceed to further treatment with substantially reduced tonnage and relatively small loss of values.

In the recovery of diamonds, optical sorters remove diamonds because they are lighter in color than most waste minerals, while X-ray sorters remove them because of their fluorescent re-

Table 4. Costs of preconcentration at Daping mine*

Items	Cost basis, $	Cost per month, $	Cost per ton of ores, $
Wages			
3 operators	16.6	49.65	0.04
18 hand-sorting workers	16.6	297.9	0.22
Electricity, 770 kW · h per month	0.025	57.75	0.04
Compressed air, 17.8 m^3 per ton of feed	0.0028	66.0	0.05
Depreciation charge for equipment, 5% per year	525 (three sorters)	23.85	0.02
Consumption of materials, 4% of investment per year		19.12	0.02
Total		514.27	0.38

* Plant works three-shift basis, 550 h per month, eleven months per year; 1350 t in 30–20-mm size range per month, 20% of which is quartz.

sponse. Sorting is applied to gravity concentrates in which diamonds form a small minority of particles. Feeds of particles 4–8 mm in size are treated at rates of 300–350 kg per channel per hour. Diamond recovery is 98–99%, and the product is only 0.3–0.5% of the feed mass. This product is sorted further by hand.

3. Economic Aspects

With wide ranges of particle sizes, different types of applications, and variations in local cost factors, general validity cannot be assigned to the somewhat scarce published costs of sorting. Capital costs of sorters appear to range from ca. $ 11 000 to $ 3000 per ton · hour capacity for 6-mm feed and 150-mm feed, respectively. Approximate costs of installation and commissioning range from 50 to 150% of machine costs, depending on convenience (i.e., distance from supplier and ease of access) and climate (cold climate requires housing, heating, and lighting; hot climate requires airconditioning, dust extraction, etc.).

Among the factors contributing to operating costs, electric power ranges from 0.2 to 3.0 kW · h/t and compressed air from 5 to 30 m^3 per ton of feed, or ca. 50–60 m^3 per ton ejected. Costs of spares and maintenance are generally claimed to be very low, around $ 0.07–0.14 per ton treated. If the ore must be washed, water consumption appears to be around 100 L per ton of feed. A quotation of operating costs for sorting limestone (70–130 mm) in Finland in 1970 gives $ 0.05 per ton for power, labor, and maintenance [11].

Operator costs probably constitute the largest and most variable component of operating charges. In general, operator attention appears to range from 0.1 to 2.0 min per ton of feed, i.e., a throughput of 200 t/h could require one to six operators.

An interesting example of cost is available from the Daping tungsten mine in China [9]. The ore contains wolframite associated with quartz veins in metamorphic host rocks of generally dark color, and it is sorted for preconcentration in a size range of 20–30 mm. Magnetic detectors are used to eject wolframite, which is dark, but has higher magnetic susceptibility than the host rock. At the same time, optical detectors are used to eject white quartz (20% of the feed) because this contains unliberated wolframite. Machine sorting (three machines with one operator each) is followed by hand sorting of both products (18 operators), which permits a substantial reduction in operators compared with hand sorting alone. The feed grade of 0.11% wolframite is raised to 0.20% by machine sorting and to 0.49% by subsequent hand sorting, an enrichment ratio of 4.5 at 91.9% recovery. The costs of this preconcentration during the period July 1981 to November 1982 are shown in Table 4.

4. References

General References

[1] N. L. Weiss (ed.): *SME Mineral Processing Handbook*, Society of Mining Engineers, AIMMPE, New York 1985.

[2] B. A. Wills: *Mineral Processing Technology*, 3rd ed., Pergamon Press, Oxford 1985.

[3] G. Tarjan: *Mineral Processing*, Akademiai Klado, Budapest 1981.

[4] P. Somasundaran (ed.): *Advances in Mineral Processing*, SME, Colorado 1986.

[5] L. White (ed.): *E & MJ Second Operating Handbook of Mineral Processing*, McGraw-Hill, New York 1980.

Specific References

[6] R. G. Weavind: "The treatment and recovery of refractory diamonds," *J. Chem. Metall. Min. Soc. S. Afr.* (1952) 243–264.

[7] A. Balint: "Ore sorting according to electrical conductivity," *J. S. Afr. Inst. Min. Metall.* Special Issue, October 1975, 40–42.

[8] D. Collier et al.: "Ore sorters for asbestos and scheelite," 10th International Mineral Processing Congress, London 1973, pp. 1007–1022.

[9] Shi Fengnian: "Magnetophotometric sorter and its application at a wolframite mine," Mineral Processing and Extractive Metallurgy IMM International Conference, Kunming 1984, pp. 589–597.

[10] D. G. Kidd, N. P. G. Wyatt: "Radiometric sorting of ore," 12th CMMI Congress, Johannesburg 1982, pp. 645–651.

[11] Gunson's Sortex Ltd., Personal Communication.

19. Magnetic Separation

WILLIAM J. BRONKALA, Applied Magnetic Systems, Inc., Greenfield, Wisconsin 53 220, United States

1. History

In the mid-1800 s, FARADAY demonstrated in a series of experiments that any substance placed in a magnetic field will either increase or decrease, to a varying extent, the magnetic field passing through it. This variation in the magnetic susceptibilities of different materials makes possible the phenomenon of magnetic separation (see Chap. 2).

Based on its magnetic susceptibility, a material can be classified as ferromagnetic (strongly attracted magnetically), paramagnetic (attracted magnetically to some degree), or diamagnetic (repelled by a magnetic field).

All materials may be termed magnetic, although the relative value of the induced magnetism might be very small. *Ferromagnetic materials* have magnetic properties similar to iron, and some ferromagnetic materials will retain their magnetic properties in the absence of an applied magnetic field. This phenomenon is known as *remanence*. *Paramagnetic materials* have a positive magnetic susceptibility (showing a weak attraction to a magnet), and *diamagnetic materials* have a negative magnetic susceptibility (showing a weak repulsion from a magnet). Ferromagnetic materials are amenable to *low-intensity separation* and paramagnetic materials to *high-intensity separation*.

Since the work of FARADAY, numerous magnetic devices have been developed that utilize differences in magnetic susceptibility to obtain physical separation of mineral products. The first successful commercial application of magnetic separators was the separation of iron from brass in the late 1860 s, followed by the separation and concentration of the ferromagnetic mineral magnetite.

Low-intensity electromagnetic separators of various configurations were developed after 1880 for wet and dry magnetic separation of ferromagnetic materials; these were followed in the early 1900 s by high-intensity magnetic separators and permanent magnet separators. The first *permanent magnet material* was an aluminum–nickel–cobalt–iron alloy designated ALNICO, which was followed in the 1950 s by ceramic barium and strontium ferrite permanent magnets, and in the 1980 s by rare-earth magnets (→ Magnets and Magnetic Materials, treated in the A series).

High-intensity magnetic separators of several types were developed to separate paramagnetic materials of lower positive magnetic susceptibility. The first high-intensity separators were largely confined to dry separation, with subsequent development of high-intensity wet-type separators in the early 1960 s and the high gradient magnetic separator (HGMS) in the mid-1980 s.

2. Principles of Magnetic Separation

Magnetic separation is a physical separation of discrete particles with different permeability or susceptibility, based on a three-way competition among tractive magnetic forces; gravitational, frictional, and inertial forces; and attractive interparticle forces. The feed to magnetic separators is split into two or more components. If the separator is to produce a magnetic concentrate, then a weakly paramagnetic or diamagnetic material in practice simply called nonmagnetic, constitutes the tailings (or reject) product. If liberation is incomplete, a less magnetic (middling) product can result. Each product must be transported into, through, and out of the magnetic separator. The magnetic and competing forces (gravitational, frictional, hydrodynamic, or inertial) tend to reduce the degree of separation.

Definitions (for a more comprehensive treatment, → Magnets and Magnetic Materials, treated in the A series). A *magnetic field* is defined by its *magnetic field strength H* (sometimes also called *field intensity*). The field strength of a homogeneous magnetic field in a solenoid with n turns per meter is given by

$$H = nI$$

where I is the current per turn. (The SI unit for magnetic field strength is A/m.)

The *magnetic flux density B* is a measurement of the *magnetic forces* acting on particles in a magnetic field. In a vacuum, magnetic flux density and magnetic field strength are related by

$$B_0 = \Phi/A = \mu_0 H$$

where Φ is the *magnetic flux* (SI unit weber, $1\ \text{Wb} = 1\ \text{m}^2\ \text{kg}\ \text{s}^{-2}\ \text{A}^{-1}$), A is the cross section of the coil, and μ_0 is the *permeability of the vacuum* ($\mu_0 = 4\pi \times 10^{-7}\ \text{N/A}^2$). The SI unit for magnetic flux density is the tesla ($\text{T} = \text{kg}\ \text{s}^{-2}\ \text{A}^{-1}$), $1\ \text{T} = 1\ \text{Wb/m}^2$.

If a material is placed in a magnetic field, the flux density of this material is given by

$$B = \mu H$$

where μ (SI unit N/A^2) is the permeability of the material.

Another method of calculating B for a material is

$$B = \mu_0 (H + M) \qquad (1)$$

where M (SI unit A/m) is the *induced magnetization* of the substance observed whenever it is placed in a magnetic field.

The *magnetic susceptibility* of a material, χ (volume susceptibility), is dimensionless and is defined as the ratio of induced magnetization to magnetic field strength:

$$\chi = M/H \qquad (2)$$

Substitution in Equation (1) gives

$$B = \mu_0 H(1 + \chi)$$

The specific magnetic susceptibility ψ is given by

$$\psi = \chi/\varrho$$

where ϱ is the density of the material. Table 1 lists the magnetic susceptibility of various minerals.

Table 1. Magnetic susceptibility of minerals*

Mineral	Susceptibility	Mineral	Susceptibility
Magnetite	0.12–3.07(a)	Graphite	2.2×10^{-6}
Franklinite	3.7×10^{-3}(a)	Fluorite	-2.85×10^{-7}(b)
Ilmenite	1.5×10^{-3}(a)	Aragonite	-3.92×10^{-7}
Magnetic pyrite	$0.337–5.75 \times 10^{-3}$(a)	Calcite	-3.63×10^{-7}
Siderite	8.4×10^{-4}	Ruby	4.7×10^{-7}(b)
Hematite	$0.11–1.1 \times 10^{-3}$(a)	Topaz	4.2×10^{-7}(b)
Zircon	-1.7×10^{-7}	Beryl	8.26×10^{-7}
Limonite	$7–8 \times 10^{-4}$(a)		3.86×10^{-7}(b)
Corundum	3.4×10^{-7}(b)	Epidote	2.38×10^{-5}
Pyrolusite	$6.2–7 \times 10^{-4}$(a)	Augite	2.66×10^{-5}
Manganite	4.9×10^{-4}(a)	Adularia	-4.27×10^{-7}
Garnet	3.75×10^{-4}(a)		-3.84×10^{-7}
Quartz	$1.75–4.38 \times 10^{-4}$(a)		-3.17×10^{-7}(b)
Rutile	1.96×10^{-6}	Diopside	8.8×10^{-6}(b)
Pyrite	$0.2–1.5 \times 10^{-4}$(a)	Sapphire	5.7×10^{-6}(a)
Zincblende	-2.64×10^{-7}(b)	Cobaltite	5.8×10^{-6}(a)
Dolomite	1×10^{-6}(b)	Feldspar	1.6×10^{-6}(a)
Apatite	2.64×10^{-6}(b)	Limestone	6×10^{-6}(a)
Willemite	1.9×10^{-4}(a)	Red serpentine	4.1×10^{-5}(a)
Chalcopyrite	8.5×10^{-7}(b)	Green serpentine	3.5×10^{-4}(a)
Spinel	0.2×10^{-7}(b)	Antimonite	-8.5×10^{-7}(a)
Galena	-3.5×10^{-7}(b)	Mica, transparent	$0.8–1.2 \times 10^{-5}$(a)
Halite	-5.0×10^{-7}(b)		
Celestine	-3.42×10^{-7}		
	-3.14×10^{-7}		
	-3.59×10^{-7}(b)		
Tourmaline	1.12×10^{-6}(b)		

* (a) Volume susceptibility; (b) specific susceptibility.

Substances with a positive susceptibility (corresponding to $\mu > 1$) are called paramagnetic; substances with a negative susceptibility (corresponding to $0 < \mu < 1$) are called diamagnetic. As can be shown in Equation (2), substances with a positive susceptibility have a positive magnetization and augment the flux density of a magnetic field; substances with a negative susceptibility have a negative magnetization and weaken the flux density.

A magnetic field exerts a force on each of the two poles of a *magnetic dipole*, making it align with the lines of the magnetic field. Because these forces are exerted in opposite directions, they are equal in a uniform magnetic field. Therefore, the net force on the dipole is zero. However, if the field has a *gradient*, i.e., it varies in space, the force on the dipole will be greater in the direction of the higher field and will be proportional to the magnetic dipole moment and the magnitude of the magnetic field gradient.

Process of Magnetic Separation. Magnetic separation relies on the different behavior of individual mineral particles under the influence of a magnetic field. Ferromagnetic or paramagnetic materials are attracted along the lines of the magnetic force from areas of lower magnetic field strength to higher field strength. Diamagnetic particles are repelled from areas of higher magnetic field strength to those of lower field strength.

The magnetization of various materials is directly dependent on their degree of magnetic susceptibility and the strength of the applied magnetic field (Eq. 2).

Ferromagnetic materials are quickly saturated magnetically, and increasing the magnetic field strength will not increase the magnetization beyond a certain point. For paramagnetic materials that are difficult to magnetize, the induced magnetization is proportional to the magnetic field strength applied, and some materials, practically speaking, cannot be saturated.

The efficiency of magnetic separation may be expressed by both the recovery (i.e., the ratio of magnetic material in the magnetics relative to that in the feed) and the grade (i.e., the fraction of magnetic material in the magnetic concentrate).

3. Types and Basic Application of Magnetic Separators

Magnetic separators are used in two basic areas:

1) tramp iron removal—equipment protection, and
2) mineral concentration and material purification.

3.1. Tramp Iron Removal

Tramp iron magnetic separators are used to protect material handling and process equipment such as crushers, pulverizers, and screens. These applications usually involve dry material or material with only surface moisture. Iron coarser than 3 mm is usually defined as tramp iron.

The size and shape of the tramp iron, together with the material handling system in use or proposed, must be considered in selecting magnetic separators suitable for tramp iron removal.

Magnetic equipment that has been developed for tramp iron removal includes

1) magnetic head pulleys—permanent and electromagnetic pulleys used as the head pulley in belt conveyor systems;
2) suspended magnets—magnetic units installed over belt conveyors or feeders;
3) plate magnets, which are used at the bottom of chutes or launders; and
4) grate magnets employed in discharge hoppers.

Magnetic Pulleys. Magnetic pulleys are elongated cylinders supported by a shaft, used as the head pulley for conveyors that transport material in a plant. Magnetic pulleys have magnets installed in them to provide a magnetic field of the

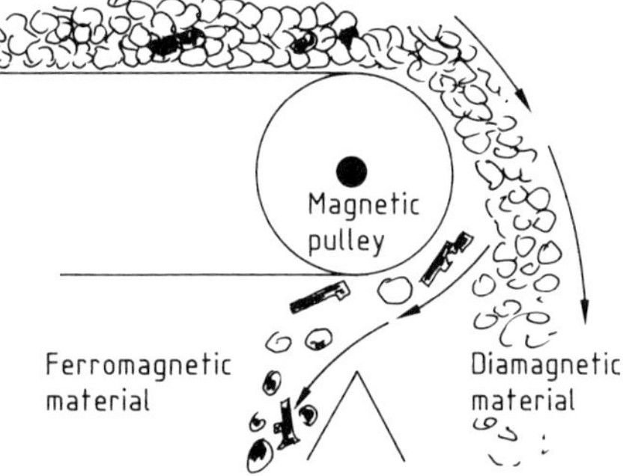

Figure 1. Operating principle of a magnetic pulley

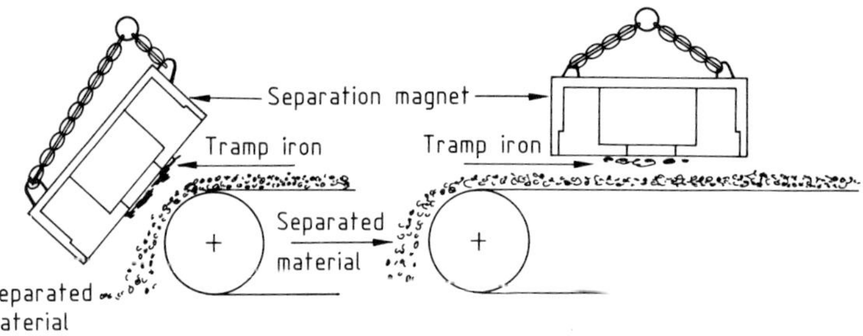

Figure 2. Two ways of mounting suspended magnets to remove core nails from foundry sand on a conveyor belt

same strength around the pulley circumference, which removes any tramp iron contained in the product. Magnetic pulleys are easy to install, have low initial cost, and accomplish continuous and automatic removal of tramp iron. A typical installation is shown in Figure 1. Magnetic pulleys are available in diameters from 0.2 to 6.8 m and in widths to match the conveyor belt width.

Suspended Magnets. Suspended magnets are rectangular steel boxes that contain permanent or electromagnets and provide a magnetic field to remove tramp iron from material carried on belt conveyors.

These magnets are suspended over conveyor belts or feeders and develop a deep magnetic field through which the material on the conveyor must move. The term deep means that the magnetic field of a suspended magnet can extend to 760 mm (a pulley's effective magnetic field only extends to 130 mm). Any tramp iron contained in the material being transported will be intercepted and attracted to the suspended magnet face.

Suspended magnets are available in both permanent and electromagnetic types, with the electromagnetic unit applied to the deeper loads carried on belt conveyors. The depth of material carried on the conveyor belt, the belt speed, and the clearance required over the material on the belt will determine the size of the suspended magnet. Suspended magnets can be made self-cleaning by installing a conveyor belt that runs over the face of the magnet. A typical application of a suspended magnet installation is shown in Figure 2.

Tramp Iron Magnetic Drums. Magnetic drum separators are used for tramp iron removal when installation of magnetic pulleys or suspended magnets is not feasible. The tramp iron magnetic drum incorporates a magnet assembly held in a fixed position inside a rotating drum cylinder or a shell rotated around this magnet assembly. The active separation zone (within a few centimeters of the surface of the drum) draws the ferromagnetic particles onto the surface of the drum; here, the motion of the drum carries them to a region outside the influence of the magnetic assembly. The ferromagnetic particles then drop or are scraped off the edge of the drum.

Feed material can be introduced on the top vertical center line of the rotating shell (*overfeed* arrangement) or under the drum for clean pickup of magnetic material (*underfeed*). These two feed arrangements are illustrated in Figure 3. Over-

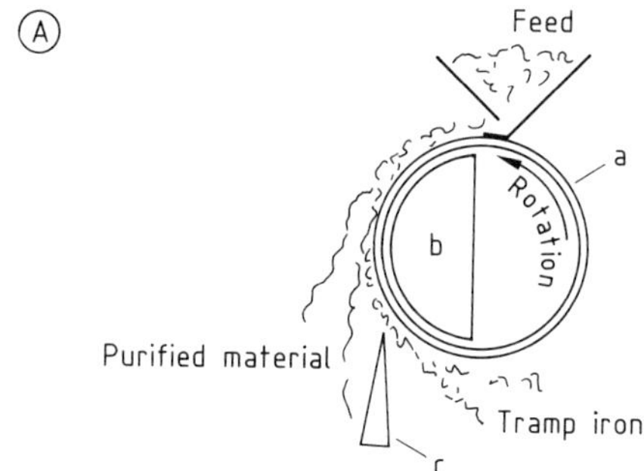

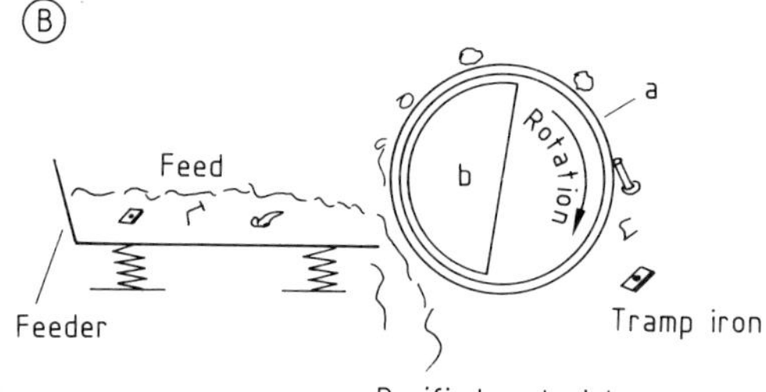

Figure 3. Feed arrangement for magnetic drums
A) Overfeed; B) Underfeed
a) Drum shell; b) Magnet; c) Adjustable nonmagnetic (diamagnetic) splitter

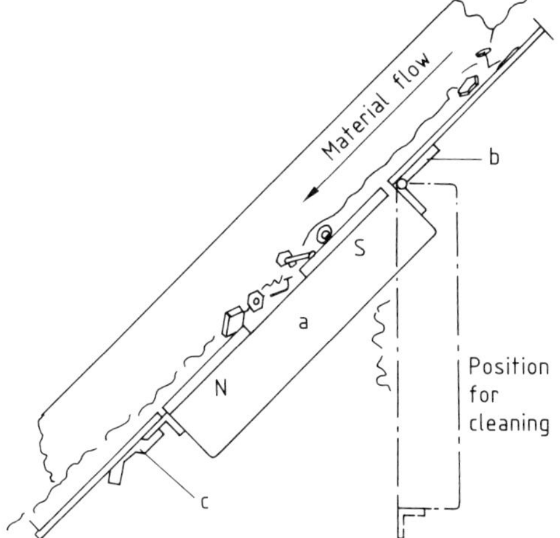

Figure 4. Operating principle of a plate magnet
a) Plate magnet; b) Hinge; c) Latch

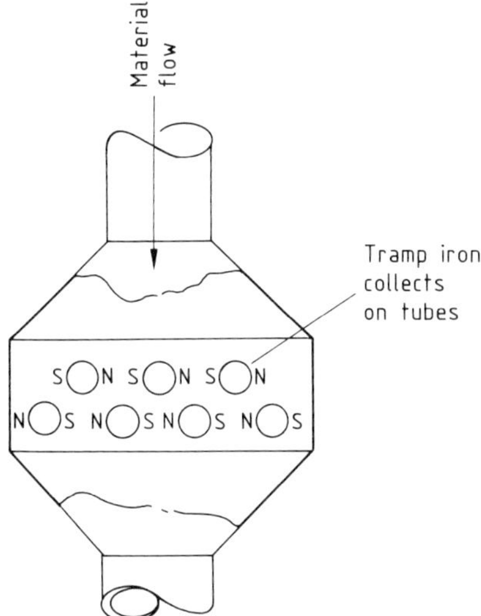

Figure 5. Operating principle of a grate magnet

feed magnetic drums provide the most complete magnetic removal because gravity is working with the magnetic field rather than against it as in the underfeed arrangement.

Plate Magnets. Plate magnets (see Fig. 4) are installed at the bottom of chutes or launders carrying raw materials. They develop a magnetic field which attracts and holds any tramp iron that might be present. Tramp iron is attracted and held on the face of the plate magnet from which it must be removed periodically by hand.

A chute angle of 45° or less is recommended, and the maximum depth of material flowing over the face of the plate magnet is 115 mm. The plate magnet should be installed as close to the feed point as possible.

Grate Magnets. A grate magnet consists of a set of magnetized tubes installed in a support frame through which feed material is allowed to flow. These tubes, containing the magnet elements, are installed on 25- or 37.5-mm centers, and the feed material is directed through these openings. The magnetic material collected on the magnetized tubes must be cleaned periodically by hand to remove collected ferromagnetics. Grate magnets can also be used to remove tramp iron from slurries or liquids. A typical grate magnet installation is shown in Figure 5.

3.2. Mineral Concentration and Product Purification

The magnetic responsiveness of minerals provides an effective means of concentrating naturally occurring ores.

Magnetic equipment used in *wet mineral concentration* includes

1) wet magnetic drum separators,
2) magnetic filters,
3) wet high-intensity magnetic separators (WHIMS), and
4) high-gradient magnetic separators (HGMS).

Magnetic separators used in *dry mineral concentration* include

1) alternating polarity magnetic drum separators,
2) induced roll magnetic separators,
3) high-intensity crossbelt magnetic separators, and
4) high-intensity disk-magnetic separators.

3.2.1. Wet Magnetic Separators

Wet magnetic drum separators incorporate a stationary electro- or permanent magnet assembly held in a fixed position within a revolving drum shell. The magnetic drum is mounted in a nonmagnetic tank arrangement.

The tank arrangement has a feed entry point as well as collection hoppers for removal of the weakly paramagnetic or diamagnetic ("non-

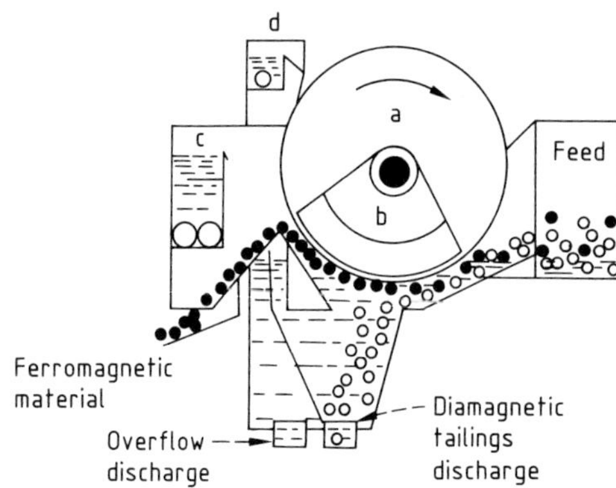

Figure 6. Wet drum magnetic separator
a) Rotation drum; b) Stationary magnet assembly; c) Repulping box; d) Drum wash

magnetic") and ferromagnetic products of the magnet assembly. A typical arrangement is shown in Figure 6.

The feed to a wet drum separator is typically a slurry in which the products to be separated are carried, with water as a medium.

Wet drum separators have generally been used to separate ferromagnetic particles such as magnetite and ferrosilica from diamagnetic silica and coal fines.

Magnetic Filters. A magnetic filter is a simple device in which a direct-current electric coil or permanent magnets are used to inductively magnetize a steel grid. The steel grid is arranged in such a way that the feed slurry or liquid passes through it. The grid has a large number of magnetized edges that serve to collect any ferromagnetic particles present in the slurry or liquid feed.

Periodically, the filter assembly must be cleaned of the ferromagnetic particles collected, which means that the feed must be stopped or bypassed to permit washing or back flushing of the filter element. Magnetic filters are used to clean oil, paint, clay slip, and other liquids or slurries containing fine iron contamination.

Wet High-Intensity Magnetic Separators. The WHIMS magnetic separator employs a rotating carousel in which a matrix of vertically grooved plates, enclosed wedge-shaped bars, expanded metal mesh, or steel balls is fixed within a ring. Figure 7 shows the operation of a carousel WHIMS unit. At each feed point, the slurry to be separated is introduced and the magnetics are collected on the matrix with the diamagnetic "nonmagnetics" passing into a "nonmagnetic" collection hopper below the separator. As the matrix rotates, it is washed to remove any physically entrapped "nonmagnetics."

Continued rotation of the matrix brings the collected ferro- and paramagnetics into an essentially unmagnetized zone where they are washed off by a high-pressure water jet. These WHIMS separators have high magnetic field strength (intensity) and high magnetic field gradient and can separate very weakly paramagnetic particles.

High-Gradient Magnetic Separators. The HGMS magnetic separator also has a high magnetic field strength and a high magnetic field gradient, but it employs a static magnetized matrix canister through which the feed slurry is passed.

Typically, this canister has a matrix of packed stainless steel wool which acts as a collecting element for the magnetic particles. Copper coils

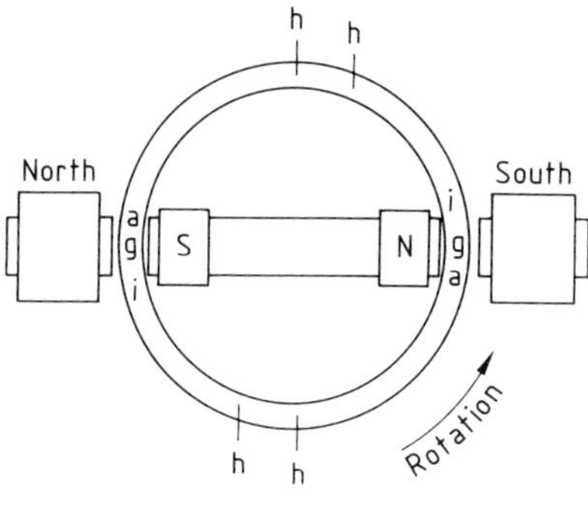

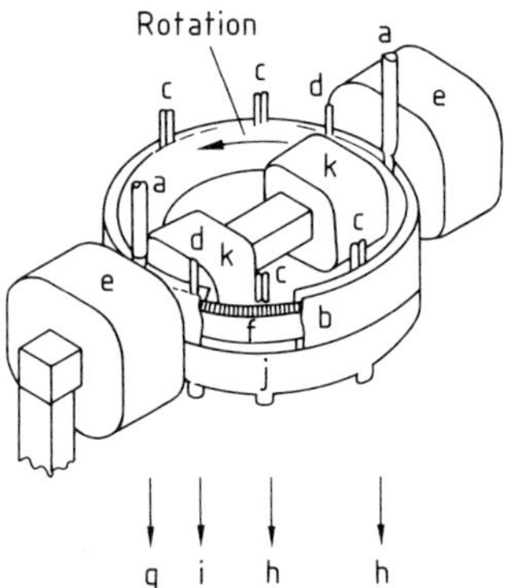

Figure 7. Wet high-intensity magnetic separator (WHIMS)
a) Feed pipe; b) Rotor; c) High-pressure water jet; d) Low-pressure water jet; e) Outer coil; f) Matrix; g) "Nonmagnetics" discharge; h) Magnetics discharge; i) Middlings discharge; j) Trough; k) Inner coil

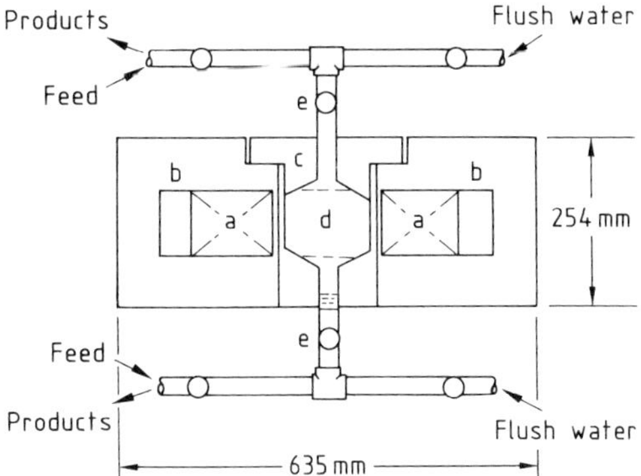

Figure 8. Typical flow through a high-gradient magnetic separator (HGMS)
a) Coil; b) Mild steel circuit; c) Nonmagnetic stainless steel matrix housing; d) Matrix area; e) Feed control for overfeed or retention time control for underfeed operation

which contain hollow conductors energize the canister and are cooled by circulating water or liquid helium circulating through these hollow conductors to permit the development of magnetic flux densities as high as 5 T. This extremely high flux density will extract very weakly paramagnetic impurities. A typical application is in kaolin cleaning. The separation of paramagnetic biomaterials is described elsewhere (→ 11. Biochemical Separations, **B3**, p. **11**-22).

When the matrix is saturated with particles of a positive magnetic susceptibility, water is introduced to wash the matrix of diamagnetics. The magnet is then deenergized, and the matrix is flushed with high-velocity water to remove the paramagnetics. Processing of the slurry is resumed after reenergization of the magnet. Typical flow through an HGMS magnetic separator is shown in Figure 8.

3.2.2. Dry Magnetic Separators

Magnetic Drum Separators. Two types of magnetic drum are used in mineral concentration or product purification. In the first type, as illustrated in Figure 9, the poles of the stationary magnet assembly *alternate across the face width* of the drum. This presents a strong pattern for holding the ferromagnetics to be removed and permits use with coarse material (up to 305-mm diameter).

The second type of magnetic drum, as illustrated in Figure 10, uses a magnet assembly in which the magnetic poles *alternate around the circumference* of the drum. This type of magnet assembly usually has a greater length of magnet arc (that part of the drum cylinder—expressed in degrees—under which the magnet is located) and develops considerable agitation and reorientation of the ferromagnetic particles as they are carried from the feed to the magnetic discharge point. A magnetic drum is usually fed by a vibrating or other type of spreading feeder *on the top vertical center line* of the drum. Ferromagnetic particles are attracted and held to the drum shell until they pass beyond the end of the mag-

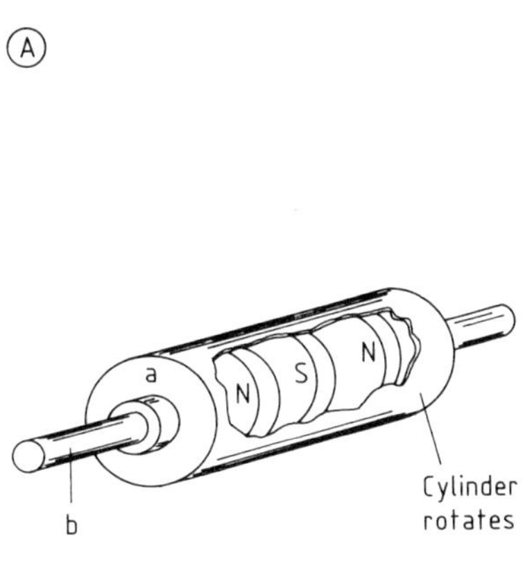

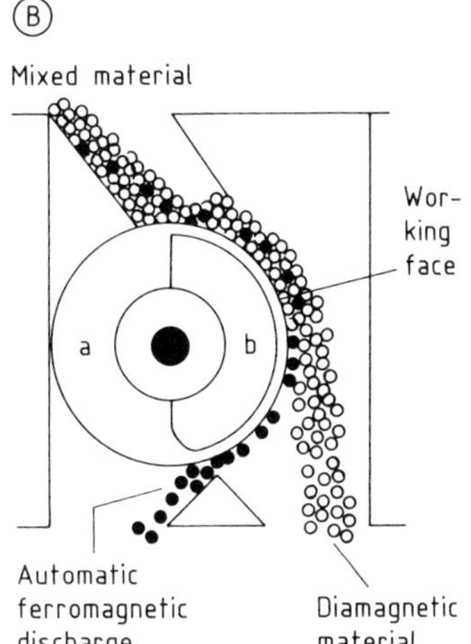

Figure 9. Magnetic drum separator with radial poles alternating across the face width of the drum
A) Arrangement of poles: a) Drum heads; b) Stationary shaft
B) Principle of operation: a) Revolving cylinder; b) Stationary magnet assembly

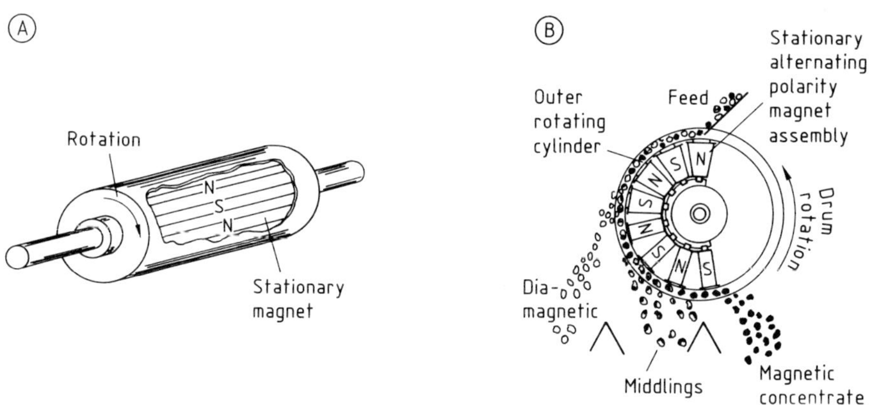

Figure 10. Magnetic drum separator with axial poles alternating around the circumference of the drum
A) Arrangement of poles; B) Principle of operation

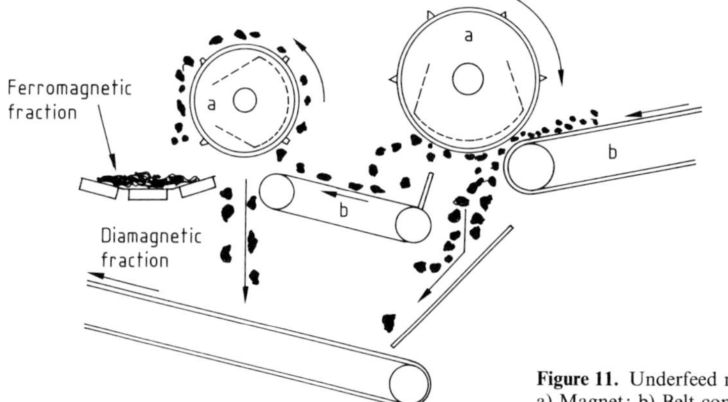

Figure 11. Underfeed magnetic drum separator
a) Magnet; b) Belt conveyor

net assembly, where they are discharged into a magnetic product collection hopper.

In other applications where maximum purity of the ferromagnetic product is desired, such as in municipal waste treatment, an *underfeed* such as illustrated in Figure 11 is used. In this application, the ferromagnetics must be picked up against the influence of gravity to reach the magnetic drum face.

Induced Roll Magnetic Separators. The induced roll magnet separator is a high-field-strength magnetic separator that achieves a high field gradient and can remove paramagnetic minerals from a dry granular feed. Because of the narrow magnetic gaps used on this type of separator, the feed is usually less than 3 mm in diameter, and because of the surface activity of very fine material, all particles ≤ 74 μm (≤ 200 mesh) in diameter should be removed.

The induced roll separator has one or more rotating rolls, made up of alternating steel and diamagnetic disks that are inductively magnetized by an electro- or permanent magnet source. The principle of operation is shown in Figure 12.

Feed is introduced from a hopper or feeder on the top vertical center line of the roll, and the highly magnetized edges of the steel disks serve to attract and hold the weakly magnetic mineral to the roll surface. A change in polarity occurs as the roll rotates, and the collected magnetics are discharged after the roll reaches its null point.

Various adjustments, such as air gap, roll speed, splitter setting, feed rate, and field strength, can be used to control induced roll separators.

High-Intensity Crossbelt Magnetic Separators. For very selective concentration of weakly magnetic (paramagnetic) minerals, the high-in-

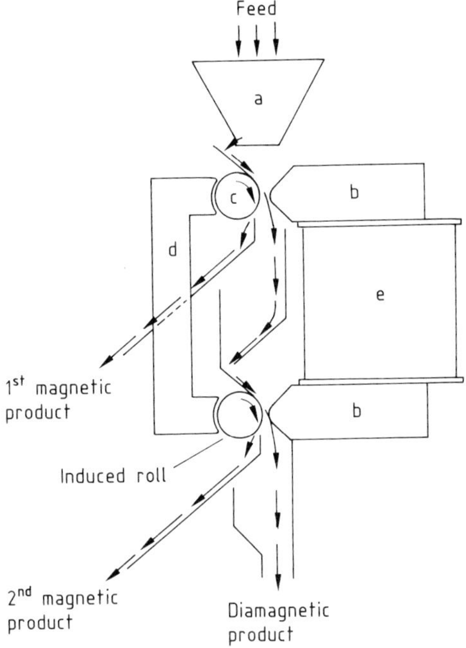

Figure 12. Induced roll magnetic separator
a) Feeding hopper; b) Pole piece; c) Induced roll; d) Bridge bar; e) Coil

tensity crossbelt separator has been used. This separator employs a feed belt on which a thin layer of the feed material is introduced to a high-strength magnetic field. Paramagnetic materials are lifted to the upper pole of this field and transferred by a crossbelt to a collecting hopper. To obtain the high field gradient required for separation of weakly magnetic minerals, the upper pole is shaped to a point (or series of points) for improved capacity, while the bottom pole is flat. When various weakly magnetic minerals are to be concentrated, a series of high-intensity poles is used, with coil strength increasing at each pole, or a variation in the field strength of the air gap may be employed. The operating principle of the high-intensity crossbelt magnetic separator is shown in Figure 13.

High-Intensity Disk-Type Magnetic Separators. The basic design and operation of a high-intensity, disk-type, rotor separator are shown in Figure 14. The induced magnet is a steel ring arranged to rotate between the feed belt and the primary magnet in such a way as to discharge para- and ferromagnetic products to the sides of the belt. This separator is used to very selectively concentrate such weakly paramagnetic materials as columbite, tantalite, wolframite, monazite, uexenite, and other high-value minerals. The use of a magnetized ring permits narrower separating air gaps and develops magnetic fields with a high flux density at each air gap.

4. Applications

The many types of magnetic separators result in a wide range of applications for these units. They are currently used in tramp iron removal, mineral processing, and iron recovery:

Removal of tramp iron:

Chemicals	Pharmaceuticals
Cooling fluids	Scrap metals
Food processing	Miscellaneous materials
Minerals	(water, glass, cork, textiles)

Mineral beneficiation:

Aluminum	Molybdenum
Barium	Nickel

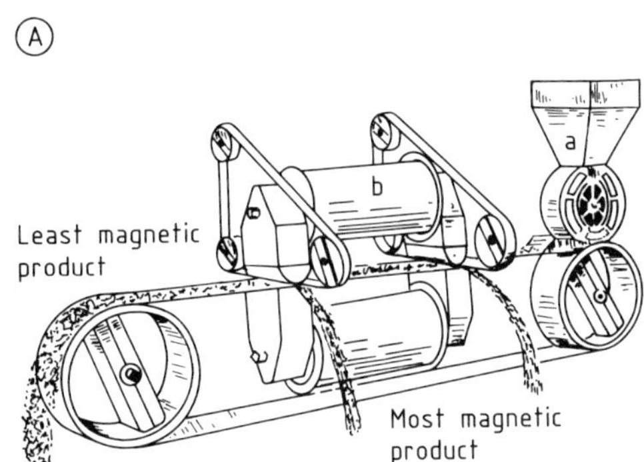

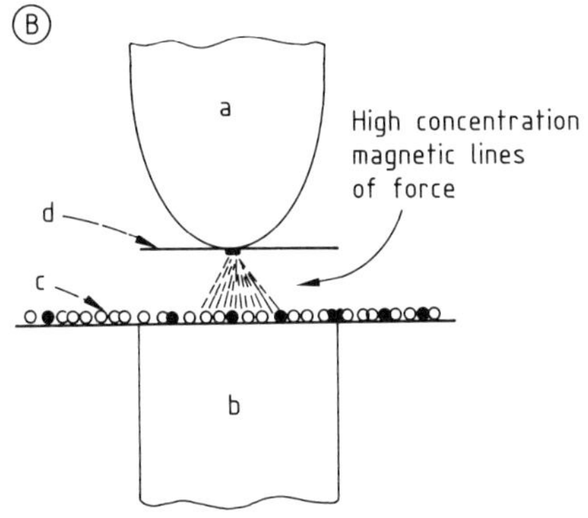

Figure 13. High-intensity crossbelt magnetic separator
A) Side view: a) Feeding hopper; b) Coil
B) Typical cross section and crossbelt arrangement: a) Parabolic curved nose of upper pole magnet; b) Flat lower pole; c) Feed belt; d) Crossbelt

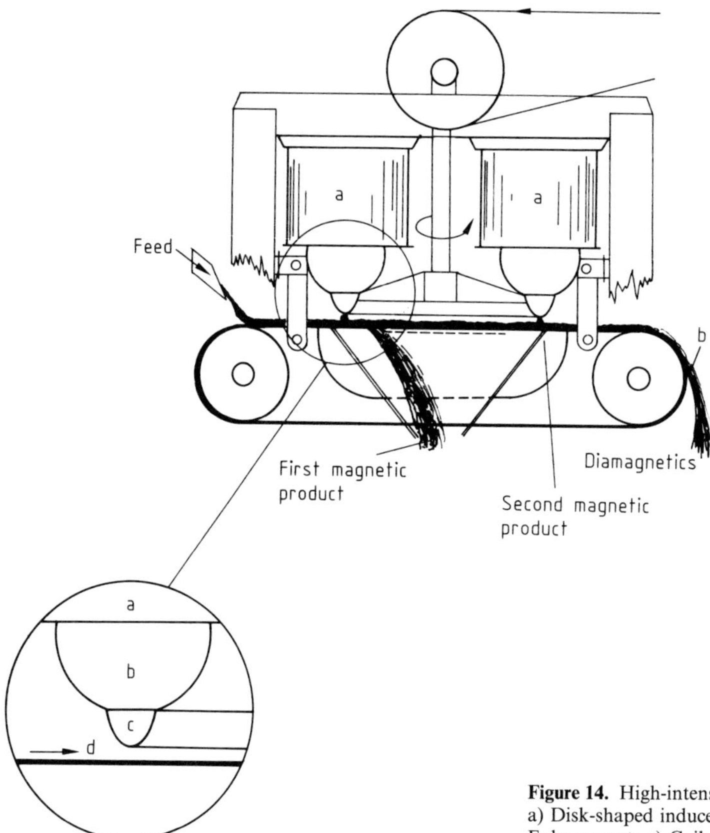

Figure 14. High-intensity disk-type magnetic separator
a) Disk-shaped induced magnet; b) Feed belt
Enlargement: a) Coil; b) Pole; c) Ring; d) Belt

Chromium	Niobium
Clay	Rare earths
Diamond	Rhenium
Garnet	Talc
Germanium	Tantalum
Hafnium	Tin
Iron	Titanium
Kyanite	Tungsten
Manganese	Yttrium

Recovery of iron:
Solid waste
Heavy media (ferrosilicon)

The newer, high-intensity, high-gradient magnetic separators open up other areas of magnetic treatment that have been pursued in varying degrees to date. Some of these potential applications are indicated in the following material:

Mineral beneficiation in addition to those systems listed above:

Antimony	Magnesium
Arsenic	Mercury
Asbestos	Mica
Beryllium	Platinum-group metals
Bismuth	Radium
Cadmium	Scandium
Coal	Silicon
Cobalt	Silver
Copper	Sulfur
Feldspar	Tellurium
Gallium	Thallium
Gold	Thorium
Graphite	Uranium
Indium	Vanadium
Lead	Zinc
Lithium	Zirconium

Water treatment (magnetic filtration):
Diamagnetic suspended solids (by magnetic seeding and flocculation)

Dissolved solids (by magnetic seeding and flocculation)
Paramagnetic suspended solids
Oils

Waste treatment for recovery of ferromagnetic materials:
Coal and oil ash
Ore tailings
Smelter and furnace dust

Removal of paramagnetic particulate impurities:
Chemicals
Fluids
Minerals
Pharmaceuticals

Chemical processing:
Recovery of paramagnetic fine precipitates: deposition and recovery of substances on ferromagnetic particles.

The selection of a particular magnetic separator will be influenced by several factors such as

1) size of feed material,
2) tonnage or capacity to be handled,
3) relative magnetic responsiveness (i.e., susceptibility) of the material to be separated magnetically,
4) condition of feed material (wet, dry, or liquid slurry),
5) purity required in the ferro- or paramagnetic concentrate or the diamagnetic product,
6) temperature of the feed material and point of application,
7) material handling system involved,
8) operating cost parameters, and
9) level of magnetic strength required.

5. References

All references given are general references

[1] J. D. Kraus: *Electromagnetics*, McGraw-Hill, New York 1953.
[2] D. M. Hopstock: "Fundamental Aspects of Design and Performance of Low Density Dry Magnetic Separators", *Trans Soc. Min. Eng. AIME* **258** (1975) 222–227.
[3] *Permanent Magnet Handbook*, Crucible Steel Company of America.
[4] R. L. Sanford: *Permanent Magnets*, National Bureau of Standards C448, 1944.
[5] J. E. Lawver, D. M. Hopstock: "Wet Magnetic Separation of Weakly Magnetic Minerals", *Miner. Sci. Eng.* **6** (1974) no. 3, 154–172.
[6] R. S. Dean, C. W. Davis: *Magnetic Separation of Ores*, Bulletin 425, US Bureau of Mines, 1942.
[7] A. Nussbaum: *Electronic and Magnetic Behavior of Materials*, Prentice-Hall, Englewood Cliffs, N.J., 1967.
[8] V. G. Derkatsch: *Die magnetische Aufbereitung schwachmagnetischer Erze*, VEB Deutscher Verlag für Grundstoffindustrie, Leipzig 1960.
[9] J. A. Oberteuffer: "Magnetic Separation: A Review of Principles, Devices, and Applications", *IEEE Trans. Magn.* **MAG 10** (1974) no. 2, 223–238.
[10] P. W. Selwood: *Magnetochemistry*, 2nd ed., Interscience, New York 1956.
[11] J. E. Lawver, D. M. Hopstock: "Electrostatic and Magnetic Separation", *SME Mineral Processing Handbook*, Sect. 6, Society of Mining Engineers, New York 1985.
[12] I. S. Wells: "Wet Separation of Paramagnetic Minerals", *Chem. Eng. (Rugby, England)*, 1982.
[13] G. Clark (ed.): "Magnetic Separation", *Ind. Miner.*, 1985.
[14] J. E. Forciea, L. G. Hendrickson, O. E. Palasvirta: "Magnetic Separation for Mesabi Magnetite, Taconite", *Min. Eng. (Littleton, Colo.)* **10** (1958) 339–345.
[15] J. E. Forciea, R. W. Salmi: "Primary Magnetic Separation Specifications", *Trans. Soc. Min. Eng. AIME* **232** (1965) 339–345.
[16] B. Skold: "Progress in Magnetic Separation", *Mineral Processing Meeting of Swedish Mining Society*, Skellerftea, January 1970.
[17] G. H. Jones, *Proc. Int. Miner. Process. Congr.*, Institute of Mining and Metallurgy, 1960, pp. 717–732.
[18] J. Iannicelli, *Clays Clay Miner.* **24** (1976) 64–68.
[19] E. J. Tenpas: "Magnetic Separators—Types and Applications", *Rock Prod.* , 1971.
[20] E. Laurilla: *An Approach to the Theoretical Treatment of Magnetic Concentration*, ser. A, vol. 6, Physica-Helsinki 1958.
[21] H. W. Buus: "How to Select Magnetic Separation Equipment", *Foundry*, December 1960.
[22] R. K. Singhal: "Magnetic Separators for Mineral Processing", *Min. Mag.*, **113** (1965) no. 5, 356–365.
[23] E. J. Roberts, P. Stavenger, J. P. Bowersox, A. K. Walton, M. Melita: "Solids Concentration", *Chem. Eng. (N.Y.)*, June 29, 1970, pp. 52–68.
[24] W. J. Bronkala: "Purification: Do It With Magnets", *Chem. Eng. (N. Y.)* **95** (1988) no. 3, 133–138.
[25] P. B. Sherwood: "Cross Belt and In-Line Primary Magnets", *N.S.A. Operators Meeting*, Salt Lake City, Utah, February 4–6, 1976.
[26] R. B. Jacob, J. A. Selvaggi: "Operating Characteristics of the Newly Designed High Gradient Magnetic Separator (HGMS)", *AIME Annual Meeting*, Atlanta, Ga, March 6–10, 1983.
[27] H. Harley-Smith: "Magnetic Separation in Mineral Beneficiation", Metals and Minerals International. *Perry's Chemical Handbook*: "Solid–Solid Systems and Liquid–Liquid Systems", 6th ed., Chapter 21, McGraw Hill, New York 1984.
[28] M. Gupta: "A New Method of Magnetic Separation of Ferro and Para-Magnetic Minerals from Feebly Magnetic and Non-Magnetic Minerals", *Proc Australas. Inst Min Metall*, No. 274, June 1980.
[29] G. Clary, (ed.): "Magnetic Separation–4", *Industrial Minerals*, May 1985.

[30] L. A. Roc: "Magnetic Separation of Ores–History and New Foreign Developments", *Annual AIME Meeting*, New York, February 16–20, 1958.
[31] R. F. Merwin: "How to Remove Tramp Iron from Burdens Carried on Conveyor Belt", *Annual AIME Meeting*, Tampa, Florida, October 13–15, 1966.
[32] C. K. McArthur: P. R. Porath, "Concentrating Iron Ore in a Mobile Plant", *Min. Congr. J.* **33** (1965) October 28.
[33] J. A. Oberteuffer: *Magnetic Separation: A Review of Principles, Devices and Applications*, Institute of Electrical and Electronics Engineers Inc., Copyright, 1977.

20. Electrostatic Separation

Frank S. Knoll, Carpco, Inc., Jacksonville, Florida 32206, United States
James E. Lawver, Consultant, Lakeland, Florida 33803, United States
Joseph B. Taylor, Carpco, Inc. Jacksonville, Florida 32206, United States

1. General Principles

Millions of tonnes of titanium ores, iron ores, salts, and other minerals are processed every year by electrostatic separation. Over the past 10 years, major advances in electrostatic separation and its application have occurred. New processes include dewatering clay slurries, separating halite and sylvite, and shape separation of vermiculite and gangue minerals.

The common factor for all electrostatic separations of particles is that opposite charges attract one another. How these charges occur distinguishes one method of electrostatic separation from another. When processing granular mixtures, electrostatic separation can be effected due to differences in conductivity, triboelectric effect, and polarizability. How charges occur when a particular separation occurs will influence the form of electrostatic separation. Electrostatic separation in which charge is transferred to or from a particle is called *electrophoresis*. Electrostatic separations of this type are based on conductivity or triboelectric differences in the particles comprising a mixture. Electrostatic separation where there is no external transfer of charge but where polarization is induced is called *dielectrophoresis*. Electrostatic separations by dielectrophoresis are based on differences in polarizability of the particles comprising a mixture. Polarizability of particles, depending on their environment, can be due to differences in dielectric constant, shape factors, and material structure. In both forms of electrostatic separation—electro- or dielectrophoresis—the mechanism is the same; opposite charges attract one another. A comprehensive introduction into the theory of electrostatic separation is given in [1].

1.1. Electrophoresis

Electrophoresis is the motion of a charged body or particle under the influence of an electric field. Simply put, like charges repel and unlike charges attract one another (see Fig. 1). An electrostatic mineral separation uses electrophoresis when the force acting on a particle is due to the interaction of an electric field and a charged particle. The electric field can either be from a high voltage source or from the charged particle's own electric field. Particle charging can occur in three forms—ion bombardment, conductive induction, or contact charging.

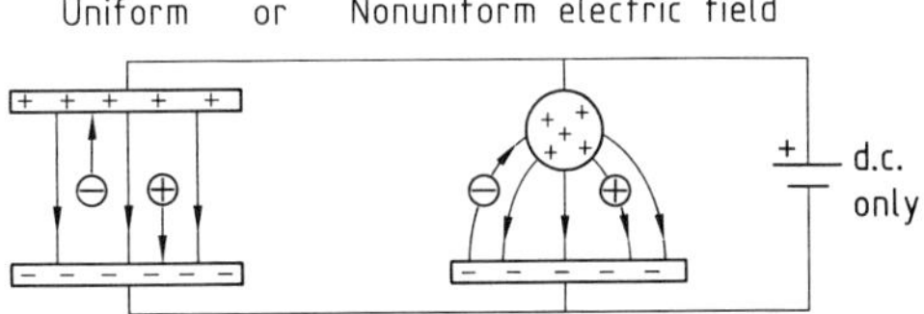

Figure 1. Electrophoretic force on charged particles

1.1.1. Charging by Ion Bombardment

Charging and separating dry minerals by ion bombardment is the most common form of electrostatic separation. Millions of tonnes of minerals are processed each year by this method (see Table 1).

In an ion bombardment separation, granular material is fed onto a grounded metal cylinder (or roll) and charged by a corona-producing electrode placed above the roll's surface (see Fig. 2 a). Typical equipment to generate an ion or corona discharge includes a high-voltage d.c. power supply and a specially made electrode to create corona. While both conductors and nonconductors become charged, only the conductor is able to lose its charge. The charged nonconductor, as it rests on the roll's surface, "sees" an oppositely charged image of itself in the metal surface. It is attracted to the image charge, becomes electrostatically pinned to, and moves with the roll's surface. The conductive particle also sees an image and is attracted to it. But upon touching the roll's surface, it discharges rapidly to the grounded surface and is thrown free from the roll's surface with a projectile motion.

Shape separation is a new form of ion bombardment separation. Unlike conventional ion bombardment where separation is based on conductivity, shape separation uses advanced roll construction and controlled charging techniques to separate materials based on the particle's shape and density, without consideration to the constituents' inherent conductivities (see Fig. 3) [3]. The separation characteristics of large particles can be determined by the simple calculation of the "flatness" coefficient.

The flatness coefficient for a typical particle is determined by considering the particle resting on a horizontal surface in its most stable position. The maximum dimensions of the particle in the direction parallel and perpendicular to the surface on which it is resting are taken as particle length L, and thickness T. The flatness coefficient K is determined by the ratio of L to T. The greater the coefficient, the flatter the particle. Effective separations can be carried out if the ratio of flatness coefficients K_A/K_B of two particulate materials is >2. The efficiency of separation will increase with the ratio of flatness coefficients. A recent installation using this process is the recovery of coarse vermiculite from gangue (silicious rock) whereby the overall existing plant throughput was increased by rejecting a barren gangue fraction early in the process. Figure 3 illustrates the flatness coefficient calculation for typical vermiculite and gangue particles. Operating particle size ranges for shape separation as well as other electrostatic techniques are shown in Figure 4.

Table 1. Application of electrostatic separators

Process	Application	
	Industrial installation	Pilot or laboratory testing
Electrophoresis:		
Ion bombardment	beach sands and alluvial tin ores	coal cleaning * electrostatic assaying
	silica from iron and chromite ores	
	production of iron ore super-concentrates	
	chopped wire and plastic separation	
	removal of stained particles from glass sands	
	total metallic removal from non-metallic materials (ceramics, plastics, etc.)	
	shape separation: vermiculite – mica from rock	
Conductive induction	final rutile cleaning final zircon cleaning impurities from foodstuffs	
Contact charging	halite – sylvite * separation clay dewatering *	barite – quartz fluorspar coal cleaning feldspar – quartz phosphate – quartz *
Dielectrophoresis:		
Fluid media	catalyst fines from petroleum products	Dielectric separation of various minerals, i.e. rutile from quartz
Air media	fiber – foodstuffs	vermiculite and mica fiber separations paper – plastics fiber – foodstuffs

* incorporates some form of chemical surface conditioning

1.1.2. Charging by Conductive Induction

Conductive induction is similar to that of ion bombardment in that large differences in particle surface conductivity are needed to effectively separate materials. Particle charging is less.

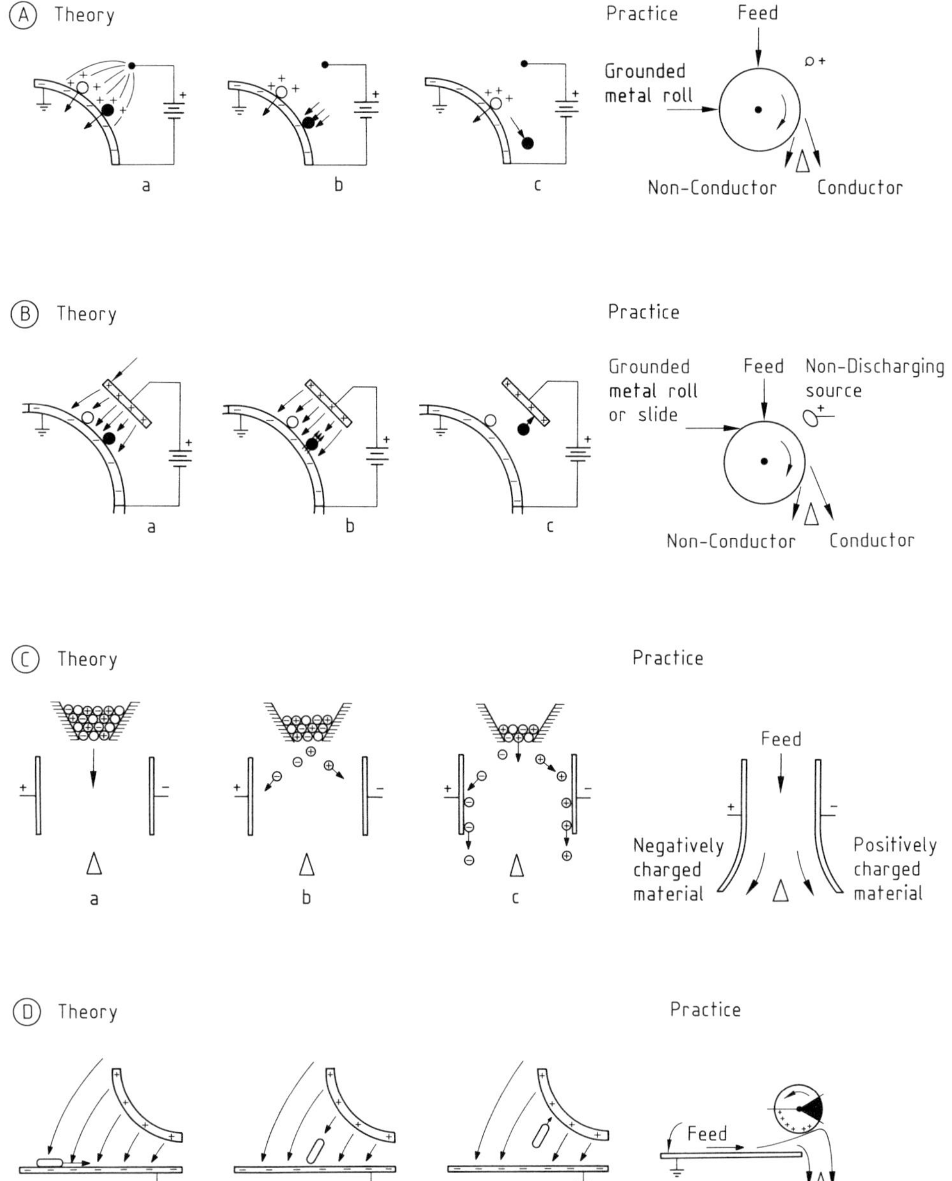

Figure 2. Forms of electrostatic separation
A) Ion bombardment
a) Both particles charged by corona cling to grounded metal roll; b) The conductor allows charge to dissipate; c) Discharged particle falls free
B) Conductive induction
a) Both particles rest on grounded plate under electrode; b) Charge conducts through and charges conductive particle; c) Charged particle lifts from surface attracted to charged electrode
C) Contact charging
a) Material previously charged by contact enters separator; b) Particles, while falling, are attracted to electrode of opposite potential; c) Particles fall on either side of a divider
D) Dielectrophoresis
a) Particle moves into nonuniform electric field; b) Particle polarizes and aligns itself with electric field lines; c) Particle lifts and moves in the direction of the highest intensity

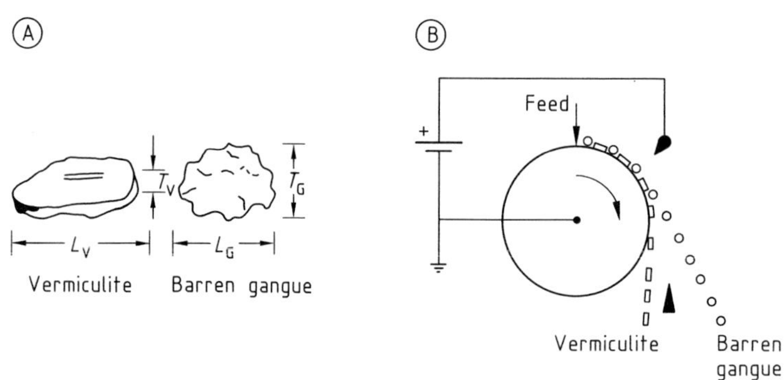

Figure 3. Electrostatic shape separation of vermiculite and barren gangue
A) Flatness coefficient calculation; ratio of "flatness" coefficients $= \frac{K_V}{K_G} = 3.70 \geq 2$

Vermiculite: $T = 5$ mm, $L = 25$ mm, flatness ratio $K = L/T = 5$
Gangue: $T = 17$ mm, $L = 23$ mm, flatness ratio $K = L/T = 1.35$
B) Industrial practice

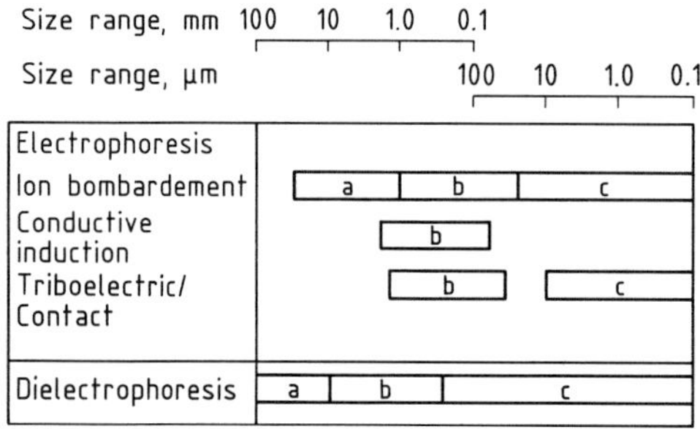

Figure 4. Electrostatic separation vs. size range
a) Dry methods–extended range with special materials; b) Dry methods—standard separation range; c) Wet methods—standard separation range

Therefore, the forces generated by this type of separation are generally weaker than that of ion bombardment, and per-stage efficiency is decreased.

In a conductive induction separation, a granular mixture is fed onto a plate or grounded roll above which a high voltage static or non-ionizing electrode is placed (Fig. 2 B). Conductive particles in contact with a grounded surface, when passed under an active nondischarging electrode, become charged to the polarity of the grounded surface. Under the influence of the static electrode, the inductively charged particle is attracted to and lifted away from the particulate flow by a static (nondischarging) electrode of opposite polarity. In case larger and smaller particles have the same density, attraction is a function of size with smaller particles being deflected the furthest due to their low mass. Less conductive particles unable to gain or conduct charge pass through the separation zone unaffected. Industrial conductive-induction separators, referred to as screen and plate-type separators, are used for the final electrostatic cleaning of industrial minerals such as rutile and zircon.

1.1.3. Charging by Contact Electrification

Contact electrification is one of the oldest methods used to separate nonconductive minerals electrostatically. Contact charging occurs when two minerals of different composition are repeatedly brought into contact with one another. A rule of thumb often used for predicting the sign of the surface charge resulting from contact electrification is Cohen's rule. It states that "when two dielectric materials are contacted and separated, the material with the higher dielectric constant becomes positively charged." Once charged, the material is allowed to free-fall between two oppositely charged electrodes (see Fig. 2 C) and separated. The naturally occurring charges that are generated by contact charging are usually weak and easily dissipated by particle collisions and fluctuations in relative humidity [1].

These limitations have been overcome for certain mineral separations by the introduction of surface conditioning techniques prior to electrostactic separation using free-fall methods. Wet or gas phase conditioning methods, combined with improvements in separator design today, permit the separation of sodium from potassium salts on a large scale. Other laboratory and pilot scale tests that incorporate forms of preconditioning show promise in the fields of pebble phosphate and coal (see Table 1).

1.2. Dielectrophoresis

Dielectrophoresis is the movement of neutral particles in a nonuniform electric field (see Fig. 5). Dielectrophoresis differs from electrophoresis in [1], [2]:

1) Dielectrophoresis produces motion on bodies suspended in a fluid medium. The direction of that motion is independent of the sign of the field. Accordingly, either a.c. or d.c. fields can be used. Electrophoresis on the other hand produces a motion of the suspended particle where the direction of the resultant path depends on both the sign of the charge on the particle and the sign of the field direction. Reversal of the field reverses the direction of travel.

2) Dielectrophoresis gives rise to an effect that is proportional to the particle volume and is, therefore, more easily observed on coarse particles. It is observable at the molecular level only under special conditions. Contrary to that electrophoresis is observable with particles of any size—atomic ions, molecular ions, charged colloidal particles, or charged macroscopic bodies.

3) Dielectrophoresis usually requires quite divergent fields for strong effect. Electrophoresis operates in both uniform and divergent fields.

4) Dielectrophoresis requires relatively high field strengths. In media of low dielectric constant K (e.g., $K = 2-7$), this is usually some 10^4 V/m. In media of high dielectric constant (e.g. water, 80), lower fields may be used successfully (500 V/m). Contrary to that electrophoresis with previously charged particles (e.g. ions, charged sols) requires relatively low fields.

5) Dielectrophoresis usually requires a substantial difference in the relative dielectric constants of the particle and the surrounding medium, e.g. $(K_2\text{-}K_1) > 1$. Electrophoresis can be appreciable even when the free charge per unit weight of the particle is quite small.

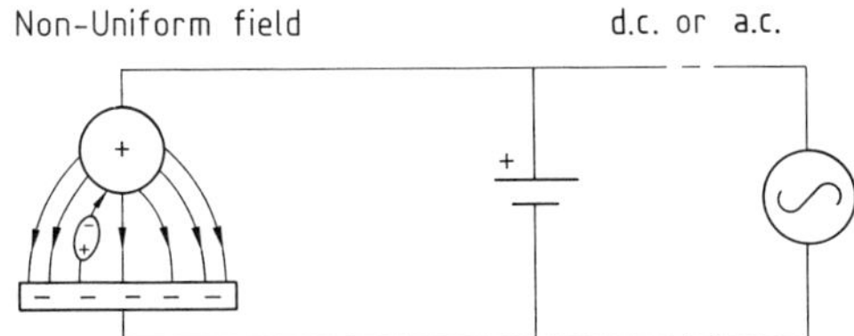

Figure 5. Dielectrophoretic force on neutral particle

Presently (in the 1980s), dielectrophoretic separations are largely a laboratory and pilot plant phenomenon. Industrial applications are limited to filtration of dielectric fluids, such as petroleum, and separation of impurities from foodstuffs. Dielectrophoresis is an emerging technology that will provide new solutions in the area of physical processing.

The properties of a dielectrophoretic separation vary with the medium in which separation occurs. In a dielectric field medium of an intermediate dielectric constant to the constituents, a separation can be performed based on differences in the dielectric constants of the feed materials. Commercial use of such a dielectric separation is hindered, though, by the cost and hazard associated with various intermediate dielectric fluids. Examples of intermediate dielectric fluids include nitrobenzene, kerosene, xylene, ethanol, and propanol. When air is used as a medium, separation is largely based on differences in shape, density, and dipole moment. Potential commercial applications using air as a medium, include fiber–foodstuff separation, and separation of vermiculite–mica from gangue mineral (see Table 1).

2. Applications of Electrophoresis

2.1. Ion Bombardment

Separation of Minerals. One of the largest ongoing ion bombardment electrostatic applications is at Wabush Mines in Labrador, Canada. This 1 kt/h installation uses a two-stage circuit (rougher–scavenger) equipped with high-tension roll separators to reduce the silica content of a wet gravity concentrate from 6.5% down to 2.5% SiO_2 at yields of 92%. This hard rock iron ore plant incorporated the first major use of electrostatic separation in minerals outside the beach sands industry (see Fig. 6).

The electrostatic separation of conductive titanium minerals from less conductive heavy minerals is one of the most quoted applications of this technology. Ion bombardment roll separators are used worldwide throughout the beach sands industry to isolate TiO_2-bearing minerals from a gravity concentrate containing 20 or

Figure 6. Installation for ion-bombardment electrostatic separation at Wabush Mines (Canada)

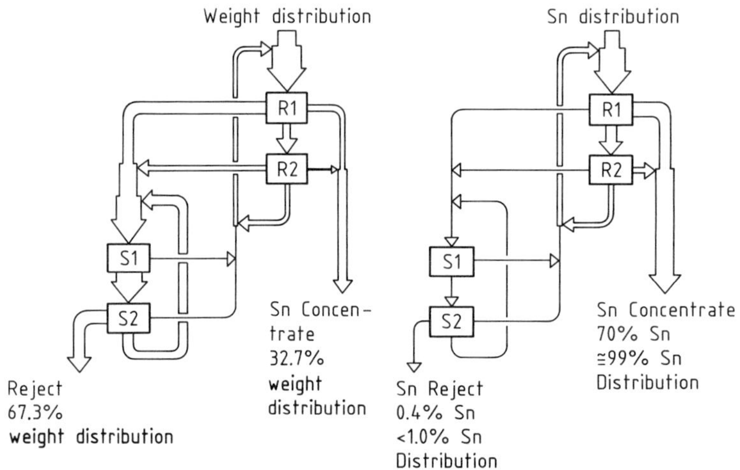

Figure 7. Electrostatic separation results using ion bombardment techniques
Materials: Hard rock, tin ore; feed grade ≃ 22.5 % Sn

more other minerals. Production of the nonconductive mineral, zircon, also necessitates the use of both ion bombardment and conductive induction techniques (roll and plate separators) to remove TiO_2 as a contaminant.

The widespread use of electrostatic separation in the beach sand field has spawned a number of improvements in equipment design in the 1970s. Advances in feeding, electrode design, and high voltage controls have led to simplified process circuitry, reduced circulating loads, and greater stage efficiency. Machine improvements center around the use of a new velocity feed system which presents particles to a rotating roll at near to the surface speed of the roll. Electrode design has been improved by the combination of charging and attracting electrodes working together to separate materials. Automated electrode cleaning reduces dust buildup on ionizing electrodes and maintains maximum operating efficiency of the separator. These improvements can often be retrofitted to existing separators at considerable savings. Three-stage circuits (rougher, scavenger, and cleaner) for recovering conductive minerals of tin and titanium typically achieve recovery of more than 98 % while eliminating virtually all other nonconductor minerals. A material balance is shown in Figure 7 for a new Canadian hard rock tin plant in operation since 1985/1986.

Recovery of Values from Electronic Scrap. With the rapid growth of integrated circuit technology, vast amounts of electronic scrap have been generated from production and assembly operations. Significant amounts of precious metals such as gold and silver are often incorporated as part of the design of integrated circuits or printed circuit board manufacture. Recovery of these values can be accomplished by grinding or granulating the chips and circuit boards to liberate the metal values and processing electrostatically to concentrate the metal values by ion bombardment techniques. Semiprecious metals encapsulated in ceramic materials can also be concentrated after granulation by using ion bombardment techniques to strip out nonferrous metals. Smelting or leaching is then used to further refine the precious metal values contained in the conductor fraction or nonconductor fraction.

Recovery of Values from Chopped Wire and Cable. Thousands of tonnes of wire and cable scrap are generated each year from off-specification in-house production and power company maintenance. The metal values, copper and aluminum, are usually recovered with granulation followed by an air–gravity separation. The most frequent problems associated with a wire granulation line are the recovery of fine metallics from stranded wire, and reusable plastics (PVC, polyethylene, nylon etc.) free of metallics.

Ion bombardment separation enhances conventional air–gravity separation in two ways:

Middling Circuit. Excessive fines are generated and lost when fine liberated wire is regranulated. These values can be recovered by electrostatic separation, leaving unliberated particles for rechopping. Typical middlings product is composed of 60 % metallics and 40 % plastics. After an electrostatic separation step, a metallic concentrate (>90 % recovery of the free metal) would consist of 99 % metallics and 0.5 % plastics. The remaining unliberated plastic and metal concentrate would then go to regranulation.

Plastics Recovery. As the value of petroleum and plastics made from it increases, so will the need to recycle these materials. Thermosetting plastics such as nylon, PVC, and polyethylene can be recovered from chopped wire and cable for reuse. A typical air–gravity reject product would consist of 5 % fine metallics and 95 % plastics. After electrostatic processing, a plastic concentrate (>90 % recovery of plastic) would contain 99.9 % plastics and <0.1 % fine metallics. The cleaned plastic is then recompounded, screened, and pelletized for reuse.

Electrostatic Assaying. A roll-type electrostatic separator with modifications has been successfully developed as a control technique to effect the removal of trace amounts of metallic particles from large quantities of granular nonconductive materials. With this rapid technique a single metallic particle can be recovered from a 136-kg sample of sand-like material (6×10^{18} grains) with a confidence level of 99.2 %.

Improved Electrode Systems. A new electrode system is now available to improve the performance of separators known as high-tension roll separators. The Carpco IC Electrode System combines ion bombardment and conductive induction techniques to improve the recovery of conductive materials, especially fine conductors. When applied to conventional ion bombardment separations, the IC Electrode System has proven, both in the laboratory and in industrial practice, gains in grade/recovery up to 25 % over conventional electrodes. This increase in per-separation-stage efficiency will alter present day plant design by reducing middling recirculation load and possibly eliminating some scavenging machine stages. Most conductor/nonconductor roll-type electrostatic separations can be improved when retrofitted with the new IC Electrode System.

Electrostatic Shape Separation. Laboratory test work in the late 1970s led to the development and installation of electrostatic shape separation of vermiculite and barren gangue in 1981 in South Africa. A single unit processes material at a rate of 7.5 t/h of vermiculite. The unit is a modified Carpco roll-type ion bombardment electrostatic separator [3].

Pilot-scale tests with a feed containing 19–20 % vermiculite and 80–81 % rock of a particle size range of 2–25 mm gave a concentrate with a vermiculite content of 93 % at recovery >94 %.

Continuing development had shown the process sucessful for other shaped materials such as mica–rock and flat–round metal separations.

2.2. Contact Charging

Recent laboratory work with contact (triboelectric) charging of minerals using a lined air cyclone prior to electrostatic separation has met with success on a laboratory or pilot scale [4]. Table 2 shows results to date for various minerals.

Table 2. Results of separations using contact charging

Material	Preconditioning	Test results	Notes
Barite–quartz	acrylic lined air cyclone	concentrate: 97–98% $BaSO_4$ recovery >95%	two-stage separation
Fluorspar	stainless steel air cyclone	concentrate: 97% CaF_2 recovery >90%	one cleaning stage followed by two scavenging stages
Coal (Federal Republic of Germany)	stainless steel air cyclone	starting: 0.82% inorganic sulfur 20.3% ash concentrate: 0.33% inorganic sulfur 5.0% ash	one or two stages of cleaning
Feldspar–Quartz	stainless steel air cyclone, feed and cyclone heated to 100 °C	quartz product grade +90% SiO_2	
Phosphate	stainless steel air cyclone, feed and cyclone heated to 150–200 °C	grades higher than 75% $Ca_3P_2O_8 \cdot H_2O$	increase of 10% with respect to run-of-mine feed grade

2.2.1. Contact Charging with Chemical Surface Conditioning

Salt Separation. The problems associated with the disposal of brine solution from sylvite–halite separation by flotation or leaching has led to the development of a dry electrostatic technique for concentrating sylvite and halite [5], [6]. The first commercial installation (in the Federal Republic of Germany in 1973/74) was processing 1.8 to 2×10^6 t/a in 1978 [7].

The process involves two stages:

1) Vapor phase preconditioning of the feed material with either trichloroacetic or salicylic acid in an enclosed vessel (Fig. 8). Preconditioning enhances the naturally occuring charge differences of the sylvite and halite.
2) Free-fall type electrophoretic separators are used to separate the halite–sylvite mixture. The free-fall electrostatic separator used consists of two parallel electrodes made of tubes placed side by side. This arrangement allows the surfaces of the electrodes to be rotated and cleaned continuously, preventing subsequent particle buildup on the electrodes and possible loss of efficiency. The free-fall separator installation is reported to be 40 m long and 46 m high and can handle 30 t m^{-1} h^{-1} of feed material.

Typical results are: particle size up to 1.5 mm; separator type free-fall, 4 kV/cm, pretreatment temperature, 70° to 160 °C, vapor phase; feed grade, 12.8% sylvite, 7.6% carnallite, 17.1% native magnesium sulfate, and 61.2% rock salt (K_2O content, 9.38%).

Final concentrate grade is 18% to 20% K_2O at 67% to 92% recovery, 21% to 43% native magnesium sulfate at 74% to 98% recovery, and 2.5% to 4.4% rock salt at <4% of total distribution.

Other references on electrostatic separation of salts are given in [1], [8], [9].

Purification of Coal. In the beginning of the 1980s a new process has been developed for the pretreatment and electrostatic separation of coal from inorganic sulfur compounds (pyrite) and ash. Pilot plant and laboratory work has been performed with good results. The process is, reportedly, to be used at the power plant where the coal will be pulverized, pretreated, separated electrostatically, and the concentrate burned as fuel [10], [11]. The benefits of such a process are:

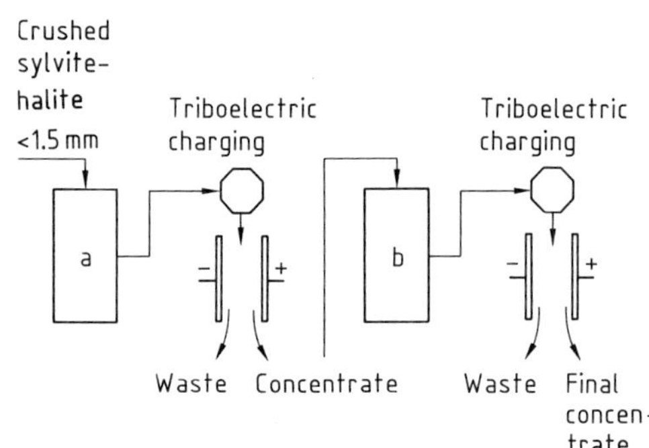

Figure 8. Electrostatic separation of sylvite–halite
a) Vapor phase conditioning with salicylic acid;
b) Vapor phase conditioning with trichloroacetic acid

Table 3. Performance of coal cleaning system on various coals (sulfur and ash contain in kg per kJ of fuel value)

Coal	Feed grade, kg/10^6 kJ		Product grade, kg/10^6 kJ			Yield	
	Sulfur	Ash	Sulfur	Ash		wt%	Fuel value, %
				<0.053 mm (−270 mesh)	>0.053 mm (+270 mesh)		
Pittsburgh (2.38 mm) washed	1.38	5.37	0.92	4.47	2.9	93.0	97.0
Picway blend washed and unwashed	1.54	5.16	1.04	3.74	1.98	85.4	90.0
Pittsburgh (2.38 mm) unwashed	2.14	14.7	1.03	7.14	1.9	61.0	75.0
Lower Freeport (3.36 mm) unwashed	1.24	16.04	0.40	8.08	1.98	70.0	87.0
Hagy unwashed	0.90	9.46	0.57	3.91	1.5	72.0	85.0
Freeport–Kittanning pond tailings	0.77	10.1	0.44	7.61	3.5	78.3	80.0

1) Coal is treated in fine sizes (106 mm, i.e., <140 mesh) to assure complete liberation of constituents. In-plant processing and separation minimize hazards of shipping fine pulverized coal.
2) Pretreatment alters surface conductivity of pyrite and ash particles. The practice of using gaseous ammonia to alter the particle conductivity of electrostatic precipitator feed has been successfully adapted to fine coal.
3) Coal is treated in a low-oxygen (flue gas) environment, thus eliminating the explosion hazard due to electrostatic discharge in coal dust.
4) The process allows power plants to purchase lower grade high-sulfur coal.

From storage, the coal is retrieved, pulverized, and sized to less than 100 μm (170 mesh). Pretreatment consists of vapor phase conditioning with ammonia along with an a.c. neutralization–deagglomeration step followed by separation on a modified roll-type ion bombardment electrostatic separator. Results of separations performed on various coals are shown in Table 3.

Electrofiltration. Recent work in the area of solid–liquid separations has led to the development of electrically augmented vacuum filtration [12], [13]. The unit is similar to an electrolytic cell where two parallel plate electrodes are placed in a dispersed clay slurry. When a potential is applied across these plates (Fig. 9), clay particles in the slurry which normally have a net negative charge are attracted to the positive electrode and collected. This cake is then harvested as the solid phase product from electrofiltration. Filtered liquid is then drawn out of solution from both the positive and negative electrodes by vacuum methods. This type of unit has been installed on an industrial scale in a Georgia kaolin plant. Some guidelines given for the application of electrofiltration are:

Dewaterability. The electrofiltration process is capable of increasing the solids content from >50% up to >70% solids as with other more power intensive filtration methods. Therefore, this method should be applied to feeds that are difficult or impossible to dewater by conventional means (e.g. vacuum filtration) or extremely costly to dewater by pressure filtration.

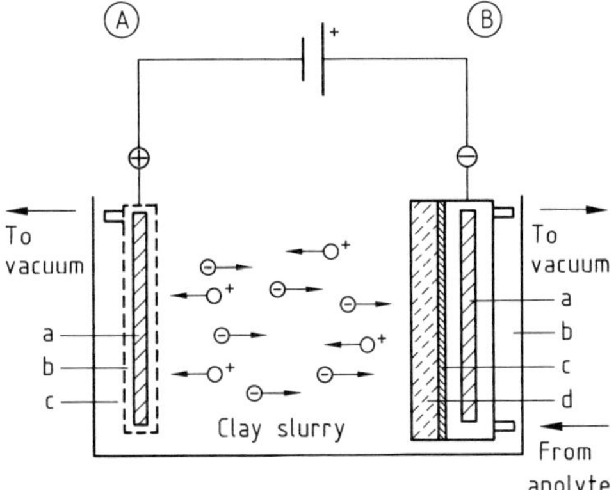

Figure 9. Electrically augmented vacuum filtration
A) Cathode procedure
a) Cathode; b) Filtrate; c) Filter cloth
B) Anode procedure
a) Anode; b) Anolyte; c) Electrolyte; d) Cake

Particle Size. A high content of fine particles is desirable. As *a general* guideline, the particles should be smaller than 10 μm (1250 mesh). Work has shown that finer particles yield a drier cake.

Dispersion. Electrofiltration requires a well dispersed suspension.

Conductivity. The background conductivity of the suspension should be low to maximize the power introduced into the dispersion at any specific current. Dewatering is a function of power input. As conductivity increases, power drops at constant current.

While industrial practice has been limited to kaolin slurries, other materials that have been tested include bentonite and hectorite clays, calcium carbonate, iron ore, and fine coal refuse.

3. Applications of Dielectrophoresis

3.1. Dielectric Separation

In the area of dielectrophoresis, work performed at the Tuscaloosa Research Center [14] has led to the development of a laboratory or pilot-scale continuous dielectric separator for mineral beneficiation (Fig. 10). The process uses a drum with fine wires placed along its surface, parallel to its axis. A screen electrode is in solution beneath the drum, and an a.c. potential is placed across the electrode and drum assembly. Particles of a higher dielectric constant are fed onto and attracted to the region of high electric field gradient (the wires) on the drum's surface. Lower dielectric constant material is repelled to a region of low field gradient by an intermediate dielectric fluid and falls through the screen electrode. The high dielectric constant material releases after rotating out and away from the screen electrode. Table 4 gives a partial list of minerals separated by this unit.

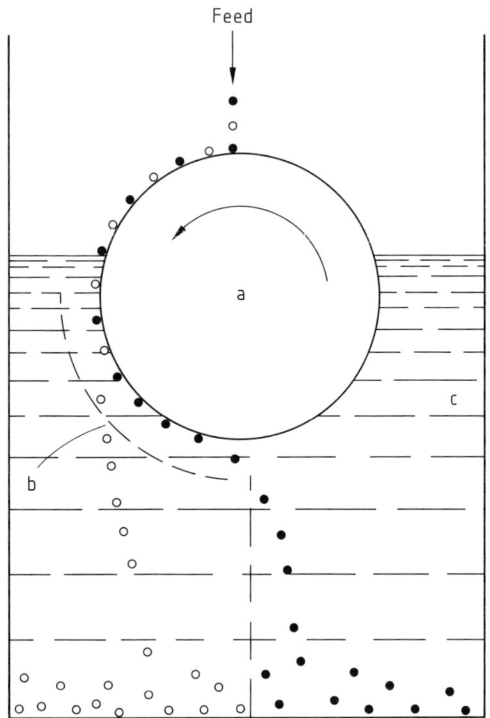

Figure 10. Dielectric mineral separator
a) Drum electrode; b) Screen electrode; c) Intermediate dielectric liquid of constant K
○ Low-dielectric-constant particles $<K$
● High-dielectric-constant particles $>K$

Table 4. Results of continuous dielectric separation of mineral mixtures

Feed composition, wt % and mineral name		Components of dielectric liquids*	Dielectric constant of separation	High-dielectric concentrate		
High-dielectric mineral(s)	Low-dielectric mineral(s)			Grade, %	Recovery, %	Grade ratio
10 Rutile	90 quartz	nitrobenzene and kerosene	4.5	62	94	6.2
11 Zircon	89 quartz	amyl alcohol and xylene	4.5	61	54	5.5
43 Chromite	57 olivine and talc	ethanol and carbon tetrachloride	6.0	72	90	1.7
10 Barite	90 quartz	nitrobenzene and kerosene	4.5	48	53	4.8
9 Corundum	91 quartz	amyl alcohol and xylene	4.5	45	54	5.0
10 Tourmaline	90 quartz	amyl alcohol and xylene	4.5	46	68	4.6
9 Monazite	91 quartz	nitrobenzene and kerosene	4.5	47	74	5.2
10 Witherite	90 quartz	amyl alcohol and xylene	4.5	29	67	2.9
8 Garnet	92 quartz	nitrobenzene and kerosene	4.5	38	62	4.7
9 Celestite	91 quartz	nitrobenzene and kerosene	4.5	51	72	5.7
10 Beryl	90 quartz	ethanol and carbon tetrachloride	4.5	32	51	3.2
10 Ilmenite	90 quartz and feldspars	nitrobenzene and kerosene	5.0	52	94	5.2

* intermediate dielectric fluid

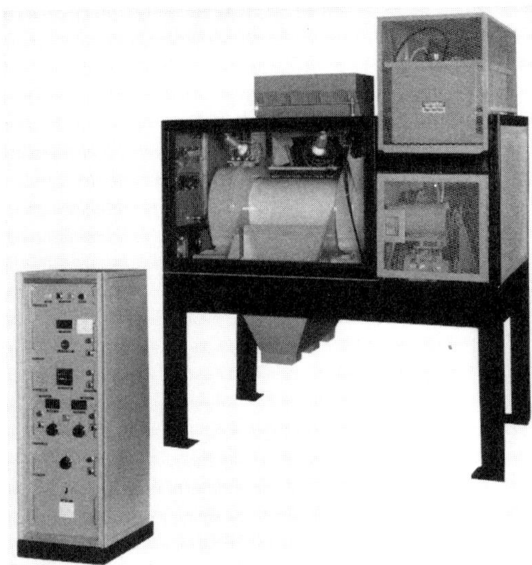

Figure 11. Multifield electrostatic separator (Carpco)

3.2. Multifield Electrostatic Separator

The multifield electrostatic separator (see Fig. 11) has been developed to separate materials that can be polarized through differences in shape, particle structure, and dielectric properties [15]. Actual separation of material is achieved by attracting highly polarizable particles such as fibers from less polarizable granular material.

The multifield separator uses a selective electrostatic field produced by a high voltage source. Feed material is transported under the multifield separation zone by means of a vibratory feeder or other conveyance (Fig. 2 D) Upon entering the active area, particles experience an attractive force toward the roll, based on their physical and electrical properties. Particles that are highly polarizable are lifted and become attached to the surface of the roll. As the multifield roll rotates, it carries particles out of the separation zone over a divider to a discharge zone.

Several unique features of the multifield separator include:

1) Mass-acting process which is not sensitive to changes in temperature and humidity that typically limit the effectiveness of other surface–acting electrostatic processes
2) Selective roll charging and discharging
3) Isolated nondischarging high voltage operation
4) Ability to separate lightweight materials, that are not readily treated by other methods (air–gravity, conventional electrostatic separation).

The unit has been used for separation of the following on a pilot (laboratory) scale: vermiculite–mica from rock, fiber from foodstuffs, and paper from plastic. The particle size is up to 12.5 mm of granular free-flowing materials. The capacity is up to 1785 kg $h^{-1}m^{-1}$ based on bulk density of material. High voltage power consumption is <200 W.

4. References

General References

[1] A. D. Moore: *Electrostatics and its Applications*, Wiley & Sons Inc., New York 1973.
[2] H. A. Pohl: *Dielectrophoresis*, Cambridge University Press, Cambridge 1978.

Specific References

[3] F. S. Knoll, US 4 247 390, 1981.
[4] M. Carta et al.: "Triboelectric Phenomena in Mineral Processing. Theoretic Fundamentals and Applications," *J. Electrost.* **10** (1981) 177–182.
[5] A. Singewald: "Trennen von Kalium–und Magnesiummineralien im elektrischen Hochspannungsfeld," *Kali und Steinsalz* **8** (1982), 252–260.
[6] G. Fricke: "The use of Electrostatic Separation Processes in the Beneficiation of Crude Potassium Salts," *Phosphorus Potassium* (1977) no. 90 July/August.
[7] A. Singewald et al., US 4 276 154, 1981.
[8] G. Samsel, CA 665 730, 1963.
[9] F. Fraas: "Electrostatic Separation of Granular Materials," Bull. No. 603, US Bureau of Mines, 1962.
[10] S. R. Rich: "Pilot Plant Program for the AED Advances Coal Cleaning System," Phase II, Contact No. 80–69, State of Ohio, Department of Energy, Aug. 1980.
[11] S. R. Rich, US 4 260 394, 1981.
[12] R. A. Adams: "The Application of Electrofiltration to the Minerals Industry, Hydrometallurgy, Research, Development and Plant Practice" *Proc. Int. Symp. Hydrometall. 3rd* 1983.
[13] M. P. Freemann: "Electrically Augmented Vacuum Filtration in Dewatering Kaolin Slurries," *The Clay Minerals Society Meeting*, Macon, GA, Aug. 1979.
[14] C. E. Jordan et al.: "A Continuous Dielectric Separator for Mineral Beneficiation," *US Bureau of Mines, Report of Investigations* 8437, 1979.
[15] F. S. Knoll et al, US 4 363 723, 1982.

21. Gravity Concentration

RICHARD O. BURT, Tantalum Mining Corporation of Canada Ltd., Lac du Bonnet, Manitoba, Canada REO 1 AO

Gravity concentration is the separation of one or more minerals, usually of different density, from others by relative movement in response to the force of gravity and other forces.

Next to handpicking, gravity concentration is the most ancient mineral-processing method, with early practices described over 2000 years ago [1]. One of the classic works on the subject is the 6th century book by AGRICOLA [2]. In the 20th century, the importance of gravity concentration declined somewhat as other processes, such as flotation or magnetic and electric separation methods, were developed (→ 23. Flotation, → 19. Magnetic Separation, → 20. Electrostatic Separation). Nevertheless, it remains one of the major recovery processes in both mineral processing and, especially, coal preparation (→ Coal, **A7**, p. 178). Furthermore, the development of high-capacity equipment, along with heightened awareness of the negative environmental impact of many chemicals used in flotation and leaching, has resulted in renewed interest in gravity concentration for a wide range of products.

Today gravity concentration is used for a diverse range of minerals: from andalusite to zircon, from coal to diamond, from mineral sands to metal oxides, and from industrial minerals to gold.

1. Principles

Important factors in determining the relative movement of a particle in a fluid include relative density, mass, size, and shape of the particle.

If, in a hypothetical two-mineral separation, only one of these factors is significantly different, separation is relatively easy. In nearly every case, however, the variety of minerals present have different relative densities and a range of particle masses, sizes, and shapes. The ease or difficulty of separating one species from another depends on the relative differences between these factors and whether such differences assist or oppose separation. Some idea of the ease of separating two minerals can be obtained from the *concentration criterion* (CC), which is defined as follows:

$$CC = \frac{d_h - d_f}{d_l - d_f}$$

where d_h is the relative density of the heavy mineral; d_l the relative density of the light mineral; and d_f the density of the suspending fluid.

In general, if CC > 2.5, then separation is relatively easy; however, if CC < 1.25 separation is not likely to be effective. For developmental test work, the concentration criterion is not sufficiently accurate, and tests involving heavy-liquid separation should be carried out (→ 22. Dense-Medium Separation, p. **22**-5).

Gravity concentration is effected by the relative movement of minerals of different density in response to gravity and one or more other forces, which often include the resistance to motion offered by a viscous fluid. One characteristic of all separation devices is that, for the particles to move relative to one another, at some stage of the process the particles must be kept slightly apart, or *dilated*, by the superimposed force.

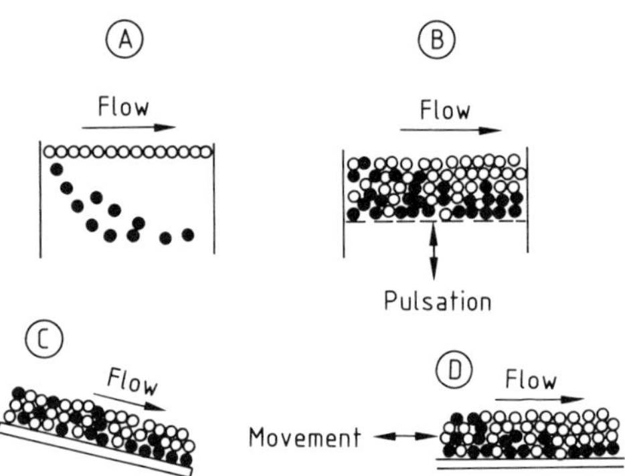

Figure 1. Mechanisms in gravity concentration
A) Density; B) Stratification; C) Flowing film; D) Shaking
○ = light; ● = heavy

No single mechanism can satisfactorily explain the behavior of any particular device; a combination of two or more mechanisms comes closer to describing the operation (but rarely does so completely). Conversely, however, various mechanisms have been proposed that are common to apparently diverse items of equipment and gravity concentration processes, and these are a useful means of classifying the various devices (see Fig. 1) [3].

Density (Fig. 1 A). The viscous fluid used has a density, or apparent density, between that of the minerals to be separated. Thus, one mineral(s) has a net positive buoyancy and floats, whereas the other(s) has a net negative buoyancy and sinks. This classification includes dense-medium separation (→ 22. Dense-Medium Separation), one of the most important gravity concentration processes, and magnetohydrostatic separation.

Stratification (Fig. 1 B). An intermittent fluidization caused by the pulsation of the fluid in a vertical plane stratifies the various mineral constituents. This classification applies to a wide range of jigs.

Flowing Film (Fig. 1 C). The constituents are separated by their relative movement through a stream of slurry, which flows down a plane by the action of gravity. Flowing-film concentration is the oldest process used in gravity concentration and remains of major importance, with such units as the sluice, pinched sluice, Reichert cone, and, through inclusion of lateral movement caused by the addition of centrifugal force, the spiral.

Shaking (Fig. 1 D). The mineral components are stratified by superimposing a horizontal shear force on the flowing film. This force can be oscillating, as in a shaking table, or orbital, as in the Bartles–Mozley separator or crossbelt concentrator.

The theoretical aspects of these and other mechanisms involved in gravity separation are discussed in [4].

2. Choice of Equipment

A wide range of gravity concentration equipment is available. Factors that determine choice of the correct unit include the size range of the particles to be separated, the duty required, the desired throughput and efficiency, and—all other things being equal—the unit cost, both capital and operating, of the equipment.

Size Range. The overall particle-size range that can be separated by using gravity concentration is larger than for any other process. Particles can be separated from each other as long as there is a suitable density differential between them; however, the top size may be limited by the mechanical handling ability of the device, 500 mm is generally accepted as the largest size for separation. The practical lower cutoff limit for fine gravity concentration is ca. 6 μm.

No device is capable of treating the entire size range amenable to gravity concentration. In addition, few devices operating as a single unit can efficiently treat the complete range which they are designed to handle. Therefore, some form of feed preparation is normally required. Figure 2 indicates the optimum size range for some of the more common equipment types [4]. Jigs, Reichert cones, and sluices are essentially mass classification devices and are capable of treating a fairly wide range of sizes, but some loss of efficiency occurs at the top, bottom, and middle of the size range. Treatment of material in *series circuits* is practical with these units (Fig. 3).

Other types of equipment (e.g., the table or Bartles–Mozley separator, concentrating by reverse classification, or the spiral concentrating by hindered settling) operate significantly better on sized or classified feed, and treatment in *parallel circuits* is required (Fig. 4).

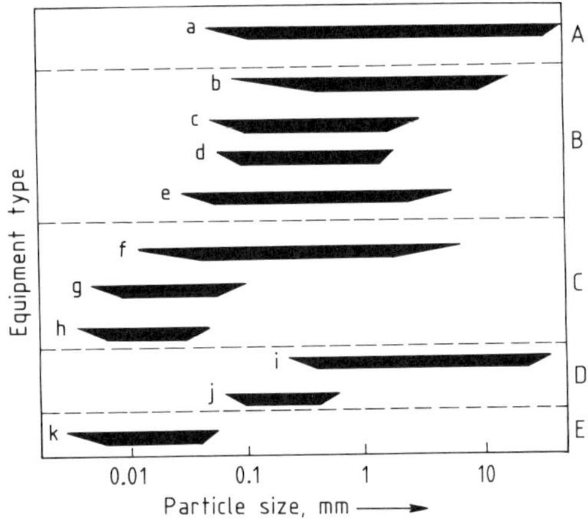

Figure 2. Operating range of gravity concentration equipment
A) Stratification devices: a) jig
B) Flowing-film devices: b) Sluice box; c) Reichert cone; d) Pinched sluice; e) Spiral
C) Shaking devices: f) Shaking table; g) Bartles–Mozley; h) Crossbelt
D) Air devices: i) pneumatic jig; j) Air table
E) Miscellaneous devices: k) centrifugal separators

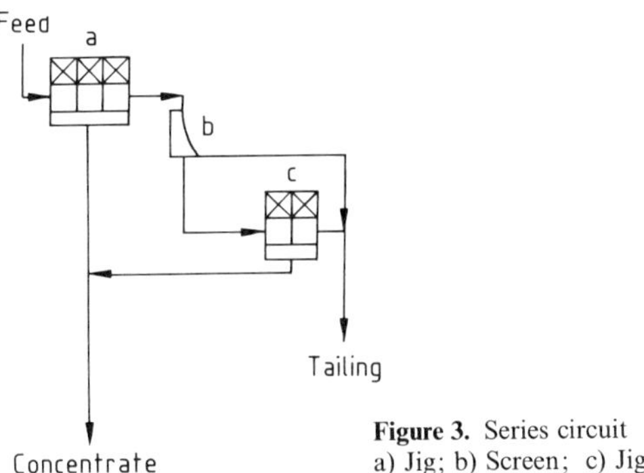

Figure 3. Series circuit
a) Jig; b) Screen; c) Jig

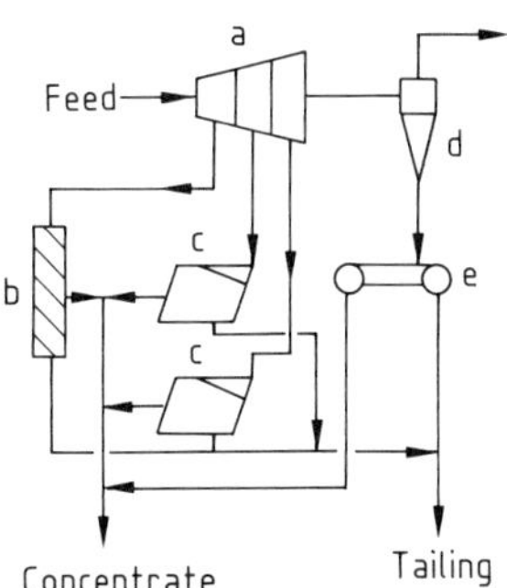

Figure 4. Parallel circuit
a) Hydrosizer; b) Spiral; c) Shaking table; d) Cyclone; e) Crossbelt concentrator

Table 1. Process applicability—density of separation

Wt % of feed within ±0.10 kg/L of density of separating medium	Degree of difficulty expected	Type
0–7	simple	sluices, jigs
7–10	relatively simple	cones
10–15	moderately difficult	tables spirals, DMS*
>15	difficult	DMS*

* Dense-medium separation

Generally, the two types of circuit are combined. For example, a series circuit of jigs treating unsized feed may be followed by a sizing step, sized fractions being heated on a parallel circuit of tables and film sizing units.

Capacity. The capacity range of gravity concentration equipment is enormous (from a few kilograms per hour to ca. 1000 t/h per unit). Whereas sluices and jigs are manufactured in a variety of sizes suited for a range of capacity, other items are produced in one or two sizes only, and high throughputs require multiple units. Typical capacities for gravity concentration equipment are as follows (in tons per hour):

Jig	≤800
Sluice	≤500
Reichert cone	60–70
Spiral	1–12
Table (ore)	0.2–2
Table (coal)	8–12
Bartles–Mozley Separator	2–2.5
Crossbelt Concentrator	0.3–0.5
Centrifugal	1–15

Efficiency. Most devices can efficiently separate materials containing only very heavy and very light fractions. However, most materials consist of particles with a range of densities. The more efficient a device, the more capable it is of effectively separating near-density particles [5].

Equipment efficiency classified by this measure is shown in Table 1.

3. Stratification Devices

Jigs sort by movement of a bed of particles, which are intermittently fluidized by pulsation of the fluid in a vertical plane. The particles become arranged in layers, with density increasing from top to bottom. This particle arrangement (stratification) results from several continuously varying forces which act on the particles. Although many types of jigs are manufactured, their construction principles are similar (Fig. 5).

Theories postulated to describe the action of the jig include differential acceleration, hindered settling, and interstitial trickling (Fig. 6). Differential (initial) acceleration at the start of particle movement is related to particle density, rather than particle size or mass. Hindered settling occurs once the particles have reached their terminal velocity; movement is related to mass. In the interstitial trickling phase, movement is related to particle size. An alternative theory proposes that the mechanism of concentration is the attainment of minimum potential energy in the heterogeneous mass [6]. Each of these mechanisms undoubtedly plays some part in the stratification process; however, which (if any) is the most important is open to question. Because the design of the pulsation cycle (i.e., duration of the cycle and the intensity of both upward and downward pulsation) can be varied, different manufacturers have developed units that accentuate different mechanisms.

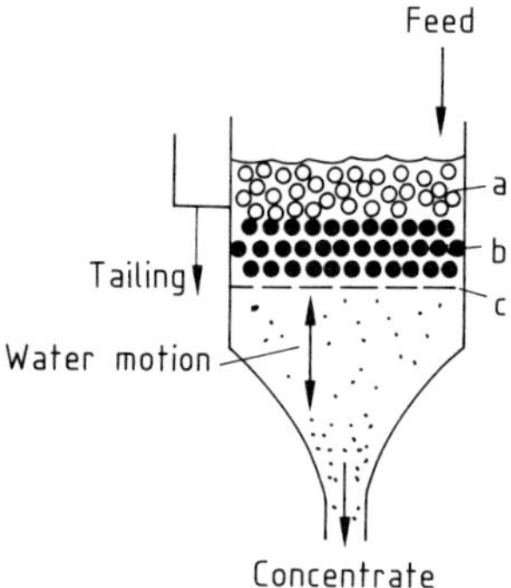

Figure 5. Basic construction of a jig
a) Jig bed; b) Ragging; c) Jig screen

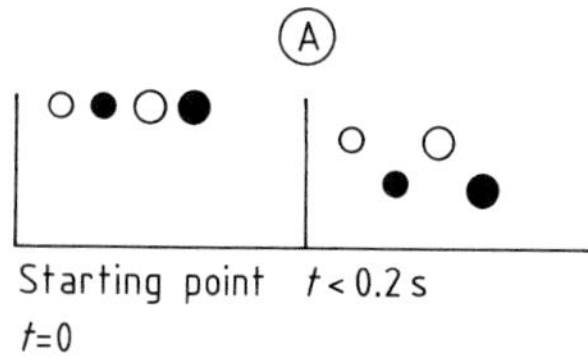

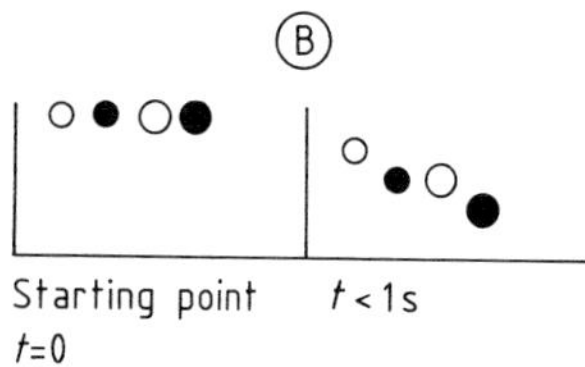

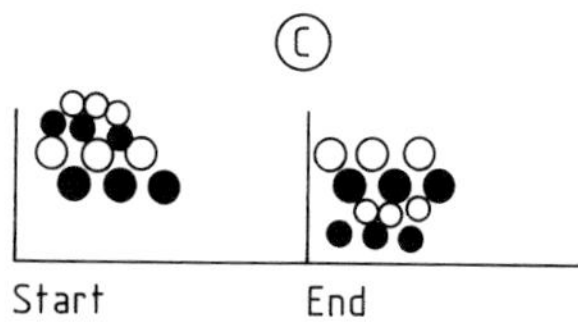

Figure 6. Three mechanisms of jigging
A) Differential acceleration; B) Hindered settling; C) Interstitial trickling
○ = light; ● = heavy

Jigs that rely on a long pulsation phase and a short suction phase, such as the Baum, OPM (Fig. 7 A), and Batac coal-washing jigs, tend to have a high capacity at the expense of recovery of finer particles. Although less efficient than heavy-medium separation, the high capacity and low operating cost of these jigs ensure that they remain the primary means of concentrating thermal coal as well as metallurgical coal where the amount of near-density shale is not excessive.

Units with a longer suction phase, such as the Denver, Bendelari, and IHC (Fig. 7 B–D), exhibit a higher recovery of fines with some compromise to capacity. Therefore, these units are more commonly used for mineral recovery, such as alluvial tin or hard-rock gold ores.

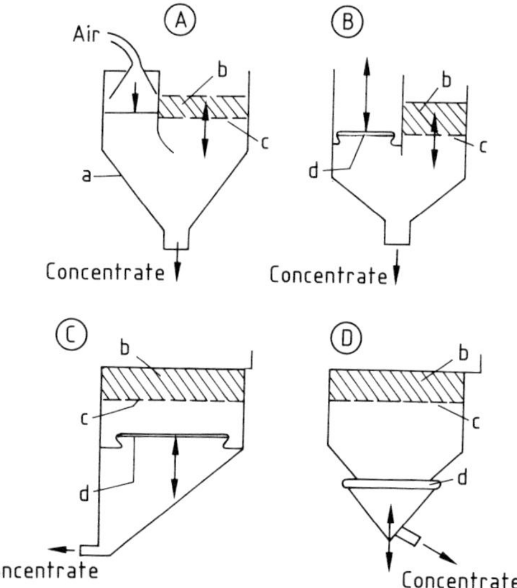

Figure 7. Typical jigs
A) Baum and OPM; B) Denver; C) Bendelari; D) IHC
a) Air box; b) Jig bed; c) Screen; d) Diaphragm

4. Flowing-Film Devices

When a film of water flows down an inclined surface under laminar conditions, the velocity distribution is parabolic, being zero at the bottom of the film and maximal at the top. When mineral particles of different size and relative density are introduced into this laminar film, they arrange themselves in the general downslope sequence shown in Figure 8. A variety of devices use this mechanism for separation.

Sluice Box. A sluice box is essentially an inclined trough or launder, usually of rectangular cross section, through which a wide size range of feed is washed by a large volume of water. A series of riffles is arranged at the bottom of the sluice box so as to provide some turbulence between each riffle and allow concentration of heavies.

These units are relatively inefficient and generally used only for alluvial mining in areas where power is not readily available.

Pinched Sluice and Reichert Cone. The principle of operation of a typical *pinched sluice* is shown in Figure 9.

Pulp, at a high solids density, enters the channel in a relatively thin stream at the upper or wider end. As this

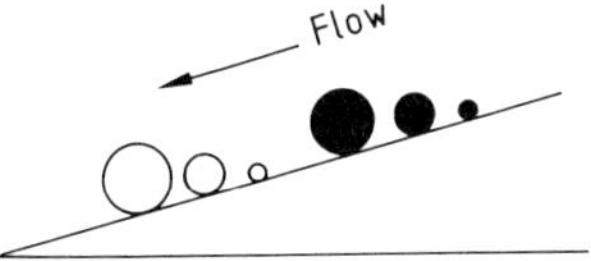

Figure 8. Particle arrangement on an inclined surface
○ = light; ● = heavy

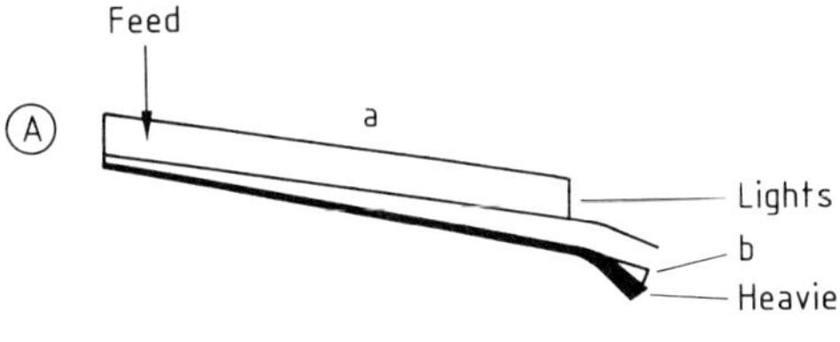

Figure 9. Pinched sluice
A) Cross section: a) Concentrating segment; b) Splitter
B) Plan view

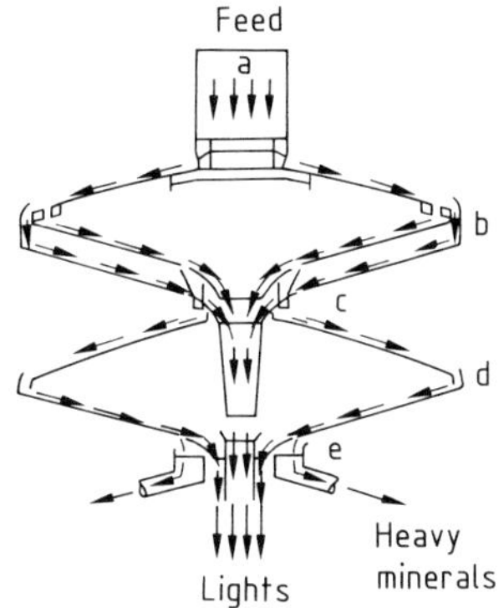

Figure 10. Reichert cone
a) Feed distributor; b) Double-cone assembly; c) Adjustable insert; d) Single-cone assembly; e) Variable insert

stream flows down the sluice, the finer and heavier particles concentrate (a) in the lower levels of the stream and increase in depth as the width of the sluice decreases. At the end of the channel, the slower moving lower level of the stream, enriched in heavy minerals, is peeled away from the main stream by a splitter (b).

A variety of pinched sluice designs are used, especially in Australia. Elsewhere, the more common variant is the *Reichert cone* (Fig. 10). The inverted cone concentrating surface of this device may be viewed as a number of pinched sluices placed adjacent to one another to form a circle, without any sidewalls to interfere with stratification [7]. The cone concentrator consists of a number of stages in series and in parallel; it can thus achieve multiple stages of separation in one unit. This is important because neither cone nor sluice is particularly efficient in a single stage.

The advantages of the Reichert cone are its high capacity, low capital cost per unit throughput, light weight, compactness, and metallurgical performance. Cones have two disadvantages: one is the need for a feed with 60% solids because the device is sensitive to variations in pulp density; the other is that cones are unsuitable for operations with a new feed rate of less than ca. 50 t/h.

Spiral Concentrator. Spiral concentrators are used widely in mineral processing and, more recently, in coal preparation. The basic design of the spiral has remained virtually unchanged for many years (Fig. 11).

Recently, however, spiral technology has developed rapidly, more so perhaps than that of

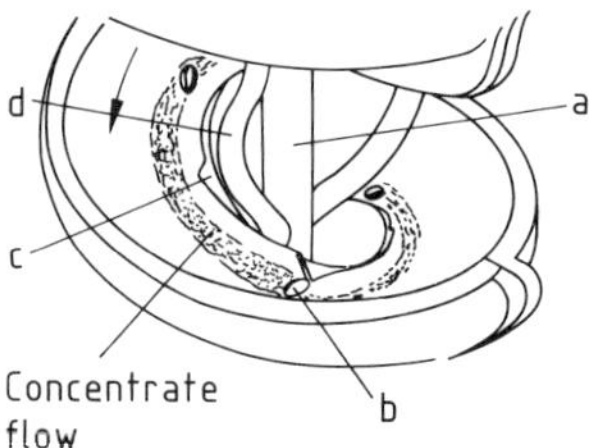

Figure 11. Spiral concentrator
a) Center column; b) Splitter; c) Wash-water channel; d) Concentrate pipe

any other gravity concentrator. Developments, mainly in Australia, the Soviet Union, the Peoples Republic of China, and the United States, have included the introduction of a range of trough profile, pitch, and diameter; the introduction of limited offtake spirals; and the use of spirals that do not require wash water [8]. The correct choice of these parameters for a specific duty is probably the most important factor in efficient spiral flow sheet design.

Concentration in a spiral is a complex blend of a variety of forces: hindered settling classification, interstitial trickling, Bagnold shear, and stream cross-sectional rotation. Many gravity concentration devices use the first two forces, several employ the first three, but the addition of cross-sectional rotation makes the spiral unique as a self-compensating device.

5. Shaking Separators

Shaking Table. The shaking table (Fig. 12) remains one of the workhorses of the mineral-processing industry [9].

The forces by which the shaking table separates the heavy particles from the lights are complex and, to some extent, opposing. Particles are stratified into discrete layers behind a series of riffles (Fig. 13) by superimposing, on the flowing film, a horizontal, assymmetrically reciprocating

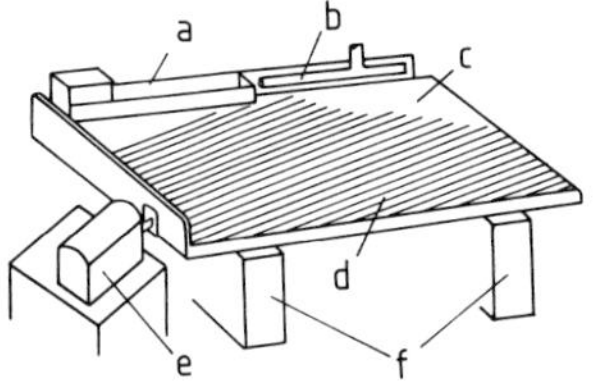

Figure 12. Shaking table
a) Feedbox; b) Wash water; c) Deck; d) Riffles; e) Shaking mechanism; f) Supports

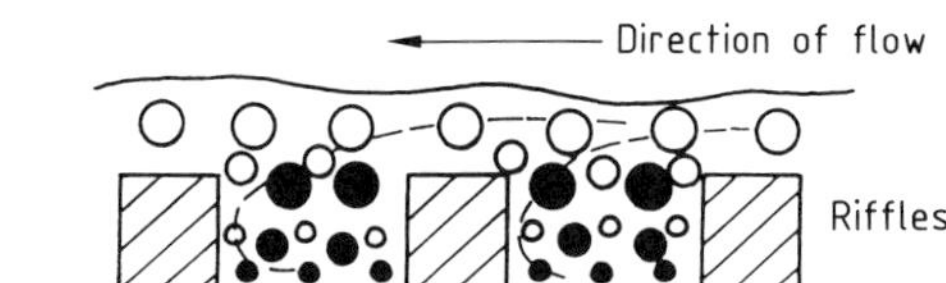

Figure 13. Stratification behind riffles

shear force acting laterally to the flow. The stratified layers are separated from each other by a lateral movement caused by the asymmetry of the shear force.

Two types of tables are currently available. The multidecked Concenco (USA) and SKO (USSR) units are ideally suited for roughing duties in larger plants and for coal preparation; the latter application is by far the most important in terms of volume. Single-deck units with mechanical head motions, such as the Holman (UK), Deister (USA), SKM (USSR), Wilfley (UK and USA), and Yun Tin (Peoples Republic of China) types, are better suited for smaller plants or for cleaning duty, mainly in mineral processing. In such operations, careful feed preparation and treatment of table middlings on separate equipment are essential for optimum performance.

6. Slime Separators

The inherent difficulty of recovering very fine particles (< 50 μm) by gravity has led to the development and, in many cases, obsolescence of a more diverse range of equipment than for any other particle-size range.

The Endless Belt concentrator and Johnson barrel are used in South Africa to recover fine gold in milling circuits. Likewise, the Rocking–Shaking vanner is used in China for the treatment of tin, tungsten, and iron ores. The more widely used Bartles–Mozley separator and the Bartles Crossbelt concentrator superimpose, an orbital shear motion on a slightly inclined surface, and are capable of recoveries as fine as 6 μm. The former (Fig. 14) is a semibatch device that discharges concentrates discontinuously; the latter is a continuous device that uses a slowly moving belt to remove concentrates from the pulp stream.

Spiral sluices are popular in both the Soviet Union and the Peoples' Republic of China. Similar in principle to the spiral separator, the cross section of the helix is much flatter, enabling recoveries as fine as 15 μm to be achieved.

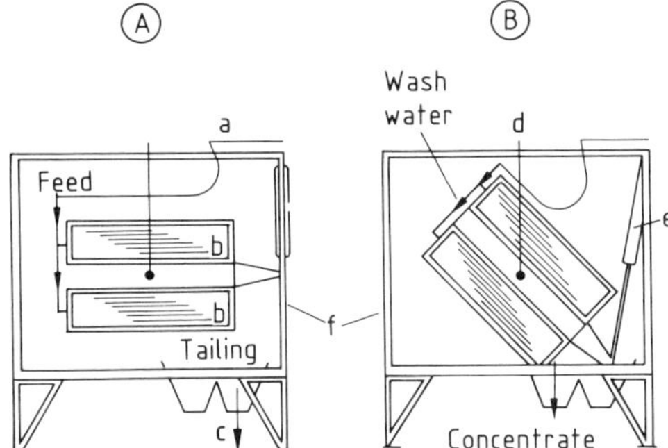

Figure 14. Bartles–Mozley separator
A) Collecting step: a) Feedwater valve assembly; b) Two sandwiches; c) Product launders
B) Discharging step: d) Suspension cables; e) Tilt cylinder; f) Main frame

The GEC (General Electric Co.) Duplex concentrator uses side shear rather than orbital shear. It is a semibatch device, with the decks tilting sideways rather than lengthways as on the Bartles–Mozley separator.

Centrifugal separators are growing in popularity. The Knelson concentrator consists of a riffled basket rotating on a vertical axis with a continuous supply of back-pressure water in the riffles to keep the bed of ore (usually gold) mobile. The Yunnan separators used in China for iron, gold, and metal oxide recovery are smooth-walled drums rotating on a horizontal axis. Both units are semibatch devices.

7. Dry Gravity Concentration

The majority of commercially available, dry gravity concentration processes are similar in principle to their wet gravity concentration counterparts. The latter include dense media separation, jigging, pinched sluices, and tabling. Although the mechanisms of separation of the wet and dry units may not be the same, the wet units already described make a reasonable point of reference for their dry counterparts. Other dry gravity concentration devices, which do not have a direct counterpart in wet concentration, have been developed. These include the fluidized-bed separator and the zigzag.

All dry concentrating equipment, except the dry jig, uses a constant controlled upflow of fluidizing air. The dry jig, however, uses a pulsating air flow; pulsation effects stratification in a manner comparable to its counterpart in wet concentration.

Dry separators generally require closely sized feed and are less efficient than wet separators. Consequently, they are used only where water is not available, or for "cleanup" of rough concentrates already dried for other processes.

8. Flow Sheet Example

Apart from sample separations, plants rarely incorporate only one type of device. This is especially true of metal oxide plants. Figure 15 is an example of the complex flow sheet that may be

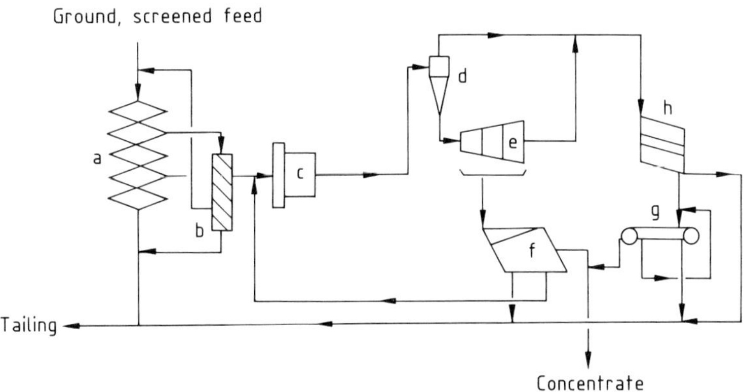

Figure 15. Example of gravity concentration circuit
a) Reichert cone; b) Spiral; c) Grinding mill; d) Cyclone; e) Classifier; f) Shaking table; g) Crossbelt concentrator; h) Bartles–Mozley separator

required to maximize both recovery and purity of the valuable minerals.

Acknowledgment. Material from this section is drawn mainly from the book *Gravity Concentration Technology*, written by the author (assisted by C. Mills) and published by Elsevier Science Publishers BV, Amsterdam, who retain the copyright. Their permission to utilize this material is duly acknowledged.

9. References

[1] C. P. S. Pliny: *Natural History Book 33*, ca. 70 A.D.

[2] G. Agricola: *De Re Metallica*, translated by H. C. Hoover, L. H. Hoover, Dover Publications, New York 1950, Book XIII.

[3] E. G. Kelly, D. J. Spottiswood: *Introduction to Mineral Processing*, Wiley Interscience, 1982.

[4] R. O. Burt (assisted by C. Mills): *Gravity Concentration Technology*, Elsevier, Amsterdam 1984.

[5] J. W. Leonard, D. R. Mitchell: *Coal Preparation*, 3rd ed., AIME, New York 1968, Section 4, pp. 23–30.

[6] F. W. Mayer: *Proc. 7th Int. Mineral Processing Cong.*, Gordon and Breach, New York 1964, pp. 75–97.

[7] J. J. Ferree, *Min. Eng.* **25** (1973) no. 3, 29–31.

[8] R. O. Burt, A. V. Yashin, *Mineral Processing & Extract. Metall.*, IMM. London 1984, pp. 117–128.

[9] R. O. Burt, *XV. Int. Mineral Processing Cong.*, Cannes 1985, pp. 272–281.

[10] A. V. Yashin, M. F. Anekin, V. A. Skrepko, *Spiral Concentrator Operation*, Nedra Press, Moscow 1984, pp. 54–65.

22. Dense-Medium Separation

RICHARD O. BURT, Tantalum Mining Corporation of Canada Ltd., Lac du Bonnet, Manitoba, Canada ROE 1AO

Dense-medium separation (DMS), also known as heavy-medium separation or sink–float separation, is one of the most widely applied and most efficient gravity separation processes in both mineral processing and coal preparation. Essentially, it is a process in which particles of different relative densities are separated in a liquid or in a fairly stable suspension of predetermined density chosen such that the liquid density is higher than that of the lighter particles but lower than that of the heavier particles.

The range of operating density of separation is 1300–3800 kg/m^3. The medium can be any suitable fluid; however, in practice, it is invariably a suspension of fine, normally magnetic particles in water. Dense-medium separation effectively separates particles with density differentials of <100 kg/m^3 in the size range of 500–0.5 mm. The largest size is dictated by the mechanical considerations of the separating vessel; the smallest, by medium rheology.

Uses. Dense-medium separation has two main areas of application. It is used for preconcentration to obtain an economically acceptable waste product that does not warrant further treatment, such as in the processing of diamonds, sulfides, and metal oxide ores. This allows either (1) a significant reduction in the quantity of ore that requires grinding prior to other separation methods or (2) the treatment of a lower grade ore than would otherwise be economically viable. Alternatively, DMS is used to produce a salable end product, as in coal preparation and industrial mineral separation.

1. The Process

Unlike most other gravity concentration techniques, dense-medium separation is a system rather than a unit process [1]. To work effectively and economically, the system must consist of a series of interconnected phases:

1) feed preparation,
2) feed and medium presentation,
3) separation of heavies and lights,
4) product recovery and dewatering, and
5) medium recovery and recycling.

These stages are illustrated in a typical dense-medium separation circuit in Figure 1.

In operation, the feed must be screened (a) to remove fine ore and slimes before it is fed, with reconstituted medium, to the separatory vessel (b) which separates the sinks (heavies) from the floats (lights). The vessel may be either static or dynamic; generally, the type of vessel chosen does not change the overall design concept of the circuit.

Floats and sinks are withdrawn separately and are drained of most of the medium on static or vibratory screens (c) and (d). The medium is either returned directly to the system or cleaned prior to return. Next, the floats and sinks are washed on the vibratory screens to remove the remaining adhering heavy medium.

The undersize products from the washing screens, consisting of medium, wash water, and fines, are too dilute and contaminated to be returned directly as medium to the separatory vessel. They are treated individually or together by magnetic separation (g) to separate the magnetic ferrosilicon or magnetite from the nonmagnetic fines. Reclaimed, cleaned medium is thickened to the required density by a suitable densifier (h), and continuously returned to the DMS circuit. The densified medium discharge passes through a demagnetizing coil (i) to ensure a nonflocculated, uniform suspension in the separatory vessel [2].

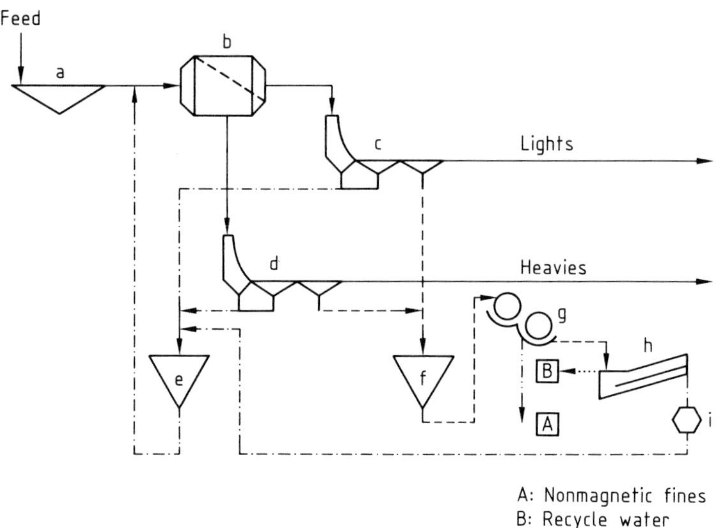

Figure 1. Typical dense-medium separation circuit
a) Feed preparation screen; b) Heavy-medium separation; c) Lights drain–wash screen; d) Heavies drain–wash screen; e) Heavy-medium sump; f) Dilute heavy-medium sump; g) Magnetic separator; h) Densifier; i) Demagnetizing coil
——— Ore, heavies, lights; –·–·– Heavy medium; ····· water

One of the essential components of the heavy-medium separation process is the medium itself (see Chap. 3). The correct choice of medium and its effective control, in terms of both consistency and physical parameters, are essential for efficient operation of the system.

2. The Separatory Vessel

Separatory vessels can be either static or dynamic. In *static vessels*, separation (generally of particles coarser than 2 mm) is carried out at normal gravity, whereas in *dynamic vessels*, finer particles (generally limited to the range of 20–0.5 mm) are separated at elevated gravitational force.

Static DMS Vessels. Static vessels can be subdivided into cone, drum, trough, and combination types (Fig. 2). The feed is generally introduced at or near the top of the separatory vessel. Lights float on the surface and are removed over a weir, with or without the assistance of paddles. Removal of the sinks varies with type of vessel.

Cone separators (Fig. 2 A) contain the greatest depth of medium, which should be slightly agitated to minimize the density gradient from top to bottom of the vessel. Sinks are removed from the bottom by an internal air lift, external pump, or bucket elevator, with drained medium being recycled directly to the vessel. Cone separator circuits tend to be exceptionally stable be-

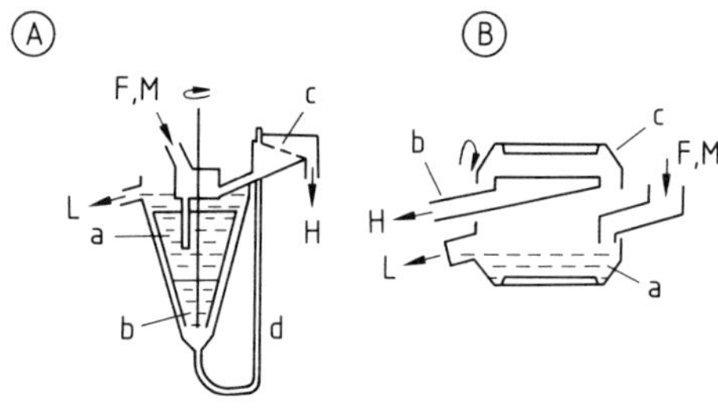

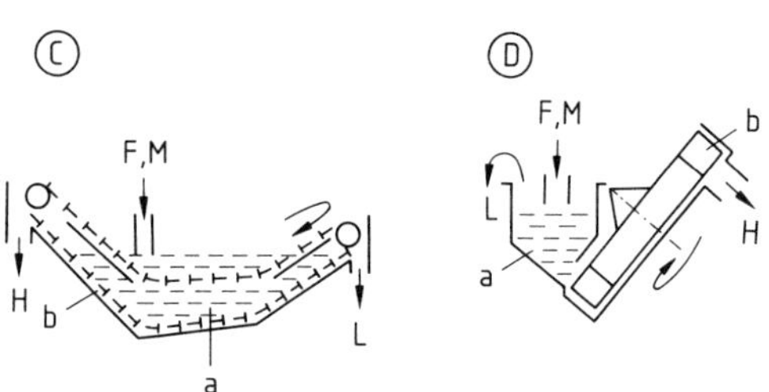

Figure 2. Types of static separatory vessels
A) Cone: a) Bath; b) Paddles; c) Screen; d) Airlift
B) Drum: a) Bath; b) Launder; c) Drum
C) Trough: a) Bath; b) Chain
D) Combination (Drewboy): a) Bath; b) Elevator
M = Medium; F = Feed; H = Heavies; L = Lights

cause of the relatively large quantity of medium employed; they are especially suitable for separation of materials having similar densities.

Drum separators (Fig. 2 B) consist essentially of a rotating cylindrical drum fitted with internal lifters, which elevate the heavies out of the medium bath into the sinks discharge launder. The lifters also cause some agitation of the medium, overcoming any tendency for it to settle. Drum separators are common in both coal preparation and mineral separation.

Trough separators (Fig. 2 C) consist of a shallow bath or trough for the medium and a moving product removal system, which typically is a single-chain conveyor that moves the two products to opposite ends of the trough. Trough separators are used mainly in North American coal preparation plants.

Combination separating vessels (Fig. 2 D), such as the Drewboy and Norwalt, are classified as shallow-bath vessels. Unlike other static vessels, the heavies removal system is not located within the medium bath itself but, like the deep cones, is a separate part of the system. These vessels, therefore, combine the advantages of the shallow bath (large heavies discharge capacity, low medium requirement, and low density gradient) with the advantages of the deep cone (division of the functions of product separation and removal). Therefore, these separators are among the most efficient available.

Dynamic DMS Vessels. Dynamic dense-medium separation was first studied by DRIESSEN at the Dutch State Mines Laboratories with a standard hydrocyclone [3]. The higher forces involved in centrifugal separation permit much finer particles to be separated successfully. The high shear within the vessel enables finer medium, essential for stability, to be used without resulting in excessive viscosity. In addition to the cyclone, a variety of modified cyclone vessels have been developed (Fig. 3).

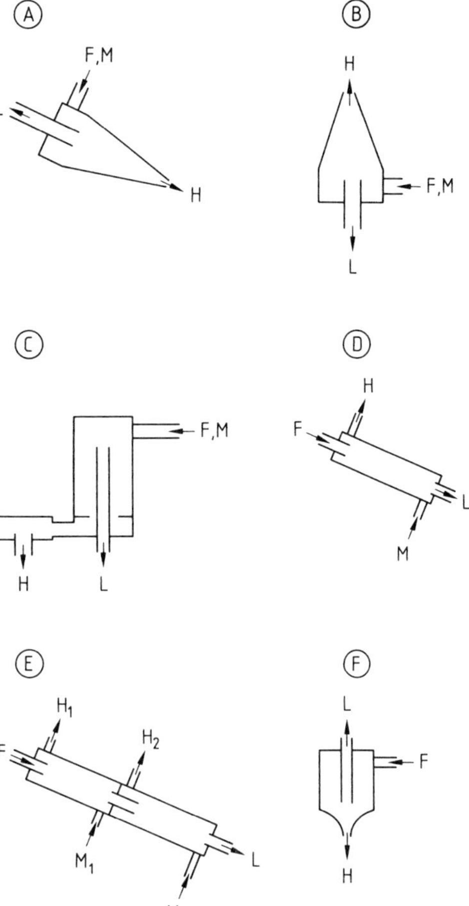

Figure 3. Types of dynamic dense-medium separating vessels
A) Cyclone; B) Swirl; C) Vorsyl; D) Dyna whirlpool; E) Triflo; F) Water-only cyclone
M = Medium; F = Feed; H = Heavies; L = Lights

The smallest particle that can be effectively separated by dynamic dense-medium systems is rarely related to the separatory vessel used; rather, it is generally related to the medium recovery system. The accepted smallest size for efficient separation is 0.5 mm.

The principle of operation of the dense-medium cyclone is similar to that of the conventional cyclone (→ 13. Dust Separation, pp. **13**-4 – **13**-8), except that separation is based on particle density rather than size. Only a limited size classification effect occurs, on large or small particles or those with a relative density close to the separating density. A typical cyclone separator is operated on an axis that is close to horizontal (Fig. 3 A), whereas the Japanese swirl cyclone (Fig. 3 B) operates upside down. The Vorsyl separator (Fig. 3 C) is a modified cyclone with a cylindrical body; heavies leave the body via an annular throat to a secondary densifying cylinder.

In the separators shown in Figure 3 A – C, medium and solids enter the separator together at high pressure (210 – 275 kPa) through the feed port; this results in high wear of feed pump and piping. In the Dyna whirlpool separator

(Fig. 3 D) and the Triflo separator (Fig. 3 E) however, medium is pumped only into the lower tangential inlet; the feed enters by gravity at lower pressure through an axial pipe. Heavies pass through the cycloning medium and leave via the tangential sinks pipe. Lights remain on the axis of the vessel and exit through the axial discharge pipe. The Dyna Whirlpool is a single-stage unit. The Triflo separator has two stages; it can be set to produce two sink products at the same density of separation or at two different densities.

The water-only cyclone (Fig. 3 F) can be regarded as an autogenous dense-medium cyclone, in that fines in the ore recirculate within the cyclone to form the medium. These cyclones are employed only in the coal preparation industry, and usually in series, because they are relatively inefficient.

3. The Medium

Although the correct choice of separatory vessel is obviously important in terms of capital and operating costs, as well as efficiency of separation, the control and efficiency of the operation are, to a large extent, dependent on the medium.

Ideally, the medium should be a true heavy liquid. However, although such liquids exist and some are used for laboratory simulations, their toxicity and cost preclude them from commercial application.

Practically, a commercial medium should

1) be available in quantity at or near the site,
2) be comparatively inexpensive at the site,
3) be noncorrosive,
4) be nontoxic,
5) form stable mixtures in water,
6) be physically and chemically stable
7) mix readily with water,
8) rinse from the product with high-pressure water sprays,
9) be capable of regeneration, and
10) be adjustable over a range of densities [4].

Dense media used for commercial separation are almost exclusively suspensions of fine insoluble particles in water. The most commonly utilized solids are *magnetite* [*1309-38-2*] ($d = 5.0-5.2$) and *ferrosilicon* [*8049-17-0*] ($d = 6.7-6.9$). Magnetite is most commonly used in coal preparation; ferrosilicon is generally chosen for preconcentration of metalliferous ores, although to reduce costs it is often mixed with magnetite. Magnetite and ferrosilicon are preferred because they can be regenerated almost completely in the medium recovery circuit by use of low-intensity magnetic separators as both recovery and densification units.

Ideally, the medium should have low yield stress, low viscosity, and high stability. Unfortunately, however, stable suspensions tend to have high yield stress and high viscosity, and vice versa. The classic work on the rheology of dense media was carried out by DeVaney and Shelton [5]. These workers showed that, for any suspension, *viscosity* (and hence apparent density) increases with increasing solids content. Above a critical value, the viscosity increases rapidly to a limiting point at a volumetric concentration of ca. 40 % solids for all suspensions (Fig. 4). Conversely, the *settling rate*, which is the reciprocal of stability, decreases rapidly with increasing volumetric solids content for equisized particles (Fig. 5).

The coarser the *particle size* of the medium, the faster the solids settle, producing a less stable slurry but a lower viscosity effect at higher concentration. Conversely, the finer the medium particle size, the more stable is the suspension,

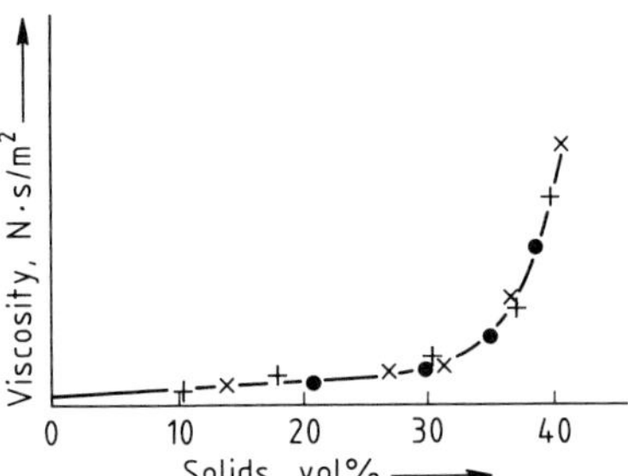

Figure 4. Effect of volumetric solids content on viscosity of three media
● Quartz; × Magnetite; + Ferrosilicon

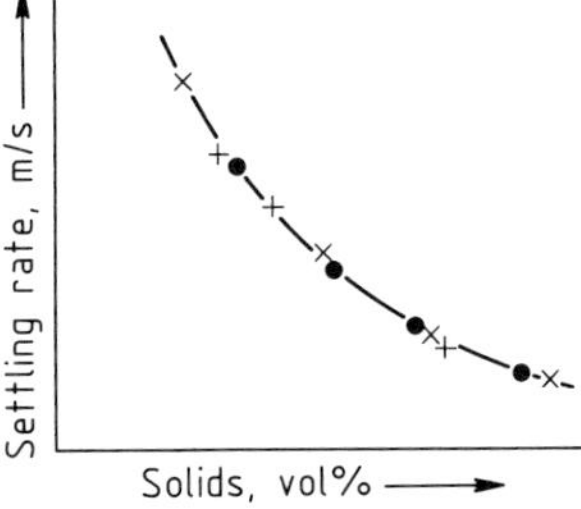

Figure 5. Effect of volumetric solids content on settling rate of three media
● Quartz; × Magnetite; + Ferrosilicon

although the viscosity effect is higher at higher slurry concentrations.

A more stable, and consequently finer, medium is required in dynamic separation because the higher forces involved increase the tendency of the medium to segregate.

The more difficult the separation, the greater is the necessity to have a stable, low-viscosity medium at the operating density and to minimize variation in the medium density during the process.

Magnetohydrostatic separation, in which the apparent density of a magnetic fluid is controlled by a non-uniform magnetic field, is a potential extension of dense-medium separation. It has been studied in the Soviet Union, Israel, South Africa, Great Britain, and the United States; however, it is not yet a fully commercialized process.

4. Testing for Dense-Medium Separation

Complete separation, even of particles very close to the separating density, can be achieved in the laboratory by using true heavy liquids and allowing sufficient time for such separation. The most common heavy liquids used are carbon tetrachloride [*56-23-5*] ($\varrho = 1590$ kg/m^3), 1,1,2,2-tetrabromoethane [*25167-20-8*] ($\varrho = 2970$ kg/m^3), and thallium malonate formate (Clerici's solution) ($\varrho = 4200$ kg/m^3). All can be diluted to the desired density, the first two with white spirits or acetone, the last with water. Both cost and toxicity increase with increasing density.

Sequential heavy-liquid separation, in which successive separations occur in a series of liquids of increasing (or decreasing) density, generates either washability curves or partition curves.

A *washability curve* is used to determine the response of an ore to gravity separation. A typical set of curves is shown in Figure 6. These curves allow prediction of the mass and assay of the sink and float; the anticipated recovery at any predetermined density of separation; and the ease or difficulty of the separation proposed. Alternatively, the curves can be used to determine the density of separation required to achieve a desired weight split, product assay, recovery, etc.

In the example shown, preconcentration of an industrial mineral (M) is to be carried out at a density of 2640 kg/m^3, equivalent to the minimum (point A) on the tolerance curve (g), thereby assuring maximum efficiency of a reasonably difficult separation. At this density, on the line AB, the weight split is 68% to sinks, equivalent to a floats assay of 0.3% (point C) and a sinks assay of 5.1% (point D). Line DE then shows that recovery of M is 93% (point E).

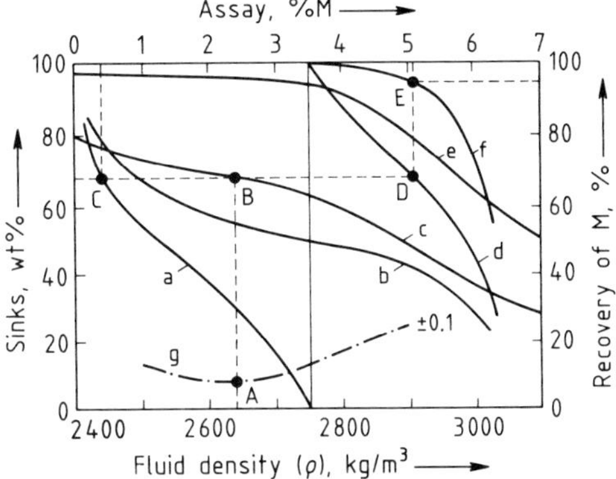

Figure 6. Typical set of washability curves
a) Cumulative assay of floats; b) Characteristic assay curve; c) Density curve—the weight percent of material reporting to the sinks at the density of separation; d) Cumulative assay of sinks; e) Recovery: density curve—the percent of heavy mineral M reporting to the sinks (i.e., recovery) at the density of separation; f) Grade: recovery curve—the recovery to sinks of heavy mineral M, at a specific assay of the sinks (%M); g) Tolerance curve—the weight percent of material within a specified density range at the density of separation

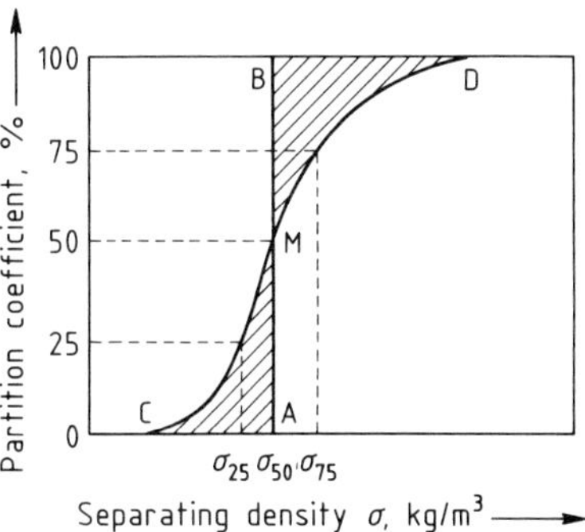

Figure 7. Partition curve

Partition curves are used to determine the efficiency of a particular separation process [6]. They are constructed by carrying out sequential heavy-liquid analyses of both products from the separation, and then plotting at each density the proportion of the calculated feed that reports to the sinks or floats. A typical partition curve is shown in Figure 7.

The separating density, σ_{50}, is the density at which a particle has an equal chance of sinking or floating or, con-

Table 1. Anticipated separator results in treating any coal of size range 1.5 to 0.5 cm in the same DMS unit at density 1600 kg/m^3

Liquid density, kg/m^3	Heavy-liquid test on head sample of a raw coal			Partition coefficient at assumed cut point	Calculated sink–float analysis for separator					
					Clean coal (floats)			Rejects (sinks)		
	wt%	% ash	ash units	$d = 1.6$	wt%	% ash	ash units	wt%	% ash	ash units
Floats 1300	52.8	2.6	1.373	100.0	52.80	2.6	1.373		2.6	
1300–1400	12.3	8.7	1.070	96.5	11.87	8.7	1.033	0.43	8.7	0.037
1400–1500	4.2	18.8	0.790	84.5	3.55	18.8	0.667	0.65	18.8	0.122
1500–1600	2.1	30.0	0.630	63.0	1.32	30.0	0.396	0.78	30.0	0.234
1600–1700	2.9	39.4	1.143	34.5	1.00	39.4	0.394	1.90	39.4	0.749
1700–1800	1.0	47.1	0.471	13.3	0.13	47.1	0.061	0.87	47.1	0.410
1800–1900	1.1	53.6	0.590	4.0	0.04	53.6	0.021	1.06	53.6	0.568
1900–2000	1.0	60.2	0.602	0.5	0.01	60.2	0.006	0.99	60.2	0.596
Sinks 2000	22.6	85.1	19.233			85.1		22.60	85.1	19.233
Total	100.0	25.9	25.902		70.72	5.59	3.951	29.28	75.0	21.949

versely, at which it has an equal chance of being misplaced. This is equivalent to the 50% point, M, on the partition curve. A perfect separation is represented by the vertical line AMB, i.e., all lighter material floats and all heavier material sinks. The actual separation is represented by the curve CMD. The difference between this curve and the vertical represents the amount of error of separation; hence, the total shaded area is the *error area.*

The most common guide to efficiency of separation is the *probable error* E_p. This is defined as half the differential in density between the 25 and 75% partition coefficients on the partition curve:

$$E_p = 0.5\,(\sigma_{75} - \sigma_{25})$$

When plotted on probability paper, the partition curve can approximate a straight line. However, significant "tails" occur at the top and bottom of the line. Modern computer-modeling techniques permit the shape of the generated curve to be described mathematically, although the model most likely to fit the data is still in question [7].

The partition curve, error area, probable error, and more recently, the computer-generated curve can be used not only to monitor plant performance and acceptance testing, but also to design and predict dense-medium circuit operation. By starting with washability data generated as described earlier and applying the partition coefficients, related to E_p (which are known for a specific separator), the partition curve, and hence the characteristics of both products from the separator, can be predicted. A typical predicted separation of coal is shown in Table 1. In this example, at a cut point of 1600 kg/m^3 coal can be obtained with 5.59% ash and a yield of >70%.

Dense-medium separation processes have been modeled for many years by means of washability data, because separation is almost entirely dependent on the relative density of particles and medium. Recently, considerable developments have been made in the use of washability data to model most other types of gravity concentration devices; hence, the importance of heavy-liquid analysis in industry is growing. Within a few years, sufficient data will likely be available for design of complete gravity concentration circuits by computer.

Acknowledgment. Material from this section is drawn mainly from the book **Gravity Concentration Technology**, written by the author (assisted by C. Mills) and published by Elsevier Science Publisher BV, Amsterdam, who retain the copyright. Their permission to utilize this material is duly acknowledged.

5. References

[1] R. O. Burt, C. Mills: *Gravity Concentration Technology*, Elsevier, Amsterdam 1984, pp. 139–183.
[2] R. J. Gochin, M. R. Smith, *Min. Mag.* (1983, Dec.) 453–460.
[3] M. G. Driessen, *J. Inst. Fuel* **19** (1945) 33–45.
[4] C. G. Wilson, Mineral Processing Plant Design, AIME 1978, pp. 520–538.
[5] F. D. DeVaney, S. M. Shelton U.S. Bureau of Mines (USBM), Report of Investigations 3469R, 1940.
[6] K. F. Tromp, *Colliery Guardian* **154** (1937) 995–999.
[7] K. J. Reid, Lu Maixi, Zhang Shenggui, *Int. J. Miner. Process.* **14** (1985) 291–299.

23. Flotation

Baki Yarar, Colorado School of Mines, Golden, Colorado 80401, United States

Symbols

C = equilibrium concentration of surfactant in solution
E = electric field between electrodes applying the field
k = Boltzmann constant
p = pressure inside a bubble
p_o = pressure of the bulk of the liquid
P_a = probability of particle-bubble adhesion
P_c = probability of particle-bubble collision
P_f = probability of flotation
P_s = probability of formation of a stable particle-bubble aggregate
S = spreading coefficient for air on solid
T = temperature
V_e = electrophoretic velocity of particle
W = work of adhesion
Γ = surface excess concentration of surfactant measured in quantity per unit area (e.g., molecules/cm^2)
γ = specific surface energy of liquid film
γ_0 = specific surface energy of an infinitely thick film of a liquid
γ_{LG} = liquid–gas interfacial tension
γ_{SG} = solid–gas interfacial tension
γ_{SL} = solid–liquid interfacial tension
ε = dielectric constant
ζ = zeta potential
η = viscosity of aqueous medium
θ = contact angle
π_D = disjoining pressure

Flotation is one of the well-established physicochemical separation processes, widely applied in mineral concentration and recovery operations; other applications include water treatment and purification, the recycling of secondary materials, and the recovery of ionic and colloidal materials from aqueous solutions.

Various forms of this process are known as froth flotation, ion flotation, precipitate flotation, piggyback flotation, flotoflocculation, skin flotation, column flotation, and two-liquid flotation [1]–[8]. The recently introduced gamma flotation is based on the control of liquid–vapor interfacial tension to facilitate the particle–bubble adhesion process [9].

Because most flotation processes are applied with water as the principal liquid, and the widest area of applications is in the field of mineral processing, mostly systems containing water and minerals are discussed here.

1. Introduction

1.1. Interfacial Energies and Contact Angle

The flotation process relies primarily on the fact that hydrophilic particles are wetted by water, whereas hydrophobic particles are wetted by oils and air bubbles; therefore, if air bubbles are introduced into an aqueous slurry, the bubbles adhere to the hydrophobic solid particles. As a result, air–solid aggregates are carried to the surface, forming a froth layer; this explains the name froth flotation. The froth layer can be removed manually or mechanically; the result is the separation of hydrophobic from hydrophilic particles. For further information on detergency, see also →Detergents, **A8,** p. 320.

The principal processes occurring in a slurry-containing vessel known as a *flotation cell* are shown in Figure 1. Equation (1), known as the Young-Dupre equation, and Equations (2) and (3) are used to describe the relationship of physicochemical variables to particle–bubble adhesion.

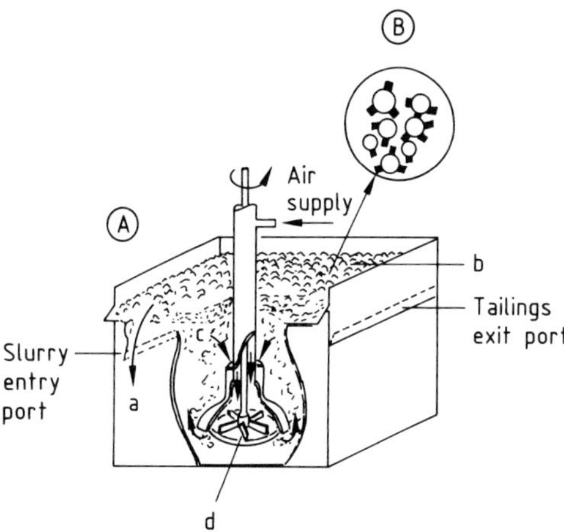

Figure 1. Processes occurring in a flotation cell (Schematic)
A) Flotation cell a) Froth overflow; b) Froth layer; c) Pulp; d) Rotor for pulp agitation
B) Mineralized air bubbles within flotation cell

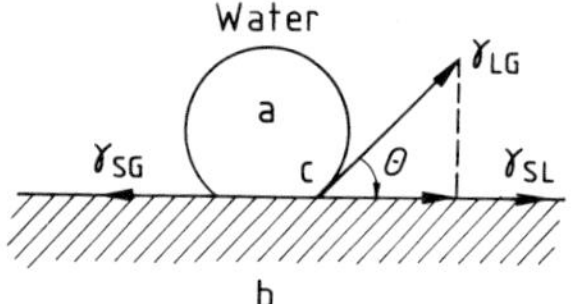

Figure 2. Equilibrium contact angle
a) Captive bubble in water; b) Hydrophobic solid $\cos\theta = \frac{\gamma_{SG} - \gamma_{SL}}{\gamma_{LG}}$

Table 1. Contact angles of solids in water and various aqueous media

Solid	Contact angle (degrees)	Solution conditions
Colemanite	43	5×10^{-3} M oleate* solution
Copper metal	93	in 50 mg/L oleate* solution
Fluorite	91	in 10^{-5} M oleate*, pH = 8.1
Galena	60	in 10^{-3} M ethyl xanthate solution
Graphite	96	water
Ilmenite	80	treated with oleate*, pH = 8, T = 75 °C
Colorado oil shale with 28 % organic carbon	59.5	water
Paraffin wax	108	water
Silica	81	2.5 mg dodecylammonium chloride per L solution, pH = 10
Teflon	160	water
Teflon	0	methanol–water solution with surface tension <20 mN/m

* Sodium oleate.

$$\gamma_{SG} - \gamma_{SL} = \gamma_{LG} \cos\theta \tag{1}$$

$$W = \gamma_{LG}(1 - \cos\theta) \tag{2}$$

$$S = \gamma_{SL} - (\gamma_{LG} + \gamma_{SG}) \tag{3}$$

where γ_{SG} = solid–gas interfacial tension, γ_{SL} = solid–liquid interfacial tension, γ_{LG} = liquid–gas interfacial tension, θ = contact angle, S = spreading coefficient for air on solid, and W = work of adhesion. For particle–bubble adhesion the conditions $W > 0$, $S > 0$, and $\theta > 0$ are satisfied.

Figure 2 describes the mechanical equilibrium among the three interfacial tensions indicated as vectors applied to point C, known as the three-phase contact point; the contact angle θ is measured in the liquid phase. The two variables $\cos\theta$ and γ_{LG} are readily accessible by experiment. Contact angles on minerals in various media are given in Table 1. The significance of the contact angle arises from the fact that it is a measure of surface wettability; $\theta > 0$ indicates a hydrophobic solid. Contact angle values of $\theta > 10°$ usually indicate that particles can form bubble–particle contacts strong enough to resist turbulence in a conventional flotation cell, enhancing the probability of flotation as decribed in Chapter 4.

1.2. Hydrophobic and Hydrophilic Solids

Although most naturally occurring minerals are hydrophilic, some are hydrophobic. Examples of the latter include graphite, sulfur, antimonite (Sb_2S_3), molybdenite (MoS_2), talc, and high-rank coals such as anthracite. Many polymers, such as Teflon and Nylon, are also hydrophobic.

Hydrophilic materials can be made hydrophobic by the adsorption of chemicals. For example, calcite ($CaCO_3$) can be readily made hydrophobic by treating it with low concentrations of sodium oleate ($C_{17}H_{33}COONa$). Silica (SiO_2) can be made hydrophobic by treatment with dodecylamine ($C_{12}H_{25}NH_2$). A solid made hydrophobic by chemical treatment can revert to the hydrophilic state on further chemical change. For example, galena (PbS), in contact with 25 mg/L potassium ethyl xanthate (potassium *O*-ethyldithiocarbonate) [*140-89-6*] (C_2H_5OCSSK) solution, is hydrophobic at pH below 10.8, but hydrophilic at higher pH. Teflon is highly hydrophobic in most aqueous media, but behaves like a hydrophilic solid if the surface tension is reduced below 20 mN/m by the addition of methanol [10]. Chemicals used to affect the hydrophobic–hydrophilic properties of solids are known as flotation reagents.

2. Flotation Reagents

Flotation reagents can be classified according to their function or their chemistry.

2.1. Functional Classification of Flotation Reagents

Collectors. These are used primarily to make solids hydrophobic and promote adhesion to air bubbles or oil droplets. Common examples are fatty acids, sulfonates, xanthates (dithiocarbonates), amines, and dithiophosphates.

Frothers. Frothers promote the formation of a metastable froth phase that facilitates the removal of particles carried by air bubbles to the top of the flotation cell. They also reduce the induction time; that is, they allow colliding parti-

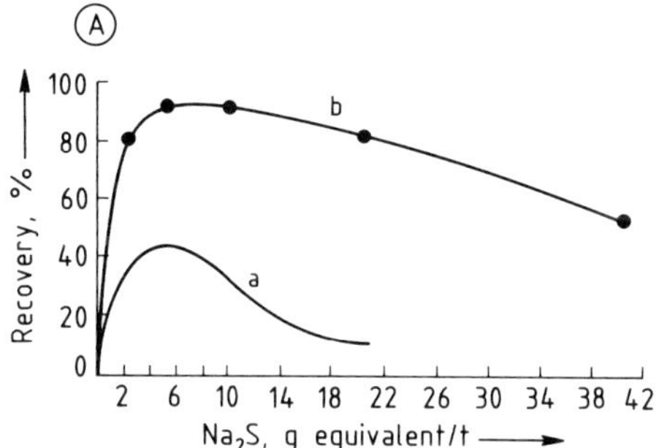

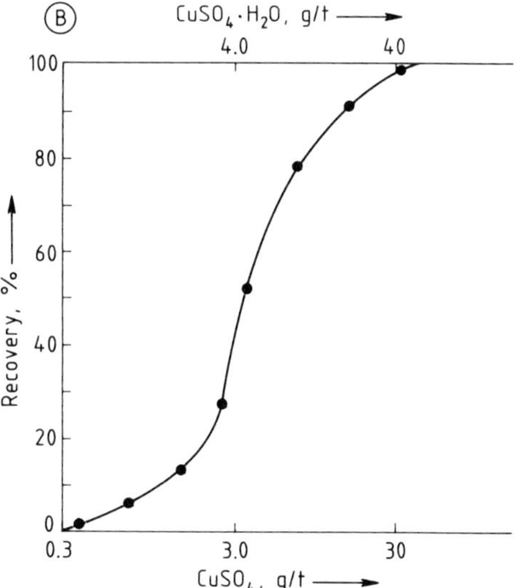

Figure 3. Activation Processes
A) Activation of malachite [$CuCO_3 \cdot Cu(OH)_2$] by sodium sulfide a) In the presence of 0.43 mol/L ethyl xanthate; b) In the presence of 0.43 mol/L isoamyl xanthate[a, b]
B) Activation of ZnS by $CuSO_4$ in the presence of 25 mg/L potassium ethyl xanthate[c]
[a] With permission of [7]. [b] Excessive quantities of sulfide reduce the recovery. [c] Redrawn from [11].

cles and bubbles to establish contact more rapidly. The usual induction times in sulfide flotations are of the order of milli- or microseconds [12]. Examples of frothers are pine oil, long-chain alcohols, and polyoxypropylene derivatives.

Auxiliary Reagents. These reagents include *depressants,* which are used to prevent solids from becoming hydrophobic, and *activators,* which promote the adsorption of reagents onto selected solids. For example, high pH impedes the flotation of sulfide minerals by xanthates. Thus, reagents used to increase pH, e.g., NaOH or $Ca(OH)_2$, would be depressants in this context. Sulfide ions similarly impede the flotation of sulfides in the same system, and Na_2S used as a source of sulfide ions would act as a depressant. Other depressants include starch and quebracho.

On the other hand, sphalerite (ZnS) does not readily become hydrophobic in the presence of xanthates unless copper (II) ions are introduced into the aqueous pulp. In this case $CuSO_4$, used to supply copper (II) ions, is an activator. Other examples of activation are provided by divalent ions such as barium or calcium in the flotation of silica by fatty acids. In this case, silica is inert to these reagents unless a divalent cation is introduced into the pulp at alkaline pH. Thus, $CaCl_2$ or $Ba(NO_3)_2$ used as sources of these divalent cations and NaOH used to raise the pH would be activators of silica for flotation by fatty acids. The terms *collector* and *frother* do not precisely describe the action of such reagents in all systems.

Although oleic acid acts primarily as a collector in the flotation of solids such as scheelite ($CaWO_4$) it also exhibits excellent frothing properties when present in sufficient quantities. On the other hand, nonanol [*143-08-8*], a well-known frother, is the sole reagent required for the flotation of talc or high-rank coals, where it acts simultaneously as frother and collector. The same is true for activators and depressants. For example, sodium sulfide was cited previously as a depressant for sulfide minerals in the presence of xanthate reagents. In fact, heavily oxidized (tarnished) sulfide minerals are treated with sodium sulfide before the introduction of xanthates for flotation, making Na_2S an activator in this type of flotation pulp (see Fig. 3).

2.2. Chemical Classification

Flotation reagents may be inorganic or organic. The former are largely used as auxiliary reagents (Table 2), whereas the latter (Tables 3 and 4) fall into all categories of the functional classification. The organic reagents can be classified as follows:

Polar:

anionic: polar group is an anion, e.g., oleate, $CH_3(CH_2)_7CH{=}CH(CH_2)_7COO^-$

cationic: polar group is a cation, e.g., RNH_3^+

amphoteric: anionic or cationic, depending on pH, e.g., alkylaminocarboxylic acids, $RNH(CH_2)_xCOOH$

Table 2. Inorganic auxiliary flotation reagents

Compound	Composition	Common Applications
Lime	CaO	pH regulator, depressant
Sodium carbonate (soda ash)	Na_2CO_3	pH regulator, dispersant
Sodium hydroxide (caustic soda)	NaOH	pH regulator, dispersant
Sodium sulfide	Na_2S	sulfide depressant and ore sulfidizer
Sodium bisulfide	NaHS	sulfide depressant and ore sulfidizer
Sulfuric acid	H_2SO_4	pH regulator
Sodium cyanide	NaCN	sulfide depressant
Calcium cyanide	$Ca(CN)_2$	sulfide depressant
Sodium dichromate	$Na_2Cr_2O_7$	PbS depressant
Cupric sulfate	$CuSO_4$	ZnS, FeAsS, Sb_2S_3 activator
Lead acetate	$Pb(CH_3COO)_2$	Sb_2S_3 activator
Sodium ferrocyanide	$Na_4Fe(CN)_6$	depressant in Cu–Mo sulfide circuits
Potassium permanganate	$KMnO_4$	FeS_2 depressant in FeAsS flotation
Sulfur dioxide	SO_2	activated ZnS depressant
Sodium thiosulfate	$Na_2S_2O_3$	SO_2 source in acid circuits
Sodium silicate	Na_2SiO_3	siliceous gangue dispersant
Sodium fluosilicate	Na_2SiF_6	depressant in iron-flotation circuits
Sodium polyphosphates	e.g., $(NaPO_3)_6$	dispersant
Sodium fluoride	NaF	activator in silicate flotation
Nokes reagent	complex mixture of P_2S_5, As_2O_3, Sb_2O_3, NaOH, etc.	general depressant in molybdenite flotation circuits except for MoS_2

Table 3. Organic reagents commonly used as flotation collectors

Compound	Active component(s)	Area of application
Primary amine salts	$RNH_3^+Cl^-$	silica, silicates, sylvite
Quaternary ammonium salts	$RN(CH_3)_3^+Cl^-$ $R = C_{10}-C_{16}$	silicates, oxides, clays
p-Tolylarsonic acid	$H_3C-C_6H_4-AsO(OH)_2$	cassiterite
Sodium salts of carboxylic acids[a]	RCOONa	oxides, carbonates, apatite, iron ores, chromite, scheelite
Alkyldithiocarbamates	$R_2N-C(=S)-S^-\ Na^+(K^+)$	sulfides, metallic minerals
Dixanthogens	$S=C(OR)SSC(OR)=S$	sulfides, metallic minerals
Hydrocarbon oils[b]	C_nH_{2n+2}	coal, molybdenite, colemanite with sulfonates
Sodium alkylhydroxamates	$R-C(OH)=N-O^-\ Na^+$	iron ores, wolframite, cassiterite
Naphthenic acids	$R_m-C_5H_8-(CH_2)_nCOOH$	fluorapatite, colemanite
Oximes	$R-CH(OH)-C(=NOH)-R$	chrysocolla, cassiterite
Alkylsulfates and -sulfonates	$R-O-SO_3^-\ M^+$ $R-SO_3^-\ M^+$	iron ores, beach sand cleaning, borates, carbonates, fluorite
Sodium 2-(Methyloleylamino) ethylsulfonate	$C_{17}H_{33}-N(CH_3)-CH_2CH_2SO_3Na$	celestite
O-Ethyl isopropyl thionocarbamate	$C_3H_7(H)N-C(=S)-OC_2H_5$	copper sulfides
Thionocarbanilide	$C_6H_5-NH-C(=S)-NH-C_6H_5$	sulfide minerals
Alkyldithiocarbonates (xanthates)	$R-O-C(=S)-S^-Na^+(K^+)$	sulfide minerals, gold
Xanthogen formates[c]	$R-O-C(=S)-S-C(=O)-OR$	sulfide minerals
Dialkyl-dithiophosphates[d]	$(R^1-O)(R^2-O)P(=S)S^-M^+$	sulfide minerals, native gold, copper

[a] R may be saturated or unsaturated. [b] Vapor oils, kerosene, fuel oils. [c] Trade name: Minerec. [d] Trade name: Aerofloat.

Table 4. Organic chemicals used in flotation as depressants, slime depressants, and flocculants

Common name	Active components	Application
Sodium isopropyl-naphthalene sulfonate Trade name: Aerosol OS	$(CH_3)_2CH$, SO_3Na (naphthalene ring structure)	wetting agent, defoamer, dispersant, emulsifier
Sodium dioctylsulfo-succinate Trade name: Aerosol OT	$CH_2COOC_8H_{17}$ $NaO_3SCHCOOC_8H_{17}$	wetting agent, defoamer, dispersant, emulsifier
Poly(ethylene oxide)	$-(CH_2CH_2O)_n-$	flocculant, dewatering aid
Sodium polyacrylate	$\left(\begin{array}{c}CH-CH_2\\ \vert \\ COONa\end{array}\right)_n$	flocculant, dewatering aid
Starch	CH_2OH, CH_2OH, O, O, OH, OH, O, O, O, OH, OH (structure)	slime depressant, iron ore flocculant, slime control agent
Tannic acid	OH, OH, OH, OH, HO, O, OH, O, COOH (structure)	depressant for fluorite, carbonates, and non-sulfides
Quebracho (trade name) extracted from shinopsis trees	HO COOH, HO, OH, OH — Quinic acid [*77-95-2*] COOH, HO, OH, OH — Shikimic acid [*138-59-0*] COOH, HO, OH, OH — Gallic acid [*149-91-7*] O, O — Flavone [*525-82-6*] O — Coumaran [*496-16-2*] OH, HO, O, OH, OH, OH — Catechin [*154-23-4*]	depressant for carbonates, fluorite, and nonsulfides

nonionic: polar group is not ionized, e.g., *n*-hexanol, $CH_3(CH_2)_4CH_2OH$
Nonpolar: e.g., saturated hydrocarbons

Organic flotation reagents are also classified by functional group, including fatty acids, carboxylic acids, thiols, xanthates, sulfates, sulfonates, and amines.

The following organic reagents are commonly used as flotation frothers:

Long-chain aliphatic alcohols	$CH_3(CH_2)_nCH_2OH$ $n = 3-6$
Camphor oil Active component: Safrole [*94-59-7*]	O, O (ring structure), $CH_2CH{=}CH_2$
Dimethyl phthalate [*131-11-3*]	$COOCH_3$, $COOCH_3$ (benzene ring)
Dimethylphenylcarbinol (2-Phenyl-2-propanol) [*617-94-7*]	OH, $H_3C-C-CH_3$ (phenyl)

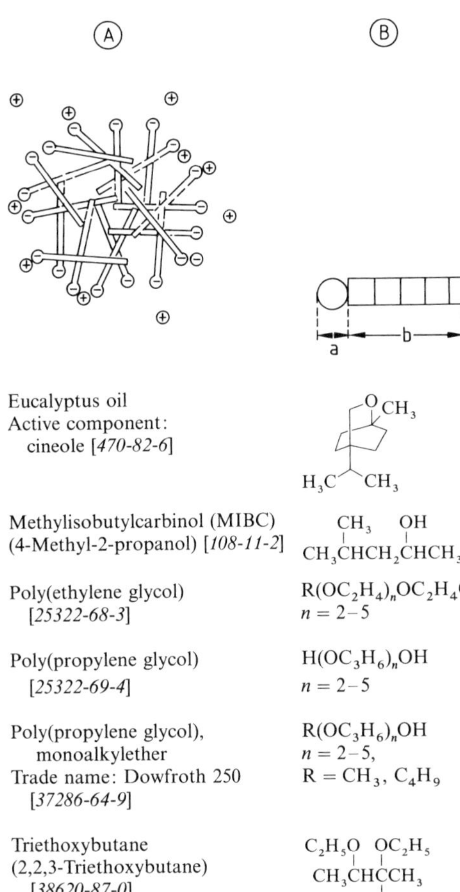

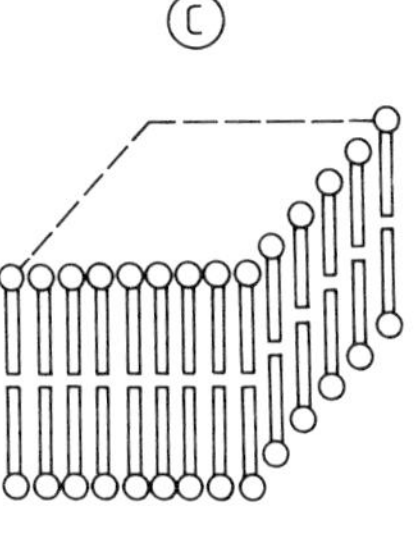

Figure 4. Micelle formation
A) Spherical micelle
B) Straight-chain surfactant (amphipatic) molecule
a) Polar end; b) Nonpolar end
C) Laminar micelle

Eucalyptus oil Active component: cineole [*470-82-6*]	(structure: bicyclic ether with O, CH_3, H_3C, CH_3)
Methylisobutylcarbinol (MIBC) (4-Methyl-2-propanol) [*108-11-2*]	$CH_3CH(CH_3)CH_2CH(OH)CH_3$
Poly(ethylene glycol) [*25322-68-3*]	$R(OC_2H_4)_nOC_2H_4OH$ $n = 2-5$
Poly(propylene glycol) [*25322-69-4*]	$H(OC_3H_6)_nOH$ $n = 2-5$
Poly(propylene glycol), monoalkylether Trade name: Dowfroth 250 [*37286-64-9*]	$R(OC_3H_6)_nOH$ $n = 2-5$, $R = CH_3, C_4H_9$
Triethoxybutane (2,2,3-Triethoxybutane) [*38620-87-0*]	$CH_3CH(OC_2H_5)C(OC_2H_5)_2CH_3$

2.3. Surface and Solution Chemistry

2.3.1. Solubility

Inorganic reagents used in flotation are always highly soluble in water whereas organic reagents may not be. Nonpolar oils, for example, are insoluble in water and must be added to the flotation cell or the preceding conditioning tank, at high shear created by intense agitation of the pulp. Alternatively, these reagents can be prepared as emulsions or sometimes dissolved in an alcohol such as methanol before addition to the flotation system. Straight-chain alcohols containing more than eight carbon atoms are insoluble in water and are used as frothers.

Most ionic or polar flotation reagents are sufficiently soluble in water. Solubility is limited mainly by the tendency of such compounds to form molecular aggregates in the aqueous medium known as micelles when present in high concentrations (Fig. 4). The concentration at which micelles form is known as the *critical micelle concentration* (cmc).

2.3.2. Surface Activity

The ability of reagents to alter the liquid–vapor interfacial tension, γ_{LG}, of aqueous solutions is called surface activity (see also →Surfactants). For example, although the surface tension of distilled water is about 72 mN/m at room temperature, the introduction of a flotation reagent such as sodium dodecyl sulfate at a concentration of 0.005 mol/L reduces this value to 48.5 mN/m. Such organic reagents contain two chemical

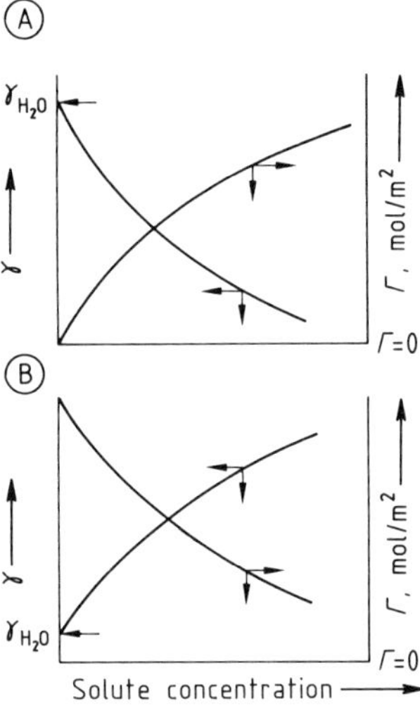

Figure 5. Variation of solution surface tension by solutes according to the Gibbs adsorption equation (Schematic)
A) Effect of organic surfactants on γ_{LG} and Γ. Condition of surface excess concentration
B) Effect of inorganic solutes on γ_{LG} and Γ. Condition for surface deficiency.
γ_{H_2O} = Surface tension of pure water; γ = Surface tension of solution; Γ = Surface excess concentration

groups: a nonpolar moiety, usually a saturated hydrocarbon, and a polar moiety, which may be ionic (e.g., $-COO^-$, $-SO_3^-$, $-SO_4^-$) or nonionic (e.g., $-OH$, $-NH_2$, $>C=O$).

The quantitative relationship between the concentration of an amphipatic molecule and its effect on the surface tension at concentrations below the cmc is described by the Gibbs adsorption equation:

$$\frac{d\gamma_{LG}}{dC} = -\Gamma k \frac{T}{C} \tag{4}$$

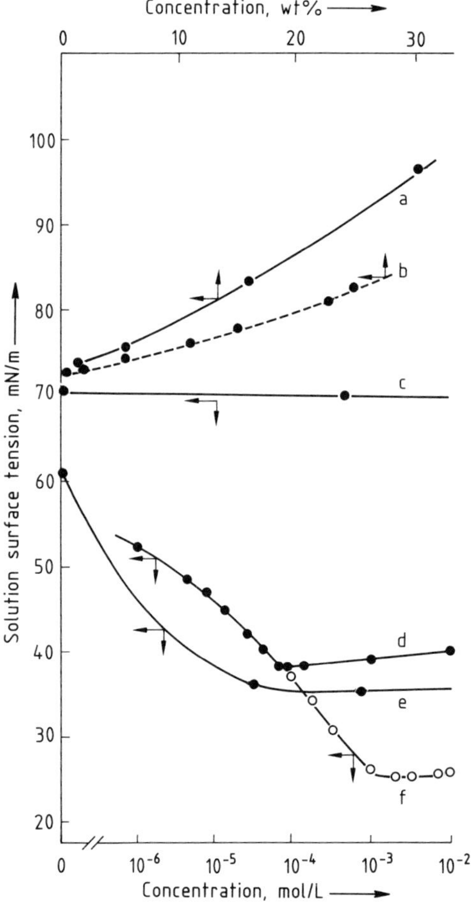

Figure 6. Effects of inorganic salts and organic flotation reagents on the surface tension of water
a) Sodium hydroxide[a]; b) Sodium chloride[a]; c) Potassium ethyl xanthate[b] d) Polyoxyethylene *n*-decanol with 30 ethylene oxide groups[b]; e) Cetyltrimethylammonium bromide[c]; f) Sodium oleate[d];
[a] Drawn from data in [13]. [b] With permission from [14]. [c] Drawn from data in [15]. [d] Redrawn from data in [16].

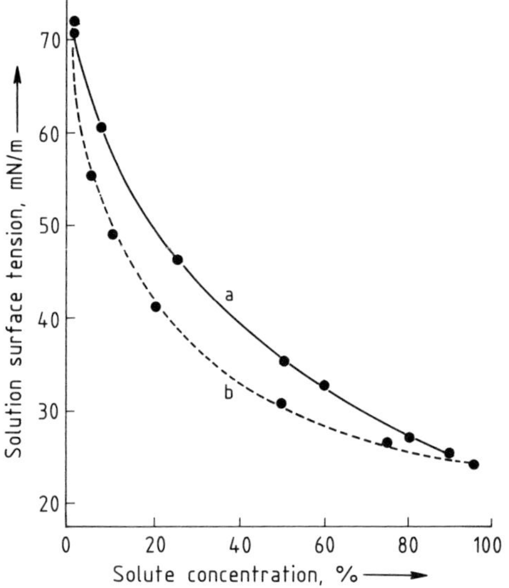

Figure 7. Surface tension of aqueous solutions of acetone and methanol
a) Acetone; b) Methanol

where γ_{LG} = solution surface tension, C = equilibrium concentration of surfactant in solution, k = Boltzmann constant, Γ = surface excess concentration of surfactant measured in quantity per unit area (e.g., molecules/cm^2), and T = temperature.

Based on Equation (4), the variation of solution surface tension can be discerned to be as shown in Figure 5, which predicts that some substances ought to increase surface tension, creating a surface deficiency, as opposed to surface excess concentration. This condition is met by a number of inorganic compounds shown in Figure 6.

Not all amphipatic flotation reagents exhibit pronounced surface activity, as demonstrated by Figure 6. Compounds that lower the surface tension of water are not necessarily surface active, although their structures may indicate amphipatic properties. Short-chain alcohols, acetone, and simple ethers fall into this category (see Fig. 7).

2.4. Interactions between Flotation Reagents and Solids

As a general rule, for alteration of the hydrophobic–hydrophilic properties of solids in a flotation pulp, adsorption or desorption at the surface is a prerequisite. Thus, for a solid to become hydrophobic or hydrophilic it must first take up

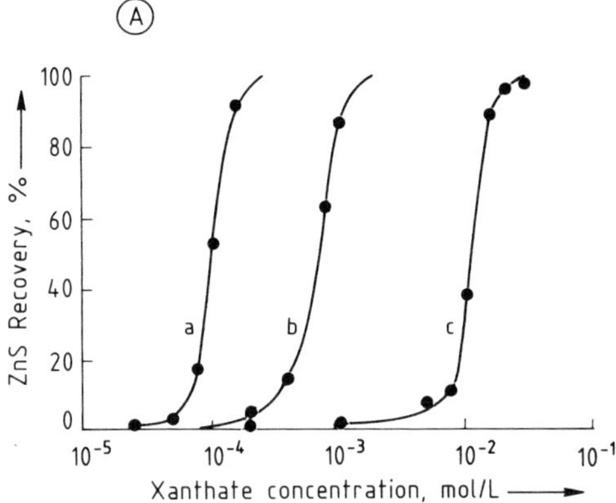

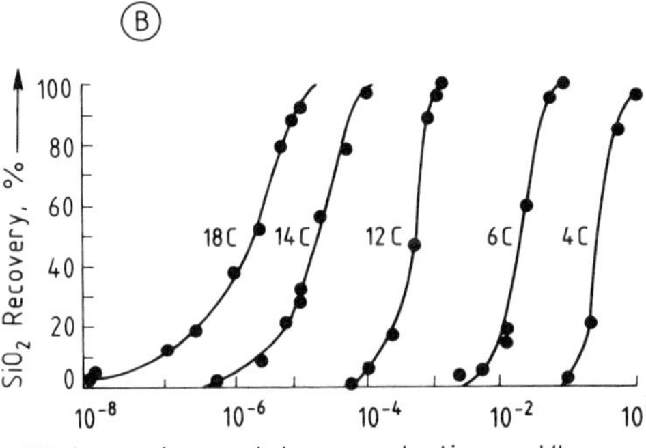

Figure 8. Role of carbon chain length in flotation
A) Sphalerite (ZnS) in the presence of various chain-length xanthates at pH = 3.5*
a) 6 Carbons, hexyl xanthate; b) 4 Carbons, isopropyl xanthate; c) 2 Carbons, ethyl xanthate
B) Quartz (SiO_2) flotation by alkylammonium acetate solutions; chain lengths indicated at their natural pH**
* With permission [18]. ** With permission [19].

ions or molecules from the medium or release them into it. Alternatively, a solid made hydrophobic by the adsorption of a reagent changes only by the action of other species which alter the surface properties. This may or may not require the desorption of the species originally present. For example, at a given pH, a given sulfide mineral requires a minimal concentration of flotation collector before stable bubble–particle contact (flotation) can occur [8] (Fig. 8).

Silica, treated with trimethylchlorosilane ($Si(CH_3)_3Cl$) shows hydrophobic–hydrophilic transitions that depend on time, pH, and temperature [17]. In all the experimental cases reported for this system, whether the solid was hydrophilic or hydrophobic, the organic reagent was present at the solid surface. The variation of hydrophobic–hydrophilic properties was attributed to the tenacity of the aqueous layer at the solid–liquid interface; the desorption of species that confer hydrophobic characteristics is not always essential for a solid to exhibit hydrophilic properties.

Naturally hydrophobic solids need to be considered a special case of this general rule, because they may require no surfactant uptake from the medium to become hydrophobic, but can be made hydrophilic by adsorbed reagents. For example, most wetting agents, e.g., Aerosol OT, (sodium dioctylsulfosuccinate) [*577-11-7*], eliminate the natural hydrophobicity of graphite.

2.4.1. Adsorption–Desorption of Inorganic Reagents

In general, the chemical reactions of inorganic flotation species in aqueous solution can be described in terms of solubilities. The classical equilibrium constant (solubility product) applies:

$$MA_2 \rightleftharpoons M^{2+} + 2A^- \qquad (5)$$

$$K_{sp} = [M^{2+}][A^-]^2/[MA_2] \qquad (6)$$

where M^{2+} indicates the cation (metal) concentration in solution, A^- the anion concentration, and K_{sp} the solubility product. Figure 9 shows the relationships of hydrolysis and precipitation that occur in the aqueous phase for various ions in accordance with the solubility product princi-

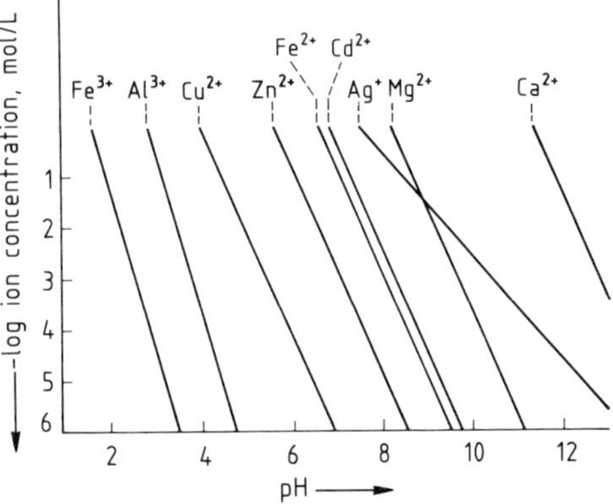

Figure 9. Cation concentrations derived from the dissolution of some metal hydroxides at different pH [21]*
* Calculations made assuming that the solubility product expression given in Equation (6) applies; more elaborate calculations including other hydrolytic species can also be made [22], [23]

ple. Traces of ions or their hydrolysis products many play important activator roles in flotation [20].

2.4.2. Adsorption – Desorption of Organic Reagents

Organic reagents and solids suspended in a flotation pulp interact by one or more of the mechanisms given below. Some processes, such as hydrogen bonding, are also important in the solubilization of polar substances in aqueous media.

Hydrogen Bond Formation. This bond (strength ca. 21 – 29 kJ/mol), usually associated with the physical properties of ice and water, is favored when a group contains hydrogen attached to a highly electronegative element such as F, O, S, or N. Thus, the adsorption of an amide group, present in proteins or nonionic polyacrylamide (a depressant), on fluorite or silica is due to this bond [24], [25]. The interaction of elemental sulfur and alcohols probably also takes place by hydrogen bond formation.

Electrostatic Interactions. Such interactions are nonspecific, occurring between groups that carry electrostatic charges of opposite sign. Bond formation occurs by charge attraction. A typical example in flotation systems is the adsorption of amines that carry cationic polar groups on silica, which carries a net negative electric charge at pH > 2. Many treatments of reagent – solid interactions in flotation systems tend to exaggerate the importance of electrostatic interactions, overlooking other mechanisms.

Chemical Bond Formation. Chemical bonds are characterized by their strength (typically greater than ca. 42 kJ/mol) and usually lead to the formation of saltlike structures at solid – liquid interfaces. A surface compound is not necessarily identical to a compound formed in the bulk of the solution [26], [27]. Examples of such bonds include the interaction of fatty acids, which carry the polar carboxylate group ($-COO^-$), with sparingly soluble solids such as calcite ($CaCO_3$) or fluorite (CaF_2). Infrared spectroscopy shows that the calcium oleate formed by the interaction of oleic acid with calcite or fluorite is not identical to the calcium salt of oleic acid, $Ca(C_{17}H_{33}COO)_2$, obtained from $CaCl_2$ and sodium oleate [28].

Hydrophobic Bonding. Association of the hydrocarbon ends of amphipatic substances leads to the formation of *micelles* (Sect. 2.3.1). Similarly, many saturated hydrocarbons, such as kerosene or fuel oils, are readily adsorbed on naturally hydrophobic solids, e.g., graphite or freshly cleaved bituminous coal. Such interactions are mainly a result of London – van der Waals forces. Furthermore, the hydrocarbon ends of surfactants, such as dodecylbenzenesulfonate, and dodecylbenzenesulfate, associate at the solid – liquid interfacial region, resulting in low-energy structures known as hemimicelles [29].

The strength of association is a function of the hydrocarbon chain length and is of the order of 1 kT per $-CH_2$-group (where k = Boltzmann constant and T = temperature). The unique bond that leads to such adsorptions and molecular associations is called the *hydrophobic bond.* Although the term has been criticized [30], it has been generally accepted.

Multiple Dipole Interaction Bonds (Crystal-field Interactions). The adsorption of tannins (depressants in flotation) or polyacrylamide on fluorite has been attributed to the formation of a bond in which an active group of the reagent is held in the crystal field of the adsorbing solid [31]. Little information on such bonds is available.

Collector Fitting into Solid Lattices. The separation of similar salts, such as NaCl and KCl, by the use of amines is a well-established indus-

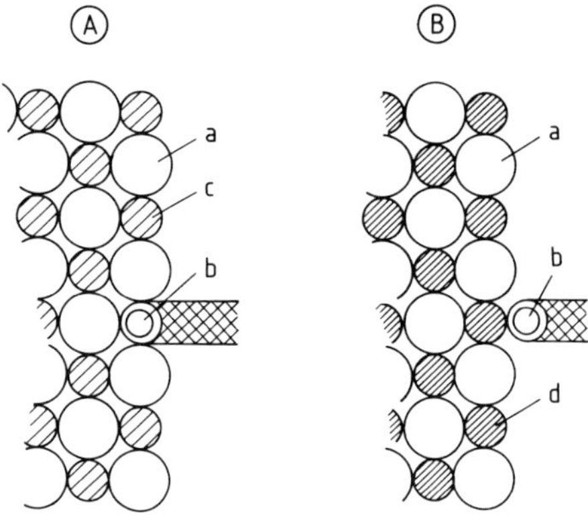

Figure 10. Schematic representation of the selective adsorption of amine ions on sylvite (KCl) [32]
A) Potassium chloride (KCl)
B) Sodium chloride (NaCl)
a) Cl^- b) NH_3^+; c) K^+; d) Na^+

trial practice. Although a number of mechanisms have been suggested for the selective interaction between KCl and primary amines, the most widely accepted is that the amine fits into spaces on the KCl lattice vacated by K^+ because the ionic sizes of $-NH_3^+$ and K^+ are comparable. Thus, KCl is made selectively hydrophobic, whereas NaCl is not (Fig. 10).

In many systems more than one mechanism may act simultaneously and synergistically. The term adsorption is used here for the accumulation of a dissolved species on a solid particle. The term sorption is preferred by some authors. The terms physical adsorption or physisorption and chemisorption are also used to distinguish between bonds with strengths less than 20 kJ/mol and those greater than 42 kJ/mol, respectively.

3. Electrical Phenomena

The main phenomena to be considered in this connection are those associated with the solid–liquid interfacial region. On contact of the solid with the aqueous medium, ionization, ion dissolution, or ion adsorption occurs, resulting in charging of the particle surface. Hydrolysis in the aqueous medium may also alter the character of the ionic species passing into it. The predominant phenomena here relate to the fact that cations are hydrated more than anions and that thermal energy promotes ion mobilization in the system. Ruptured bonds and the lattice parameters also affect the properties of the solid–liquid interface. The cleavage of fluorite, graphite, and chalcopyrite are shown in Figure 11.

Within short periods of contact, a double layer (usually referred to as the electric double layer), consisting of the charged surface of the solid and counter ion, is established. Although earlier theories considered this charge arrangement to be in the form of a simple plate capacitor, current colloid science shows that the double layer consists of a surface with fixed charge whereas the ionic cloud countering it is diffuse [33]. The overall system, consisting of the solid and the ionic medium surrounding it, conforms to the principle of electroneutrality.

Zeta potential is widely used to describe the role played by the electrical double layer in flotation. It corresponds to the potential that exists at the so-called shear plane, which is created when a charged particle moves under the influence of an external electric field.

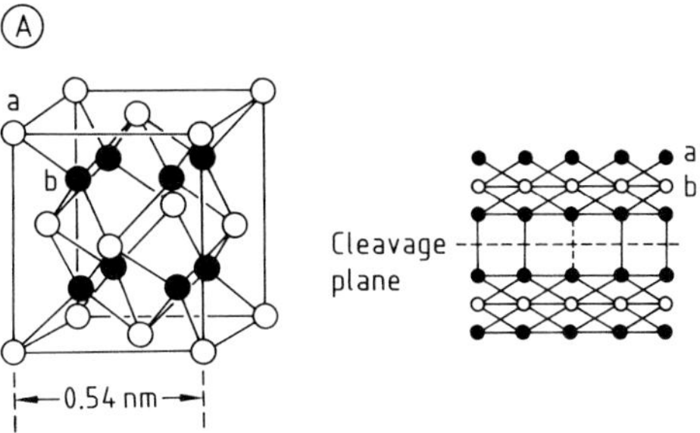

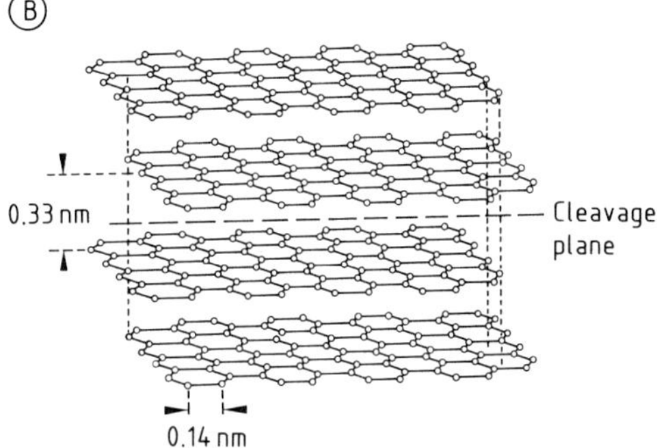

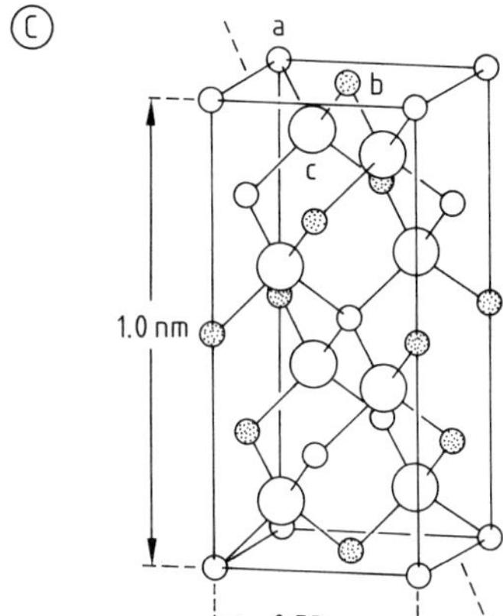

Figure 11. Crystal structures and mode of cleavage of minerals
A) Fluorite (CaF_2) and its cleavage plane
a) Calcium; b) Fluorine
B) Graphite
C) Chalcopyrite ($CuFeS_2$) which cleaves irregularly, sometimes at the (011) plane
a) Copper; b) Iron; c) Sulfur

Zeta potential can be measured by electrophoresis, sedimentation potential, streaming potential, and electroosmosis. In the first two of these methods the shear plane is created by the motion of the particle; in the other two, the particle (more precisely, a porous bed consisting of many charged particles) is fixed [34], [35]. Equation (7) correlates

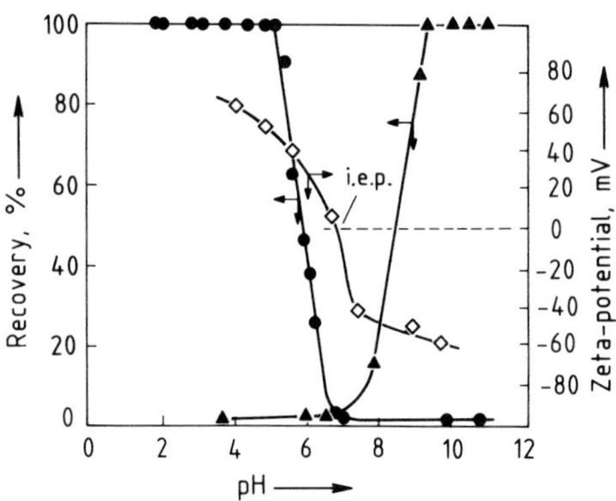

Figure 12. Relationship between iep and flotation properties of goethite (FeO (OH)) [36]*
● = Sodium dodecyl sulfate or -sulfonate; ▲ = Dodecylammonium chloride; ◇ = iep
* Courtesy USBM; zeta potential measurements made in 10^{-4} M NaCl; note that minimum flotability occurs at the pH of iep and anionic reagents act as collectors at $\zeta > 0$ whereas cationic reagents collect goethite at $\zeta < 0$.

the zeta potential with the electrophoretic mobility of the particles under the influence of an external electric field:

$$\zeta = \frac{4\pi\eta}{\varepsilon} \frac{V_e}{E} \qquad (7)$$

where
ζ = zeta potential
η = viscosity of the aqueous medium
ε = dielectric constant
V_e = electrophoretic velocity of particle
E = electric field between electrodes applying the field

At 25 °C and $\eta = 0.895$ mPa · s, Equation (7) reduces to $\zeta = 12.8\ V_e = 12.8\,u$
where u = electrophoretic mobility.

The significance of the zeta potential becomes apparent where electrostatic interactions play a major role in flotation reagent adsorption, as shown in Figure 12. The condition at which $\zeta = 0$ is known as the isoelectric point (iep), sometimes, less accurately, the zero point of charge (zpc). In principle, although an iep can be defined for each solid for numerous conditions, the variable most frequently used to describe the ionic conditions at which iep occurs is pH.

Adsorption mechanisms correlate with zeta potential; that is, physical adsorption affects the magnitude of the zeta potential. A change of sign in the presence of an appropriate concentration of adsorbate (see Fig. 13) is taken as an indication of a chemisorption-type reagent uptake [38]. As seen from Figure 13, the iep does not necessarily indicate zero potential at the surface of the particle, only that the potential at the shear plane is equal to zero.

Depending on the mechanism of reagent–solid interaction, many systems exhibit correlations between system variables and zeta potential (Fig. 14).

The following isoelectric points ($\zeta = 0$) are observed in the absence of other substances.

AgCl	$pAg^+ = 4$
AgBr	$pAg^+ = 5.4$
Ag_2S	$pAg^+ = 10.2$
$CuSiO_3 \cdot 2H_2O$	$pCu^{2+} = 4$, at pH = 7
CaF_2	$pCa^{2+} = 3$
Fe_2O_3	pH = 6.5 and 8.2
FeS_2	pH = 7
FeS_2	pH = 5 (30 mg NaCN/L)
PbS	pH = 3.5
SiO_2	pH = 2–3.7
SiO_2	neutral pH in the presence of 3×10^{-3} mol dodecylammonium acetate/L

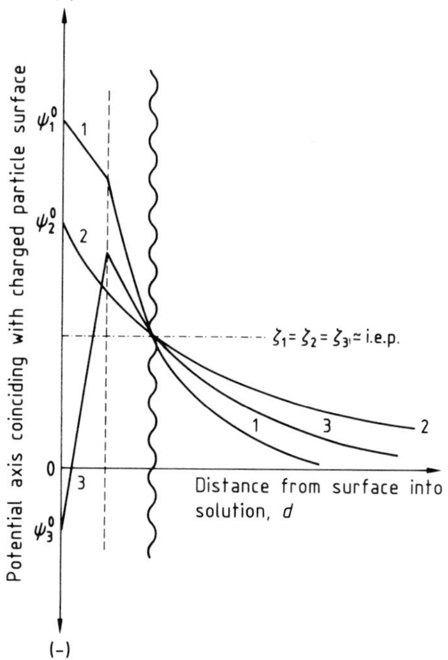

Figure 13. Representation of the fact that zero zetapotential does not necessarily indicate zero surface potential (ψ^o) [35, p. 221]*
---- = Inner Helmholz plane; ~~ Shear plane
* Although ψ_1^o and ψ_2^o are positive and ψ_3^o is negative, all conditions indicate $\zeta = 0$; inner Helmholz plane passes through the centers of unhydrated, specifically adsorbed ions.

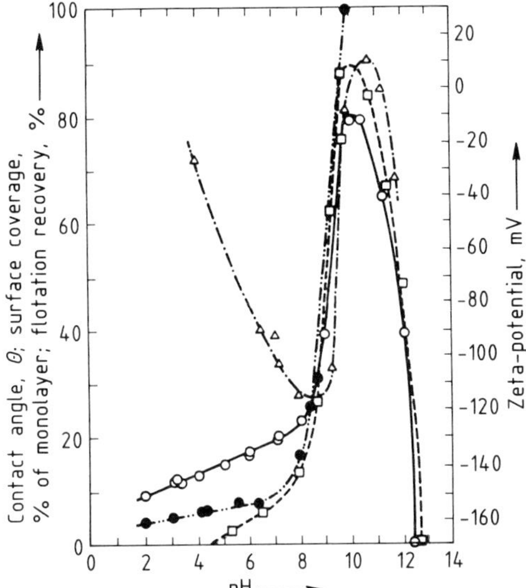

Figure 14. Correlation of commonly measured flotation and surface chemical variables in the quartz-dodecylammonium acetate system*
□–□ = % Recovery; △–·–△ = Zeta potential; ○–○ = Contact angle; ●–··–● = Monolayer
* With permission [37].

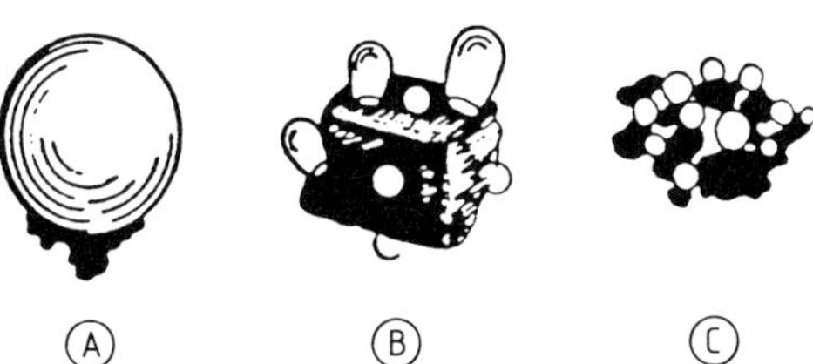

Figure 15. Different forms of particle–bubble aggregates [7], [41]
A) Particle attached to bubble
B) Multiple bubble attached to large particle (common in vacuum flotation)
C) Particle–bubble flocculation in pulps containing fine particles

4. Bubble–Particle Interactions

The ultimate objective of a flotation process is the selective removal of solid particles from the aqueous medium, which is accomplished by the adhesion of air bubbles to the hydrophobic particles. Particle flotability can be treated as a probability [39]:

$$P_f = P_c \cdot P_a \cdot P_s \qquad (8)$$

where
P_f = probability of flotation
P_c = probability of particle–bubble collision
P_a = probability of particle–bubble adhesion
P_s = probability of formation of a stable particle–bubble aggregate

In some methods, such as vacuum flotation, where dissolved gases become the bubble source, or in situ bubble-generation processes, where acids generate bubbles of carbon dioxide from carbonate-containing pulp, the probability treatment needs to be modified. Equation (8), however, is widely applied, because most flotation systems rely on extraneously introduced air bubbles. Particle–bubble aggregates are shown in Figure 15.

Solids to be floated carry an aqueous film, which may have a thickness as low as tens of nanometers [40]. The stability of this film determines whether a particle will adhere to a bubble on collision within the slurry: if the film thins, ruptures, and recedes, adhesion is facilitated. The time taken for the film to undergo these processes is known as the induction time.

Film stability has been quantified in terms as disjoining pressure [27].

$$\pi_D = p - p_o \qquad (9)$$

$$\gamma = \gamma_0 \int_h^\infty \pi_D \, dh \qquad (10)$$

where
π_D = disjoining pressure
p = pressure inside bubble
p_o = pressure of the bulk of the liquid
γ = specific surface energy of the liquid film
γ_0 = specific surface energy of an infinitely thick film of the same liquid

The conditions for film stability are shown in Figure 16.

For stable films: $\left(\frac{\partial \pi_D}{\partial h}\right)_T < 0$

For unstable films: $\left(\frac{\partial \pi_D}{\partial h}\right)_T > 0$

Disjoining pressure depends on the summation of electrical double-layer effects (π_E), solvation effects (π_s), and London–van der Waals forces (π_v); i.e., $\pi_D = \pi_E + \pi_s + \pi_v$, as well as on the heterogeneity of the solid surface [42]. The role of frothers is to make π_D more positive, i.e., reduce film resistance to rupture.

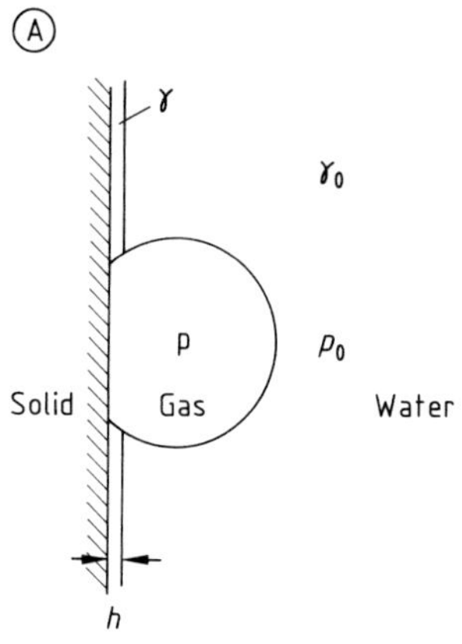

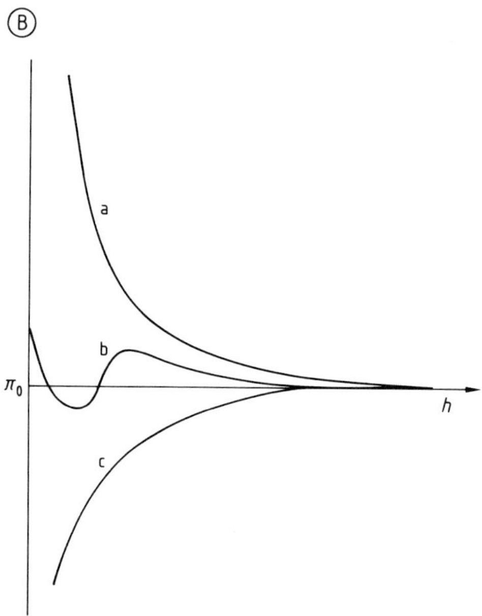

Figure 16. The concept of disjoining pressure [40]
A) Schematic representation of a disjoining film between a particle and a bubble
B) Relation of film stability to disjoining pressure
a) Stable film; b) Metastable film; c) Unstable film

5. Flotation Processes

The phenomena described so far relate to the froth flotation process. The following variations and techniques are also used:

Collectorless flotation:
froth flotation using the natural hydrophobicity of some minerals

Two-liquid flotation:
variant of froth flotation, using oil droplets instead of air bubbles and an oil layer instead of froth

Column flotation:
variant of froth flotation, employing columns 3–9 m high with 0.3–1.5 m cross-section

Microbubble flotation:
variant of column flotation, employing bubbles 10–50 µm in diameter

Dissolved-air flotation:
air bubbles are generated by applying vacuum to air-saturated slurries

Gamma flotation:
liquid–vapor surface-tension control is used to separate hydrophobic particles

Piggyback flotation:
also known as carrier flotation: uses the mutual coagulation of solids; when one is floated, the other (usually more difficult to float on its own) is also collected

Skin flotation:
hydrophobic particles are removed without froth, but at a layer approximately one particle thick

5.1. The Gamma Flotation Process

The gamma flotation process is characterized by emphasis on the control of solution surface tension. The critical surface tension of wetting of

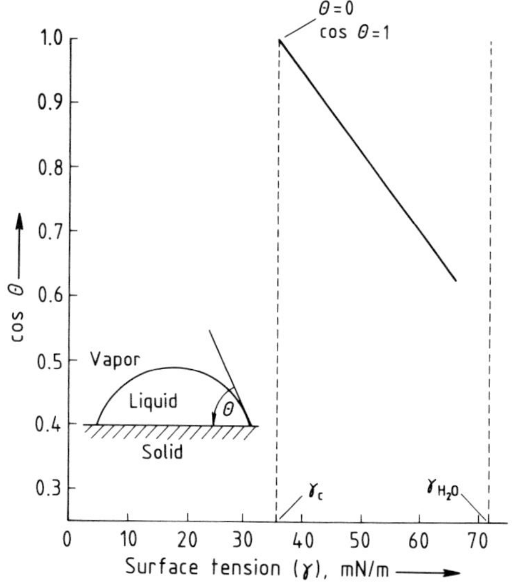

Figure 17. Schematic of critical surface tension of wetting by contact-angle measurement

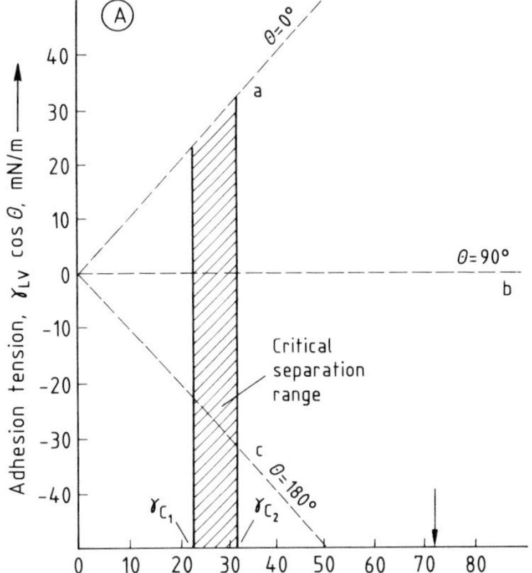

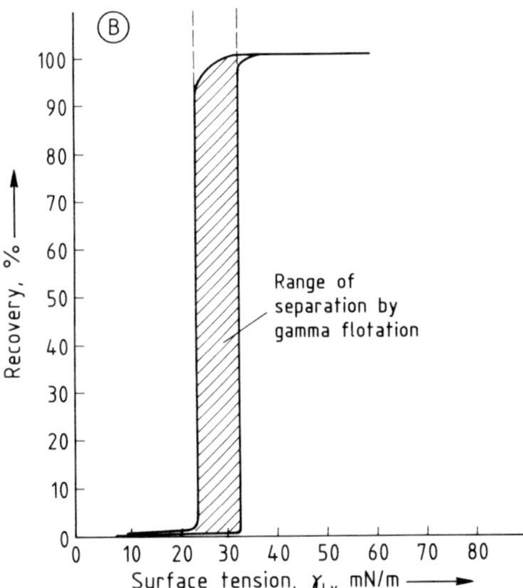

Figure 18. Schematic of physical basis of the gamma flotation process [9]
A) Adhesion tension vs. solution tension for two solids with different critical surface tension or wetting
B) Relation of flotation recovery to solution surface tension for two solids with different critical surface tension or wetting
γ_{c_1} = Critical surface tension of wetting of solid 1;
γ_{c_2} = Critical surface tension of wetting of solid 2;
* Lines a, b, and c obtained by substituting values into the adhesion tension equation, e.g., [$\gamma_{LG} \cos\theta$].
** In the shaded area solid 1 with γ_{c_1} floats, whereas solid 2 with γ_{c_2} is completely wetted by the aqueous solution.

solids is based on a plot of the cosine of contact angle θ against the surface tension of the solution in contact with a solid [43]. Each hydrophobic solid exhibits $\cos\theta = 1$ (zero contact angle) at a given solution surface tension as shown in Figure 17. Difficulties in contact-angle measurements on powdered solids, as used in flotation practice, can be overcome by a technique where flotation recoveries are plotted against solution surface tension [10], [44], [45]. The γ_{LG} at which flotation recovery is equal to zero is taken as the critical surface tension of wetting, γ_c, of the solid under investigation. The two techniques [43]–[45] are equivalent [10]; the proof of this constitutes an independent demonstration of the thermodynamic basis of the flotation process [46].

Figure 18 shows the use of critical surface tension differences of two hydrophobic solids by controlling the surface tension of the solution in which they are suspended. This method is called gamma flotation [9]. The value of the process has been demonstrated for a number of flotation systems including oil shale, sulfide minerals, and others [48]–[50] (see Figures 19 and 20). Salt flotation, which has been used in coal processing, is

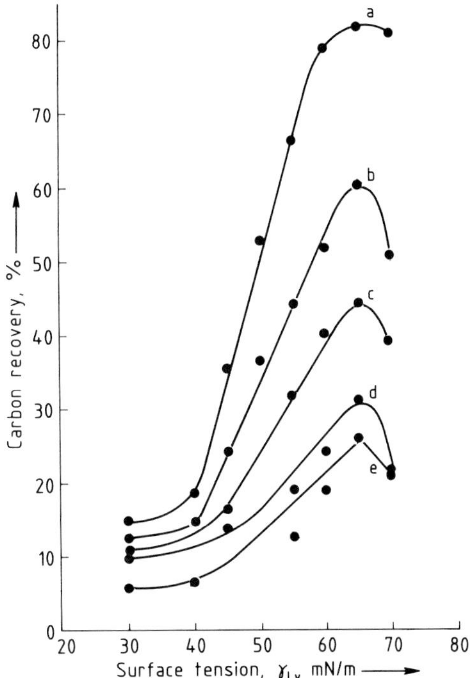

Figure 19. Separation of Colorado oil shale sizes fractions by exploiting the kinetic aspects of gamma flotation [47]
a) <13 µm; b) 18–13 µm; c) 26–18 µm; d) 38–26 µm; e) 48–38 µm

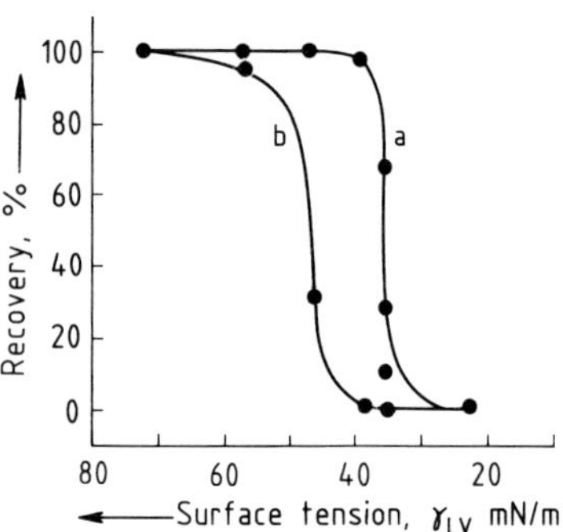

Figure 20. Separation of a mixture of silica and magnetite by gamma flotation*
a) SiO_2; b) Fe_3O_4
* Powders have been initially made hydrophobic by treatment with appropriate surfactants and mixed after filtration [49]

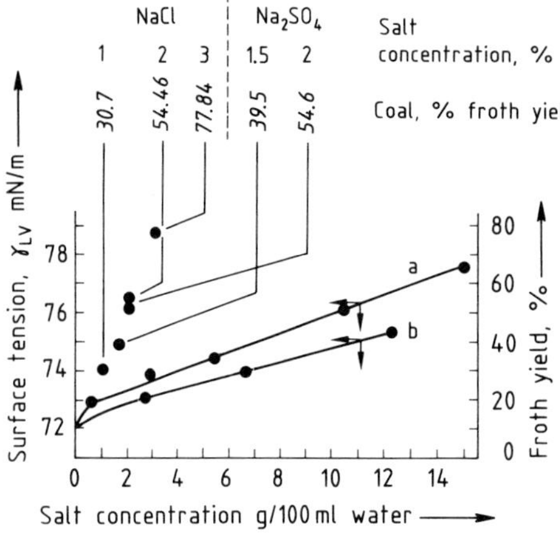

Figure 21. Flotation in high surface tension solutions* [9]
a) NaCl; b) Na_2SO_4
* This method is also known as salt flotation [7, p. 338, 339], [13].

a variant of gamma flotation in which the natural hydrophobicity of coal is exploited and the surface tension of the solution is controlled by salts. Figure 6 shows that some salts raise the surface tension of water to values above the usual 72 mN/m. Figure 21 demonstrates the relationship between solution surface tension modified by salts and flotation results obtained with various coals [9].

5.2. Design of Flotation Operations

Flotation testing is generally preceded by geological, engineering, and mineralogical work.

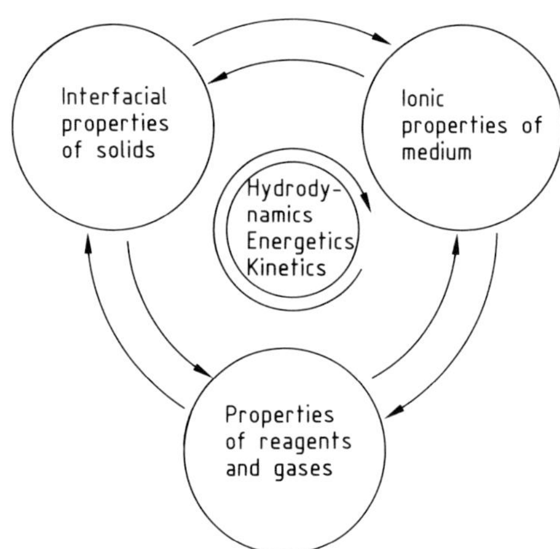

Figure 22. System variables that affect the flotation process [9]

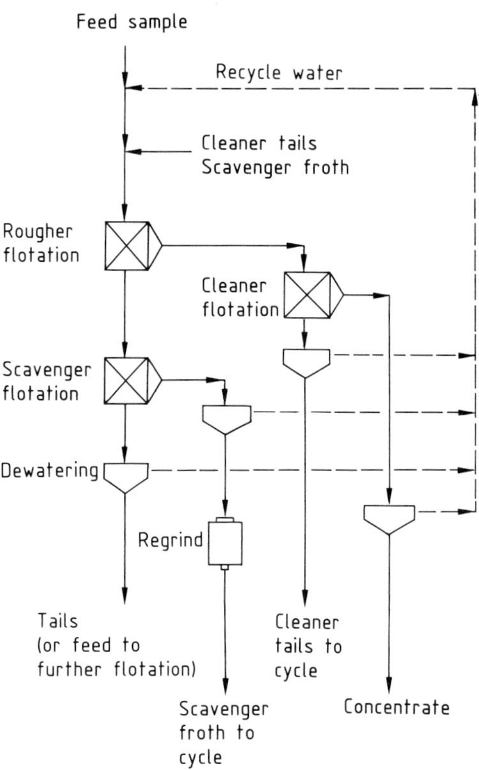

Figure 23. Schematic of locked-cycle test [51]

The overall flotation system is a complex combination of variables outlined in Figure 22, but laboratory testing is based on mineral characteristics and reagent properties. Recovery is "the percentage of total values contained in the ore which is recovered in the concentrate"; grade refers to "assay of values in the concentrate." Recovery and grade tend to vary inversely.

Laboratory tests are designed to establish crushing and grinding conditions and optimize such variables as pulp density, reagent addition, pH, and auxiliary reagent concentration before the pilot plant stage. An important laboratory test is the "locked-cycle test", which simulates pilot plant conditions (Fig. 23) [1].

Flotation machine variables, such as volume, area, throughput, aeration rate, and cell layout are considered at the plant-design stage. Simulation studies using computer modeling techniques have gained increasing acceptance [1], [52]. The final plant layout provides for product handling, filtration, dewatering, tailings disposal, and plant-water recycling.

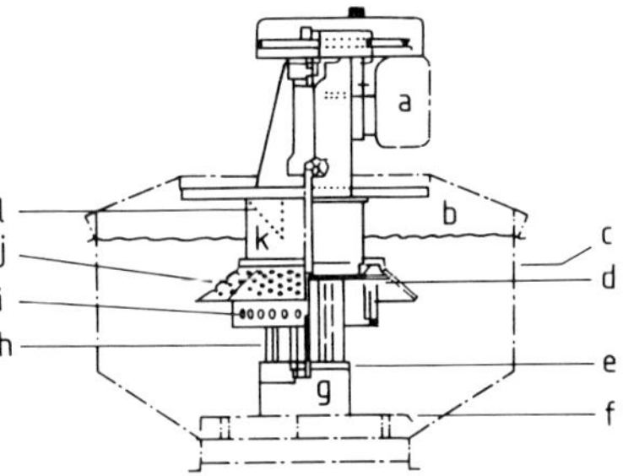

Figure 24. Wemco flotation machine*
a) Motor; b) Weir level; c) Rotor submergence; d) Rotor top; e) Adjustable collar; f) False bottom; g) Draft tube; h) Rotor; i) Disperser; j) Hood; k) Stand pipe; l) Air inlet duct
* With permission [53]

5.3. Flotation Machines

Flotation machines consist of a chamber (cell) equipped with entry and exit ports and an agitation assembly that keeps the pulp in constant motion during the separation. These ma-

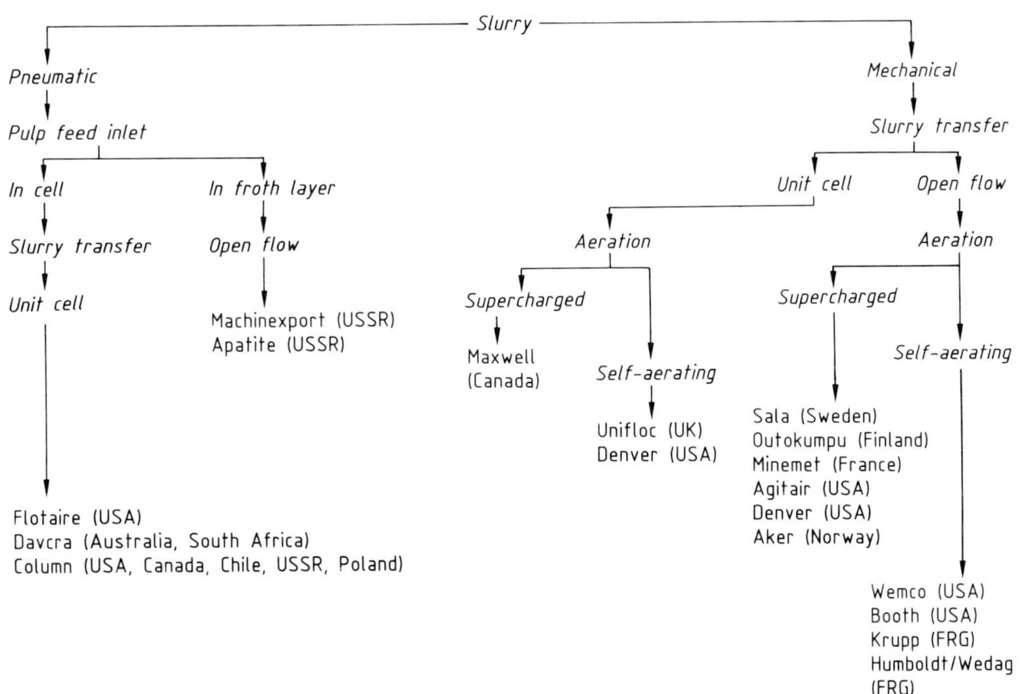

Figure 25. Types of flotation machines, their modes of operation, and representative manufacturers [54]–[56]

chines fulfill the following functions: (1) mixing and agitation of the slurry, (2) aeration and promotion of particle–bubble collisions, (3) formation of a froth layer and its removal for product recovery, and (4) continuous discharge of tailings, consisting of components not collected in the froth layer [53].

The design of a flotation machine (or cell) is largely empirical and aims at maintaining high throughput subject to the previously stated criteria; the Wemco flotation machine is shown in Figure 24. Flotation cells can reach volumes of 20–25 m^3 [54]. Figure 25 is an outline of contemporary flotation machine types; Figure 26 shows cell geometries and paddle designs in mechanically agitated flotation machines. Flotation machines used in water treatment are not always mechanically agitated but may rely on the quieter conditions provided by the generation of air bubbles in the system.

Such methods include vacuum flotation, in which air-saturated water is subjected to a vacuum for bubble generation. Pressure flotation, on the other hand, does not require the application of vacuum, since air bubbles are released after saturation of water under pressure.

Electroflotation has also been used to a limited extent. In this method hydrogen and oxygen bubbles are generated by electrolysis [57].

A new version of an old flotation cell has led to the development of column flotation (see Fig. 27) [58]–[60]. Column cells seem to perform best with highly hydrophobic solids. They are used for the recovery of sulfides and coal, and to recover valuable particles rejected in the tailings of conventional, mechanically agitated, batteries.

Column flotation offers the following advantages:

1) reduction of gangue entrainment in froth,
2) increase of bubble residence times in the pulp,

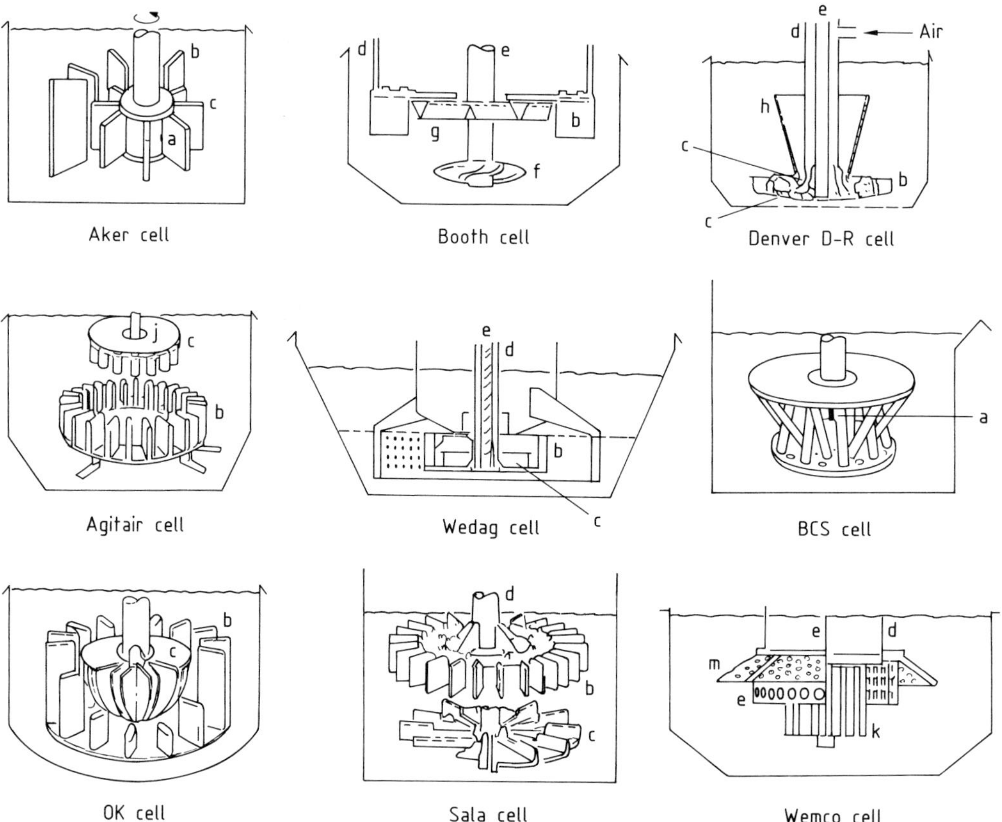

Figure 26. Schematic of impeller and tank geometries of mechanically agitated flotation cells [55]
a) Air port; b) Diffuser; c) Impeller; d) Standpipe; e) Shaft; f) Propeller; g) Aerating impeller; h) Recirculation well; j) Pulp entry; k) Rotor; m) Disperser hood

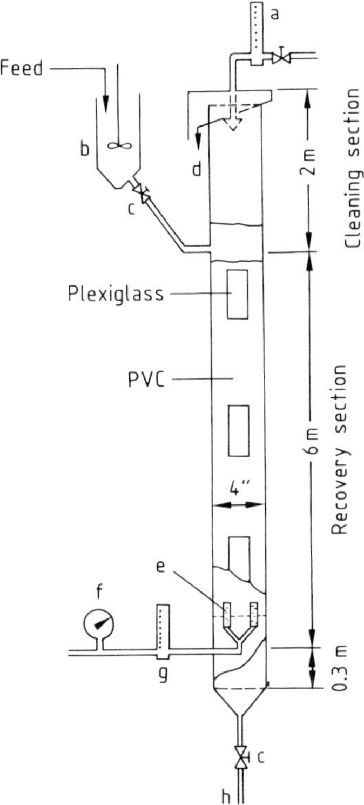

Figure 27. Pilot plant size flotation column
a) Wash water flow meter; b) Conditioner; c) Valve; d) Concentrate discharge; e) Porous glass diffuser; f) Air regulator; g) Air flow meter; h) Tails discharge

3) improved selectivity,
4) energy economy per unit weight of values recovered, and
5) improved control and instrumentation.

The following relationships are used in the design of flotation machines [56]:

Geometrical
 Effective volume: $V \propto L^{n}$ (n ~ 2–3)
 Froth area: $A \propto L^{2}$
 Impeller/tank aspect ratio: $D/L \propto D^{r}$ (r ~ 0–0.5)

Impeller speed
 Peripheral speed: $S = \pi ND \propto D^{a}$ (a ~ 0–0.5)

Power (constant viscosity and density)
 Net power consumption: $P \propto N^{3}D^{5}/N^{c}D^{2c}$ (c ~ 0–0.33)

Aeration
 Air-flow number: $N(Q) = Q/AND \propto D^{f}$ (f ~ 0)

Impeller diameter is the parameter of machine size
 Geometry: $L \propto D^{1-r}$; $A \propto D^{2(1-r)}$; $V \propto D^{n(1-r)}$
 Power: $P \propto D^{(a+1)(3-c)-1}$
 Aeration: $Q \propto D^{a+f+2(1-r)}$
 Power intensity or specific power:
 $P/V \propto D^{(a+1)(3-c)-n(1-r)-1} \propto D^{b}$
 Nominal residence time of air:
 $V/Q \propto D^{(2-n)(r-1)-(a+f)} \propto D^{q}$

N = Impeller rotational speed in rpm; D = Impeller diameter; L = Tank length. Exponents are experimental parameters characteristic of specific machine design.

5.4. Industrial Applications

Various methods of separation exploit differences in size, relative density, and magnetic and electrostatic properties. Flotation, however, exploits two unique controllable properties of particulates (Table 5): (1) it operates at very fine particle sizes, and (2) it is based on inherent or modified surface properties of solids.

Since its discovery in the late 19th century, the flotation process has been used principally in the minerals industry. At present, it is used to treat several thousand million tonnes of ore annually. Most of the minerals-related applications aim at recovering ore values, rejecting the rest as tailings.

Applications of flotation in the minerals industry are diverse and include recovery of sulfide minerals, upgrading phosphates ores, ash and sulfur removal from coal, separation of soluble minerals such as halite (NaCl) and sylvite (KCl) from each other, and recovery of fine gold [6]–[8], [61].

Water-treatment applications include the removal of oily components and colloidal matter from industrial waters, water recycling, and the removal of color and metal ions.

Mineral Flotation. Minerals containing valuable elements can be recovered economically by flotation (Table 6) in the following stages:

1) *Crushing and grinding*. These steps reduce particle size and mechanically liberate the desired minerals. Some deposits, such as beach sands or placers, do not require crushing, which is usually effected by jaw or gyratory crushers. For grinding, equipment such as ball mills and rod mills is employed.
2) *Classification and sizing*. As can be seen from Table 5, the optimal size range for flotation is 10–500 µm. Coal particles can be floated at the upper limit, and copper sulfide particles at the lower. Particles of nonsulfide minerals finer than 50 µm, however, interfere with flotation. Such fines are separated by hydrocyclones, sedimentation tanks, and other classi-

Table 5. Physical properties of particulates and size ranges suitable for separation technology

Physical property	Physical separation technique	Approximate particle size range, mm
Color or radiation	sorting	10–100
Size	dry or wet screening	0.01–10
Size, shape, density	wet or dry cyclones	0.005–1
	sedimentation and hydraulic classification	0.003–3
	centrifugation	0.001–0.05
Density	heavy media	1–10
Size, density	jigging	5–50
	tabling	0.005–3
Magnetic susceptibility	wet or dry magnetic separation	0.001–5
Surface or bulk conduction	electrostatic separation	0.05–5
Surface chemistry	flotation	0.001–0.5
	spherical agglomeration	0.01–0.5
	selective coagulation and selective flocculation	0.05–0.001

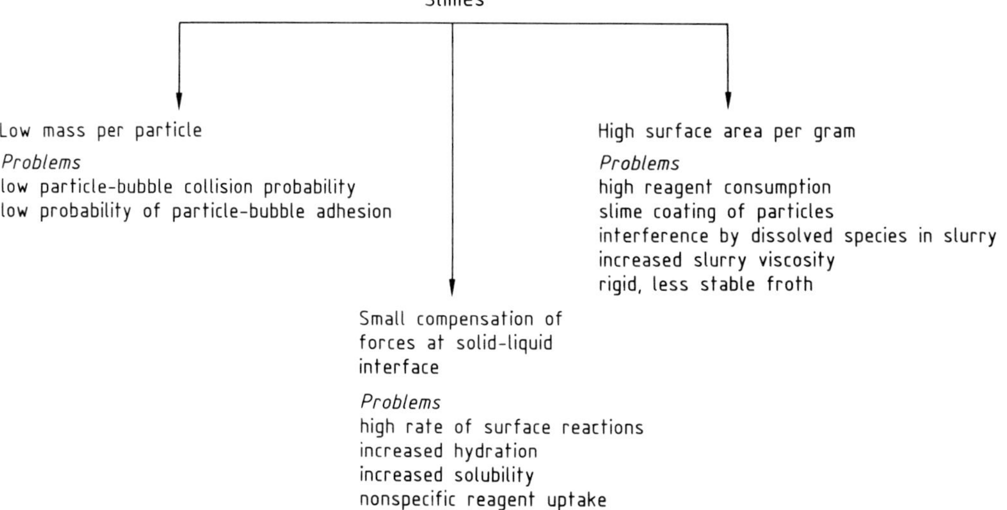

Figure 28. Characteristics of slimes [7]

fication equipment; they are discarded as slimes (Figure 28).

3) *Conditioning*. In the conditioning operation the ore powder is slurried with water and flotation reagents, forming the flotation slurry. In the flotation step, air is introduced to transfer hydrophobic particles to a froth layer, which is skimmed away mechanically by paddles.

The solids content of the pulp, known as pulp density, averages 20–40%, but may be 3–10% in some systems, such as coal flotation. In the preparation of potable water, the solids content may be well below 1%.

The mineral flotation process is carried out in interconnected batteries of flotation cells, where slurry flow occurs by gravity and the level of the slurry is controlled by gates or weirs.

The arrangement of flotation cells, sometimes with intermittent grinding of the products, leads to flotation circuits of which five basic types have been identified [62] (Fig. 29).

5.4.1. Sulfide Mineral Flotation

Sulfide minerals such as those of copper ($CuFeS_2$), lead (PbS), and zinc (ZnS) are com-

Table 6. Elements contained in minerals recovered by flotation

Element	Mineral	Composition	Element	Mineral	Composition
Al	bauxite	$Al_2O_3 \cdot xH_2O$ (Fe,Mn,Ti)	Mn	pyrolusite	MnO_2
Al	kaolinite	$Al_2Si_2O_5(OH)_4$	Mo	molybdenite	MoS_2
As	arsenopyrite	$FeAsS_2$	Na	halite	NaCl
Au	native gold		P	fluorapatite	$Ca_5F(PO_4)_3$
B	colemanite	$Ca_2B_6O_{11} \cdot 5H_2O$	S	native sulfur	
Ba	barytes	$BaSO_4$	Si	quartz	SiO_2
C	diamond	carbon	Sn	cassiterite	SnO_2
C	graphite	carbon	Sr	strontianite	$SrCO_3$
Ca	calcite	$CaCO_3$	Th	fergusonite	(Y,Ce,U,Th,Ca) (Nb,Ta,Ti) O_4
Cr	chromite	$FeO \cdot Cr_2O_3$	Ti	rutile	TiO_2
Cu	chalcopyrite	$CuFeS_2$	U	carnotite	$K_2O \cdot 2U_2O_3 \cdot V_2O_5 \cdot 2H_2O$
F	fluorite	CaF_2	V	vanadinite	$Pb_5ClV_3O_{12}$
Fe	hematite	Fe_2O_3	W	scheelite	$CaWO_4$
Hg	cinnabar	HgS	Y	monazite	(Ce,Y) PO_4
K	sylvite	KCl	Zn	sphalerite	ZnS
Li	spodumene	$LiAl(SiO_3)_2$	Zr	zircon	$ZrSiO_4$
Mg	magnesite	$MgCO_3$			

monly found in the same ore. Xanthates (dithiocarbonates) and dithiophosphates are used as collectors in sulfide mineral flotation; frothers include pine oil and synthetic commercial compounds such as Dowfroth 250. Some qualified generalizations apply to sulfide mineral flotation:

1) Xanthates and dithiophosphates are collectors for sulfides.
2) The chemisorption of xanthates and the presence of oxygen are essential for the flotation of sulfide minerals such as galena [63]–[65]. This correlates well with the electrochemistry of the system and the stability of species that collect at the solid surface (Fig. 30).
3) Other variables being constant, an increase in chain length for a homologous series of reagents strengthens them as collectors; this rule also applies to other minerals; see Figure 14.
4) At a given constant collector concentration, a critical pH for each sulfide mineral determines the boundary of flotation–no flotation conditions. The ratio of concentration of xanthate to hydroxide for such a critical transition usually conforms to the Barsky relationship: $[X^-]/[OH^-] = \text{constant}$
5) The ions OH^-, S^{2-}, CN^-, $Cr_2O_7^{2-}$ are common sulfide mineral depressants for selectivity control in flotation; Cu^{2+} is an activator for ZnS in flotation by xanthates. Figure 31 is a flow diagram of a sulfide flotation plant.

5.4.2. Nonsulfide Flotation

Nonsulfide ores include those of iron, phosphate, fluorite, magnesite, and beach sand. Some ores may contain sulfur-bearing minerals as secondary components. Collectors in these systems are generally fatty acids and their salts, sulfonates, and sulfates. Amines are also used, especially in iron-bearing ores and in silicate flotation. Figure 32 shows the separation of lead carbonate, barium sulfate, and calcium fluoride in a 100 t/d plant; the main gangue is siliceous material.

Phosphate-ore concentration plant flow sheets are shown in Figures 33 and 34. Their difference lies in the run-of-mine grades and the complexity of gangue minerals.

An interesting application of nonsulfide mineral flotation is the separation of soluble salts from brines (Fig. 35). Primary amines are well-established selective collectors and have long been used in the production of KCl from the Dead Sea. In some cases, clays interfere with the flotation and need to be depressed by the addition of starch, lignin, or cellulose derivatives.

5.4.3. Coal Flotation

Coal is cleaned by gravity separation to reduce ash and pyrite content [69]. These methods include jigging, tabling, and heavy-media separation. The fines (ca. 400 μm, 35 mesh) are treated by flotation. To liberate ash and sulfur-bearing components, the run of mine coal is sometimes ground to ca. 600 μm (28 mesh) or finer, and the whole mass subjected to flotation without recourse to primary gravity separation. The natural hydrophobicity of freshly ground coal is an asset which minimizes the use of collectors. A small quantity of MIBC or pine oil (50–100 g

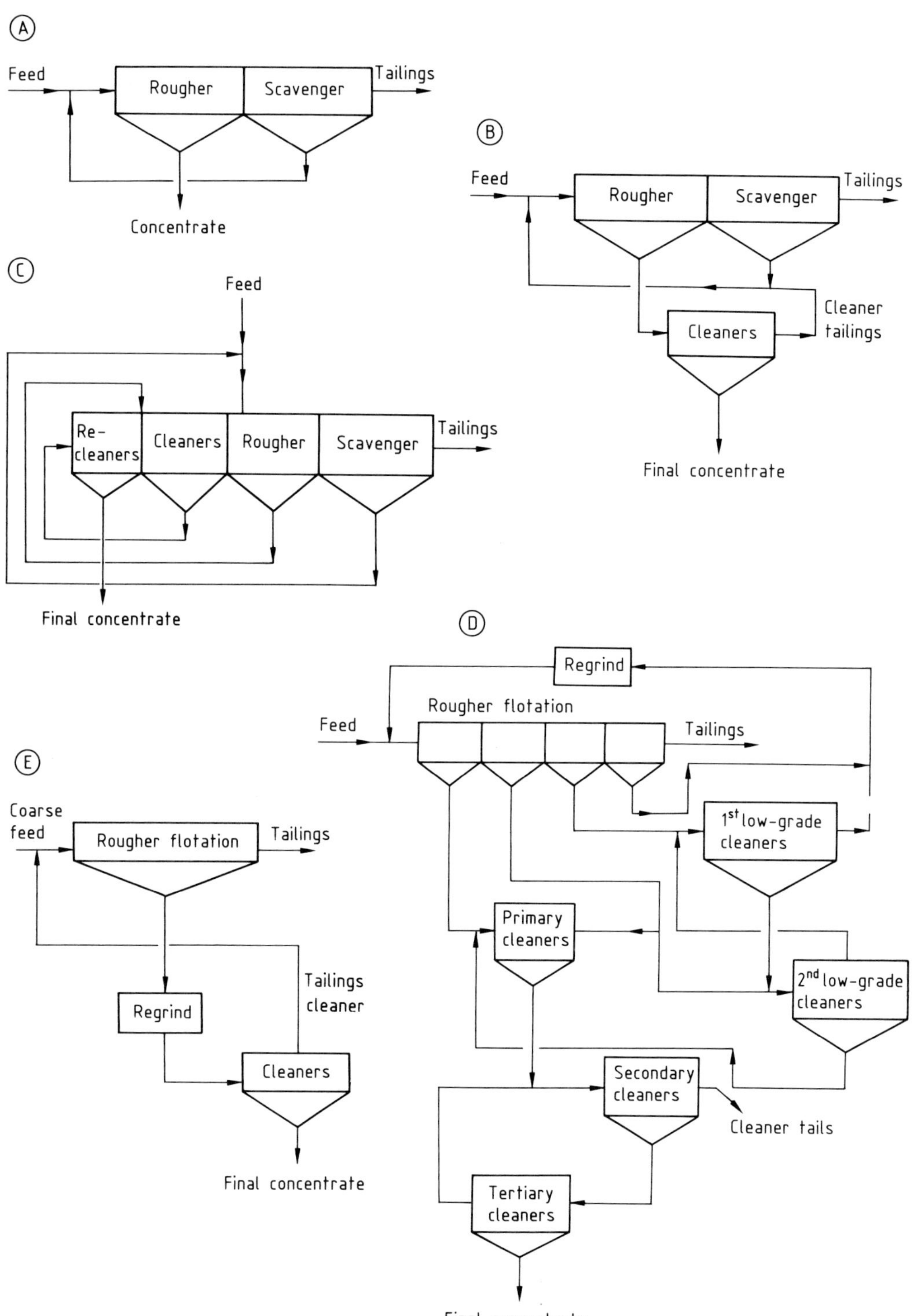

Figure 29. Flotation circuit configurations*
A) Type 1; B) Type 2; C) Type 3; D) Type 4; E) Type 5 * With permission [62].

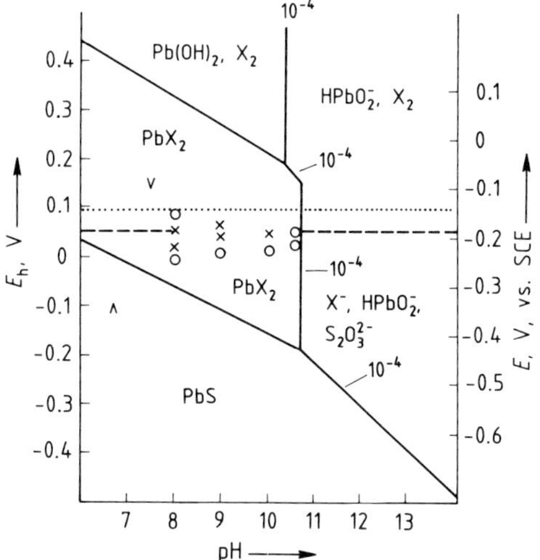

Figure 30. E_h-pH Diagram of thermodynamically stable species and associated experimental data in the galena–oxygen–xanthate flotation system [42]
x = Strong bubble contact; o = Feeble bubble contact;
v = Oxidizing depression by potassium chromate;
∧ = Reducing depression by sodium sulfite-ferrous mixture; ····· = Xanthate (10^{-3} M)–dixanthogen (saturated);
----- = Galena–dixanthogen system in equilibrium with 10^{-3} M xanthate ion; E = Electrical potential

per tonne of mineral) acts as both collector and frother.

On extended storage, or if the coal seam has been in contact with air for geological time periods, coal loses its hydrophobicity and is referred to as "weathered coal". In such cases, MIBC, low pH, and the addition of hydrocarbon oils such as kerosene or fuel oil make coal matter floatable [42], [70].

Sometimes, pyrite is first removed by conventional sulfide mineral flotation using xanthates as collectors. This is followed by the flotation of coal matter and the rejection of ash-forming materials such as silica, silicates, and carbonaceous shale.

Unlike other minerals, coal exhibits a high degree of variation with respect to origin, age, rank, moisture content, degree of weathering, physical and chemical structure, and nature of gangue. As a result, it is difficult to develop a unified flotation strategy from sample to sample. Even samples of the same origin undergo surface chemical changes to different extents on prolonged storage.

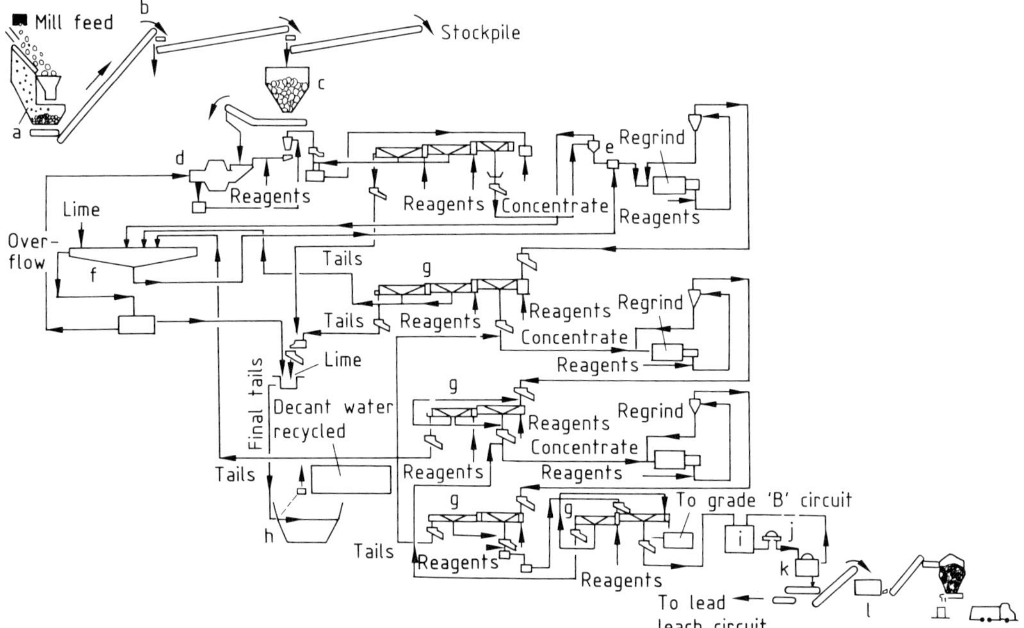

Figure 31. Flow sheet of plant treating 30 000 t/d of molybdenite ore, Henderson Mill, Colorado*
a) Gyratory crusher; b) Self-cleaning magnet; c) Ore storage bin; d) Semiautogenous mill; e) Pump box; f) Thickener; g) Cleaner; h) Tailing pond; i) Concentrate holding tank; j) Pump; k) Disc filter; l) Dryer
* Courtesy of Denver Equipment Co.

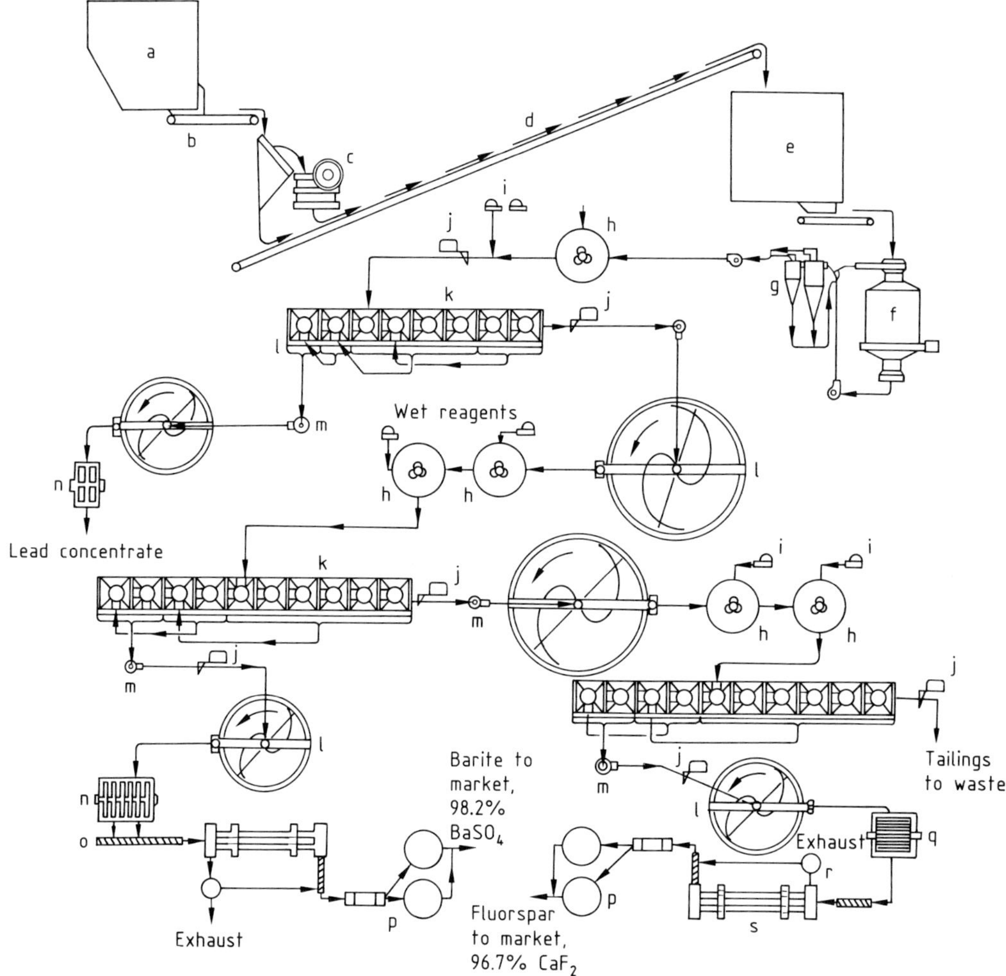

Figure 32. Flow sheet for $PbCO_3$, CaF_2, and $BaSO_4$ concentration *
a) Coarse ore bin; b) Apron feeder; c) Jaw crusher; d) Belt conveyor; e) Fine ore bin; f) Ball mill; g) Cyclone; h) Conditioner; i) Reagent feeders; j) Sampler; k) Flotation machine; l) Concentrate thickener; m) Pump; n) Disc filter; o) Screw conveyor; p) Elevator and storage bins; q) Fluorspar filter; r) Dust cyclone; s) Dryer
* Courtesy of Denver Equipment Co.

Nonetheless, generalized strategies are possible (Table 7). Figure 36 shows a flow sheet of coal flotation [71].

5.4.4. Flotation of Precipitates and Ions

Precipitate flotation is based on the same principles as mineral flotation. For example, in the preparation of potable water, the adjustment of pH to precipitate ferric ions as hydroxides eliminates most of the iron and colloidal matter by a subsequent flotation step. In this case, the water contaminants render the hydrolysis product sufficiently hydrophobic for bubble–particle adhesion to permit flotation under quiescent conditions.

Alternatively, dissolved ions can be electrostatically attached to selective cations, which in turn can be collected in a foam layer (Fig. 37). The recovery of Au, Pd, Pt, and Ir by such foam fractionation methods, as well as the removal of uranium from aqueous solutions by the use of amines, has been demonstrated [73]–[75].

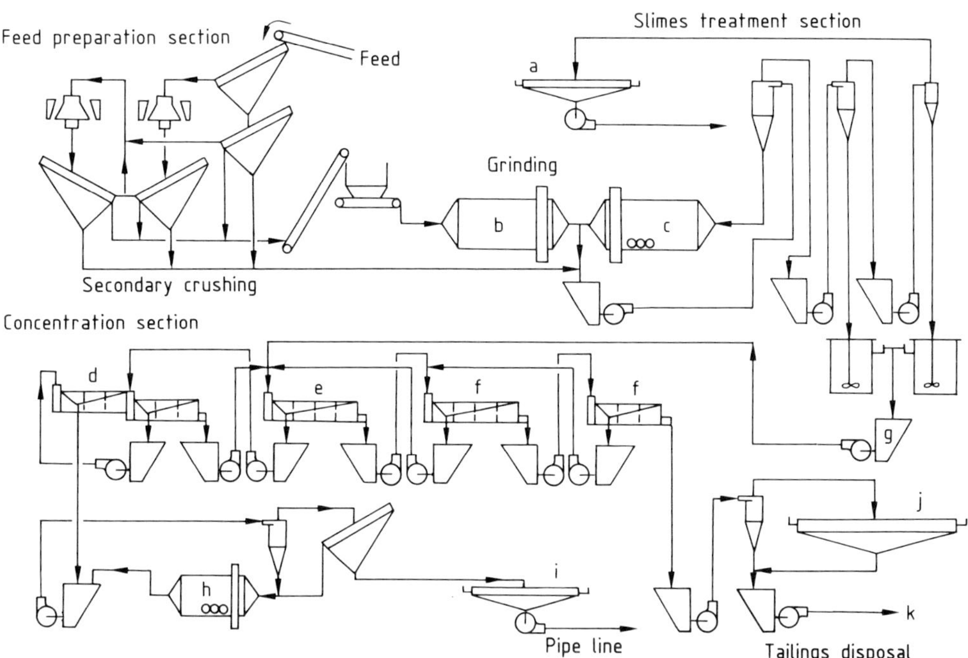

Figure 33. Flow sheet for phosphate ore flotation in Brazil*
a) Slime thickener; b) Rod mill; c) Ball mill; d) Cleaner; e) Rougher; f) Scavenger; g) Conditioner; h) Concentrate regrinder; i) Concentrate thickener; j) Tailing thickener; k) Tailing pond
* The main mineral of value is apatite; the gangue contains barytes, iron oxides, and silicates; fatty acid flotation generates concentrates assaying an average of 36% P_2O_5, with recoveries of 62–63%; with permission [66].

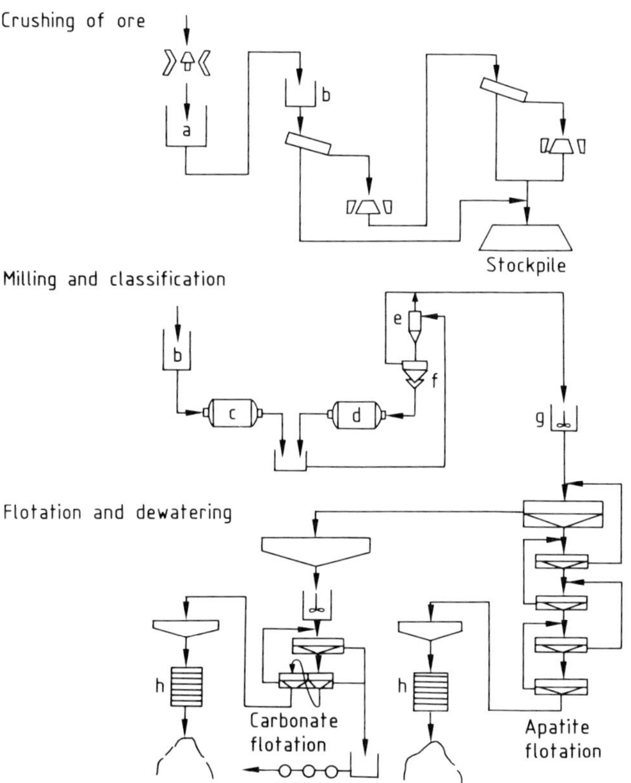

Figure 34. Flotation flow sheet for Finnish low grade apatite ore* [67]
a) Silo; b) Feed bin; c) Rod mill; d) Ball mill; e) Hydrocyclone; f) Cone classifier; g) Conditioner; h) Pressure filter
Head ore = 3.9% P_2O_5; Concentrate grade = 36.5% P_2O_5;
Recovery = 75–79%
* Gangue minerals include calcite, dolomite phlogopite $[(KMg_3(AlSi_3)O_{10}(OH,F)_2]$, and mica.

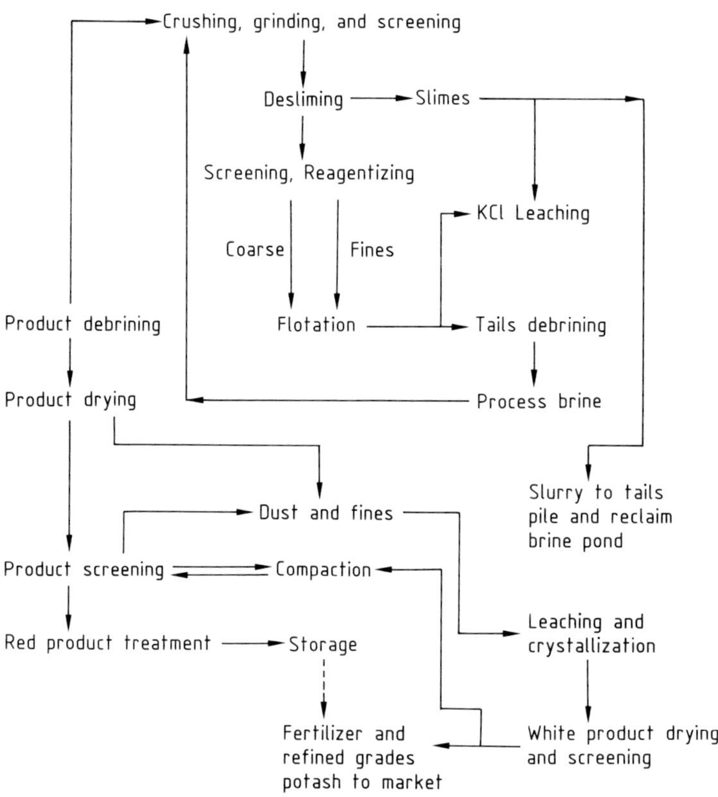

Figure 35. Potash treatment flow sheet as practiced by the Potash Corporation of Saskatchewan, Canada [68]

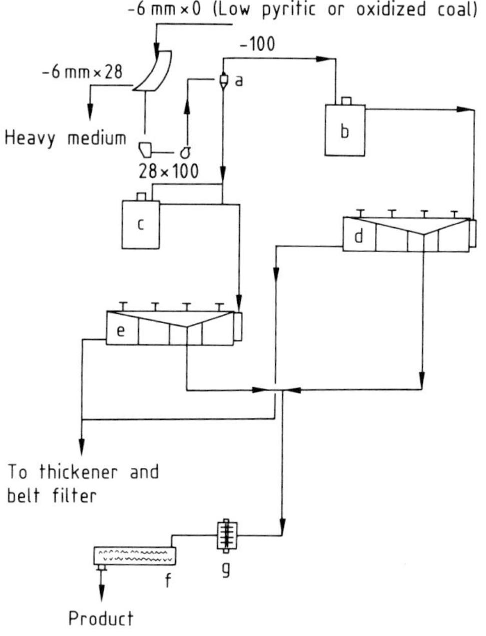

Figure 36. Flotation flow sheet applicable to oxidized coals or low pyrite-content coal*
a) Cyclone; b) Conditioner, 6%; c) Conditioner, 25%; d) Fine rougher float, 6%; e) Coarse rougher float, 15%; f) Drier; g) Disc filter
* With permission [71].

Figure 38 shows a foam fractionation (ion flotation) assembly. At times of increasing environmental concern, the removal of heavy metal ions and toxic substances such as PCBs (polychlorinated biphenyls) seems a promising area for further application of ion flotation and foam fractionation techniques.

5.4.5. Use of Oil Droplets instead of Air Bubbles

Surfaces to which air bubbles can adhere are also readily wetted by hydrocarbon oil droplets in preference to water. This has been demonstrated thermodynamically and experimentally.

Table 7.

Flotation circuit*	Yield, wt%	Product, ash%	Tailings, ash%
Single-stage flotation (A)	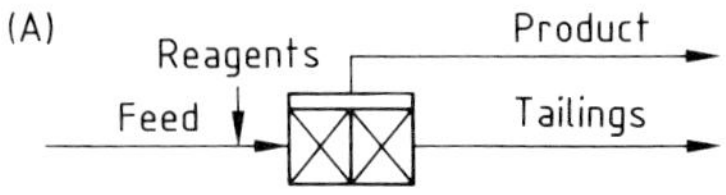 59.6	11.6	27.0
Desliming (B)	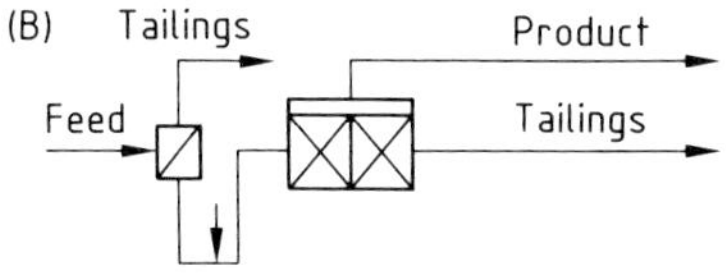 60.0	8.2	31.3
Separate conditioning of the coarse and fine size fractions (C)	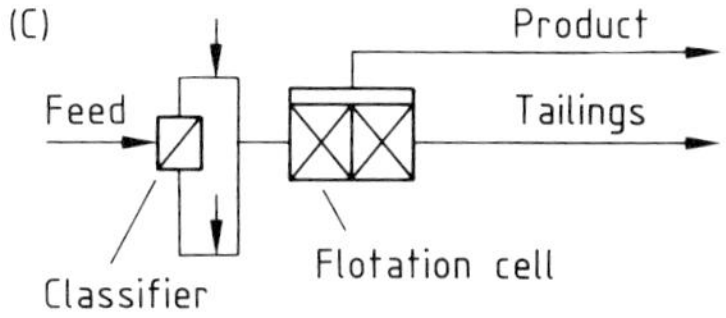 51.0	13.0	23.5
Two-stage conditioning (D)	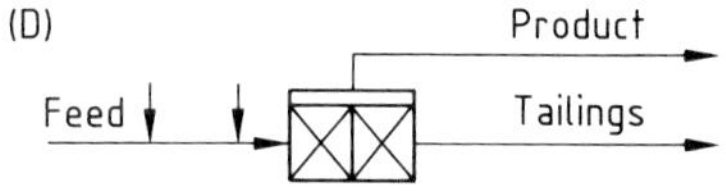 65.0	12.8	28.6
Split feed flotation (E)	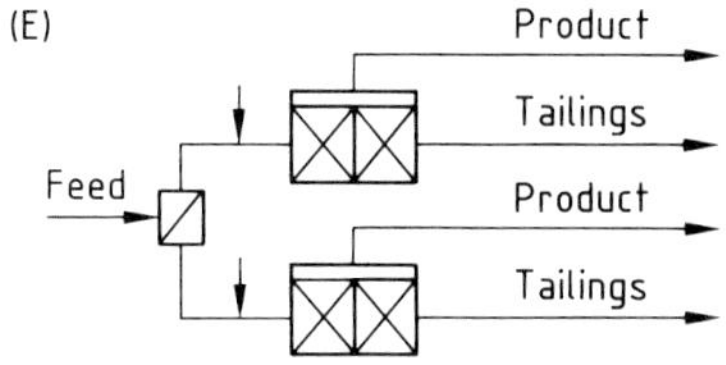 75.2	11.2	39.5
Two-stage reagent addition (F)	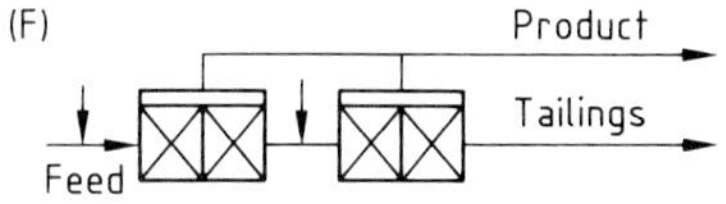 79.6	11.0	43.6
Reflotation of classified tailings (G)	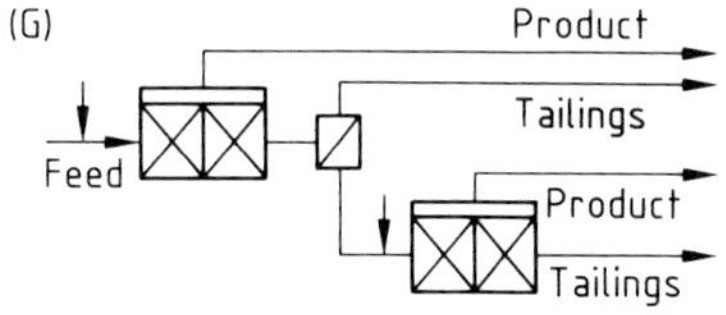 81.0	9.9	58.0

*Symbols defined in (c)

Although micrometer-size solid particles may adhere nonselectively to air bubbles, the replacement of air by hydrocarbon oils provides better selectivity, as proved experimentally. These considerations have given rise to the development of two-liquid flotation for extraction of finely divided solids into an oil phase (oil extraction) in assemblies similar to those used in conventional flotation technology.

This method has been used for alumina (Al_2O_3), hematite (Fe_2O_3), cassiterite (SnO_2), and artificial diamonds, by first making the powders hydrophobic with an amine or a sulfonate, followed by extraction into an oil phase [76],

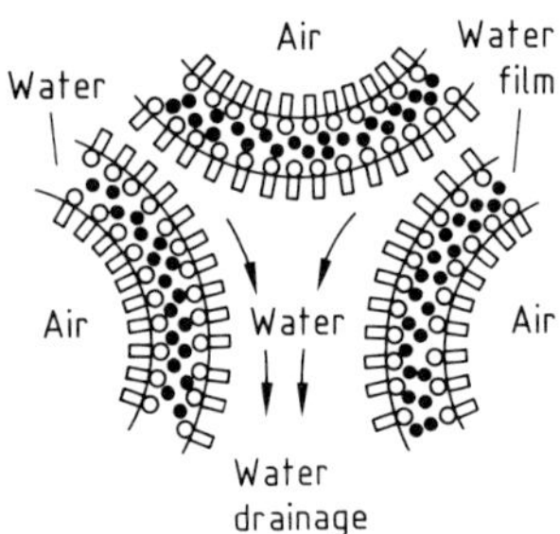

Figure 37. Scheme of selective ion flotation by foam fractionation
○▭ = Surfactant; ● = Ion removed

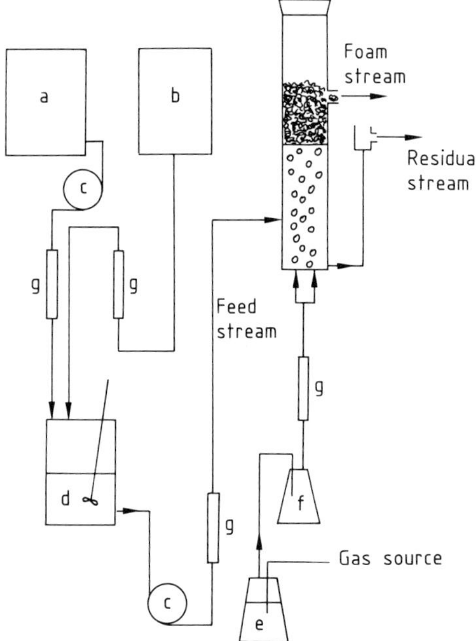

Figure 38. Schematic of continuous-flow foam fractionation column [72]
a) Ion storage; b) Surfactant storage; c) Pump; d) Mix tank; e) Saturator; f) Moisture trap; g) Flow meter

[77]. The thermodynamics of the system apply equally well to the now well-established process of coal cleaning by oil agglomeration.

6. Future Trends

The recent decline in the mining and metallurgical industries has shifted emphasis away from conventional flotation technology. Nonetheless, understanding of the fundamentals of flotation is by no means complete and development of sophisticated instrumentation is leading to combined research on materials science and on surface fundamentals of flotation [78].

As new applications, such as control of the so-called pulp potential, control of redox conditions, E_h, use of critical surface tension of wetting, and flotation of ions and dissolved species, such as toxic and hazardous materials, become more widespread, froth flotation will be employed more and more.

Computer technology and instrumental control now make larger plants possible with higher-grade products and more complete recoveries. Furthermore, pyrometallurgy and hydrometallurgy are now used together with flotation for recovery of metals from slags or melts [79].

The flotation of precious metal-bearing ores and native gold is widely applied despite the apparent fate of sulfide mineral flotation. Less familiar systems such as rare-earth minerals are finding increased interest. Column flotation promises wider application. The custom-design of flotation chemicals is also under development and wider application of flotation in recycling industries is at its early stages.

7. References

[1] B. Yarar, Z. Dogan (eds.): *Mineral Processing Design,* Martinus-Nijhof, Dordrecht, The Netherlands 1987.

[2] B. Yarar: "Investigation of the Recovery of Cellulose from Difficult to Process Sources, Using Surface Chemical Principles," *Proc. 4th Recycling World Congress,* New Orleans, La., 1982.

[3] B. Yarar, D. J. Spottiswood (eds.): *Interfacial Phenomena in Mineral Processing,* Engineering Foundation Publ., New York 1982.

[4] B. Yarar, S. R. Rao: "Flotation," in J. J. McKetta (ed.): *Encyclopedia of Chemical Processing and Design,* vol. 23, Marcel Dekker, New York 1985, pp. 454–508.

[5] P. Somasundaran (ed.): *Advances in Mineral Processing,* Society of Mining Engineers (SME), New York 1986.

[6] A. M. Gaudin: *Flotation,* McGraw-Hill, New York 1957.

[7] V. I. Klassen, V. A. Mokrausov: *An Introduction to the Theory of Flotation,* (Transl. J. Leja, G. W. Poling), Butterworths, London 1963.

[8] K. L. Sutherland, I. W. Wark: *Principles of Flotation,* Australian Institute of Minerals and Metals, Melbourne 1955.

[9] B. Yarar: "Gamma Flotation: A New Approach to Flotation, Using Liquid–Vapor Surface Tension Control," *Proc. 2nd Latin American Flotation Congress,* vol. 1, Concepcion, Chile, Aug. 19–23, 1985, pp. 2.1–2.36.

[10] B. Yarar, J. Kaoma, *Colloids and Surfaces* **11** (1984) no. 3, 429–437.
[11] D. W. Fuerstenau, "Activation and De-activation in the Flotation of Sulfide Minerals," in P. E. Richardson et al. (eds.): *The Physical Chemistry of Mineral-Reagent Interactions in Sulfide Flotation,* USBM, Information Circular no. 8818, 1980, pp. 100–120.
[12] J. Leja: *Surface Chemistry of Froth Flotation,* Plenum Publ., New York 1982, p. 587.
[13] R. C. Weast (ed.): *Handbook of Chemistry and Physics,* 60th ed. CRC Press, New York 1979, p. F–43.
[14] H. Lange in M. J. Schick (ed.): *Nonionic Surfactants,* Marcel Dekker, New York 1967, p. 459.
[15] M. H. Buckenham, J. H. Schulman, *Trans. Soc. Min. Eng. AIME* **226** (1963) 1–5.
[16] Y. Zimmels: ref. [3], p. 135.
[17] J. Laskowski, J. A. Kitchener, *J. Colloid Intface Sci.* **29** (1969) 670–679.
[18] M. C. Fuerstenau, K. L. Clifford, M. C. Kuhn, *Int. J. Min. Process.* **1** (1974) 307–318.
[19] D. W. Fuerstenau, T. W. Healy, P. Somasundaran, *Trans. Soc. Min. Eng. AIME,* **229** (1964) 321–325.
[20] M. C. Fuerstenau, J. D. Miller, M. C. Kuhn: *Chemistry of Flotation,* Society of Mining Engineers (SME), New York 1985.
[21] W. Stumm, J. J. Morgan: *Aquatic Chemistry,* J. Wiley & Sons, New York 1970, p. 164, p. 171.
[22] R. M. Garrels, C. L. Christ: *Solutions Minerals and Equilibria,* Freeman, San Francisco 1965.
[23] J. N. Butler: *Ionic Equilibrium, a Mathematical Approach,* Addison-Wesley, Reading, Mass., 1964.
[24] O. Griot, J. A. Kitchener, *Trans. Faraday Soc.* **61** (1965) no. 509, 1026–1031.
[25] B. Yarar: "Polymeric Flocculants and Selective Flocculation: An Overview," in K. L. Mittal, E. J. Fendler (eds.): *Solution and Surfactant Chemistry, Theoretical and Applied Aspects,* Plenum Publ. New York 1982, pp. 1333–1364.
[26] B. Yarar: "The Surface Chemical Mechanism of Calcite-Colemanite Separation by Flotation" in J. M. Barker, S. J. Lefond (eds.): *Borates: Economic Geology and Production,* AIME, New York 1985, pp. 221–233.
[27] O. Mellgren, *Trans. Soc. Min. Eng.-AIME* **235** (1966) 46–60.
[28] A. S. Peck, M. E. Wadsworth: "Infrared Studies of Oleic Acid and Sodium Oleate Adsorption on Fluorite, Barite and Calcite," U.S. Bureau of Mines (USBM) Report of Investigation no. 6202, 1963.
[29] S. Chander, D. W. Fuerstenau, D. Stigter: "On Hemimicelle Formation at Oxide–Water Interfaces," in R. H. Otewill et al. (eds.): *Adsorption from Solution,* Academic Press, London 1983, pp. 197–210.
[30] H. E. Garrett: *Surface Active Chemicals,* Pergamon, London 1972, p. 150.
[31] R. W. Slater, J. P. Clark, J. A. Kitchener: "Chemical Factors in the Flocculation of Mineral Slurries with Polymeric Flocculants," *Proc. 8th Int. Mineral Processing Congress,* Leningrad 1968.
[32] J. Laskowski: *Physical Chemistry in Mineral Processing Technology,* (Polish Text), Wydawnictwo "Slask", Katowice, Poland 1969, p. 215.
[33] H. R. Kruyt (ed.): *Colloid Science,* Elsevier, Amsterdam 1952, p. 126.
[34] A. Sheludko: *Colloid Chemistry,* Elsevier, Amsterdam 1966, p. 141.
[35] R. J. Hunter: *Zeta Potential in Colloid Science,* Academic Press, London 1981, p. 11.
[36] I. Iwasaki, S. R. B. Cooke, A. F. Colombo: "Flotation Characteristics of Goethite," USBM, Report of Investigation, no. 5593, 1960.
[37] D. W. Fuerstenau, *Trans. Soc. Min. Eng.-AIME,* **208** (1957) 834–836.
[38] A.Yucesoy, B. Yarar, *Trans. Inst. Min. Metallo.* **83** (1974) C 96–C 100.
[39] J. Laskowski, *Miner. Sci. Eng.* **6** (1974) no. 4, 223–235.
[40] A. D. Read, J. A. Kitchener: "The Thickness of Wetting Films," *Wetting, Soc. Chem. Ind. Monograph no. 25,* Gordon-Breach, New York 1967, pp. 300–317.
[41] J. H. Schulman, J. Leja, *Kolloid-Z.* **136** (1954) 107–119.
[42] B. Yarar, J. Leja, *Trans. Soc. Min. Eng.-AIME,* **272** (1983) 1978–1983.
[43] W. A. Zisman: "Relation of the Equilibrium Contact Angle to Liquid and Solid Constitution," in K. J. Mysels et al. (eds.): *20 Years of Surface and Colloid Chemistry, The Kendall Award Addresses,* Amer. Chem. Soc., Washington, D.C., 1973, pp. 109–157.
[44] B. Yarar, J. Kaoma, *Trans. Soc. Min. Eng.-AIME* **276** (1985) 1875–1878.
[45] J. Kaoma, B. Yarar: "Correlation of the Critical Surface Tension of Wetting and the Non-stoichiometric Composition of Molybdenum Sulfide," in H. F. Barry, P. C. H. Mitchell (eds.): *Proc. 4th Conference on the Chemistry and Uses of Molybdenum,* Climax Molybdenum Co., Ann Arbor, Mich., 1982, pp. 117–122.
[46] J. Laskowski: "The Relationship between Flotability and Hydrophobicity," ref. [5], pp. 189–208.
[47] B. Yarar, G. P. Hemphill: "Size-Surface Energy Relationships in the Flotation-Upgrading of Oil Shale," *Proc. 1983 Eastern Oil Shale Symp. Institute of Mining and Minerals Research-Kentucky,* Lexington, Ky., 1983, pp. 241–248.
[48] G. P. Hemphill, B. Yarar: "Use of Critical Surface Tension of Wetting in the Beneficiation of Oil Shales," in F. A. Curtis (ed.): *Proc. Energy-84,* Pergamon, Toronto 1984, pp. 111–116.
[49] J. A. Finch, G. W. Smith, *Can. Metall. Q.* **14** (1975) no. 1, 44–51.
[50] D. T. Hornsby, J. Leja, *Colloids and Surfaces* **7** (1983) 339–349.
[51] M. R. Smith, R. J. Gochin: ref. [1], pp. 166–201.
[52] A. L. Mular, M. A. Anderson: *Design and Installation of Concentration and Dewatering Circuits,* Society of Mining Engineers (SME), Littleton, Colo., 1986, p. 588.
[53] G. W. Poling: "Selection and Sizing of Flotation Machines," in A. L. Mular, R. B. Bhappu (eds.): *Mineral Processing Plant Design,* 2nd ed., Society of Mining Engineers (SME)-AIME, New York 1980, pp. 887–906.
[54] I. S. Blagov et al.: "State and Development of Coal Flotation in the USSR," *Proc. 9th Int. Coal Preparation Congress,* New Delhi, India, 1982, pp. C 1–C 5.
[55] G. Barbery: "Engineering Aspects of Flotation in the Minerals Industry: Flotation Machines, Circuits and their Simulation," *Scientific Basis of Flotation, NATO-Advanced Study Institute on Scientific Basis of Flotation, Preprints,* Cambridge, UK, 1982.
[56] C. C. Harris: ref. [35], p. 753.

[57] A. E. Zhangozhina et al., *Chem. Abstr.* **102** (1985) 133410e.
[58] D. A. Wheeler: "Column Flotation: The Original Column," ref. [9], pp. c.1–c.28.
[59] G. S. Dobby, J. A. Finch: " Flotation Column Scale-Up and Simulation," *Proc. 17th Annual Meeting of Canadian Mineral Processors,* Jan. 22–24, Can. Inst. Min. Metall., Ottawa 1985, pp. 614–638.
[60] K. S. Narasimhan et al., *Eng. Min. J.* **173** (1972) no. 5, 84–85.
[61] M. C. Fuerstenau (ed.): *Flotation,* (A. M. Gaudin Memorial vol.) AIME, Vols. 1 and 2, New York 1976.
[62] S. G. Malghan: "Typical Flotation Circuit Configurations," ref. [70], pp. 76–92.
[63] B. Yarar, B. C. Haydon, J. A. Kitchener, *Trans. IMM (London)* **78** (1969) C 181–C 184.
[64] D. Toperi, R. Tolun, *Trans. IMM (London)* **78** (1969) C 191–C 197.
[65] R. Woods, P. E. Richardson: "The Flotation of Sulfide Minerals: Electrochemical Aspects," ref. [5], pp. 154–170.
[66] E. W. Betz: "Beneficiation of Brazilian Phosphates" in J. Laskowski (ed.): *Proc. 12th International Mineral Processing Congress,* vol. 2 B, Elsevier, Amsterdam 1981, pp. 1846–1874.
[67] K. Kiukkola et al.: "Selective Flotation of Apatite from Carbonatite Glimmerite Ore at the Silinjarvi Mine of Kemira Oy, Finland," *Proc. 14th Int. Mineral Processing Congress,* vol. 5, CIM, Toronto 1982, pp. 7.1–7.15.
J. Li: "The Practice of Concentration and Ways to Increase Grade and Recovery of the Graphite Concentrate in Washu Graphite Mine, Shandong, China," ref. [75], pp. 9.1–9.10.
[68] G. G. Strathdee et al.: "The Processing of Potash Ore by PCS," ref. [75], pp. 12.1–12.20.
[69] J. W. Leonard (ed.): *Coal Preparation,* AIME, New York (1979) p. 9-1.
[70] G. Ozbayoglu: "Coal Flotation," ref. [1], p. 76.
[71] T. M. Plouf, *Min. Eng.* **32** (1980) 1218–1224.
[72] R. B. Grieves: "Surfactant–Anion Interactions in Foam Fractionation," ref. [3], pp. 357–374.
[73] R. Lemlich (ed.): *Adsorptive Bubble Separation Techniques,* Academic Press, London 1972.
[74] D. W. Downey, E. W. Berg: "Ion Flotation of Platinum Group Metals," ref. [3] pp. 375–394.
[75] D. J. Wilson: "Topics in Precipitate and Floc Flotation," ref. [3], pp. 395–414.
[76] H. L. Shergold, O. Mellgren, *Trans. IMM (London)* **78** (1969) C 121–C 132.
[77] H. L. Shergold, R. Stratton-Crawley, *Colloids and Surfaces* **3** (1981) 253–265.
[78] F. D. Schowengerdt, D. J. Spottiswood, R. Sen, B. Yarar: "Surface Behavior of Minerals in Flotation: An Auger Electron Spectroscopic Analysis," *Proc. 15th Int. Mineral Processing Congress,* vol. 1, Cannes 1985, pp. 49–58.
[79] D. R. Gaskell et al. (eds.): *The Reinhardt Schuhmann International Symposium on Innovative Technology, and Reactor Design in Extraction Metallurgy,* The Metallurgical Society Inc., Warrendale, Pa., 1986, p. 159.

24. Mixing, Introduction

MARKO ZLOKARNIK, Bayer AG, Leverkusen, Federal Republic of Germany

Mixing, stirring, and kneading have been used by man since time immemorial in the preparation of food and drink, and belong to the classical unit operations used in the most diverse branches of industry—production of building materials, glass, chemicals, food processing, etc. In the chemical industry, this technology is of special importance: to carry out a chemical reaction, the solutions and mixtures must be prepared in such a way that the reagents have a suitable form for reaction. During the reaction itself, the mixing operation must ensure that the reagents stay in close contact with each other; this is particularly important when the reagents have a different aggregate state.

The operations used for mixing must first be classified according to the state of aggregation of the predominant component; as a rule, the resulting mixture has this state, too. From the process engineering point of view, stirring a gas into a liquid is not the same as mixing a liquid into a gas stream. Even the term liquid state must be made more specific, since low-viscosity liquids place drastically different demands on mixing equipment than viscous, pastelike substances. Bearing these considerations in mind, mixing operations can be divided into the groups listed in Table 1.

Mixing of Homogeneous Miscible Substances. A mixture that is homogeneous at the molecular level results only when mixing gases or miscible liquids. Here the mixing operation can be divided into two steps: (1) pre-mixing taking place through the large-scale convective transport of eddies and (2) elimination of concentration differences through purely diffusive transport. The former is governed by the physical properties as well as by the type of mixing device and the mixing energy expenditure; it is scale-dependent. The better the pre-mixing, i.e., the smaller the eddies produced, the faster the following mixing process can proceed. The latter is governed by those material properties of the system that influence diffusion (viscosity, diffusion coefficient) and is only rate-limiting if the premixing was not intensive enough. Good mixing devices (nozzles, stirrers causing a pronounced bulk circulation) usually function so well that the process of diffusion has no effect on the homogenization time (no effect of the Schmidt number *Sc* on the homogenization characteristic, → "Homogenization of Liquid Mixtures" in Stirring, p. **25**-6).

In the last few years, *micromixing* has received special attention; this refers to the production of extremely small eddies with the purpose of mixing the reaction components as fast as possible. For example, micromixing plays a key role in competitive reactions of the type

$$A + B \longrightarrow C \tag{1}$$
$$A + 2\,B \longrightarrow D \tag{2}$$

In this case, the selectivity of the desired Reaction (1) can be increased through a fast reduction of the reagent concentrations. If the reaction, however, is of the competitive-consecutive type,

$$A + B \longrightarrow C \tag{3}$$
$$A + C \longrightarrow D \tag{4}$$

then it is also important to avoid backmixing of the reagents with the product C. In such cases a

Table 1. Classification of mixing operations according to the aggregate state of the predominant component

Aggregate state	Operation	Most common mixing equipment
Gaseous*	mixing	mixing chamber, nozzle
Liquid	stirring	stirrer
Paste	kneading	kneader, screw extruder
Solid (granular)	mixing	mixer

* The mixing of gases is treated under Continuous Mixing of Fluids in Volume B4.

plug-flow reactor is used, and mixing is performed with a nozzle or an in-line mixer.

Mixing in Heterogeneous Systems. The main objective of the mixing operation in the spraying of liquids in a gas and in the dispersion of immiscible liquids or gases in a liquid is to create as large an interfacial area as possible. In these operations, however, fine primary droplets or gas bubbles often have the tendency to coalesce; i.e., a part of the work done in forming the interfacial area is wasted. During the production of fine bubbles or droplets in stirred vessels, the size distribution of the bubbles or droplets is the result of a continuous process of dispersion and coalescence.

The *coalescence process* depends to a great extent on the state of flow in the mixing device, thus requiring care in equipment scale-up. To allow reliable conclusions, design guidelines for mixing devices for these operations must be based on experiments in several different scales. The exact physical properties that determine coalescence are still not known: some substances suppress coalescence strongly, even in extremely small concentrations (a few ppm), while others increase the coalescence tendency. Especially in the dispersion of gases in liquids, the coalescence behavior of the system must be determined through appropriate experiments in the laboratory.

Mixing of Pastes and Granular Materials. Pastes and powders or granular materials are usually mixed to achieve the best possible blending. Here, however, only a stochastically homogeneous mixture is possible, a mixture that depends strongly on the physical properties of the substances to be mixed (viscosity and rheological behavior for pastes, size and morphology for granular solids). In the mixing of granular materials, *mixture separation,* determined by the state of flow in the device and therefore scale-dependent often presents a problem. Since the sampling, analysis, and the statistical evaluation of experimental data are problematic, there are comparatively few reliable design and scale-up guidelines for these mixing devices. In the future, significantly more intensive research activity in this field is necessary.

25. Stirring

Marko Zlokarnik, Helmut Judat, Bayer AG, Leverkusen, Federal Republic of Germany

In addition to the standard symbols defined in the front matter of this volume (p. IX), the following symbols are used:

A	area	m^2
a	interfacial area per unit volume	m^{-1}

c	concentration	kg m^{-3}
c_p	specific heat capacity	J kg^{-1} K^{-1}
D	vessel diameter	m
d	stirrer diameter	m
E	sorption efficiency	kg O_2 kW^{-1} · h^{-1}
G	rate of mass transfer	kg s^{-1}
g	acceleration due to gravity	m s^{-2}
H	liquid height	m
H^*	liquid height above stirrer	m
h	stirrer clearance from bottom	m
k	heat-transfer coefficient	W m^{-2} K^{-1}
k_L	liquid-side mass-transfer coefficient	m s^{-1}
k_s	mass-transfer coefficient	m s^{-1}
L	length	m
M_d	torque	N m
n	speed of rotation (rotational frequency)	s^{-1}
P	power	W
p	pressure	Pa
Q	rate of heat transfer	W
q	flowrate	m^3 s^{-1}
q'	stirrer pumping capacity	m^3 s^{-1}
R	universal gas constant	J mol^{-1} · K^{-1}
T	temperature	K
V	liquid volume	m^3
v	superficial velocity	m s^{-1}
α	film heat-transfer coefficient	W m^{-2} K^{-1}
$\dot{\gamma}$	rate of shear	s^{-1}
$\mathfrak{D}$	diffusion coefficient	m^2 s^{-1}
δ	mean particle or droplet diameter	m
ε	power dissipation per unit mass	W kg^{-1}
η	dynamic viscosity	Pa s
θ	mixing time	s
λ	thermal conductivity	W m^{-1} K^{-1}
μ	scale factor	
ν	kinematic viscosity	m^2 s^{-1}
ϱ	density	kg m^{-3}
σ	interfacial tension	N m^{-1}
τ	shear stress	Pa
Φ	mass fraction	
Φ'	volume fraction	
ω	angular velocity	rad s^{-1}

Dimensionless groups

Ar	Archimedes number
E^*	sorption efficiency number
Fr	Froude number
Ga	Galileo number
Ne	Newton number
Nu	Nusselt number
Pr	Prandtl number
Q	gas flowrate number
Q'	pumping capacity number
Re	Reynolds number
Sc	Schmidt number
Sh	Sherwood number
We	Weber number
X	gas dispersion number
Y	sorption number

1. Stirring Operations, Stirring Vessels, and Stirrer Types

1.1. Stirring Operations

When the liquid component predominates in the mixture, the mixing operation is termed *stirring*, and a stirrer is used to perform it. The following five stirring operations may be defined:

1) Homogenization, i.e., equalization of differences in concentration and temperature
2) Intensification of heat transfer between the liquid and the heat-exchange surface
3) Suspension (and possibly dissolution) of a solid in the liquid, or slurry formation
4) Dispersion (possibly emulsification) of two immiscible liquids
5) Dispersion or sparging of a gas in the liquid (gas–liquid contacting)

Homogenization refers to the formation of a uniform phase out of several miscible liquids or to the elimination of concentration and temperature gradients during a chemical reaction in the liquid phase. (In the food industry, the same term is also used for emulsification under extreme shear conditions, e.g., the homogenization of milk.)

Intensification of heat transfer in mixing vessels—especially in highly viscous liquids—can be an important stirring operation, particularly in the case of a strongly exothermic chemical reaction, for example, block polymerization. In this case, stirring aims to reduce the boundary layer at the vessel wall, as well as to provide for the motion of the liquid to and from the heat-exchange surface.

When particulate materials are to be dissolved in a liquid, or when a solid-catalyzed reaction is to take place, the stirrer must make a *suspension* of the particles in the liquid to allow their entire surface to take part in the process. For continuous processes, a quasi-uniform distribution of the solids in the entire liquid volume is required; this allows the solid particles to be transferred from stage to stage together with the liquid (e.g., in crystallization cascades). In this case, the solids are exposed to high mechanical stress, which can cause their abrasion.

In order to enhance mass transfer between two immiscible liquids, the phase with the smaller volume is dispersed in the other, i.e., a *liquid–liquid dispersion* is made. When surfactants are added, a more or less stable emulsion can be produced.

Stirrers are also used to perform gas–liquid contacting to make *gas–liquid dispersions*. The process is also called *sparging*. Different stirring operations must often be performed simultaneously; an example is solid-catalyzed hydrogenation, in which the stirrer must both distribute the gas (hydrogen) in the liquid, as well as suspend the solid catalyst (e.g., Raney nickel).

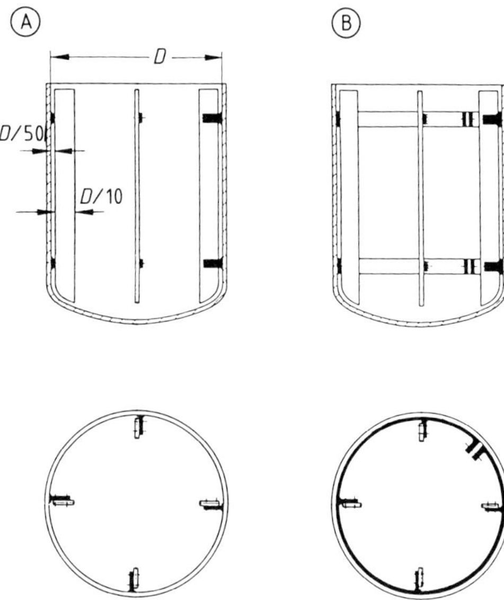

Figure 1. Baffle designs
A) Normal design; B) Design for glass and coated vessels (baffle basket with pressure-fitted ring)

1.2. Stirring Vessels and Auxiliary Equipment

Mixing operations are mainly performed in agitated vessels, but storage tanks and pipes can also be employed (cf. Section 2.6.). The design of mixing vessels is governed by standards (DIN 28 136, ASME Code Section VIII). The interior auxiliary equipment consists of baffles, feed and drain pipes, coils, and probes (e.g., thermometers, level indicators). All of these inserts can affect stirring.

If an axially positioned stirrer is operated in a *vessel without inserts*, the liquid is set into rotation, and a vortex is formed. The vortex can reach the stirrer and cause gas to be entrained in the liquid. This is generally undesirable because it results in an extraordinarily high mechanical stress in the stirrer shaft, as well as in its bearings and seal, due to the absence of the liquid bearing, often leading to the destruction of the stirrer. Even when vortex formation causes no gas entrainment, bulk rotation of the liquid is always undesirable when a two-phase system with different densities is concerned since the centrifugal force counteracts stirring.

Bulk rotation of the liquid in cylindrical vessels is prevented by the installation of baffles. Fully effective baffling is achieved with four baffles having a width of $D/10$ (D = inside diameter of vessel), which are vertically arranged along the entire vessel wall. In order to avoid dead volume behind the baffles, the baffle width is reduced to $D/12$ and they are set at a clearance of $D/50$ from the wall. The baffles are usually attached to the vessel wall by means of welded brackets (Fig. 1 A). In enamel-coated vessels, they are attached to the cover. When this is not possible (glass vessels, wooden vats), they are made in the form of a basket with pressure-fitted rings (Fig. 1 B).

Baffles are not necessary when stirring is carried out in vessels with rectangular cross section (pits), or when the stirrer is introduced into the vessel laterally. When the stirring is weak, liquid rotation can be prevented even in cylindrical vessels by installing the stirrer off-center and/or at an angle to the axis. In this case, however, uneven mechanical stress in the stirrer shaft must be accepted.

In order to supply or remove large amounts of heat, stirring vessels are equipped with *coils*. A spiral coil (Fig. 2 A) is only effective with stirrers that generate an axial flow pattern, since these produce good liquid circulation between the coil and the wall. In contrast, the liquid circulation produced by radial stirrers is strongly deflected

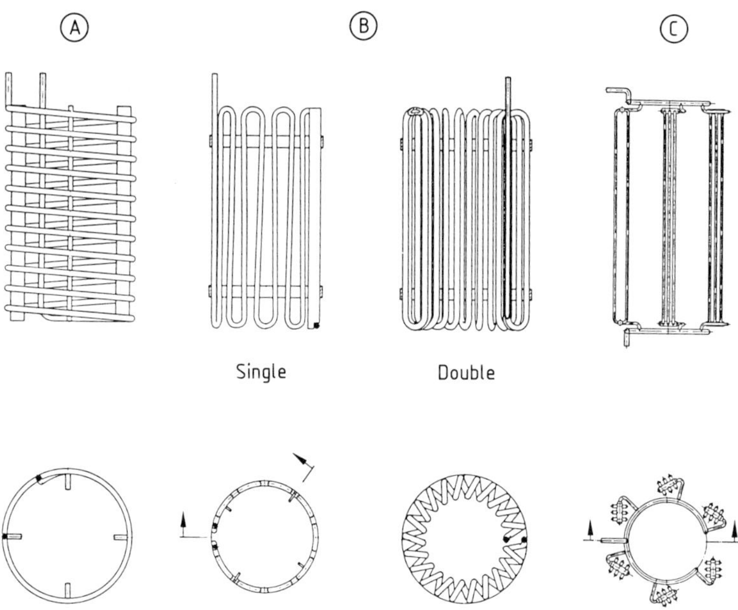

Figure 2. Tube coil designs
A) Spiral coil; B) Meander coils (cooling basket); C) Tube bundles

by the spiral coil, so that the flow through the annulus between the coil and wall is insufficient. For these types of stirrers, it is better to arrange the coil vertically along the vessel wall (meander coil, Fig. 2 B). This arrangement does not deflect the radial flow pattern, but prevents bulk rotation of the liquid to such an extent that baffles are often unnecessary. The heat-exchange tubes can also be arranged into bundles and installed instead of baffles (Fig. 2 C).

1.3. Operation of Individual Types of Stirrers

The mixing operations described in Section 1.1 naturally cannot all be performed with a single stirrer type; therefore, a wide variety of stirrers exists. There are stirrers suitable to a specific mixing operation and a given material system. The following discussion is limited to those stirrer types which are most widely used in the chemical industry, and for which established design guidelines exist.

These stirrer types are arranged in Figure 3 according to the predominant flow pattern they produce, as well as to the range of viscosities over which they can be effectively used. The flow patterns generated by stirrers causing radial and axial fluid motion are shown in Figure 4.

The *turbine stirrer* (Rushton turbine, six blades on a disk) is the only high-speed stirrer that sets the fluid in *radial motion*—or, at higher viscosities, in *tangential motion*. This type is only effective with low-viscosity liquids and baffled vessels. In this case, the diameter ratio D/d (D = vessel diameter; d = stirrer diameter) ranges from 3 to 5. During rotation, the turbine stirrer causes high levels of shear and is well suited to dispersion processes. The *impeller stirrer* was developed for use in enamel-coated vessels, and thus has rounded stirring arms. It is used in conjunction with small clearances from the bottom and with a ratio $D/d = 1.5$, either with or without baffles. It can also operate with strongly fluctuating fill levels (e.g., during vessel discharging) because it mixes even small amounts of liquid very well. The *cross-beam, grid,* and *blade stirrers* belong to the group of low-speed stirrer types and are used with $D/d = 1.5-2$. They can operate with baffles or, for viscous fluids, without, and are especially well suited to homogenization. As a rule, the low-speed *anchor stirrer* is operated at very small clearances from the wall ($D/d \simeq 1.05$) and is only appropriate for enhancing heat transfer in highly viscous liquids.

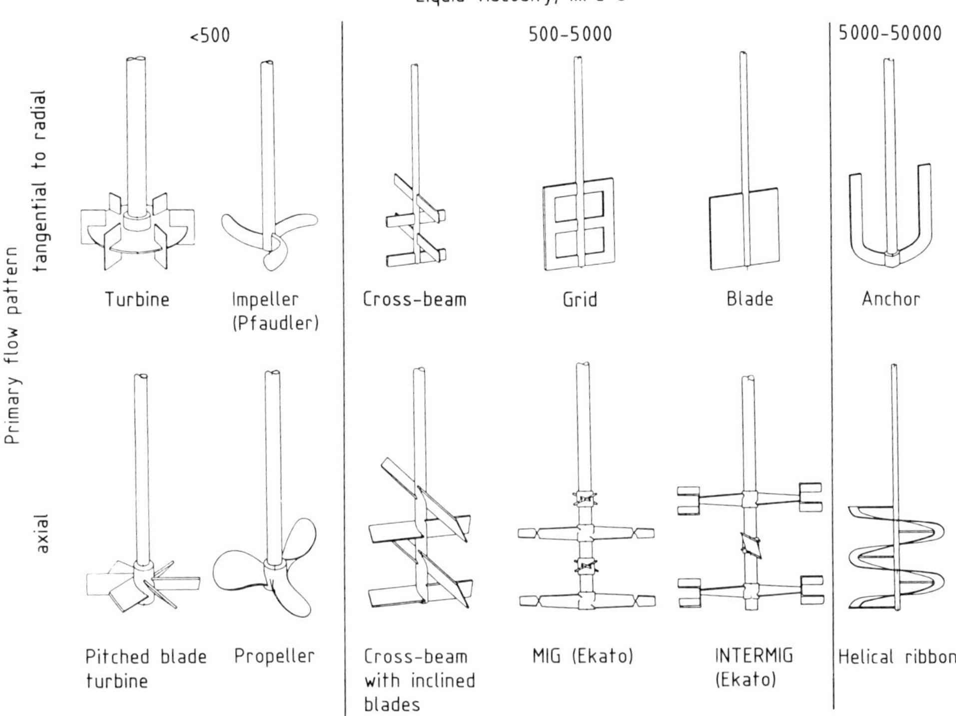

Figure 3. Common stirrer types

The *paddle stirrer* with pitched blades (pitched blade turbine) and particularly the *propeller stirrer* belong to the group of high mixers that generate an *axial flow pattern*. As a rule, both are used with low-viscosity liquids and in baffled vessels. They are well suited to homogenization and suspension of solids. In order to enhance the axial flow component in more viscous media and/or for $H/D > 1$ (e.g., in fermenters; H = liquid height) multistage stirrers with pitched stirring surfaces are required. This category includes the *cross-beam stirrer with pitched beams* and the *MIG* and *INTERMIG* stirrers of the Ekato company, Schopfheim, Federal Republic of Germany. These stirrers are operated at low speeds, with $D/d \simeq 1.5$ (with baffles) or $D/d = 1.1$ (without baffles). Mixing operations for these types include homogenization, suspension of solids, and dispersion. The very low-speed *helical ribbon stirrer* is used with small wall clearances ($D/d \simeq 1.05$) and operated in such a way that it drives the liquid downwards along the wall. Under these conditions, it is the best-suited of all stirrer types for the homogenization of highly viscous liquids.

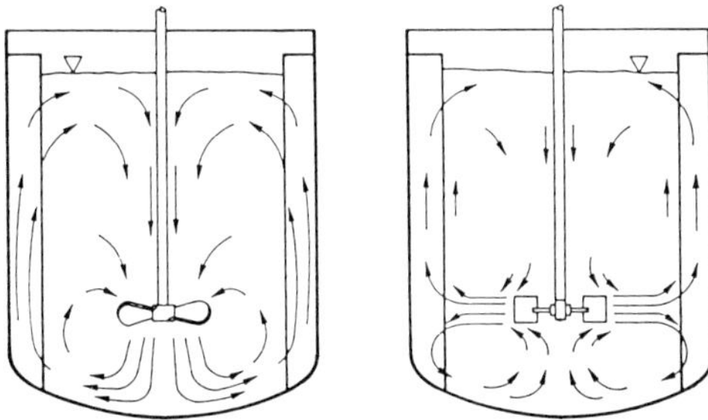

Figure 4. Flow patterns in a baffled tank, generated by an axial-flow propeller and a radial-flow turbine stirrer

In addition to these often used types, a large number of special designs also exists. Of these, only three should be mentioned: the stirrer operating on the rotor–stator principle, the sawtooth disk, and the hollow stirrer.

In *rotor–stator stirrers* (Fig. 5), the rotor consists of a blade or paddle stirrer enclosed by a ring of baffles (stator). As a result, high levels of shear are exerted on an extremely small volume. When the stirrer consists of a planar *sawtooth disk* (Fig. 6), the liquid is accelerated radially in a thin ring away from the center, and then

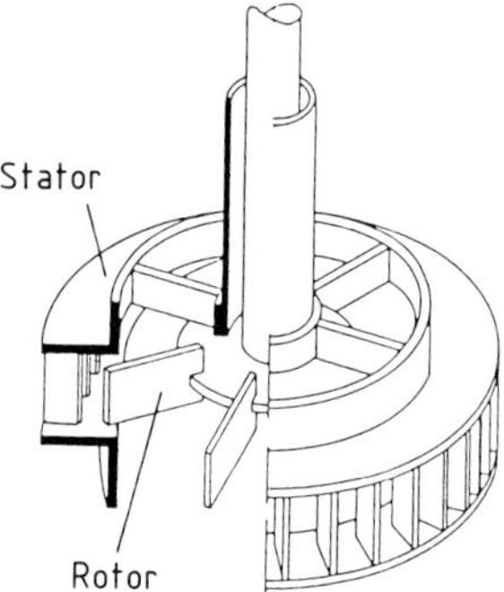

Figure 5. A rotor–stator stirrer

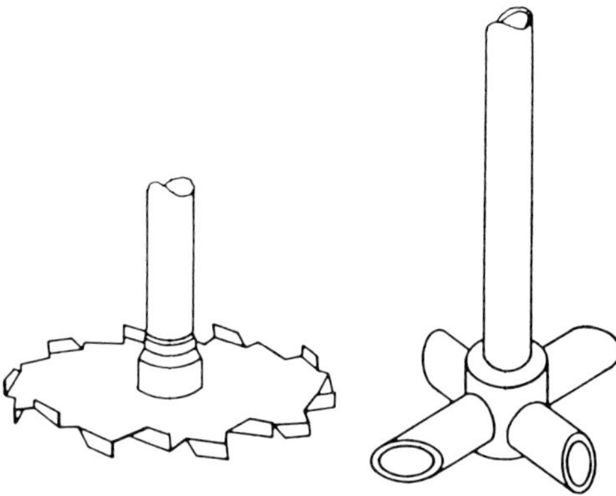

Figure 6. Disperser disk

Figure 7. Hollow stirrer (tube type)

quickly decelerated. High levels of shear can thus be achieved, even without a stator ring or baffles. Both of these stirrer types are particularly well suited to emulsification and dispersion over a wide range of viscosities, e.g., in the production of paint pigments.

In *hollow stirrers*, the stirrer head is hollow and is connected through a hollow shaft to the gas-filled space above the liquid surface. The suction generated behind the stirrer edges during rotation can thus be used to supply a gas to the liquid. As a single unit combining both stirrer and gas supply, hollow stirrers are well suited for enhancing mass transfer in gas-liquid systems. Figure 7 shows the so-called *tube stirrers*, a simple, effective hollow stirrer. All hollow stirrers operate at high speeds and are used in baffled vessels with $D/d = 3-5$.

2. Homogenization of Liquid Mixtures

2.1. Problem Definition

Homogenization deals with the mixing of miscible liquids to obtain the desired degree of homogeneity, or with the maintenance of homogeneity for the purpose of carrying out a reaction under well-defined conditions. The question here is what type of stirrer and what mixing conditions to adopt, so the mixing operation can be performed with the minimum effort (inexpensive stirrer, low work of mixing). In order to answer this question, the mixing time and power characteristics must be known for the stirrer.

The term *characteristic* means a relationship between dimensionless groups that describes the dependence of a target quantity (mixing time, power, etc.) on geometrical, physical, and process parameters. In terms of fluid mechanics, mixing is a complicated operation. Process-related questions are almost exclusively answered by experiments, which are best evaluated using dimensional analysis. This allows the number of parameters to be reduced, and ensures a reliable scale-up in combination with the theory of models.

2.2. Mixing-Time Characteristic

2.2.1. Definition and Measurement Method

The mixing time θ is the time required for the mixer to achieve the desired degree of homogeneity. A statement of the mixing time is therefore only meaningful in conjunction with the degree of homogeneity. The most common methods for determining the mixing time are the schlieren method and the chemical decolorization method. In the *schlieren method*, two liquids having different refractive indices are mixed; the complete homogenization of the liquids is indicated by the disappearance of the streaks (schlieren). In the *chemical decolorization method*, the first liquid, which contains one reaction component, is colored with an indicator; the second liquid, containing the other reaction component, is then added as the stirring begins; and the time at which the color disappears is measured. (Common reaction systems: sulfuric acid/alkali solution with phenolphthalein indicator; thiosulfate/iodine with starch indicator.) The degree of homogeneity at the point of decoloration depends on the excess of the added reaction component (Danckwerts method [1]). The principles and applications of this method are described in [2]. Chemical decolorization methods fail for extremely short mixing times (micromixing). In such cases, a consecutive dye reaction (diazotization) is used instead; the degree of mixing can be determined from the selectivity. Details of both methods can be found in Mixing, Introduction (page **24**-1).

2.2.2. Mixing-Time Characteristic at $\Delta\varrho \simeq 0$ and $\Delta\nu \simeq 0$

In the homogenization of liquids having the same density ϱ and kinematic viscosity ν, the mixing time θ for a given stirrer type and vessel inserts depends on the rotational speed n, the stirrer diameter d, and the kinematic viscosity. In terms of dimensional analysis, this can be expressed as $n\theta = f(Re)$. In this equation, known

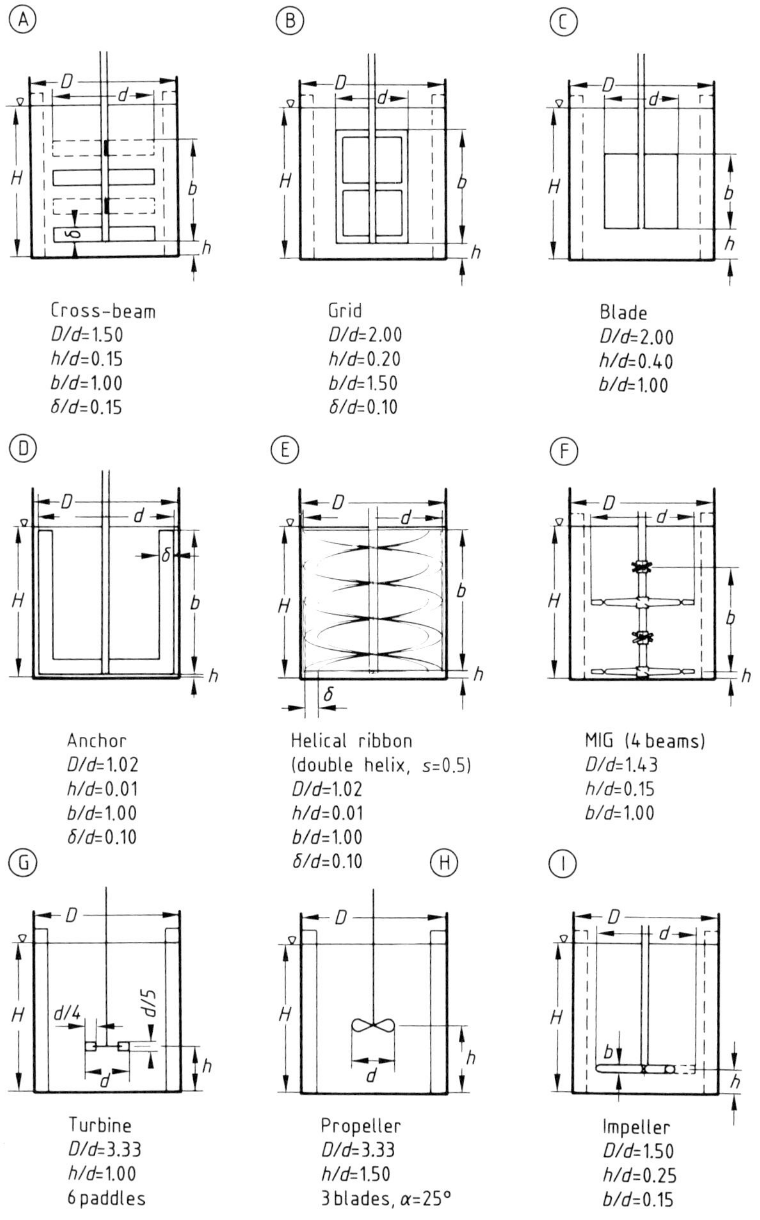

Ⓐ Cross-beam
D/d=1.50
h/d=0.15
b/d=1.00
δ/d=0.15

Ⓑ Grid
D/d=2.00
h/d=0.20
b/d=1.50
δ/d=0.10

Ⓒ Blade
D/d=2.00
h/d=0.40
b/d=1.00

Ⓓ Anchor
D/d=1.02
h/d=0.01
b/d=1.00
δ/d=0.10

Ⓔ Helical ribbon
(double helix, s=0.5)
D/d=1.02
h/d=0.01
b/d=1.00
δ/d=0.10

Ⓕ MIG (4 beams)
D/d=1.43
h/d=0.15
b/d=1.00

Ⓖ Turbine
D/d=3.33
h/d=1.00
6 paddles

Ⓗ Propeller
D/d=3.33
h/d=1.50
3 blades, α=25°

Ⓘ Impeller
D/d=1.50
h/d=0.25
b/d=0.15

Figure 8. Dimensions and installation conditions of stirrer types for vessel with $H/D = 1$
(Baffles drawn with dotted lines indicate that the stirrer can be used in baffled or unbaffled vessels.)

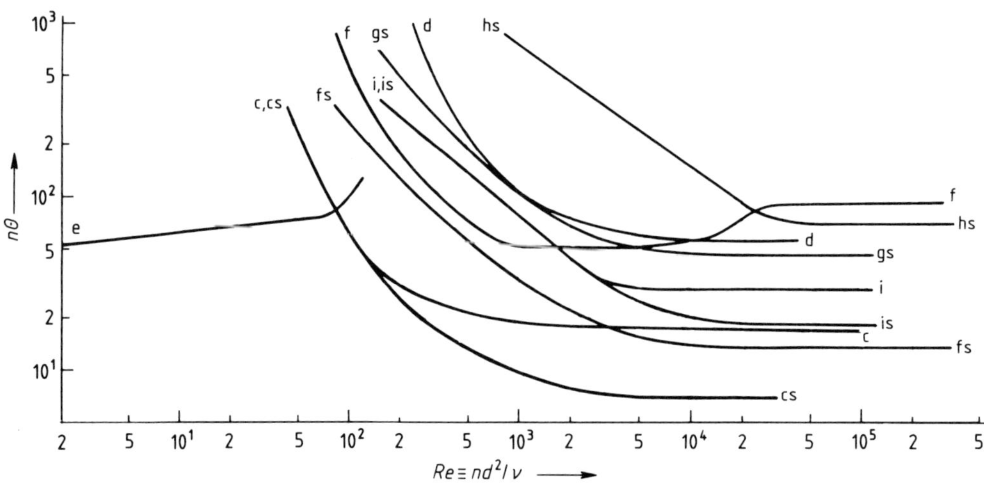

Figure 9. Mixing-time characteristics of the stirrer types in Figure 8 [3]
Cross-beam stirrer a = 1.8 · c (i.e., the mixing time is 1.8 times that of Curve c); as = 1.8 · cs; grid stirrer b = 1.25 · c, bs = 1.25 · cs; blade stirrer c, cs; anchor stirrer d; helical ribbon stirrer e; MIG-stirrer f, fs; turbine stirrer gs; propeller stirrer hs; impeller stirrer i, is. In each case s indicates the presence of baffles.
Example: What value of θ is obtained when a liquid with $\nu = 10^{-4}$ m^2/s (= 1 St) is mixed in a baffled vessel using a cross-beam stirrer (as = 1.8 · cs) with $d = 1$ m and $n = 1\ s^{-1}$?
$Re = nd^2/\nu = 10^4$
$n\theta = 1.26 \times 10$
$\theta \approx 13$ s

as the mixing-time characteristic, $n\theta$ represents the mixing number and *Re* the Reynolds number, $Re = nd^2/\nu$.

Since each mixing-time or power characteristic is only valid for the particular geometry, the dimensional proportions of the stirrer types treated are given in Figure 8.

Figure 9 shows the mixing-time characteristics for these stirrer types as determined by the chemical decolorization method. With an excess of ca. 0.01 N of acid, the decolorization indicated an essentially complete molecular homogenization [3]. Figure 9 allows the determination of the mixing conditions required to achieve a desired mixing time. It provides the number of stirrer rotations ($n\theta$) needed to homogenize the mixture under given flow conditions (*Re*). Depending on the range of Reynolds numbers, the following stirrers can produce homogenization with the minimum number of rotations: helical ribbon mixer, blade stirrer, and the MIG stirrer, the last two in baffled vessels.

As the ratio *H*/*D* increases, the mixing time becomes longer; for a baffled vessel with a cross-beam stirrer, the following relationship is valid over the range $10^3 < Re < 10^5$ [3]:

$$n\theta = 16.5\,(H/D)^{2.6} \tag{1}$$

(The number of stirrer beams is assumed to increase with vessel height in accord with Figure 8 A).

Mixing-time characteristics for 25% and 10% deviation from the final mixed state have been obtained by C. J. Hoogendoorn and A. P. den Hartog [4]. Homogenization and power characteristics for enamel-coated stirrers with rounded edges can be found in [5]; a number of other mixer types and insert configurations are treated in [6].

2.2.3. Mixing-Time Characteristic at $\Delta\varrho \neq 0$ and $\Delta\nu \neq 0$

In the homogenization of liquids with different densities and viscosities, coarse mixing is affected by the specific weight difference of the two components, $g\Delta\varrho$; however, the elimination of the final concentration gradients takes place under conditions characterized by the properties of the homogeneous mixture, $\bar{\varrho}$ and $\bar{\nu}$. In this case, the mixing time depends on the following parameters: $\theta = f\,(n, d, \bar{\varrho}, \bar{\nu}, g\Delta\varrho)$; in terms of dimensional analysis, this can be expressed as $n\theta = f(Re, Ar)$, where $Re \equiv nd^2/\bar{\nu}$ is the Reynolds number and $Ar \equiv d^3 g\Delta\varrho/(\bar{\nu}^2\bar{\varrho})$ the Archimedes number. The mixing-time characteristic of a cross-beam stirrer, as in Figure 8 A, operating in a baffled vessel with $H/D = 1$ is as follows [7]:

$$n\theta = 51.6\ Re^{-1}(Ar^{1/3} + 3) \quad (2)$$

This relationship is valid for $10 < Re < 10^5$ and $10^2 < Ar < 10^{11}$.

2.3. Power Characteristic

2.3.1. Definition and Measurement Method

In addition to the mixing power P, the motor drive of a stirrer must also account for the power losses in the the gearbox and the seals. In order to determine the mixing power P, the torque M_d, and the rotational speed n of the stirrer must be known: $P = M_d\,\omega$, with $\omega = 2\,\pi n$ as the angular velocity. The torque can be measured using, for example, a torsion shaft with strain gauges, electrically with eddy current torque transducers, or mechanically using a swiveling motor. The speed of rotation can be measured using mechanical, electrical (photocell), or optical (stroboscope) instruments.

2.3.2. Power Characteristics in Homogeneous Systems

The power required by a given stirrer type in a given vessel configuration depends on the speed of rotation n, the stirrer diameter d, the density ϱ, and the kinematic viscosity ν of the medium. In vessels without baffles, the liquid vortex, and therefore the acceleration due to gravity g, is immaterial, as long as no gas is entrained in the liquid. Thus, $P = f(n, d, \varrho, \nu)$ applies, and in accordance with dimensional analysis $Ne = f(Re)$. Here, $Ne \equiv P/(\varrho n^3 d^5)$ is the Newton or power number, and $Re \equiv n d^2/\nu$ the Reynolds number. The relationship $Ne = f(Re)$ is called the power characteristic of a stirrer.

The power characteristics of the stirrer types discussed in this section are presented in Figure 10. Three flow regimes can be identified:

1) In the *laminar regime* ($Re \leq 10$; for stirrers with very small wall clearances, such as the anchor or helical ribbon mixer, $Re \leq 100$), the effect of inertial forces (density) is suppressed by the viscous forces; therefore, baffles are unnecessary. In this regime, $Ne \propto Re^{-1}$, or $Ne\,Re$ = constant; the power is therefore given by

$$P = k_1 n^2 d^3 \eta \quad (3)$$

η being the dynamic viscosity, and k_1 being the constant

2) In the *turbulent flow regime*, the viscosity, and therefore the Reynolds number, ceases to have an effect. In this case Ne = constant applies, and the power is given by

$$P = k_2 n^3 d^5 \varrho \quad (4)$$

This regime begins at $Re \geq 10^2$ for baffled vessels and at $Re \geq 5 \times 10^4$ for vessels without baffles. In this regime baffles are fully effective and increase the required stirring power by up to a factor of ten. For example, the power characteristics of the blade stirrer in a vessel with baffles (cs) and without (c) can be compared.

3) The *transition regime* arises only in vessels without baffles and for $Re \simeq 10 - 5 \times 10^4$; in this case, both viscosity and density have an effect. The relationship $Ne \propto Re^{-1/3}$ applies here.

The stirring power can be determined from Equations (3) and (4) using the constants k_1 and k_2 from Table 1. In general, the power can be calculated using Equation (5). The Reynolds number is determined first, and the corresponding value of Ne read from Figure 10; using this value, the power can now be evaluated by:

$$P = Ne\,n^3 d^5 \varrho \quad (5)$$

Table 1. The proportionality constants for Equations (3) and (4)

Stirrer type	Curve	k_1 Ne ($Re = 1$)	k_2 Ne ($Re = 10^5$)
Cross-beam stirrer	a	110	0.4
	a s	110	3.2
Grid stirrer	b	110	0.5
	b s	110	5.5
Blade stirrer	c	110	0.5
	c s	110	9.8
Anchor stirrer	d	420	0.35
Helical ribbon stirrer	e	1000	0.35
MIG stirrer	f	100	0.22
	f s	100	0.65
Turbine stirrer	g s	70	5.0
Propeller stirrer	h s	40	0.35
Impeller stirrer	i	85	0.20
	i s	85	0.75

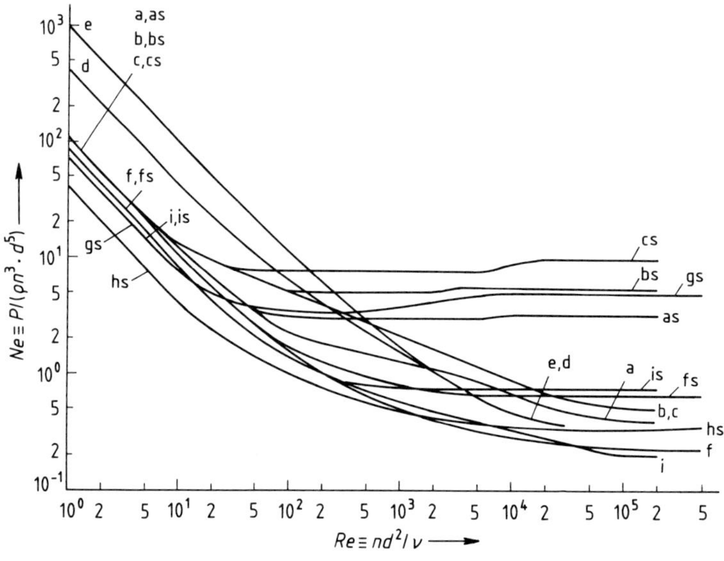

Figure 10. Power characteristics of the stirrer types in Figure 8 [3] (except g [11] and h [9])

The power characteristic for vessels without baffles decreases monotonically over a wide range of flow conditions; this offers the possibility of determining the viscosity of a reaction mixture and therefore the conversion, or the degree of polymerization, as a function of time by measuring the mixing power.

If large differences in density or viscosity exist at the start of mixing, it is advisable to calculate the mixing power using the larger of the two values.

Of the extensive literature on stirring power, the basic works of W. Büche [8], J. H. Rushton and co-workers [9], as well as S. Nagata and co-workers [10] should be mentioned. Power characteristics for various designs of turbine and paddle stirrers are given in [11]. Many power characteristics can also be found in the literature on mixing time [3]–[6].

2.4. Optimization: Minimum Work of Mixing

The optimum stirrer type and stirring conditions for the homogenization of a liquid mixture are those that require the least work of mixing. These conditions can be found quickly with Figure 11, which shows the relationship between the modified power number $Ne\,Re^3\,(D/d) \equiv P\,D\,\varrho^2/\eta^3$ and the modified mixing number $n\theta\,Re^{-1}\,(D/d)^{-2} \equiv \theta\eta/D^2\varrho$. This representation is based on the information in Figures 9 and 10. The stirrer whose curve lies lowest in Figure 11 gives the desired mixing time with the least power. For the sake of clarity, Figure 11 includes only the curves for the optimum stirrers along with their corresponding Reynolds number scales.

The use of the figure is as follows: first the modified mixing number is determined, and then the best stirrer type can be read off the abscissa. The abbreviations are explained in Table 1. The point of intersection of this value with the curve provides the rotational speed via the Reynolds number, $Re \equiv nd^2\varrho/\eta$. The ordinate of this point gives the modified power number, and through this the mixing power. A specific example is worked out in Figure 11.

To obtain a value of the mixing number lower than $\theta\eta/D^2\varrho \simeq 10^{-4}$ is practically impossible. For liquids with viscosity similar to that of water and for large vessels, the following approximation is valid:

$$\theta_{min} \simeq 10^{-4}\,D^2\,\varrho/\eta$$

Following a suggestion in [12], the pairs of values $(n\theta, Re)$ can be selected from Figure 9 in which both $n\theta$ and Re are lowest; using the corresponding Ne values from Figure 10, the following relationship can be derived from Figure 11:

$$\frac{P\,D\,\varrho^2}{\eta^3} = a\left(\frac{\theta\eta}{D^2\,\varrho}\right)^{-3} \quad \text{or} \quad \frac{P\,\theta^3}{\varrho\,D^5} \mathrel{\hat{=}} \frac{P\,\theta^3}{V\varrho\,D^2} = \text{constant} \qquad (6)$$

where $a \simeq 700$. Two observations may be drawn from this relationship, which is valid in the turbulent regime (η irrelevant):

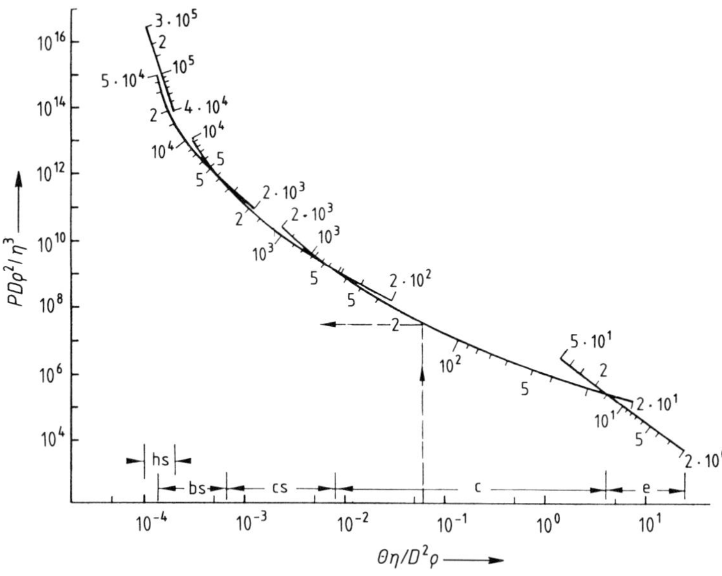

Figure 11. Curves for determining the optimum stirrer type and the operating conditions that minimize the mixing work [3]
The numbers along the curves are the Reynolds number, $Re = nd^2\,\varrho/\eta$
Example: $D = 1.0$ m, $\eta = 1$ kg m s^{-1} = 1000 mPa · s, $\varrho = 1000$ kg/m^3,

$\theta = 60\ \mathrm{s} \quad \dfrac{\theta\eta}{D^2\varrho} = 6 \times 10^{-2}$

Blade stirrer, without baffles (Fig. 8 C)
$D/d = 2$; $d = 0.5$ m
$Re = 170$; $n \approx 41$ min^{-1}
$PD\,\varrho^2/\eta^3 = 3 \times 10^7$; $P = 30$ W
The dotted arrows show this example in the figure itself

1) at constant P/V, θ increases as $D^{2/3}$
2) at constant D, θ decreases as $(P/V)^{-1/3}$
This confirms that the constancy of P/V in the turbulent regime does not constitute a scale-up criterion for the homogenization of liquid mixtures.

If the corresponding values for the helical ribbon mixer in the laminar regime are inserted in the same relationship [12], then

$$\frac{P\theta^2}{D^3\eta} = 5.8 \times 10^5 \qquad (7)$$

This indicates that the constancy of P/V is a scale-up criterion only in the laminar regime. It also shows that $\theta \propto (P/V)^{-1/2}$.

2.5. Homogenization in Storage Tanks

Stirrers are rarely used for homogenization in storage tanks since the liquids involved are of low viscosity and relatively long mixing times are generally acceptable. In such cases, mixing is carried out with side-entering propeller stirrers, liquid jets, or rising gas bubbles.

In homogenization using *propeller stirrers* [13], [14], one or more are introduced laterally through the wall of the tank at an angle of 7–10° to the radius. At this angle, liquid circulation is intensive without significant rotation of the contents. Since the mixing time θ and the liquid throughput q' of the stirrer are related through $\theta \propto V/q'$ (V = liquid volume), the operating conditions are chosen to maximize the liquid throughput at a given power. This is achieved through large stirrer diameter and low stirrer speed (Section 2.7). At $H/D = 1$ and P = constant, the following is valid:

$$\theta \propto (D/d)^{2.3} \qquad 10 < D/d < 30$$

At a given d, the mixing time is inversely proportional to the power, $\theta \propto P^{-1}$, and increases rapidly with density differences, $\theta \propto (\Delta\varrho/\varrho)^{0.9}$ [13].

Two conditions can generally arise during the mixing:

1) At high mixing intensities, homogenization mainly depends on convection and turbulent diffusion
2) At lower mixing intensities, which is commonly the case, small density differences lead to the formation of layers, which are subsequently broken up through interfacial waves. This process is much slower and depends to a great extent on the tank geometry and physical properties of the system.

In the second case, the boundary between the two flow regimes is defined by the value of the critical Froude number, $Fr'_{crit} \equiv n^2 d\varrho / g \Delta\varrho$ [15]:

$$\log Fr'_{crit} = 1.40 + 0.04\,(D/d) \tag{8}$$

At $Fr' \gg Fr'_{crit}$, $n\theta$ is independent of Fr', but depends strongly on D/d:

$$n\theta = 90\,(D/d)^{1.3} \tag{9}$$

On the other hand, at $Fr' \ll Fr'_{crit}$, the following applies:

$$n\theta = 90\,(D/d)^{1.3}\,(Fr'_{crit}/Fr')^{1.5} \tag{10}$$

The most favorable installation geometry for the propeller stirrer (also according to [16]) is $\alpha = +\,10°$ and $\beta = +\,10°$ (where α is the angle between the stirrer shaft and the horizontal, and b the angle between the shaft projected on the horizontal plane and the radius). The power characteristic of side-entering propeller stirrers is approximately 1/3 lower than that of a vertical stirrer [16].

In the homogenization with *liquid jets*, the mixing time can be estimated relatively accurately under the assumption $\theta \propto V/q'$ [14]. The total liquid throughput q' consists of the jet stream itself and the entrained liquid. For calculation of the jet stream, see [17], [18]; for the estimation of liquid entrainment, see [19].

The liquids with relatively low viscosity ($n < 100$ mPas) can be homogenized effectively with *rising gas bubbles* from a distributor at the bottom of the tank (sintered or perforated plate, two-fluid nozzle). Their effectiveness is due to their buoyancy, and therefore mixing ability, remaining constant over the entire height of the tank. For liquids similar to water and without significant density or viscosity differences, the following relationship applies [20]:

$$\theta\,(g/D)^{1/2} = 4.75\,Fr^{-1/4}\,(dH^2/D^3)^{1/4} \tag{11}$$

where $Fr \equiv q^2/(D^5/g)$. After transformation, this becomes

$$\theta/D = 4.75\,(H/q)^{1/2}\,(d/g)^{1/4} \tag{12}$$

q is the gas flowrate, and d the diameter of the gas distributor. For the estimation of the gas flowrate for lower surface tensions or higher viscosities than that of water, refer to [20].

Homogenization with rising gas bubbles is also effective for the mixing of water reservoirs or lakes. Model measurements [21] yield the following relationship for the mixing time θ,

$$\theta = 250\,[H^4/(qg)]^{1/3} \tag{13}$$

which can be expressed in dimensionless form as

$$\theta\,(g/D)^{1/2} = 250\,Fr^{-1/6}\,(H/D)^{4/3} \tag{14}$$

2.6. Homogenization in Pipelines

In continuously operating large facilities, miscible liquids are often simply mixed in the pipelines themselves. The mixing effect of a turbulent flow is not particularly intense: for liquid mixtures without density differences, a pipe length of ca. 100 tube diameters is required to reduce the relative concentration fluctuations to 1% [22]. The mixing effect is enhanced by feeding the secondary component through an injector or a Venturi nozzle.

A rotary pump installed in the pipeline acts as an excellent mixing device [23]. One or more stirrers may be installed on the same shaft over a short length of pipe (in-line blending) [24].

Viscous liquids are mixed in pipelines using so-called static mixers. Examples of these are the "Static Mixer" of the Kenics Corp. (Figure 12) and the Sulzer mixer [25]. The deflecting elements in a static mixer divide the stream into two

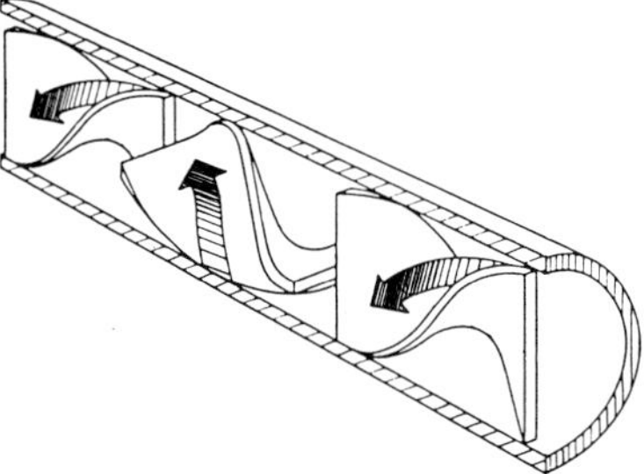

Figure 12. Kenics mixer, an example of a static mixer used in in-line mixing

streams and turn each through 180 °, so after passing through N elements the stream has been blended 2^N times. The Sulzer inserts, on the other hand, take advantage of the mixing effect of flow bending around an oblique web.

Comparative studies of various static mixers have been carried out by PAHL and MUSCHELKNAUTZ [26, 27]. Pressure drop, residence time distribution, and homogenization characteristics can be obtained from these studies. The homogenization characteristics are presented in the form of a relationship between the variation coefficient and the required number of pipe diameters (L/D).

Homogenization in pipelines is treated in more detail under → Continuous Mixing of Fluids, Volume B4.

2.7. Liquid Throughput of Stirrers

The liquid flowrate that is displaced through the area swept by the stirrer is termed the liquid throughput of the stirrer or its pumping capacity q'. This area is given by $\pi d^2/4$ for axially operating stirrers, and by πdb for radial flow stirrers, where b is the paddle height. For axial flow stirrers, q' is determined from the circulation time of particles with $\Delta \varrho \simeq 0$; for radial flow stirrers, it is obtained from the integral value of the measured velocity distribution of the flow.

In continuously stirred tanks, homogenization can only be achieved if $q' \gg q$, where q is the liquid flowrate through the vessel. Under these conditions, the mixing time is $\theta \propto V/q'$, where V is the liquid volume. In the case of propeller stirrers, knowledge of q' enables calculation of the flow velocity along one or more spiral coils concentric to the stirrer. The installation of such coils may be unavoidable in the case of extremely exothermic reactions.

The throughput characteristic of a given stirrer is provided by $Q' = \mathrm{f}(Re, D/d, h/d)$, where $Q' \equiv q'/(nd^3)$ represents the dimensionless throughput number; $Re \equiv nd^2/\nu$ the Reynolds number; and D/d, h/d the geometric parameters of the stirrer installation. For propeller stirrers with pitch $s = 1$ (pitch s = tangent of the blade angel α) operating in baffled vessels with $H/D = 1$, the following relationship is valid in the turbulent regime ($Re > 10^3$) [28]:

$$Q' = 0.654\,(D/d)^{-0.16} \quad D/d > 3 \tag{15}$$

For paddle stirrers with six pitched paddles ($\alpha = 45°$, $b/d = 0.2$), the relationship is as follows [29]:

$$Q' = 1.29\,(D/d)^{-0.20};\ 2 < D/d < 3.3 \tag{16}$$

For turbine stirrers with six paddles, as in Figure 8 G, the throughput is given by [30]:

$$Q' \simeq 0.75 \tag{17}$$

Other stirrer configurations and paddle numbers are treated in [31].

2.8. Vortex Depth in Unbaffled Stirred Tanks

All calculations for the design of stirrers in unbaffled vessels are valid only under the assumption that the vortex does not reach the stirrer head, otherwise, gas entrainment alters the physical properties of the system. The highest allowable rotation speed is given by the following relationships [32].

Propeller stirrer with $D/d = 3.33$ and $h/d = 1$ (Fig. 8 H):

$$Fr_{\max} = 0.072\,(H'/d)^{1.33}/(0.25 - Ga^{-0.10}) \tag{18}$$

Turbine stirrer with $D/d = 3.33$ and $h/d = 1$ (Fig. 8 G):

$$Fr_{\max} = 0.016\,(H'/d)^{1.16}/(0.10 - Ga^{-0.18}) \tag{19}$$

These expressions are valid over the range $2.3 < H'/d < 5.7$ and $10^6 < Ga < 10^{10}$ In these equations, $Fr_{\max} \equiv n_{\max}^2\, d/g$ is the Froude number; H' the height of liquid above the stirrer; and $Ga \equiv d^3 g/\nu^2$ the Galileo number. For other stirrer types and geometries, see [32]–[34].

3. Intensification of Heat Transfer

3.1. Problem Definition

Stirrers are often used to intensify the heat transfer between the liquid and the heat exchange surface in a vessel. The question is what type of stirrer to use and under what conditions so that the desired heat flux can be effected at a minimum expense (inexpensive stirrer, low power requirements). In the case of cooling, this question involves an optimization problem: high rotational speeds result not only in an increase of the heat flux, but also in the dissipation of more heat of agitation.

3.2. Heat-Transfer Characteristics

3.2.1. Definitions

Heat transfer is described through the equation $Q = k\, A\, \Delta T$, where Q is the heat flowrate, A the heat-exchange surface area, ΔT the temperature difference, and k the heat-transfer coefficient, which is composed of the inner and outer film heat-transfer coefficients, α_i and α_o, the thermal conductivity of the wall material λ, and its thickness δ, according to the relationship

$$1/k = 1/\alpha_o + \delta/\lambda + 1/\alpha_i$$

Stirring affects only the inner-side heat-transfer coefficient α_i. This coefficient is defined through the relationship $\alpha_i \equiv Q/(A\, \Delta T)$, where the characteristic temperature difference ΔT is generally taken to be the difference between the average wall temperature T_w and the average temperature of the vessel contents T, i.e., $\Delta T = |T_w - T|$.

For a given stirrer type and a fixed geometry, α_i is a function of the vessel diameter D, the stirrer diameter d, and the rotational speed n, as well as the physical properties of the liquid (specific heat capacity c_p, heat conductivity λ, density ϱ, dynamic viscosity in the bulk η and at the wall η_w): $\alpha_i = \mathrm{f}(D, d, n, c_p, \lambda, \varrho, \eta, \eta_w)$. Dimensional analysis yields the following expression: $Nu = f\,(Re, Pr, \eta/\eta_w, D/d)$, where $Nu \equiv \alpha_i\, D/\lambda$ is the Nusselt number, $Re \equiv nd^2\, \varrho/\eta$ the Reynolds number, $Pr \equiv c_p \eta/\lambda$ the Prandtl number, η/η_w the viscosity ratio, and D/d the diameter ratio. The relationship between these dimensionless numbers is termed the heat transfer characteristic.

3.2.2. Heat-Transfer Characteristics in the Range Re > 200

In this flow regime, heat-transfer characteristics are generally of the form $Nu = \mathrm{c}\; Re^{2/3}\; Pr^{1/3}\; (\eta/n_w)^m$, where the constant c is a function of the stirrer type and the nature of the heat exchange surface (baffled or unbaffled tank wall, spiral or meander coil). The exponent of the viscosity term, m, which enables both heating and cooling data to be presented together, has been found to be 0.14, 0.18, or 0.24 in several investigations.

Table 2 lists the values of the constant c for various stirrer types for heat transfer through the tank wall (W) or a spiral coil (S); these values are taken from [35], in which 77 original works are evaluated and compared. A further collection of such expressions is given in [36].

The constant of the heat-transfer characteristic c has the following values:

1) Unbaffled vessels, $c = 0.35-0.4$, independent of stirrer type and diameter ratio
2) Baffled vessels, $c = 0.5$ for axial flow mixers (e.g., propeller stirrers) and $c = 0.75-0.80$ for radial ones (e.g., blade and turbine stirrers)
3) Tanks with spiral coils, $c = 0.70-0.90$ for radial flow stirrers, independent of whether baffles are present or not, whether inside or outside the coil.

3.2.3. Heat-Transfer Characteristics in the Range Re < 200

In the laminar flow regime, the exponents a and b in the heat-transfer characteristic $Nu \propto Re^a\, Pr^b\, (\eta/\eta_w)^m$ have different values than in the turbulent regime.

For the *anchor stirrer*, for example, the following relationship has been established [37]:

$$Nu = 1.4\, Re^{0.43}\, Pr^{1/3}\, (\eta/\eta_w)^{0.18} \tag{20}$$

In contrast, extensive measurements [38] with low-clearance anchor stirrers have shown that the effect of Re and Pr on Nu decreases steadily as the Reynolds number decreases. In the range $0.5 < Re < 200$, the relationship $Nu \propto Re^{0.1}$ was found to apply for cooling. The results can be expressed in the usual form by including a second constant:

$$Nu = k_3\, (Re\, Pr^{1/2} + k_4)^{2/3} \tag{21}$$

At high Reynolds numbers ($Re \gg k_4$), this expression approaches the usual form, i.e., $Nu \propto Re^{2/3}\, Pr^{1/3}$.

For cooling, heat-transfer measurements with wall-scraping elements [39] show that the effect of the Reynolds number in the regime $Re < 200$ is drastically reduced.

For the *helical ribbon mixer*, the following expression is given in [40]:

$$Nu \propto (Re\, Pr)^{1/3}\, (\eta/\eta_w)^{0.20} \tag{22}$$

Table 2. Values of c in the relationship $Nu = c\ Pr^{1/3}\ Re^{2/3}\ (\eta/\eta_w)^{0.14}$

Stirrer Type and Installation	c	Re Range	Reference [d]
Cross-beam stirrer with one beam, $D/d = 1.67$, $\delta/D = 0.16$, no baffles, spiral coil	W 0.36 S 0.78 [a]	$3\times10^2 - 3\times10^5$	[53]
Two paddle stirrers with 6 paddles, arranged one above the other, $D/d = 2.5$, $\delta/d = 0.16$, no baffles, spiral coil	W 0.40 S 0.70 [a]	$2\times10^2 - 1\times10^6$	[32]
Blade stirrer $D/d = 2.0$, $b/d = 1.42$, no baffles, spiral coil	W 0.36 S 0.70 [a]	$1\times10^3 - 1\times10^6$	[54]
Blade stirrer, 4 baffles ($D/15$ wide), $D/d = 2.0$, $b/d = 1$, $d/D = 0.25 - 0.60$, $b/D = 0.10 - 0.60$	W 0.80 W 1.31 $(d/D)^{0.39}(b/D)^{0.34}$	$3\times10^2 - 5\times10^5$	[57]
Turbine stirrer, 4 baffles ($D/10$ wide) $D/d = 3.33$, $h/d = 1$ (Fig. 8 G) $h/D = 0.08 - 0.033$, $H/D = 0.66 - 1.0$	W 0.75 [b] W 1.15 $(h/D)^{0.4}(H/D)^{-0.56}$	$3\times10^1 - 4\times10^4$	[42]
Turbine stirrer, 4 baffles ($D/10$ wide) $D/d = 3.33$, $h/d = 1$ (Fig. 8 G) h/D = 0.05 − 0.70; $d/D = 0.16 - 0.75$	W 0.76 W 1.01 $(h/d)^{0.12}(d/D)^{0.13}$	$5\times10^4 - 8\times10^5$	[34]
Turbine stirrer (Fig. 8 G), spiral coil $d/D = 0.3$, $\delta/D = 0.03$, $D_s/S = 0.7$ $d/D = 0.25 - 0.58$, $\delta/D = 0.018$-0.036 (D_s = coil diameter, δ = tube diameter) Baffles, within or outside the coil, have no effect	S 0.87 S 0.17 $(d/D)^{0.1}(\delta/D)^{-0.5}$	$4\times10^2 - 1\times10^6$	[52]
Propeller stirrer, 4 baffles ($D/10$ wide) pitch $s = d$, $h = d$ $s/d = 0.4 - \infty$, $h = d$	W 0.50 W 0.64 $(s/d)/(0.285 + s/d)$	$2\times10^5 - 9\times10^5$	[60]
Anchor stirrer (Fig. 8 D), $D/d = 1.04$	W 0.38	$3\times10^2 - 4\times10^4$	[59,71]
Anchor stirrer (Fig. 8 D), $D/d = 1.02$ heating cooling	W 0.40 [c] W 0.26 [c]	$2\times10^2 - 1\times10^5$	[69]

[a] Based on $Re^{2/3}$
[b] With (η/η_w)
[c] Without (η/η_w) term
[d] The reference numbers refer to the list of references in [35].
W wall; S spiral coil

This work is further evidence that a considerably reduced effect of the Reynolds number must be expected for $Re < 200$.

3.3. Stirrer Power During Heat Transfer

The relationships in Section 2.3. also apply to the stirrer power in heat transfer. If wall clearances are small, the viscosity at the wall temperature should be used to calculate the Reynolds number.

3.4. Optimum Stirring Conditions for the Removal of Reaction Heat

Heat removal is a particularly severe problem in the case of strongly exothermic reactions in viscous mixtures. Figure 13 illustrates the prob-

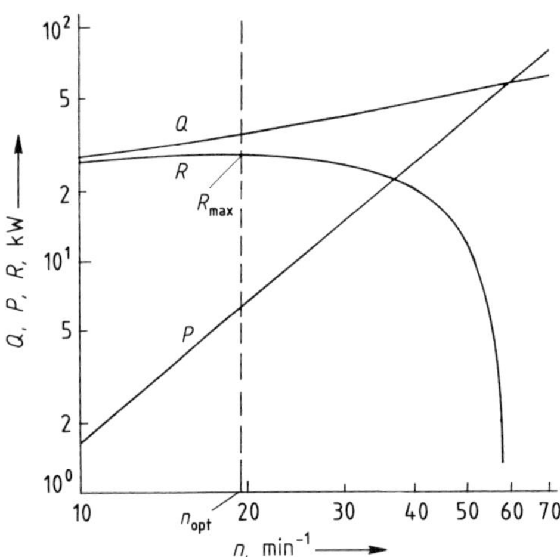

Figure 13. Graphical representation of the relationship between Q, P, $R = Q - P$, and the rotational speed n [41]
Anchor stirrer in a vessel with
$D = 2.0$ m, $V = 5.86$ m^3, $d = 1.8$ m ($D/d \simeq 1.10$),
$\varrho = 1000$ kg/m^3, $\eta = 50$ kg/(m · s), $\lambda = 2$ W/(m · K),
$c_p = 2000$ J/(kg · K), $A = 14$ m^2, $\Delta T = 25$°C, $k \approx \alpha$

lem with an example. The values of the rate of heat transfer Q through the wall, as well as the stirrer power P, are plotted as functions of the speed of rotation n. These values are calculated using the heat-transfer and power characteristics of the anchor stirrer (Fig. 10 and [38]). In the appropriate range of Reynolds numbers ($5 \leq Re \leq 50$), stirrer power increases much more rapidly with rotational speed ($P \propto n^2$) than the heat flux ($Q \propto \alpha \propto n^{1/2}$). The thick curve represents the difference between the two straight lines on the log-log plot and corresponds to the heat of reaction $R = Q - P$ that can be removed through the wall: R goes through a maximum at $n \simeq 20$ min^{-1} and reaches zero at n = 59 min^{-1}. At this point, the heat removed is that generated by the stirrer itself.

In order to predict the optimum conditions (n_{opt}, R_{opt}) the following procedure is applied [41]: using $Q = \alpha A \Delta T$ (in the laminar regime, $k \simeq \alpha$) and $B \equiv R/V$, the relationship $R = Q - P$ yields:

$$B = \alpha A \Delta T/V - P/V$$

In dimensionless form, this relationship can be formulated as follows:

$$\Pi_2 = Nu - (D/d)\,\Pi_1^{-1}\,Re^3\,Ne$$

where

$$\Pi_1 \equiv \frac{D^2 \varrho^2 \lambda \Delta T}{\eta^3} \cdot \frac{A}{D^2} \quad \text{and} \quad \Pi_2 \equiv \frac{B D^2}{\lambda \Delta T} \cdot \frac{V}{DA}$$

To find the optimum conditions, the known expressions for $Nu = Nu(Re,Pr)$ and $Ne = Ne(Re)$ are substituted into the equation above, the resulting expression differentiated with respect to Re, and the derivative set to zero.

The determination of these optimum conditions is facilitated by the graph in Figure 14, which applies to two anchor stirrers (as in Fig. 8 D) with different wall clearances ($D/$

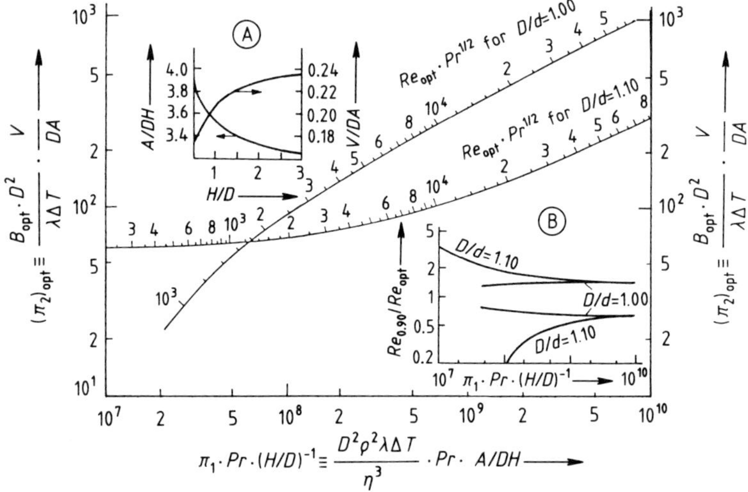

Figure 14. Curves for determining the optimum operating conditions for two anchor stirrers (Fig. 8 D) with different clearances (D/d = 1.00 and 1.10) in the laminar flow regime [4]

$d = 1.00$ and 1.10) in the laminar regime ($Re < 100$). The geometric parameters A/DH and V/DA for vessels with elliptical bottoms can be obtained as functions of the H/D ratio from auxiliary Diagram A. However, the optimum rotation speed can only rarely be realized in practice; therefore, auxiliary Diagram B gives the range of Re (and hence n) over which B does not fall below 90 % of B_{opt}.

The determination of the optimum stirring conditions has also been treated by PENNEY and KOOPMAN [42]. They found that in the laminar regime the maximum amount of reaction heat can be removed when the stirrer power is approximately 20 % of the heat of reaction; in the turbulent regime, this figure is 5–20 %.

Example. The conditions are the same as those given in the caption of Figure 13, including the condition $D/d = 1.10$. For $Pr = 5 \times 10^4$ and an abscissa value of $\Pi_1\, Pr\, (H/D)^{-1} = 2.82 \times 10^8$, the optimum condition obtained from Figure 14 is $Re_{opt}\, Pr^{1/2} = 4.8 \times 10^3$, which gives the ordinate $(\Pi_2)_{opt} = 8.0 \times 10$, resulting in $n_{opt} \simeq 20\ \mathrm{min}^{-1}$ and $R_{max} = B_{opt}\, V = 28.5$ kW (dashed optimum operating line in Fig. 13). At this rotational speed, the stirrer power is approximately 6 kW; this is roughly 20 % of the maximum heat of reaction that can be removed. The auxiliary Diagram B shows that the range of rotational speeds with which 90 % of this maximum heat flow can be achieved, i.e., $R_{0.90} = 25.6$ kW, lies between 8 and 32 min^{-1}.

If more heat than the calculated R_{max} must be removed, the anchor stirrer with $D/d \simeq 1.00$ can be used instead of the one with $D/d = 1.10$; this mixer allows the removal of $R_{max} \simeq 60$ kW at a somewhat lower optimum rotation speed ($n = 17\ \mathrm{min}^{-1}$).

An alternative method to increase R_{max}, which is simpler to implement, is to select a more slender vessel. With $H/D = 2$, the tank diameter corresponding to the volume in the example ($V = 5.86\ \mathrm{m}^3$) is $D = 1.57$ m. (For a given V and H/D, the desired value of D can be obtained from auxiliary Diagram A by forming the product $(A/DH)\,(V/DA)\,(H/D) = V/D^3$ in this case $V/D^3 = 1.52$). With the new abscissa value, $\Pi_1\, Pr\,(H/D)^{-1} = 1.7 \times 10^8$, R_{max} is found to be $R_{max} \simeq 39$ kW at $n_{opt} = 19.7\ \mathrm{min}^{-1}$. At $H/D = 3$ ($D = 1.37$ m, $d = 1.25$ m), it is possible to remove $R_{max} = 46.5$ kW at $n_{opt} = 20.6\ \mathrm{min}^{-1}$. In this example, R_{max} increases with $\sqrt{H/D}$. The value of $R_{max} = 60$ kW calculated for the anchor stirrer with $D/d \simeq 1.00$, however, is reached for $D/d = 1.00$ only at $H/D \simeq 6$.

4. Suspension of Solids in Liquids

4.1. Problem Definition

In order to quickly dissolve solids in liquids, or to allow them to be fully effective in solid-catalyzed reactions, the entire surface of the solid must be accessible to the liquid; in other words, the solid particles must be suspended in the liquid. From the point of view of stirring, the quantities of interest are the critical rotational speed—the speed at which this state is just reached—as well as the other conditions (stirrer type and installation geometry) that allow this state of flow to be maintained with minimum power.

4.2. Suspension Characteristic

For a given stirrer type and installation geometry, the critical speed of rotation n_{crit} is generally a function of the stirrer diameter d, the mean particle diameter δ, the mass fraction of solids in the mixture ϕ, and the physical properties of the system (liquid density ϱ, liquid kinematic viscosity ν, specific weight difference between solid and liquid $g\,\Delta\varrho$): $n_{crit} = \mathrm{f}(d, \delta, \varrho, \nu, g\,\Delta\varrho, \phi)$. Dimensional analysis leads to the relationship $Re_{crit} = \mathrm{f}(Ar, \delta/d, \phi)$ where $Re_{crit} \equiv n_{crit}\, d^2/\nu$ is the Reynolds number based on the critical speed of rotation and $Ar \equiv d^3 g\,\Delta\varrho/\varrho\,\nu^2$ is the Archimedes number.

The suspension characteristic obtained by KNEULE and WEINSPACH [43] can be divided into two regimes:

1) $Re_{crit} \propto Ar^{0.4}\,(\phi\, d/\delta)^{0.2}$
$$Ar\,(\delta/d)^2\,\phi^{0.5} < 5 \times 10^4 \tag{23}$$

2) $Re_{crit} \propto Ar^{0.5}\,\phi^{0.25}$
$$Ar\,(\delta/d)^2\,\phi^{0.5} > 5 \times 10^4 \tag{24}$$

Since $Ar \equiv Re^2/Fr'$, Fr' being $n_{crit}^2\, d\varrho/g\,\Delta\varrho$, Regime 2 can also be described by

$$Fr'_{crit} = b\,\phi^{0.5} \tag{25}$$

which defines Fr' as the scale-up criterion and leads to the conclusion that $(P/V) = (P/V)_M \sqrt{\mu}$ (μ = scale factor, M (subscript) = model).

Later work by others has failed to confirm these relationships. When the more recent findings are expressed by the P/V relationship, the values obtained for the exponent of the scale factor μ range from -1 to 0.

According to VOIT and MERSMANN [44], there are two limiting laws that describe suspension, both represented by modified Froude numbers. A force balance on a suspended particle yields $Fr'\, d/\delta$ = constant; this condition applies to small particles ($\delta/D \leq 10^{-3}$), which are unlikely to reach the tank floor. The criterion $Fr'\, d/D$ = constant can be derived from an energy balance of the suspension process and applies to large particles ($\delta/D \geq 10^{-3}$), which have a higher probability of reaching the floor. The rotational speeds required for suspension can be well reproduced with these two criteria. These observations are also supported by experimental results of several authors.

KNEULE and WEINSPACH [43] have measured the suspension characteristics of numerous stirrer types with various installation geometries. They found the optimum stirrer diameter d and distance from the bottom h to be given by $D/d = 3.0-3.5$ and $h/d = 0.3-0.5$. The optimum shapes for the vessel bottom are hemispherical and elliptical; a flat vessel bottom is unsuitable for particle suspension. For a vessel with an elliptical bottom, baffles, and a propeller stirrer installed at $h/d = 0.2-0.8$ pumping the liquid towards the floor, the constant b in Equation (25) has the value $b = 3.06$. For a turbine stirrer with six paddles and $h/d = 0.3$, the value is $b = 1.21$. In order to keep the particles in the same material system in suspension, the propeller stirrer must therefore operate at a rotational speed $(3.06/1.21)^{1/2} = 1.59$ times higher than a turbine stirrer of the same size.

In suspending solids with simultaneous gas dispersion, the suspension characteristic corresponds to that in the two-phase system [45].

Reference [46] gives design guidelines for cases where the stirrer must distribute the solid quasi-homogeneously in the vessel; for example, this may be required to allow the transport of solids from stage to stage in continuous processes. Multi-stage stirrers are better suited for this operation than single mixers, with the cross-beam stirrer (beam angle $\alpha < 45°$) being especially effective [47].

Suspension characteristics for enameled stirrers with rounded edges are listed in [48].

One of the most difficult mixing operations is to stir into the liquid a nonwettable solid that is lighter than the liquid. The usual method consists of using a high-speed stirrer (propeller, turbine, sawtooth disk) in an unbaffled vessel and mixing in the solid through the vortex that extends to the stirrer. Another possibility is offered by self-aspirating rotor–stator systems, although design guidelines for these operations are not available yet.

4.3. Stirrer Power in the Suspension of Solids

Experiments [49] have shown that the flow behavior of suspensions with solid volume fractions ϕ' up to 0.25–0.30 approximates that of a Newtonian fluid. In this range of ϕ' values, the stirrer power required for suspension can be calculated from the corresponding power characteristics $Ne = f(Re)$ for homogeneous fluids (cf. Fig. 10); only the physical properties ϱ and η must be replaced with the effective values ϱ_s and η_s for the suspension:

$$\varrho_s = \varrho + \phi' \Delta\varrho$$

and

$$\eta_s = \eta[1 + 1.25\phi'/(1 - \phi'/\phi'_{max})]^2 \quad (26)$$

For spherical particles of the same size, $\phi'_{max} = 0.74$ and therefore $\eta_s = \eta[1 + 0.926\,\phi'/(0.74-\phi')]^2$. In the turbulent regime with Ne = constant, therefore, the stirrer power is given by $P = Ne\,n^3\,d^5\,\varrho_s$.

The optimum stirrer, in terms of energy, is the one that requires the least power at the critical speed of rotation. In terms of dimensional analysis, this is expressed as $Ne\,Fr_{crit}^{3/2}$ = minimum. For a propeller stirrer with Ne = 0.35 and a turbine stirrer with Ne = 5.0 (Table 1), and with the values of b aready given for the two stirrers, the propeller stirrer requires only 28 % of the power needed by the turbine stirrer.

4.4. Scale-up of the Solids Suspension Process

According to the suspension characteristics for the turbulent regime, the scale-up rule for a material system ($\Delta\varrho/\varrho$, δ, and ϕ = constant) is either that the tip speed u is constant or Fr is constant, depending on the value of the ratio δ/D. This results in the following rules for the critical speed of rotation in an industrial-scale facility:

u = constant, $n = n_M\,\mu^{-1}$
Fr = constant, $n = n_M\,\mu^{-1/2}$

For a scale factor of $\mu = 5$, the critical speed of rotation for the larger stirrer in the first case is only 20 % of that required on a laboratory scale, and in the second case 44.6 %.

In the mixing of fluid systems (liquid–liquid, gas–liquid), the power per unit volume P/V is of prime importance for the interfacial area per unit volume. Since solids are often suspended in these systems (e.g., hydrogenation with a suspended catalyst), the behavior of P/V with respect to both scale-up criteria must be examined. In the turbulent regime, $P \propto n^3 d^5$; for scale-up under geometric similarity ($V \propto d^3$), this results in $P/V \propto n^3 d^2$. Expressing the relationship between n and d in terms of the tip speed $u \propto nd$ yields the proportionality $P/V \propto \mu^3 d^{-1}$. In the second case, $n^3 d^2$ has to be replaced by

$(n^2 d)^{3/2} d^{1/2}$, and the relationship $P/V = (P/V)_M \mu^{1/2}$ is the result for $Fr \propto n^2 d$ = constant. The two regimes describing solid suspension are characterized by extremely different mixing intensities, and knowledge of the appropriate operating regime is therefore particularly important.

4.5. Mass Transfer in Solid–Liquid Systems

Mass transfer is described with the equation $G = k_s A \Delta c$, where G is the mass transferred in a given time, k_s the mass-transfer coefficient, A the interfacial area, and Δc the driving concentration difference. The mass-transfer coefficient k_s is obtained from the mass-transfer characteristic expressed by the general equation $Sh = f(Ar, Re, Sc, \delta/d)$ for solids suspension. In this expression, $Sh \equiv k_s d/\mathfrak{D}$ is the Sherwood number, $Ar = d^3 g \Delta\varrho/(\varrho \nu^2)$ the Archimedes number, $Re = n d^2/\nu$ the Reynolds number, $Sc \equiv \nu/\mathfrak{D}$ the Schmidt number, and $\mathfrak{D}$ the diffusion coefficient.

Figure 15 shows the first experimentally determined mass-transfer characteristic, which was obtained by Hixson and Baum [50]. The critical Reynolds number, and therefore n_{crit}, is clearly indicated by the change in slope of the curve. At rotational speeds lower than n_{crit} not only k_s but also the effective interfacial area A of the heap of solids lying more or less at the bottom of the vessel changes. Since only the total interfacial area A is known and used in the equation to determine k_s, only *apparent* values of k_s are measured in this regime. Above the critical rotation speed, only k_s changes as the liquid boundary layer at the particle surface is reduced with increasing turbulence.

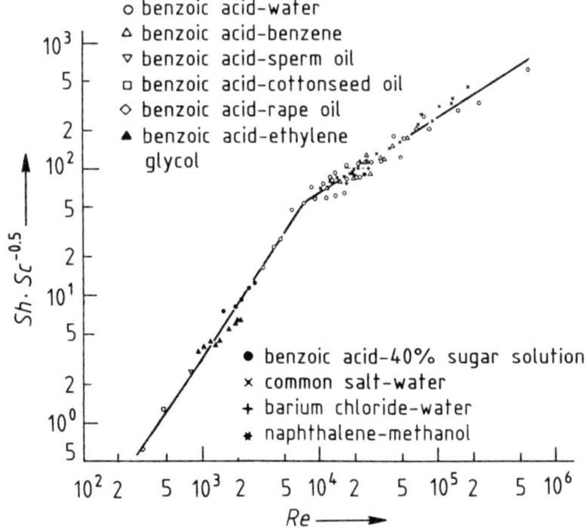

Figure 15. Mass-transfer characteristic for various solid–liquid systems [50]
The 4-blade propeller stirrer had the following dimensions: $H/D = 1$; $D/d = 3$; $h/d = 0.5$; $\alpha = 45°$. The height of the blades was 0.25 d

In the energetically favorable state of flow given by the critical Froude number, the turbulent and gravitational forces are in equilibrium. Under these conditions, the measured mass-transfer rates are the same as for particles moving freely under the effect of gravity. The value of the mass-transfer coefficient depends neither on the stirring conditions nor on the particle size: it is exclusively determined by the physical properties of the system (Section 5.5 and [63]). Weinspach [51] found that the relationship $Sh/Ar^{1/3} = 7.2 \times 10^{-2}\ Sc^{0.56}$ exactly matches his experimental mass-transfer characteristic for a heap in a vertical tube obtained at the point of incipient fluidization. From this relationship, it follows that $k_s \propto (\mathfrak{D}/\nu)^{0.44} (\Delta\varrho\, g\, \nu/\varrho)^{1/3}$.

In the regime $n > n_{crit}$, mass transfer in solid–liquid systems is correlated with the power dissipation per unit mass ε for various types and arrangements of stirrers:

$$Sh - 2 = a\,(\varepsilon \delta^4/\nu^3)^{\alpha}\, Sc^{1/3} \quad (27)$$

Various authors [52–54] have found $a = 0.4$–0.55 and $\alpha = 0.21$–0.26. The parameter $\Delta\varrho/\varrho$ is irrelevant in this case. The term in parentheses is the Reynolds number, as derived from Kolmogoroff's theory of isotropic turbulence. For high values of the Reynolds number, the particle diameter δ is irrelevant. Equation (27) was also found to apply to bubble columns over a wide range of viscosities ($100 < Sc < 50\,000$) [54]. The following relationship can therefore be recommended for mass-transfer calculations:

$$Sh - 2 = 0.55\,(\varepsilon \delta^4/\nu^3)^{0.25}\, Sc^{1/3} \quad (28)$$

5. Dispersion in Liquid–Liquid Systems

5.1. Problem Definition

Liquid–liquid dispersion aims at increasing the interfacial area between two immiscible liquids, in order to enhance the rate of processes such as extraction or chemical reaction (saponification, bead polymerization, etc.). Only the increase in interfacial area due to stirring is considered here; when the increase is produced by the addition of surfactants, the process is termed emulsification, for which completely different laws apply.

In general, the phase with the smaller volume is stirred into the phase with the larger volume (continuous phase) and broken up into droplets (dispersed phase). When the volume fractions of both phases are nearly the same, however, phase inversion can occur: which of the two phases becomes the continuous one depends on the starting conditions, as well as on the physical properties of the system [55].

In dispersion in liquid–liquid systems, the quantities of interest are the minimum speed of rotation required to disperse one phase completely into the other, as well as the interfacial area and the mass-transfer coefficient.

5.2. Dispersion Characteristic

In liquid–liquid dispersion a steady state is established, in which the formed droplets continuously coalesce into larger ones and are broken up by the stirrer into new droplets. The droplet size distribution of a dispersion generated by stirring is characterized by a mean droplet size δ, $\delta \equiv \Sigma n_i \delta_i^3 / \Sigma n_i \delta_i^2$ (n_i = number of droplets in class i; in the English-language literature, δ is referred to as the Sauter mean diameter d_{32} [56]). According to this definition, a simple relationship exists between δ and the interfacial area per unit volume of the dispersed phase $a \equiv A/V = \Sigma \pi \delta^2 / \Sigma (\pi \delta^3/6)$, namely $a = 6/\delta$.

For a given stirrer type and installation geometry, the steady-state values of δ and a are functions of the following variables: δ, $a = \mathrm{f}(n, d, \Delta\varrho g, \varrho', \varrho, \nu', \nu, \sigma, \phi')$. In this expression, n is the speed of rotation of the stirrer, d the stirrer diameter, $g\Delta\varrho$ the specific weight difference between the two phases, ϱ' and ϱ being the densities, and ν' and ν the kinematic viscosities, σ the interfacial tension, and ϕ' the volume fraction of the dispersed phase (the prime indicates the dispersed phase). In terms of dimensional analysis, the above relationship can be stated as follows: $\delta/d = \mathrm{f}(Re, Fr', We, \varrho'/\varrho, \nu'/\nu, \phi')$, where $Re \equiv nd^2/\nu$, $Fr' \equiv n^2 d\varrho / g\Delta\varrho$ (modified Froude number), and $We \equiv \varrho n^2 d^3/\sigma$ is the Weber number.

The dispersion characteristic of the turbine stirrer in Figure 8 G is

$$\delta/d = a\, We^{-0.6} (1 + b\phi') \tag{29}$$

where the numerical value of a has been generally found to lie between 0.05 and 0.08 [57], [58].

The minimum rotational speed required to mix in the dispersed phase is given by completely [59] as

$$Fr' = 36.1\, (Re\, We)^{-0.2} (1 + 3.5\phi')^{2.34} \tag{30}$$

and by [60] as

$$Fr'^{\,0.4} = C_s^2\, (\eta_c/\eta_d)^{2/9}\, (\Delta\varrho/\varrho)^{0.1}\, We^{-0.6} \tag{31}$$

where the values of the constant for various stirrer types can be found in [60].

5.3. Stirrer Power in Dispersion

The power characteristics are the same for homogeneous systems and liquid–liquid dispersion as long as the liquid density is replaced by the mean density of the dispersion, $\varrho_d = \varrho + \phi' \Delta\varrho$, and the viscosity is replaced by the weighted geometric mean of the viscosities of the two phases, $\eta_d = \eta'^{\phi'} \eta^{(1-\phi')}$ or $\eta_d = [\eta/(1-\phi')]\,[1 + 1.5\phi' \eta'/(\eta' + \eta)]$, the latter more easily interpreted physically [61], [62].

5.4. Scale-up of the Dispersion Process

The dispersion characteristic $\delta/d \propto We^{-0.6}$ yields the following scale-up criterion:

$$(\delta/d)\, We^{0.6} = \delta\, (\varrho/\sigma)^{0.6} (n^3 d^2)^{0.4} = \text{constant} \tag{32}$$

This results in the scale-up rule $n = n_M \mu^{-2/3}$ for the speed of rotation. Replacing the term $n^3 d^2$ with the equivalent expression $P/\varrho V$, which is valid in the turbulent regime, produces the relationship

$$\delta\, (\varrho^{0.2}/\sigma^{0.6})\, (P/V)^{0.4} = \text{constant} \tag{33}$$

Therefore, the mean droplet size, and the volumetric interfacial area a, remains unaltered if the same power per unit volume (P/V) is used in the scale-up.

5.5. Heat and Mass Transfer in Dispersion

In analogy to Section 4.5, the mass-transfer coefficient k in the interfacial boundary layer of the continuous phase can be described by the mass-transfer characteristic $Sh = f(Ar, Re, Sc, \delta/d)$, and the heat-transfer coefficient α can be described by the heat-transfer characteristic $Nu = f(Ar, Re, Pr, \delta/d)$. CALDERBANK [63] has shown that the characteristics are the same for all disperse systems; therefore, the following discussion (based on [63]) applies to both liquid–liquid and gas–liquid systems.

As the speed of rotation is increased, more power is dissipated in the liquid volume. As a result, smaller droplets or bubbles are formed, and the interfacial area per unit volume a increases. Under these conditions the mass-transfer coefficient k remains practically constant. As soon as the droplet or bubble size becomes less than 2.5 mm, the k values begin to fall rapidly. In the transition range $0.8\ \text{mm} < \delta < 2.5\ \text{mm}$, the value of k falls by a factor of approximately 4 and then remains constant, although further increase in the stirrer power results in smaller droplets. The reason for the large drop in k values in the regime $\delta < 2.5$ mm is the increasing hindrance of turbulence in the boundary layer. Under these conditions, the droplets or bubbles behave as near rigid spheres.

This description is supported by the following experimental relationships:

$\delta > 2.5$ mm. This gas bubble regime generally arises in the dispersion of gases in pure liquids (coalescing system) in stirred tanks or bubble columns. In the ranges $1 < Sc, Pr < 10^5$. the following relationships apply:

$$\mathrm{Sh}/Ar^{1/3} = 0.42\, Sc^{1/2}$$

$$Nu/Ar^{1/3} = 0.42\, Pr^{1/2}$$

When expressed in terms of k, this gives

$$k \propto (\mathfrak{D}/\nu)^{1/2}\,(\nu g \Delta\varrho/\varrho)^{1/3}$$

i.e., k is independent of both the flow conditions and δ, and is determined exclusively by the physical properties of the system.

$\delta < 2.5$ mm. This regime includes droplets formed during liquid–liquid dispersion, as well as gas bubbles produced in liquid mixtures or in the presence of surfactants (noncoalescing systems). Within the ranges $1 < Sc, Pr < 10^5$, the following relationships apply:

$$Sh/Ar^{1/3} = 0.31\, Sc^{1/3}$$

$$Nu/Ar^{1/3} = 0.31\, Pr^{1/3}$$

The mass-transfer characteristic in terms of k becomes

$$k \propto (\mathfrak{D}/\nu)^{2/3}\,(\nu g \Delta\varrho/\varrho)^{1/3}$$

Therefore,

$$k \propto \mathfrak{D}^{2/3}$$

In the extended range $0.1 < Ar\, Sc < 10^{12}$, the mass-transfer characteristic for $\delta < 2.5$ mm can be more precisely expressed as

$$Sh = 2 + 0.31\,(Ar\, Sc)^{1/3}$$

This relationship indicates that the Sherwood number takes the value 2 for $\Delta\varrho \simeq 0$, as is the case in natural convection.

6. Dispersion in Gas-Liquid Systems

6.1. Problem Definition

The mixing operation generally termed *gas–liquid contacting*, and most often meaning simply *aeration*, deals with increasing the interfacial area between liquid and gas. In the chemical industry this operation is employed in hydrogenation, chlorination, oxidation with air, etc. In fermentations and in aerobic wastewater treatment, the introduction of O_2 into the liquid is one of the most important process-engineering operations.

Two alternatives exist for the introduction of a gas in a liquid: through the surface or into the bulk of the liquid. The difference between these two methods plays a significant role in the choice of the equipment, as well as the absorption process itself. A distinction is therefore made between surface aerators and bulk aerators. Bulk

Table 3. Classification of aeration eqipment

Type of aeration	Equipment designation	Adjustable process parameters	
		No.	Type
Surface	*surface aerators*		
	turbine aerator	1	rotational speed
	brush aerator	1	rotational speed
	plunging jet aerator	1	water flowrate
	immersed jet aerator	1	water flowrate
Bulk	*bulk aerators*		
	stirrers		
	hollow stirrer	1	rotational speed
	stirrer with separate gas supply	2	rotational speed and gas flowrate
	gas distributors		
	Frings aerator	1(2)	rotational speed and gas flowrate
	sintered plates	1	gas flowrate
	static mixers	1	gas flowrate
	two-component nozzles	2	water and gas flowrates

aeration equipment must be subdivided into two classes. One subclass comprises the various stirrer types that directly determine the hydrodynamic state of the system, irrespective of the gas throughput. The second subclass consists of various gas distributors whose action is almost exclusively limited to the generation of primary gas bubbles. In this case, the hydrodynamic state is determined by the bubble swarm alone. This classification of aeration equipment is shown in Table 3.

6.2. Surface Aerators

Common to all four types in this category is the one free adjustable process parameter: for the first two types, the aerator's speed of rotation; for the last two, the water flowrate.

Mass transfer in gas–liquid systems is described by the general relationship

$$G = k_L A \Delta c \tag{34}$$

where G is the mass-transfer rate across the interface, k_L the liquid-side mass-transfer coefficient, A the interfacial area, and Δc the characteristic concentration difference. In surface aeration, Δc is given by

$$\Delta c = c_s - c \tag{35}$$

For example, the O_2 concentration in the liquid is assumed to be spatially uniform. The O_2 saturation concentration c_s, which is a function of the temperature and the partial pressure of O_2, can also be assumed to be constant.

The performance of a surface aerator is characterized by the mass-transfer coefficient $k_L A$ defined by Equation (34):

$$k_L A = G/\Delta c \tag{36}$$

6.2.1. Turbine-Type Surface Aerator

This surface aerator has been in use for decades in the biological wastewater treatment in basins with $H < 4$ m. Extensive measurements are available in the literature concerning O_2 uptake and power requirements with this stirrer type [64], [65]. By measuring the mass transfer in the same material system (water–air) with various sizes of turbine stirrers whose disks were positioned exactly in the surface of the liquid, the following sorption characteristic is obtained:

$$Y_1 = 1.41 \times 10^{-4} Fr^{1.2} Ga^{0.115} \tag{37}$$

$$0.02 \leq Fr \leq 0.34, \quad 1.5 \leq Ga \leq 200, \quad h/d = 0.2$$

In the above expression, $Y_1 \equiv [G/(d^3 \Delta c)] (\nu/g^2)^{1/3}$ represents the sorption number, $Fr \equiv n^2 d/g$ the Froude number, $Ga \equiv d^3 g/\nu^2$ the Galileo number, and h the immersed height of the stirrer blades.

The power characteristic of a turbine-type surface aerator is given by the relationship:

$$Ne \equiv P/(\varrho n^3 d^5) = \mathrm{f}(Fr) \tag{38}$$

as confirmed by measurements with geometrically similar turbine stirrers with diameter d from 180 to 350 mm [64].

By combining the sorption and power characteristics, the efficiency $E \equiv G/P$ of a surface aerator, in units of kg O_2 kW^{-1} h^{-1}, can be obtained in the form of the efficiency characteristic [64]:

$$(Y_1 Ne^{-1} Fr^{-1.5}) Ga^{-0.115} = \mathrm{f}(Fr) \tag{39}$$

where the expression in parentheses represents the dimensionless efficiency number E^*. An evaluation of the available experimental results with respect to Equation (39) shows that the relationship $E^* \propto Fr^{-0.35}$ applies. The decrease of E^* with Fr suggests that the Froude number should be kept as low as possible in industrial-scale equipment. If the scale-up is carried out at $Fr =$ constant, then Equation (39) predicts the following change in the efficiency:

$$E^* Ga^{-0.155} = G d^{0.155}/(P \Delta c) = \text{constant} \tag{40}$$

For constant Δc,

$$E = E_M \mu^{-0.155} \tag{41}$$

At a scale factor of $\mu = 4$ (or 10), the efficiency for large scale is 80% (or 70%) of that for the model. This relation has been confirmed by measurements with various turbine-type surface aerators, both on model and on industrial scales [65].

6.2.2. Plunging Water Jet Aerator

With few exceptions, these surface aerators have not been used industrially. Extensive laboratory experiments also show that this surface aerator type is inefficient.

E. van de Sande and J. M. Smith [66] have measured the absorption rate and penetration depth for plunging water jets with varying jet velocity and nozzle diameter (4 mm $< d <$ 12 mm). The distance between the nozzle and the water surface was kept constant ($H = 20$ cm). The evaluation of experimental material for vertical jets yields the following dimensionless relationship:

$$Y_1 = 0.15\, Fr'\, Ga^{-0.19} \qquad 2\times 10^1 < Fr' \equiv q^2/d^5 g < 4\times 10^3 \tag{42}$$

where the Froude number is formed with the liquid flowrate q [65]. Through transformation of Equation (42) and for $\Delta c =$ constant, the relationship $G \propto q^2/d^{2.57}$ is obtained. The power of a water jet is given by $P = q\Delta p$, where Δp can be replaced by $\varrho v^2/2$ for Bernoulli flow, giving $P \propto q^3/d^4$. Consequently, the efficiency follows the relationship $E \equiv G/P \propto d^{1.43}/q = d^{0.1}/p^{1/3}$. For a constant nozzle diameter, the efficiency of O_2 uptake drops with increasing liquid flowrate, thus also with increasing power consumption. For further details, see [65], where two surface stirrers (brush and immersed jet aerators) are also described.

6.3. Bulk Aerators

The stirrers are used in agitated tanks, while gas distributors are used in bubble columns and in wastewater treatment in pits, basins, and tanks.

Under the assumption of a quasi-uniform material system, mass transfer is best related to the volume of the liquid:

$$G/V = k_L a \Delta c \qquad a \equiv A/V \tag{43}$$

In bulk aeration, the characteristic concentration difference cannot be represented with $\Delta c = c_s - c$ in most cases because the saturation concentration c_s can hardly be expected to be spatially uniform.

On one hand, the system pressure (= atmospheric + hydrostatic pressures) varies with liquid depth, while on the other hand the mole fraction of the gas drops steadily, e.g., as a result of O_2 depletion in the air and the additional dilution of the gas by CO_2 from biological processes. All this can be taken into account by replacing Δc with the logarithmic mean concentration difference Δc_m:

$$\Delta c_m = \frac{c' - c''}{\ln\left(\dfrac{c' - c}{c'' - c}\right)} \tag{44}$$

The symbols are defined as follows:

$c' = c^\circ x' p'$	O_2 saturation cocentration at gas inlet
$c'' = c^\circ x'' p''$	O_2 saturation concentration at gas outlet
x', x''	O_2 mole fraction at gas inlet and outlet, respectively
p', p''	system pressure at gas inlet and outlet, respectively
c°	O_2 saturation concentration at the operational temperature, mg O_2/(L bar O_2)
c	O_2 concentration in the liquid

In particularly slender and/or tall aeration vessels (deep shaft, Tower Biology, etc.) the liquid cannot be assumed to be perfectly backmixed, which must be taken into account by modification of Equation (44).

The performance of a bulk aerator is characterized with the mass-transfer coefficient $k_L a$ as defined by Equation (43), with the mean logarithmic concentration difference given by Equation (44):

$$k_L a = G/(V \Delta c_m) \tag{45}$$

6.3.1. Stirrers as Bulk Aerators

Two basically different stirrer types are used as aerators: (1) self-aspirating hollow stirrers, which create suction by rotation and thus pump the gas and distribute it into the liquid; and (2) paddle stirrers with a radial flow pattern, in which the gas is introduced through a separate inlet under the stirrer. When hollow stirrers are used as aerators, their gas throughput and power consumption are important, i.e., the optimum operating conditions under which the desired gas flowrate can be obtained with the minimum power. When the gas is supplied to the bulk of the liquid separately, the mixing conditions needed to distribute this gas throughput and the power needed to achieve this must be determined. Since the goal of gas–liquid contacting is to produce a large interfacial area for gas absorption, the most important aspect is the dispersion characteristic, which is given here by the so-called sorption characteristic.

Hollow Stirrers [67–70]. Hollow stirrers operate on the principle of water jet ejectors. The less hydrodynamic their shape, the greater is the suction generated at the trailing edge, and therefore the gas throughput. The tube stirrer (Fig. 7) has proven to be the most effective hollow stirrer, as well as the simplest to construct. The stirrer can either draw external air or recirculate the gas of the tank's gas head, depending on whether the inlet holes in the shaft are located above or below the vessel cover. Recirculation is particularly useful since bubbling the gas only once through the liquid does not use it up completely. Hollow stirrers are particularly suitable in operations under pressure since no pumping is required for gas recirculation in chemical reactions at constant pressure.

In liquid–liquid–gas systems with large density differences between the liquids (bromina-

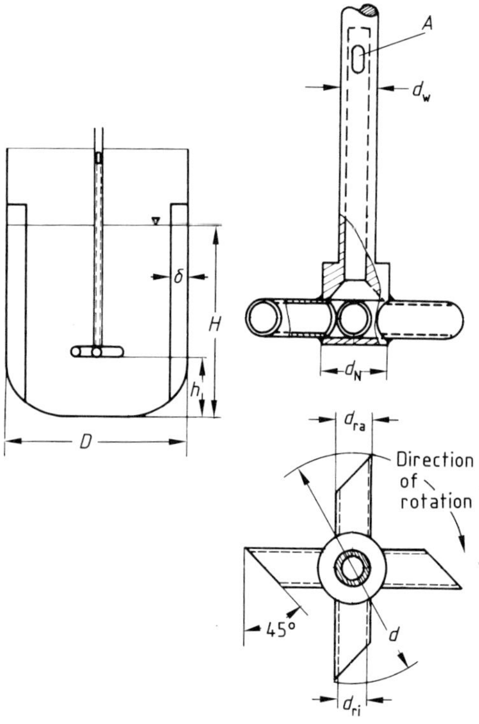

Figure 16. Installation and dimensions of a tube stirrer
Dimensions: $h/d = 1$; $H/D \approx 1$; $D/\delta = 10$;
$A = 1.5\,d_w^2$; $d/d_N = 10$; $d/d_N = 3$; $d/d_{ri} = 7.5$; $d/d_{ra} = 6$

tion, reactions with mercury), or in aeration in very slender autoclaves, the tube stirrer is designed so that it draws gas with two arms and recirculates the liquids with the other two [69]. In liquid–gas–solid systems, the (radial-flow) tube stirrer is unable to suspend heavy solid particles, e.g., catalyst particles. In such cases, it is combined with a propeller stirrer [71].

Being aspirated through a hollow stirrer, the gas must overcome the hydrostatic pressure of the liquid above the stirrer, $H^*\varrho g$. The liquid head over the stirrer $H^* \equiv (H - h)$ therefore has a significant influence on the gas flowrate q. A uniform aeration of the vessel contents can be achieved at $h/d = 1$, where h is the stirrer clearance from the bottom. Above $D/d \simeq 3$, the D/d ratio has no effect on aeration. To minimize the pressure drop in the hollow shaft, the shaft and the inlet openings are generously dimensioned (Fig. 16).

The surface tension σ does not affect gas flowrate. The same is true for the dynamic viscosity of the liquid η, as long as it is relatively low. For viscosities greater 0.5 Pa · s, however, the use of hollow stirrers for aeration becomes senseless. Since aeration deals with a material system with large density differences, the operation is strongly influenced by the parameter $g\,\Delta\varrho/\varrho \simeq g$ at normal pressures, and therefore by the Froude number.

The *gas throughput characteristic* of a hollow stirrer generally has the form $Q = f(Fr\,d/H^*, Ga, D/d, h/d)$, where $Q \equiv q/(nd^3)$ is the dimensionless flowrate number, $Fr \equiv n^2\,d/g$ the Froude number, and $Ga \equiv d^3\,g/\nu^2$ the Galileo number. For liquids with viscosities close to that of water and for $h/d = 1$, the gas throughput characteristic for a tube stirrer like that in Figure 16 are as follows:

$$Q^{-1} = 33\,(Fr\,d/H^*)^{-1.47} + 180\,(D/d)^{-4.1} \quad Fr\,d/H^* \leq 1.80 \tag{46}$$

$$Q^{-1} = 15.5\,(Fr\,d/H^*)^{-0.19} + 180\,(D/d)^{-4.1} \quad Fr\,d/H^* \geq 1.80 \tag{47}$$

The tube stirrer begins to draw in the gas at $Fr\,d/H^* = 0.156$. Its *power characteristic* is as follows:

$$Ne = [0.42 + \exp(-0.317\,Fr - 0.616)]\,(H^*/d)^{0.25} \tag{48}$$

The optimum operating conditions under which the desired gas flowrate is attained at the minimum stirrer power (per unit of gas flow) can be found by combining the dimensionless number Q, Ne, and Fr. The appropriate dimensionless group in this case is $Ne\,Fr/Q \equiv P/(q\varrho g d)$, and is likewise a function of $Fr\,H^*/d$. The optimum operating condition for the tube stirrer is given by $Fr\,d/H^* = 1.80$.

Turbine Stirrers. Stirrer power and maximum gas throughput are the two important factors.

Stirrer Power [72], [73]. As the flowrate of the gas supplied to the stirrer from below increases, the effective density of the gas-liquid mixture is reduced, and power consumption drops. In contrast to the tube stirrer, where the gas flowrate and the stirrer power are coupled through the speed of rotation, the gas flowrate here is an independent variable. The power characteristic consequently has the form $Ne = f(Q, Fr, Ga, D/d)$. This function is plotted in Figure 17 for a turbine stirrer (Figure 8 G). The flowrate number Q exerts the dominant influence on the power number Ne. The fitted curve is valid for $D/d \geq 3.33$ and $Ga \geq 10^7$, and corresponds to

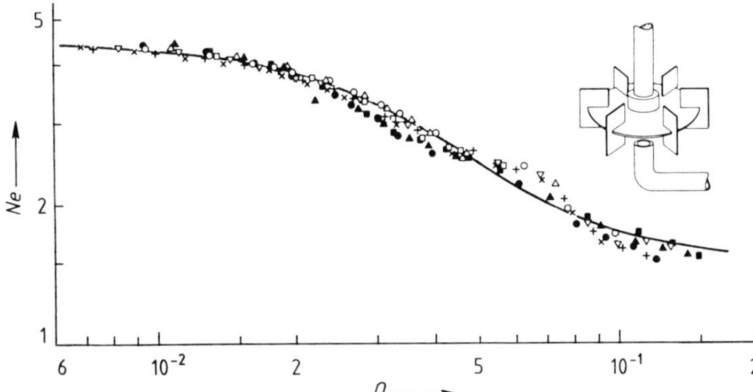

Figure 17. Power characteristic for a turbine stirrer in an aerated liquid [72]
The various symbols correspond to various pairs (Fr, D/d)

the expression:

$$Ne = 1.5 + (0.5\,Q^{0.075} + 1.6\times 10^3\,Q^{2.6})^{-1} \quad (49)$$

which is valid for $Re > 10^4$ and $Fr \geq 0.65$. The Froude number begins to have a slight effect for $Fr < 0.5$. The smallest values of Ne are observed shortly before flooding; under otherwise identical conditions, the stirrer power then is only about 40% of that in a nonaerated liquid.

Maximum Gas Throughput [69]. For a given speed of rotation, every stirrer can thoroughly distribute a gas only up to a well-defined maximum flowrate. When this limit is exceeded, the stirrer is said to be *flooded*, i.e., it is completely surrounded by gas and ceases to work. For the turbine stirrer in Figure 8 G, the maximum gas flowrate that can be dispersed well is given by:

$$Q = 0.19\,Fr^{3/4};\; 0.1 < Fr < 2.0 \quad (50)$$

The power requirements under these conditions can be calculated from the relationship $Ne = 1.36\,Fr^{-0.56}$. Equation (50) is valid for $D/d = 3.33$; for other values of D/d, see [73]. The power consumption and the maximum gas flowrate for two turbine stirrers on the same shaft are given in [74].

Sorption Characteristics. The dependence of the mass-transfer coefficient k_La (Eq. (45)) on physical properties and process parameters is given in a dimensionless form by the sorption characteristics. The name stems from the fact that both *absorption* and *desorption* are governed by the same laws [75].

Determination of k_La. According to whether the concentration of the dissolved gas c changes with time or remains constant, a distinction is made between steady-state and non-steady-state experimental conditions.

Non-steady-state methods often measure the concentration of dissolved oxygen versus time during absorption or desorption with oxygen electrodes. Such methods are consequently limited to O_2 or air in water or aqueous solutions. The so-called manometric method, on the other hand, can be used with any liquid and any gas, which, however, must be very pure [75].

Steady-state methods use reagents that instantaneously bind the dissolved O_2 and thus guarantee c = constant.

The oxidation of hydrazine [75], [76], $N_2H_4 + O_2 \xrightarrow{Cu^{2+}} H_2O + N_2$, functions for salt concentrations $\ll 1$ g/L. Under these conditions, the system coalesces. The O_2 concentration, however, must be $c \gg 0$ = constant, and be monitored with an O_2 electrode. The sulfite method, $Na_2SO_3 + 1/2\,O_2 \longrightarrow Na_2SO_4$, works satisfactorily with salt concentrations of > 10 g/L; under these conditions, the system is noncoalescing. The reaction must be catalyzed to proceed fast enough to give $c = 0$. It is only moderately catalyzed by Cu^{2+}. Although the chemical reaction is complete in the liquid bulk, physical absorption occurs in the presence of a concentration gradient of O_2 in the liquid boundary layer. Salts of Co^{2+} catalyze the reaction so strongly that complete reaction is accomplished within the liquid boundary layer (chemisorption). Thus, diffusive resistance is eliminated, and k_L is a constant that can be calculated from the kinetic parameters of the reaction.

Effect of Physical Properties on k_La. Of the usual physical properties, only the viscosity of the liquid has a strong influence on the sorption process. On the other hand, substances that affect the coalescence [77], [78] of gas bubbles also have a strong effect on k_La in low-viscosity liquids ($\eta < 50$ mPa · s).

Finely dispersed bubbles in pure liquids have the tendency to collect into larger ones. This process, called *bubble coalescence*, is caused by thinning of the liquid film (lamellae) between two neighboring bubbles until it collapses. This process is very fast in pure liquids and extremely slow in solutions that tend to foam. The reason for the thinning of the liquid film lies in the unequal capillary pressures between the film and the bulk of the liquid due to different radii of curvature and van der Waals forces. The pressure difference Δp is always in equilibrium with the difference in surface tension σ:

$$\delta\,\Delta p = 2\,\Delta\sigma; \quad \delta = \text{film thickness} \tag{51}$$

In solutions, the increase in surface area changes the surface concentration, which in turn causes the surface tension to change. This is described by Gibbs' law:

$$\Gamma_i = -\frac{c_i}{RT}\frac{d\sigma}{dc_i} \tag{52}$$

where Γ_i is the concentration excess or deficiency in the surface.

A positive $d\sigma/dc$ results in a deficiency of the solute in the surface, while a negative value results in an excess. However, according to the coalescence theory of MARUCCI [79], the substantial surface change taking place during an individual coalescence event is always accompanied by a large increase in the surface tension: the lamellae thin out to a few hundred ångströms before collapsing, the surface area per unit volume thus reaching up to $10^6\ cm^{-1}$. Therefore $d\sigma/dc$ greatly influences the dynamics of the film and is expected to be a suitable measure for the prediction of the coalescence behavior of a solution. In pure liquids, $d\sigma/dc = 0$, and no increase in surface tension, which could slow down or even stop the coalescence process, occurs in the thinning lamellae.

In the aeration of low-viscosity pure liquids such as water, methanol, or acetone, a stable bubble diameter of 3–5 mm results, irrespective of the type of the gas distributor. This state is reached immediately after the tiny primary bubbles leave the area of high shear forces. The generation of fine primary gas bubbles in pure liquids is therefore uneconomical.

In salt solutions or mixtures of miscible liquids, the coalescence of tiny primary gas bubbles is suppressed significantly: the higher the concentration of the solution, the better the size of the primary gas bubbles is preserved. The stable bubble size in this case is 0.2–0.5 mm, an order of magnitude smaller than in pure liquids.

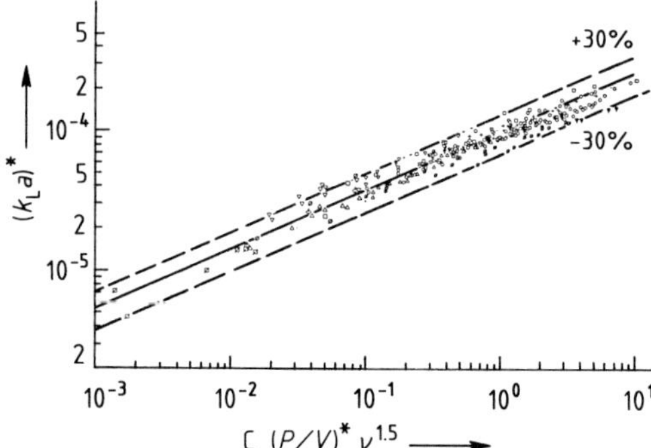

Figure 18. Sorption characteristics for aqueous coalescing systems in mixing vessels with a turbine stirrer [83]
$(k_La)^* \equiv k_La\,(\nu/g^2)^{1/3}$ = dimensionless mass-transfer coefficient;
$(P/V)^* \equiv (P/V)/[\varrho\,(\nu g^4)^{1/3}]$ = dimensionless mixing power per unit volume
where C is a constant

Due to *coalescence suppression*, the enhancement factor of physical sorption $m = (k_La)_{sol}/(k_La)_{solv}$ rises to 7 or 8, which has been confirmed by measurements of k_La as a function of the concentration of various inorganic salts (both strong and weak electrolytes) as well as normal aliphatic alcohols (methanol to octanol) [80], [86].

On the other hand, some substances, even in minute concentrations, promote coalescence. Some of the nonionic surfactants often used as antifoaming agents are so effective that a concentration of 3 ppm is sufficient to halve the k_La of pure water [87]. Clearly the use of surfactants for foam prevention should be avoided whenever possible, and an efficient mechanical foam breaker used instead [81].

Effect of Process Parameters in k_La. Traditionally, k_La values measured in a given system have been correlated as a dimensional quantity in the form

$$k_La = f(P/V,\ v) \tag{53}$$

where P/V is the stirrer power per unit volume and v the superficial velocity. After evaluating 76 original works, K. VAN'T RIET [82] arrived at two numerical equations (SI units).

For coalescing systems:

$$k_La = 2.6 \times 10^{-2}\,(P/V)^{0.4}\,v^{0.5} \tag{54}$$

For noncoalescing systems:

$$k_La = 2.0 \times 10^{-3}\,(P/V)^{0.7}\,v^{0.2} \tag{55}$$

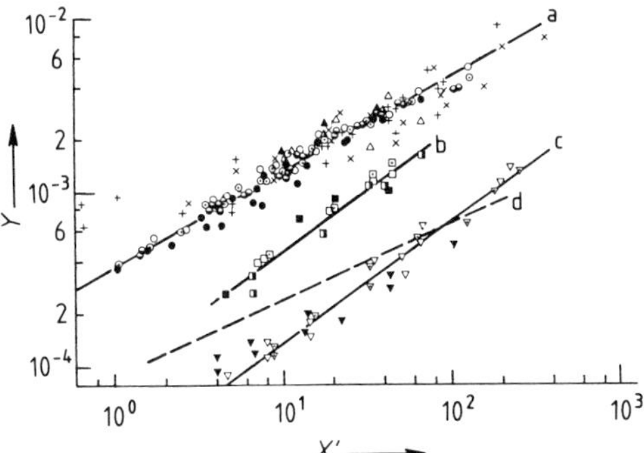

Figure 19. Sorption characteristics for systems with various rheological behaviors [84]
a) Aqueous carboxymethylcellulose (CMC) solution; b) Millet jelly; c) Aqueous glucose solution; d) Water

JUDAT [83] has compared the results of nine experimental works on the coalescing air–water system made in tanks with turbine stirrers with the volume varying from 2.5 L to 906 m^3 (0.15 m $\leq D \leq$ 12 m). His correlation is

$$k_L a \propto (P/V)^{0.4} v^{0.6} \qquad (56)$$

which is shown in dimensionless form in Figure 18.

HENZLER [84] was able to correlate the same experimental results by using the dimensionless group $Y \equiv (k_L a/v)\ (v^2/g)^{1/3}$ as ordinate and $X' \equiv P/(V \varrho g v)$ as abscissa (Fig. 19). The dimensionless group Y is the determining sorption number for bubble columns, while X' represents the ratio of the mechanical power of the stirrer to the hydraulic power of the gas throughput. The determining sorption number Y also allows comparison between mixing tanks and bubble columns [85].

The effect of viscosity seems to be appropriately taken into account by Y. The curves for different systems, however, do not coincide since each system exhibits a different coalescence behavior.

ZLOKARNIK [75] has measured the sorption characteristics of various hollow stirrers for the scale-up range $\mu = 5{:}1$, using 1 N aqueous sodium sulfite solution and air with Cu^{2+} as catalyst. These conditions give rise to a system where coalescence is completely suppressed. As a result, the parameter P/V has the dominating influence, while the gas flowrate plays no role at all:

$$k_L a \propto (P/V)^{0.8} \qquad (57)$$

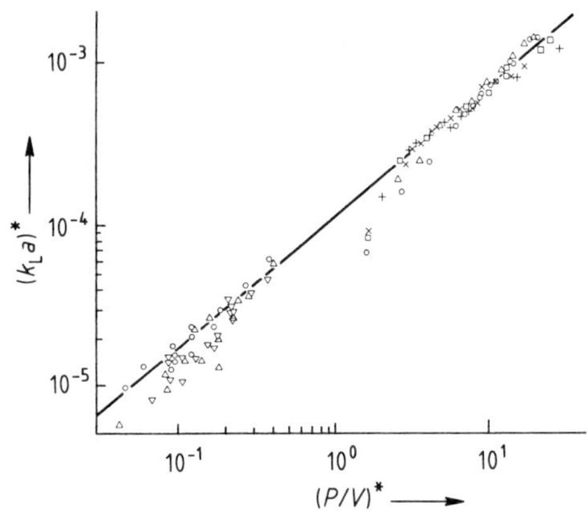

Figure 20. Sorption characteristics for noncoalescing systems [75]
$(k_L a)^* \equiv k_L a\,(v/g^2)^{1/3}$ is a dimensionless mass-transfer coefficient.
$(P/V)^* \equiv (P/V)[\varrho\,(v g^4)^{1/3}]$ is a dimensionless mixing power per unit volume. The points along the curve cover a fivefold scale-up factor.

The experimental results are presented in Figure 20 in dimensionless form.

6.3.2. Gas Distributors

The effect of gas distributors is limited almost exclusively to the production of primary gas bubbles so that the hydrodynamic state of the liquid volume is determined only by the rising bubble swarm. Vessels equipped with gas distributors can therefore be treated as bubble columns.

Mass transfer in vessels with gas distributors (perforated pipes, sintered bodies, static mixers) whose only free process parameter is the gas flowrate can be described in the same way as in bubble columns [86]:

$$k_L a \propto v \qquad (58)$$

where v is the superficial velocity.

Replacing $k_L a$ with $G/(V \Delta c_m)$ (Eq. (45)) and v with q/A (A = cross section of the vessel) yields the following relationship:

$$\frac{G}{V \Delta c_m} \cdot \frac{A}{q} = \frac{G}{H \Delta c_m q} = \text{constant} \qquad (59)$$

since $V = AH$. Therefore, the mass transfer rate G through the interface is directly proportional to the gas flowrate q, the liquid depth H, and the concentration difference Δc_m: $G \propto q H \Delta c_m$.

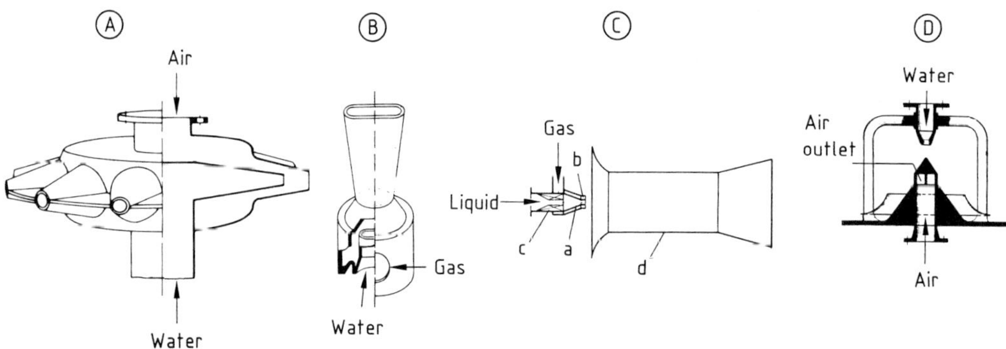

Figure 21. Two-component nozzles in biological wastewater treatment
A) EMJA-D cluster of Penberthy-Houdaille; B) Slot injector of Bayer AG; C) Jet nozzle of BASF; D) Radial flow nozzle of Hoechst AG
a) Propulsion jet nozzle; b) Air nozzle; c) Twisting element; d) Momentum tranfer tube

The performance of the Frings aerator, sintered plates, and Kenics-type static mixers as gas distributors in coalescing systems (e.g., in activated sludge basins of a wastewater treatment plant) is discussed in [86].

Two-component nozzles (injectors, jet ejectors, etc.) are devices in which a gas continuum is dispersed into bubbles by a high-energy liquid jet. Figure 21 shows some of the two-component nozzles used in the aeration of activated sludge basins.

In addition to the gas flowrate q, two-component nozzles (injectors) also have the liquid flowrate q_L as a second freely adjustable process parameter. In this case, $k_L a/v$ is not a constant but is a function of q/q_L. In order to consistently follow the sorption characteristics for stirrers, the $k_L a$ values for injectors are related to the liquid jet power ($P_L = q_L \Delta p_L$) per unit gas flowrate [87]:

$$k_L a = f(P_L/q) \qquad (60)$$

Design guidelines have only been published for the slot injector and the radial flow nozzle [80], [87]. Figure 22 shows the sorption characteristics of the slot injector in this dimensionless form. The coalescence behavior of the system is determined by the NaCl concentration in water [87]. These sorption characteristics, along with the pressure-drop characteristic of the slot injector, allow the calculation of the liquid flowrate q_L required for a desired oxygen uptake G for any gas flowrate and for given values of H and Δc. A desired oxygen uptake G can be achieved with any number of pairs of q and q_L, but there is only

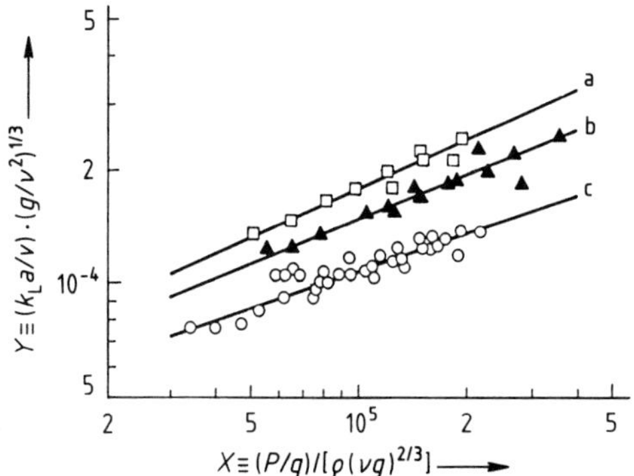

Figure 22. Sorption characteristics for the slot injector (Fig. 21 B) as a function of the coalescence behavior of the system [87]
NaCl concentration, g/L: a) 10; b) 5; c) 0 (water)

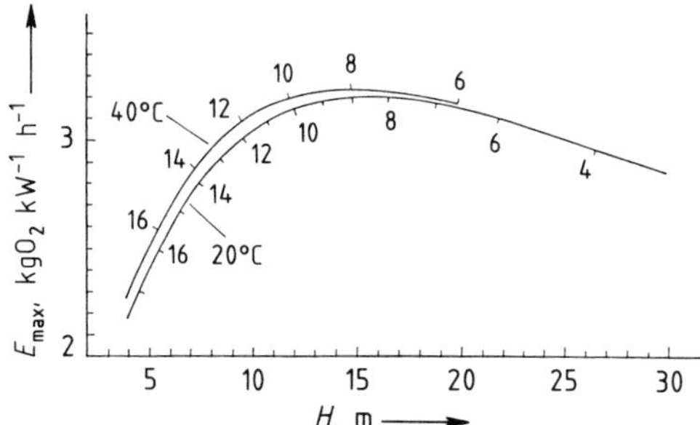

Figure 23. Maximum attainable efficiency for the slot injector: E_{max} as a function of the liquid height H for water–air [78]
$G/V = 0.1$ kg O_2 m^{-3} h^{-1}, 1 injector per 10 m^2; $c =$ 1 mg O_2/L
The numbers along the two curves are the residual oxygen, vol%, in the off-gas.

one pair (q, q_L) for which ΣP is the minimum, and thus $E = G/\Sigma P$ is a maximum.

In determining $\Sigma P = P + P_L$ an efficiency of $\eta = 0.60$ is assumed for the blower and $\eta = 0.75$ for the pump.

This maximum attainable efficiency for the slot injector is shown in Figure 23 as a function of the liquid depth H. The maximum value E_{max} occurs for $H = 15$ m, with the off-gas still containing 8 vol% O_2. If the utilization of O_2 must be increased to minimize the amount of purged gas, the liquid depth must be increased: at $H = 26$ m the utilization of O_2 is 80%, and the off-gas contains only 4 vol% O_2. These considerations are the basis for the design of tower-shaped bioreactors [80].

Two-component nozzles disperse the gas using only the liquid jet, while a stirrer must keep the entire contents of the vessel in motion to achieve the same result. This makes injectors more efficient than stirrers, as long as the objective is only to disperse the gas and not to achieve a simultaneous fast homogenization of the liquid or an intensive heat transfer.

At high aeration rates, some gas bubbles adhere to the wall and insulate it, leading to heat-transfer coefficients approximately 10% lower than in nonaerated liquids [88].

7. Mixing of Non-Newtonian Liquids

7.1. Basic Definitions

The design guidelines described in the previous sections are strictly valid only for liquids whose viscosities obey Newton's law of friction, $\tau = \eta \dot{\gamma}$ (η = viscosity, τ = shear stress, $\dot{\gamma}$ = shear rate), and are therefore only functions of temperature. In the last decades, however, an increasing number of liquids have come into use whose viscosity is also a function of the shear rate; such liquids are termed *non-Newtonian*. These substances have an entirely different viscosity behavior illustrated by plots of the function $\tau(\dot{\gamma})$, called flow curves. When the viscosity decreases as $\dot{\gamma}$ increases, the liquid is termed pseudoplastic. In the opposite case, the liquid is called dilatant. If a log–log plot of the flow curve is a straight line, the liquid is said to obey the so-called power law by Ostwald–de Waele, $\tau = K\dot{\gamma}^m$. Since $\eta = \tau/\dot{\gamma}$, this leads to $\eta = K\dot{\gamma}^{m-1}$, where K and m are rheological constants.

Pastes are characterized by a lower flow limit, which distinguishes them from other non-Newtonian liquids. The friction law in this case is $\tau - \tau_0 = \eta_{plas}\dot{\gamma}$; η_{plas} = constant corresponds to a so-called Bingham plastic.

A further class of non-Newtonian fluids consists of long-chain polymers (e.g., polyacrylamide, polyacrylonitrile), which exhibit viscoelastic flow behavior: the fluid strands resist mechanical shear by contracting like elastic bands. This generates secondary flow patterns [89] opposite to the flow pattern caused by inertial forces, e.g., the creeping of a fluid up a stirrer shaft, the so-called Weissenberg effect.

Current research in stirring aims to express non-Newtonian behavior in such a way that the relationships obtained for Newtonian fluids can also be applied to non-Newtonian fluids.

7.2. Stirrer Power in Non-Newtonian Fluids

METZNER and OTTO [90] were the first to examine the power consumption of a turbine stirrer (Fig. 8 G) in liquids whose viscosities are adequately described by the power law. They found the mean shear rate to be proportional to the speed of rotation of the stirrer $\dot{\gamma} = kn$, with $k = 13$. Therefore, the Reynolds number becomes $Re' = n^{2-m} d^2 \varrho / K$, and the power characteristic $Ne = \mathrm{f}(Re', m)$.

An analogy between the drag characteristic for pipe flow of a non-Newtonian fluid and the power characteristic suggests the function $\mathrm{f}(Re', m)$ is

$$\mathrm{f}\left[\frac{n^{2-m} d^2 \varrho}{K} 8 \left(\frac{m}{6m+2}\right)^m\right] \tag{61}$$

For $K = \eta$ and $m = 1$, the expression reduces to $\mathrm{f}(Re = n d^2 \varrho / \eta)$. CALDERBANK [91] determined the value of the rheological constant to be $K = 1.25$ for $m < 1$. He extended the measurements for blade and propeller stirrers and established the relationship $\dot{\gamma} = 10\, n$ for pseudoplastic fluids and Bingham plastics; however, for dilatant fluids he found that $\dot{\gamma} = 12.8\, n\, (D/d)^{0.5}$. For anchor stirrers [92] operating in the range $Re < 100$, the relationship $Ne = 82\,[Re'/\mathrm{f}(m)]^{-0.93}\,(1 - d/D)^{-1/4}$ was found to be valid, where $\mathrm{f}(m) = [a(1-m)]^{m-1}$ and a is a function of D/d.

The power characteristic for viscoelastic fluids [93], [94] can also be described accurately enough as $Ne(Re')$. Additional parameters to account for viscoelastic properties are needed only when the ratio of tangential to normal stresses is less than 2 [93].

Power characteristics for the turbine stirrer in aerated non-Newtonian fluids (carboxymethylcellulose (CMC) and polyacrylamide (PAA) solutions) have been presented [95].

7.3. Homogenization of Non-Newtonian Liquids

Homogenization characteristics of the form $n\theta = \mathrm{f}(Re_{\mathrm{eff}} \equiv n d^2 \varrho/\eta_{\mathrm{eff}})$ have been published for various stirrer types and a pseudoplastic liquid with power-law behavior [96], [97]. The homogenization time is compared to that for Newtonian fluids [96], and the homogenization properties of a propeller stirrer in agitated tanks are compared with those in loop reactors [97].

7.4. Heat Transfer in Non-Newtonian Liquids

The heat transfer between pseudoplastic fluids and the tank wall with a paddle stirrer (four paddles at 45°) can be well characterized with the usual relationship $Nu = a[Re'\,\mathrm{f}(m)]^{2/3}\,Pr^{1/3}$, where a = 3.41 for heating and 1.43 for cooling. The expression given by Metzner [90] must be used for $Re'\,\mathrm{f}(m)$ while the Prandtl number $Pr \equiv c\eta/\lambda$ must be calculated with the constant viscosity value at high shear rates. The heat-transfer rates, however, can be better correlated by defining the Prandtl number in a way analogous to $Re'\,\mathrm{f}(m)$: corresponding relationships for various stirrer types and non-Newtonian fluids can be found in the review [98].

7.5. Scale-Up with Non-Newtonian Fluids

The rheological properties of many different materials can be adequately described by the Ostwald–de Waele power law. Nevertheless, this representation is unsatisfactory because it fails to describe fluids with more complex rheological properties and because the dimensions of the parameter K depend on the exponent m of the power law, thus making it a physically inconsistent quantity. The development of better rheological equations of state, however, involves great experimental difficulties. From the process-engineering viewpoint, therefore, the generalized dimensional analysis [99], [100] of this problem is of special importance.

An analysis [99] of the rheological parameters shows that they can be reduced to two parameters having dimensions, even for fluids with most complex behavior. These are H, with dimension of viscosity, and θ, with dimension of time. In addition to these two parameters, a number of dimensionless parameters Π'_k representing physical properties may be involved.

To obtain governing relationships of non-Newtonian fluids as extensions to those already available for Newtonian liquids, the relationship between the dimensionless process equation for a Newtonian liquid, $\mathrm{f}(\Pi_1, \Pi_2, \ldots, \Pi_n) = 0$, and the equation for the non-Newtonian fluid, $\mathrm{f}'(\Pi_1^*, \Pi_2^*, \ldots, \Pi_j^*) = 0$ is of interest. In each case, Π represents a dimensionless group.

The correspondence between the two complete Π sets can be obtained by dividing the Π groups of the Newtonian case into two subsets: those that contain at least one process parameter, and those that consist of physical properties only. When making the extension from a Newtonian to a non-Newtonian fluid, the following must be observed:

1) All Π groups in the Newtonian set also appear in the non-Newtonian set, with η being replaced by H in groups containing the viscosity.
2) The subset of dimensionless process groups is extended by a maximum of one additional group involving θ.
3) All dimensionless groups Π'_k belong to the subset of physical property groups.

For experimental measurements dealing with one rheological substance, as is usually the case in model studies (i.e., Π'_k = constant), the dimensional-analysis framework is extended by a single additional dimensionless group. The power characteristic for a non-Newtonian fluid, for example, has the form $\mathrm{f}'[P/(\varrho n^3 d^5), nd^2\varrho/H, n\theta] = 0$; the heat-transfer characteristic is given by $[\alpha D/\lambda, nd^2\varrho/H, n\theta] = 0$, where the Prandtl number $Pr \equiv c_{\mathrm{p}}H/\lambda$ is also constant.

The dimensional analysis of stirring problems involving non-Newtonian fluids makes it possible to develop design guidelines for industrial-scale equipment by conducting experiments with the same fluid in smaller models.

8. References

Chapter 2

[1] P. V. Danckwerts: "Measurement of molecular homogeneity in a mixture," *Chem. Eng. Sci.* **7** (1957) 116–117.
[2] J. W. Hiby: "Homogenisation," *Fortschr. Verfahrenstec. B* **17** (1979) 137–155.
[3] M. Zlokarnik: "Eignung von Rührern zum Homogenisieren von Flüssigkeitsgemischen," *Chem. Ing. Tech.* **39** (1967) nos. 9/10, 539–548.
[4] C. J. Hoogendoorn, A. P. den Hartog: "Model studies on mixers in the viscous flow region," *Chem. Eng. Sci.* **22** (1967) 1689–1699.
[5] H. Gramlich, S. Lamadé: "Vergleich emaillierter Rührsysteme zum Homogenisieren von Flüssigkeitsgemischen," *Chem. Ing. Tech.* **45** (1973) no. 3, 116–122.
[6] H.-J. Henzler: "Untersuchungen zum Homogenisieren von Flüssigkeiten oder Gasen," *VDI-Forschungsh.* **587** (1978).
[7] M. Zlokarnik: "Einfluß der Dichte- und Zähigkeitsunterschiede auf die Mischzeit beim Homogenisieren von Flüssigkeitsgemischen," *Chem. Ing. Tech.* **42** (1970) no. 15, 1009–1011.
[8] W. Büche: "Leistungsbedarf von Rührwerken," *VDI-Z.* **81** (1937) no. 37, 1065–1069.
[9] J. H. Rushton, E. W. Costich, H. J. Everett, : "Power characteristics of mixing impellers," *Chem. Eng. Prog.* **46** (1950) no. 8, 395–404, and no. 9, 467–476.
[10] S. Nagata, K. Yamamoto, T. Yokoyama: "Studies on the power requirement of mixing impellers," *Mem. Fac. Eng. Kyoto Univ.* **19** (1957) no. 3, 274–290.
[11] R. L. Bates, P. L. Fondy, R. R. Corpstein: "An examination of some geometric parameters of impeller power," *Ind. Eng. Chem. Process Des. Dev.* **2** (1963) no. 4, 310–314.
[12] A. Mersmann et al.: "Auslegung und Maßstabsvergrößerung von Rührapparaten," *Chem. Ing. Tech.* **47** (1975) no. 23, 953–964.
[13] J. Y. Oldshue, H. E. Hirschland, A. T. Gretton: "Blending of low-viscosity liquids with side-entering mixers," *Chem. Eng. Prog.* **52** (1956) no. 11, 481–484.
[14] J. G. von de Vusse: "Vergleichende Rührversuche zum Mischen löslicher Flüssigkeiten in einem 12000 m^3-Behälter," *Chem. Ing. Tech.* **31** (1959) no. 9, 583–587.
[15] J. A. Wesselingh: "Mixing of liquids in cylindrical storage tanks with side-entering propellers," *Chem. Eng. Sci.* **30** (1975) 973–981.
[16] H. Blasiński, J. Nowicki, E. Rzyski: "Mixing power and mixing times of propeller agitators introduced laterally," *Int. Chem. Eng.* **20** (1980) no. 1, 92–97.
[17] H. Fosset, L. E. Prosser: "The application of free jets to the mixing of fluids in bulk," *J. Instn. Mech. Engr. London* **160** (1949) 224–232.
[18] L. S. Harris: "Jet eductor mixers," *Chem. Eng. N.Y.* **73** (1966) no. 21, 216–222.
[19] R. G. Folsom, C. K. Ferguson: "Jet mixing of two liquids," *Trans. ASME* **71** (1949) 73–77.
[20] M. Zlokarnik: "Homogenisieren von Flüssigkeiten durch aufsteigende Gasblasen," *Chem. Ing. Tech.* **40** (1968) no. 15, 765–768.
[21] J. M. Smith, L. H. J. Goosens: "The mixing of ponds with bubble columns," *Proc. 4th Eur. Conf. Mixing,* NL-Leeuvenhorst 1982, 71–80.
[22] J. W. Hiby: "Mischung in turbulenter Rohrströmung," *Verfahrenstechnik (Mainz)* **4** (1970) no. 12, 538–543.
[23] O. Bolzern, J. R. Bourne: "Rapid chemical reactions in a centrifugal pump," *Chem. Eng. Res. Des.* **63** (1985) no. 5, 275–282.
[24] H.-J. Henzler, "Eignung von kontinuierlich durchströmten Mischern zum Homogenisieren," *Chem. Ing. Tech.* **51** (1979) no. 1, 1–8.
[25] F. Streiff: "Statische Mischer mit großer Anpassungsfähigkeit," *Chem. Ing. Tech.* **52** (1980) no. 6, 520–522.
[26] M. H. Pahl, E. Muschelknautz: "Einsatz und Auslegung statischer Mischer," *Chem. Ing. Tech.* **51** (1979) no. 5, 347–364.
[27] M. H. Pahl, E. Muschelknautz: "Statische Mischer und ihre Anwendung," *Chem. Ing. Tech.* **52** (1980) no. 4, 285–291.
[28] I. Fort: "Pumping capacity of propeller mixer," *Collect. Czech. Chem. Commun.* **32** (1967) 3663–3678; **34** (1969) 1094–1097.
[29] I. Fort, J. Podivinska, R. Baloun: "The study of convective flow in a system with a rotary mixer and baffles," *Collect. Czech. Chem. Commun.* 34 (1969) 959–974.
[30] R. G. Cooper, D. Wolf: "Velocity profiles and pumping capacities for turbine type impellers," *Can. J. Chem. Eng.* **46** (1968) 94–100.
[31] Y. Sano, H. Usui: "Interrelations among mixing time, power number and discharge flow rate number in baffled mixing vessels," *J. Chem. Eng. Japan* **18** (1985) no. 1, 47–52.
[32] M. Zlokarnik: "Trombentiefe beim Rühren in unbewehrten Behältern," *Chem. Ing. Tech.* **43** (1971) no. 18, 1028–1030.
[33] A. Le Lan, H. Angelino: "Etude du vortex dans les cuves agitées," *Chem. Eng. Sci.* **27** (1972) 1969–1978.
[34] F. Rieger, P. Ditl, V. Novák: "Vortex depth in mixed unbaffled vessels," *Chem. Eng. Sci.* **34** (1979) 397–403.

Chapter 3

[35] R. Poggemann, A. Steiff, P.-M. Weinspach: "Heat Transfer in agitated vessels with single-phase liquids," *Ger. Chem. Eng. (Engl. Transl.)* **3** (1980) 163–174.
[36] M. F. Edwards, W. L. Wilkinson: "Heat transfer in agitated vessels, Part I: Newtonian fluids," *Chem. Engr. (London)* (1972) no. 264, 310–319.
[37] V. W. Uhl, H. P. Voznick: "The anchor agitator," *Chem. Eng. Prog.* **56** (1960) no. 3, 72–77.
[38] M. Zlokarnik: "Wärmeübergang an der Wand eines Rührbehälters beim Kühlen und Heizen im Bereich $10^{\circ} < Re < 10^5$," *Chem. Ing. Tech.* **41** (1969) no. 22, 1195–1202.
[39] H. Judat, to be published.
[40] M. Kuriyama et al.: "Heat transfer and temperature distributions in an agitated tank equipped with heli-

cal ribbon impeller," *J. Chem. Eng. Japan* **14** (1981) no. 4, 323–330.

[41] J. Pawlowksi, M. Zlokarnik: "Optimieren von Rührern für eine maximale Abfuhr von Reaktionswärme," *Chem. Ing. Tech.* **44** (1972) no. 16, 982–986.

[42] W. R. Penney, R. N. Koopman: "Prediction of net heat removal capabilities for agitated vessels," *AIChE Symp. Ser.* **68** (1972) no. 118, 62–73.

Chapter 4

[43] F. Kneule, P. M. Weinspach: "Suspendieren von Feststoffpartikeln im Rührgefäß," *Verfahrenstechnik (Mainz)* **1** (1967) no. 12, 531–540.

[44] H. Voit, A. Mersmann: "General statement for the minimum stirrer speed during suspension," *Ger. Chem. Eng. (Engl. Transl.)* **9** (1986) 101–106.

[45] J. A. Wiedmann, A. Steiff, P. M. Weinspach: "Fluid dynamics of stirred three-phase reactors," *Ger. Chem. Eng. (Engl. Transl.)* **4** (1981) 125–136.

[46] W.-D. Einenkel: "Beschreibung der fluiddynamischen Vorgänge beim Suspendieren im Rührwerk," *VDI-Forschungsh.* **595** (1979).

[47] W. Müller, D. Pysall: "Das Suspendieren von Feststoff mittels Balkenrührern in Stufenanordnung," *Chem. Ing. Tech.* **58** (1986) no. 6, 508–509.

[48] S. Lamadé: "Auswahl und Auslegung emaillierter Rührer für das Aufwirbeln von Feststoffen in Flüssigkeiten," *Verfahrenstechnik (Mainz).* **11** (1977) no. 2, 72–81.

[49] P. M. Weinspach: "Hydrodynamisches Verhalten von Suspensionen im Rührgefäß," *Chem. Ing. Tech.* **41** (1969) nos. 5/6, 260–265.

[50] A. W. Hixson, S. J. Baum: "Mass transfer coefficients in liquid–solid agitation systems," *Ind. Eng. Chem.* **33** (1941) no. 4, 478–485.

[51] P. M. Weinspach: "Der Lösevorgang im Fließbett und im Rührgefäß," *Chem. Ing. Tech.* **39** (1967) nos. 5/6, 231–236.

[52] D. M. Levins, J. R. Glastonbury: "Particle-liquid hydrodynamics and mass transfer in a stirred vessel," *Trans. Inst. Chem. Eng.* **20** (1972) 132–146.

[53] Y.Sano, N. Yamaguchi, T. Adachi: "Mass transfer coefficients for suspended particles in agitated vessels and bubble columns," *J. Chem. Eng. Japan* **7** (1974) no. 4, 255–261.

[54] P. Sänger, W.-D. Deckwer: "Liquid-solid mass transfer in aerated suspensions," *Chem. Eng. J. (Lausanne)* **22** (1981) no. 3, 179–186.

Chapter 5

[55] A. H. Selker, C. A. Sleicher: "Factors affecting which phase will disperse when immiscible liquids are stirred together," *Can. J. Chem. Eng.* **53** (1965) no. 4, 298–301.

[56] J. Sauter: "Die Größenbestimmung von Brennstoffteilchen,"*Forschungsarbeiten,* **279** (1926).

[57] C. A. Coulaloglou, L. L. Tavlarides: "Drop size distributions and coalescence frequencies of liquid–liquid dispersions in flow vessels," *AIChE J.* **22** (1967) no. 2, 289–297.

[58] A. Mersmann, H. Großmann: "Dispergieren in flüssigen Zweiphasensystemen," *Chem. Ing. Tech.* **52** (1980) no. 8, 621–628.

[59] J. W. van Heuven, W. J. Beek: "Vergrotingsregels voor turbulente vloeistof-vloeistof dispersies in geroerde vaten," *Ingenieur (Utrecht)* **82** (1970) no. 44, Ch 51-Ch 60.

[60] A. H. P. Skelland, R. Seksaria: "Minimum impeller speeds for liquid–liquid dispersion in baffled vessels," *Ind. Eng. Chem. Process Des. Dev.* **17** (1978) no. 1, 56–61.

[61] T. Vermeulen, G. M. Williams, G. E. Langlois: "Interfacial area in liquid–liquid and gas–liquid agitation," *Chem. Eng. Prog.* **51** (1955) no. 2, 85–94.

[62] D. S. Laity, R. E. Treybal: "Dynamics of liquid agitation in the absence of an air–liquid interface," *AIChE J.* **3** (1957) no. 2, 176–180.

[63] P. H. Calderbank, M. B. Moo-Young: "The continuous phase heat and mass-transfer properties of dispersions," *Chem. Eng. Sci.* **16** (1961) no. 2, 39–54.

Chapter 6

[64] M. Zlokarnik: "Scale-up of surface aerators for waste water treatment," *Adv. Biochem. Eng.* **11** (1979) 157–180.

[65] M. Zlokarnik: "Eignung und Leistungsfähigkeit von Oberflächenbelüftern für biologische Abwasserreinigungsanlagen," *Korrespondenz Abwasser* **27** (1980) no. 1, 14–21.

[66] E. van de Sande, J. M. Smith: "Mass transfer from plunging water jets," *Chem. Eng.* J. *(Lausanne)* **10** (1975) 225–233.

[67] M. Zlokarnik: "Auslegung von Hohlrührern zur Flüssigkeitsbegasung. Teil I: Bestimmung des Gasdurchsatzes und der Wellenleistung," *Chem. Ing. Tech,* **38** (1966) no. 3, 357–366.

[68] M. Zlokarnik: "Auslegung von Hohlrührern zur Flüssigkeitsbegasung. Teil II: Ermittlung des erreichbaren Stoff- und Wärmeaustausches," *Chem. Ing. Tech.* **38** (1966) no. 7, 717–723.

[69] M. Zlokarnik, H. Judat: "Rohr- und Scheibenrührer—zwei leistungsfähige Rührer zur Flüssigkeitsbegasung." *Chem. Ing. Tech.* **39** (1967) no. 20, 1163–1168.

[70] M. Zlokarnik: "Rohrrührer zum Ansaugen und Dispergieren großer Gasdurchsätze in Flüssigkeiten," *Chem. Ing. Tech.* **42** (1970) no. 21, 1310–1314.

[71] M. Zlokarnik, H. Judat: "Rohr- und Propellerrührer—eine wirkungsvolle Rührkombination zum gleichzeitigen Begasen und Aufwirbeln," *Chem. Ing. Tech.* **41** (1969) no. 23, 1270–1273.

[72] M. Zlokarnik: "Rührleistung in begasten Flüssigkeiten," *Chem. Ing. Tech.* **45** (1973) no. 10a, 689–692.

[73] H. Judat: "Das Dispergieren von Gasen mittels schnellaufender Rührertypen," *Fortschr. Verfahrenstech. B* **15** (1977) 141–159.

[74] P. Kurpiers, A. Steiff, P.-M. Weinspach: "Zum Überflutungsverhalten ein- und zweistufiger Rührbehälter," *Chem. Ing. Tech.* **57** (1985) no. 1, 62–63.

[75] M. Zlokarnik: "Sorption characteristics for gas–liquid contacting in mixing vessels," *Adv. Biochem. Eng.* **8** (1978) 133–151.

[76] P. Weiland, R. Sick, C. Osorio, U. Onken: "Oxidation of hydrazine—a reliable method for the determination of volumetric mass transfer coefficients in gas/liquid systems," *Ger. Chem. Eng. (Engl. Transl.)* **9** (1986) 143–148.

[77] R. R. Lessard, S. A. Zieminski: "Bubble coalescence and gas transfer in aqueous electrolytic solutions," *Ind. Chem. Fundam.* **10** (1971) no. 2, 260–269.

[78] T. O. Oolman, H. W. Blanch: "Bubble coalescence in air-sparged bioreactors," *Biotechnol. Bioeng.* **28** (1986) 578–584.

[79] G. Marucci: "A theory of coalescence," *Chem. Eng. Sci.* **24** (1969) 975–985.

[80] M. Zlokarnik: "Tower-shaped reactors for aerobic biological waste water treatment," *Biotechnology* vol. 2, Chap. 23, VCH Verlagsgesellschaft, Weinheim 1985.

[81] M. Zlokarnik: "Design and scale-up of mechanical foam breakers," *Ger. Chem. Eng. (Engl. Transl.)* **9** (1986) 314–320.

[82] K. Van't Riet: "Review of measuring methods and results in nonviscous gas–liquid mass transfer in stirred tanks," *Ind. Eng. Chem. Process Des. Dev.* **18** (1979) no. 3, 357.

[83] H. Judat: "Gas/liquid mass transfer in stirred vessels—a critical review," *Ger. Chem. Eng. (Engl. Transl.)* **5** (1982) 357–363.

[84] H.-J. Henzler: "Verfahrenstechnische Auslegungsunterlagen für Rührbehälter als Fermenter," *Chem. Ing. Tech.* **54** (1982) no. 5, 461–476.

[85] H.-J. Henzler, J. Kauling: "Scale-up of mass transfer in highly viscous liquids," *Proc. 5th Europ. Conf. Mixing*, Würzburg, Germany, June 10–12, 1985.

[86] M. Zlokarnik: "Eignung und Leistungsfähigkeit von Volumenbelüftern für biologische Abwasserreinigungsanlagen," *Korrespondenz Abwasser* **27** (1980) no. 3, 194–209.

[87] M. Zlokarnik: "Sorption characteristics of slot injectors and their dependency on the coalescence behaviour of the system," *Chem. Eng. Sci.* **34** (1979) no. 10, 1265–1271.

[88] A. Steiff, P.-M. Weinspach: "Heat transfer in stirred and non-stirred gas–liquid reactors," *Ger. Chem. Eng. (Engl. Transl.)* **1** (1978) 150–161.

Chapter 7

[89] H. Giesekus: "Sekundärströmungen in viskoelastischen Flüssigkeiten bei stationärer und periodischer Bewegung," *Rheol. Acta* **4** (1965) no. 2, 85–101.

[90] A. B. Metzner, R. E. Otto: "Agitation of non-Newtonian fluids," *AIChE J.* **3** (1957) no. 1, 3–10.

[91] P. H. Calderbank, M. B. Moo-Young: "The prediction of power consumption in the agitation of non-Newtonian fluids," *Trans. Inst. Chem. Eng.* **37** (1959) 26–33.

[92] J. L. Beckner, J. M. Smith: "Anchor-agitated systems: power input with newtonian and pseudoplastic fluids," *Trans. Inst. Chem. Eng.* **44** (1966) T 224–T 236.

[93] E. O. Reher: "Rühren nicht-Newtonscher Flüssigkeiten. 1. Mitteilung: Der Leistungsbedarf beim Rühren viskoelastischer Flüssigkeiten," *Chem. Techn. (Leipzig)* **21** (1969) no. 1, 14–22.

[94] P. Schümmer: "Das Rühren viskoelastischer Flüssigkeiten im Übergangsbereich laminar-turbulent," *Chem. Ing. Tech.* **42** (1970) no. 5, 322–327.

[95] H. Höcker, G. Langer: "Zum Leistungsverhalten begaster Rührer in Newtonschen und nicht-Newtonschen Flüssigkeiten," *Rheol. Acta* **16** (1977) 400–412.

[96] M. Opara: "Homogenisieren von nicht-Newtonschen Flüssigkeiten im Rührgefäß," *Verfahrenstechnik (Mainz)* **9** (1975) no. 9, 446–449.

[97] K. H. Tebel, P. Zehner. G. Langer, W. Müller: "Homogenisieren strukturviscoser Flüssigkeiten in Schlaufenreaktoren und Rührkesseln," *Chem. Ing. Tech.* **58** (1986) no. 10, 820–821.

[98] M. F. Edwards, W. L. Wilkinson: "Heat transfer in agitated vessels, part II, non-Newtonian fluids," *Chem. Eng. (London)* (1972) no. 265, 328–335.

[99] J. Pawlowski: "Zur Theorie der Ähnlichkeitsübertragung bei Transportvorgängen in nicht-Newtonschen Stoffen," *Rheol. Acta* **6** (1967) no. 1, 54–61; abstract in *Chem. Ing. Tech.* **38** (1966) no. 11, 1202–1203.

[100] J. Pawlowski: *Die Ähnlichkeitstheorie in der physikalisch-technischen Forschung*, Springer Verlag, Berlin-Heidelberg-New York 1971.

26. Mixing of Highly Viscous Media

DAVID B. TODD, APV Chemical Machinery Inc., South Plainfield, NJ 07080, United States

Thick mixtures with viscosities >10 Pa s do not lend themselves to simple mixing with a small turbine or propeller stirrer in a large pot. The high viscosity may arise from a high concentration of solids in a slurry, the high viscosity of the matrix fluid itself, or by interactions between the ingredients.

With high viscosity fluids, the mixing Reynolds number ($Re = D^2 N/\mu$) is likely less than 100. Thus, mixing occurs by viscous forces alone, and turbulence plays no part. The viscous forces cause either shear flow or elongational flow. Relative motion of an agitator stretches and deforms the material between itself and the vessel wall. As a layer of fluid gets stretched into thinner layers, and then undergoes reorientation, the striations become ever thinner until homogeneity is achieved. Likewise, the shear forces tear away at solid agglomerates and cause their disintegration.

Many mixers have been independently developed for specific process applications, and then their usage has been extended to other processes involving somewhat similar, but difficult mixing tasks. The agitated tubs used for kneading bread dough evolved into horizontal vessels that could more readily be tilted for emptying. The design evolved into double arm mixers, which in turn led to a variety of mixing blades tailored for other end uses as these double arm mixers began to be used for adhesives, gums, putties, rubber solutions, pigment dispersions, and a host of other products.

The very difficulty in defining precisely the consistency of these various doughs, pastes, and thick slurries has fostered a wide art of mixer design, rather than a sound scientific approach. Consequently, mixers are generally classified more by their mechanical features than by the process function for which they may best be suited.

With solids present, producing a homogeneous distribution is frequently only part of the mixing task. It may also be necessary to reduce the particle size of the solids, even down to submicrometer size, to achieve the desired end product, such as pigment dispersion. However, not all products need such fineness; for example, the fiber glass that is introduced into molten plastic should be uniformly distributed with minimum breakup.

High viscosity mixers usually have a limited high shear zone to minimize overall heat buildup, and depend upon the impeller also circulating all of the mixer contents past the high shear zone. Stagnant zones can exist, material may ride on blades and be kept out of the mixing region, and pockets of unmixed components may be carried through continuous process equipment since there is a sharp drop off of transmitted shear away from the shear creating device. High velocity impellers may be ineffective since they may create an isolated hole in the mass without producing any circulation, particularly with pseudoplastic materials.

When mixing a minor component (of either a solid or another viscous fluid) into a continuous phase viscous fluid, two kinds of mixing action may be required. The first is distributive mixing where interfaces are maintained intact, but the minor component is distributed uniformly throughout the mass. The second is dispersive mixing where the interfaces are broken up. With two-phase fluid systems, the differences between distributive and dispersive mixing has been described by OTTINO and coworkers [1]–[3].

ERWIN [4], [5] has measured distributive mixing in single screw extruders. Dispersive mixing has been evaluated for a variety of flow conditions by RUMSCHEIDT and MASON [6], [7], GRACE [8], and HAN [9]. ERWIN [10] concluded that extensional flow is the most efficient of plain strain laminar flows.

The degree of distributive mixing for a two-phase mixture can be characterized by the relative increase in interface area or number of striations per unit length. ERWIN and coworkers [2], [3] have shown that mixing increases linearly with the average total shear strain applied. They have also shown that mixing efficiency in single screw extruders was increased significantly by re-orientation of the interface, such as with mixing pins. Also, poorer mixing occurs in laminar shear flows when the ratio of the viscosity of the disperse phase to the continuous phase increases.

For a given shear field, there is a maximum viscosity ratio beyond which a liquid droplet cannot be broken up. KARAM and BELLINGER [11] have shown that breakup is easier when the interfacial tension (σ) is low, the viscosity of the continuous phase (μ_c) is high, and the radius (r) of the droplet is large. Their plot of reduced shear rate vs. viscosity ratio ($\mu_r = \mu_d/\mu_c$) is reproduced here as Figure 1. Thus, when extremes in viscosity ratio are encountered, real fine dispersions cannot be achieved in simple shear fields alone.

For viscous fluid–fluid mixtures, the efficiency of a mixer in theory can be determined from a combination of three factors: critical capillary number at which breakup occurs, the time required for this breakup, and the number and size of the resulting drops. The capillary number is the ratio of the viscous forces to the interfacial forces. The critical capillary number in shear flow goes through a minimum at a viscosity ratio of 1, and above 4, simple shear flow is no longer effective.

With a hyperbolic flow, the critical capillary number is lowered. The time for breakup increases as the viscosity ratio is increased. As the capillary number is increased above the critical value, the number of fragments produced increases sharply at first, but then becomes constant. Thus, the relative dispersion efficiency of different mixing devices should be compared on-

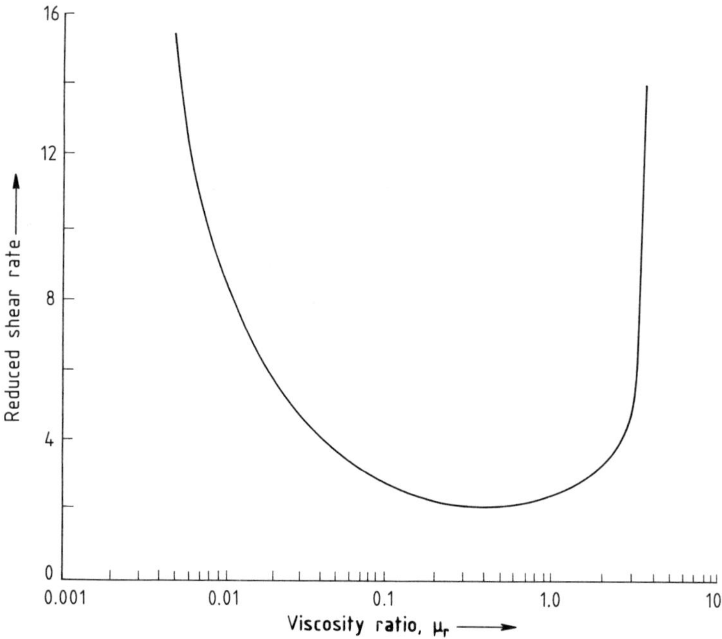

Figure 1. Effect of viscosity ratio on dispersion [11]

Reduced shear rate $\frac{\gamma_b \mu_c r}{\sigma}\left(\frac{1.19\mu_r + 1}{\mu_r + 1}\right)$

where σ = interfacial tension; μ_c = viscosity, continuous phase; μ_d = viscosity, dispersion phase; $\mu_r = \mu_d/\mu_c$; r = drop radius; γ_b = shear rate at breakup

ly under similar parameters of viscosity ratio, initial drop size, continuous phase viscosity, and interfacial tension. The number and size of the fragments produced versus time would be a reasonable measure of dispersion efficiency.

Equipment for viscous mixing is characterized by small clearances between impeller and vessel walls, high power per unit volume, and relatively small volume. Because of probable heat buildup, impeller speeds may have to be very slow. Intermeshing blades or stators may also be required to keep the material from cylindering on the rotating impeller.

A smearing blade profile may be required where dispersion is important, whereas a scraping profile may be desired if heat transfer is critical. A mixing action which causes material to work against itself can be used to advantage.

Ease of discharge, and ease of cleaning between batches are important requirements for paste mixers. Mere bottom drainage is frequently inadequate, in which case the vessel may need to be dumped.

Mixing in general has been treated in the two-volume edition *Mixing—Theory and Practice*, and in particular for viscous mixing in the chapter by IRVING and SAXTON [12]. Similar topics are also covered by PAHL [13]. The triennial European Conferences on Mixing [14], although concentrating more on multiphase dispersion and reaction in turbulent mileau, do also have some papers on viscous and non-Newtonian mixing phenomena.

Power. For viscous mixers with agitators that sweep near the vessel wall, the power drawn depends primarily on the viscous drag of the agitator rather than on the pumping effect incurred in circulating the material to the high shear zone. For scale up, the following approach may be taken:

Shear rate (DN/t)

Shear area (DL)

Shear stress = (shear rate) · (viscosity)

$$\propto \left(\frac{DN}{t}\right)\mu$$

Force = (shear stress) · (shear area)

$$\propto \left(\frac{\mu DN}{t}\right)(DL)$$

Torque = (force) · (radius)

$$\propto \left(\frac{\mu D^2 LN}{t}\right)D$$

Power = (torque) · (speed)

$$\propto \frac{\mu D^3 LN^2}{t}$$

for dimensional similarity, with both clearance (t) and agitator length (L) proportional to diameter D:

$$\text{Power} = \propto \mu D^3 N^2$$

Actually, for most very viscous mixtures, the viscosity μ is not constant, and is usually shear thinning, with an effective viscosity depending upon the $(n-1)$ power of speed, where n is the power index number of the fluid (the slope of shear stress vs. shear rate on logarithmic coordinates).

1. Batch Mixers

Single Stirrer Mixers. Thick pastes can be handled in a variety of simple batch mixers in which the agitator is in close proximity to the vessel wall. With an anchor blade (Fig. 2), the proximity of the blade to the vessel wall improves heat transfer while providing gentle agitation. Use of single or double helical blades (Fig. 3) provides not only the scraping action at the wall but also an end-to-end turnover of vessel contents. A single helical auger may be used inside of a draft tube with reliance on flow back up the annulus caused by the pumping action of the central auger. Some mixers also employ stators attached to the vessel wall that are in close proximity to the multi-arm central agitator, such as shown in Figure 4.

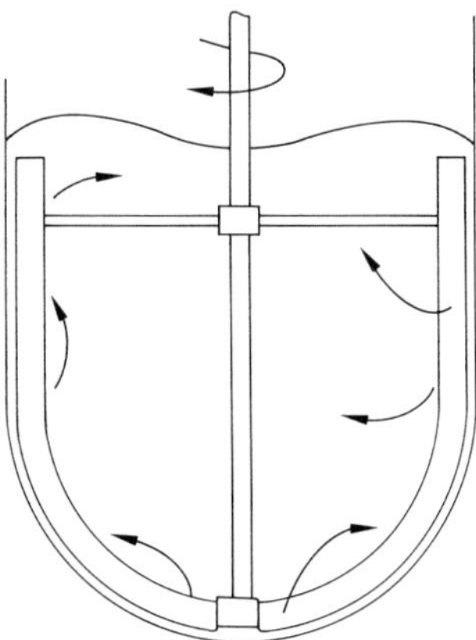

Figure 2. Anchor mixer

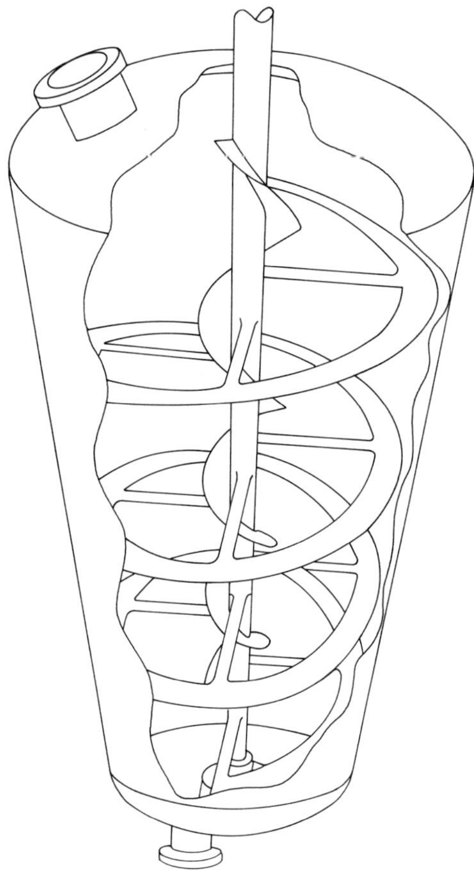

Figure 3. Helical blade mixer

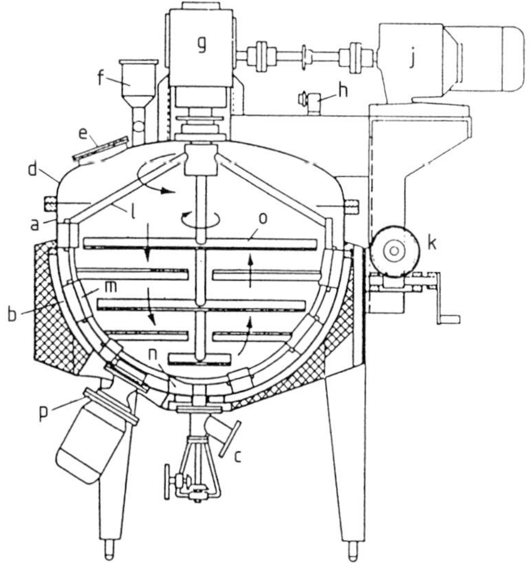

Figure 4. Multi-arm shearbar mixer
a) Body; b) Insulating heating mantel; c) Outlet; d) Cover; e) Observation port; f) Inlet; g) Counter-rotating drive; h) Tachometer; j) Motor; k) Cover mechanism; l) Outer stirrer; m) Wall scrapper; n) Discharge scrapper; o) Inner stirrer; p) Homogenization mill

Change Can Mixers. Change can mixers are vertical batch mixers in which the mixing vessel can readily be separated from the agitator. Agitation is usually provided by two planetary mixing blades (Fig. 5). Separate cans permit accurate weighing of the ingredients prior to mixing, easy transport of the mixture to the next process operation, and easier cleaning of the mixer between batches.

With most smaller change can mixers, such as used in ink and pigment manufacture, the mixing head is raised from the can either by tilting or vertically. Change can mixers also have permanently mounted agitator elements with the can raised for mixing. The latter type (Fig. 6) is used extensively for mixing solid propellants.

A variety of blade shapes has evolved to provide for thorough mixing of the can contents. The blades are usually intermeshing and pitched to provide some axial movement. The can may

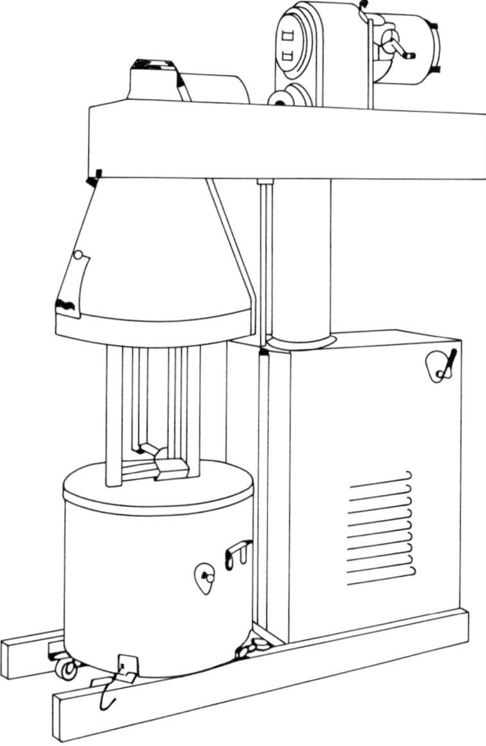

Figure 5. Change can mixer (Chas. Ross & Son)

Figure 6. Vertical mixer (Baker Perkins)

Figure 7. Intensive mixer with withdrawable agitator and tilting discharge (Eirich)

be rotated on a turntable to bring all material past the mixing head, or the planetary gear mounting may rotate around a fixed can. The can may also tilt for dumping, as shown in Figure 7.

Change can mixers are available generally in the 1–500-L size, with the propellant mixers up to 4000 L.

Double Arm Kneading Mixers. These mixers consist of two tangential or overlapping horizontally mounted mixing shafts rotating in a rectangular trough with the bottom curved in two half cylinders to match the sweep of the mixing blades (Fig. 8). Several mixing blade shapes are available (Fig. 9). Most common is the Sigma Blade (Fig. 9 A), with unequal arms angled and unequal rotating speeds to impart randomness to the mixing action. Material is swept arount the bottom and divided as it is brought down over the saddle ridge between the half cylinders.

Dispersion of agglomerates is improved with blades that provide compressive shear forces against the trough shell with a wedged shape face (Fig. 9 B) rather than a scraping blade profile. Heat transfer is promoted by using a cored blade and a scraping profile on the blade front (Sigma Blade, Fig. 9 A). Shredded action can also be imparted (Fig. 9 C), or cutting action (Fig. 9 D).

If the mixed contents are free draining, or if a heel of the batch can be left behind without impairing the quality of succeeding batches, the mixer may be emptied through one or two gates or valves on the bottom. However, the usual procedure is to tilt the mixer for emptying. The mixer can also be provided with a screw discharge mounted in the saddle section (Fig. 10). The screw is reversed during the mixing stage to assist mixing.

Double arm mixers are available in size ranges from 1 L to 5 m^3. As size is increased, rotational speeds are generally reduced. Power inputs range from 0.02–0.5 kW/kg. A typical 500-L mixer might run with a front blade speed of 30 rpm and be supplied with a 40-kW drive motor.

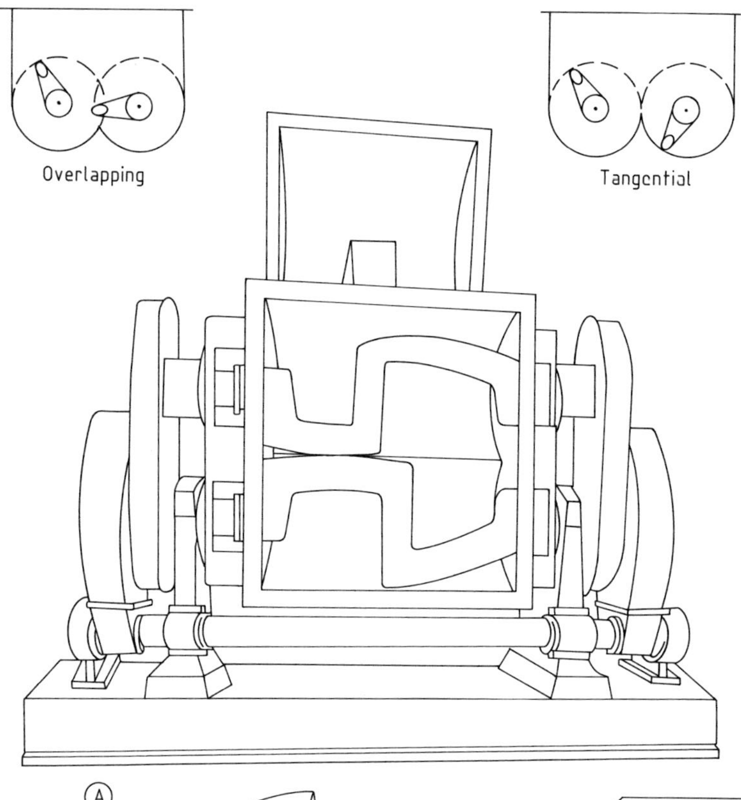

Figure 8. Double arm kneading mixer (Baker Perkins)

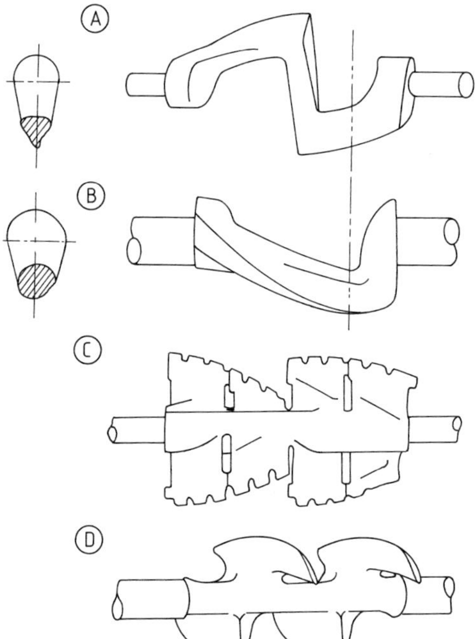

Figure 9. Agitator blades for double arm kneaders (Baker Perkins)
A) Sigma; B) Dispersion; C) Multiwing overlap; D) Double naben

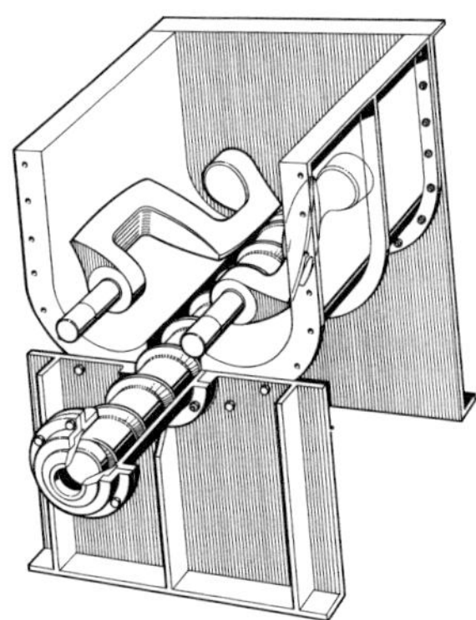

Figure 10. Screw discharge batch mixer (Werner & Pfleiderer)

Ram Mixers. The Banbury type mixer (Fig. 11) is capable of extremely high power inputs, up to 6 kW/kg, in size ranges from small laboratory to 500 kg. A compression ram is used to confine the mixture between the tight fitting rotors and vessel walls. The masticating action is most useful for mixing elastomers and plastics. The mixed product is dumped through a front or bottom discharge door. The relatively large

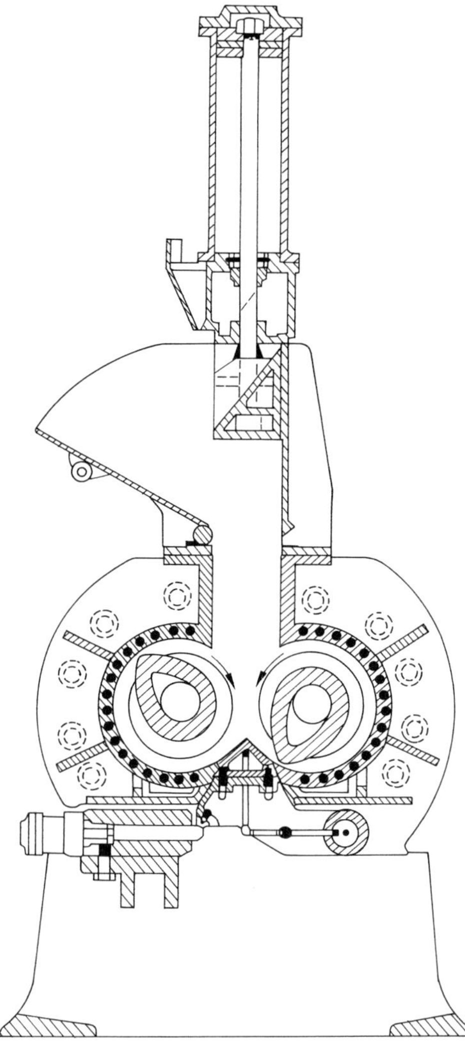

Figure 11. Banbury mixer (Farrel)

shafts permit very high torques. As with all very viscous mixing operations, batch temperature rise may be very rapid and may be the main factor limiting cycle times. At 1 kW/kg, temperature rise per minute could readily reach 30°.

High Intensity Mixers. High intensity mixers (Fig. 12) utilize a high speed bottom-mounted blade to impart high shear and to establish a vortex flow in the vessel. The bottom blade scoops the batch upward at peripheral speeds of 40 m/s. Blade impact is effective in reducing agglomerates and aids intimate dispersion. Energy input is high (up to 200 kW/m^3), and these high intensity mixers are frequently paired with a slower speed, larger batch cooler of similar design. These mixers are particularly suited for rapid mixing of powders and granules with minor amounts of liquids, such as when plasticizer is added to poly(vinyl chloride) (PVC) resin. The dry PVC granules are heated by the blade action within a few minutes, after which the plasticizer is added and rapidly penetrates the pores of the PVC particles to provide a homogeneous feed mixture for downstream processing. Other uses are for removing volatiles from pastes under vacuum, and for dissolving solids in liquids. Sizes range from 1 – 1100 L.

A somewhat similar mixer, shown in Figure 13, utilizes a horizontal impeller. Incorporated into the mixer is an optical pyrometer to indicate when the desired energy has been imparted by sensing the inside product temperature. Since a high level of energy can be supplied, the mix cycle time is quite short.

Plow Mixers. Plow mixers have plow elements on a horizontal shaft rotating in a cylindrical vessel (Fig. 14). After charging, the smooth fitting cover is closed, and rapid shaft rotation hurls material into the free space. Additional blending is provided by the impellers plowing through the solids bed. If agglomerates are present, an additional small high speed chopper may be added.

Cone and Screw Mixers. Cone and screw mixers (Fig. 15) are used for a variety of low-energy blending applications ranging from dry powders to pasty materials. An orbiting screw rotates about its own axis and also around the center line of the cone-shaped vessel near the wall to provide bottom to top circulation. The screw is reversed to aid discharge of pasty materials. Screw and cone mixers, available in capacities from 40 L to 20 m^3, work equally well when only partially filled.

Mullers. Mullers (Fig. 16) can be used for powders and pastes if the latter are not too soupy or too sticky. The mullers smash agglomerates as they rotate in a circular patch inside the round pan. An attached plow brings material progressively into the path of the mullers. Typical applications are for mixing putty and clay pastes, battery paste, and chocolate coatings. A continuous muller variant consists of two rotating sets of

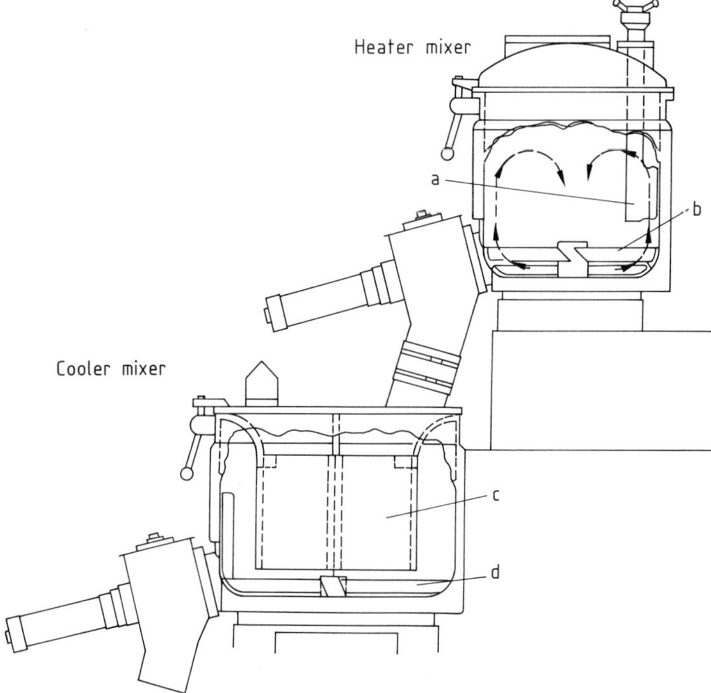

Figure 12. High density non-fluxing mixer (Baker Perkins)
a) Baffle, b) Impeller; c) Cooling ring; d) Impeller

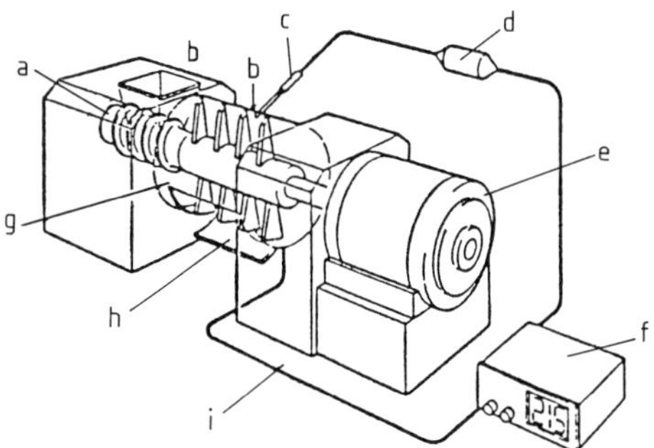

Figure 13. High intensity Gelimat fluxing mixer (Draiswerke)
a) Integrated screw feed; b) Sapphire lens; c) Optical fiber infrared sensor; d) Converter; e) Drive; f) Adjustable temperature control panel; g) Mixing chamber; h) Discharge door; i) Signal to discharge door

Figure 14. Plow mixer (Littleford Bros.)

muller wheels operating with some overlap in a figure-eight pan.

Roll Mills. Roll mills provide a very high localized shear by drawing the mixture into the nip of closely spaced parallel rolls. The viscosity of the matrix can be held high to aid dispersion because of the extensive surface area for heat transfer. The material passing through the nip is returned to the feed point by rotation of the rolls. A temperature differential between the rolls frequently causes the material to selectively adhere to one of the rolls, as it does when blending rubber stocks. Although there is little distributive mixing as the material passes through the nip,

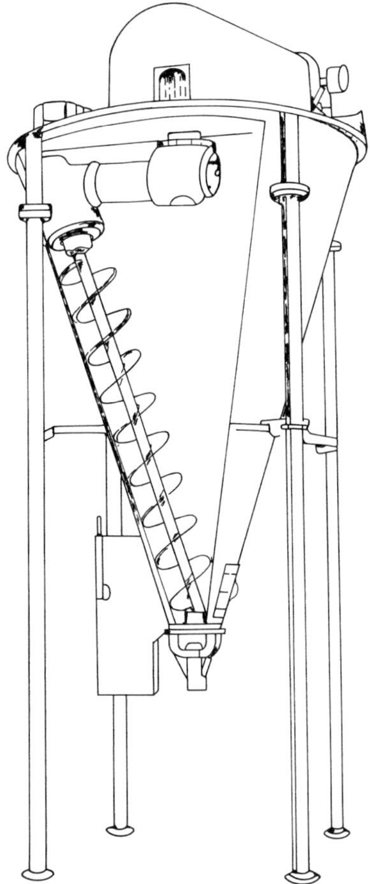

Figure 15. Cone and screw mixer (Day Mixing)

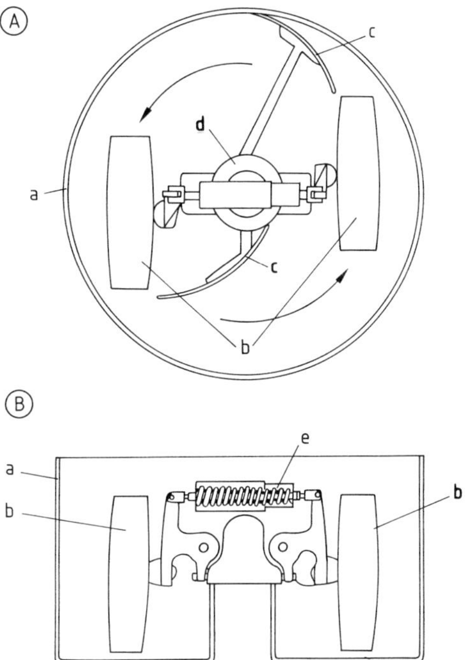

Figure 16. Muller (National Engineering)
a) Crib wall; b) Mullers; c) Plows (inner and outer); d) Stationary turret; e) Screw

there is a rotary motion of the bead of material immediately upstream of the nip. In addition to blending rubber stocks and dispersing carbon black, roll mills are used extensively in grinding and dispersing inks and pigment pastes.

2. Continuous Mixers

Although some of the batch mixers can be modified for continuous processing, the generally broad residence time is usually not conducive to the best product uniformity. However, where the component feed streams can be adequately and accurately metered, several mixers have been developed specifically for continuous processing. Continuous mixers generally consist of a closely fitted rotating element within a stationary housing with flow through the unit controlled to prevent bypassing.

2.1. Motionless Mixers

Longitudinal and transverse distribution of components can be obtained in flowing systems by placement of stationary flow redistribution devices, called motionless or static mixers, in the flow stream. Many proprietary designs have evolved, all of which divide and displace a laminar streamline from one portion of the conduit cross section to another. Examples of static mixers are shown in Figures 17–19. The Kenics mixer shown in Figure 17 consists of alternate right and left handed twisted ribbons arranged in sequence. The Ross ISG (Fig. 18) uses four drilled holes through solid sections between axially displaced cavities to redirect flow streamlines both radially and peripherally. The Sulzer SMV mixer (Fig. 19) has stacks of corrugated lamellae with open crossing channels. Most other static mixers involve variations of these three basic types with features that may permit easier fabrication or installation.

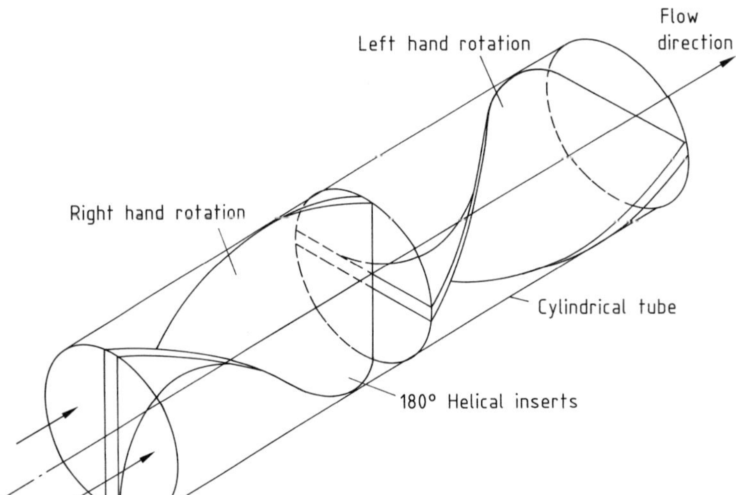

Figure 17. Kenics static mixing features

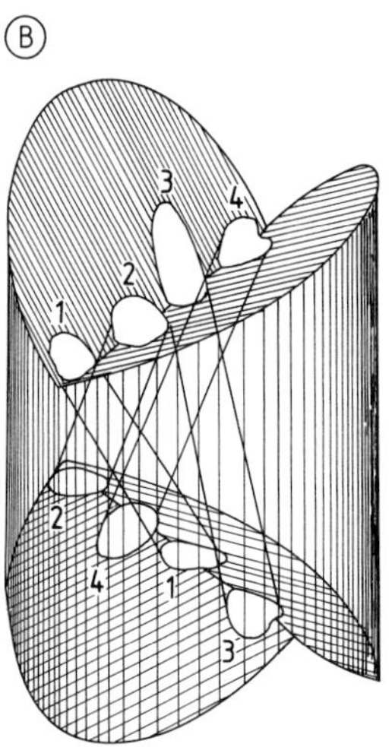

Figure 18. Ross ISG mixer
A) Assembly; B) Individual element

PAHL and MUSCHELKNAUTZ [15] have presented both mixing and pressure drop comparisons for these and other static mixers. In general, the more divisions, the better the homogenization. However, the smaller the passageways and the larger the fraction of conduit cross section occupied by metal, the higher the pressure drop. Choice of the best static mixer for a particular application involves a compromise based on energy and space available. Static mixers are of most value for distributive, rather than dispersive, fluid mixing applications. With no moving parts, there are low investment, operating, and maintenance costs.

2.2. Single Screw Mixers

The use of continuous mixers has increased considerably in recent years with the expansion of the plastics industry. End products of greater usefulness depend upon the uniform incorporation of a multitude of additives, stabilizers, antioxidants, fillers, reinforcing agents, as well as alloys and blends of differing base polymers. The base polymer resin needs to be in the melt stage to permit uniform distribution of the additives. Since the thermal conductivity of polymers is generally very low, melting is best accomplished by work energy rather than by heat transfer through a vessel wall. The screw extruder shown in Figure 20 consists of a cylindrical barrel with a helical screw. The root diameter of the screw is smaller in the feed zone. The dry ingredients, sometimes having been premixed in a batch mixer, are conveyed by the rotation of the screw. As the root diameter of the screw is increased, the

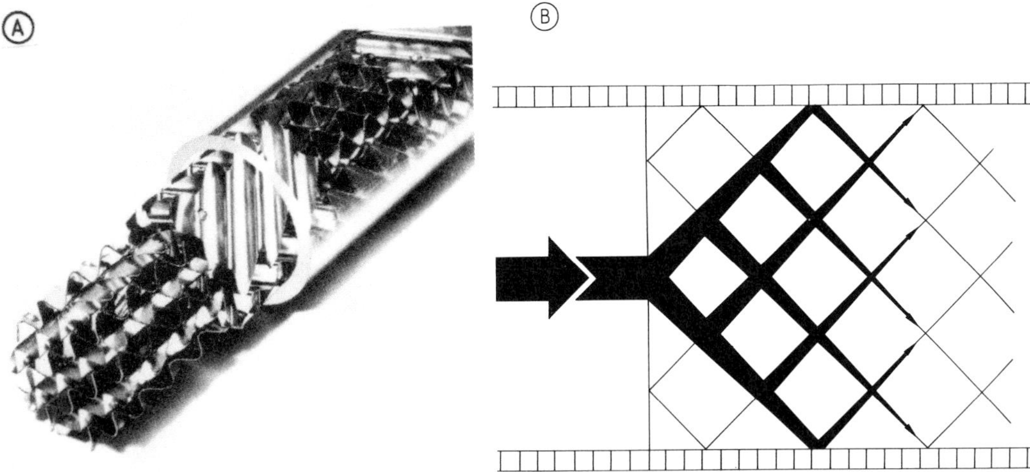

Figure 19. The Sulzer mixer
A) Sulzer mixing elements
B) Operating principle of the Sulzer mixing element

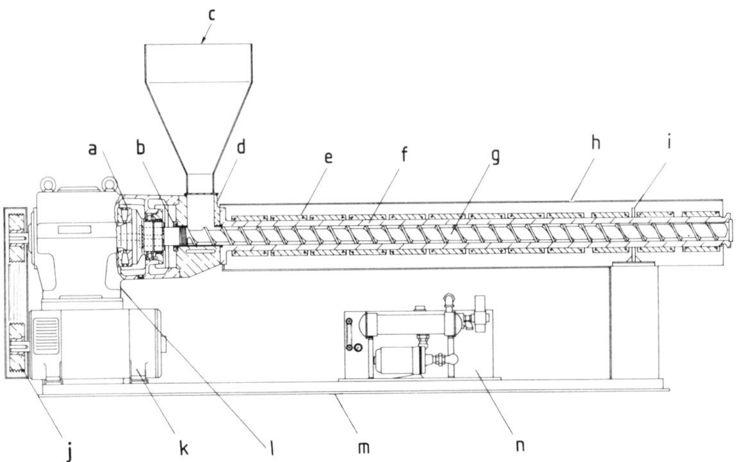

Figure 20. Extruder (Sterling Extruder)
a) Throat bearing assembly; b) Powder seal; c) Hopper; d) Throat; e) Water-cooled heaters; f) Barrel; g) Screw; h) Insulated barrel guard; i) Barrel support; j) Belt and sheave assembly; k) Motor; l) Reducer; m) Base; n) Closed-loop cooling system

solids are compressed and caused to melt primarily by friction against each other and against the barrel wall. The homogenized product is forced through some shaping element, such as dies for film, sheet, profile, or strands for subsequent pelletizing. Since the final mixture is extremely viscous (10 – 5000 Pa s effective viscosity), the pressure drop through the die is high; consequently, a pressure flow back toward the feed occurs down the screw channel, which partially offsets the drag flow caused by the conveying action of the screw. Cross-channel flow occurs, and most of the mixing is accomplished by the time the last bit of polymer has melted. A further homogenization can be achieved by using a mixing head at the end of the screw, such as shown in Figure 21. The mixing heads operate by forcing all of the product over confined flow paths, or reorienting the elongated laminar layers. The capacity of an extruder is determined primarily by the power which can be supplied. Single screw extruders can be equipped with large gears and

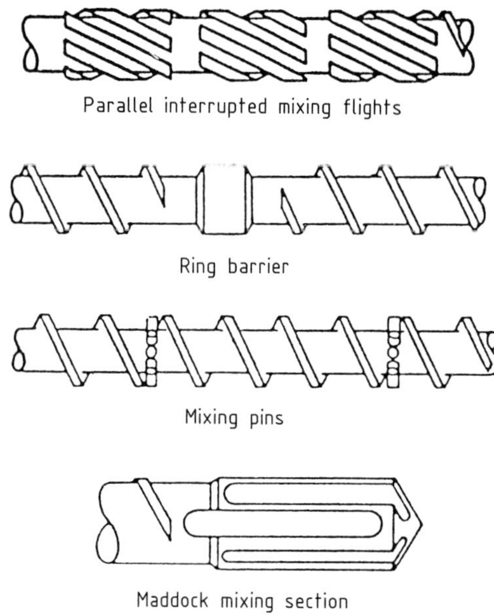

Figure 21. Mixing promoters for single screw extruders

thrust bearings to operate with high torque. Although the majority of extruders are in the 25–200 mm diameter range, large extruders do exist for tasks such as polyethylene homogenization: these are considerably larger and are equipped with up to 4000 kW drives. Comprehensive extruder texts are RAUWENDAAL [16], BERNHARDT [17], and TADMOR and GOGOS [18].

Some variants of single screw extruders are offered which enhance mixing by greater reorientation of the product as it progresses down the barrel. The KoKneader (Fig. 22) achieves this by three rows of teeth projecting inward from the barrel wall, which intermesh with the interrupted screw flight as the screw reciprocates once each rotation. The passage of the teeth, as shown in Figure 23, provides an intimate kneading action [19].

Reorientation of the laminar striations to improve distributive mixing is accomplished in the Transfermix (Fig. 24) and the Cavity Transfer mixer (Fig. 25) as material being conveyed in the

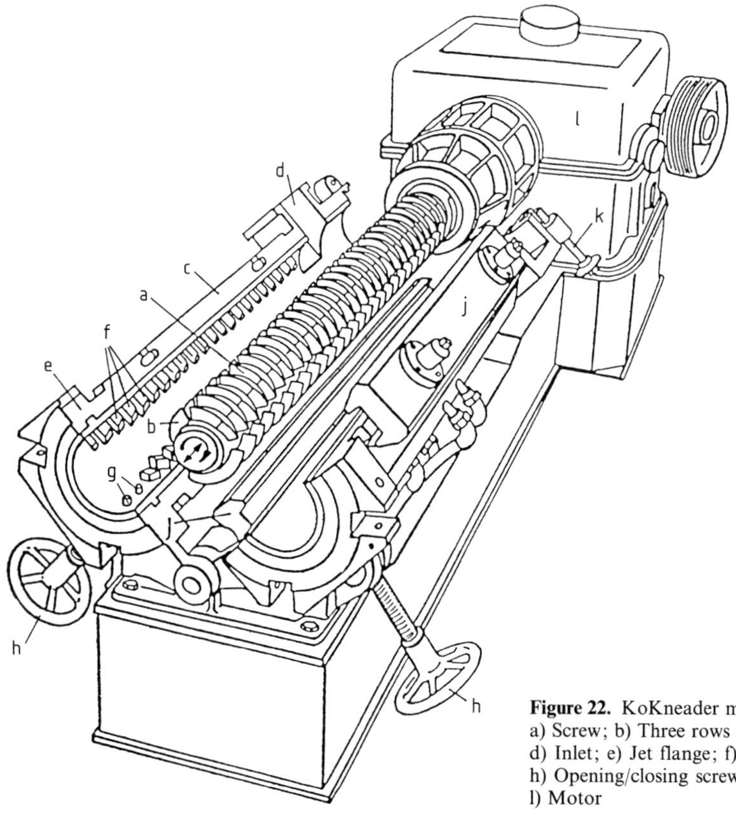

Figure 22. KoKneader mixer (Buss AG)
a) Screw; b) Three rows of teeth; c) Barrel, demountable; d) Inlet; e) Jet flange; f) Cylinder teeth; g) Mixing pins; h) Opening/closing screw; j) Clamps; k) Locking screws; l) Motor

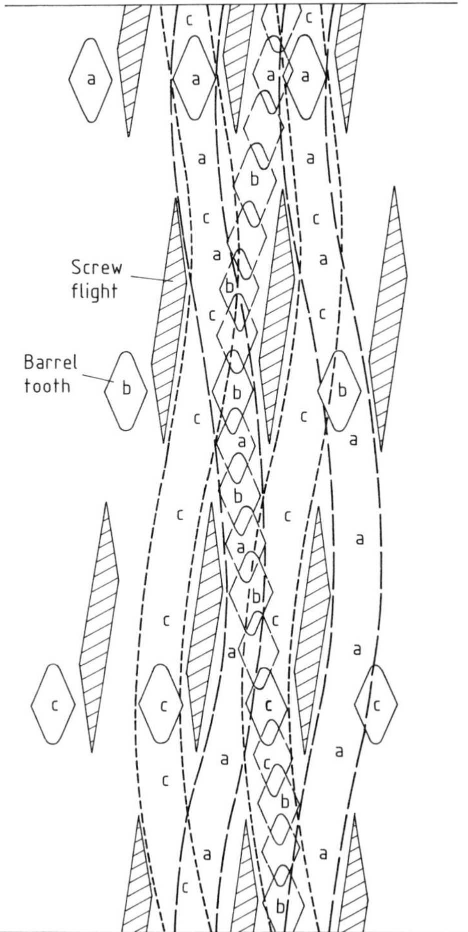

Figure 23. Paths of barrel teeth through KoKneader screw flights
a)–c) Barrel teeth

screw flights or pockets is repositioned into complementary flights or pockets in the barrel.

When starting with solid resin to be melted and mixed, the overall conversion of mechanical energy of a single screw extruder into useful work is improved by segregating the melt as it is formed by use of some type of barrier screw, as is shown in Figure 26. A channel of slightly decreased outer diameter and increasing width between the flights of the main channel enables the melt to be accumulated free of the unmelted solids and enables the solids to be more effectively compressed and fused between the decreasing width screw channel and barrel.

2.3. Twin Screw Mixers

With two screws inside a figure-eight barrel, a variety of mixing effects can be created depending on whether the screws co-rotate or counter-rotate, and whether they are tangential or they intermesh. Twin-screw extruders are used for a broad variety of products, but their most extensive usage is in polymer processing. They lend themselves to good control of temperature and residence time, and for staging of sequential events such as degassing or further additive addition at the appropriate stage down the barrel. Like the single-screw extruders, twin-screw extruders are used for generating the pressure required for discharging the product through forming dies for profile or stranding.

With the co-rotating intermeshing twin screw depicted in Figure 27, there is complete wiping of all surfaces. Intense mixing action can be cre-

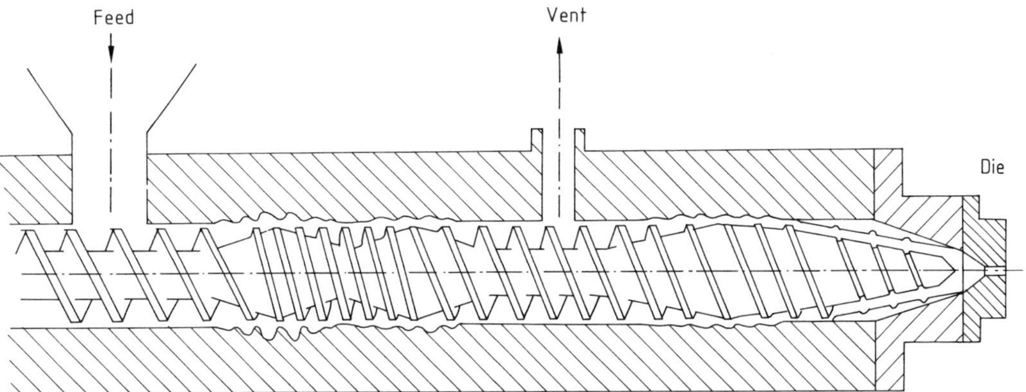

Figure 24. Transfermix (Sterling Extruder)

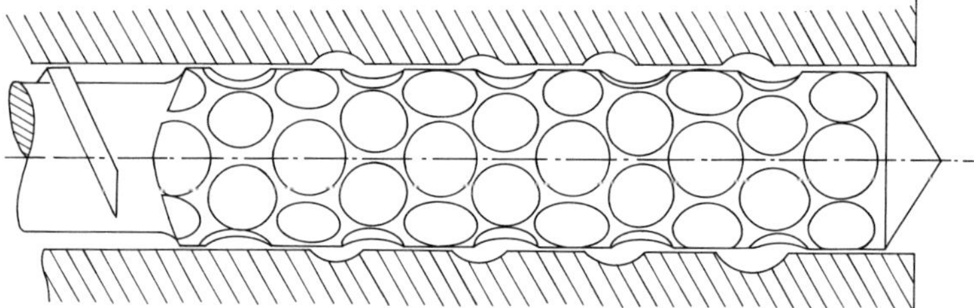

Figure 25. Cavity transfer mixer (RAPRA)

Figure 26. Barrier screw for single screw extruder (Sterling Extruder)

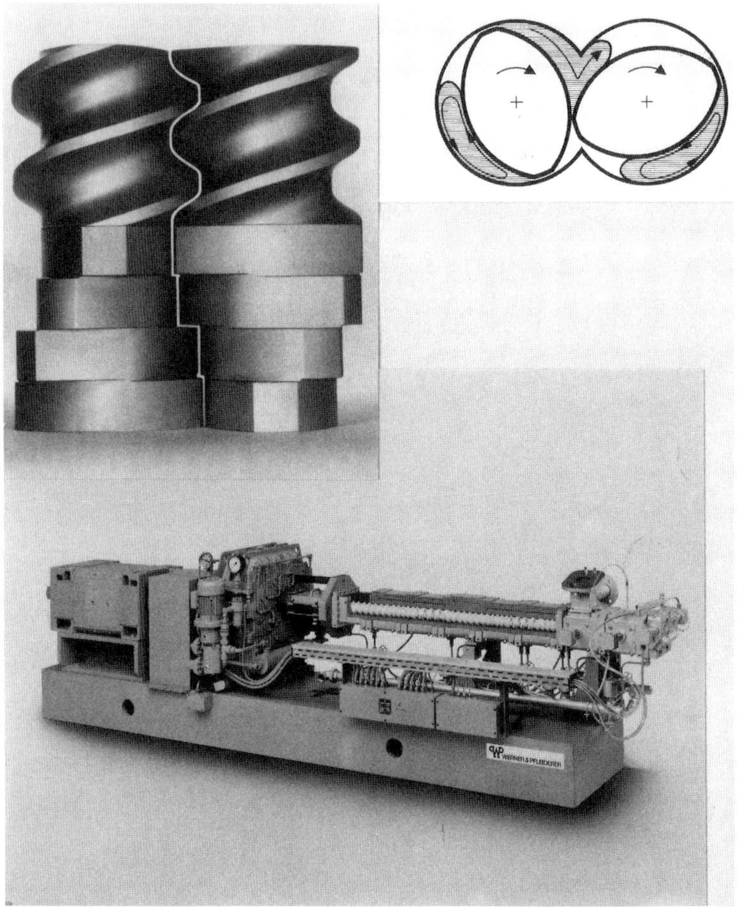

Figure 27. Co-rotating intermeshing twin screw extruder (Baker Perkins)

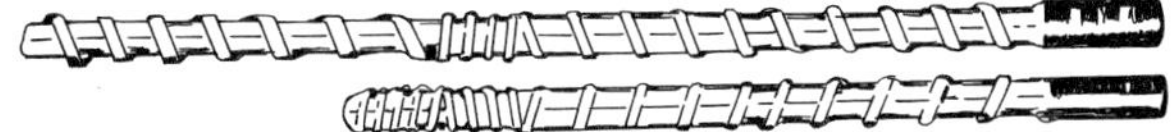

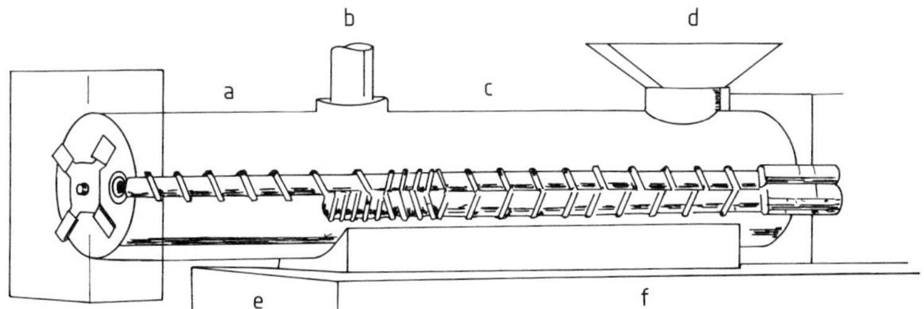

Figure 28. Counter-rotating tangential twin screw extruder (Welding Engineers)
a) Extrusion area; b) Vent area for volatile and air removal; c) Final mixing; d) Feed throat; e) Transfer area; f) Melt down and initial mixing

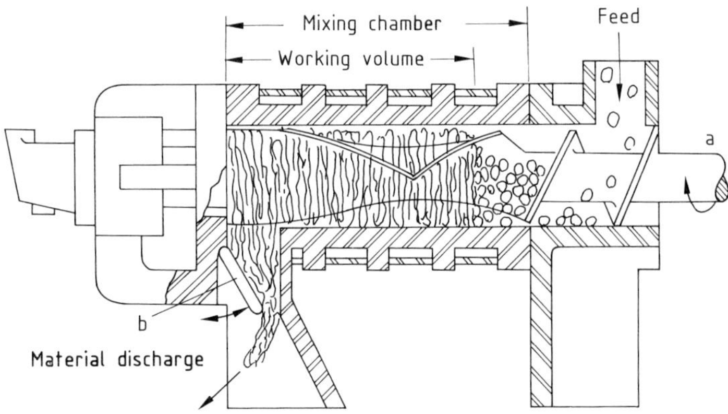

Figure 29. Farrel continuous mixer (Farrel)
a) Rotor drive; b) Discharge orifice gate

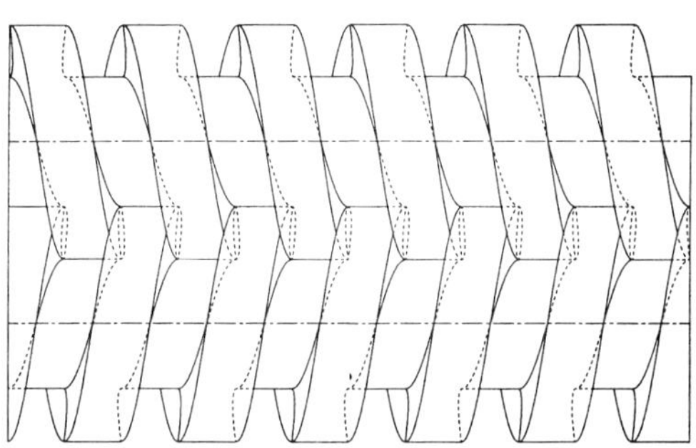

Figure 30. Counter-rotating intermeshing screws

ated with the use of kneading paddles of similar cross section, but arranged axially in an offset fashion. The kneading paddles create a series of compression–expansion zones [20] while forwarding the mixture down the barrel. Such twin-screw mixers have been built from 15 mm diameter laboratory size to huge commercial machines of 889 mm diameter.

With counter-rotating tangential screws (Fig. 28) the cross mixing between screws provides much distributive mixing and striation reorientation [21]. The large surface exposure is also of benefit in devolatilizing applications.

Robust rotors can be used in short barrels to provide a continuous variant (Fig. 29) of the Banbury-type intensive-fluxing mixer. The mixting action is a combination of intensive shear material between the rotors, and a rolling action of the material itself.

Counter-rotating intermeshing twin-screw extruders (Fig. 30) are very effective in pressure generation since axial back flow is severely restricted by the intermeshing.

3. References

[1] J. M. Ottino, R. Chella, *Polym. Eng. Sci.* **23** (1983) 357.

[2] R. Chella, J. M. Ottino, *Ind. Eng. Chem. Fundam.* **24** (1985) 170.

[3] D. V. Khakhar, R. Chella, J. M. Ottino, *Proc. Int. Cong. on Rheology* Mexico **2** (1984) 81.

[4] D. W. Gailus, L. Erwin, *Proceedings SPE ANTEC* **27** (1981) 639.

[5] R. A. Strasser, L. Erwin, *Adv. Polym. Sci.* **4** (1984) 17.

[6] F. D. Rumscheidt, S. G. Mason, *J. Colloid Sci.* **16** (1961) 238.

[7] F. D. Rumscheidt, S. G. Mason, *J. Colloid Sci.* **17** (1962) 260.

[8] H. P. Grace, *Chem. Eng. Commun.* **14** (1982) 738.

[9] C. D. Han: *Multiphase Flow in Polymer Processing*, Academic Press, New York 1981.

[10] L. Erwin, *Polym. Eng. Sci.* **18** (1978) 738.

[11] H. J. Karam, J. C. Bellinger, *Ind. Eng. Chem. Fundam.* **7** (1968) 576.

[12] H. F. Irving, R. L. Saxton in V. W. Uhl, J. B. Gray (eds.): *Mixing–Theory and Practice*, "Mixing of High Viscosity Materials," vol. 2, chap. 8, Academic Press, New York 1967.

[13] M. H. Pahl: *Mischen bei der Herstellung und Verarbeitung von Kunststoffen*, VDI-Verlag, Düsseldorf 1986.

[14] *Proc. Eur. Conf. Mixing,* **4** (1982), **5** (1985).

[15] M. H. Pahl, E. Muschelknautz, *Chem. Ing. Tech.* **52** (1980) 285; *Int. Chem. Eng.* **22** (1982) 197.

[16] C. Rauwendaal: *Polymer Extrusion*, Hanser Verlag, Macmillan 1986.

[17] E. C. Bernhardt (ed.): *Processing of Thermo-Plastic Materials*, Reinhold Publ. Co., New York 1967.

[18] Z. Tadmor, C. G. Gogos: *Principles of Polymer Processing*, Wiley-Interscience, New York 1979.

[19] D. B. Todd, *Proceedings SPE ANTEC* **33** (1987) 128.

[20] D. B. Todd, E. A. Leach, 2nd Int. Conf., "Reactive Processing of Polymers", Pittsburgh 1982.

[21] R. J. Nichols, J. C. Golba, B. C. Johnson, Proceedings Polymer Processing Society, Montreal 1986.

27. Mixing of Solids

Karl Sommer, Technische Universität München, Lehrstuhl für Maschinen- und Apparatekunde Weihenstephan, Freising–Weihenstephan, Federal Republic of Germany

Symbols

In addition to the standard symbols defined in the front matter of this volume (p. IX), the following symbols are used:

D	diameter in the mixing drum, m
D	diffusion coefficient, m^2/s
D^*	dispersion coefficient, m^2/s
g	mean particle mass, kg
G	sample mass, kg
L	mixer length, m
m	degree of freedom (d.f.)
$\dot{M}$	flow, m/s
n	constant particle number
P	nominal concentration
r	given position, m
s^2	empirical variance
T	transport coefficient, m/s
X	property and sample concentration
ξ	integral variable for R
Δz	displacement, m
σ^2	variance of mixing
χ^2	chi-square variable

Mixing of solids is the distribution of at least one solid component through another. The mixing of granular solids is a widely employed process in chemistry, pharmacy, food manufacturing, the ceramics industry, and ore beneficiation.

Mixing solids is much more difficult than mixing liquids (stirring), where in most cases just a few material constants are enough to characterize the components of the mixture. The mixing process with solids is influenced by disperse properties such as grain shape, grain size, particle-size distribution (gradation), and friction in the bulk material. A comprehensive survey of equipment and methods was published by Ries [1].

Basic mixing mechanisms are not sufficiently well known. This is manifest in the variety of existing mixer types and in the continual appearance of new and novel mixers on the market. In principle, there are two limiting cases:

1) The mixing of powders depends so heavily on product qualities that every product requires its own mixer design.
2) Mixing is a universal process and can be carried out in any apparatus, provided that the material is placed in motion.

Limiting conditions are often more important in mixer selection than the actual mixing process. The following points are often crucial:

1) Explosion safety
2) Cleanability
3) Simultaneous use of the mixer as a silo or interim storage bunker
4) Controlled addition of liquid

1. Mixedness

To assess a mixing process, it is necessary to define a suitable measure of mixedness; and this measure must be kept in mind in all tests, also in regard to the sampling conditions.

Considerable space in the literature on solid–solid mixing has been devoted to the definition and determination of mixedness. Definitions differ so much that any attempt to comprehensively present the state of the art would necessarily end in confusion. The reviews on mixedness are, unfortunately, of little assistance in clarifying the situation [2]–[5].

One of the reasons for the differences is the measure of mixedness presented in a particular review is usually only appropriate for the mixing problem described in the review. Thus FAN et al. introduce elements of packing structure into the definition of mixedness by considering the number of contact points between particles of the two components [6]–[12]. Another reason is that theoreticians often find it easier to develop their own statistical measure of mixedness on the basis of some reasonable assumption than to follow up the lines that had been pursued by others.

The first critical evaluation of methods describing the state of a mixture appeared in 1965 [13]. An analysis of 37 statistical expressions describing the state of a mixture is presented in [14]. Although this description of mixedness is unchallenged, it frequently is too complex for the practicing chemical engineer or industrial chemist.

1.1. Theoretical Variance

All mixing operations, whatever their purpose, have in common that they attempt to even out a nonuniformity in the composition or other properties (e.g., color). Such mixing always involves relative motion between particles, for which energy must be expended.

Total equalization within a mixture leads to ideal mixing, which is symbolized by, say, a checkerboard. In powder mixtures, ideal mixing occurs only in the rarest of cases. The best possible result in technical powder mixing is "random mixing." Particles are randomly mixed when the probability with respect to composition or some property is equal in all regions of the mixture or samples from it. In real mixtures, the properties always differ from the ideal value, i.e., nominal concentration, P. This departure is used in the definition of mixedness. Since it is not critical whether the real values depart upward or downward from the nominal value, the mean absolute value of the deviation can be introduced as a measured value. But the absolute value is difficult to handle mathematically, and so the mean-square deviation, the *variance*, has come to be used.

The variance characterizes the mixture as a whole and thus requires a knowledge of the characteristic property everywhere in the mixture. Such knowledge can be gained only by making theoretical assumptions about the state of mixing or by determining the property everywhere in the mixture. Theoretical variances σ^2 must therefore be distinguished from variances determined on the mixture in practice, which are based on a finite number of samples. To prevent confusion, empirical variances are represented by s^2 (see Section 2.2).

The property X can normally be determined only by a measurement procedure of finite precision. Even when the properties really are completely identical in the samples, the measured values thus depart from the nominal value and contribute to the variance [15]:

$$\sigma^2_{\text{tot}}(t) = \sigma^2_{\text{M}} + \sigma^2_{\text{Z}} + \left(1 - \frac{g}{G}\right) \cdot \sigma^2_{\text{syst}}(t) \qquad (1)$$

The variance is thus composed of three terms: σ^2_{M}, the variance of the precision of measurement; σ^2_{Z}, the variance of random mixing, which represents the optimal case for granular materials; and σ^2_{syst}, the variance characterizing the mixture, which itself can be used to assess the mixedness.

$$\sigma^2_{\text{syst}}(t) = \frac{1}{1 - \dfrac{g}{G}} \{\sigma^2_{\text{tot}}(t) - \sigma^2_{\text{M}} - \sigma^2_{\text{Z}}\} \qquad (2)$$

The factor g/G is the ratio of the particle size (e.g., mean particle mass g) and the sample size (e.g., sample mass G). If the sample and particle sizes G and g are assumed to be equal, the measured property depends only on the particle component, not on the state of mixing. The parameter σ^2_{syst} can no longer be measured in the selection of a sampling method. Evaluation of Equation (2) requires a knowledge of σ^2_M and σ^2_Z.

1.2. Variance of the Precision of Measurement

The precision can usually be determined only through preliminary tests. The variance of the precision must be estimated on the basis of a sufficiently large number of measurements of the selected sample quantity G. Care must be taken that the measurement technique corresponds closely to reality; for example, preparation steps must be included. What is more, a check must be made to see how strongly the precision of measurement depends on the value of the property itself. Often all that is known for a measurement technique is the maximum error. In such cases, a rough estimate for use as a starting point is $\sigma_M \approx \Delta X_{max}/3$.

1.3. Variance of Random Mixing

If the property to be measured correlates directly with a concentration measure of the mixture, then the variance of random mixing can be calculated from the sampling condition [16], [17]. There are a number of special cases:

Constant particle number n; number concentrations x, p

$$\sigma^2_Z(x) = \frac{p(1-p)}{n}$$

Constant particle number n; mass concentrations X, P

$$\sigma^2_Z(X) = \frac{P(1-P)}{E(G)} \cdot (P \cdot g_y + (1-P)\, g_x)$$

where g_x is the mean grain mass of component (X) weighted with the distribution q^x_3 of component (X); g_y is the mean grain mass of component (Y) weighted with the distribution q^y_3 of component (Y), and $E(G)$ is the mean value of sample mass G.

Constant sample property G, mass concentrations X, P

$$\sigma^2_Z(X) = \frac{Q^2 P}{G} \cdot \int_0^1 \frac{g(Q^x_3)}{(Q - P Q^x_3)^2}\, \mathrm{d}Q^x_3$$

$$Q = 1 - P$$

where $g(Q^x_3)$ gives the individual grain property of component (X) as a function of the undersize Q^x_3. The smallest grain of component (X) with gradation function Q^x_3 is larger than the largest grain of component (Y).

Constant sample property G, mass concentrations X, P

$$\sigma^2_Z(X) = \frac{P}{G} \cdot \int_0^1 g(R)\,\mathrm{d}R - \frac{2P^2}{G} \int_0^1 (1-R) \int_0^R \frac{g(\xi)}{(1-\xi)^2}\,\mathrm{d}\xi\,\mathrm{d}R$$

where $g(R)$ gives the individual grain properties of components (X) and (Y) as functions of the oversize R. Components (X) and (Y) have the same gradation, $Q^x_3 = Q^y_3 = 1 - R$.

1.4. Mixedness versus Time

The random variance σ^2_Z and, if the precision is dependent of concentration, the variance σ^2_M are independent of time. The time dependence of the total variance $\sigma^2_{tot}(t)$ is then given by the time variation of the systematic deviations alone.

At time $t = t_0$ let the components in a mixer be completely separated. With the passage of time, the spatial concentration distribution will become uniform. The changes in the system can be described in terms of variations along one coordinate (Fig. 1). In a practical mixing device, the material particles should not undergo any predetermined changes of position. Changes of position should be stochastic. This type of motion is treated by the theory of random processes. MÜLLER [18] has shown that mixing, as a stochastic process without aftereffect, can be described by Kolmogorov's second differential equation. Equation (3) is valid if there is no selec-

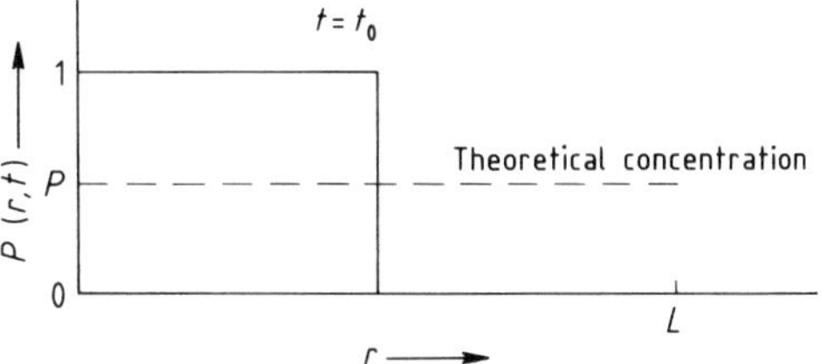

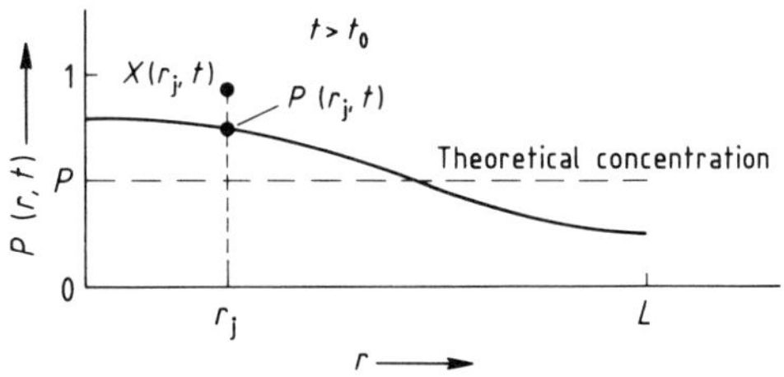

Figure 1. Axial concentration profile in a batch mixer at times t_0 and $t > t_0$. L = length of the mixer, P = theoretical concentration, $P(r_j, t)$ = expected concentration in position r_j at time t

tive transport of the constituents:.

$$\frac{\partial P(r,t)}{\partial t} = \frac{\partial^2}{\partial r^2}\left[D^*(r,t)\cdot P(r,t)\right] \tag{3}$$

where $D^*(r,t)$ is the dispersion coefficient, initially dependent on position and time. If we further assume that $D^*(r,t) = D^* =$ constant, then Equation (3) becomes Fick's second law of diffusion:

$$\frac{\partial P(r,t)}{\partial t} = D^*\cdot\frac{\partial^2 P(r,t)}{\partial r^2} \tag{4}$$

If only batch mixers are considered, the solution can be given as a Fourier series [18]:

$$P(r,t) = P + \frac{2}{\pi}\sum_{\mu=1}^{\infty}\frac{\sin\mu\cdot\pi}{\mu}\cos\mu\cdot\pi\frac{r}{L}\cdot\exp\left(-\mu^2\left(\frac{\pi}{L}\right)^2\cdot D^*\cdot t\right) \tag{5}$$

The system variance is again calculated as the average of the concentration deviations:

$$\sigma_{syst}^2 = \frac{1}{L}\int_0^1 [P(r,t) - P]^2\,\mathrm{d}r$$

and we obtain

$$\sigma_{syst}^2 = \frac{2}{\pi^2}\sum_{\mu=1}^{\infty}\frac{\sin^2\mu\pi P}{\mu^2}\cdot\exp\left(-2\mu^2\left(\frac{\pi}{L}\right)^2 D^*\cdot t\right) \tag{6}$$

The series can be evaluated for two limiting cases:

1) At time $t = 0$ [19]:

$$\sigma_{syst}^2(t=0) = P(1-P) = \sigma_0^2 \tag{7}$$

The limiting case of the initial variance is embodied in Equation (6).

2) For long times, the higher-order terms ($\mu > 1$) can be neglected:

$$\sigma_{syst}^2(t) = \frac{2}{\pi^2}\cdot\sin^2(\pi\cdot P)\cdot\exp\left(-2\left(\frac{\pi}{L}\right)^2 D^*\cdot t\right) \tag{8}$$

Equation (8) shows the desired concentration dependence of the system variance. For $t\to\infty$,

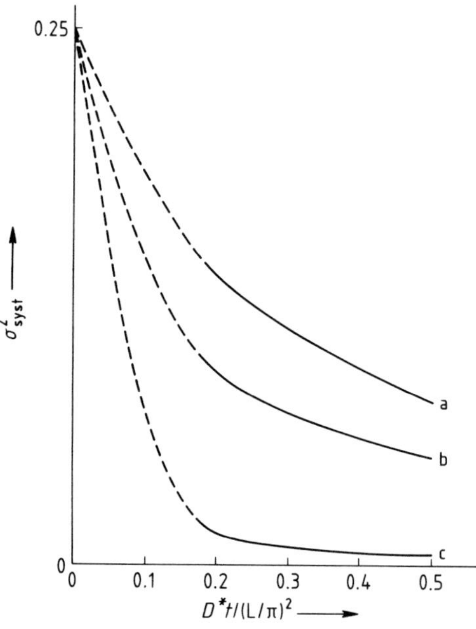

Figure 2. Systematic variance $\sigma_{syst}^2(t)$ as a function of dimensionless time, for pure dispersion. D^* is the dispersion coefficient; L, the mixer length. Parameter: nominal concentration P: a) 0.5; b) 0.3; c) 0.1

$\sigma_{syst}^2\to 0$ (Fig. 2). The limiting case of infinitely long times is likewise embodied in Equation (6).

Limit of Low Concentrations *P*. If component (X) is present only in small quantities, so that $P \ll 1$, the sine in Equation (8) can be expanded into a series, and all but the first term dropped:

$$\frac{\sigma_{syst}^2(t)}{P^2} = 2\cdot\exp\left(-2\left(\frac{\pi}{L}\right)^2 D^*\cdot t\right) \tag{9}$$

Here σ_{syst}/P is called the system coefficient of variation. From Equation (9) follows that, for small admixtures $P \ll 1$, the system coefficient of variation is invariant with respect to both sample size and mixture composition, satisfies the requirements of SCHMAHL [20], and is suitable as a measure of mixedness.

Limit of Equal Proportions. If the two components to be mixed are present in roughly equal quantities, so that $P\approx 0.5$, Equation (8) simplifies to

$$\frac{\sigma_{syst}^2(t)}{P^2} = 2\cdot\exp\left(-2\left(\frac{\pi}{L}\right)^2 D^*\cdot t\right) \tag{10}$$

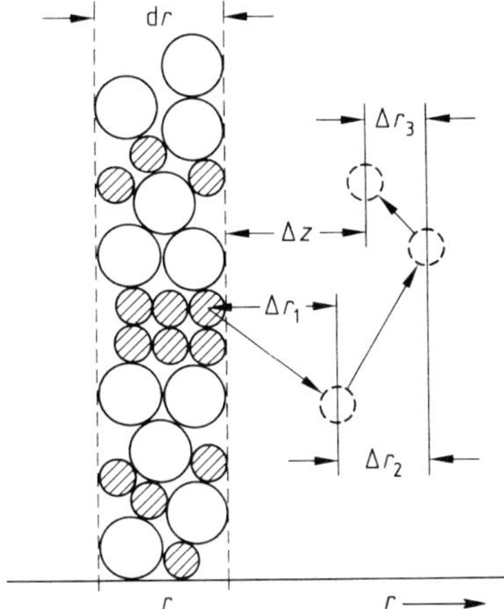

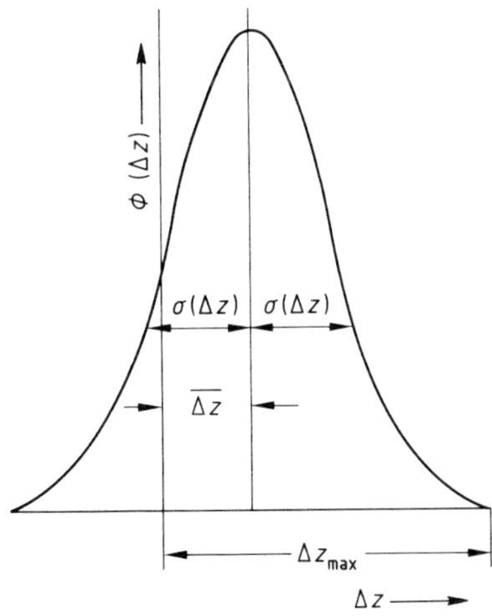

Figure 3. Schematic diagram illustrating the derivation of the Fokker–Planck differential equation: Δz is the displacement in time τ; $P(r)$, the concentration at point r; $\phi(\Delta z)$, the distribution function of Δz

In the region of almost equal proportions, the system variance itself is independent of sample size and mixture composition, satisfies the requirements of SCHMAHL, and is an appropriate measure of mixedness.

If the components of powder mixtures differ in grain shape, grain size, or density, stochastic mixing processes [21]–[28] often exhibit segregation that cannot be accounted for by the simple diffusion model with positive dispersion coefficient. The more fundamental MARKOV theories of random processes must be introduced in these cases; they lead to the so-called Kolmogorov equations. These equations were first written for statistical dynamics by FOKKER [29] and PLANCK [30] (Fig. 3).

Suppose there are a large number of spheres, mutually independent and having identical surface properties. A model system comprises small glass spheres in a matrix of large glass spheres. Let each of the spheres be set in motion. At a given time t, the concentration at a given position r has some value $P(r)$. Each of the small spheres experiences very small, rapid, stochastic collisions. Now the question is how the concentration $P(r)$ at point r changes in an interval of time. In this interval, each sphere experiences several collisions r in which the direction of motion may change. The sum of all these changes of position is called the displacement Δz. Each sphere will exhibit a different displacement Δz, even if the spheres are located at the same point r. The displacements Δz are statistically distributed with distribution function $\phi(\Delta z)$, where Δz_{max} is the greatest displacement that occurs. It should always be small in comparison with the mixer dimensions. Here $\overline{\Delta z}$ is the mean displacement of the spheres, that is, a mean convective transport.

Besides this convective motion, a broadening of the distribution density results from the stochastic nature of the process. The broadening is described by the mean square of the displacement $\overline{\Delta z^2}$:

$$\overline{\Delta z^2} = \sigma^2(\Delta z) + (\overline{\Delta z}^2) \tag{11}$$

where $\sigma^2(\Delta z)$ is the variance of the distribution density $\phi(\Delta z)$.

The quadratic quantity $\overline{\Delta z^2}$ is, of course, always positive. FOKKER [29] and PLANCK [30] calculated the change in concentration $P(r)$ in segment dr from the difference between the number of particles leaving dr in time τ and the number entering dr from adjacent layers in the same time. They stated the result in differential equation form:

$$\frac{\partial P(r)}{\partial t} = \frac{\partial}{\partial r}\left[P(r)\lim_{\tau\to 0}\frac{\overline{\Delta z}}{\tau}\right] + \frac{\partial^2}{\partial r^2}\left[P(r)\lim_{\tau\to 0}\frac{\overline{\Delta z^2}}{2\tau}\right] \tag{12}$$

As was already mentioned, $\overline{\Delta z}$ is the mean convective transport of the small spheres at position r in the time interval. Then $\overline{\Delta z}/\tau$ is the mean convective transport velocity. If the limit $\lim_{\tau\to 0}\overline{\Delta z}/\tau$ exists, it is called the transport coefficient T and has the units of velocity. If the limit $\lim \overline{\Delta z^2}/2\tau$ exists, it is called the *dispersion coefficient D^**; it has the units of area per unit time, the same as the diffusion coefficient D. The dispersion coefficient, however, is not a pure adjustment factor but a statistically derived quadratic quantity, and as such it is always positive. Then

Equation (12) is written as

$$\frac{\partial P(r)}{\partial t} = -\frac{\partial}{\partial r}[P(r)\,T(x)] + \frac{\partial^2}{\partial r^2}[P(r)\,D^*(r)] \quad (13)$$

The form of this equation, which describes the time variation of concentration at positive r, matches Fick's second law of diffusion.

The analog of Fick's first law of diffusion is the mass flow rate through a unit cross-sectional area at position r at time t:

$$\dot{M}(r) = P(r)\,T(r) - \frac{\partial}{\partial r}(P(r)\,D^*(r)) \quad (14)$$

Thus the mass flow $\dot{M}$ is composed of the convective transport $\dot{M}_T$ and the dispersive transport $\dot{M}_D$.

Let us make the simplifying assumption that $D^*(r) = D^*$ independent of the coordinate r. If in this case the transport coefficient $T(r) = 0$, Equation (14) becomes Fick's first law of diffusion, and the dispersion coefficient D^* is identical with the diffusion coefficient D. If in the mixture under consideration the components are completely separate from each other at time t, the diffusive equalization already described takes place. The variance σ^2 of the sample concentration declines steadily from the initial state σ_0^2 until it reaches the value of the sampling error σ_z^2.

When a transport flow is present, there are two possibilities: (1) flows initially in the same direction (Fig. 4, third curve from top) and (2) flows initially in opposing directions (Fig. 4, bottom curve). If the two flows are initially in the same direction, transport at first aids the equalization of concentration, but then the diffusion flow reverses and opposes the transport until both flows are equal in magnitude. This state, which is steady, is a condition of partial mixing. At first the mixedness rapidly improves, but as demixing increases the mixedness becomes worse again. An experimental example of such qualitative segregation is found in the trials reported in [18] (see Fig. 5). The differing transport properties of iron (630–750 µm) and quartz (100–200 µm) and of iron (300–400 µm) and limestone (40–60 µm) give rise to segregation.

The empirical coefficients of variation reflect the prediction from the model; the diffusion and transport flows are initially in the same direction in Test 1, opposing in Test 2. For this reason, Test 1 (Fig. 5, Curve a) leads to homogeneous mixing at first but then to segregation. In Test 2, (Fig. 5, Curve b) the stochastically homogeneous state is not reached at any time.

The segregations at constant dispersion coefficient, as described in the preceding section, always occur when movement of internal parts or inclination of the drum causes selective transport that opposes diffusive mixing. Segregations are seen in drum-type mixers even when no external convective transport processes are taking place. However, there is always an equalization of concentration when the transport coefficient vanishes and the dispersion coefficient is a constant. It should therefore be checked whether a position-dependent or concentration-dependent dispersion coefficient can account for the observed segregations without convective transport. For $T(r) = 0$, Equation (14) becomes

$$\frac{\partial P(r)}{\partial t} = \frac{\partial^2}{\partial r^2}[P(r)\,D^*(r)] \quad (15)$$

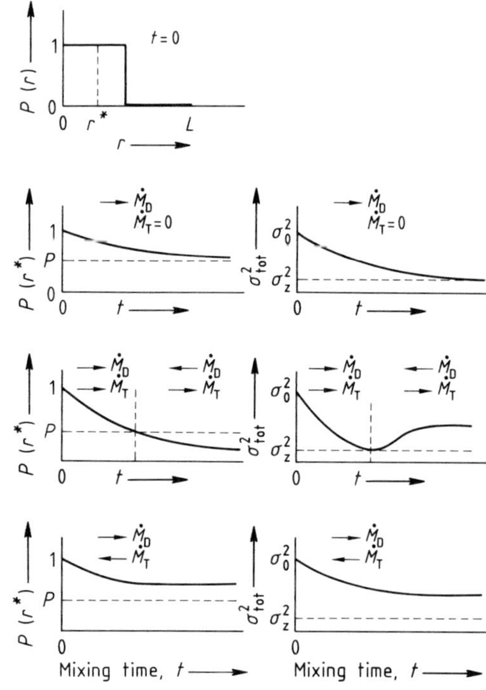

Figure 4. Concentration $P(r^*)$ at position r^* and mixedness σ^2_{tot} as function of mixing time [18]

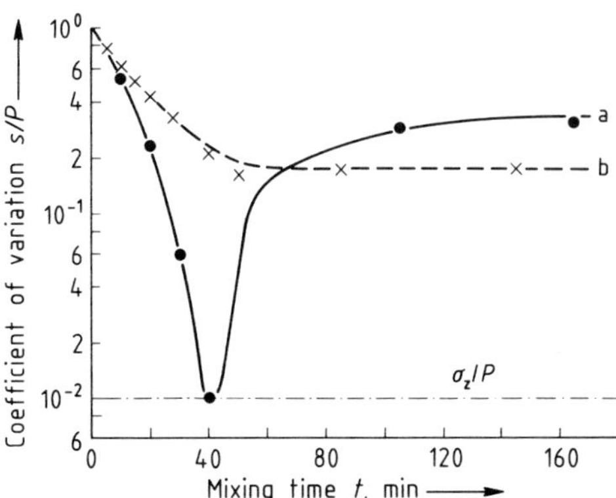

Figure 5. Coefficient of variation s/P versus mixing time t
a) Quartz, 100–200 µm, and iron, 630–750 µm (indicator); b) Limestone, 40–60 µm, and iron, 300–400 µm (indicator), [18]

If the product is first differentiated with respect to concentration and the concentration is then differentiated with respect to r, a form comparable with Fick's first law of diffusion is obtained:

$$\dot{M}(r) = -\frac{\partial (P\,D^*)}{\partial P}\frac{\partial P(r)}{\partial r} \tag{16}$$

If the dispersion coefficient is taken to depend only on the concentration, the product rule can be used in differentiation:

$$\dot{M}(r) = -\left(D^* + P\frac{\partial D^+}{\partial P}\right)\frac{\partial P(r)}{\partial r} \tag{17}$$

The expression in parentheses corresponds exactly to the Fick diffusion coefficient:

$$D = \left(D^* + P\frac{\partial D^*}{\partial P}\right) \tag{18}$$

While the dispersion coefficient D^* is always positive, by hypothesis, the diffusion coefficient D can become negative when the concentration dependence of D^* supplies a sufficiently large negative derivative. Then there exists a material flow in the direction of higher concentrations; that is, the components separate.

To investigate these processes, glass spheres were mixed in a glass drum 205 mm long and 114 mm in diameter. First, dark and light spheres of the same size (d = ca. 2 mm) were used, with a fillage of 38 %; the drum rotated at 45 rpm. The mixture ratio was 1:1. Because the spheres were equal in size and made of the same material, the dispersion coefficient could be taken as constant independent of the concentration. The contents of the drum thus became mixed to stochastic homogeneity in a short time ($\approx$20 min).

If small glass spheres (d = 1 mm) are mixed with large ones (d = 2 mm), the mobility of the small particles in the matrix of larger particles is estimated to be greater than in an environment of equal-sized spheres. Thus the dispersion coefficient decreases as the concentration increases. If the decline is great enough, the diffusion coefficient may become negative and the components, completely separate at first, will not mix. In the practical experiment, mixing was not observed even over very long mixing times. The contents of the drum remained separate.

If the small spheres are layered on top of the large ones, the concentration is initially uniform in the axial direction. Diffusion flow is theoretically impossible in this instance. But the equilibrium is labile if the diffusion coefficient is assumed to be negative. Because of irregularities in charging and the boundary conditions, violations can always occur, leading to concentration shifts and thus making segregation possible. Such upsets were seen close to the right and left end walls in the experiment. Radial disks near these walls showed an increased concentration of small spheres after five minutes of mixing, while the concentration in the other zones decreased. The observed segregations remained stable even over longer mixing times. The number of violations depends on the length of the mixing drum.

Numerous strips could be generated by varying the drum geometry [21]–[23]. Symmetry in the stated form can be expected only when the initial conditions are symmetrical. If violations are imposed in the form of local increases on concentration, individual strips can even be programmed to appear at various places. Exact observation reveals that the quantity of small spheres concentrated in the stripes does not make up half the mixture contents, as at the beginning. Thus a large fraction of the small spheres must still be present in the matrix of the large spheres.

This fact can also be explained on theoretical grounds. Equation (17) shows that, for concentrations $P \to 0$, the diffusion coefficient becomes positive in every case, even when the derivative $\partial D^*/\partial P$ is large and negative. At low concentrations, then, the components are always miscible. If it is assumed, for example, that a linear relationship holds between the dispersion coefficients of the small spheres, D_1^* for concentration $P \to 0$ and $D_{2_i}^*$ at concentration $P = 1$, then the limit of segregation lies at $P_0 = 0.5$ in the first case (D_2^*) and at a higher concentration P_0 in the second case (D_2^*) (Fig. 6). If the concentration is below the limit P_0, the diffusion coefficient is positive, so that equalization of the concentration takes place by diffusion. At a higher concentration, segregation occurs. If the dispersion coefficients are in the ratio of $D_1^*:D_2^* = 1:2$, the concentration limit is near $P_0 = 1$; that is, at this ratio and all ratios nearer to unity, no segregation takes place.

The linear concentration dependence is merely an assumption, which could not be applied to the experimental system because of the high concentration limit of $P_0 = 0.5$. On the other hand, if the dependence resembled a normal distribution, the segregation limit P_0 lies at the steepest part of the curve, so that it may occur at much lower concentrations (Fig. 6).

Practical Consequences. The postulated concentration dependences of the dispersion coefficients are quite arbitrary. The first attempts to determine these dispersion coefficients for solids

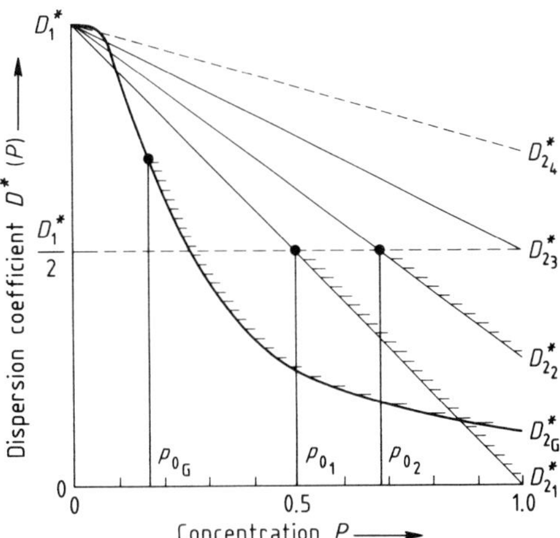

Figure 6. Estimated dispersion coefficient as a function of concentration P

were made by CAHN and FUERSTENAU [31], [32]. Unfortunately, these results cannot be applied generally. The analysis of mixing processes in terms of stochastic processes does, however, have qualitative implications. If there is a tendency to segregation because of convective transport, then according to [18], [24] the selective transport behavior can be altered by even slight changes such as the addition or removal of mixing elements.

Convective segregation tendencies should be offset, whenever possible, by a suitable choice of grain size and density [25]. If it is known that two components tend to segregate on account of negative diffusion coefficients, then the concentration dependence of the dispersion coefficient must be reduced. This is accomplished in practice by agitation in fluid-bed and centrifugal mixers. The mean free spacing between particles is large in comparison to the particle size, so that the mobility of individual particles is less dependent on the size of the surrounding particles. The success of powder mixers using the fluid-bed and centrifugal principles is due not only to the absolutely larger dispersion coefficients, but certainly also to the weaker concentration dependence. There is, however, a great danger that the initially homogeneous mixtures may separate again during slow discharging and charging, because of the concentration dependence of the dispersion coefficient.

A second way to homogenize a mixture that tends to segregate is to add small amounts of liquid [33]–[35]. The liquid causes water to form bridges between the spheres. Because of the lesser mobility, the dispersion coefficients are much lower, but the difference in environment and thus the concentration dependence no longer play a dominant role. Diffusive mixing is thus possible. The addition of about 1 wt% water to a charge of glass spheres 1 and 2 mm in diameter leads to a stochastically homogeneous final state after 10 min of mixing.

2. Practical Determination of Mixedness

For the theoretical mixedness (σ^2_{tot} or σ^2_{syst}) to be determined, the property must be known everywhere in the mixture. In practice, however, only a finite number of samples can be taken from the mixture. A variety of devices can be used for this purpose.

2.1. Sampling

The simplest methods of sampling are manual ones employing sampling shovels or ladles. Sampling shovels have the top and one side open, and thus are suitable for pulling material out of a stationary bed. The sampling ladle is open only on top and is used for removing granular material from a falling stream of material. Sampling forks are good only for coarse materials such as coke.

Thieves and sampling drills are used for sampling from mixers and silos. These devices do not just sample the surface, as shovels do, but reach to depths of several meters. The thief, in simplest

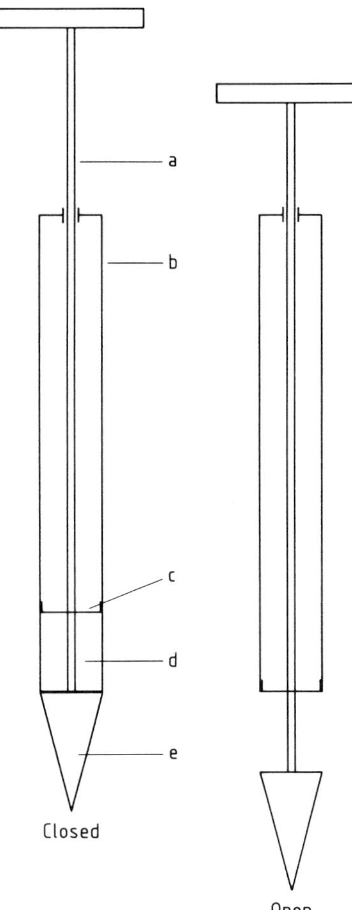

Figure 7. Thief for taking samples of free-flowing solids (Design: Lehrstuhl für Maschinen- und Apparatekunde [Department of Machinery and Apparatus], Freising–Weihenstephan, Federal Republic of Germany)
a) Guide rod; b) Tube; c) Cover; d) Sample chamber; e) Point

form, consists of a steel tube with an inside diameter of 30–50 mm, a wall thickness of 2–4 mm, and a length of 1.50 m (up to 4 m in special cases). The thief in Figure 7 is especially well suited to taking samples from free-flowing powders [36], [37]. The thief is closed and pushed into the material; sliding the tube axially along the guide rod opens and fills the chamber; the thief is then reclosed and withdrawn. For cohesive material such as moist soils, the sampling drill is appropriate. It consists of two fixed, curved plates mounted on a crosspiece. The drill is advanced into the pile by a screw motion.

The best-known sampling device for fluidized solids is the channel sampler. Normally, the channel sampler is advanced across the falling stream of material. Where the material under study is being conveyed by a belt, trough or screw conveyor, hand sampling is more and more being replaced by automatic sampling. A simple way to take samples from a belt is to use a fixed stripper, which intercepts an exact cross-sectional area of material transverse to the conveying direction (Fig. 8).

Pendulum samplers (Fig. 9) and impact samplers (Fig. 10) find use in automatic sampling at the discharge from a belt or trough conveyor. Material is acquired at regular intervals, either transverse or parallel to the discharge direction. The pendulum sampler consists of a slotted vessel suspended from a pendulum arm. A drive rod causes it to move through the stream of material. When the sampler returns to its rest position, the material is dumped by opening the bottom gate [38], [39]. In the impact sampler, a concave impactor knocks the material out of the falling stream in the conveying direction.

A simple slot (Fig. 11) takes samples out of a trough conveyor. The upright slot diverts a sub-

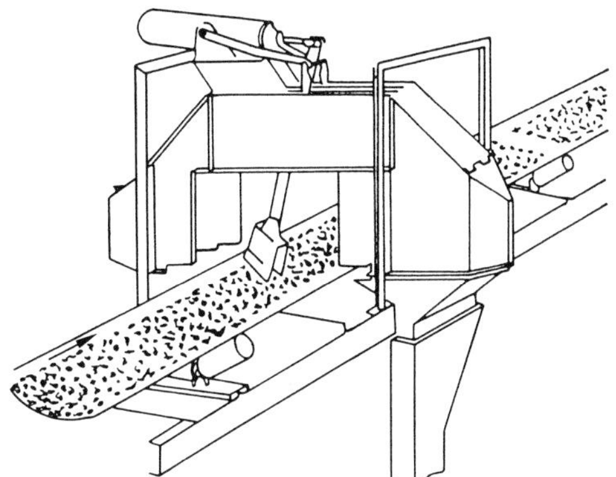

Figure 8. Sample stripper

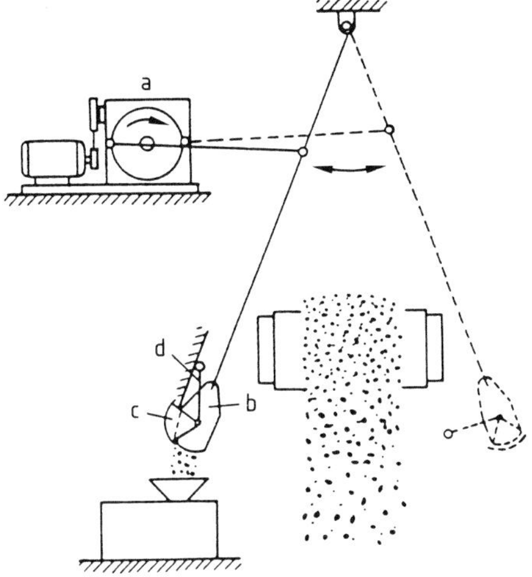

Figure 9. Pendulum sampler
a) Sabo timer drive; b) Slotted vessel; c) Bottom gate; d) Actuating lever with roller

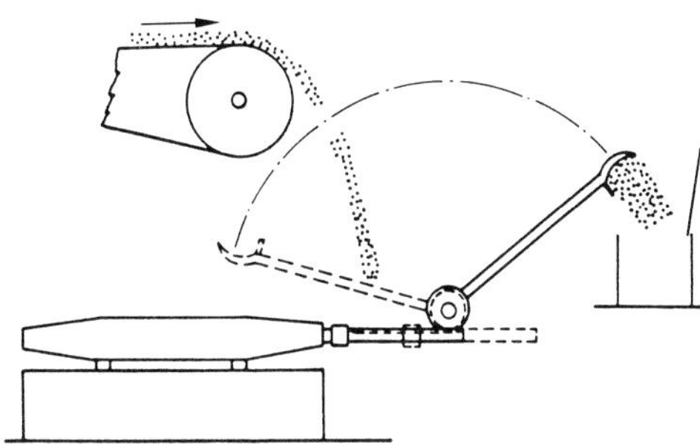

Figure 10. Impact sampler

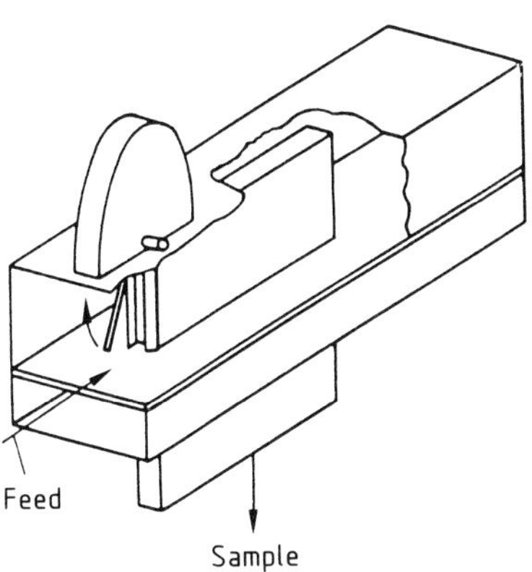

Figure 11. Slot arrangement for sampling from a trough conveyor

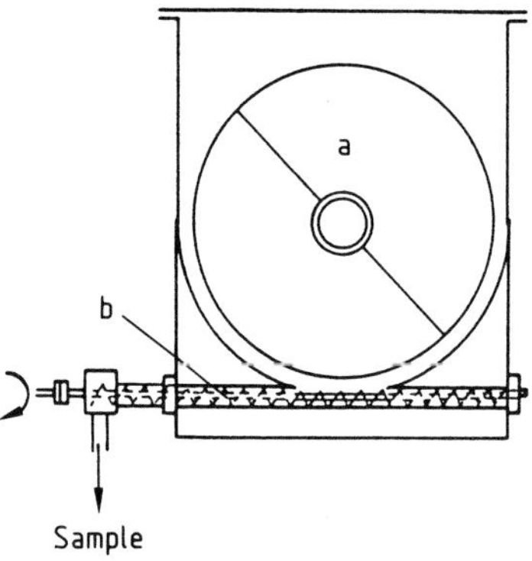

Figure 12. Screw sampler (b) mounted on screw conveyor (a)

stream of material, the sample, out of the middle of the trough. The substream remains proportional to the material flow rate as the layer depth and conveying speed vary. With a star-wheel device or a screw sampler with several slots and holes on top, fluidized material can be diverted from discharge ducts and conveying pipes [40]. The screw-type sampler can also remove a substream from a larger screw conveyor (Fig. 12). If the sampling screw rotates at a constant speed, the quantity sampled is independent of the quantity of material moved by the large screw.

In many cases, the samples thus acquired are too large for measurement. The samples are remixed in the laboratory, and a sample for analysis is then taken. This procedure should be used with caution. The laboratory mixing process must itself be monitored, since there is always a danger of segregation. It is better to use a sample divider to reduce the sample to a size that can be analyzed. Errors are also possible with this technique, but they can be determined. If, however, the particle size is small in comparison with the sample for analysis, these errors are usually negligible. Otherwise, the starting material must be comminuted until this requirement is satisfied before the sample is divided. Given these conditions, the analytical value can be assigned to the starting sample.

2.2. Empirical Variance

2.2.1. Frequency Distribution and Chi-Square Distribution

If there are analytical values from k samples, the *mean-square deviation* or *variance* s^2 is

$$s^2(X) = \frac{1}{k-1} \sum_{i=1}^{k} (X_i - \bar{X})^2 \tag{19}$$

$$s^2(X) = \frac{1}{k} \sum_{i=1}^{k} (X_i - P)^2 \tag{20}$$

The second formula is used when the mixture composition P is known. If the analytical values X_i are normally distributed (an assumption that can always be made as a first approximation), the quantity $\chi^2 = m \frac{s^2(x)}{\sigma^2(x)}$ has a *chi-square distribution*; the *number* of *degrees of freedom* (d.f.) is $m = k - 1$ when Equation (19) is used or $m = k$ when Equation (20) is used. The chi-square distribution is an asymmetrical one-dimensional probability distribution (Fig. 13). It allows a statistical assessment of the quality of the estimate s^2 in relation to σ^2, where σ^2 is the expectation of the statistically varying quantity s^2 (Fig. 14). The more d.f. and thus the greater the number of samples, the more frequently s^2 will

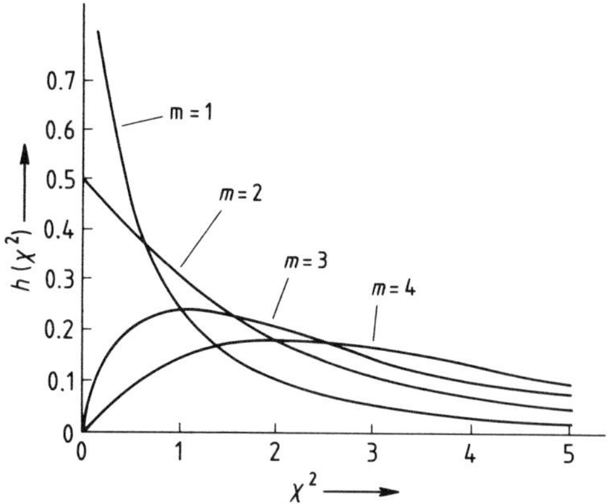

Figure 13. Chi-square distributions for degrees of freedom $m = 1$, 2, 3, and 4

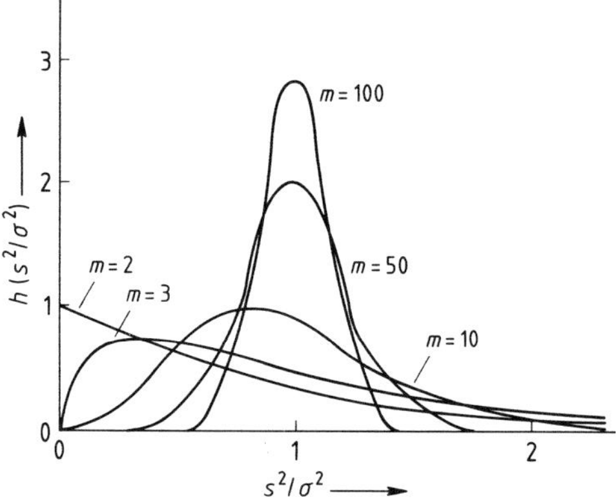

Figure 14. Frequency distributions of s^2/σ^2 for degrees of freedom $m = 2$, 3, 10, 50, and 100

be found close to the expected value σ^2, and the more symmetrical the frequency distribution will be.

2.2.2. Confidence Intervals

If a certain probability is stated (in engineering usually 95–99%, but often higher in the pharmaceutical field), the variance can be found from the chi-square distribution (Fig. 15). For example, it is possible to say with 95% confidence that

$$\frac{s^2}{\sigma^2} \geqq \frac{\chi_u^2}{m}$$

Conversely, it can be asserted with the same degree of confidence that the true value of σ^2 (the mixedness) satisfies the relation

$$\sigma^2 \leqq \frac{m}{\chi_u^2} \cdot s^2 \qquad (21)$$

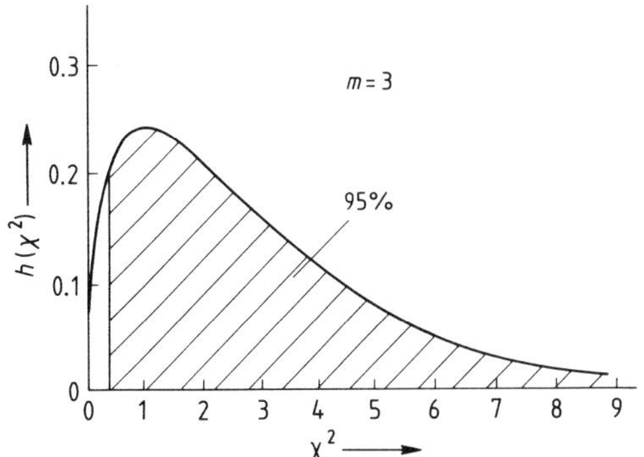

Figure 15. One-sided 95% confidence interval of the chi-square distribution, $m = 3$, $\chi_u^2 = \chi_{0.05}^2$

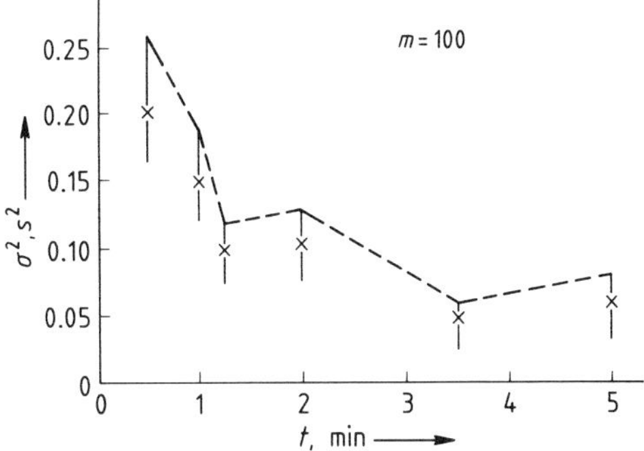

Figure 16. Example of a curve of mixedness versus time. Vertical bars represent confidence intervals

In the diagram of mixedness versus time, Equation (21) gives the confidence interval for the true mixedness (Fig. 16). With 95% confidence, the true value is lower than the stated limit.

2.2.3. Determination of Mixing Time

More difficult to determine than the confidence interval is the mixing time required to achieve a given mixedness σ_{goal}^2. The frequency distributions of s^2 (Fig. 14) show that, for few d.f., the number of s^2 values smaller than the true value of σ^2 is greater than the number of s^2 values greater than the true σ^2. For $m = 2$ d.f., the frequency is actually greatest for $s^2 = 0$.

The measured values of s^2 thus give the illusion of far too good a result. For the determination of the mixing time, as in the case of the mixedness, it is more correct to use the upper limit of the confidence interval.

Two cases must be distinguished in practice:

1) The mixedness σ_E^2 in the mixer is clearly better than the prescribed mixedness σ_{goal}^2 (Fig. 17 A). The mixing curve distinctly inter-

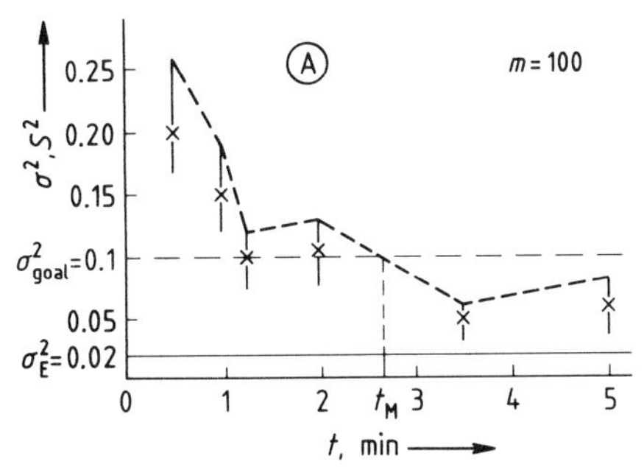

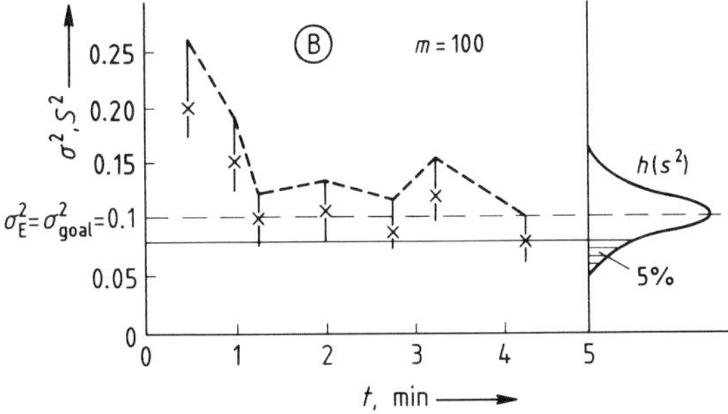

Figure 17. Mixedness versus time curves, where the final mixedness σ_E^2 is (A) very much smaller than the goal value σ_{goal}^2 or (B) roughly equal to σ_{goal}^2. Vertical bars represent confidence intervals.

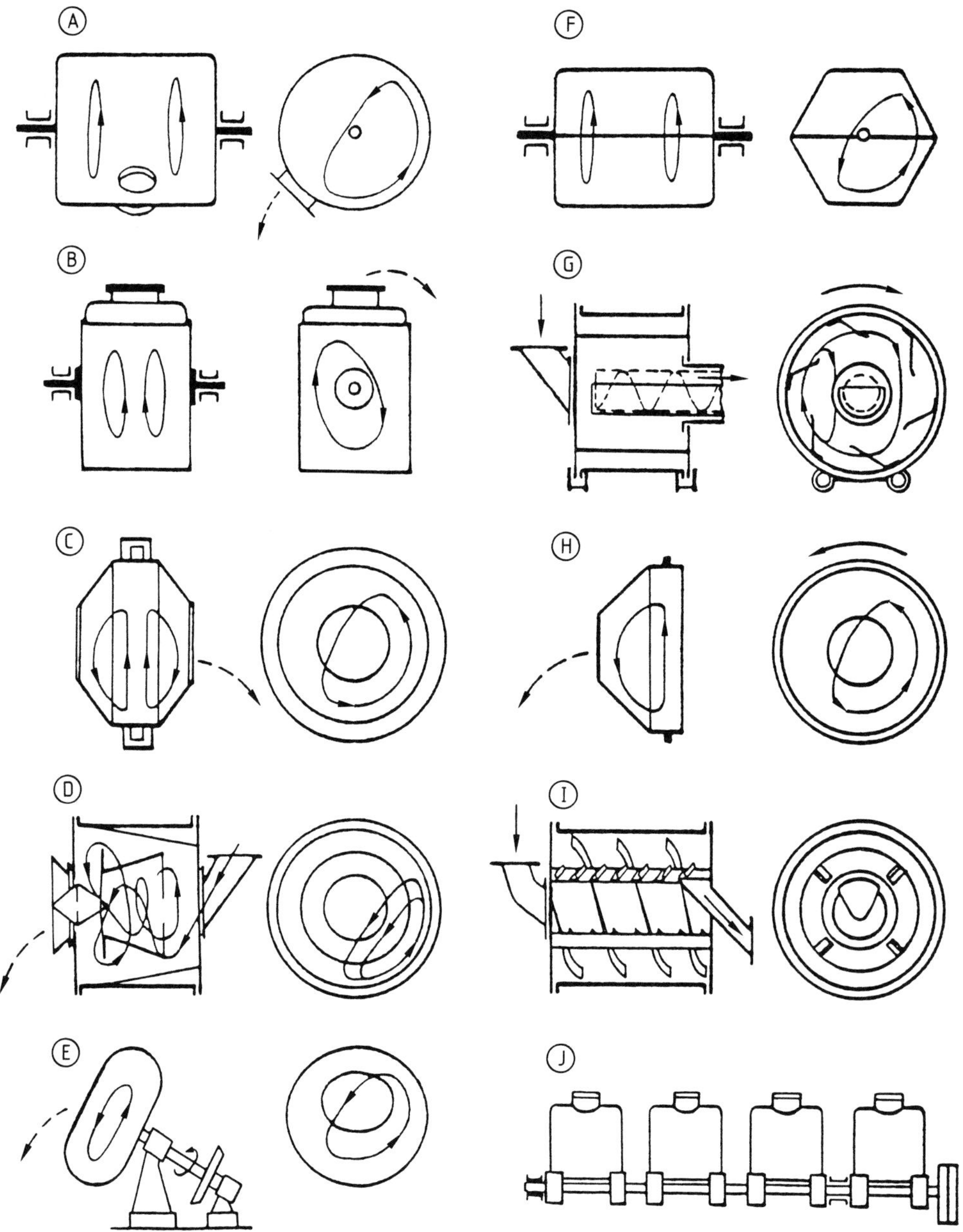

Figure 18. Tumbler mixers
A) Cylindrical drum; B) Barrel mixer; C) Cone mixer; D) Drum with adjustable baffles; E) Pear-shaped drum; F) Hexagonal drum; G) Drum with screw discharge; H) Single-cone mixer; I) Drum with spiral flights; J) Laboratory device for mixing vessels

sects the goal value within two time intervals, and the mixing time can be determined accurately.

2) The mixedness in the mixer σ_E^2 just reaches the prescribed mixedness σ_{goal}^2 (Fig. 17 B).

When the final mixedness has been reached, the measured values of s^2 fluctuate about this value in accord with their frequency distribution. In most cases, the upper limit of the confidence interval is above the mixedness target, and no clear intersection can be expected. Only when a measured value is smaller than the value s_u^2 from the chi-square distribution χ_u^2 does the upper limit of the confidence interval lie below the target value. This happens $(1 - P)$ times, where P is the selected probability; thus, if the confidence interval is 95%, this happens 5 times in 100 cases.

To keep from choosing too long a mixing time, the time intervals for sampling must be held short, even though this increases the cost.

If the final mixedness is prescribed not by the mixer (σ_{syst}^2) but by the material being mixed and the precision of measurement,

$$\sigma_E^2 = \sigma_M^2 + \sigma_Z^2$$

then minimizing the final variance σ_E^2 is often a better way to reduce the cost in the determination of mixing time. This can be done by improving the sample preparation method and the precision of measurement. If σ_Z^2 is the only possible approach for improvement, the ratio of particle size to sample size can be reduced (larger samples or smaller particles).

3. Designs of Solid–Solid Mixers

Coarse mixing in solids mixers is accomplished by the continuous movement of material in the process volume. It must be possible for adjacent particles to change places. Depending on the flowability of the powders being mixed, this can be achieved through a variety of mechanisms [41] and in a variety of geometries [42]. The large number of mixers on the market can be classified under several headings [43], [44].

3.1. Tumblers

A tumbler mixer is typically a vessel in which the mixture is rotated (Fig. 18). A tumbling motion may be superimposed on the rotational motion, or components mounted on the inside of the mixing drum (internals) may force the loose material into motion. A quite simple and well-known mixer of this type is the concrete mixer. It has a capacity of 2 m^3/h on average, with mixing times of some 1–3 min. When rotation speeds are low, the processes of importance for mixing occur in relatively thin layers at the surface of the material bed (*cascade motion*). The slow relative motion promotes any segregation effects that occur. As the rotation speed n increases, the fraction of solid material in motion becomes greater. Some individual particles escape from the bed and fall freely through the empty volume onto the bed below (*cataract motion*). If a critical rotation speed n is exceeded, the particles are held against the drum wall by centrifugal force, and free movement and thus mixing are prevented. The scaling law for model experiments is

$$Fr = n/n_{crit} = n^2 D/g = \text{constant}$$

where D is the diameter of the mixing drum. Fr is the dimensionless Froude number. Tumblers have drum volumes of 2–100 000 L and drive powers of 0.2–33 kW [43].

3.2. Screw Mixers

Another class of mixers operating at relatively low Froude numbers comprises screw mixers, in which the pan or drum is stationary and a screw or screw-shaped flights or ribbons move inside it (Fig. 19). The flights move through the bed of material at tool speeds of up to 2 m/s, loosening the material and moving portions of it along. This process creates voids, which can be filled by free-flowing material. Ribbon-type impellers, usually made of flat steel strip, rotate about a horizontal axis in horizontal drums or about one or more vertical axes in pans.

Another widespread form is the screw mixer in a conical vessel (Fig. 20). One or two orbiting screw conveyors cause the material not only to rotate in the cone but also to move upward at the same time. The dual motion causes continuous mixing. Entrop [44] has developed design diagrams for this type of mixer, giving both the mixing time and the power consumption.

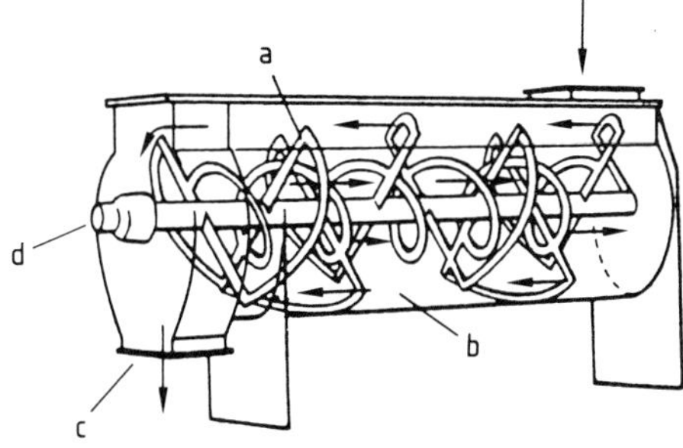

Figure 19. Ribbon mixer (Drais)
a) Helical ribbon; b) Vessel; c) Discharge; d) Drive

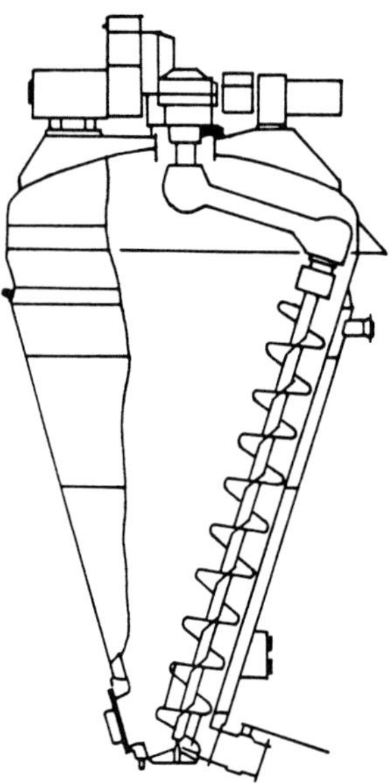

Figure 20. Hosokawa – Nauta mixer

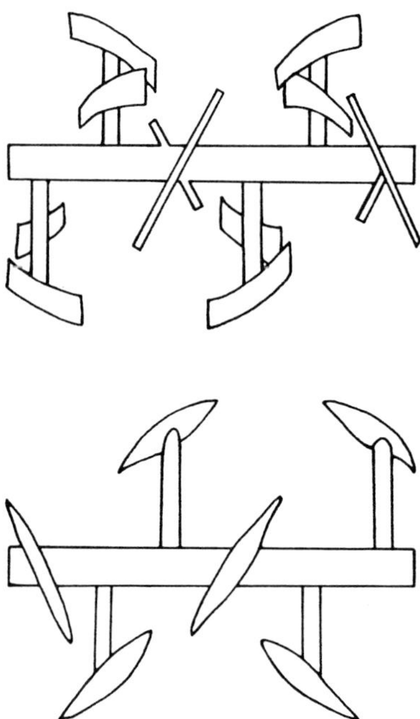

Figure 21. Paddle shafts

3.3. Paddle Mixers

Paddle or spreader mixers operate at higher rotation speeds; the peripheral speed of the mixing tool is 10 – 40 m/s. The shaft, usually vertical, bears paddles or shovels on short end pieces (Fig. 21); plowshares or partial screw flights are also used. The mixing elements separate particles from the bed of material, setting them in motion in the air space above the charge and allowing them to mix. Angled paddles or plowshares allow an axial movement that promotes axial mixing. Because the cylinder wall and the end surfaces are accessible, liquids can be metered in with little or no crust formation. Agglomerates can be broken either with cutter heads rotating at high speed in the bed of material or with pin devices. The Eirich mixer features a pan and shovels rotating in opposite directions, thus efficiently combining the advantages of the drum mixer and the paddle mixer. The solid material moves in continually overlapping trajectories and is thus comminuted, mulled and, if necessary, moistened.

3.4. Pneumatic Mixers

In pneumatic mixers, the mixing action is due to the injection of air. The simplest mixer of this type is the fluid-bed mixer.

The material bed is located above a horizontal or slightly inclined perforated plate. Air flows upward through the plate and bed until the gas velocity exceeds the value required to loosen and expand the bed. Gas bubbles form and carry the particles along as they rise. At the surface, the bubbles burst, distributing the particles in the air space above the fluidized bed.

Mixers now on the market differ in the air delivery pattern, the bottom plate being divided into quadrants or segments. As mixing proceeds, one quadrant or segment after another is supplied with extra air, and in this way the bed material is caused to circulate.

Fluid-bed mixers can be used only when the individual components are not too different in sinking velocity, since otherwise segregation will occur. Another class of pneumatic mixers includes air-jet mixers of free-flowing solids. These devices consist of vessels into which gas is injected through one or more nozzles at velocities up to supersonic (Fig. 22 A). The air-jet mixer has the advantage that the mixing time is short and there is not much danger of segregation. It has the disadvantages that a high-pressure compressor must be used and that the particles are subjected to high loads in the driving jet.

Much gentler in operation are rotating mixers (Fig. 22 B) with internal or external rotation.

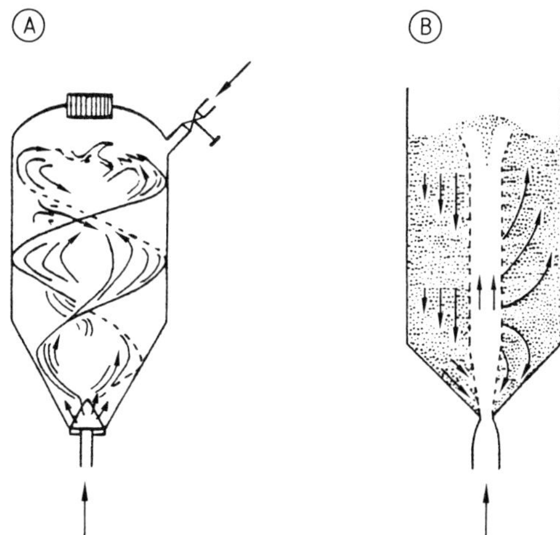

Figure 22. Pneumatic mixers
A) Airmix-system fluid-bed mixer; B) Rotating mixer

These are often used to homogenize granular resins. Air is blown into an ascending pipe and conveys the granular material upward to a separating zone. Material is admitted to the central ascender through, for example, mixing slots in various regions of the bed, so that compulsory mixing takes place.

3.5. Silo Mixers

The principle of the pneumatic rotating mixer, where material portions from various regions of the bed are brought together, is used so that large silos can be used as continuous mixers, especially for large-tonnage products. The gathering of material from various parts of the silo can be monitored with, for example, concentric floor discharge openings at various heights.

3.6. Pile Mixing

When the quantities to be mixed are so large that they can no longer be stored in a single silo, mixing must be done by appropriate piling and reclaiming techniques. This form of solid–solid mixing is important for the ore and beneficiation industries [45].

4. References

[1] H. B. Ries, *Aufbereit. Tech.* **20** (1979) no. 1, 1–25, no. 2, 78–90.

[2] L. T. Fan, S. J. Chen, C. A. Watson, *Ind. Eng. Chem.* **62** (1970) no. 7, 53–69.

[3] H. B. Ries, *Aufbereit. Tech.* **17** (1976) no. 1, 16–27.

[4] H. B. Ries, *Int. Chem. Eng.* **18** (1978) no. 3, 426–442.

[5] L. T. Fan, *Powder Technol.* **24** (1979) 73–89.

[6] Y. Akao, *Powder Technol.* **15** (1976) 207–214.

[7] Y. Akao, *Powder Technol.* **21** (1976) 267–277.

[8] H. Shindo, *Powder Technol.* **21** (1978) 105–111.

[9] L. T. Fan, *Powder Technol.* **22** (1979) 205–213.

[10] J. R. Too, *Powder Technol.* **23** (1979) 99–113.

[11] S. H. Shin, L. T. Fan, *Powder Technol.* **19** (1978) 137–146.

[12] R. H. Wang, *Powder Technol.* **21** (1978) 171–182.

[13] T. Yano, Y. Sano, *Kagaku Kogaku Abr. Ed. Engl.* **3** (1965) no. 2, 199–203.

[14] J. Boss, *Bulk Solids Handling* **6** (1986) no. 6, 1207–1215.

[15] K. Sommer, *Aufbereit. Tech.* **23** (1985) no. 5, 266–269.

[16] K. Sommer: *Sampling of Powders and Bulk Materials*, Springer-Verlag, Heidelberg 1986.

[17] K. Sommer: "Solid-Solid Mixing," *Fortschr. Verfahrenstech.* **19** (1981) 189–208.

[18] W. Müller, *Untersuchungen über Mischzeit, Mischgüte und Arbeitsbedarf in Mischtrommeln mit rotierenden Mischelementen*, Dissertation, Karlsruhe 1966.

[19] I. S. Gradshteyn, I. M. Ryzhik: *Table of Integrals, Series and Products*, Academic Press, New York-London 1965.

[20] G. Schmahl, *VDI-Forschungsh.* **533** (1969).

[21] M. B. Donald, B. Roseman, *Br. Chem. Eng.* **7** (1962) 749–753.

[22] B. Roseman, M. B. Donald, *Br. Chem. Eng.* **7** (1962) 823–827.

[23] M. B. Donald, B. Roseman, *Br. Chem. Eng.* **7** (1962) 922–923.

[24] Fr. Müller, *Aufbereit. Tech.* **7** (1966) 274–285.

[25] M. Ullrich, *Chem. Ing. Tech.* **41** (1969) 903–907.

[26] H. Mattke, *Keram. Z.* **23** (1971) 282–286.

[27] J. C. Williams, M. J. Kahn, *Chem. Eng. (N.Y.)* **80** (1973) 19–25.

[28] R. N. Rowe, A. W. Nierrow, *Powder Technol.* **15** (1976) 141–147.

[29] A. D. Fokker, *Ann. Physik (Leipzig)* **43** (1914) 810–820.

[30] M. Planck, *Sonderber. preuß-Akad. Wiss. physik math. Kl.* (1917) 324–341.

[31] D. S. Cahn, D. W. Fürstenau, *Powder Technol.* **1** (1967) 174–182.

[32] D. S. Cahn, D. W. Fürstenau, *Powder Technol.* **2** (1968/69) 215–222.

[33] N. Hoffmann, K. Schönert, *Aufbereit. Tech.* **12** (1971) 513–518.

[34] A. Z. M. Abouzied, D.-W. Fürstenau, *Powder Technol.* **23** (1979) 261–271.

[35] M. C. Coelho, N. Harnby, *Powder Technol.* **23** (1979) 209–217.

[36] *Analyse der Metalle, Probenahme*, vol. 3, Springer Verlag, Berlin–Heidelberg–New York 1975.

[37] L. T. Fan, *Powder Technol.* **12** (1975) 139–156.

[38] R. Köhling, *Aufbereit. Tech.* **12** (1972) 689–704.

[39] E. Block, *Glückauf* **101** (1965) no. 4, 255–264.

[40] M. Hilbig, *Aufbereit. Tech.* **11** (1972) 705–712.

[41] H. Schubert: *Mechanische Verfahrenstechnik II*, VEB-Verlag, Leipzig 1979.
[42] Z. Sterbacek, P. Tausk: *Mixing in the Chemical Industry,* Pergamon Press, Oxford 1965.
[43] M. Pahl: *Mischen beim Herstellen und Verarbeiten von Kunststoffen*, VDI-Verlag, Düsseldorf 1986.
[44] W. Entrop: "Scaling up Solid-Solid Mixers," *Int. Symp. Mixing*, Mons 1978.
[45] Ch. G. Schofield: *Homogenisation/Blending System Design and Central of Minerals Processing*, Trans. Tech. Publications, Clausthal 1980.